The Coconut Palm (*Cocos nucifera* L.) - Research and Development Perspectives

K. U. K. Nampoothiri
Editor-in-Chief

V. Krishnakumar • P. K. Thampan
M. Achuthan Nair
Editors

The Coconut Palm (*Cocos nucifera* L.) - Research and Development Perspectives

Springer

Editor-in-Chief
K. U. K. Nampoothiri
ICAR-CPCRI
Kasaragod, Kerala, India

MS Swaminathan Research Foundation
Chennai, Tamil Nadu, India

Nigerian Institute for Oil Palm Research
Benin City, Nigeria

FAO Consultant
Thiruvananthapuram, Kerala, India

Editors
V. Krishnakumar
ICAR-Central Plantation Crops Research
Institute, Regional Station
Kayamkulam, Kerala, India

M. Achuthan Nair
Retired Professor of Agronomy
and Director (Academic)
Kerala Agricultural University
Thrissur, Kerala, India

P. K. Thampan
Retired Chief Coconut Development Officer
Coconut Development Board,
Government of India
Kerala, India

ISBN 978-981-13-2753-7 ISBN 978-981-13-2754-4 (eBook)
https://doi.org/10.1007/978-981-13-2754-4

Library of Congress Control Number: 2018965416

This Springer imprint is published by the registered company Springer Nature Singapore Pte Ltd.
The registered company address is: 152 Beach Road, #21-01/04 Gateway East, Singapore 189721, Singapore

Foreword

The coconut palm is rightly eulogized as "Kalpavriksha" – the tree of life due to the multifarious uses it offers to human kind. Apart from the fascinating fact that all the parts, right from its roots to leaf, have something or the other to offer, the magnanimity of this eco-friendly palm in accommodating a large number of crops in its ambit, giving scope for intercropping, multiple cropping, high-density multispecies cropping and mixed farming, paving the way for additional sustainable income, is adorable. It is not surprising that the crop has spread to 94 countries giving employment opportunities and livelihood options to 64 million families across the world. However, it is a paradox that the small landholders of coconut are now in a dilemma faced with many challenges. The wide gap between the potential yield and the realized farm yield should not go unnoticed. It has to be examined whether the scientific and political worlds have given the attention the crop deserves.

This is the most opportune moment, when coconut research has completed 100 years in 2016, to have an introspection of what has been done so far in the research and development field relating to coconut. Progress is not possible unless we have such information on hand in a most understandable form, and that is what this book, *The Coconut Palm (Cocos nucifera L.) – Research and Development Perspectives*, aims at. A sincere attempt has been made to capture all the available information in one publication. It is in fact a bit surprising that there has been very little effort to comprehensively make available to the world the tremendous amount of knowledge accrued on this crop during the past 100 years. The book, in its 17 chapters, covers all aspects of coconut right from origin to cultivation, breeding, physiology and value addition apart from subjects of topical interest like biotechnology, health and nutrition as well as climate change and carbon sequestration. The publication reminds us that the "bio-happiness" of the small coconut landholders is the ultimate aim of the scientists and developmental agencies and outlines certain important strategies to make coconut farming more remunerative globally. Although there is a preponderance of Indian literature in the publication, which is understandable in view of the prime position the county holds in the field, work in other major coconut growing countries has as well been described adequately. The chapters have been well written by authors who have vast experience in their respective fields and cover the entire spectrum of coconut sector. I congratulate the editors for taking the initiative in bringing out this compilation, which is bound to be of utmost value to researchers, students, extension workers, developmental agencies, progressive farmers as well as all those who love this versatile palm.

M. S. Swaminathan Research Foundation
Chennai, India

Ex-Member, Parliament (Rajya Sabha)

M. S. Swaminathan
founder@mssrf.res.in
swami@mssrf.res.in,

Preface

The coconut palm, being a perennial heterozygous and solely seed-propagated crop and requiring large area and enormous resources for field experimentation, had been hard to research upon and obtain the expected scientific backup. However, exemplifying tribute to the scientific fraternity, most of whom have spent their whole professional career towards fine-tuning research outcome of coconut, a large quantum of information has been accrued over the years, but somehow the coconut researchers have been shying away from a comprehensive compilation of the information after the commendable book- *The Coconut Palm: A Monograph* by Menon and Pandalai in 1960 apart from the one by Child (*Coconuts*, 1974), Ohler (*Coconut, Tree of life*, 1984), Freemond (*le Cocotier*, 1966) and other publications on individual aspects.

It is appropriate to make available the scientific information on hand collated for the next generation to avoid duplication of efforts and form a strong basis for the scientific pursuit and, hence, warranted this publication, when coconut research and developemnt has just celebrated its century.

Farmers in the coconut growing areas have evolved their own ingenious ways of selecting the planting material and managing the palms. A structured research on coconut started during 1916 in India. Now research work is being undertaken in almost all the coconut growing countries especially India, Philippines, Indonesia, Côte d'Ivoire and Sri Lanka. The research outcome on various aspects available from the research laboratories has been dealt in detail in the 17 chapters of this book.

It is a matter of great apprehension that the coconut sector scenario does not present a rosy picture as on today, with decline in productivity and inadequate remuneration at the farm gate. These coupled with unexpected price fluctuation has been a worrisome factor for farmers who are not able to obtain a sustainable income from the crop. One of the factors for low income from coconut is its sole dependence on vegetable oil price which obviously fluctuates with one or the other oils in the market. Fortunately, coconut is quite versatile to counter this auguring many valuable health products apart from coconut oil. Inspired optimism has been expressed in various fora that by popularizing tender coconut water, *kalparasa*, toddy, coconut sugar, coir, shell products, etc. and projecting them in supply value chain, there could be a turnaround in product diversification scenario. However, the impact due to these has been marginal. Though there have been controversies on the health aspects of coconut oil, copra oil as well as the virgin coconut oil are now being accepted to be of high nutritive value with many health benefits attributed, which can also be taken advantage of through effective market interventions.

In the emerging scenario on climate change and consequent upsurge of pests and disease problems, greater emphasis on strengthening quarantine, systematic surveillance and constant vigilance in transborders are the need of hour to counter biosecurity risks. It is the experience in all the coconut growing countries that the technologies emanating from research organizations are not adequately transformed to field realities and level of adoption is very low, as is adequately explained in Chapter 14 of this book. The reasons could be many such as the lack of knowledge, practical difficulties in implementation, slackness on the part of the implementing agencies and financial constraints. Very effective transfer of technology mechanism supported by strong policy decisions is necessary to help the farmers to come out of this crisis. The fact that coconut is not in the priority list of agricultural crops in many countries indicates the inadequate attention it attracts from governments.

Area under coconut has not been increasing globally, and this situation can be expected to continue in the wake of growing demand for land for nonagricultural purposes. So the only option is to increase the productivity per unit area, which on the contrary has been dwindling due to the large proportion of neglected and senile palms raised under rain-fed condition as well as poor management. The problem is worsened further with fragmentation of land resulting in coconut cultivation resting mostly with the small holders who have their own special problems in managing a profitable coconut garden. Cooperative farming ably aided by government support is one way to help the comparatively resource poor coconut farmers. It is possible to bring back the glory of coconut sector through government support and with the fortitude of the farming community. The few international agencies related to coconut research and development have also become defunct or nonfunctional, APCC being the only exception. Increased international cooperation is necessary to revive this crop considering its unique problems and the voluminous ecosystem services it provides to mankind.

A publication of this nature cannot be expected to be entirely perfect in spite of the best efforts taken by the editors, which means that it has to be necessarily revised from time to time, incorporating the latest research findings as well as considering the appropriate suggestions for improvement from well meaningful readers.

Thiruvananthapuram, India K. U. K. Nampoothiri
Kayamkulam, India V. Krishnakumar
Kochi, India P. K. Thampan
Thiruvananthapuram, India M. Achuthan Nair

Acknowledgements

We wish to place on record our respectful regards to Professor M. S. Swaminathan for providing the foreword and for his valuable advice from time to time.

The editors are indebted to Dr. P. Chowdappa, Director, ICAR-CPCRI, Kasaragod, for his encouragement at various stages during the preparation of this book.

We are grateful to M. Arivalagan, S. V. Ramesh, S. R. Voleti, B. Champakam, K. Shivashankar and (late) A. Ramadasan for their contribution in collecting very useful information on physiological and biochemical aspects in the initial stage of preparation of the book. The contribution of (late) S. Robert Cecil in drafting the chapter on soil fertility and nutrition during the first stage is also gratefully acknowledged.

The editors wish to thank the authors of all the chapters for their sincere efforts in collecting the large volume of information and compiling them within the time frame. We have received immense and valuable support from many, and we are happy to acknowledge their help. Our special thanks go to Mr. S. Thajudin, Asst. Chief Technical Officer, and Mr. E. R. Asokan, Technical Officer of ICAR-CPCRI (Regional Station, Kayamkulam) for their sincere help in literature collection and preparation of figures.

Any mistakes which might have occurred inadvertently are regretted. It is our pleasure to thank the publishers, M/s Springer Nature for the excellent printing and time-bound publication in a commendable manner.

A compilation of this nature would not have been possible without the effort of many individuals, whose names have not been possible to be individually mentioned. We greatly appreciate their professionalism and encouragement, and we are indebted for their support and would like to profusely thank all of them. We are personally indebted to our family members, but for whose support and forbearance, it would not have been possible to bring out this publication in time.

Contents

About the Editors

Dr. K. U. K. Nampoothiri is a former Director of the Central Plantation Crops Research Institute (CPCRI), Kasaragod, Kerala, Indian Council of Agricultural Research, Government of India. He served as the Head of the Division of Plant Breeding at the Nigerian Institute for Oil Palm Research, Benin City, Nigeria, for 5 years. He was also the Director of M. S. Swaminathan Research Foundation, Chennai, India, and an FAO consultant. He is the recipient of the Platinum Jubilee Award for coconut research and development from CPCRI. His major field of specialization is agriculture: cytogenetics and plant breeding. unampoothiri@gmail.com

Dr. V. Krishnakumar is currently a Principal Scientist (Agronomy) and Head, ICAR-Central Plantation Crops Research Institute (Regional Station), Kayamkulam, Kerala. He served as Editor of the *Journal of Plantation Crops*. As an agronomist, he has edited 10 books and published more than 100 research and technical articles as well as contributed chapters in various books and technical bulletins. dr.krishnavkumar@gmail.com

Mr. P. K. Thampan is the (retired) Chief Coconut Development Officer of the Coconut Development Board, Government of India. Specializing in agronomy, he was involved in the implementation of numerous international COGENT (Coconut Genetic Resources) and APCC (Asian Pacific Coconut community) projects. His field activities cover agricultural extension, agricultural education, and integrated development of the coconut industry. Sri Thampan is the founder President of Peekay Tree Crops Development Foundation, a nongovernmental organization. He is the recipient of the Dr. C. S. Venkat Ram Memorial Award. He has authored 18 books as well as 200 scientific and technical essays. thampan.pk@yahoo.com

Dr. M. Achuthan Nair served as a Professor of Agronomy and Director (Academic), Kerala Agricultural University. He is an FAO (UN) consultant. His fields of specialization are plant nutrition, agronomy, and agroforestry, and he has published over 75 scientific articles. He has 25 years of teaching experience both at under and post graduate levels. manair46@gmail.com

Editors and Contributors

Editor-in-Chief

K. U. K. Nampoothiri ICAR-CPCRI, Kasaragod, Kerala, India
MS Swaminathan Research Foundation, Chennai, Tamil Nadu, India
Nigerian Institute for Oil Palm Research, Benin City, Nigeria
FAO Consultant, Thiruvananthapuram, Kerala, India

Editors

V. Krishnakumar ICAR-Central Plantation Crops Research Institute, Regional Station, Kayamkulam, Kerala, India

P. K. Thampan Coconut Development Board, Peekay Tree Crops Development Foundation, Kochi, Kerala, India

M. Achuthan Nair Kerala Agricultural University, Thiruvananthapuram, Kerala, India

Contributors

U. Archana Westvest 233, Delft, The Netherlands

S. Arulraj ICAR-IIOPR, ICAR- CPCRI, Krishnagiri, Tamil Nadu, India

Merin Babu ICAR-CPCRI, Regional Station, Kayamkulam, Kerala, India

Shameena Beegum ICAR-CPCRI, Kasaragod, Kerala, India

R. Dhanapal Palaniappa Nagar, PN Pudur, Coimbatore, Tamil Nadu, India

M. Gunasekharan Bharathiyar University, Coimbatore, Tamil Nadu, India

H. Hameed Khan Optic Craft, RS Puram, Coimbatore, Tamil Nadu, India

K. B. Hebbar ICAR-CPCRI, Kasaragod, Kerala, India

Vinayaka Hegde ICAR-CPCRI, Kasaragod, Kerala, India

Rohini Iyer Navasakti Trust, Karunagappally, Kollam, Kerala, India

D. Jaganathan ICAR-CTCRI, Thiruvananthapuram, Kerala, India

S. Jayasekhar ICAR-CPCRI, Kasaragod, Kerala, India

B. A. Jerard Division of Horticulture and Forestry, ICAR-CIARI, Port Blair, Andaman and Nicobar Islands, India

A. Josephrajkumar ICAR-CPCRI, Regional Station, Kayamkulam, Kerala, India

Anitha Karun ICAR-CPCRI, Kasaragod, Kerala, India

K. V. Kasturi Bai ICAR-CPCRI, Kasaragod, Kerala, India

M. R. Manikantan ICAR-CPCRI, Kasaragod, Kerala, India

A. C. Mathew ICAR-CPCRI, Kasaragod, Kerala, India

B. Mohan Kumar College of Forestry, Kerala Agricultural University, Thrissur, Kerala, India

Chandrika Mohan ICAR-CPCRI, Regional Station, Kayamkulam, Kerala, India

C. P. R. Nair ICAR-CPCRI, Regional Station, Kayamkulam, Kerala, India

P. K. Ramachandran Nair School of Forest Resources and Conservation, University of Florida, Gainesville, FL, USA

R. V. Nair ICAR-CPCRI, Kasaragod, Kerala, India

T. Nalinakumari Kerala Agricultural University, Thiruvananthapuram, Kerala, India

S. Naresh Kumar Centre for Environment Science and Climate Resilient Agriculture, ICAR-IARI, New Delhi, India

N. M. Nayar TBGRI, Thiruvananthapuram, Kerala, India

V. Niral ICAR-CPCRI, Kasaragod, Kerala, India

R. Pandiselvam ICAR-CPCRI, Kasaragod, Kerala, India

V. A. Parthasarathy ICAR-IISR, Kozhikode, Kerala, India

P. S. Prathibha ICAR-CPCRI, Kasaragod, Kerala, India

V. Rajagopal ICAR-CPCRI, Kasaragod, Kerala, India

Society for Hunger Elimination, Tirupati, Andhra Pradesh, India

T. Rajamohan Coconut Research and Development Centre, Thiruvananthapuram, Kerala, India

M. K. Rajesh ICAR-CPCRI, Kasaragod, Kerala, India

Rajkumar ICAR-CPCRI, Kasaragod, Kerala, India

D. V. Srinivasa Reddy ICAR-ATARI, Main Research Station, Bengaluru, Karnataka, India

P. Rethinam ICAR, Coimbatore, Tamil Nadu, India

APCC, Jakarta, Indonesia

C. V. Sairam ICAR-CIBA, Chennai, Tamil Nadu, India

M. Shareefa ICAR-CPCRI, Regional Station, Kayamkulam, Kerala, India

L. Geetha Nellithara House, Oachira, Kollam District, Kerala, India

J. J. Solomon Porkudam Apartments, SBI officers colony, Madurai, Tamil Nadu, India

C. Thamban ICAR-CPCRI, Kasaragod, Kerala, India

George V. Thomas ICAR-CPCRI, Vakayar, Kerala, India

Regi J. Thomas ICAR-CPCRI, Regional Station, Kayamkulam, Kerala, India

Abbreviations

AC	Activated carbon
AD	Alzheimer's disease
ADOT	Andaman ordinary
AGT	Andaman giant tall
AIFTA	ASEAN-India Free Trade Agreement
AIS	Alien invasive species
APCC	Asian and Pacific Coconut Community
API	Annual productivity index
ATARI	Agricultural Technology Application Research Institute
BGD	Brazilian green dwarf
BSR	Basal stem rot
BUROTROP	Bureau for the Development of Research on Tropical Perennial Oil Crops
CABI	Coconut for Agro-Based Industry Indonesia
CAD	Coronary artery disease
CBFS	Coconut-Based Farming System
CBOs	Community-based organizations
CBT	Cambodia tall
CCRI	Cocoa and Coconut Research Institute
CCV	Complex coconut varieties
CDKA	Cyclin-dependent kinase
CDM	Clean development mechanism
CFDV	Coconut foliar decay virus
CFTRI	Central Food Technological Research Institute, Mysore
CGD	Chowghat green dwarf
CGRD	Coconut Genetic Resources Database
CHD	Coronary heart disease
CHRC	Chumphon Horticulture Research Centre
CIAE	Central Institute of Agricultural Engineering
CIBA	Central Institute of Brackish Water Aquaculture

CICY	Centro de Investigacion Cientifica de Yucatan Merida
CIRAD	Centre de Cooperation Internationale en Recherche Agronomique pour le Development
CKP	Coconut kernel protein
CLB	Coconut leaf beetle
CLP	Coconut leaf pruning
CMA	Coconut monoethanolamide
CMC	Carboxymethyl cellulose
COD	Chowghat orange dwarf
COGENT	International Coconut Genetic Resources Network
CPCRI	Central Plantation Crops Research Institute
CPE	Cumulative pan evaporation
CRD	Cameroon red dwarf
CRI	Coconut Research Institute, Lunuwila, Sri Lanka
CRICATAS	Chinese Academy of Tropical Agriculture Sciences
CSA	Climate-smart agriculture
CSM	Coconut skim milk
CSMP	Coconut skim milk powder
CVD	Cardio vascular disease
CYD	Coconut yellow decline
DC	Desiccated coconut
DFR	Differential fertilizer requirement
DNA	Deoxyribonucleic acid
DRIS	Diagnosis and Recommendation Integrated System
DSC	Differential scanning calorimetry
DUS	Distinct, Uniformity and Stability
ECT	East coast tall
EFA	Essential fatty acid
EPN	Entomopathogenic nematodes
ERP	Eppawala rock phosphate
ET	Evapotranspiration
ETo	Estimated actual evapotranspiration
FFA	Free fatty acid
GALD	Galas green dwarf
GBGD	Gangabondam green dwarf
GHGs	Greenhouse gases
GI	Glycaemic index
GMP	Good manufacturing practices
HDL	High-density lipoprotein
HDMSCS	High-density multispecies cropping system
HI	Harvest index
HPLC	High-performance liquid chromatography
HRI	Horticulture Research Institute
HRM	High-resolution melt
IARI	Indian Agricultural Research Institute

ICAR	Indian Council of Agricultural Research
ICECRD	Centre for Estate Crops Research and Development
ICOPRI	Indonesian Coconut and Other Palmae Research Institute
INM	Integrated nutrient management
IRHO	Institut de Recherches Pour les Huiles et Oleagineux
ISSR	Inter simple sequence repeats
ISTR	Inverse sequence-tagged repeat
JAT	Jamaica tall
KAU	Kerala Agricultural University
LAGT	Lakshadweep green tall
LAMP	Loop-mediated isothermal amplification
LAW	Liquid albumen weight
LCFA	Long-chain fatty acid
LCT	Laccadive ordinary
LDL	Low-density lipoprotein
LYD	Lethal yellowing disease
MAD	Malayan dwarf
MAP	Modified atmosphere package
MARDI	Malaysia Agriculture Research and Development Institute
MARI	Mikocheni Agricultural Research Institute
MAWA	Malayan dwarf x West African Tall
MCFA	Medium-chain fatty acid
MCW	Mature coconut water
MGD	Malayan green dwarf
MIT	Mineralization and immobilization turnover
MRD	Malayan red dwarf
MSP	Minimum support price
MYD	Malayan yellow dwarf
MZT	Mozambique tall
NAM	No apical meristem
NFDM	Non-fat dry milk
NGT	New Guinea tall
NIST	National Institute for Interdisciplinary Science and Technology
PBCk	High-potential buffering capacities of potassium
PBR	Plant breeders' rights
PCA	Philippine Coconut Authority
PGPR	Plant growth-promoting rhizobacteria
PHILCORIN	Philippine Coconut Research Institute
PHOT	Philippines ordinary
PNT	Panama tall
PNUE	Photosynthetic nitrogen use efficiency
PPVFRA	Plant Varieties and Farmers' Rights Authority
RCA	Revealed comparative advantage
RDF	Recommended dose of fertilizers
REE	Rare earth elements

RGA	Resistance gene analogue
RI	Refractive index
RIT	Rennell Island tall
RPW	Red palm weevil
RSW	Rugose spiralling whitefly
RWD	Root (wilt) disease
SAW	Solid albumen weight
SBTN	Snow ball tender nut
SCoT	Start codon-targeted polymorphism
SEM	Scanning electron microscopy
SGD	Sri Lanka green dwarf
SMCFA	Short- and medium-chain fatty acid
SNT	Solomon tall
SSAT	Strait Settlements
SSR	Simple sequence repeats
SV	Saponification value
TAG	Tagnanan tall
TAT	Tahiti tall
TCW	Tender coconut water
TDM	Total dry matter
TNAU	Tamil Nadu Agricultural University
TTPT	Tutupaen tall
VCO	Virgin coconut oil
VRD	Vanuatu red dwarf
VTT	Vanuatu tall
WAT	West African tall
WCLWD	Weligama coconut leaf wilt disease
WCT	West Coast tall
WTA	World Trade Agreement

Chapter 1
Introduction

N. M. Nayar

Abstract This chapter introduces the coconut palm to the reader. The coconut, *Cocos nucifera* L. (family Arecaceae), has been the most useful plant to the humans since every part of it has been finding an active economic use. Its importance has been diminishing in the present era. Coconut palm is the hallmark of the tropical beaches with its often slanting trunks and symmetrical crown. The members of the family Arecaceae are unique among plants because they are the longest living plants, since stem cells of several kinds remain throughout the life of palms.

For about a century from the dawn of the Industrial Revolution, the coconut oil was the most traded among all the vegetable oils. Its importance began to diminish with end of the Second World War. When the FAO began to publish area-production figures in 1961, the rank of the coconut oil had come down to 4th out of 14 traded vegetable oils in the world, and in 2011, it ranked as low as 11th. The reasons for the downslide have been analyzed. There are certain unique aspects of the coconuts in the world in matters of production and consumption that have been recorded for general information.

1.1 Introduction

The coconut palm (*Cocos nucifera* L., family Arecaceae) has been the most useful plant species to the humans (Nayar 2016). Every part of the palm was being put to active economic uses from very ancient times. The setting in of the Anthropocene Age has been marked in the coconut including those who habit the regions where the coconut is the "staff of life," but the status of coconut palm has continued largely as before in most of the regions of the world where the coconut matters. The coconut palm is the most ubiquitous plant species in the more than 30,000 islands that jot the tropical and subtropical seas of the Old World and in their littoral regions.

N. M. Nayar (✉)
TBGRI, Thiruvananthapuram, Kerala, India
e-mail: nayar.nm@gmail.com

K. U. K. Nampoothiri et al. (eds.), *The Coconut Palm (*Cocos nucifera *L.) - Research and Development Perspectives*,
https://doi.org/10.1007/978-981-13-2754-4_1

It has been an integral component of the legends, lores, ethnobotany, and lives of the people of these vast regions from prehistoric times (Nayar 2016).

The coconut oil was the most traded vegetable oil for about a century from the mid-1850s until the end of the Second World War (1950s). Some of the European imperial powers had then set up large coconut plantations in their overseas colonies. In the years following the Second World War, the preeminent position began to slip down steadily, and presently, it is in a pathetic 11th position among the 14 oil crops for which the FAO reports production figures. There are several reasons for this. This may find an analysis in some of the following chapters.

1.2 The Uniqueness of the Palms

John Dransfield, the best known authority on palm taxonomy, has observed about the coconut palm that "the often slanting stems and graceful crowns of the coconut are largely responsible for the palms being considered the hallmark of the tropics" (Dransfield et al. 2008). Peter Tomilinson, the doyen of palm biology, has stated that "the palms are emblematic organisms of the tropics." They are the world's longest living trees, because stem cells of several kinds remain active in different tissues, throughout the life of the palms. Palms are distinctive from the other groups of organisms in that they can make tall and long-lived trees entirely by primary developmental processes, i.e., all the tissues are the direct result of continuously active root and shoot apical meristems. These explain why we are able to successfully transplant palms of any age at any new location. These are the attributes that make the palms in general and the coconut in particular unique and different from other plant species (Tomlinson 2006).

1.3 Development Perspectives of the Coconut Sector

The present status and the outlook for the coconut do not present a rosy picture. In the early 1960s, the coconut accounted for nearly ¼ of the total oil crops production and ranked 4th during the year 1961 in nut production among all the 14 vegetable crops recognized by the FAO – the first year from which the FAO began to maintain production statistics. In 2011 AD, after 50 years, it had come down to less than 1/7th (Tables 1.1 and 1.2) of the total oil seeds production. Presently, in 2018 the coconut is grown – or it occurs naturally – in 94 out of 284 countries and territories of the world (FAO STAT 2018).

1.4 The Poor State of Affairs of the Coconut

This is obvious from the data given in Tables 1.1 and 1.2. During the 50-year period, 1961–2011, the areas under coconut and total oil seeds increased by 2 $^{1/2}$ times and of rice by about 2 times. During this period, while the production of rice and oilseeds increased by more than five time each, that of the coconut increased by 2 $^{1/2}$ times only.

Table 1.1 Area and production of coconut, oil seeds, and rice 1961–2011

Commodity particulars		Year		
		1961	2011	Increase (%)
Coconut	Area M ha	5.3	12.0	251.2
	Production Mt	22.9	57.2	249.8
Oil seeds	Area M ha	113.6	280.1	246.6
	Production Mt	104.8	550.9	525.7
Rice[a]	Area M ha	115.3	215.7	187.1
	Production Mt	162.3	738.2	455.5

Source: Calculated from data downloaded from FAOSTAT, 11 July 2014
[a]Rice data given for comparison

Table 1.2 Production of vegetable oils in the world[a]

Crops	Production 10^6 t and ranking	
	1961	2011
Coconut oil	1.6 (4): 8.6%	3.1 (9): 1.95%
Cotton seed oil	2.2 (3)	5.2 (7)
Groundnut oil	2.5 (2)	5.7 (6)
Maize germ oil	1.5 (6)	2.3 (10)
Olive oil	1.3 (7)	3.6 (8)
Palm kernel oil	1.1 (8)	6.0 (5)
Rape and mustard	0.1 (12)	22.9 (3)
Rice bran oil	0.4 (10)	1.1 (11)
Sesame oil	0.4 (10)	1.1 (11)
Soybean oil	3.0 (1)	41.9 (2)
Sunflower oil	1.0 (5)	13.4 (4)
Other oil crops	2.2 (Not applicable)	16.1 (Not applicable)
Palm oil	0.4 (9)	48.5
Vegetable oil, total	18.6	159.2

[a]Calculated from data downloaded from FAOSTAT, 11 July 2014

The total production of oil has remained stagnant in 4/14 oil crops, coconut, cotton seed, groundnut, and maize germ; the production has increased 2–6 times in another four oil crops – olive, palm kernel, rice bran, and sunflower – those of oil palm, rape and mustard, and sunflower have increased phenomenally.

Each success story has been a different one: in a nutshell, in oil palm through low pricing and deft marketing, in cotton seed and rape and mustard through the removal of anti-nutritional factors, and, in sunflower, by improving its adaptability and expanding cultivation. In the coconut, no effort whatsoever has been done to either remove the stigma, about its consumption, or expand its uses. Hybrid coconuts give enhanced yields, but its impact on overall yield may be still insignificant.

1.5 Causes for Decline of Coconut

Let us now analyze the reasons for the decline of coconut from the 1950s. From this time, the soybean lobby in the USA mounted a strong lobbying against the palm-based vegetable oils (meaning coconut and oil palm oils highlighting the presence of anti-nutritional factors in them). There was then a prevailing suspicion that the polysaturated fatty acids present in the palm oils can cause cardiac problems. They managed even to get the American Medical Association to promote their stand. The palm oil industry undertook massive R&D efforts, to overcome the problem. They soon managed to develop a technique to fractionate the polysaturated fatty acid and market a liquid form of palm oil (palm olein) and market it at highly lower prices. No similar efforts were done in the case of coconut oil, even though it had the advantage that most of the saturated fatty acids contained in the coconut oil consist of medium- and short-chain fatty acids (Nayar 2016). For this reason, coconut oil continues to carry the stigma of having high levels of saturated fatty acids, while the case of palm oil is hardly ever raised because of its great price advantage also. This situation has had an adverse effect on the use and consumption of coconut and its oil (Tables 1.3 and 1.4).

Notwithstanding the above, the local populations of the countries of the world continue to use significantly high quantities of the coconut, its oil, and/or both

Table 1.3 Changes in production and the use of coconut and coconut oil, 1961 and 2011

	1961			2011		
Crops	Production 10^6 t	Food %	Processing %	Production 10^6 t	Food %	Processing %
Coconut	22.9	31.0	62.8	57.2	35.8	40.1
Oil seeds total	104.8	14.4	70.6	550.9	8.7	76.7
Coconut oil	16.0	88.8	0.12	3.1	67.8	0.42
Vegetable oil total	18.6	77.1	0.35	159.2	50.6	0.28

Source: Data from FAOSTAT, 11 July 2014

Table 1.4 Changes in food supply of coconut and coconut oil, 1961, 2011[a]

	Food supply capita^{-1}			
	(kg year^{-1})		Kcal day^{-1}	
Item	1961	2011	1961	2011
Coconut/copra	2.3	3.0	9.0	11.0
Oil seeds (total)	4.9	7.0	38.0	57.0
Coconut oil	0.5	0.3	11.0	7.0
Vegetable oils total	4.7	11.7	113.0	280.0

[a]Domestic supply, and not production, is used for calculation

Table 1.5 Countries of the world where the food supply of coconut and coconut oil is among the highest, 2011

	Food supply			
	Coconut		Coconut oil	
Countries	(kg cap^{-1}year^{-1})	(Kcal cap^{-1}day^{-1})	(kg cap^{-1}year^{-1})	(Kcal cap^{-1}day^{-1})
Fiji	62.9	190.0	3.2	77.0
Kiribati	123.2	62.1	5.0	120.0
Philippines	3.4	10.0	3.4	82.0
Samoa	173.8	530.0	3.7	91.0
Sao Tome and Principe	136.7	348.0	1.4	34.0
Solomon Island	143.0	226.0	0.8	22.0
Sri Lanka	66.3	272.0	2.2	55.0
Vanuatu	136.4	374.0	3.7	50.0
World	3.0	11.0	0.3	7.0

Source: FAOSTAT, downloaded 11 July 2014

(Table 1.5). Incidentally, there are no reports of any higher incidence of any coronary health problems in these countries.

1.6 Major Coconut-Growing/Coconut-Using Countries

The major coconut-growing countries of the world are listed in Table 1.6. Generally, such lists give advantage to the larger countries. To provide a better perspective, the countries are ranked with four criteria. They provide revealing insights. For instance, the Philippines, which ranks first or second in 3/4 criteria, is ranked last in the terms of yield ha^{-1}.

Further, there are some countries that have more area under coconut than the reported net cropped area of the respective country (Table 1.7). This appears to happen because coconut occurs naturally in such countries.

Table 1.6 Leading coconut-producing countries of the world, 2013[a]

Country	Area (M ha)[b]	Area (% net cropped area)[b]	Production (Mt)[b]	Yield (t/ha)[b]
Brazil	0.26 (6)	0.35 (11)	2.82 (4)	11.0 (1)
India	2.16 (3)	1.38 (9)	11.93 (3)	5.5 (5)
Indonesia	3.00 (2)	12.77 (3)	18.3 (1)	6.1 (4)
Malaysia	0.11 (11)	3.67 (7)	0.61 (10)	5.4 (7)
Mexico	0.17 (9)	8.50 (4)	1.10 (8)	6.5 (3)
Philippines	3.55 (1)	65.02 (2)	15.35 (2)	4.3 (10)
Sri Lanka	0.42 (5)	8.00 (5)	2.2 (5)	5.2 (8)
Thailand	0.20 (8)	1.26 (10)	1.01 (9)	4.8 (9)
Tanzania	0.68 (4)	4.69 (6)	0.58 (11)	0.9 (12)
Vietnam	0.14 (10)	2.14 (8)	1.31 (6)	9.6 (2)
Papua New Guinea	0.22 (7)	73.44 (1)	1.20 (7)	N.A. (6)

[a]Source: FAOSTAT downloaded 21 October 2014

[b]Figures in parenthesis give relative ranking in the respective item

Table 1.7 Coconut-growing countries of the world having relatively large areas under coconut

Countries	Country area (000 ha)	Arable area (000 ha)	Coconut area (000 ha)	% area under coconut
Fiji	1827.0	165.0	65.0	39.4
French Polynesia	400.0	2.5	22.0	880.0
Kiribati	81.0	2.0	30.0	1500.0
Maldives	30.0	3.0	1.1	36.7
Marshall Islands	18.0	2.0	6.5	325.0
Micronesia	70.0	2.0	17.0	850.0
Papua New Guinea	46,284.0	300.0	220.0	73.3
Samoa	284.0	8.0	27.0	337.5
Solomon Islands	2890.0	19.0	53.0	278.9
Sri Lanka	6561.0	1250.0	420.0	33.6
Tonga	75.0	16.0	9.3	38.1
Vanuatu	1219.0	20.0	98.0	490.0

Source: Calculated from data sourced from FAOSTAT, downloaded 21 October 2014

1.7 History of the Taxonomy of Coconut

The coconut is taxonomically *Cocos nucifera* L. (Sp. pl.1188 (1753)) (Dransfield et al. 2008). The classification of the species from the family downward is given below.

Family – Arecaceae/Palmae

Subfamily – Arecoideae (one of the five subfamilies)

Tribe – Cocoseae (1 of the 14 tribes)

Subtribe – Attaleinae (one of the three subtribes)

Genus – *Cocos* (1 of the 12 tribes)

Species – *nucifera* (monospecific)

Linnaeus used as type specimen the figure of the coconut palm and its parts given in the chapter on coconut – "Thengu" in the 10 volumes on the herbals of Malabar Coast, *Hortus Malabaricus* (1678–1693) written in Dutch language and published from Amsterdam. The figure of the type specimen and chapter on coconut are reproduced as annexure to the chapter.

1.8 Brief History of *Cocos* Classification

The genus *Cocos* is now monospecific. But it has been so only since 1966. Prior to this, new species were being added to the coconut. Beccari (1916) transferred all the more than 90 *Cocos* species and then included *Cocos* to other genera (except *nucifera*), mostly to Butia and *Syagrus*, (refer also IPNI list 2015). This list has 180 extant names of *Cocos nucifera*.

Annexure I: Description of Coconut (*Cocos nucifera*) Given in Linnaeus C (1753)

HORTUS MALABARICUS

ON VARIOUS KINDS OF TREES
AND PODDED FRUITS

Described in Latin, Malayalam, Arabic and Brahmanic characters and names, with the addition of a true delineation of flowers, fruits and seeds in their natural size and with an accurate description of (their) colours and properties

Adorned by
The Most Noble and Generous Lord
HENRY VAN RHEDE TOT DRAAKESTEIN,
Chief ruler in Mydrecht, once the Governor of the Supreme Council of the Malabar Kingdom, Extraordinary Senator among the Indian Belgians but now one among the Renowned and Most Valiant Nobles of Utrajactine Province under the name of Equestrian Order

and
THEODORE JANSON of ALMELOVEEN, M.D.

Enriched with notes and illustrated with commentaries by
JOANNES COMMELINUS

Printed at Amsterdam
with the expense met by: Widow of John van Someren, heir of John van Dyck, Henry and Widow of Theodore Boom
in the year 1678.

PART ONE

OF

The Malabar Gardens

ON

Trees

Tenga

TENGA, in the language of the Brahmins *Mado*, is a tree with erect stem, rising high and growing in sandy soil. The root with thick bark (which is) reddish and turns dark is internally with soft wood, spreading (its) fibres copiously and transversely.

The stem erect, one foot thick, with the maximum thickness coming up to two feet in one part with leafy branches which arise only from the apex growing taller and taller (higher and higher) surrounded at the tip in a decussate manner like a crown, and in the bark on rind turning dark, which cannot be peeled off, grooved (furrowed) with semicircular rings and marks by which the base of the branches were attached, and consists of less hard wood, and interwoven with thick filaments and becoming red or reddish, inwardly soft and more hard towards the bark, when older, is more hard and more dark, which when exposed to open air, is liable to rooting within two or three years, is more durable under water, the apex of which is surrounded by tender branches, while becoming taller and taller tapers into a cone and consists of a soft and whitish pulp or core which is called *Palmita* in Portuguese language, which on the outside is covered with various mutually intertwining coating

Fig. 1.1 The type specimens of coconut *Cocos nucifera* L .
Figures givenin Linnaeus C (1753)

and inside is solid. When *Tenga* is young and one cubit in height, this is sweet and is of pleasing taste, when older, of less pleasing and astringent taste, and with no sweetness at all, when taller and older is of more sweet and most pleasing taste, emulating our walnuts, and is edible, and is much sought after by elephants, which because of its extreme liking, uproots these trees. The foliage branches are simple, not divided into others, in the base by which they tightly cling to them, interiorly plane and concave and from above interiorly reduced to a pointed curving protuberance, externally on the whole convex, broader in the lowest part of its base and thence gradually becoming narrow and becoming thin at the edges, and with a red-brown, reticulate, smooth rind the filaments of which arise from the margins of the base on both sides, first clothe all around the stem and mutually cover themselves, and are as if with a yellow green bark and inside consisting of hard woody filaments covering soft, whitish flesh. Those that are older droop down from the stem and are curved, then fall down one after another at fixed time, while others arise anew from the apex of the stem.

The leaves which cover the upper part of the branches up to their flat part where they are thinned down to the belly and arise obliquely in the margins and are placed opposite to each other, and are narrow, oblong, more or less six cubits in length, three transverse fingers in breadth, from the origin (base) gradually contracted to a narrow point, with a very flat surface, grooved (marked) lengthwise on the lower side, with parallel and subtle veins, towards the base with a whitish light yellow suture curving to an angle, are inserted and towards the exterior part somewhat closed from the angle of the suture and in the interior which meets with the interior surface of the branch is a rib standing out very much, which marks the exterior part with a furrow, situated in the middle, is of dark and shining greenness, when they wither they turn yellow, are of no taste and smell except that of forest (wild).

From above the origin of the foliaceous branches from the stem, there arise oblong capsules in which tender flowers and fruits inserted to stalks are enclosed; these capsules are round and somewhat smooth, three or four spans long, more narrow and more yellow towards the base and at the tip constructed into a cone, almost half a finger in breadth, the thick rind striated lengthwise with veins and woody filaments, externally green and marked with furrows and layers which deepen towards the tip, internally and smooth.

The fruit – bearing stalks, while enclosed within the case are smooth and shining and are glandular towards the base, (are) angular and curved with undulating hollow spaces, some (are) more straight others proceed in a somewhat zig-zag fashion (and) are placed around a more thick middle branch, and tapers to a point towards the tip of the case (capsule) and are closely packed together and on rupture of the capsule they spread out obliquely from the middle branch in all directions and are covered with a thicker rind, at first when soft and tender and enclosed within the capsules are light yellow or whitish yellow and then becoming more yellow and more red and finally turning green. On these stalks within the case, the flowers and fruits arise in due order that fruits come out from parts lower than the flower buds.

The flower buds are whitish – slight yellow spikes which later become more and more yellow, smooth and shining, some are more flat while others are more round (they are) tightly held in small, trifoliate, whitish yellow calices and (are) closely packed together when they are within the capsule, as they increase in size they break out of the capsule along with their stalks, (are) loose (not compact) and are placed sideways (transversely) or a little anteriorly or bent back with its tip, and they open into three leaves, which along one margin are more round and cuspidate at the tip, are hard, rigid and thick and have six yellow stamens in the middle, which are provided with oblong thick, first erect and then drooping apices and arise from the axil of leaves and with these (is) a short thick and whitish style which is trifid at the top, and are of pleasing smell, like that of lily though weak, the flowers immediately fall away from the stalk after they had broken forth from the capsule.

The fruits which arise together with the flowers in the oblong capsules and are tightly (attached to the stalk below) inserted on the stalk below the flowers, are quite large when they reach maturity and are like nuts consisting of woody covering which internally towards the interior surface of the covering or testa are covered with a cavity which is filled with water or a liquid, these when they are tender and break forth from the case, with the spikes, are round in shape and whitish yellow in colour and are tightly and wholly covered with a whitish yellow calyx which is formed of round leaves mutually rolled round, inside under the covering by which they are wrapped around (are) with thick whitish flesh, having no cavity, later, as they grow in size with age are oblong to angular in shape, increased greatly in length, less in breadth, growing green and shiny in the outer covering the leaves of the calyx from which the fruits protrude, becoming red from reddish, in these (fruits) a narrow oblong mark of a cavity to be introduced later with more growth (as they increase in size) (is seen) in the bottom part or towards the base of the fruit, before the cavity is formed, these tender fruits are called *Belleca* by the Malabari, *Coquinhos* by the Portuguese, when the cavity already formed the fruits are first filled with water and these are called *Caricu* by the Malabari, *Carica* by the Portuguese; when the growth of the cavity is over, after each one has

been filled with its water, each one is covered lined by a kernel along the internal surface of the covering.
What is to be noted about the cavity is that it is very oblong and narrow in the tender fruits and along with the fruit first it increases more in length than in width, is limited by an oval shape, the more acute angle of which is situated towards the top of the fruit, in older fruits however it is approaching the elliptical or round shape: the kernel of the fruits which grows for certain interval of time is sought after in eatables, at first is tender, white and light azure, with pleasing and sweet taste, not at all oily, later becomes white and more and more hard, when hard (it is) with the taste of nut and the more older and the more harder (it is) of the taste of oil, when ground and crushed it gives out a milky liquid which is of the taste of milk, but is somewhat oily, when very old, most oily and becomes azure blue because of oil, emitting strong smell of oil which is not quite pleasing.

According to the different names, when it is tender and similar to milk and soft it is called *Corumba* by Malabari, when it is complete (full grown in size) and does not enlarge any more, it is called *Elani* by the Malabari and *Lania* by the Potuguese, when it is somewhat hard, is called *Malinga* by both Malabari and the Portuguese, when hard, is called *Tenga* by the Malabari and *Coquo* by the Portuguese, when still more hard, (it is called) *Barettu Tenga* and when really hard, (is called) *Cotta Tenga* by Malabari and *Coco Sicco* by the Portuguese. As for the water inside the cavity, this, in a tender fruit known as *Caricu,* (is) slightly biting, a little salty and a bit acidic, less sweet, then as the fruit gets older, it becomes more and more sweet; in an old fruit like *Barettu Tenga,* (it is) at once very sweet and salty, before the formation of the kernel, it is clear and very hispid like water. As the kernel is being formed it is turbid or less hispid and becomes pale white; from the fruits known as *Caricu,* which is already filled with all its water, when a small narrow hole is made (it) gushes forth with great force in sprays, less (water) (comes forth) from the fruits which is filled with kernal, like *Corumba* or *Lania.* The water in *Malinga* has this property that when the liquid called *Suri* is notably shaken by fermentation towards the eye, when the kernel is hard and begins to dry within the fruits, is poured into a vessel and exposed to air, loses its sweet taste immediately and is rendered slightly acidic, if it is shaken by wind, without any fermentation, lest this way the water of the fruit *Malinga* loses all its fermentation.

Regarding the covering surrounding the water and the kernel, it is first soft and in the fruit *Corumba* is still tender and edible and is of the taste of walnut, when smeared with lime, it first becomes yellow then is tinged with green-dark colour, the more the fruit is old and the kernel is the more hard, it also becomes hard, and when is of the maximum hardness, it is less liable to cutting, is striated in length by three protruding sutures on the outside and in the lower part in between the sutures is provided with three closed eyes, of which one is larger and less hard, is situated between sutures which stand more apart, (it) is in use with the goldsmiths because of the hot flames it gives.

The exterior rind surrounding the covering is first whitish, as in the fruits *Coquinhos* when they are very tender, then it turns green, and when they get older, (with more age) it turns yellow from green, when the fruit is more old and begins to dry up and is finally dry, it is ashen dark and in the beginning it is soft and this, then consists inside of soft, white and thick flesh which in *Lania* is tender and is still soft and easily pervious to cutting, and when smeared with lime is tinged with yellow colour, later becomes more and more hard, is interwoven in the flesh, with woody fibres suitable for making ropes, which closely adhere to the woody covering towards the outside.

As regards the external shape, these fruits are round when very tender, oblong round as they grow older and with three obtuse angles projecting towards the part of sutures, when more old less oblong, and when very old the shape is more contracted (narrow) than *Lania*; in old coconuts, towards their basal part, inside the hard kernel arises an oblong – round whitish eye, which later becomes yellow, it is the young sprout of the fruit, which when laid in soil gives out a new shoot from the upper part through the less hard and more broad eye of the covering and form the lower part it gives out 'piram' called .... by Mal., *Pongo* in Portuguese, which grows larger and larger as the water in the meantime gets dried up and fills the whole cavity; the new shoot coming out of the eye, springs out of the covering, a little bent, bores through the external covering and gives out rootlets from the lower part, which first running obliquely through the exterior covering and finally rupturing it, fix themselves here and there: the *piram (pongo)* which arises in the cavity is white and in the constricted part, by which it comes out of the eye, is light yellow, is externally striated in length, with wrinkles, and is granulated and besmeared with oil, inside is fungus-like, humid soft and white, is edible, is with a taste more sweet than the kernel but rather disgusting (fastidious).

From this tree also comes the liquor *Suri* which is a drink like wine having inembriating power, is of pleasing taste, is a bit sweet, slightly saltish and slightly acidic, when first taken out, is with greater sweetness, when gets old is more acidic, is of a colour from whitish to azure blue and pallid, covered with bubbles or froth on the surface, and within a short time ferments and during fermentation producing a hissing noice and throwing out in air tiny drops like strong and vigorous wine, having also the power of fermenting a mass of flour, especially when it is more sweet, is a liquor that serves with great advantage (can be easily put to many uses) from which is formed a strong wine which has the power of intoxicating vinegar, sugar, called *Iagra*, this liquid is tapped by those called *Chegues* thus:-

They cut the tip of the fruit-bearing or flower-bearing capsules called '*Mammam*' of the tree and hang a vessel to it and at a distance of four fingers down the tip, they cut the rind of the capsule obliquely, they lift the cut part towards the tip forming what is called "beard" so that the liquid from the cut tip flowing through the rind of the capsule, may fall into the suspended vessel from the lip, they cover the cut tip with certain mud or with leaves of *Vetti Tali* or with the shavings of the green outer rind, in order to close the holes situated at the tip, lest the liquor suri oozing from the shoots (stalks) which are inside the capsule should spread within the capsule itself: they tie round the capsule at the incision with rind that is detached from the foliaceous branches, lest it should break; the *Mammam* or capsule, after the tip has been cut for the first time, gives *Suri* after five days, and is a liquid which first exudes with a reddish colour; they take away the vessels with *Suri* twice a day i.e. in the morning and in the evening according to the common usage, thrice with also at noon; the vessels before being suspended are well washed and purified from any liquor, whenever they are suspended, the tip is cut and smeared in the manner mentioned above and further the *Mammam* is beaten in four parts near to the margin (mouth) (edge) but not in the mouth itself nor in between two edges, with a bone filled with butter, this beating is lest the liquid is intercepted and get stuck but may remain more easily and freely and so much copiously, this liquid called *Suri* which is taken at noon in vessels hung in the morning, is sweet what is taken in the evening slightly acidic, (that which is taken) the next day, turns sour the third day, the sweet liquid is rendered acidic, losing all its sweetness; the liquid which is taken from a tender (young) tree is less strong and vigorous than that from an older and taller tree, and less spirited so much so that it gives less of strong wine when distilled, emits the smell of the soil in which it grows; the liquid which the trees that are quite old give, (flows) in less quantity and is more sweet and more intoxicating; strong wine (spirit) is extracted by distilling after adding some drops of cocos oil from this liquid after it has been kept for a day in the vessels which have been removed; in order to make vinegar from it, they put the vessels in which the liquid is collected, in lime for fifteen days, by which provoked to vehement fermentation.

It is changed to vinegar forming much froth and depositing in the bottom of a vessel a whitish material which is of ashen colour; sugar called *Iagra* is made from this liquid thus into the vessels which they hang to the *Mammae* of trees, they put a little lime just enough to tinge the liquid collected in, with red colour. For, if the lime is superfluous, the liquid is more whitish and milky by the colour of the lime if less lime is added the liquid turns white by its own colour, thus they cook this liquid *Suri* thus mixed and sufficiently steeped with lime. Shaking it continuously with a spoon, till it is thickened then red sugar is obtained; it is to be noted that from the liquid *Suri* collected drop by drop to which less lime is given, sugar cannot be made, if more of lime is added, to that it settles down, mixed together, at the bottom, in order to make sugar. First the remaining lime has to be removed; in making white sugar or *Iagra*, this besides is observed that they transfer the liquid *Suri* mixed with lime into different vessels for the separation of lime, changing the vessels thrice, namely if it is taken in the morning at noon it is poured into a new vessel, leaving the lime that has settled down at the bottom, thus again at 2 o'clock and in the evening is poured into a new vessel and boiled.

This tree is very voraceous, especially that which grows in salty soil and near the sea. That which grows in mountainous regions is less prolific and gives less fruits or *Cocos*. The water of which is of less sweet taste and less pleasing and gives the liquid *Suri* in less quantity, but is more strong; is of the greatest vigour from the twenty first to the thirtieth year of age, producing larger and more fruits or *Cocos*. So also bigger capsules or *Mammas* in which they are enclosed, and pouring down the liquid *Suri* more abundantly, from the time it gives fruits, always produces, new fruits. The mature ones falling down, when is of old age, (fruits are of) smaller size and are less abundant; the liquid *Suri* which an older one gives is also in small quantity, though more strong and more sweet. Ordinarily lives for hundred years; while is languishing and is close to death, all the foliaceous branches turn yellow and look as if burnt.

It is observed about this tree that in the third year after sowing it is clothed with branches which have grown to the right size, and from that year branches begin to fall and stand up to the height of one and half or two men and the stem which is completely clothed around and covered by branches is not more than one cubit in height, with the thickness equalling one foot: ordinarily every month a new branch comes out, in the meanwhile the old ones fall down, the falling of branches is more frequent during the summer than in the rainy season, within the space of three months the branches grow to the right size, for example those which come out after the third year of planting; before the appearance of capsules or fruit bearing *Mammae*, branches are of greater length than after, in a big tree ordinarily more or less twenty eight branches are found; they produce capsules with their fruits at times in the third, or fifth or eighth or tenth year of planting, according as the soil is more or less good, cobs which come out in between each of the seven branches which come out from the tip of the tree, reach their right size within the space of three months and opens up on the fifteenth day; flowers fall away on the fifth or sixth day after the opening of the cob. In a tree, ordinarily there are twelve branches of *Cocos*, one bunch bears dry *Tenga*, *Cocos* called *Barettu Tenga*, or *Cotta Tenga*, another (bunch) bears mature fruits called *Palupen Tenga*, which are with flesh more hard than in *Maninga* less hard than Tenga; third branch (which) bears immature fruits called *Belacca - Caricu* fourth (bunch of) fruits called *Maninga*, fifth (bunch) fruits called *Elani*, sixth fruits called *Curimba*, besides two opened *Mammae* and four closed (ones).

From the tender fruits and from those which first break forth from the *Mammae*, ordinarily fifteen or sixteen fall away from the branches, the others growing into *Cocos*, in the third month after the opening of the *Mamma*, fruits acquire a little of water. in the sixth month they are filled with all their water, then in the space of three months they are filled with all their kernel which becomes hard after three months and then coco falls from the tree, however one month before falling the water is diminished and the eye appears in the kernel; one month after the fruit has fallen down, the *Pongo* appears which within the space of one month, fills the whole cavity; within which time, if the coco is put in the soil, a short white new branch arises from the eye to the height of one cubit.

Its Powers

The root boiled in water with dry ginger and drunk is useful in warm fever, the same well crushed and decoction made with the oil of the fruit or *Coco* is employed in washing the mouth for blisters arising in the gum; the juice squeezed out from tender branches and mixed with a little honey is applied to eyes in lotion to soothe pain, juice of leaves with oil of the fruit or coco, given in decoction helps in haemorrhages, flowers eaten with *jaggery* helps those urinating purulent matter; juice pressed out from flowers and given in cow's milk is good in gonorrhoea; the cob of the flowers crushed and boiled with oil and besmeared is useful in ulcers caused by burning, juice expressed out from tender fruits helps in blisters of the mouth in infants when washed with it, as also for sore –eyes used in the same way.

Canjee water drunk with flowers is helpful in extinguishing burning of liver; the same in repeated lotion of the head is conducive for the redness of eyes;

the kernel of the fruit or coco chewed is good for blisters of the mouth; oil extracted from the kernel and smeared dries out blisters of head; the same employed in the same way is also useful for haemorrages; the *Pongo* which grows from the water in the cavity, fried with rice is given for coughs originating from melancholy, from the bark of the tree an oil is pressed out, with which is smeared the part affected by skin disease caused by worms.

Cocos nucifera L., Sp. Pl. 1188. 1753; Moore & Dransfield, Taxon 28: 64. 1979; Manilal & Sivar., Fl. Calicut 300. 1982; Sasidh. & Sivar., Fl. Thrissur 481. 1996; Sivar. & Philip, Fl. Nilambur 747. 1997 [ARECACEAE].

Tenga Rheede, Hort. Malab. 1: 1-8, tt. 1-4. 1678.

Thenga (Malayalam script on tt. 3, 4) is the coconut fruit and thengu (t. 1) is the coconut palm. Thengumpookkula (Malayalam script on t. 2) means coconut (thengu) inflorescence (pookkula), the subject of t. 2. Poo is flower and kula bunch. The names of the fruits at their different stages of growth, which are still in use today, are given in the text.

Coconut is one of the most important cash crops of Malabar and is abundant in coastal areas. All parts of this plant are used by natives, for various purposes. Fermented sap (phloem exudation), called toddy, is used as an intoxicant; fresh toddy is a health drink. The sap is extracted from the inflorescence (spadix) after various pre-treatments [described generally in the text]. The word Thenga means honey-fruit; "then" is honey and "nga" (or ka) is fruit in local language Malayalam. The allusion is to the sweet water inside the nut.

Acknowledgement: Permission accorded by Professor M R Haridas (Formerly) Professor of Botany, Calicut University,to reproduce the chapter is gratefully acknowledged.

Annexure II Relevant pages from Linnaeus (1753) Sp. Plantarum (page 1188) enumerating the latin binomial of the coconut.http//bioversity library.org/35209 downloaded on 02 January, 2016

1188 PALMÆ PENNATIFOLIÆ.

PENNATIFOLIÆ.

CYCAS.

circinalis. 1. CYCAS frondibus pinnatis circinnalibus: foliolis planis.
Cycas frondibus pinnatis, foliolis lineari-lanceolatis, ſtipitibus ſpinoſis. *Hort. cliff.* 482. *Fl. zeyl.* 393. *Roy. lugdb.* 5.
Palma indica, caudice in annulos protuberante diſtincto. *Raj hiſt.* 1360.
Arbor Zagoe amboinenſis. *Seb. theſ.* 1. *p.* 39, *t.* 25. *f.* 1.
Teſſio. *Kæmpf. jap.* 897.
Olus calappoides *Rumpf. amb.* 1. *p.* 86. *t.* 22. 23.
Todda-pana ſ. Mouta-panna. *Rheed. mal.* 3. *p.* 9. *t.* 13. -- 21.
Habitat in India.
Foliatio circinalis more Filicum peragitur.

COCOS.

nucifera. 1. COCOS frondibus pinnatis: foliolis replicatis.
Coccus frondibus pinnatis: foliolis enſiformibus margine villoſis. *Hort. cliff.* 483. *Fl. zeyl.* 391. *Roy. lugdb.* 4.
Palma indica coccifera anguloſa. *Bauh. pin.* 502.
Palma indica nucifera. *Bauh. hiſt.* 1. *p.* 375.
Calappa. *Rumpf.* 1. *p.* 1. *t.* 1. 2.
Tenga. *Rheed. mal.* 1. *p.* 1. *t.* 1. 2. 3. 4.
Habitat in Indiæ *paludoſis, umbroſis.*
Foliola omnia (excepto utrinque infimo) retro-plicata ſunt, contra ac in ſequente.

PHOENIX.

dactylifera. 1. PHOENIX frondibus pinnatis: foliolis complicatis.
Phœnix frondibus pinnatis: foliolis alternis enſiformibus baſi complicatis, ſtipitibus compreſſis dorſo rotundatis. *Hort. cliff.* 482. *Hort. upſ.* 306. *F. zeyl.* 390. *Roy. lugdb.* 5.
Palma major. *Bauh. pin.* 506.
Palma dactylifera major vulgaris. *Sloan. jam.* 174.
Palma. *Bauh. hiſt.* 1. *p.* 351. *Dod. pempt.* 819. *Raj. hiſt.* 1352.

Palma hortenſis mas. *Kæmpf. amœn.* 688. *t.* 1. 2. *f.* 1. 2.	Palma hortenſis femina. *Kampf. exot.* 668. 686. *t.* 1. 2. *f.* 2. 16. 11.

Habitat in India.

ARE-

References

Beccari O (1916) It genera *Cocos* Linn. Ele Palmae affini. L'*Agricultura Coloniale*, 10, 435–471 489–532, 585–623

Dransfield J, Uhl NW, Asmussen CB, Baker WJ, Harley MM, Lewis CE (2008) Genera Palmarum: the evolution and classification of palms. Kew Publishing, Richmond

FAOSTAT (2018) Production data. downloaded 06032018

Linnaeus C (1753) Cocos. p 1188. http://bioversitylibrary.org/. page 359 209

Nayar NM (2016) Does the coconut have a future? Keynote address. In: Chowdappa P et al. (eds) Abstracts of third international symposium on coconut research and development. CPCRI, Kasaragod, India, 10–12, December 2016

Tomlinson PB (2006) The uniqueness of palms. Bot J Linn Soc 1451:5–14

Dr. N. M. Nayar holds a PhD in Botany/Plant Breeding, LSU, USA; served as Director of four ICAR institutes, Humboldt Fellow and Fellow of several societies; and recently published two books by Elsevier, Origins and Phylogeny of Rices and The Coconut: Phylogeny, Origins, and Spread. He has also edited six volumes. nayar_nm@yahoo.co.in

Chapter 2
International Scenario of Coconut Sector

P. Rethinam

Abstract Coconut, known as the 'tree of life', *Kalpavriksha*, 'tree of abundance' and 'nature's supermarket', cultivated in more than 94 countries worldwide, provides livelihood security to millions of people in the Asia and Pacific regions. The coconut industry, which traditionally relied upon copra and coconut oil and to some extent coir, is experiencing tremendous transformation towards product diversification, high-value product development, by-product utilization and more importantly as a health drink. This chapter provides an account of global market dynamics of organic coconut water, virgin coconut oil, functional foods and health drinks from coconut including neera, coconut sugar, cosmeceuticals, oleochemicals and biofuel/bio-lubricants in the consumers' market. India's domestic consumption and global export and import of coconut products such as coconut oil, copra meal/copra cake, fresh coconut, desiccated coconut, coconut milk, cream, milk powder, coconut shell charcoal, activated carbon, coir and coir products are also elaborated. Causes for low productivity in coconut across the producing countries and various strategic solutions for increasing production as well as approaches needed to face future global challenges to transform coconut industry into a successful venture are outlined. A brief description of global organizations promoting coconut industry such as APCC, COGENT and BUROTROP is also provided in this chapter.

2.1 Introduction

Coconut (*Cocos nucifera* L.), the 'tree of life', *Kalpavriksha*, 'tree of abundance' and 'nature's supermarket', has geographically spread around the world to even very distant islands from Asia to America aided by waves as well as through mariners migrating and trading between various countries. Coconut palms can be seen

P. Rethinam (✉)
ICAR, Coimbatore, Tamil Nadu, India

APCC, Jakarta, Indonesia
e-mail: palms02@hotmail.com; dr.rethinam@gmail.com

K. U. K. Nampoothiri et al. (eds.), *The Coconut Palm (*Cocos nucifera *L.) - Research and Development Perspectives*,
https://doi.org/10.1007/978-981-13-2754-4_2

along the coast and in interior regions of almost all the tropical countries between the tropics of Cancer and Capricorn. Its wide distribution has been favoured by its usefulness, adaptability to different ecological conditions and ability to float in sea water and germinate on the coastal soils when washed ashore. Coconut has been used by human and their immediate ancestor species for at least half a million years as tender nut, dry fruit, source of food, drink, oil and wood as well as for shelter and aesthetic purposes (Foale 2003).

Health-promoting qualities of coconut milk and oil have been recognized in Ayurveda system of medicine for over 4000 years. There is a detailed mention of the functional properties of coconut in all classical Ayurveda texts called Samhitas. There are lot of ancient and present evidences to show that coconut is good for the health. But, unfortunately, the soya bean lobby defamed the coconut industry with false propaganda to promote soya oil. Now that the true facts have been realized, the demand for various products in the nonproducing countries has enormously increased which made the sunset coconut industry to a sunrise coconut industry.

Coconut water and neera are emerging as world health drinks, and the coconut water industry is growing as a multimillion dollar industry. Demand for coconut products as functional foods, functional drinks, nutraceutical, pharmaceutical and cosmeceutical products, etc. is growing at a fast rate. Coconut value-added products such as coir, coir pith, coir pith briquettes, grow bags, husk chips, geotextiles, shell charcoal, activated carbon, etc. are gaining importance and are now in huge demand because of their eco-friendly nature.

2.2 Global Coconut Scenario

As per the statistics of 2015, coconut is cultivated in more than 94 countries in the world in 11.988 million ha producing 67.04 billion nuts with a productivity of 5592 nuts ha^{-1}. Coconut is one of the major crops which provides livelihood security to millions in the Asia and Pacific regions, which together occupy more than 89.60% of global coconut area and 85.91% of copra production earning more than 1.08 billion US$ as export income.

Coconuts have, for a very long time, been an important crop in these regions and play an important part in the local economy and culture of not only for large producers such as the Philippines, India and Indonesia but also the Pacific Islands, where coconut palms are integral part to the livelihoods of many smallholders.

The area under coconut and its production have increased almost two times from 1969 to 2015. The overall global productivity remained mostly less than 1 mt copra ha^{-1} $year^{-1}$ which has to be substantially increased to provide sustainability. Over a period of about five decades, there was considerable increase in area and production from the initial period with no improvement in the productivity. Though the world area under coconut cultivation increased from 5.23 million ha in 1969 to 11.988 million ha in 2015, there is a gradual decline in total area from 2012 onwards. More or less the same trend was also observed in APCC countries which have a share of

Table 2.1 Area, production and productivity of coconut during various years

Country	2011	2012	2013	2014	2015
Area(million ha)					
World	12.035	12.241	12.225	12.195	11.988
APCC countries	10.793	10.988	10.975	10.928	10.733
Production of whole nuts					
World	65,381,223	71,851,535	69,909,092	68,505,812	67,041,674
APCC countries	56,307,021	63,835,154	61,736,420	60,180,632	58,783,273
Productivity (nuts ha^{-1})					
World	5433	5870	5707	5618	5592
APCC countries	5216	5810	5625	5507	5477
Productivity (Copra mt ha^{-1})					
World	0.910	1.001	0.995	0.952	0.958
APCC countries	0.846	0.970	0.960	0.910	0.916

89.5% of area and 87.7% of production (APCC 2015). The trend in area, production and productivity during various years is given in Table 2.1.

The trend in production of coconut as whole nut as well as copra equivalent is presented in Tables 2.2 and 2.3. Production of coconut in copra equivalent for top three countries (India, Indonesia and the Philippines) of the world is depicted in Fig. 2.1.

World area under coconut (Table 2.4) clearly indicates that no significant area expansion has taken place during 2010 to 2015 period. Therefore, any future efforts should focus on vertical expansion, by consolidation of coconut plantations, filling up of gaps, replanting and productivity increase. The trend in area under coconut cultivation in the world and APCC countries during 2011 to 2015 is given in Fig. 2.2.

Although coconut is widely dispersed in most of the tropical regions and grown in 93 countries in the world, out of 11.988 million ha of global area under this crop, close to ten million ha is contributed by only 4 countries, viz. Indonesia, the Philippines, India and Sri Lanka. These countries together contribute nearly 80% of the total area under coconut and its production in the world (Fig. 2.3).

As per the data available from APCC Coconut Statistical Yearbook 2015, Indonesia is the largest coconut-producing country with an area of 3.57 million ha and production of 2.96 million t of copra equivalent, followed by the Philippines with an area of 3.52 million ha and production of 2.26 million mt of copra equivalent. India, with 1.98 million ha and production of 2.73 million mt copra equivalent, occupies third place in area and second place in production (Fig. 2.4.).

The productivity of copra equivalent is more than 1 mt ha^{-1} in India, Malaysia, Papua New Guinea, Sri Lanka, Thailand and Vietnam. Vietnam recorded fast improvement in productivity. Countries like China, Myanmar, Brazil and Venezuela have shown an average productivity level of around 2.27 mt ha^{-1} of copra. Under normal ecological conditions, the global productivity ranges from 900 to 1500 kg copra equivalent ha^{-1} only. The productivity of coconuts has not made much progress over a period of five decades because the bulk of the existing coconut palms is

Table 2.2 Production of coconuts as whole nuts during various years

Country	2011	2012	2013	2014	2015
A. APCC countries	56,307,021	63,835,154	61,736,420	60,180,632	58,783,273
F.S. Micronesia	40,000	40,000 r	45,000	59,000 r	59,000
Fiji	160,000	179,500	148,000	200,000	165,000
India	16,943,000	23,351,000	22,680,000	21,665,000	20,440,000
Indonesia	16,189,000	16,060,000	15,563,000	15,330,000	14,804,000
Jamaica	95,500	95,700	95,700	98,500	99,200
Kenya	181,041	185,024	246,416	258,737	265,206
Kiribati	57,500	58,000	56,200	54,600	43,900
Malaysia	563,000	624,000	501,000 r	528,000 r	538,000
Marshall Islands	29,500	33,000	35,000	35,000	35,000
Papua New Guinea	1,495,000	1,495,000	1,482,592	1,483,000	1,483,000
Philippines	15,245,000	15,864,000 r	15,354,000	14,696,000	14,735,000
Samoa	180,000	262,000	267,000	267,000	267,000
Solomon Islands	100,000	100,000	100,000	100,000	100,000
Sri Lanka	2,707,000	2,927,000	2,513,320	2,870,000	3,056,000
Thailand	845,000	806,000	838,000	800,000 r	809,000
Tonga	86,100	81,600	81,762	75,100	71,698
Vanuatu	450,000	447,000	493,980	415,110	378,269
Vietnam	940,380	1,226,330	1,235,450	1,245,585	1,434,000
B. Other countries	9,074,202	8,016,381	8,169,672	8,325,180	8,258,401
Continents					
Asia	956,779	919,021	951,183	984,935	955,709
Pacific	278,950	251,255	248,132	245,274	236,342
Africa	1960,735	1,973,332	2,017,843	2,029,370	2,034,715
America	5,877,738	4,872,773	4,952,514	5,065,601	5,031,635
World total	65,381,223	71,851,535	69,906,092	68,505,812	67,041,674

Source = APCC Coconut Statistical Yearbook (2015)

old and almost one third are senile. Very often, coconut plantations are grown under suboptimal management conditions. The plantations are also affected by serious diseases and pests. The varieties and hybrids developed in various research institutes have made only very little impact in terms of area expansion except probably in India especially in areas where coconut is cultivated under irrigation. Since bulk of the coconuts is grown under rainfed conditions, farmers prefer local talls than hybrids. Coconut-producing countries should aim to reach a much higher yield level which is possible with adoption of adequate management practices, because many farmers are able to get more than 100 to 150 nuts palm^{-1} year^{-1}. Indian average productivity of coconut is 10,614 nuts ha^{-1} as in 2015–2016. In some of the countries, proper harvesting is not being done, and even the fallen nuts are not collected fully, leading to incorrect low production figures.

Table 2.3 Production of coconuts in copra equivalent during various years (mt)

Country	2011	2012	2013	2014	2015
A. APCC countries	91,32,898	10,652,936	10,535,164	9,942,086	9,832,948
F.S. Micronesia	6500	7275	7373	7373	7373
Fiji	26,667	29,833	24,568	38,161	27,500
India	2,259,000	3,113,000	3,024000	2,889,000	2,725,000
Indonesia	3,174,378	3,189,895	3,051,585	3,005,916	2,960,851
Jamaica	15,917	15,950	15,950 r	16,417	16,500
Kenya	13,926	14,233	18,956	19,904	20,402
Kiribati	9594	9708	9418	9134	7349
Malaysia	112,511	124,830	120,240	126,720	129,120
Marshall Islands	5405	6046	7000	5000	5000
Papua New Guinea	299,000	299,000	299,000	299,000	299,000
Philippines	2,077000	2,633,000	2,715,000	2,216,000	2,258,000
Samoa	65,000	52,000	52,920	53,000	53,000
Solomon Islands	20,000	20,000	20,126	20,126	20,126
Sri Lanka	517,000	559,656	480,497	548,757	584,321
Thailand	218,000	219,000	202,000 r	219,000	202,000
Tonga	17,000	16,325	17,100	15,699	14,988
Vanuatu	61,000	89,400	98,796	79,204	72,175
Vietnam	235,000	253,785	370,635	373,675	430,243
B. Other countries	1,815,050	1,603,275	1,633,936	1,665,036	1,651,080
Continents					
Asia	191,356	183,805	190,236	196,987	191,142
Pacific	55,790	50,251	49,627	49,055	47,268
Africa	392,147	394,666	403,569	405,874	406,943
America	1,175,757	974,553	990,504	1,013,120	1,006,327
World total	10,947,948	12,256,211	12,169,100	11,607,122	11,484,628

Source: APCC Coconut Statistical Yearbook (2015)

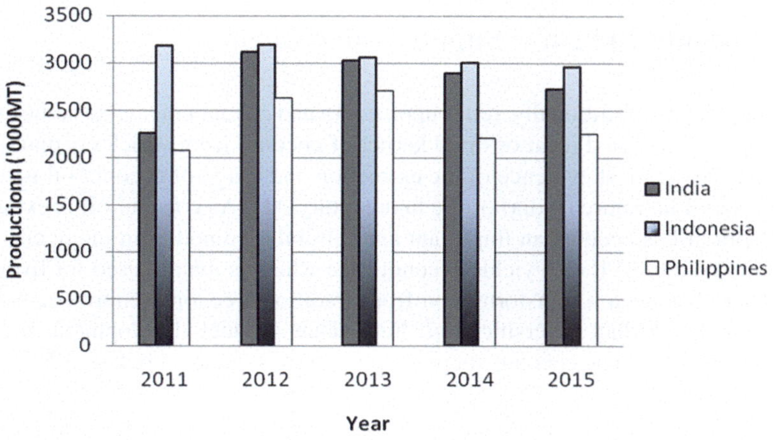

Fig. 2.1 Production of coconut in copra equivalent for top three countries

Table 2.4 Area under coconut cultivation during various years (in 1000 ha)

Country	2010	2011	2012	2013	2014	2015
A. APCC countries	10,778	10,793.0	10,988	10,975	10,927	10,733
F.S. Micronesia	17.0	17.0	18	18	18	18
Fiji	60.0	60.0	65	60	60	64
India	1895.0	1896.0	2071	2137	2141	1975
Indonesia	3739.0	3768.0	3782	3654	3610	3571
Jamaica	15.4	15.6	15.8	15.8	15.9	15.9
Kenya	78.0	82.0	86	177	177	177
Kiribati	19.0	19.0	19	20	20	20
Malaysia	106.0	106.0	101	88	88	82
Marshall Islands	8.0	8.0	8	8	8	8
Papua New Guinea	221.0	221.0	221	221	221	221
Philippines	3576.0	3562.0	3574	3551	3502	3517
Samoa	102.0	104.0	97	99	99	99
Solomon Islands	38.0	38.0	38	38	38	38
Sri Lanka	395.0	395.0	395	395	440	440
Thailand	232.0	216.0	214	209	208	202
Tonga	34.0	34.0	33.8	33.8	31	31
Vanuatu	96.0	96.0	92	92	92	92
Vietnam	147.0	155.0	157	158	158	162
B. Other countries	1243	1242.0	1253	1250	1268	1255
Continents						
Africa	389.0	388.0	397	394	455	451
America	620.0	620.0	622	622	581	580
Asia	130.0	130.0	130	130	186	184
Pacific	104.0	104.0	104	104	46	40
World total	12,021.0	12,035.0	12,241	12,225	12,195	11,988

Source = APCC Coconut Statistical Yearbook (2015)

2.3 Coconut Industry: Supply–Value Chain

Coconut industry, traditionally, relies upon copra and coconut oil and to some extent coir. Copra is the dried meat or dried kernel of coconut from which coconut oil is extracted. The earliest evidence of the extraction and usage of coconut oil is found in early Tamil literature (India) of the first century AD. As coconut oil is extracted from copra, it has become an important agricultural commodity in many coconut-producing countries. It also yields coconut cake which is mainly used for livestock feed. There has been a transformation from traditional coconut-copra-coconut oil production to product diversification, high-value product development and by-product utilization. The coconut supply–value chain is depicted in Fig. 2.5.

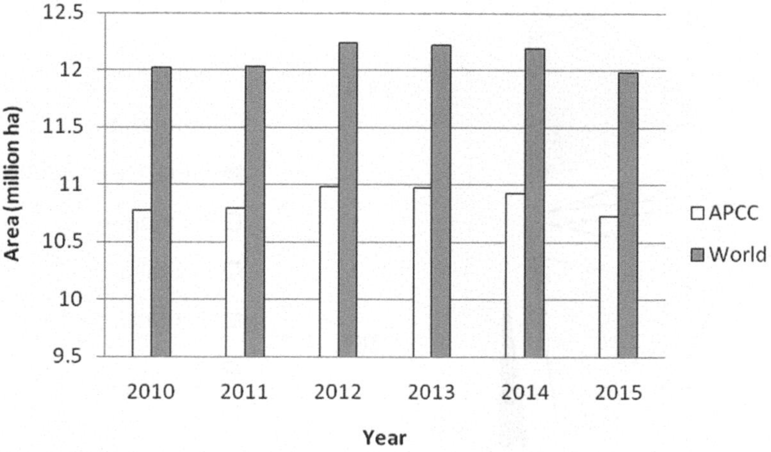

Fig. 2.2 Trend in area under coconut during 2011 to 2015

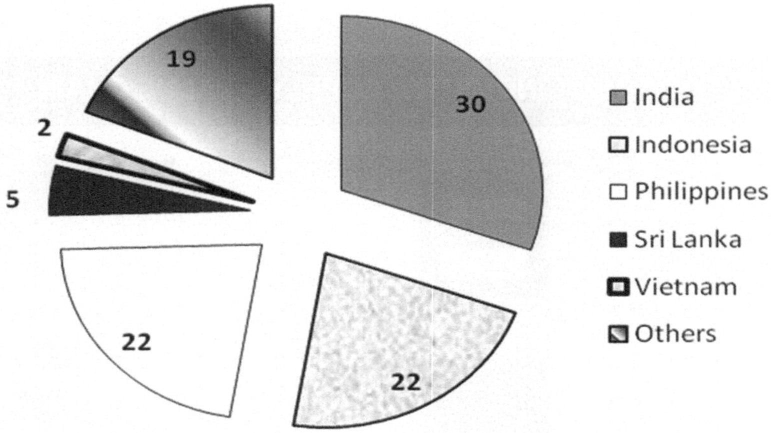

Fig. 2.3 Percentage share in world coconut area during 2015

2.4 Opportunities for Product Diversification, Value Addition and By-Product Utilization

Coconut being a versatile crop, it is possible to use each and every part of coconut palm for making value-added products as given below.

1. Inflorescence: Sugar (jaggery), *neera*, toddy (fermented sap – coconut wine), vinegar, beverage, confectionery, jelly.
2. Coconut meat: Tender coconut in syrup, coconut jam, puddings/ice cream.

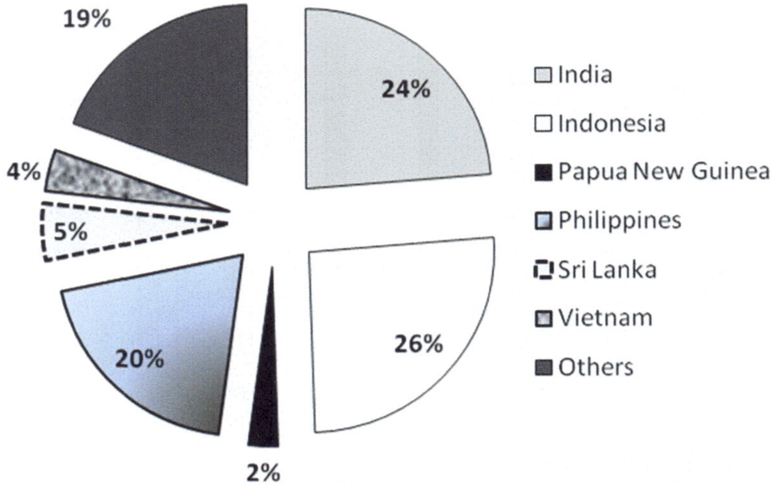

Fig. 2.4 Percentage share in world coconut production in copra equivalent

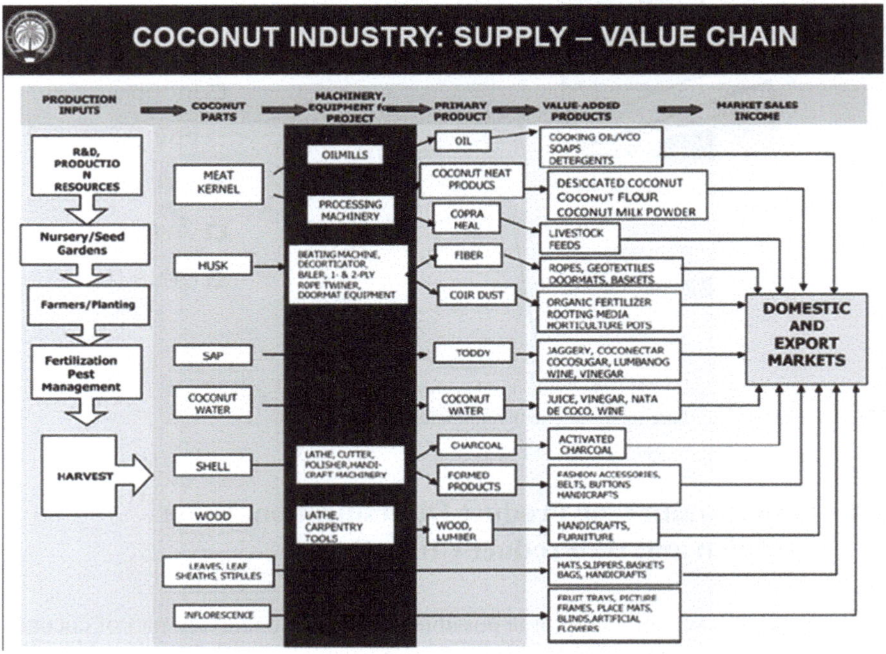

Fig. 2.5 The coconut industry: supply–value chain. (Source: Arancon 2013)

3. Mature coconut: Desiccated coconut, partially defatted coconut powder, roasted coconut paste, coconut chips, ball copra, cup copra, dehydrated edible coconut meat, fresh coconut gratings, coconut milk, milk powder, coconut oil, coconut flour, coconut meal.
4. Coconut milk: Coconut milk powder (dehydrated milk), coconut milk syrup, coconut spread/cheese, coconut honey/candy, coconut skim milk beverage, coconut protein, coconut curd, ice cream, low-fat/high-fat jam, virgin coconut oil, coconut flour, coconut cake.
5. Coconut water: Tender nut water and meat shake, pouched/tinned coconut water, coconut water spray dried powder, water concentrate/syrup, vinegar, *nata de coco*, carbonated coconut water, toddy.
6. Husk: Coir, fibre, rubberized coir fibre, coir yarn, coir mats, high-value coir wall panelling, rope, mattresses, coir rugs, geotextiles, coir dust, coir dust briquettes, grow bags, husk chips, coir dust compost, coir dust-based flowering pots, fibre-based ornaments, etc.
7. Coconut shell: Shell powder, charcoal, activated charcoal, ice cream cups, shell ornaments, spoons, and many eco gift products.
8. Coconut leaves: Leaf midribs for broomstick, roofing, flower vase and drawing mats, coconut leaf-based vermicompost.
9. Coconut stem: Furniture, doors, window and door frames, wall panels, floor tiles, packing cases, stem chips, etc.

For more details, please see Chap. 13.

2.5 Demand for Coconut

Demand for coconut has grown upwards up to 500% in the last decade, so much so the producers are not able to keep pace with. European markets have, therefore, taken a number of steps to curb their demand. Specifically, the European Union has proposed levies on vegetable oil imports to the EU, and they have promoted the use of alternate vegetable oils such as palm, canola and soya and put stricter aflatoxin regulations into place within the copra production market. Indeed, numerous foreign firms are looking to invest in the supply side of coconut production, especially in places such as Sri Lanka's ultra-productive 'Coconut Triangle' region. Today, the top coconut suppliers are struggling to meet the increasing demands of the global economy. Coconut has been a cash crop for decades and even with stiff competition from other vegetable oils, it promises to continue to be a profitable venture in the future. Nonetheless, the major coconut producers must be aware of the current situation and take steps to ensure that their farms are sustainable enough to stand the test of time.

2.6 Domestic Consumption of Major Coconut Products

The domestic consumption of major coconut products is given in Table 2.5.

Among the various products, the consumption of desiccated coconut has increased considerably (the highest being 549,574 mt during 2013), while no spectacular variations are observed in terms of domestic consumption of copra, coconut oil and copra meal.

2.7 Global Export of Coconut Products

Global export of coconut products exceeds US$ 1.2 billion annually. Though more than 50 products of coconut are being exported, only 14 products, viz. copra, coconut oil, desiccated coconut, coconut milk, milk powder, cream, coconut water, neera, coco sugar, coco chemicals, virgin coconut oil, shell charcoal, activated charcoal and coir-based products, are being exported on a larger scale. The export market of coconut and coconut products is highly concentrated with less than half a dozen exporting countries accounting for over 80% of the total quantity traded in most cases.

The Philippines is the largest exporter of coconut products earning US$ 841 million year^{-1}, followed by Indonesia, Sri Lanka, Malaysia, India and Thailand. India exports mainly coir and coir products including geotextiles and coir pith. World export of coconut products is presented in Table 2.6.

Among the products exported, coconut oil remains as the major one. But in terms of growth rate, it shows a fluctuating trend. The copra export reduced considerably because the copra-producing countries themselves have started processing coconut oil and other value-added products which is a recent development. Global export of desiccated coconut also showed fluctuation over the years, and it was 3.04 lakh mt during 2015. India imports desiccated coconut from Sri Lanka. In the past 5-year period, export of activated carbon increased from 152,490 t to 189,938 mt. The export of coir and coir products also showed fluctuations, and it decreased from 331,221 mt during 2014 to 286,671 mt during 2015. India and Sri Lanka are the major coconut fibre-producing countries.

Table 2.5 Domestic consumption of major coconut products during various years

Products	2011	2012	2013	2014	2015
Copra (million t)	5.482	5.845	5.442	5.430	5.341
Coconut oil (million t)	3.493	3.712	3.418	3.359	3.297
Copra meal (million t)	1.779	1.906	1.752	1.690	1.662
Desiccated coconut (mt)	269,275	254,190	549,574	397,426	394,064

Source: APCC Coconut Statistical Yearbook (2015)

Table 2.6 World export of various coconut products during 2011 to 2015

Products	2011	2012	2013	2014	2015
Fresh coconut(mt)	445,837	373,179	491,835	791,551	682,018
Copra (mt)	161,584	167,866	106,794	186,926	154,130
Copra meal (mt)	611,643	1,060,417	1,184,585	851,428	696,469
Coconut oil (mt)	1,862,669	2,142,817	2,228,404	2,100,013	2,096,558
Desiccated coconut (mt)	386,286	360,916	379,881	440,500	304,280
Coconut milk/cream and coconut milk powder (mt)	25,618	42,556	42,724	50,208	63,395
Charcoal (mt)	275,905	261,749	308,964	360,525	414,269
Activated carbon(mt)	152,490	256,720	209,311	239,594	189,938
Coconut shell (mt)	275,905	261,749	308,964	355,288	414,269
Coir yarn (mt)	6339	6588	6362	6392	6630
Coir and coir products (mt) [a]					
APCC countries	304,305	295,421	290,458	331,025	286,475
Other countries	200	196	196	196	196
Total	304,505	295,617	290,654	331,221	286,671

Source: APCC Statistical Yearbook (2015)

[a]Mostly coir yarn, coir mattings, coir mats, rugs, carpets, rubberized coir and coir rope (from India) and mattress fibre, bristle fibre, coir yarn, twisted fibre and coir twine (from Sri Lanka). Other countries include Indonesia, Malaysia, the Philippines and Thailand

Table 2.7 Global coconut oil production (1000 mt)

Country	2010	2011	2012	2113	2014
Philippines	1746	1137	1353	1511	1360
Indonesia	857	840	830	868	980
India	413	407	401	376	563
Malaysia	45	51	44	53	43
Vietnam	33	34	34	34	2
Papua New Guinea	52	55	55	36	22
Mexico	132	130	131	129	127
Other countries	310	311	311	322	315
Total	3616	2991	3288	3358	3412

Source: Uron Salum (2017)

2.7.1 Coconut Oil

Coconut oil production has increased from 1.95 million mt in 1960 to 3.41 million mt in 2014 (Table 2.7). It can be seen that the coconut oil production has been ranging between 2.99 (during 2011) and 3.61 million mt (during 2010).

The world coconut oil export from 1999 to 2015 ranged from 1.86 million mt in 2011 to a maximum of 3.24 million mt in 2005 (Fig. 2.6). Coconut oil continues to remain the major product of export. Coconut oil was the largest source of lauric oil

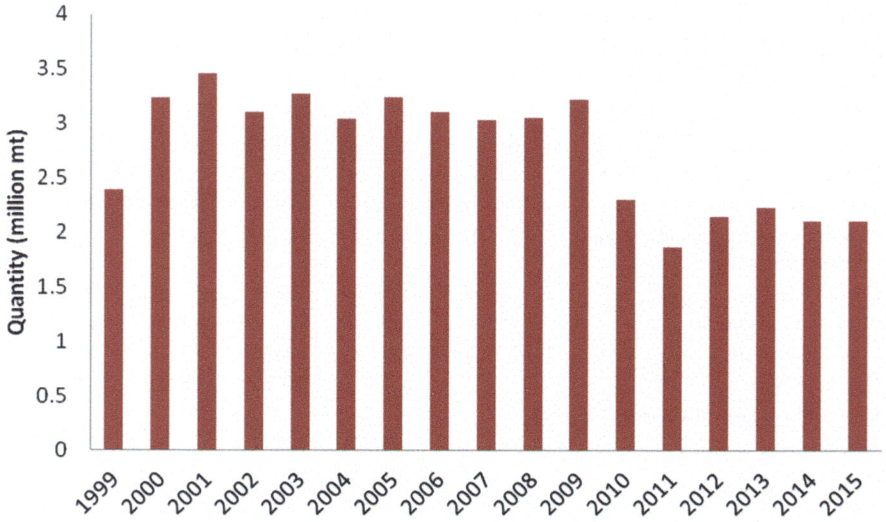

Fig. 2.6 World coconut oil export: 1999–2015

Fig. 2.7 Percentage share of coconut oil exports in the world

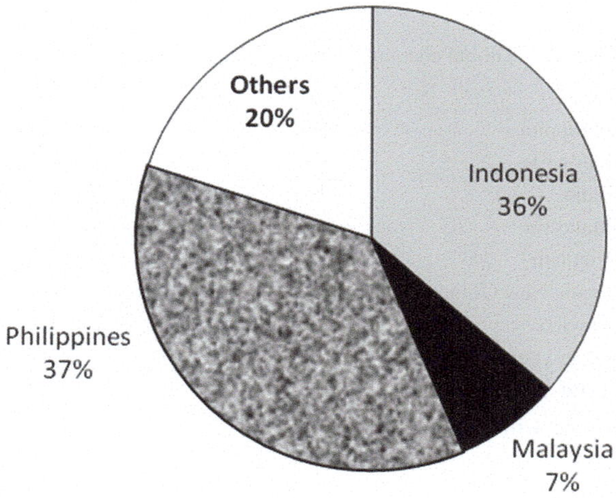

in the world. In the recent years, palm kernel oil production has exceeded that of coconut oil, and this has contributed to a slight reduction in export of coconut oil. Therefore, the price has to be competitive with other oils in the global market.

The Philippines, Indonesia and Malaysia are the leading coconut oil exporters, together contributing about 80% of the total world export (Fig. 2.7). Malaysia may be exporting coconut oil after importing from other countries. A comparison of export of coconut oil in the world vis-a-vis APCC countries and among the top three

Table 2.8 World export of coconut oil during various years (mt)

Country	2011	2012	2013	2014	2015
A. APCC countries	1,546,241	1,845,859	1,880,730	1,821,156	1735,108
Fiji	1328	3794	1494	1630	1794
India	4251	6552	6829	7067	7725
Indonesia	540,050	799,973	630,568	771,419	760,072
Kenya	600	553	38	612	161
Malaysia	143,611	136,783	131,068	187,665	152,091
Marshall Islands	–	3956	3330	124	–
Papua New Guinea	54,349	19,847	13,466	11,068	18,467
Philippines	781,411	852,236	1,080,836	815,018 r	765,558
Samoa	2509	3935	1428	1450	1020
Solomon Islands	470	172	196	238	1163
Sri Lanka	1931	2499	3821	11,254	8679
Thailand	1200	366	651	1960	15
Tonga	2531	3961	1428	1452	1020
Vanuatu	12,000	10,011	5535	9208	6570
Vietnam	–	1221	42	991	10,773
B. Other countries	316,428	296,958	347,674	278,857	361,450
Continents					
Asia	7700	8636	7204	6553	8110
Pacific	–	7333	6423	5287	7527
Africa	10,928	13,571	11,210	7930	8312
America	36,900	14,057	17,268	27,368	60,877
Europe	260,900	253,361	305,569	231,719	276,624
World total	1,862,669	2,142,817	2,228,404	2,100,013	2,096,558

Source: APCC Coconut Statistical Year book (2015)

APCC countries is presented in Table 2.8. APCC countries contributed around 83% of the total world export of coconut oil during 2015.

India and Sri Lanka are not able to compete since the domestic price is always higher than the world export price. The share of India is quite insignificant (0.3%). Virgin coconut oil (VCO) has received much attention globally in the recent times. The popularity of VCO is growing among consumers in all the continents due to its myriad properties including potential health benefits. The fast-developing and high-value niche global market for virgin coconut oil offers a good prospect for improvement of the income of coconut farmers. The USA is the largest importer of VCO in the world. The European market for VCO has also grown significantly over the past few years, driven by the increasing attention that European consumers are paying to healthier diets. Globally, the Philippines is the largest exporter of VCO. Besides the Philippines, other leading VCO exporters are Indonesia, India, Malaysia and Papua New Guinea. India exported 818 mt VCO in the year 2015–2016, and this is 3.8% higher than the quantity exported during the previous year. The USA is India's major export destination.

2.7.2 *Copra Meal/Copra Cake*

Copra meal is an important commercial by-product of copra while extracting coconut oil and has significant use in the large-scale production of commercial cattle feed. The bulk of coconut meal is exported from APCC countries. The world export market of copra meal almost falls into the category of oligopoly, wherein Indonesia and the Philippines together contribute more than 95% of the total exports (Table 2.9). India never developed an efficient export chain of copra meal, chiefly due to the huge absorption of this by-product into the domestic market as cattle feed.

2.7.3 *Fresh Coconut*

Fresh coconut is mainly used for domestic consumption and coconut-based industries (Table 2.10). Future potential exists for export of fresh nut for various purposes, which can be exploited.

Table 2.9 Export of copra meal during various years (mt)

Country	2011	2012	2013	2014	2015
A. APCC countries	599,628	1,035,224	1,166,255	835,226	689,670
Fiji	54	209	0.14	75	30
India	3	14	418	37	–
Indonesia	182,833	356,237	256,392	281,336	281,482
Kiribati	578	314	74	216	–
Malaysia	1495	1779	2064	2206	2088
Marshall Islands	2000	487	940	482	–
Papua New Guinea	22,630	10,195	5084	5250	8471
Philippines	373,135	646,044	892,281	536,186	381,873
Samoa	3061	3908	1121	2094	6531
Sri Lanka	6522	9685	187	2556	4184
Thailand	–	4	2	2	–
Vanuatu	6917	6082	7692	4786	5011
Vietnam	400	266	–	–	–
B. Other countries	12,015	25,193	18,330	16,202	6799
Continents					
Asia	100	103	1098	588	177
Pacific	–	365	431	503	867
Africa	3815	18,084	10,151	13,588	5505
America	–	247	308	1003	169
Europe	8100	6394	6342	521	81
World total	611,643	1,060,417	1,184,585	851,428	696,469

Source: APCC Coconut Statistical Yearbook (2015)

Table 2.10 World export of fresh coconut during 2011–2015 (mt)

Country	2011	2012	2013	2014	2015
A. APCC countries	424,372	322,829	414,149	687,470	599,345
Fiji	–	267	236	74	64
India	21,717	38,088	51,466	52,720	14,896
Indonesia	304,911	159,503	154,231	417,042	420,561
Jamaica	75	25	88	77	29
Kenya	–	436	–	–	–
Malaysia	55,588	18,498	53,127	48,247	44,906
Philippines	7318	2316	40,020	1753	3670
Samoa	555	453	673	1311	609
Sri Lanka	7910	19,510	21,246	33,073	13,204
Thailand	25,077	37,017	32,620	49,239	68,255
Vietnam	1221	65,214	60,442	83,934	33,151
B. Other countries	21,465	50,511	77,709	104,081	82,673
Continents					
Asia	955	1997	3491	8201	3172
Pacific	50	26	41	49	23
Africa	3486	4322	32,090	11,969	13,098
America	16,974	18,144	18,577	39,215	40,522
Europe[a]	–	25,861	23,487	44,647	25,858
World total	445,837	373,179	491,835	791,551	682,018

[a]re-export Source: APCC Coconut Statistical Yearbook (2015)
Conversion factor: 1 mt of husked nuts = 1250 whole nuts

2.7.4 *Desiccated Coconut*

Desiccated coconut (DC) is prepared from coconut kernel. It is rich in healthy saturated fats and is an excellent source of dietary fibre. The export of desiccated coconut (Table 2.11) is also increasing every year, and there is a lot of potential for this product since it is being used in confectioneries, chocolate industries, etc. With 30% of the world's imports, the European Union (EU) remains as the largest importer of DC coconut in the world.

The Philippines and Indonesia are the major global exporters of DC, accounting for more than 50% of the exports (Fig. 2.8.). Though India is the largest producer of raw coconut in the world, DC export is only to the tune of about 1% of the global exports. Nevertheless, during the year 2015–2016, India exported 4261 mt of DC worth Rs 52.60 crores. Since the demand is likely to increase by 10% annually, coconut-growing countries can exploit the situation.

Table 2.11 World export of desiccated coconut during various years (mt)

Country	2011	2012	2013	2014	2015
A. APCC countries	260,553	238,809	2617,387	305,749	232,883
India	4190	5173	3004	5166	2606
Indonesia	51,665	61,511	75,930	86,797	85,902
Malaysia	8194	7762	7137	9800	11,246
Philippines	111,868	94,877	130,669	109,099	69,548
Sri Lanka	45,135	40,224	28,202	51,132	36,131
Samoa	560	2747	38	12.4	–
Thailand	3941	3063	1839 r	3441 r	2311
Vietnam	35,000	23,452	14,568 r	40,302	25,139
B. Other countries	125,733	122,107	118,494	134,895	71,397
Continents					
Asia	24,409	20,364	21,800	24,761	21,874
Pacific	391	155	80	144	174
Africa	22,778	27,966	18,500	31,808	287
America	41,811	35,790	35,017	37,372	6564
Europe[a]	36,344	37,832	43,097	40,636	42,498
World total	386,286	360,916	379,881	440,500	304,280

[a]re-export Source: APCC Coconut Statistical Yearbook (2015)

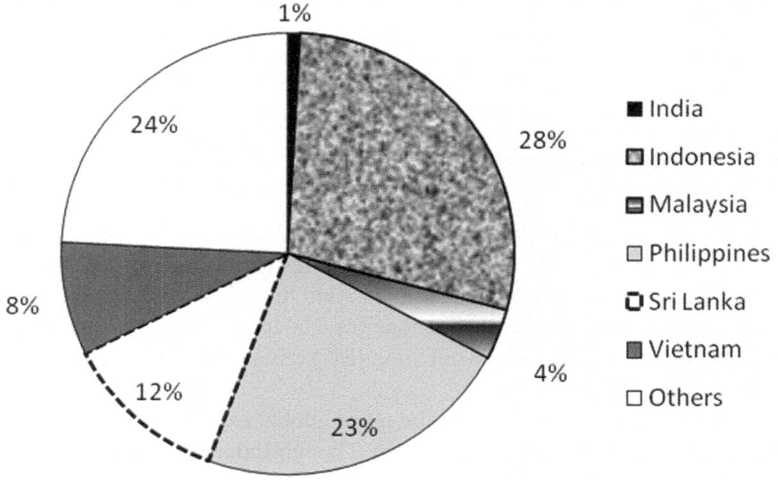

Fig. 2.8 Percentage share of DC exporting countries in the world

2.7.5 Coconut Milk/Cream and Milk Powder

Coconut milk/cream and milk powder are high-value healthy, commercial products for the higher-end consumer market across the world. They are in great demand as could be seen from Table 2.12. These are being used in various functional and

Table 2.12 World export of coconut milk/cream and milk powder during various years (mt)

Country	2011	2012	2013	2014	2015
Coconut milk/cream	12,808	32,823	34,033	39,497	49,139
Indonesia	1394	18,297	19,212	19,440	20,229
Philippines	2918	3103	306	2544	4001
Samoa	400	348	354	100	77
Sri Lanka	8096	11,075	14,161	17,413	24,832
Coconut milk powder	12,810	9733	8691	10,711	14,256
Malaysia	3925	3493	3600	3575	6358
Philippines	2467	1599	389	1301	2256
Sri Lanka	4446	4641	4702	5835	5642

Source: APCC Coconut Statistical Yearbook (2015)

Fig. 2.9 Percentage share of export of coconut milk/cream by various countries

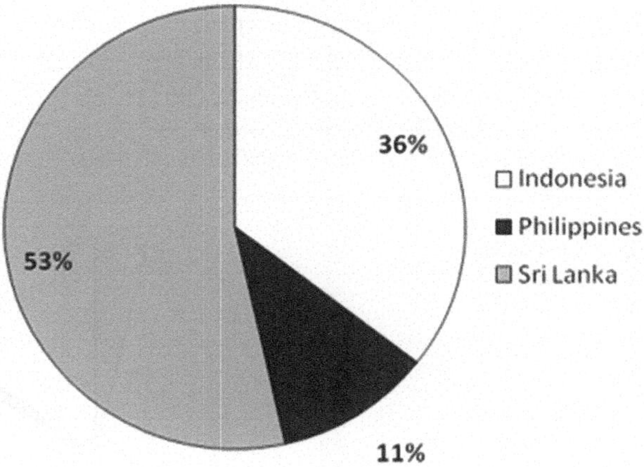

nutritional food preparations. The demand shows a steady increase due to the increased awareness of the health benefits of these products. Coconut milk is an excellent substitute for the dairy milk, and of late the health-conscious consumer segment is much aware of the comparative benefits of this product. Sri Lanka and Indonesia together contribute more than 90% of total export share of coconut milk/cream (Fig. 2.9.). India's share in this segment is insignificant.

2.7.6 Coconut Shell Charcoal and Activated Carbon

Coconut shell charcoal and activated carbon are the by-products of coconut. The coconut shell which was being wasted for a long time is now being utilized for making valuable products, which are highly useful in many purification processes. The demand will be increasing in the years to come, and the coconut-producing countries should take advantage of the situation. Their export details are given in Table 2.13. Indonesia has almost monopolized the shell charcoal export market in

Table 2.13 Export of coconut shell charcoal and activated carbon from selected countries during various years (mt)

Country	2011	2012	2013	2014	2015
Shell charcoal	275,905	261,749	308,964	360,525	414,269
Philippines	24,634	47,926	92,505	77,334	60,373
Sri Lanka	4957	6916	6359	8736	8853
Indonesia	230,452	180,671	178,500	249,682	330,012
India	15,862	26,236	31,600	24,773	15,031
Activated carbon	152,490	256,720	209,311	239,594	189,938
Philippines	41,449	128,246	62,362	60,060	54,561
Sri Lanka	31,359	30,271	36,629	37,230	31,747
Indonesia[a]	13,548	25,225	20,208	21,724	25,713
Malaysia	20,557	16,066	16,222	15,197	15,311
Thailand	9096	5475	6170	8822	8261
India	36,481	51,437	67,720	96,561	54,345
Total[b]	428,395	518,469	518,275	600,119	604,207

Source: APCC Coconut Statistical Yearbook (2015).
[a]*Includes wood–/coal-based activated carbon*
[b]*Aggregate of coconut shell charcoal and activated carbon on shell charcoal basis*

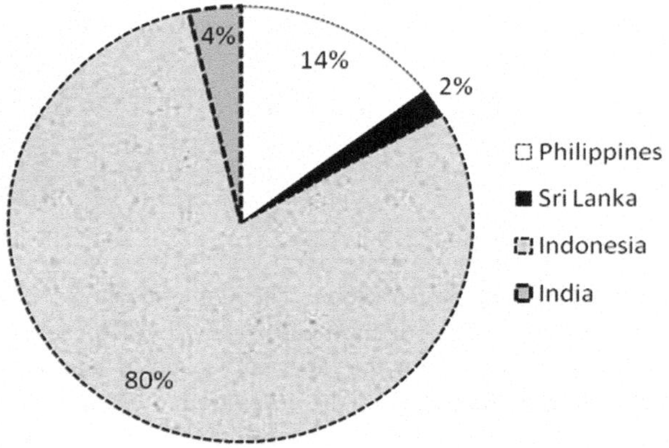

Fig. 2.10 Percentage share of export of shell charcoal by various countries

the world with around 80% contribution of the total exports, while the Philippines follows with 14.6% share in the world exports (Fig. 2.10). India too shows positive signals in the export segment of shell charcoal with a share of about 4% in the total exports, while the remaining share is that of Sri Lanka (2.1%).

2.7.7 *Coir and Coir Products*

India and Sri Lanka are the pioneers in exporting coir products throughout the world (Fig. 2.11), and till recently, they were the only coir-producing and coir-exporting countries. In the recent past, coir industry was established in other countries such as the Philippines, Indonesia and Thailand. Quantity of coir and coir products exported in the year 2015 was 0.29 million tonnes (Table 2.14). Around 80% of the export is in the form of coir fibre. China is the major buyer of coir fibre (90%), and its requirement is expected to increase 10–20% every year. At present, there is a deficit of nearly 20% in supply of coir fibre in the world. Since there is a preference for eco-friendly products globally, natural fibre-based products will have a great demand in the future.

2.8 Price of Coconut Products

Global coconut oil price has been highly fluctuating. The international price of coconut oil was US$ 1110 in 2015 which is 11.7% less compared to the 2014 price (Uron Salum 2017). In major coconut-producing countries like India, the domestic price is much higher (Table 2.15). There is a growing competition for fresh coconut since it is being increasingly used for other coconut products.

Coconut water, which is being mostly wasted, fetches US$ 800/t, the price for coconut milk is US$ 800 /t and those of milk powder is US$ 2800 /t, fatty alcohol US$ 1125 /t, coir bristle fibre US$ 460/t and activated carbon US$ 945/t (Table 2.16) (Rethinam 2004).

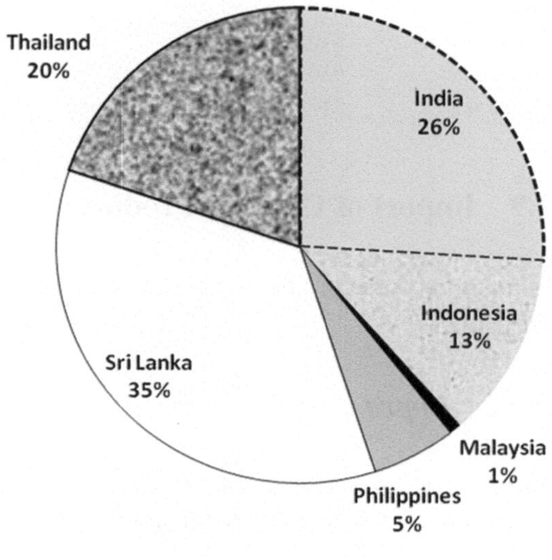

Fig. 2.11 Percentage share of export of coir and coir products by various countries

Table 2.14 Export of coir and coir products during various years (mt)

Country	2011	2012	2013	2014	2015
APCC countries	304,305	295,421	290,458	331,025	286,475
India	75,733	71,592	75,652	81,738	74,225
a. Coir yarn	5572	4479 r	4360	4187	4013
b. Coir mattings	1434	1461	2834	2270	1550
c. Coir mats	63,510	61,139	63,384	69,132	62,471
d. Coir rope	754	405	468	632	525
e. Rugs and carpets	418	78	68	114	363
f. Rubberized coir	393	390	283	993	955
g. Others	3652	3640	4255	4410	4348
Indonesia	46,173	43,089	24,109	31,972	36,171
Malaysia	12,570	13,963	14,027	11,877	2399
Philippines	7635	8201	29,407	27,834	16,085
Sri Lanka	120,616	119,376	112,529	137,900	100,863
a. Mattress fibre	78,174	57,924	53,142	36,320	18,090
b. Bristle fibre	9450	8160	10,251	16,409	11,62
c. Coir yarn	767	1025	2002	2205	2617
d. Twisted fibre	26,808	45,119	41,406	74,744	62,584
e. Coir twine	5417	7148	5728	8222	5960
Thailand	41,578	39,200	34,734	39,704	56,732
Other countries	200	196	196	196	196
Total	304,505	295,617	290,654	331,221	286,671

Source: APCC Coconut Statistical Yearbook (2015)

Table 2.15 Coconut oil price in major coconut-growing countries (US$/mt)

Countries	1997	1999	2001	2010	2014	2015
India	1448	1265	649	1263	2316	NA
Indonesia	570	628	267	944	1267	1069
Philippines	636	775	282	1271	1396	1043
Sri Lanka	946	945	680	1923	2043	1951

2.9 Import of Coconut Products

Import of coconut products is either for domestic consumption or for re-export.

2.9.1 Copra

Since coconut-producing countries are utilizing copra to produce coconut oil, its availability in the international market has considerably reduced (Table 2.17).

Table 2.16 Annual average price for selected coconut products during various years (US$/mt)

Product	1997	2000	2010	2014
Copra	434	305	732	834
Coconut oil	657	450	1116	1257
Coconut meal	NA	NA	108	253
Desiccated coconut	1149	791	1637	2650
Fatty alcohol	1225	1082	1500	NA
Fatty acid	923	833	335	654
Coconut milk – liquid	1502	1361	1771	1996
Coconut milk powder	3010	3097	2715	4301
Coconut water	807	870	1019	1270
Nata de coco	1301	1114	1104	720
Coconut vinegar	813	779	431	1555
Mattress fibre	183	199	312	212
Bristle fibre	619	524	517	396
Coir rope	689	599	615	1009
Rubberised fibre	2134	1526	2626	2657
Coir dust	194	195	280	316
Coir rope	NA	NA	1326	1009
Shell charcoal	282	253	402	438
Activated carbon	1109	994	1376	1554

Value addition and utilization of by-products give more income than coconut oil as could be seen from Table 2.14

Table 2.17 Import of copra during various years (mt)

Country	2011	2012	2013	2014	2015
European countries	441	395	619	426	1611
African countries	17	1790	50	73	1979
American countries	–	157	108	–	6
Asian countries	169,024	143,297	76,604	38,290	121,159
Bangladesh	28,765	13,500	13,500	–	–
China	4	52	–	–	40
India	1600	1600	1600	84	–
Indonesia	15	66	190	448	–
Malaysia	33,090	18,601	7140 r	19,499	24,018
Pakistan	10,869	18,519	17,804	18,100	20,135
Philippines	91,465	64,712	33,274	–	75,351
Singapore	3165	3150	3000	–	–
Sri Lanka	–	23,097	–	–	–
Others	51	–	96	159	1435
Pacific countries	–	2560	–	44	2023
World total	169,482	147,647	76,654	38,833	126,778

Source: APCC Coconut Statistical year book (2015)

2.9.2 Coconut Oil

Coconut oil import by European and American countries is continuing to remain high (Table 2.18). Out of the 21 million mt of import of coconut oil during 2015, around 45% was to European countries, followed by 28% by American countries.

2.9.3 Copra Meal

Asian countries are the major producers, importers and consumers of copra meal (Table 2.19), and among such countries, South Korea leads with around 60%. Coconut meal as cattle feed has always a good demand in China, Vietnam and India.

2.9.4 Desiccated Coconut

Import of desiccated coconut continues to increase every year, and European countries are importing large quantities of this product (36.2% during 2015) followed by American (22.8%), African (19.3%), Asian (18.1%) and Pacific (3.6%) countries (Table 2.20). This trend is likely to continue since desiccated coconut is largely being used in confectionery and chocolate industries.

Table 2.18 Import of coconut oil (mt)

Country	2011	2012	2013	2014	2015
European countries	1,024,500	1043,336	1,089,250	931,464	948,779
American countries	517,600	539,525	588,250	585,124	586,326
African countries	25,200	18,915	288,153	18,359	31,876
Asian countries	526,362	401,715	476,319	514,629	530,667
Pacific countries	15,400	14,950	20,209	15,868	18,449
World total	2,109,062	2,018,441	2,462,181	2,065,444	2,116,097

Source: APCC Coconut Statistical Yearbook (2015)

Table 2.19 World import of copra meal during various years (mt)

Country	2011	2012	2013	2014	2015
European countries	18,000	18,330	18,330	8226	7364
American countries				3488	4976
African countries	3400	3210	3210	8366	1538
Asian countries	563,955	623,705	623,255	259,032	638,863
Pacific countries	10,900	13,300	13,300	10,342	11,484
Others	18,300	16,900	16,900	2154	17,464
World total	614,555	675,445	674,995	291,608	681,689

Source: APCC Coconut Statistical Year Book (2015)

Table 2.20 World import of desiccated coconut during various years (mt)

Country	2011	2012	2013	2014	2015
European countries	127,573	136,549	133,255	240,804	139,441
American countries	71,069	70,519	69,639	101,398	87,858
African countries	21,284	18,730	279,599	61,428	74,252
Asian countries	66,080	64,877	54,778	85,980	69,610
Pacific countries	11,479	11,179	11,026	12,538	13,842
Others	11,290	11,254	–	–	–
World total	308,775	313,108	478,658	502,148	385,003

Source: APCC Coconut Statistical Yearbook (2015)

Table 2.21 World import of coir fibre during various years (1000 mt)

Country	2011 r	2012 r	2013	2014	2015
European countries	244.265	250.947	244.556	313.61	295.920
American countries	98.204	94.025	58.191	126.89	159.85
Others	529.396	588.629	602.791	1697.33	1832.88
World total	871.865	933.601	905.538	2138.13	2288.65

Source: APCC Coconut Statistical Year Book (2015) r – revised

2.9.5 Coir Fibre

Import of coir fibre is also increasing since there is good demand for natural fibres in the West. European countries import large quantities of coir fibre (Table 2.21), and among the other countries, India and China are the leaders. China makes value-added products like rubberised mattress, while European countries make high-value coir products like wall panel.

2.9.6 Coir Yarn

European countries import coir yarn (Table 2.22), for blending with other fibres to produce high-value coir products. Among the countries in the Asian continent, China imports sizeable quantity of coir yarn, and it touched 612.09 mt during 2015.

2.10 Production of Major Oilseeds

The production of major oilseeds is given in Table 2.23. The increase in production of coconut oil is only 1.78%, while the same for palm oil, soya bean and rapeseed mustard oils were 35.0%, 26.65% and 16.07%, respectively. In the case of palm kernel oil, another source of lauric oil, growth is 3.78% which is much more than that of coconut oil. There is vast scope to increase production of virgin coconut oil

Table 2.22 World import of coir yarn (1000 mt)

Country	2011	2012	2013	2014	2015
European countries	3.99	2.724	8973	342.05	167.98
American countries	0.44	0.606	1114	132.84	166.07
Others	1.40	0.676	3495	1735.10	2016.95
World total	5.83	4.006	13,582	2209.99	2351.00

Source: APCC Coconut Statistical Year Book (2015)

Table 2.23 Production of nine major vegetable oils (million mt)

Vegetable oils	1997	2001	2005	2009	2013	2014	2015	% share 2015
Palm	17.62	23.96	33.59	45.27	53.83	59.74	62.51	35.96
Soybean	20.82	27.84	33.55	36.11	43.18	45.25	48.81	28.08
Rape seed	11.83	13.74	16.06	21.72	23.80	27.00	26.27	15.11
Sunflower	9.32	8.16	9.73	13.04	13.75	16.16	15.11	8.69
Cotton	3.95	4.04	5.01	4.70	5.28	4.91	4.71	2.71
Palm kernel	2.17	2.95	3.95	5.24	6.25	6.54	6.85	3.94
Groundnut	4.21	5.14	4.51	4.16	5.29	3.92	3.69	2.13
Coconut	3.43	3.25	3.50	3.26	3.52	3.02	2.95	1.70
Olive	2.68	2.77	2.92	3.02	2.87	3.39	2.92	1.68
World production	76.03	91.85	112.82	136.52	157.77	169.93	173.82	100

Source: Uron Salum (2017)

and other value-added products such as coconut milk, milk powder, desiccated coconut, etc., which fetch higher price. It is obvious that a fair price for coconut cannot be ensured if used solely as a vegetable oil source.

2.11 Emerging New Applications

It has been realized that coconut oil with health benefits and possessing good flavour and taste is a very versatile product. In the past 20 years, it has been projected as one of the forbidden ingredients, by vested lobbies, that people wanted to avoid because of the fat content. But with the evidences from clinical studies and with awareness, coconut oil is getting accepted rapidly. Indeed, looking at the many coconut products available in the market, coconut, formerly maligned as being too fatty, has become some kind of a wonder product.

2.11.1 Organic Foods

The global organic food market is growing due to increased consumer awareness about overall health benefits of organic foods. But, the demand for organic foods is concentrated in countries where consumers have high purchasing power. The largest organic markets in terms of global revenue distribution are European Union (53% market share) and North America (40%). In the USA, the growth rate for the sale of organic products is 17%. Consumption of organic oil is 95,000 mt which is only 0.1% of the total oil consumption. Compared to other oils, organic coconut oil and organic virgin coconut oil can easily be produced since the bulk of coconut plantings are managed without the use of inorganic fertilizers and pesticides (Rethinam 2008). Coconut organic foods as niche products, coming from certified organic farms, processed by authorized firms and certified by registered certification bodies, can have a cutting edge in the market.

2.11.1.1 Organic Coconut Water: Global Market Dynamics

The global market for organic coconut water is expected to be steered by its taste and many health benefits it offers. Moreover, the presence of antioxidants and micronutrient content could prompt consumers to prefer it over other conventional soft drinks. The drink is also regarded as having curative properties for disorders such as vomiting and diarrhoea and in restricting cholesterol levels. For details, please refer to Chap. 15.

Emerging evidence about the skin and hair benefits of coconut water and its consumption in yoga institutes and health clubs are all expected to serve as an opportunity for companies in the organic coconut water market. Coconut water is by far the leading plant-based water available for sale worldwide. In 2016, coconut water accounted for 96% of the volume share in the global sale of all plant-based water with over 700 million litres sold and with a market value of about 2.2 billion US dollars. In the USA, the industry for coconut water has experienced a continued growth in market size with a projected 1.98 billion US dollars, with the forecasted generated revenue for 2019 being 612.5 million.

In order to meet the challenges associated with the relatively short shelf life of coconut water, several producers are now offering frozen organic coconut water. This enhances its shelf life, allowing it to be transported to regions where coconut water is not readily available. Packaging is proving to be an especially important product differentiator for companies competing in the global organic coconut water market. It is now available in small, easy-to-carry tetra packs, cans and plastic bottles. Organic coconut water is now available in powder form also.

Latin America, North America, Asia-Pacific, Europe, Japan, the Middle East and Africa are major global markets for organic coconut water. South Africa has been identified as the fastest-growing regional market for packed coconut water. In the United Kingdom, consumption of organic coconut water approximates 25–26 million litres. The consumption of coconut water is also remarkably high in tropical countries, such as Sri Lanka, India, Thailand and Brazil, where coconuts grow in abundance.

2.11.1.2 Virgin Coconut Oil

Virgin coconut oil (VCO) processed from coconut kernel and coconut milk with very low free fatty acid (FFA) is gaining popularity as healthy nutritive oil. The oil is found to be good for the immune system as it contains lauric acid, which the body converts to a monoglyceride that has antiviral, antibacterial and antifungal properties. Please see Chap. 15 for details. Marketing opportunities of natural VCO and organic all-natural VCO may be exploited in niche health markets to obtain premium prices. A large number of small-scale units of VCO have been set up in many countries which use different processing methods. It is all the more necessary to strictly follow quality standards in order to sustain credibility and demand. The market for VCO seems to be a fast-growing one with its growing demand as functional food and pharmaceutical, nutraceutical and cosmeceutical agent (Rethinam 2006a, b; Rethinam and Amrizal 2006).

The Philippines is the major producer of VCO followed by most of the Asian coconut cultivating countries such as Thailand, Indonesia and Sri Lanka with Indian production also increasing considerably over the years. Export of VCO from the Philippines has increased from a meagre 103 mt during 2003 to 1693 mt during 2008 and then to 4914 mt during 2011 and to 36,332 mt during 2015. It is being exported to countries such as the USA, Korea, Japan, Netherlands, Singapore, Malaysia, South Africa and Australia. Countries in the Pacific region like Fiji and Samoa are also exporting VCO to Australia. Indonesia is producing VCO involving more than 200 small and medium manufacturers in the country, most of the production being consumed domestically. Virgin coconut oil export market itself is of recent origin and expanding at a faster rate. Organic VCO fetches a very high price compared to normal coconut oil.

2.11.2 Functional Foods

The market is attracting health-conscious groups with functional foods. While the conventional food sector has an expected growth rate of 1–3%, functional foods are catching up with a growth rate of 7–8%. Since coconut milk, milk powder and desiccated coconut provide lauric acid which can help to build up resistance/immunity against viral, fungal and bacterial diseases, these can find a market as components of functional foods.

2.11.2.1 Functional Drinks from Coconut

The global functional food drinks market, defined as 'soft drink with added health benefits', shows a continuously growing trend especially as a sports drink, energy drink, wellness drink and welcome drink. Beverages showed strong growth between 1998 and 2003, expanding by a compound annual growth rate of almost 11%. One reason behind the success of functional beverages is convenience and portability.

Tender nut water (young coconut water) and mature coconut water in pure form and also with mineral and vitamin fortification can have a wider market – both domestic and international – if supported by a well-directed marketing network. Apart from the drink's credibility as a lower calorie alternative to other juices, there are other factors which make it so popular. Coconut water contains many vitamins and minerals, including niacin, pantothenic acid, riboflavin, thiamine, potassium, magnesium, manganese, phosphorus, selenium, zinc, calcium and vitamins C, A, E and K. It is being marketed as an alternative to sports drinks. From athletes to an increasingly health-conscious middle class, it is credited as an answer to the growing demand for natural, healthy products.

The most important attributes of this drink are that it is slightly sour tasting, cholesterol free, fat free, and packed with electrolytes and potassium. There are over 100 brands manufacturing coconut water, and over 250 companies have beverages with some form of coconut water in them. There is an increasing demand for the coconut drink in the European Union countries and America. Coconut water currently represents an annual turnover of US$ 2 billion which is expected to reach US$ 4 billion in the next 5 years. Brazil is marketing most of the coconut water in cartons. Ten per cent of coconuts available are used for industrialized coconut water. Proper packaging and labelling with good shelf life will definitely increase its demand in the world market.

2.11.2.2 Neera

Coconut nectar juice, from the inflorescence, in the fresh form, is known as *neera* which becomes toddy on fermentation. *Neera* is used as a health and nutritive drink which is popularized by some of the government agencies such as in India.

2.11.2.3 Coconut Sugar

The coconut palm syrup can be crystallized to produce fine granules of sugar, which is well accepted by the global market. Nature created this product such that it cannot be processed in factories. Coconut sugar is produced in palm habitats by local people, which is a good livelihood option. In countries like Indonesia, around 50,000 mt of coconut sugar is produced per month, which has a good local market in the country. The greatest advantage of the coconut sugar industry is that it can be produced in villages by small- or medium-scale enterprises and cooperatives involving women groups.

Organic-certified supply chain now encompasses over 7000 smallholder farmers specific to coconut sugar (Benjamin-Ripple 2012). The global market for coconut sap sugar has been increasing and dominated by Indonesia, Thailand and the Philippines. There is an ever-increasing demand for coconut sugar both in the local and international markets due to its health benefits. It can be a most suited alternative sweetener, especially when agave sugar is being rejected owing to the high fructose content. Coconut sugar has tremendous market potential owing to its low glycemic index (35 per serving compared to 65–100 of cane-based sugar) and high nutrient content. Thus, coconut sap sugar is safe even for diabetic patients. It was in 2007 that the Philippines first exported coconut sap sugar to the USA. Afterwards, the Philippines supplied it to the Middle East, South Korea, Hong Kong, Norway, Canada, Switzerland, Australia, France and New Zealand. This alternative sugar industry is estimated to be a US$ 1.3 billion industry, and hence the market prospects are enormous.

2.11.3 Coconut Shell Charcoal and Activated Carbon

Global Coconut Shell Activated Carbon Industry 2014 Market Research Report indicates that the major markets are the regions including North America, Europe and Asia and countries such as the USA, Germany, Japan and China.

2.11.4 Cosmeceuticals

Production of cosmetics and personal products in Asia and Pacific countries is just developing with emerging popularity as skin whitening products. Coconut oil, which is rich in C12 and C14 fatty acids (lauric and myristic), is good for skin care when applied pure or as cleaning products. With a growth of 10% for cosmetics and personal care products and 5 to 19% for soap in Asia and Pacific, the requirement is enormous. Many of the Pacific countries are importing all these products now.

2.11.5 Oleochemicals

Industrially useful natural fatty acids known as 'lauric oils' (capric acid, lauric acid, myristic acid, palmitic acid) and 'fatty alcohol' (lauryl alcohol and myristyl alcohol) which are key ingredients for common oleochemical products such as detergents, soaps and cosmetics are currently derived from coconut oil, palm kernel oil and also babassu seed. The long-term market trend for oleochemicals is favourable with world capacity expected to rise to 12 million mt and production to 10.8 million mt. The global market for natural fatty acids (primarily derived from coconut, palm and palm kernel oil) is expected to reach $ 13 billion by 2018. The market for lauric acid alone was estimated at about $ 1.4 billion in 2008. The market for myristic acid is estimated at about $ 600 million. Consumer trend is increasing towards application of oleochemicals in detergent, soap and personal care products, and hence there is good scope for coconut-based oleochemicals.

2.11.6 Biofuel/Bio-Lubricants

Energy security perspectives have become a driving force for the use of vegetable oil-based biodiesel fuels. Numerous countries are in the process of making biofuel. Three challenges the biofuel sector must overcome are price considerations, lack of awareness about the fuel and negative impact on the glycerin supply to existing markets. Bio-lubricants are functional fluids made from vegetable oils and downstream esters. Coconut oil as a bio-lubricant has been used in India for three-wheelers. Overall the global usage of renewable raw material in lubricants and related functional fluid applications is about 250,000 mt. The Philippines is moving forward in this aspect followed by Thailand, Vanuatu and Marshall Islands. Marshall Islands uses the double filtered coconut oil directly to run car, fish boat, truck, etc. The Philippines is using a mixture of diesel and methyl ester from CNO at 99:1 ratio and planning to increase it to 95:5. These developments of the coconut oil in biofuel industry indicate a bright future for the use of natural renewable resources to produce fuels.

It is an irony that the Clean Development Mechanism (CDM) of the Kyoto Protocol currently excludes tree crops like coconut from carbon sequestration credits. These perennial plantations are very similar to forest plantations in the carbon benefits they offer. In addition to timber, coconut yields nuts regularly, providing renewable energy. Moreover, coconut palms are mostly owned by smallholders, a fact which complies with the CDM goals like poverty alleviation and sustainability.

2.11.7 Biomass as Alternate Source for Fuel

Coconut biomass like coconut shell and coconut petioles can be used for producing alternate sources of energy. Coconut shell-based gasifiers are now becoming popular. The leaf petioles along with other biomasses are being used for generating energy.

2.11.8 Premium Grade Monolaurin and HIV/AIDS

Over a period of 22 years, 42 million people in the world would have been affected by HIV/AIDS. Coconut oil, with 48% lauric acid, is a potential source for producing monolaurin (lauricidin) which has been experimentally found to reduce the virus. Though Dr. Jon Kabara, a US scientist, has done preliminary work, pilot-scale testing with a large number of AIDS patients has to be carried out. The Philippines has done certain basic studies, and further elaborate research is planned. For details on nutrition and health, please refer to Chap. 15.

2.11.9 Coir and Coir Pith

Currently, the global annual production of coir fibre is 350,000 mt. Yet, this renewable resource is underutilized, with local coir mills processing only a fraction of the available husk, which accrue more or less year round as a waste during coconut processing. Historically, Sri Lanka has been the world's largest exporter of various fibre grades, whereas India exports largely value-added products – yarn, mats and rugs. Global trade volume for coir fibre, value-added products now stands at about \$ 140 million year^{-1} with India and Sri Lanka, respectively, accounting for about 70 and 60 million dollars. Multifarious uses of coir fibre include coir wood for furniture, doors, wall panel, wardrobe, coir veneer board, medium density coir board, coco lawns, handmade paper, other eco-friendly products, low-cost wall panels, rubberized coir for car seats, biodegradable pots in horticulture sector, natural fibre pith, geotextile for soil conservation and natural fibres as reinforcements in industrial products. Moreover, coir milling and value addition, especially spinning and weaving, are important regional employers, particularly in rural Southern India and coastal Sri Lanka, providing work to more than half a million people, mostly women working part-time. Other coconut-growing countries, including the Philippines, Thailand and Vietnam, are now expanding their production with a view to exporting coir fibre.

2.12 Yield Gap in Coconut

A maximum yield of 480 nuts palm^{-1} year^{-1} was reported in a palm in Thazhava, Kerala, India, with an estimated yield of 13.84 mt copra year^{-1} (Iyer et al. 1979). A yield of 6 mt copra year^{-1} was recorded in the high-yielding hybrids and varieties in farmers' fields in certain areas. A yield of 3–4 mt ha^{-1} of copra for hybrid varieties under ideal management has been achieved at research institutes around the world. But the realized average yield is less than 1 mt ha^{-1} (Fig. 2.12) indicating the tremendous scope for increasing the productivity.

2.12.1 Causes for Low Productivity

The low productivity of coconut around the world in general is due to various reasons, viz. the presence of old and senile palms in more than two third of the area; overcrowding in some countries; sparse populations in some other countries; planting nondescriptive varieties without proper selection of either mother palms or seedlings; non-adoption of recommended management practices; natural disasters like hurricane, cyclone, etc.; pest infestation; disease problems; and improper harvesting. In addition, coconut predominantly being a smallholders' crop, the resource-poor

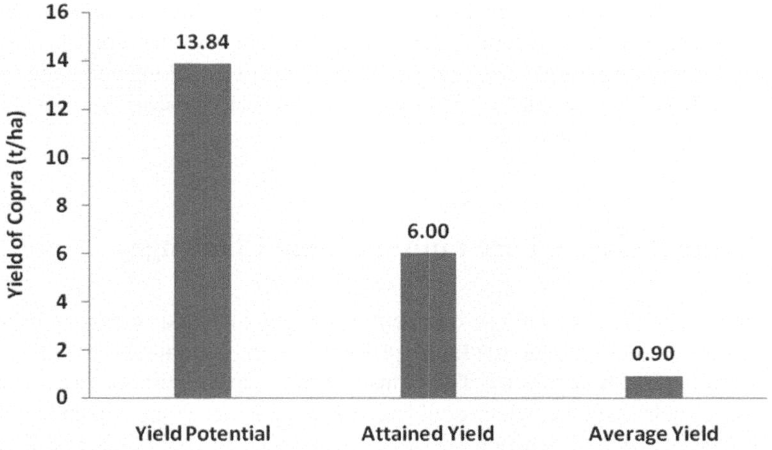

Fig. 2.12 Yield gaps in coconut

farmers are not able to take up replanting and provide proper management due to financial constraints. The political will was also either inadequate or totally absent. Many countries started feeling that coconut industry is dying, and some of the countries have reduced the area. Now they have realized the export potential of various coconut products stimulating attention to coconut (Rethinam 2008).

2.12.2 Strategic Solutions for Increasing Production

Sustainable productivity will be necessary to meet the growing demand of raw material for the processing industries at cost-effective prices. Production can be increased through area expansion and productivity increase. Since land has become a scarce commodity due to diversified uses, urbanization and industrialization, the only option left is productivity increase (Rethinam 2005).

2.12.3 Integrated Approach to Increase Productivity

Over a period of time, the research conducted in different parts of the coconut-growing countries has helped to develop technologies to increase farm income and coconut productivity as well as farm-level processing technologies to increase the income and employment (Rogonon 2004). Productivity of coconut plantations can be increased through rejuvenation of existing plantations by gap filling with high-yielding varieties and hybrids; optimum input management through organic recycling; nutrient and water management introducing micro-irrigation, fertigation, soil and moisture conservation, replanting/under-planting to eliminate old senile and unproductive plantations; adopting integrated farming system approach (inter/mixed/multiple/multi-storied cropping and mixed farming); taking adequate care to maintain soil health; and taking up proper plant protection measures. For details, please see Chaps. 7, 10, and 12.

2.13 Approaches to Face Future Global Challenges

In the wake of mounting intense competition in global markets arising from the liberalized trade environment, the future prospects of the coconut industry lie solely on its overall competitive ability. The industry has to tap the full potential of coconut as a renewable resource, which could be used to generate a range of environment-friendly, natural products, with a wide variety of end uses and applications. Likewise, producing countries need to exploit to the fullest their individual comparative advantages in cultivation, processing and marketing of coconuts and various products. The moot factors in achieving a sustainable degree of overall competitiveness in the coconut industry are listed hereunder:

1. *Encouragement for farm-level processing*: Encourage community processing at farm level engaging farm youth and women to go for farm-level processing such as dehusking, shelling, collecting coconut water and making copra, virgin coconut oil or coconut products like chips, vinegar, charcoal from shell and handicrafts from shell, coir fibre and coir pith. Farm-level processing, on a farmer private entrepreneur participatory approach, will provide large quantity of raw materials at one place.
2. *Organic farming*: Converting the weakness of non-application of inorganic fertilizers into strength by adopting organic farming to produce niche organic products to derive better price and income. Organic farming associations can be encouraged to take up organic farming in larger areas.
3. *Linking farmers to the market*: If farmers are linked to the market, it will increase their profits encouraging them to produce more coconut as well as take up processing at farm level as indicated below:

    (i) Diversifying products/developing products for niche market: Large-scale product diversification and value addition from coconut and its by-products.

    (ii) Increasing processing efficiency and quality standards matching with international standards: Precision processing for making quality products, maintaining international quality standard, proper labelling and packaging with suitable branding.

    (iii) Intensive market promotional efforts, market study and market forecasting, conducting trade/investment mission and participation in international trade fairs and exhibitions.

(iv) Promotion of nontraditional coconut products like biofuel, bio-lubricants, etc. and coir-based products like geotextile, coir pith composites, grow bags, etc.
(v) Promoting coconut-based eco-tourism in island countries/establishment of coconut world.
(vi) Effective transfer of technology through training by setting up farmer's field school, farmer's participatory demonstration, etc.
(vii) Improving international/national transport facilities particularly water transport to reduce the freight charges.

2.14 Global Organizations Promoting Coconut Industry

The major organizations are Asian and Pacific Coconut Community (APCC), COGENT and BUROTROP.

2.14.1 Asian and Pacific Coconut Community (APCC)

The Asian and Pacific Coconut Community (APCC) is an intergovernmental organization of coconut-producing countries organized in 1969 under the aegis of the United Nations Economic and Social Commission for Asia and the Pacific (UN-ESCAP). The APCC Secretariat is located in Jakarta, Indonesia, and is headed by an executive director.

The APCC has 18 coconut-producing member countries accounting for over 90% of world coconut production and export of coconut products. The APCC member countries include Federated States of Micronesia, Fiji, India, Indonesia, Kiribati, Malaysia, Marshall Islands, Papua New Guinea, the Philippines, Samoa, Solomon Islands, Sri Lanka, Thailand, Tonga, Vanuatu and Vietnam. Jamaica and Kenya are associate member countries of the APCC.

For executing its functions, the APCC Secretariat coordinates with member countries through a network of national liaison officers. Their responsibilities include the development of the coconut industry and national focal points that deal with coconut in their respective countries. The Asian and Pacific Coconut Community maintains close contacts with the United Nations Organization and its special agencies. The objectives of the Asian and Pacific Coconut Community (APCC) are to promote, coordinate and harmonize all activities of the coconut industry.

Its vision is to improve the socio-economic conditions of all stakeholders of the coconut industry in the member countries, particularly the small coconut farmers. APCC's mission is to assist the member countries to develop their coconut industries to increase productivity; reduce cost of production and adopt integrated coconut-based farming systems; encourage organic farming; promote farm-level processing; promote product diversification, value addition and by-product utilization; improve quality standards; intensify market promotional activities and research; harmonize trade-related issues; and develop human resources for effective transfer of technology.

2.14.2 International Coconut Genetic Resources Network (COGENT)

COGENT's goal is to strengthen international collaboration in conservation and use of coconut genetic resources, to improve coconut production on a sustainable basis and to boost livelihoods and incomes of coconut stakeholders in developing countries, 39 producing countries with their own research centres and 2 regional organizations working on global conservation. COGENT is a scientific network built for mutual interest and not a formal organization. Decisions are taken during the steering committee meetings, expected to be held annually.

The achievements are as follows:

COGENT multisite International Coconut Gene banks have been established to conserve 200 important accessions in 5 regions: Brazil, Côte d'Ivoire, India, Indonesia and Papua New Guinea.

Development of the International Coconut Genetic Resources Database (CGRD).

Protocols for in vitro embryo culture, cryopreservation, morphometric and molecular marker-based methods, pest risk assessment and germplasm health management, collection, characterization and ex situ and in situ conservation.

Evaluation of 38 high-yielding hybrids in a multilocation trial involving four African and three Latin American/Caribbean countries.

Evaluation of farmers' varietal preferences in 15 countries.

From 1994, the network was involved in more than 288 projects, trainings, meetings and workshop activities. One thousand ninety coconut researchers shared information and technologies. Inadequacies of resources have unfortunately forced COGENT to slow down its activities.

2.14.3 Bureau for the Development of Research on Tropical Perennial Oil Crops (BUROTROP)

BUROTROP is a global network for the development of research on tropical perennial oil crops started in 1990. It is a French-based international organization. The general objective is to strengthen the capabilities of member countries in research and development of coconut and oil palm.

The target beneficiaries are the NARS (35 countries representing 90% of the production for coconut). Through these, the poor smallholders in developing countries who live below poverty level, consumers in developing countries, processors and traders, commercial sector in Europe, development agencies, policy-makers and donors are benefitted.

BUROTROP had 94 active members, 63 individual members and 31 corporate members (11 private companies, 19 public organizations and 1 NGO). In addition to the active members, BUROTROP has many associated members. BUROTROP is not itself a research organization; however, it initiates and coordinates research team groups and networks. It is unfortunately non-functional now.

2.15 Future Strategy

Though the growth of coconut industry is slow, it holds promise in the future as it is such a versatile crop. Availability of quality planting material to meet the increasing demand and good management practices will help in increasing the productivity. Efforts made in product diversification, value addition and by-product utilization, in the recent past, with innumerable high-value products, will make it more competitive and will go a long way in providing food, nutritional, health and social security. Full advantage has to be taken on the positive effects of monolaurin derived from coconut on health effects. There will be growing demand for coconut since its shortage is already felt in many countries like Thailand, Malaysia and Indonesia. International cooperation in research, development and marketing has to be necessarily strengthened. Apart from providing environmental sustainability, coconut can provide livelihood security in the future, provided there is willingness from farmers and strong commitment and political will to develop this industry as a whole and make coconut industry as a sunrise industry.

References

APCC (2015) Coconut statistical yearbook, 2015. Published by Asian and Pacific Coconut Community, Jakarta, Indonesia, p 288

Arancon Romulu N Jr (2013) Market and trade of coconut products. Paper presented at Expert Consultation on Coconut Sector Development in Asia and the Pacific, Bangkok, 30 Oct-1 Nov. 2013

Benjamin-Ripple (2012) Coconut palm sweetners –tripple bottom line sustainability. Paper presented at 45th Cocotech meeting of the Asian & Pacific Coconut Community (APCC) countries held in Kochi from July 2 to 6, 2012

Foale M (2003) The coconut odyssey: the bounteous possibilities of the tree of life. ACIAR Monograph No. 101, p 132

Iyer RD, Rao EVVB, Govindankutty MP (1979) Super yielders in coconut. Indian Fmg 28(10):3–5

Rethinam P (2004) Strategies for enhancing productivity and income of coconut farmers. In: Proceedings of 41st Cocotech meeting of the Asian & Pacific Coconut Community (APCC) countries held during July, 2004, Santo, Vietnam. APCC Publication p 200

Rethinam P (2005) Increase coconut productivity through soil moisture conservation in coconut plantations. Coco Info Int 12(2):5–9

Rethinam P (2006a) Virgin coconut oil – healthy oil for all. APCC Publication p 628

Rethinam P (2006b) Strategic solutions to improve coconut production, processing and value addition. Paper presented in 42nd Cocotech meeting of the Asian & Pacific Coconut Community (APCC) countries in Manila, Philippines from August 21 to 25, 2006

Rethinam P (2008) Current status of organic farming in coconut in the Asia and Pacific countries. Invited paper presented at the national conference on organic farming in horticulture crops, Kasaragod, 14, September 2008

Rethinam P, Idroes A (2006) Generation of creative income and development opportunities through coconut industry in Pacific countries. Paper circulated at the special body on Pacific islands developing countries, 62nd ESCAP session and Asia – Pacific Business Forum 2006, Jakarta, Indonesia, 4–5 April 2006

Rogonon F (2004) Strategies for enhancing productivity and income of coconut farmers. In: Proceedings of 41st Cocotech meeting of the Asian & Pacific Coconut Community (APCC) countries held during July, 2004, Santo, Vietnam. APCC Publication, pp 79–85

Salum UN (2017) Global scenario on production, productivity and marketing of coconut commodities. In: Chowdappa P, Niral V, Jerard BA, Samsudeen K (eds) Coconut. Daya Publishing House, New Delhi, p 440

Dr. P. Rethinam has served as Project Coordinator (Palms); Director, National Research Centre for Oil Palm; Chairman, Coconut Development Board (Government of India); and Assistant Director General (Plantation Crops), ICAR. He has contributed in releasing coconut hybrids and containing root (Wilt) and Tatipaka diseases of coconut. As Executive Director of Asia and Pacific Coconut Community, he intensified the value addition activities of coconut. He has published 215 articles, coedited 2 volumes and edited 37 books. His awards include Konda Reddy Gold Medal and Rolling Shield, Dr. Nathanael Gold Medal, Kalpavriksha Award, Dr. M H Marigowda National Award, Udyan Ratan Award and Lifetime Achievement Awards. palms002@yahoo.com; dr.rethinam@gmail.com

Chapter 3
Botany, Origin and Genetic Resources of Coconut

V. Niral and B. A. Jerard

Abstract The coconut palm, *Cocos nucifera* L., native of the humid tropics, is a versatile, multipurpose palm with slender unbranched stem crowned with a cluster of long green pinnate leaves producing fruit bunches borne on the leaf axils. The stem lacks bark, cambium, and with no secondary growth features, volume once formed remains unaltered. Similarly, the fibrous roots never expand in thickness after formation. The palm neither produces branches nor forms vegetative buds on the stem, the only vegetative buds being those at the apex of the stem and if this bud gets killed, the palm dies. The palm is non-tillering and monoecious, the spadix bears distinct male and female flowers and the fruits take about 10–12 months to mature. *C. nucifera*, though a monotypic species with no known wild/domesticated relatives, still presents considerable intraspecific diversity and heterozygosity. Systematic classification of the genetic diversity in the crop, using the distinct morphological features of the palm, has been attempted by many. The genetic resources in this crop are conserved mainly in field gene banks, in the coconut-growing countries. Various aspects covering botany, morphology, phylogeny, development, varieties, their classification, the variability reported and the genetic resources in the crop are discussed.

V. Niral (✉)
ICAR-CPCRI, Kasaragod, Kerala, India
e-mail: niralv@yahoo.com; niralvittal@gmail.com

B. A. Jerard
Division of Horticulture and Forestry, ICAR-CIARI, Port Blair, Andaman and Nicobar Islands, India
e-mail: jerardba@yahoo.com; jerardba@gmail.com

K. U. K. Nampoothiri et al. (eds.), *The Coconut Palm (*Cocos nucifera *L.) - Research and Development Perspectives*,
https://doi.org/10.1007/978-981-13-2754-4_3

3.1 Introduction

Coconut, *Cocos nucifera* L., a monoecious perennial palm, belongs to the lower group of flowering plants called monocotyledons. It is a monotypic species under the genus *Cocos* (Dransfield et al. 2008) and is placed in the family Arecaceae (formerly Palmaceae) and the subfamily Arecoideae (one of five subfamilies) which includes a total of 27 genera and 600 species. Formerly, the genus Cocos included over 30 species, distributed mainly in Central and South America. Later, Beccari (1916) distinguished *Cocos* as a monotypic genus with *Cocos nucifera* as the sole species and categorised the rest of the species into nine other genera. This systematic classification of *Cocos* was justified by Tomlinson (1961) based on detailed anatomical examination of these genera. The taxonomic information of the coconut palm is listed below.

Kingdom: Plantae
Subkingdom: Tracheobionta
Super division: Spermatophyta
Division: Magnoliophyta
Class: Liliopsida
Subclass: Arecidae
Order: Arecales
Family: Arecaceae
Genus: *Cocos*
Species: *C. nucifera* L

3.2 Origin and Distribution

The name *Cocos*, for *coconut* and its relatives in other European languages, goes back to Spanish word *coco* meaning 'spectre or goblin', with reference to the three marks on the nuts which make it look like an eerie face. The botanical species' name *nucifera* is a neo-Latin formation meaning 'bearing nuts' (*nux* = nut and *ferre* = bring/carry/bear).

In India, coconut is referred to as *narikela* in Sanskrit, and almost all names of coconut in the Indic languages are based on this, e.g. Hindi *nariyal*, Urdu *nariyal*, Marathi *nara*, Parsi *nargil* and in Telugu *Kobbarikaya/nalikeram*. Further, the first element in the Indic languages resembles the several Austronesian names of coconut, e.g. Tagalog *niyog*, Malaysian *nyiur* and Hawaiian *niu*. The Polynesian and Melanesian term *niu* and the Philippine and Guamanian term *niyog* are said to be based on the Malay word *nyiur*. Some of the other vernacular names for coconut include the following: tengu and kobbari in Kannada; nalikeram and thenga in Malayalam; tengai in Tamil; kokosneut in Afrikaans; coco in Catalan, Spanish; kokoyashi in Japanese; kelapa and nyiur in Indonesian/Malay; kelapa in Portuguese; pol in Sinhalese; cot dua in Vietnamese; niu in Hawaiian; maprao in Thai; and so on.

Coconut is considered to be an ancient species with a long history of domestication and cultivation. It is presumed that early humans, while domesticating habitats in coastal areas, started domesticating the coconut palms, resulting in the absence of wild forms in the present-day coconut population. The origins of the coconut palm are a subject of controversy, with two divergent points of thought: most authorities claiming South Asian origin, while others claim its origin in West Coast of Central America and in the adjacent islands of the Pacific. The various arguments and evidences presented in support of the origin of coconut are given hereunder.

Martius (1850) considered the West Coast of Central America as the centre of origin of coconut. This was supported by Cook (1901). On the other hand, De Candolle (1886), in the second edition of *Origin of Cultivated Plants*, opined the origin of coconut in the Indian archipelago. Asiatic origin of coconut has been advocated by Beccari (1917), Mayuranathan (1938) and Nayar (1978). Child (1974) reviewed the literature and postulated the likely origin of coconut in the Old World and that it would not be possible to assign a precise habitat in the Pacific Ocean as the centre of origin of the coconut. Purseglove (1968, 1985) also supported a Pacific origin of coconut. Harries (1978) opined that speculation on the origin, without extra evidence, is futile, considering the widespread distribution of the palm and the divergent opinions implicating a large area from Africa to South America. He postulated that the evolution of the cultivated coconut (niu vai) occurred from the primordial *Cocos*, which he designated as niu kafa. He speculated that the large fruited coconut might have originated from a smaller fruited progenitor and further suggested that the coconut can be compared to the small fruited *Nypa* palm, *Nypa fruticans* Wurmb, and contrasted it with the large fruited coco de mer (double coconut), *Lodoicea maldivica* (Gimel.) Pers.

An origin for the whole Cococeae tribe in western Gondwanaland seems most compatible with the present-day distribution. It has been hypothesised that the tribe probably differentiated shortly before the break up of that supercontinent. Members radiated and became very diverse in the Americas; some rafted on the African and Madagascar Plates, where they survive to the present day; others rafted on the Indian plate, where they are now extinct. With its ability to float, the coconut became independent of plate tectonics for its dispersal. The wild type dispersed by floating between the islands and atolls fringing the continental plates. It is further postulated that the coasts and islands of the Tethys Sea could have been the ancestral home of the coconut, from where it dispersed by floating to other islands in the Indian Ocean and from there into the Pacific (but not into the Atlantic) Ocean (Harries 1990, 1992, 1995).

Gunn et al. (2011) based on extensive genetic analysis, reported genetic differentiation between coconut populations of Indian Ocean and Pacific region and absence of substantial genetic admixture between the two major genetic subpopulations, in spite of the absence of any known reproductive isolating barriers and also extensive movement of coconuts by humans between these oceanic basins, concluded independent origins of coconut cultivation in the Pacific and Indian Ocean basins.

Considering the absence of wild types in coconut (Baudouin et al. 2014) and the fact that *Cocos nucifera* L. is a monotypic species, the origin of this palm has been hypothesised on the basis of historical accounts, fossil records and current status of dissemination and distribution of the palm. Coconut fossils have been discovered in India [10 to 13 cm long *Cocos*-like fruits in the Deccan Plateau (Prasad et al. 2013;

Srivastava and Srivastava 2014) and 3 to 16 cm long *Cocos*-like fruits in west central region (Kaul 1951; Shukla et al. 2012)], Madagascar [coconut shell fragments (Crowther et al. 2016)], New Zealand [3 to 5 cm long *Cocos*-like fruits (Hayward et al. 1960; Endt and Haywar 1997)], Australia [10 to 13 cm long *Cocos*-like fruits (Rigby 1995)], French Polynesia [whole coconuts (Lepofsky et al. 1992)], Vanuatu [coconut endocarp fragments (Spriggs 1984)] and Columbia, South America [25 cm long *Cocos*-like fruits with only 4 mm thickness (Gomez-Navarro et al. 2009)]. Nayar (2017a) observed that among the fossils of *Cocos*-like fruits recovered so far, some of the fossils from India, Australia and Columbia in South America come within the size and morphological characteristics of present-day coconuts, especially natural stands of coconut observed in Nicobar Islands and Lakshadweep (both in India; Balakrishnan and Nair 1979; Samsudeen et al. 2006; Jerard et al. 2017), eastern Philippines (Greuzo 1990) and New Guinea (Dwyer 1938), while those from New Zealand and some material of India appear to be of some other taxa related to *Cocos*. He also hypothesised that the process of domestication or ennoblement of the coconut could have begun in the islands and coastal regions bordering northern Indian Ocean (Arabian Sea, Bay of Bengal), Malesia and the westernmost region of the Pacific Ocean. Nayar (2017b) has reviewed elaborately on coconut origin and domestication in his book *The Coconut: Phylogeny, Origins, and Spread*. Phylogenetic studies in the last few decades do not support an American origin of coconut, in spite of the fact that members of the subtribe Attaleinae are predominantly distributed in South America (Baudouin et al. 2014; Gunn et al. 2015), except for two genera (*Parajubaea*, *Voanioala*) in Madagascar and one (*Jubaeopsis*) in South Africa (Dransfield et al. 2008). The plausible origin of the coconut palm encompasses a vast region, the Indian archipelago, northern Indian Ocean, the Philippines in Southeast Asia, New Guinea and the Western Pacific, covering Northern Indian Ocean, Malesia, Melanesia and Micronesia. The progenitor of present-day coconut is also not known. Phylogenetic studies have indicated some affinity of *Cocos* with that of *Attaleina*, *Syagrus* and *Parajubaea* (Baker et al. 2009; Meerow et al. 2009, 2015).

However, the results have not been consistent, differing between studies (Gunn 2004; Baker et al. 2009; Meerow et al. 2009, 2015; Baker and Couvreur 2013a, b). It has also been postulated that the nearest relative might even have become extinct, since much of this region had been subject to severe geological and climatic perturbations in the past (Prebble and Dowe 2008). Regardless of its origin, coconut has spread across much of the tropics, probably aided by sea currents and also in many cases by seafaring people. *Cocos nucifera* is, hence, regarded as a semi-domesticated species with evolution of local populations having varying degrees of dependence upon humankind.

3.3 Varieties of Coconut

The coconut palm, though a monotypic species with no known wild/domesticated relatives, exhibits considerable intraspecific variability, widely differing from each other in morphological characters, particularly with respect to fruit characters and plant habit.

Considerable diversity is observed in the fruit size, fruit shape and fruit colour of coconut palm. The fruit colour varies from yellow, shades of green and brown to orange. Shapes of the coconut fruit are broadly classified as round, oblong or elliptic. Further, based on the equatorial view, the shape of coconut fruits can be classified as angled, round or flat based on the curvature of the fruit and the presence of ridges on the fruit. Variations are recorded in shape of the nut inside the fruit, and these are broadly categorised as round, oval and oblong.

Based on relative plant habit and a group of morphological characteristics, coconut palms are broadly categorised into two groups, viz. the talls and the dwarfs (Narayana and John 1949). Tall palms grow to a height of 20–30 m, have a sturdy stem, commence flowering 5–7 years after planting and continue bearing up to the age of 80–100 years. Tall palms are normally cross-pollinated and hence highly heterozygous. The fruits are generally medium to large in size and produce higher quantity and good quality copra with fairly high oil content. Among the indigenous Indian tall cultivars, West Coast Tall, East Coast Tall, Benaulim Tall, Tiptur Tall, Andaman Ordinary Tall and Laccadive Ordinary Tall are predominantly grown in different areas. Some popular exotic tall cultivars elsewhere are Fiji Tall, Rennell Tall, Philippines Ordinary Tall, Sri Lankan Tall, West African Tall, Panama Tall, Malayan Tall, Jamaican Tall and San Ramon Tall.

Dwarf palms have gained commercial importance due to their early bearing, short stature, tender nut qualities and resistance to certain diseases. They are of shorter stature, 8–10 m high when 20 years old, and start bearing about 3–4 years after planting and have a short productive life of about 40–50 years. The dwarf palms are more homozygous than talls, due to a high degree of self-pollination. They produce nuts, which are generally small to medium in size. The dwarfs are presumed to have originated from talls either through mutation or by inbreeding (Swaminathan and Nambiar 1961; Purseglove 1968). The popular dwarf cultivars grown in India are Chowghat Green Dwarf, Chowghat Orange Dwarf, Kenthali Orange Dwarf and Gangabondam Green Dwarf. Among the exotic dwarf cultivars, Malayan Yellow Dwarf, Malayan Orange Dwarf and Malayan Green Dwarf have become popular in most of the coconut-growing countries of the world. Further, selection and hybridization in nature resulted in intermediate types.

The tall and dwarf types have been utilised for development of hybrids, combining the early flowering trait of dwarfs with the hardiness and high-yielding character of tall parents, simultaneously exploiting hybrid vigour.

3.3.1 Native Varieties and Classification Systems

The earliest documented reference to the native varieties of coconut was in Rumphius' *Herbarium Amboinense* (1741), which listed 13 varieties from the Netherlands Indies. Subsequently, numerous attempts have been made to document the different forms of coconut. Miquel (1855) listed 18 varieties with Latin names and descriptions, which was used by Dutch workers to refer to the different coconut populations in Dutch colonies. Different varieties of coconut in the Philippines, along with a few Latin varietal names, have been reported. In the 1950s, FAO initiated a questionnaire-based survey and documented the different coconut types in 18 countries (Mao 1959). In 1970, FAO published a list of descriptors of coconut on the basis of information obtained from 30 independent sources (Harries 1970).

Narayana and John (1949), acknowledging the need for categorisation as essential for proper understanding of the variations and forms of coconut, attempted the first systematic classification of the known coconut varieties (Table 3.1). They categorised coconut varieties into two groups, viz. talls and dwarfs. The talls were further differentiated into three varieties and nine forms, while the dwarfs were differentiated into two varieties and two forms. These varieties and forms were named mostly after the countries from which they were originally obtained or where they are largely grown or after a particular distinguishing character.

Table 3.1 Varietal classification system proposed by Narayana and John (1949)

Variety	Forms	Important features
Talls		
Spicata	–	Inflorescence unbranched (spikeless), with numerous female flowers and few male flowers
Androgena	–	Male tree, inflorescence contains numerous male flowers, does not produce female flowers and nuts
Typica	Ramona	Large nuts and high copra content
	Kappadam	Robust with high tender nut content
	Gigantea	Robust with majestic appearance, large fruits, high copra content, less nut yield
	Nova-guineana	Robust with stout trunk, massive crown with large number of long leaves and bunches, susceptible to fungal disease/insect pests
	Cochin-chinensis	Robust with stout trunk, large number of leaves and bunches, thin kernel, susceptible to fungal disease/insect pests
	Malayensis	Stout trunk, poor yield, green nuts, less copra/oil content, sweet tender nut water and with peculiar aroma, pink colour at base of buttons
	Siamea	Comparatively short with stout trunk, compact crown, medium-yield, green nuts, tender nut water sweet, high copra oil content
	Laccadive	Regular bearer, high yield, medium-size fruits, good quality and quantity of copra, good toddy yielders
	Pusilla	High female flower production, high setting percentage, small size fruits, good quality copra, very high copra oil content, suitable for ball copra production
Dwarfs		
Nana	Nana	Dwarf, short trunk, high yield, good quality and quantity of copra, suitable for tapping
	Maldiviana	Dwarf, short thin trunk, small crown, short leaves, sweet tender water, poor quality of copra
Javanica	–	Intermediate between tall and dwarf, robust, short trunk, prolific and early bearer

This method of classification was quoted by Gangolly et al. (1957) in their review of literature on varieties of coconuts, which also formed the basis of the chapter on varieties by Menon and Pandalai (1960), but being cumbersome was not used for classifying the coconut varieties further. Since the varieties from the Pacific Islands, Africa or America were not included, this classification was not considered comprehensive.

Liyanage (1958) adopted the classification of Narayana and John (1949), primarily based on fruit size and appearance, to classify the varieties of Sri Lanka. He reduced the coconut varieties to three – typica, nana and aurantiaca – by not including androgena and spicata varieties. The new variety, aurantiaca, was considered to be semi-tall, distinct from either the tall or the dwarf groups. The forms corresponded with some of those in the Indian system but had different names. The variety typica had eight forms, viz. typica, kamandala, bodiri, navasi, ran thembili, gon thembili, pora pol and dikri pol; variety nana had three forms, viz. pumila, eburnea and regia; and aurantiaca had two forms, namely, King Coconut/rath thembili and navasi thembili. Carlos (1963) identified four botanical varieties, viz. typica, nana, javanica and spicata in the Philippines. McPaul (1962) grouped the coconuts of Fiji under two groups, talls and dwarfs. Rattanapruk (1970) classified the coconuts in Thailand into three varieties based on the nut characteristics and age at fruiting.

Fremond et al. (1966) put forth another system of classification, on the basis of pollination characteristics. They divided the varieties into two groups, viz. autogamous or self-pollinating as in the case of most of the dwarfs and allogamous or cross-pollinating as in the talls. However, this classification also has limitations as the dwarfs are easily cross-pollinated when surrounded by talls, and talls are also capable of self-pollination (Whitehead 1965a; Rognon 1976). Further, this method of classification requires testing of pollination behaviour of the local forms from different countries under controlled condition, which presents a lot of practical difficulties.

Harries (1978) put forth a method for practical identification of coconut varieties to enable comparison of varieties irrespective of their country of origin and the conditions under which they grow. He advocated the use of fruit component analysis for characterising and classifying varieties. According to Harries, the fruits of a palm being the physiological sinks are not only the most interesting but also the most uniform despite the exposure of palms to variations in the growing conditions. He further reasoned that by considering the relationship between the fruit components rather than absolute values, the effects of fruit size and fruit number are reduced. Based on these assumptions, Harries (1978) recommended a system of classification based on contrasting the proportion of husk in the fruit with the fresh weight of the fruit. He introduced the concept of coconut variety identification based on the niu kafa-niu vai-introgression method. According to Harries, the niu kafa type represents the coconuts, which have evolved through natural selection in uninhabited islands and coral atolls. These coconuts are named after a variety characterised by long angular, thick-husked fruits. It has the capability of slow germination, which facilitates its survival under natural conditions. The niu kafa type being large fruited is suited for copra production and coir processing and has therefore been introduced into inland areas for commercial cultivation. As a result the identity of its distinctive natural habitat has been lost.

The niu vai type, on the other hand, derives its name from a variety used as a source of sweet uncontaminated water. According to Harries (1979), the coconut palm was first used by humans as a source of water. Therefore, this selection pressure led to an increase in the volume of the liquid endosperm in the immature fruit. This caused a change in the fruit characters, with these being spherical with larger inner cavity and lesser husk. The competition for light and space resulted in these nuts developing the ability to germinate quickly and producing vigorous seedlings. Harries did not include the dwarf varieties under niu vai and niu kafa types as these evolved much later and can survive only under cultivation. The two ancestral types maintain their distinctive characteristics in the Central American isthmus, where these populations are geographically isolated (Richardson et al. 1978). However, in Asia and Pacific, human intervention through migration and cultivation has brought these forms together, and the resulting opportunity for cross-pollination has allowed the development of intermediate forms through introgression. This introgressive hybridization has resulted in the development of many of the cultivated types available today.

Harries (1981) has studied and classified a few of the present-day coconut populations into niu vai and niu kafa types. According to him, the West African Tall and the coconuts along the Atlantic and Caribbean coast of America, such as Jamaica Tall, belong to the niu kafa type, while the Malayan Tall belongs to the niu vai type. He considered the Tahiti Tall to be a variable population, as a result of greater introgression between niu vai and niu kafa types. The coconut populations in most islands in the Pacific Ocean are introgressed forms similar to the Tahiti Tall but sufficiently different to warrant distinct names. Within these introgressed types, he observed that niu vai and niu kafa types occur as minor variants. Harries (1981) concluded that the techniques such as the electrophoresis of polymorphic enzymes would show the relationship between varieties and anticipated that these studies would support the niu kafa-niu vai-introgression theory.

Satyabalan (1997) utilised the then available information on known varieties/cultivars of coconut to comprehensively classify them under different groups. On the basis of the ratio of fruit component traits, tall coconut varieties were categorised into five groups, while dwarf varieties were categorised into three groups under each of the colour groups, viz. green, yellow and red/orange: Group I corresponding to small nuts with less copra content, Group II with medium to large fruits with medium copra content and Group III with medium to large fruits and more copra content (Table 3.2). This classification did not consider the brown dwarfs and the Niu Leka Dwarf which has a much larger fruit with higher copra and shell content showing similarity to palms of the tall variety and unlike the normal green, yellow and orange/red dwarfs.

Therefore, the widely used method of classifying coconut varieties is on the basis of their morphology and growth habit. The coconut varieties are generally classified as tall and dwarf, prefixed by the name of the country of origin.

Table 3.2 Varietal classification by Satyabalan (1997)

Variety	Groups	Important features	Regional distribution/example of varieties
Talls			
Talls	Group 1	Very large fruits, spherical or ovoid shape, thin husk; husked fruit large, spherical, with thin meat, more water and thin shell; copra content >300 g, <30% of husked fruit weight	Southeast Asia, Oceania and America, Thailand Tall, Bali Tall, San Ramon Tall, Malayan Tall, Rennell Tall, Rotuma Tall, Tahiti Tall, Panama Tall, Ecuadorian Tall
	Group 2	Large fruits, spherical or ovoid shape, thin husk; husked fruit large, spherical, with thin meat, more water and thick shell; copra content 200–300 g, <30% of husked fruit weight	Southeast Asia, Oceania and America, Africa; Thailand Klarng, Borneo Tall, Tenga Tall, Philippines Laguna Tall, Philippines Lono, Solomon Tall, Fiji Tall, Natava Tall, Rangiroa Tall, Surinam Tall, Nigerian Tall
	Group 3	Large fruits, spherical or ovoid shape, thin husk; husked fruit large, spherical, with thick meat, less water, thick shell; copra content 200–300 g, >30% of husked fruit weight	All coconut-growing regions Park Choke Tall, Guam Tall, Polynesian Tall, Jamaican Tall, St. Vincent Tall, Atlantic Tall, West African Tall, Zanzibar Tall, Ceylon Tall
	Group 4	Medium-sized fruits, spherical or ovoid shape, thick husk; husked fruit medium, spherical, with thick meat, less water, thick shell; copra content <200 g, >30% of husked fruit weight	All coconut-growing regions Standard Kudat Tall, Philippines Dalig Tall, Solomon Tall, New Hebrides Tall, Blanchisseuse Tall, West African Tall, Mozambique Tall, Kenya Tall, Gon thembili, Laccadive Ordinary Tall, Laccadive Micro Tall, Benaulim Tall
	Group 5	Medium-sized fruits, spherical or ovoid shape, thick husk; husked fruit medium, spherical, with thin meat, more water, thick shell; copra content <200 g, <30% of husked fruit weight	All coconut-growing regions Park Choke Tall, Kong Thein Yong, Klapawangi, Fiji Tall, Niu Ui, Solomon Tall, Kiriwana Tall, Kavieng Tall, Surinam Tall, Seychelles Tall, Indian East Coast Tall, Indian West Coast Tall
Dwarfs			
Green Dwarfs	Group I	Small fruits; husked fruits with high percentage of shell (26–39) and copra (32–43)	Chowghat Green Dwarf, Ayiramkachi, Pumilla
	Group II	Medium-sized fruits; husked fruits with lesser percentage of copra (23) and slightly lesser shell (23–27)	Nam Hom (aromatic coconut), Nok Koom
	Group III	Medium-sized fruits, husked fruits with slightly higher percentage of copra (25–35) and lesser shell (20–25)	Equatorial Green Dwarf, Guinea Green Dwarf, Mu-se-keo, Thailand Green Dwarf, Thungkhled, Pathiu Green Dwarf; Malayan Green Dwarf, Gangabondam Green Dwarf

(continued)

Table 3.2 (continued)

Variety	Groups	Important features	Regional distribution/example of varieties
Yellow Dwarfs	Group I	Medium small fruits; husked fruits with high percentage of shell (20–29) and copra (24–35)	Malayan Yellow Dwarf
	Group II	Medium-sized fruits; husked fruits with lesser percentage of copra (21–30) and lesser shell (16–19)	Ghana Yellow Dwarf, Malayan Yellow Dwarf
	Group III	Medium-sized fruits; husked fruits with slightly higher percentage of copra (24–31) and slightly lesser shell (21–27)	Malayan Yellow Dwarf, Nari-kay
Red/Orange Dwarfs	Group I	Medium small fruits; husked fruits with high percentage of shell (20–29) and copra (24–35)	Malayan Red Dwarf, King Coconut, Mapro fire, Thalai Roi
	Group II	Medium-sized fruits; husked fruits with lesser percentage of copra (21–30) and lesser shell (16–19)	Malayan Red Dwarf
	Group III	Medium-sized fruits; husked fruits with slightly higher percentage of copra (24–31) and slightly lesser shell (21–27)	Cameroon Red Dwarf, Malayan Red Dwarf, Chowghat Orange Dwarf

In spite of the absence of wild relatives, the present-day population of the coconut palm presents a wide range of variability broadly grouped into two groups – talls and dwarfs – on the basis of a few important characters like stature, growth characteristics of the palm, precocious nature in flowering and nut and copra characters (Table 3.3). This is the widely accepted classification used for distinguishing coconut cultivars. The talls are the most commonly cultivated ones for commercial production in all coconut-growing regions, while the dwarfs are increasingly being grown for their aesthetic value, as parents for production of high-yielding hybrids as well as for tender nuts.

To conclude, plant habit, fruit colour and other fruit characteristics are presently the most convenient for grouping of the varieties. In most cases, these traits are genetically determined and not solely the effect of environment. However, further refinements are possible for a more fool-proof classification system. Based on molecular marker studies, the present-day coconut populations have been classified into two major groups: the Pacific group with five subgroups (Southeast Asia, Melanesia, Micronesia, Polynesia and the Pacific coast of Central and South America) and the Indo-Atlantic group. The Pacific group includes the domesticated coconut, while the Indo-Atlantic group includes niu kafa coconut types. However, human intervention through migration and cultivation has brought the different forms together and the resulting opportunity for cross-pollination has allowed the development of intermediate forms through introgression. Presently, there is no

Table 3.3 General characteristics of tall and dwarf coconuts

Trait	Tall	Dwarf
Stem height	Relatively more at a given age	Relatively less at a given age
Stem circumference	Sturdy, with bole at base	Thin, without bole at base
Leaf	Longer, wider leaf base, strong attachment to the stem	Shorter, narrow leaf base, weak attachment to the stem
Leaflets	Longer and broader	Shorter with relatively lesser width
Initiation of flowering	Late (5–7 years)	Early (3–4 years)
Mode of pollination	Predominantly cross-pollinated	Predominantly self-pollinated
Colour of fruits and petioles	Generally mixtures of greens and browns in the population	Green, yellow, red-yellow (orange) or brown in the population
Arrangement of leaf scars on the trunk	Widely spaced	Closely spaced
Fruit size	Very small to very big	Small to medium
Copra	High quantity and better quality	Lesser quantity and poor quality
Tender nut water	Low to medium	Medium to high, sweet water
Phenotypic variation	High	Low
Within cultivar	High	High
Between cultivars		
Leaf and bunch attachment	Very strong	Fragile
Root distribution	Generally more dense and plentiful	Less dense and few
Productive life span	About 60 years	About 40 years

single method of classification that can account for all the variability observed in the global coconut populations.

In addition, to the variation described above, variation in coconut endosperm texture and quality, varieties with aromatic water, thick-shelled nuts, thin-husked fruits, pink-husked fruits, persistent inflorescence and leaves have been reported in different natural coconut populations which are dealt under Sect. 3.6. Besides these, a number of abnormalities in stem, vegetative parts, inflorescence and fruit shape have been reported from different coconut-growing regions. These abnormalities, elucidated under Sect. 3.6, are freaks of nature and not distinct varieties, from the botanical point of view.

3.4 Botany and Morphology

The coconut palm is a monocot. The single stem of the palm lacks bark, cambium and secondary growth features, characteristic of gymnosperms and dicotyledons, and hence the stem once formed never alters in thickness, except for a slight shrinkage when the stem gets old. Similarly, the root lacks a taproot, and the roots once formed never grow in thickness. The palm is unbranched and does not form vegetative buds on the stem. The features of the different parts of the palm are described below.

3.4.1 Root Growth and Development

The palm being a monocot has an adventitious root system. Roots emerge from the base of the stem (bole) and continue to be produced throughout the life of the palm. The number of roots varies with age of the palm, girth of the bole, soil fertility and management. It ranges from 1500 to 7000 (Sampson 1923; Copeland 1931; Patel 1938) and in rare instances as many as 11,360 roots have been recorded (Menon and Pandalai 1960). The main roots form a number of secondary roots which branch profusely forming a large absorptive surface through which the palm takes in nutrients from the soil. However, these rootlets are short-lived and are frequently replaced. The roots do not have root hairs. From the main roots and rootlets, numerous pneumatophores develop and serve as breathing organs, facilitating gaseous exchange between the roots and the atmosphere. The tender growing tip of the roots is protected from injury by a root cap.

The main roots of the palm are uniform in size and long-lived. They generally measure around 5–10 metres in length and 8 mm in diameter. However, occasionally, longer roots, up to 25 m in length, have also been reported (Menon and Pandalai 1960). The growing root is initially yellowish-white in colour and gradually turns light red and subsequently becomes reddish-brown with age. Thus in a fairly old and growing root, gradations in colour are generally visible. Coconut roots cannot indefinitely grow into water or withstand continuous water stagnation and submerged portions of the root get decayed. However, when the water table recedes, these roots produce new branch roots and/or rootlets (Sampson 1923; Menon and Pandalai 1960; Ohler 1984). The roots of the coconut palm, in general, can live for many years. However, the rootlets have a very short life span, as influenced by the ecological conditions in the root zone.

The coconut root, both the main roots and rootlets, has three regions, distinct in their internal structure, viz. the root cap, the growing root tip and the absorbing region, a little distance away from the root tip. However, the tissues in rootlets are much smaller and less defined and carry less conducting tissues, especially xylem vessels.

The tip of the root contains the root cap (8–26 layers of dead tissue), covering the tender growing portion of the root and the meristem called calyptrogen, which is responsible for the formation of the root cap. The growing root tip, immediately behind the root cap, is undifferentiated and includes the epidermis and uniform meristematic cells which later on differentiate into various tissues. The cells in the absorbing region of the root are differentiated, with three distinct layers, viz., the outer single-layered epidermis, the inner central stele and the thick cortical region in between the epidermis and the stele. The epidermis is the main absorbing tissue, and the cells are much larger, multinucleate (two to four nuclei), thin-walled and short-lived and wither away as soon as the hypodermis attains an impervious sclerenchymatous condition.

The central stele accounts for about one-fifth the diameter of the root and comprises of conducting tissues bordered by a single layer of living cells, the pericycle.

The rootlets and pneumatophores emerge from the pericycle. Immediately beneath the pericycle are 30–55 radically arranged groups of small dividing cells called the procambial strands, which differentiate into the xylem and phloem. The xylem vessels (are large) alternate with the phloem vessels (are inconspicuous). Often, two xylem vessels coalesce to form a bigger vessel, and a mature root may contain about 25 large metaxylem vessels. The companion cells of the vascular bundles, which are initially thin walled, later develop into thick-walled sclerenchymatous bundle sheath and contribute to the tensile strength of the stele. In the centre of the stele is the narrow pith, comprising of living cells, which subsequently disintegrates as the root becomes old. In between two rows of phloem and xylem, there is a narrow ray of cells, which connects the inner ground tissue bordering the central hollow with the pericycle (Patel 1938).

The cortical region accounts for the major portion of the root and comprises of the hypodermis/exodermis, endodermis and mesodermis. The hypodermis lies immediately below the epidermis and comprises of several layers of small and thin-walled cells, which subsequently becomes thick walled, and the epidermis loses its absorbing capacity. The hypodermis protects the central stele by its thick and pliable covering. The central mesodermis consists of large-sized and loose parenchyma enclosing many air spaces and includes tannin cells as well as some empty cells of unknown function. The endodermis has a single row of small cells, whose distinct identity is maintained even in older roots. The inner wall of the endodermal cells is thick as compared to the lateral walls, while the outer wall is thin. A few thin-walled passage cells exist in this layer (Menon and Pandalai 1960).

3.4.2 Stem

The coconut palm has a single, straight stem, greyish in colour, topped by a crown of leaves. It has a single terminal bud, also called the 'cabbage', the death of which results in the death of the palm. The stem is marked by leaf scars. The thickness of the stem is determined by its vigour and soil conditions, in addition to varietal differences. In certain varieties, the base of the stem referred to as the bole is swollen. The stem of the coconut palm becomes visible once the bole reaches the full stage of its development. Under favourable conditions, the trunk in a young palm is formed within 3–4 years. In the initial years, the stem gradually becomes thick, and once the maximum size is reached, there is not much change in the stem girth with age. In the tall variety, the base of the trunk (bole) is up to 0.8 m in diameter, tapering quickly to about 0.4 m (Child 1974). However, the stem becomes thinner as the palm grows old as well as under unfavourable growth conditions.

The length of the stem is determined by the age of the palm, variety and ecological conditions. Dwarf varieties have shorter trunks than tall varieties. Palms under unfavourable management practices, including under planting with excessive shade and very close planting, exhibit rapid stem elongation. Stem growth is fastest at early stages, with annual height increment of as much as 1.5 m $year^{-1}$. The incremental growth reduces and the stem tapers down over the years and levels off as the palms grow old, with annual stem elongation of about 10–15 cm from the 40th year. In rare instances, palms with branching, due to damage to the terminal bud, are observed, and up to five branches have been reported (Davis 1969). The stem, being predominantly fibrous, combines stiffness with adequate suppleness and is tough enough to withstand considerable lateral strain, particularly when exposed to severe winds/gales, due to the presence of the numerous vascular bundles (as much as 18,000 in a mature palm).

The stem is derived from the terminal bud, which is approximately 0.5 mm × 0.5 mm in size, visible as a small protuberance at the apex of the trunk and well protected by the leaves in various stages of development. The growing point comprises of a mass of minute cells with dense cytoplasm and large multinucleolate (four to ten nucleolate) nuclei. This region is three to four layered and has meristematic activity. Immediately below this region, in the tender stem, the cells are large and contain numerous procambial strands along with large starch granules and sugar reserves. These procambial strands later differentiate into xylem and phloem bundles. The bundles derived from the procambial strands are closed ones and are scattered throughout the stem. The first formed bundles are comparatively short-lived but bigger than those formed subsequently from the ground meristem. The first formed protoxylem vessels are smaller than the subsequently formed metaxylem. In a young stem, three to nine xylem vessels are observed, and these form a V-shaped tissue, with the base towards the centre of the stem. Small groups of xylem vessels are seen outside or above the xylem, and both are enclosed within a layer of thin-walled cells that subsequently thicken with age and form the fibrous sheath. However, a few cells in certain patches of the fibrous sheath as well as the parenchyma close to the xylem and phloem are not thickened even in old stems. This is mainly confined to the apex of the bundle as a broad round tissue, and the base of the bundle is practically without the fibres. The xylem tissue, V-shaped when young, changes shape with age. The xylem may be found in two or three groups. However, big bundles with many xylem vessels are rare. The most common type of bundle in the old stem is the one with only one big xylem vessel, a small phloem tissue above and a broad fibrous tissue lying at the outermost of the bundle. The vascular bundles towards the centre are more widely spaced than towards the periphery. The number of vascular bundles in the periphery is also few or negligible, instead large numbers of small groups of fibres derived from the ground meristem are observed. Between the central part of the stem and the periphery, the vascular bundles are closely packed with few layers of intervening cells. Occasionally, the separating parenchyma is absent, and therefore, two bundles coalesce into a double bundle. The size of the vascular bundles increases considerably with the age of the palm. The xylem and phloem are enclosed by cells that are initially thin walled but subsequently thicken and form the fibrous sheath.

The cortex comprises of several layers of thin-walled parenchyma and a single layer of epidermis with thicker outer wall. In the parenchyma, tannin cells and air passages occur, mostly towards the periphery. Both the stem and the root lack the periderm or corky tissue seen in most of the plants. However, the periphery of the stem contains a peculiar storied type of cork cells, called rhytidome.

There is no secondary thickening of stem in coconut palms as they lack bark and cambium (the living and growing tissue between the bark and wood). However, the stem of a seedling is thin, while that of an adult palm is stout. The thicker stem as the seedling grows older is the result of increased meristematic activity of the growing point resulting in the formation of more and more cells and vascular bundles, thereby forming the bole, which appears as an inverted cone. Once the bole is formed, the stem of almost uniform girth begins to appear.

3.4.3 *Leaf*

The first few leaves of a growing coconut seedling have the pinnae fused together and appear as entire leaves. After about six to eight leaves have been formed, the subsequent ones tend to split into leaflets. By about 3–4 years, the trunk becomes visible with a single terminal growing point, from where new leaves develop.

The adult coconut palm bears a crown of leaves at the apex, comprising of the opened leaves and those surrounding the bud in various stages of development. The number of leaves in the crown varies with the variety and ecological and cultural conditions. Generally, the crown of an adult palm carries, in addition to the opened leaves, a similar number of leaf primordia, in different stages of differentiation. These leaves belong to four distinct sets. The first set comprises of the oldest 10–12 leaves, from the axils of which fruit bunches have been harvested. The next set comprises of the next older 10–14 leaves, supporting fruit bunches in different stages of development. The third set includes the younger 10–12 opened leaves, with spadices in various stages of development in their axils. The last set comprises of leaves in the cabbage with the outermost ones in different stages of unfurling and the remaining which have not yet emerged.

The leaves are long and vary from 3 to 6 m in length depending on the variety, age of the palm, soil fertility and vigour of the palm. The individual leaf consists of a strong petiole, extending to form a rachis with numerous leaflets (150–250) inserted on it. The leaflets on either side of the rachis are not exactly paired, with one-half having about two to ten leaflets more than the other side. The leaflets are also long (60–150 cm), narrow, linear, tapering and lanceolate. The leaves when young retain the stipules at their bases, forming a fibrous tissue that more or less surrounds the whole stem. As the leaf becomes older, the stipules dry and fall away. In young palms, the stipules persist till the leaf dies.

The progression of leaf development from the minute bud to the adult leaf takes several years. On an average, the time taken from inception to final abscission of the leaf is almost 5 years. Different workers have explained the stages in development of the leaf (Patel 1938; Padmanabhan 1963), formation of leaflets (Venkatanarayana 1957) and sequence of events involved in differentiation of leaf primordia into adult leaf (Periaswamy 1965). The leaf primordium is differentiated almost 28–32 months prior to its emergence from the leaf sheath. Under favourable conditions, the leaves of good bearing palms remain on the crown for 3–3.5 years after emergence. Generally, the life span of leaves is lesser in poor bearers than medium and heavy bearers (Patel 1938). In addition, the season and also soil conditions affect leaf shedding. Satyabalan (1993) observed that high yielders bear higher number of leaves on the crown indicating the longevity of the leaf as well as the drought tolerance of the palm. In regions where the seasons are more marked, there is a considerable cyclic variation in the rate of leaf opening which appears to be more dependent on the temperature rather than the rainfall. Generally, 1-year-old seedlings contain seven to nine leaves, with majority of the seedlings having eight leaves. The number of leaves on the crown increases to 30–35 during maturity, and an adult palm on an average bears about 30 leaves.

The length of the leaf varies from 4 m in dwarfs up to 7 m in talls, depending on variety, growing condition and age of the palm. The petiole accounts for about one-fourth of the total length of the leaf but varies with the variety. The petiole continues as the midrib of the leaf. A short and stout petiole is able to better withstand vertical pressure exerted by the developing bunch in its axil. the leaflets are borne on either side of the midrib and have differential length based on their position in the leaf. The first leaflets at the base are short, followed by a gradual increase in length of the subsequent leaflets with the maximum length being achieved at about one-third of the midrib followed by a graded decline in length towards the top of the leaf. The smallest leaflets at the tip are about 25 cm in length, while the largest leaflets measure about 80–120 cm. The number of leaflets in a mature palm leaf ranges from 200 to 250.

The leaves in a coconut palm are so arranged as to ensure maximum light availability to each leaf. They are arranged in five spirals, running either in the clockwise or anticlockwise direction, and spirality in a palm can either be right-handed (bunches hang towards the right of the petiole) or left-handed but remains the same throughout the life of the palm. Based on the phyllotaxy, the 6th leaf is positioned over the first leaf, the 11th leaf over the 6th and so on. The two types of spirals are distributed almost equally in a population, with a slight preponderance of the left spiral (Davis 1962; Louis and Chidambaram 1976).

Patel (1938) reported that the number of leaves in the unopened cabbage is 1 to 1½ times the opened leaves on the crown of the palm. Leaf primordium differentiation occurs 28–32 months prior to the emergence from the leaf sheath as a small indistinct protuberance on the side of the growing point. Within a month, it undergoes repeated periclinal and anticlinal cell divisions and develops into a seven to ten cell-layered fingerlike structure having three layers of primary meristem, viz. the outermost dermatogen, the middle three cell-layered periblem and innermost layer plerome. In the succeeding month, the rapidly growing protuberance assumes the shape of a hollow cone enclosing the next younger leaf and the growing point. At this stage, procambial initials, which subsequently form the vascular bundles, appear in the plerome. The developing leaf initially contains only the petiole portion, and differentiation of the leaflets occurs 8 months later, in approximately the seventh rudimentary leaf from the growing point. Venkatanarayana (1957) observed the formation of 18-layered rectangular mass of cells united at the margin that develops into leaflets, followed by subsequent longitudinal split in the middle of this tissue, forming the sides of the leaflets. The epidermis of the leaflets, both upper and lower, is derived from only the plerome tissue and not from the dermatogen. The inception of plications and the separation of leaflets in palm leaves is on the basis of vertical and transverse growth stresses (in the expanding and elongating lateral faces of the rachis) acting at the base of the nonplicate lamina which bears alternating lines of weaker parts between parallel rows of provascular strands whose cells elongate at right angles to the forces of vertical stress and resist them (Padmanabhan 1984).

The procambial strands of the leaflets are formed after the xylem vessels of the first formed vascular bundles of the petiole and the main rachis have thickened. Within the leaflet, the procambial strands of the midrib are formed first, followed by those of the sides 2 months later. This is followed by the formation of the cross veins. In a mature leaflet, there is a strong central midrib from which slant down the two sides of the leaflet. The epidermis is observed on either side with a thick outer cuticle. The lower epidermis is thinner than the upper epidermis. Below the upper epidermis are two layers of hypoderm that serves as water storage tissue, while beneath the lower epidermis is a single broken layer of hypoderm. Below the upper hypoderm and spreading almost to the lower hypoderm is the broad palisade tissue comprising of thin walled, elongate and closely packed cells. Below the palisade tissue is a scanty spongy parenchyma. The chloroplasts are formed in the mesophyll just prior to the emergence of the leaf. Each leaflet contains about 20–25 vascular bundles running all along its length and occupying the entire thickness of the leaflet from the upper to lower hypoderm. Of these, five to six bundles are big while the rest are narrow. In between two large vascular bundles and just above the lower hypoderm are three small vascular bundles made up of a few tracheids, three to four phloem elements and a small group of fibres. The top of the petiole contains as many as 500–1000 diminutive vascular bundles alternating with small group of fibres, which provide additional mechanical strength to the leaf. The upper epidermis is highly cutinized. On the lower epidermis, numerous multicellular, dark brown, short-stalked scales containing tannin (trichomes) occur at regular intervals in small depressions. In addition, bundles of crystals of raphides are seen all over the lower epidermis. The stomata are first formed about a year prior to the emergence of the leaf and are initially confined to the margin of the leaflets away from the midrib, while the vast majority of stomata are formed just about 3 months prior to the emergence of the leaf. The stomata are confined to the lower surface of the leaflets and are distributed all along its length in two to four longitudinal rows in between every two vascular/fibre bundles. The stomata are elliptic with two guard cells containing large starch grains and a small opening. The guard cells are bordered by subsidiary cells, and a small square cell at either end of the stoma joins these subsidiary cells. The stomata are fairly large with a dimension of 38 $\mu \times 19\ \mu$, extending to 38 $\mu \times 40\ \mu$ along with the subsidiary cells. Normally, a leaflet contains about 170–220 stomata mm^{-2}. However, stomatal density is a varietal character. Dwarf varieties, in general, have higher stomatal density than talls (Satyabalan

1993; ICAR-CPCRI 2015). The mature stomata are tetracytic, with both the polar and the lateral subsidiary cells cut off by the neighbouring protodermal cells. Each lateral subsidiary cell is the result of a longitudinal division of a trapezoid cell formed by two oblique divisions of a lateral protodermal cell. The ontogeny of stomata conforms to the perigynous type with nonintersecting oblique division (Ghose 1979).

The midrib of the leaflet is a very strong structure with a central ring of seven to eight vascular bundles enclosed within a fibrous sheath formed by two to three rows of thickened cells. The outer epidermis is highly cuticularised and is continuous with that of the sides of the leaflets. Between the epidermis and the fibrous sheath are two layers of hypoderm (continuous with the upper hypoderm of the sides of the leaflet) followed by two to three layers of parenchyma. The attachment of each half of the leaflet to the midrib is narrow and tapering, but the place of attachment is strengthened by a group of fibres. There is also a special motor tissue bordering the inner epidermis that is continuous with the lower epidermis of the leaflet. Behind the motor tissue are a vascular bundle and a patch of sclerenchyma. In young leaflets, this motor tissue is nonfunctional and comprises of smaller, much compressed, thin-walled elongated cells. By the time the leaflets are about to open, the motor tissue develops considerably and consists of two rows of big cells. These cells absorb water supplied by the vascular bundle behind it and become turgid, thereby causing a lateral push to be exerted along the sides of the leaflet, which gets separated from its oppressed position in the bud and remains at an angle. The bundle of fibres at the base of the attachment of the sides act as a pivot on which the side moves. However, as the epidermal folds are highly cutinised, sudden movements are checked. The outer margins of the sides of the leaflets are peculiarly folded somewhat like a hook; one margin is folded on the upper epidermis and another on the lower so that both point towards the same direction. At the bend of the fold, the epidermis is in short folds, and immediately behind the fold is a motor tissue similar to the one at the base of the sides of the leaflets. The cells in the motor tissue are initially narrow, elongated and much compressed but are developed at the time of unfolding of the tissue. When the cells become big and turgid, they exert an outward pressure, and as a result, the fold is straightened, and the interlocked margins are released, thereby helping the leaflets to unfold. This motor tissue at the margin of the leaflets persists in some of the older leaves, while in others it is lost and the epidermal folds straighten out and therefore the margin becomes a truncate structure instead of a bent one. In addition to these two motor tissues, two other motor tissues are seen at the base of the leaflet. Of these two, one is located at the angle made by the inner or upper half of the leaflet on the main rachis and helps to spread out the leaflet as a whole, while the other located on the inner side of the outer half of the leaflet at the place of attachment on the main rachis helps keep the outer half of the leaflet in position. Thus, there are four different motor tissues in the leaflet – one pair running the whole length of the midrib, one at the margins and two at the base.

The petiolar sheath is first visible as a soft wing on either side of the petiole, about a year after the leaf is differentiated. The young sheath is made up of a mass of parenchyma forming the ground tissue in which are scattered a few procambial strands. As the leaf matures, these strands develop into vascular bundles of considerable size and length with a broad mass of sclerenchyma, like those in the stem. In addition to the vascular bundles, isolated groups of fibres occur all over the sheath, especially towards the periphery. A narrow cortex is also visible in the young sheath. The old sheath consists of closely woven fibres and vascular bundles with a mass of sclerenchyma and without much of the thin-walled ground parenchyma. The importance of the mechanical tissue of the petiole sheath is evident, as at an early stage it encircles the stem and partly bears the weight of the leaf in the developing bud, and as the petiole of the other young leaves enclosed in the sheath develops, it is gradually torn up. The thickness of leaf sheath, fibre of weft and warp strands has relevance to adaptation, geographical affinity, pollination system and taxonomic forms (Arunachalam et al. 2005).

3.4.4 *Inflorescence*

The coconut palm is monoecious with distinct male and female flowers borne on the same inflorescence. The inflorescence emerges from the leaf axils, and in adult palms under favourable conditions of growth, one inflorescence is produced every month from successive leaf axils. The age at initial flowering varies with the variety as well as the growing conditions and ranges from 3 to 7 years after planting the seedlings in the main field. Dwarfs in general commence flowering earlier than the Talls. The inflorescence, referred to as a 'spadix', is initially visible as an oblong flat structure enclosed by a double sheath called spathe and when fully mature becomes more cylindrical. Due to the pressure exerted by the growing inflorescence, the spathe ruptures along a longitudinal groove and exposes the inflorescence.

The length of the inflorescence ranges from 60 to 200 cm and has about 8000–10,000 male flowers and 0–400 female flowers, depending on the variety, cultural conditions, season and age of bearing (Patel 1938; Menon and Pandalai 1960; Niral et al. 2008). The central inflorescence axis, referred to as rachis (peduncle), bears about 30–35 branches, the spikelets. The spikelets carry numerous male flowers, with few female flowers (generally 1 to 2, sometimes none and occasionally up to 5) borne near the base of the spikelets. Occasionally a few hermaphrodite flowers are seen alongside the female flowers in some cultivars. All the flowers are sessile/subsessile.

The flower primordium is formed about 4 months after the initiation of the leaf primordium and about 32 months before the opening of the spathe. The male and female flowers are differentiated about 12 months prior to the opening of the fully grown spadix. Generally, the male flowers begin to form a month later than the females and mature a month before the stigma is receptive. The ovary is differentiated 6–7 months before the opening of the spathe. About 75% of the total growth in the length of the inflorescence occurs during the period of about 6 months before the opening of the spathe. Soon after the opening of the spadix, the male flowers complete their life cycle, while the female flowers have a longer history as it takes another 12 months for the nuts to fully ripen.

The development of the inflorescence has been studied and described by Juliano (1926) and Patel (1938). The inflorescence arises in every leaf axil, and the rudimentary inflorescence primordium is formed at almost the same time as the subtending leaf, and therefore the inflorescence initials are present in the axils of the young rudimentary leaves near the growing point. The inflorescence primordium is a minute cone-like protuberance about 77 μ × 107 μ found in the axil of the fourth leaf from the growing point, about 32 months prior to the opening. The whole inflorescence cluster consists of a central axis (the rachis), from which the rachillae arise in spiral succession. The rachillae at their apices bear the male flowers (either singly or in pairs) in the axils of tertiary bracts (floral bracts), with female flowers at the base. The inflorescence initially begins as a minute protuberance at the end of two clasping bracts, one of which is situated towards the cabbage and the other between the petiole and the inflorescence initial. The first primary bract envelops the second primary bract, and these later outgrow the whole floral cluster and form the outer and inner spathes enveloping the growing inflorescence. The enclosed inflorescence primordium has at its apex a ring of actively dividing cells, which shows signs of formation of the primary inner bract which later develops into a persistent inflorescence envelope. After the spathes have completely enveloped the growing point of the inflorescence, secondary bracts make their appearance as minute lateral protuberances. The lower secondary bracts tend to elongate more rapidly than those at the apex, so that their tips are nearly as long as the apex of the main rachis of the inflorescence.

The next stage in the development of the inflorescence is marked by the appearance of the primordia of the rachillae, in the axils of the secondary bracts. These primordia rapidly elongate vertically and lateral to the main rachis leaving behind the subtending secondary bracts. The main rachis always terminates with a single rachilla almost similar to the lateral branches. In the rachillae, tertiary or floral bracts appear, and in the axil of these floral bracts, there is a zone of actively dividing cells, which form the initials of the flower primordium.

3.4.4.1 Floral Morphology

Male Flowers The male flowers outnumber the female flowers in a spadix. The number of male flowers varies depending on the length of spikelet-bearing portion and the number and length of spikelets in an inflorescence. The male flower comprises of three sepals, three petals and six stamens, which are about 8 mm in length and are arranged in a single whorl with a rudimentary/abortive pistil in the centre. Anthers are yellowish in colour and attain a bluish-green tinge on maturity. Opening of the male flowers commences from the apex of the spikelets and extends downwards and occurs throughout the day, with maximum blooming during 8–10 AM. The interval between the opening of the first male flower and the shedding of the last male flower is termed as the male phase, which lasts for 18–22 days.

The number of male flowers varies with the variety, age, as well as environment. Nampoothiri (1970) reported that the proportion of male flowers at the distal, middle and proximal portion of the inflorescence was constant from tree to tree in the case of tall varieties, while in the case of dwarf varieties, the ratios of male flowers in the three positions were found to vary from palm to palm.

Female Flowers The female flowers are comparatively few in number, as compared to the male flowers. They are larger having a diameter of 13–25 mm, globular in structure and are bracteolate. They contain six rounded, concave, imbricate perianth with a staminodal ring at the base and a short style with three stigmas at the centre. Ovary is tricarpellary, syncarpous with a single anatropous ovule in each carpel. However, only one ovule is fertile. Two to five male flowers, referred as accessory or axillary male flowers, accompany each female flower. Generally, the female flowers become receptive 3–4 weeks after the opening of the spathe, by which time the male flowers in the spadix would have shed their pollen. Each female flower remains receptive for 1–3 days. The interval between the receptivity of the first female flower and the last female flower is termed as the female phase, and it lasts for 5–7 days, depending on the variety and growing conditions.

3.4.5 Pollen, Pollination and Fruit Set

The male and female flowers being separate, transmission of the pollen from the male flower to the female flower is through wind or insects (Sampson 1923; Patel 1938; Louis and Chelladurai 1984). The possibility of both self and cross-pollination exists in coconut, depending on the time interval between the opening of the male flower and the receptivity of the female flowers, and varies with the variety as well as cultural and environmental conditions. In general, there is a gap of at least 2–3 days between the end of the male phase and commencement of the female phase in an inflorescence (Ratnambal et al. 1995, 1999), and hence pollination is effected through pollen from neighbouring palms (cross-pollination). However, in majority of the dwarf populations, self-pollination is observed (Jack and Sands 1922), due to overlapping of the female and male phases in an inflorescence (Ratnambal et al. 1995, 1999). Self-fertilisation can also occur through overlapping of the male and female phases of successive inflorescences, during certain seasons like summer months in India (Patel 1938), as well as rainy season in the Philippines (Copeland 1931) and in Sri Lanka (Petch 1913).

The pollen sacs burst and shed their pollen before the opening of the male flower or simultaneously with splitting of the perianth lobes. The fresh pollen grains are smooth and spherical in shape and measure about 0.063 mm in length and 0.02 mm in breadth. The coconut pollen retains its viability for up to a week at room temperature. The viability period is enhanced when pollen is stored at low temperatures and/or in the desiccators. Nampoothiri (1970) observed differential fertility of pollen within an inflorescence, with pollen from male flowers at the proximal end of the inflorescence being the most sterile, while at the distal end, the pollen sterility was the least.

3.4.6 Fruit Development: Endosperm, Embryo

The fruit starts developing upon fertilisation of the female flowers. Even though an inflorescence produces many female flowers, only a few develop into mature fruits, while the rest are shed during the course of development. The coconut fruit contains an internal endosperm (kernel) with embryo embedded in it and protected externally by a thick pericarp. The fruit of the coconut palm, commonly referred to as the 'nut', is botanically a drupe. The pericarp has three distinct regions, the exocarp/epicarp (tough fibrous outermost layer, with varying shades of green, brown, red or yellow colour, depending on the variety), mesocarp (husk portion immediately beneath the epicarp) and endocarp (shell). In the tender fruit, husk is fleshy and has an astringent taste, which becomes more fibrous as the fruit matures. The thickness of the mesocarp is dependent on the variety and varies from 2 to 15 cm. The shell, on its basal side, has three pores (eyes) representing the three carpels of the ovary. One of the eyes is soft, while the other two are quite hard. the embryo is protected by the shell and is present beneath the soft eye. The thickness of the endosperm ranges from 0.8 to 2.0 cm, depending on the variety. In between the endocarp and the albuminous endosperm is a thin layer of testa/seed coat. The testa is brown in colour and adheres to the endosperm. in the middle of the endosperm is a cavity filled with sweet water, also referred to as the liquid endosperm. In the immature fruit, this central cavity is completely filled with water. However, the quantity of liquid endosperm reduces gradually during development and on storage for a few months after harvest. If the nut water is completely exhausted, the fruit loses its ability to germinate (Menon and Pandalai 1960).

The young embryonic fruits are initially yellowish in colour, but turn green, yellow or red (depending on the variety) on exposure to light. On reaching complete maturity (10–12 months after fertilisation), the fruits turn brown. Juliano (1926) observed that after fertilisation, the fruit develops with the pericarp developing more rapidly at the basal region which appears whitish and soft till almost maturity. The endocarp is differentiated even before fertilisation and appears as a soft creamy white structure. As the fruit develops, the embryo sac increases in size leaving a large vacuole at the centre. The young fruits initially grow more in length than width, and later, there is a greater increase in width rather than length, and finally the fruits in most of the varieties are wider than long.

The developing nut attains its maximum size and weight around 8 months after fertilisation and remains so for another 2 months. Subsequently, there is a drastic reduction in nut weight along with a slight decrease in the size. In the final stages of ripening, there is loss of water from the liquid endosperm. Abraham and Mathew (1963), based on their studies on cytology of coconut endosperm, reported that the development of the solid endosperm (kernel) in the fruit becomes visible 6 months after fertilisation as a thin jellylike coating around the periphery of the large embryo-sac cavity. The endosperm tissue is thicker at the antipodal end. The thickness of solid endosperm reaches its maximum around ninth month. The rate of oil deposition in the endosperm peaks at this stage and continues well after the fruits have turned brown (Menon and Pandalai 1960).

Jayasuriya and Perera (1985), based on their studies on the growth, development and dry matter accumulation in coconut fruits, reported four distinct growth phases, viz. initiation, pre-pollination development, post-pollination development and finally maturation and senescence. They reported commencement of husk, shell and endosperm growth in the first, fifth and sixth months, respectively, with simultaneous growth of all the three components from 5 to 8 months after fertilisation. The rapid growth phase of the husk, shell and endosperm is extended from 3 to 7 months, 5 to 9 months and 6 to 10 months, respectively. The weight of the nut and the individual nut component traits are influenced by the variety as well as growth conditions (Patel 1938; Harries 1978; Satyabalan 1993; Niral et al. 2009).

3.4.6.1 Fruit Maturity

The fruits mature 10–12 months after fertilisation. In tall varieties, it generally takes 11–12 months for the fruits to mature, whereas in dwarf varieties, nuts will mature in 10–11 months after emergence of the inflorescence. On maturity, the outer surface of the fruit (the exocarp) starts turning brown. At this stage, the fruits, in certain varieties, tend to detach from the fruit stalk and fall. However, in certain varieties, especially some of the dwarf varieties, the attachment to the fruit stalk is strong, and the mature fruits do not shed from the palm even after the fruits have turned completely brown. In such cases, if mature fruits are not harvested regularly, the endosperm tends to dry and get spoilt inside the nut.

3.4.6.2 Morphology

Structure of the Fruit A mature embryo is cylindrical in shape and approximately 0.8 mm in length. The embryo is in the endosperm just below the germ pore, seen as a dark circular spot on the endocarp. The plumule and radicle can be distinguished within the proximal end of the embryo. In cross sections, the plumule shows a central meristematic zone surrounded by the scaly leaf primordia, which in turn are enclosed by the coleoptiles. The radicle is situated opposite to the plumule and within the apical mass of meristematic cells. The proximal part of the embryo is separated by a small constriction from the cotyledon which develops into the haustorium (apple) during germination of coconut seed nuts.

On germination, the embryo simultaneously develops in two directions, from proximal end, the apical part forces its way out through the germ pore and the plumule and radicle then grows outside the endocarp and from distal end of the embryo, the cotyledon expands to form a pear-shaped haustorium inside the central cavity of the fruit. Haustorium is mainly an absorptive and storage organ which supports the growth of the embryo by providing the products of endosperm hydrolysis till the seedling becomes self-sustaining. Compared with various other tissues within a haustorium, the surface tissues are markedly different, having an undulating structure which is closely attached to the degrading endosperm, with starch grains and oil droplets. It accumulates relatively high amounts of sucrose and starch, with considerably higher activities of phosphoglucomutase and phosphoglucose isomerase, and plays a key role in absorption of oil reserves released from degraded endosperm as well as in the conversion of sugars (Sugimura and Murakami 1990). López-Villalobos et al. (2001) studied the changes in germinating coconut over a four month period and reported continuous and proportionate growth of haustorium, plumule and radicle during coconut seed germination, with haustorium increasing to 45 g nut^{-1} and weighing four to five folds higher than the other two tissues. The vascular bundles run in parallel to the haustorium surface, extending from shoot apex to the distal tip of the haustorium, and the combined strength of these strands of vascular tissue passing through the soft eye prevents the breakage of the growing seedlings' attachment to the haustorium. Various biochemical changes occur in the developing fruits. For details, please see Chap. 9 on Physiology and Biochemistry.

3.4.7 Propagation

Coconut palms are propagated only by seeds. The mature, husk dried nuts containing the seeds germinate on sowing, and the time taken for germination is influenced by the variety and season (Whitehead 1965b; Nampoothiri et al. 1972; Satyabalan 1993; Niral et al. 2006). Natural regeneration also occurs widely in favourable environments if the fallen nuts are left in the field as such. Although the seed nuts are sown either vertically or horizontally keeping the broader surface of fruit facing up (as in commercial nurseries), the seeds can germinate mostly in all directions. In a natural fall from the palm, mostly the seeds settle on the ground with the broader surface touching the ground. Generally, seed nuts of tall varieties germinate from 60 to 200 days after sowing and those of dwarf varieties germinate from 30 to 95 days after sowing. There are some exceptional varieties too with still earlier germination and also showing viviparous germination.

3.4.7.1 Seed Germination, Growth and Development of Seedlings

During germination, the sprout appears as a spear out of the nut and the lamina unfurl upon growth and development after producing a couple of scaly leaves without lamina. The laminae of the first few leaves are fused. The length of the leaves increases from few centimetres in scale leaves and up to 2 m in sixth leaf stage depending on the cultivar and growing conditions. Generally, the seedlings are taken for planting at six to seven leaf stage with at least one leaf showing splitting to leaflets.

3.5 Cytology

Santos (1929) was the first to study the cytology of *C. nucifera* who reported the chromosome number as n = 16. In India, the chromosome number (n = 16) was first reported by Janaki Ammal (1945) and Venkatasubban (1945) followed by other workers (Ninan et al. 1960; Abraham et al. 1961). These studies and those of several others (Nambiar and Swaminathan 1960; Swaminathan and Nambiar 1961; Raveendranath and Ninan 1973) have confirmed the somatic chromosome number of 2n = 32.

3.5.1 Karyomorphology

The gross features of chromosome complements of tall (WCT) and dwarf (CGD, COD) varieties have been studied in India. Raveendranath and Ninan (1973) observed the presence of secondary constrictions on the long arm of chromosome VI in talls and long arm of chromosome III in dwarfs. However, these differences were not consistent, and additional satellites were observed on chromosome II (long arm), chromosome I (short arm), short arm of chromosome XII (Raveendranath and Ninan 1973), long arm of chromosome XII (Thankamma Pillai et al. 1983) and IX (Nambiar and Swaminathan 1960) in talls and in chromosome VI (long arm) in dwarfs (Raveendranath and Ninan1973). Nambiar and Swaminathan (1960) observed that in talls, majority of the chromosomes had submedian centromeres, with two pairs of chromosomes much longer and three pairs relatively short. On the other hand, Raveendranath and Ninan (1973) observed that talls as well as dwarfs had a preponderance of chromosomes with median centromeres, with four submedian chromosomes (II, IV, VII, XIV) in WCT, three (chromosome II, VII, XII) in COD and only one (chromosome II) in CGD. In higher plants, karyotypic evolution has been from complete symmetry to asymmetry (Stebbins 1950). From this angle, WCTs show a more evolved karyotype than COD and CGD. Total chromatin content is found to be greater in CGD than WCT (Raveendranath and Ninan 1973). The total chromatin content is more in wild species than cultivated ones. Therefore, CGD appears to be the most primitive among the three accessions studied. However, evidences from morphology, breeding system and meiotic behaviour support the possible evolution of dwarfs from talls.

3.5.2 Meiotic Studies

The different varieties of talls and dwarfs (both open pollinated and inbred populations) show significant differences in their meiotic behaviour. The dwarfs are reported to show less stable meiosis than talls, and it has been proposed that ancestral types show more stable meiosis (Lindquist 1960). In general, microsporogenesis is more regular in open pollinated than inbred progenies. Nambiar et al. (1970) studied cytological behaviour in five tall accessions, viz. Laccadive Ordinary (LCT), Philippines Ordinary (PHOT), Andaman Ordinary (ADOT), New Guinea (NGT) and Cochin China (CCNT), and observed that microsporogenesis was relatively regular in both inbred and open-pollinated progenies of LCT, while comparatively higher frequencies of chromosome aberrations and pollen sterility were observed in inbred as well as open-pollinated progenies of CCNT and NGT and inbred progenies of PHOT and ADOT. The lack of inbreeding depression only in LCT could either be due to differences in intensity of inbreeding and selection between these geographically distinct varieties or due to the LCT being comparatively less sensitive to inbreeding.

Nambiar and Swaminathan (1960) observed many meiotic irregularities in Apricot from Strait Settlements (SSAT) and Dwarf Red forms, which are derived from the Dwarfs, while meiosis was regular in LCT. Consequently, higher pollen sterility occurred in these two dwarf derivatives in comparison with LCT. Thankamma Pillai et al. (1983) studied meiosis in nine cultivars and hybrids and indicated that the percentage of abnormalities was highest in CGD and COD, while chromosome abnormalities and sterility were very low in D × T and T × D hybrids. They concluded that the higher degree of inbreeding in dwarfs might be the reason for higher chromosome aberrations and sterility in them. Cytological studies on *Spicata* palms (Ninan et al. 1960; Ninan and Satyabalan 1963) indicated irregular meiosis with inversions, translocations and many other abnormalities. *Spicata* palms, being predominant outbreeders, are believed to have arisen from talls through mutation. Further, cytological studies have been undertaken on abnormal palms, bulbiferous palms and root (wilt) disease-affected palms.

Nambiar and Prasannakumari (1964) studied microsporogenesis in root (wilt) disease-affected palms and reported low frequency of cytological aberrations, high pollen fertility and seed set. Thankamma Pillai and Vijayakumar (1972) studied the course of microsporogenesis in a palm bearing defective nuts (self-pollinated progeny of NGT) and observed that the aberrant meiosis and sterility in this palm was attributed to inbreeding. Raveendranath et al. (1975) found no appreciable karyological differences between the normal (Talls) and abnormal palms producing bulbils (in the place of normal inflorescences) and opined that cryptic structural changes or genetic mutations might be responsible for this abnormality.

3.5.3 Cytology of Endosperm and Embryo

Abraham and Thomas (1962) reported free nuclear divisions in coconut water (liquid endosperm). Abraham and Mathew (1963) and Abraham et al. (1965), based on their studies on 6-month-old fruits, observed that size of nuclei varied considerably in the developing endosperm. They found that the tissues adjacent to endothelium were normally triploid ($3x = 48$), less frequently hexaploid ($6x = 96$) and still less frequently dodecaploid (12x = 192) and proposed that higher ploidy levels arise by C-mitosis. They also recorded an inverse relationship between ploidy and percentage oil content, with the inner part of the endosperm having the highest ploidy level and lowest oil content (Abraham 1963; Abraham et al. 1965). In tall variety, the percentage of oil content in the outer, middle and inner layers of endosperm was 75.7, 54.1 and 41.4, respectively. Abraham et al. (1965) recorded higher ploidy levels (48x and above) in buttery endosperm (Philippine makapuno coconuts), which they felt arose through amitosis and nuclear fusion. Unlike the endosperm, the young coconut embryos are diploids and divide by normal mitosis. Raveendranath and Ninan (1973) studied karyomorphological features of somatic chromosomes from 6-month-old embryos and observed an essential uniformity in relative chromosome length from root tip (Nambiar and Swaminathan 1960) and embryo cells of WCT palms. Ninan and Raveendranath (1965) reported occurrence of a haploid embryo in a WCT palm.

3.6 Genetic Resources

Before, we deal with the genetic resources in coconut, it is important to understand some of the terms commonly used to describe coconut diversity. There are diverse perspectives in the understanding of taxonomical terminology used in coconut. Most non-scientific observers and stakeholders do not differentiate between the terms in vogue, which confuses the readers. Even in many scientific papers, the terms cultivar, variety, ecotypes, variants, types and forms are loosely used. The traditional taxonomists consider variety as the last entity in classification, which leads to the indiscriminate coining and use of different terms with no uniformity and specific definition.

After many attempts in earlier years, the coconut breeders seem to have come to the conclusion that it is almost impossible to classify coconuts perfectly, in view of the range of variability for most of the characters. While largely agreeing to this, there is a necessity to remove the chaos existing in the nomenclature among the coconut workers based on some consensus. Taking in the wisdom of the present and earlier coconut workers, a coconut nomenclature is suggested below with the hope that it will guide future workers.

3.6.1 Variety

A variety can be defined as a group of palms having similar traits that can be reproduced from generation to generation. From the taxonomical point of view, the term variety is ranked just below the level of species, but above subvariety and form and hence in coconut, *typica* and *nana* can be considered as botanical varieties. Variety *typica* would represent varieties with tall plant habit with all the associated features of typical talls such as in Indian West Coast Tall, West African Tall, Malayan Tall, Rennell Tall, etc. Variety *nana* would represent varieties with dwarf plant habit and all the associated features of typical dwarfs such as in the Malayan Yellow Dwarf, Chowghat Green Dwarf, Malayan Orange Dwarf, Rangiroa Red Dwarf, etc.

However, this taxonomical grouping does not include *intermediate* varieties having certain features of *nana* and some features of *typica.* It will be desirable to have at least four varietal groups in coconut, as has been suggested by Bourdeix and coworkers (Bourdeix et al. 2017). Therefore, in addition to dwarfs *nana*, which are predominantly self-pollinating, there could be another varietal group of dwarfs, viz. *Javanica*, to accommodate the compact dwarfs, which are predominantly cross-pollinating as in the case of *typica* but with dwarf plant habit as in the case of *nana.* The Niu Leka Dwarf, with dwarf plant habit but has nonoverlapping male and female phases, facilitating cross-pollination, will fall in this category. The *Aurantica* could be the fourth group, encompassing coconut palms with semi-tall plant habit and with variable reproduction modes (cross/self-pollination) as is seen in the tall × dwarf and dwarf × tall hybrids. The King Coconut and Navasi Thembili of Sri Lanka are classical examples of the fourth varietal class.

3.6.2 Cultivar

It is an assemblage of palms which is distinct, comparatively uniform and stable, having been purposely selected for a particular character or a combination of characters, cultivated for centuries, and is capable of reproducing its characters from one generation to the next in nature. The name "cultivar" comes from a combination of two words: cultivated and variety. Cultivars, therefore, represent cultivated varieties that have evolved due to purposive and continued selection. Examples include the Malayan Yellow Dwarf, Malayan Orange Dwarf, Chowghat Green Dwarf, San Ramon Tall, Laguna Tall, etc. Some coconut researchers suggest that all conserved coconut accessions can be referred to as cultivars, arguing that almost all the coconut palms are selected and planted by mankind, the only exceptions being coconut palms growing naturally in remote/uninhabited islands.

3.6.3 Ecotype

Ecotypes are groups of palms or populations within the same variety that are adapted to certain climatic and edaphic conditions and survive as a distinct group due to environmental selection or geographic isolation. Ecotypes are generally named or known by the location where they are grown. However, it is very difficult to identify distinct ecotypes in coconut, except in the instances of geographical isolation, as in isolated atolls. Coconuts in the Cocos (Keeling) Islands of the Indian Ocean and along the Northern Coast of Australia are some such examples. Vanuata Tall, with resistance to foliar decay virus, is an example having ecotype status (Bourdeix et al. 2005a). Ecotypes, having developed adaptive features for particular environmental conditions over a period of time, can serve as a rich source of valuable genes for coconut breeding. In India, the Laccadive and Andaman ecotypes evolved because of their isolation have provided better sources for coconut breeding such as Lakshadweep Ordinary Tall, Lakshadweep Micro Tall, Andaman Ordinary Tall and Andaman Giant Tall which resulted in release of improved varieties. Despite the involvement of human selection process, Kuttiyadi coconut can also be considered as an ecotype, as it got adapted and evolved in the hilly tracts of Malabar region of Kerala, India.

3.6.4 Form or Morphotype

These can be used to refer to smaller groups of palms within the cultivar/ecotype, which are different from each other for a few distinct characters. The variation could be for colour/size of fruits, taste of tender nut water, etc., such as the Thailand Aromatic Dwarf. Purposive selection of these forms/morphotypes, over a period of time, could result in their progression into the domain of cultivars.

3.6.5 Variant

The term variant can be applied to refer to a group of plants with some special morphological features found in different ecotypes or cultivars. This would include variants arising due to single mutations in a cultivar and are mostly considered abnormal. A classical example of variant is the *Spicata*, wherein the inflorescence is unbranched, with numerous female flowers borne directly on the central axis and with very few male flowers. This variant has been encountered in many cultivars and from countries as distant as the Philippines, India and Samoa and referred to as Spicata Tall, Spicata Red Dwarf, Standard Kudat, etc. Another typical example is the soft endosperm coconut, found in different regions and referred to as makapuno/kopyor/coco gras/dikiri/thayiru thengai. The male coconut tree would also fit into this category as would many other freaks observed in natural populations.

3.6.6 Population

Population refers to any subgroup located in a restricted location, such as an island, atoll or continuous strip of coastline. It could be a mixture of diverse varieties/cultivars/forms. Hence, population could also refer to the coconut palms in a location, whether highly heterozygous, as are most tall populations or homozygous, as in the self-pollinating dwarfs. For instance, in India, coconuts in the west coast will be one population, and the ones in the east coast will be another population. The term population is also commonly used to refer to a group of individuals obtained from a cultivar. For example, WAT 06, West African Tall *Ouidah* from Benin, is a population of the West African Tall cultivar, and MYD 01 refers to a population of Malayan Yellow Dwarf collected from Kulasekharam in Tamil Nadu, India.

3.6.7 Subpopulation

A subset of a population that shares one or more characters is referred to as a subpopulation. In coconut, subpopulations within a coconut population can be delimited based on morphological traits as well as genetic markers. For example, a group of palms producing high number of medium-sized nuts within a population of Indian West Coast Tall would represent a subpopulation.

3.6.8 Accession

Accession refers to the basic working unit of conservation in the gene banks and is not a taxonomic entity. The conserved genetic resources could represent a cultivar or form/morphotype, etc. Accessions are generally assigned numbers for their precise identity and are unique to a particular gene bank and tend to continuously increase numerically, as new collections are added to the gene bank. For example, in the case of coconut, "IND006" refers to the accession Andaman Giant Tall (AGT) planted in 1940 at the ICAR-Central Plantation Crop Research Institute in India and "SMD NJM R2" designates Malayan Yellow Dwarf planted in 1981 at the Marc Delorme Research Station in Côte d'Ivoire. The list of coconut accessions conserved in the gene banks of COGENT country members has been documented (Bourdeix et al. 2010) and is available at COGENT website.

In order to create public awareness about coconut genetic resources and to facilitate dissemination of information among the various stakeholders, COGENT/Bioversity International initiated a project to develop an international coconut genetic resources database (CGRD), in the year 1994. The CGRD (Hamelin et al. 2005) was released into the public domain in 1999 and is hosted on the COGENT website. In the CGRD, data on coconut accessions are divided into two main parts – passport data and priority characterisation and evaluation data – taking into consideration the standardised descriptors for coconut palms and the methods detailed in the STANTECH manual. Presently 725 accessions are listed in the CGRD.

With advances in biotechnology, the term accession is also used in the context of conserving DNA/genomic resources. Considering that coconut is a cross-pollinated crop with inherent heterozygosity, the DNA of individual coconut palms is conserved separately, and hence the term 'accession' in a DNA gene bank refers to the DNA of a single palm.

3.6.9 Genotype

A group of palms can be referred to as a genotype if their specific parents are clearly known, for example, a hybrid between two known palms or offspring of a single known palm.

3.6.10 Plant Variety

This is a non-taxonomic term, applied to an intraspecific rank, usually a cultivar or hybrid. In legal parlance, the term plant variety is used to refer to varieties of plants for which patent protection has been applied for. Plant breeders' rights (PBR), also known as plant variety rights (PVR), are rights granted to the breeder for a new variety of plant that provides for exclusive control over propagating the material of a new variety for a specified number of years. In India, guidelines for the conduct of test for distinctiveness, uniformity and stability (DUS) on coconut have been developed by the Protection of Plant Varieties and Farmers Rights' Authority (PPVFRA 2011), in collaboration with ICAR-Central Plantation Crops Research Institute, and so far six varieties, viz. Kalparaksha, Kalpa Sankara, Kalpa Mitra, Kalpa Dhenu, Kalpa Pratibha and Kalpasree, have been registered in the country.

3.6.11 Naming Coconut Cultivars/Ecotypes

The national researchers working on coconut genetic resources are responsible for naming coconut accessions/cultivars/ecotypes of the respective countries. However, considering the need to guide the researchers, avoid confusion and streamline the naming process, COGENT has developed guidelines towards having a standardised procedure for naming coconut accessions (Baudouin et al. 2010).

The international naming of a coconut cultivar or ecotype comprises of two parts: a cultivar name and an abbreviation. An important consideration in naming is to avoid duplicating any previously recorded name (including synonyms). The cultivar name is to be in two parts, written in English (except possibly for the first part) and not exceeding 30 characters. The first part of the name may include either a vernacular name (Agta Tall – Agta being vernacular for blackish necrotic pericarp), place/region/country of origin (West African Tall), a prominent biological trait of the cultivar (Andaman Giant Tall), an ethnological trait linked to the history of the cultivar (Raja Tall), fruit colour (only if a cultivar is homogeneous for colour, such as Chowghat Green Dwarf) or a combination of those (Raja Brown Dwarf – a combination of an ethnological trait and a colour).

An international abbreviation of a cultivar should consist of three to four letters, followed by two digit numerical, avoiding duplication of any previously recorded abbreviation. The first part of the abbreviation is linked to the cultivar name, such as EAT for East African Tall, WCT for the Indian West Coast Tall. In giving abbreviation for dwarf cultivars, showing homogeneity in fruit colour, it is suggested to include a letter corresponding to the colour of the fruit (G, Y, R, O or B), with a provision to optionally omit the letter G in the case of green fruits: MYD for Malayan Yellow Dwarf, CGD for Chowghat Green Dwarf and CATD for Catigan Green Dwarf. The abbreviation should indicate the plant habit, using the letters T or D, as applicable: PRD for Pemba Red Dwarf, WAT for West African Tall and RTB for King Coconut (synonym Rath Thembili, semi-tall).

It is suggested that specific populations or variants within a cultivar be referred to by using a population name after the cultivar name and by a number after the abbreviation, such as West African Tall *Ouidah* (WAT06), a population of the West African Tall cultivar from Benin; Malayan Red Dwarf Fiji (MRD04), a population of the Malayan Red Dwarf cultivar from Taveuni; Malayan Red Dwarf Kulasekharam (MRD01), a population of the Malayan Red Dwarf cultivar from Tamil Nadu, India; and so on.

3.7 Germplasm Collection and Conservation

In due recognition of the importance of coconut genetic resources to enhance coconut productivity and to mitigate genetic erosion in native habitats, collection and conservation of coconut germplasm have received considerable attention in the major coconut-growing countries, and ex situ field gene banks have been established especially in the Philippines, India, Indonesia and Sri Lanka in the early part of the twentieth century. Subsequently, considering the dwindling productivity of coconut in traditional coconut communities and the fact that about 96% of the coconut farmers worldwide are smallholders, the International Coconut Genetic Resources Network (COGENT) was formed in 1991 to facilitate research on coconut. COGENT started with 15 coconut-growing countries as members and currently has 39 member countries that are divided into five subnetworks: Southeast and East Asia (China, Indonesia, Malaysia, Myanmar, the Philippines, Thailand and Vietnam), South Asia and Middle East (Bangladesh, India, Pakistan, Sultanate of Oman and Sri Lanka), South Pacific (Cook Islands, Fiji, Kiribati, Papua New Guinea, Samoa, Solomon Islands, Tonga and Vanuatu), Africa and the Indian Ocean (Benin, Côte d'Ivoire, Ghana, Kenya, Madagascar, Mozambique, Nigeria, Seychelles and Tanzania) and Latin America and the Caribbean (Brazil, Colombia, Costa Rica, Cuba, Guyana, Haiti, Honduras, Jamaica, Mexico and Trinidad and Tobago). To provide double security for conserved germplasm in national gene banks and to promote effective access and safe germplasm movement, five multisite International Coconut Genebanks (ICG) have been established, one in each region: Southeast and East Asia (Indonesia), South Asia and Middle East (India), South Pacific (Papua New Guinea), Africa and the Indian Ocean (Côte d'Ivoire) and Latin America and the Caribbean (Brazil). COGENT in partnership with research institutes has developed numerous research methodologies such as varietal identification techniques using standardised morphological descriptors (IPGRI 1995), safe movement of coconut germplasm (Frison et al. 1993), creation and development of the International Coconut Genetic Resources Database (CGRD) and other dedicated software (Hamelin et al. 2005), strategies for germplasm survey and conservation (Bourdeix et al. 2005b) and the publication of catalogues of coconut varieties (Bourdeix and Batugal 2005; Bourdeix et al. 2010). Presently 24 gene banks from 23 COGENT member countries are conserving 725 unique populations with 1837 living accessions.

In India, germplasm collection began in 1924 with the introduction of cultivars from Fiji, Indonesia, Malaysia, the Philippines, Sri Lanka and Vietnam. The germplasm collection was intensified in 1952, and in 1958 the first indigenous germplasm survey and collection were started. The Central Plantation Crops Research Institute Kasaragod is actively involved in the collection and conservation of coconut biodiversity in the field gene bank for utilisation in the coconut improvement programme. The institute has undertaken exotic collections of coconut germplasm from Pacific and Indian Ocean Islands and from the South Asian countries of Sri Lanka and Bangladesh with funding from ADB/IPGRI. Further extensive prospection and collection of indigenous coconut germplasm from different coconut-growing regions of the country have been undertaken, and distinct accessions have been conserved in the National Active Germplasm Site at the institute.

The Indian coconut population harbours a wide range of diversity. The islands of Lakshadweep and Andaman and Nicobar with their natural coconut populations, some of which have established with no human interference, are reservoirs of vast genetic diversity. Majority of the native populations in the traditional coconut-growing zones belong to the wild type with greater proportion of husk, while some of the evolved types have lesser husk and more endosperm. Further, introgression between the niu kafa and niu vai types in the nature has resulted in the emergence of intermediate types (Thomas and Niral 2010).

Presently, ICAR-CPCRI has the largest collection of coconut germplasm with 455 accessions from 28 countries, representing coconut germplasm of south and Southeast Asia, Caribbean Islands, Indian Ocean Islands, Pacific Ocean Islands and African countries and India. The indigenous coconut germplasm comprises of collections from 12 states as well as from Lakshadweep and Andaman and Nicobar Islands. Germplasm characterisation is being undertaken using the IBPGR descriptor (Anon 1978). The institute has brought out descriptors for 74 coconut accessions (Ratnambal et al. 1995, 1999), and a coconut germplasm database has been developed (Rajagopal et al. 2005). ICAR-CPCRI has also contributed to the development of the world catalogue of conserved coconut germplasm and catalogue of farmers' varieties brought out by COGENT/Bioversity International (Bourdeix et al. 2010).

3.7.1 Alternate Conservation Strategies

Plant genetic resources are among the most essential of the world's natural resources, and during the last three to four decades, major advances have been made in conserving them. During the past two decades, there have been increasing efforts to develop improved in situ conservation methods which would permit dynamic conservation of plant populations (Sthapit and Jarvis 1999).

Bourdeix proposed a new experimental approach to coconut genetic resources conservation termed 'Polymotu' concept, as an alternate conservation strategy, combining the ancestral practice of the Polynesians, wherein they planted distinct coconut varieties separately in small islands or insulated valleys (Bourdeix et al. 2011). The Polymotu concept cannot be a substitute to conventional field gene banks but is intended to complement efficiently these gene banks. It is envisaged herein that several coconut accessions could be planted, each in a distinct isolated site, and the geographical remoteness will ensure the reproductive insulation required for true to type breeding of each of the accessions through natural pollination. This in addition to conservation will also facilitate germplasm access to people, considering that conventional field gene banks are mostly used by researchers and institutions wherein the main limiting factor is the huge cost incurred on maintenance of the gene bank and reproduction of the accessions.

In coconut, other complementary conservation strategies include conservation of embryos and pollen, which being amenable for long-term conservation can provide a viable backup to field gene banks. The possibility of long-term conservation of zygotic embryos was first suggested by Bajaj (1984), when embryos were observed to resume growth after freezing at -196 °C. Assy-Bah and Engelmann (1992) reported successful cryopreservation of immature embryos of coconut 7–8 months after pollination as well as mature zygotic embryos. In India, cryopreservation of mature embryos after desiccation pretreatments has been reported by Karun et al. (2005). Sisunandar et al. (2010a) reported an improved cryopreservation protocol for a wide range of coconut cultivars from Indonesia. In India, cryopreservation of mature coconut embryos through vitrification has been attempted by Sajini et al. (2011). Coconut plumule (apical dome with three to four leaf primordia) extracted from mature embryos (11–12 months after pollination) was also utilised for cryopreservation, by encapsulation techniques (Nan et al. 2008).

Pollen cryopreservation has been successfully used in a variety of plant species (Towill and Walters 2000). Coconut pollen has a life span of few days (Patel 1938) and can be stored for a few days at room temperature in desiccators. Freeze-drying of coconut pollen for short-term storage has also been reported (Whitehead 1963; Rognon and Nucé de Lamothe 1978). Karun et al. (2014) reported long-term cryopreservation of coconut pollen in liquid nitrogen. For details, please refer Chap. 6 on Coconut Biotechnology.

3.8 Novel Traits and Abnormalities

In addition to the varieties and forms of coconut discussed earlier, a number of variants have been reported by different workers from different coconut-growing regions. These include abnormalities, which are freaks of nature and not distinct varieties, from the botanical point of view. Davis has observed and reported most of the abnormalities in coconut palms, while Menon and Pandalai (1960) have compiled these in their coconut monograph.

3.8.1 Novelties Associated with Stem

3.8.1.1 Polyembryony

This is a phenomenon wherein a single nut produces more than one seedling. This has been reported sporadically in different places, and up to five seedlings have been reported to emerge from a single nut. Furtado (1927) was the first to conclude that this phenomenon is due to the development of many embryos from one ovule and not due to polycarpy as was previously believed. Davis (1979) reviewed polyembryony in coconut.

3.8.1.2 Branching

This is a rare phenomenon and has been reported by a number of workers (Ridley 1907; Burkill 1923–25; Furtado 1923–25; Quisumbing 1926–27; Jacob 1935; Davis 1947, 1950, 1956a, Indiresh 1992; Mao and Lai 1993). Branching is found to take place at all stages of growth and from various regions of the stem. Branching is a commercially important phenomenon as it can help obtain increased yields with less planting space.

Branching in coconut has been attributed to a number of causes. In a young palm, when the growing point is injured, adventitious buds develop and form new growing points from the meristematically active ground tissue of the bole. Branching is also induced when the terminal bud is injured by lightening, fire, storm or as a result of some pest attack. Davis (1950) and Jacob (1935) have reported instances of branching due to dichotomy and stem fascinations, respectively. Davis (1968) and Balaga (1975) were able to induce branching in coconut seedlings but with a very limited success. They observed formation of neomeristems of adventitious origin on making a longitudinal incision in the shoot of young seedlings. Fisher and Tsai (1979) observed formation of a twin shoot, emerging from a single growing tip, while growing isolated coconut embryos in vitro. They concluded that this was a case of branching, possibly arising due to an injury to the embryo at the time of dissection and culture.

3.8.1.3 Suckering

This is another rare phenomenon, synonymous to branching, but restricted to the underground portion of the stem (bole). Shortt (1885) reported six shoots in a seedling. In India, Patel (1938) reported as many as 40 suckers arising from the base of a palm. Aiyadurai et al. (1959) reported 18 suckers. Davis (1956b) and Chatterjee (1959) have reported a number of instances of suckering in coconut palms. In India, suckering has been observed in three seedlings belonging to three different varieties as well as in a seedling in a farmer's garden. Suckering is an economically interesting phenomenon as it offers the possibility for vegetative propagation of the palms through stimulation of the suppressed basal buds. However, what exactly stimulates the development of suckers is still unknown. Davis (1960) and Michael and Varghese (1963) were able of induce two to three suckers by causing mechanical injury to the growing point. However, as the number of suckers produced is very much less, this cannot be used for commercial propagation of palms.

3.8.2 Novelties Associated with Leaf

3.8.2.1 Plicate Palms

Abnormality in the leaves is rare. Patel (1938) reported fused leaflets in adult coconut palms, wherein the leaflets are fused and do not separate, as in ordinary palms, and attributed it to genetic and environmental factors. Zuniga et al. (1970) grouped these palms under the name 'plicata'. The palms with fused leaflets are referred to as plicata in the Philippines and India and as niu yabia in the Fiji Islands (Satyabalan 1997). Subsequent research workers (Sugimura et al. 1994a; Arunachalam et al. 2001) have observed that plicata palms have reduced leaf length and are poor bearers with a long juvenile period. Moa and Lai (1993) reported sterility in plicata palms. Boron deficiency can also cause the leaflets to remain fused, but in contrast to palms suffering from boron deficiency, the plicata palms do not respond to borax application.

Davis (1956c) described some other abnormalities like forked leaves, twin leaves, fused leaves, multi-leaf, etc. Lilly (1962) observed formation of secondary midribs in some of the leaflets in a leaf of a 10-year-old palm. The secondary midribs were smaller in size than the normal midribs. However, she concluded that this could also be considered as a case of fused leaflets.

3.8.2.2 Chimeras

In palms showing chimera, green and yellow variegations are observed in fruits/leaves. This is due to somatic or bud mutation and has been reported from India and Sri Lanka (Satyabalan 1997). Variations in colour of the fruits can occur within a bunch or in different bunches of the same palm. A young coconut seedling showing chimera was observed in India wherein part of a leaf was observed to be albino (white/yellowish white) while the rest were dark green. Subsequently produced new leaves also showed this chimeric pattern.

3.8.2.3 Albinos

In coconut nurseries, a few albino seedlings are observed at very low frequencies (less than 1%). These seedlings do not respond to manurial treatment and generally die after producing four to six leaves. Patel (1938) opined that albinism is, due to genetic factors, while Furtado (1926–29) believed that albinism in coconut seedlings was the result of chlorosis due to the absence of ferrugineous products in the endosperm. Pandalai and Pillai (1959) based on their experiments with albino seedlings concluded that inadequate availability of iron (due to the inability of the plant to utilise the iron in the leaves) results in the albinic condition. The mobilisation of iron appears to be controlled by recessive gene/genes, since albinism is an inherited character.

3.8.3 Novelties Associated with Inflorescence

3.8.3.1 Midget Palm

The terminal inflorescence or hapaxanthic spadix has been reported from India and Indonesia. Here, the inflorescence emerges from the terminal portion and then the palm dies. Davis and Menon (1953) (quoted by Menon and Pandalai 1960) and Davis (1955) first reported the curious phenomenon of coconut seedlings producing inflorescences within 12 months of sowing and referred to them as midget palms.

3.8.3.2 Bulbils

Bulbiferous coconut palms, producing numerous vegetative shoots from spadices, were reported from India by Davis (1948a, b). Venkataraman (1928) recorded a rare instance wherein the buttons/young nuts in a spadix, instead of developing into normal nuts, grew into bulbils somewhat resembling miniature seedlings. This transformation was visible at a very early stage in the development of the inflorescence, even before the spathe had fully opened, as even the unfertilised ovules in the tender inflorescences had started to produce growth primordia. Venkataraman (1928) also referred to this as a case of parthenogenesis. Jerard et al. (2014a) reported a naturally occurring, rare bulbiferous coconut palm among West Coast Tall population, which produced only bulbil shoots in leaf axils in place of normal inflorescence. The identified palm happened to be twins in which both of them produced only bulbil shoots instead of floral parts. Morphological and molecular studies on these palms revealed that the main palm and their bulbil progenies are genetically uniform. The genetic uniformity of the twin mother palms and their bulbil progenies was confirmed through microsatellite analysis using ten polymorphic SSR primer pairs specific to coconut and has been well demonstrated to differentiate coconut cultivars. The primary and secondary bulbil shoots were found to be capable of growing into independent plants making it possible to use them as propagules to develop a homogeneous clonal population hitherto unavailable in coconut. The bulbils showed axillary growth in 6th–12th leaf axil, which further develop as secondary bulbils indicating the complete vegetative state of the palm. Comparison of shoot apices of a normal seedling with bulbil shoot revealed variation in cell growth pattern. Conservation of bulbiferous palms as a unique genetic resource needs to be taken up to utilise these rare sources for future breeding programmes, provided their seed fertility can be restored.

3.8.3.3 Double Spadix in Leaf Axils

Davis (1957) first reported the occurrence of a double inflorescence. Each of the two spadices was independent with a fully developed spathe and normal spikelets bearing male and female flowers. However, the outer spathe was common to both the spadices. A palm with double spadices in each of the leaf axils was reported in the Indian WCT population indicating the possibility of natural occurrences of such palms.

3.8.3.4 Multi–Spatheate Inflorescence

In a normal inflorescence, only four bracts are visible. The third bract enlarges to form a spathe that encloses the spadix. At the base of spathe and extending a third of its length is the second bract that has transformed to a fibrous sheath to protect the spathe during the early stages of development. At the base of the outer spathe is the first/outer bract, which is about 25 mm long and scaly. The fourth bract is a scaly bract at the base of the spadix. Davis and Menon (1952-quoted by Menon and Pandalai 1960) reported a bispatheate palm wherein all the inflorescences in the palm had two fully developed spathes. In these palms, the fifth/innermost bract develops to the size of the original spathe, and both these spathes completely cover the spadix till its emergence. Subsequently many workers from various coconut-growing countries have reported the occurrence of bispatheate palms. Thomas and Mathew (1960) reported a trispatheate coconut palm. Michael (1963), however, observed a multispatheate coconut palm wherein the inflorescence had five fully developed spathes, the outer spathe enclosing four closely set inner spathes. Here again, the first and second spathes originated from the third and fourth bracts of the inflorescence. However, the origin of the third, fourth and fifth spathes was unclear and was hypothesised to have originated from the rudimentary bracts generally found at the base of the spadix. The higher number of the spathes offers greater mechanical protection for the spadix and also more protection against insect attacks. Moa and Lai (1993) have reported multispatheate palms at Hainan Island, China.

3.8.3.5 Hermaphrodite Flowers

Normally, the spikelet in a coconut inflorescence has a few female flowers at the base with numerous male flowers towards the top. In between the female and male flowers, occasionally some hermaphrodite flowers are observed (Gopal Rao 1948). Hermaphrodite flowers are bigger than the male flowers, but smaller than the females. Hermaphrodite flower resembles the female flower but shows development of the staminodal ring into stamens. Davis et al. (1954) reported quantitative data on four palms of the tall variety from Kerala, which produce appreciable numbers of hermaphrodite flowers in all the spadices examined. Smit (1970) reported high percentage of hermaphrodite flowers ranging from 10% to 55% between palms of Nias Yellow Dwarf at the Manado seed garden (Davis et al. 1981). Higher occurrence of hermaphrodite flowers in S_1 population of Chowghat Green Dwarf has been reported from India (CPCRI 2014).

3.8.3.6 Male Palms

The male coconut palm, also referred to as Androgena, produces only male flowers and was first reported from India (John and Narayana 1942). Subsequently, male coconut palms have been reported from China (Mao and Lai 1993), Jamaica and Markham Valley, New Guinea (Whitehead 1966). Ninan et al. (1960) observed meiotic abnormalities such as aneuploidy in a few cells in Androgena palms, but no in depth cytological investigations have been undertaken. Male palms are characterised by robust inflorescences with numerous spikelets bearing large number of male flowers. The inflorescence of the male coconut palm is much larger than those of the ordinary palms and also has a much greater number of male flowers per inflorescence than in the ordinary palm. Davis et al. (1955) observed branching of spikes in spadices of male trees and secondary spadices, as well as numerous spathes or highly developed bracts.

3.8.3.7 Spicata Inflorescence

Palms with unbranched inflorescences, referred to as spicata (Beccari 1916; Boldingh 1920; Jacob 1941; Davis 1980), though very rare, are reported in about all coconut-growing regions. The inflorescences of the Spicata palm have a central spike which is unbranched or rarely with one or two branches, unlike the normal inflorescences that are branched with 30–35 spikelets. The inflorescences of the Spicata palms bear a large number of female flowers (125–130) with very few male flowers (50), in stark contrast to the normal inflorescences, with innumerable male flowers and few female flowers at the base of each spikelet (0–4). However, the fruit setting is low in Spicata, with Jacob (1941) observing 10% setting, while Sugimura et al. (1993) reported 4% setting.

The Spicata palms cross freely with other coconut varieties. On selfing, John and Narayana (1949) found that only 50% of the progenies breed true to the mother. Whitehead (1965a) on crossing a tall parent with pollen of the dwarf spicata observed the resultant hybrids to be intermediate in form and precocity. Whitehead (1966) reported that these hybrids had lesser inflorescence branches and therefore, advocated that the spicata character is incompletely dominant or pleiotropic. Ninan and Nambiar (1974) observed that the hybrids from crosses of tall and spicata show a low proportion of male flowers per inflorescence, a character associated with spicata.

Spicatas are also locally referred to as KelapaBrol/KelapaBrodjol in Indonesia, Spicata-Maure in the Philippines, Niu Tuave in Samoa, Niu Yalewa in Fiji Islands, Ngohard Tapala and Niu Toga in Solomon Islands, Loholohotahia in Tonga Islands and Ma Praew in Thailand. Whitehead (1966) noted spicata palms in many islands of Pacific, with the first report of spicata trait in dwarf palms (small red fruited dwarf) in New Guinea. Whitehead also reported the absence of such palms in Jamaica. In India, wide variation has been reported among selfed spicata tall progenies (CPCRI 2014), and a yellow spicata dwarf population has been developed.

Ninan and Satyabalan (1963) observed meiotic abnormalities and a high degree of pollen sterility in spicata palms. Subsequently, Ninan and Raveendranath (1972) undertook further cytological studies in these palms and observed that the 15th pair of chromosomes is heteromorphic (one member has an additional terminal knob). Barring this, the chromosomes of this variety are similar to those of the typical tall variety. This heterozygous condition accounts for segregation following open-pollination of spicata into about 50% tall palms having the normal branched inflorescence and 50% spicata palms.

Sankaran et al. (2015) identified a rare form of multiple *spicata* from South Andaman, India. Unlike the normal *spicata*, the inflorescence of this peculiar palm contains, in addition to the central spike, 10–15 spikelets of 30–32 cm length, with 23–26 female flowers in each spikelet/rachillae. On an average, each inflorescence produces 330–345 female flowers, but only one or two fruits are produced per bunch, due to high percentage of button shedding.

3.8.3.8 Partial Suppression of Spikelets

This was reported by Davis (1957) wherein about half the spikes towards the basal end of the inflorescence are reduced and in extreme cases represented by one or two flowers only. However, the spikes at the distal half develop normally and bear male flowers. The main axis is extended and bears numerous female and male flowers.

3.8.3.9 Secondary Branching of Spikes

This is generally reported in the male inflorescences of the male coconut tree, wherein the inflorescences bear a larger number of spikelets most of which are branched (John and Narayana 1942). In addition, numerous spathes or highly developed bracts are also seen in such inflorescences. Sugimura et al. (1994b) observed secondary branching of spikes in two of the five inbred lines of Markham Valley Tall but with a normal pre-bearing period.

3.8.3.10 Varied Number of Pistillodes, Astamens, and Perianth

The male flowers normally have three valvate petals and three smaller triangular and valvate sepals, while the female flowers have three huge sepals and three smaller and thinner petals and a bract and a bracteole. Davis observed certain flowers with a larger number of perianths, with one flower having 13 perianths. In one particular female flower, Davis and Menon (as quoted by Menon and Pandalai 1960) observed seven perianth parts and concluded that the seventh perianth was a developed pistillode. They also observed a male flower with six sepals and six large petals with a single ring of 12 stamens with well-developed anthers and six pistillodes. Certain abnormal male flowers having 7/8/10 stamens and 4/5 pistillodes have also been recorded (Menon and Pandalai 1960).

3.8.4 Novelties Associated with Fruits

3.8.4.1 Polycarpy

In a normal coconut fruit, with tricarpic ovary, each one has an ovule, but only one embryo develops and produces a shoot. In rare cases, polycarpy has been observed. Jacob (1940) described a two-seeded coconut, wherein the two seeds were separated by a leathery septum while the third ovule was aborted. Davis (1948c) and Daniel Sundararaj (1952) observed a three-seeded coconut with three fertile eyes giving rise to three shoots, one each from each eye. Forbes (1879) reported a nut with 14 carpels, embryos in all of which germinated to produce a palm with 14 stems united at the base.

3.8.4.2 Varied Number of Carpels

A normal female flower has three syncarpic carpels with an embryonic ovule. In rare instances, mono-, bio- and tetracarpic nuts have been reported with the nuts being one-seeded (Davis 1948c). Davis and Menon (1953) (quoted by Menon and

Pandalai 1960) observed that some flowers in the midget palm were bicarpellary. Davis has also observed young coconut fruits having four to ten carpels (Menon and Pandalai 1960).

3.8.4.3 Horned Coconut

A number of workers reported the occurrence of coconut fruits bearing flat horns ranging from one to six. Davis (1965) regarded the production of horns in palm fruits as atavistic. The development of horns in coconut fruits has been attributed to various reasons, like development of the staminodes (Furtado 1926), duplication of the segments of the gynoecium (Petch 1924), apocarpic nature of the ovary (Costerus and Smith 1923; Gadd 1924; Davis and Menon 1953 – quoted by Menon and Pandalai 1960) and enlargement of some perianths (Masters 1869); Nair and Sadanandan (1976) studied anatomically the development of hornlike structures in some coconut fruits from initiation to maturity. The accession, viz. 'Andaman Horned Cocos', a coconut germplasm collected from South Andaman, India, and conserved at ICAR-CPCRI, has been registered for this distinct trait of horny nuts (Jerard et al. 2014b).

3.8.4.4 Vivipary

Thomas (1960) described a case of vivipary in coconut and reported viviparous seedlings to be healthier and more vigorous than the normal seedlings. Das and Thakur (1996) in their article on the production and prospects of coconut in Assam have reported the occurrence of vivipary in certain pockets within the state. At ICAR-CPCRI, vivipary has been reported in Malayan and Andaman dwarfs, with fruits germinating on the crown (CPCRI 2013; Shareefa et al. 2014). Viviparous coconuts have been reported from Indonesia also (Satyabalan 1997). Vivipary reported in Andaman Green Dwarf cultivar (Sankaran et al. 2012) was interpreted as an adaptive reproductive strategy that enables seedlings to establish more rapidly and subsequent dispersal by water or other means.

3.8.4.5 Sweet Husk

Menon and Pandalai (1960) have reported certain palms wherein the young buttons have higher sugar content and less tannin content in their pericarp. In normal cases, due to the higher percentage of tannin, the sweetness in the pericarp of young buttons is masked resulting in an astringent taste. In these rare instances, the husk is sweet with less fibre and can therefore be eaten. The nuts of these palms yield very fine white-coloured fibre. The retting period required for these husks is also comparatively less. Tampake et al. (1982) have reported the occurrence of palms with sweet edible husk in Indonesia. The variety Kaithathali from Lakshadweep, India, is reported to have sweet husk and is chewed by the seafarers as an antidote to seasickness. Sweet-husked varieties have been reported from other coconut-growing countries: Navasi Thembili (Sri Lanka); Kalpa Tebu (Indonesia); Kalpa Logi (Malaysia); Cuyamis, Caumanis, Tabal, Tamisan and Mais (the Philippines); Uta (Rotuma, Fiji Islands); Preug-wan (Thailand); Nu wa or Cocos sucre (New Caledonia); and Cay dua bong (Vietnam). In India, sweet husked-type variants were reported from Lakshadweep as well as Nicobar Group of Islands, wherein the young buttons were sweet to taste, while the mature nuts had less of pale-coloured hard fibres and more of pale or white-coloured coir dust as compared to normal mature coconuts (Niral and Jerard 2017).

3.8.4.6 Soft Husk

Soft-husked types, with less hard husk fibre and more pale-coloured fibres, have been infrequently observed in coconut populations. Niral and Jerard (2017) reported soft-husked types among coconut populations of Lakshadweep Islands and Nicobar Group of Islands, wherein the attachment of the husk to the nut in the dry, mature fruits was observed to be loose and they produce a dull sound when tapped. Such nuts, on complete drying, also tend to show splitting of husk. These are very similar to sweet husk types in fibre colour and texture.

3.8.4.7 Pink–Husked Fruits and Anther Filaments

A few coconut palms bearing pink-ringed fruits, with pink-coloured mesocarp in developing fruits, have been reported (Bourdeix et al. 2010; Niral and Jerard 2017) from different regions of India (in west coast tall population, East Coast tall population, yellow dwarf), the Philippines (Pilipog Green dwarf) and French Polynesia (Tahitian Red Dwarf). A green dwarf with attractive pink mesocarp colour in tender fruits from Fiji has also been reported. At ICAR-CPCRI, Jerard et al. (2016) reported palm-bearing fruits with pink-coloured mesocarp from San Ramon tall population. While the pink colour appeared on all the female flowers (at the base of the tepals) and also in tender fruits, the pink colour was exhibited only in a proportion of male flowers. This is the first report on differently coloured anther filaments in coconut male flowers, one with dark pink and another with normal yellow filament which has the potential in marker-assisted selection, towards development of tender nut varieties with attractive pink husk. Although no morphological differences could be seen on size of male flowers and anthers, the flowers with pink filaments could be easily identified even at unopened stage since the pink tinge can be seen at the bottom of tepals.

3.8.4.8 Sweet Endosperm

A variant with sweet endosperm, referred to as ‘Mohachao Narel’, has been reported from India (Samsudeen et al. 2010, 2013). The sweet endosperm has very little fibre content, making the kernel palatable as a salad for table purpose. Only a small percentage of the nuts in a bunch express this character. Studies on nut component traits of sweet and normal fruits of this population indicated slightly lesser fruit and husked fruit weight as compared to normal nuts. Though endosperm thickness was similar, endosperm weight, copra weight and copra recovery were more in normal nuts at similar age of maturity. The development of a large population bearing sweet kernel fruits is a challenging task as the inheritance and genetics need further understanding through progeny testing. Among the progenies of Mohachao Narel, conserved at ICAR-CPCRI, the percentage of sweet kernel fruits in a palm varied from 4.5% to 60% (ICAR-CPCRI 2017).

3.8.4.9 Soft or Buttery Endosperm

The endosperm (kernel) of a mature coconut is normally hard, but in certain palms, few of the nuts have a thicker and softer flesh of a buttery consistency with less liquid endosperm which is viscous. Such coconuts with buttery kernel are popularly referred to as makapuno in the Philippines and are used in preparation of specific coconut delicacies and fetch a premium prize. In India, Patel (1938) recorded rare instances wherein the internal cavity of the shell was filled with a jelly-like substance of the consistency of thick curd and locally referred to as ‘thairuthengai’ (curd coconut). Satyabalan (1953) observed rugged/warty kernel in 10% of the fruits in a Laguna tall palm at ICAR-CPCRI, India. The abnormal kernel was not sweet and had an oily taste/flavour but with lesser fibre content, while the embryos were normal in size and shape. Soft kernel types were collected from Andaman Islands (Jerard et al. 2013) and Kerala (CPCRI 2014), belonging to Andaman and West Coast Tall coconut populations, respectively. The fruit component studies in the soft kernel types indicated that the fresh kernel has lesser fibre, oil and sweetness. The tender nut water from soft kernelled fruits was more viscous due to suspended kernel tissue and recorded higher TSS. Adriano and Manahan (1931) distinguished three different types of endosperm in soft kernel fruits: nuts with a hard outer layer, a soft viscous middle layer and a semi-liquid inner layer of endosperm, nuts with a hard outer layer and a soft viscous middle layer and nuts with one layer of hard endosperm, filling the whole central cavity.

Makapuno is reported to be a single recessive gene mutation (Torres 1937) and has been successfully exploited through controlled pollination among makapuno-bearing palms followed by embryo culture to develop elite populations giving higher percentage of makapuno nuts. Mutants similar to the makapuno type have been reported from other coconut-growing regions, such as Coco Gra of Seychelles, Kopyor of Indonesia, Thairu Thengai or Nei Thengai or Ghee Thengai of India, Dikiri Pol of Sri Lanka, Mapharao Khati of Thailand, Sap of Vietnam, Niu Garuk of Papua New Guinea and Pia of Polynesia.

Zuniga (1953) obtained the expected Mendelian ratios in the progeny of makapuno. However, Cruz and Ramirez (1968) advocated a more complex behaviour, considering the different types of makapuno being found and the differences in nuclear behaviour between various makapuno types. The embryos though normal in size do not germinate under natural conditions, but germinate in vitro when cultured in appropriate nutrient media. Cedo et al. (1984) based on their observations from controlled pollination studies, with makapuno palms derived from embryo culture, concluded that makapuno coconuts are homozygous and that pollen produces a xenia effect. Mujer et al. (1983, 1984a, b) based on his biochemical studies on normal and makapuno coconut concluded that the makapuno endosperm occurs due to a deficiency in the activity of the enzyme alpha-D-galactosidase. Samonte et al. (1989) observed that the activity of three galactomannan-degrading enzymes, viz. alpha-D-galactosidase, beta-mannanase and beta-mannosidase, was consistently low during endosperm maturation in the makapuno endosperm as compared to normal endosperm.

3.8.4.10 White Testa

The testa in a normal coconut is brownish in colour, while in rare instances, the testa has been observed to be white (Tampake et al. 1982).

3.8.4.11 Aromatic Types

Aroma, which is released when a tender coconut is cut open, is a novel trait. A few exceptional palms have been reported to produce fragrant endosperm in different countries, such as Klapawangi, aromatic green dwarf, Nam-Hom, etc. The liquid endosperm (tender nut water) in aromatic coconut is characterised by a pleasant 'pandan-like' aroma, similar to the aroma of aromatic rice, which adds palatability in tender nut consumption and fetches a premium price in the market. In Thailand, Saensuk et al. (2016) concluded that the aroma in the liquid endosperm of the popular aromatic green dwarf coconut is due to the presence of 2-acetyl-1-pyrroline (2AP), and this qualitative trait is controlled by a single recessive gene.

3.8.5 Novelties Associated with Chromosome Number

3.8.5.1 Haploidy

Whitehead and chapman (1962) reported a diploid-haploid twin (seedling) in coconut, the haploid being the weaker member of the pair. At ICAR-CPCRI also such diploid-haploid twin (seedlings) has been observed in the coconut nursery, in different varieties, at very low frequencies. Ninan and Raveendranath (1965), during the course of extensive investigations on cytology of embryo and endosperm in coconuts, encountered a haploid embryo with chromosome number of 16 and nuclear volume of only about a third of that of the diploid nuclei (2n = 32), in a West Coast Tall palm. They concluded that the haploid embryo probably resulted from haploid parthenogenesis.

3.8.6 Novelties Associated with Gross Morphology

3.8.6.1 Compact Crown

Compactness is an important trait, in development of improved varieties, especially as it is amenable for high density planting. In Fiji, compact dwarfs have been reported in natural populations, possibly having arisen from introgression between the local types. The compact dwarfs are allogamous, but with a very low rate of trunk extension as in the case of dwarfs, but with trunk diameter and flowering behaviour similar to talls but with the crown being more compact than those of talls or the slender dwarfs. The first compact dwarf described in coconut is the Niu Leka Dwarf. Bourdeix et al. (2017) reported the presence of compact dwarfs in French Polynesia and the Cook Islands, some with green, brown, red-orange and yellow fruits, including a compact red dwarf in French Polynesia, with pink mesocarp colour in young developing nuts and hypothesised that some of the compact dwarfs presently available in Polynesia are progenies of the Niu Leka × Malayan Red Dwarf crosses made in Fiji by Marechal (1926). In India (CPCRI 2010), a dwarf mutant identified in seedling progenies from Lakshadweep coconut population showed extreme dwarfism as compared to the other conserved dwarf accessions. Besides, compact dwarf hybrids of MYD × NLAD have been developed in India (CPCRI 2013).

3.9 Future Strategy

Coconut, with a long history of domestication and cultivation, has evolved, through the ages, from being an important oil-producing crop to an important food crop, with high nutritive value today. The crop now finds a place in the list of 49 priority crops of global importance in food and livelihood security of the International Treaty on Plant Genetic Resources for Food and Agriculture, wherein country signatories to this treaty agree to make their genetic diversity and related information about the crops stored in their gene banks available to all, to facilitate research, innovation and exchange of information to stimulate sustainable crop production.

The fact that coconut belongs to a monotypic genus with no known wild relatives has been a major handicap in unravelling the origin as well as the evolution of the present-day cultivated coconut. In the absence of experimental studies, our understanding of the origin and domestication has been based on historical accounts, fossil records and some phylogenetical studies. Comprehensive phylogenetic studies using all related genera, viz. *Attalea*, *Syagrus*, *Parajubaea*, *Jubaeopsis caffra* and *Voanioala*, as well as the species earlier classified under the genus *Cocos* may help to throw light on the evolutionary relationship of *Cocos nucifera*, under the subtribe Attaleinae. Wide hybridisation between these related genera and geographically isolated/remote coconut populations can also be attempted to facilitate better understanding of the evolution of the present-day coconut.

In view of the basic need for a broad genetic base for varietal improvement, especially for developing resistant planting material to combat serious diseases which is a major constraint limiting coconut production in all the coconut-growing countries, intensified efforts are required for collection, conservation and in-depth studies on coconut populations of remote/uninhabited islands in the Pacific Ocean as well as the Indian Ocean, including the Andaman and Nicobar Islands and the Lakshadweep Group of Islands. Activities of COGENT need to be strengthened for this purpose. Simultaneously, new approaches and synchronised global efforts are required to conserve the available variability in the crop, both in situ and ex situ in field gene banks, cryo-genebanks as well as DNA banks, to mitigate loss of biodiversity through genetic erosion, from loss of native habitats, sea level rise, growing urbanisation and replanting of existing coconut plantations.

References

Abraham A (1963) Chromosome constitution and oil content in the coconut endosperm. J Indian Bot Soc 42A:1–3

Abraham A, Mathew PM (1963) Cytology of coconut endosperm. Ann Bot 3:505–512

Abraham A, Thomas KJ (1962) A note on the *in vitro* culture of excised embryo. Indian Cocon J 15:84–88

Abraham A, Mathew PM, Ninan CA (1961) Cytology of *Cocos nucifera* L. Cytologia 26:327–332

Abraham A, Ninan CA, Gopinath P (1965) Cytology of development of abnormal endosperm in Philippine makapuno coconuts. Caryologia 18:395–409

Adriano FT, Manahan M (1931) The nutritive value of green, ripe and sport coconut (buko, niyog and makapuno). Philippine Agric 20:195–198

Aiyadurai SG, Mohammed Ibrahim PA, Ramanathan T (1959) Multiple seedlings from a germinating coconut. Sci Cul 24:339

Anonymous (1978) Consultation on coconut genetic resources; report. International Board for Plant Genetics Resources, Roma, p 24

Arunachalam V, Jerard BA, Elangovan M, Ratnambal MJ, Dhanapal R, Rizal SK, Damodaran V (2001) Unexploited diversity in coconut palm (*Cocos nucifera* L.). Plant Gen Resour Newsl 127:39–43

Arunachalam V, Jerard BA, Damodaran V, Ratnambal MJ, Kumaran PM (2005) Phenotypic diversity of foliar traits in coconut germplasm. Gen Resour Crop Evol 52(8):1031–1037

Assy-Bah B, Engelmann F (1992) Cryopreservation of immature embryos of coconut (*Cocos nucifera*). Cryo Letters 13:67–74

Bajaj YPS (1984) Induction of growth in frozen embryos of coconut and ovules of citrus. Curr Sci 53:215–216

Baker WJ, Couvreur TLP (2013a) Global biogeography and diversification of palms shed light on the evolution of tropical lineages. I. Historical biogeography. J Biogeogr 40:274–285

Baker WJ, Couvreur TLP (2013b) Global biogeography and diversification of palms sheds light on the evolution of tropical lineages. II. Diversification history and origin of regional lineages. J Biogeogr 40:286–298

Baker WJ, Savolainen V, Asmussen-Lange CB, Chase MW, Dransfield J, Forest F, Harley MM, Uhl N, Wilkinson M (2009) Complete generic level phylogenetic analyses of palms (Arecaceae) with comparisons of supertree and super-matrix approaches. Syst Biol 58(2):240–256. https://doi.org/10.1093/sislro/sypo.21

Balaga HY (1975) Induction of branching in coconut. Kalikasan 4:135–140

Balakrishnan NP, Nair RB (1979) Wild populations of *Areca* and *Cocos* in Andaman and Nicobar Islands. Indian J For 2(4):350–363

Baudouin L, Bourdeix R, Harries H, George MLC (2010) Coconut palm nomenclature. In: Bourdeix R, Batugal P, Oliver JT, George MLC (eds) Catalogue of Conserved Coconut Germplasm. International Coconut Genetic Resources Network (COGENT), Bioversity International, Regional Office for Asia, the Pacific and Oceania, Serdang, Selangor Darul Ehsan, Malaysia, pp 1–3

Baudouin L, Gunn BF, Olsen KM (2014) The presence of coconut in southern Panama in pre-Columbian times: clearing up the confusion. Ann Bot 113:1–5

Beccari O (1916) *Cocos nucifera* Linn *forma palmyrensis* Becc. In: Rock JF (ed) Palmyra Island with a description of its flora. College of Hawaii Bull, Honolulu: 1–58, 1–80

Beccari O (1917) The origin and dispersal of *Cocos nucifera*. Philippine J Sci Series C: Bot 12:27–43

Boldingh I (1920) Over de veelvormigheid van der klapper, Mededlingen Afdeling Zaadteelt, Department van Landbouw, Nijverheiden Handel 1. Kolff, Batavia, pp 1–20

Bourdeix R, Batugal P (2005) Standardized catalogues of coconut germplasm: catalogue of conserved germplasm and farmers' varieties. In: Batugal P, Ramanatha Rao V, Oliver J (eds) Coconut genetic resources. International Plant Genetic Resources Institute, Serdang, pp 456–462

Bourdeix R, Santos G, Labouisse JP, Baudouin L (2005a) Annex 1. Useful definitions of terms and nomenclature. In: Batugal P, Ramanatha Rao V, Oliver J (eds) Coconut genetic resources. International Plant Genetic Resources Institute, Serdang, pp 9–10

Bourdeix R, Guarino L, Rao R, Baudouin L (2005b) Status, gaps and strategy in coconut germplasm collecting. In: Batugal P, Ramanatha Rao V, Oliver J (eds) Coconut genetic resources. International Plant Genetic Resources Institute, Serdang, pp 44–64

Bourdeix R, Batugal P, Oliver JT, George MLC (2010) Catalogue of conserved coconut germplasm. International Coconut Genetic Resources Network (COGENT), Bioversity International, Regional Office for Asia, the Pacific and Oceania, Serdang, Selangor Darul Ehsan, Malaysia

Bourdeix R, Johnson V, Baudouin L, Tuia VS, Kete T, Planes S, Lusty C, Weise S (2011) Polymotu: a new concept of island-based germplasm bank based on an old Polynesian practice. Ogasawara Res (37):33–51

Bourdeix R, Perera L, Rivera RL, Saena-Tuia V, Masumbuko L (2017) Global coconut communities – status and strategies in *in situ* diversity management and utilization. In: Chowdappa P, Niral V, Jerard BA, Samsudeen K (eds) Coconut. Daya Publishing House, A Division of Astral International Pvt. Ltd, New Delhi, pp 407–441

Burkill IH (1923–25) The fertility of branched coconut palms. Gard Bull. Straits Settlement 3:1–2

Carlos T (1963) Agronomic studies in some cultivars of coconut in Philippines M.Sc Thesis, UPLB College, Laguna, Philippines

Cedo LO, de GEV, Rimando TJ (1984) Controlled pollination of embryo-cultured makapuno coconut (*Cocos nucifera* L.). Philippine Agric 67:100–104

Chatterjee D (1959) On multiple seedling of coconut. Sci and Cult 25:370

Child R (1974) Coconuts, 2nd edn. Longman, London, p 335

Cook OF (1901) The origin and dispersal of the cocoa palm. Contrib US Natl Herbarium 7:257–293

Copeland EB (1931) The coconut. Macmillan and Co, London

Costerus JC, Smith JJ (1923) Studies in tropical teratology. Ann Jordanian Bot Buitenz 33:95

CPCRI (2010) Annual report 2009–2010. Central Plantation Crops Research Institute, Kasaragod, Kerala, India, p 150

CPCRI (2013) Annual report 2012–2013. Central Plantation Crops Research Institute, Kasaragod, Kerala, India, p 142

CPCRI (2014) Annual report 2013–2014. Central Plantation Crops Research Institute, Kasaragod, Kerala, India, p 139

Crowther A, Lucas L, Helm R, Horton M, Shipton C, Wright HT, Walshaw SM, Pawlowicz M, Radimilahy C, Douka K, Picornell-Gelabert L, Dorian Q, Fuller DQ, Boivin NL (2016) Ancient crops provide first archaeological signature of the westward Austronesian expansion. Proceedings Natl Acad Sci U S A 113(24):6635–6640. https://doi.org/10.1073/pnas.1522714113

Cruz SSD, Ramirez DA (1968) The cytology of the developing endosperm of the normal and the mutant (makapuno) coconut. Philippine Agric 52:72–81

Daniel Sundararaj D (1952) Morphology of the normal and abnormal fruits in coconut. Indian Cocon J 5:149–152

Das RP, Thakur AC (1996) Production and prospects of coconut in Assam. In: Proceedings of the seminar on problems and prospects of agricultural research and development in North-East India, 27–28 November 1995. Assam Agricultural University, Jorhat, pp 165–171

Davis TA (1947) A three crowned coconut palm. J Bombay Nat Hist Soc 47:398–400

Davis TA (1948a) Proliferation in two coconut palms (*Cocos nucifera* L). Indian Cocon J 2:11–17

Davis TA (1948b) Abnormal palms of Travancore III.A bulbiferous coconut palm. J Bombay Nat Hist Soc 47:527–529

Davis TA (1948c) Abnormal palms of Travancore IV. Polycarpy in a coconut (*Cocos nucifera* L.). J Bombay Nat Hist Soc 48:704–706

Davis TA (1950) Branching in some Indian palms. Indian Cocon J 3:135–145

Davis TA (1955) Meet the smallest member of the family. Cocon Bull 8:244–246

Davis TA (1956a) Branching in coconut palm. Cocon Bull 9:197–200

Davis TA (1956b) Freaks of nature. I. Suckering in coconut palms. Cocon Bull 9:179–180

Davis TA (1956c) Forked and fused leaves of the coconut palm. Cocon Bull 10:51–53

Davis TA (1957) Spadices with secondary branches or none at all. Cocon Bull 10:163–165

Davis TA (1960) Clonal propagation in coconuts. Curr Sci 29:273

Davis TA (1962) Asymmetry and yield in *Cocos nucifera*. Experentia 18:321–322

Davis TA (1965) Addition to the theories on the morphology of horns in coconut fruits. Ceylon Cocon Qtly 16:99–101

Davis TA (1968) Difficulties in the genetic improvement of the coconut: a promising alternative method. Indian J Gen 28A:154–164

Davis TA (1969) Prospects of the clonal propagation of the coconut. Ceylon Cocon Planters Rev 6:1–5

Davis TA (1979) Some unusual formations in palms. Principles 23:80–83

Davis TA (1980) Super abundant female flower production in a coconut: a report. Philippine J Cocon Stud 5:23–30

Davis TA, Menon KPV (1952) A bispatheate coconut palm.2. Indian Cocon J 6:30–34

Davis TA, Menon KPV (1953) A midget coconut palm. Indian Cocon J 6:139–144

Davis TA, Anandan AP, Menon KPV (1954) Hermaphroditism in *Cocos nucifera* L. Indian Cocon J 7:133–142

Davis TA, Anandan AP, Menon KPV (1955) An instance of dimorphic sex expression and compound spadices in a coconut palm. Indian Cocon J 8:150–157

Davis TA, Sudasrip H, Azis H (1981) Significance of hermaphrodite flowers in coconut breeding. Proceedings Indian Natl Sci Acad 47:527–535

de Condolle A (1886) Origin of cultivated plants, 2nd edn. Reprint. Hafner Publishing Company, New York

De Jayasuriya VUS, Perera RKIS (1985) Growth, development and dry matter accumulation in the fruit of *Cocos nucifera* L. var *nana* form *pumila*. COCOS 3:16–21

Dransfield J, Uhl NM, Asmussen CB, Baker WJ, Harley MM, Lewis CE (2008) Genera palmarum: the evolution and classification of palms. Royal Botanic Gardens, Kew

Dwyer REP (1938) Coconut improvement by seed selection and plant breeding. New Guinea Agric Gazette 4:24–102

Endt R, Hayward BW (1997) Modern relatives of New Zealand's fossil coconut in high attitude South America. Geol Soc N Z Newsl 113:67–71

F'urtado CX (1923-25) Branched coconut palms and their fertility. Gdns' Bull Singapore, 3:274–279
Fisher JB, Tsai JH (1979) A branched coconut seedling in tissue culture. Principes 23:128–131
Forbes HO (1879) Notes on *Cocos nucifera*. J Bot 17:193–194
Fremond Y, Ziller R, de Nuce de Lamothe M (1966) The coconut palm. International Potash Institute, Berne
Frison EA, Putter CAJ, Diekmann M (1993) FAO/IBPGR Technical guidelines for the safe movement of coconut germplasm. FAO/IBPGR, Rome
Furtado CX (1926–1929) Teratological notes. A. Abnormalities in coconut palms, Gdns' Bull, Dermatol Sin, 4:78–84
Furtado CX (1927) Abnormalities in coconut palms. Gdns' Bull 4:78–86
Gadd CH (1924) An abnormal inflorescence from a nut. Tear Book, Dept. Agric Ceylon 21–23
Gangolly SR, Satyabalan K, Pandalai KM (1957) Varieties of the coconut. Indian Cocon J 10:3–28
Ghose M (1979) Ontogenetic study of stomata and trichomes in some palms. Phytomorphology 29:26–33
Gomez-Navarro C, Jaramillo C, Herrera F, Wing SL, Callejas R (2009) Palms (Arecaceae) from a Paleocene rainforest of northern Colombia. American J Bot 96:1300–1312
Gopal Rao T (1948) A note on the occurrence of a hermaphrodite flower in coconut (*Cocos nucifera*). J Indian Bot Soc 27:208–211
Gruezo WS (1990) Fruit component analysis of eight "wild" coconut populations in the Philippines. Philippine J Cocon Stud 15:6–15
Gunn BF (2004) Phylogeny of the Cocoseae (Arecaceae) with emphasis on *Cocos nucifera*. Ann Mo Bot Gard 91:505–525
Gunn BF, Baudouin I, Olsen KM (2011) Independent origins of cultivated coconut in the old world tropics. PLoS One 6(6):e21143. https://doi.org/10.1371/journal.pone.0021143
Gunn BF, Baudouin I, Buele T, Ilbert P, Duperray C, Crisp M, Issali A, Konan J, Alain Rival A (2015) Ploidy and domestication are associated with genome size variation in palms. American J Bot 102:1625–1633
Hamelin C, Bourdeix R, Baudouin L (2005) The international coconut genetic resources database. In: Batugal P, Ramanatha Rao V, Oliver J (eds) Coconut genetic resources. International Plant Genetic Resources Institute, Serdang, pp 427–438
Harries HC (1970) The Malayan dwarf supersedes the Jamaica tall coconut: 1Reputation and performance. Oléagineux 25:527–531
Harries HC (1978) The evolution, dissemination and classification of *Cocos nucifera*. Bot Rev 4:265–320
Harries HC (1979) Nuts to the garden of Eden. Principes 23:143–148
Harries HC (1981) Practical identification of coconut varieties. Oleagineux 36:63–72
Harries HC (1990) Malesian origin for a domestic *Cocos nucifera*. In: Baas P et al (eds) The plant diversity of Malesia. Springer, Dordrecht, pp 351–357
Harries HC (1992) Biogeography of the coconut, *Cocos nucifera* L. Principes 36:155–162
Harries HC (1995) Coconut, palmae. In: Smartt J, Simmonds NW (eds) Evolution of crop plants, 2nd edn. Longman, Burnt Mill, Harlow, pp 389–394
Hayward BW, Moore PR, Gibson RW (1960) How warm was the late Oligocene in New Zealand? Geol Soc N Z Newsl 90:39–41
ICAR-CPCRI (2015) Annual report 2014–15. ICAR-Central Plantation Crops Research Institute, Kasaragod, Kerala, India, p 126
ICAR-CPCRI (2017) Annual report 2016–17. ICAR-Central Plantation Crops Research Institute, Kasaragod, Kerala, India, p 236
Indiresh KM (1992) Branching in coconut. Indian Cocon J 22:14–15
IPGRI (1995) Annual report. International Plant Genetic Resources Instt, Rome
Jack HW, Sands WN (1922) The dwarf coconut in Malaya. Malayan Agri J 10:4–12
Jacob KC (1935) Stem fasciation in the coconut palm. J Bombay Nat Hist Soc 37:901
Jacob KC (1940) A bicellular coconut (*Cocos nucifera* L.). J Bombay Nat Hist Soc 41:905–906
Jacob KC (1941) A new variety of coconut palm (*Cocos nucifera* Linn.) var. Spicata. J Bombay Nat Hist Soc 41:906–907

Janaki Ammal EK (1945) In: Darlington CD, Janaki Ammal EK (eds) Chromosome atlas of cultivated plants. George Allen & Unwin Ltd, London

Jerard BA, Damodaran V, Niral V, Samsudeen K, Rajesh MK, Sankaran M (2013) Conservation and utilization of *ThairuThengai*– soft endosperm coconut accession from Andaman Islands. J Plantn Crops 41(1):14–21

Jerard BA, Rajesh MK, Elain SE, Sajini KK, Rahman S, Fayas TP, Karun A (2014a) Scope of novel and rare bulbiferous coconut palms (*Cocos nucifera* L.). Genetic Resour and Crop Evol 61:1–6

Jerard BA, Niral V, Dhanapal R, Damodaran V, Arunachalam V, Rajesh MK, Devakumar K, Samsudeen K, Nair RV, Kumaran PM, Thomas GV (2014b) IND 221–Andaman horned Cocos (IC0598221; INGR13063), a coconut (*Cocos nucifera*) germplasm with distinct character of horny nuts. Indian J Plant Gen Resour 27(1):66–89

Jerard BA, Niral V, Samsudeen K, Gayathri UK (2016) Pink husked coconut selection- a trait of promise. In: Chowdappa P, Muralidharan K, Samsudeen K, Rajesh MK (eds) Abstracts – 3rd International Symposium on Coconut Research and Development, 10–12 December, 2016. ICAR-CPCRI, Kasaragod, p 21

Jerard BA, Rajesh MK, Thomas RJ, Niral V, Samsudeen K (2017) Island ecosystems hosts rich diversity in coconut (*Cocos nucifera*): evidences from Minicoy islands. India Agri Res 6:214–226

John CM, Narayana G (1942) The male coconut tree. Madras Agri J 31:75–77

John CM, Narayana G (1949) Varieties and forms of the coconut (*Cocos nucifera* Linn). Indian Cocon J 2:209–226

Juliano JB (1926) Origin, development and nature of the stony layer of the coconut (*Cocos nucifera*). Philippine J Sci 30:187–200

Karun A, Sajini KK, Parthasarathy VA (2005) Cryopreservation of mature zygotic embryos of coconut by desiccation methods. CORD 21(1):13–19

Karun A, Sajini KK, Niral V, Amarnath CH, Remya P, Rajesh MK, Samsudeen K, Jerard BA, Engelmann F (2014) Coconut (*Cocos nucifera* L.) pollen cryopreservation. Cryo-Letters 35(5):407–417

Kaul KN (1951) A palm fruit from Kapurdi (Jodhpur, Rajasthan desert), *Cocos sahnii sp. nov*. Curr Sci 20:138

Lepofsky D, Harries HC, Kellum M (1992) Early coconuts in Moorea, French Polynesia. J Polynesian Soc 101:299–308

Lilly VG (1962) An abnormal coconut leaf. Cocon Bull 16:333–335

Lindquist K (1960) Cytogenetic studies in the seriola group of *Lactuca*. Hereditas 46:75–157

Liyanage DV (1958) Varieties and forms of the coconut palm grown in Ceylon. Ceylon Cocon Qtly 9:1–10

López-Villalobos A, Dodds PF, Hornung R (2001) Changes in fatty acid composition during development of tissues of coconut (*Cocos nucifera* L.) embryos in the intact nut and *in vitro*. J Exp Bot 52(358):933–942

Louis IH, Chelladurai M (1984) Nature and frequency of insects pollinating the coconut palm (*Cocos nucifera* L.). Indian Cocon J 15:12–17

Louis IH, Chidambaram A (1976) Inheritance studies on the phyllotaxy of coconut palm. Ceylon Cocon Qtly 27:22–24

Mao YT (1959) A preliminary survey of types of coconut palms. FAO Crop Production and Improvement Branch, Rome

Mao ZS, Lai YT (1993) The coconut germplasm of Hainan Island, China. Plant-Gen Resour Newsl 91–92:53–57

Marechal H (1926) Observations and preliminary results on the coconut palm with a view to developing improved seed nuts for Fiji. Agric J Fiji 1:16–45

Masters MT (1869) Vegetable teratology. Ray Society, London, pp 428–429

Mayuranathan PV (1938) The original home of the coconut. J Bombay Nat Hist Soc 40:174–187 776

McPaul JW (1962) Coconut growing in Fiji. Bulletin 38, Dept. Agric, Fiji

Meerow AW, Noblick L, Borrone JW, Couvreur TLP, Mauro-Herrera M, Kuhn DN, Nakamura K, Olcas NH, Schnell RJ (2009) Phylogenetic analysis of seven *WRKY* genes across the palm subtribe Attaleinae (Arecaceae) identifies *Syagrus* as sister group of the coconut. PLoS One 4(10):e7353. https://doi.org/10.1371/journal.pone.0007353

Meerow AW, Noblick L, Salas-Leiva DE, Sanchez V, Francisco-Ortega J, Jestrow B, Nakamura K (2015) Phylogeny and historical biogeography of the cocosoid palms (Arecaceae, Arecoideae, Cocoseae) inferred from sequences of six WRKY gene family loci. Cladistics 31:1–26 First published: 18 October 2014. https://doi.org/10.1111/cls.12100

Menon KPV, Pandalai KM (1960) The coconut palm, a monograph. Indian Coconut Committee, Ernakulum, p 384

Michael KJ (1963) A "multi-spatheate" coconut palm. Indian Cocon J 16:78–80

Michael KJ, Verghese EJ (1963) Production of suckers in coconut. Indian Cocon J 16:106–108

Miquel FAW (1855) Flora van Nederlandsch Indie. III. C. G. van der Post, Amsterdam, pp 64–72

Mujer CV, Arambulo AS, Mendoza EMT, Ramirez DA (1983) The viscous component of the mutant (makapuno) coconut endosperm. I. Isolation and characterization. Kalikasan. Philippine J Bio 12:42–50

Mujer CV, Ramirez DA, Mendoza EMT (1984a) Alpha-D-galactosidase deficiency in coconut endosperm: its possible pleiotropic effects in makapuno. Phytochemistry 23:893–894

Mujer CV, Ramirez DA, Mendoza EMT (1984b) Coconut alpha-Dgalactosidase isoenzymes: isolation, purification and characterization. Phytochemistry 23:1251–1254

Nair RB, Sadanandan AK (1976) A note on horns in coconut fruits. J Plantn Crops 4:81–82

Nambiar MC, Prasannakumari (1964) Effect of root (wilt) disease on the microsporogenesis in coconut (*Cocos nucifera* L.). Indian Cocon J 27:93–100

Nambiar MR, Swaminathan MS (1960) Chromosome morphology, microsporogenesis and pollen fertility in some varieties of coconut. Indian J Gen 20:200–211

Nambiar MC, Thankamma Pillai PK, Vijayakumar G (1970) Cytological behaviour of first inbred generation of coconut. Indian J Gen 30:744–752

Nampoothiri KUK (1970) Pollen studies in coconut (*Cocos nucifera*) with special reference to a sampling procedure. Indian J Agri Sci 19:457–460

Nampoothiri KUK, Mathew J, Sukumaran CK (1972) Variation in germination pattern of coconut cultivars and hybrids. J Plantn Crops 1(Suppl):24–27

Nan ON, Hocher V, Verdeil JL, Konan JL, Ballo K, Mondeil F, Malaurie B (2008) Cryopreservation by encapsulation dehydration of plumules of coconut. Cryo-Letters 29:339–350

Narayana GV, John CM (1949) Varieties and forms of coconut. Madras Agric J 36:349–366

Nayar NM (1978) Coconut research and development. Principes 22:70–74

Nayar NM (2017a) Origin and domestication. In: Chowdappa P, Niral V, Jerard BA, Samsudeen K (eds) Coconut. Daya Publishing House, Astral International Pvt. Ltd, New Delhi, pp 39–72

Nayar NM (2017b) The coconut: phylogeny, origins, and spread. Academic, London, p 240

Ninan CA, Nambiar MC (1974) Cytology and origin of spicata coconuts (*Cocos nucifera* L. var. *spicata*). In: Kachroo P (ed) Advancing Frontiers in cytogenetics. Hindustan Publishing Corporation, New Delhi, pp 175–188

Ninan CA, Raveendranath TG (1965) A naturally occurring haploid in the coconut palm (*Cocos nucifera* L). Caryologia 18:619–623

Ninan CA, Raveendranath TG (1972) Why spicata palms do not breed true? Chromosome studies provide the answer. Cocon Bull 3:2–4

Ninan CA, Satyabalan K (1963) Cytogenetic studies in the genus *Cocos*. II. Some observations on the spicata character in coconuts (*Cocos nucifera* L). Indian Cocon J 16:109–114

Ninan CA, Pillai RV, Joseph J (1960) Cytogenetic studies in the genus *Cocos*. 1. Chromosome number in *C. australis* and *C. nucifera* L. varieties. Spicata and androgena. Indian Cocon J 13:129–134

Niral V, Jerard BA (2017) Botany, growth and development. In: Chowdappa P, Niral V, Jerard BA, Samsudeen K (eds) Coconut. Daya Publishing House, New Delhi, pp 73–130

Niral V, Arunachalam V, Augustine Jerard B, Samsudeen K, Kumaran PM (2006) Planting material production in coconut through seeds as influenced by genetic variability. In: Book of abstracts

of lead papers and poster papers presented at the National Symposium on improving Input Use Efficiency in Horticulture, August 9–11, 2006, Bangalore, pp 64
Niral V, Jerard BA, Kavitha KV, Samsudeen K, Nair RV (2008) Variability and association among floral traits and pollen recovery in coconut (*Cocos nucifera* L.). J Plantn Crops 36(3):186–191
Niral V, Nair RV, Jerard BA, Samsudeen K, Ratnambal MJ (2009) Evaluation of coconut germplasm for fruit component traits and oil yield. J Oilseeds Res 26(Special issue):668–670
Ohler JG (1984) Coconut: tree of life. FAO plant production and protection paper 57, food and agriculture organization, Rome, Italy, pp 446
Padmanabhan D (1963) Leaf development in palms. Curr Sci 32:537–539
Padmanabhan D (1984) Ontogeny of plications in palm leaf – a new theory. J Swamy Bot Club 1:59–67
Pandalai KM, Pillai RV (1959) Albinism in coconut seedlings. Nature 184:1163
Patel JS (1938) The coconut: a monograph. Government Press, Madras, p 311
Periasamy K (1965) Morphological and ontogenetic studies in palms II. Growth pattern of the leaves of *Cocos nucifera* and *Borassus flabellifer* after the initiation of plications. Aust J Bot 13:225–234
Petch T (1913) The flower of the coconut. Trop Agric 41:449–455
Petch T (1924) Horned Coconut. Year Book, Dept. Agric, Ceylon, p 20–21
PPVFRA (2011) Guidelines for the conduct of test for distinctiveness, uniformity and stability (DUS) on coconut. Plant Variety J India 5(6)
Prasad M, Khare EG, Singh SK (2013) Plant fossils from the Deccan intertrappean sediments of Chhindwara district, MP, India. J Palaentol Soc India 58:229–250
Prebble M, Dowe J (2008) The late quaternary decline and extinction of palms on oceanic Pacific islands. Quaternary Sci Rev 27:2546–2567
Purseglove JW (1968) The origin and distribution of the coconut. Trop Sci 10:190–199
Pursglove JW (1985) Tropical crops: monocotyledons, 5th edn. Longman, London
Quisumbing E (1926–27) Branching in coconut. Philippine Agric 15:3–11
Rajagopal V, Manimekalai R, Devakumar K, Rajesh MK, Karun A, Niral V, Gopal M, Aziz S, Gunasekaran M, Kumar MR, Chandrasekar A (2005) A database for coconut crop improvement. Bioinformation 1(2):75–77
Ratnambal MJ, MK Nair, Muralidharan K, Kumaran PM, Rao EVVB, Pillai RV (1995) Coconut descriptors. Part. I. CPCRI, Kasaragod, Kerala, India, p 198
Ratnambal MJ, Niral V, Krishnan M, Ravikumar N (1999) Coconut descriptors part II, CD- ROM, CPCRI, Kasaragod, Kerala, India
Rattanapruk V (1970) Thailand. In: FAO Coconut Breeding. pp 31–33
Raveendranath TG, Ninan CA (1973) A study of somatic chromosome complements of tall and dwarf coconut (*Cocos nucifera* L.) and its bearing on intervarietal variation and evolution in coconuts. J Plantn Crops 1(1):17–22
Raveendranath TG, Nair PN, Ninan CA (1975) Chromosomes of a bulbiferous palm. Cocon Bull 6:2
Richardson DL, Harries HC, Balsevicius E (1978) Coconut varietés in Costa Rica. Variedades de cocoterosen Costa Rica. Turrialba 28:87–90
Ridley HN (1907) Branching in palms. Ann Bot 21:415–422
Rigby JF (1995) A fossil *Cocos nucifera* L. fruit from the latest Pliocene of Queensland, Australia. In: Pant DD (ed) Birbal Sahni centennial volume. Allahabad University, Allahabad, pp 379–381
Rognon F (1976) Floral biology of the coconut. Duration and sequence of male and female phases in different types of coconut. Oleagineux 31(1):13–18
Rognon F, De Nucé de Lamothe M (1978) Récolte etconditionnement du pollen pour la pollinisation des champs semenciers de cocotier. Oléagineux 33:17–21
Rumphius GE (1741) *Herbarium amboinense* Vol 1, Amsterdam. [The Ambonese Herbal, Volumes 1–6, Georgius Everhardus Rumphius; Translated, annotated, and with an introduction by Beekman EM, Yale University Press 2011, ISBN:9780300153767
Saensuk C, Wanchana S, Choowongkomon K, Wongpornchai S, Kraithong T, Imsabai W, Chaichoompu E, Ruanjaichon V, Toojinda T, Vanavichit A, Arikit S (2016) *De novo* transcrip-

tome assembly and identification of the gene conferring a "pandan-like" aroma in coconut (*Cocos nucifera* L.). Plant Sci 252:324–334
Sajini KK, Karun A, Amarnath CH, Engelmann F (2011) Cryopreservation of coconut (*Cocos nucifera* L.) zygotic embryos by vitrification. Cryo-Letters 32(4):317–328
Samonte JL, Mendoza EMT, Ilag LL, de la Cruz NB, Ramirez DA (1989) Galactomannan degrading enzymes in maturing normal and makapuno and germinating normal coconut endosperm. Phytochemistry 28:2269–2273
Sampson HC (1923) The coconut palm. John Bale, Sons and Danielsson Ltd, London
Samsudeen K, Jacob PM, Niral V, Kumaran PM, Salooja R, Moosa H (2006) Exploration and collection of coconut germplasm in Kadmat and Amini Islands of Lakshadweep, India. Genet Resour Crop Evol 53:1721–1728
Samsudeen K, Nagwaker DD, Karun A, Niral V, Jerard BA, Ajithkumar P, Devadas K, Nair RV (2010) Exploration and collection of sweet endosperm coconut 'Mohachao Narel' from Maharashtra, India. In: Thomas GV, Krishnkumar V, Jerard BA, Niral V, Josephrajkunmar (eds) Abstracts. International conference on "coconut biodiversity for prosperity" 25–288 October, 2010. CPCRI, Kasaragod, p 19
Samsudeen K, Rajesh MK, Reshmi R, AjithKumar P, Devadas K, Anitha K (2013) Diversity in Mohachao Narel: a sweet endosperm coconut (*Cocos nucifera* L.) population from Maharashtra, India. Natl Acad Sci Lett 36(3):319–330
Sankaran M, Damodaran V, Singh DR, Jerard BA (2012) Vivipary in *Cocos nucifera* L. var. Andaman green dwarf. Curr Sci 103(10):1139–1040
Sankaran M, Damodaran V, Jerard BA, Abirami K, Roy DS (2015) Multiple *Spicata* coconut (MSC): a rare type of coconut in Andaman Islands. Transcriptomics 3:123
Santos J (1929) Cytological study of *Cocos nucifera* Lin. Philippine J Sci 37:417–437
Satyabalan K (1953) A note on the occurrence of 'buttery' kernel in coconut. Indian Cocon J 6:152–154
Satyabalan K (1993) The coconut palm-botany and breeding. Asian Pacific Coconut Community, Jakarta, p 214
Shareefa M, Thomas RJ, Nampoothiri CK, Karun A (2014) Studies on vivipary in dwarf coconut cultivars. Indian J Hort 71:464–468
Satyabalan K (1997) Coconut varieties and cultivars – their classification. Asian Pacific Coconut Community, Jakarta, p 105
Shortt J (1885) A monograph on the coconut palm or *Cocos nucifera* L., Madras Agric Coll Misc. Publ. Govt. Press, Madras
Shukla A, Mehrotra RC, Guleria JS (2012) *Cocos sahni* Kaul, a *Cocos nucifera*-like fruit from the early Eocene rainforest of Rajasthan, western India. J Biosci 37:769–776
Sisunandar PAS, Yohannes MS, Rival SA, Adkins SW (2010a) Dehydration improves cryopreservation of coconut (*Cocos nucifera* L). Cryobiology 61:289–296
Smit EHD (1970) Morphological and anatomical studies of the coconut. Thesis, Wageningen, p 107
Spriggs M (1984) Early coconut remains from the South Pacific. J Palynological Soc 93:71–76
Srivastava R, Srivastava G (2014) Fossil fruit of *Cocos* L. (Arecaceae) Maastrichtian-Danian sediments of Central India and its phytogeographical significance. Acta Palaeobotanica 54:67–75
Stebbins GL (1950) Variation and evolution in plants. Oxford & IBH Publ Co, Bombay
Sthapit BR, Jarvis D (1999) Participatory plant breeding for on farm conservation. LEISA 15(40):41
Sugimura Y, Murakami T (1990) Structure and function of the haustorium in germinating coconut palm seed. Japan Agri Res Q 24:1–14
Sugimura Y, Kamata N, Rocat DA, Salud CD, Ceniza MS (1993) Cultivarietal characteristics of 'Spicata' coconut palm. Japan J Trop Agric 37:264–268
Sugimura Y, Rocat DA, Salud CD, Kamata N (1994a) Characteristics of coconut palm with plicate leaves. Japan J Trop Agric 38:119–123
Sugimura Y, Rocat DA, Salud CD, Kamata N (1994b) An inflorescence mutant appeared in the inbred line of coconut. Japan J Trop Agric 38:145–147

Swaminathan MS, Nambiar MC (1961) Cytology and origin of the dwarf coconut palm. Nature 192:85–86
Tampake H, Kuswara T, Davis TA (1982) Coconut germplasm survey of Nusa Tenggara, Timur Province, the initial step towards inducing drought resistance in coconut strains. Indonesian Agri Res Devt J 4:52–61
Thankamma Pillai PK, Vijayakumar G (1972) A note on the cytology of abnormal palm. Indian J Gen 32:303–305
Thankamma Pillai PK, Vijayakumar G, Ravindran PN (1983) Cytology of coconut. In: Nayar NM (ed) Coconut research and development. Wiley Eastern Ltd., New Delhi, pp 62–70
Thomas CA (1960) A viviparous coconut palm. Cocon Bull 14:335–340
Thomas CA, Mathew C (1960) A tri-spatheate coconut palm. Bull Indian Central Cocon Committee 7:256–259
Thomas GV, Niral V (2010) Coconut biodiversity for prosperity. Indian Hort 55(6):54–58
Tomlinson PB (1961) Anatomy of the monocotyledons II palms. Cocos, Cocoid Palms, pp 195–198
Torres JP (1937) Some notes on Makapuno coconut and its inheritance. Philippine J Agric 8:27–37
Towill LE, Walters C (2000) Current Research Progress and Applications. In: Engelmann F, Takagi H (eds) Cryopreservation of Tropical Plant Germplasm. JIRCAS/IPGRI, Tsukuba/Rome, pp 115–129
Venkatanarayana G (1957) On certain aspects of the development of the leaf of *Cocos nucifera*. Phytomorphology 7:297–305
Venkataraman K (1928) Parthenogenesis in coconut. Yearbook, Deptartment of Agriculture Madras, pp 29–32
Venkatasubban KR (1945) Cytological studies in Palmae: chromosome numbers in a few species of palms of British India and Ceylon. Proceedings Indian Acad Sci B 22:193–207
von Martius CFP (1850) (1823–50) Historia Naturalis Palmarumleipsig. 3, Munich
Whitehead RA (1963) The processing of coconut pollen. Euphytica 12:167–177
Whitehead RA (1965a) Flowering in *Cocos nucifera* L. in Jamaica. Trop Agric (Trinidad) 42:19–29
Whitehead RA (1965b) Speed of germination, a characteristic of possible taxonomic significance. Trop Agric (Trinidad) 42:369–372
Whitehead RA (1966) Sample survey and collection of coconut germplasm in the Pacific Islands (30 May–5 September 1964). Ministry of Overseas Development, HMSO, London
Whitehead RA, Chapman GP (1962) Twinning and haploidy in *Cocos nucifera*. Nature 195:1228–1229
Zuniga LC (1953) The probable inheritance of the makapuno character of the coconut. Philippine Agric 36:402–413
Zuniga LC, Villegas LG, Penaflorida G (1970) Collection of coconut cultivars in the Philippines. In: Emata RG (ed) Coconut production in the Philippines, pp 9–24

Dr. V. Niral is a Plant Geneticist by qualification. She has more than 22 years of research experience in coconut genetic resources and breeding and is the curator of the gene bank. She has 177 research publications including 3 books, 7 technical bulletins and 38 research papers on coconut. She has also contributed towards the development of trait-specific genetic stocks and 13 coconut varieties including hybrids. She is presently a Principal Scientist of the Division of Crop Improvement at ICAR. CPCRI. niralv@yahoo.com; niralvittal@gmail.com

Dr. B. A. Jerard has a doctoral degree in Horticulture. He has 21 years of research experience ICAR-CPCRI and ICAR-CIARI on coconut genetic resources management. His contributions include collection and conservation of coconut accessions, release of 20 coconut varieties and the use of molecular markers in coconut improvement. He has served as the Coconut Expert to the Ministry of Agriculture, Government of Fiji. He has published 170 technical articles, 14 books and 36 research papers. He is presently serving as the Head of the Division of Horticulture and Forestry, ICAR-CIARI, Port Blair, Andaman and Nicobar Islands, India. jerardba@yahoo.com; jerardba@gmail.com

Chapter 4
Varietal Improvement

K. U. K. Nampoothiri and V. A. Parthasarathy

Abstract Considerable progress has been made in evolving high-yielding coconut cultivars and hybrids with desirable traits, despite critical constraints in breeding due to heterozygous, monotypic and perennial nature of the crop with a long gestation period and requiring extensive land area for experimentation. Comprehensive efforts in prospection and international cooperation have helped in the assemblage of 1837 valuable accessions maintained in the five international gene banks and various national gene banks. The significance of selection at mother palm level and seedling stage has been well established. The most significant stride came with the advent of hybrids especially between Dwarf and Tall varieties which led to concerted attention on production of high-quality hybrids. However, the area covered under hybrids is only 3% globally, mainly because of poor accessibility and availability of adequate quality assured planting material. Possibility of mitigating this limitation through seed gardens is indicated. Since no single method is infallible, a comprehensive breeding strategy based on the principles of reciprocal recurrent selection has been suggested, irrespective of the considerable time and resources required for such an initiative. Coconut breeding programmes in the major coconut-growing countries are described in the chapter. The main challenge today is in the development of biotic and abiotic tolerant materials to alleviate the negative impact of serious diseases and moisture stress conditions. Strides in biotechnology are the way ahead to complement future breeding strategies in coconut.

K. U. K. Nampoothiri (✉)
ICAR-CPCRI, Kasaragod, Kerala, India

MS Swaminathan Research Foundation, Chennai, Tamil Nadu, India

Nigerian Institute for Oil Palm Research, Benin City, Nigeria

FAO Consultant, Thiruvananthapuram, Kerala, India
e-mail: unampoothiri@gmail.com

V. A. Parthasarathy
ICAR-IISR, Kozhikode, Kerala, India
e-mail: vapartha@gmail.com

K. U. K. Nampoothiri et al. (eds.), *The Coconut Palm (*Cocos nucifera *L.) - Research and Development Perspectives*,
https://doi.org/10.1007/978-981-13-2754-4_4

4.1 Introduction

Availability of planting material, with high yield potential and other desirable traits, is a basic requirement for profitable coconut cultivation. Though the first record on coconut breeding became available about 275 years ago from the Indonesian publication *Herbarium Amboinense* (Harries 1991a), the palm has a very long history as stated by Foale (2005) 'Through a process of natural selection, over a period of perhaps a few million years, the coconut developed the means to disperse across vast expanses of ocean and take hold firmly on the perilous boundary between land and sea, adapting to fierce windstorms and periodic inundation, thriving unassisted by any other fauna and flora and delivering its fruit in turn to the ocean vehicle for further dispersal'.

Selection of palms based on field experience and raising their progenies for generations has been in vogue ever since the crop was in cultivation for some 2000–3000 years ago (Purseglove 1968; Anon 2016). There is evidence for exchange of coconut varieties between Papua New Guinea and Solomon Islands from very early times. German Government in Papua New Guinea introduced *Niu vai* and *Niu iu* from Samoa in 1911 (Dwyer 1938). Formal and scientific breeding started in 1916 in India (Nair et al. 1991). The special features like long gestation period, height, heterozygosity, large nut size, requirement of considerable space and other resources, poor multiplication ratio, lack of vegetative propagation and monotypic nature of the genus *Cocos* are all factors discouraging scientists to undertake coconut breeding work. In spite of all these odds, considerable strides have been made in evolving high-yielding coconut varieties, which is briefed in this chapter.

4.2 Genetic Diversity

Realising that wide genetic base is the basic requirement for breeding, prospection and collection of various coconut types were given the first priority. This is all the more important, because there are no known wild-/domesticated-related species, to be used as gene pool sources. The first attempt for coconut germplasm collection was made between 1901 and 1910 in Indonesia and Madagascar. But, unfortunately, the First World War put an end to these activities in both the countries (Ohler 1984). A Samoan tall planted in 1912 at Solomon Islands is the oldest accession registered in CGRD (Bourdeix et al. 2005). In India, collections were made right from 1920, mostly from domestic sources. Introductions of a few coconut types were also made from Malaysia and the Philippines. The progenies of these were planted at the Government Farms in Kerala, India. Added emphasis was given to exotic introductions from 1927, and accessions from Vietnam, Indonesia, Thailand, Malaysia, Sri Lanka, Fiji, Papua New Guinea and the Philippines were raised at the Coconut Research Station, Pilicode (now under the Kerala Agricultural University).

The first organised germplasm bank was established in 1940 at the erstwhile Central Coconut Research Station at Kasaragod, Kerala, India (presently ICAR-Central Plantation Crops Research Institute- ICAR-CPCRI) using open pollinated progenies of the introduced accessions maintained at Pilicode (Rao et al. 1993). The collections were conserved as separate accessions in a centralised germplasm bank. Similar work was later taken up in most of the other coconut-growing countries. For details, please refer Sect. 4.11 on global scenario of this chapter.

A prospection for coconut germplasm was conducted in the Pacific islands in 1981 by ICAR-CPCRI, with the financial assistance of International Bureau of Plant Genetic Resources Institute (now Bioversity International). While only a few nuts were collected in the earlier cases, 100 nuts per each accession were systematically collected which were planted in an offshore World Coconut Germplasm Centre in Andaman Islands (Rao and Koshy 1981). With the initiative under Coconut Genetic Resources Network (COGENT), international germplasm exchange was made possible between the member countries. ICAR-CPCRI now has the largest germplasm collection with 455 accessions (323 indigenous and 132 exotic) from 28 countries (ICAR-CPCRI 2016). The institute also hosts the International Coconut Genebank for South Asia (Nampoothiri 1999). The other ICGs are in Indonesia for Southeast and East Asia, Côt d'Ivoire for Africa and Indian Ocean islands, Papua New Guinea for South Pacific region and Brazil for Latin America and Caribbean Islands (Nampoothiri 2003). Table 4.1 gives a list of 1837 coconut accessions in 24 countries.

The strategies for collecting coconut genetic diversity are fairly well established (Guarino et al. 1998). Care has to be exercised to collect only distinct types since many of them seen in different locations and various countries may be only duplicates of the existing ones, with different local names. Adequate attention should be given to collect and keep permanent records of passport data of the collections (from the site of collection itself), seedling characters in the nursery, juvenile traits (till fifth year after planting) as well as yield, various yield attributes and nut analysis data of the adult palms. Cumulative yield of first eight annual harvests gives a fairly good indication of the yield potential of accessions. It has been found that girth at collar, number of days for germination and number of leaves of seedlings are important characters to discriminate coconut accessions. Based on nursery characters, an index has also been developed by Rao and Mathew (1981).

The movement of plants or plant parts between countries or continents entails the risk of introducing exotic plant pests or pathogens. Less-developed countries often lack adequate plant quarantine and diagnostic facilities and are especially vulnerable to the damaging effects of newly introduced diseases. It is extremely important that the risk is recognised and that a minimum risk transfer form of germplasm is chosen, such as in vitro plantlets instead of nuts (Batugal et al. 2005). In view of the importance in following strict quarantine regulations, COGENT has published a manual on the germplasm movement involving seed nuts, embryos and pollen (Ikin and Batugal 2004).

In a perennial crop like coconut, with no means of vegetative propagation, in situ conservation is of special significance. This refers to maintenance of coconut

Table 4.1 Global accessions in gene banks (as in 2012)

Countries and gene banks	Number of accessions
Benin – CRC Sémé Podji	4
Mexico – CICY Yucatan	20
Pakistan	32
Tonga – Ministry of Agriculture	7
China – Wenchang Coconut Research Institute	17
Western Samoa RS	9
Fiji – Taveuni Coconut Centre	11
Ghana – Oil Palm Research Institute	16
Bangladesh – BARI	40
Malaysia – Department of Agriculture, Sabah	45
Solomon Islands – Yandina Research Centre	21
Brazil – EMBRAPA, Brazilian Agricultural Research Corporation	29
Jamaica – Coconut Industry Board	60
Vietnam – Dong Go Experimental Centre	31
Malaysia – Malaysian Agriculture Research and Development Institute (MARDI)	44
Vanuatu – Saraoutou Research Centre	79
Papua New Guinea – Stewart Res. Centre and Cocoa and Coconut Research Institute	60
Thailand – Chumphon Horticultural Research Centre	52
Tanzania – National Coconut Development Programme	72
Cote d'Ivoire – CNRA Marc Delorme Research Station	149
Indonesia – Indonesian Palm Research Institute	203
Sri Lanka – Coconut Research Institute	157
Philippines – Philippines Coconut Authority	224
India – CPCRI (as in March, 2016)	455
Total	1837

Updated to 2012 from Nampoothiri (2003)

diversity in its natural habitat. In situ genetic diversity is lost due to many factors like land use changes, adoption of high-yielding varieties, urban area development, natural disasters and lethal diseases. Farmers conserve coconut varieties based on their own perceptions, disease resistance, quality of tender nut water, etc. Such conservation efforts are to be encouraged, and the palms are to be regularly monitored. Added importance is now given to promote 'conservation beyond gene banks' which includes in situ as well as on-farm conservation (Bourdeix et al. 2017). Duplicate germplasm banks are necessary to safeguard against any untoward catastrophe threatening the collections. In India, Agricultural Universities take the responsibility to collect and maintain local collections as well as duplicate samples of the central germplasm bank.

The coconut population in the Cocos (Keeling) Islands of the Indian Ocean displays fruit characters thought to be consistent with the wild-type form. This population remains unaffected by human intervention unlike that of the main atoll. The genetic uniqueness of the North Keeling population holds possible value for improvement of atoll populations elsewhere (Leach et al. 2003).

Cocos is a monotypic genus with only *nucifera* as the sole species. It is a diploid with a chromosome number of 32 (2n = 32). Coconuts are generally classified into two varieties, viz. tall (*typica*) and dwarfs (*nana*). Talls generally attain a height of 20–30 m, flower only by 5 or 6 years and are cross-pollinated and hence heterozygous. Dwarfs flower early (from 3 years), have a lower height increment (reaching 8–10 m height) and are homozygous because of their self-pollinating nature. There are also various intermediate types which do not show the typical characteristics of either tall or dwarf. Dwarfs are hypothesised to have originated from talls through continuous inbreeding and mutations (Swaminathan and Nambiar 1961; Nampoothiri 1974). According to Perera et al. (2016), SSR allele frequency differences between dwarf and tall accessions and ethnobotanical and geographic information indicate that dwarf coconut originated from a typical domestication event in Southeast Asia. The commercial cultivation of dwarf coconut is limited in the world, representing about 5% of global coconut population. However, dwarfs are now receiving added attention, especially for tender nut water and as a source of disease resistance. For details, please see Chap. 3.

Rao and Pillai (1982) attempted characterisation of coconut germplasm based on fruit component analysis. Descriptors are now available for 74 of the Indian accessions. COGENT also facilitated characterisation of collections and creation of a Coconut Germplasm Data base (CGRD). The catalogue of conserved coconut germplasm provides comprehensive information on 116 accessions (Bourdeix et al. 2010). The accessions which are found to be better performers are identified, their progenies raised and tested in multilocation trials. Based on the results, varieties are recommended for cultivation in a particular ecological region or at a national level. This type of simultaneous evaluation is necessary to reduce the time lag between germplasm collection and its direct use by coconut farmers. Through this process, 29 varieties have been released in India for cultivation (Niral et al. 2014). Details of a few important ones are given in Table 4.2. San Ramon, a Philippine variety introduced to Sri Lanka, was found to have the potential to yield 51% more copra nut^{-1} than the Sri Lankan Tall and 50% more copra than the improved CRIC 60. The mean copra yield for San Ramon (selfed), Sri Lankan Tall and CRIC 60 was 371 g, 240 g and 246 g, respectively. Based on these, Fernando (1998) recommended the use of San Ramon and its crosses for producing copra and oil in Sri Lanka.

It could be seen that these varieties are released not only for yield but also for some of the desirable traits like better tender nut water, stress tolerance, etc. The south Pacific region contains a large genetic resource for the genetic improvement of coconut palms. Considerable diversity for fruit morphology was found that ranged from populations exhibiting wild-type characters in the central pacific to populations displaying domesticated characters in Rennell Island, Papua New Guinea, the Sikaiana and Marquesas Islands. Many populations exhibited continuous variation in fruit morphology between these two extremes which are presumed to have arisen from introgressive hybridisation between wild and domesticated

Table 4.2 Performance of certain varieties released in India

Variety	Germplasm	Important traits	Nut yield (number ha^{-1} $year^{-1}$)	Copra yield (mt ha^{-1} $year^{-1}$)	States for which recommended
Chandra Kalpa	LCT	Drought tolerant, high oil content – 72%	17,700	3.12	Kerala, Tamil Nadu, Karnataka, Andhra Pradesh, Maharashtra
Pratap	Banawali Green Round Tall	High yield	20,826	3.60	Maharashtra
Kamarupa	Assam Green Tall	High yield, cold tolerant	17,877	2.90	Assam
ALR(CN)1	Aliyar Nagar Tall	High yield	21,240	2.89	Tamil Nadu
Kera Keralam	WCT	High yield	26,019	3.53	Kerala, Tamil Nadu, West Bengal
Kalparaksha	MGD	Semi-tall, high copra yield, RWD resistant	13,260	2.85	Root(wilt)disease prevalent areas of Kerala
CARI-C2 (Surya)	Hari Papua Orange Dwarf	Ornamental, orange nuts	24,072	1.77	Andaman and Nicobar islands
Kalyani coconut	Jamaica Tall	High yield	14,240	3.90	West Bengal
Chowghat Orange Dwarf	COD	Dwarf, tender nut, ornamental	19,824	2.78	All regions

Source: Regi J. Thomas et al. (2016)

populations (Duhamel 1993). The wild and domesticated populations were found in disjunctive pockets throughout the area which did not form a part of the clines. This suggests the worthiness of further exploration to find potentially useful populations in the region, before this variability is completely lost (Ashburner et al. 1997).

Some rare types of coconuts also happen to be collected, which are worthy to be preserved. The case of palms, known as ‘sweet husk’ or SWH reported by Batugal et al. (2009) and reproduced here, is a typical example. ‘The husk of unripe fruits, which is usually tough and astringent, is tender, edible and sweet. A sufficiently ripe fruit can be husked by hand, which is impossible in the case of an ordinary coconut. The only place these rare types are found in any appreciable quantities is the islet of Onoiki in the Ha’apai group of Tonga. The islet is so small that it does not figure in most of the maps. Coconuts having a circumference of more than 60 cms have been found in another islet of Niu-foi in Tonga (Henry Teuira 1928). This islet has even taken the name of the newly located coconut variety, as its name -“Niu-foi” meaning “new coconut”. Planting in islets seems to have played some role in varietal creation and conservation. The geographical remoteness of the islets has been a reproductive barrier which has enabled new coconut varieties to become fixed’.

Standardisation of zygotic embryo culture technique has helped to increase the scope of collection, especially those from other countries, circumventing the disadvantage of nut size and ensuring better safety from a quarantine point of view. Koshy and Kumaran (1997) collected 15 accessions from the Indian Ocean islands of Mauritius, Madagascar and Seychelles. Parthasarathy (2001) used this technique to collect four accessions from Sri Lanka. Protocols developed for coconut embryo culture in the Philippines, India and France are very promising with 31–80% of inoculated embryos developing into plantlets in vivo. COGENT has published a manual on germplasm health management indicating the operational management of germplasm movement involving seed nuts, embryos and pollen (Ikin and Batugal 2004).

For long-term conservation of the coconut germplasm in vitro, a working protocol for efficient cryopreservation of embryo (i.e. storage at ultra-low temperature, − 196 °C, in liquid nitrogen) is necessary to be developed. Refinement and standardisation of the current technique, using various genotypes on a large scale, are essential (Engelmann 1999). Molecular tools when perfected would be helpful in further refining the characterisation of accessions.

4.3 Selection

Coconut is, by and large, a cross-pollinated crop and hence heterozygous, throwing out lot of segregants among its progenies. These inter- and intrapopulation variability gives ample scope for selecting the best ones for planting. In fact this was the first option available to growers, much before scientific breeding was started. Most of the populations available at present are the result of informal mass selection process. Selection can be practised at mother palm, seed nut and seedling stages.

4.3.1 Mother Palm Selection

The method, adopted for long, is mass selection wherein open pollinated seed nuts from high-yielding palms are pooled and used as planting material. The advantage of the method is its simplicity. The possibility of such a method is immense since palms yielding as much as 471 nuts palm^{-1} year^{-1}, under rain-fed condition, have been identified as in India (Iyer et al. 1979). Quite a few studies have been made in the earlier years to ascertain the efficacy of mother palm selection. There were always controversies about the efficiency of mass selection, because varied results were obtained, partially because of the cross-pollinating nature of tall variety. Liyanage (1967) observed that in a two hectare block in Sri Lanka, the yield of nuts varied from 15 to 148 nuts palm^{-1} year^{-1}. Tammes (1955) estimated that the best palms in a plantation yield about twice as much as the average. Adansi (1970) could identify only a meagre five palms qualifying as mother palms from a group of 4620 palms in Ghana. Ohler (1984) found that by selecting the best three palms from a population with an average yield of 40 nuts $palm^{-1}$ $year^{-1}$, a yield of 160–175 nuts could be obtained. Some contradictory results were reported from Sri Lanka where a trial was conducted to compare the performance of progenies of high-yielding palms and low-yielding palms and from a heap of bulked nuts. The difference in yielding capacity between the three groups was very small and insignificant (Harland 1957; Liyanage 1958). However, these findings were criticised by Sakai (1960) for the reason that the parent palms for the trial belonged to different populations.

Certain criteria are generally to be adopted for mother palm selection. The palm should have attained stabilised yield. The yield should not be less than 80 nuts $palm^{-1}$ $year^{-1}$ (based on at least 4 years' yield data), and copra yield has to be at least 20 kg $palm^{-1}$ $year^{-1}$. Apart from yield, higher number of female flowers, number of functional leaves on the crown, leaf production and time taken for flowering are important selection criteria. Disease affected or pest prone palms as well as those which show tendency of alternate bearing, barren nut production and bunch buckling are to be totally avoided (Menon and Pandalai 1960).

In such a selection procedure, there are two disadvantages. One is that the yield observed phenotypically is due to both genetic and environmental factors. It is therefore necessary to avoid the non-genetical effect to the extent possible. One way to overcome this is to avoid palms in an obviously site-specific favourable environment. The general practice is to select the best palms in a coconut garden raised in a more or less uniform environment so that the difference in yield can be presumed to be due to genetic factors. Charles (1961) observed that selection of 5% best palms from a group of 128 palms yielded 70% more nuts than the average. Liyanage (1967) reported that progenies of the best 5% palms yielded 14.4% more yield than the parents. Experiments conducted in Côte d'Ivoire have shown that by selecting 7–8% of the best families, a genetic gain ranging between 15% and 30% is achieved (Bourdeix et al. 1989).

The second short fall, in selection of mother palms based on phenotypic yield, is that weightage is given only for the potential of mother palms, the progenies being randomly mated ones, with no idea about the male parent. This defect is minimised if there are a high proportions of desirable palms in the concerned garden. It is therefore evident that, on a long-term basis, if mother palms are selected based on progeny testing, the efficiency of selection can be increased. However, the possibility of pure line breeding is questionable in view of the inbreeding depression reported from many experiments. Dwyer (1938), therefore, recommended 'plant-to-row method' in which seedlings of individual selected palms are grown in separate lines enabling rejection of rows with inferior seedlings. Vigorous seedlings from the selected rows are then planted side by side so that the potential of the mother palms can be assessed based on the performance of their progenies from the beginning. Liyanage (1967) who analysed open pollinated progenies of 104 Sri Lankan Talls found 32% higher yield could be obtained than the average when six best families were selected among them. It is generally accepted that the best 10% of the palms in a population are eligible to be selected for seed nut production, exercising caution to avoid environmental effects. In Sri Lanka, planting materials were produced through open pollination of a pool of about 50,000 selected palms and used in the national replanting programme in operation for four decades (Peries 1998). In Vanuatu, four mass selection cycles by open pollination/intercrossing were conducted from 1962 to 2002, using two Vanuatu Tall populations (Labouisse et al. 2004). There was no increase in the number of nuts palm^{-1} because the selection was mainly based on copra content nut^{-1} which is strongly and negatively correlated with the number of nuts.

The efficiency of mother palm selection can be increased by considering the heritability of characters and correlations between them. Results available across the coconut-growing countries show the significance of selection method in coconut improvement. This is not surprising since studies show comparatively high heritability for many characters of economic importance. The heritability for yield varies from 0.48 to 0.63, whereas the same for yield attributes like number of spathes, female flower production, setting percentage, etc. are higher, indicating the advantage of selection based on these characters (Nair et al. 1991). The high heritability value of 0.67 and 0.95 for nut weight and copra reported by Liyanage and Sakai (1960) meant that these characters could be reliable selection criteria. In West Coast Tall (WCT), (Fig.4.1) studies on the relationship between yield of nuts, copra content per nut, total yield of palm and yield of oil indicated the importance of exercising selection pressure for copra per nut and oil percentage in addition to number of nuts (Bavappa and Sukumaran 1976). For details, please see Sect. 4.8 on genetic studies in this chapter.

Selection is further refined by using the concept of prepotency (Harland 1957) which is defined as a phenomenon wherein the gene combination responsible for high yield potential, in certain mother palms, tends to be transmitted en bloc to progenies even under random mating. This means the high yield potential of such

Fig. 4.1 West Coast Tall. (Photo: MA Nair)

mother palms is transmitted to the progenies, irrespective of the pollen parent. Satyabalan and Mathew (1984) argued that prepotents are palms with good general combining ability. Based on progeny studies, five prepotent palms which gave 35–40% more copra than the population mean were identified in Sri Lanka (Peries 1993a). In India, Iyer et al. (1981) studied seedling progenies of 18 elite palms (bearing above 200 nuts palm^{-1} year^{-1}) and found that some progenies gave a close resemblance to those of prepotent palms. In India, eight prepotent palms could be identified through progeny studies of 30 West Coast Tall palms (Nampoothiri 1991). However, the number of prepotent palms which could be identified from a population is very few, that too after a laborious and time-consuming process of progeny testing. The selection at mother palm stage can be further refined through inter se mating of selected palms. This appears to be more effective, as it allows for a strict selection of pollen parents while retaining the potential for large seed nut production. However, experimental results are not available to assess the yield progress that could be realised from this method.

4.3.2 Seed Nut Selection

Having selected the mother palm, the next stage at which variation can be expected is among the seed nuts, since the nuts harvested from the same palm show variable nut characters for nut size, nut shape, colour, copra content, oil percentage, etc. Selection practised at the seed nut stage is to ensure that seedlings are raised from good and healthy nuts, by discarding all small, malformed and abnormal nuts. There is a preferential selection towards medium-sized nuts preferred by farmers as market price in most of the cases is based on the number of nuts. There are not many studies to prove or disprove the need/efficacy of seed nut selection. Seed nut selection is, however, practised more from an agronomic consideration.

4.3.3 Seedling Selection

Due to the heterozygous nature of the palms, high variability is observed for various characters – right from the germination in the nursery which gives scope for rejection of 'undesirables'. Nursery selection is based on characters which are correlated with high yield of adult palms, such as early sprouting, faster growth rate, early splitting of the unexpanded leaf into leaflets, seedling vigour in terms of girth at the collar, height, number of leaves and colour of petioles.

Correlations between seedling characters and adult palm performance have been reported by many workers (Charles 1959; Nampoothiri et al. 1975). Satyabalan and Mathew (1983) reported high and positive correlation of growth characters like collar girth and leaf production of the seedlings from the fifth month with those of the later months. This finding enabled them to identify palms of superior genetic value based on the growth characters of the progeny even from the fifth month. From studies conducted on seedlings of West Coast Tall variety of coconut in India, it was found that the most important criteria to be used are speed of germination, vigour, girth at collar, number of leaves and splitting of leaves (Thomas and Gopimony 1991a and b). Selection for leaf number and leaf production rate at early stages proved to be a rapid method for genetic improvement (Prabhakaran et al. 1991). In Nigeria, Akpan (1994) found that selection for seedling height and girth resulted in earlier inflorescence initiation in the local Badagry Tall. However, selection pressure is, by and large, arbitrary. Liyanage (1953) found that the advantage of seedling selection is 12% and advocated selection of 50% of the seedlings. In general, 35% rejection is considered satisfactory from a farmers' point of view. Coconut management all over the world adopt culling of weak seedlings from the nursery. Needless to emphasise that very strict selection criteria are to be used when the material is used for further breeding. The selection criteria and selection pressure are left to the discretion of the breeder, the results being better when there is stringent selection for as many characters as possible. Rajesh et al. (2013) suggested use of molecular markers in identifying the good progenies.

4.4 Hybridisation

Ever since the report of coconut hybrids in 1932 by Patel in India (Patel 1937), the attention was completely turned to hybridisation between the two major coconut varieties, the talls and dwarfs. Though the first hybrid was produced by Marechal in 1928, by crossing Niu Leka and Malayan Dwarf in Fiji, the work was discontinued due to financial crisis (Marechal 1928; Ohler 1984), and Patel was the first to report hybrid vigour in tall × dwarf hybrids (Menon and Pandalai 1960). This was followed some 15 years later by Sri Lanka and in the early 1990s by IRHO and its partners. Tall × Dwarf (T × D) as well as its reciprocal (D × T) have been found to be early bearing intermediate in height and expressing hybrid vigour resulting in higher yield (Fig.4.2).

The major problem faced was the highly variable performance between hybrids involving different parental palms (Harries 1991b). This was minimised to a certain extent by identifying parents on the basis of combining ability tests (Line × Tester, diallel) for selection of parents. This was well evidenced in the production of MAWA hybrids (Malayan Dwarf × West African Tall) in Côte d'Ivoire. Bourdeix et al. (2016) estimated a 20–30% genetic gain when seven to 8% of parents are selected. Dwarf × Tall crosses are many a times preferred because of the advantage of using comparatively homozygous dwarfs as mother palms. The feasibility of identifying genetically pure mother palms through the use of molecular markers like SSRs has also been indicated (Regi et al. 2015). In India, diallel analysis involving 16 parental lines indicated Gangabondam Green Dwarf (GBGD) and Laccadive Ordinary Tall (LCT) to be good general combiners, and LCT × GBGD was identified as the combination with the best specific combining ability for copra and nut yield (Nair et al.

Fig. 4.2 D × T coconut hybrid. (Photo: KUK Nampoothiri)

2016). A number of hybrids have been produced by crossing different dwarf and tall cultivars originating from various geographical regions. In India, West Coast Tall × Chowghat Orange Dwarf (COD) and Chowghat Green Dwarf (CGD) and their reciprocals also showed their superiority in precocity and yield. Although both T × D and D × T hybrids are high yielding, D × T has certain advantages such as short stature of mother palms facilitating easier emasculation and pollination as well as easiness in identifying hybrid seedlings, especially if COD is used as the mother palm.

It may, however, be pointed out that farmers often express some concern about these hybrids in terms of bunch buckling, high input requirement, as well as vulnerability to moisture stress and pests and diseases (Batugal et al. 2009). Since the performance of hybrids vary with the environment, they are usually tested in various locations within the country using local cultivars as control before their release for cultivation. COGENT facilitated hybrid multilocation trials involving three African countries and three Latin American and Caribbean countries resulting in the identification of 19 early bearing and high-yielding coconut hybrids (Batugal et al. 2009). Hybridisation involves selection of parent palms, emasculation of mother palms, pollen collection from male parent, pollination and bunch-wise harvest. For details, please see Menon and Pandalai (1960) and Ohler (1984).

In India, 19 hybrids with a potential for high yield and possessing other desirable characters have been released. The best among them yield as much as 167 nuts palm^{-1} year $^{-1}$equivalent to 6.28 mt copra ha^{-1} year^{-1} as compared to 80 nuts palm^{-1}year $^{-1}$ and 1.5 t copra ha^{-1} year^{-1} realised from the local WCT. Details of certain hybrids released in India are given in Table 4.3.

Apprehension is often raised on the longevity of the hybrids. It is now known that yield of hybrids does not decline even after 80 years as is evidenced from the performance of hybrids planted as early as in 1935 in India. The hybrids can be further improved using information on the individual combining ability and by exploiting within population variability (Gascon and Nuce de Lamothe 1976). It was possible to improve the yield of PB121 hybrid from 15% to 25% in just one generation of half-sib mating (Batugal et al. 2009).

Table 4.3 Performance of released hybrids in India

Hybrid	Parents	Important traits	Nut yield (number ha^{-1} year^{-1})	Copra yield (mt ha^{-1} year^{-1})	States for which recommended
Chandra Sankara	COD × WCT	High yield	20,532	4.27	Kerala, Karnataka, Tamil Nadu
Kera Sankara	WCT × COD	High yield, drought tolerant	19,116	3.78	Kerala, Karnataka, Maharashtra, Andhra Pradesh
Chandra Laksha	LCT × COD	High yield, drought tolerant	19,293	3.76	Kerala, Karnataka
Kalpa Sreshta	MYD × TPT	Dual purpose variety, High yield	29,227	6.28	Kerala, Karnataka
Kalpa Sankara	CGD × WCT	Tolerant to root (wilt) disease, high yield	14,868	3.2	Root (wilt) disease prevalent tracts of Kerala
VHC-4	LCT × CCNT	High yield	28,497	4.27	Tamil Nadu
Vasista Ganga	GBGD × PHOT	High yield	22,125	3.88	Andhra Pradesh, Karnataka
Ananta Ganga	GBGD × LCT	High yield	22,656	3.84	Andhra Pradesh, Karnataka
Konkan Bhatye coconut hybrid 1	GBGD × ECT	High yield	20,532	3.47	Maharashtra

Source: Regi J Thomas et al. (2016)

Though most commercial hybrids are D × Ts, and T × Ds, certain T × T hybrids have a comparable potential. They are more suitable as a main crop for intercropping and offer better prospects for long-term genetic progress. So far, Thailand is the only country where T × T hybrids have been released to farmers (Anupap et al. 1992). D × D crosses are not very popular among coconut breeders, but more than 40 crosses are being evaluated in Côte d'Ivoire and Malaysia since 1993. An important feature of D × D hybrids is their high genetic homogeneity, since the two dwarf parents are close to being pure lines (Batugal et al. 2009). Hybrids have been recovered even from open pollinated progenies. Komadan, a variety identified and perpetuated by a farmer family in Kerala, India, is a typical example. These are rare recombinants (10–15%) arising due to the natural crossing of Chowghat Orange Dwarfs with the neighbouring talls and hence called NCDs (natural cross dwarfs), identified at the seedling stage itself, primarily by their typical bronze colour of the petiole and high vigour. Studies have shown that Komadan has maintained its genetic identity over generations with respect to economically important characters like copra and oil content which reflects its prepotent nature (Satyabalan 1956; Ninan and Satyabalan 1964).

The PB series of hybrids were produced in Côt d'Ivoire. Comparisons made between plantations consisting of PB121, PB111 and 141 with local West African Tall (WAT) showed that the hybrids were superior for precocity, number of nuts and copra content (80% higher) than WAT (de Taffin et al. 1991). In Sri Lanka CRIC 60 (Tall × Tall) and CRIC 65 (Yellow Dwarf × Tall), yielding higher number of nuts with higher copra content, have been recommended for cultivation. A large number of hybrid combinations have been field tested in India, the Philippines, Indonesia, Sri Lanka, Côt d'Ivoire and Jamaica using local talls and dwarfs. More than 400 hybrids were developed till 1993, worldwide under established coconut improvement programmes using various breeding strategies (Batugal et al. 2009). The technology gap between average yield in farmers' fields and that of research stations in 15 coconut-growing countries, in terms of nuts or copra yield, was estimated to be 33–84% (Batugal 2005). However, these results have to be viewed in the light of the environmental influence on yield of coconut. Recent developments in biotechnological tools have resulted in considerable advancements in the identification of parental lines and hybridity testing (Preethi et al. 2016).

4.4.1 Seed Gardens

As hybrids became popular among the farmers, the demand for planting material increased phenomenally, and it was not possible to meet even a portion of the demand, which leaves the farmer with no other alternative than to plant materials of uncertain quality from various sources. In India, for example, the annual production of quality planting material from authentic institutions is only 4.2 million as against a demand of around 14 million coconut seedlings (Jayasekhar et al. 2016). The reasons for limited hybrid seed production are many, such as inadequacy of mother palms, cumbersome procedure for hand pollination of individual palms and low multiplication rate. This is one of the primary reasons for the poor coverage of area (0.1–14%) under hybrids. It is estimated that on an average, the hybrids represent only <3% of the 10 to 20 million coconut palms planted yearly worldwide. Establishment of accredited nurseries involving registration of mother palms as well as certification of seedlings is one of the ways to improve the situation. This should be read with the fact that the estimated coconut technology gap in terms of nut and copra yield ranges from 33% to 84% (Batugal and Bourdeiux 2005).

Establishment of seed gardens was suggested to circumvent the issue of inadequacy of hybrid planting material. The first isolated seed garden in the world was established in Sri Lanka at Ambakalle in 1955, for producing seed nuts of local talls. Seed gardens using inter se mated progenies of selected talls and pure dwarfs were later established in that country during 1959 (Peries 1993b). In Côt d'Ivoire, seed gardens were established with West African Tall and Malayan Yellow Dwarf in a 1:5 ratio, talls being considered only as a pollen source. But it did not have scope for any flexibility, restricting production of only D × T hybrids. Therefore, it was replaced by a more flexible system. Seed gardens came to be planted with only mother palms, isolated to avoid pollen contamination. Pollen from the desired male parents is dusted on to the inflorescences of the mother palms, which has already been emasculated. Several combinations with dwarfs as female parent can be produced in this system. In India, seed gardens were established in different states, including the isolated one in a forest where selected progenies of West Coast Tall and Laccadive Ordinary were planted in two separate blocks. Chowghat Orange Dwarf and Gangabondam were planted in adjoining rows of the talls in such a way that the two rows of talls and dwarfs form a double hedge with every dwarf coming in the centre of the talls. It is possible to produce D × T, T × D or T × T hybrids by adjusting the emasculation of mother palms as per the requirement (Bavappa 1973). Absence of male sterile lines in coconut is a handicap in this effort.

Following the success of crossing inbred lines to evolve high-yielding hybrids in annual crops, efforts were made in coconut also to produce inbreds with very little success due to inbreeding depression. Four generations required to create 95% homozygous structures will need 25–60 years. Studies on first- and second-generation inbreds of 18 WCT palms showed that the selfed progenies were inferior to their grandparents and sibs, indicating depression from S_2 generation (Sukumaran Nair and Balakrishnan 1991). Results from inbreeding populations in Mapanget Tall for three generations in Indonesia showed that inbreeding depression was evident in all vegetative characters except plant height and number of leaflets (Novarianto and Miftahorachman 1991). Same was the experience with Markham valley Tall by Sugimera et al. (1994). Thus, the possibility of producing homozygous lines through inbreeding was dismissed.

4.5 Breeding Strategy

As none of the above-mentioned methods are foolproof, it is worthwhile to incorporate all the possible methods in evolving a hybrid. A general scheme, following the reciprocal recurrent selection (RRS) method, is indicated in Fig. 4.3 (Bavappa and Nampoothiri 1974; Nampoothiri 2016).

Here a procedure is described for the improvement of the D × T hybrids, as an example. Dwarf and tall coconut varieties (the base populations) show complementarities and good reciprocal combining ability. Hence, the tall population is improved in respect to the dwarf population, and this improvement is reciprocal for both. At each stage, mother palms in both the populations are selected carefully, and their progenies are produced (preferably by inter se mating). The best seedlings in the nursery are then selected to raise the parental palms for the next stage. For stage 2, parental palms are selected not only based on physical attributes but also based on the combining ability assessed on the basis of the performance of hybrid progenies. Inter se mated progenies of these selected parents are raised from which the best seedlings are taken forward to raise progenies. At each stage D × T hybrids can be distributed to farmers, in view of the time lag of 10–15 years for each stage. These steps are continued stage after stage so that the potential of the hybrids distributed

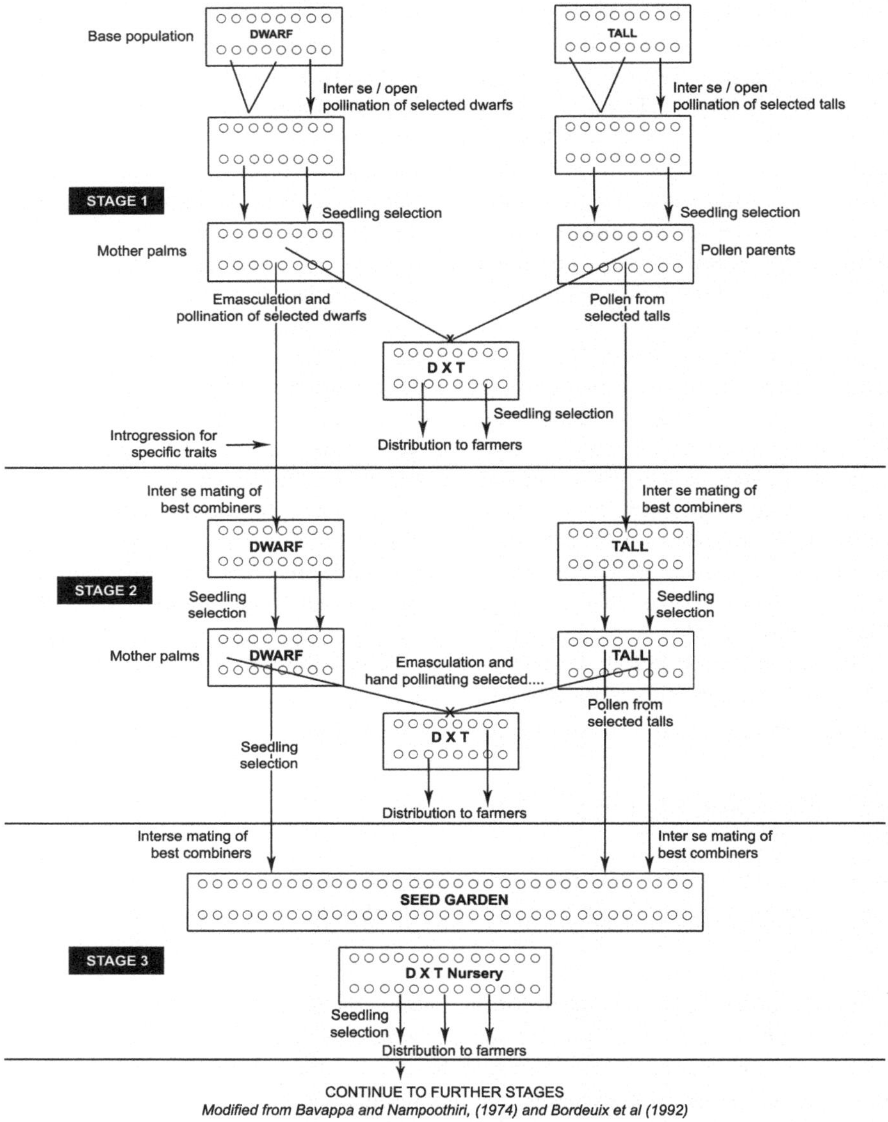

Fig. 4.3 Breeding strategy for D × T coconut hybrid production

gets higher and higher as the process continues due to the additional homozygosity as well as the improved genetic base of both the parents. To meet the increased demand of these improved hybrids, proper seed gardens have to be raised at stage 3. It is an open-ended system providing scope for introgression from desired sources identified during the programme. This only indicates a general scheme which can suitably be modified based on the local situations, additional information becoming available and the nature of populations available. Similar method can be adopted for other hybrid combinations, the major difference being in the selection of concerned base populations. Bourdeix et al. (1993) followed a comprehensive breeding scheme in Côt d'Ivoire for improving T × T and D × T hybrids. To be fully efficient, the RRS method requires the evaluation of at least 100 half-sib families per year. Including the hybridisation test, the selfing and the intercrossing of the parent palms, this would require planting of at least 100 ha each year for field experiments. This level of activity was reached in Côte d'Ivoire in the 1980s and in the Philippines in the 1990s. Currently, there is no research programme allocating such levels of resources to coconut breeding, which is a matter of concern (Batugal et al. 2009).

4.6 Complex Hybrid

A coconut hybrid with another hybrid or a variety is called a complex hybrid. While F_1 hybrids have helped in varietal improvement to a great extent, the method limits a complete exploitation of polygenic recombinations of crosses. Therefore, efforts were made for testing multiple crosses to develop varieties with desirable multiple characters. Harries (1991a) suggested the production of three-way hybrids using D × D, D × T or T × T crosses. Some plantations with composite characters could be exploited for nuts or for hybrid seed nut production depending on the demand. A three-way cross MYD × (WAT × RAT) planted in Côte d'Ivoire in 1976, under average management, yielded only 77% more than the WAT. In the 1990s, three-way hybrids (D × T) × T and (T × T) × D were produced in Thailand, which is not known to have reached a logical conclusion (Petchpiroon and Thirakul 1994). The first experiment testing of complex hybrids was established in Côte d'Ivoire in 1976 (Bourdeix et al. 1993), essentially searching for exceptional coconut palms that could be propagated rapidly, when a method of vegetative propagation would become operational. However, successful coconut in vitro propagation technique has not been developed hampering the aim of this experiment (Batugal and Bourdeiux 2005).

Complex hybrids between dwarf varieties were produced in Côte d'Ivoire in the lethal yellowing disease-resistant programme (Batugal et al. 2009). However, no breeding programme went further than the second-generation (D × T) × (D × T) or (D × D) × (D × D) since the 1990s, except in the Philippines which aimed to develop Makapuno Dwarf autogamous varieties involving Makapuno Tall and normal dwarf varieties (Nunez and de Paz 2004). One could hope to have hybrids with genes from many accessions in the future.

4.7 Synthetic Variety

Synthetic variety is an open pollinated variety purposely created to be more adaptable and stable over many generations due to its wide genetic base. Breeding for a synthetic or a composite coconut variety (CCV) has been first proposed in India and Sri Lanka. Subsequently, more attention was given to the development of CCV by Santos (1990) and Santos and Rivera (1994) in the Philippines. Philippine Coconut Authority (PCA) started work on synthetic variety in 1979. A base population of palms having a high degree of balanced heterozygosity was produced, and a breeding scheme was developed that would allow individual palms to mate at random and maintain high heterozygosity from generation to generation. Single crosses of six tall coconut cultivars and varieties, four local and two exotic ones, which had good combining ability were used. The intercrossed or open pollinated nuts from the first-generation constituted the second-generation synthetics, with no considerable loss in yield (Anon 2006). This hybrid as well as its open pollinated progenies was distributed to farmers as Complex Coconut Varieties (CCV).

San Ramon Orgullo Tall is a synthetic variety developed from a combination of 15 different coconut hybrids in the Philippines. (Please see Sect. 4.11 Global scenario in this chapter.) In spite of its simplicity, this method has some disadvantages. Though synthetic varieties are expected to have more stability and wider adaptability due to the involvement of several selected parents, compared to single cross hybrids, one major constraint in developing synthetic varieties in coconut is that the poly genes controlling the many desirable characters may attain an equilibrium only after many cycles of inter mating since inbreeding to purify parental lines is not possible to be resorted to (Batugal 2005). They are likely to be less productive than the best F_1 hybrid that could be made from the parental varieties. Though planting material can be reproduced by open pollination by the farmers themselves, the progenies will progressively become less productive because of pollination from the many coconut varieties planted in the garden, as is the usual practice of most of the farmers. Currently, there are very few countries evaluating multiple crosses to develop varieties with desirable multiple traits. Induction of mutation and polyploidy has also been experimented in India, though on a very limited scale, without any worthwhile result.

4.8 Genetic Studies

Genetic studies, which are very helpful in designing breeding strategies, have been few in coconut, for inherent reasons already explained.

4.8.1 Genetic Variability

Genetic variability among individuals of a variety or population of a species results from the many genetic differences between individuals and may manifest in differences in DNA sequence, in biochemical characteristics (e.g. in protein structure or isoenzyme properties) or in physiological properties (Ramanatha Rao and Hodgkin 2001). Nut yield, copra weight, copra yield, whole nut weight and husked nut weight have shown high heritability and genetic advance (Ganesamurthy et al. 2002; Thomas and Gopimony 1991b; N'Cho et al. 1993). Generally, economic characters showed higher genotypic coefficients of variation (16–22%) compared to vegetative and reproductive characters. Among the many characters, heritability was the highest for petiole length (52%), followed by 45% for both oil content and nut yield (Sindhumole and Ibrahim 2000a). Since such characters may be controlled by additive genes, pure line selection could improve them. On the other hand, characters like oil content, diameter of husked nut (polar as well as equatorial) and thickness of meat had high heritability coupled with low to moderate genetic advance as percentage over mean. This indicates nonadditive gene action which could be exploited by resorting to heterosis breeding. Discriminant function analysis showed husk: nut ratio, followed by thickness of meat and oil content to have more positive weight. The relative efficiency of selection based on discriminant function analysis was more rewarding compared to direct selection. Repeatability coefficients obtained by the variance analysis and structural analysis (correlation) showed the smallest values for fruit and solid albumen production. Considering the repeatability estimates by the main components method (covariance), five and three evaluations could be recommended for fruit and solid albumen production, respectively (Farias Neto et al. 2003). Prabhakaran et al. (1991) reported that selection for number of leaves and leaf production rate at early stages would quicken genetic improvement. They indicated that selection for nut yield can be effectively done, as early as in sixth year after planting, on the basis of number of functioning leaves. Varagas and Blanco (2000) in Costa Rica, evaluated tall coconut cultivars from Pacific coast of Costa Rica and the Philippines (San Ramon, Tagnanan and Laguna) for fruit characteristics. He found that most of the introduced cultivars showed high heterogeneity. A cluster analysis, using the Ward method, classified the palms into four groups with high internal homogeneity. Some of the palms from the Costa Rican Pacific area had nut characters similar to the San Ramon and Tagnanan groups but not with Laguna group. At the association level used (semi-partial $R^2 = 0.10$), another group was constituted which included the remaining palms sampled from the area. Jayalekshmy and Rangasamy (2002) did cluster analysis and grouped 30 genotypes into six clusters, based on 20 morphological traits, including vegetative, inflorescence and fruit characters. All the dwarf cultivars were grouped into one cluster. Talls fell into three clusters. T × D hybrids came along with the Tall cultivars, and the D × T hybrids were clustered with dwarf cultivars. Laccadive Micro formed a separate cluster, where the intercluster distance was the highest. Andaman Giant and San Ramon which show distinguishable nut characters constituted a separate cluster. The nut characters were efficient in assessing genetic divergence. Their analysis indicated the possibility of obtaining promising progenies from the parents belonging to divergent clusters.

4.8.2 Correlations

From coefficient of correlations calculated involving 19 floral, vegetative and yield traits, Kalathiya and Sen (1991) suggested that the number of spadices and duration of female phase, which are correlated significantly with nut yield, could be used as selection criteria for nut yield improvement. A number of leaves and length of spadix were significantly correlated with number of spadices. The number of female flowers per spadix was also correlated with spadix length. The copra yield in coconut is reported to be strongly and positively correlated with nut yield, copra weight, kernel weight, whole nut weight and dehusked nut weight. The direct effects of dehusked nut weight, percentage of husk to whole nut weight, percentage of kernel to whole nut weight, copra weight and nut yield on copra yield were positive and high. These characters are to be given emphasis while selecting for copra yield improvement (Ganesamurthy et al. 2002). In a study of yield and its morphological and chemical components in low-, medium- and high-yielding types, Narayanan Kutty and Gopalakrishnan (1991) found that yield was correlated with total number of leaves on the palm ($r = 0.69$), leaf length ($r = 0.68$), N and K content of leaves ($r = 0.42$ and 0.61, respectively), total chlorophyll content ($r = 0.63$) and sugar content ($r = 0.37$). The correlation between yield and total phenolic constituents of the leaves was negative ($r = -0.55$). There was a marked direct effect of leaf length, K content and chlorophyll content on yield and a yield prediction model based on these traits showed 74% efficiency. The morphology of the palms in relation to nut yield has also been reported. In a study on WCT palms, the number of leaves in mother palms showed positive correlation with number of bunches, whereas number of nuts per bunch showed negative correlation with most of the nut and seedling characters (Thomas and Gopimony 1991a). High nut yield was associated with high stem circumference, many closely spaced short leaves with short petioles and a few broad leaflets as in the case of Pratap, a high-yielding variety released for cultivation in Maharashtra.

A total of 43 coconut germplasm accessions were characterised for nut yield and fruit component traits by Geethanjali et al. (2014). Fruit length, fruit breadth, fruit weight, nut weight, kernel weight and copra weight nut^{-1} were found to be positively correlated with each other but showed significant negative correlation with the number of nuts produced palm^{-1} year^{-1}; shell thickness and husk thickness were not correlated with any of the fruit components. Nut yield and copra content nut $^{-1}$ had positive direct effect on the total copra yield palm^{-1}. They argued that equal consideration should be given for both nut yield and copra content nut^{-1} while selecting elite genotypes.

Nampoothiri et al. (1975) reported phenotypic and genotypic correlations of nine characters with yield. Girth at collar of seedlings, time taken for flowering, spathe production and number of female flowers were found to be correlated with nut yield. There was significant genetic correlation between yield and number of leaves at seedlings stage. In a study on dwarf coconuts, Aragao (2000) worked out correlations between 15 pairs of phenotypic, genotypic and environmental characters. Fruit weight (FW), fibre weight (FiW), nut weight (NW), copra weight (CW), solid albumen weight (SAW) and liquid albumen weight (LAW) were the characters evaluated. Phenotypic and genotypic correlations, in general, were significant and positive (at one per cent level) between FW and FiW (0.97 and 0.98), FW and NW (0.74 and 0.76), FW and SAW (0.74 and 0.83), FW and LAW (0.53 and 0.52), FiW and NW (0.55 and 0.60), FiW and CW (0.42 and 0.38), FiW and SAW (0.60 and 0.75), FiW and LAW (0.36 and 0.37), NW and SAW (0.87 and 0.82), NW and LAW (0.77 and 0.0.79), CW and SAW (0.55 and 0.36) and SAW and LAW (0.41 and 0.31). The correlations were negative between CW and LAW (−0.47** and − 0.88**) and NW and CW (−0.14**and − 0.30). The results showed a strong association between these pairs of characters due to genetic effects. The strong and positive association noted between majority of these pairs of characters in the dwarf coconut should favour selection for these characters. Association between FW and CW (0.67**) and NW and CW (0.76**) was due to environmental effects. If selection is performed with the aim of increasing fruit and nut weight, the outcome is uncertain with regard to copra weight.

Path analysis has been used by some workers to understand the pattern of relationship between characters. Relationship of 12 traits with oil yield palm^{-1} was evaluated. A number of nuts palm^{-1}, fruit weight, kernel weight, copra weight, fruit quality value and oil content had marked effects on oil yield. A number of nuts palm^{-1}, fruit weight and kernel weight had considerable direct effects. Oil yield could be increased by direct selection for these three traits or by simultaneous progeny selection for nut palm^{-1} and fruit weight or oil content. In another study in India by Sindhumole and Ibrahim (2000b), it was revealed that reproductive characters exerted less influence on nut yield compared to vegetative characters. The effect of trunk height was the most prominent but was highly modified due to indirect effects through trunk girth and female flowers. Among the reproductive characters, female flowers spadix^{-1} had the highest direct effect, whereas, the direct effect of spadix production on nut yield was the lowest.

4.8.3 *Stability*

Coconut, being a perennial crop, should perform stably to get regular yield and hence the importance of studies on stability. Balakrishnan et al. (1991) studied stability in 32 cultivars grown at Pilicode, India, by non-parametric procedures using

20 year's yield data. The two non-parametric measures of stability used were mean of the absolute rank differences of the cultivar over the environments and variance among the ranks over the environments. Significant differences were found in the stability of cultivars. West Coast Tall, Java, Fiji, Laccadive Ordinary, Philippine Ordinary, Andaman Ordinary and Godavari were stable. Cochin China, S. S. Green and Laccadive Small were among the highly unstable cultivars, mainly as a result of their biennial (alternate) bearing tendency. When cumulative yield for 2 years was considered, varieties like Cochin China, S. S. Green, Andaman Giant, Kappadam and Basanda were found stable although they were not stable in annual yield owing to irregular bearing. In another study in the Konkan region of India, the mean square due to environment + (genotype × environment) was found to be highly significant contributed by the sib components, environment (linear) and genotype × environment (linear). The population studied by them was free from the unpredictable component of interaction. On the magnitude of stability parameters, variety Pratap was the best genotype for general cultivation, whereas T × D (orange) and Benawali Yellow Round were better adapted for cultivation under fluctuating environmental conditions. Morpho-physiological variation and phenotypic plasticity in Mexican populations of coconut was reported by Daniel-Zizumbo and Colunga-GarciaMarin (2001). They studied the pattern of variation in 19 morphological and physiological characteristics of leaves and their phenotypic plasticity in 18 Mexican coconut populations experimentally grown under similar conditions and in the presence of lethal yellowing disease. They found five ecotypes, viz. Atlantic Tall; Pacific Tall 1, 2 and 3; and Malayan Yellow Dwarf, were differentiated by means and plasticity for these characteristics. The traits that best differentiate these ecotypes were the leaf length, number of leaves per unit time and percentage of proximal rachis in leaf. The ecotypes corresponded to the population groups previously observed in a study of morphological characterisation of fruit in situ. Diallel analysis in India with 36 crosses involving nine coconut cultivars of diverse origin showed that girth, height, total leaf production and leaf production during the year were significantly different among the characters studied. Laccadive Ordinary × amaica and Fiji × Jamaica exhibited, high specific combining ability (sca) effects for total leaf production and leaf production during the year. Laccadive, Jamaica, Java, Fiji and San Ramon cultivars proved as useful tall parents (Sukumaran 1982). Different values and relationships are reported by different workers, which leave the breeder with no other option than exercising selection for as many characters as possible which have shown positive correlation with yield in one study or the other.

4.8.4 Field Experiments

Timely collection of data and safe keeping of records for a long period are primary requirements for deriving meaningful results from coconut breeding experiments. The data once missed is lost forever. In the case of germplasm collection, passport data have to be recorded at the time of collection itself. Problems faced in coconut breeding trials are different from those of annual crops because of the long gestation period, height, perennial nature, larger area required, lack of uniformity in the experimental material, time lag between treatments and manifestation of effects, differences in pre-bearing age as well as yield realisation from harvests throughout the year. Measurements are usually taken right from the time of planting. The characters recorded on seedlings are the height, girth at collar region, leaf splitting, number of leaves and leaf area. Thereafter annual measurements are made on juvenile characters, and the date of emergence of the first inflorescence is noted when the palms come to bearing. Performance of palms is measured based on number of bunches, number of nuts, size of nut, copra content, copra weight, other nut characters and oil percentage (based on laboratory estimates). The annual yield increases progressively as the palms grow and therefore the actual assessment is possible only when the yield stabilises, i.e. in 8–10 years. Coconut palms are generally harvested 8 to 12 times a year, the number of nuts obtained being different in different harvests. Therefore the cumulative figure from all the harvests in a year is considered as nut yield. While it is possible to monitor each of the harvests and get data of annual yield in research stations and well-maintained farms, it is not feasible when palms are to be evaluated in cultivators' gardens or when an assessment has to be made in an unknown population (for germplasm collection, etc.). In such cases a reliable estimate can be made by using the regression eq. $Y = -0.527 + 0.914\ x$ where y is the estimated annual yield and x is the count at the time of observation. One another problem is that many populations show annual/biennial bearing habit, giving high and low nut yields in the successive years. Analysis of mean yield for two consecutive years is recommended to reduce this error. Keen observations are also to be made on resistance to biotic and abiotic stress as well as special characters like nut water quality, etc., depending on the experimental objectives. In view of the seasonal differences in nut characters, a sample of two nuts is to be analysed six times a year for 4 years on 30 palms to get a reliable estimate of nut characters.

The most common design followed for breeding trials is randomised block design. All over the world, the number of palms used has fluctuated from <10 to over >140, depending on various local factors (Batugal et al. 2009). Marshall and Brown (1973) have estimated that 60–100 palms could represent a natural population of allogamous plants. The STANTECH manual recommends a minimum number of 96 palms for hybrid evaluation (Santos et al. 1996), which calls for resources unaffordable by many research organisations. Small plots with more number of replications are now preferred. Jacob Mathew (1991) found that a plot size of six to eight palms as adequate, though larger plots understandably, increase the accuracy. The number of replications is generally decided based on available experimental area, residual degrees of freedom required and the efficacy required for estimates. In addition to standard statistical considerations, effects of competition between palms in the adjacent plots should also be considered, which is overcome by providing at least one nonexperimental guard row.

4.9 Breeding for Special Characters

Although coconut is known for its multifarious uses, its potential has not been fully exploited except for yield of nuts, copra and oil. An expanded product range will help the crop in moving away from the vegetable oil sector, especially in view of the growing competition from other vegetable oils. There are varieties which possess special attributes. Their identification and use in breeding programme deserve breeder's attention. Selection of the trait will depend on the market demand and availability of the local variability.

4.9.1 Tender Coconut

Tender coconut water is recognised as a health drink for a very long time, and it is becoming more popular with the present increased awareness on health. It is possible to use 20–25% of the fresh nuts to satisfy the domestic demand, the tender nuts often receiving a better price than mature nuts (Nampoothiri 2017). This is apart from the scope for exports, Sri Lanka, Malaysia, Thailand and the Philippines taking a lead in this. Suitable varieties will be those with better volume (say 350 ml nut^{-1}), sweetness (7 g 100 ml^{-1} of water) and good flavour of water as well as taste and consistency of kernel and mineral content, at the 7–8 months stage of the nut. Chowghat Orange Dwarf was released in India for large-scale cultivation as a tender nut variety (ICAR-CPCRI 2014). Galas Green Dwarf (Philippines), Nam Wan and Nam Hom-aromatic coconut (Thailand), Green Dwarf-BGD (Brazil) and Malayan Dwarfs and King Coconut (Sri Lanka) are some of the examples (Thampan and Gopalakrishnan 2010). Such varieties and even exceptional individual palms could be multiplied, and gardens could be established to produce inter se material for establishing further improved generations. In view of the increased demand for tender nut, certain private nurseries now concentrate their efforts in producing hybrids which are suitable for tender nut purposes, mainly using selected Malayan Dwarfs as mother palms. 'Pandan' an Aromatic Green Dwarf, in Malaysia, is preferred for fresh beverage, due to its special aromatic flavour. Brazil Green Dwarf (BGD) is also renowned for its very sweet tender nut water. Under proper management, BGD produces around 150 nuts palm^{-1} ha^{-1}, and about 59,000 ha of this cultivar are planted in Brazil annually. Both Aromatic and Brazil Green Dwarf are currently being improved in their respective countries (Batugal et al. 2009).

4.9.2 Toddy and Sugar

There are cultivars which yield 1.5–2.0 l of toddy palm^{-1}daily with a sugar recovery of 15%. Neera is the unfermented inflorescence sap, with good market scope. In Sri Lanka, the hybrids give 792 l of toddy year^{-1}. Thailand varieties such as Tha-le-Ba,

Suricha, Sai Bua, Theung Bong, Kathi and Khi Kai yield up to 4 litres of toddy. These cultivars, if used for toddy tapping and sugar production, can give 7 to 10 times more income than from nut production. The employment opportunities are also increased due to this activity. Malayan Dwarfs yield up to 3.4 l of toddy daily. The toddy is reported to be sweeter than the one from talls. Dwarf varieties are at an advantage because of the easiness in tapping due to the small stature and the possibility of high density planting. It is also possible to use the same palm for tapping and nut yield. This can be done by using alternate spathes for tapping or tapping only the top portion of the spathe, leaving the lower portion for nut production (Thampan and Gopalakrishnan 2010). The low glycaemic index (35) of coconut sugar is an added advantage for health conscious consumers. Considering these facts, breeding varieties for toddy and sugar production and studies on economic feasibility of establishing gardens, especially with dwarf varieties, solely for this purpose deserve attention.

4.9.3 Husk

Production and marketing of coir and its products help to derive additional income and thus enable better stability to the coconut industry. In some of the countries like India and Sri Lanka, where coir industry has recorded substantial progress, production of husk attains special significance. There are varieties and hybrids which produce nuts with a higher proportion of up to 54% husk (Malayan Yellow Dwarf × Jamaica Tall). But unfortunately there is a proportionate reduction in the kernel component, to say 24%, in these cases. There has not been any concerted effort to evolve 'high husk varieties'. This has to be read with the fact that the available husk is not being fully utilised especially because of the predominance of scattered small holdings. It appears more logical to advocate breeding for varieties which produce nuts with a comparatively high husk (40%) as well as kernel components (30%).

4.9.4 Dwarfness

Dwarf coconut types have the advantage of reduced harvest cost and feasibility for high density planting. Developing compact dwarfs is another possibility in coconut breeding for which, apart from the identified dwarf cultivars, every dwarf mutants from natural populations can be used. Bourdeix et al. (2017) reported the presence of compact dwarfs in French Polynesia and the Cook Islands. One mutant identified in India from Lakshadweep coconut populations has shown extreme dwarfism compared to the dwarf cultivars available in the germplasm (Nair et al. 2016). Compact dwarf hybrids of MYD × NLAD have been developed in India (CPCRI 2013). The use of such dwarf progenies would help in changing the plant type enabling high density planting of coconut.

4.9.5 Makapuno Palms

Some palms exhibit unique features, though with limited distribution. Makapuno type (gelatinous mutant coconut) of the Philippines is a typical example. In this, a few tall palms produce characteristic nuts with soft- and jelly-like endosperm filling the entire cavity. While these are not good as mature nuts, they are considered as delicacies and have high demand in ice cream industry. This character is governed by a single recessive factor, and the palm which occasionally bears Makapuno nuts is heterozygous for the character. Two issues in their propagation are that only few nuts in a bunch show this special trait and that the nuts do not germinate. Its propagation is now possible through embryo culture. As they are found only in the Laguna Tall variety which are late bearers and highly cross-pollinated, embryo-cultured homozygous Makapuno palms are developed from these heterozygous Makapuno-bearing sources. At the National Coconut Research Centre in the Visayas, 4 Dwarf × Makapuno crosses have been developed. Homozygous Makapuno palms of the second and third filial generations have been derived from those hybrids. The homozygous palms were precocious (flowering in 26 months) and self-pollinated and gave high percentage of Makapuno nuts (Nunez and de Pas 2004). Although comparable varieties are known in many other countries, such as Sri Lanka, India, Cambodia, Thailand and Vanuatu, the Philippines' Makapuno is the most popular and economically important among the soft-endosperm coconuts. Kopyor coconut of Indonesia is a similar type, occurring to the extent of 24–38% of the harvested nuts. In India also tall coconuts with similar features, called 'Thairu Thengai' (curd coconut), have been identified since the 1930s.

4.9.6 Unique Forms

Sweet husked coconut palms (SWH) of Sri Lanka produce a special type of husk which is tender, edible and sweet. A fairly mature nut can be de husked with bare hands. Various surveys have shown that while most islanders are familiar with these SWH coconut palms, the palms are becoming increasingly scarce. It is currently very difficult to obtain any of their seed nuts, as immature fruits of these palms are eaten by children. Such types are seen in many Pacific islands also. Another interesting Thailand variety called Maphrao or fiddle sound box produces very thick nuts with three lobes. The shell of nuts from this variety is very thick and dense with three lobes making it suitable to be used to make sound boxes of the local fiddles. The fruits are sold at exorbitant prices. Unfortunately only a very few palms of this type are now in existence (Thampan and Gopalakrishnan 2010). The other types of importance are Kaithathali (soft, fleshy husk) and ornamental coconuts (Maphrapo Pradap of Thailand).

Some of these types possessing special characters may not be of immediate economic importance. But these varieties should be conserved, propagated and improved. It is the breeders' prime responsibility to see that these bio resources are preserved in nature foreseeing their economic importance in the future.

4.9.7 *Disease Resistance*

Unfortunately coconut production is seriously impaired in most of the major coconut-producing countries due to very serious diseases, whether it is root (wilt) disease in India, LYD in Central America and Africa, Cadang Cadang in the Philippines or Weligama Coconut Leaf Wilt Disease (WCLWD) in Sri Lanka. In the absence of any chemical cure, the only option left is evolving and planting resistant/ tolerant varieties.

The most popular variety-West Coast Tall (WCT) grown in India is highly susceptible to root (wilt) disease. A comprehensive screening of 84 cultivars and 68 hybrids during 1961–1988 did not yield any source for resistance (Regi et al. 2010). However, individual disease-free palms with high yield have been located in the midst of heavily disease-affected palms. Selected palms among these were inter se mated to generate more homogeneous tall parents. Very strict criteria such as freedom from root (wilt) disease, yield of 80 nuts palm^{-1} and absence of phytoplasma in the sieve tubes, among others, were considered for selection.

Later, comparatively higher level of resistance was observed in the Chowghat Green Dwarf (CGD). Field testing of CGD × WCT hybrids was therefore undertaken. The criteria used for selecting the CGD mother palms, identified from heavily disease-affected gardens (with at least 80% disease affected palms), were good yield (>100 nuts palm^{-1} year^{-1}), freedom from the disease and age (more than 15 years). Individual inflorescences were artificially pollinated using pollen from selected disease-free WCT palms. These hybrids were found to be early bearing and semi-tall in nature giving a 10-year cumulative average yield of 84 nuts palm $^{-1}$ year^{-1} equivalent to 2.5 mt of copra ha^{-1}. Though after 18 years of planting, 67.7% of the palms showed root (wilt) disease symptoms, the disease-free hybrids gave an average yield of 107 nuts palm^{-1} year^{-1} compared to 72 nuts from disease-affected hybrid palms. Considering the high yield and resistance/tolerance to root (wilt) disease, three coconut varieties, viz. Kalparaksha (MGD), Kalpasree (CGD) and Kalpa Sankara (CGD × WCT), were released for cultivation in root (wilt) affected areas. Genomics approach to amplify putative RGAs from the coconut root (wilt) disease-resistant cultivar, Chowghat Green Dwarf, by Rachana et al. (2016) indicates the scope of using biotechnological tools in resistance breeding (Nair et al. 2016).

Warwick et al. (1993) found none of the eight varieties tested to be resistant to Lixas and leaf blight in Brazil, though Polynesian Tall, Rotuma Tall and Tonga Tall performed better. Lethal yellowing disease is a very serious disease devastating large areas of plantations in many countries, and identification of possible resistance at field level was started from early years (Whitehead 1968). Maypan hybrid (Malayan Dwarf × Panama Tall) claimed to possess resistance to LYD later broke down. A hybrid between Vanuatu Tall and Sri Lankan Green Dwarf is reported to be resistant to LYD in Ghana (Wikipedia 2016). One recently selected cultivar in the Philippines – the 'Maypan' – is identified as resistant to lethal yellowing disease. Brazil Green Dwarf × Panama Tall (Brapan hybrid) and Malayan Yellow Dwarf (Maybraz hybrid) as well as hybrids between Malayan Dwarf Yellow with many local varieties are being field tested against LYD in Jamaica (Coconut Industry Board 2016). 1500 lethal yellowing-resistant coconut plantlets from Mexico are reported to have been added to Grenada's 'coconut stock' under the coconut rejuvenation programme (Harries 2016). In all the countries the dwarfs, especially the Dwarf Green, have been found to be the source of resistance and the need for in depth investigations on this variety with respect to disease resistance needs no over emphasis.

4.9.8 Pest Resistance

Among the many pests attacking coconut, only a very few such as rhinoceros beetle, leaf-eating caterpillar, eriophyid mite and red palm weevil cause major economic losses. No pest-resistant variety or hybrid has so far been developed, maybe because there are satisfactory control measures available. However differences in the severity of pest attack have been reported in a few varieties. Preliminary screening of cultivars and hybrids against leaf-eating caterpillar (*Opisina arenosella*) and rhinoceros beetle (*Oryctes rhinoceros*) indicated variations in susceptibility among cultivars, though no resistant cultivar could be observed (Kapadia 1981; Sumangala Nambiar 1991). Greater tightness between the nuts and the petals seen in round rather than elongated and angled nuts offers better resistance to the eriophyid mite, *Aceria guerreronis* (Moore and Alexander 1987). In India, Chowghat Orange Dwarf and Kulasekharam Green Tall (derived from Malayan Green Dwarfs), were found to be less prone to mite infestation, whereas WCT and Laccadive Micro were highly infested (Nair 2000). Molecular studies have shown positive indications on the possibility of selecting varieties resistant to *Aceria* mite infestation.

All the coconut cultivars are prone to damage by rhinoceros beetle, the hybrids with Chowghat Orange Dwarf as pollen parent being more susceptible (Nambiar 1988). Sosamma et al. (1988) reported Java Tall, Klapawangi Tall, Kenthali and Andaman Giant Tall as more tolerant to burrowing nematodes. For details, please see Chap. 5 on varietal resistance.

4.9.9 Drought Tolerance

With coconut being, by and large, a rain-fed crop, long dry spells adversely affect the yield to the extent of 50%. Under this circumstance, developing drought-tolerant varieties/hybrids are of great importance to increase coconut production. Screening coconut germplasm and its evaluation in drought-prone areas would take a very long time. A better option would be to use phenotypic and the physiological parameters for selection, as has been demonstrated by Naresh Kumar et al. (2006). A cultivar with more roots and a fine root density is less affected by drought (Cintra et al. 1993). Physiological traits such as leaf stomatal frequency, stomatal index, chlorophyll fluorescence, epicuticular wax content, lipase activity and proteases are reliable parameters for the identification of drought-tolerant coconut cultivars (Rajagopal et al. 1988; Repellin et al. 1994). Rajagopal et al. (2005) found that talls and hybrids (with talls as mother palms) had better drought tolerance than the dwarfs and the hybrids (with dwarfs as mother palms). Federated Malay States, Java Giant, Fiji, Laccadive Ordinary, Andaman Giant, Laccadive Ordinary Tall × Chowghat Orange Dwarf and Malayan Yellow Dwarf × West Coast Tall were identified as drought-tolerant varieties in India (Kasturi Bai et al. 2006; Rajagopal et al. 2005). For details, please see Chap. 9 on Physiology and Biochemistry.

4.9.10 Cold Tolerance

Coconut prefers a typical tropical climate. However it is planted even in regions like Assam in India where comparatively cold climates are experienced. Kamrupa, a selection from Assam Green Tall, is considered as a suitable variety for such regions. It is characterised by a yield of 16.34 kg copra palm^{-1} year^{-1}, 64.5% oil and tolerance to high rainfall (>2000 mm year^{-1} and cold weather – where temperatures goes below 8 °C in winter (Chaudhury et al. 2001). There are varieties which can thrive in cooler temperatures in the island of Hainan, China. However, work on cold tolerance aspect is very limited, obviously because of the smaller area under cultivation in this situation.

4.9.11 Ideotype

Idiotype refers to an ideal coconut palm possessing all the desirable characters apart from the yield and yield attributes, as a result of increased physiological efficiency. Palms should have a good architecture possessing an open crown with a horizontal orientation of leaves rather than the one with drooping or erect crown (Iyer et al. 1981). They are high regular yielding palms with good nut characters and possessing other favourable characters such as dwarfness, early bearing and tolerance to unfavourable conditions such as drought, diseases and pests. Moreover it should be able to perform in varied situations (universal/plastic variety) with steady production predictable with least fluctuation (Wikramaratne 1993). Such an ideal palm is a dream of every breeder. While the theoretical concept is appealing, it is easily said than done as there are no donor varieties available for developing such idiotypes and combining as many good characters as possible is the only alternative, no matter the large number of years which may be required for such an effort.

4.10 Biotechnology

There are five major areas of biotechnology which are directly useful for coconut breeding, viz. micropropagation through tissue culture, embryo culture, molecular markers, cryopreservation and genome sequencing. The primary goal of tissue culture is production of uniform elite palms. Plantlets have been regenerated using plumular explants and successfully established in the field (Raju et al. 1984; Karun et al. 2016). Eeuwens and Blake (1978) obtained initial success using inflorescence tissue for callus production. But a repeatable and commercial protocol using adult palm explants is yet to be developed, since coconut palm is highly recalcitrant to in vitro manipulations (Iyer and Parthasarathy 2000).

A simple protocol from field collection of embryos to field establishment was standardised at CPCRI (Karun et al. 2004). This has been used for germplasm collection from the Indian Ocean islands of Mauritius, Madagascar, Seychelles and Sri Lanka. Karunaratne et al. (1991) studied the feasibility of developing an in vitro technique for screening drought tolerance. PCR-based molecular markers are being used in screening parents for the Weligama Coconut Leaf Wilt Disease (WCLWD) resistance programme in Sri Lanka (Lalith Perera 2014). Molecular markers are useful in differentiating varieties and genuine hybrids. This would therefore help in characterisation of varieties and eliminate duplicates in germplasm collections (Rajesh et al. 2012, 2014). Genetic mapping will be of immense value in selection of coconut palms for conventional breeding (Rohde et al. 1999). In vitro conservation techniques such as cryopreservation of both coconut zygotic embryos and pollen have been standardised, and these would enable long-term storage of coconut germplasm at a lower cost and reduced space (Karun and Rajesh 2007; Sajini et al. 2006). Only a few research institutes worldwide such as ICAR-CPCRI, Kasaragod; CCRI, Sri Lanka; PCA, the Philippines; and CIRAD, France are still engaged in serious biotechnological research. The breeding programmes will get a boost if and when the whole genome sequencing of coconut becomes available (Jaccoud et al. 2001). Please see Chap. 6 for details on Coconut Biotechnology.

4.11 Global Scenario of Coconut Breeding

The coconut breeding work in India has been adequately detailed in the foregoing sections. The work in other major coconut-growing countries is briefed below, though they all follow more or less the same pattern.

4.11.1 Sri Lanka

As per legends coconut palms were introduced to Sri Lanka as early as in 137 B.C. Selection from tall coconut palm population right from 1930 resulted in the identification of elite palms producing 100 nuts palm^{-1} year^{-1} with 220 g copra nut^{-1}, under average conditions. The tall form of coconut, San Ramon, was introduced from the Philippines and was established in the north western province. Open pollinated seeds of this form the base material for the current genetic improvement of San Ramon. This cultivar has the potential to produce 51% more copra nut^{-1} than the local talls. Tall × Tall (CRIC 60) and Dwarf × Tall (CRIC 65) were released for planting during the 1960s. Seed gardens were established to produce larger number of hybrids (Liyanage 1996). Two more hybrids, viz. Sri Lankan Tall × San Ramon (CRISL 98) and Sri Lankan Green Dwarf × San Ramon (Kapruwana), were produced in 2004. A seed garden has been established to mass produce the former hybrid. A new variety of dwarf with brown nuts, petioles and inflorescence has been identified, named Sri Lanka brown dwarf, which is superior to other forms of dwarf in the country in terms of kernel weight and tolerance to moisture stress (Perera et al. 2002). Crosses between Brown Dwarf × Sri Lanka Tall, its reciprocal as well as Brown Dwarf × San Ramon were made in 2003. Their performance till the ninth year in multilocation trials indicates that they are promising (Gopalakrishnan 2012). Added emphasis is now given to produce varieties resistant to Weligama Coconut Leaf Wilt disease (WCLWD).

4.11.2 Bangladesh

BARI Narikel-1 and BARI Narikel-0.2 (selections from local talls), producing 65 to 70 nuts palm^{-1}, have been recommended for cultivation. Hybrids involving CRI 60 and Malayan Dwarf are being produced by BARI (Bangladesh Agricultural Research Institute) (Nazrul and Amzad Hussain 2000).

4.11.3 China

In China, because of the cold weather, coconut palms can grow only in the country's southern parts, mainly in Hainan Province, parts of Yunnan Province and in Guangdong Province. A few coconut palms are sporadically distributed in Guangxi Province in the southwest and Fujian Province in the southeast. The only recommended hybrid is Malayan Yellow Dwarf (MYD) × Local Hainan Tall (HAT). This hybrid (WY78F1) is early flowering (3–4 years) and gives three- to fourfold increase in nut yield (80 palm $^{-1}$ year^{-1}) and copra (4 t ha^{-1} year^{-1}). The Wenchang Coconut Research Institute is the agency for taking up coconut research in the country (Tang and Ma 2005).

4.11.4 Philippines

Coconut breeding is undertaken at the Philippine Coconut Research Institute (PHILCORIN). Large quantity of planting materials becomes necessary when millions of palms are destroyed by typhoons like Haiyan. Important planting materials used are Genetically Multi-Ancestored Farmers' Composite Variety, Tacunan Green Dwarf (TAGD), Tutupaen Tall (TTPT), Tangnanan Tall (TAG), Bago Oshiro Tall (BAO, PCA 15-9), Tacunan Green Dwarf × Tagnanan Tall (TACD × TAGT) , PCA 15-4, Catigan Green Dwarf × Tagnanan Tall (CATD × TAGT) and Makapuno (embryo-cultured seedlings). Out of the nine hybrids recommended for cultivation by the Philippine Coconut Authority (PCA), MRD × TAGT (PCA 15–2) and MRD × BAYT (PCA15–3) have been outstanding, giving the highest number of nuts (144 to 155 palm^{-1}) and copra yield (6 mt ha^{-1}). The local tall BAYT was also comparatively good, producing 114 nuts palm^{-1} with a copra yield of 5 mt ha^{-1}. Galas Green Dwarf (GALD) is recommended for tender coconut purpose.

Dwarf × Tall F_1 coconut hybrids are popular because they bear fruits early, give high yield and recover quickly from stress. But due to various constraints, the hybrid did not become very popular, and therefore breeders have developed the Genetically Multi-Ancestored coconut farmers' variety (GMA farmers' composite variety) combining the agronomic qualities of four local tall varieties (Laguna, Bago Oshiro, Baybay and Tagnanan) and two foreign tall varieties (West African and Rennell). This composite variety combines high yield, precocity, vigour and genetic stability. It has a potential to yield two to four tonnes of copra ha^{-1} year^{-1}. Unlike the F_1 hybrids, coconut farmers can use this composite variety as a source of seeds for successive generations (PCA 2016). Seed gardens are maintained to support the long-term coconut planting/replanting programme. San Ramon is a tall variety (native to one of the islands known by the same name) characterised by very big nuts, giving up to 500 g copra nut^{-1}. San Ramon Orgullo Tall is a synthetic variety developed from a combination of 15 different coconut hybrids. It starts flowering in 3.5–5 years and produces 17 bunches in a year (Santos and Rivera 1994). One ha can yield 3.2 to 6.7 mt of copra, which is much more than that of local talls. Other aspects of work undertaken are embryo culture multiplication of Makapuno, establishment and management of coconut field gene bank, cryopreservation, biotechnology and tissue culture.

4.11.5 Indonesia

The first survey for coconut germplasm was conducted in the beginning of 1930. Unfortunately the activities were discontinued till 1973 due to financial constraints. From intensive surveys carried out later, 98 collections were made which are maintained at the Coconut Research Institute. Several dwarfs, talls and hybrids were also introduced from Côte d'Ivoire (Yohannes and Sumardjan 1993). Pati Dwarf Kopyor coconut originated from Pati (Central Java) is one of the local coconut populations in Indonesia. Three varieties, viz. Kopyor Green Dwarf, Kopyor Brown Dwarf and Kopyor Yellow Dwarf, have been released in 2010 as superior varieties. The estimated yield of nuts palm^{-1} year^{-1} is much higher than that of Kopyor Tall coconuts. Four improved intra-varietal hybrids (Kelapa Baru or KB1, KB2, KB3 and KB4) produced by crossing selected tall cultivars have been released. The hybrids started flowering in the sixth year, with 16 bunches year^{-1}, 96 to124 nuts palm^{-1} year^{-1}, 3.88

to 4.66 mt copra ha^{-1} and oil content of 67–71%. Three D × T hybrids evolved from the local material: Khina 1 (Nias Yellow Dwarf × Tenga Tall), Khina 2 (Nias Yellow Dwarf × Bali Tall) and Khina 3 (Nias Yellow Dwarf × Palu Tall) have been released. These hybrids have been found to be superior to the parental types with respect to precocity for bearing and high production of copra (Maskromo et al. 2013).

4.11.6 Thailand

The most common variety is the local tall although hybrids are being produced since 1982. The Chumphon Horticulture Research Centre (CHRC) of the Horticulture Research Institute of Thailand recommends Chumphon hybrid No.60 (Thai Tall × WAT) and Chumphon No.2 (MYD × Thai Tall) for cultivation since 1987 and 1995, respectively. These hybrids give nut and copra yields ranging from 80 to 126 nuts palm^{-1} and 3.4 and 4.2 mt copra ha^{-1}, respectively (Chulapan Petchpiroon 2000). Aromatic Green Dwarf is another popular special variety preferred for tender nuts having unique smell, sweet water and meat (Batugal 2005).

4.11.7 Malaysia

Studies from 1920s found that Malayan Green Dwarfs are robust and resistant to adverse conditions and produce 1.3 kg copra palm^{-1} year^{-1}, nearly as much as the average tall palm. MAWA hybrids (Malayan Yellow, Orange or Red Dwarf × West African Tall combinations) are less variable in terms of nuts palm^{-1}, fruit weight and copra weight and produce more copra than other hybrids. These MAWA hybrids have been used in the National Planting Programme since 1978. The Malayan Red and Yellow Dwarf × West African Tall have been found to be high yielding (25.82 and 24.98 kg copra palm^{-1}, respectively) and suitable for planting in the imperfectly drained but highly fertile coastal plains (Rethinam et al. 2005).

4.11.8 Vietnam

Ta is the most widely cultivated (79%) traditional tall variety, Dau and Giay are the other two tall varieties being commercially cultivated. Breeding work was initiated in 1984, with the introduction of 14 varieties and seven hybrids from Côte d'Ivoire, Sri Lanka, the Philippines and Fiji (Long 1993). The Vietnam Oil Plant Institute (OPI) is recommending seven introduced high-yielding hybrids, viz. PB111, PB121, PB 132, PB 141, JVA 1, JVA2 and CRIC 65 which have significantly outyielded the Ta local tall. Nut production was 48 to 69 palm^{-1} in 1996, as against 31–35 nuts by the Ta variety. Six local hybrids (Eo × Ta; Tam Quan × Ta; Tam Quan × BAOT; MYD × Rennel Tall; MYD × Palu Tall and MYD × Ta) are under field testing (Batugal et al. 2009). Among the local varieties, Ta, Dau and Giay have been identified as the best for copra making. A seed garden of 120 ha has also been established at Trang Bang, Tay Ninh Province for producing hybrids and seed material of local varieties.

4.11.9 Côte d'Ivoire

Coconut was introduced from Mozambique by Portuguese to West Africa in the early sixteenth century. Now it is cultivated in Côte d'Ivoire in about 50,000 hectares. Generally an average Ivorian coconut grove is composed of 48% hybrids and 52% West African Tall (WAT), which is the only local ecotype in the country (Konan et al. 2000). Coconut research was initiated in Port Bouet in 1950, by collection of accessions from republic of Benin. The outstanding hybrids identified are PB 213 (WAT × RIT), PB 214 (WAT × VTT), PB121 (MYD × WAT), PB 132 (MRD × TAT- Tahitian Tall), PB123 (MYD × RIT) and PB111 (CRD or Cameroon Red Dwarf × WAT). These hybrids flower very early (40–57 months after field planting) under Côte d'Ivoire conditions. They produce 100 to 132 nuts palm^{-1} year^{-1}which is 34–138% higher compared to WAT. Further, their copra yields ranged from 3.15 to 4.8 mt ha^{-1} or 86–135% more compared to WAT (Bourdeix et al. 2016). Two breeding schemes (Dwarf × Tall and Tall × Tall), using the recurrent reciprocal selection method, are adopted. The creation of complex hybrids and their clonal exploitation are expected to increase yield of copra considerably (De Taffin et al. 1991). The crop is under threat by Lethal Yellowing Disease especially since 2013. Therefore priority is now for production of hybrids involving tolerant varieties like Vanuatu Tall, Panama Tall, Sri Lanka Green Dwarf, MYD, MRD and Niu Leka Dwarf.

4.11.10 Benin

Most of the plantations in Benin are made up of 50- to 70-year-old West African Tall (WAT) variety. Breeding programme is handicapped due to poor genetic base available. PB 121 and PB 111 hybrids were found to outyield the local talls, though the yields are low, in general, due to poor environment. In 1989, promising varieties from Côte d'Ivoire were identified for collection and introduction, namely, Polynesian Tall (PYT), Rennell Tall (RT), Vanuatu Tall (VTT) and Malayan Red Dwarf (MRD) (Sanoussi 2016).

4.11.11 Nigeria

Since the vegetable oil demand is largely met from palm oil in the southern states (oil palm being native to the area) and ground nut oil in the northern states of the country, coconut had not been a popular crop in Nigeria, and obviously much attention has not been given to coconut research. A few introductions were made from India, Malaysia and Cameroon in 1966 and maintained in the germplasm. A comprehensive prospection was made in the southern states of the country which enriched the germplasm collection to 60 indigenous and 25 exotic accessions. Coconut research was intensified with the establishment of a substation at Badagry, Lagos state in 1976 (Nampoothiri 1982). Now hybrid seedlings are produced from Nigerian Tall and a mixture of Nigerian Dwarf, Cameroon Dwarf and Malaysian Dwarf. Evolving high-yielding varieties resistant to Awka wilt disease is a priority area in the breeding programme. The West African Tall (WAT) is the most extensively grown tall variety, giving up to 80 to 170 nuts palm^{-1} per year^{-1}. Two hybrid seed gardens have been established, one in Badagry and the other one near Benin city, to produce hybrid planting materials. Two rows of Malaysian, Cameroon and Nigerian Dwarf cultivars are planted to one row of tall palms in these gardens (Akpan 2016).

4.11.12 Ghana

All coconut cultivars in Ghana are considered to be at risk from the Cape St. Paul Wilt Disease (CSPWD), a lethal yellowing type of disease. Hence, the coconut breeding programme in the country is geared towards developing hybrids resistant or highly tolerant to CSPWD. There are 6 cultivars and 21 hybrids being tested in four locations. These varietal resistance trials are still under observation although some of the test materials were already infected by the CSPWD.

4.11.13 Tanzania

The Mikocheni Agricultural Research Institute (MARI) is currently testing six hybrids with the local East African Tall (EAT) as sole pollinator. Mother palms involved are from Malayan Green Dwarf (MGD), CRD, Pemba Red Dwarf (PRD), MYD, MRD and improved EAT populations. In addition to determining their yield performance, the F_1 progenies are also being evaluated for their resistance to lethal disease and tolerance to drought stress.

4.11.14 Mexico

Coconut research at the Instituto Nacional de Investigacion Agropecuaria Y Forestal is focused on developing hybrids resistant to lethal yellowing disease. Initial hybrids were mainly derived from crosses between MYD and improved Pacific Tall populations. Intra-population crosses of selected Pacific Tall were also done, and these are currently being tested.

4.11.15 South Pacific Region

Though composed of tiny islands, they all put together a substantial area (600,000 ha) with coconut. In Western Samoa and Tuvalu, there are seed gardens to produce hybrids of MYD as well as RMD × RLT to meet the demand for replanting programme. Cocoa and Coconut Research Institute (CCRI) carries out national research on coconut in Papua New Guinea. A good collection of varieties have been made, and it also hosts the International Gene Bank for the South Pacific. The tall and dwarf varieties in ICG are being used for breeding work. Malayan Red Dwarf and Malayan Yellow Dwarf × Rennell Tall are the ruling planting materials. One of the main programmes is production of Malayan Dwarf × Rennell Tall hybrids for replanting. There is a seed garden in Omuru for mass production.

In Fiji, a seed garden to produce MYD or RMD and RLT or Rotuma Tall (RTT) is functional. In Vanuatu, work started from 1962. The Vanuatu Research and Training Centre has produced hybrids involving the local cultivars, viz. Vanuatu Tall (VTT) and Vanuatu Red Dwarf (VRD), with the introduced varieties Rennell Island Tall (RIT) and Brazilian Green Dwarf (BGD). The Malayan Red Dwarf (MRD) × RIT was also produced. These hybrids as well as exotic introductions, though slightly better in yield, were found to be susceptible to coconut foliar decay (CFD), whereas the local varieties are tolerant. The VRD × VTT hybrids had lower copra yields (3.3 to 3.7 mt ha^{-1}) but were found to be more tolerant against CFD. So emphasis is laid on selection of local material. The local talls are found to produce 2.5 mt copra ha^{-1} $year^{-1}$, whereas the dwarf × local talls produce 4.0 mt copra ha^{-1} $year^{-1}$ (Duhamel 1993). In the Solomon Islands, where controlled pollination began in the 1960s, a replanting programme is being carried out with the Malayan Red Dwarf × Rennell Tall hybrid.

4.12 Future Strategy

1. Increasing the availability and accessibility of quality planting materials.
2. Emphasis on evolving resistant/tolerant varieties to biotic and abiotic stresses specific to the concerned area.

3. Evolving coconut idiotypes suitable for inter/mixed cropping.
4. Strengthening international germplasm exchange strictly adhering to the quarantine protocols. Sustaining/strengthening the existing national and international coconut gene banks.
5. Adoption of a comprehensive breeding strategy.
6. Diffusion of information to farmers and steps to narrow down the potential and realised performance of coconut varieties.
7. Developing micropropagation protocol using adult palm tissues and genome sequencing of coconut to complement the breeding efforts.
8. Establishment of an International Coconut Research Institute (ICRI).

References

Adansi MA (1970) The present position of selection and introduction. In: Chona BL, Adansi MA (eds) Coconut in Ghana. Crops Research Institute, Accra

Akpan EEJ (1994) Evaluation of tall coconut (*Cocos nucifera* L.) genotypes within the Nigerian coconut germplasm bank. Oleagineux 49:13–20

Akpan EEJ (2016) The Nigerian coconut breeding programme – Africa. Bioversity international. www.bioversityinternational.org/fileadmn/bioversity/publications/web.ch5.htm

Anonymous (2006) Development of PCA SYN VAR 001: synthetic variety of coconut. Retrieved in October, 2016 from agris.fao.org/agris-search/search.do?recordID=PH2007001709

Anonymous (2016) A brief look at the history of coconuts. www.be-healthy-with–coconuts.com/history-of-coconuts-html

Anupap T, Petchipiroon C, Rattanapruk M (1992) Existing status of coconut genetic resources of Thailand. In: Report of a working session for implementation of the International Coconut Database held at CIRAD, Montpellier, France, 19–22 May 1992

Aragao WM (2000) A importancia do coqueiro – anao verde (Coletaneas Rumos & Debates). Embrapa, Petrolina

Ashburner GR, Thompson WK, Halloran GM (1997) RAPD analysis of South Pacific coconut palm populations. Crop Sci 37:992–997

Balakrishnan PC, Devadas VS, Unnithan VKG (1991) Phenotypic stability of coconut (*Cocos nucifera* L.) cultivars. In: Silas EG, Aravindakshan M, Jose AI (eds) Coconut breeding and management. Kerala Agricultural University, Thrissur

Batugal P (2005) Performance of coconut hybrids in some countries of Asia, Africa and Latin America. In: Batugal P, Ramanatha Rao V, Jeffrey O (eds) Coconut genetic resources. International Plant Genetic Resources Institute. Regional office for Asia, the Pacific and Oceania, Serdang, Selangor D.E., Malaysia, pp 302–308

Batugal P, Bourdeiux R (2005) Conventional coconut breeding. In: Batugal P, Ramanatha Rao V, Jeffrey Oliver T (eds) Coconut genetic resources. International Plant Genetic Resources Institute, Regional office for Asia, the Pacific and Oceania, Serdang, Selangor D.E., Malaysia

Batugal P, Ramanatha Rao V, Oliver J (2005) Coconut genetic resources. In: International Plant Genetic Resources Institute – Regional Office for Asia, the Pacific and Oceania (IPGRI-APO), Serdang, Selangor DE, Malaysia

Batugal P, Bourdeix R, Baudouin L (2009) Coconut breeding. In: Breeding plantation tree crops: trop species. Springer, New York, pp 327–375

Bavappa KVA (1973) A coconut hybrid seed garden for the west coast. Cocon Bull 4(1):5–6

Bavappa KVA, Nampoothiri KUK (1974) Recent trends in coconut breeding. Indian J Gen 34A:58–65

Bavappa KVA, Sukumaran CK (1976) Coconut improvement by selection and breeding- a review in the light of recent findings. In: International symposium on coconut research and development, December, 28–31 Kasaragod, India, p 78
Bourdeix R, Sangare A, Le Saint JP (1989) Effectiveness of individual combining ability tests on hybrid coconuts. Initial results. Oleagineux 44(5):209–214
Bourdeix R, Sangare A, Le Saint JP, Meunier J, Gascon JP, Rognon F, de Nuce LM (1993) Coconut breeding at IRHO and its application in seed production. In: Nair MK, Khan HH, Gopalasundaram P, Rao EVVB (eds) Advances in coconut research and development. Oxford and IBH publishing Co, New Delhi, p 759
Bourdeix R, Guarino L, Ramanatha Rao V, Baudouin L (2005) Status gap and strategy in coconut germplasm collecting. In: Coconut genetic resources. International Plant Genetic Resources Institute – Regional Office for Asia, the Pacific and Oceania (IPGRI-APO), Serdang, Selangor DE, Malaysia
Bourdeix R, Batugal P, Oliver JT, George MLC (2010) Catalogue of conserved coconut germplasm. International Coconut Genetic Resources Network (COGENT). Selangor, Malaysia. www.bioversityinternational.org/fileadmin/bioversity/publications/Web.../ch5.htm
Bourdeix R, N'Cho YP, Sangare A (2016) Coconut breeding programme. Côted'Ivoire. www.bioversityinternational.org/fileadmin/bioversity/publications/Web./ch5.htm
Bourdeix R, Perera L, Rivera RL, Saena-Tuia V, Masumbuko L (2017) Global coconut communities – status and strategies in *in situ* diversity management and utilization. In: Chowdappa P, Niral V, Jerard BA, Samsudeen K (eds) Coconut. Daya Publishing House, New Delhi, pp 407–440
Charles AE (1959) Nursery selection of coconut seedlings. Papua New Guinea Agric J 12:116–180
Charles AE (1961) Selection and breeding of the coconut palm. In: Symposium on tropical crops improvement. South Pacific Commission, Nouméa, pp 51–55
Chaudhury D, Nath JC, Mohan NK (2001) 'Kamrupa'– A newly released coconut variety by Assam Agricultural University. Indian Cocon J 31(9):12–13
Chulapan Petchpiroon (2000) Coconut research and development in Thailand. COGENT report. www.bioversity International.org/fileadmin/bioversity/publications/Web.../ch5.htm
Cintra FLD, Passos EEM, De Leal LS (1993) Evaluation of root system distribution in tall coconut cultivars. Oleagineux 48:453–461
Coconut Industry Board (2016) Coconut farming techniques: Jamaica. www.coconutindustryboardjm.org/research. Accessed 15 Jan 2017
CPCRI (2013) Annual report 2012–2013. Central Plantation Crops Research Institute, Kasaragod, p 142
Daniel Zizumbo V, Colunga-GarciaMarin P (2001) Morpho-physiological variation and phenotypic plasticity in Mexican populations of coconut (*Cocos nucifera* L.). Genet Resour Crop Evol 48(6):547–554
De Taffin G, Zakra N, Bonny CP (1991) Performance des hybrides de cocotiers Nain × Grand en conditions industrielles en Côte d'Ivoire. Oléagineux 46:187–195
Duhamel G (1993) Crop improvement programmes in South Pacific region. In: Nair MK, Khan HH, Gopalasundaram P, Rao EVVB (eds) Advances in coconut research and development. Oxford and IBH publishing Co, New Delhi, p 759
Dwyer REP (1938) Coconut improvement by seed selection and plant breeding. New Guinea Agric Gaz 4:24–102
Eeuwens CJ, Blake J (1978) Culture of coconut and date palm tissue with a view to vegetative propagation. Acta Hortic 78:277–286
Engelmann F (1999) Cryopreservation of coconut germplasm. In: Orepez C, Verdeil J, Ashburner GR, Gardena R, Samantha JM (eds) Current advances in coconut biotechnology. Kluwer Academic Publishers, London, pp 289–296
Farias Neto JT, de Lins PMP, Muller AA (2003) Pesquisa Agropecuaria Braziliera. Brasília 38:1237–1241

Fernando WMU (1998) Genetic improvement of the coconut cultivar san Ramon. COCOS 13:30–39
Foale M (2005) An introduction to the coconut palm. In: Batugal P, Ramanatha Rao V, Oliver J (eds) Coconut genetic resources. International Plant Genetic Resources Institute. Regional Office for Asia, the Pacific and Oceania (IPGRI-APO), Serdang, Selangor DE, Malaysia, p 762
Ganesamurthy C, Natarajan S, Rajarathinam S, Vincent S, Khan HH (2002) Genetic variability and correlation of yield and nut characters in coconut. J Plantn Crops 30(2):23–25
Gascon JP, Nuce de Lamothe M (1976) Coconut improvement: Methods and suggestions for international cooperation. Oléagineux 31(11):479–482
Geethanjali S, Rajkumar D, Shoba N (2014) Correlation and path analysis in coconut (*Cocos nucifera* L.). J Plant Breed 5(4):702–707
Gopalakrishnan R (2012) Coconut breeding initiatives in India and Sri Lanka. Indian Cocon J 55(3):45–48
Guarino L, Ramanatha Rao V, Batugal P (1998) Collecting coconut diversity: elements of a strategy for COGENT germplasm collecting activities. IPGRI 1998. Strategy paper for COGENT
Harland SC (1957) The improvement of coconut palm by breeding and selection. Circulation paper No. 7157 Coconut Res Inst Bull (Ceylon) 15
Harries HC (1991a) Two hundred and fifty years of coconut research. Coconut Commun XX1(19):11
Harries HC (1991b) The promise, performance and problems of F_1 hybrid coconut. In: Silas EG, Aravindakshan M, Jose AI (eds) Coconut breeding and management. Kerala Agricultural University, Trichur, pp 380–390
Harries HC (2016) Plant-varieties revitalise-coconut-industry. www.nationnews.com
Henry Teuira (1928) Ancient Tahiti. Kraus, Honolulu, p 651
ICAR- CPCRI (2014) Varieties and hybrids developed. Retrieved from www.cpcri.gov.in/index.php/2014-07-03-05-53-45. October, 2016
ICAR-CPCRI (2016) Annual report 2015–16. ICAR- Central Plantation Crops Research Institiute, Kasaragod, p 210
Ikin R, Batugal P (2004) Germplasm health management for COGENT's multisite International gene banks. IBPGR-APO, Serdang, p 159
Iyer RD, Parthasarathy VA (2000) Biotechnology of coconut, oil palm and cacao. In: Chadha KL, Ravindran PN, Sahijram L (eds) Biotechnology in horticultural and plantation crops. Sahijram Malhotra Publishing House, New Delhi, pp 286–302
Iyer RD, Rao EVVB, Govindankutty MP (1979) Super yielders in coconut. Indian Fmg 28(10):3–5
Iyer RD, Rao EVVB, Sukumaran CK, Jacob PM (1981) Towards an ideal plant type in coconut. In: Visheswara et al (eds) Proceedings of PLACROSYM IV. Indian Society for Plantation Crops, Kasaragod, pp 29–37
Jaccoud D, Peng K, Feinstein D, Kilian A (2001) Diversity arrays: a solid state technology for sequence information independent genotyping. Nucleic Acids Res 29(4):e25
Jacob M (1991) Statistical techniques for experiments with coconut. In: Silas EG, Aravindakshan M, Jose AI (eds) Coconut breeding and management. Kerala Agricultural University, Thrissur, pp 232–242
Jayalekshmy VG, Rangasamy SR (2002) Cluster analysis in coconut. J Plantn Crops 30(2):18–22
Jayasekhar S, Chandran KP, Krishnakumar V (2016) The demand-supply dynamics of coconut seedlings- a glance through the socio –political economy frame. In: Chowdappa P, Shamsudeen K, Thamban C, Rajesh MK (eds) Planting material production in coconut. Today and Tomorrow Printers and Publishers, New Delhi, pp 121–125
Kalathiya KV, Sen NL (1991) Correlation among floral and yield characteristics in coconut, variety Dwarf Green. In: Silas EG, Aravindakshan M, Jose AI (eds) Coconut breeding and management. Kerala Agricultural University, Thrissur, pp 116–117
Kapadia MN (1981) Susceptibility of three cultivated coconut varieties to *Nephantis serinopa* Meyrick. Indian Cocon J 12(4):4–5

Karun A, Rajesh MK (2007) Biotechnological options for varietal improvement in coconut. In: Thampan PK, Vasu KI (eds) Coconut for rural welfare. APCC, Jakarta, pp 295–304
Karun A, Sajini KK, Radha E, Parthasarathy VA (2004) Efficacy of CPCRI embryo culture protocol in germplasm expedition. J Plantn Crops 32(Suppl):139–143
Karun A, Muralikrishna KS, Sajini KK, Rajesh MK (2016) Biotechnological applications for quality planting material production. In: Chowdappa P, Samsudeeen K, Thamban C, Rajesh MK (eds) Planting material production in coconut. Today and Tomorrow Printers and Publishers, New Delhi, pp 39–53
Karunaratne S, Santha S, Kovoor A (1991) An *in vitro* assay for drought tolerant germplasm. Euphytica 53:25
Kasturi Bai KV, Rajagopal V, Arunachalam V (2006) Assessment of diversity in coconut varieties for drought responsive physiological traits. J Plantn Crops 34:118–120
Konan JL, Cho YPN, Kullaya A, Bourdeix R, Batugal P (2000) West African tall. COGENT report. www.biodiversityinternational.org/fileadmn/bioversity/publications/web.../ch5.htm
Koshy PK, Kumaran PM (1997) Exotic germplasm of coconut collected through embryo culture techniques. ICAR News 3(3):2 Indian Council of Agricultural Research, New Delhi
Labouisse JP, Sileye T, Morin JP, Hamelin C, Baudouin L, Bourdeix R, Rouzière A (2004) Coconut (*Cocos nucifera* L.) genetic improvement in Vanuatu: overview of research achievements from 1962 to 2002. OCL 11(4):354–361
Lalith Perera (2014) Some progressive steps in coconut research and development in Sri Lanka through utilization of molecular markers. J Plantn Crops 42(1):1–5
Leach BJ, Foale MA, Ashburner GR (2003) Some characteristics of wild and managed coconut palm populations and their environment in the Cocos (Keeling) Islands, Indian Ocean. Genet Resour Crop Evol 50:627–638
Liyanage DV (1953) Report of the botanist. Annua report of the coconut research Board for 1953. Coconut Research Institute, Sri Lanka, p 55
Liyanage DV (1958) Varieties and forms of coconut palm grown in Ceylon. Ceylon Cocon Qtly 9(3–4):1–10
Liyanage DV (1967) Identification of genotype of coconut suitable for breeding. Exp Agric 3:205–210
Liyanage M d S (1996) Coconut Research in Sri Lanka. In: Thampan PK (ed) Coconut for prosperity. Peekay Tree Crops Development Foundation, Kochin, pp 221–237
Liyanage DV, Sakai KI (1960) Heritabilities of certain yield characters of coconut palm. J Genet 57:245–252
Long VV (1993) Coconut selection and breeding programme in Vietnam. In: Nair MK, Khan HH, Gopalasundaram P, Rao EVVB (eds) Advances in coconut research and development. Oxford and IBH publishing Co, New Delhi, p 759
Marechal H (1928) Observation and preliminary experiments on the coconut palm with a view to developing improved seed for Fiji. Fiji Agric J 1:16–45
Marshall DR, Brown AHD (1973) Definition of an optimum sampling strategy for genetic conservation. Conference technique de la FAO sur les ressources génétiques végétales, Rome, Italy
Maskromo I, Novarianto H, Sukma SD, Sudarsono (2013) Productivity of three Dwarf Kopyor coconut varieties from Pati, Central Java, Indonesia. Cord 29(2):19–28
Menon KPV, Pandalai KM (1960) The coconut palm – a monograph. Indian Central Coconut Committee, Ernakulam, p 384
Moore D, Alexander L (1987) Aspects of migration and colonization of the coconut palm by the coconut mite *Eriophyes guerreronis* (K.) (Acari: Eriophyidae). Bull Entomol Res 63:285–288
N'Cho YP, Sangare A, Bourdeix R, Bonnot F, Baodouin L (1993) Assessment of a few coconut ecotypes- a biometric approach. 1. Study of tall populations. Oleagineux 48:121–132
Nair CPR (2000) Status of coconut eriophyid mite, Aceria guerreronis Keifer in India. In: Proceedings of international workshop on coconut eriophyid mite. January 2000. Coconut Research Institute, Sri Lanka, pp 13–31

Nair MK, Nampoothiri KUK, Dhamodaran S (1991) Coconut breeding- past achievements and future strategies. In: Silas EG, Aravindakshan M, Jose AI (eds) Coconut breeding and management. Kerala Agricultural University, Thrissur, pp 17–25
Nair RV, Jerard BA, Regi JT (2016) Coconut breeding in India. In: Al-Khayri JM et al (eds) Advances in plant breeding strategies: agronomic, abiotic and biotic stress traits. Springer International Publishing, Cham
Nambiar SS (1988) Susceptibility of hybrid coconut varieties to *Oryctes rhinoceros* L. under rainfed conditions at Pilicode. In: Silas EG, Aravindakshan M, Jose AI (eds) Coconut breeding and management. Kerala Agricultural University, Thrissur, pp 158–160
Nampoothiri KUK (1974) Biosystematic studies in *Cocos nucifera*. Ph.D. thesis submitted to IARI, New Delhi
Nampoothiri KUK (1982) Coconut industry in Nigeria. Indian Cocon J XII(13):1–7
Nampoothiri KUK (1991) Impact of prepotency in coconut productivity. CORD 7:2
Nampoothiri KUK (1999) Strengthening the capability of the international coconut gene bank for South Asia. Project report, COGENT, 1999
Nampoothiri KUK (2003) Conservation of coconut genetic resources. In: Mandal BB, Chaudhury R, Engelmann BM, Tao KL, Dhillion BS (eds) Conservation biotechnology of plant germplasm. Published By NBPGR, New Delhi, p 293
Nampoothiri KUK (2016) Coconut breeding- a synthesis. In: Chowdappa P et al. (eds) Abstracts of third international symposium on coconut research and development. CPCRI, Kasaragod, India, 10-12, December 2016, pp 4–6
Nampoothiri KUK (2017) Perspectives in global coconut research. In: Chowdappa P, Niral V, Jerard BA, Samsudeen K (eds) Coconut. Daya Publishing House, New Delhi, pp 19–38
Nampoothiri KUK, Satyabalan K, Mathew J (1975) Phenotypic and genotypic correlation of certain characters with yield in coconut. In: 4th FAO technical working party. Coconut production, protection and processing, Kingston, Jamaica
Narayanan Kutty MC, Gopalakrishnan PK (1991) Yield components in coconut palms. In: Silas EG, Aravindakshan M, Jose AI (eds) Coconut breeding and management. Kerala Agricultural University, Thrissur, pp 94–98
Naresh Kumar S, Rajagopal V, Kasturi Bai KV (2006) Criteria for identification of coconut palms *in situ* tolerant to abiotic stresses in farmers' fields. Indian Cocon J 37:5–7
Nazrul I, Amzad Hussain AKM (2000) National coconut programme in Bangladesh. www.bioversity international.org/fileadmin/bioversity/publications/web/ch5.htm
Ninan CA, Satyabalan K (1964) A study of natural self and cross (Dwarf × Tall) progenies of dwarf coconuts in the west coast of India and its bearing on genetics of dwarfs and the putative hybridity of their off -type progenies. Caryologia 17(1):77–91
Niral V, Augustine Jerard B, Samsudeen K, Thomas RJ, Thomas GV (2014) Coconut Varieties and Hybrids, CPCRI Technical Bulletin No. 87. Central Plantation Crops Research Institute, Kasaragod, p 40
Novarianto H, Miftahorachman L (1991) Effect of inbreeding on some characters of the mapanget tall coconut. Ind Crops Res J 3(2):15–17
Nunez TC, de Paz VM (2004) Growth and development of F2 pure Makapuno palms. Philippine J Crop Sci 19(S.1):69
Ohler JG (1984) Coconut-the tree of life. FAO, Rome, p 446
Parthasarathy VA (2001) Deputation report. Central Plantation Crops Research Institute, Kasaragod, p 55
Patel JS (1937) Coconut breeding. Proceedings of Assoc Econ Biol 5:1–16
PCA (2016) Genetically multi-ancestored farmers' composite variety- a sustainable option for the farmers. www.pca.da.gov.Ph./pdf/techno/gmasynvar
Perera L, Peris RRA, Padmasriri MHL (2002) Sri Lanka Brown Dwarf (SLBD) coconuts: a potential coconut variety for future breeding. Plant Gen Resour Newsl 131:70–72

Perera L, Baudouin L, Mackay L (2016) SSR markers indicate a common origin of self-pollinating dwarf coconut in South-East Asia under domestication. Sci Hortic 211:255–262. Retrieved from https://doi.org/10.1016/j.scienta.2016.08.028

Peries RRA (1993a) Seedling selection in coconut based on juvenile and adult palm correlations. In: Nair MK, Khan HH, Gopalasundaram P, Rao EVVB (eds) Advances in coconut research and development. Oxford and IBH publishing Co, New Delhi, pp 71–77

Peries RRA (1993b) Improving the efficiency of the low input coconut agro ecology, through conservation and utilization of biodiversity. In: Proceedings of the 10th IFOAM conference. Lincoln University, New Zealand - 16 December, 1994, p 122

Peries RRA (1998) Coconut breeding in Sri Lanka. In: Coconut breeding. Workshop on standardization of coconut breeding research techniques, June 1994, Port Bouet, Côte d'Ivoire. IPGRI-APO, Serdang, Selangor, Malaysia

Petchpiroon C, Thirakul A (1994) Coconut breeding programme in Thailand. In: Workshop on standardization of coconut breeding research techniques, Marc Delorme Coconut Station, Port Bouet, Côte d'Ivoire, June 20–25, 1994. IPGRI/COGENT, GTZ

Prabhakaran PV, Balachandran Nair GK, Padmanabhan Thampi A, Sivasankara Pillai K, Ramachandran Nair V, Madhavan Nair G, Pushpangadan K, Sukumaran KM, Raghavan Pillai G, Koshy EP (1991) Forecast of coconut yield using biometric characters of young palms. In: Silas EG, Aravindakshan M, Jose AI (eds) Coconut breeding and management. Kerala Agricultural University, Thrissur, pp 243–246

Preethi P, Rajesh MK, Rahul CU, Jerard BA, Samsudeen K, Regi JT (2016) Identification and utilisation of informative EST-SSR markers for genetic purity testing of coconut hybrids. J Plantn Crops 44(2):77–84

Purseglove JW (1968) The origin and distribution of the coconut. Trop Sci 10:191–199

Rachana KE, Naganeeswaran S, Fayas TP, Thomas RJ, Rajesh MK (2016) Cloning, characterization and expression analysis of NBS-LRR type resistance gene analogues (RGAs) in coconut. Acta Bot Croat 75(1):1–10

Rajagopal V, Shivashankar S, Kasturibai KV, Voleti SR (1988) Leaf water potential as an index of drought tolerance in coconut. Plant Physiol Biochem 15:80–86

Rajagopal V, Kasturi Bai KV, Kumar N (2005) Breeding for drought tolerance in coconut. In: Batugal P, Rao VR, Oliver J (eds) Coconut genetic resources. IPGRI, Rome, pp 282–301

Rajesh MK, Regi JT, Rijith J, Shareefa M, Jacob PM (2012) Genetic purity assessment of D × T hybrids in coconut using SSR markers. Indian J Gen Plant Breed 72(4):472–474

Rajesh MK, Jerard BA, Preethi P, Regi JT, Fayas TP, Rachana KE, Karun A (2013) Development of a RAPD- derived SCAR marker associated with tall- type palm trait coconut. Sci Hortic 150:312–316

Rajesh MK, Jerard BA, Preethi P, Regi JT, Karun A (2014) Application of RAPD markers in hybrid verification in coconut. Crop Breed Appl Biotech 14:36–41

Raju CR, Prakash Kumar P, Chandra Mohan M, Iyer RD (1984) Coconut plantlets from plant tissue cultures. J Plantn Crops 12:75–78

Ramanatha Rao V, Hodgkin T (2001) Genetic diversity and conservation and utilization of plant genetic resources. Plant Cell Tissue Org Cult 68:1–19

Rao EVVB, Koshy PK (1981) Coconut germplasm collection in Pacific ocean islands. IBPGR/FAO/CPCRI project. CPCRI, Kasaragod, p 47

Rao EVVB, Mathew J (1981) An index for evaluation of coconut germplasm in the nursery. In: Vishveshwara S et al (eds) Proceedings of PLACROSYM IV. Indian Society for Plantation Crops, Kasragod, pp 200–205

Rao EVVB, Pillai RV (1982) Characterisation of coconut germplasm based on fruit component analysis. In: Bavappa KVA et al (eds) Proceedings of PLACROSYM V. Indian Society for Plantation Crops, Kasragod, pp 112–124

Rao EVVB, Pillai RV, Ratnambal MJ (1993) Current status of coconut genetic resources research in India. In: Nair MK, Khan HH, Gopalasundaram P, Rao EVVB (eds) Advances in coconut research and development. Oxford and IBH publishing Co, New Delhi, p 759

Regi JT, Jacob PM, Nair RV (2010) Kalpa Sankara, a coconut hybrid (CGD × CT) suitable for the root (wilt) disease prevalent tract of Kerala. In: Bose A (ed) In a nutshell- essays on coconut. Coconut Development Board, Publication No. 151, pp 243

Regi JT, Rajesh MK, Jacob PM, Jose M, Nair RV (2015) Studies on genetic uniformity of Chowghat Green Dwarf and Malayan Green Dwarf varieties of coconut using molecular and morphometric methods. J Plantn Crops 2(43):89–96

Regi JT, Shareefa M, Karun A (2016) Enhancing productivity through varietal improvement. In: Kalavathi S, Regi JT, Abdul Haris A, Josephrajakumar, Krishnakumar V, Chowdappa P (eds) Souvenir- gleanings of coconut. ICAR-CPCRI. Central Plantation Crops Research Institute, Kasaragod, Kerala, India, p 72

Repellin A, Daniel C, Zuily-Fodil Y (1994) Merits of physiological tests for characterizing the performance of different coconut varieties subjected to drought. Oleagineux 49:155–168

Rethinam P, Batugal P, Rognon F (2005) Performance evaluation of coconut varieties and farmers' varietal preferences. In: Batugal P, Ramanatha Rao V, Oliver J (eds) Coconut genetic resources. International Plant Genetic Resources Institute – Regional Office for Asia, the Pacific and Oceania (IPGRI-APO), Serdang, Selangor DE, Malaysia

Rohde W, Becker D, Kullaya A, Rodriguez J, Herran A, Ritter E (1999) Analysis of coconut germplasm biodiversity by DNA marker technologies and construction of a genetic linkage map. In: Oropeza C, Verdeil JL, Ashburner GR (eds) Current advances in coconut biotechnology. Kluwer Academic Publishers, Dordrecht pp 99–120

Sajini KK, Karun A, Kumaran PM (2006) Cryopreservation of coconut zygotic embryos by pre-growth desiccation. J Plantn Crops 34(3):576–581

Sakai KI (1960) Method of breeding coconut palm. Trop Agric 116:185–189

Sanoussi A (2016) Coconut genetic improvement programme in Benin. www.bioversityinternational.org/fileadmin/bioversity/publications/Web.../ch5.htm

Santos GA (1990) Activities in coconut genetic resources collection, conservation and genetic improvement in the Philippines. Philippine J Coconut Stud 15:16–20

Santos GA, Rivera RL (1994) Coconut breeding programme of the Philippines. In :Workshop on standardization of coconut breeding research techniques, Marc Delorme Coconut Station, Port Bouet, Côte d'Ivoire, June 20–25, 1994. IPGRI/COGENT, GTZ

Santos G, Batugal P, Oathman A, Baudoin Labouisse JP (1996) Manual on standardised research techniques in coconut breeding. IPGRI-COGENT Singapore, p 46

Satyabalan K (1956) A note on the performance of the natural cross dwarf (dwarf female) × Tall. Indian Cocon J 9:166–173

Satyabalan K, Mathew J (1983) Identification of pre potent palms in West Coast Tall coconut based on growth of the progeny in the nursery. In: Nayar NM (ed) Coconut research and development. Wiley Eastern Ltd, New Delhi, pp 15–22

Satyabalan K, Mathew J (1984) Identification of prepotent palms in W C T coconuts based on the early stages of growth in the nursery. In: Nayar NM (ed) Coconut research and development. Wiley Eastern Limited, New Delhi, p 225

Sindhumole P, Ibrahim KK (2000a) Path analysis of nut yield in cultivars of coconut (*Cocos nucifera* L.). South Indian Hortic 48(1/6):160–162

Sindhumole P, Ibrahim KK (2000b) Genetic variability in nine cultivars of coconut (*Cocos nucifera L.*). Trop Agric 38(1/2):18–20

Sosamma VK, Koshy PK, Rao EVVB (1988) Response of coconut cultivars to the burrowing nematode, *Radopholus similis*. Indian J Nematol 18(1):136–137

Sugimera Y, Rocat DA, Salud CD, Kamata (1994) An inflorescence mutant appeared in the inbred line of coconut "Markham". Japan J Trop Agric 38(2):145–147

Sukumaran CK (1982) Assessment of specific and general combining abilities of nine coconut cultivars. In: Bavappa KVA et al (eds) Proceedings of PLACROSYM V. Indian Society for Plantation Crops, Kasaragod, pp 131–137

Sukumaran Nair S, Balakrishnan PC (1991) Inbreeding depression in coconut (*Cocos nucifera* L.). In: Silas EG, Aravindakshan M, Jose AI (eds) Coconut breeding and management. Kerala Agricultural University, Thrissur, pp 51–54

Sumangala Nambiar S (1991) Susceptibility of hybrid coconut varieties to *Oryctes rhinoceros* under rain fed conditions at Pilicode. In: Silas EG, Aravindakshan M, Jose AI (eds) Coconut breeding and management. Kerala Agricultural University, Thrissur, pp 158–160

Swaminathan MS, Nambiar MC (1961) Cytology and origin of the dwarf coconut palm. Nature 192:85–86

Tammes PML (1955) Review of coconut selection in Indonesia. Euphytica 4:17–24

Tang L, Ma Z (2005) Status of coconut genetic resources research in China. In: Batugal P, Ramanatha Rao V, Oliver J (eds) Coconut genetics, Selangor DE, Malaysia, pp 648–653

Thampan PK, Gopalakrishnan R (2010) Suitable coconut varieties for multipurpose uses. In: Bose A (ed) In a nutshell- essays on coconut. Coconut Development Board, Publication No. 151, p 243

Thomas M, Gopimony R (1991a) Influence of seed nut characters on seedling vigour in coconut. In: Silas EG, Aravindakshan M, Jose AI (eds) Coconut breeding and management. Kerala Agricultural University, Thrissur, pp 99–102

Thomas M, Gopimony R (1991b) Heritability and correlations in West Coast Tall coconut palms. In: Silas EG, Aravindakshan M, Jose AI (eds) Coconut breeding and management. Kerala Agricultural University, Thrissur, pp 103–105

Varagas A, Blanco FA (2000) Fruit characterisation of *Cocos nucifera* L. (Arechaceae) cultivars from the Pacific coast of Costa Rica and the Philippines. Genet Resour Crop Evol 47:483–487

Warwick DRN, Leal EC, Bezerra APT, Tupinamba EA (1993) Field resistance of coconuts to 'Lixas' and Leaf blight in Brazil. In: Nair MK, Khan HH, Gopalasundaram P, EVVB R (eds) Advances in coconut research and development. Oxford and IBH publishing Co, New Delhi, p 759

Whitehead RA (1968) Selecting and breeding coconut palms (*Cocos nucifera* L.) resistant to lethal yellowing disease. A review of recent work in Jamaica. Euphytica 17:81

Wikipedia (2016) www.cirad.fr. Research results

Wikramaratne MRT (1993) The quest for the ideal palm. In: Nair MK, Khan HH, Gopalasundaram P, Rao EVVB (eds) Advances in coconut research and development. Oxford and IBH publishing Co, New Delhi, p 759

Yohannes S, Sumardjan H (1993) Current status of germplasm collection in Indonesia. In: Nair MK, Khan HH, Gopalasundaram P, Rao EVVB (eds) Advances in coconut research and development. Oxford and IBH publishing Co, New Delhi, p 759

Dr. K. U. K. Nampoothiri is a former Director of the Central Plantation Crops Research Institute (CPCRI), Kasaragod, Kerala, Indian Council of Agricultural Research, Government of India. He served as the Head of the Division of Plant Breeding at the Nigerian Institute for Oil Palm Research, Benin city, Nigeria for 5 years. He was also Director of M. S. Swaminathan Research Foundation, Chennai, India and an FAO consultant. He is the recipient of the Platinum Jubilee Award for coconut research and development from the CPCRI. His major field of specialization is agriculture: cytogenetics and plant breeding. unampoothiri@gmail.com

Dr. V. A. Parthasarathy Former Director and Emeritus Scientist, Indian Institute of Spices Research, Kozhikode, India, has made significant contributions in the field of biotechnology and breeding of a range of horticultural crops. His major contributions are in the tribal region of North East at ICAR Research Complex for NEH Region, Barapani, India, for which he received the Lifetime Achievement Award. He has published 200 research papers and 100 reviews/chapters in books. He has edited/authored 50 books. vapartha@gmail.com

Chapter 5
Varietal Resistance in Coconut

Regi J. Thomas, M. Shareefa, and R. V. Nair

Abstract Host plant resistance offers long-term and sustainable solution in biotic stress management in coconut. Among the diseases affecting coconut, phytoplasmal diseases reported from various countries are a major concern. Furthermore, bud rot, basal stem rot and stem bleeding are also very serious diseases limiting coconut production. Replanting with resistant varieties will go a long way in mitigating the adverse effects of these biotic factors, especially in cases where effective control measures are not available. The best and viable option in the future will be the management of coconut diseases mostly relying on integrated management practices centred on the exploitation of sources of disease resistance. Anticipating a rapid advancement in density and damage by sucking pests in the present era of climate change, efforts should be oriented to develop coconut varieties that can resist/tolerate coreid bug, eriophyid mite and emerging invasive pests as well. Besides, the focus to breed coconut varieties with resistance to red palm weevil, the most dreaded pest of coconut, should continue. It would be appropriate if the breeding programme in each coconut-growing country is fine-tuned to develop resistant varieties to combat pests/diseases of regional importance fully utilising the vast genetic resources available in those countries.

5.1 Introduction

Diseases and pests of various types are all the time a major production constraint in coconut. Cultivation of resistant varieties is considered as the most ideal and stable method for combating diseases and pests. Resistance or susceptibility is relative

---

R. J. Thomas (✉) · M. Shareefa
ICAR-CPCRI Regional Station, Kayamkulam, Kerala, India
e-mail: regijacob@yahoo.com; hishareefa@gmail.com

R. V. Nair
ICAR-CPCRI, Kasaragod, Kerala, India
e-mail: rvncpcri@gmail.com

K. U. K. Nampoothiri et al. (eds.), *The Coconut Palm (*Cocos nucifera *L.) - Research and Development Perspectives*,
https://doi.org/10.1007/978-981-13-2754-4_5

which is used in parlance only when it is transmitted from parents to offspring. Host plant resistance may be defined as the relative amount of heritable qualities possessed by the plants which influence the ultimate degree of damage caused by the pathogen or the insect pest (Painter 1951). Definitions of the common terms used in resistance breeding are given below to give a clear idea about the terminologies used.

5.2 Terminologies Used in Resistance Breeding

Resistance: Resistance is a state of less disease with immunity (no disease) representing the extreme (Simmonds 1983). All varieties that fall in between the extremes of the resistance scale, that is, immunity and extreme susceptibility, are partially resistant and partially susceptible (Kulkarni and Chopra 1989).

Immunity: 'Exempt from infection'. Immunity can be defined as absolute resistance to a pathogen (Robinson 1969).

Susceptibility: It is the sum total of qualities which make a plant a 'fit host' for a pathogen. Susceptibility is the opposite of resistance and is also inversely propositional to it (Robinson 1969).

Tolerance: A cultivar will be termed tolerant if it has the same yield and quality as another cultivar but supports more of the pathogen (Politowski and Browning 1978).

5.3 Disease Resistance

Breeding of disease-resistant varieties is one of the focus areas of coconut research. The major thrust of resistance breeding programme in coconut across the world has been to develop varieties with resistance/tolerance to phytoplasmal diseases. However, efforts have also been made for developing varieties with resistance to major fungal diseases as well. Among the phytoplasmal diseases affecting coconut, the major ones are root (wilt) disease and lethal yellowing disease. Of these, root (wilt) disease is restricted within India, whereas lethal yellowing is a global disaster especially in Carribean and African regions.

5.3.1 Phytoplasmal Diseases

5.3.1.1 Root (Wilt) Disease

Root (wilt) is a serious, non-lethal, debilitating malady of coconut palm which was first reported from Kottayam District, India, as early as in 1882 (Butler 1908). Systematic investigations on the etiology have ruled out the role of any physiological and nutritional disorders and also the role of biotic agents such as fungi, bacteria, viruses and nematodes (Nair et al. 1996). Investigations carried out at the Central Plantation Crops Research Institute under the Indian Council of Agricultural Research (ICAR-CPCRI), Regional Station, Kayamkulam, India, on the etiology of the disease have suggested the association of phytoplasma (Solomon et al. 1983a) transmitted through *Stephanitis typica* (lace bug) (Mathen et al. 1987) and *Proutista moesta* (plant hopper) (Anon 1991). Srinivasan (1991) reported that leaf rot disease gets superimposed on 65% of root (wilt) affected palms. For details, please refer to Chap. 11 on Phytoplasmal Diseases.

Donors for Resistance: In the absence of any effective control measure, planting of disease-resistant/disease-tolerant varieties is the only option to combat this disease. The search for resistance to coconut root (wilt) disease started with Butler (1908) who suggested that there could be resistance to coconut root (wilt) disease in the local cultivars grown in the diseased tract. Varghese (1934) initiated the studies to identify coconut palms resistant/tolerant to root (wilt) disease and surveyed about 10 km^2 in and around Kayamkulam, but in vain. Davis (1953) reported occurrence of high yielding palms among the heavily diseased palms.

Studies at ICAR-CPCRI (the erstwhile Central Coconut Research Station) Kayamkulam, India, showed that progenies of palms from disease-free areas also develop the disease when planted in the disease-affected areas during the course of 15 years (Anon 1966). Davis (1970) suggested that Malayan Dwarf could be a probable answer to root (wilt) disease of coconut, a surmise based on the performance of Malayan Dwarf in Jamaica where it replaced Jamaican Tall, which is susceptible to Lethal Yellowing Disease.

Rawther and Pillai (1972) studied the average yield in both healthy and diseased coconut palms and also disease incidence in Dwarf × Tall (natural cross), Tall × Dwarf, Dwarf Orange, Dwarf Green and West Coast Tall (WCT) as well. They found that Dwarf × Tall (natural cross) gave higher yield in both healthy and diseased palms compared to other varieties. Dwarf × Tall (natural cross) had the lowest percentage of disease incidence (4.6), and WCT had the lowest yield and also the highest disease incidence (48.5%). All the other varieties and hybrids studied were superior to WCT with regard to disease resistance and yield. Considering the higher level of resistance and yield, Dwarf × Tall (natural cross) was suggested for large-scale propagation in the root (wilt) diseased tract. However, this could not be put into practice since the specific male parents of these hybrids were not known and the quantity of natural cross progenies realisable from open-pollinated Chowghat Orange Dwarf (COD) was limited.

Ninan (1978) reported from a survey involving different varieties like WCT, Chowghat Green Dwarf (CGD), COD and their hybrids that CGD palms were the most resistant to root (wilt) disease with 92.2% palms being free from the disease. He also advocated further detailed study of hybrids of CGD with WCT for yield and disease resistance. Iyer et al. (1979) identified nine elite WCT palms in the disease-affected tract and monitored them for yield and reaction to root (wilt) disease. Among the nine elite palms under observation, three palms were subsequently affected by root (wilt) disease but still gave nut yields ranging from 200 to 471 nuts palm^{-1} year^{-1} (Anon 1988). However, all the open-pollinated (OP) progenies of these elite palms planted at ICAR-CPCRI, Regional Station, Kayamkulam, India, during 1980 have contracted the disease with a disease incidence ranging from 40 to 100%.

All the field trials on disease resistance were conducted in the disease hot spots of Kerala, where the disease incidence was the highest with an average incidence of 45–50%. As leaf rot was found to be closely associated with root (wilt) disease, resistance to this disease was also simultaneously screened in most of the trials. Mathai (1978), from the Kerala Agricultural University, reported that out of 75 hybrids and 14 varieties tested for susceptibility to root (wilt) disease, Laccadive Dwarf and its hybrid progenies showed some tolerance. Mathai (1980) also studied the resistance to root (wilt) disease on 7-year-old palms belonging to Tall × Dwarf, Tall × Gangabondam, Tall × Yellow Dwarf, Tall × Nyiorgading, Tall × Strait Settlement, Tall × Laccadive Dwarf, Tall × Andaman Dwarf, Tall × Tall and WCT. All the hybrids contracted root (wilt) disease with varying intensity. WCT was the most susceptible followed by Tall × Nyiorgading. Tall × Laccadive Dwarf was the most resistant. Similarly for leaf rot, Tall × Dwarf was the most susceptible followed by Tall × Laccadive Dwarf. Tall × Nyiorgading was the most resistant.

Attempts to screen the available germplasm from ICAR-CPCRI, Kasaragod (a disease-free area), were made as early as in 1961 at CPCRI, Regional Station, Kayamkulam. The cultivars tested were Andaman Ordinary, Andaman Giant, Cochin China, Ceylon Tall, Laccadive Dwarf, Laccadive Ordinary, New Guinea, the Philippines, Strait Settlement Apricot, Strait Settlement Green, St. Vincent and Spicata. All these cultivars evaluated in the field developed typical symptoms of root (wilt) disease (Menon et al. 1981).

Mathai et al. (1985 and 1991) reported the reaction of 15 coconut cultivars and Dwarf × Tall (NCD) hybrid to root (wilt) and leaf rot disease of coconut evaluated in two field trials. All the tested varieties showed varying degrees of intensity to root (wilt) disease. The intensity of the disease was significantly lower in Andaman Ordinary, San Ramon and Guam cultivars. Java and WCT were found to be highly susceptible to the disease with a higher disease index. The results also showed that leaf rot disease was the highest in WCT, whilst Saint Vincent, San Ramon, Cochin China and Andaman Ordinary had the lowest incidence of leaf rot. The result on Andaman Ordinary should be taken with caution since the number of palms screened under this cultivar was limited.

Large-scale field experiments were undertaken during 1972 at ICAR-CPCRI, Kayamkulam, as well as in cultivators' gardens in different soil types in 63 villages of Alappuzha, Kollam and Kottayam districts of Kerala State. All the open-pollinated progenies of 36 cultivars and 43 hybrid combinations planted, in gardens that had 40–70% incidence of root (wilt) disease, contracted the disease. It was also observed that the percentage of disease incidence was comparatively high in alluvial and reclaimed soil types (Jacob and Rawther 1991).

Data collected for 10 years from 21 cultivars and 15 hybrids (Tables 5.1 and 5.2), along with WCT as control, planted during 1972 revealed that none of the cultivars and hybrids were resistant to root (wilt) disease. The percentage of disease incidence ranged from 33.3 to 100. Andaman Ordinary and FMS recorded the lowest percentage of disease incidence (33.3), yielding 18.4 and 18.7 nuts $palm^{-1}$ $year^{-1}$, whereas WCT recorded 61% disease incidence with 29 nuts $palm^{-1}$ $year^{-1}$. Laccadive Ordinary yielded 40 nuts with 58.3% disease incidence in the 16th year of planting. The average annual yield was the maximum in WCT × COD hybrids with 52.5 nuts $palm^{-1}$ $year^{-1}$ though the disease incidence was as high as 91.6% (Jacob and Rawther 1991). Another field experiment, laid out in 1972, in cultivator's garden, indicated that hybrid combination COD × WCT gave an average yield of 80 nuts $palm^{-1}$ $year^{-1}$ with 50% disease incidence followed by its reciprocal hybrid yielding 70.3 nuts with 94% disease incidence. WCT × GB (Gangabondam) yielded 52.5 nuts showing 87.5% disease incidence compared to 70 nuts and 37.5% disease incidence in WCT (Jacob and Rawther 1991). Trials involving 27 cultivars, 10 hybrids, F_2 (OP) of Dwarf × Tall and Tall × Dwarf progeny, progenies of elite palms of high-yielding WCT palms and prepotent WCT were laid out in cultivator's gardens

Table 5.1 Reaction of coconut cultivars to root (wilt) disease

Sl. No.	Cultivars	Percentage of disease incidence	Annual yield (nuts $palm^{-1}$)
1.	Andaman Giant	100.0	4.5
2.	Andaman Ordinary	33.3	18.4
3.	B.S.I.	100.0	9.7
4.	Car Nicobar	92.0	22.1
5.	Chowghat Orange Dwarf	66.6	30.7
6.	Cochin China	50.0	14.4
7.	F.M.S.	33.3	18.7
8.	Gangabondam	92.0	13.3
9.	Java Giant	92.0	25.4
10.	Jamaica	66.6	24.1
11.	Kulasekharam Green Dwarf	66.6	22.0
12.	Kulasekharam Orange Dwarf	66.6	24.1
13.	Laccadive Micro	75.0	36.7
14.	Laccadive Ordinary	58.3	40.0
15.	M_2 irradiated	100.0	28.2
16.	Malayan Green Dwarf	83.0	31.4
17.	Malayan Orange Dwarf	66.6	15.4
18.	Malayan Yellow Dwarf	100.0	29.2
19.	Philippines Ordinary	41.6	21.9
20.	S.S. Apricot	92.0	19.9
21.	S.S. Green	83.3	14.1
22.	West Coast Tall	61.0	29.0

Note: No. of palms – 12 in all except West Coast Tall, which had 36 palms
Source: Jacob and Rawther (1991)
Age at observation: 10 years
Year of planting: 1972

Table 5.2 Reaction of coconut hybrids to root (wilt) disease

Sl. No.	Hybrids	Percentage of disease incidence	Annual yield (nuts $palm^{-1}$)
1.	COD × WCT	100.0	32.8
2.	Fiji × Gangabondam	66.6	16.4
3.	Fiji × SS Green	75.0	31.1
4.	Jamaica × Gangabondam	75.0	12.2
5.	Java Giant × KGD	83.3	43.0
6.	Java Giant × KOD	83.3	29.5
7.	Java Giant × KYD	33.3	38.2
8.	KGD × Java Giant	50.0	35.0
9.	KOD × Java Giant	75.0	37.0
10.	KYD × Java Giant	83.3	35.5
11.	Laccadive Ordinary × San Ramon	66.6	11.7
12.	San Ramon × Gangabondam	66.6	13.1
13.	WCT × COD	91.6	52.5
14.	WCT × Gangabondam	91.6	24.4
15.	WCT × MYD	66.6	13.1

Source: Jacob and Rawther (1991)
Age of observation: 10 years
Year of planting: 1972
No. of palms: 12

during 1982. Among the 27 cultivars tested, all except Kenthali took up the disease, and the disease incidence ranged from 7.1% to 55.6%. The disease also affected other hybrid combinations with disease incidence ranging from 12.5% to 66.7% in the fifth year of planting itself (Jacob and Rawther 1991).

Breeding Programme: A comprehensive breeding programme for evolving resistant/tolerant variety to root (wilt) disease is in progress at ICAR-CPCRI, Kayamkulam, since 1987 (Nair et al. 1996). In an intensive survey in a large number of farmers' plots in the hot spot districts of Kerala involving 200 Chowghat Green Dwarf (CGD) palms revealed that 75% of the observed palms were disease-free indicating that CGD had a higher level of resistance to root (wilt) disease compared to other varieties. Besides, a screening trial involving ten varieties initiated in 1988 revealed that CGD had the highest level of resistance (75%) 6 years after planting followed by Philippines Lono (70.8%), Zanzibar (70.8%) and FMS (66.7%). Please refer Table 5.3. WCT had intermediate level of resistance (62.5%). King Coconut (8.3%) and Kappadam (33.7%) were the most susceptible (Nair et al. 2004). Since the studies involving field survey as well as screening trials confirmed the higher level of resistance of CGD to root (wilt) disease, it was identified as a source of resistance for the breeding programme.

Breeding for resistance to root (wilt) disease was initiated based on the following observations: (i) CGD possesses higher level of resistance compared to other varieties. (ii) High-yielding disease-free WCT palms can be found in the disease-endemic

Table 5.3 Resistance of coconut varietal progenies to root (wilt) disease

		Disease incidence (per cent)		
Sl. No.	Variety	Self	Inter se	Average
1.	Karkar	50.0 (6.97)	66.7 (8.17)	58.3 (7.57)
2.	SS Apricot	58.3 (7.62)	58.3 (7.45)	58.3 (7.53)
3.	Kappadam	58.3 (7.45)	75.0 (8.65)	66.7 (8.05)
4.	CGD	25.0 (4.93)	25.0 (4.93)	25.0 (4.93)
5.	FMS	25.0 (4.93)	41.7 (6.42)	33.3 (5.68)
6.	Zanzibar	33.3 (5.58)	25.0 (4.93)	29.2 (5.25)
7.	Philippines Lono	25.0 (4.93)	33.3 (5.58)	29.2 (5.25)
8.	Fiji Rotuma	58.3 (7.62)	41.7 (6.42)	50.0 (7.02)
9.	King Coconut	100.0 (10)	83.3 (9.08)	91.7 (9.54)
10.	WCT	41.7 (6.42)	33.3 (5.77)	37.5 (6.10)
	General mean			6.69

Source: Nair et al. (2004)
Palms per treatment: 12
Year of planting: 1988
Year of observation: 1994

areas, in the midst of heavily disease-affected palms. Such palms were identified as the base material for the breeding programme. (iii) In the disease-endemic areas, a number of disease-free and high-yielding Chowghat Orange Dwarf (COD) palms were also found. These were also included in the programme to exploit the reported occurrence of high yield and tolerance to the disease in COD × WCT hybrids.

Artificial pollination of mother palms was carried out in farmers' plots for producing progenies which can be used either for developing a root (wilt) resistant/tolerant variety or for the production of parental materials for establishing seed gardens (Nair et al. 2000). The mother palms were serologically tested to ascertain their freedom from disease (Solomon et al. 1983b). WCT (inter se), WCT (self), WCT × CGD, CGD (self/inter se), CGD × WCT and COD (self/inter se) cross combinations were produced for field planting.

In the absence of an artificial screening technique for systematic evaluation of resistance, the seedlings produced were tested for field resistance under natural conditions. For evaluation of resistance, 2725 seedlings of the above cross combinations were underplanted since 1991 with a spacing of 7.5 m × 7.5 m inside the experimental farm. Care was taken to see that the old diseased palms were retained at least for 6–7 years after underplanting of the seedlings to provide sufficient inoculum for natural infection under field conditions. Heavy incidence of root (wilt) disease was observed in all the underplanted seedlings. Observations on root (wilt) disease recorded 12 years after planting from various progenies of WCT (WCT inter se/self/mixed pollen) revealed that the disease incidence varied from 55.0 to 58.7%, indicating that there was no significant variation in the susceptibility to the disease among the WCT progenies (Nair et al. 2003). However, in the case of the open-pollinated progenies of WCT, there was 80% disease incidence, showing the superiority of artificially pollinated seedlings (full sibs) over the open-pollinated seedlings (half sibs) with regard to their resistance to the disease.

Observations on 31 CGD × WCT hybrids during 1991 revealed that hybrids came to flowering in 30–40 months after planting. 67.5% of these palms took up the disease 14 years after planting (Nair et al. 2006). Though both the parents involved were deficient in some of the desirable characters like disease resistance, palm height, average yield of nuts, early bearing, nut size and shape, the hybrids were better in the overall performance. Even though majority of CGD × WCT hybrids were diseased, they gave a 10-year cumulative average yield of 84 nuts $palm^{-1}$ $year^{-1}$ indicating that this hybrid is tolerant to root (wilt) disease. This yield is quite attractive compared to the average productivity of Kerala State (35 nuts $palm^{-1}$ $year^{-1}$). Considering the performance of CGD × WCT in the root (wilt) disease prevalent tract, it was released under the name 'Kalpa Sankara' (Table 5.4).

Root (wilt) disease incidence and intensity of selected CGD was further evaluated and compared with WCT in the disease prevalent tracts of Kerala. The studies confirmed the high yield and low incidence of root (wilt) disease in CGD in hot spots of root (wilt) disease. Considering the high yield and low incidence of root (wilt) disease, selection made from Chowghat Green Dwarf was released under the name 'Kalpasree' for cultivation in homesteads of the root (wilt) prevalent tracts.

Studies carried out during 2004 involving five dwarf varieties of coconut, viz. Malayan Green Dwarf (MGD), Malayan Yellow Dwarf (MYD), Malayan Orange Dwarf (MOD), Chowghat Green Dwarf (CGD) and Chowghat Orange Dwarf (COD), resulted in identification of another promising variety, viz. Malayan Green Dwarf (MGD) (Fig. 5.1), as resistant to root (wilt) disease (Nair et al. 2007).

West Coast Tall (WCT), in an adjacent plot, showed 84% disease incidence indicating the availability of sufficient inoculum for evaluation of resistance of the dwarf varieties. CGD showed maximum resistance with disease incidence of 19.9% followed by MGD with a disease incidence of 22.4%. MGD gave the highest nut yield (88.8 $palm^{-1}$ $year^{-1}$) followed by WCT (48.7 $palm^{-1}$ $year^{-1}$). Considering the high yield and resistance to root (wilt) disease, this MGD selection was released under the name 'Kalparaksha' for cultivation in the root (wilt) disease prevalent tracts of India.

Table 5.4 Root (wilt) disease incidence and yield of 'Kalpa Sankara' in comparison to its parents

Sl. No.	Variety	Disease incidence (per cent)	Average nut yield ($palm^{-1}$ $year^{-1}$)	Copra content (g nut^{-1})	Copra (kg $palm^{-1}$)
1.	Kalpa Sankara (CGD × WCT)	67.7	82.5	170	14.00
2.	CGD	25.0	90.0	104	9.36
3.	WCT	80.0	48.7	177	8.62

Source: Nair et al. (2006)
*Period of observation: 1999–2004
Year of planting: 1991

Fig. 5.1 Malayan Green Dwarf (MGD). (Photo: V. Krishnakumar)

Crossing programme for the development of an improved WCT variety, by inter se mating/selfing of selected disease-free WCT palms from the hotspots, was initiated during 1990–1991. The selected disease-free palms were selfed and inter se mated to produce two sets of first-generation progenies (viz. 100 self and 450 inter se mated progenies). Observations recorded on these progenies after 18–19 years of planting revealed that the disease incidence in inter se mated progenies was only 47% compared to 63% in selfed progenies. The inter se mated progenies recorded 35–40% higher nut yield compared to selfed progenies and were also superior with regard to nut characters. A total of 40 high-yielding and disease-free palms belonging to the first generation were selected during 2009–2010 to produce the second-generation material. The inter se mated progenies were superior to selfed progenies with respect to nut yield and resistance to the disease (Thomas et al. 2016). Analysis of population structure of the mother palms and first-generation progenies using SSR markers indicated higher values for inbreeding coefficient and homozygozity in subsequent generations.

In order to meet the acute shortage of quality planting materials, five seed gardens were established in the root (wilt) prevalent tract, by planting selfed or inter se progenies of selected WCT, CGD and COD mother palms, spread over nearly 40 ha (Nair et al. 2001). These seed gardens are expected to produce nearly 2,50,000 seedlings annually.

Molecular Markers: Considering the long life cycle of coconut, selection of resistant varieties through conventional methods will be time consuming and laborious. Molecular markers offer numerous advantages over markers traditionally used in plant mapping and selective breeding. Rajesh et al. (2002) identified AFLP primer pairs detecting variations between resistant and susceptible palms.

Devakumar et al. (2011) studied the population structures among the root (wilt) disease-resistant and susceptible coconut palms from 12 locations in the 3 disease-endemic districts of Southern Kerala using 9 microsatellite markers. Two major populations and a subpopulation cluster were identified among the resistant palms. The analysis of genetic relatedness between the resistant mother palms showed that most of the palms located in a single locality shared sib relationship among them.

Rajesh et al. (2015) used RNA-Seq to generate the transcriptome of leaf samples of coconut root (wilt) disease-resistant cultivar CGD. Analysis using bioinformatic tools identified 243 resistance gene analog (RGA) sequences, comprising 6 classes of RGAs. Domain and conserved motif predictions of clusters were performed to analyse the architectural diversity. Phylogenetic analysis of deduced amino acid sequences revealed that coconut NBS-LRR-type RGAs were classified into distinct groups based on the presence of TIR or CC motifs in the N-terminal regions. Furthermore, qRT-PCR analysis validated the expression of randomly selected NBS-LRR-type RGAs. The results of this study provide a sequence resource for the development of RGA-tagged markers in coconut, which would aid mapping of disease-resistant candidate genes.

Rachana et al. (2016) subsequently used a comparative genomics approach to amplify putative RGAs from the coconut root (wilt) disease-resistant cultivar, Chowghat Green Dwarf (CGD), by using primers designed based on conserved motifs of the NBS-LRR domain of the date palm. The amplified sequences were cloned, sequenced and characterised. The coconut RGAs had high identity to monocot NBS-LRRs. Real-time quantitative polymerase chain reaction analysis indicated that the isolated coconut NBS-LRR class RGAs were expressed more in root (wilt) disease-resistant genotypes than in susceptible ones.

5.3.1.2 Lethal Yellowing Disease (LYD)

Lethal yellowing was first reported nearly 100 years ago from Jamaica. Subsequently, it was reported both from East and West Africa including Kenya, Tanzania, Nigeria, Ghana and Mozambique, the Caribbean, the Natuna Islands of Indonesia and most recently from Madang on the north coast of Papua New Guinea. The disease is caused by phytoplasma which is transmitted by adult plant hopper (*Myndus crudus*). Because of its ability to spread rapidly and destroy coconut populations, lethal yellowing disease (LYD) is considered as one of the most serious threats to coconut cultivation globally (Harries 1978). In Jamaica, five million Jamaican Tall palms were destroyed in 20 years following the spread of this disease (Romney 1983). Please see Chap. 11 on Phytoplasmal Diseases for details.

Sources of Resistance: Most of the exotic coconut varieties introduced and tested in Jamaica were susceptible to LYD (Whitehead 1968; Harries 1973). However, resistance of Malayan Dwarf variety (red, yellow and green colour forms) to LYD in Jamaica was first recognised in the midst of the 1950s (Nutman and Robert 1955). Harries (1974a) also reported that all the three colour forms of the Malayan Dwarfs have good resistance to LYD in Jamaica, and they also outyielded Talls when grown under suitable conditions. Large-scale planting of Malayan Dwarf in Jamaica led to the control of the disease. Subsequently F_1 hybrids were introduced in large scale, combining the high resistance of the Malayan Dwarf with more attractive agronomic characteristics of the more susceptible Tall parents such as Jamaican Tall, Rennel Tall and Panama Tall. Each parent was deficient in at least one of the characteristics considered such as productivity, leaf and bunch production, disease resistance, number and weight of nuts and crop production. The nature of resistance to LYD has not been determined. In an effort to locate additional sources of resistance, some evidence has been obtained about resistance in certain cultivars. Most of the hybrids have intermediate degree of resistance between that of parents. The degree of resistance of Panama Tall (PNT) nearly approaches the resistance of Malayan Dwarf (Harries 1974b). However, Malayan Dwarf × Panama Tall (Maypan hybrids) was found to be the most promising, and it was widely planted in Jamaica during the 1970s (Harries and Romney 1974).

Malayan Dwarf and Maypan, as well as crosses involving Cameroon Dwarf, showed high resistance to lethal yellowing disease in Jamaica. However, observations in Ghana suggested that Malayan Dwarf and Cameroon Dwarf are more susceptible to LYD in that country than in Jamaica (Anon 1976). The Malayan Dwarf variety was highly resistant to LYD in West Africa, Caribbean Islands and Florida (USA), and the only means of prevention in affected areas was to plant resistant varieties. However, studies conducted during 1982–2001 showed a 70% crop loss in over 19 years for Malayan Dwarf coconuts and 83% for Maypan coconuts. Results indicate that these cultivars cannot be considered resistant to lethal yellowing caused by phytoplasmas, as claimed in other studies (Broschat et al. 2002). The PNT is the pollen parent of the Maypan hybrid, which used to be planted in Jamaica to control lethal yellowing. The main source of contamination was the susceptible Jamaica Tall, thus increasing the susceptibility in the resulting Maypan progeny. The incidence of genetic contamination seems to be insufficient to be the only cause of the latest outbreak of the disease in the country. Hence, Maypan and its parents cannot be classified as resistant in the present context of Jamaica (Baudouin et al. 2008).

Fiji Dwarf (also known as Niu Leka) has shown variable resistance to LYD in Florida (USA). Flores (2008) reported that not even a single Fiji Dwarf died of the disease at Subtropical Horticultural Research Station (SHRS), Miami, USA, during the period 2000–2008. However, further confirmatory work was not carried out at SHRS. It was noted that recovery from lethal yellowing has been observed only in Malayan Dwarf, in crosses of Malayan Dwarf and in one unspecified cross not involving Malayan Dwarfs. Some Tall varieties also showed variable resistance (Sherman and Maramorosch 1977). Maypan hybrid was also found to be resistant to LYD in Florida (USA), Belize and other countries (Eden-Green 1997). In pure stands, LY resistance was estimated at 90% for Maypan (Table 5.5) compared to 96% for Malayan Dwarf (Been 1981). Resistance was also recognised in certain populations of Panama Tall and a progeny population of Malayan Dwarf × Fiji

Table 5.5 Incidence of lethal yellowing disease (percentage of incidence)

Variety/hybrid	Average	Fair prospect	Kildare	Caenwood	Rodney Hall	Plantain Garden
Jamaica Tall (JAT)	90	90	88	83	98	97
Malayan Dwarf (MAD)	4	1	10	11	3	4
Panama Tall (PNT)	44	–	38	–	–	67
Fiji-Malayan × JAT	5	4	5	–	–	–
Fiji-Malayan × self	8	8	–	–	–	–
JAT × Cambodia Tall (CBT)	53	53	–	–	–	–
JAT × Cameroon Red Dwarf (CRD)	14	8	20	–	–	–
JAT × Cuban Dwarf	40	–	40	–	–	–
JAT × MAD	36	49	9	–	–	–
JAT × Mozambique Tall (MZT)	53	53	–	–	–	–
JAT × PNT	46	54	29	–	–	–
JAT × Rennel Tall (RLT)	95	92	–	–	–	97
JAT × Solomon Tall (SNT)	52	–	–	–	–	–
JAT × Tahiti Tall (TAT)	79	79	–	–	–	–
Malayan Dwarf × CBT	11	0	–	50	–	–
Malayan Dwarf × CRD	15	15	–	–	–	–
Malayan Dwarf × JAT	23	24	15	34	5	20
Malayan Dwarf × MZT	30	17	–	50	–	–
Malayan Dwarf × PAT (Maypan)	10	13	4	21	13	7
Malayan Dwarf × RLT	26	29	30	44	23	25
Malayan Dwarf × SNT	20	14	–	24	–	–
Malayan Dwarf × TAT	38	0	–	100	–	–
PNT × CRD	28	28	–	–	–	–
PNT × JAT	38	58	17	–	–	–
PNT × MAD	11	–	11	–	–	–
PNT × RLT	50	54	–	–	–	42
PNT × SNT	17	17	–	–	–	–

+ Fair Prospect, Kildare, Caenwood, Rodney Hall and Plantain Garden are the experimental locations
Source: Been (1981)
*Year of planting: 1962–1970
– Data not available
Year of observation: 1979

Dwarf (F_1 hybrid). Promising source of resistance was also identified in Sri Lankan Dwarf, Indian Dwarf and King Coconut (Table 5.6) and intermediate resistance in a few of the tall varieties (Been 1981).

An assessment made in 1977 suggested that coconut genetic resources in Africa were vulnerable to loss from disease and replacement by improved varieties was considered necessary. Since then, losses due to diseases in Tanzania, particularly

Table 5.6 Incidence of lethal yellowing disease (percentage of incidence)

Variety	Average	Fair prospect	Plantain Garden	Kildare	Caenwood	Orange River
Indian Dwarf	0.0	0.0	–	–	–	–
King Coconut	0.0	0.0	–	–	–	–
Malayan Dwarf	1.1	1.1	–	–	0.0	–
Sri Lanka Dwarf	3.2	–	–	–	–	–
Malayan Dwarf (Local)	6.0	1.0	6.7	16.7	13.8	0.0
Rotuma Tall	26.3	60.0	14.0	–	–	
Bougainville Tall	31.0	35.0	28.0	–	–	–
Cambodia Tall	35.4	35.4	–	–	–	–
Thailand Tall	36.0	31.9	–	–	57.1	21.8
Malayan Tall (Solomon Is)	39.5	54.3	26.5	28.6	–	–
Malayan Tall	39.8	44.1	–	–	52.1	31.0
Yap Tall	45.7	76.5	26.2	4.9	61.5	–
Markham Valley Tall	48.0	58.0	44,0	–	–	–
Peru Tall	48.7	48.7	–	–	–	
Karkar Tall	50.0	19.5	44.0	–	–	–
Panama Tall	50.6	–	70.5	44.7	–	–
Niu Leka Dwarf	53.9	31.6	62.1	100.0	–	–
Sarawak Tall	55.8	56.0	–	58.0	–	52.8
Fiji Tall	66.7	60.7	–	73.4	–	–
Rennel Tall	67.7	72.9	67.3	–	–	–
Solomon Tall	70.2	80.0	58.9	–	–	–
Rangiroa Tall	71.8	76.1	100.0	40.3	–	–
Seychelles Tall	72.5	78.9	88.8	27.7	79.7	–
Samoa Tall	74.7	74.7	–	–	–	–
Tahiti Tall	77.4	83.7	71.1	73.1	–	–
Tonga Tall	86.6	83.0	79.0	100.0	–	–
Rangiroa Dwarf	93.5	100.0	86.5	–	–	–
Vanuatu Tall	94.0	94.0	–	–	–	–
Sri Lanka Tall	95.8	100.0	95.0	100.0	100.0	–
India Tall	96.9	100.0	–	–	96.0	–
Jamaica Tall	100.0	100.0	100.0	100.0	100.0	100.0

+ Fair Prospect, Plantain Garden, Kildare, Caenwood and Orange River are the experimental locations
Source: Been (1981)
–Data not available
Year of planting: 1962–1970
Year of observation: 1979

those caused by phytoplasma, have proved higher than anticipated as several introduced varieties with resistance in their countries of origin were not resistant under African conditions. Also, with reference to Tanzania, foreign introductions were not as numerous as anticipated because of the disease problems encountered and also because many potential introductions had very similar counterparts among the native varieties. However, the need to conserve coconut genetic resources was emphasised. Though germplasm collections existed in Côte d'Ivoire and Tanzania, it was recommended that every African country should have its own collection of native and standard varieties (Harries 1991).

The results of studies conducted by Sangare et al. (1992) during 1981–1992 on the occurrence of LYD in West, Central and East Africa revealed resistance in both Dwarf and Tall varieties, although West African Tall and its hybrids with both a Dwarf and a Tall were highly susceptible. Losses of 81% were reported in imported cultivars at Kifumangao, the tall cultivars being the most severely affected (Mpunami et al. 1992). Local palms were also susceptible, with the exception of East African Tall from Tanga. Some resistance to lethal disease was found in a hybrid between East African Tall and Pemba Red Dwarf.

Lethal disease of coconut in Tanzania is very similar to lethal yellow disease in the Caribbean Islands and West Africa. Schuiling et al. (1992) reported that varieties that show resistance to lethal yellowing in Jamaica are highly susceptible to lethal disease in Tanzania. The dwarf varieties are only marginally less susceptible than the Talls though hybrids are equally susceptible as their corresponding Tall parents. However, all the ecotypes appear to be more susceptible to disease in Tanzania. Promising resistance to lethal disease in Tanzania was observed only in two sub-populations of the local East African Tall. Resistance to lethal yellowing in different regions may be influenced by factors such as temperature, vector preference or sub-optimal growing conditions (Harries 1998), or it could be due to different strains of phytoplasma (Eden-Green 1997). No permanent cure has yet been found for LY, and to date, the use of resistance is the only effective means of coping with it to some extent (Been 1998).

Mariau et al. (1996) observed that very few cultivars are absolutely resistant/tolerant, necessitating testing of several most promising hybrids on a larger scale. A successful coconut rehabilitation scheme, developed in Jamaica in the 1960s–1980s, replaced the lethal yellowing-susceptible Jamaica Tall variety, first with the resistant Malayan Dwarf and then with its Maypan hybrid. The hybrid, which is still being produced and planted, is the product of a breeding programme that showed very promising performance in field exposure trials. Yet a rehabilitation programme, using introduced hybrids against lethal disease in Tanzania in the 1980s–1990s, did not succeed, and field exposure trial results were disappointing. Similar trials laid out with introduced material in the early 1980s in Mexico also did not give any encouraging result (Harries 1998). An alternate strategy for field exposure trials, based on the Jamaican experience and formulated in 1975, was partially implemented in Tanzania and Mexico after 1990, and positive results were reported. It was argued that the coconut palm-phytoplasma disease relationship calls for a pragmatic strategy in which simple, block plantings generate both the test material for field exposure trials and the seed nuts and pollen of resistant selections for direct use in the rehabilitation programmes. The more complicated strategy of small, replicated trial plots does not screen resistance realistically and cannot provide source material directly to rehabilitation programmes. The limitations of the replication method have been overlooked though, superficially, it appears to be more scientific. It was concluded that disease screening must be done under conditions that match the farming systems that will apply to the rehabilitation programme (Harries 1998).

Zizumbo-Villarreal and Arellano-Morín (1998) studied 20 coconut populations and suggested a correlation between precocity and resistance. The general perception is that cultivars from Malaysia and South East Asia are resistant to lethal yellowing disease. However, Caribbean and Atlantic Coast coconut cultivars are susceptible to LY, but hybrids of parents which originated in South East Asia have maintained excellent levels of resistance (Harries 2001). Sporadic outbreaks of non-epidemic phytoplasma disease in South East Asia may be linked to the introduction of susceptible cultivars from the Caribbean and West Africa (Maramorosch and Harries 1998).

Considerable efforts have been devoted throughout the world to screen varieties often involving international cooperation, which though was a lengthy and difficult task. Baudouin et al. (2009) opined that although no variety so far has been proven fully and permanently resistant, treating resistance level as a threshold trait makes it possible to demonstrate significant differences among varieties, which can be exploited effectively to make genetic improvement a component of an integrated control strategy. They have also made a few suggestions to increase the diversity of resistance sources and increase the level and the sustainability of resistance to LY in coconut.

The most viable alternative for control of lethal yellowing in Mexico and the Caribbean was to develop hybrids of crosses of dwarf coconut palms with tall landraces, combining disease resistance, earliness and production of large number of nuts per cluster of dwarf palm, with the hardiness and greater fruit size of the tall palms of the Pacific. The Donají hybrid, which is resistant to lethal yellowing, was produced at the experimental field of Oaxaca Coast of INIFAP, by crossing the Malayo Enano Amarillo cv. Acapulco as the female parent and the Landrace Alto Pacífico cv. Escondido as the male parent. This dual-purpose hybrid is recommended for all coconut-growing regions of Mexico, as it is high yielding and precautious (Serrano et al. 2011).

In the 1990s, bunch and nut yield dominated the coconut breeding programmes in Nigeria. With the outbreak of Awka wilt, focus was shifted to resistance breeding for LYD due to decline in yield and the high susceptibility of West African Tall (WAT). Odewale et al. (2013) evaluated the performance of different coconut varieties surviving under natural field conditions in an LYD endemic area of Nigeria. Five coconut varieties (viz. Malayan Green Dwarf (MGD), Malayan Yellow Dwarf (MYD), Malayan Orange Dwarf (MOD), West African Tall (WAT) and a hybrid derived from WAT and Malayan Dwarf) were evaluated for bunch and nut production for a period of 7 years. The results indicate that the surviving palms in the respective varieties are vigorous with the dwarf varieties showing more resistance to LYD compared to WAT. However, WAT recorded the highest yield for bunch (8.1) and nut production (63.7 nuts palm^{-1} year^{-1}). Evidence of biennial rhythm was revealed among the palms across the years. Despite the relatively poor performance of the varieties, the high yielding palms could be used in crossing programme for the production of nucleus planting material for further testing whilst serving as a germplasm base for resistance breeding.

Arroyo-Serralta et al. (2012) isolated a number of metabolites from coconut leaf cuticular waxes and explained the role of the wax layer in host plant-vector and host plant-pathogen interactions. Chromatographic analysis of the leaf cuticular wax from five coconut ecotypes showed that three main components (I, II and III) can be used as chemotaxonomical markers for classification of varieties either as resistant or susceptible. However, the results obtained from the antifeedant assay did not conform to the positive correlation found between the metabolites in the wax obtained from the various ecotypes and their resistance or susceptibility to the lethal yellowing disease. However, their study indicates that the full composition of the cuticular wax may have some role in host plant-insect vector interaction.

Molecular Markers: Breeding coconuts for any desirable trait is hindered by the long life span, low multiplication rate and ineffective clonal propagation. Additionally, the lack of adequate genotypes for identifying markers linked with LY resistance demands alternative approaches. Cardeña et al. (2003) identified three coconut populations which could be used for this purpose comprising of the susceptible West African Tall (WAT), the resistant Malayan Yellow Dwarf (MYD) and a resistant population of Atlantic Tall (AT) palms. AT was closely related to WAT, and both of them were distantly related to MYD. The objective of this work was to use those populations for identifying RAPDs associated with LY resistance. A total of 82 RAPDs could differentiate the DNA pools from MYD and WAT, and 12 of them appeared at frequencies ≥ 0.85 in MYD and ≤ 0.15 in WAT.

Zizumbo-Villarreal et al. (2006) reported that the Mexican genetic pool had high genetic diversity ($pl = 94$; $H_T = 0.34 \pm 0.02$) similar to that of the populations imported from the world's main gene pools ($pl = 94$; $H_T = 0.36 \pm 0.01$). Both molecular variance and Wright's index of differentiation indicated strong differentiation among Mexican ecotypes ($F_{ST} = 0.32$) despite significant gene flow ($Nm = 1.4$ to 5.6). UPGMA analysis and exact tests of differentiation suggested that the Indo-African gene pool is found along the Caribbean and Gulf of Mexico coasts, whilst the Asia-Pacific pool is found on the Pacific coast. High positive correlations were found between genetic distance and LY mortality percentages under severe incidence conditions during 9 and 14 years ($r^2 = 0.80$; $P = 0.02$; $r^2 = 0.78$; $P = 0.04$), suggesting that genetic distance may be useful for the estimation of the potential LY mortality in regions as yet unaffected as well as identification of potential parents for LY resistance breeding.

Konan et al. (2007) evaluated the genetic differences between the tolerant genotypes and the susceptible ones using 12 microsatellite markers. This work aimed to use identified materials as reference to select suitable parents for gene mapping studies. A total of 58 alleles were detected at the 12 microsatellite loci. The number of alleles varied from three to seven, with an average of 4.83 alleles. The F_{st} index revealed that 59.7% of the total allele variability explained differences between the three accessions. Genotypes of West African Tall, susceptible to the lethal yellowing disease, were less genetically clustered with the genotypes of the two resistant/tolerant accessions. This differentiation was based on specific alleles and frequency variation of shared alleles in the three accessions. This molecular typology was useful as reference for large molecular screening of coconut genetic resources and the identification of suitable parents for the development of mapping populations for tagging the lethal yellowing resistance genes.

There are several possible causes for the devastation of the Maypan hybrid due to an epidemic outbreak of LYD in Jamaica. Studies revealed that the LY affected planting material in Jamaica is genetically the same as the material shown to be resistant. Lebrun et al. (2008) compared the deoxyribonucleic acid (DNA) of MYD sampled from four locations in Jamaica with a reference DNA of the same cultivar collected in five different countries. The results of analyses showed more variation at 34 simple sequence repeat loci in Jamaica than in the rest of the world providing clear evidence for the presence of about 16% of alleles that do not match with the typical MYD genotype. These alleles appear to have already been present in the introduced germplasm. The observed heterogeneity might have caused some loss of resistance though insufficient to explain a massive outbreak of the disease.

Puch-Hau et al. (2015) used degenerate primers to amplify nucleotide-binding site (NBS)-type sequences from coconut ecotypes which were either resistant or susceptible to lethal yellowing. Genomic DNA fragments of approximately 500–700 bp were obtained and sequenced. Phylogenetic analysis of these fragments revealed that they clustered in seven different clades. All CnRGC sequences were grouped within the non-TIR-NBS-LRR subclass of NBS-LRR genes. The expression analysis revealed changes in expression profiles in response to salicylic acid (SA) and a constitutive expression profile in plants untreated with SA. This is the first large-scale analysis of NBS-type sequences in coconut palm.

5.3.1.3 Cape St. Paul Wilt

The Ghanaian form of LYD of coconut is known as the Cape Saint Paul Wilt Disease (CSPWD). The disease is caused by phytoplasmas and has been active in the country since 1932.

Varietal Screening: Danyo and Dery (2011) conducted studies to update the disease situation among Malayan Yellow Dwarf × Vanuatu Tall (MYD × VTT) hybrids in the Western and Central Regions of Ghana. Studies also compared the level of LYD susceptibility in the local West African Tall (WAT) variety and the MYD × VTT hybrid recommended for replanting. Losses were higher in the Central Region than in the Western Region. Overall, only 4.8% of total area under MYD × VTT coconut plantings had been affected by the disease as in 2009. LYD incidence was significantly higher in the WAT variety than the MYD × VTT hybrids. There was variation in disease incidence and severity in MYD × VTT hybrids between the different coconut growing areas. The susceptibility of the MYD × VTT hybrids under intense disease pressure calls for screening of more coconut varieties, to identify truly resistant types to the lethal yellowing disease.

Quaicoe et al. (2009) reported the results of screening trials on CSPWD carried out on 38 varieties since 1956. Two varieties, viz. Sri Lanka Green Dwarf (SGD) and Vanuatu Tall (VTT), have shown high resistance to the disease, and their hybrid (SGD × VTT) is under observation to determine its performance. A programme to rehabilitate the CSPWD-devastated areas was started in 1999. Emerging results indicate that the MYD × VTT hybrid used for the programme succumbs to the disease under intense disease pressure.

Johnson and Harries (1976) after a visit to Ghana and Togo in 1975 reported that there are great similarities between Cape St. Paul Wilt in Ghana, maladie de kaincope in Togo and Dahomey, Kribi disease in Cameroon and lethal yellowing in the Caribbean. The situation is more serious because both Malayan Dwarf, which was planted as a resistant variety in the Caribbean and the Cameroon Dwarf, which withstands the disease in Cameroon, succumbed to the disease in Ghana. Information on the susceptibility of various Dwarf and Tall varieties includes evidence which questions the resistance of Malayan Dwarf and Cameroon Dwarf, which have shown resistance to lethal yellowing in the Caribbean area and to Kribi disease in Cameroon, respectively, but have succumbed to Cape St. Paul Wilt in Ghana.

Molecular Markers: Swarbrick et al. (2013) isolated eight putative receptor-like kinase (RLK) genes from coconut using oligonucleotides designed against kinase subdomains of RLKs of other plant species, and the intron sequence of one of these was further analysed. Three single nucleotide polymorphisms (SNPs) were identified within this intron that could be used as a traceable marker to differentiate two distinct genotypes which could be differentiated using high resolution melt curve analysis. Analysis of different varieties of coconut used in the breeding programme included promising hybrids such as Sri Lankan Green Dwarf × Vanuatu Tall. F_1 crosses between these palms were self-pollinated to generate F_2 populations. Genotyping of palms at the RLK marker suggested that some F_2 progenies of parent F_1 palms might have been sired via cross-pollination from neighbouring palms, a possibility that would bear significance for such breeding programmes.

5.3.2 Fungal Diseases

5.3.2.1 Bud Rot

Bud rot is a fatal disease of coconut. The disease is characterised by rotting of the terminal bud and surrounding tissues with a foul smell and can even lead to mortality of the palm. Franqueville et al. (1989) observed the existence of *Phytophthora* resistance characters in a wide range of coconut varieties and hybrids in Côte d'Ivoire and suggested that coconut performance with respect to *Phytophthora* can be improved. The West African Tall proved sensitive to bud rot, but resistant to nut fall, whereas the Malayan Yellow Dwarf × West African Tall hybrid reacted vice versa. Some hybrids are sensitive to both forms of diseases, whilst others presented good tolerance. Within the Malayan Yellow Dwarf × West African Tall hybrid (PB

121), there is considerable variability, and some crosses are sensitive, whereas others are highly resistant. These studies reveal that a compromise can be found, both between the different hybrids and within the same hybrid to accommodate lower resistance to *Phytophthora* whilst retaining high yields.

With a view to improving PB 121 hybrid, two combining ability trials of WAT parents with a MYD tester were undertaken in 1978 and 1982 (Bourdeix et al. 1992). Results of the first trial showed that three progenies were significantly superior to PB 121. In terms of copra palm^{-1}, the difference amounted to 19% in young palms (4–8 years old) and 15% in adult palms, during which these progenies exceeded 4.8 mt ha^{-1} of copra, whereas PB 121 levelled off at 4.2 mt. Two out of three progenies had an excellent level of tolerance to nut-fall caused by *Phytophthora katsurae*, whilst the third was too productive for disease to adversely affect it.

Brahmana et al. (1993) conducted a study on the resistance of eight varieties to bud rot (*Phytophthora palmivora*), using the extent of disease attack on the nuts and the content of phenolic compounds in the husk as indicators of resistance. The varieties Nias Yellow Dwarf × Palu Tall and Bali Tall were more resistant, with high content of phenolic compounds and low disease attack. Whilst fungicide application is recommended for control, planting of resistant varieties is the most sustainable and practical method to combat this malady.

5.3.2.2 Basal Stem Rot

Basal stem rot (BSR) caused by *Ganoderma lucidum* is a major limiting factor in coconut production in Tamil Nadu, India. An experiment was laid out during 1989 in disease-endemic area to evaluate the performance of East Coast Tall (ECT) crossed with BSR-resistant ECT in comparison with nine other coconut cultivars, i.e. San Ramon, Lakshadweep Ordinary, British Solomon Islands, Java Giant, Straight Settlement Green, WCT × COD, COD × WCT, VHC-1 and ECT. Observations recorded after 15 years of planting revealed that ECT × BSR-resistant ECT recorded higher survival percentage, better growth characters, higher nut yield and lower disease index compared to the other cultivars (Karthikeyan et al. 2005).

5.3.2.3 Stem Bleeding

Stem bleeding disease starts with bleeding from the stem and later causes wilting and drying of leaves. Radhakrishnan and Balakrishnan (1991) conducted a trial with six coconut hybrids involving Gangabondam (GB) for resistance to stem bleeding caused by *Ceratocystis paradoxa*, and the correlation between disease index and mean yield during 1975–1984 was also worked out. There was a negative correlation between the disease index and yield. The lowest percentage of infection (29%) was found in Cochin China × Gangabondam (GB). Ramanujam et al. (1998) studied the reaction of 26 coconut cultivars involving 16 Talls, 6 Dwarfs and 4 hybrids against stem bleeding disease using a detached petiole inoculation method. All cultivars tested were susceptible to *C. paradoxa* to different degrees. The maximum lesion size was recorded in Malayan Green Dwarf, and the minimum was recorded in Banawali Green Round.

5.3.2.4 Leaf Spot and Leaf Blight

Leaf spot and leaf blight are minor diseases of coconut. During 2000–2002, more than 14,000 coconut palms were destroyed in Brazil by leaf spot caused by *Bipolaris incurvata*. Experiments were carried out in 2003 at Sococo plantations located in Para (Brazil) on 1.5-year-old juvenile coconut palms using three coconut hybrids, viz. PB 121, PB 123 and PB 132. Less number of leaf spot was observed in PB 121, whilst the hybrids PB 123 and PB 132 were more susceptible to the foliar spots (Gomes et al. 2009).

Ray and Kaiser (1990) reported that an acceptable level of resistance to leaf blight (*Pestalotia palmarum*) was found in 11 cultivars (Andaman Ordinary, Local Tall, Laccadive Micro, Laccadive Ordinary, Strait Settlement Green, Philippines Ordinary, West Coast Tall, Dwarf, Tall × Dwarf, Dwarf × Tall and Malayan Dwarf × West Coast Tall) out of 13 tested. In these, resistance was found to be dominant over susceptibility. Govindan et al. (1991) reported that none of the 8 varieties and 12 hybrids evaluated in Kerala proved resistant to leaf blight, but Malayan Yellow Dwarf showed low disease incidence, based on percentage of leaves and leaflets affected, as reflected with a lower disease index of 4.44. This was followed by CGD (6.66). In both these varieties, the area of infection was low.

Performance of eight coconut cultivars, viz. Tiptur Tall, Malayan Orange Dwarf, Malayan Yellow Dwarf, Sakhigopal Tall, Chowghat Orange Dwarf, Chowghat Green Dwarf, West Coast Tall and East Coast Tall, was evaluated against grey leaf spot in coastal Odisha, India. Tiptur Tall was found to be the most tolerant among the eight cultivars with 26% mean annual disease incidence, and Chowghat Orange Dwarf was the most susceptible with 35.4% disease incidence. The disease occurrence was probably influenced by high rainfall and drop in temperature (Ghose et al. 2006).

Studies on reaction of 28 different coconut hybrids/varieties against grey leaf spot or leaf blight caused by *Pestalotiopsis palmarum* revealed that no hybrid/variety was resistant to the disease (Suriachandraselvam et al. 2000). The intensity of leaf spot ranged from 23.7 (Malayan Yellow Dwarf) to 57% (Philippines × San Blas). The hybrids were more susceptible showing an intensity range of 42–57% compared to varieties which recorded an intensity range of 24–44%. This study proved that coconut hybrids are more susceptible to grey blight than varieties.

Five coconut hybrids, viz. PB 121, 141, 111, 231 and 132, introduced to Brazil were evaluated for resistance to *Botryodiplodia theobromae*. Ten plants were selected at random from each hybrid, and data were collected monthly for 34 months on the number of healthy leaves, number of infected leaves, lesion size and number of new lesions. The PB 141 hybrid was more leaf blight tolerant than Brazilian Tall and PB 231 hybrid (Warwick et al. 1991).

Helminthosporium halodes causes serious damage to introduced coconut varieties in Côte d'Ivoire. Inoculation techniques were used in the nursery to investigate entry site, susceptibility and progression of damage. Coconut varieties were classified as resistant, susceptible or intermediate (Quillec and Renard 1975). The red and yellow forms of Malayan Dwarf tended to be more susceptible to *Helminthosporium incurvatum* than the green form (Anon 1976).

5.3.2.5 Small and Big Verrucosis

Small and big verrucosis are important diseases affecting Brazilian coconut palms. It produces necrotic lesions on the palm leaflets. Five coconut hybrids introduced from Côte d'Ivoire were evaluated for their susceptibility to small verrucosis and big verrucosis in Sergipe (Brazil). The hybrid PB-141 showed low susceptibility to both the verrucoses. PB 132 and PB 121 were less susceptible to the small verrucoses and PB 111 to the big verrucoses. The climatic factors had no effect on the occurrence of the diseases. Higher disease incidence was observed from May to June and from November to January (Leal et al. 1996).

The dwarf coconut germplasm accessions planted in Brazil were evaluated in relation to the incidence of small verrucosis (*Phyllachora torrendiella*) and big verrucosis (*Sphaerodothis acrocomiae*). The incidence of verrucosis disease was observed in a 5-year-old dwarf coconut plantation. Of the six varieties assessed, the lowest incidence of small verrucosis was detected in Malaysian Yellow Dwarf, Gramame Yellow Dwarf and Cameroon Red Dwarf. The varieties Malayan Red Dwarf, Gramame Red Dwarf and Jiqui Green Dwarf were susceptible to big verrucosis disease. However, since the yellow dwarfs have a greater foliar emission, they might be able to tolerate the disease better (Leal et al. 1997).

Field experiments were conducted in Brazil during 1989 to 1992, to determine the severity of these two diseases on eight tall coconut genotypes under field conditions. The cultivars studied were GPY (Polynesia Tall), GRL (Rennell Tall), GOA (West African Tall), GRT (Rotuma Tall), GTG (Tonga tall), GML (Malaysia Tall), GVT (Vanuatu Tall) and GBR-PF (Brazil-Praia do Forte Tall). A number of leaves and verrucosic stromata were counted from selected palms. GBR-PF, GML and GOA had the highest number of stromata. Cultivar GPY, although not significantly different in stromatal count from the others, was the one with the lowest average number of stromata for both diseases (Leal et al. 1998).

5.3.3 Protozoan Diseases

Protozoa of the genus *Phytomonas* (Trypanosomatidae) have been implicated as the causal agent of several lethal diseases of coconut palms in South America and islands of the southern Caribbean. Parthasarathy et al. (1976) recorded these flagellated organisms in the phloem of ‘hartrot’ diseased coconut palms in Surinam.

Infected coconut palms exhibit inflorescence necrosis, rapid dieback of foliage and extensive root loss. Alexander (1980) reported that Surinam Tall was the most resistant to hartrot.

5.4 Pest Resistance

The coconut palm is damaged by a number of pests. It includes the larvae of Lepidopteran pests like leaf eating caterpillar (*Opisina arenosella*), which feeds on the lower epidermis and mesophyll tissues of coconut leaves. The nut is damaged by eriophyid mite (*Aceria guerreronis*). This mite can reduce up to 90% of coconut production. In addition, borers like Rhinoceros beetle and Red Palm weevil also damage coconut. Please refer to Chap. 12 on Pest Dynamics and Suppression strategies for details.

Management strategies have been developed against major and minor pests of coconut. However, emerging and invasive pests always pose a threat to coconut cultivation. Chemical and biological control measures have, at times, proved ineffective and ecologically undesirable. Breeding for resistance/tolerance to pest is an alternative sustainable and effective method for controlling major pests. Among the many pests, eriophyid mite is a serious pest in almost all coconut growing countries. Hence, breeding for resistance/tolerance to eriophyid mite has received more attention in coconut breeding programmes.

5.4.1 Eriophyid Mite *(Aceria guerreronis)*

Eriophyid mite develops in the meristematic regions of the immature nuts, which is covered by perianth (tepals). Their feeding causes scarring and distortions of the nuts and may cause their premature drop. Based on laboratory observations, the tightness of the perianth to the nut was identified as a key factor in determining susceptibility or resistance to attack by *A. guerreronis*. Penetration tests showed that as the nuts grew, it became increasingly larger in proportion to the perianth. Tepal aestivation in female flowers, shape of the developing nut, growth rate and pattern of nut enlargement are some of the traits identified as contributing to a lesser mite attack. Biophysical traits of habitat of the mite in the plant provide selection indices for tolerance. However, a conclusive test to determine resistance is still elusive.

The entry of mites depends on the tightness of tepals to the fruits at the early stages of fruit development. Greater tightness is achieved in round rather than elongated and angled fruits (Moore 1986; Aratchige et al. 2007). Varadarajan and David (2002) measured the gap by estimating the ratio of length of nut to radius of tepal. A large gap could also allow the predatory mites and hence is not congenial for herbivorous mite. In Malayan Yellow Dwarf, the space developed between the coconut surface and the perianth was large enough to allow mite to penetrate to the meristematic tissue (Howard and Rodriguez 1991). Differences for the gap in uninfested nuts are significantly different among the three varieties, Sri Lankan Green Dwarf, Sri Lankan Tall and their hybrids. Sri Lankan Green Dwarf palms are susceptible to mites as the nuts of this variety are small with an elongated shape. The gap in the nuts of this variety before infestation is large enough for the eriophyid mite to enter, but too small for the predatory mites. However, this perianth-fruit-rim gap in infested nuts does not differ significantly and hence accessible to predatory mites. Access of predatory mites long after the eriophyid mites reach a sufficient population is insufficient to keep the pest population below normal level. Hence, the gap between the nut and tepal is important and needs breeders' attention.

A good ideotype for resistance/tolerance to mite needs to consider the size, shape and perianth-fruit-rim gap of 2-month-old nuts before the mite could infest. Aestivation of tepals in coconut is of two types: *contortions* (with regular twisting or overlapping of tepals at one end) and *imbricate* (with irregular twisting or non-overlapping of inner tepals or one tepal overlapping at both ends). Aestivation also decides the tightness of the tepals (Moore 1986) and consequently the population of different herbivorous and predatory mites (Lawson-Balagbo et al. 2007). Tall varieties possess a higher percentage of contorting tepals than dwarf varieties, whereas Dwarf × Tall hybrids are intermediate (Davis et al. 1990). Round and dark green fruits show better tolerance against the eriophyid mite than the elongated fruits and those of other colours (Moore and Alexander 1987). Drought is also a predisposing factor which makes coconut palms susceptible to mite attack, since the growth rate of nuts is slow because of the lack of available soil moisture (Mariau 1986).

Thirty one varieties in India were evaluated in 1999 for eriophyid mite damage using a one to five scale. Under conditions of natural infestation, Laccadive Ordinary, Cochin China, Andaman Ordinary and Gangabondam recorded minimum nut damage, whereas Seychelles, St. Vincent and Nigerian Tall were highly susceptible (Muthiah and Bhaskaran 1999). Screening of coconut varieties for tolerance to coconut mite resulted in identification of some accessions such as Kenthali (Ramaraju et al. 2000) and Chowghat Orange Dwarf (Nair 2000) with lower incidence of mite. Muthiah and Rajarathinam (2002) screened 33 coconut cultivars for eriophyid mite resistance for 3 years during 1999–2002. Chowghat Orange Dwarf, Siam, British Solomon Island (BSI), Ayiramkachi, Philippines Ordinary and Spicata were found to be moderately tolerant. The cultivars Seychelles and St. Vincent were found to be highly susceptible to mite attack. Among 34 coconut hybrids screened during 2001, 4 hybrids, viz. Java Giant × East Coast Tall, Ayiramkachi × West Coast Tall, Cochin China × Philippines Ordinary and West Coast Tall × Chowghat Orange Dwarf, were found to be moderately tolerant.

Levin and Mammooty (2003) reported that the Spicata mutant showed a fair level of tolerance to eriophyid mite. They also reported that the genotype BSI recorded the highest percentage of nut damage by mites followed by Philippines Lono (81.1%). Laksha Ganga (Laccadive Ordinary × Gangabondam) recorded the minimum incidence (19.4%) compared to the maximum mite damage (30.0%) in Ananda Ganga. The cultivars Ayiramkachi (90.2%) and Andaman Dwarf (85.3%)

were the most susceptible to mite damage among the indigenous cultivars, whereas the genotypes Bombay (6.4%), Laccadive Micro (7.4%), Chowghat Orange Dwarf (8.8%) and Spicata (9.5%) were the least susceptible.

Muthiah and Natarajan (2004) conducted a field experiment in Tamil Nadu, India, during 1999–2001 to evaluate the resistance of 33 coconut cultivars and 34 hybrids to eriophyid mite. Four cultivars, viz. BSI, Chowghat Orange Dwarf, Philippines Ordinary and Spicata, and two hybrids, viz. Philippines Ordinary × San Blas and Cochin China × Philippines Ordinary, were moderately resistant, whereas all the other materials tested were moderately susceptible or highly susceptible to the pest.

The cross between Sri Lankan Yellow Dwarf × Sri Lankan Tall has been identified as tolerant to *Aceria* mite, in an evaluation of five commercially cultivated coconut cultivars in Sri Lanka in a severely mite-affected area (Perera 2005, 2006). Sri Lankan Yellow Dwarf and Gonthembili have also been identified as tolerant cultivars to coconut *Aceria* mite (Perera 2006).

Eleven coconut cultivars were screened against eriophyid mite in Andhra Pradesh (India) during September 2000 to January 2005 (Raju et al. 2006). None of the cultivars were resistant to *A. guerreronis*. However, Java Giant and Ceylon Green Tall were moderately resistant, and the rest were susceptible. The maximum percentage of infested nuts was recorded by Fiji (83.65), followed by Chowghat Orange Dwarf (80.13).

Girisha and Nandihalli (2009) screened ten coconut varieties, viz. West Coast Tall, Arsikere Tall, Laccadive Ordinary, Gangabondam, Philippines, 'Green Dwarf', Andaman Dwarf, Laccadive Dwarf, 'Green Tall' and Spicata, against eriophyid mite. Significantly less mite population was recorded in Gangabondam (28.96/28.28 mm^2 area of perianth) which was found significantly superior over other varieties followed by West Coast Tall, whereas varieties Laccadive, Green Dwarf, Arsikere Tall and Green Tall recorded more mite population. The superiority of Gangabondam might be due to the tight attachment of perianth to nut surface. Gangabondam also recorded least damage grade (1.40) and damaged nuts (16.00) with the highest number of healthy nuts (78 nuts $palm^{-1}$).

Sujatha et al. (2010) screened 8 tall coconut varieties (45 years old) and 17 hybrids (15 years old) for resistance against the mite for 4 years during 2004 to 2007 under natural conditions of coastal ecosystem of Andhra Pradesh, India. The lowest mite damage index was recorded in Laccadive Ordinary and the highest in Laccadive Micro. Among the 17 coconut hybrids screened, ECT × GB (Godavari Ganga) recorded the lowest mite damage, whereas LM × GB recorded the highest damage among various cross combinations. Badge et al. (2016) screened coconut genotypes for their level of susceptibility to coconut eriophyid mite in India. Based on mean damage score, none of them were found to be resistant to eriophyid mite. Among the 26 coconut cultivars screened, minimum infestation was observed in the genotypes, viz. Jamaica, BSI, Philippines Lono, Guam and Orange Dwarf.

Molecular markers: Molecular analysis of coconut accessions from different parts of South India was done using 32 simple sequence repeats (SSRs) and 7 RAPD primers. In single-marker analysis, nine SSR and four RAPD markers associated with mite resistance were identified. In stepwise multiple regression analysis of SSRs, a combination of six markers showed 100% association with mite infestation. Stepwise multiple regression analysis for RAPD data revealed that a combination of three markers accounted for 83.9% of mite resistance in the selected materials. Combined stepwise multiple regression analysis of RAPD and SSR data showed that a combination of five markers explained 100% association with mite resistance in coconut (Shalini et al. 2007).

5.4.2 Rhinoceros Beetle *(Oryctes rhinoceros)*

Although all coconut cultivars are prone to damage by rhinoceros beetles, the hybrids developed with Chowghat Orange Dwarf as pollen parent was reported to be more susceptible (Nambiar 1988). Coconut varieties were screened in Tamil Nadu, India, for reaction to rhinoceros beetle. Average leaf damage over 3 years recorded from different hybrid combinations and varieties revealed that Laccadive Ordinary × Cochin China and Gangabondam × East Coast Tall had significantly minimum damage. Among the 12 varieties evaluated, West Coast Tall (WCT) and East Coast Tall (ECT) recorded less damage. The damage was more in hybrids involving dwarf genotypes (Muthiah and Bhaskaran 2000).

5.4.3 Red Palm Weevil *(Rhynchophorus ferrugineus)*

Red palm weevil (RPW) is known to cause serious damage to the crop and has attained key pest status. A preliminary survey on the damage of coconut cultivars in different districts of Tamil Nadu, India, showed that Andaman Giant, Java Giant, East Coast Tall, West Coast Tall, Federated Malay States × Laccadive Ordinary, East Coast Tall × Malayan Green Dwarf, West Coast Tall × Gangabondam, Java Giant × San Blas and Laccadive Ordinary × Cochin China were more susceptible to red palm weevil attack (Sadakathulla and Ramachandran 1993). Faleiro and Rangnekar (2001) studied the ovipositional preference of RPW to different coconut cultivars and reported that the highest cumulative egg lay was in CGD, COD and Benaulium, which recorded an average of 31.3, 30.9 and 27.4 eggs, respectively.

5.4.4 Red Spider Mite *(Oligonychus velascoi)*

Red spider mite feeds on coconut leaves. Large-scale attack causes yellowing and early dieback of the leaves of young palms and can retard growth severely. Five biological parameters were evaluated in populations of *O. velascoi* growing on three coconut varieties in the Philippines. All parameters indicated Baybay Tall as the most susceptible variety. On the basis of total developmental period, fecundity and adult longevity, MAWA (Malayan Yellow Dwarf × West African Tall) was the most resistant hybrid, followed by Malayan Orange Dwarf (MOD). However, mortality of immature mites was significantly higher in MOD than in the other varieties, suggesting the presence of an antibiotic component in its sap (Capuno and de Pedro 1982).

5.4.5 Nematodes

The most important nematode found in coconut root is burrowing nematode, *Radopholus similis*. General decline symptoms like stunting, yellowing, reduction in number and size of the leaves and leaflets, delay in flowering, button shedding and reduced yield are exhibited by burrowing nematode infested palms. Sosamma et al. (1980) reported that the cultivar Java had the least root lesion index and root population of burrowing nematode (*Radopholus similis*; 7 g^{-1} of roots), whereas cultivar Jamaica Tall (259 g^{-1} of roots) and Fiji Tall (137 g^{-1} of roots) were most severely infested. Java Tall, Klapawangi, Kenthali and Andaman Giant showed fair level of tolerance to burrowing nematode (Sosamma et al. 1988). But Sudha (1998) reported that all 25 coconut cultivars screened against *R. similis* were susceptible to the nematode. Philippines Ordinary was the least susceptible with an average population of 48.4 nematodes 10 g^{-1} root, whereas COD was highly susceptible recording 953.4 nematodes 10 g^{-1} root.

5.4.6 Rodents

Rodents damage tender coconuts throughout the year, but with a higher intensity during July–October (Sadakathulla 1996). Evaluation of 28 cultivars and 9 hybrids for rodent (*Rattus rattus wroughtoni* and *Funambulus palmarum*) damage for 1 year in India indicated that Ayiramkachi, Laccadive Micro and East Coast Tall × Ayiramkachi were highly susceptible, followed by Siam × Ayiramkachi. The varieties Jamaica Tall, Andaman Giant, Andaman Ordinary and Federated Malay States were moderately resistant to rodents (Sadakathulla and Kareem 1994).

5.4.7 Ash Weevil

A study was carried out in Kerala, India, on 37 coconut cultivars and hybrids to record the infestation of ash weevil (*Myllocerus curvicornis*) and the relationship between infestation by this pest and incidence of root (wilt) disease. The highest infestation rate (69.7%) was observed on Malayan Green Dwarf and the lowest (7.29%) on Fiji × S.S. Green. Weevil damage was related more to the high nitrogen content of diseased palms (Ponnamma et al. 1985).

5.5 Breeding Coconut with Their Microbiome

The current concept of 'holobiont' theory, where the plant is considered as a meta-organism (i.e. plant and its microbiome present in the rhizosphere and endophytic compartment), is giving a new dimension to the plant breeding programme (Zilber-Rosenberg and Rosenberg 2008; Theis et al. 2016; Gopal and Gupta 2016). The plant microbiome is known to govern many of the positive traits in plants (Vandenkoornhuyse et al. 2015). In coconut, basic research on rhizosphere microbiota have shown differences in microbial community profile, particularly with relation to arbuscular mycorrhizae in drought-tolerant varieties (Thomas et al. 1993) and in palms located in the high-yielding tracts in Kerala, India (Rajeshkumar et al. 2015). A preliminary study of rhizosphere microbiota of coconut root (wilt) diseased and field resistant palms indicated differences in plant-beneficial bacterial communities (Gopal et al. 2005) that need further detailed investigation using genomic protocols. Thus, the use of arbuscular mycorrhizae and endophytic microbiota has already emerged as an innovative technology for phytoplasma disease management (Musetti et al. 2012; Bianco et al. 2013) that could possibly lead to tailored microbiome for plant disease management (Gopal et al. 2013).

5.6 Future Strategy

From the data so far available, it appears that MGD, CGD and CGD × WCT hybrids can be very important in the management of coconut root (wilt) disease. The development of an improved WCT variety, by inter se crossing or selfing of selected WCT palms, seems to be a promising line of research irrespective of the long time and resources required. Considering the long gestation period required for the large-scale production of planting materials, the approach of establishing nucleus seed gardens in different districts of root (wilt) prevalent tracts, concurrent with the varietal improvement programmes, is a pragmatic approach to reduce the acute shortage of quality planting materials in the diseased tract. Control of coconut root (wilt) disease, in the long run, will most likely rely on integrated management practices centred on the exploitation of source of disease resistance.

Collective effort for sharing coconut germplasm to breed varieties with resistance to phytoplasmal diseases should be accorded maximum priority. In addition, research on tagging disease-resistant genes should employ molecular markers which are stable and dependable, and such work should utilise next-generation sequencing technology so that the long-cherished objective of identification of resistant progenies in nursery stage itself materialises. Among other diseases affecting coconut, bud rot, basal stem rot and stem bleeding are the most devastating, and developing resistant varieties will go a long way to combat the diseases by taking up replanting programmes. Since, Talls have certain specific advantages over Dwarfs, recurrent selection to improve the level of disease resistance in tall populations should be envisaged, involving sib (inter se) mating between highly divergent coconut populations so as to simultaneously increase the yield of the resultant progenies.

In the present era of climate change, a rapid increase in density and damage of sucking pests should be anticipated. Hence, efforts should be oriented to develop coconut varieties that can resist/tolerate sucking pests like coreid bug, eriophyid mite and – recently emerged – invasive spiralling whitefly. Besides, the focus to breed coconut varieties with resistance to red palm weevil, the most dreaded pest of coconut, should continue. It would be appropriate if the breeding programme in each coconut-growing country is fine-tuned to develop resistant varieties to combat pests/diseases. Since growing coconut as a monocrop is no more profitable, it is suggested to develop coconut varieties which can fit well into high-density multispecies cropping systems with the additional trait of resistance to major pests/diseases of specific coconut-growing tracts. In view of the regional specificity of the biotic factors, priority attention should be given for location-based breeding programmes with necessary international cooperation wherever appropriate.

References

Alexander VT (1980) Varietal resistance studies for 'hart rot' disease of coconuts. Oleagineux 35(6):302–303

Anonymous (1966) Annual report for 1966. Central Coconut Research Station, Kayamkulam, Kerala, India, pp. 64

Anonymous (1976) 15th Report of the Research Department, July 1974-Dec 1975. Coconut Industry Board, Jamaica, pp. 55

Anonymous (1988) Annual report for 1987. Central Plantation Crops Research Institute. Kasaragod, Kerala, India, pp 174

Anonymous (1991) Annual report for 1990–91. Central Plantation Crops Research Institute. Kasaragod, Kerala, India, pp. 154

Aratchige NS, Sabelis MW, Lesna I (2007) Plant structural changes due to herbivory: do changes in Aceria infested coconut fruits allow predatory mites to move under the perianth? Exp Appl Acarol 43:97–107

Arroyo-Serralta GA, Zizumbo-Villareal D, Escalante Erosa F, Peña-Rodríguez LM (2012) Cuticular wax composition of coconut palms and their susceptibility to lethal yellowing disease. J Mex Chem Soc 56(1):67–71

Badge AS, Patil PD, Pashte VV (2016) Screening of coconut genotypes against coconut eriophyid mite (*Aceria guerreronis* K). Int J Bioresource Stress Manag 7(1):18–23

Baudouin L, Lebrun P, Berger A, Myrie W, Been B, Dollet M (2008) The Panama Tall and the Maypan hybrid coconut in Jamaica: did genetic contamination cause a loss of resistance to Lethal Yellowing? Euphytica 161(3):353–360

Baudouin L, Philippe R, Quaicoe R, Dery S, Dollet M (2009) General overview of genetic research and experimentation on coconut varieties tolerant/resistant to Lethal Yellowing. Oléagineux 16(2):127–131

Been BO (1981) Observations on field resistance to lethal yellowing in Jamaica. Oleagineux 36(1):9–12

Been BO (1998) Breeding for resistance to lethal yellowing. In: Proceedings of International cashew and coconut conference: trees for life - the key to development, Dar es Salaam, Tanzania, 17–21st February 1997, pp 380–384

Bianco PA, Marzachi C, Musetti R, Naor V (2013) Perspectives on endophytes as biocontrol agents in the management of phytoplasma diseases. In: Torres E, Laviña A, Jarausch W, Bertaccini A (eds) New Perspectives in Phytoplasma Disease Management. COST Action FA0807, Barcelona, pp 69–72

Bourdeix R, N'Cho YP, Sangare A, Baudouin L, Lamothe MW d N d (1992) The improved PB121 coconut hybrid, a cross between the Malayan Yellow Dwarf and selected West African Tall parents. Oléagineux 47(11):619–633

Brahmana J, Sipayung A, Purba P (1993) Content of phenolic compounds as an index of resistance in coconut fruit to premature nutfall disease. Bull Pusat Penelitian Kelapa Sawit 1(1):81–87

Broschat TK, Harrison NA, Donselman H (2002) Losses to lethal yellowing cast doubt on coconut cultivar resistance. Palms 46(4):185–189

Butler EJ (1908) Report on Coconut Palm Disease in Travancore. Bulletin No. 9. Imperial Agricultural Research Institute, Pusa, pp 1–23

Capuno MB, de Pedro LB (1982) Varietal reaction of coconut to *Oligonychus velascoi* Rimando using five mite-based biological parameters. Ann Trop Res 4(4):274–280

Cardeña R, Ashburner GR, Oropeza C (2003) Identification of RAPDs associated with resistance to lethal yellowing of the coconut (*Cocos nucifera* L.) palm. Sci Hortic 98(3):257–263

Danyo G, Dery SK (2011) Overview of the epidemiology of coconut Lethal Yellowing Disease (LYD) in Malayan yellow dwarf × Vanuatu tall hybrids in the Western and central regions of Ghana. J Ghana Sci Assoc 13(1):30–38

Davis TA (1953) Breeding in coconut for disease resistance. Indian Cocon J 6(3):95–100

Davis TA (1970) The Malayan Dwarf, probable answer to the root (wilt) in Kerala. Cocon Bull 1(3):3–7

Davis TA, Sudasrip H, Darwis SN (1990) Development of coconut perianth and fruit. Coconut Research Institute, Mandao, Indonesia

de Franqueville H, de Taffin G, Sangare A, le Saint JP, Pomier M, Renard JL (1989) Detection of *phytophthora heveae* tolerance characters in coconut in Côte d'Ivoire. Oléagineux 44(2):93–103

Devakumar K, Thomas RJ, Nair RV, Jerard BA, Rajesh MK, Jacob PM, Jayadev K, Parthasarathy VA (2011) Analysis of population structure and genetic relatedness among root (wilt) disease resistant and susceptible coconut palms (*Cocos nucifera* L.) cv. West Coast Tall using microsatellite markers. Indian J Agric Sci 81(5):487–493

Eden-Green SJ (1997) History, world distribution and present status of lethal yellowing disease of palms. In: Eden-Green SJ, Offori F Elmina (eds) Proceedings of International Workshop on Lethal Yellowing Like disease of Coconut. Ghana, November 1995, pp 9-26

Faleiro JR, Rangnekar PA (2001) Ovipositional preference of red palm weevil *Rhynchophorus ferrugineus* Olive. to coconut cultivars. Indian Cocon J 32(6):22–23

Flores A (2008) Fiji Dwarf sets new durability standards in coconuts. Agri Res (May-June): 17

Ghose S, Mishra BD, Rout MK (2006) Varietal evolution and occurrence of coconut grey leaf spot disease in coastal Orissa. J Mycopathol Res 44(1):105–107

Girisha RC, Nandihalli BS (2009) Seasonal abundance and varietal reaction of coconut perianth mite, *Aceria guerreronis* Keifer in Dharwad area. Karnataka J Agric Sci 22(3):606–608

Gomes BEG, Miranda VS, Poltronieri LS, do NMM, Farias PRS (2009) Assessment of incidence of three hybrids of coconut palm against leaf spot in the field conditions. Revista de Ciências Agrárias 51:161–170

Gopal M, Gupta A (2016) Microbiome selection could spur next-generation plant breeding strategies. Front Microbiol 7:1971. https://doi.org/10.3389/fmicb.2016.01971

Gopal M, Gupta A, Nair RV (2005) Variations in hosting beneficial plant-associated microorganisms by root (wilt) diseased and field tolerant coconut palms of West Coast Tall variety. Curr Sci 89(11):1922–1927

Gopal M, Gupta A, Thomas GV (2013) Bespoke microbiome therapy to manage plant diseases. Front Microbiol 4:355. https://doi.org/10.3389/fmicb.2013.00355

Govindan M, Radhakrishnan TC, Sathiarajan PK (1991) Reaction of certain varieties and hybrids of coconut to natural infection of leaf blight caused by *Pestalosphaeria elaeidis*. In: Silas EG, Aravindakshan M, Jose AI (eds) Proceedings of national symposium on coconut breeding and management. Kerala Agricultural University, Thrissur, pp 156–157

Harries HC (1973) Selection and breeding of coconuts for resistance to diseases such as lethal yellowing. Oleagineux 28(8–9):395–398
Harries HC (1974a) Selection and breeding of coconuts for resistance to diseases such as lethal yellowing. Indian J Gen 34A:66–74
Harries HC (1974b) The performance of F_1 hybrid coconuts in Jamaica. J Plantn Crops 2(2):15–20
Harries HC (1978) Lethal yellowing diseases of coconut in global perspective. Philippine J Cocon Sci 3:1–4
Harries HC (1991) The vulnerability of the coconut genetic resources of Africa- Working towards a better future for the African coconut farmer. In: Proc. First African Coconut Seminar, 4–8th February 1991, Dar es Salaam, Tanzania, pp 77–81
Harries HC (1998) Breeding phytoplasma disease resistant coconuts: alternative strategies for field exposure trials. In: Proceedings of International Cashew and Coconut Conference: trees for life – the key to development, Dar es Salaam, Tanzania, 17–21st February 1997, pp 375–379
Harries HC (2001) Coconut varieties and lethal yellowing: a regional perspective for the Americas. Palms 45(3):148–150
Harries HC, Romney DH (1974) Maypan: An F_1 hybrid coconut variety for commercial production in Jamaica. World Crops 26:110–111
Howard FW, Rodriguez EA (1991) Tightness of the perianth of coconuts in relation to infestation by coconut mites. Fla Entomol 74(2):358–361
Iyer RD, Rao EVVB, Kutty MPG (1979) Super yielders in coconut. Indian Fmg 28(10):3–5
Jacob PM, Rawther TSS (1991) Varietal resistance. In: Nair MK, KKN N, Koshy PK, Jayasankar NP (eds) Coconut root (Wilt) disease. Central Plantation Crop Research Institute, Kasaragod, pp 67–72
Johnson CG, Harries HC (1976) A survey of Cape St. Paul Wilt of coconut in West Africa. Ghana J Agri Sci 9(2):125–129
Karthikeyan G, Karpagavalli S, Natarajan C, Arulraj S (2005) Evaluation of coconut hybrid, ECT × BSR tolerant ECT in basal stem rot sick soil. Indian Cocon J 36(5):12–14
Konan KJN, Koffi KE, Konan JL, Lebrun P, Dery SK, Sangare A (2007) Microsatellite gene diversity in coconut (*Cocos nucifera* L.) accessions resistant to lethal yellowing disease. African J Biotech 6(4):341–347
Kulkarni RN, Chopra VL (1989) Breeding for disease resistance. In: Chopra VL (ed) Plant breeding. Oxford & IBH Publishing Co. Pvt. Ltd, New Delhi, India, pp 285–308
Lawson-Balagbo LM, Gondim MGC Jr, de Moraes GJ, Hanna R, Schausberger P (2007) Refuge use by the coconut mite *Aceria guerreronis*: fine scale distribution and association with other mites under the perianth. Biol Control 43:102–110
Leal EC, Ram C, Warwick DRN, Leal M de L da S (1996) Behavior of hybrid of coconut *Cocos nucifera* in relation to small-verrucose *Phyllachora torrendiella* and big-verrucose *Sphaerodothis acrocomiae*. Fitopatol Bras 21(3):325–327
Leal EC, Silva Leal M d L d, Ram C, Tupinambá EA (1997) Evaluation of dwarf coconut germplasm to the incidence of small-verrucosis (*Phyllachora torrendiella*) and big-verrucosis (*Sphaerodothis acrocomiae*) in Sergipe. Agrotrópica 9(1):13–18
Leal EC, Warwick DRN, Silva Leal M d L d, Tupinambá EA (1998) Evaluation of tall coconut germplasm against the verrucoses *Phyllachora torrendiella* and *Sphaerodothis acrocomiae* in Sergipe State. Brazil Fitopatología 33(4):220–223
Lebrun P, Baudouin L, Myrie W, Berger A, Dollet M (2008) Recent lethal yellowing outbreak: why is the Malayan Yellow Dwarf coconut no longer resistant in Jamaica? Tree Genet Genomes 4(1):125–131
Levin L, Mammooty KP (2003) Incidence of coconut eriophyid mite *Aceria guerreronis* Keifer (Eriophyidae: Acari) in different coconut cultivars and hybrids. J Trop Agric 41(1/2):59–62
Maramorosch K, Harries H (1998) Prevention of lethal yellowing disease of coconut palms. Intl J Trop Plant Dis 16(2):189–194
Mariau D (1986) Comportement de *Eriophyes guerreronis* Keifer a l'egard de diferentes varietes de cocotier. Oleagineux 41:499–505

Mariau D, Dery SK, Sangaré A, N'Cho YP, Philippe R (1996) Coconut lethal yellowing disease in Ghana and planting material tolerance. Plantations Recherche Développement 3(2):105–112

Mathai G (1978) Search for root (wilt) disease resistance in coconut. Indian Cocon J 9(6):7–8

Mathai G (1980) Seasonal and varietal variations on the incidence of leaf rot of coconut in the root (wilt) affected tracts. Indian Cocon J 11(6):1–3

Mathai G, Mathew AV, Kunju UM (1985) Reaction of coconut (*Cocos nucifera* L.) cultivars to root (wilt) and leaf rot disease. Indian Phytopathol 38(3):561–562

Mathai G, Mathew J, Balakrishnan B (1991) Reaction of exotic cultivars of coconut (*Cocos nucifera* L.) to root (wilt) disease of Kerala. In: Silas EG, Aravindakshan M, Jose AI (eds) Proceedings of National Symposium on Coconut Breeding and Management. Kerala Agricultural University, Thrissur, India, pp 161–162

Mathen K, Solomon JJ, Rajan P, Geetha L (1987) Electron microscopic evidence on the role of *Stephanitis typica* (Distant) as the vector of coconut root (wilt). Curr Sci 56(23):1239–1240

Menon KS, Thomman KJ, Sukumaran AS (1981) Varietal resistance studies. In: Mathen K, MPG K, Joseph T, Mathew C (eds) Review of research on root (wilt) disease. Central Plantation Crop Research Institute, Kayamkulam, pp 78–82

Moore D (1986) Bract arrangement in the coconut fruit in relation to attack by the coconut mite *Eriophyes guerreronis* Keifer. Trop Agric 63:285–288

Moore D, Alexander L (1987) Aspects of migration and colonization of the coconut palm by the coconut mite, *Eriophyes guerreronis* Keifer. (Acari: Eriophyidae). Bull Entomol Res 77:641–650

Mpunami A, Schuiling M, Kaiza D (1992) The lethal disease of coconut palm. Tanzania Res Train Newslett (Dar es Salaam) 7(1–3):12–15

Musetti R, Farhan K, Grisan S, Polizzotto R, De Maroc F, Ermacora P (2012). Fungal endophytes as innovative tools for phytoplasma disease control in abstract- Integrated Management of Phytoplasma Epidemics in Different Crop Systems. COST Action 0807. San Michele: European Cooperation in Science and Technology, 5

Muthiah C, Bhaskaran R (1999) Screening of coconut genotypes and management of eriophyid mite *Aceria guerreronis* in Tamil Nadu. Indian Cocon J 30(6):10–11

Muthiah C, Bhaskaran R (2000) Screening of coconut hybrids/varieties and management of rhinoceros beetle on coconut. Indian Cocon J 31(3):58–59

Muthiah C, Rajarathinam S (2002) Screening of coconut genotypes/hybrids and management of eriophyid mite on coconut. Proceedings of 15th Plantation Crops Symposium PLACROSYM-XV, Mysore, India, 10–13 December, 2002, pp 583–587

Muthiah C, Natarajan C (2004) Varietal reaction and nutrient management of coconut eriophyid mite. Planter 80(936):159–169

Nair MK, Koshy PK, Jacob PM, Nair RV, Rao EVVB, Nampoothiri KUK, Iyer RD (1996) A root (wilt) disease resistant coconut hybrid and strategy for resistance breeding. Indian Cocon J 27(1):2–5

Nair CPR (2000) Status of coconut eriophyid mite, *Aceria guerreronis* Keifer in India. In: Proceedings of International Workshop on coconut eriophyid mite. 6–8 January 2000. Coconut Research Institute, Sri Lanka, pp 13–31

Nair RV, Rajesh MK, Jacob PM (2000) Chowghat Green Dwarf - a suitable variety for the coconut root (wilt) diseased tract. Indian Cocon J 31(3):55

Nair RV, Jacob PM, Ajithkumar R (2001) Production of quality planting material for root (wilt) disease prevalent tracts. Indian Cocon J 32(3):6–8

Nair RV, Jacob PM, Rajesh MG, Thomas RJ (2003) Incidence of root (wilt) disease in various cross combinations of West Coast Tall variety of coconut. CORD 19(1):45–50

Nair RV, Jacob PM, Ajithkumar R (2004) Screening of coconut varieties against root (wilt) disease. J Plantn Crops 32(1):50–51

Nair RV, Jacob PM, Thomas RJ, Pillai SP, Mathews C (2006) Performance of CGD × WCT hybrid in the root (wilt) disease prevalent tract. J Plantn Crops 34(1):15–20

Nair RV, Thomas RJ, Aravazhi E, Sasikala M, Nair CPR (2007) Malayan Green Dwarf, a root (wilt) resistant coconut variety with high yield suitable for the disease prevalent regions. Proceedings of 94th Indian Science Congress, Annamalai University, Chidambaram, Tamil Nadu, January 2007, pp 140

Nambiar SS (1988) Susceptibility of hybrid coconut varieties to *Oryctes rhinoceros* Linn. (Scarabaeidae) under rainfed conditions at Pilicode. In: Silas EG, Aravindakshan M, Jose AI (eds) Proceedings of National Symposium on Coconut Breeding and Management. Kerala Agricultural University, Thrissur, India, pp 17–18

Ninan CA (1978) Exploitation of dwarf germplasm in coconut breeding. Indian Cocon J 8(11):1–4

Nutman FJ, Robert FM (1955) Lethal Yellowing: 'The unknown disease of coconut palm in Jamaica. Empire J Agri 23:257–267

Odewale JO, Okoye MN, Odiowaya G, Ahanon JM, Agho CI (2013) Bunch and nut production of surviving coconut palms in lethal yellowing disease endemic area of Nigeria. Intl J Plant Breed Gen 7(1):57–64

Painter RH (1951) Insect resistance in crop plants. Macmillan, New York, p 520

Parthasarathy MV, Van Slobbe WG, Soudant C (1976) Trypanosomatid flagellate in the phloem of diseased coconut palms. Science 192:1346–1348

Perera L (2005) Report of the genetics and plant breeding division. Annual Report of the Coconut Research Institute of Sri Lanka, Lunuwila

Perera L (2006) Report of the genetics and plant breeding division. Annual Report of the Coconut Research Institute of Sri Lanka, Lunuwila

Politowski K, Browning JA (1978) Tolerance and resistance to plant diseases: an epidemiological study. Phytopathol 68:1178–1185

Ponnamma KN, Cecil SR, Kurian C, Abraham VA, Nambiar PTN (1985) Relation between incidence of *Myllocerus curvicornis* (F.) and intensity of root (wilt) disease in coconut cultivars and hybrids. J Plantn Crops 13(1):69–73

Puch-Hau C, Oropeza-Salín C, Peraza-Echeverría S, Gongora-Paredes M, Córdova-Lara I, Narvaez-Cab M, Zizumbo-Villareal D, Sáenz-Carbonell L (2015) Molecular cloning and characterization of disease resistance gene candidates of the nucleotide binding site (NBS) type from *Cocos nucifera* L. Physiol Mol Plant Pathol 89:87–96

Quaicoe RN, Dery SK, Philippe R, Baudouin L, Nipah JO, Nkansah-Poku J, Arthur R, Dare D, Yankey EN, Pilet F, Dollet M (2009) Resistance screening trials on coconut varieties to Cape Saint Paul Wilt Disease in Ghana. Oléagineux 16(2):132–136

Quillec G, Renard JL (1975) Helminthosporiosis of the coconut palm. Preliminary studies. Oleagineux 30(5):209–213

Rachana KE, Naganeeswaran S, Fayas TP, Thomas RJ, Rajesh MK (2016) Cloning, characterization and expression analysis of NBS-LRR type resistance gene analogues (RGAs) in coconut. Acta Bot Croat 75(1):1–10

Radhakrishnan TC, Balakrishnan PC (1991) Evaluation of intensity of stem bleeding disease of coconut. In: Silas EG, Aravindakshan M, Jose AI (eds) Proceedings of national symposium on coconut breeding and management. Kerala Agricultural University, Thrissur, pp 163–164

Rajesh MK, Jayadev K, Chandrasekar A, Anuradha Upadhyay, Devakumar, K, Manimekalai R, Nair RV, Parthasarathy VA (2002) Improved protocol for AFLP analysis as a base for tagging root (wilt) resistance genes in coconut. In: Proceedings of 15th Plantation Crops Symposium PLACROSYM XV, Mysore, India, pp 204–208

Rajesh MK, Rachana KE, Naganeeswaran S, Shafeeq R, Thomas RJ, Shareefa M, Babu M, Karun A (2015) Identification of expressed Resistance Gene Analogue (RGA) sequences in coconut leaf transcriptome and an evolutionary analysis of palm RGAs. Turkish J Agri Forestry 39(3):489–502

Rajeshkumar PP, Thomas GV, Gupta A, Gopal M (2015) Diversity, richness and degree of colonization of arbuscular mycorrhizal fungi in coconut cultivated along with intercrops in high productive zone of Kerala, India. Symbiosis 65(3):125–141

Raju NS, Raghavaiah G, Rao KM (2006) Reaction of different coconut genotypes against eriophyid mite (*Aceria guerreronis*) in Andhra Pradesh. Indian Cocon J 36(10):18–19

Ramanujam B, Nambiar KKN, Ratnambal MJ (1998) Screening of coconut cultivars/hybrids against *Thielaviopsis paradoxa* (de Seynes) Hohnel, using petiole inoculation technique. In: Mathew NM, KuruvillaJacob C (eds) Proceedings of PLACROSYM XII. Allied Publishers, New Delhi, India, pp 284–286

Ramaraju K, Natarajan K, Sundara Babu PC, Palaniswamy S, Rabindra RJ (2000) Studies on coconut eriophyid mite *Aceria guerreronis* Keifer in Tamil Nadu, India. In: Proceedings of International Workshop on Coconut Eriophyid Mite. 6–8 January 2000. Coconut Research Institute, Lunuwila, Sri Lanka, pp 13–31

Rawther TSS, Pillai RV (1972) Note on field observations on the reaction of coconut varieties to root (wilt). Indian J Agric Sci 42(8):747–749

Ray SKD, Kaiser SAKM (1990) Observations on some elite genotypes (cultivars) of coconut palm for resistance to leaf blight (grey leaf spot) disease. J Mycopathol Res 28(1):9–11

Robinson RA (1969) Disease Resistance Terminology. Rev Appl Mycol 48(11–12):593–606

Romney DH (1983) Brief review of coconut lethal yellowing. Indian Cocon J 13:1–8

Sadakathulla S (1996) Evaluation of coconut genotypes for rodent damage. Pest Management Horti Ecosyst 2(1):37–44

Sadakathulla S, Ramachandran TK (1993) Field screening of coconut cultivars to red palm weevil at Tamil Nadu. Ann Plant Res Insects Newsl 19:43

Sadakathulla S, Kareem AA (1994) Relative rodent damage in coconut genotypes. Ann Plant Res Insects Newsl 20:46

Sangare A, de Taffin G, de Franqueville H, Arkhust ED, Pomier M (1992) Coconut lethal yellowing in Ghana. Initial results on planting material performance in the field. Oléagineux 47(12):699–704

Schuiling M, Mpunami A, Kaiza DA, Harries HC (1992) Lethal disease of coconut palm in Tanzania. III. Low resistance of imported germplasm. Oleagineux 47(12):693–698

Serrano AV, Cortazar Ríos M, Ovando Cruz ME (2011) Donají: new coconut hybrid resistant to lethal yellowing in Mexico. Revista Mexicana de Ciencias Agrícolas 2(5):773–778

Shalini KV, Manjunatha S, Lebrun P, Berger A, Baudouin L, Pirany N, Ranganth RM, Prasad DT (2007) Identification of molecular markers associated with mite resistance in coconut (*Cocos nucifera* L.). Genome 50(1):35–42

Sherman KE, Maramorosch K (1977) Present status of lethal yellowing disease of the coconut palm. J Plantn Crops 5(2):75–83

Simmonds NW (1983) Strategy of disease resistance breeding. FAO Plant Prot Bull 31:2–10

Solomon JJ, Govindankutty MP, Nienhaus F (1983a) Association of mycoplasma-like organisms with the coconut root (wilt) disease in India. Z Pflkrankheiten Pflschutz 90:295–297

Solomon JJ, Sasikala M, Shanta P (1983b) A serological test for the detection of root (wilt) disease of coconut. In: Nayar NM (ed) Coconut Research and Development. Wiley Eastern Ltd., New Delhi, pp 401–405

Sosamma VK, Koshy PK, Rao EVVB (1980) Susceptibility of coconut cultivars and hybrids to *Radopholus similis* in field. Indian J Nemato 10(2):250–252

Sosamma VK, Koshy PK, Rao EVVB (1988) Response of coconut cultivars to the burrowing nematode, *Radopholus similis*. Indian J Nemato 18(1):136–137

Srinivasan N (1991) Occurrence of coconut leaf rot in relation to root (wilt) disease. Indian Cocon J 21(10):14–18

Sudha S (1998) Screening of coconut cultivars against the burrowing nematode, *Radopholus similis*. In: Proceedings of First National Symposium on Pest Management in Horticultural Crops: Environmental implications and thrusts, Bangalore, India, 15-17 October 1997, pp 332–333

Sujatha A, Chalam MSV, Kalpana M (2010) Screening of coconut germplasm against coconut eriophyid mite, *Aceria guerreronis* Keifer in Andhra Pradesh. J Plantn Crops 38(1):53–56

Suriachandraselvam M, Bhaskaran R, Ramachandran TK (2000) Evaluation of coconut hybrids/varieties against grey leaf spot diseases. Madras Agric J 87(7/9):497–498

Swarbrick PJ, Yankey EN, Nipah JO, Quaicoe R, Dickinson MJ (2013) Identification of receptor like kinase genes in coconut and development of a marker for validation of breeding materials resistant to a phytoplasma disease in Ghana. African J Biotech 12(45):6347–6357

Theis KR, Dheilly NM, Klassen JL, Brucker RM, Baines JF et al (2016) Getting the hologenome concept right: an eco-evolutionary framework for hosts and their microbiome. *BioR*. https://doi.org/10.1101/038596

Thomas GV, Rajagopal V, Bopaiah BM (1993) VA-Mycorrhizal associations in relation to drought tolerance in coconut. J Plantn Crops 21:98–103

Thomas RJ, Shareefa M, Rajesh MK, Nair RV, Jacob PM, Sasikala M, Karun A (2016) Studies on improvement of West Coast Tall variety of coconut for yield and resistance to root (wilt) disease through recurrent selection. Acta Hortic. https://doi.org/10.17660/ActaHortic.2016.1127.23

Vandenkoornhuyse P, Qaiser A, Duhamel M, Le Van A, Dufrense A (2015) The importance of the microbiome of the plant holobiont (Tansley Rev.). New Phytol. https://doi.org/10.1111/nph.13312

Varadarajan MK, David PMM (2002) Population dynamics of the coconut mite and associated arthropods in Tamil Nadu, India. Insect Sci Appl 22:47–59

Varghese MK (1934) Diseases of coconut palms. Department of Agriculture and Fisheries, Travancore Government Press, Trivandrum, p 105

Warwick DRN, Bezerra APO, Renard JL (1991) Reaction of coconut hybrids to leaf blight (*Botryodiplodia theobromae* Pat.). 1. Field observations. Oléagineux 46(3):100–106

Whitehead RA (1968) Selecting and breeding coconut palms resistant to lethal yellowing disease. Euphytica 17:81–101

Zilber-Rosenberg I, Rosenberg E (2008) Role of microorganisms in the evolution of animals and plants: the hologenome theory of evolution. FEMS Microbiol Rev 32:723–735

Zizumbo-Villarreal D, Arellano-Morín J (1998) Germination patterns in coconut populations (*Cocos nucifera* L.) in Mexico. Genet Resour Crop Evol 45:465–473

Zizumbo-Villarreal D, Ruiz-Rodriguez M, Harries HC, Colunga-GarcíaMarín P (2006) Population genetics, lethal yellowing disease, and relationships among Mexican and imported coconut ecotypes. Crop Sci 46(6):2509–2516

Dr. Regi J. Thomas is presently a Principal Scientist (Horticulture) at ICAR-CPCRI Regional Station, Kayamkulam, Kerala, India. His areas of interest are in breeding for resistance to root (wilt) disease of coconut, coconut germplasm characterization and establishment of coconut seed gardens. He has published 34 research papers, presented 43 papers in conferences/symposia, written 10 book chapters and edited 2 books. He was involved in the release of three coconut varieties and received the Best Scientific Team Research Award. He is also a recipient of the Netherlands Fellowship Programme. regijacob@yahoo.com

Dr. M. Shareefa has a PhD in Horticulture. She started her career in ICAR as a Scientist and has been working on coconut improvement, especially breeding for resistance to root (wilt) disease of coconut, quality planting material production and coconut tissue culture. She has published 6 research papers, 2 book chapters and 25 popular articles. She has received the Best Oral Paper Presentation Award at the 3rd International Symposium on Coconut Research and Development. hishareefa@gmail.com

Dr. R. V. Nair superannuated from the Agricultural Research Service of ICAR as Head of the Crop Improvement Division of CPCRI. He is a Plant Breeder by qualification, and breeding for resistance to coconut root (wilt) disease has been his main area of interest. He has won the Rashtriya Gaurav Award. He has partnered in developing and releasing 15 varieties in plantation crops. He has published 150 research papers and a book. rvncpcri@gmail.com

Chapter 6
Coconut Biotechnology

M. K. Rajesh, Anitha Karun, and V. A. Parthasarathy

Abstract The major hurdles for genetic improvement of coconut are attributed to its perennial nature, long juvenile period and high heterozygosity, which make conventional coconut breeding time-consuming, resource demanding and laborious. Biotechnological tools have permitted researchers to overcome some of these impediments. Embryo culture has been successfully utilised for international germplasm exchange. Cryopreservation techniques have been standardised for zygotic embryos and pollen, which enable safe and long-term conservation of coconut genetic resources. A repertoire of molecular markers has been utilised for comprehending and exploiting the genetic diversity existing among coconut populations world-wide. Molecular markers have also been employed to generate linkage map for QTLs involving yield attributes. The limited availability of genomic resources/information, a restraint to genomics-assisted crop improvement in coconut, is gradually being surmounted with the research on coconut genomics and transcriptomics gaining impetus during the last few years. This chapter amply elaborates the efforts that have been instilled and the success accomplished with respect to coconut biotechnology.

6.1 Introduction

Conventional genetic analysis has been difficult in coconut, given its perennial nature, extended juvenile period, long breeding and selection cycle, difficulties in raising F_2 progenies and high inherent heterozygosity, especially of tall cultivars. Therefore, genetics of most of the agronomically important traits have been

M. K. Rajesh (✉) · A. Karun
ICAR-CPCRI, Kasaragod, Kerala, India
e-mail: mkraju.cpcri@gmail.com; anithakarun2008@gmail.com

V. A. Parthasarathy
ICAR-IISR, Kozhikode, Kerala, India
e-mail: vapartha@gmail.com

K. U. K. Nampoothiri et al. (eds.), *The Coconut Palm (Cocos nucifera L.) - Research and Development Perspectives*,
https://doi.org/10.1007/978-981-13-2754-4_6

inadequately investigated in coconut. These intrinsic limitations in traditional coconut breeding have made the prospects of exploitation of biotechnological tools attractive.

Non-availability of quality planting materials has been a major constraint in coconut productivity. Since the palm has a single apical meristem, without branches or suckers, its propagation is restricted to seed propagation exclusively. Without any other known method of propagating the coconut palm through vegetative means, much hope was placed on use of tissue culture to vegetatively propagate high yielding and disease-free coconut genotypes. In spite of concerted efforts in a number of laboratories world-wide, a commercial in vitro regeneration protocol has not been standardised till date. The difficulties of cloning coconut through tissue culture have been addressed in a number of research laboratories around the globe since the late 1970s. Many of the earlier efforts were concentrated on procedures for reversion of floral meristem into vegetative ones; the lack of success in these initiatives diverted research efforts to induction of somatic embryos from a variety of explants. Unfortunately, success has been quite limited due to a number of limitations (Fernando et al. 2010). The frequency of regeneration obtained has been quite low, and only a small number of tissue culture-raised plantlets have been successfully established in the field. The success obtained in embryo culture and its use in germplasm collection and exchange and also embryo rescue have been the major achievements with regard to use of tissue culture techniques in coconut.

In vitro conservation has been used as an adjunct approach to facilitate conservation of coconut genetic resources and ease safe international exchange of germplasm. These techniques need minimum space for storage of vast germplasm collections and enable supply of valuable genetic resources. Success in in vitro conservation of coconut genetic resources has been achieved by cryopreservation of zygotic embryo, pollen and plumular tissues.

Application of molecular markers in coconut commenced in the early 1990s, and their uses have been varied, ranging from assessment of genetic diversity to generating linkage maps. Initial investigations were focused on evaluation of coconut genetic diversity and genetic relatedness at the protein and isozyme levels and afterwards at the DNA level using a gamut of molecular markers, with many of the recent research utilising the codominant microsatellites for characterisation of germplasm, for hybridity testing and for construction of genetic linkage maps.

Knowledge derived from recent genome and transcriptome studies will facilitate discovery of candidate genes that govern key developmental and agronomical traits in coconut, in addition to providing important resources for comparative genomic studies in Arecaceae.

This chapter comprehensively covers the major developments in coconut biotechnology during the past few decades emphasising tissue culture methodologies and their applications, molecular markers, gene discovery and, finally, genomics and transcriptomics.

6.2 In Vitro Techniques

6.2.1 Embryo Culture

6.2.1.1 Germplasm Collection

The natural distribution and transcontinental dispersal across land/sea routes since time immemorial and scientific collection and exchange of coconut genetic resources have solely been based on seed nuts since they represent the only planting material for propagation in coconut with no means of vegetative propagation. However, the bulky nature of the seed nuts, its short dormancy period, presence of nut water, stringent phytosanitary requirements and high cost for transportation are major impediments encountered during the course of germplasm collection and storage. Therefore, according to the technical guidelines of the Bioversity International, collection and transportation of coconut for the safe movement of coconut germplasm is advocated through embryo culture (Diekmann 1997). Besides overcoming formalities of quarantine regulations which include treatments of the nuts with chemicals, collection as embryos also lessens the cost of transport to a great extent, and therefore coconut zygotic embryo culture has immense practical value with respect to collection and exchange of germplasm between countries.

Though many efforts in standardising zygotic embryo culture have been reported in the past (Cutter and Wilson 1954; Abraham and Thomas 1962; Ventura et al. 1966), the ICAR-CPCRI developed for the first time a successful procedure for culturing coconut zygotic embryos from 8- to 11-month-old nuts, which has been effectively utilised in the germplasm expeditions of the institute since then (Karun et al. 1999, 2004). The protocol utilises an effective artificial medium which sustains the extracted embryo to grow and to form entire plantlets under in vitro conditions. The field collection technique includes inoculation of the sterilised zygotic embryos, excised from the nuts, on to the nutrient medium in vitro (Assy-Bah et al. 1987; Sossou et al. 1987; Karunaratne 1988; Rillo and Paloma 1991; Karun et al. 1993; Ashburner et al. 1995). Storage of zygotic embryos is essential when the collection sites are remote. Assy-Bah et al. (1987) reported that endosperm plugs, scooped from mature nuts, could be stored in KCl solution for up to 2 weeks. In a later study, Karun and Sajini (1994) established that sterile water could also be made use of as medium for storing zygotic embryos for 2 months; this approach has been utilised in all the germplasm collection expeditions carried out by the ICAR-CPCRI. Germination of embryos was inhibited in media free of sucrose and activated charcoal (De Guzman et al. 1971; Karunaratne et al. 1985; Assy-Bah et al. 1987; Rillo and Paloma 1990).

The embryo culture protocol developed at the Central Plantation Crops Research Institute under the Indian Council of Agricultural Research (ICAR-CPCRI) by Karun et al. (1993) was successfully employed for the first time during 1994 for transferring six Pacific Ocean accessions, maintained at the World Coconut Germplasm Centre, Andaman Islands, to the mainland. Afterwards, 5 international expeditions were conducted by the institute during 1997–2001 for the collection of coconut genetic diversity (Karun et al. 2002), wherein a total of 4182 embryos of 45 accessions were collected from 8 countries (Mauritius, Madagascar, Seychelles, Maldives, Comoros, Reunion, Sri Lanka and Bangladesh). Variations were recorded for the per cent retrieval of embryos among locations and accessions with the percentage of germination varying between 54 (Sri Lanka) and 82.2 (Bangladesh). All the exotic accession collections made as embryo cultures have been planted in the International Coconut Genebank (ICG-SA) and have commenced flowering since 2004. From the results obtained from studies on in vitro retrieval of embryos and their ex vitro establishment, it has been concluded that about 300–400 embryos (i.e. three to four times the actual requirement) are needed for successful establishment of 100 palms in a field gene bank. The diverse collections, after evaluation, are being exploited for evolving new varieties in coconut.

6.2.1.2 Embryo Rescue

Embryo culture technique has been successfully utilised in coconut for obtaining plantlets from embryos which fail to germinate naturally (De Guzman et al. 1971; Gosal and Bajaj 1983; Thomas and Pratt 1981). A commercial application of embryo culture has been the rescue and culture of Makapuno variety, with jelly-like endosperm, in the Philippines (De Guzman 1970; De Guzman et al. 1971; Del Rosario and De Guzman 1976). Successful embryo rescue of a similar variety from Sri Lanka ('Dikiri Pol') has been also achieved (Vidhanaarachi et al. 1998). In India, embryos from sweet kernelled nuts from the Konkan region of Maharashtra have been excised out and plantlets regenerated successfully through the embryo culture protocol (CPCRI 2011). Horned coconut from Andaman Islands, India, produces multiple ovaries which result in development of hornlike structures over the mature fruits, thus delaying germination. Such embryos were cultured in vitro, and plantlets could be retrieved using embryo culture technique. Field evaluation of these plantlets has revealed that the trait is inherited (CPCRI 2012).

6.2.2 *Cryopreservation*

Palm genetic resources are traditionally conserved ex situ as whole plants in field gene banks which require large area, besides huge investments in terms of financial, infrastructural and manpower resources. Cryopreservation of coconut pollen and embryo is an excellent option for long-term conservation of genetic resources, which can provide a viable backup to field gene banks. Coconut, being a recalcitrant crop, is sensitive to desiccation and can thus be conserved only for short periods even under optimal moisture conditions (Assy-Bah and Engelmann 1992b). This necessitates pretreatments before conserving in liquid nitrogen at −196 °C; pretreatments include simple desiccation (using laminar air current and use of silica gel), the use of high concentration of sucrose, various cryoprotectants in various combinations (glycerol, propylene glycol, DMSO, sorbitol, formamide) and encapsulation with 3% sodium alginate and dehydration techniques.

6.2.2.1 Embryo Cryopreservation

Bajaj (1984) proposed the prospects of long-term conservation of coconut zygotic embryos when the embryos resumed growth after freezing at −196 °C. In the experiments carried out, immature embryos of West Coast Tall cultivar were partially dehydrated and cut into transverse halves, treated with a cryoprotectant solution (7% DMSO and 7% sucrose in MS liquid medium), blotted dry and wrapped in a single layer of sterile aluminium foil. It was then frozen by gradually lowering into liquid nitrogen and kept for 5 min. The frozen samples were thawed in warm water (35 °C to 40 °C), washed and cultured on MS medium containing 2,4-D (0.2 mg l^{-1}), NAA (0.5 mg l^{-1}) and kinetin (0.1 mg l^{-1}). The retrieved embryos showed a lag period of up to 4 months without any sign of growth. In a few of these cultures, the embryos subsequently showed an overall swelling and elongation. The survival of a single coconut embryo, 15 months after freezing, was reported by Chin et al. (1989) using a classical protocol, i.e. cryoprotection with DMSO and slow freezing. Hornung et al. (2001) followed an encapsulation-dehydration protocol for the cryopreservation of plumular tissues of coconut in which the encapsulated plumular tissues were pre-cultured for 72–96 h in medium with 0.75 M sucrose and then desiccated with silica gel to around 30% moisture content. Callus growth was observed from the plumular tissues after freezing in liquid nitrogen.

Assy-Bah and Engelmann (1992a) could successfully establish rooted plantlets from coconut embryos from the coconut hybrid PB 121, frozen in liquid nitrogen. Immature embryos of coconut (7–8 months after pollination) were placed for 4 h in petri dishes on standard medium containing 600 g l^{-1} glucose. Pregrowth on this medium was compared with pregrowth on medium supplemented with the cryoprotectants glycerol, sorbitol or polyethylene glycol (PEG) 6000 at 5, 10 or 15%. Thereafter the embryos were immersed rapidly in liquid nitrogen. Thawing was carried out by immersion of the cryotubes for 30 seconds in a water bath at 40 °C. After freezing in liquid nitrogen, survival was obtained in three conditions only: pregrowth with 10% and 15% glycerol (25% and 10% survival, respectively) and 10% sorbitol (43% survival). PEG showed no cryoprotective effect at the concentrations tried. However, only one rooted plantlet could be obtained from embryos pretreated with 15% glycerol after 2.5 months.

Assy-Bah and Engelmann (1992b) reported cryopreservation of mature embryos of four varieties of coconut (Hybrid PB 121, Cameroon Red Dwarf, Indian Tall and Rennell Island Tall). The embryos were pretreated in the laminar air current for 4 h and subsequently incubated in a medium containing 600 g l^{-1} glucose and 15% glycerol for 11–20 h. After rapid freezing and thawing, a recovery rate of 33% and 93% was observed depending on the variety. Cryopreservation of mature embryos of West Coast Tall variety of coconut after desiccation pretreatments has been reported. Maximum retrieval of healthy plantlets was achieved from embryos subjected to 18 h silica gel or 24 h laminar airflow desiccation treatment. When the moisture content of the embryo was reduced to below 20%, irreversible damage of shoot meristem was noticed (Karun et al. 2005).

Cryopreservation of coconut plumular tissues (apical dome with three to four leaf primordia) extracted from mature zygotic embryos of Malayan Yellow Dwarf cultivar has also been attempted. The plumular tissues were first pre-cultured on standard medium with 0.12 M sucrose for 3 days and subsequently suspended in standard medium containing 3% (v/v) Na-alginate and 0.15 M sucrose for encapsulation. After beads were made in 0.1 M calcium chloride containing 0.15 M sucrose, which was subjected to sequential pretreatment for 2–3 days in standard medium containing various sucrose concentrations (0.5 M, 0.75 M, 1 M). Thereafter, the beads were dried for 6–24 h on sterile filter paper over 40 g silica gel in 125 ml airtight boxes. After freezing, regrowth of plumules was obtained for plumules dehydrated for 14 h (21%) and 16 h (20%). Pregrowth of encapsulated plumule beads in 1 M sucrose for 16 h resulted in 20% leafy shoot production from the cryopreserved samples. These observations were consistent with the histological data showing similarity with control cells without any treatments (Nan et al. 2008).

Sisunandar et al. (2010a) put forth a modified and improved cryopreservation protocol for a number of Indonesian coconut cultivars. The protocol is comprised of four steps, viz. rapid dehydration, rapid cooling, rapid warming and recovery in vitro and acclimatisation and soil supported growth. For rapid dehydration, the embryos were positioned in a glass jar fitted with a stainless steel mesh (1–2 mm) platform and an autoclavable fan placed below the platform and housed in a polycarbonate tube. Activated silica gel (680 g) was placed into the lower portion of the glass jar in two separate zones. Following this protocol, 20% (when cryopreserved 12 days after harvesting) and 40% (when cryopreserved at the time of harvest) of embryos cryopreserved could be returned to normal seedlings. Differential Scanning Calorimetry (DSC) studies revealed that this procedure induced a drop in embryo fresh weight to 19% and considerably reduced the amount of water remaining that could produce ice crystals (0.1%). Of the 20 cultivars tested, 16 were found to produce between 10% and 40% normal seedlings, while 4 cultivars generated between 0% and 10% normal seedlings after cryopreservation. In a subsequent study, Sisunandar et al. (2010b) carried out morphological, cytological and molecular studies in coconut plantlets recovered after cryopreservation. The embryos from four cultivars were subjected to rapid dehydration in a drying chamber containing activated silica gel for 8 h to decrease the moisture content from 78% to 80% to 19% to 20%. The plants recovered from cryopreservation showed no morphological variation in terms of shoot elongation rates, production of opened leaves and number and total length of primary roots. In karyotype analysis, there was no variation in chromosome number (2n = 32), type of chromosomes, the length of the long and short arms, the arm length ratio and centromeric index in all studied cultivars independently of cryopreservation as revealed by karyotype analysis. Genetic and epigenetic fidelity of coconut plants recovered from cryopreservation was assessed through microsatellite (SSR) analysis and global DNA methylation rates; these studies revealed that there were no significant differences between seedlings originating from cryopreserved embryos and respective controls.

Bandupriya et al. (2007) studied the effect of abscisic acid (ABA) on encapsulation-dehydration procedure for cryopreservation of coconut plumules. Supplementation of ABA (40 μM) to the sucrose pretreatment medium contributed significantly to the survival and recovery rates of cryopreserved plumules with 84% survival and 39% recovery. In a subsequent study, Bandupriya et al. (2010) investigated the most appropriate technique to transport and store mature zygotic embryos of coconut (for excision of plumules) for cryopreservation studies. The conditions tested include transportation as solid endosperm cores containing embryos (refrigerated for 10 days) and embryos in solidified agar or KCl solution (stored at 27 °C in dark for 10 days). Subsequent to encapsulation-dehydration, the plumule excised from embryos stored in KCl and solidified agar showed significantly higher rate of recovery than embryos in albumen cores in unfrozen samples. In frozen plumular tissues, there was no noteworthy difference in recovery under the conditions experimented. In frozen plumules pretreated with 1.0 M sucrose, the rate of recovery was significantly higher (40%) in the ones excised from embryos and stored in solidified agar in comparison to other conditions.

A protocol for cryopreservation of mature coconut embryos through vitrification has been developed by Sajini et al. (2011) after detailed studies on the effect of preculture conditions, vitrification and unloading solutions on survival and regeneration of coconut zygotic embryos after cryopreservation. PVS 3 solution, consisting of equal proportion of glycerol and sucrose, was most effective for regeneration of cryopreserved embryos among the seven plant vitrification solutions tested. The most efficient procedure included pre-culture of embryos for 3 days on medium with 0.6 M sucrose, followed by PVS 3 treatment for 16 h and subsequently cooling rapidly in liquid nitrogen and rewarming and unloading in 1.2 M sucrose liquid medium for 1.5 h. With this protocol, survival rates of 70% to 80% (corresponding to size enlargement and weight gain) and recovery of 20% to 25% (showing normal shoot and root growth) could be obtained from cryopreserved embryos. Sisunandar et al. (2014) examined the effect of maturity on the outcome of cryopreservation in four cultivars, viz. 'Nias Yellow Dwarf', 'Tebing Tinggi Dwarf', 'Takome Tall' and 'Bali Tall'. About 28% of plantlet recovery was achieved with 11-month-old nuts, which was considerably higher compared to younger nuts.

6.2.2.2 Pollen Cryopreservation

As viability of coconut pollen is only a few days (Patel 1938), alternative methods such as freeze drying was explored for short-term storage (Rognon and Nucé de Lamothe 1978; Whitehead 1963). Based on the results obtained by Karun et al. (2005) and Karun and Sajini (2010), it was concluded that storage of coconut pollen in liquid nitrogen for 24 h did not influence in vitro germination. Karun et al. (2014) have described procedures for long-term cryopreservation of coconut pollen of West Coast Tall and Chowghat Orange Dwarf cultivars. Pollen of both cultivars retained their viability and fertility even after a storage period of 4 years in liquid nitrogen and gave normal nut set when used for pollination. One hundred per cent germination was observed in embryos extracted from hybrid nuts produced with cryostored pollen, and plantlet development was normal confirming the feasibility of setting up pollen cryobank in coconut.

Although field gene bank is the preferred means of conservation, embryo and pollen cryopreservation can also be taken up as complementary conservation strategies in coconut. Even when valuable genetic materials conserved in field gene banks are irrevocably lost, there is always a chance for their retrieval from the cryopreserved germplasm.

6.2.3 Tissue Culture

Mass multiplication of elite coconut palms, endowed with desirable traits like high yield, resistance to biotic and abiotic stresses, is of paramount importance in coconut. Research on coconut tissue culture started in the 1980s after success was reported in oil palm tissue culture. It was initially presumed that application of these techniques would be successful in coconut also. But the culture media developed for oil palm was indubitable for coconut, and it was later proved that the coconut palm is highly recalcitrant to in vitro manipulations and every stage of the procedure brought its share of problems (Verdeil et al. 1998). Poor in vitro responses, irrespective of explants, cultivars or culture conditions, have been attributed to the cumulative effect of several factors such as influence of genotype and explant maturity, adsorption of nutrients and hormones by activated charcoal making culture conditions undefined, production of compact calli, poor plantlet regeneration, underperformance of regenerated plantlets and very slow rate of growth during in vitro culturing (Fernando et al. 2010). Optimisation of composition of culture media, age and type of explants, plant growth regulators and their concentrations, period of subculturing and alternate additives, therefore, are contemplated for standardising a repeatable tissue culture protocol.

Selection of explants is a crucial factor in the success of in vitro experiments. After Eeuwens (1976) initial standardisation of media (Y3) and successful report of callus induction from various explant sources like stem, leaf and inflorescence, a few laboratories around the world initiated intensive research in this area. With respect to coconut, one of the foremost impediments is the lack of formation of friable embryogenic calli and formation of abnormal tissues. Many tissues, viz. leaves, inflorescence, ovary, anthers and zygotic embryos, have been utilised in tissue culture experiments. Coconut leaf explants from juvenile palms were made use of in some initial studies, and callus induction could be achieved (Pannetier and Buffard-Morel 1982; Raju et al. 1984). However, since the embryogenic capacity of leaf explants lasts only for a short duration, its use as an explant is limited (Karunaratne et al. 1991).

Immature inflorescence explants contain numerous meristematic points and therefore are a potential source in coconut, the important criteria being selection of inflorescence of correct maturity stage. Branton and Blake (1984), utilising immature inflorescence explants, observed that sections of inflorescence rachillae proliferated in medium to form calli, which they termed 'colloids'. Verdeil et al. (1994) reported formation of somatic embryos possessing functional bipolar organisation and completely differentiated shoot meristems from immature inflorescence explants. Vidhanaarchchi and Weerakoon (1997) reported that basal media composition influenced the responses of immature inflorescence explants, apart from its size. Raju (2006) reported that prolonged incubation of immature inflorescence explants in auxin-cytokinin combination media resulted in conversion of floral primordia to vegetative primordia, albeit at a very low frequency.

The earliest reports of using an in vitro anther culture approach in coconut are by Iyer (1981), Thanh Tuyen and De Guzman (1983) and Monfort (1985). Iyer (1981) reported formation of multicellular embryoids from coconut anthers. Stages of coconut inflorescence at which anthers and microspore are to be sampled were standardised and defined by Perera (2003). Perera et al. (2007, 2008) have reported formation of somatic embryo-like structures, possessing root and shoot apices, via anther culture. Success has been achieved in anther culture in a culture medium developed by Karunaratne and Periyapperuma (1989) supplemented with 9% sucrose (Perera et al. 2008, 2009). Addition of 0.1% activated charcoal was indispensable to lessen callus necrosis. Callus induction from microspores could be achieved in dark conditions in a medium supplemented with 2,4-D (100 μM), TDZ (9 μM) and NAA (100 μM). Maturation of the somatic embryos could be accomplished in a medium supplemented with 5 μM ABA and 10 μM $AgNO_3$ (Perera et al. 2007). Transfer of callus to a hormone-free medium and then to a medium supplemented with 5 μM BAP promoted germination of somatic embryos. Supplementations of 0.35 μM gibberellic acid along with 5 μM BAP further enhanced the germination frequency of mature somatic embryos (Perera et al. 2008, 2009). Ploidy analysis of anther-derived plantlets indicated that half of the regenerated plantlets were haploid and rest was diploid (Perera et al. 2008).

Unfertilised ovary has been reported as a promising source of explant with high regeneration potential – efficient callogenesis has been reported in CRI72 medium supplemented with 2,4-dichlorophenoxyacetic acid (2,4-D) (Perera et al. 2007). Perera et al. (2009) reported 76% callusing frequency and 70% regeneration frequency from Sri Lankan Tall palms. A major advantage is that unfertilised ovaries can be extracted with minimal injury to the palm. Immature embryos of coconut (6–7 months post anthesis) cultured in medium supplemented with 2,4-D produced callus with 50% of them turning in to globular embryos. Around 22% of germination was observed in these cultures. Age of the embryo was considered as an important factor determining callus proliferation and subsequent embryogenesis (Karunaratne and Periyapperuma 1989).

The focus of a majority of recent studies has been the use of plumular tissues as explants, with promising and consistent results (Chan et al. 1998; Lopez-Villalobos 2002; Fernando et al. 2003; Saenz et al. 2006; Pe'rez- Nu'n˜ez et al. 2006; Rajesh et al. 2005, 2014b). Multiplication of the embryogenic calli is the key to scaling up protocols; Perez-Nunez et al. (2006) could achieve rapid multiplication of embryogenic calli by subdividing and repeated subculturing of calli. Rajesh et al. (2005) described a regeneration protocol from plumular explants of West Coast Tall cultivar. Induction of callus was achieved in Y3 media supplemented with either an auxin alone (2,4-D; 74.6 μM) or in combination with a cytokinin (TDZ; 4.54 μM). Subculturing of calli at monthly intervals to media containing lower levels of 2,4-D and a constant level of either cytokinins (BA and TDZ) or polyamines (spermine and putrescine) resulted in formation of somatic embryos. Enhanced frequency of embryogenic calli, somatic embryoids and meristemoids were obtained in Y3 media supplemented with either spermine or BA. Evidences for development of shoot buds (organogenesis) and typical bipolar embryoids (somatic embryogenesis) were provided through histological studies. Albeit plantlets have been regenerated and successfully established in the field, a commercial scale protocol has not been accomplished, conversion of somatic embryos into plantlets remaining a major bottleneck. Various efforts have been made in coconut tissue culture to refine the protocol such as use of polyamines by Rajesh et al. (2014b) working with dwarf cultivars. Besides multiplication, maintenance of embryogenic callus for lengthy periods could ensure year-round supply of embryogenic calli. Bhavyashree et al. (2015) reported that coconut embryogenic calli obtained from plumular tissues could be maintained for 21 weeks without compromising on the embryogenic potential by subculturing the calli from lower 2,4-D concentration (74.6 μM/l) to higher levels (90.4 μM/l).

One of the major problems encountered during coconut tissue culture is browning, brought about by the release of secondary plant products especially polyphenols. Activated charcoal has been extensively used to avert browning (Samosir 1999). The supplementation of activated charcoal in the culture medium, however, hinders the action of the exogenously applied plant growth regulators and other media supplements, leading to uncertainties in the exact medium composition (Pan and van Staden 1998). The incidence of callogenesis and formation of somatic embryos have been reported to be influenced by variations in particle size and the potency of the different types of activated charcoal (Samosir 1999).

6.3 Molecular Studies

Knowledge of the extent of genetic diversity of a crop is a key element in any breeding programme. Traditionally, coconut genotypes have been characterised using morphological traits (Balakrishnan and Nair 1979; Panda 1982; Sugimura et al. 1997; Zizumbo-Villarreal and Piñero 1998). Assessment of genetic diversity using morphological markers has been difficult due to dearth of phenotypic markers, long juvenile phase of the crop and requirement of large-scale field trials, which is expensive. Molecular markers have helped to surmount these limitations as they are abundant, highly polymorphic and independent of environmental influences, and the characterisation can be performed rapidly (Perera et al. 1998). In coconut, effective use of molecular markers has helped in characterisation and management of germplasm, estimation of extent of genetic diversity and population structure, linkage mapping, identification of QTLs for marker-assisted selection (MAS) and association analysis. Molecular markers used in coconut include biochemical markers (proteins, isozymes and polyphenols) and DNA-based markers [restriction fragment length polymorphism (RFLP), randomly amplified polymorphic DNA (RAPD), amplified fragment length polymorphism (AFLP), inverse sequence-tagged repeat (ISTR), simple sequence repeats (SSR), inter simple sequence repeats (ISSR), start codon targeted polymorphism (SCoT) and single-nucleotide polymorphism (SNPs)]. The use of these molecular methods has opened up new avenues for phylogenetic analysis and provided new tools for the efficient conservation and utilisation of coconut genetic resources.

6.3.1 Biochemical Markers

Initial studies on characterisation of genetic diversity in coconut involved proteins, isozymes, carotenoid or polyphenol markers (Benoit 1979; Jayasekera 1979; Carpio 1982; Jay et al. 1989; Fernando 1995). Work on isozyme analysis was first initiated in IRHO in 1978, with initial studies undertaken with pollen with nine enzyme systems (Benoit and Ghesquiere 1984). Four systems were selected and used to compare eight ecotypes; however, they demonstrated only a weak enzyme polymorphism, few polymorphic loci per system and not more than two alleles per locus. White et al. (1987) analysed six tall populations from Papua New Guinea and Solomon Islands using 15 enzymatic systems; polymorphism could be obtained only in four enzymatic systems with low diversity. Meunier (1992) analysed 6 tall populations from different origins with 17 isozyme systems. Although this method initially uncovered a clear distinction between ecotypes from Southeast Asia, Africa and Pacific regions (Bourdeix et al. 1993), its use was eventually abandoned due to marked environmental effects, low genetic diversity and low polymorphism. Fernando and Gajanayake (1997) utilised horizontal starch gel electrophoresis to characterise Sri Lankan coconut germplasm using six enzyme systems. Only two polymorphic loci, viz. peroxidases and esterases, could be recognised, which however are systems which have been frequently reported to be influenced by environment. Electrophoretic patterns of leaf peroxidase, endopeptidase and proteins were analysed in four cultivars and two hybrids by Cardena et al. (1998). The results obtained were utilised in coconut cultivar identification, testing progeny legitimacy, pollen contamination and studying breeding systems. Zizumbo-Villarreal et al. (2002) estimated the diversity in 22 Mexican and imported coconut populations using 15 enzymatic systems and reported low polymorphism. The fixation indices indicated low total heterozygosity and low heterozygosity within populations suggesting endogamy and genetic drift and a high diversity among populations due to differentiation between Pacific and Gulf of Mexico coastal populations. Forty coconut cultivars from different geographical regions and 6 hybrids were analysed using 11 isozyme systems by Parthasarathy et al. (2004). Genetic diversity studies with enzyme systems came up against technical problems, as numerous systems proved monomorphic or not very active.

Other groups working with proteins (White et al. 1987; Canto-Canche et al. 1992) found that leaf extracts were easily oxidised, giving low enzyme activities and inconsistent results. On the other hand, analysis of leaf polyphenol polymorphism, using high-performance liquid chromatography (HPLC), provided a picture of variability that matched geographical origins and a clear distinction between talls and dwarfs (Jay et al. 1989). Chempakam and Ratnambal (1993) reported significant differences in the levels of leaf polyphenols in 36 cultivars from 8 different geographical origins. But the sensitivity of the polyphenol banding patterns to ecological conditions limits its applications. Presently, the characterisation of genetic diversity in coconut germplasm at the DNA level (Ashburner 1999) has largely replaced these strategies.

6.3.2 DNA-Based Markers

6.3.2.1 Diversity Studies

Restriction Fragment Length Polymorphism (RFLP): Lebrun et al. (1998) utilised RFLP technique to examine the genetic diversity in 100 palms, representing 10 tall and 7 dwarf coconut populations, from diverse geographical locations. Twenty cDNA probes from oil palm, rice, maize and coconut and one cytoplasmic probe from wheat were hybridised on digested DNA using four restriction enzymes. Based on molecular polymorphism, two genetically distinct groups could be defined – one group comprising of ecotypes from the Far East and the South Pacific and the other group comprising of ecotypes from Indian sub-continent and West Africa. All the dwarfs (except Niu Leka) formed a highly homogenous group related to the first group of talls. The results were in tune with earlier reports on the historical dispersion of coconut. The RFLP studies revealed substantial diversity in coconut populations from the Pacific and Far East regions, which are considered as putative regions of origin of the coconut palm. Furthermore, tall ecotypes, as expected, exhibited higher polymorphism compared to dwarf ecotypes.

Randomly Amplified Polymorphic DNA (RAPD): Seventeen distinct South Pacific populations displayed a moderate level of genetic diversity when analysed using 14 RAPD primers (Ashburner et al. 1997). Variability within the populations was over 60%. The low interpopulation diversity among observed coconut populations in this study was attributed to genetic drift and a possible bottleneck in the history of the species. A few RAPD markers unique to specific populations were also identified in the same study. RAPD technique has also been utilised to access the genetic diversity of coconut populations from Sri Lanka, the Philippines and Brazil as reported by Everard et al. (1996), Rodriguez et al. (1997) and Wadt et al. (1999), respectively.

Genetic diversity of four dwarf populations from East Java was analysed using RAPD (Hayati et al. 2000). Variability of coconut population grown outside East Java was reported higher than that at East Java since those coconut populations were raised from seeds of open pollinated palms.

Genetic diversity of 15 Indian and 5 exotic accessions of coconut was tested using 8 polymorphic RAPD primers (Upadhyay et al. 2004). The results revealed higher heterozygosity, higher proportions of polymorphic bands and higher genetic diversity in tall accessions compared to dwarf accessions. Likewise, exotic accessions displayed higher variation in comparison to Indian accessions. Dwarf accessions, from geographically distant regions, were found to group together. Genetic diversity studies within dwarf populations with yellow fruits, viz. Malayan Yellow Dwarf (MYD), Kulashekaram Yellow Dwarf (KYD) and Andaman Yellow Dwarf (AYD), analysed using 16 highly polymorphic RAPD primers, indicated the presence of greater diversity within accessions (Ritto et al. 2008).

Amplified Fragment Length Polymorphism (AFLP): AFLP analysis of 42 indigenous Sri Lankan genotypes, carried out using 8 primer pairs, revealed more variation in tall varieties (*typica*), rather than intermediate (*aurantiaca*) and dwarf (*nana*) varieties. *Aurantiaca* group was more similar to *nana* group rather than the typica group. In addition, putative duplicate accessions were identified in the a*urantiaca* group (Perera et al. 1998). Teulat et al. (2000) used AFLP markers in combination with SSR markers to analyse genetic diversity of 31 palms from 14 coconut populations from different ecological regions.

Inverse Sequence-Tagged Repeat (ISTR): Utilising inverse sequence-tagged repeat (ISTR) approach (primers complementary to repetitive copia-like sequences in the coconut genome), Rohde et al. (1992) could amplify a large number of genetic loci with an abundance of polymorphisms occurring among a set of selected coconut genotypes representing different regions of the world. Duran et al. (1997) extended this technique to analyse East African Tall coconut populations. It was concluded that ISTR markers constitute robust tools for genotype identification, analysis of germplasm variability and breeding purposes in coconut. These studies also provided evidence for the existence of truncated, copia-like repetitive sequences in the coconut genome indicating that retro-elements may have played a putative role in the generation of genetic diversity in coconut.

Inter Simple Sequence Repeats (ISSR): Thirty-three coconut accessions, comprising of coconut accession from different geographical regions of the world, maintained at the International Gene Bank in India, were analysed using 19 ISSR primers (Manimekalai and Nagarajan 2006). A total of 199 ISSR markers were generated, out of which 154 were polymorphic. Least similarity was found between Nicobar Tall and Chowghat Orange Dwarf, both accessions from India. Coconut accessions from Southeast Asia, South Asia and South Pacific formed distinct clusters, in accordance with their origin and dispersal.

Simple Sequence Repeats (SSR): The use of polymorphic microsatellites has gained popularity as a powerful tool for assessment of genetic diversity in coconut populations because of their high information content and codominant nature. Rivera et al. (1999) isolated 38 informative SSR markers from an enriched genomic library. A preliminary screening of 20 coconut samples from the Philippine Coconut Authority (PCA) gene bank indicated the capability of these SSRs to detect substantial polymorphism with an average of 5.2 alleles per microsatellite. Additional confirmation using 40 coconut samples using 8 of the SSRs revealed an average of 8 alleles per SSR. Dwarfs grouped separately from talls and showed less genetic diversity, commensurate with their autogamous breeding behaviour. Using a precloning enrichment procedure, Perera et al. (1999) isolated eight coconut microsatellites, which were used to study the levels and patterns of genetic diversity of Sri Lankan coconut populations. The results showed that the Sri Lankan Tall coconuts (*typica*) exhibit higher levels of diversity than the dwarfs (*nana)* and intermediates (*aurantiaca*) and the intermediate coconuts are more similar to dwarfs than talls. This was in agreement with the results obtained using AFLPs in the same set of genotypes in an earlier study (Perera et al. 1998). Perera et al. (2000) used 8 pairs of SSR primers to analyse the genetic diversity in 130 individuals of coconut comprising 75 tall and 55 dwarf individuals representing 94 ecotypes from different geographical regions. A phenetic tree based on genetic distance clustered individuals into five groups, each mainly composed of either talls or dwarfs. Thirty-three tall coconut populations from Sri Lanka were subjected to microsatellite assay with eight SSR primer pairs in order to study the levels and distribution of genetic variation for formulating future collection strategies and selecting parents for breeding programmes (Perera et al. 2001). A high level of genetic diversity was detected in all the populations. A coconut microsatellite kit was developed by the CIRAD in collaboration with the COGENT, and it consists of 14 microsatellite markers with sufficient discriminating power for practical identification of coconut cultivars (Baudouin and Lebrun 2002). Standard protocols, without the use of radioactive probes, as well as dedicated statistical software, GeneClass2, were developed which could be adapted to use in developing countries.

Merrow et al. (2003) utilised 15 simple sequence repeat (SSR) microsatellite DNA loci to analyse genetic variation within coconut germplasm collections maintained at two locations in South Florida, USA, representing 8 cultivars. Parentage analysis of 'Fiji Dwarf' cultivar was also carried out using these loci. The Red Malayan Dwarfs were found to be genetically distinct from green and yellow ones. Also, genetic identity of 'Red Spicata' was found to be closer to 'Fiji Dwarf'.

A detailed assessment of the extent of genetic diversity in 21 Indian and 24 exotic coconut accessions was undertaken employing 8 SSR primers (Devakumar et al. 2006). The microsatellite loci distinguished a total of 48 alleles, with a mean of 6 alleles per locus. Genetic diversity values were low for most of the dwarfs and high for the tall accessions, which is in accordance with their breeding behaviour. However, Kulasekharam Orange Dwarf, an indigenous dwarf, showed genetic diversity higher than that of many of the tall populations. Within population variation (58%) was found to be higher than among population variation (42%).

Microsatellite analysis of lethal yellowing disease tolerant genotypes (Vanuatu Tall and Sri Lankan Green Dwarf) and susceptible genotype (West African Tall) was carried out by Konan et al. (2007). A total of 58 alleles were detected by the 12 microsatellite loci utilised for the study. Genotypes of susceptible West African Tall cultivar were found to be less genetically clustered to the genotypes of the two tolerant cultivars. The fingerprinting based on microsatellites aided in identification of suitable parents to be used in crossing programmes for developing a segregating mapping population for marker-assisted selection of lethal yellowing resistant genes. Maypan coconut populations, a hybrid of Malayan Yellow Dwarf (MYD), and Panama Tall, earlier considered highly resistant, were devastated by an outbreak of lethal yellowing disease in Jamaica. Lebrun et al. (2007) used 34 SSR markers to compare the MYD sampled from 4 locations in Jamaica along with a reference DNA of MYD collected from 5 different countries, viz. Ghana, Malaysia, the Philippines, Mexico and India, to ensure whether the affected planting material in Jamaica was genetically similar to the material earlier shown to be resistant to lethal yellowing disease. The results revealed more variation at 34 simple sequence repeat loci in MYD samples from Jamaica than from the rest of the world. About 16% of alleles in Jamaican MYD samples did not match with the typical MYD genotypes indicating that Jamaican MYD palms were only partially true to type and this heterogeneity might have had an undesirable effect on its degree of resistance to lethal yellowing disease.

Rajesh et al. (2008a) studied the extent of genetic diversity in 26 coconut accessions from the Andaman and Nicobar Islands, India, utilising 14 SSR markers. A total of 103 alleles were obtained with an average of 7.35 alleles per locus. The average observed and expected heterozygosities were 0.29 and 0.66, respectively. Mean fixation index (F_{ST}) of 0.49 indicated a high level of population differentiation. More number of rare alleles was observed in tall accessions from the Nicobar Islands. Rajesh et al. (2008b) estimated the pattern of diversity in 10 landraces from 3 coconut-growing communities of India using 14 SSR markers. A total of 90 alleles were recorded with an average of 6.42 alleles per locus. Heterozygosity was highest for the tall landraces compared to dwarf ones.

Sixteen SSR markers were used by Dasanayaka et al. (2009) to detect genetic relationships of 43 coconut accessions conserved ex situ in the field gene banks of the Coconut Research Institute of Sri Lanka. These markers could uncover, without any ambiguity, the genetic relationships of Sri Lankan coconut populations. As expected and also reported in numerous earlier studies, gene diversity and polymorphism information content were relatively higher in allogamous tall coconut populations studied than in autogamous dwarf populations. Genetic lineages unveiled, based on evolutionary mechanisms, revealed a narrow genetic base of coconut germplasm, with most of the diversity restricted to the tall populations.

SSR markers were used to characterise two ecotypes of coconut from Kerala, India, namely, Annur and Bedakam. Clustering analysis shows their distinct nature as compared to local West Coast Tall (WCT) populations. However Annur ecotype was comparatively closer to WCT (Rajesh et al. 2014c). Rajesh et al. (2014d) studied the genetic and phylogenetic relationships of coconut populations selecting Laccadive Ordinary Tall (LCT) and Laccadive Micro Tall (LMT) from Amini and Kadamat Islands from Lakshadweep, India, using 20 polymorphic SSR markers. The variation was observed to be large among the distinct types of these cultivars. For example, the elliptical type from Amini emerged as a distinct type and was found to be related to round type from Amini. The pear-shaped nuts from both islands showed affinity and seem to have formed as a result of introgression between elliptical and round types. The round form of LMT from Kadamat Island was found to be distinct according to SSR studies. The extent of genetic diversity and the population structure of four distinct coconut accessions (giant, ordinary, micro and mini micro types) of Minicoy Island, India, were carried out by using 19 SSR markers by Jerard et al. (2017). A total of 70 alleles were detected with a mean of 3.68 alleles per locus. The study revealed Laccadive Mini Micro Tall to be a genetically distinct accession.

Diversity and genetic relationships among 2 tall Brazilian coconut accessions ('Praia do Forte' and 'Merpe') and 7 accessions introduced from different geographic regions of the world ('Rennell Island Tall', 'Polynesian Tall', 'West African Tall', 'Malaysian Tall', 'Vanuatu Tall', 'Rotuman Tall' and 'Tonga Tall'), maintained at International Coconut Genebank for Latin America and the Caribbean (ICG-LAC), Brazil, were studied employing 19 polymorphic microsatellite primers. The primers could detect between four and ten alleles per locus, with an average of 6.57. The analyses unveiled the genetic relationships between Brazilian and African accessions and among the Southeast Asian and the South Pacific accessions, confirming the common origin of the accessions (Loiola et al. 2016).

Using 14 SSR markers, Rasam et al. (2016) characterised 5 indigenous accessions ('Banawali', 'Gangabondam Green Dwarf', 'Pratap', 'Konkan Bhatye Coconut Hybrid-I' and 'East Coast Tall') from the Konkan region of Maharashtra, India. The levels of polymorphism detected using SSR markers among the five accessions ranged from 85.7% to 100%. Existence of genetic variation, both between and within samples of each of the five accessions, was observed.

Diagnostic SSR markers have also been identified for use in hybridity testing in coconut in cases where morphological assessment, based on leaf petiole colour, is cumbersome (Perera, 2010; Rajesh et al. 2012). With the advances made in next-generation sequencing (NGS), the cost of development of molecular markers from non-model organisms has come down drastically. A few recent studies have reported the development and validation of SSR markers from expressed sequence tags (EST-SSRs) in assessment of genetic diversity in coconut (Xiao et al. 2013; Preethi et al. 2014) and genetic purity of coconut hybrids (Preethi et al. 2016).

Start Codon Targeted Polymorphism (SCoT): This novel marker system was described first by Collard and Mackill (2009) and is based on detection of short conserved region flanking the ATG translation start codon in plant genes. Genetic diversity in 23 coconut accessions (10 talls and 13 dwarfs), representing different geographical regions, was tested using SCoT marker technique. Results indicated the potentiality of SCoT markers in detection of DNA polymorphism in coconut accessions (Rajesh et al. 2015b).

WRKY Loci: Mauro-Herrera et al. (2006) utilised sequences of WRKY transcription factors, comprising of single-nucleotide polymorphisms (SNPs) and one microsatellite repeat, to generate ten informative markers. Fifteen genotypes, corresponding to six coconut cultivars, were analysed using these markers, and a total of two to four alleles could be detected. As an extension of this study, the population used by Meerow et al. (2003) for SSR analysis was analysed by 13 WRKY markers. In spite of the lesser number of alleles detected using WRKY markers (37 alleles) compared to SSR markers (67 alleles), clustering of coconut populations using data derived from both these markers was comparable (Mauro-Herrera et al. 2007). Meerow et al. (2009) carried out detailed analysis of DNA sequences of 7 WRKY loci across 72 samples of Arecaceae tribe Cocoseae subtribe Attaleinae. Based on the results obtained, genus *Syagrus* was recognised as sister to *Cocos*.

Pesik et al. (2017) sought to evaluate nucleotide sequence diversities of *WRKY* gene in a collection of Kopyor (a mutant where endosperm is detached from its shell and forms endosperm crumbs in the shell) coconut germplasm in Indonesia. SNP-specific primers were designed based on eight informative SNPs identified, and duplex PCR could be successfully utilised for differentiating Banten Tall, Jember Tall, Kalianda Tall, Pati Dwarf and Sumenep Tall Kopyor coconuts.

6.3.2.2 Non-linkage Association of Markers with Phenotypic Traits

Attempts were made to identify RAPD markers associated with resistance to lethal yellowing disease using three coconut populations, viz. susceptible West African Tall (WAT), resistant Malayan Yellow Dwarf (MYD) and a resistant population of Atlantic Tall (AT) palms (Cardena et al. 2003). A particular RAPD band was regarded as associated with LY resistance if its frequency were high in MYD and surviving AT and low in WAT. Based on this criteria, a total of 12 markers were identified, and their possible use in marker-assisted breeding (MAS) was suggested. Shalini et al. (2007) carried out a detailed study to identify markers associated with resistance to coconut mite (*Aceria guerreronis* Keifer). A group of mite-resistant and mite-susceptible accessions from India were genotyped with 32 SSRs and 7 RAPD markers. Nine SSR and four RAPD markers, associated with mite resistance, were identified. Based on stepwise multiple regression analysis of each marker, a combination of six SSR showed 100% association with mite infestation, while three RAPD markers accounted for 83.86% of mite-resistant genotypes. A combination of five markers (three SSRs and two RAPD) could explain 100% of the association with mite resistance. In order to identify markers associated with palm habit, RAPD technique was employed to differentiate tall, dwarf and dwarf x tall hybrids using a bulked DNA approach (Rajesh et al. 2014a). The RAPD primer OPBA3 could clearly differentiate both the tall and dwarf bulks. The utility of the primer was validated in individual tall and dwarf coconut palms representing different geographic regions. This RAPD primer was also used to screen the parents and validate hybrids of dwarf x tall crosses. Purity of the dwarf x tall hybrids was tested using RAPD-derived sequence-characterised amplified region (SCAR) marker (Rajesh et al. 2013a).

6.3.2.3 Linkage Mapping and QTL Identification

An important step in genetic analysis is to produce genetic linkage maps which represent the relative order of genetic markers and their relative genetic distances from one another, along each chromosome of an organism. Co-segregation of markers with QTLs identified for agronomic traits can offer prospects for marker-assisted selection in coconut breeding programmes. Four DNA markers, viz. AFLP, ISSR, ISTR and RAPD, were utilised to generate the first linkage map in coconut using 52 progenies of a cross between Malayan Yellow Dwarf x Laguna Tall. A total of 382 markers generated 16 linkage groups for each parent with a total map length of 2226 cM. Six QTLs for early germination were identified; these QTLs were correlated with early germination and yield and represent characters which are important in coconut breeding (Herran et al. 2000). Lebrun et al. (2001) made use of 227 AFLP and SSR markers to construct a linkage map, consisting of 16 linkage groups and total map length of 1971 cM, utilising progenies of a cross between Cameroon Red Dwarf and Rennell Island Tall. Nine QTLs were detected for yield characters, viz. number of bunches and number of nuts. Using the same mapping population, Baudouin et al. (2006) investigated genetic factors which control fruit components in coconut. Out of the 52 putative QTLs identified for the 11 traits studied, 34 got grouped in 6 small clusters, which were presumed to correspond to single pleiotropic genes. Interestingly, QTLs for fruit component weight, endosperm humidity and fruit production were found at different locations in the genome, which implies the requirement for selection of QTLs for individual traits for efficient marker-assisted selection for yield. Waxes have been implicated in plant defences to different stress. Riedel et al. (2009) mapped QTLs related to cuticular wax on a linkage map derived from a population of 94 progenies of a cross between East African Tall and Rennell Island Tall using AFLP and SSR markers and COS clones. A total of 704 markers were placed onto the integrated map resulting in a total map length of 2739 cM. A total of 46 QTLs, relating to cuticular wax composition, could be mapped onto the coconut linkage map.

6.3.2.4 Association Analysis

Geethanjali et al. (2017) estimated the genetic diversity and population structure of 79 genotypes, representing accessions from around the globe, utilising 48 SSR loci. The number of alleles ranged from two to seven among the genotypes, with a mean of 4.1 alleles per locus. The genotypes exhibited reasonably high amount of genetic diversity, which was strongly structured based on their place of origin. Based on hierarchical clustering, the genotypes were grouped into two major clusters, each with two subclusters, consistent with their geographic origins. The first cluster encompassed tall genotypes from Indo-Atlantic and South Asia regions, while the second cluster consisted of dwarf genotypes and a few tall genotypes from Indo-Pacific and Southeast Asia. The occurrence of distinct genetic structuring in the accessions, with two major populations ($K = 2$) and four subpopulations ($K = 4$), was corroborated by model-based clustering by STRUCTURE analysis. Only a low proportion of SSR locus pair (2.4%) was found to be in linkage disequilibrium. Association analysis attempted in a subset of 44 genotypes uncovered a SSR locus, viz. CnCir73, present in chromosome 1, putatively linked with fruit yield component traits (fruit breadth, kernel weight, nut weight and copra content), which was consistent with the results obtained in an earlier study (Baudouin et al. 2006).

6.4 Genomics and Transcriptomics

6.4.1 Gene Discovery

6.4.1.1 Somatic Embryogenesis

Various studies have been undertaken to understand and decipher the molecular basis of recalcitrance exhibited by different tissues in vitro. One of the first such studies was carried out by Perez-Nunez et al. (2009), who obtained the complete coding sequence (2240 nucleotides) of *somatic embryogenesis receptor-like kinase* (*SERK*), working with plumular explants. In situ RT-PCR studies revealed the spatial expression of *CnSERK* in meristematic centres among the embryogenic structures of calli. Expression of *CnSERK* could be detected in embryogenic calli even before somatic embryo formation could be observed, whereas it could be barely detected in non-embryogenic calli. From the results of this study, the use of *CnSERK* expression as a potential marker of competence of somatic cells to form embryos in coconut tissues cultured in vitro was suggested.

Montero-Cortés et al. (2010) isolated a putative cyclin-dependent kinase (*CDKA*), linked to cell cycle control, and carried out its expression analysis during somatic embryogenesis from plumular explants. Expression of *CDKA* was reported to be enhanced during embryogenic callus formation phase and progressively decreased as somatic embryos developed. Also, it was suggested that *CDKA* could be utilised as a marker to ascertain the meristematic capability and embryogenic competence of in vitro cultured tissues. For obtaining clues on genes expressed during somatic embryogenesis in coconut from transcriptome data of embryogenic calli, Rajesh et al. (2016) carried out detailed qRT-PCR analyses of 14 identified genes in 6 developmental stages of regeneration from plumular explants. The results revealed differential expression pattern of the 14 transcripts among different developmental stages: enhanced expression of *CLAVATA1* (CLV) was observed during the initial callogenesis; *germin-like protein* (GLP), *glutathione S-transferase*, *PICKLE* (PKL), *WUSCHEL* (WUS) and *WRKY* were expressed more in somatic embryos, while expression of *SERK*, *mitogen-activated protein kinase* (MAPK), *APETALA2/ethylene-responsive factor*, *SAUR family protein*, *embryogenic cell protein* (ECP), *late embryogenesis-abundant protein* (LEA), *arabinogalactan protein* (AGP) and *AINTEGUMENTA* (ANT) was higher in the embryogenic calli in comparison to initial callogenesis and somatic embryos. Bhavyashree et al. (2015)

attempted to correlate embryogenic potential of long-term maintained calli through gene expression studies. While enhanced expression of *ECP*, *GST*, *LEAFY* and *WUS* was recorded in long-term embryogenic calli (21 weeks old), expression of genes such as *SERK*, *GLP*, *WRKY* and *PK*L was higher in initial embryogenic calli (21 days old). From the results obtained, it was concluded that coconut plumular calli could be maintained for extended periods without loss in its embryogenic potential. In a later study, Bhavyashree et al. (2016) also carried out comparative studies of gene expression patterns during different stages of in vitro regeneration in two coconut cultivars, viz. a tall (WCT) and a dwarf (COD). The results of the study uncovered noteworthy differences in both the regeneration potential (higher somatic embryogenesis was recorded in WCT as compared to COD) and gene expression patterns (enhanced expression of *SERK*, *PKL* and *WUS* in embryogenic calli and GLP and *GST in* somatic embryos of WCT compared to COD), signifying genotypic differences of cultivars to in vitro regeneration.

6.4.1.2 Transcription Factors

An AINTEGUMENTA-like gene from coconut (*CnANT*), encoding two APETALA2 (AP2) domains and a conserved linker region, was cloned and characterised from a Sri Lankan Tall cultivar by Bandupriya et al. (2013). Comparison of genomic sequence of CnANT in tall and dwarf cultivars uncovered the presence of one single-nucleotide polymorphism and one indel in the first exon and first intron, respectively. A SSR marker, designed from within the indel sequence, could distinguish the tall and dwarf cultivars. Expression analysis by qRT-PCR indicated higher expression of CnANT in in vitro grown tissues in comparison to vegetative tissues. As an extension of this study, Bandupriya et al. (2014) reported enhanced expression of *CnANT* in mature zygotic embryos and also in embryogenic callus in comparison to other phases of somatic embryogenesis. Overexpression of *CnANT* in *Arabidopsis* resulted in induction of hormone-free regeneration of explants. Moreover, ectopic expression of CnANT improved in vitro regeneration, implying the role of *CnANT* in cellular proliferation in the course of in vitro culture. Sun et al. (2017) carried out detailed characterisation of Wrinkled (CoWRI1), an AP2/EREBP domain-containing transcription factor, which is an important regulator of oil accumulation, from coconut endosperm tissues. Transcriptional activities CoWRI1 and its interaction with acetyl-CoA carboxylase promoter were established by the yeast two-hybrid and yeast one-hybrid approaches, respectively. Ectopic expression of *CoWRI1*, undertaken through seed-specific expression in *Arabidopsis* and endosperm-specific expression in rice, appreciably enhanced oil content in the seeds of transgenic *Arabidopsis* and *rice*.

6.4.1.3 Endosperm Development

Knutzon et al. (1995) cloned the coding region of lysophosphatidyl acyltransferase (LPAAT), a pivotal enzyme controlling the metabolic flow of lysophosphatidic acid into different phosphatidic acids, from a coconut endosperm cDNA library. The upstream region of *LPAAT* from coconut genome was amplified employing chromosome walking, and a number of putative promoter elements were described, including TATA-box, CAAT-box and Skn1-motif. The expression pattern of coconut *LPAAT* promoter, based on stable genetic transformation in transgenic rice plantlets, revealed that the expression was confined to rice endosperm (Xu et al. 2010).

Li et al. (2009) compared the expression profiles of miRNAs at two developmental stages in coconut endosperm development [immature (8 months) and mature (12 months)], using miRNA microarray assays, to categorise miRNAs involved in endosperm tissue development and compound anabolism. A total of 179 miRNAs were recorded in immature (95 expressed miRNAs) and mature tissues (176 expressed miRNAs), based on the annotation in miRBase. A total of 32 miRNAs displayed differential expression (23 were upregulated and 9 downregulated in mature endosperm), thereby associating the role of miRNAs in development of coconut endosperm. The differential expression patterns of a few of these miRNAs were validated using qRT-PCR. Also, the target genes of 32 miRNAs, with differential expression patterns, were computationally predicted. Liang et al. (2014), utilising suppression subtractive hybridisation, identified 737 unigenes which were differentially expressed during 3 developmental stages of coconut endosperm. Out of these, 103 were found to encode enzymes implicated in fatty acid and carbohydrate biosynthesis and metabolism. Stage-specific expression of selected transcription factors and other relevant genes were established by qRT-PCR.

6.4.1.4 Resistant Gene Analogues

Transcriptome data of leaf samples of coconut root (wilt) disease-resistant cultivar (Chowghat Green Dwarf) were mined, and 243 resistance gene analogue (RGA) sequences were identified, encompassing 6 classes of RGAs, by detailed domain and conserved motif predictions by computational analysis (Rajesh et al. 2013b, 2015a, 2017). Phylogenetic analyses grouped coconut nucleotide binding site and leucine-rich repeat (NBS-LRR) class of *R* genes into discrete clusters, based on the occurrence of either TIR or CC motifs in the N-terminal regions. In addition, the expression pattern of a few NBS-LRR type RGAs was validated through qRT-PCR analysis. Using degenerate primers, Puch-Hau et al. (2015) amplified NBS-type RGAs from coconut accessions displaying either resistance or susceptibility to the lethal yellowing disease. Based on phylogenetic analysis of the sequences obtained, seven distinct clades were obtained, with all sequences belonging to the non-TIR-NBS-LRR subclass of NBS-LRR type RGAs. Variations in expression profiles of a few of these transcripts were observed in response to exogenous application of salicylic acid. In a later study, Puch-Hau et al. (2016) reported species-specific evolution of resistance gene candidates in coconut palm.

Utilising primer pairs based on conserved motifs of the NBS-LRR domain of the date palm, Rachana et al. (2016) amplified putative RGAs from Chowghat Green Dwarf. The isolated RGAs were homologous to monocot NBS-LRRs. Detailed structural analysis and 3-D modelling of the NBS domain of these RGAs were also carried out.

6.4.1.5 Aroma

The presence of 2-acetyl-1-pyrroline (2AP) imparts a pleasant 'pandan-like' aroma to liquid endosperm of Aromatic Green Dwarf from Thailand. Saensuk et al. (2016) identified an ortholog of the rice aromatic gene in coconut (*CnAMADH2*). Comparison of coding sequences of *CnAMADH2* of Aromatic and non-Aromatic Green Dwarf palms uncovered a modification of guanine (G) in non-aromatic accession to cytosine (C) in the aromatic accession in the exon 14, which caused a non-synonymous amino acid change from alanine (A) to proline (P) at position 442 (P442A) of CnAMADH2. This substitution was linked to 2AP content in the aromatic palms. Based on this sequence variation, a PCR marker, capable of detecting aromatic and non-aromatic alleles of the gene, was developed and validated.

6.4.2 Estimation of Genome Size

Sandoval et al. (2003) carried out flow cytometer analysis of different tissues (immature leaves, shoot meristems extracted from zygotic embryos and culture in vitro and slow- and fast-growing calli) and estimated the average nuclear DNA content of coconut to be 5.6 ± 0.2 pg DNA/2C ($\simeq 5.4 \times 10^9$ bp). Utilising results obtained from flow cytometry studies of 23 coconut cultivars (comprising of talls, dwarfs and a hybrid), Gunn et al. (2015) estimated the average genome size to be around 5.96 pg, using isolated nuclei from young palm leaves. They concluded that the coconut genome is large and displayed intraspecific variation linked to domestication, with variations among talls significantly more in comparison to dwarf cultivars. In a later study, Neto et al. (2016) estimated the genome sizes of 14 coconut accessions, comprising of 8 tall and 6 dwarf cultivars, using flow cytometry. The 2C DNA content varied from 5.72 to 5.48 pg for talls and from 5.58 to 5.52 pg for dwarf accessions, and the mean genome size was estimated to be 5.59 and 5.55 pg for the tall and dwarf accessions, respectively.

6.4.3 Chloroplast Genome

Huang et al. (2013) described the chloroplast genome sequence of a dwarf coconut palm and provided detailed account of its gene content and organisation, inverted repeat fluctuations and repeated sequence structure, in addition to occurrence of RNA editing. Chloroplast genome of coconut, about 154,731 bp in length, was predicted to encode 130 genes and 4 pseudogenes. Albeit the smallest reported so far in palms, chloroplast genome of coconut shared the same overall organisation, gene content and repeat structures with chloroplast genome of other palms. Pseudogenisation of *rps19*-like gene and an unusual high number of RNA editing sites are a couple of unique features encountered in the coconut chloroplast genome.

6.4.4 Mitochondrial Genome

The assembly of complete coconut mitochondrial (mt) genome of an Oman Local Tall cultivar was reported by Aljohi et al. (2016). The mt genome is around 678.65 kbp in length, possessing a GC content of 45.5%, and was found to encode 72 proteins, 9 pseudogenes, 23 tRNAs and 3 ribosomal RNAs. The chloroplast (cp)-derived regions accounted for 5.07% of the total assembly length in contrast to the date palm mt genome, where 93.5% of the genome sequence was found to be cp derived (Fang et al. 2012). In coconut, the cp-derived regions included 13 proteins, 2 pseudogenes and 11 tRNAs. The mt genome of coconut has a relatively large fraction of repeat content (17.26%), including both forward (tandem) and inverted (palindromic) repeats. Sequence variation analysis shows that the transition/transversion ratio of 0.3 in the mt genome was much lower (2–2.1) when compared to that of the nuclear genome.

6.4.5 Transcriptome Analysis

6.4.5.1 Host–Pathogen Interactions

The first genome-wide study of coconut using transcriptome analysis was carried out by Rajesh et al. (2013b) to ascertain the intricate host-pathogen interactions with respect to root (wilt) disease in coconut. Reads obtained from RNA-Seq analysis of healthy and diseased Chowghat Green Dwarf (CGD) palms, utilising an Illumina HiSeq 2000 platform, were assembled into 59, 282 transcripts, with a mean size of 987 bp (Rajesh et al. 2017). Based on sequence similarity searches, 39,665 of assembled transcripts had at least 1 significant hit in the Uniprot and date palm proteome databases. Overall, 2718 transcripts were differentially expressed in diseased samples (fold change of 2 and above with a p value $< = 0.05$) in comparison to healthy ones. The differentially expressed transcripts could be sorted out into pathways like formation of cell wall, plant-pathogen interactions, primary and secondary metabolism, biosynthesis of hormones and signalling cellular transport. The expression patterns of a set of genes, both upregulated and downregulated, were validated by quantitative real-time PCR (qRT-PCR), the results of which were comparable to those observed by RNA-Seq analysis.

Nejat et al. (2015) undertook RNA-Seq analysis to distinguish the global transcriptome responsive to coconut yellow decline disease by comparing the RNA-Seq profiles of naturally infected and healthy Malayan Red Dwarf palms. Illumina sequencing generated a total of 72,019,264 and 70,935,896 reads, which could be assembled into 108,994 and 148,264 contigs from the healthy and infected leaf transcriptomes, respectively. A total of 18,013 transcripts were upregulated, and 21,860 transcripts downregulated in infected leaves in comparison to healthy ones. The results obtained revealed reprogramming of many of the biological and cellular processes due to phytoplasma infection. A number of genes associated with the production of defence-related proteins, reactive oxygen species, ABC transport protein family, hydrolase and kinases, flavanol synthesis, auxin-induced protein, no apical meristem (NAM) gene family and ethylene were upregulated as a result of phytoplasma infection. Upregulation of *GA-2ox* (gibberellin 2-oxidase) was presumed to downregulate levels of gibberellins in infected coconut palms; this could be the possible reason for stunting, inflorescence necrosis and premature nut fall, which are the characteristics of phytoplasma infection in coconut. Genes involved in photosynthesis were downregulated and could constitute diversion of cellular machinery towards defence mechanisms in response to biotic stress. It was also implied that phytoplasma might modify the expression of genes involved in carbohydrate metabolism in infected palms. The upregulation of ABC protein family transcripts during phytoplasma infection could be construed as a mechanism by phytoplasma to import sugars, which are their main source of energy, through the ABC transporter systems.

6.4.5.2 Fatty Acid Biosynthesis and Metabolism

Fan et al. (2013) undertook transcriptome analysis combining RNA isolated from juvenile leaves and fruit flesh of Hainan Tall cultivar. A total of 57,304 unigenes could be obtained after assembly of the RNA-Seq data, with an average length of 752 bp. Out of these, 23,168 unigenes could be mapped into 215 KEGG pathways, mainly galactose metabolism, hormone signal transduction and plant-pathogen interaction pathways. Additionally, 347 unigenes involved in fatty acid synthesis and metabolism were also reported. These unigenes could be assigned to five steps of the fatty acid biosynthesis pathway (fatty acid biosynthesis, unsaturated fatty acid, citrate cycle, fatty acid metabolism and fatty acid elongation), thus aiding elucidation of the molecular mechanisms involved in fatty acid biosynthesis in coconut. Of these, 20 unigenes were predicted to be related to fatty acyl-ACP thioesterase, which is a key enzyme for terminating the elongation of carbon chains and therefore regulating the length of fatty acids. From the leads obtained in this study, it was proposed that expression of fatty acyl-ACP thioesterase might be associated with the observed accumulation of medium chain fatty acids (i.e. lauric acid) in coconut.

6.4.5.3 Embryogenesis

De novo assembly and analysis of global transcriptome of coconut embryogenic calli, derived from plumular explants of West Coast Tall cultivar, was carried out using Illumina paired-end sequencing. From the assembled reads, transcripts known to be involved in SE, namely, protein kinases like receptor-like kinases (*SERK* and *CLV1*), mitogen-activated protein kinase (*MAPK*), transcription factors (*WUS*, *AP2/ERF*, *PKL*, *ANT* and *WRKY*), extracellular proteins (*AGP*, *GLP*, *ECP* and *LEA*) and *GST*, were mined (Rajesh et al. 2016). Bandupriya and Dunwell (2016) carried out a comprehensive study of ESTs by exploring the transcriptome data of immature embryo, mature embryo, microspore-derived embryo and mature leaves to categorise key embryo-specific genes. Transcripts with putative roles in embryogenesis, viz. chitinase, β-1,3-glucanase, ATP synthase CF0 subunit, thaumatin-like protein and metallothionein-like protein, were identified.

6.4.5.4 RNA-Directed DNA Methylation

Huang et al. (2014) carried out transcriptome analysis of maturing gelatinous endosperm, mature embryo and young leaf of a fragrant dwarf coconut. After assembly of the sequencing data, a total of 58,211, 61,152 and 33,446 unigenes could be identified from embryo, endosperm and leaf tissues, respectively. Putative homologues of factors required for RNA-directed DNA methylation in coconut were identified. The results imply the importance of RNA-directed DNA methylation, especially small RNA-mediated epigenetic regulation during seed development, particularly in maturing endosperm, in coconut.

6.4.5.5 Aroma

De novo assembly of transcriptome from the liquid endosperm of Aromatic Green Dwarf coconut of Thailand was undertaken by Saensuk et al. (2016) to identify the gene(s) responsible for biosynthesis of 2-acetyl-1-pyrroline (2AP) biosynthesis, which imparts a ‘pandan-like’ aroma. From the assembled whole-transcriptome data, derived through RNA-Seq of 7-month-old endosperms of Aromatic and a non-Aromatic Green Dwarf coconut, size differences were observed in transcripts encoding 2AP in aromatic (2371 bp) and non-aromatic (1921 bp) palms. These transcripts were orthologous to rice 2AP.

6.4.5.6 Whole Genome Sequencing

Recently, the whole genome sequence of coconut (cultivar Hainan Tall) was made available, and a total of 419.08 gigabases (Gb) of clean data was obtained (Xiao et al. 2017). This study generated a scaffold length of 2.2 Gb (with a scaffold N50 of 418 kb) that represents over 91% of the calculated genome of coconut. The study further predicted that coconut genome harbours 28,039 protein-coding genes compared to *Phoenix dactylifera* (PDK30 variety: 28,889), *Phoenix dactylifera* (DPV01 variety: 41,660) and *Elaeis guineensis* (34,802). Around 72.75% of the genome comprise transposable elements with long-terminal repeat elements (LTRs) contributing to a little over 92% of transposable elements in coconut (Xiao et al. 2017). In addition, the completeness of genome, analysed using BUSCO, showed that 90.8% of the 1440 expected plant genes were identified as complete. Bayesian molecular clock analysis showed that the divergence time between coconut and oil palm is about 46.0 (25.4–83.3) million years ago. Comparative analysis of coconut and *Arabidopsis thaliana* encoded antiporter and ion channels revealed that coconut has acquired significant gene family expansions including Na+/H+ antiporters, carnitine/acylcarnitine translocases, potassium-dependent sodium-calcium exchangers and potassium channels. Further, it was also inferred that the expansion of these gene families could be ascribed to the adaptation of coconut to salt stress, fatty acids and potassium accumulation.

6.5 Transformation Studies

Samosir (1999) attempted genetic transformation studies for the first time in coconut utilising microprojectile bombardment for targeting *GUS* gene into embryogenic calli and juvenile leaf tissues. It was reported that two constitutively expressed promoters, viz. actin (*Act1*) and ubiquitin (*Ubi*), produced the strongest transient expression in these tissues. Andrade-Torres et al. (2011) described *Agrobacterium*-mediated transformation of a wide range of tissues such as immature anthers, excised zygotic embryos, plumule-derived embryogenic calli and somatic embryogenesis-derived roots and leaves. A number of reporter genes and procedures for antibiotic selection of transformants were evaluated. Since embryogenic calli displayed endogenous *GUS*-like activity, other genes (e.g. green/red fluorescent protein) were experimented with. From the results obtained, it was concluded that successful gene transfer could be achieved by a combination biolistics (to generate micro-wounds in explants) and *Agrobacterium*-mediated transformation (to introduce genes into explants).

6.6 Future Strategy

Despite its importance as a crop in the tropics, the genetics of coconut has not received much attention in comparison with many other crops. The genetic basis of inheritance of important traits has to be understood. Coconut remains a difficult crop to manipulate in vitro. A viable protocol for micropropagation of desired coconut hybrids/selections has to be developed to enable disseminating the benefits of various breeding programmes. The technique, when perfected, could also be used for the mass multiplication of the disease-resistant/tolerant types especially, in the context of the epidemic and devastating nature of lethal yellowing and root (wilt) diseases. Embryo culture, embryo rescue and cryopreservation techniques have helped in collection and preservation of germplasm. The important advances in biotechnological techniques which have been made particularly during the last decade, such as identification of molecular markers associated with polymorphism and potentially linked to specific traits, should be streamlined to benefit the coconut breeder in practice.

The use of new-generation sequencing techniques has witnessed large-scale generation of ESTs in coconut, which has provided insights into plant-pathogen interactions and facilitated elucidation of genes involved in somatic embryogenesis. The available genome sequence forms a potential resource to be utilised in genetic enhancement of coconut palm. It is anticipated that a large volume of genomic resources, with respect to transcriptome sequencing and whole genome sequencing using next-generation sequencing platforms, would be made available soon in the public domain due to the initiatives undertaken by numerous laboratories around the globe. The obvious challenge to unravel and integrate the genomic knowledge into appropriate methodologies should be taken up which would revolutionise future coconut breeding programmes, especially in terms of palm productivity and resistance to biotic and abiotic stresses, and lead to deciphering in vitro recalcitrance.

References

Abraham A, Thomas KJ (1962) A note on *in vitro* culture of excised coconut embryos. Indian Cocon J 15:84–88

Aljohi HA, Liu W, Lin Q, Zhao Y, Zeng J, Alamer A, Alanazi IO, Alawad AO, Al-Sadi AM, Hu S, Yu J (2016) Complete sequence and analysis of coconut palm (*Cocos nucifera*) mitochondrial genome. PLoS One 11(10):e0163990. https://doi.org/10.1371/journal.pone.016399

Andrade-Torres A, Oropeza C, Sáenz L, González-Estrada T, Ramírez-Benítez JE, Becerril K, Chan JL, Rodríguez-Zapata LC (2011) Transient genetic transformation of embryogenic callus of *Cocos nucifera* L. Biologia 66(5):790–800

Ashburner GR (1999) The application of molecular markers to coconut genetic improvement. Curr Plant Sci Biotechnol Agri 35:33–44

Ashburner GR, Faure DR, Tomlinson, Thompson WK (1995) A guide to the zygotic embryo culture of coconut palms (*Cocos nucifera* L.). ACIAR Technical Reports, p 16

Ashburner GR, Thompson WK, Halloran GM (1997) RAPD analysis of South Pacific coconut palm populations. Crop Sci 37:992–997

Assy Bah B, Engelmann F (1992a) Cryopreservation of immature embryos of coconut (*Cocos nucifera* L.). Cryo-Letters 13:67–74

Assy Bah B, Engelmann F (1992b) Cryopreservation of mature embryos of coconut (*Cocos nucifera* L.) and subsequent regeneration of plantlets. Cryo-Letters 13:117–126

Assy Bah B, Durand-Gasselin T, Pannetier C (1987) Use of zygotic embryo culture to collect germplasm of coconut (*Cocos nucifera* L.). Plant Gen Resour Newsl 71:4–10

Bajaj YPS (1984) Induction of growth in frozen embryos of coconut and ovules of citrus. Curr Sci 53:1215–1216

Balakrishnan NP, Nair RB (1979) Wild populations of Areca and Cocos in Andaman & Nicobar Islands. Indian J For 2(4):350–363

Bandupriya HD, Dunwell JM (2016) Transcriptome analysis for discovering candidate genes involve in embryogenesis in coconut (*Cocos nucifera* L.) through 454 pyrosequencing. J Natl Sci Found Sri Lanka 43:4

Bandupriya HDD, Fernando SC, Verdeil JL, Malaurie B (2007) Effect of abscisic acid on survival and recovery of cryopreserved plumule explants of coconut (*Cocos nucifera* L.). COCOS 18:58–68

Bandupriya HDD, Fernando SC, Verdeil JL, Malaurie B (2010) Cryopreservation of encapsulated plumules of coconut: effect of transport/store conditions. Asia Pac J Mol Bio Biotechnol 18(1):135–137

Bandupriya HDD, Gibbings JG, Dunwell JM (2013) Isolation and characterization of an AINTEGUMENTA-like gene in different coconut (*Cocos nucifera* L.) varieties from Sri Lanka. Tree Genet Genomes 9:813–827

Bandupriya HDD, Gibbings JG, Dunwell JM (2014) Over expression of coconut AINTEGUMENTA-like gene, CnANT, promotes in vitro regeneration in transgenic Arabidopsis. Plant Cell Tissue Org Cult 116(1):67–79

Baudouin L, Lebrun P (2002) The development of microsatellite kit and dedicated software for use with coconuts. Burotrop Bull 17:16–20

Baudouin L, Lebrun P, Konan JL, Ritter E, Berger A, Billotte N (2006) QTL analysis of fruit components in the progeny of a Rennell Island Tall coconut (*Cocos nucifera* L.) individual. Theoretic Appl Gen 112:258–268

Benoit H (1979) Preliminary studies on polymorphism of coconut isozymes. 5th FAO Technical Working Party on Coconut Production, Protection and Processing, Manila, Philippines

Benoit H, Ghesquiere M (1984) Electrophore'se, compte renducocotier. IV. Déterminisme génétique, Rapport interne, IRHO-CIRAD, France

Bhavyashree U, Lakshmi Jayaraj K, Rachana KE, Muralikrishna KS, Sajini KK, Rajesh MK, Karun A (2015) Maintenance of embryogenic potential of calli derived from shoot meristem of West Coast Tall cv. of coconut (*Cocos nucifera* L.). J Plantn Crops 43(2):105–116

Bhavyashree U, Rachana KE, Lakshmi KJ, Rajesh MK, Karun A (2016) Comparative gene expression profiling during *in vitro* regeneration in two coconut cultivars. Emirates J Food Agri 28(8):572–580

Bourdeix R, Sangare A, Le Saint JP, Meunier JP, Gascon F, de Rognon N d LM (1993) Coconut breeding at IRHO and its application in seed production. In: Nair MK et al (eds) Advances in coconut research and development. Indian Society for Plantation Crops, Kasaragod, pp 79–94

Branton RL, Blake J (1984) Development of organized structures in callus derived from explants of *Cocos nucifera* L. Ann Bot 52:673–678

Canto-Canche B, Quintal-Salazar E, Villanueva MA (1992) Biochemical markers of variety in *Cocos nucifera* L. from Yucatan. Turrialba 42(3):375–381

Cardeña R, Oropeza C, Zizumbo-Villarreal D (1998) Leaf proteins as markers useful in the genetic improvement of coconut palms. Euphytica 102:81–86

Cardena R, Ashburner GR, Oropeza C (2003) Identification of RAPDs associated with resistance to lethal yellowing of the coconut (*Cocos nucifera* L.) palm. Sci Hortic 98:257–263

Carpio CB (1982) Biochemical studies of *Cocos nucifera*. Kalikasan. Philippine J Biol 11:319–338
Chan JL, Saenz L, Talavera C, Hornung R, Robert M, Orepeza C (1998) Regeneration of coconut (*Cocos nucifera* L.) from plumule explants through somatic embryogenesis. Plant Cell Rep 17:515–521
Chempakam B, Ratnambal MJ (1993) Variation for leaf polyphenols in coconut cultivars. In: Nair MK et al (eds) Advances in coconut research and development. Indian Society for Plantation Crops, Kasaragod, pp 51–54
Chin HF, Krishnapillay B, Hor YL (1989) A note on the cryopreservation of embryos from young coconuts (*Cocos nucifera* var. Mawa). Pertanika 12(2):183–186
Collard BCY, Mackill DJ (2009) Start codon targeted polymorphism: a simple novel DNA marker technique for generating gene targeted markers. Plant Molecular Biol Reporter 27:86–93
CPCRI (2011) Research highlights 2010–11. Central Plantation Crops Research Institute, Kasaragod, Kerala, India, p 36
CPCRI (2012) Annual Report 2011–12, Central Plantation Crops Research Institute Kasaragod, Kerala, India, p 128
Cutter VM, Wilson KS (1954) Effect of coconut endosperm and other stimulants upon the development *in vitro* of embryos of *Cocos nucifera*. Bot Gaz 115:234–240
Dasanayaka PN, Nandadasa HG, Everard JMDT, Karunanayaka EH (2009) Analysis of coconut (*Cocos nucifera* L.) diversity using microsatellite markers with emphasis on management and utilisation of genetic resources. J Natl Sci Foundn of Sri Lanka 37(2):99–109
De Guzman EV (1970) The growth and development of coconut Makapuno embryos *in vitro*. The induction of rooting. Philippine Agric 53:65–78
De Guzman EV, Del Rosario AG, Eusebio EC (1971) The growth and development of Makapuno coconut embryo in vitro. III. Resumption of root growth in high sugar media. Philippine Agric 53:566–579
Del Rosario AG, de Guzman EV (1976) The growth of coconut Makapuno embryos as affected by mineral composition and sugar level of the medium during the liquid and solid cultures. Philippine J Sci 105:215–222
Devakumar K, Jayadev K, Rajesh MK, Chandrasekhar A, Manimekalai R, Kumaran PM, Parthasarathy VA (2006) Assessment of genetic diversity of Indian coconut accessions and their relationship to other cultivars using microsatellite markers. Plant Gen Resour Newsl 145:38–45
Diekmann M (1997) The use of biotechnology for the safe movement of coconut germplasms. Paper presented at International Symposium on Coconut Biotechnology, Merida, December 1–5
Duran Y, Rohde W, Kullaya A, Goikoetxea P, Ritter E (1997) Molecular analysis of east African tall coconut genotypes by DNA marker technology. J Gene Breed 51:249–288
Eeuwens CJ (1976) Mineral requirements for growth and callus initiation of tissue explants excised from mature coconut palms (*Cocos nucifera*) and cultured in vitro. Physiol Plant 36(1):23–28
Everard JMDT, Katz M, Gregg K (1996) Inheritance of RAPD markers in the coconut palm, *Cocos nucifera* L. Trop Agri Res 8:124–135
Fan H, Xiao Y, Yang Y, Xia W, Mason AS, Xia Z, Qiao F, Zhao S, Tang H (2013) RNA-Seq analysis of *Cocos nucifera*: transcriptome sequencing and de novo assembly for subsequent functional genomics approaches. PLoS One 8(3):e59997. https://doi.org/10.1371/journal.pone.0059997
Fang Y, Wu H, Zhang T, Yang M, Yin Y, Pan L, Yu X, Zhang X, Hu S, IS A-M, Yu J (2012) A complete sequence and transcriptomic analyses of date palm (*Phoenix dactylifera L.*) mitochondrial genome. PLoS One 7(5):e37164. https://doi.org/10.1371/journal.pone.0037164
Fernando MTN (1995) Economic analysis of coconut-based farming systems in Sri Lanka. Scottish Agri Econ Rev 8:107–120
Fernando WMU, Gajanayake G (1997) Patterns of isozyme variations in coconut (*Cocos nucifera* L.) populations used for breeding improved varieties. Plantations, Recherche, Développement 4(4):256–263

Fernando SC, Verdeil JL, Hocher V, Weerakoon LK, Hirimburegama K (2003) Histological analysis of plant regeneration from plumule explants of *Cocos nucifera*. Plant Cell Tissue Org Cult 72:281–283

Fernando SC, Vidhanaarachchi VRM, Weerakoon LK, Santha ES (2010) What makes clonal propagation of coconut difficult? Asia-Pac J Mol Biol Biotech 18:163–165

Geethanjali S, Anitha Rukmani J, Rajakumar D, Kadirvel P, Viswanathan P (2017) Genetic diversity, population structure and association analysis in coconut (*Cocos nucifera* L.) germplasm using SSR markers. Plant Gen Resour:1–13. https://doi.org/10.1017/S1479262117000119

Gosal SS, Bajaj YPS (1983) *In vitro* hybridization in an incompatible cross – blackgram x green gram. Curr Sci 52:556–557

Gunn BF, Baudouin L, Beulé T, Ilbert P, Duperray C, Crisp M, Issali A, Konan JL, Rival A (2015) Ploidy and domestication are associated with genome size variation in palms. American J Bot 102(10):1625–1633

Hayati DPK, Hartana A, Suharsono, Aswidinnoor H (2000) Genetic diversity of Jombang coconut populations based on RAPD markers. International Conference on Science and Technology for Managing Plant Genetic Diversity in the 21st Century, 12–16 June, 2000, Kuala Lumpur, Malaysia p 11

Herran A, Estioko L, Becker D, Rodriguez MJB, Rhode W, Ritter E (2000) Linkage mapping and QTL analysis in coconut (*Coconut nucifera* L.). Theoret Appl Gen 101:292–300

Hornung R, Domas R, Lych PT (2001) Cryopreservation of plumular explants of coconut (*Cocos nucifera* L.) to support programmes for mass clonal propagation through somatic embryogenesis. Cryo-Letters 22:211–220

Huang YY, Matzke AJM, Matzke M (2013) Complete sequence and comparative analysis of the chloroplast genome of coconut palm (*Cocos nucifera*). PLoS One 8(8):e74736. https://doi.org/10.1371/journal.pone.0074736

Huang YY, Lee CP, Fu JL, Chang BCH, Matzke AJ, Matzke M (2014) *De Novo* transcriptome sequence assembly from coconut leaves and seeds with a focus on factors involved in RNA-directed DNA methylation. G3: Genes Genomes Genetics 4:2147–2157

Iyer RD (1981) Embryo and tissue culture for crop improvement, especially of perennials, germplasm conservation and exchange. In: Rao AN (ed) Tissue culture of economically important plants. Costed, ANBS, Singapore, pp 229–230

Jay M, Bourdeix R, Potier F, Sanslaville C (1989) Premiers resultats de l'etude du polymorphisme des polyphenols foliares du cocotier. Oleagineux 44:151–161

Jayasekara C (1979) Annual Report of the Coconut Research Institute, Sri Lanka (ARCRI) for 1979 pp 38–39

Jerard BA, Rajesh MK, Thomas RJ, Niral V, Samsudeen K (2017) Island ecosystems host rich diversity in coconut (*Cocos nucifera*): evidences from Minicoy Island, India. Agric Res 6(3):214–226

Karun A, Sajini KK (1994) Short-term storage of coconut embryos in sterile water. Curr Sci 67(2):118–120

Karun A, Sajini KK (2010) Cryopreservation of coconut zygotic embryos and pollen, Techical bulletin no. 63. ICAR-CPCRI, Kasaragod

Karun A, Shivasankar S, Sajini KK, Saji KV (1993) Field collection and *in vitro* germination of coconut embryos. J Plantn Crops 21:291–294

Karun A, Sajini KK, Shivashankar S (1999) Embryo culture of coconut: the CPCRI protocol. Indian J Hort 56(4):348–353

Karun A, Sajini KK, Parthasarathy VA (2002) Increasing the efficiency of embryo culture to promote germplasm collecting in India. In: Engelmann F, Batugal P, Oliver J (eds) Coconut embryo *in vitro* culture: Part II. International Plant Genetic Resources Institute, Regional Office for Asia the Pacific and Oceania (APO), Serdang, pp 7–26

Karun A, Sajini KK, Radha E, Parthasarathy VA (2004) Efficacy of CPCRI embryo culture protocol in germplasm expedition. J Plantn Crops 32:139–143

Karun A, Sajini KK, Parthasarathy VA (2005) Cryopreservation of mature zygotic embryos of coconut by desiccation methods. CORD 21(1):13–19

Karun A, Sajini KK, Niral V, Amarnath CH, Remya P, Rajesh MK, Samsudeen K, Jerard BA, Engelmann F (2014) Coconut (*Cocos nucifera* L.) pollen cryopreservation. Cryo-Letters 35(5):407–417

Karunaratne S (1988) Short-term *in vitro* preservation of coconut seed material: a method to facilitate field collection and transport of coconut germplasm. CORD 4:40–47

Karunaratne S, Periyapperuma K (1989) Culture of immature embryos of coconut, *Cocos nucifera* L.: callus proliferation and somatic embryogenesis. Plant Sci 62:247–253

Karunaratne S, Kurukularachchi C, Gamage C (1985) The culture of embryos of dwarf coconut (*Cocos nucifera* L. var.Nana.) *in vitro*. COCOS 3:1–8

Karunaratne S, Santha S, Kovoor A (1991) An *in vitro* assay for drought-tolerant coconut germplasm. Euphytica 53:25

Knutzon DS, Lardizabal KD, Nelsen JS, Bleibaum JL, Davies HM, Metz JG (1995) Cloning of a coconut endosperm cDNA encoding a 1-acyl-sn-glycerol-3-phosphateacyltransferasethatacceptsmedium-chain- length substrates. Plant Physiol 109:999–1006

Konan JNK, Koffi KE, Konan JL, Lebrun P, Dery SK, Sangare A (2007) Microsatellite gene diversity in coconut (*Cocos nucifera* L.) accessions resistant to lethal yellowing disease. African J Biotech 6:341–347

Lebrun P, N'Cho YP, Seguin M, Grivet L, Baudouin L (1998) Genetic diversity in coconut (*Cocos nucifera* L.) revealed by restriction fragment length polymorphism (RFLP) markers. Euphytica 101:103–108

Lebrun P, Baudouin L, Bourdeix R, Konan JL, Barker JHA, Aldam C, Herran A, Ritter E (2001) Construction of a linkage map of the Rennell Island tall coconut type (*Cocos nucifera L.*) and QTL analysis for yield characters. Genome 44:962–970

Lebrun P, Baudouin L, Myrie W, Berger A, Dollet M (2007) Recent lethal yellowing outbreak: why is the Malayan yellow dwarf coconut no longer resistant in Jamaica? Tree Genet Genomes 4:125–131

Li D, Zheng Y, Wan L, Zhu X, Wang Z (2009) Differentially expressed microRNAs during solid endosperm development in coconut (*Cocos nucifera* L.). Sci Hortic 122:666–669

Liang Y, Yuan Y, Liu T, Mao W, Zheng Y, Dongdong L (2014) Identification and computational annotation of genes differentially expressed in the pulp development of *Cocos nucifera* L. by suppression subtractive hybridization. BMC Plant Biol 14:205. https://doi.org/10.1186/s12870-014-0205-7

Loiola CM, Azevedo AON, Diniz LREC, Aragão WM, de O, Azevedo CD, Henrique P, Santos AD, Ramos HCC, Pereira MG, Ramos SRR (2016) Genetic Relationships among Tall Coconut Palm (*Cocos nucifera* L.) Accessions of the International Coconut Genebank for Latin America and the Caribbean (ICG-LAC), Evaluated Using Microsatellite Markers (SSRs). PLOS Mar 08 2016 – Mar 15 2016

Lopez-Villalobos A (2002) Role of lipids in coconut (*Cocos nucifera* L.) embryogenesis. University of London, Lopez-Villalobos

Manimekalai R, Nagarajan P (2006) Assessing genetic relationships among coconut (*Cocos nucifera* L.) accessions using inter simple sequence repeat markers. Sci Hortic 108:49–54

Mauro-Herrera M, Meerow AW, Borrone JW, Kuhn DN, Schnell RJ (2006) Ten informative markers developed from WRKY sequences in coconut (*Cocos nucifera*). Mol Ecol Notes 6(3):904–906

Mauro-Herrera M, Meerow AW, Borrone JW, Kuhn DN, Schnell RJ (2007) Usefulness of WRKY gene-derived markers for assessing genetic population structure: an example with Florida coconut cultivars. Sci Hortic 115(1):19–26

Meerow AW, Noblick L, Borrone JW, Couvreur TLP, Mauro-Herrera M, Hahn WJ, Kuhn DN, Nakamura K, Oleas NH, Schnell RJ (2009) Phylogenetic analysis of seven WRKY genes across the palm subtribe Attaleinae (Arecaceae) identifies Syagrus as sister Group of the Coconut. PLoS One 4(10):e7353. https://doi.org/10.1371/journal.pone.0007353

Merrow AW, Wisser RJ, Brown JS, Kuhn DN, Schnell RJ, Broschat TK (2003) Analysis of genetic diversity and population structure within Florida coconut (*Cocos nucifera* L.) using microsatellite DNA, with special emphasis on the Fiji dwarf cultivar. Theoret Appl Gen 106:715–726

Meunier J (1992) Genetic diversity in coconut a brief survey of IRHO's work. In: Papers of the IBPGR Workshop on Coconut Genetic Resources, International Crop Network series 8, IBPGR, Rome, pp 59–61

Monfort S (1985) Androgenesis of coconut: embryos from anther culture. Zeitschrift fur Pflanzenzuchtung= J Plant breed 1:113

Montero-Cortés M, Rodríguez-Paredes F, Burgeff C, Pérez-Nuñez T, Córdova I, Oropeza C, Verdeil JL, Sáenz L (2010) Characterisation of a cyclin-dependent kinase (CDKA) gene expressed during somatic embryogenesis of coconut palm. Plant Cell Tissue Organ Cult 102(2):251–258

Nan ON, Hocher V, Verdeil JL, Konan JL, Ballo K, Mondeil F, Malaurie B (2008) Cryopreservation by encapsulation dehydration of plumules of coconut. Cryo-Letters 29:339–350

Nejat N, Cahill DM, Vadamalai G, Ziemann M, Rookes J, Naderali N (2015) Transcriptomics-based analysis using RNA-Seq of the coconut (*Cocos nucifera*) leaf in response to yellow decline phytoplasma infection. Mol Gen Genomics 290(5):1899–1910

Neto FM, Pereira TNS, Geronimo IGC, Azevedo AON, Ramos SRR, Pereira MG (2016) Coconut genome size determined by flow cytometry: tall versus dwarf types. Gen Molecular Res 15(1). https://doi.org/10.4238/gmr.15017470

Pan MJ, Van Staden J (1998) The use of charcoal in in vitro culture–a review. Plant Growth Regul 26(3):155–163

Panda KC (1982) Coconut types of Orissa. Indian Cocon J 13:10–18

Pannetier C, Buffard-Morel J (1982) Production of somatic embryos from leaf tissues of coconut (*Cocos nucifera* L.) In: Proceedings of the 5th international plant tissue culture Congress, July11–16, 1982. Tokyo, Japan

Parthasarathy VA, Geethalakshmi P, Niral V (2004) Analysis of coconut cultivars and hybrids using isozyme polymorphism. Acta Bot Croat 63:69–74

Patel JS (1938) The coconut: a monograph. Government press, Madras, p 313

Perera L (2010) Hybrid testing and variety identification of coconut (*Cocos nucifera* L.) in Sri Lanka using microsatellite markers. CORD 26(1):39–43

Perera L, Russel JR, Provan J, McNicol JW, Powel W (1998) Evaluating genetic relationships between indigenous coconut (*Cocos nucifera* L.) accessions from Sri Lanka by means of AFLP profiling. Theor Appl Genet 96:545–550

Perera L, Russell JR, Provan J, Powell W (1999) Identification and characterization of microsatellites in coconut (*Cocos nucifera* L.) and the analysis of coconut population in Sri Lanka. Mol Ecol 8:344–346

Perera L, Russell JR, Provan J, Powell W (2000) Use of microsatellite DNA markers to investigate the level of genetic diversity and population genetic structure of coconut (*Cocos nucifera* L.). Genome 43:15–21

Perera L, Russell JR, Provan J, Powell W (2001) Levels and distribution of genetic diversity of coconut (*Cocos nucifera* L, var. *Typica* form *typica*) from Sri Lanka assessed by microsatellite markers. Euphytica 122:381–389

Perera PIP (2003) Cytological examination of pollen development for microspore and anther culture of coconut (*Cocos nucifera* L.) cv Sri Lanka tall. Cocos 15:53–59

Perera PIP, Hocher V, Verdeil JL, Doulbeau S, Yakandawala DMD, Weerakoon LK (2007) Unfertilized ovary: a novel explant for coconut (*Cocos nucifera* L.) somatic embryogenesis. Plant Cell Rep 26:21–28

Perera PIP, Hocher V, Verdeil JL, Bandupriya HDD, Yakandawala DMD, Weerakoon LK (2008) Androgenic potential in coconut (*Cocos nucifera* L.). Plant Cell Tissue Organ Cult 92(3):293–302

Perera PIP, Yakandawala DMD, Hocher V, Verdeil JL, Weerakoon LK (2009) Effect of growth regulators on microspore embryogenesis in coconut anthers. Plant Cell Tissue Organ Cult 96(2):171–180

Perez-Nunez MT, Chan JL, Saenz L, Gonzalez T, Verdeil JL, Oropeza C (2006) Improved somatic embryogenesis from *Cocos nucifera* (L.) plumule explants. In Vitro Cell Develop Biol Plant 42(1):37–43

Perez-Nunez MT, Souza R, Saenz L, Chan JL, Gonzalez T, Zuniga JJ, Oropeza C (2009) Detection of a SERK-like gene in coconut in vitro cultures and analysis of its expression during the formation of embryogenic callus and somatic embryos. Plant Cell Rep 28:11–19

Pesik A, Efendi D, Novarianto H, Dinarti D, Sudarsono S (2017) Development of SNAP markers based on nucleotide variability of WRKY genes in coconut and their validation using multiplex PCR. Biodiversitas J Biol Diversity 18(2):465. https://doi.org/10.13057/biodiv/d180204

Preethi P, Rajesh MK, Naganeeswaran S, Shafeeq Rehaman, Anitha Karun (2014) Identification of EST-SSRs in coconut (*Cocos nucifera* L.) by deep transcriptome sequencing. In: Book of Abstracts of National Seminar on 'New Horizons and Challenges in Biotechnology and Bioinformatics', 9–10 October, 2014, ICAR-CPCRI, Kasaragod, p 101

Preethi P, Rajesh MK, Rahul CU, Jerard BA, Samsudeen K, Thomas RJ, Karun A (2016) Identification and utilization of informative EST-SSR markers for genetic purity testing of coconut hybrids. J Plantn Crops 44(2):77–84

Puch-Hau C, Oropeza-Salín C, Peraza-Echeverría S, Gongora-Paredes M, Córdova-Lara I, Narvaez-Cab M, Zizumbo-Villareal D, Sáenz-Carbonell L (2015) Molecular cloning and characterization of disease-resistance gene candidates of the nucleotide binding site (NBS) type from *Cocos nucifera* L. Physiol Mol Plant Pathol 89:87–96

Puch-Hau C, Oropeza C, Góngora-Paredes M, Córdova I, Tun-Suárez J, Sáenz L (2016) New insights into the evolutionary history of resistance gene candidates in coconut palms and their expression profiles in palms affected by lethal yellowing disease. Genes Genom 38(9):793–807

Rachana KE, Naganeeswaran SA, Fayas TP, Thomas RJ, Rajesh MK (2016) Cloning, characterization and expression analysis of NBS-LRR-type resistance gene analogues (RGAs) in coconut. Acta Bot Croat 75(1):1–10

Rajesh MK, Radha E, Sajini KK, Karun A, Parthasarathy VA (2005) Plant regeneration through organogenesis and somatic embryogenesis from plumular explant of coconut. J Plantn Crops 33(1):9–17

Rajesh MK, Nagarajan P, Jerard BA, Arunachalam V, Dhanapal R (2008a) Microsatellite variability of coconut a ccessions (*Cocos nucifera* L.) from Andaman and Nicobar Islands. Curr Sci 94:1627–1631

Rajesh MK, Arunachalam V, Nagarajan P, Lebrun P, Samsudeen K, Thamban C (2008b) Genetic survey of ten Indian coconut landraces by simple sequence repeats (SSRs). Sci Hortic 118:282–297

Rajesh MK, Thomas RJ, Rijith J, Shareefa M, Jacob PM (2012) Genetic purity assessment of D x T hybrids in coconut with SSR markers. Indian J Gen Plant Breed 72(4):472–474

Rajesh MK, Jerard BA, Preethi P, Thomas RJ, Fayas TP, Rachana KE, Karun A (2013a) Development of a RAPD-derived SCAR marker associated with tall-type palm trait in coconut. Sci Hortic 150:312–316

Rajesh MK, Rachana KE, Babu M, Thomas RJ, Karun A (2013b) Characterization of the global transcriptome responsive to root (wilt) disease in coconut using RNA-seq. In: National Symposium on 'Pathogenomics for Diagnosis and Management of Plant Diseases', ICAR-CTCRI, October, 2013, Thiruvananthapuram, India, pp 24–25

Rajesh MK, Jerard BA, Preethi P, Thomas RJ, Karun A (2014a) Application of RAPD markers in hybrid verification in coconut. Crop Breed Appl Biotech 14:36–41

Rajesh MK, Radha E, Sajini KK, Karun A (2014b) Polyamine-induced somatic embryogenesis and plantlet regeneration *in vitro* from plumular explants of dwarf cultivars of coconut (*Cocos nucifera*). Indian J Agri Sci 84(4):527–530

Rajesh MK, Samsudeen K, Rejusha P, Manjula C, Rahman S, Karun A (2014c) Characterization of Annur and Bedakam ecotypes of coconut from Kerala state, India, using microsatellite markers. Intl J Biodiversity 2014:1. https://doi.org/10.1155/2014/260895

Rajesh MK, Samsudeen K, Jerard BA, Rejusha P, Karun A (2014d) Genetic and phylogenetic relationships of coconut populations from Amini and Kadmat Islands, Lakshadweep (India). Emirates J Food Agri 26(10):898–906

Rajesh MK, Rachana KE, Naganeeswaran SA, Shafeeq R, Thomas RJ, Shareefa M, Merin B, Karun A (2015a) Identification of expressed resistance gene analog sequences in coconut leaf transcriptome and their evolutionary analysis. Turk J Agri For 39:489–502. https://doi.org/10.3906/tar-1409-75

Rajesh MK, Sabana AA, Rachana KE, Rahman S, Jerard BA, Karun A (2015b) Genetic relationship and diversity among coconut (Cocos nucifera L.) accessions revealed through ScoT [start codon targeted polymorphism] analysis. 3 Biotech 5:999–1006. https://doi.org/10.1007/s13205-015-0304-7

Rajesh MK, Fayas TP, Naganeeswaran S, Rachana KE, Bhavyashree U, Sajini KK, Karun A (2016) *De novo* assembly and characterization of global transcriptome of coconut palm (*Cocos nucifera* L.) embryogenic calli using Illumina paired-end sequencing. Protoplasma 253(3):913–928

Rajesh MK, Rachana KE, Kulkarni K, Sahu BB, Thomas RJ, Karun A (2017) Comparative transcriptome profiling of healthy and diseased Chowghat Green Dwarf coconut palms from root (wilt) disease hot spots. Eur J Plant Path:1–21. https://doi.org/10.1007/s10658-017-1365-8

Raju CR (2006) Direct *in vitro* shoot development from immature rachilla explants of coconut (*Cocos nucifera* L.). CORD Intl J Cocon Res Dev 22(1):51–58

Raju CR, Kumar PP, Chandramohan M, Iyer RD (1984) Coconut plantlets from leaf tissue cultures. J Plantn Crops 12:75–78

Rasam DV, Gokhale NB, Sawardekar SV, Patil DM (2016) Molecular characterisation of coconut (*Cocos nucifera* L.) varieties using ISSR and SSR markers. J Hort Sci Biotechnol 91:347–352

Riedel M, Riederer M, Becker D, Herran A, Kullaya A, Arana-López G, Peña-Rodríguez L, Billotte N, Sniady V, Rohde W, Ritter E (2009) Cuticular wax composition in *Cocos nucifera* L. physicochemical analysis of wax components and mapping of their QTLs onto the coconut molecular linkage map. Tree Genet Genomes 5(1):53–69

Rillo EP, Paloma MBF (1990) Comparison of three media formulations for *in vitro* culture of coconut embryos. Oleagineux 45(7):319–323

Rillo EP, Paloma MBF (1991) Storage and transport of zygotic embryos of *Cocos nucifera* L. for *in vitro* culture. Plant Gen Resour Newsl 86:14–18

Ritto P, Jiji G, Rajesh MK, Jerard BA, Niral V (2008) Genetic diversity in yellow dwarf populations of coconut (*Cocos nucifera* L.) assessed using RAPD markers. Indian J Hort Res 65:60–64

Rivera R, Edwards KJ, Barker JHA, Arnold GM, Ayad G, Hodgkin T, Karp A (1999) Isolation and characterization of polymorphic microsatellites in *Cocos nucifera* L. Genome 42:668–675

Rodriguez MJB, Estioko LP, Namia TI, Soniega JA (1997) Analysis of genetic diversity in coconut by RAPD. Philippine J Cocon Stud 22:1–7

Rognon F, De Nuce de Lamothe M (1978) Recolte et conditionnement du pollen pour la pollinisation des champs semenciers de cocotier. Oleagineux 33:17–21

Rohde W, Salamani F, Ashburner R, Randles JW (1992) An *Eco* RI repetitive sequence family of the coconut palm (*Cocos nucifera* L.) shows sequence homology to copia-like elements. J Gen Breed 46:391–394

Saensuk C, Wanchana S, Choowongkomon K, Wongpornchai S, Kraithong T, Imsabai W, Chaichoompu E, Ruanjaichon V, Toojinda T, Vanavichit A, Arikit S (2016) *De novo* transcriptome assembly and identification of the gene conferring a "pandan-like" aroma in coconut (*Cocos nucifera* L.). Plant Sci 252:324–334

Saenz L, Azpettia B, Chuc-Armendariz Chan JL, Verdeil HV, Oropeza C (2006) Morphological and histological changes during somatic embryo formation from coconut plumule explants. In Vitro Cell Develop Biol Plant 42:19–25

Sajini KK, Karun A, Amarnath CH, Engelmann F (2011) Cryopreservation of coconut (*Cocos nucifera* L.) zygotic embryos by vitrification. Cryo-Letters 32(4):317–328

Samosir YMS (1999) Optimisation of somatic embryogenesis in coconut (*Cocos nucifera* L.). PhD thesis, The University of Queensland, Australia

Sandoval A, Hocher V, Verdeil JL (2003) Flow cytometric analysis of the cell cycle in different coconut palm (*Cocos nucifera* L.) tissues cultured in vitro. Plant Cell Rep 22:25–31

Shalini KV, Manjunatha S, Lebrun P, Berger A, Baudouin L, Pirany N, Ranganath RM, Prasad DT (2007) Identification of molecular markers associated with mite resistance in coconut (*Cocos nucifera* L.). Genome 50:35–42

Sisunandar Novarianto H, Mashud N, Samosir Y, Adkins SW (2014) Embryo maturity plays an important role for the successful cryopreservation of coconut (*Cocos nucifera*). In Vitro Cell Develop Biol Plant 50:688–695

Sisunandar, Peter AS, Yohannes MS, Samosir, Rival A, Adkins SW (2010a) Dehydration improves cryopreservation of coconut (*Cocos nucifera* L.). Cryobiology 61:289–296

Sisunandar RA, Turquay P, Samosir Y, Adkins S (2010b) Cryopreservation of coconut (Cocos *nucifera* L.) zygotic embryos does not induce morphological, cytological or molecular changes in recovered seedlings. Planta 232:435–447

Sossou J, Karunaratne S, Kovoor A (1987) Collecting palm: *In vitro* explanting in the field. Plant Gen Resour Newsl 16:7–18

Sugimura Y, Itano M, Salud CD, Otsuji K, Yamaguchi H (1997) Biometric analysis on diversity of coconut palm: cultivar classification by botanical and agronomical traits. Euphytica 98(1–2):29–35

Sun R, Ye R, Gao L, Zhang L, Wang R, Mao T, Zheng Y, Li D, Lin Y (2017) Characterization and ectopic expression of CoWRI1, an AP2/EREBP domain-containing transcription factor from coconut (*Cocos nucifera* L.) endosperm, changes the seeds oil content in transgenic *Arabidopsis thaliana* and Rice (*Oryza sativa* L.). Front Plant Sci 8:63. https://doi.org/10.3389/fpls.2017.0006

Teulat B, Aldam C, Trehin R, Lebrun P, Barker JHA, Arnold GM, Karp A, Baudouin L, Rognon F (2000) An analysis of genetic diversity in coconut (*Cocos nucifera L.*) population from across the geographical range using sequence tagged microsatellites (SSRs) and AFLPs. Theoret Appl Gen 106:411–422

Thanh-Tuyen NT, De Guzman EV (1983) Formation of pollen embryos in cultured anthers of coconut (Cocos nucifera L.). Plant Sci Lett 29(1):81–88

Thomas BR, Pratt D (1981) Efficient hybridization between *Lycopersicon esculentum* and *L. peruvianum* via embryo callus. Theoretical and Appl Gen 59:215–219

Upadhyay A, Jayadev K, Manimekalai R, Parthasarathy VA (2004) Genetic relationship and diversity in Indian coconut accessions based on RAPD markers. Sci Hortic 99:353–362

Ventura PF, Zuniga LC, Figueroa JE, Lazo FD (1966) A progress report on the development of coconut embryos in artificial media. Philippine J Plant Indust 3:81–87

Verdeil JL, Huet C, Grosdemange F, Buffard Morel J (1994) Plant regeneration from cultured immature inflorescences of coconut (*Cocos nucifera* L.): evidence for somatic embryogenesis. Plant Cell Rep 13:218–221

Verdeil JL, Hocher V, Triques K, Lyakurwa R, Rival A, Durand-Gasselin T, Engelmann F, Sangare A, Hamon S (1998) State of research on coconut embryo culture and acclimatization techniques in the IDEFOR (Côte d'Ivoire) and ORSTOM/CIRAD laboratories (France). Coconut Embryo:17–28

Vidhanaarachch VR, Weerakoon LK, Fernando SC, Gamage CKA, Santha ES (1998) Status of research on coconut embryo culture and acclimatization techniques in Sri Lanka. In: Batugal PA, Engelmann F (eds) Coconut embryo in vitro culture. IPGRI, APO, Serdang, pp 85–88

Vidhanaarchchi VRM, Weerakoon LK (1997) Callus induction and direct shoot formation in *in vitro* cultured immature inflorescence tissues of coconut. COCOS 12:39–43

Wadt LHO, Sakiyama NS, Pereira MG, Tupinambra EA, Riberio FE, Aragao WM (1999) RAPD markers in the genetic diversity study of the coconut palm. In: Oropeza C, Verdeil JL, Ashburner GR, Cardena R, Santamaria JM (eds) Current plant science and biotechnology in agriculture – current advances in coconut biotechnology. Kluwer Academic Publishers, Dordrecht, pp 89–97

Whíte T, Moran G, Knox RB (1987) Estimation of genetic distance in coconut. ACIAR Project -Coconut improvement. Reports 1 and 2. School of Botany, The Univ of Melboume, Australia

Whitehead RA (1963) The processing of coconut pollen. Euphytica 12:167–177

Xiao Y, Luo Y, Yang Y, Fan H, Xia W, Mason AS, Zhao S, Sager R, Qiao F (2013) Development of microsatellite markers in *Cocos nucifera* and their application in evaluating the level of genetic diversity of *Cocos nucifera*. Plant Omics 6(3):193–200

Xiao Y, Xu P, Fan H, Baudouin L, Xia W, Bocs S, Xu J, Li Q, Guo A, Zhou L, Li J (2017) The genome draft of coconut (*Cocos nucifera*). Giga Sci 6(11):1–11

Xu L, Ye R, Zheng Y, Wang Z, Zhou P, Lin Y, Li D (2010) Isolation of the endosperm-specific *LPAAT* gene promoter from coconut (*Cocos nucifera* L.) and its functional analysis in transgenic rice plants. Plant Cell Rep 29:1061–1068

Zizumbo Villarreal D, Cardena-Lopez R, Pinero D (2002) Diversity and phylogenetic analysis in *Cocos nucifera* L. in Mexico. Genet Resour Crop Evol 49(3):237–245

Zizumbo-Villarreal D, Piñero D (1998) Pattern of morphological variation and diversity of *Cocos nucifera* (Arecaceae) in Mexico. American J Bot 85:855–865

Dr. M. K. Rajesh is a Principal Scientist (Biotechnology) in the Division of Crop Improvement, ICAR-CPCRI. He has 20 years of experience in molecular characterization of coconut, cocoa and areca nut germplasm, the use of molecular markers for hybrid authentication, in vitro culture and in vitro conservation of coconut and gene expression studies in coconut. He has published 100 research papers, written 20 book chapters and edited 5 books. His current area of research includes genome and transcriptome analysis of coconut. mkraju.cpcri@gmail.com

Dr. Anitha Karun has a doctorate degree in Horticulture. She is presently the Head of Crop Improvement Division of ICAR-CPCRI, Kasaragod, Kerala, India. Her areas of specialization are palm tissue/organ culture and cryopreservation. She has been awarded 'ICAR Award for Team Research' for her contribution in germplasm exchange of coconut in the form of embryos. She has published 100 research papers, 12 book chapters and 3 technical bulletins. anithakarun2008@gmail.com

Dr. V. A. Parthasarathy Former Director and Emeritus Scientist, Indian Institute of Spices Research, Kozhikode, India, has made significant contributions in the field of biotechnology and breeding of a range of horticultural crops. His major contributions are in the tribal region of North East at ICAR Research Complex for NEH Region, Barapani, India, for which he received the Lifetime Achievement Award. He has published 200 research papers and 100 reviews/chapters in books. He has edited/authored 50 books. vapartha@gmail.com

Chapter 7
Agro-management Practices for Sustainable Coconut Production

George V. Thomas, V. Krishnakumar, R. Dhanapal, and D. V. Srinivasa Reddy

Abstract Improved agro-techniques have been standardized, through research conducted over several decades, to achieve sustainable productivity and profitability in coconut farming. Adoption of refined nursery techniques enables production of quality planting material. Poly bag nursery technique with bio-priming of biofertilizer formulations helps in production of superior quality seedlings. The good management practices validated to improve the productivity in adult coconut palms include integrated nutrient management, green manuring/cover cropping, soil and water conservation measures, weed management, irrigation, fertigation and cropping/farming system approach. Fertigation helps to increase the fertilizer use efficiency, saves fertilizer costs, reduces labour requirement and ensures continuous nutrient supply in tune with crop requirement. Sustainable cropping system models are evolved to optimize utilization of natural resources and to enhance the economic viability. Integrated farming involving cultivation of fodder grass in the interspaces of coconut and integration of animal husbandry enterprises offer significant ecological and economic benefits. Effective formulations of agriculturally important microorganisms such as nitrogen fixers, plant growth-promoting rhizobacteria and arbuscular mycorrhizal fungi have been developed as valuable inputs for sustainable crop production. Lignocellulosic residues from coconut plantations can be con-

---

G. V. Thomas (✉)
ICAR- CPCRI, Vettimoottil House, Vakayar (P.O.), Kerala, India
e-mail: georgevthomas@yahoo.com

V. Krishnakumar
ICAR-Central Plantation Crops Research Institute, Regional Station, Kayamkulam, Kerala, India
e-mail: dr.krishnavkumar@gmail.com

R. Dhanapal
B-5, Palaniappa Nagar, PN Pudur, Coimbatore, Tamil Nadu, India
e-mail: ramaswamydhanapal@yahoo.co.uk

D. V. Srinivasa Reddy
ICAR-ATARI, Main Research Station, Bengaluru, Karnataka, India
e-mail: reddydvs@yahoo.com

K. U. K. Nampoothiri et al. (eds.), *The Coconut Palm (*Cocos nucifera *L.) - Research and Development Perspectives*,
https://doi.org/10.1007/978-981-13-2754-4_7

verted into brown, granular vermicastings using earthworms. Organic farming practices with focus on building soil biological fertility foundations through integrated application of organic and bio-inputs including recycling of waste biomass, in situ cultivation and incorporation of leguminous cover crops and biofertilizers of *Azospirillum* and *Bacillus* and other cultural practices are combined with micro-irrigation techniques to obviate moisture stress and enable sustainable coconut production, in an environment-friendly way.

7.1 Introduction

The coconut palm, a native of humid tropics, is grown extensively in coastal areas between the latitudes 26°N and 26°S of the equator. The main coconut-growing areas are located in Asia, Oceania, West Indies, Central and South America and West and East Africa, and the major production comes from the Asia-Pacific region. Though the palm grows up to 1000 m above mean sea level, most of the commercial coconut cultivation is up to an altitude of 600 m (Ohler 1999). Conditions near the sea are ideal for vigorous growth and fruiting of the palm, but it comes up well even hundreds of kilometres in the inland away from the sea. The coconut palm is also planted widely on banks between fields of other crops. As the palm has perennial growth habit with a prolonged reproductive phase of 44 months (from the initiation of the inflorescence primordium to full maturity of the nuts), the weather conditions at different stages of the development cycle influence the growth and production of nuts.

Temperature is an important weather factor that has great influence on the growth and productivity of the palm. The palm prefers less diurnal variation between day and night temperatures and does not tolerate extremes of temperature. The aberrations in weather such as long spells of hot and dry weather, severe winters and extremes of temperatures are not favourable for coconut growing. The ideal temperature requirement for vigorous growth and yield of the palm is in the range of 27–32 °C with a diurnal variation of not more than 7 °C. Abnormalities in fruit development and reduction in yield of palms occur when temperature drops to less than 15 °C (Child 1974). High temperature during the flowering stage leads to drying up of the developing inflorescences and results in a decrease in production. The palm requires plenty of sunlight, and shaded condition does not favour proper development of the palm. Sunshine in excess of 2000 h in a year, with 120 h of sunshine per month, is ideal for the better growth of the palm (Menon and Pandalai 1960; Nair 1979). Maintenance of proper level of moisture in coconut basin either through well-distributed rainfall or irrigation and provision of sufficient drainage in waterlogged areas is essential for proper growth of the palm. The palm comes up well in areas receiving well-distributed rainfall ranging from 1000 mm to 3000 mm annually. Persley (1992) reported that coconut requires an annual rainfall of at least 1800 mm, evenly distributed throughout the year for optimum production. Ideal relative humidity (RH) for best growth and production is 80–90% (Child 1974). RH below 60% results in stomata closure that restricts transpiration, while RH higher than 90% predisposes the palm to a number of diseases.

The highly adaptable nature of the palm makes it an ideal crop to be cultivated under diverse soil conditions ranging from littoral sands to clayey soils, ill-drained low-lying areas to well-drained hill slopes and strongly acidic peaty soils to alkaline calcareous soils (Khan et al. 1979). In the littoral (coastal) sand which has many unfavourable features for successful cultivation of many other crops, the coconut palm can be cultivated provided special care is taken from the early stages of growth with organic manuring and watering. In the west coast of India, the major soil types where coconut is grown extensively are laterites, littoral sands and red sandy loams, while deltaic alluvium, red soils and littoral sand are the major soils along the east coast of the country. The soil type for the best growth of the palm is found to be deltaic alluvial followed by red sandy loam. A well-drained soil with a 1.0–1.5 m depth, rich in organic matter and having good fertility and water-holding capacity, is well suited. Though the palm tolerates salinity up to a pH of 8.5, a pH range of 5.0–7.5 supports better growth of the palm (John 1952; Nair 1979; Khan et al. 1979).

Coconut farming at the global level is constrained by the low productivity which is mainly due to lack of proper management, senility, applications of low level of inputs and poor health of soil. In order to make the farming profitable in the scenario of worsening climatic conditions and severe constraints on the available natural resource base, it is necessary to adopt scientific management practices to attain higher yield per unit area of land, time and inputs. The technological advancements made so far are of great relevance to achieve sustainable productivity and to enhance economic viability of coconut farming.

7.2 Quality Planting Material Production

Being the foundation for successful coconut production, the quality of planting material largely determines the ultimate returns from coconut. As it is a perennial crop, the yield performance of the crop will be known only after several years of planting and any mistake in the selection of seedlings leads to considerable economic loss to the farmer for several decades. In the absence of a viable technology for vegetative method of propagation, coconut is propagated only through seeds. The seedling selection is very important as the palm does not breed true to itself due to the cross-breeding nature. The fact that certain seedling characters are strongly correlated with earliness in flowering, nut yield and copra production makes the seedling selection highly relevant for sustainable coconut farming (Ramadasan et al. 1993). It is possible to produce quality seedlings through a series of selections at different stages. Nursery techniques to produce quality seedlings have been standardized, which involves several steps, viz. selection of seed gardens, selection of mother palms, collection of seed nuts, storage of seed nuts, site selection and raising of nursery with proper care and management, as well as selection of seedlings. Please refer to Chap. 4 on Varietal Improvement for details.

7.2.1 Collection of Seed Nuts

Suitable time for seed nut collection will vary from country to country and region to region within the country according to seasonal factors. Maturity and development of nuts for satisfactory germination and the period of storage required (which depends on the preparedness for sowing in the nursery) are to be considered. Generally seed nuts are harvested during February to May in the west coast of India so that the rains of south-west monsoon can be taken advantage of to sow the nuts (Menon and Pandalai 1960).

Nuts harvested during April to July and sown during June to September germinated earlier, the average time being 125 days after sowing. However, there is no harm in collecting seed nuts in other months, if there is a peak coconut yield. But nuts harvested from August to December and sown during October to February required a longer time (about 190 days) for germination. However, the total germination percentage was almost the same for nuts harvested in all months, indicating that the nuts germinate satisfactorily irrespective of the month of harvest, if proper nursery techniques are adopted (Nelliat et al. 1976). In the case of East Coast Tall variety, nuts collected from February to August and stored for 1 or 2 months gave high percentage of germination (Sundaresan et al. 1974).

Though Nelliat et al. (1976) reported that there is no difference in performance between palms raised from nuts which are 11 to 14 months old, the generally accepted practice is to sow 11–12-month-old nuts. While conducting studies on hybrid seed collection in Côte d'Ivoire, Wuidart and Nuce de Lamothe (1981a) suggested to collect the bunches when one or two nuts just begin to turn brown. Nuts under 11 months germinated slowly than 11–12-month-old nuts and gave a large number of abnormal sprouts. From India, Chattopadhyay and Hore (2012) observed early and maximum germination and better seedling vigour when seed nuts with higher weight (1100 g) was used when compared to seed nut with 600 g weight. Wickramaratne et al. (1987) from Sri Lanka warned that seed nuts should not be selected or rejected on the basis of size, quantity of nut water or shape, as they are not strongly correlated with germination. Only immature, empty, exceptionally small or outsized nuts should be rejected, and all others could be used as seed nuts.

7.2.2 Storage of Seed Nuts

Though the seed nut season starts in February in the west coast of India, the nuts may have to be stored for about 4 months because the nuts can be sown in the nursery only with the start of south-west monsoon in June. During this hot period of storage, care has to be taken to avoid drying of nuts (Menon and Pandalai 1960).

According to Fremond et al. (1966), storage of seed nuts in the shade for 1 month is essential for breaking the dormancy. Kailash Rao and Srirama Rao (1968) advocated storage of seed nuts for 2 months for the West Coast Tall variety of coconut.

Marimuthu and Natarajan (2005) also found that to get more quality seedlings, the seed nuts are to be stored for 1 month in open shade followed by sand curing for 2–3 months. The storage time should be shorter in the case of seed nuts of dwarf varieties of coconut.

Nampoothiri et al. (1973) studied the variation in germination pattern of five coconut cultivars and one hybrid by harvesting 12-month-old seed nuts at monthly intervals from February to June and sowing in June (to give storage periods of 4 months, 3 months, 2 months, 1 month and no storage, respectively). They found that when the number of days taken for germination from the date of harvest was considered, storage does not appear to be necessary. Prolonged storage of cultivars such as Laccadive Ordinary and Gangabondam adversely affected germination. Nuts of Straight Settlement Apricot was found to germinate 50–80 days after harvest, and sprouts emerge if stored for longer than about 2 months.

7.2.3 Pre-sowing Treatment

Among the many pre-sowing treatments recommended for reducing the period of germination, soaking of seed nuts in water for a period up to 15 days has been found to result in quicker and better germination, but in case the period is prolonged, germination and quality of seedlings will be adversely affected. In an experiment in Tanzania, water soaking recorded the lowest period for germination (81.1 days) as against 142.9 days for untreated seed nuts (Thomas 1974). Germination process was hastened by soaking the nuts in water or in a solution of 0.01 M potassium nitrate and 0.02 M sodium carbonate for 48 h (Thomas 1974), chopping the husk from both ends of the nuts as well as injecting different hormones (Liyanage 1952; Deshpande and Kulkarni 1962) and major and minor nutrients (Menon and Pandalai 1960; Sumathykutty Amma 1964).

A practice that can be followed to improve sprouting is the slicing off of a part of the husk ridge at the stalk end of the nut above the germination eye. Removal of the exocarp facilitates penetration of water through the husk, and the moisture will reach the sprout earlier, providing optimum moisture conditions in the husk for germination and sprouting. Sometimes a slice of husk is also removed on the opposite side of the nut, permitting easier penetration of the roots through the husk into the soil (Fremond et al. 1966; Borah 1991; Ugbah and Akpan 2003). Peries (1984) compared a pre-nursery system (slicing seed nuts and transplanting sprouted seed nuts from pre-nursery beds to nursery beds, soon after sprouting) with the conventional nursery. Though the rate of sprouting at the end of 20 weeks in the pre-nursery system was higher (76%) as compared to 67% in the conventional nursery, the number of quality seedlings produced was more or less the same. The cost of production of seedlings in the pre-nursery system was more than over 50% of the conventional nursery. Though slicing the seed nut facilitated sprouting, transplanting sprouted seed nuts from the pre-nursery caused a setback to the development of the seedlings.

7.2.4 Nursery Raising Techniques

Coconut seed is not to be sown directly into the field. The nursery practices are intended to provide optimum conditions for the germination of seed nuts and subsequent growth of seedlings so as to get high percentage of quality seedlings from the nuts sown in the nursery. It also makes watering as well as pest and disease control easier and more efficient. Selection of good quality seedlings for field planting is possible only by raising them in a nursery.

Site Selection and Raising Nursery For raising the nursery, the site selected should be well-drained, with coarse-textured soil and near dependable water source for irrigation. Raising nursery in open condition is the best; however, it can be raised anywhere if there is not too much shade as in the coconut garden itself. The seedlings become lean and lanky and have an unhealthy appearance under heavy shade. The layout of the nursery is to be designed according to the irrigation system used. Beds, which are long and narrow, are to be prepared with provision for walking space or drains in between, as found necessary.

To facilitate access to the seedlings, these are sometimes planted in double rows, keeping the spacing at about 15 cm within the row and about 45 cm between rows. In general, spacing of about 60 cm is kept between the beds in the nursery. When the seed nuts are sown too close, the emerging seedlings will become etiolated and lanky. Muliyar and Pillai (1989) suggested keeping a distance of 80 cm between beds. This will help irrigating the nursery and examination of seedlings from the sides without getting inside the beds.

Sowing of Seed Nuts The positioning of the soft eye is important, as it is believed that this affects sprouting (germination). In the Philippines, trimming the edge (below which the functional eye through which the sprout emerges) and positioning the nut with the flat surface downward and ridge pointing upward were found to facilitate sprouting and normal development of seedlings (Anon 1975).

Laying the nuts with the soft eye towards the upper side helps the sprout to emerge faster, as it would be closer to the surface. Placing the nuts with the soft eye closer to the lower surface may ensure that the soft eye, and therefore the embryo, remains in contact with the nut water for a longer period, which might enhance germination. Since there is no method to determine the position of the soft eye of the seed nut, it is not possible to position it in the nursery such that early sprouting is facilitated. Practically, the best planting position of the nuts is horizontal, irrespective of which side is lowermost. This will increase the speed of sprouting and reduce nursery costs considerably (Wickramaratne and Padmasiri 1986). As the roots grow rather quickly, the position of the soft eye in relation to contact with the soil will be of only minor importance, provided the nuts are adequately watered. Romney et al. (1968) opined that seedlings from horizontally placed nuts are less likely to be damaged at transplanting because the attachment between shoot and nuts is much better protected by the husk. In Papua New Guinea, the rate of germination and subsequent growth of seedlings were much faster in horizontal sowing than vertical sowing. In 8 weeks from transplanting, 47% germination was recorded for the former method against 39% in the latter (Kenman 1973). Remison and Mgbeze (1988) and Chattopadhyay et al. (2004) also reported beneficial effect of horizontal sowing in comparison with vertical sowing.

In some countries, nuts are planted in a vertical position, permitting the roots to grow through the husk for a longer time, which is an advantage at transplanting. But this can be done only with nuts that have enough water to fill almost the entire cavity, such that the haustorium does not lose its contact with the nut water. This method of sowing will be ideal when the seedlings are to be transported to distant places.

Results of a large-scale trial conducted by Wuidart and Nuce de Lamothe (1981b) showed that MYD germinates very rapidly and WAT slowly. MYD, RLT and WAT took 51, 81 and 126 days, respectively, to attain 50% germination. It was concluded that late germination was influenced by husk moisture, water content of nuts and the distance separating the haustorium from this water. The differences between varieties are, thus, determined mainly by the time the haustorium takes to reach the water in the nut's cavity.

Borah (1991) observed that seed nuts that floated vertically upright in water were more vigorous than those floating horizontally. The strong and stout seedlings from such nuts resulted from early germination and less exposure to drought. The coconut water may be absorbed completely within a period of about 5 months, and by this time about half the kernel may still be left in the nut, available to the developing seedling.

The seed nuts can be sown in flat beds if the soil is well-drained. Otherwise, they should be sown in raised beds at a spacing of 40 cm between rows and 30 cm between the nuts during May–June, either horizontally or vertically in 20–25 cm deep trenches. Spacing of seed nuts in the nursery is determined based on the duration for which the seedlings are to be retained in the nursery. In Odisha and Godavari areas of India, where much older seedlings are preferred for planting, it is usual to transplant the seedlings when they are 1 year old to a secondary nursery with a spacing of 90–150 cm between seedlings. In Sri Lanka, in the conventional nursery, seed nuts are laid in the nursery beds spaced at 45 cm between rows and 15 cm within row and remain there for 7 to 9 months until seedlings are selected for field planting (Anon 1971). The depth of sowing is to be adjusted such that the husk appears just above the surface of soil.

7.2.5 Care and Management of Nursery

The seedlings are to be properly maintained with adequate watering, weeding, fertilizer application and plant protection measures. The nursery beds should also be kept weed-free by frequent weeding. The most popular method of weeding in coconut nurseries is hand weeding, though it is labour intensive and time consuming (Remison and Mgbeze 1987).

During dry and hot weather, the nursery beds should be mulched and shaded with dry coconut leaves or any other suitable materials. Mulching with coconut leaves has been found to promote early and better germination, good growth of seedlings and recovery of high percentage of good seedlings (Liyanage 1952; Verghese et al. 1953; Aiyadurai 1954). Remison and Mgbeze (1987) observed that mulching helped to reduce competition from weeds and to conserve moisture, leading to increased plant height, plant girth and dry matter of seedlings.

Manuring of Nursery Seedlings Inorganic manures are usually not applied to the seedlings in the nursery as the intrinsic quality of seedlings is likely to be masked to some extent making a proper selection of seedlings difficult (John 1952). The seed nut contains adequate nutrients to meet the needs of the growing plant, at least up to the field planting stage. Maravilla et al. (1978) pointed out that the non-responsiveness of seedlings to manure application in the early nursery stages could be due to the availability of nutrients at sufficient levels to meet its growth requirements from the endosperm. However, as the nutrient reserve from the endosperm decreases from the fourth month after germination (Foale 1968b), the need for fertilizer application to the nursery has been stressed to produce healthy and vigorous seedlings, facilitating better establishment, faster growth and early bearing in the main field (Ziller and Fremond 1961; Fremond et al. 1966; Nelliat 1972; Mathew and Ramadasan 1964; Santiago 1978; Maravilla 1986; Srinivasa Reddy et al. 1998a).

The effect of N on the growth of coconut seedlings was more significant than P and K (Mathew and Ramadasan 1964). On the other hand, Bachy et al. (1962) observed in West Africa that excess N has a depressing effect on the growth of the seedlings. They reported that the application of P-K-Mg fertilizers gave a highly significant improvement in the vigour of the seedling. In Davao, it was observed that chloride application significantly increased the girth of the seedlings (Magat and Prudente 1974). It was also reported that application of KCl and NaCl to coconut seedlings influenced the growth and increased their resistance to diseases particularly leaf spots (Magat et al. 1977; Abad et al. 1978). Application of nutrients (N, P, K, Ca and Mg at 120:60:120:90:60 kg ha^{-1} month^{-1}) to the seedlings in the nursery bed has been found to be beneficial to produce seedlings with proper vigour and quality as indicated by higher chlorophyll content and nutrient concentration in leaves (Nelliat et al. 1976). In the Philippines, application of N and K fertilizers in coconut nursery produced taller seedlings with higher girth and greater vigour index (Almaden and Santiago 1980). Veloso and Ly (1982) also reported application of ammonium sulphate and NaCl at 30 g seedling^{-1} produced taller seedlings with greater girth and lesser degree of leaf spot/blight disease infection compared to unfertilized seedlings. The collar girth, leaf area, dry matter production and number of roots were superior with the application of farmyard manure at 25 tonnes ha^{-1} and 160 kg each N and K fertilizers as soil application in three splits at fifth, seventh and ninth months after sowing (Srinivasa Reddy et al. 1998a). Embryo-cultured plantlets were significantly taller and had higher girth and more total leaves than unfertilized Laguna Tall and Makapuno seedlings, with application of NaCl at a total dose of 18–54 g seedling^{-1} and chicken manure at 250–750 g seedling^{-1} (at 2,4,6,8 and 10 months stage) (Ubaldo et al. 2006).

In order to develop a fertilizer mixture that supports the plant growth with simultaneous improvement of soil health in Sri Lanka, Ranaweera et al. (2010) evaluated inorganic fertilizers, organic manures (cattle manure and compost) and biofertilizers on 8-month-old coconut seedlings of CRIC 65 cultivar. They observed significant improvement in growth of seedlings as well as the soil microbial activity with application of Biogold® and compost. Application of organic manure (neem seed powder and leaves of gliricidia) and inorganic fertilizers also resulted in production of quality seedlings of Sri Lanka Tall variety in Pakistan (Solangi and Iqbal 2012). Application of NaCl at 50–60 g seedling^{-1} was recommended in the Philippines, and it has become a common practice for growing seedlings in the nursery (Magat et al. 1977).

According to Ho et al. (1978), monthly application of NPK fertilizers from the second month improved the girth of seedlings. Oguis et al. (1979) found that Cl significantly increased the girth of seedlings, while S greatly increased its height. Of the Cl sources, NaCl at 30 g Cl (60 g NaCl seedling^{-1}) gave the best result followed by KCl. They also found that 60 g Cl from KCl or higher rates of NaCl significantly reduced leaf spot incidence compared with higher rates of NH_4Cl. Maravilla (1986) also reported a beneficial effect of application of fertilizers for production of poly bag seedlings. Under Indian conditions, Ratnambal (1995) suggested application of 20 g $(NH_4)_2$ SO_4 and 25 g KCl 2 months after germination and 45 g $(NH_4)_2$ SO_4 and 45 g KCl bag^{-1} 4 months after germination as recommended in Indonesia. It is suggested that nursery manuring is not necessary if seedlings are removed 9 months after sowing. After this period, the seedling responds to manuring, especially if soil is of low fertility.

Irrigation of Nursery Irrigation is an important input for promoting early establishment and growth of freshly planted seedlings. The available moisture on the seed bed has been found to be the crucial factor facilitating germination (Wuidart 1981b). The nursery is to be watered twice a week or more often according to necessity during rainless periods. Peries and Everard (1995) noticed that the vigour of seedlings was enhanced by a good soil water supply during the early nursery stages combined with a higher level of solar radiation in the nursery site. Wuidart (1981b) suggested the following quantities of water every other day: 0–2 months 8 mm, 2–4 months 10 mm, 4–6 months 12 mm and > 6 months 15 mm. From 6 months onwards, the requirement will be about 75 m^3 of water day^{-1} ha^{-1} so that the hourly discharge of water should be about 10 m^3 ha^{-1}. The traditional method practised by farmers in some parts of India is to place earthen pots of 20 l capacity at a distance of 75 cm on either side of the seedling and filling the pots periodically with water, to provide sufficient moisture for establishment and vigorous growth of seedlings.

A modified system of raising seedlings is followed in some of the seed garden as follows:

Germination beds are used to obtain rapid germination and healthy, well-formed sprouts (Wuidart 1981a). The germinated ones are transferred to the nursey periodically. These beds should be as close to the nursery beds as possible, to reduce transport and the risk of drying out during transplanting. Placing nuts in a germination bed before planting out in the nursery has several advantages. In the germination bed, nuts can be planted much closer than in the nursery bed, saving space, water and labour. The soil should be friable and well-drained to facilitate the lifting of the seedlings for transplantation to the nursery and to reduce the risk of termite attack. Before placing the nuts in the germination bed, the soil should be weeded thoroughly. Spacing could be 5 cm between nuts in the row and about 20 cm between rows, using a density of about 16 seed nuts m^2. There should be more than 80% germination in 5 months with optimum management, and seed nuts that do not germinate by such time are to be discarded as failures (Harries 1983). The percentage recovery can be fixed at the sprouting stage itself, and all the ungerminated seed nuts can be disposed of (Pillai 1994).

Seedling Selection in the Nursery When the environmental conditions are favourable, seed nuts of tall cultivars commence to germinate 11 to 12 weeks after sowing. The percentage of germination reaches the maximum between the 17th and 18th week and then commences to decline. In India, nuts which do not germinate within 20 weeks from the date of sowing and dead sprouts are removed from the nursery.

The period between sowing and sprouting depends very much on the variety and may range from 3 to 6 months, the dwarf nuts sprouting earlier than those from tall variety. Early germinating and sprouting of seed nuts are related to high leaf production, early flowering and high yield. Rigorous roguing out of inferior seedlings in the nursery is as important as the selection of mother palms and seed nuts. Liyanage (1955), while analysing the results of a field trial in Sri Lanka to study the comparative effects of selection of seedlings, came to the conclusion that selection of seedlings alone will increase the crop yield by 10%. Later on Liyanage and Abeyawardena (1957) opined that in coconut, seedling vigour is highly correlated with adult palm characters such as early flowering, nut yield and copra production.

Fernando et al. (1993) studied the variation in seedling characters of 3 different coconut cultivars and their use in identification in the nursery. In general, the colour of the petiole and vigour of the seedlings can be used as selection criteria for dwarfs and hybrids. Seedlings of tall varieties usually grow tall with long leaves and long and broad leaflets showing late leaf splitting. The percentage of seedlings to be selected for planting differs depending on whether heterogeneous or homogeneous nuts are used for sowing. Where a comparatively homogeneous planting material is used, 65–70% good quality seedlings can be expected from a well-managed nursery (Nampoothiri et al. 2000). Details on selection of quality seedlings are given in Chap. 4.

7.2.6 Poly Bag Nursery for Superior Seedlings

Production of coconut seedlings in poly bags with suitable potting media was first introduced in 1969 at Côte d'Ivoire to produce vigorous seedlings with the advantage of early production (Wuidart 1981c). The method has many advantages such as facility for applying homogeneous and regular fertilizer doses leading to better seedling growth, easiness in handling the plants and providing seedlings with intact root system without damage for field planting (Foale 1968a; Chang 1978; Harries 1983; Srinivasa Reddy et al. 1996). The improved water-holding capacity of the potting medium would also help to maintain required moisture for early germination.

To derive the best benefits from the poly bag system, a pre-nursery is essential, wherein seed nuts are sown very closely and transplanted in poly bags when the sprouts are 8–10 cm long. The germinated nuts are taken out at weekly intervals until about 80% of the nuts are germinated or up to 5 months from sowing whichever is earlier. The germinated nut is placed in the half-filled bags with the sprout planted vertically in the centre of the bag and enough potting mixture is added to fill the bag up to two-third portion. Care must be taken to see that the collar of the young seedling is not covered by the potting medium.

The poly bag also facilitates prolonging the nursery period until the environment is conducive for field planting although no particular advantage has been obtained by planting older seedlings (Ho et al. 1978). In studies with different methods of raising seedlings in coconut nursery at the ICAR-CPCRI by Srinivasa Reddy (2000), better growth with higher number of roots and leaf splitting was observed in poly bags (14.5 roots and 9.1 split leaves per seedling, respectively) compared to the conventional bed nursery system (9.5 roots and 4.6 split leaves, respectively). The above-ground dry matter production of 12-month-old seedlings was also high (151.6 g seedling^{-1}) compared to 110 g seedling^{-1} in the case of bed seedlings. Unless seedlings are raised in close proximity to the planting site, the cost of transportation will be high. The extra labour and material costs will also be major constraints in production of poly bag seedlings. It is possible to accommodate 25,000 seedlings in a 1 hectare nursery area spaced at 60 cm × 60 cm in triangles.

Potting Media Preparation and Sowing of Seed Nuts Poly bags used for raising seedlings are made of 0.2 mm-thick black polyethylene (500 gauge), which are resistant to ultraviolet rays. Sufficient holes should be provided to drain excess water applied in the bags, but without the risk of the soil drying out. The poly bag usually takes about 16–18 kg of topsoil (Wuidart 1981c). On coastal clays, topsoil mixed with sand in the ratio of 3:1 is a suitable mixture for the poly bags (Ho et al. 1978). Researchers in the Philippines (Cano et al. 1989) have recommended the use of loose friable soil for potting coconut seedlings. The general potting mixture for raising coconut seedlings in poly bags in Sri Lanka is a 1:2:3 mixture of topsoil, cow dung and coir dust (Peries and Everard 1991). However, subsequent study by Perera et al. (1996) indicated that a mixture containing river sand, cow dung and coir dust (3:2:1) is the best. Among the materials which could be suitable for potting mixture, coir dust appears to be the best, considering the weight of roots, root anchorage and easiness in refilling the bags during different stages of seedling growth (Bandaranayake et al. (1997).

The study at ICAR-CPCRI, Kasaragod, India, indicated that sowing in potting mixture medium containing 1:1:1 mixture of red earth, cow dung and sand either in poly bag or cement tank was beneficial in producing vigorous seedlings (Srinivasa Reddy 1998). Since the potting mixture is not only costly, but its availability is also limited, the study on alternative media indicated that sand + vermicompost mixture in 1:1 proportion and sand + P + K + biofertilizer were similar in response to potting mixture media in terms of seedling growth, physiological parameters and final seedling vigour (Srinivasa Reddy et al. 2001). The seedlings raised in poly bags responded to salt at low rates of about 2 g as evidenced from increased girth, height and leaf number (Remison and Iremiren 1990). The plants tolerated higher levels of salt, but there was no significant increase in any of the growth parameters.

Laying Out of Poly Bag Nursery With the increasing size of the seedlings, spacing between bags is also to be increased. In a trial conducted in Nigeria, Iremiren (1986) found no difference in growth parameters between seedlings raised in poly bags and seedlings raised in the soil. In this case poly bag seedlings may not have shown any advantage, because the size of the bags was only 40 cm × 40 cm. Growth parameters were also similar at spacing treatments of 30 cm × 30 cm, 45 cm × 45 cm, 60 cm × 60 cm and 75 cm × 75 cm, except for height and leaf area index, which decreased significantly with wider spacing. The size of the nursery bed can be 3 m × 6 m with about 1.5 m spacing between beds. Each bed can accommodate 115 poly bag seedlings arranged in a triangular fashion with 60 cm between bags. Wuidart (1981b) recommended, as a general rule, spacing of 60 cm × 60 cm (up to 6 months), 80 cm × 80 cm (6–9 months) and 100 cm × 100 cm (9–12 months).

The seedlings removed from the nursery should be planted in the main field as early as possible, preferably within 10 days (Aiyadurai 1954). Seedlings in poly bags can be retained in the nursery for longer compared to seedlings planted in the bed nursery. Pruning of roots is not harmful in younger seedlings which are 7–12 months old, but may cause some delay in establishment and retard growth in older ones (Patel 1938).

7.2.7 Age of Seedlings for Transplanting in the field

Menon and Pandalai (1960) recommended that as a general rule, under normal conditions, seedlings of 9–18 months of age can be considered suitable for field planting in many places. At this stage, they will have about 60% of the kernel till left in the nuts as an available food reserve for the growing seedling, and this will facilitate their quicker establishment when transplanted out in the field. These seedlings do not wither quickly and being light in weight, can also be transported easily to other places. The optimum time for transplanting seedlings from the nursery to the field in Solomon Islands is 4–5 months after germination or when the seedlings have produced a total of 3–5 leaves (Anon 1967). In Fiji Islands also, seedlings with 3–4 leaves in the nursery are generally selected for transplanting in the field (Satyabalan 1976). Early transplanting of coconut seedlings from the nursery to the field has been recommended in almost all the coconut-growing countries to avoid the shock sustained during the removal of older seedlings from the nursery (Foale 1968c; Jenkin and Foale 1968; Sumbak 1968; Sumbak 1970; Anon 1977; Satyabalan 1983; Romney et al. 1968).

Thampan (1981) reviewed the different practices and norms adopted by the major coconut-growing countries and concluded that 6–9-month-old seedlings should be preferred for transplanting under normal conditions, since at this stage, the seedlings would have developed 3–5 fully opened leaves and a few roots with adequate availability of stored food in the nut. Satyabalan and Mathew (1983) and Satyabalan (1984) reported high and positive correlation of growth characters like collar girth and leaf production of the seedlings from the fifth month of growth of seedlings with those of the later months. This finding enabled them to identify palms of superior genetic value based on the growth characters of the progeny even from the fifth month. Remison (1987) recommended that seedlings should be transplanted after 9 months in the nursery in Nigeria. In India, for practical considerations, it is found to be ideal to plant 1-year-old seedlings, especially as it coincides with the monsoon. The coconut growers in Karnataka (India) prefer 3-year-old seedlings commonly known as *geppe* for planting in the main field rather than 9–12-month-old *molake* (Hanumanthappa et al. 2000). The steps involved for raising *geppe* seedlings include transplantation of 12-month-old seedlings from primary nursery to secondary nursery, where it is planted at a spacing of 1 m in trenches of 0.1 m^2. Farmers' preference to *geppe* seedlings is due to their better establishment and early bearing.

7.3 Field Planting

Adequate attention needs to be paid in land preparation like clearing, leveling, terracing and fencing before planting is taken up.

7.3.1 Field Planting Systems

The spacing to be adopted for planting coconut will depend upon the planting system, soil type and the type of cropping. Triangular, square, single hedge and double hedge are the common systems that could be followed. The hedge row system is based on narrow spacing between palms in the row and large spacing between rows. In single hedge system, the spacing could be 6.5 m within the row and 9 m between rows, whereas the double hedge row system is based on alternation of a narrow spacing (6.5 m) between two rows with a very wide spacing (9 m) between pair of rows.

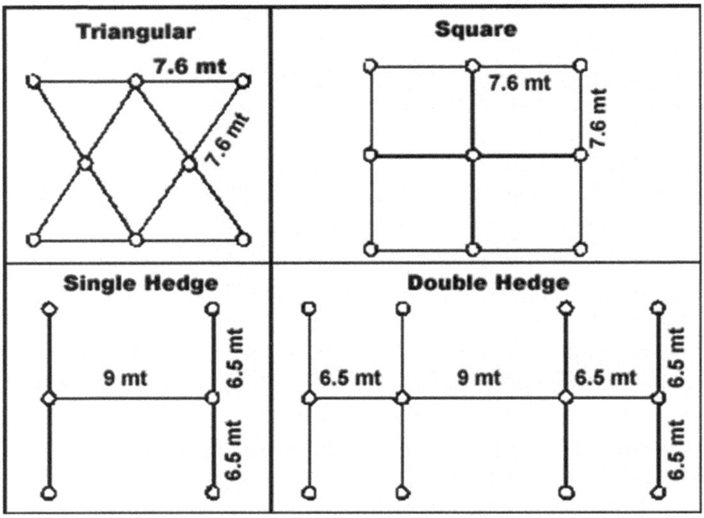

The planting distance is generally in the range of 7 m × 7 m to 10 m × 10 m for tall cultivars, with an average planting density of around 130–180 palms ha^{-1}. A spacing as close as 5.5 m × 5.5 m is recommended for dwarf palms giving a density of 400 palms ha^{-1} depending on the planting system adopted (Friend 1990). Adoption of hedge (rectangular) system of planting coconut with wider row spacing and rows oriented in east-west direction would increase light availability and would facilitate growing annuals and perennials right from the time of planting coconuts. Triangular system of planting (equilateral triangle) accommodates 30 palms more than square system of planting. Multiplication of the number of palms obtained by square planting (at the same planting distance) by a factor 1.155 gives the approximate number of palms, which can be planted in triangular system (Abeyawardana 1954).

Whitehead and Smith (1968) in Jamaica, using Jamaica Tall variety on square system of planting at 6.6 m, 7.5 m, 9.0 m and 10.5 m, observed increase in nut yield palm^{-1} from the closest (135 palms ha^{-1}) to the widest (85 palms ha^{-1}) spacing. Coomans (1974), from an experiment in Côte d'Ivoire with West African Tall coconuts planted at 26 densities ranging from 123 to 278 palms ha^{-1}, noticed effects of competition at higher densities, about 2 years before the palms came into bearing; leaf production was greater and flowering was earlier at the lower densities. The nut and copra yields palm^{-1} decreased with increasing palm density. In these conditions, the optimal plantation density was 150 palms ha^{-1}. Comparison of the yield of West Coast Tall coconut palms planted at 6.6 m, 7.5 m and 9.0 m in triangular method to have densities of 150,200 and 250 palms ha^{-1} in Kerala, India revealed that the yield was the highest from plots with 250 palms ha^{-1} (Kannan et al. 1977). In Sri Lanka, significant differences were observed by Manthriratna and Abeywardena (1979) in the mean yield palm^{-1} for both between-row spacings and within-row spacings. They opined that as far as yield per unit area is concerned, the plant density matters rather than the 'rectangularity' of the system of planting. The number of nuts palm^{-1} decreased with increasing density with 83 nuts palm^{-1} at a density of 128 palms ha^{-1} to 54 nuts palm^{-1} at a density of 239 palms ha^{-1}. From a long-term trial on spacing and manuring (6.1 m × 6.1 m, 7.6 m × 7.6 m and 9.1 m × 9.1 m) with normal and double the recommended dose of NPK fertilizers in coconut palms (East Coast Tall) in Tamil Nadu, India, Louis et al. (1981) reported that though the growth characters were not affected by various spacings tried, there was forced vertical growth in the closest spacing tried. Though this also resulted in reduction of yield palm^{-1}, the yield ha^{-1} was increased due to the compensating effect of number of palms ha^{-1}. Studies by Fernando and Bandaranayake (1996) on the effect of planting density (128–239 palms ha^{-1}) on coconut yield in Sri Lanka indicated that nut yield palm^{-1} decreased with increasing density, but the yield maximized (2.26 tonnes ha^{-1}) at a density of 171 palms ha^{-1} and decreased beyond this. They concluded that density ranging from 171 to 179 palms ha^{-1} is the optimum for planting coconut in dry-intermediate zone.

Wickramaratne (1987) suggested that a rectangular system is the best for intercropping, whereas triangular system is ideal for coconut monocropping. The Coconut Research Institute of Sri Lanka (Anon 1987) recommended planting densities according to agroclimatic zone (and rainfall) with higher densities in the dry zone where intercropping is not recommended. Liyanage and Dassanayake (1993) suggested avenue planting system (e.g. 10 m × 5 m, 12 m × 5 m or 15 m × 5 m) with wider rows in east-west direction to facilitate better light penetration. In Indonesia, Darwis (1988) recommended new planting at 5 m × 12 m to have population of 160 palms ha^{-1} and wide interrow areas for cultivating other crops. Darwis (1990) indicated that the direction of coconut rows from east to west enables optimal use of sunlight, thereby minimizing shading of the intercrops.

Ovasuru (1988) reported that in Papua New Guinea, while coconuts were traditionally spaced at 7 m × 7 m to 9 m × 9 m, some new plantings have been taken up with wider spacing at 12 m × 12 m to accommodate intercrops such as cocoa. Intercropping experiments on cocoa with coconut conducted at Thailand, Dootson et al. (1987) indicated that triangular planting at 8.5 m may be a compromised option between the optimum for monoculture and the lower density appropriate for intercropping cocoa. In Tonga, traditionally coconut palms were planted on a 10 m × 10 m grid pattern (Thompson 1988). Some farmers adopt spacings of 5 m × 15 m to allow more light into the interiors to enable more crops to be grown beneath the coconuts. Opio (1990) discussed about 'hedge planting' which is based on wider spacing between the rows and closer spacing between the palms. Under such a system, though there is no significant difference in plant population per unit area compared to the conventional spacing of 9 m × 9 m, yields from hedge planting are generally higher, sometimes by as much as 25%. In Indonesia, the distance between the rows is normally much more than the distance between the palms in each row, creating adequate space for intercropping. This system could be developed with the same palm population as the traditional approach but with an option for intercropping. Research done by the Indonesian Coconut and Other Palmae Research Institute (ICOPRI) has shown that the combination of 6 m by 16 m (within and between rows, respectively) is an appropriate arrangement to allow successful intercropping (Novarianto and Warokka 2006).

7.3.2 Size of Pits and Planting of Seedlings

The type of soil decides the depth of pits to be taken for plating. In areas with laterite soil and rocky substratum, deeper and wider pits (1.2 m × 1.2 m × 1.2 m) are to be dug and filled up with topsoil, powdered cow dung and ash up to a depth of 60 cm before planting. Addition of 2 kg of common salt helps to loosen the soil in such areas. Arranging two layers of coconut husk, with concave surface facing up at the bottom of the pit before it is filled up, helps to conserve soil moisture in the pits. For planting the seedling, take a small pit at the centre of the filled pits (up to 3/4 depth of pit), keep the seedlings, apply fertile topsoil around it, press firmly and apply water around the seedling. Nelliat (1968) found that irrigation with 45 l of water once in 4 days combined with application of 0.15 m^3 of red earth in planting pits prior to planting in littoral sandy soil resulted in quick and vigorous growth of young palms. In loamy soils with low water table, planting in pits (1.0 m × 1.0 m × 1.0 m) filled up to 50 cm depth is generally recommended. However, in places where the water table is high, planting at the surface or even on mounds becomes necessary. Even under such cases, digging pits and filling have to be done. If planting is taken up in littoral sandy soil, application of 0.15 m^3 of red earth helps to improve the physical characteristics of soil. In areas subject to water logging, proper drainage is to be provided by making drainage channels. In Kuttanad region of Kerala, India, where below sea-level cultivation is done, planting of coconut seedlings is carried out on raised bunds.

7.3.3 Time of Planting

Planting the seedlings with the onset of pre-monsoon rains is ideal. In Kerala (India) and Sri Lanka, planting is undertaken in May–June with the onset of south-west monsoon. In places where assured irrigation is available, planting can be done at least a month before the monsoon sets in to allow seedlings to establish well before the onset of heavy rains. In areas liable to water logging, planting is taken up towards the end of monsoon. In Tamil Nadu and Andhra Pradesh (India), planting is taken up during north-east monsoon in October–November.

7.4 Care of Young Plantation

Availability of sunlight is one of the cardinal principles involved in farming and coconut is no exception. For enhancing the growth and photosynthetic efficiency of palms, which leads to early flowering, the seedlings should be planted in the open space. Ample sunlight, moisture and well-aerated soil are prerequisites for good establishment of coconut palms. The seedlings, after field planting, are to be protected from heavy wind by staking and from sunlight by proper shading. The palm base is to be kept mulched using coconut husk, dry leaves, etc. during the dry period to conserve soil moisture. Irrigation during summer months is necessary. In case soil happens to be covering the collar region and leaf axils, its frequent removal is necessary. The pits are to be widened every year before the application of manure and gradually filled up as the young plants grow. Necessary remedial measures should be taken up against diseases and pests as and when required. Please refer to Chaps. 10 and 12 for details.

7.5 Management of Adult Palms

The management practices to improve the productivity and to achieve sustainable production from coconut gardens include fertilizer application, green manuring/cover cropping, soil and water conservation measures, weed management, irrigation/fertigation, cropping/farming system and organic farming practices.

7.5.1 Fertilizer Application

Coconut is a crop that exports nutrients to the above-ground parts continuously from a limited volume of soil throughout its existence. The production phenology of coconut is such that once the palm starts flowering, it continues for decades (producing a spadix in the axil of each leaf every month), and therefore, the yield in coconut depends largely on the number of leaves produced per year. During these years, considerable quantity of nutrients is removed from the soil. It is, therefore, essential that a nutritionally rich environment is provided in the root zone of coconut all round the year to realize adequate yields. Soil test-based nutrient application will ensure judicious supply of nutrients.

Application of fertilizer would be necessary 3 months after transplanting (Sumbak 1970). ICAR-CPCRI has recommended that one-tenth of the dosage recommended for adult palms should be applied after 3 months of planting of seedlings in the main field, 1/3 after 1 year of growth, 2/3 after 2 years and full dosage from fourth year onwards. The general recommendation from the institute for fertilizing the adult palms is 500 g N, 320 g P_20_5 and 1200 g K_2 O palm^{-1} year^{-1} (Nelliat 1972).

Under average management of coconut gardens, a minimum of 340 g N, 170 g P_2O_5 and 680 g K_2O palm^{-1} year^{-1} is to be given (Anon 1989). Fertilizers are to be applied within a radius of 1.8 m and forked in immediately. In the case of rainfed coconut cultivation, application of fertilizers, in general, is to be made in two splits every year, 1/3 of the recommended dose after the pre-monsoon showers and the remaining 2/3 applied in September when the south-west monsoon rains recede. However, in irrigated areas, application in more split doses is better.

Nutrient indexing of West Coast Tall (WCT), Chowghat Orange Dwarf (COD) x WCT and WCT x COD palms indicated that hybrid palms do not require higher N, P and K inputs for higher productivity. Studies on rationalization of P application to coconut palms indicated that if the soil test value is >20 ppm P, its application could be skipped (Khan et al. 1992). Application of magnesium at 500 g palm^{-1} is advantageous for root (wilt) disease-affected palms to improve the palm vigour (Cecil 1991). Boron deficiency causes characteristic malformation of leaves like hook leaves, nut cracking, drying of the female flowers and a number of other symptoms. Soil application of borax at 50 g palm^{-1} twice at monthly intervals after appearance of the first symptom corrects the deficiency (Baranwal et al. 1989). Please refer to Chap. 8 for details.

7.5.2 Green Manuring/Cover Cropping

Coconut is predominantly grown in sandy soils which often lack adequate quantity of organic matter and other major, secondary and micronutrients essential for growth and yield of coconut. The organic matter status of the soil can be maintained by the addition of green leaves, compost or farmyard manure (FYM). However, in many coconut gardens, farmers are unable to apply the required quantities of compost or FYM. Hence, cultivation of green manure crops in situ and their incorporation which is a very convenient and economic method of enhancing the organic matter status in the soil, is a viable alternative.

Green manuring involves growing of crops, mainly leguminous crops, and incorporation of green biomass when they attain the maximum vegetative growth. Leguminous green manure crops improve the soil structure and N status of the soil. In addition, they help in releasing plant nutrients, reducing leaching, regulating soil temperature and enhancing the activity of soil microflora. The nitrogen-fixing potential of legumes – *Rhizobium* symbiosis – can be exploited in the interspaces of coconut through intercropping of green manures, leguminous cover crops and forage legumes. Among the six green manure crops evaluated by Vijayaraghavan and Ramachandran (1989) in Tamil Nadu, India, *Desmodium tortuosum* was the best which produced 12.8 kg of green leaf matter basin^{-1} followed by cowpea (*Vigna unguiculata*) and sunn hemp (*Crotalaria juncea*). The dry matter production was also found to be superior in *Desmodium* compared to others.

In trials on alley cropping of *Leucaena* as one or two double-row hedges between two rows of coconuts in Western Ghats of India, dry matter and organic N from seven prunings of *L. leucocephala* could meet the requirements for green manure and mulching materials for coconut gardens (Vijayakumar et al. 1986). A series of experiments on *L. leucocephala* conducted in Sri Lanka showed that the highest biomass yield (13.4 t ha^{-1} $year^{-1}$) was obtained in the dry zone on entisols with a pH of 6.0. Application and incorporation of 30 kg coconut $palm^{-1}$ of fresh loppings of *Leucaena* as a green manure around the palm provided the entire N and 20% of P and K requirements of an adult palm (Liyanage et al. 1993a). Application of fresh *Leucaena* loppings in a quarter circle trenches around low-yielding palms on degraded ultisols resulted in 29% increase in nut production and 51% increase in copra yield.

Gliricidia sepium lopping is an ideal green manure for coconut palms that supply significant amounts of N. It is moderately shade tolerant and performs well in acid soils where *Leucaena* does not. A well-established gliricidia intercrop is capable of producing about 8–10 mt ha^{-1} of fresh loppings from three prunings $year^{-1}$. Application of at least 30 kg lopping around each palm can completely replace inorganic N input and about 20% of P and K requirement of coconut (Liyanage 1994). In addition, *G. sepium* can enrich the subsoil through N fixation and mining nutrients from subsoil with its deep root system. *G. sepium* could be successfully grown as an intercrop in the coconut garden in littoral sandy soil also for green manure production (Subramanian et al. 2000). Three rows of *G. sepium* in between two rows of coconut palms with three prunings $year^{-1}$ resulted in the best biomass yield (7970 kg ha^{-1}). Results showed that application of *G. sepium* prunings to the coconut palms could meet a major portion of N (90%) and part of the P and K requirements (25 and 15%, respectively) of coconut palms.

Solangi et al. (2010) studied the usefulness of *G. sepium* in coconut plantations of Pakistan and found that providing about 30 kg of leaves on the surface as mulch provided sufficient N, P and K for coconut. Ilangamudali et al. (2014) reported accumulation of soil organic matter due to incorporation of *G. sepium.* Higher total N, exchangeable K and Mg as well as higher soil microbial activity observed under such cases will be helpful to improve soil fertility of degraded coconut lands in intermediate and dry zones of Sri Lanka. Lekadou et al. (2012) studied the effect of spatial arrangement and population density of *Acacia auriculiformis* trees with coconut palms grown in littoral quaternary sands in Côte d'Ivoire. Though the initial growth of coconut palms (up to 3 years) was better in sole coconut gardens, its subsequent growth (5 years after planting) improved when pruning of *Acacia* and application of leaves were done. This also improved the N, P and K status. However, *Acacia* pruning should be done only 3 years after planting to reduce the depressive effects on coconut palms.

In cover cropping a semi-permanent vegetation of leguminous creepers is maintained in the interspaces of perennial tree crops. It helps in preventing soil erosion and weed growth and adds organic matter and acts as thick mulch which in turn improves soil fertility and water-holding capacity. The choice of cover crop depends on climate, soil as well as age of palms (Bourgoing 1990). Among the four leguminous crops (*Atylosia scarabaeoides, Pueraria phaseoloides, Centrosema pubescens* and *Calopogonium mucunoides*) evaluated for their microbial indices and relationships in a 19-year-old coconut plantation in South Andaman Island (India), Dinesh (2004) found that the latter two were better suited as cover crops due to their positive contribution to soil organic carbon, N and microbial activity. From another study, Pandey and Begum (2010) also observed that growing *Pueraria* as cover crop in coconut plantation increased the soil N mineralization rate, mineral N pool and microbial biomass carbon by 37, 46 and 41%, respectively. In areas where P levels are low, in order to get optimum benefits out of *Pueraria phaseoloides* (effective ground cover and increased nodulation for N fixation), it is necessary to apply phosphate fertilizer (Wijebandara 2010).

The technique for utilization of leguminous cover crops as green manures to supply biologically fixed nitrogen and easily decomposable biomass to coconut was standardized (Thomas and Shantaram 1984). It involves cultivation of leguminous creepers such as *Pueraria phaseoloides*, *Mimosa invisa* and *Calopogonium mucunoides* in coconut basins during the monsoon period from June to October and incorporation of the legume biomass in respective basins. During a growth period of 140–150 days, the legumes yielded 15 to 28 kg of biomass and 102 to 197 g of N in the basin of a coconut palm. Legumes such as cowpea, sunn hemp, etc. can also be cultivated in coconut gardens to generate large quantities of biomass for recycling.

7.5.3 Soil and Moisture Conservation

The coconut palm, which exhibits simultaneous vegetative and reproductive phases of growth, requires a regular supply of water to realize and maintain its potential growth and nut production. As it is generally cultivated under rainfed conditions, soil moisture stress during the non-rainy seasons is an acute problem, particularly in sandy or gravelly soils. During extended periods of soil moisture stress, the cells of the absorption zone of coconut roots were found to be inactive by suberization and dehydration, thereby adversely affecting the water and nutrient absorption processes (Vidhana Arachchi 1996a, Vidhana Arachchi et al. 2000). Soil moisture stress is known to affect the growth of young palms, delay initiation of flowering, increase shedding of buttons and immature nut fall as well as reduce the number and size of nuts (Abeywardena 1981).

For reducing surface evaporation and improving water retention under rainfed conditions and reducing ill effects of soil erosion, various conservation methods such as (i) mulching with coconut husk, coir dust, green leaves, dried coconut leaves, etc., (ii) addition of organic manures such as FYM or green leaf manure, (iii) coconut husk burial (effect lasting for 7 years), (iv) inter-cultivation and (v) bunding/terracing are to be adopted.

Mulching is a simple agro-technique of practical significance to reduce soil moisture loss and to create suitable agroclimatic conditions for proper growth of plant roots and soil flora and fauna. Organic materials from plantations having high moisture holding capacity are ideal for spreading in coconut basins before the withdrawal of monsoon when sufficient moisture is available in soil. Mulches decompose over a period of time and add to the soil organic reserves.

Coir dust, which is a major by-product in coconut fibre industry, could be used to maintain and improve the organic matter content of depleted soils (Vidhana Arachchi and Jayasekara 1988). Coir dust, being a spongy material, absorbs ample quantity of water compared to its weight (Liyanage 1988; Vidhana Arachchi and Jayasekara 1988; Vidhana Arachchi and Somasiri 1993; Van Holm 1993). In gravelly soils, 20% increase in nut yield and 15% increase in copra yield were observed as a result of burying coir dust. Vidhana Arachchi and Somasiri (1997) suggested application of 21 tonnes of coir dust ha^{-1} (6.3% or 1:15 coir dust/sand; vol/vol) in sandy soil to improve moisture and nutrient retention capacities and physical properties.

Coconut husk, a biodegradable material and because of its plentiful availability in tropical and subtropical regions, has very good potential as resource for environmentally friendly agricultural purposes. Das et al. (1991) reported 50% increase in yield after incorporating coconut husks as mulch to the basin of coconut palms. Manufacture of soil erosion control materials from coconut husk fibre was introduced in the middle of nineteenth century (Ziegler and Sutherland 1997). Sutherland and Ziegler (2007) reported that natural fibre rolled erosion control systems are applied on bare slopes as they are biodegradable, less costly, environmentally friendly and equally effective in reducing erosion and generally provide a favourable microclimate for biomass production. Utilization of coconut fibre mat or geotextile as buffer zones on bare soil is found to be highly effective in reducing runoff and mitigating soil losses.

Liyanage et al. (1993b) and Abeygunawardena et al. (1995) concluded that soil moisture conservation, using coconut husk and coir dust in lateritic gravel and sandy soils, is an economically viable proposition. Vidhana Arachchi (1996a, b) found that 75–80% of effective roots of adult coconut palm was localized in a depth range of 20 cm to 80 cm. Neutron probe study also corroborated the result indicating that such roots were more responsible for extraction of water from the soil profile. Only about 5% of roots went beyond 100 cm depth. The maximum absorption took place within a distance of 1 m away from the palm, and therefore, the placement of any soil moisture conservation measure should target the effective root zone.

Integration of vegetative (intercropping) and engineering measures was found to be effective for soil and water conservation in coconut plantations located in sloppy areas (Dhanapal et al. 2002). Contour trench (4 m length × 0.5 m width × 0.5 m depth) filled with coconut husk with two lines of pineapple in the interspaces of coconut proved to be the best for soil and moisture conservation in a laterite soil having a slope of 14–16%. This also minimized loss of soil. The inter-cultivation of CO-3 grass (hybrid of Bajra x Napier grass) not only reduced runoff but also produced fodder at 100 tonnes ha^{-1} $year^{-1}$ in eight harvests. The highest increase in coconut yield (162%) was observed in the same treatment where the annual nut yield was increased from 35 nuts $palm^{-1}$ to 93 nuts $palm^{-1}$. Among the various soil and water conservation technologies such as half-moon bund, coconut husk burial, filling trenches with coconut husk, mulching, providing cover crops and catch pits adopted by farmers, mulching coconut basins with leaves, or coir pith was more advantageous compared to other soil and water conservation technologies (Thamban et al. 2014).

Coastal sandy soils suffer from poor retentive capacity for water and nutrients. Besides, they also show excessive infiltration due to the porosity of sand, easy leachability and low inherent fertility status. These problems could be overcome by adopting certain agro-techniques such as preparing trenches or pits in between coconut rows and filling with coconut husk or raw coir pith to 5 cm height (Subramanian et al. 2006).

7.5.4 Irrigation

One of the critical resources required in coconut production is the availability of water. Though most of the coconut-growing regions are endowed with high rainfall, the rainy period is confined to a few months during the monsoon season. The palm experiences moisture stress and drought conditions for varying periods extending up to 7 months in a year. The adverse effect of moisture stress on the productivity of coconut has been well established. Utilization of the available water in the most effective manner by optimizing irrigation schedules and by adopting soil moisture conservation practices and water harvesting techniques assume particular significance in coconut cultivation. According to Prasada Rao (1986), drought produces injuries to leaves of coconut palm and reduces the yield for several months. Proper irrigation management of coconut palms leads to maximum productivity and continuous nut harvest.

In coconut, initiation and differentiation of vegetative and reproductive primordia and enlargement of cells are very sensitive to moisture stress. Severe drought results in drooping of leaves, breaking of petioles and even death of palms. Even in the well-managed fields, drought affects the nut yield up to 30% in the succeeding year. Rao and Vamadevan (1982) have shown that moisture stress period varies between 14 to 15 weeks in southern parts and 18 to 21 weeks in the northern parts of Kerala, India. Yusuf and Varadan (1993) summarized the results of water management studies conducted on coconut by various research workers in India.

7.5.4.1 Water Requirement and Response to Irrigation

In order to maintain the optimum level of water in the plant biomass and to determine the frequency of irrigation, the assessment of water requirement is essential. Crop water requirements are expressed by the rate of evapotranspiration (ET) in mm per day. The evapotranspiration (Eo) together with the crop coefficient (Kc) gives the water requirement of the crop. Joshi et al. (1988) reported that the water requirement of coconut for optimum growth is 20 mm of water at IW/CPE ratio of 1.0. Rajagopalan et al. (1988) also observed the same in the case of young West Coast

Tall x Gangabondam (T x G) coconut hybrids, when the palms were irrigated at 30 mm cumulative pan evaporation (CPE). The evapotranspiration rates of 5-year-old coconut palms (cv. West Coast Tall) grown in an Oxisol on the west coast of Kerala (determined by soil moisture depletion studies and lysimetric measurements) increased from 2.9 mm day^{-1} in December to 5.5 mm day^{-1} in April and decreased to 2.3 mm day^{-1} in June following the onset of monsoon rain (Rao 1989). Jayakumar et al. (1988) measured the consumptive use of water in 6-year-old irrigated palms (*cv*. West Coast Tall; leaf area index 2.4) over a 6-month dry period from November 1986 to May 1987. They reported the consumptive use of water to range from 2.7 to 4.1 mm day^{-1}. The crop coefficient values were 0.54, 0.73, 0.60 and 0.65 by the Penman, Blaney-Criddle, radiation and US Class-A pan methods, respectively. Saseendran and Jayakumar (1988) computed the mean yearly consumptive use of coconut to be 1126 mm (37 1 palm^{-1} day^{-1} for a basin area of 1.2 m^2). The yearly irrigation requirement was estimated to be 4656 1 palm^{-1} spread over the non-monsoon months of December to May. A stepwise multiple linear regression equation fitted to the coconut productivity and monthly rainfall data for Kerala state for the period from 1956–1957 to 1989–1990 indicated that coconut yield increased as a result of summer rains in March and May (Babu et al. 1993). The irrigation requirement of coconut worked out (using the reference evapotranspiration) varied from 1106 l in December to 13,91 l palm^{-1} month^{-1} during May and the total irrigation requirement for 6 months being 7807 l (Rao 1994).

Soil- and climate-based irrigation schedule study for coconut in Kerala, India, indicated that requirement of water varied according to the type of soil (Salam and Mammen 1990). Based on yield trends and irrigation water consumption, irrigation at 50 mm CPE (cumulative pan evaporation) with 50 mm water was suggested as the best schedule for irrigating West Coast Tall coconuts during dry spells in the west coast in Kerala (Jose Mathew et al. 1996). Yields became stable with adequate irrigation showing minimum fluctuation among harvests during different periods of the year. A comparison of different irrigation treatments with mature tall palms growing in a sandy clay loam/clay soil with a low water table (below 5 m throughout the period) indicated that the best results in terms of yield and economic water use were obtained with 50 mm water at 50 mm Epan (Mathew et al. 1993a). The annual irrigation and water requirements, during the non-rainy period, were determined as 538 mm and 1093 mm, respectively. The consumptive use during this period was estimated at 272 mm with irrigation: CPE ratio of 1.02.

In Sri Lanka, Vidhana Arachchi (1998b) formulated criteria for the design of a drip irrigation system for coconuts. Roots at a distance of 0.5 to 1.0 m away from the base of the palm were responsible for most of the water absorption, and the highest moisture extraction was observed at 1 m distance. The maximum flow rate recommended was 30 l ha^{-1} for 2.5 h from each of the four drippers placed equidistant in the circumference of a circle of radius 1 m around the base of the trunk. The evapotranspiration rate of 15-year-old coconut palms (cv. CRIC 60) during the dry period was 2.52 ± 1.12 mm day^{-1}, and therefore, irrigation frequency for coconut grown in drought-prone gravelly soils of the Andigama series (Red Yellow Podzolic) was determined to be 8 days.

Using a soil water balance approach, Azevedo et al. (2006) estimated actual evapotranspiration (ETc) of 6-year-old dwarf green palms grown on sandy soil in Northeast Brazil over a 2-year period. The irrigation treatments were 50, 100 and 150 l palm^{-1} day^{-1} equivalent to 1.0, 2.0 and 3.0 mm day^{-1}, respectively, applied using two sprinklers positioned at 0.8 m apart from each palm, which works out to 2.5, 2.9 and 3.2 mm day^{-1} with cumulative annual totals of 900 mm to 1100 mm at a planting density of 205 palms ha^{-1} (triangular system at 7.5 m). These are equivalent to 120 l to 160 l palm^{-1} day^{-1}. There were no yield differences between treatments in terms of the number of bunches palm^{-1} or the number of fruits bunch^{-1}, but extra irrigation water increased the volume of water nut^{-1} by about 16%. When yield was expressed as the number of nuts ha^{-1}, there was a significant 12% yield loss from applying 1.0 mm day^{-1} compared with 2 mm day^{-1} (equivalent to a reduction in the number of nuts palm^{-1} from 93 to 82). Water-use efficiency (WUE) values decreased with increasing irrigation water level for all productivity parameters.

In a detailed experiment in Vanuatu, Roupsard et al. (2006) monitored water use of Vanuatu Red Dwarf x Vanuatu Tall Hybrid over a 3-year period in a typical coconut plantation, displaying a constant leaf area index (LAI = 3) and a grass understorey. The eddy flux method was used to estimate actual evapotranspiration (ET) from the palms and grass understorey and the sap flow method to measure transpiration (T) from the palms alone. The annual transpiration was 642 mm (ranging monthly between 1.3 and 2.3 mm day^{-1}), amounting to around 68% of E. ET rates varied seasonally between 1.8 and 3.4 mm day^{-1} and ETo (Penman-Monteith) from 2.4 to 5.8 mm day^{-1}. At a density of 144 palms ha^{-1}, these ET values equate to 93 to 160 l palm^{-1} day^{-1}. The crop coefficient Kc values averaged 0.79 and 0.59 in the cool and warm seasons, respectively.

From the studies in the coastal region of Ceara, Brazil, Miranda et al. (2007) derived Kc values for micro-sprinkler irrigated dwarf green coconut palms over a 32-month period, commencing 11 months from planting. Using the water balance approach (based on tensiometers), ETc increased from a minimum of 0.52 mm day^{-1} (25 l palm^{-1} day^{-1}), at the 11th month after planting, to a maximum value of 5.01 mm day^{-1} (244 l palm^{-1}day^{-1}), at the 36th month after planting. Over the same period, ETo (Penman-Monteith) varied between 3 and 6 mm day^{-1}. During the canopy development phase, Kc increased linearly from 0.63 (11 months after planting) to 1.0 (23 months, when the palms were flowering). During the flowering and fruit development stage, the average Kc value was 1.02.

Madurapperuma et al. (2009b) used the 'compensation heat pulse method' (which has been successfully evaluated on palms in Australia by Madurapperuma et al. (2009a)) to measure actual water use of mature palms (20 years old) of two cultivars grown on two contrasting soils in Sri Lanka planted in square system with a spacing of 8.3 m × 8.3 m (145 palms ha^{-1}). Peak rates of water use differed between the two cultivars, reaching 13 to 14 l palm^{-1} h^{-1} for CRIC 60 (a tall x tall hybrid) but only 9 to 10 l palm^{-1} h^{-1} for CRIC 65 (a dwarf x tall hybrid). Total daily water use averaged 120 l palm^{-1}day^{-1} (ranging from 105 to 135 l palm^{-1} day^{-1}) or 1.74 mm day^{-1} for CRIC 60 and 25% less at 90 l palm^{-1} day^{-1} (ranging from 75 to 97 l palm^{-1} day^{-1}) or 1.31 mm day^{-1} for CRIC 65. The mean daily ET rate over the period of measurement was 3.5 mm giving a Kc value of 0.37 to 0.50. Palms growing on the water-retentive soil had larger leaf areas and trunk diameters (and hence more stem water storage) than the corresponding palms grown on the second soil. In Sri Lanka, CRIC 65 is known for its sensitivity to water stress, while CRIC 60 is recognized as being drought tolerant.

Yield responses to irrigation have been recorded in field experiments in Kerala, (India), Sri Lanka and Brazil. Dhanapal et al. (2000a) reported that irrigation increased the root production in coconut palms in India, which is considered important from the point of view of increased WUE as well as better nutrient absorption. More main roots were produced from one-fourth of the basin area in irrigated palms (1149–1212) compared to that in rainfed palms (429).

The irrigation requirement of young palms (from 5 to 7 years; cv. WCT) was reported by Nelliat and Padmaja (1978) in Kerala. The best treatment in terms of yield of nuts and water use efficiency was the application of 40 mm of irrigation water during December to May (IW/CPE ratio of 0.75). In this way, an average total of 680 mm of water was applied in the summer months, yielding a total of 157 nuts palm^{-1} over the 3 years after the palms came to bearing. By comparison, when the IW/CPE ratio was 0.5, the total yield was significantly less, at 126 nuts palm^{-1}. Bhaskaran and Leela (1978) conducted a trial in red sandy loam soil for 12 years with cv. WCT. Four categories of palms, based on yield per palm (poor, < 20 nuts; low, 20 to 40 nuts; medium, 40 to 60 nuts; and high, 60 to 80 nuts), were selected for the study. Water was applied at 800 l palm^{-1} once in a week (equivalent to 2 mm day^{-1}) in 2 m radius basins during the summer months (December to May). The maximum yield increase of 25.9 nuts in transition period (initial 3 years of the study) was recorded by 'low yield group' closely followed by 'medium yield group' recording an increase of 23.4 nuts. 'High yield group' recorded comparatively low increase in yield (12 nuts). This yield increase was mainly attributed to high setting and high female flower production as nuts from the newly formed bunches will be ready for harvest only after 3 years. Further assessment of 8 years' yield revealed an annual yield increase of 9, 13, 8 and 12 nuts for 'poor', 'low', 'medium' and 'high' yield groups, respectively. The average increase over 11 years from the commencement of irrigation was also maximum in 'low yield group' (38.3 nuts) followed by 'medium group' (32 nuts).

Shanthamalliah et al. (1978) reported that irrigation at 80–100% or 60–100% of available soil moisture, providing 15 cm-thick coir dust mulch, resulted in maximum number of leaves and increased girth of stem. The total water requirement was 1591 mm and 1533 mm year^{-1} for maintaining 80–100% of field capacity and 60–100% of field capacity, respectively. Mulching with coir dust reduced water requirement by about 40–55%.

Nair et al. (1988) reported the results of an irrigation trial for cv.WCT in which water was applied at 500 l palm^{-1} in a sandy clay loam soil at different intervals during the summer months (December to May) over a 5-year period. Compared to the control (rainfed), significant increase in yield was obtained from the third year on irrigating palms when the CPE totalled 50 mm. The B/C ratio was the highest (1.50) in drip irrigation at 100% Eo followed by 66% Eo (1.42), and the water saving of the latter was 34% over the former and 43% over basin irrigation (IW/CPE ratio of 1). The annual yield increase of 30–40 nuts palm^{-1} was reported from mature coconut palms (cv. WCT) over a 6-year period as a result of irrigation (Naresh Kumar and Kasturi Bai 2009). In Aliyar Nagar, India, basin irrigation at 4 cm depth and drip irrigation at 100% of Eo generally gave the highest values for total leaf production, number of spadices, number of female flowers and higher nut yields in hybrid coconut than the other irrigation treatments (Venkitaswamy et al. 1997). Studies on the effect of mulches and irrigation on young coconut plants in coastal Karnataka, India, indicated better growth due to drip irrigation and coir pith mulch (Uthaiah et al. 1993). Hybrids (D xT) planted in a dry climate with supplementary irrigation and NPK fertilizers gave copra yields of 4.1–4.3 tonnes ha^{-1} year^{-1} during 8–16 years compared to 3.4–3.6 tonnes ha^{-1} year^{-1}, when management input was lower (Daniel et al. 1991).

Irrigation with 20 l of water applied twice a week or two earthenware pots buried on either side of the plant and filling them twice a week (27 l week^{-1}) resulted in better establishment and vigorous growth of the young seedlings (cv. CRIC 60) in the dry zone of Sri Lanka (Liyanage and Mathes 1989). From a 3-year trial with 3 coconut cultivars, viz. Malayan Yellow Dwarf, Malayan Green Dwarf and Malayan Red Dwarf (13–16-year-old palms), it was reported that irrigation on alternate days was highly effective to enhance the yield compared to irrigation at fortnightly intervals (Louis et al. 1980).

More than four decades of research in Sri Lanka has indicated that the cv. CRIC 65 was capable of producing a sustained higher yield than cv. CRIC 60 in the absence of adverse soil water deficit (Peries 1995). The cumulative yields of nuts and copra were 50% higher in CRIC 65 than in CRIC 60 over a period of 32 years. Development, precocity and production of dwarf coconut palms varied under different irrigation frequencies varying from 6 to 28 l palm^{-1} day^{-1} (Miranda et al. 1999).

7.5.4.2 Methods of Irrigation

Irrigation methods commonly adopted in coconut gardens are flooding, basin irrigation, sprinkler or perfo spray and drip irrigation.

Flood Irrigation In this method of irrigation, water is allowed to flood the land surface under coconut. Flood irrigation was the most common method practised before the micro systems of irrigation were introduced. However, this method of irrigation is still in use only in some of the coconut-growing areas. There are many problems associated with this type of irrigation such as (1) large quantity of water required for meeting the water requirement of the palms; (2) deep percolation of water and leaching of the nutrients away from the root zone, especially in areas with highly porous soil; (3) difficulty in farm operations especially if the soil is too heavy as in clay soil; (4) non-uniform wetting; (5) increased cost for electricity; and (6) requirement of special land preparation like land leveling and provision of irrigation channels. The quantity of water required will be around 0.75 l to 1.00 lakh l day^{-1} ha^{-1}

which is generally given only once in a week or 10 days. In littoral sand, significant increase in functional leaves and yield were recorded under flood irrigation (Bhaskaran and Leela 1978), with an estimated cost-benefit ratio of 1:3.

Basin Irrigation In basin irrigation, water is applied to the basins of 1.8 to 2.0 m radius around the trunk which is the active root zone of coconut. Irrigation channels are provided in between two rows of coconut, and each basin is connected with the channel. In this method, there will be loss of water due to deep percolation, seepage and evaporation. This loss can be reduced by irrigating the basin through hose pipes. However, there is no control over the flow of water. The main advantages are (1) saving in quantity of water compared to flood irrigation as only limited area is irrigated and (2) application of more uniform quantity of water. Irrigation with 200 l of water once in 4 days in the basin of coconut grown in red sandy loam soil has been recommended for adult palms (Nelliat and Padmaja 1978). Nair et al. (1988) reported that irrigating coconut with 500 l of water in basins at cumulative pan evaporation (CPE) value of 50 mm at an interval of 12 days was economical. A 7-year study with basin irrigation during dry periods at 373 l palm^{-1} weekly or fortnightly, or at 745 l palm^{-1} fortnightly, indicated that total copra yield increased from 1.59 tonnes ha^{-1} to 2.44 tonnes ha^{-1} by weekly applications in drier years and from 2.28 tonnes ha^{-1}to 2.87 tonnes ha^{-1} in wetter years (Abeywardena 1981). The disadvantages of this system are the following: (1) more water is required compared to drip irrigation; (2) there is a chance of deep percolation and wastage of water, especially in porous sandy soil; and (3) there is increased weed growth in the basin area.

Sprinkler or Perfo Irrigation Sprinkler irrigation or perfo sprays are most suited for inter or mixed cropping systems where the entire surface requires wetting. Perfo irrigation is a kind of sprinkler irrigation where small holes are formed throughout the pipe through which water is forced out in small sprinkle and wets the gardens. To wet a 1 hectare garden, approximately 200–300 m of aluminium pipes are required.

The advantages are as follows: (1) an ideal microclimate is maintained in the garden, (2) uniform wetting of the soil, (3) intercrops irrigated along with the main crop and (4) best suited for high-density multispecies cropping system. Irrigation efficiency is also high in this method compared to basin and flood irrigation. However, this method requires higher initial investment; sediment in the water may clog the small pores and affect the irrigation; it is difficult to create the required moisture stress for some of the crops, which are grown along with coconut (coffee, pepper), since the system works on high pressure and the pump should work continuously; there is a chance of water being carried away in drift form if the area is open and experiences heavy wind; and there will be more weed growth as the whole garden gets wet.

Liyanage et al. (2008) tested girdle sprinkler system, which is a new approach for irrigation of coconut. It comprises of 1 kg cm^2 inlet pressure with sub-mainline having 20 mm PVC laterals. While it had 94% distribution uniformity, similar system with 16 mm conduit laterals showed 91% distribution uniformity, which could be considered as an excellent system. This system helps to conserve water as well as labour compared to hose irrigation. It also has less clogging problems compared to drip irrigation so that even harvested rain water or water with silt particles could be used to irrigate with just a screen filter.

Drip Irrigation (Trickle Irrigation) Drip irrigation, a micro-irrigation system, is an efficient method of providing irrigation water directly to the root zone of plants facilitating watering close to the crop as per its requirement. The system applies water under pressure at a low rate to keep the soil moisture within the desired range. The system has an overall application efficiency of 90% as compared to 25–30% for surface irrigation. It is ideally suited for widely spaced crops like coconut as it saves water, energy and labour and WUE is high. It can be adopted in any type of soil especially in very porous soils and land with undulated topography where any other type of irrigation can lead to lot of wastage of water and energy.

The characteristic features of drip irrigation (trickle irrigation) method are as follows: (1) low rate of application, (2) application of water over a long period to meet the water requirement of the crop, (3) application of water at frequent intervals to suit the infiltration rate avoiding wastage of water and (4) application of water near or at the root zone of palms.

A study conducted in Kerala indicated that yield of coconut with drip irrigation at 30 l palm^{-1} day^{-1} during January to May was comparable to basin irrigation at 600 l palm^{-1} week^{-1}, achieving 66% saving of water (Varadan and Chandran 1991). Experiments to evaluate the influence of drip irrigation in comparison with surface (basin) irrigation in Tamil Nadu, India, revealed that irrigation methods had a significant effect on nut yield (Subramanian et al. 1997). Drip irrigation at 40 l palm^{-1} resulted in a saving of 40% of water compared with surface irrigation with nut yields comparable to those produced with surface irrigation. In a low rainfall area in Tamil Nadu, India, the monthly water requirement of a coconut palm irrigated with drip irrigation ranged from 55 l day^{-1} in December to 115 l day^{-1} in June (Kulandaivelu 1990). A trial conducted to evaluate the economic viability of drip irrigation in a full-bearing coconut plantation in Gujarat, India, indicated that it can save 45–50% water over surface irrigation without any significant reduction in yield (Kapadiyal et al. 1998).

In drought-prone gravelly soils of the Andigama series (Red Yellow Podzolic) in Madampe, Sri Lanka, irrigation through four drippers placed equidistant in the circumference of a circle of radius 100 cm around the base of the palm and discharging water at 30 l h^{-1} for 2.5 h wetted a large volume of soil in the effective root zone (Vidhana Arachchi 1998a). Nainanayake et al. (2008) evaluated the response of 20-year-old palms (grown on a shallow (0.6 m) sandy clay loam soil) to drip irrigation over a period of 2–4 years. Irrigation lowered the temperature of the canopy microclimate and the nut surface temperature, thereby reducing the possibility of empty nut formation during dry spells. Irrigation also increased female flower production and reduced immature nut fall. Over a 2-year period, application of 80 l of water palm^{-1} day^{-1} resulted in a 45% yield increase over the control. Reducing the quantity of water applied by 50% (40 l palm^{-1}; 0.65 mm day^{-1}) reduced the yield benefit to 20%.

Drip irrigation along with mulching will be a useful practice with regard to both soil moisture conservation and soil temperature regulation in case of littoral sandy soil. It has been reported from India that the available soil moisture was higher by 22.2 to 28.8% in the drip irrigated basins under mulch compared to drip without mulch. Similarly, there was a reduction in the soil temperature under irrigated, mulched plots by 1.6–1.7 °C compared to un-mulched rainfed plots at 15 cm depth (Maheswarappa et al. 1998b). The mean yield under drip irrigation was 73.3 nuts palm^{-1} under mulching, whereas it was only 58.8 nuts palm^{-1} without mulching indicating the beneficial effect of mulching along with drip irrigation. The pooled data on nut yield for 4 years did not show any significant difference between drip and basin irrigation treatments in littoral sandy soil. The nut yield under different irrigation treatments was on par with each other but was significantly superior to that of rainfed control (26.8 nuts palm^{-1} year^{-1}) (Dhanapal et al. 1998). The benefit/cost ratio in the drip irrigated coconut garden was 2.02 as compared to 1.68 under basin irrigation and 1.07 in rainfed gardens. In addition, the payback period for this investment on the drip irrigation system was only 2 years, thus confirming the economic viability of drip systems for coconut gardens. Through adoption of drip irrigation system, 80 man-days of labour ha^{-1} could be saved compared to the conventional basin irrigation system. Bastine and Palanisami (1998), based on the analysis of irrigation investments in existing and new plantations of coconuts in mixed cropping system, also reported the economic viability of investment in irrigation.

A comparative study of irrigation at 66,100 and 133% of Eo along with basin irrigation at 100% Eo and no irrigation with and without coconut leaf mulch was undertaken with WCT palms in Kerala, India (Dhanapal et al. 2000b). Water was applied during December to May through six emitters placed 0.75 m away from the bole of the palm. The nut yield obtained in drip (at all levels) or basin irrigation was not different, and they suggested adopting drip irrigation at 66% Eo. The quantity of water to be applied at this level will be 27 l palm^{-1} day^{-1} during December to February and 35 l palm^{-1} day^{-1} during March to May. The number of dripping points should be six for sandy soils and four for other soil types (Dhanapal et al. 1999). The rate of water application should be 2–3 l hour^{-1} emitter^{-1} (Fig. 7.1).

More emitters are required for the sandy soil because the horizontal spread is only about 50% of that of the laterite and red sandy loam soil and the wetted volume of active root is only 10.2% for four emitters. The wetted volume with 6 emitters is 15.1% (Dhanapal et al. 2000a). Dhanapal et al. (2003) studied the effect of drip irrigation in COD x WCT hybrid coconut palms in Kerala, India. For drip irrigation, water was applied at 2 l h^{-1} through 4 emitters placed 1 m away from the bole at equidistance with the help of 4 mm LDPE micro tubes, while water was applied in basins of 1.8 m radius under basin method of irrigation. The annual leaf production and leaf nutrient status improved due to irrigation. It was found that drip irrigation equal to 66% Eo was economically efficient with 34% water saving. They recom-

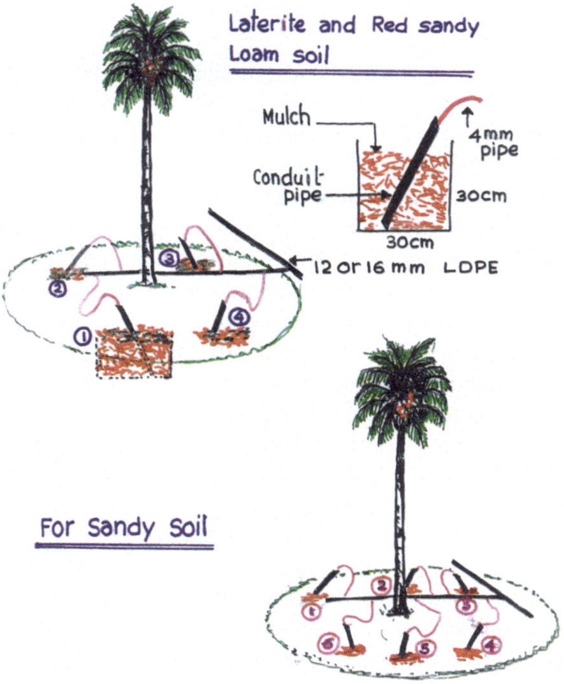

Fig. 7.1 Drip irrigation layout in the coconut basin. (Dhanapal et al. 1999)

mended, under Northern Kerala condition, to irrigate coconut palms through drip method at 27 l water palm^{-1} day^{-1} during December–January and 32 l water palm^{-1} day^{-1} during February–May. Similar results were reported in the case of WCT palms also (Dhanapal et al. 2004).

Factors which generally differentiate the soil water regime for drip irrigation from other irrigation systems are as follows: (i) the flow regime is two- or three-dimensional rather than vertical, (ii) water is added at high frequency and (iii) soil water is maintained within a relatively narrow range. A minimum of 15–20% of the active root zone should be wetted to absorb the water required by the palms. As the effective root zone of coconut palm is confined to 0.75–1.25 m from the bole, it is recommended to place the emitter or micro tubes 1 m away from bole. The quantity of water applied influenced the vertical and horizontal movement of water as well as the volume of root zone wetted in the coconut palm basin. The soil moisture content in various soil layers also was directly proportional to the rate of water application (Dhanapal et al. 1995). The per cent volume of active root zone wetted was 13.6 and 18.2, respectively, in surface and subsurface placement of emitters. The subsurface placement of emitters covered 35% more volume of the basin and maintained higher moisture than surface placed emitters. Thus, evaporation loss can be prevented by allowing water to drip at 30 cm depth by making a pit of 30 cm^3. A conduit pipe of 40 cm may be placed diagonally and the water allowed dripping in that pipe. The pit can be filled with locally available mulch or coir pith. Considering the yield ha^{-1}, cost/benefit ratio and water saving, drip irrigation with 30 l water $palm^{-1}day^{-1}$ during October to January and with 40 l water $palm^{-1}$ day^{-1} during February to May with four drippers placed at distance of 1 m away from the bole was recommended by Nagwekar et al. (2006) for WCT palms grown in sandy soil in Konkan region of India. The productive and economic advantages of drip systems, designed for areas with limited water supply compared to standard drip systems designed for full irrigation to meet potential evapotranspiration in Kerala, have been described by Keller et al. (1992).

7.5.4.3 Irrigation with Saline Water

Pomier and Brunin (1974) in Côte d'Ivoire studied the effect of irrigation with water having salinity (brackish water) equal to half the seawater (ECw = 23 dS m^{-1}) in coconut grown in soil with predominance of coarse sand. No damage was noticed on the palms, and the yield of palms under brackish water irrigation increased by 30%. Nagarajan et al. (1975) used seawater, fresh water and their mixtures in the ratios of 2.5:1 and 1:2.5 for irrigating coconut palms grown in sandy loam soil during summer months (December–May) by pot watering at 45 l of water $palm^{-1}$ $irrigation^{-1}$ twice a week in shallow basins. The palms did not suffer salt injury even though saline water was used for summer irrigation for more than a decade. Although salinity build-up in the soil as a result of these irrigations was much pronounced, highly permeable nature of the soil with low water table, leaching effect of rains (annually 3500 mm) and high salt-tolerant nature of the palms appear to be the favourable factors offsetting injurious effect, if any, of salinity on the palms.

A study on 1-year-old seedlings planted in sandy soil in Maharashtra, India, and irrigated with seawater, seawater + fresh water and fresh water indicated the adverse effect of undiluted seawater on the survival, growth and tertiary root production of seedlings (Patil et al. 1998). Higher levels of salinity had adverse effect on the growth of two coconut cultivars, Nigerian Dwarf Green (NGD) and Nigerian Red Dwarf (NRD) (Amalu 1998). For all parameters, chloride salinity was the most deleterious to growth followed by sulphate. Lower salinities (<4 mmho cm^{-1}) enhanced growth of seedlings, whereas higher concentrations (>12 mmho cm^{-1}) reduced dry matter production up to 50%.

The quality of tender nuts in relation to salinity of irrigation water was studied by Ferreira Neto et al. (2002). The increase in salinity of irrigation water (from 0.1, 5.0, 10.0 and 15.0 dSm^{-1}) increased the total soluble solids (° Brix) and the electrical conductivity of coconut water, especially when irrigated with water EC 15 dS m^{-1}. Increasing the salinity of irrigation water, besides reducing the weight and volume of coconut water, impaired the shape (appearance) of the nut. The chloride and potassium ions were found to be present in higher proportions. It was concluded that irrigating with water up to 10 dS m^{-1} can satisfactorily produce tender nut for marketing.

7.5.5 Fertigation

Fertigation is an efficient method of fertilizer application through drip irrigation system. Drip fertigation increases the fertilizer use efficiency, saves fertilizer cost, reduces labour requirement and supplies nutrients according to the crop demand besides helping in application of fertilizers exactly and uniformly to the wetted root zone (Dhanapal et al. 2005). Water soluble fertilizers like urea, diammonium phosphate (DAP) and muriate of potash (MOP) can be combined and supplied through drip irrigation. The other possibility is to use liquid fertilizers which are highly soluble and therefore do not cause any interference or clogging though it involves high cost. Through this method, fertilizers can be applied in six equal splits from December to May at monthly intervals in areas such as in Kerala.

Miguel et al. (2011) evaluated the effect of various doses of N (urea) and K_2O (potassium chloride) under fertigation in development and the production of Anão Verde coconut palm in Brazil. The number of leaves, canopy diameter, plant height and other parameters were improved with 2910 g palm^{-1} year^{-1} of K_2O and 2353 g palm^{-1} year^{-1} of N producing the highest production in sixth year of the crop. However, during the seventh year, the dose of 1540 and 1539 g palm^{-1} year^{-1} of K_2O and N, respectively, promoted maximum production. Though in the east coast region of Tamil Nadu, India, 100% of recommended dose of fertilizers (RDF) (0.56: 0.32: 1.20 kg N, P_2O_5, K_2O palm^{-1} year^{-1}) was the best for obtaining higher values of yield attributes and nut yield, in addition to effective build-up of soil available NPK as well as economic viability. Thiruvarassan et al. (2016) suggested that the fertilizer application through micro-irrigation may be restricted to 75% of RDF (0. 42: 0.24: 0.90 kg) for obtaining sustainable coconut yields. Similar results were obtained by Basavaraju et al. (2014) in Karnataka, India, indicating the possibility of saving 25% of recommended NPK by adopting fertigation.

For West Coast Tall coconut palms, application of 50% of the recommended dose of fertilizers (0.50:0.32:1.20 kg N, P_2O_5, K_2O palm^{-1} year^{-1}) through drip fertigation was sufficient to produce yield equivalent of 100% of the RDF through soil application (Subramanian et al. 2012a) (Table 7.1). Fertigation resulted in increased availability of soil N, P and K, higher annual leaf production and higher photosynthetic activity and production of more female flowers. However, Khandekar et al. (2016) from Maharashtra reported that application of 100% recommended fertilizer through drip (1.0 kg N + 0.5 kg P_2 O_5 + 1.0 kg K_2O palm^{-1}) in eight splits from October to May and was found suitable to increase yield (113.78 nuts palm^{-1}) with maximum net returns (Rs.108,000 ha^{-1}) for 28-year-old West Coast Tall variety.

Fertigation with polythene mulching increased the productivity in coconut, besides ensuring higher efficiency of water, nutrients and profitability in coconut (Jayakumar et al. 2017). They used different doses of fertilizers (100, 80 and 60% of recommended dose of 0.50:0.32:1.20 kg of N, P_2O_5, K_2O palm^{-1} year^{-1}) with and without polythene mulch (100 μm thickness) along with basin irrigation without mulch in Tamil Nadu, India. All the plant growth characters (plant height, canopy development, etc.) and yield attributes (spathe length, number of inflorescences,

Table 7.1 Coconut yield as influenced by fertigation (palm^{-1} year^{-1})

Treatments	Pre-treatment yield		10 years' mean data	
	No. of female flowers	Nut yield	No. of female flowers	Nut yield
No fertilizer	262	102	226	81
25% NPK (drip)	254	111	280	97
50% NPK (drip)	257	106	306	121
75% NPK (drip)	239	108	294	123
100% NPK (drip)	248	108	326	131
100% NPK (soil)	272	114	276	114

number of bunches palm^{-1} year^{-1}) as well as number of nuts bunch^{-1} were maximum with fertigation using 100% of the RDF with polythene mulching. The number of nuts was also the highest (292 nuts palm^{-1} year^{-1}) in the same treatment. Besides, gradual reduction in weed population, superior water and nutrient use efficiency and high profitability were observed in all the polythene mulch combination treatments when compared with other treatments without mulch.

7.5.6 Weed Management

Coconut is generally planted at a wider spacing resulting in growth of a wide range of annual and perennial weed species. Such weeds compete with coconut for soil moisture and nutrients, thereby affecting the growth and yield of coconut as well as obstructing routine cultural practices. Hence, it is necessary to adopt appropriate weed control measures in coconut plantations.

A variety of weed control methods are available, but Abad (1980) reported that hand weeding is the most common method of weed control in coconut gardens. However, if the area is too large and labour is costly, chemical control may become necessary. Use of kerosene or petrol-operated weed cutters is also common (Dhanapal et al. 2017). Gunathilake (1985) described the management practices including raising of cover crops, intercropping with shade-loving crops, cattle grazing or mulching with coconut fronds, husks or slashed weeds around palms in coconut plantations for weed management in Sri Lanka.

In new coconut plantations, weed control was carried out by establishing cover crops in furrows and spraying with Paracol (paraquat + diuron) at 1.2 kg or diuron 2.4 kg and paraquat 0.55 kg hectare^{-1} between the furrows (Abad 1980). In trials at the Davao Research Center of the Philippine Coconut Authority, winged bean (*Psophocarpus tetragonolobus* L.) was found to be a suitable cover crop in coconut plantations where it suppressed weeds and improved soil fertility (San Juan and Abad 1981). Sanico and Laguna (1989) also found that cover cropping with *P. phaseoloides* was the most effective method in controlling weeds. Rolling between coconut rows with a light-weight wooden roller at the time of cover crop sowing controlled *Imperata* in young coconut plantations (Bourgoing and Boutin 1987). The highest average yield year^{-1} was obtained with clean scraping of the soil surface with inter-cultivation in mature WCT coconuts over a period of 6 years in Kerala (Mathai 1979).

Senarathne et al. (2003) compared different weed management systems and their effects on yield of coconut plantations in Sri Lanka. Controlling weeds with glyphosate at 1.44 kg ai ha^{-1} resulted in a 25% increase in nut yield over the uncontrolled weed plots and was found to be the most effective and economical method of weed control. Senarathne and Gunathilake (2010) evaluated tractor slashing (3 times per year), cover cropping with *P. phaseoloides* and buffalo grazing (once a month) for weed management in mature coconut plantations in Sri Lanka. Though both the latter treatments could effectively reduce the weed biomass, the yield was the highest in the buffalo grazing plots, which also gave the highest BC ratio of 1.86. Later on Senarathne and Perera (2011) reported that cover cropping with *P. phaseoloides* and application of glyphosate at 1.44 kg ai ha^{-1} were very effective resulting in 20% and 23% increase in nut yield over the uncontrolled weed plots, respectively.

7.5.7 Leaf Pruning in Coconut

Coconut leaf pruning (CLP) involves the removal of coconut leaves in the lower whorls of canopy so as to allow adequate sunlight for the normal development and high yield of perennial and annual crops, grown along with coconut. According to Sampson (1923), 6–8 lower leaves of a coconut palm which have past their prime and are of little use to the palm can be removed.

Coconut palms of five age groups, viz. 5, 10, 20, 30 and over 50 years, were studied to determine the effect of leaf growth in relation to age and the relation between leaf size and bearing status of such palms by de Silva and Abeywardena (1970). Leaf growth parameters (leaf length, petiole length and total number of leaflets) of 10-year-old nonbearing coconut palms were more than those of bearing palms of the same age. Marar and Padmanabhan (1970) measured the effects of CLP for a period of 4 years and found that removing old coconut leaves from axils of which mature nuts had been harvested did not harm palm growth or productivity. On the other hand, where all opened leaves on one side of the palms were removed additionally, there was significant decline in average production (45.6 coconuts palm^{-1} year^{-1} compared to 68.6 nuts before treatment). Bailey et al. (1977) recorded significant decline in coconut yield due to increased premature fruit shedding following pruning treatments and concluded that defoliation above 40% has long-term negative effects on the health of palms.

No change in yield was noticed in 55-year-old WCT coconut palms, in which 3–10 of the lower leaves were removed during the 5-month dry season of each year (January to May) for a period of 5 years (Sudhakara et al. 1989). Magat et al. (1994) tried maintaining CLP for a longer period of time and got different results. Although there was once again no decrease in yield during the first year, trees retaining the

13 younger leaves showed a 29% loss in production of nuts in the second year followed by a further 20% reduction in the third year, i.e. a nearly 50% decline in nut production after 3 years. Treatments retaining the 18 youngest leaves caused no significant decline and even showed an improved yield in the third year. Thus, it was concluded that maintaining 18 functional leaves in the crown was sufficient to provide optimum yield in coconut.

According to Dauzat and Eroy (1997), a possible alternative to choosing a lower density for intercropping purposes can be the pruning of the coconuts. Their simulations showed that limiting the frond numbers to 18 in coconut stands at regular densities is quite effective to enhance the light transmission. Pruning seemed to be a very flexible and cost-effective means to modify the light competition in a coconut-based farming system. Aterrado and Abad (1998) pointed out that no changes in yield occurred within the first year of pruning. The effect of pruning was felt a year after implementation when yield for 95% pruning dropped to 54.4 nuts palm^{-1} year^{-1} compared to 25% and 50% prunings with 86.8 and 77.1 nuts palm^{-1} year^{-1}, respectively. Eroy et al. (2001) reported that CLP did not significantly affect the yield in the first year, but nut and copra yield tree^{-1} had been significantly reduced by 21% and 17%, respectively, after 2 years.

In areas with a distinct dry period of 3–6 months with a monthly rainfall of < 1000 mm, leaf pruning during nut harvest before the onset of dry season can minimize the adverse effects of drought on the fruit set (Anon 2000a). In cases where damages of pests occur on lower and older leaves (e.g. attack of *Opisina arenosella* Walker, the black-headed caterpillar), pruning of these leaves would serve as a mechanical control measure. CLP is not recommended for very tall coconut palms due to the excessive transmission of sunlight under coconut and the difficulty usually encountered in pruning tall palms (Anon 2000a).

Padrones et al. (2000) from the Philippines reported that higher number of leaves was produced under pruned coconuts than those under non-pruned coconuts. However, nut and copra production of bearing palms were not significantly affected by leaf pruning. From a 9-year study on coconut leaf pruning (CLP) in the Philippines, involving local tall Laguna variety (19–28 years old, 9 m × 9 m triangular planting) intercropped with selected annual food crops (corn, peanut and sweet potato), Magat et al. (2002) reported reduction in nut and copra yield palm^{-1} with CLP starting at leaf ranks (LR) 19 or 23 and onwards. However, 2.3–2.8% increase was noted in copra weight nut^{-1}. There was better growth and yield (34% and 45% increase, respectively) for sun-loving annual intercrops (corn and peanut) with CLP, whereas the yield was not affected in the case of shade-tolerant sweet potato. Consequently, CLP contributed to higher income, higher net returns and benefit/cost ratio (BCR), thus, offsetting any reduction in coconut yield due to CLP practice eventually achieving optimized total farm productivity and maximum farm profitability.

Canja et al. (2003) carried out pruning of coconut palm from leaf no.19 (maintaining 18 younger leaves in the crown) and found that the CLP did not affect the yield and nutrition of coconut palms. Though lower number of nuts and copra yield palm^{-1} were observed on palms with CLP, there was improvement in copra weight nut^{-1}.

From an 8-year experiment conducted at the Davao centre, the Philippines, Secretaria et al. (2003) reported 20–25% decline in the yield of palms pruned to retain 18 youngest leaves. Based on yield reduction coefficients, they have developed yield predicting models in areas with well-distributed rainfall for local tall and hybrid coconuts under coconut-based farming system at two pruning levels (retaining 19 and 23 leaves).

Rosenfeld (2009) reviewed the research work done on pruning on the health of palms including coconut. He summarized the following points relevant to coconut:

(a) Pruning increased the rate of production of new leaves, but size of new leaves decreased as a result of higher levels of pruning.
(b) Leaf nutrients were not affected much by leaf pruning in healthy palms, but in nutrient-deficient palms, the symptoms became worse.
(c) There were significant negative effects on nut production when fewer than 18 younger leaves were retained.

The studies taken up so far on leaf pruning in coconut indicate that pruning retaining 18 functional leaves on the crown is advisable.

7.5.8 Underplanting/Replanting in Coconut

Coconut palms commence full production at the age of 10–16 years which continues at an increasing rate between 30 and 40 years and thereafter starts to decline (Nelliat et al. 1974). The useful bearing life of tall varieties is estimated to be up to around the age of 60 years. The main factor that leads to low income in coconut production systems in most of the coconut producing countries is the declining productivity of coconut palms due to old age and senility (Mwinjaka et al. 1994; Aguilar and Benard 1993). Replacing such palms is necessary to ensure that production and profitability are maintained so that the future of the industry is safeguarded (Ollivier et al. 2001). Magat (1993) reported that 15% of the palms in the Philippines were senile by 1990 and cautioned that unless replanting measures were promptly taken, there would be a 2% decline per annum in coconut production.

Coconut farmers are generally reluctant to uproot a palm even when it attains senility and becomes less productive, mainly because coconut has a long juvenile phase and therefore, the benefit from coconut is derived after a long time. Replanting, which is the complete removal of old coconut palms before new seedlings are planted, is an agronomically superior method in replacing a senile coconut plantation. This results in early flowering, early bearing and development of uniform plantation compared to the underplanting system (Perera and Fernando 2002).

In the underplanting system, new seedlings are planted along with the old palms, which are removed in stages over a period of 5–6 years. Even though most of the coconut farmers are aware that replanting is superior to underplanting with respect to the performance of the new plantation, the underplanting system is still being practised by the farmers, because the continued income from the old palms is necessary till the underplanted ones start yielding especially in small holdings.

Ollivier et al. (2001) suggested, considering variety, residual density, productivity and condition of the palms as the factors to decide replanting. Farmers who would like to take up underplanting should decide when to replace their palms judiciously. In Tanzania, under very different conditions of low rainfall and sandy soil, Romney (1987) observed that competition of adult tall palms is likely to substantially reduce the performance of the young palms. Sometimes, the old palms tend to show yield improvement due to the care given to the newly planted seedlings, which leads to reluctance of farmers to remove the old palms which is detrimental to the growth of underplanted seedlings.

Pordesimo and Noble (1989), in the Philippines, recommended 'strip replanting' based on a simulation model. Perera and Fernando (2002) suggested that if the plantation is over 60 years and is producing less than 2400 nuts ha^{-1} $year^{-1}$, replanting could be undertaken. The factors such as the height of the old palms that makes harvesting difficult, existing number of vacancies and severity of damage due to drought or long-term negligence are also to be considered along with the age and yield in deciding when to replace old plantation effectively.

In Samoa, Opio (1987) suggested that the yields of mature local tall varieties are not economically viable beyond 40 years, unless they are intercropped with other cash crops. According to Mwinjaka et al. (1999), there is a recommendation to coconut growers in Tanzania to replant their coconut fields when the palms are 66 years old. The gradual replacement of existing coconut palms should be undertaken at planting densities of 15–17 m × 10 m. This density, if maintained, will allow for various intercropping systems to be established. As the tendency is to replace old tall palms with more precocious and higher yielding D x T hybrid varieties, the density should often be higher than in the original design.

In the Philippines, vegetative pith or 'ubod' of coconut is an edible food item, which finds use in many food preparations. A system or strategy of underplanting seedlings among adult stand of palms was developed to provide an alternative source of 'ubod', thus, preventing the indiscriminate cutting of existing productive coconut palms for the purpose (Anon 2000b). This system involves planting of seedlings in the interspaces of full-bearing coconut in a 3 m × 3 m triangular pattern in two rows about 2 m away from the row of coconut-bearing palms. The seedlings are planted in pits either with 1 seedling per pit (742 seedlings ha^{-1}) or 2 seedlings per pit at 60 cm apart (1484 seedlings ha^{-1}). The 'ubod' can be harvested at 3 years from field planting. This system can also be used in replanting old existing palms with some remaining young palms arranged in a 9 m × 9 m triangular planting distance. However, this technology cannot be applied to full-bearing palms with close planting (<8 m), and, therefore, it is limited to tall-bearing palms and not for dwarf varieties with close planting. Padrones et al. (2000) from the Philippines recommended underplanting two young coconuts in each pit taken at 3 m × 3 m distance between spaces of coconut-bearing palms, which according to them is more profitable.

7.6 Coconut-Based Farming System

Very often, the coconut farmers in many parts of the world find difficulties in sustaining their families' livelihoods from the income of coconut alone. One of the best options to overcome this problem is to effectively utilize the space available in coconut gardens for cultivation of other compatible crops. Such a system will also offer considerable scope for increasing productivity per unit area, time and inputs by more efficient utilization of resources like sunlight, soil, water and labour. Production alternatives for intercropping in coconut plantations can take the form of a single intercrop, a mixture of crops or a crop/livestock combination which are compatible with each other and other environmental factors. One of the most common farming systems practised by coconut-growing traditional farmers is the coconut-based farming system (CBFS).

According to Aguilar and Benard (1993), CBFS is the integration of complementary enterprises in coconut farming (e.g. intercropping, livestock, processing integration linked with marketing) to increase productivity unit area^{-1}, increase employment opportunities and provide a buffer against low and fluctuating copra prices. An example of a working definition of CBFS was given by Magat (1999) as: 'a system or practice in coconut production in which the available farm resources like coconut trees, soil and water/rainfall, farm labour, agricultural inputs (seeds, livestock, fertilizers and other agro-chemicals), and farm tools are utilized to produce nuts, food and non-food agricultural produce from the farm, in a productive and profitable way'. In any CBFS, integrated crop management practices of the main as well as intercrops should be followed to achieve optimum productivity, profitability and sustainability of coconut palms and to maximize the total productivity and economic benefits of the system.

Research programmes in the 1960s and 1970s enabled the development of many coconut-based cropping systems to increase the productivity and income from unit area of plantations without decreasing coconut yield. According to Liyanage and Gunathilake (1998), CBFSs have been recognized as a strategy for optimizing the productivity and augmenting the economic viability of coconut lands, particularly in the wet and intermediate agroclimatic zones in Sri Lanka, which comprise of nearly 80% of the land occupied by coconut. The success of cropping system in coconut gardens greatly depends on the canopy architecture and rooting pattern of crops; extent of availability of sunlight in the plantation; selection of crops adaptable to local climatic and soil conditions; arrangement of crops in relation to air space and soil utilization; shade tolerance, growth pattern and duration of the crops planned to be cultivated; irrigation facilities available; sociocultural factors of the farmers; marketing facilities for the products; as well as economic competitiveness of the crops.

7.6.1 *Amenability of Coconut Palm to Cropping Systems*

Coconut palm utilizes the natural resources only to a very limited extent producing less than 10% of the potential for dry matter production in the tropics (Nelliat et al. 1974). Mialet-Serra et al. (2001a) are also of the opinion that in mature coconut plantations, significant quantities of light, water and nutrients may not be captured by the palms and therefore can be utilized by intercrops without reducing main crop yield.

7.6.1.1 Space Availability

In gardens, with monocropping as little as 25% of the land is only effectively used (Ontolan 1988; Darwis and Tarigans 1990; Magat 1990). Though the potential (maximum) annual biological productivity of a cropping system under optimum conditions (Loomis and Williams 1973) is to the extent of 280.5 tonnes ha^{-1}of dry matter (770 kg ha^{-1} day^{-1}), according to Magat (1990), for coconuts, even at high nut yields of 100 nuts palm^{-1} and 200 nuts palm^{-1}, the annual productivity is only 18.70 tonnes ha^{-1} and 35.5 tonnes ha^{-1} of dry matter or 6.6 and 12.6%, respectively, of the potential biological productivity. Due to the unique morphological features (with a terminal crown of leaves, growing to a height of 20–30 m), the coconut palms may occupy less than 30–40% of the available air space between canopy and ground during the major part of their life span of 80–100 years (Bavappa 1975).

7.6.1.2 The Rooting Pattern of Coconut

Though under favourable conditions, as many as 4000–7000 roots are found in the middle-aged coconut palms (Menon and Pandalai 1960), about 74% of the roots produced by a palm under good management do not go beyond the 2 m lateral distance. IAEA (1975) reported the depth and distance for coconut root to be 0.6 m and 4.0 m, respectively, in the Philippines, whereas the corresponding figures were 0.6 m and 3.0 m in Sri Lanka. In Bali, Steel and Humphreys (1974) showed that palm roots were still very dense at 3 m from the trunk and that some laterals occurred. Studies using P isotopes (Anilkumar and Wahid 1988) confirmed that more than 80% of the root activity was confined to an area of 2 m radius around the palm and to the 25–60 cm depth of the soil. Thus, on surface area basis, the area occupied by the palm is 56.25 m^2 (7.5 × 7.5 m); and the area of active root zone is 12.57 m^2 (πr^2 where r = 2 m). Therefore, the fraction of total area effectively utilized by the palm is only 22.24%. Thus, in a pure stand of coconuts at normal planting density and management conditions, about 75% of the total area is not being effectively utilized to the fullest extent by coconut roots.

7.6.1.3 The Availability of Solar Radiation

The venation structure of the coconut crown and the orientation of leaves allow part of the incident solar radiation to pass through the canopy and fall on the ground. According to Nair and Balakrishnan (1976), Liyanage (1985) as well as Liyanage and Martin (1987), on an average some 56% of solar radiation is available for intercrops, although this will vary with age of the coconut stand and planting density. The diffused sunlight facilitates growing a number of shade-tolerant crops in the interspaces. The leaves in a coconut palm crown are not randomly distributed, but clumped around few widely spaced growing points. This non-random distribution will also lead to low extension coefficient of around 0.65 for photosynthetically active radiation (PAR). Light penetration through the canopy is influenced by age, spacing, soil fertility, varietal characteristics, leaf area and time of the day. The amount of light transmitted ranges from 5% in a 5–10-year-old D x T hybrid at a density of 650 palms ha^{-1} to about 90% in a 60–70-year-old plantation at a density of 120 palms ha^{-1} (Reynolds 1995). In CBFS, transmitted radiation especially the PAR regime is very important as it has a bearing on the behaviour and productivity of intercrops. Intercropping experiments under coconuts in the Philippines demonstrated that, in the absence of strong water deficit and with a proper fertilizer supply, the intercrop yields are more or less linearly related to the available PAR (Benard et al. 1996). Thus, optimizing CBFS can be achieved mainly by providing sufficient light for intercropping through the choice of coconut density.

The apparent coverage of ground by canopies of palms of various age groups varies (Nelliat et al. 1974). When the palms are about 8–10 years old, the percentage of sunlight transmitted is only 20, and this remains almost constant till about 25 years. Subsequently, the percentage of light transmission increases progressively, and the canopy coverage of ground decreases inversely. By the time the palms are 40 years old, the light transmission increases to about 50%. Based on the growth habit of the palm and the amount of light transmitted through its canopy, the life span of coconut palm could be divided into three distinct phases from the point of view of intercropping.

A. Up to about 8 years: Good transmission initially; decreasing with age; suitable for growing annuals/biennials
B. Eight to 25 years: Maximum ground coverage and low canopy; poor light availability; not suitable for multiple cropping
C. Above 25 years: Increasing trunk height; reduction in crown size, light transmission increasing with age; ideal for raising annual and/or perennial crops

Classification of some of the crops based on tolerance to sunlight/shade for intercropping is as in Table 7.2.

The major criteria for selection of component crops are:

(1) Shade tolerance of selected crops and amount of solar radiation in the coconut garden
(2) Differential rooting pattern to exploit different soil layers for moisture and nutrients

Table 7.2 Classification of crops based on tolerance to sunlight/shade for intercropping

Sl. no.	Parameter	Effect on production/remarks	Examples of suitable crops
1	Not tolerant to even slight shade	Drastic decline in yield even under slight shade. Cannot be grown successfully as intercrop	Paddy, bhendi, sweet potato, leguminous plants, groundnut
2	Not tolerant to shade	Yield declines with intensity of shade. Can be grown in border areas	Coleus, brinjal, tomato, chilli, dioscorea, Nendran banana
3	Tolerant to shade	Slight reduction in yield due to shade. Can be grown successfully when 50% sunlight is available	Colocasia, cassava, amorphophallus, banana varieties other than Nendran
4	Shade loving	Yield increases to some extent with increase in shade. More ideal as intercrop than sole crop	Ginger, turmeric, arrowroot, Kacholam

(3) Possibility of supplying biomass for recycling
(4) Lesser height than coconut and non-interference with cultural operations of the main crop
(5) Should not be alternate host for any pest or disease
(6) Lesser economic life span than coconut
(7) Adequate marketing and processing facility
(8) Suitability of climatic conditions, soil type, rainfall pattern and irrigation requirements of the crops intended for inclusion in the cropping system

Dauzat (1995) and Dauzat and Eroy (1997) modelled the architecture of the coconut palm and generated virtual coconut stands in which simulations of light transmission was carried out. Subsequently the percentage of light transmission through the coconut canopy depending on the geographical location of the site, the age of the coconut palms and the adopted planting design was calculated, and the most suitable plants for intercropping according to the light environment under coconut were recommended. It was found, by simulations, that the PAR transmission under coconuts at a given age is sensibly a linear function of the tree density, irrespective of their planting pattern. Therefore, it could be possible to adjust the palm density according to PAR requirement of intercrops. Because the light transmission is similar in triangular and square designs, the choice of a coconut planting design may be guided by practical considerations, e.g. ease in cultural practices like cross ploughing in square designs (Dauzat and Eroy 1997).

7.6.2 Relevance of Coconut-Based Farming Systems

The main relevance of CBFS includes the following: social and ecological benefits, conservation of natural resources, biodiversity conservation, supply of biomass and employment generation.

1. Social benefits: Social benefits relate to the food and nutritional functions of coconuts as well as various crops grown under the system. Growing of intercrops in coconut lands produces more food and agricultural products, ensuring food security of the people in rural and urban areas. At the same time, the practice generates jobs and livelihood, enhancing farm incomes and the purchasing power of people and thus alleviating poverty in farming communities (Magat 2004a).
2. Ecological benefits: Compared to the ecological conditions of the long-term monoculturing, those of lands under CBFS are more favourable and stable for intensive and sustainable agricultural production. This is mainly due to the more efficient utilization of various natural resources, higher biomass generation and recycling over a period of time.
3. Conservation of natural resources: The land cover minimizes the direct impact of rainfall and the separation of soil aggregates under coconut environment, which can control surface runoff and soil erosion by 70–90%, compared to bare soil or uncropped condition. An adequate ground cover can also increase rainwater infiltration and storage, eventually increasing water supply of the entire area. Because of the shade under coconut stand and full canopy coverage, evaporative demand is very much reduced, and intercropping allows a better retention of water in the soil for a longer period. The microclimate in the coconut garden maintains not only lower air temperatures (by 4–6 °C beneath the canopy) but also lower soil temperatures.
4. Biodiversity conservation: Due to the uniqueness in the growth pattern, coconut offers scope for accommodation of many crop species for inter/mixed cropping, which in turn helps plant genetic resources conservation and management. The CBFS favours diversity in the soil microflora as well.
5. Supply of biomass for recycling and soil fertility improvement: CBFS enables production of large quantities of biomass which could be effectively recycled and put back in to the system for soil fertility improvement.
6. Employment generation from farm diversification: Since various crops with different growth patterns and durations are included in the CBFS, their cultivation coupled with value addition provides opportunities for more employment generation. Besides, mixed farming also offers year-round employment to the farm family.

The main cropping systems adopted in coconut plantations are the following.

7.6.3 Coconut-Based Intercropping

Growing annuals/biennials in the interspaces of coconut is referred to as intercropping. A large variety of crops have been found suitable for growing under irrigated and rainfed conditions. Intercropping takes advantage of the nature of the coconut palm's canopy and rooting system (Reynolds 1988; Proud 2005). Apart from the advantages listed above, intercropping helps to reduce the dependence on coconut products (which often experience an unstable market) and ensures economic support during the long juvenile period of coconut.

Several reports and experimental research results are available on intercropping in coconut gardens from different parts of the world covering a number of crops. Many reviews of the experiments in India on intercropping in coconut are available (Nair 1976, 1977; Nair and Bavappa 1975; Nair and Varghese 1976; Nair et al. 1974; Nelliat et al. 1974; Nelliat and Krishnaji 1976; Gopalasundaram and Nelliat 1979a, b; Chattopadhyay et al. 1995; Srinivasa Reddy and Biddappa 2000). These reviews indicate that the practice extends to many of the cultivated crops. Notable among them are cereals, tuber crops, pulses, oilseeds, fruit crops, rhizome spices, ornamental and medicinal plants and fodder grass (Table 7.3). Coconut intercropping systems followed in Peninsular Malaysia and Sri Lanka were described by Denamany et al. (1979) and Liyanage et al. (1984), respectively. The ideas and considerations for coconut farm diversification are given by Scheewe (2003).

The details of coconut-based cropping systems adopted in the wet and wet intermediate zones of Sri Lanka, their constraints and prospects are given by Ranatunga et al. (1988). Godoy and Bennett (1991) presented an analysis of the profits of cultivating modern and traditional varieties of coconuts as a monocrop as well as intercrop, under ideal and average growing conditions, with good and average management. The results showed that intercropping generated more income than monocropping by smallholders in Indonesia.

Tropical tuber crops such as cassava, elephant foot yam, colocasia, Chinese potato, sweet potato, greater yam and lesser yam were found to be suitable intercrops in coconut gardens. The tuber crops partially meet the food requirements of a farm family and almost always find a place in the homestead gardens of Kerala (Varghese et al. 1978b). In an intercropping trial of coconut with tuber crops at ICAR-CPCRI, Kasaragod, India, there was a general reduction in the yield of coconut when the intercrop alone was manured, but no such reduction was noticed when both the intercrop and the main crop were manured (Varghese et al. 1978b). Intercropping of tuber crops, elephant foot yam and yams resulted in increased yield of root (wilt) disease-affected coconut gardens. The cost/benefit analysis showed that coconut+tapioca combination gave the highest net return per rupee invested (Sethumadhava Menon and Ramakrishnan Nayar 1978). Ginger and turmeric are the important spice crops commonly intercropped in coconut gardens (Varghese et al. 1978b; Hegde et al. 1990). Experiments at Kasaragod, India, have indicated the suitability of vegetables like snake gourd, bottle gourd, brinjal, coccinia and bitter gourd as compatible crops with coconut (Hegde et al. 1993). Intercropping with vegetables helped to generate additional employment to the tune of 215 to 365 man-days ha^{-1} $year^{-1}$. Among the different sequences tried, snake gourd-ridge gourd-amaranthus were the most remunerative (Rs. 22,217 ha^{-1} $year^{-1}$) followed by amaranths-bottle gourd-brinjal (Rs. 20,920). Intercropping of fodder grass, vegetable crops (amaranthus, pumpkin and ash gourd) and fruit crops (banana and pineapple) could be successfully taken up in coconut gardens under coastal sandy soil by adopting appropriate soil moisture conservation measures, viz. husk and coir pith application in the planting zone (Subramanian et al. 2009).

Table 7.3 Common intercrops grown in coconut gardens in India

Crops	References
1. Fruit crops	
Banana (*Musa* sp. different varieties), Pineapple (*Ananas comosus*),Papaya (*Carica papaya*), Guava (*Psidium guajava*),Lemon (*Citrus* sp.)	Nelliat et al. (1974), Nair (1977), Gopalasundaram and Nelliat (1979b), Gopalasundaram et al. (1993), Maheswarappa et al. (2003), Subramanian et al. (2009), Athmanathan et al. (2000), and Girijadevi et al. (2013)
2. Fodder crops	
Hybrid Bajra Napier, Guinea grass, Fodder cowpea, *Stylosanthes gracilis*	Ramakrishnan Nayar and Sahasranaman (1978), Jacob Mathew and Mohamed Shaffee (1979), Sahasranaman et al. (1983), Maheswarappa and Hegde (1995), CPCRI (2004), and Subramanian et al. (2007)
3. Medicinal and aromatic crops	
Chittadalodakam (*Adhatoda beddomei*), Karimkurinji (*Nilgirianthus ciliatus),* Nagadanthi (*Baliospermum montanum),* Vetiver (*Vetiveria zizanioides)*,Indian long pepper (*Piper longum*),Noni (*Morinda citrifolia*),Arrow root (*Maranta arundinacea*), Galangal (*Kaempferia galanga*),Patchouli (*Pogostemon patchouli*)	Nair et al. (1991), Rajagopalan et al. (1992), Viswanathan et al. (1992, 1993), Lalitha Bai et al. 1996), Maheswarappa and Nanjappa (2000), Maheswarappa et al. (2000, 2008), Srinivasa Reddy and Arunachalam (2002), Suneetha and Chandrakanth (2003), CPCRI (2003, 2008), Ghosh et al. (2007a), Mohandas (2011), Ravi Bhat and Krishnakumar (2011), Bari and Rahim (2012), Nagwekar et al. (2013), Khandeker et al. (2014), Thiruvarassan and Maheswarappa (2014), and Nath et al. (2015)
4. Flowering crops	
Heliconia sp., *Anthurium* sp., *Jasminum* sp.	CPCRI (2003) Arunachalam and Reddy (2007), and Nihad and Krishnakumar (2015)
5. Tree crops	
Acacia mangium, Acacia auriculiformis, Casuarina equisetifolia,Ailanthus sp., *Tectona grandis, Tamarindus indica, Erythrina indica*	CPCRI (1989)
6. Spices/tree spices	
Ginger (*Zingiber officinale*),Turmeric (*Curcuma longa*),Black pepper (*Piper nigrum*), Nutmeg (*Myristica fragrans*),Cinnamon (*Cinnamomum verum*),Clove (*Syzygium aromaticum*) Vanilla (*Vanilla planifolia*)	Nelliat et al. (1974), Nair (1977), CPCRI (1984, 2009), Jayachandran et al. (1991, 1998), Sharma et al. (1996), Manjunath et al. (1998), Srinivasa Reddy et al. (2000), Maheswarappa and Anithakumari (2002), Nagwekar et al. (2002), and Maheswarappa et al. (2003, 2012)
7. Beverages	
Cocoa (*Theobroma cacao*),Coffee (*Coffea arabica/C. robusta*)	Nelliat et al. (1974), Nair et al. (1975), Nair (1977), Abdul Khader et al. (1984), Bavappa et al. (1986), and Elain Apshara (2013)

(continued)

Table 7.3 (continued)

Crops	References
8. Other crops	
Mulberry (*Morus* sp.),Ramie (*Boehmeria nivea*)	Shanthamallaiah et al. (1982a, b), CPCRI (2002), and Manjunath et al. (2010)
9. Millets	
Varagu (*Panicum scrobiculatum*),Finger millet (*Eleusine coracana*)	Nambiar (1978)
10. Pulses	
Horse gram (*Macrotyloma uniflorum*), Cowpea (*Vigna unguiculata*),Bengal gram (*Cicer arietinum*),Soybean (*Glycine max*)	CPCRI (1975), Shanthamallaiah et al. (1982b), Joseph (1992), Lourduraj et al. (1992), Hegde and Yusuf (1993), and Jayaraman and Subramanian (1994)
11. Oilseed crops	
Groundnut (*Arachis hypogea*)	Sahasranaman (1964), Kannan and Nambiar (1976), and Leela and Bhaskaran (1978)
12. Tuber crops	
Elephant foot yam (*Amorphophallus paeoniifolius*) Cassava (*Manihot esculenta*) Sweet potato (*Ipomoea batatas*) Colocasia (*Colocasia esculenta*) White yam (*Dioscorea rotundata*)	Kannan and Nambiar (1976) Menon and Nayar (1978) Varghese et al. (1978a) Suja et al. (2003a, b, 2004a, b) Girijadevi et al. (2013) Krishnakumar et al. (2013)
13. Vegetables	
Chilli, French bean, dolichos bean, tomato, knolkhol, capsicum, brinjal, snake gourd, bottle gourd, amaranthus, coccinia, bitter gourd, ridge gourd, cucumber, cluster beans, etc.	Sahasranaman (1961), Shanthamalliah et al. (1982a), George and Nair (1987), Rethinam (1989), Patil et al. (1992), Hegde et al. (1993), Nagwekar et al. (1997), and Manjunath et al. (1998)

Intercropping trials at ICAR-CPCRI Kasaragod with ornamental, medicinal and aromatic crops in coconut gardens revealed that *Heliconia*, anthurium, *Jasminum pubescens* and marigold under ornamental crops and long pepper and patchouli under medicinal crops were compatible as intercrops in coconut garden (CPCRI 2003). A medicinal plant *Plumbago rosea* L. (known as rosy-flowered leadwort) was successfully grown as intercrop in coconut gardens and the use of bio-resources such as neem cake and FYM in the ratio of 1:4 along with microbial inoculants was suggested to achieve the highest benefit/cost ratio (Nihad et al. 2010). Intercropping of flowering plant, *Heliconia stricta*, in root (wilt) disease-affected coconut gardens enabled to enhance the profitability from such gardens (Nihad et al. 2013).

Among the fruit crops, banana is a popular, stable and marketable long-term crop that could be planted between stands of coconut palms. To be a compatible and productive intercrop, banana suckers are to be planted 2 m away from the base of coconut palms. Banana can be intercropped in gardens when the coconut palms are 1–3 years old and from the 25th year. Generally, banana and coconut do not compete for soil resources, except when grown in dry zones. Magat (2004b) and

Secretaria and Magat (2006) from the Philippines had described the agro-management practices to be followed when banana and root crops are to be inter-cropped in coconut garden.

Ennin et al. (2009) reported replanting of coconut with lethal yellowing disease (LYD) tolerant MYD x VTT hybrid intercropped with banana and cassava with minimum fertilizer application. This system showed biological compatibility in that they did not affect the vegetative growth of young coconut and produced high cassava yield (mean of 35.3 tonnes ha^{-1}) and banana yield (mean of 2.9 tonnes ha^{-1}) giving high economic returns with a B/C ratio of 5 for cassava.

A study on the effect of cropping systems, residue management and tillage practices on organic carbon sequestration in clay loam soil in Kerala, India, revealed that among the different cropping systems, coconut + pineapple cropping system maintained the highest soil organic carbon (SOC) content of 1.30% at the end of 2 years, whereas the coconut + maize system maintained only 1.21% SOC. Surface mulching with crop residues could maintain SOC carbon up to 1.37%, but when the residues were incorporated to soil, the SOC status was only 1.1%. Among tillage practices, reduced till maintained 1.29% SOC, whereas the conventional tillage could maintain only 1.22% SOC after 2 years. Improvement in soil properties, like aggregate stability, porosity, bulk density and water-holding capacity, was observed with the maintenance of SOC which was reflected on yield and returns (Sudha and Annamma 2011).

Lamanda et al. (2008) used 3D architectural modelling approach for providing indicators for assessing above-ground competition for light and below-ground competition for space, in order to optimize intercropping in 6–60-year-old coconut holdings. Intercropping with shade-tolerant species was not limited by light transmission from the 35th year after coconut planting. However, at that stage of coconut tree development, the density of primary roots in the interrow limited intercrop development, especially for root and tuber crops. Tubers such as taro, yam, cassava and kava are therefore not recommended in this type of intercropping unless the number of coconut palms is reduced. This modelling approach could be used for recommending coconut planting patterns and densities, as well as indicating intercrop potential depending on their location in the most sunlit areas with minimum root competition.

7.6.4 Coconut-Based Mixed Cropping

Growing of perennial crops with adult coconut palms is referred to as mixed cropping. A number of perennials like cacao, clove, nutmeg, coffee, black pepper, mulberry, cinnamon, mango, sapota, papaya, cardamom and other crops are successfully grown with coconut.

Mixed cropping coconut with perennials is popular in large-scale plantations. Perennials are particularly suited for mixed cropping with coconut because once they reach maturity, they continue to provide a steady flow of income with limited maintenance requirements. This is also considered important under smallholder production systems where resources are limited. Coffee is a popular mixed crop under mature coconut stands. The shade from coconut palms provides optimum conditions for coffee's growth and productivity. The crop is best planted at 2 m away from the base of coconut palms in three rows at 3 m × 3 m triangular system under coconut grown at a spacing of 10 m × 10 m square or two rows at 3 m × 3 m in a triangular pattern under coconut at a spacing of 8 m × 8 m or 9 m × 9 m square. Canja and Magat (2006) described the agro-management practices to be adopted for coconut-coffee mixed cropping in the Philippines. In this system, application of fertilizers to coconut with or without coffee fertilization increased copra yield significantly, whereas application of fertilizers to coffee, without coconut fertilization, gave low yield suggesting that coconut could not benefit from the fertilizers applied to coffee. Margate et al. (1993) suggested the need to apply fertilizers separately to both the crops. Economic analysis revealed that fertilizer application to both coconut and coffee gave the highest net return followed by that for coffee alone.

The profitability of growing cacao as mixed crop in coconut has been established in field experiments conducted at two locations in Kerala, India (Nair et al. 1975; KAU 1979). According to Creencia (1979), coconut areas producing at least 50 nuts palm^{-1} year^{-1} may be intercropped with cacao. The productivity of coconut and net return from the system were significantly higher under mixed cropping with cacao both in double hedge and single hedge systems. The double hedge system was found superior at Pilicode (Kerala) in an experiment conducted in plantations where the coconut palms were cultivated at a wider spacing of 9 m × 9 m. In Kerala, India, coconut intercropped with double rows of cacao was more profitable than that intercropped in a single row (Nair 1979; Bastine et al. 1986).

Mixed cropping with Forestero variety of cacao in 16-year-old root (wilt) disease-affected West Coast Tall coconut palms in single and double hedge systems increased the coconut yield by 27–35% as compared to that of monocropping (Kamalakshy Amma et al. 1982). There was also a build-up in the soil nutrient status which was more pronounced at 4 m than at 1.5 m away from the base of palm, where cacao was planted. The status of nutrients was low in double hedge system, perhaps due to the effective utilization of nutrients and higher yields of crops per unit area. Evaluation of 9 elite clones of cacao in double row system of planting in the plantation of Laccadive Ordinary Tall coconut cultivar revealed that VTLCP-22 and VTLCP-1 involving crosses of NA-33 and II-67 x NC-29/66 performed best with high vigour and yield under coconut (Elain Apshara 2010).

The microclimate of coconut garden (planted at 7.5 m × 7.5 m) mix cropped with cacao and cinnamon (two rows of cacao or two rows of cinnamon planted in double hedge system between coconut rows) was studied by Balakrishnan et al. (1976), and they reported considerable reduction in the evaporation during the peak period (December to May) in the mixed cropped field compared to the open surface. The diurnal variations on relative humidity and vapour pressure in the microclimate of cacao and cinnamon were also relatively much less compared to those of microclimate of coconut.

Studies of coconut-based agroforestry system on the aerial development in cacao in monoculture and intercropped with coconut by Mialet-Serra (1998) in Vanuatu and Indonesia have shown that some architectural features of young cacao can be influenced by the amount of shading, from the coconut palms, leading to differences between monoculture and intercropped conditions. In East Java, Indonesia, Karmawati et al. (2010) found that the production of cacao under the shade of coconut palms was normal and stable, having almost similar productivity as the monoculture system. Such conditions could be achieved through spacing of coconut palms at 12.0 m × 8.0 m or the density at 104 palms ha^{-1} and cacao with spacing of 3.0 m × 2.0 m and 1152 trees ha^{-1}. Utomo (2013) studied the environmental performance of cacao production from monoculture system and agroforestry system in Indonesia. The land productivity ratio (LER) of cacao-coconut agroforestry was the highest (1.36), indicating the higher level of yield advantages from this system. It also had soil fertility advantages in terms of higher content of organic carbon, C:N ratio and soil organic matter, which stimulated growth and activity of two beneficial soil microbe groups (bacteria and fungi) that exist in the cacao-coconut system. Such a system will also have the least impact on global warming, acidification and eutrophication.

Unlike in some other cacao-growing countries, cacao is not grown under the shade of coconut in Ghana. Osei-Bonsu et al. (2002) compared the merits of four cacao-coconut cropping systems with the traditional cultivation of cacao under *Gliricidia sepium* shade. Growth of cacao seedlings was not affected when grown under coconut, whereas growth and yield were considerably affected when mix cropped with *G. sepium*. Seven years' mean dry bean yield of cacao and profitability was the highest (1.23 tonnes ha^{-1}) when cacao was spaced at 2.5 m (1739 ha^{-1}) under a spacing of coconuts at 9.8 m in the triangular system (105 palms ha^{-1}).

Large coconut areas on fertile alluvial clays along the west coast of Peninsular Malaysia have been underplanted with cacao. The favourable price of cocoa beans, the unstable copra prices and the suitability of coconut shade have accounted for the success of this cropping system.

Mixed cropping of cacao under older stands of coconuts resulted in greatly improved economic returns in Malaysia (Ramadasan et al. 1978). Magat and Secretaria (2007) described the agro-management practices to be adopted for coconut-cacao mixed cropping in the Philippines and by Daswir and Dja'far (1988); Abbas and Dja'far (1989) in North Sumatra, Indonesia. Fagon and Topper (1988) found in Jamaica that close planting of cacao under coconuts was beneficial than wider spacing to get higher cocoa yield. Dootson et al. (1987) studied the effect of underplanting modern Upper Amazon cacao in 15-year-old Thai Tall coconuts and reported increase in economic returns once the cacao came into bearing. According to Mathes (1986), the relative return to a cacao-coconut mixed cropping was the second highest compared to coconut alone, coconut-coffee (*Coffea robusta*) and coconut – black pepper cropping systems.

In the study conducted in coconut stands at different densities at the Davao Research Center of the Philippines Coconut Authority, Dauzat and Eroy (1997) computed three-dimensional numerical mock-ups of coconut palms. Mialet-Serra et al. (2001b), using plant architectural models for estimation of radiation transfer in a coconut-based agroforestry system in Côte d' Ivoire, tried to validate the model computed by Dauzat and Eroy (1997), which predicted the yield of understorey corn and mung bean intercrops as a function of estimated transmitted PAR in several planting patterns and densities of coconut. They obtained good agreement between the estimated PAR through a coconut canopy and field measurements at several sites for several varieties with various planting patterns and densities. Thus, by using 3D numerical mock-ups of coconut linked to the radiation calculation modules, the distribution of transmitted radiation reaching the soil under a coconut stand could be estimated accurately for various planting patterns. However, their method has some limitations for large-scale adoption.

Nelliat et al. (1979) and Srinivasa Reddy and Thomas (2001) have reported the beneficial effect of growing cacao, black pepper, clove, nutmeg and cinnamon as mixed crops in coconut plantations (CPCRI 1975, 1979). Panniyur-1 variety of black pepper yielded 2 kg of dried berries vine^{-1}, and the maximum yield was 5 kg vine^{-1} (Anon 1977). Experiments conducted under Goa (India) conditions revealed that black pepper grows satisfactorily as mixed crop in coconut gardens and the plants started yielding from the third year onwards. The average yield obtained from a 1 hectare coconut garden was 0.76 t and 0.44 t of dry black pepper from Panniyur-1 and Karimunda varieties, respectively (Mathew et al. 1993b).

Shanthamallaiah et al. (1982b) reported that mulberry as a mixed crop increased the yield of coconut by 920 nuts and net income by Rs. 7379 ha^{-1} and doubled the employment potential. The profitability of growing perennial spice crops such as clove, nutmeg and cinnamon has been reported (Nelliat et al. 1979; Srinivasa Reddy et al. 1998b, 2000). Experiments at ICAR-CPCRI, Kasaragod, on mixed cropping with black pepper, clove, nutmeg and cacao indicated highest net returns with coconut + nutmeg cropping system (Rs. 94,300 ha^{-1}) followed by that in coconut + clove (Rs. 46,800 ha^{-1}) compared to coconut + black pepper (Rs. 26,200 ha^{-1}) and coconut + cacao (Rs. 31,400 ha^{-1}). The net return under coconut alone was only Rs. 22,300 ha^{-1} (Nair et al. 1991). Mixed cropping with spice crops such as cinnamon, clove, nutmeg, black pepper, garcinia and allspice was beneficial in Ratnagiri, Maharashtra (India), to increase the coconut yield (Patil et al. 1991).

Vanilla (*Vanilla planifolia*), a climbing orchid and a shade-loving spice crop, has been found to be a suitable mixed crop in coconut gardens. Growing vanilla in coconut gardens with the application of cow dung slurry (6 tonnes ha^{-1} in two splits or vermicompost 5 kg plant^{-1} year^{-1} in two splits along with biofertilizers of phosphate-solubilizing *Bacillus* sp. and nitrogen-fixing *Azospirillum* sp. at 25 g plant^{-1} year^{-1}) resulted in higher fresh bean yield and improvement of microbial properties in rhizosphere. There was 53% increase in coconut yield in vanilla intercropped plots compared to the pre-experimental yield (Maheswarappa et al. 2016). However, vanilla is not a preferred crop now due to very low price of the vanilla beans.

Coconut canopy was found to provide adequate shade for shade-loving cardamom (*Elettaria cardamomum*) in the mixed cropping system in Karnataka, India, which increased overall profits from the coconut gardens and the net return from the system was 2.5 times greater than from monocropping (Korikanthimath et al. 2000a, b).

Maheswarappa et al. (2008) evaluated the performance of different medicinal plants such as Orila (*Desmodium gangeticum*), Moovila (*Pseudarthria viscida*) and Coleus (*Coleus aromaticus*) (herbs of 8 months duration), Chittadalodakam (*Adhatoda beddomei*), Karimkurinji (*Nilgirianthus ciliatus*) and Nagadanthi (*Baliospermum montanum*) (shrubs of 18 months duration) as inter/mixed crops in a 30-year-old WCT coconut garden spaced at 7.5 m × 7.5 m at ICAR-CPCRI, Kasaragod, India. Among the annuals, Orila recorded the highest net return (Rs. 12,929 ha^{-1}), while among the biennials, the highest net return obtained was with Karimkurinji (Rs.1,93,049 ha^{-1}). Mohandas (2011) reported the suitability of sitharathai (*Alpinia galanga*), chotrukuthali (*Aloe vera*) and tulasi (*Ocimum sanctum*) as medicinal plants and lemon grass (*Cymbopogon flexuosus*) and patchouli (*Pogostemon patchouli*) as aromatic plants for intercropping in a 36-year-old East Coast Tall coconut garden at the Coconut Research Station, Veppankulam, Tamil Nadu, India.

Mensah and Ofosu-Budu (2012) evaluated coconut-citrus mixed cropping systems in the context of lethal yellowing disease of coconut in Ghana and found that MYD x VTT coconut hybrid planted at 9.5 m triangular offered optimal spacing for citrus mixed cropping at the convergence point of two diagonal lines linked with 4 adjacent coconut palms. The mixed cropping system did not hinder the optimal growth and yield of coconut or citrus, whereas it enabled a more efficient use of land and generated higher productivity by fitting more trees per unit area of land as compared with sole cropping. Though the cost-benefit ratio of the intercropping came next to sole coconut planting, intercropping provided 26% of fruit income as insurance against lethal yellowing disease.

From multidimensional analysis of coconut-based mixed farming systems adopted by farmers, Thamban et al. (2006) reported coconut + areca nut + black pepper + banana was the most commonly adopted model of CBFS followed by coconut + areca nut + black pepper system. The degree of crop intensification varied widely across the farms, and the economic analysis of various CBFS models indicated that they are technically feasible and economically viable. The level of profitability increased with increase in the number of the component crops.

7.6.5 Coconut-Based Agroforestry System

Coconut-based agroforestry systems hold promise as a sustainable land-use activity in areas where food and nutritional security is of concern and availability of land for expansion of cultivation is limited. The task in coconut farms is to diversify by integrating fruit trees and multipurpose trees. The numerous crop species in the homesteads serve the primary needs of the farmers' families. Apart from food, the small plots provide fuel, fodder, timber and cash.

An agroforestry experiment was conducted at ICAR-CPCRI, India, with different tree species during 1983 in an adult coconut garden. Data on coconut yield has shown that with high population of subabul, casuarina and eucalyptus, the coconut yield was adversely affected, whereas *Ailanthus* tree, though slow growing, appeared to be compatible with coconut and had not affected the yield of coconut (CPCRI 1989). Taffin et al. (1991) developed stable cropping systems combining coconut and N-fixing trees such as *Acacia mangium*, *A. auriculiformis* and *Casuarina equisetifolia. A. mangium* produced maximum biomass (457 kg ha^{-1}), while A. *auriculiformis* provided the highest volume of harvested wood (49 m^3 ha^{-1}). Liyanage et al. (1993a) studied growth performance and biomass yield of four nitrogen-fixing trees (NFT) (viz. *A. auriculiformis*, *Calliandra calothyrsus*, *G. sepium* and *L. leucocephala*) and their effect on coconut yield by planting them in double rows of 2 m × 1 m in the avenues of coconut. Due to the compatibility between coconut and NFT's, nut yield of coconut has also been increased by 15–26% in different NFT plantings. Coconut-based agroforestry systems, particularly with NFTs, offer much scope for reducing the use of fertilizer inputs through biological N-fixation, recycling of nutrients and by adding organic matter (Gunathilake 2011). Among them, *G. sepium* has been the best species. The coconut-NFT tree-based integrated system helps to minimize additional input of fertilizers in coconut plantations, thus saving on fertilizer cost (Tennakoon 2011).

Peiris et al. (2003) evaluated 26 agroforestry models developed and established in farmers' fields by Coconut Research Institute of Sri Lanka for studying their economic feasibility and biological productivity and found that all economic indicators including net present value (NPV) of agroforestry models were higher than the monocrop indicating their economic sustainability in the long run. The study of Bullecer et al. (2006) from the Philippines indicated that agroforestry systems could be improved by using more strategic plant combinations, plant densities and planting patterns for the different life stages (and related growth patterns) of the coconut palms. According to them planting of fast-growing and tall timber trees under coconut is not advisable, and more than 30% shading adversely affects its growth. Only shallow-rooted and shade-tolerant intercrops should be planted within a 3 m radius around the coconut trees. They found that planting fast-growing and timber trees under coconut such as *Leucaena*, *mahogany* and *gmelina* can adversely affect nut production and, hence, not recommendable. While environmental protection is afforded, economic benefits from coconut are reduced which could make the scheme less acceptable to farmers. They suggested more in-depth studies before potential best coconut-based agroforestry systems can be identified and extrapolated to other areas. The financial analysis conducted in Davao City by Secretaria and Magat (2004a, b) showed that the farming system combining coconut and *gmelina* is profitable even though coconut yield is reduced to a certain extent.

7.6.6 High-Density Multispecies Cropping System

High-density multispecies cropping system (HDMCS) or multistoried cropping system involves growing coconut and a combination of annual and perennial crops of different heights, rooting characteristics and canopy patterns in the same garden (between coconuts) so as to maximize utilization of solar radiation, nutrients and moisture (Nelliat et al. 1974; Bavappa et al. 1986; FSSRI-PCA 1984). This system requires more management skills, labour and other inputs than most other systems (Ohler 1992). Diversifying the farming system by intercropping cash crops, such as cacao, coffee, banana, pineapple, etc., and changing to multistoried cropping systems can generate much higher returns (Proud 2005).

Crops having varying canopy heights are selected in this intensive cropping system with the objective of greater utilization of solar energy and soil resources. The most profitable multistoried cropping system with coconut as main crop (1 ha) was established at ICAR-CPCRI, Kasaragod, India, with black pepper trained on coconut, 350–600 cacao seedlings planted between rows of coconut and 3500 pineapple suckers planted between rows of coconut and cacao. The output from the system included 17,000 coconuts, 300 kg dried beans of cacao, 60 kg dry pepper and 4 tonnes of pineapple in the single hedge system ha^{-1} $year^{-1}$ (Nelliat et al. 1974, 1979). Based on the feasibility, marketability and economic viability, among the eight different cropping models evaluated by Thiruvarassan et al. (2014) in a 25-year-old East Coast Tall (ECT) coconut garden, the one with coconut + black pepper + banana + elephant foot yam recorded the highest B/C ratio (2.16) and net income (Rs. 57,577 ha^{-1}) in the east coast region of Tamil Nadu, India.

A multistoried cropping experiment at Lampung indicated that the highest income was obtained with cropping system of cacao + cinnamon (*Cinnamomum* sp.) + black pepper + pineapple, followed by banana + maize and kapok (*Ceiba pentandra*) + cacao (Dwiwarni et al. 1987). Margate and Magat (1983) reported that planting black pepper (on the base of coconut palms) + pineapple (1 m × 1 m) + dwarf papaya (3 m × 3 m triangular)/Forastero cacao (3 m in a row), together with coconuts (planted at 9 m × 9 m square system) in a multistoried cropping system, increased nut yield and copra production $palm^{-1}$ as well as the total profitability of all the crops planted in the same area. The income generated from the intercropping system was more than double that of monocropping. However, Cabangbang et al. (1991) and De Luna (2008) reported that net incomes are lower in farms with more intercrops due to substantial additional labour costs incurred when integrating more intercrops in coconut farms. Income from the four-crop combination involving coconut + coffee + black pepper + *lanzones* is half that from the two-crop combination of coconut + coffee (Cabangbang et al. 1991). Thus, while diversification or intercropping is admittedly one good strategy to increase farmers' income, identifying the right combinations of intercrops is crucial considering the additional expenditure requirements of each intercrop.

Growing a large number of crop species in unit area of coconut plantation at high plant densities is practised to achieve maximum resource use efficiency and to meet the diverse needs of the farmer. A high-density multispecies cropping system model with many compatible annual/perennial crops (with 17 species) at higher plant density (total of 14,976 planting points ha^{-1} of coconut plantation) was established at Central Plantation Crops Research Institute, Kasaragod, India, during 1983 (Bavappa et al. 1986). A gross margin of Rs. 92,230 ha^{-1} was realized in 1996–1997 compared to Rs. 1750 ha^{-1} during 1983–1984 (Sairam et al. 1999). As the perennial crops grew and utilized more and more space, the annual crops except banana were removed so that the system consisted of nutmeg, banana and pineapple in the coconut garden. The coconut yield increased by 176% as compared to the pre-experimental yield as a response to the adoption of high-density cropping system and irrigation provided to coconut and companion crops.

This experiment was subsequently maintained with different levels of recommended dose of fertilizers. The mean productivity of crops revealed that the yields declined with reduction of fertilizers below one-third of recommended dose in the system. The coconut yield did not vary much between one-third, two-third and full dose of recommended fertilizers (147 to 157 nuts palm^{-1} year^{-1}). The component crops performed better under two-third and full dose of fertilizers in the system. Hence, only two-third level of recommended fertilizers was found necessary to sustain the yield of coconut and component crops at economically higher levels (Srinivasa Reddy et al. 2000). The biomass availability varied from 17 tonnes ha^{-1} in control to 22.58 tonnes ha^{-1} in full recommended fertilizer treatment. Coconut yield and economics of the system indicated the possibility of maintaining the system with one-third level of fertilizers complemented with biomass recycling (CPCRI 2002). In Assam, India, adoption of HDMSCS involving coconut + black pepper + banana + Assam lemon + pineapple + ginger and coconut + betel vine+ banana + Assam lemon + turmeric + colocasia had resulted in a nut yield increase of 110 and 83%, respectively, over pre-experimental nut yield (Chowdhury and Deka 1997).

In Sri Lanka, studies on canopy architecture in a multispecies cropping system involving 13 fruit crops by Jamaluddeen and Jacob (1983) indicated that the percentage increase in canopy diameter between 24 and 33 months after planting was greatest with coffee (59), followed by mango (51), coconut and jackfruit (both 40). Studies in the Mid Country Research Station, Sri Lanka, indicated the agronomic and economic potential with cash crops such as black pepper, coffee and clove and food crops such as banana and lime (Premaratne and de Silva 1991). This system was found to be financially viable in improving the income levels of upland farmers. High capital cost for establishment of the system is the only disadvantage of the system.

The results of various studies conducted in Kerala, India, indicated increased coconut yield due to introduction of HDMSCS compared to the monocropping of coconut (Maheswarappa et al. 2003, 2005; Maheswarappa 2008; Krishnakumar and Maheswarappa 2010). The increase in yield of coconut palms from the initial level of 30 nuts to 75.8 nuts palm^{-1} was reported in root (wilt) disease-affected area through adoption of HDMSCS (Maheswarappa et al. 2003). The crops like banana, pineapple, black pepper, nutmeg and tuber crops performed very well and provided additional yield and income. The higher B/C ratio of 2.28 and higher positive net present worth (Rs.1,80,106 ha^{-1}) indicated that HDMSCS is economically viable in root (wilt)-affected area. Improvement in properties like water-holding capacity, organic carbon as well as major and micronutrient status of the soil was observed due to adoption of integrated nutrient management practices and HDMSCS (Maheswarappa et al. 2005). The improvement in the yield of the palms was coupled with reduction in root (wilt) disease intensity indices due to reduction in yellowing of leaves. The percentage increase in average yield was 54.5, 52, 48.3 and 40.9, respectively, under apparent healthy, disease early, disease middle and disease advanced palms in comparison with pre-experimental yield. From the root (wilt) disease-affected gardens of Kerala, India, Krishnakumar et al. (2011) reported the economic advantage of HDMSCS over monocropping being 61% with a B/C ratio of 1.59 indicating that the coconut-based high-density cropping system is economically viable in root (wilt) disease-affected areas provided the disease is well managed by adopting integrated practices.

HDMSCS supplies large quantities of biomass for recycling and thus offers scope for reducing the need of inorganic fertilizers for coconut and component crops. Palaniswami et al. (2007) studied the possibility of reducing the quantity of inorganic fertilizer application to coconut and various component crops cultivated in HDMSCS in Kerala, India, and reported the mean yield of coconut to range from 127 nuts palm^{-1} under no fertilizer treatment to 147 nuts palm^{-1} with either 66 or 33% of recommended fertilizer dose (Table 7.4). The productivity of palms declined beyond 33% of the dosage. The yield of clove and black pepper was the highest at 66% of the dosage, while that of pineapple and banana was the highest with application of fertilizers in the full dosage. Later on Maheswarappa et al. (2013) evaluated the performance of coconut-based HDMSCS under organic and integrated nutrient management by using reduced doses of NPK fertilizers (2/3 or 1/3 of recommended dose), application of vermicompost (by recycling biomass), biofertilizers, green manuring (in basins) as well as fully organic manures. The mean nut yield, copra content and oil yield for 5 years did not differ among the treatments (ranging between 145 to 155 nuts palm^{-1} year^{-1}, 159.5 to 164.6 g nut^{-1} and 65.7 to 65.8%, respectively) indicating the beneficial effect of organic cultivation by recycling biomass, application of biofertilizers and green manuring.

Among the different HDMSCS models evaluated in a 27-year-old coconut garden of cv. Assam Tall spaced at 8.0 m × 8.0 m in India by Nath et al. (2008), the highest nut yield as well as per cent increase in nut yield was recorded with coconut + black pepper + banana + Assam lemon + pineapple + ginger. This was more profitable giving the highest net return of Rs.42,155 ha^{-1} with a B/C ratio of 1.67. Nath

Table 7.4 Output from crops under coconut-based high-density multispecies cropping system model in Kerala (1999–2005)

Fertilizer dosage (%)	Coconut (nuts palm^{-1} year^{-1})	Pineapple (kg fruit^{-1})	Clove (dry kg tree^{-1} year^{-1})	Banana (kg bunch^{-1})	Black pepper (kg dry) (bush^{-1} year^{-1})
100	145	0.89	1.44	5.76	0.87
66	147	0.70	1.55	5.43	1.66
33	147	0.57	1.25	4.70	0.90
25	137	0.43	1.12	4.36	1.06
20	129	0.48	1.00	3.91	0.42
Control	127	0.45	1.32	3.86	0.46

Full fertilizer dosage (N/P/K g plant^{-1}): coconut, 500:320:1200; pineapple, 8:4:8; clove, 300:250:750; banana, 200:200:400; black pepper, 50:50:150

and Deka (2010) also from Assam reported increase in yield of 28-year-old coconut garden spaced at 7.5 m × 7.5 m over the years after planting of various intercrops irrespective of different cropping systems, wherein the maximum increase of nut yield (68.62%) was recorded in coconut + black pepper + turmeric followed by coconut + black pepper + ginger (50%), while only 8.69% increase was observed in control (coconut + black pepper). The highest net return of Rs.1,46,549 ha^{-1} and B/C ratio 2.20, respectively, was recorded in coconut + black pepper + turmeric cropping system. Among the various multispecies cropping system models evaluated under 27-year-old ECT palm spaced at 7.5 m × 7.5 m in the east coast region of Tamil Nadu, India, coconut + black pepper + banana + elephant foot yam+ green leafy coriander) recorded the maximum net return of Rs. 57,577 with the B/C ratio of 2.16. The increase in cumulative nut yield in this model was 24.28% (Subramaniam et al. 2010).

7.6.7 *Coconut-Based Mixed Farming*

This involves integration of other subsidiary enterprises such as livestock, poultry, rabbitry, pisciculture and others along with the cultivation of fodder or pasture in the coconut garden. Coconut, due to its perennial nature, offers a unique opportunity for integration of animal husbandry enterprises such as rearing of cattle, goat or sheep to provide significant economic benefits to the farmer (Shelton 1991).

According to Darwis (1990), the coconut farming systems adopted by Indonesian smallholders can be classified into four types: farmyard, polyculture, monoculture and tidal swamp. In Java, about 79% of the coconut smallholdings fall within the polyculture pattern, where coconuts are grown with annuals, perennials or both types of intercrops and the remaining is coconut monoculture. In tidal swamp areas, coconuts are combined with fish, prawn, duck, crab or lowland rice production. In a polyculture system, involving perennial crops such as cloves, bananas, breadfruit trees (*Artocarpus altilis)* and sawo trees (*Manilkara kauki*), or pineapple and banana; the best profitable crops were pineapple and banana.

Much of the area under coconut plantations is under tall varieties which are often more than 30 years old, and therefore, light levels are high enough to support understorey vegetation such as pasture grasses. Sahasranaman and Pillai (1976) in India found that Guatemala grass (*Tripsacum laxum*), hybrid Napier (Pusa giant and NB 21) and guinea grass (*Panicum maximum*) are the ideal fodder crops which gave a yield of 50–60 mt of fresh fodder ha^{-1} $year^{-1}$ under coconut shade, while legumes such as Brazilian lucerne (*Stylosanthes gracilis*) and cowpea *(Vigna unguiculala)* yielded 30 tonnes ha^{-1} $year^{-1}$. At a feeding rate of 30–40 kg of fresh fodder in the ratio of 3:1 grasses and legumes $animal^{-1}$ day^{-1}, an area of 1 ha could support 4 milch cows.

Sahasranaman and Pillai (1976) reported 28% increase in nut yield in the root (wilt) disease-affected area by adopting mixed farming practice over a period of 5 years. Jacob Mathew and Mohamed Shaffee (1979) and Nelliat and Krishnaji (1976) indicated the incremental benefit of coconut + dairy over coconut alone. They also recorded increased coconut yield, satisfactory milk yield and employment potential for about 800–850 days as against 150 days for pure coconut.

Generation of additional employment from 150 man-days to 1000 man-days ha^{-1} by adoption of mixed farming was also reported (Sahasranaman and Pillai 1976; Sahasranaman et al.1976). Ferdinandez (1978) reviewed the studies done at the Coconut Research Institute, Sri Lanka, and reported that there was long-term beneficial effect on nut production by intercropping with certain pasture species, provided both crops are adequately fertilized and grown in favourable rainfall regions.

When grazing was introduced in coconut stands, a near doubling of coconut yield was reported by several researchers (Childs and Groom 1964; Ovasuru 1988; Moog and Faylon 1991). Moog and Faylon (1991) reported higher yield (80 to 100 nuts $palm^{-1}$ $year^{-1}$) when grazing was practised in gardens with improved pastures compared to that in gardens with natural pastures (30 to 50 nuts $palm^{-1}$ $year^{-1}$). Reynolds (1993) concluded that competition for moisture is likely to occur where annual rainfall is below 1750 mm, particularly if it is not evenly distributed.

A coconut-based mixed farming system involving 14 activities and integrating the crop and livestock systems was found to be the best in the linear programming model (Salam et al. 1991). The structural and functional diversity of the components of the system ensures a high level of resource use efficiency, meeting the multiple demands (food, fodder, fuel and timber) of the farmer. The model also recorded a B/C ratio of 1.64. Maheswarappa et al. (1998a) reported increase in nut yield under mixed farming by 39.6% and 33.5% for WCT and Laccadive Ordinary (LO) varieties of coconut, respectively. There was improvement in yield to the extent of 91.6 and 60.7% for WCT and LO, respectively, due to adoption of mixed farming (Maheswarappa et al. 2001). From an integrated nutrient management study, Subramanian et al. (2008) found that Bajra Napier hybrid (CO3) grown as intercrop in coconut garden under red sandy loam soil gave a fodder yield of 96 mt ha^{-1} $year^{-1}$ with the application of chemical fertilizers alone. In coastal sandy soil when Bajra Napier hybrid was grown as intercrop with soil moisture conservation measures, the fresh fodder yield obtained was 92 mt ha^{-1} $year^{-1}$ (Subramanian et al. 2007, 2009, 2012b).

From a comparison of the performance of 8 grass species, Smith and Whiteman (1983) from Australia recommended *Stenotaphrum secundatum* under deeper shade, with light transmission (LT) of 30–50%. In the Philippines, introduction of improved grasses or grass-legume pastures and cattle into coconut plantations resulted in total income ranging from US $ 608 to 809 compared to US $ 10 from coconuts alone (Deocareza and Diesta 1993). Stur et al. (1994) reviewed the available information on cattle rearing under coconuts, citing several examples in the Pacific Islands, and concluded that the level of production in such systems with adapted forages was comparable to that obtained in open conditions. The prevalence of the pasture-cattle-coconut systems in the different coconut-growing countries has been reviewed by Reynolds (1995). In the Philippines, CBFS integrates complimentary enterprises such as intercropping, livestock, processing of coconut products/by-products and marketing (Aguilar and Benard 1993).

The cropping systems practised in Sri Lanka include intercropping with seasonal and annual crops, mixed cropping with perennials, alley cropping with a combination of tree legumes and seasonal crops and mixed farming involving cattle and pasture (Liyanage and Dassanayake 1993). They also reported that coconut palms in the integrated system involving *Brachiaria milliformis/Pueraria phaseoloides* mixed pasture and *G. sepium* and *Leucaena leucocephala* fodder trees and crossbred cattle yielded 17% more nut and 11% more copra while maintaining the nutrient status of the palm above the critical level despite reduced application of fertilizer. Despite several factors which limit its widespread use by farmers, it was concluded that the coconut + cattle integrated system could contribute to the development of a sustainable and productive farming system in Sri Lanka (Liyanage and Dassanayake 1993). Iniguez and Sanchez (1991) estimated the percentage contribution of the cattle component in a coconut cattle system in Bali, Indonesia, to be 75.

The information available from the research on pasture species for the coconut plantations have been reviewed by Chen (1991), Reynolds (1993), Shelton et al. (1987), Stur and Shelton (1991) and Wong (1991). Various suitable pasture species have been identified, and their nutritional qualities have been ascertained (Norton et al. 1991). Intercropping coconut with nitrogen-fixing trees is a sustainable and productive land-use system. Such trees produce nutritious protein-rich fodder that supplements pasture feed, reducing grazing pressure. They also provide rich organic matter to the nutrient-poor soils of coconut plantations, besides additional farm products such as fuel wood and fodder. According to Dalla Rosa (1993a), *L. leucocephala* and *G. sepium* both perform well in the coconut understorey – yielding useful fodder, fuel wood and green manure.

Mantiquilla et al. (2000) evaluated various forages under coconut in Mindanao, the Philippines. *Brachiaria decumbens* applied with N fertilizer gave the highest DM yield of 24 t ha^{-1} $year^{-1}$ in a grazing system, while *B. decumbens* or *Setaria sphacelata* grown with legumes yielded about 15 mt. In a cut-and-carry system, *Pennisetum purpureum* gave the highest yields, but the forage quality was higher with *Panicum maximum.* Nayar and Sahasranaman (1978) observed that mixed farming had little effect on the size of soil aggregates when observed after a period of 5 years. However, the system was found to improve soil physical properties such as water-holding capacity, porosity and hydraulic conductivity and reduce bulk density both in the coconut basins and grass-cultured plots (Maheswarappa et al. 1998a).

Progressive increase in coconut production was noticed by Anitha et al. (2010) from the initial yield of 24 nuts $palm^{-1}$ to 42 nuts $palm^{-1}$ in a 10-year-old coconut plantation, spaced at 8 m × 8 m square pattern (156 palms ha^{-1}) with the introduction of a unit of 10 meat goats ha^{-1} of coconut farm utilizing the natural feeds available in the farm. According to Premaratne and Somasiri (2011), mixed farming system involving ruminants provides regular supply of organic matter to the soil as daily excretion of faeces and urine (amounting to 9.5% and 3.5% of body weight of animals, respectively). Integration of coconut/livestock could also enhance the coconut production in the system due to the increase in fertility of soils from animal manure as well as removal of weeds by animals (Sahasranaman and Pillai 1976; Bopaiah and Shetty 1991a; Biddappa et al. 1993; Maheshwarappa et al. 1998a). Higher organic carbon, N, P and K status has been reported under grass-cultured plot. The addition of slurry from the biogas plant helps to add considerable organic matter into the system, thereby increasing the various microbial and biochemical properties of the soil. Among the grass species tried, guinea grass intercropped plots had higher available N, P and K compared to coconut + hybrid Napier Bajra intercropping. Soils under mixed farming showed relatively lower values of available Ca, Mg, Mn, Cu and Zn and marginal increase in available Fe status (Maheswarappa et al. 1998a). In the integrated system involving coconut, mixed pastures and crossbred cattle, Liyanage et al. (1993c) from Sri Lanka reported that recycling of animal excreta (73 kg fresh cattle manure and 30 l urine palm^{-1} year^{-1}) reduced the cost of fertilization of coconut by 69% and improved soil fertility by providing organic carbon, total nitrogen and available phosphorus. The integrated system was also found to be economically viable when compared with monoculture coconut. The beneficial microbial community increased significantly in the root region of coconut in the farming system compared to the populations in monocrop of coconut. Intercropping of fodder hybrid Napier with coconut palms resulted in the proliferation of total bacteria and nitrogen-fixing organisms in the coconut rhizosphere in root (wilt)-diseased and apparently healthy palms. Compared to the palms in the control plot, crop mixing enhanced phosphate-solubilizing bacteria in root region of root (wilt)-affected coconut palms harbouring significantly higher numbers (Potty and Jayasankar 1976; Potty et al. 1977). The hybrid Napier + *Stylosanthes gracilis* combination proved to be the best among the combined treatments because of low level of denitrifiers and comparatively high proliferation of nitrifiers (Sahasranaman et al. 1983).

Bopaiah and Shetty (1991b) found that bacterial counts were higher in the root zone of coconut and Napier grass in mixed farming than that in monocropping of coconut. Enhanced soil biological activity was also indicated by higher levels of microbial biomass and activities of soil enzymes (phosphatase and dehydrogenase) in the farming system. The impact of intercropping fodder grass and organic recycling on microbial proliferation was evident not only at 0–25 cm depth but also in 25–50 cm and 51–100 cm depths in the basin (Thomas et al. 2010b). Aparna and Arya Nath (2014) compared intercropping systems with fodder grass, banana, vegetables and tuber crops in red loam and laterite soils and found that coconut + fodder grass improved many of the soil physical and chemical attributes as the quantum of organic matter added or recycled to the soil was more compared to the others. The microbial activity was also more under organic system of cultivation which resulted in higher enzyme activities.

According to Dalla Rosa (1993b) and Reynolds (1995), some of the potential benefits of coconut-animal production systems are:

1. Improved nutritional security of farm family. Increased overall farm income (coconut and animals) and greater employment opportunities
2. Increased stability for coconut farms through crop diversification and reduced market and financial risks

3. Increased soil fertility and reduced fertilizer costs as cow dung and cow urine serve as organic manures for the cropping system
4. Improved growth and yields of coconut and effective understorey weed control
5. A better grazing environment for cattle provided by coconut and higher relative humidity and soil water availability for pasture

7.6.8 Coconut-Based Homestead Farming System

The homestead is an operational farm unit in which a number of crops (including tree crops) are grown mostly as coconut-based farming system involving livestock, poultry or fish, mainly for the purpose of meeting the farmer's basic needs. Homestead farms or home gardens can be found in almost all tropical and subtropical ecosystems where subsistence land-use systems predominate. Such systems can be found in Sri Lanka (Jacob and Alles 1987; Nimal 1989), Mexico (Rico Gray et al. 1990), India (Nair and Sreedharan 1986; Krishnakumar 2010; Subramanian et al. 2014), Bangladesh (Leuschner and Khalique 1987), Pacific Islands (Vergara and Nair 1985), Indonesia (Michon et al. 1986) and several other countries, each one with its own characteristics.

The small and marginal farmers of Kerala place high value on their homestead farms as a source of nutritional security, additional income and for risk minimization. These homestead farms also play a significant role in maintaining the quality of microenvironment within the farm. An array of agricultural, horticultural crops and other miscellaneous plants in the homestead farms indicate a high level of crop diversity. Homestead farms are known to be repositories of biodiversity. Coconut-based homestead farms often present crowding of all kinds of plants, and when interplanted with multipurpose trees, the light availability is considerably reduced. Homestead farming satisfies the requirements of sustainability by being productive, ecologically sound, stable, economically viable and socially acceptable (John 2014). Systematic homestead farming requires careful selection of crops, and the plot layouts should be well planned for maximum sunlight utilization and also to cover the entire ground area with the identified crops. Productivity from the coconut-based homestead farms can be further enhanced by integrating livestock into the unit, which helps recycling of nutrients. Salam et al. (1991) opined that the home garden system as seen in Kerala, though not scientifically laid out, is productive and ensures better efficiency of scarce resources of land, water, nutrients and solar energy. An economic analysis of the coconut-based cropping system using data collected from 172 holdings in Southern Kerala showed that labour, manure and land area have significant influence on productivity (Job et al. 1993). Krishnankutty et al. (2013) reported that the homestead farms in medium elevation lands in Palakkad district, Kerala, are predominantly with coconut-based cropping pattern, with coconuts, areca nuts and a few tubers as intercrops.

Nair and Sreedharan (1986) reported 66 cultivated crop species, while Jose and Shanmugharatnam (1994) documented more than 130 species in a single home garden, dominated by coconut as the main crop. They opined that the crops, though at first glance appear to be haphazardly planted, have definite spatial arrangement within the farm and these subsistence oriented systems were managed mostly by family labour. From a survey of 815 smallholding coconut farms in Kerala, Krishnakumar and Reddy (2007) reported that coconut-based cropping system was the most predominant homestead cropping system followed by majority of farmers (98.0%), and the size of homestead farms ranged from 0.04 ha to 2.40 ha with the average being 0.54 ha. As the size of homestead farms increased, there was an increase in the investment made and the profit obtained on account of cultivation of coconut and other intercrops coupled with other enterprises. Later on, appropriate interventions in crop production technology and management practices in a farmer participatory approach were identified, and restructuring of homestead farms has led to increase in cropping intensity ranging from 56% to 86% in Kerala, India (Krishnakumar et al. 2007). An economic analysis indicated that the maximum income could be realized by those farmers having land holdings of more than 0.4 ha size (Krishnakumar 2010).

John and Nair (1998) developed an integrated model for small coconut-based homesteads (0.2 ha) by linear programming. One model consisted of 43 enterprises with a cropping intensity of 162%. Such a model could provide a net profit of Rs.37,426 on investing Rs.25,000 indicating a B/C ratio of 2.5. The amount of sunlight transmitted in coconut stand was 36% (John and Nair 1999), whereas when interplanted with multipurpose trees, the light availability was reduced to less than 1 %. Therefore, meticulous selection of understorey crops and their varieties, timely planting, following the temporal sharing concept and selective pruning of overstorey canopy for shade regulation could mitigate the problem of low light availability and improve overall productivity of the system. A net income of Rs.5,50,214 ha^{-1} year^{-1} could be realized from 1 hectare of coconut-based homestead farming model comprising of coconut with black pepper trailing on its trunk, banana, cows, fodder grass, poultry and aquaculture at ICAR-CPCRI (Maheswarappa et al. 2001; Subramanian et al. 2014).

Studies by Pandey et al. (2014) from South Andaman Islands of India indicated that the coconut palms extended their roots quite close to the companion crops, but companion crops extended their roots only up to a certain distance within the radius of their canopies. Thus, the main crop and its intercrops separated their niches horizontally. While there was facilitative mechanism by the main crop to its associative crops above the ground, there existed an exploitative mechanism below the ground.

The floral diversity of homestead farms of various agroclimatic zones of 14 districts of Kerala, covering 2500 farmers, was studied (Krishnakumar et al. 2010). Coconut was the base crop in most of the homestead farms surveyed, with the interspace of which a wide variety of crops ranging from annuals to perennials were cultivated so as to generate cash income after meeting the subsistence needs. John et al. (2010) found coconut+ black pepper+ nutmeg to be the most sustainable system in terms of economics and environmental consideration followed by coconut+ black pepper+ vanilla. Kalavathi et al. (2010a, b) reported significant improvement in food and nutritional security as well as farm family income of 150 small and marginal coconut homesteads of three community-based organizations (CBOs) of Kerala, through integrating interventions like intercropping, livestock rearing and product diversification. Diversification of crops and coconut-based enterprises implemented through CBOs emerged as the most effective strategy for improving the quality of life of the marginal coconut farmers – both in terms of income and food and nutritional security.

7.7 Harnessing Beneficial Microbial Resources

The beneficial effect of incorporation of legume creepers was reflected in the counts of total microflora, asymbiotic nitrogen fixers, P solubilizers, and enzyme activities. Soil spore counts and mycorrhizal infection of coconut roots were also increased in green manure applied plots (Thomas 1987). Association of legumes with efficient strains of *Rhizobium* spp. helps the nitrogen-rich green matter to be easily decomposed and the bound nutrients to be released. In order to enhance symbiotic effectiveness of rhizobia in green manure legumes, the introduced strains of rhizobia can be protected against acidic soil conditions by pelleting with alkaline substances such as lime and rock phosphate (Thomas and Ghai 1991).

Field experiment to evaluate the effect of green manure legumes on root (wilt) disease index and coconut yield indicated that basin cultivation and incorporation of *M. invisa* and *P. phaseoloides* increased coconut yield by 21.6 and 14.7%, respectively (Thomas and George 1990). Thomas et al. (2001a) reported the possibility of substituting 25–50% N fertilizer by raising either *C. mucunoides* or *M. invisa* in coconut basin.

HDMSCS promotes the proliferation of microbial diversity and biological activities to a great extent (Bavappa et al. 1986). In the HDMSCS with black pepper, banana, clove and pineapple as component crops, the root region of coconut recorded a sixfold increase in the population of bacteria and two- to threefold increase in population of fungi in the cropping system at different depths in coconut basins, compared to the monocrop of coconut (Bopaiah and Shetty 1991a). Bacteria formed the most important component followed by actinomycetes and fungi in the root region of coconut. The resource heterogeneity resulting from above-ground crop diversity might have led to greater diversity of decomposers and detrivores. The positive impact of cropping system was reflected on the soil microbial biomass carbon, soil microbial nitrogen and soil enzyme activities. Medium levels of fertilizer inputs along with organic recycling favoured higher activities of soil enzymes in the cropping system. The highest values of microbial biomass were recorded at 1/3 recommended levels of NPK (Kavitha 2009). The ratio of microbial biomass to total carbon was high in rhizosphere of coconut under the cropping system indicating higher contribution of microbial biomass to total carbon. This shows the occurrence of large portion of inactive biomass at higher doses of mineral fertilizer inputs (Thomas et al. 2016). Rohini Iyer (1983) reported that black pepper and coffee grown with 1/3 doses of fertilizer application supported the maximum mycorrhizal activity, whereas for banana, clove, and pineapple, a 2/3 dose was found to exhibit maximum activity.

Arbuscular mycorrhizal fungi belonging to the five genera, *Glomus*, *Gigaspora*, *Scutellospora*, *Serelerocystis* and *Acaulospora*, formed symbiotic association with roots of coconut seedlings (Thomas et al. 1991). A study on diversity of AM fungi associated with coconut intercropping systems of Kerala, India, revealed that mycorrhizal parameters like spore density, root colonization, species richness and relative occurrence of species varied significantly among the cropping systems (Ambili et al. 2012). The diversity of fungal species was reported to be maximum in the HDMSCS of ICAR-CPCRI, Kasaragod, India, with coconut as the main crop and banana, black pepper, pineapple and clove as component crops. From the coconut palm cultivated in crop mixed system under rainfed condition in a highly productive zone in Kerala, India, 40 AM species belonging to 10 genera were recorded indicating high level of AM richness in coconut rhizosphere (Rajeshkumar et al. 2015). A study on second generation high-density multispecies cropping system in Assam revealed the highest level of stimulation of microbial community in coconut + black pepper + turmeric cropping system followed by that in coconut + black pepper + ginger cropping system (Nath and Deka 2010), whereas in the east coast region of Tamil Nadu, the microbial proliferation was the highest in the model having coconut + black pepper + banana + elephant foot yam + greens (Subramaniam et al. 2010).

Plants exert influence on soil microbial communities through root exudates, which directly or indirectly influence plant growth (Bopaiah et al. 1987). The root exudates of coconut from mixed farming and multistoried cropping systems had significantly higher total sugar, amino acids and phenol content when compared to monocrop of coconut. Amino acids constituted an important component of organic fraction in root exudates of coconut and component crops in the cropping system (Kavitha 2009). The total amino acid content varied in relation to the fertilizer treatments in the root exudates of coconut (282.90 μ mol. g^{-1} root) and component crops, viz. pineapple, banana and clove (253.35–303.00 μ mol. g^{-1} root). The amino acids recovered from the root exudates of coconut and component crops showed various proportions of acidic and neutral amino acids.

Investigations on various cropping systems over several decades have revealed the profound influence of the cropping systems in supporting higher population of plant beneficial microbial community comprising of bacteria, fungi and actinomycetes in the rhizosphere and root region of the main crop coconut, irrespective of the crops which formed the components (Nair and Subba Rao 1977a; Potty et al.1977; Antony 1983; Rohini Iyer 1983; Thomas 1987; Bopaiah and Shetty 1991a, b; Kavitha 2009; Nath and Deka 2010).

Mixed cropping of cacao in coconut plantations improved the microbial activity in the rhizosphere of coconut to a higher level which was also attributed to an increase in organic matter content of soil due to periodic shedding of cacao (Nair and Subba Rao 1977b). Intercropping of fodder grass hybrid Bajra Napier with coconut palms resulted in the proliferation of total bacteria and N-fixing organisms in the rhizosphere of apparently healthy and root (wilt)-diseased coconut palms compared to the palms in the monocropped plot (Potty et al. 1977). Antony (1983) reported increase in soil microflora in the rhizosphere of root (wilt)-diseased coconut palms as a result of intercropping with tuber and rhizome crops.

7.7.1 Microbial Resources with Function-Specific Traits

Function-specific microorganisms which are found to be of great relevance to the growth and productivity of coconut and intercrops belong to 5 groups, viz. N-fixing bacteria in the rhizosphere, N-fixing *Rhizobium* living in symbiosis with legumes, phosphate-solubilizing bacteria, phosphate-solubilizing fungi and arbuscular mycorrhizal fungi (AMF). The constant association of beneficial microbes with the roots helps the palm in several ways contributing to the establishment, survival and growth, particularly in nutrient-poor soils and under drought conditions. Studies on mycorrhizal colonization showed that drought-tolerant genotypes were superior to the susceptible ones in harbouring a higher level of mycorrhizae colonization in roots, indicating the active role of host-fungus association under the conditions of low water availability (Thomas et al. 1993). Summer irrigation had a positive influence on AM symbiosis in coconut (Harikumar and Thomas 1991). The root region of coconut palm is inhabited by a number of free-living and associative N-fixing bacteria which possess different levels of N-fixation activity. *Beijerinckia indica* is the predominant asymbiotic N_2 fixer, capable of fixing N under acidic soil conditions in coconut soils, and is also endowed with the properties of production of polysaccharides which help in soil aggregation (Merilyn and Thomas 1992).

A study on 26 crops including plantation crops and intercrops in different cropping systems revealed the occurrence of *Azospirillum* in different levels in coconut-based farming systems (Ghai and Thomas 1989). The coconut harboured endophytic association of unique diazotrophs with N-fixation and plant growth promotion properties (Thomas et al. 2001a; Thomas and Prabhu 2003). Fungal isolates from coconut soils, particularly those of *Aspergillus* and *Penicillium* species, possess high level of phosphate-solubilizing ability solubilizing up to 72% of insoluble phosphorus supplied (Thomas et al. 1985). Inoculation of soils with efficient phosphate-solubilizing bacteria after addition of farmyard manure and rock phosphate results in the release of more available P from insoluble P sources (Thomas and Shantaram 1986). Coconut rhizosphere soils have also been found to harbour potassium-solubilizing bacteria belonging to *Acinetobacter*, *Alcaligenes* and *Micrococcus* species (Alka Gupta et al. 2016).

From an extensive study on plant growth-promoting rhizobacteria (PGPR) associated with coconut involving rhizosphere and root samples from 5 states of India, 512 PGPR isolates were collected which were found to possess various plant growth promoting and biocontrol traits (Priya George et al. 2012a, c). They observed that pseudomonas isolates from coconut rhizosphere showed antagonism towards *G. applanatum* and *T. paradoxa* and significantly inhibited both pathogens.

The dose of fertilizer applied has a direct bearing on the mycorrhizal activity, and this may vary with host-fungus combinations. A good number of AM fungi can withstand only low concentrations of soil nutrients. AM infection is negatively related to the available soil P (Harikumar and Thomas 1991). Root (wilt) disease had an adverse effect on AM symbiosis in coconut which was highly pronounced in disease advanced palms (Thomas 1988). Thomas and Ghai (1987) reported genotype-dependent variation in AM colonization in coconut, and the tall cultivars had higher level of colonization than dwarfs and hybrids.

7.7.2 Biofertilizers as Inputs in Coconut Production

Formulations containing living cells of beneficial microorganisms multiplied in a suitable carrier can form low-cost eco-friendly inputs for organic coconut production. The group of microorganisms responsible for biological N-fixation, P-mobilization, uptake of immobile elements, biological control and production of plant growth-promoting substances have been found to be closely associated with the coconut palm (Thomas and Prabhu 2003; Priya George et al. 2012a, b, c; Rajeshkumar et al. 2015). Isolations and multilevel screening of microbes from coconut ecosystem resulted in selection of efficient strains and development of microbial formulations for application in crop production. These bioinoculants enhanced root biomass and branching of secondary roots of the coconut seedlings. Inoculation of PGPR, viz., *Brevibacillus brevis* and *Bacillus coagulants*, resulted in production of coconut seedlings with high seedling quality index (Alka Gupta et al. 2006). Significant improvement in seedling growth parameters such as girth and root production was recorded when *Azospirillum brasilense* was applied to coconut seedlings raised in poly bags containing the potting media of coir pith-soil-sand mixture (Srinivasa Reddy et al. 2001).

Research efforts on harnessing the rich microbial resources associated with coconut palm resulted in the development and release of talc-based bio-formulations of plant growth-promoting rhizobacteria (PGPR), viz. 'CPCRI Kera Probio' based on *Pseudomonas* sp. for application to coconut. KerAM, a soil-based arbuscular mycorrhizal (AM) formulation, containing *Claroideoglomus etunicatum*, as the dominant AM species isolated from coconut agroecosystem with high potential to increase the growth parameters of coconut seedlings, is another resource made available as bioinoculants in coconut (Thomas et al. 2016). In Sri Lanka, Ilangamudali and Senarathne (2016) found mixing arbuscular mycorrhizal fungus-based biofertilizer at 50 g with potting medium in each poly bag at the time of planting seed nuts to be beneficial to produce quality seedlings with well-developed roots, which gave good establishment in the field.

Field experiments conducted at ICAR-CPCRI, Kasaragod, India, revealed that PGPR strains *Bacillus megaterium* TSB16 isolated from coconut and *Pseudomonas putida* KDSF23 from cacao can be used as a bioinoculant for vegetable production in organic agricultural systems indicating cross-compatible nature of PGPR. They can also serve as a single bioinoculant for coconut and vegetables in coconut-based cropping system (Khadeejath Rajeela et al. 2016).

7.8 Organic Farming as a Viable Strategy in Coconut

Organic farming is a holistic production management system which promotes and enhances agroecosystem health, including biodiversity, biological cycles and soil biological activity, as per FAO/WHO Codex Alimentarius Commission. In organic production systems, agronomic, biological and mechanical methods of management practices are adopted avoiding the use of synthetic materials. Organic farming system is defined as an integrated farming system that strives for sustainability, the enhancement of soil fertility and biological soil fertility while, with rare exceptions, prohibiting synthetic fertilizers, synthetic pesticides, antibiotics, genetically modified organisms and growth hormones (Paull 2010).

Coconut is one of the most amenable crops for organic farming, and a number of agro-techniques are available for organic cultivation (Prabhu et al. 2000; Thomas et al. 2001b, 2010a, b, 2012a). Recycling of waste biomass available in coconut plantations, cultivation of N-fixing leguminous green manure crops, biofertilizers and biopesticides are vital inputs in organic cultivation of coconut. Bio-resource-based strategy to strengthen biological foundations of soil fertility is of vital importance to achieve sustainable productivity (Thomas et al. 2012b). The diversified farming system, for which coconut is very amenable, will also enable addition of large quantities of organic matter and their effective recycling within the system helping to increase organic content of soil, improve microbial activity and make the entire production system more productive even with little or no external inputs.

7.8.1 Vermicomposting of Coconut Leaf Biomass

The technology for bioconversion of lignin-rich crop residue biomass available abundantly in coconut gardens to enrich organic resource, using local earthworm, gives a new dimension to biological soil fertility management and crop nutrition in coconut. The availability of biomass from well-managed coconut garden with 175 palms ha^{-1} is estimated as 14 tonnes ha^{-1} year^{-1} in the form of leaves, spathe, bunch stalk and husk. Though major portion of husk is used for extraction of coir fibre, coir dust (the by-product of coir processing factories) can be used as a bio-resource after biodegradation. The organic biomass can either be converted into compost by using earthworms or microbial cultures. Coconut palm residues, being very hard, are decomposed rather slowly, and hence vermicomposting has great relevance. Lignocellulosic residues from coconut plantations can be converted into brown, granular vermicastings using earthworms. At ICAR-CPCRI, a local earthworm (*Eudrilus* sp.) closely related to the African night crawler and very effective in vermicomposting of coconut palm residues has been isolated, and the methodology for vermicomposting has been standardized (Prabhu et al. 1998).

The recovery of compost is as high as 70%. The vermicompost, thus, produced has a nutrient content of 1.8% N, 0.21% P and 0.16% K with a C/N ratio of 9.9. From a 1 hectare coconut garden, it is possible to produce around 4 tonnes of vermicompost from coconut leaves alone. The vermicompost produced from coconut leaves using *Eudrilus* sp. at ICAR-CPCRI is available in the trade name 'Kalpa Organic Gold'. In coconut plantations with irrigation facilities, in situ vermicomposting can be done in basins or in trenches dug in interspaces of four coconut palms or by the heap method in plantations. This will enable disposal of coconut residues in a less expensive and eco-friendly manner with the benefit of producing high quality manure in coconut plantation.

The vermicompost increases soil fertility through addition of plant nutrients, growth hormones, increased level of soil enzymes and important microorganisms as they are rich in microbial diversity, population and activity. The agro-residues from component crops in the cropping system like banana, pineapple and others recycled through vermicomposting process produce quality organic manure (Thomas et al. 2012b). Application of entomo-pathogenic fungus, *Metarhizium anisopliae*, along with coconut leaves in vermicomposting tanks, was effective for the control of *Oryctes rhinoceros*, which multiplies in coconut leaf substrate during vermicompost production (Murali Gopal et al. 2006). Nine out of 15 microbial communities, particularly the plant beneficial ones, were enriched in the vermicompost produced from coconut leaves + cow dung mixture compared to 5 communities in vermicompost produced from cow dung alone. The coconut leaf vermicompost contained significantly high population of fungi, free-living nitrogen fixers, phosphate solubilizers, fluorescent pseudomonas and silicate solubilizers (Murali Gopal et al. 2009).

Vermiwash is a liquid organic fertilizer generated as a spin-off technology from vermicompost production. Coconut leaf vermiwash produced during vermicomposting of coconut leaves had the properties to increase crop production capacities of soil by enhancing the organic carbon content in the soil and increasing the populations of the soil microorganisms, particularly plant beneficial ones and their activities which facilitated increased uptake of the nutrients by the plants. Application of coconut leaf vermiwash (CLV) was observed to boost the biomass yield of crops such as cowpea, maize and bhendi (Murali Gopal et al. 2010, 2012).

7.8.2 Bioconversion of Coir Pith to Valuable Resources

Coir pith, with high C/N ratio and lignin and polyphenol content, is resistant to natural degradation. It can be made more amenable for earthworm/microbial activity and subsequent decomposition by various chemical and decomposition methods. The *Eudrilus* sp. of earthworm was found to be ideal for vermicomposting of coir pith when mixed with coconut leaves (Thomas et al. 2001b), producing granular vermicompost in 2 months with 1.2% N and a C/N ratio of 16.7:1. The composted biomass was found to be of superior quality with respect to content of major and micro nutrients and plant beneficial microorganisms particularly the bacteria involved in nutrient transformation and growth promotion.

Co-composting of coir pith can be done using poultry manure, lime and rock phosphate at 10 kg, 0.5 kg and 0.5 kg, respectively, for every 100 kg of coir pith. This brings about bioconversion of coir pith to a final product in 45 days, which possesses physicochemical characteristics required for quality organic manure with low C/N ratio (21.42), which is considered as a maturity index of composting process (Thomas et al. 2013). The composted coir pith can be used as manure in coconut plantations and can increase the capability of soils to store moisture and nutrients. The coir pith compost, thus, produced has been released with the trade name 'Kalpa Soil Care' by ICAR-CPCRI.

Ghosh et al. (2007b) composted coir pith fortified with edible oyster mushroom and urea in a multilayer heap fashion using perforated PVC pipes. The composted pith was an excellent organic manure, with a reduced C/N ratio of 20:1, pH of about 6.5 and electrical conductivity of 0.23 dS cm^{-1}. Tripetchkul et al. (2012) attempted multilayered heap composting of coir pith and obtained compost within 21 days. Forced aeration during composting of coir pith is not necessary if coir pith is fortified with cow manure, rice bran and coconut water and operated under low C/N ratio; however, maturity was reached only at 35 days of post-composting.

Use of microbial starter cultures having lignin degradation capacity enhances the decomposition of coir pith (Nagarajan et al. 1985). The efficacy of various biocontrol agents such as *Trichoderma* and *Chaetomium* with cellulolytic activities was tested for degrading coir pith (Ramamoorthy et al. 1999). The work done at ICAR-CPCRI, Kasaragod, resulted in the isolation of an efficient ligninolytic basidiomycete fungus, *Marasmiellus troyanus.* Studies on naturally decomposing coconut biomass resulted in the isolation of efficient ligninolytic and cellulolytic fungi such as *Lepista* sp., *Lentinus squarrosulus* and *Schizophyllum commune* with degradation potential (Thomas et al. 2001a). Lignocellulosic biomass from coconut palm including coconut leaf stalk and bunch waste could be used as substrate for cultivation of oyster mushroom with a biological conversion efficiency of 57–70%, and the spent substrate available after the cultivation can be utilized as quality organic manure for crop production (Thomas et al. 1998a). Mixing of coir pith and leaflets with leaf stalk, bunch waste and leaflets resulted in higher yield of mushrooms (Thomas et al. 1998b). Crop duration of *Pleurotus sajor-caju* (oyster mushroom) was more in lignocellulosic biomass of coconut, particularly when leaf stalk was the substrate.

7.8.3 Direct Utilization of Organic Matter

Coconut husk, coir pith and leaves are ideal materials for use as mulch in coconut basins due to their lignocellulosic nature which offer resistance to microbial decomposition. Mulching is an important cultural practice in coconut gardens not only to conserve moisture and soil but also to suppress the weed growth and supply nutrients and organic matter. One hundred husks will provide 1 kg of potash apart from 270 g N and 150 g P_2O_5 (Jothimani 1994). Effect of husk buried will be observed from third year onwards, and the beneficial effect lasts for 5–6 years. The biomass available from inter/mixed crops can be used for mulching and as a source of nutrients in coconut plantation. Cacao, as a component of multiple cropping system, adds substantial quantity of organic matter to the soil. When grown under single and double hedge system in adult coconut plantation, cacao litter fall was to the tune of 818 and 1785 kg ha^{-1} $year^{-1}$, respectively, on oven-dry basis (Varghese et al. 1978a, b). With a nutrient content of 2.84% N, 0.26% P and 1.73% K (Eernstman 1968), the cacao leaf litter in the double hedge system could contribute 50 kg N, 11 kg P_2O_5 and 35 kg K_2O every year to the soil in the coconut-cacao mixed cropping system. Effective recycling of organic materials (farmyard manure, poultry manure, cow urine, cowshed washing, etc.) from 1.2 ha coconut garden adopting coconut mixed farming could supply 125 kg N, 78 kg P_20_5 and 115 K_20 $year^{-1}$ (Maheswarappa et al. 1998a).

7.8.4 Field Validation of Organic Farming Technologies

From a 6-year study on coconut-based integrated farming system (CBIFS) maintained with only organic inputs, a mean yield of 108 nuts $palm^{-1}$ $year^{-1}$ from coconut could be obtained in addition to a fodder yield of 106 t ha^{-1} $year^{-1}$ (Table 7.5). The soil nutrient status in terms of organic carbon and available N and P was substantially improved under organic treatments. However, the K content was low due to the higher uptake of K by coconut and fodder crop. The integrated system was highly remunerative with an average net return of Rs. 2,85,512 ha^{-1} achieving economic and environmental advantages. Organic farming with integrated package of biological management practices, involving animal husbandry enterprises, avoiding

Table 7.5 Effect of nutrient management on yield of coconut and fodder grass in CBIFS

Treatments	Mean coconut productivity (nuts $palm^{-1}$)	Mean hybrid Bajra Napier (CO3) yield (fresh fodder tonnes ha^{-1})	Crude protein content of fodder (%)
Coconut monocrop + recommended fertilizers	97		
CBIFS: 50% inorganic fertilizers + 50% through organics	119	117	12.19
CBIFS: 100% through organics	108	106	12.69
CBIFS: 100 % through inorganics	106	96	11.94

all chemical inputs, enables to achieve sustainable productivity from coconut plantations. The integrated farming system under higher organic inputs recorded very high level of microbial biomass content and enzyme activities of phosphatase and dehydrogenase in the root region soils, indicating the stimulation of biological and microbial activities (Thomas 2010).

Organic farming has been found to be feasible in coconut with integrated treatments utilizing organic and bio-inputs including recycling of waste biomass by vermicomposting, in situ cultivation and incorporation of leguminous cover crops and biofertilizers of *Azospirillum* and *Bacillus* at 100 g each annually and other cultural practices including irrigation through the drip system in summer months.

A long-term study (2003–2014) conducted at ICAR-CPCRI, Kasaragod, India (Thomas et al. 2010b; Subramanian et al. 2016) has clearly indicated the possibility of organic farming in coconut under coastal ecosystem. The increase in yield for WCT variety of coconut was 65% (from the mean yield of 55 nuts palm^{-1} year^{-1} to the mean yield of 96 nuts palm^{-1} year^{-1}), while that of Chandrasankara (COD x WCT) was 55% (from the mean yield of 68 nuts palm^{-1} year^{-1} to 108 nuts palm^{-1} year^{-1}), respectively, during the initial 3 years of conversion period and during the 8 years of post-conversion period (Table 7.6). The practice of vermicomposting the recyclable biomass in trenches made in the interspaces, application of biofertilizers (Phosphobacteria and *Azospirillum* at 100 g palm^{-1}), raising cover crop in basins and its incorporation were found to be the best for improvement in soil nutrient status and enhancement of soil microbial population, as well as nut yield and copra content.

From an analysis of organic farming practices followed by 150 coconut farmers from Tamil Nadu, Kerala and Karnataka, India, Jaganathan et al. (2013) reported that growing intercrops, use of green manure crops and mulching were the main agronomic practices adopted. The major organic inputs produced or prepared at the farm were crop residues, farmyard manure, cow dung slurry and vermicompost. The mean yield and productivity of coconut was found to be 93 nuts palm^{-1} year^{-1} and

Table 7.6 Effect of organic and biological sources of nutrients on yield of WCT and COD x WCT coconut palms (nuts palm^{-1} year^{-1})

Treatments	Yield during the conversion period (2003–2006)		Mean yield (2007–2014)	
	WCT	COD x WCT	WCT	COD x WCT
Vermicomposting (in basin) + biofertilizers + cover cropping in interspace	67	74	88	106
Vermicomposting (in trenches) + biofertilizers + basin management with cover crops	64	77	96	108
Vermicomposting (in basin) + biofertilizers + intercropping vegetables	66	72	88	98
Vermicomposting (in trenches) + biofertilizers + intercropping vanilla and black pepper	68	77	91	107
Control	54	64	55	68

13,140 nuts ha^{-1}, respectively. Experimental results from long-term trials and the experience in farmers' gardens revealed that organic farming can become a viable strategy for sustainable coconut production, providing greater advantages in terms of ecological and economic benefits, with locally available resources and adoption of integrated farming involving animal husbandry enterprises.

From the experiments conducted over the years, the following practices are suggested for organic farming in coconut:

1. Use of seed nuts from organically grown plantations or from plantations where chemicals are not used.
2. Raising of seedlings without chemical inputs with bio-priming of PGPR, arbuscular mycorrhizae, *Azospirillum* and phosphobacteria.
3. Adopting a suitable spacing of coconut to facilitate inter/mixed cropping.
4. Recycling of biodegradable biomass from coconut, intercrops and animal origin by on-farm vermiculture.
5. Planting of leguminous cover crops and green manure crops in basins or interspaces.
6. Application of biofertilizers of nitrogen-fixing bacteria, P-mobilizing bacteria and plant growth-promoting rhizobacteria (PGPR).
7. Promoting biodiversity by inter/mixed cropping or high-density multispecies cropping by growing annuals, biennials and perennial crops.
8. Establishing mixed farming system with dairy unit including milch cows and raising fodder crops in interspaces. Subsidiary enterprises such as poultry, rabbitry, pisciculture, bee keeping and sericulture could be included.
9. Mulching with organic wastes from coconut to improve moisture holding capacity.
10. Irrigation to avoid moisture stress during summer months, preferably adopting drip irrigation.
11. Use of biopesticides such as microorganisms, parasites, predators and natural plant-based pesticides to manage the pests and diseases.

7.9 Future Thrust

The future strategies in coconut production should focus on higher level of productivity with cost-effective technologies based on natural resource management aimed at achieving competitiveness in coconut cultivation across all the producing countries through the following measures:

Achieving higher productivity in organic farming through integrated application of the technologies, approaches for replenishing higher export of K through appropriate means since there are reports of K becoming deficient under prolonged organic cultivation, explicit elucidation of information on the economic viability and long-term effect of adopting organic farming on soil properties and environment, research on impact of climate change on coconut cultivation, developing climate-resilient technologies which permit adaptation of the crop to climatic changes and building crop soil resilience, designing soil and crop management strategies based on conservation and effective utilization of natural resources to enhance sequestration of organic carbon, to improve the quality of soil and to prevent the loss of carbon as carbon dioxide with a view to mitigating the adverse effect of global warming, identification of constraints in coconut cultivation and standardization of precision farming, using Global Positioning System and Geographical Information System, development of site-specific management practices for optimum utilization of resources to achieve higher productivity simultaneously reducing the environmental hazards of overapplication of fertilizers and chemicals, developing cropping/production systems with higher nutrient and water use efficiency incorporating the beneficial properties of crop diversity, conservation agriculture and the rhizosphere ecology playing a crucial role in soil nutrient dynamics, plant nutrient uptake and soil health, utilization of the enormous potential existing in use of bio-resources, recycling of waste biomass, tapping nitrogen-fixing potential of legumes, utilizing plant-microbe synergy and biofertilizers, better understanding the microbiome of coconut roots and identifying novel molecules produced from rhizosphere microorganisms, and validate them for nutrient use efficiency and soil health management to achieve sustainable production in an eco-friendly manner.

References

Abad RG (1980) Weeds in coconut and their control. Weed Sci Soc Philippine Newsl 8(1):3–5

Abad RG, Prudente RL, Magat SS (1978) Incidence of leaf spots in coconut fertilized with various combinations of N, P and K. Philippine J Cocon Stud 3(1):37–44

Abbas BS, Dja'far D (1989) Sensitivity analysis of cacao cultivation: relative profitability of coconut and *Leucaena glauca* L. (*L. leucocephala)* as shade plants, based on a case study at Bah Lias Plantation, North Sumatra. Bull Perkebunan 20(2):97–103

Abdul Khader KB, Rajamony L, Balakrishnan PC (1984) Effect of raising cocoa as a mixed crop in adult coconut plantations. In: Proceedings of PLACROSYM VI, Oxford and IBH Publishing, New Delhi, pp 367–374

Abeyawardana V (1954) The density of palms in triangular and square planting. Ceylon Cocon Qtly 5(2):94

Abeygunawardena P, Abeyratne F, Fernando MTN (1995) Economics of soil moisture conservation in coconut lands using coconut husks and coir dust. COCOS 10:35–47

Abeywardena V (1981) Influence of watering on the yield of coconut. Ceylon Cocon Qtly 30(3–4):91–100

Aguilar EA, Benard GH (1993) Coconut based farming systems research and development in the Philippines. In: Nair MK, Khan HH, Gopalasundaram P, Rao EVVB (eds) Advances in coconut research and development. Indian Society for Plantation Crops, Kasaragod, pp 347–355

Aiyadurai S (1954) A note on the nursery studies on coconut seedlings. Indian Cocon J 7:156–163

Alka Gupta, Murali Gopal, Thomas GV (2006) Bioaugmentation of *Cocos nucifera* L. seedlings with plant growth promoting rhizobacteria, *Bacillus coagulans* and *Brevibacillus brevis* for growth promotion. In: Proceedings of XXXVI conference and annual meeting of European Society for new methods in agricultural research. Iasi, Romania, Sept. 10–14, 2006

Alka Gupta, Murali Gopal, Ravindran D, Thomas GV (2016) Prevalence of potassium solubilizing bacteria in the rhizosphere of coconut palms (*Cocos nucifera* L.) growing in different soil types. In: Chowdappa P et al (eds) Abstracts of 3rd International symposium on coconut research and development. ICAR-CPCRI, Kasaragod, Kerala, India, p 178

Almaden EA, Santiago RM (1980) Response of coconut seedlings to spacing and application of nitrogen and potassium. Ann Trop Res 2(2):89–95

Amalu UC (1998) Effect of salinity on growth of Nigerian dwarf coconut (*Cocos nucifera* L.) seedlings. Glob J Pure Appl Sci 4(1):15–20

Ambili K, Thomas GV, Indu P, Murali Gopal, Alka Gupta (2012) Distribution of arbuscular mycorrhizae associated with coconut and arecanut based cropping systems. Agric Res 1(4):338–345

Anilkumar KS, Wahid PA (1988) Root activity pattern of coconut palm. Oleagineux 43(3):37–342

Anitha S, Joseph L, Aravindakshan K (2010) Integration of meat goats on the productivity of coconut farm. In: Thomas GV, Krishnakumar V, Augustine Jerard B (eds) Improving productivity and profitability in coconut. Proceedings of International conference on coconut biodiversity for prosperity. CPCRI, Kasaragod, India, pp 181–182

Anonymous (1967) Coconut research in the British Solomon Islands. Ext. Pamphlet 10/67 Dept. of Agri. British Solomon Islands Protectorate pp 1–33

Anonymous (1971) Nursery management and selection of seedlings. Leaflet No 2. Revised edn. Coconut Research Institute, Sri Lanka

Anonymous (1975) The Philippine recommends for coconut 1975. The Philippine Council for Agriculture and Resources Research, p 63

Anonymous (1977) Selection of seedlings in the nursery. Reports and papers, Coconut convention, Directorate of Extension Education, Kerala Agricultural University, Thrissur, Kerala, p 44

Anonymous (1987) Planting coconut seedlings. Advisory circular no. A1, Coconut Research Institute of Sri Lanka

Anonymous (1989) Coconut- package of practices. CPCRI, Kasaragod

Anonymous (2000a) Coconut leaf pruning. A potential tool for higher farm productivity. Techno Guide Sheet No. 3. Series of 2000. Department of Agriculture, Philippine Coconut Authority, p 2

Anonymous (2000b) Under-planting of young coconuts on bearing palms for the production of edible vegetative pith or "Ubod"- Techno Guide Sheet No. 7.Series of 2000. Department of Agriculture, Philippine Coconut Authority, p 2

Antony J (1983) Microbiological activities in the root region of coconut (*Cocos nucifera* Linn.) under intercropping with tuber crops. PhD thesis, University of Kerala, Trivandrum

Aparna B, Arya Nath V (2014) A comparative study on soil chemical and biological regimes under coconut ecosystem in red loam and laterite soils of Trivandrum district- a farmer participatory active research. Indian J Sci Res Tech 2(5):14–17

Arunachalam V, Reddy DVS (2007) Foliar traits of jasmine plants intercropped in coconut. Agrofor Syst 71:19–23

Aterrado ED, Abad RG (1998) Response of coconut to simulated injury levels caused by leaf-feeding pests. Philippine J Crop Sci 23.(Suppl 1):39–40

Athmanathan U, Bhaskaran R, Karthikeyan A (2000) Banana: ideal intercrop for coconut. Coconut Research Station, Veppankulam, Tamil Nadu. Online edition of India's National Newspaper – The Hindu. Thursday, November 30, 2000

Azevedo PV, Sausa IF, Silva BB (2006) Water-use-efficiency of dwarf-green coconut (*Cocos nucifera* L.) orchards in northeast Brazil. Agric Water Manage 84:259–264

Babu KS, Asan RB, Bhaskaran C, Mohanakumaran N, Kunju UM (1993) Coconut yield response to monthly rainfall. Indian Cocon J 23(10):8–10

Bachy A, Briolle C, Villeman G (1962) Etude de la fumure des pepinieres de cocotiers en Cote d'Ivoire et au Dahmoey. Oleagineux 17(3):161–164

Bailey P, O'Sullivan D, Perry C (1977) Effect of artificial defoliation on coconut yields in Papua, New Guinea. Papua New Guinea Agric J 28(2, 3 and 4):39–45

Balakrishnan TK, Nair PKR, Nelliat EV (1976) Microclimate of coconut varieties and cacao and cinnamon grown as mixed crops with coconut. Indian J Meteorol Hydrol Geophys 27(1):71–74

Bandaranayake CK, Peries RRA, Perera L (1997) The suitability of optional material as a replacement for coir dust in raising coconut seedlings in polybags. COCOS 12:33–38

Baranwal VK, Manikandan P, Ray PK (1989) Crown choking disorder in coconut- a case of boron deficiency. J Plantn Crops 17:114–120

Bari MS, Rahim MA (2012) Economic evaluation and yield performance of some medicinal plants in coconut based multistoried agroforestry systems. Agriculturist 10(1):71–80

Basavaraju TB, Bhagya HP, Prashanth M, Arulraj S, Maheswarappa HP (2014) Effect of fertigation on the productivity of coconut. J Plantn Crops 42(2):198–204

Bastine CL, Palanisami K (1998) Irrigation in small holder coconut cultivation in Kerala, India – a financial analysis. COCOS 13:40–53

Bastine L, Khader A, Rajamoy L, Abdul K (1986) Cost-benefit analysis of intercropping of cacao in coconut gardens. Madras Agric J 73(3):170–172

Bavappa KVA (1975) Intensive cropping possibilities in plantation corps. Planters Chron 69(9):260–262

Bavappa KVA, Abdul Khader KB, Biddappa CC, Khan HH, Kasturi Bai KV, Ramadasan A, Sundararaju P, Bopaiah BM, Thomas GV, Misra LP, Balasimha D, Bhat NT, Shama Bhat K (1986) Coconut and arecanut based high density multi-species cropping systems. J Plantn Crops 14(2):74–87

Benard G, Margate RZ, Daniel C, Eroy MN (1996) Food intercrops under adult coconut palms. In: EEC-STD3-TS3- CT92-0132, 3rd technical report, p 62

Bhaskaran UP, Leela K (1978) Response of coconut to irrigation in relation to production status of palms and soil type. In: Nelliat EV (ed) Proceedings of PLACROSYM I. Indian Society for Plantation Crops, Kasaragod, pp 200–206

Biddappa CC, Bopaiah MG, Kailasam C, Shanavas M, George KM (1993) Effect of long term coconut-grass system on certain nutrients in the soil. J Plantn Crops 21(suppl):41–46

Bopaiah BM, Shetty HS (1991a) Microbiology and fertility in coconut-based mixed farming and coconut monocropping systems. Trop Agric 68(2):135–138

Bopaiah BM, Shetty HS (1991b) Soil microflora and biological activities in the rhizospheres and root regions of coconut-based multistoreyed cropping and coconut monocropping systems. Soil Biol Biochem 23(1):89–94

Bopaiah BM, Shekhara Reddy H, Nagarajan KV (1987) Biochemical characterization of root exudates of coconut palm. Curr Sci 56:832–833

Borah SC (1991) Standardization of nursery techniques in coconut. In: Silas EG, Aravindakshan M, Jose AI (eds) Coconut breeding and management. Kerala Agricultural University, Thrissur, pp 276–277

Bourgoing R (1990) Choice of cover crop and planting method for hybrid coconut growing on smallholdings. Oleagineux 45:23–30

Bourgoing R, Boutin D (1987) Method of controlling Imperata using light wooden rollers and establishing cover crops (Pueraria) in young hybrid coconut plantations in a village environment. Oleagineux 42(1):19–23

Bullecer R, Arellano Z, Stark M (2006) Participatory assessment of coconut-based agroforestry in San Isidro, Bohol. Paper presented during the 2nd National Agroforestry Congress, Pili, Camarines Sur, Philippines, 26–27 October 2005, pp 1–11

Cabangbang RP, Aguilar EA, Ilao SSL, Malisci LC (1991) Intercropping coconut with cacao. In: Success stories of farmer-managed coconut-based farming systems, Philippine Council for Agriculture, Forestry and Natural Resources Research and Development (PCARRD) Book Series, Los- BaZos, Philippines, vol. 1, pp 1–10

Canja LH, Magat SS (2006) Coconut–coffee (Robusta/Excelsa/Arabica) cropping model. Coconut intercropping guide no. 6, pp 1–10

Canja LH, Magat SS, Margate RZ (2003) Long-term coconut leaf pruning (CLP) effects on CBFS: coconut-coffee agroecosystem. CORD 19(2):39–46

Cano SB, Santos GA, Bahala RT (1989) Germination, growth and development of coconut seedlings in relation to seed nut preparation, position and polybagging. In: Training manual of short course on collection, conservation and characterization of coconut genetic resources. Philippines, 1990

Cecil SR (1991) Calcium and magnesium nutrition of the coconut palm. In: Silas EG, Aravindakshan M, Jose AI (eds) Coconut breeding and management. Kerala Agriculturtal University, Thrissur, pp 219–224

Chang AK (1978) An investigation into the optimum period of retention of MAWA coconut seedlings in polybag and field nurseries. In: Proceedings of the International conference on cocoa and coconuts, Kuala Lumpur, pp 568–578

Chattopadhyay N, Hore JK (2012) Influence of nut weight and method of planting on germination and seedling growth of coconut. J Crop Weed 8(1):98–100

Chattopadhyay PK, Mitra SK, Dasgupta MK (1995) Coconut-based farming system for sustained production. In: Ghosh DC, Das Gupta D, Majumdar DK, Chattopadhyay GN, Ganguli PK, Munsi PS, Bhattacharya D (eds) Proceedings of National symposium on sustainable agriculture in sub-humid zone, pp 171–172

Chattopadhyay N, Sharangi AB, Hore JK (2004) Effect of position and depth of planting on germination of coconut. Haryana J Hortic Sci 33(1–2):85–86

Chen P (1991) Management of forages for animal production under tree crops. In: Iñiguez L, Sánchez M (eds) Proceedings of International workshop on integrated tree cropping and small ruminant production systems, Medan, Sumatra, Indonesia

Child R (1974) Coconuts, tropical agricultural series, 2nd edn. Longman, London, pp 568–578

Childs AHB, Groom CG (1964) Balanced farming with cattle andcoconuts. East African Agric For J 29(3):206–207

Chowdhury D, Deka KK (1997) Coconut based inter/mixed cropping under Assam conditions. J Plantn Crops 25(1):106–108

Coomans P (1974) Planting densities for the coconut palm. Oleagineux 29:409–414

CPCRI (1975) Central Plantation Crops Research Institute, Annual Report 1974, Kasaragod, India, pp 24–27

CPCRI (1979) Central Plantation Crops Research Institute, Annual Report for 1977, Kasaragod, India, pp 32–34

CPCRI (1984) Annual report 1982. Central Plantation Crops Research Institute, Kasaragod, India, p 32

CPCRI (1989) Annual Report 1988. Central Plantation Crops Research Institute, Kasaragod, India, p 37

CPCRI (2002) Annual report 2001-02. Central Plantation Crops Research Institute, Kasaragod, India, pp 28–29

CPCRI (2003) Annual report 2002-03. Central Plantation Crops Research Institute, Kasaragod, India, pp 32–33, 38

CPCRI (2004) Annual report 2003-04. Central Plantation Crops Research Institute, Kasaragod, India, pp 33–34, 38

CPCRI (2008) Annual report 2007-08. Central Plantation Crops Research Institute, Kasaragod, India, pp 31–32

CPCRI (2009) Annual report 2008-09. Central Plantation Crops Research Institute, Kasaragod, India, p 40

Creencia RP (1979) Inter cropping coconut with cacao. Philippine J Cocon Stud 4(3):29–35

Dalla Rosa KR (1993a) Intercropping coconuts with nitrogen fixing tress. Agrofor Pac Technol 6:179–182

Dalla Rosa KR (1993b) Cattle under coconuts – a practical Pacific tradition. Published by the Agroforestry Information Service (AIS) of the Nitrogen Fixing Tree Association (NFTA). A publication of the Agroforestry Information Service. Number 7. Winrock International, pp 183–186

Daniel C, Adje I, Vihoundje F (1991) Performance of Dwarf x Tall coconut hybrids in a dry climate and with supplementary irrigation. Oleagineux 46(1):13–22

Darwis SN (1988) Status of intercropping on coconut lands in Indonesia. In: Proceedings of workshop on intercropping and intergrazing in coconut areas, 7–11 September, 1988, Colombo, Sri Lanka
Darwis SN (1990) Models of coconut based farming system in Indonesia. Indonesian Crop Res J 2(2):43–49
Darwis SN, Tarigans DD (1990) Management of soil in coconut based farming system. In: de Silva S (ed) Proceedings of XXVII COCOTECH meeting on coconut based farming systems, Manila, Philippines, pp 503–518
Das PK, Yusuf M, Hegde MR (1991) Reducing the risk in rainfed coconut cultivation. Indian Fmg 40(10):29–34
Daswir HAS, Dja'far D (1988) Analysis of cacao shaded with coconut (*Cocos nucifera*) compared with *Leucaena glauca* in North Sumatra. Buletin-Perkebunam 19(2):99–106
Dauzat J (1995) Coconut architecture modelling and simulation of the radioactive climate under coconut. In: Proceedings of the seminar on European research working for coconut, Montpellier, France, pp 91–99
Dauzat J, Eroy MN (1997) Simulating light regime and intercrop yields in coconut based farming systems. Eur J Agro 7:63–74
De Luna CC (2008) Socio-economic factors influencing choice of crop combinations in coconut-based agroforestry system in Makiling Forest Reserve, Philippines. MS thesis. University of the Philippines, Los Baños
de Silva MAT, Abeywardena V (1970) Leaf growth in relation to age and the bearing status of coconut palms. Ceylon Cocon Planters' Rev 6(2):59–62
Denamany G, Sharif Bin Ahmad MD, Hamid NBB (1979) Coconut intercropping systems in Peninsular Malaysia. Oleagineux 34(1):7–15
Deocareza AGF, Diesta HE (1993) Animal production on improved pasture crops under coconuts. In: Proceedings of 3rd meeting of regional working group on grazing and feed resources of Southeast Asia on srategies for suitable Forage-based Livestock production in South East Asia, pp 189–193
Deshpande BR, Kulkarni VG (1962) Studies in germination of coconut. Part I. Cocon Bull 16:336–338
Dhanapal R, Yusuf M, Bopaiah MG (1995) Moisture movement studies under drip irrigation in coconut. J Plantn Crops 23(1):28–34
Dhanapal R, Maheswarappa HP, Subramanian P (1998) Low cost water saving irrigation technology and its adaptability for coconut. Indian Cocon J 24(4):85–87
Dhanapal R, Subramanian P, Maheswarappa HP (1999) Methods of irrigation in coconut gardens. Indian Cocon J 29(11):4–7
Dhanapal R, Maheswarappa HP, Subramanian P (2000a) Response of coconut roots to the methods of irrigation in littoral sandy soil. J Plantn Crops 28(3):208–211
Dhanapal R, Maheswarappa HP, Subramanian P, Sairam CV, Gopalasundaram P (2000b) Micro irrigation technique for coconut in sandy soil. In: Muralidharan M, Rajkumar R (eds) Recent advances in plantation crops research. Allied Publishers, Chennai, pp 181–186
Dhanapal R, Palaniswami C, Mathew AC, Manojkumar C (2002) Effect of hybrid grass (CO3) in soil and water conservation in Western Ghat Region. In: Proceedings of National symposium on soil and water conservation measures and sustainable land use systems with special reference to the Western Ghat Region, held at ICAR Research Complex for Goa, pp 43–46
Dhanapal R, Maheswarappa HP, Subramanian P, Upadhyay AK (2003) Influence of drip irrigation on growth and yield of COD x WCT coconut (*Cocos nucifera* L.) hybrid. CORD 19(2):32–38
Dhanapal R, Maheswarappa HP, Sairam CV, Subramanian P, Upadhyay AK (2004) Influence of drip irrigation on growth and yield of coconut (*Cocos nucifera* L.) in laterite soil. J Plantn Crops 32(3):26–30
Dhanapal R, Subramanian P, Mathew AC, Palaniswami C (2005) Drip fertigation to increase the productivity of coconut. Indian Cocon J 31:17–19

Dhanapal R, Subramanian P, Bhat SR, Alka Gupta, Murali Gopal (2017) Production systems, nutrient dynamics and organic farming. In: Chowdappa P, Niral V, Jerard BA, Samsudeen (eds) Coconut. Divya Publishing House, New Delhi, pp 245–276

Dinesh R (2004) Long-term effects of leguminous cover crops on microbial indices and their relationships in soils of a coconut plantation of a humid tropical region. J Plant Nutri Soil Sci 167(2):189–195

Dootson J, Rattanapruk M, Suwannawuth e W (1987) Coconut development in Thailand and its stimulation by cocoa. Oleagineux 42(6):233–240

Dwiwarni I, Yuhono JT, Kemala S (1987) Multistoreyed cropping among coconut palms. Pemberitaan Penelitian Tanaman Industri Indonesia 12(3–4):67–73

Eernstman T (1968) Chemical analysis of leaves and other organs of *Theobroma cacao* L. as a means of diagnosing fertilizer requirements. Pub No. 54, PUDOC, Wageningen, p 55

Elain Apshara S (2010) Performance of elite cocoa clones intercropped in coconut. In: Thomas GV, Krishnakumar V, Augustine Jerard B (eds) Improving productivity and profitability in coconut farming. Central Plantation Crops Research Institute, Kasaragod, pp 183–187

Elain Apshara S (2013) Performance of selected cocoa (*Theobroma cacao* L.) clones under arecanut and coconut. J Plantn Crops 41(2):242–246

Ennin AS, Banful B, Andoh-Mensah E, Nuhu Issaka R, Nii Lamptey J, Aduen- ing-Manu J, Bolfrey-Arku G, Dery SK (2009) Food crop intercropping alternatives for replanting coconut farms destroyed by lethal yellowing disease. J Food Agric Environ 7(3&4):581–587

Eroy MN, Margate RZ, Magat SS, Santos GA (2001) Development of coconut leaf pruning techniques in bearing palms to increase sunlight transmission and productivity of coconut-based farming systems. Research Highlights 2000. Bureau of Agricultural Research, Department of Agriculture, Diliman, Quezon City, Philippines, p 76

Fagon HJ, Topper BF (1988) Agronomic research on cocoa in Jamaica; 1960–1980 and current trends. Trop Agric (Trinidad) 65(4):291–294

Ferdinandez DEF (1978) Integration of animal production and other crops into the coconut cropping systems in Sri Lanka. Ceylon Cocon Qtly 29:81–86

Fernando WMU, Bandaranayake CK (1996) Effect of planting density on the yield of coconut. COCOS 11:32–39

Fernando WMU, Pieris TSG, Wickramaratne MRT (1993) Variation in seedling characters of three different coconut cultivars and their use in identification in the nursery. COCOS 9:23–29

Ferreira Neto M, Gheyi HR, Holanda JS, Medeiros JF, Fernandes PD (2002) Qualidade do fruto verde de coqueiro em função da irrigação com água salina. (Quality of green coconut fruits in relation to salinity of irrigation water). Revista Brasileira de Engenharia Agrícola e Ambiental 6:69–75

Foale MA (1968a) The growth of the young coconut palm (*Cocos nucifera*).II. The influence of nut size on seedling growth in three cultivars. Australian J Agric Res 19:927–937

Foale MA (1968b) The growth of the young coconut palm (*Cocos nucifera*).1. The role of the seed and of photosynthesis in seedling growth up to 17 months of age. Australian J Agric Res 19(5):781–789

Foale MA (1968c) Seedling growth pattern. Australian J Agric Res 19(5):781–789

Fremond Y, Ziller R, Lamothe MN (1966) The coconut palm. International Potash Inst, Berne

Friend D (1990) The coconut industry, growing, harvesting and extraction. The Cocon Qtly Supplement, Dec 1990, pp 1–33

FSSRI-PCA (1984) Coconut-based farming systems: status and prospects. Farming Systems and Soils Research Institute-Philippine Coconut Authority

George S, Nair VR (1987) Effect of shade on growth, nodulation and yield of cowpea (*Vigna unguiculate* L. Walp). Agric Res J Kerala 25(2):281–284

Ghai SK, Thomas GV (1989) Occurrence of *Azospirillum* spp. in coconut based farming systems. Plant Soil 114:235–241

Ghosh DK, Bandopadhya A, Maji MK, Mahapatra S (2007a) Studies on the performance of medicinal plants under coconut plantation in West Bengal. Indian Cocon J 38(8):15–18

Ghosh PK, Sarma US, Ravindranath AD, Radhakrishnan S, Ghosh P (2007b) A novel method for accelerating composting of Coir Pith. Energy Fuel 21:822–827

Girijadevi L, Gladis R, Joseph B (2013) Benefit cost ratio analysis of coconut based cropping systems. J Plantn Crops 41(1):101–104

Godoy R, Bennett CPA (1991) The economics of monocropping and intercropping by smallholders: the case of coconuts in Indonesia. Hum Ecol 19(1):83–98

Gopalasundaram P, Nelliat EV (1979a) Intercropping in coconut. In: Nelliat EV, Bhat KS (eds) Multiple cropping in coconut and arecanut gardens. CPCRI, Kasaragod, pp 6–23

Gopalasundaram P, Nelliat EV (1979b) Intercropping in coconut gardens. Cocon Bull 16:201–206

Gopalasundaram P, Thomas Varghese P, Hegde MR, Nair MGK, Das PK (1993) Experiences in coconut based farming systems in India. In: Nair MK, Khan HH, Gopalasundaram P, Rao EVVB (eds) Advances in coconut research and development, Oxford and IBH Publishing, New Delhi, pp 383–393

Gunathilake HAJ (1985) Weed control in coconut lands. Planter 61:531–537

Gunathilake HAJ (2011) Fertilizer trees for enhancing coconut land productivity. In: Pushpakumara DKNG, Gunasena HPM, Gunathilake HAJ, Singh VP (eds) Increasing coconut land productivity through agroforestry interventions. Proceedings of a symposium held at the Coconut Research Institute of Sri Lanka, pp 28–44

Hanumanthappa M, Khan HH, Mohankumar B (2000) Production of coconut *geppe* seedlings in Karnataka. Indian Cocon J 31(8):9–10

Harikumar VS, Thomas GV (1991) Effect of fertilizer and irrigation on vesicular arbuscular mycorrhizal association in coconut. Philippine J Cocon Stud 16(2):20–24

Harries HC (1983) A ten point coconut nursery programme to avoid germination problems. Planter 59:207–241

Hegde MR, Yusuf M (1993) Performance of soybean varieties as intercrop in coconut garden. J Plantn Crops (Suppl) 21:29–31

Hegde MR, Gopalasundaram P, Yusuf M (1990) Intercropping in coconut garden. Technical bulletin no. 23, Central Plantation Crops Research Institute, Kasaragod, India, p 7

Hegde MR, Yusuf M, Gopalasundaram P (1993) Intercropping of vegetables in coconut garden. In: Advances in coconut research and development. Oxford and IBH Publishing, New Delhi, pp 407–412

Ho DT, Loh HP, Chan E (1978) Germination, nursery practices and field planting techniques for the MAWA hybrid coconut. In: Proceedings of the International conference on cocoa and coconuts, Kuala Lumpur, June 21–24

IAEA (1975) Root activity patterns of some tree crops. Technical report series no. 170. International Atomic Energy Agency, Vienna, p 154

Ilangamudali IMPS, Senarathne SHS (2016) Effectiveness of Arbuscular Mycorrhizal Fungi based biofertilizer on early growth of coconut seedlings. COCOS 22:1–12

Ilangamudali IMPS, Senarathne SHS, Egodawatta WCP (2014) Evaluation of coconut based Gliricidia sepium agroforestry systems to improve the soil properties of intermediate and dry zone coconut growing areas. Glob Adv Res J Agric Sci 3(2):67–76

Iniguez LC, Sanchez MD (1991) Integrated tree cropping and small ruminant production systems. In: Proceedings of a workshop on research methodologies in Medan, North Sumatra, Indonesia, September 9–14, 1990, p 329

Iremiren GO (1986) Response of coconut nursery seedlings to some cultural treatments in Nigeria. In: Pushparajah E, Chew PS (eds) Cocoa and coconuts. Incorporated Society of Planters, Kaula Lumpur, pp 861–868

Jacob VJ, Alles WS (1987) Kandyan Garden of Sri Lanka. Agrofor Syst 5:123–137

Jacob Mathew, Mohamed Shaffee (1979) Mixed farming in coconut garden. In: Nelliat EV, Shama Bhat K (eds) Multiple cropping in coconut and arecanut gardens, Tech Bull No 3. CPCRI, Kasaragod, pp 42–46

Jaganathan D, Thamban C, Jose CT, Jayasekhar S, Anithakumari P (2013) Analysis of organic farming practices in coconut in South India. J Plantn Crops 41(1):71–79

Jamaluddeen K, Jacob VJ (1983) Studies on canopy architecture in a multi-species cropping system. J Plantn Crops 11(1):39–46

Jayachandran BK, Meerabai M, Abdulsalam M, Mammen MK, Kunjamma M (1991) Performance of ginger under shade and open conditions. Indian Cocoa Arecanut Spices J 15(2):40–41

Jayachandran BK, Ancy J, Babu P, Nizam SA, Mridula KR (1998) Under the coconut tree: in India, ginger has it made in the shade. Agrofor Today 10(3):16–17

Jayakumar M, Saseendran SA, Hemaprabha M (1988) Crop coefficient for coconut (*Cocos nucfera* L.): a lysimetric study. Agric For Meteorol 43(3-4):235–240

Jayakumar M, Janapriya S, Surendran U (2017) Effect of drip fertigation and polythene mulching on growth and productivity of coconut (*Cocos nucifera* L.), water, nutrient use efficiency and economic benefits. Agric Water Manage 182(1):87–93

Jayaraman S, Subramanian S (1994) Lodging of soybean genotypes in understorey of coconut plantations. Madras Agric J 81:369–370

Jenkin RN, Foale MA (1968) Land resource study. An investigation of the coconut growing potential of the Christmas Island, vol. 1. The environment and the plantation. Land Resources Division. Directorate of Overseas Survey, p 12

Job E, Asan RB, Ramasubramonian PR (1993) Economic analysis of coconut-based cropping system in Kerala. Indian Cocon J 24(4):2–4

John CM (1952) Coconut cultivation. Indian Central Coconut Committee, Ernakulam

John J (2014) Homestead farming in Kerala: a multi-faceted land-use system. Rev Agrarian Stud 4(1):80–94

John J, Nair MA (1998) An integrated coconut based mixed farming homestead model for southern Kerala. Indian Cocon J 29(8):4–7

John J, Nair MA (1999) Light – a limiting factor in coconut based homegardens. Indian Cocon J 30(8):1–3

John J, Joy M, Geetha V, Sheeba U (2010) A holistic systems approach to identify sustainable perennial crop based systems – a case study. In: Thomas GV, Krishnakumar V, Augustine Jerard B (eds) Improving productivity and profitability in coconut. Proceedings of International conference on coconut biodiversity for prosperity. CPCRI, Kasaragod, pp 215–229

Jose D, Shanmugharatnam N (1994) Micro farms and house hold economy of rural Kerala: an empirical analysis with special reference to the homegarden farming system. In: Proceedings of the International Congress on Kerala studies, pp 86–88

Jose Mathew, Pillai GR, Santhakumari G, Varughese K (1996) Irrigation management on yield stabilization and annual productivity of coconut. J Trop Agric 34(1):33–35

Joseph PA (1992) Intercropping groundnut in coconut garden. Indian Cocon J 23:2–3

Joshi OP, Gopalasundaram P, Yusuf M (1988) Moisture depletion patterns in coconut plantation under rained and irrigated condition. In: Agrometeorology of plantation crops. Kerala Agricultural University, Thrissur, pp 22–26

Jothimani S (1994) Organic farming in coconut. Indian Cocon J 25:48–49

Kailash Rao T, Srirama Rao T (1968) Influence of the month of seednut maturity on germination and vigour of coconut seedlings. Andhra Agric J 15:18–20

Kalavathi S, Krishnakumar V, Thomas RJ, Thomas GV, George ML (2010a) Improving food and nutritional security of small and marginal coconut growers through diversification of crops and enterprises. J Agric Rural Dev Trop Subtrop 111(2):101–110

Kalavathi S, Krishnakumar V, Thomas RJ, Thomas GV, George ML (2010b) Diversification of livelihood for the marginal coconut farmers: a replicable model. Agric Ext Rev 22(4):10–16

Kamalakshy Amma PG, Menon KS, Patil PB, Sadanandan AK (1982) Effect of mixed cropping with cacao in coconut root (wilt) affected garden. Indian Cocon J 13(4):14–17

Kannan K, Nambiar KPP (1976) Studies on intercropping coconut garden with annual crops. Cocon Bull 5(9):1–3

Kannan K, Nambiar PKN, Nambiar KPP (1977) Studies on spacing in coconut. Indian Cocon J 8(2):1–2

Kapadiyal PK, Koradiya UA, Sarvaiya RB, Gandhi AP, Bhambhaniya LB, Lad AN, Raman S (1998) Drip irrigation for coconut in south Saurashtra. In: Mathew NM, Kuruvilla Jacob C (eds) PLACROSYM XII, Plantation crops: developments in plantation crops research. Allied Publishers, New Delhi, pp 157–159

Karmawati E, Mahmud Z, Syakir M, Munarso SJ, Ardana IK, Rubiyo (2010) Cocoa cultivation and post harvest processing. Center for Plantation Research and Development, The Ministry of Agriculture, Indonesia

KAU (1979) Kerala Agricultural University, Research Report for 1977-78. Vellanikkara, Thrissur, India, pp 90–98

Kavitha MS (2009) Soil microbial diversity and biological processes in high density multispecies cropping systems of coconut. PhD thesis. Mangalore University, Karnataka, India, p 174

Keller J, Sivanappan RK, Varadan KM (1992) Design logic for deficit drip irrigation of coconut trees. Irrig Drain Syst 6(1):1–7

Kenman ET (1973) Effect of seed nut trimming on the germination and growth of coconuts. Papua New Guinea Agric J 24(1):269

Khadeejath Rajeela TH, Murali Gopal, Alka Gupta, Thomas GV (2016) Cross- compatibility evaluation of plant growth promoting rhizobacteria of coconut and cocoa on yield and rhizosphere properties of vegetable crops. Biocatal Agric Biotechnol 9:67–73

Khan HH, Sankaranarayanan MP, Narayana KB (1979) Characteristics of coconut belts of India. 1. Morphology, some physico-chemical characteristics and taxonomy. In: Nelliat EV, Ranganathan V, Vishveshwara S, Potti SN, Krishna Murthy Rao W, Zachariah PK (eds) Proceedings of PLACROSYM-1. Indian Society for Plantation Crops, Kasaragod, pp 54–79

Khan HH, Biddappa CC, Nagarajan M, Thomas CV, Cecil SR (1992) Residual effect of long term phosphorus application on soil and plant nutrient contents, mycorrhizal association and yield of coconut. In: Nambiar KKN, Iyer RD, Rao EVVB, Krishnamurthy Rao W, Premkumar T, Hegde MR (eds) Proceedings of PLACROSYM-IX. Indian Society for Plantation Crops, Kasaragod, pp 14–18

Khandekar RG, Nagwekar DD, Sawant VS, Malshe KV, Haldankar PM (2016) Increase in productivity by fertigation technique in coconut. J Indian Soc Coastal Agric Res 34(1):88–91

Khandeker RG, Nagwekar DD, Sawant VS, Gurav SS, Haldankar PM (2014) Performance of *Morinda citrofolia* as mixed crop in coconut under Konkan Region of Maharashtra. Int J Res Emerg Sci Technol 1(4):63–66

Korikanthimath VS, Hiremath GM, Hosmani MM (2000a) Requirement of labour in the mixed cropping system of coconut and cardamom and its relationship with productivity levels. Crop Res Hisar 20(3):520–524

Korikanthimath VS, Hegde R, Mulge R, Hosmani MM (2000b) Coconut based cardamom multistoreyed cropping system compatibility and yield pattern. Crop Res Hisar 20(2):343–349

Krishnakumar V (2010) Restructuring coconut based homestead farms for sustained productivity. Indian Cocon J 53(8):15–19

Krishnakumar V, Maheswarappa HP (2010) Integrated nutrient management for root (wilt) diseased coconut (*Cocos nucifera*) palms. Indian J Agric Sci 80(5):394–398

Krishnakumar V, Reddy DVS (2007) Homestead farms in Northern Kerala – their basic characteristics and socio-economic status. J Plantn Crops 35(2):106–110

Krishnakumar V, Reddy DVS, Thamban C, Sairam CV (2007) Restructuring of homestead farms for sustainable income and employment opportunities. J Plantn Crops 35(3):146–151

Krishnakumar V, Rageena S, John J, Geetha K, Reddy DVS (2010) An assessment of species structure and crop diversity in the homestead farms of Kerala, India. In: Thomas GV, Krishnakumar V, Augustine Jerard B (eds) Improving productivity and profitability in coconut. Proceedings of International conference on coconut biodiversity for prosperity. CPCRI, Kasaragod, India, pp 211–214

Krishnakumar V, Maheswarappa HP, Jayasekhar S, Shanavas M (2011) Economic evaluation of high density multispecies cropping system in rsoot (wilt) disease affected coconut (*Cocos nucifera*) area in Kerala. J Plantn Crops 39(1):125–130

Krishnakumar V, Dhanapal R, Thomas GV (2013) Integrated nutrient management for elephant foot Yam intercropped in coconut gardens. J Root Crops 39(2):263–264

Krishnankutty J, Krishnapriya N, Shaiby KR (2013) Plant diversity and area trends in the coconut based homesteads of Kerala. Acta Biol Indica 2(2):461–466

Kulandaivelu R (1990) Crop water requirement for coconut through drip system. In: Proceedings of International congress on the use of plastics in agriculture, New Delhi, India, 26th February-2nd March 1990, pp 181–183

Lalitha Bai EK, Prasanna Kumari KT, Viswanathan TV, Rekha K (1996) Performance of pachouli in coconut gardens. South Indian Hortic 43(5–6):154–155

Lamanda N, Dauzat J, Jourdan C, Martin P, Malézieux E (2008) Using 3D architectural models to assess light availability and root bulkiness in coconut agroforestry systems. Agrofor Syst 72:63–74

Leela K, Bhaskaran UP (1978) Effect of intercropping coconut stands with groundnut on soil fertility and plantation management. In: Nelliat EV, Ranganathan V, Vishveshwara S, Potti SN, Krishna Murthy Rao W, Zachariah PK (eds) Proceedings of PLACROSYM-1. Indian Society for Plantation Crops, Kasaragod, pp 393–398

Lekadou TT, Issali AE, Konan KJL, Yao SDM, Allou K, Zakra AN, Yoro GR (2012) Effect of population density and spatial arrangement of Acacia auriculiformis trees on vegetative characters and mineral nutrition of coconut palm (*Cocos nucifera* L.) upon quaternary sands. Int J Emerg Technol Adv Engg 2(12):18–27

Leuschner WA, Khalique K (1987) Homestead agroforestry in Bangladesh. Agrofor Syst 5(2):139–151

Liyanage DV (1952) Report of the acting Botanist, annual report of the coconut research scheme for 1950. Ceylon Govt Sessional Paper 14:22–30

Liyanage DV (1955) Planting material for coconuts. Paper presented at the coconut conference, 29th August 1955, Bandirippuwa Estate, Lunuwila Coconut Research Institute, Ceylon

Liyanage LVK (1985) Rationale for intercropping in coconut lands. Cocon Bull 2(2):31–35

Liyanage MS (1988) Use of coir dust for moisture conservation. Cocon Bull 5(1):18–19

Liyanage MS (1994) Gliricidia intercropping enhances productivity of coconut lands. Cocon Bull 9:23–24

Liyanage DV, Abeyawardena V (1957) Correlations between seed nut, seedling and adult palm characters in coconut. Trop Agric 113:325–340

Liyanage MdeS, Dassanayake KB (1993) Experiences in coconut based farming systems in Sri Lanka. In: Nair MK et al (eds) Advances in coconut research and development. Proceedings of International symposium, Oxford & IBH Publishing, New Delhi, pp 357–367

Liyanage MdeS, Gunathilake HAJ (1998) Agronomy and farming systems research in coconut: Sri Lankan experience. In: Topper CP, Caligari PDS, Kullaya AK, Shomari SH, Kasuga LJ, Masawe PAL, Mpunami AA (eds) Proceedings of the International cashew and coconut conference: trees for life – the key to development. Dar es Salaam, Tanzania. Bio Hybrids International Ltd, Reading, pp 411–417

Liyanage MS, Martin MPLD (1987) Soybean-coconut intercropping. J Cocon Res Inst Sri Lanka 5:28–31

Liyanage LVK, Mathes DT (1989) Effect of irrigation on establishment and early growth of coconut (Var. CRIC 60) in the dry zone of Sri Lanka. COCOS 7:1–13

Liyanage MS, Tejwani KG, Nair PKR (1984) Intercropping under coconuts in Sri Lanka. Agrofor Syst 2(3):215–228

Liyanage MS, Jayasundara HPS, Gunasekera TGLG (1993a) *Leucaena* as a multipurpose tree for coconut plantations in Sri Lanka. J Trop For Sci 6(9):1–97

Liyanage MS, Jayasekara KS, Fernandopulle MN (1993b) Effects of application of coconut husk and coir dust on the yield of coconut. COCOS 9:15–22

Liyanage MS, Jayasundara HPS, Fernando DNS, Fernando WSMA (1993c) Integration of legume-based pasture and cattle into coconut farming systems in Sri Lanka. J Asian Fmg Syst Assoc 1:579–588

Liyanage LRMC, Purusothaman Y, Panditharathna DP, Tennakoon NA (2008) Evaluation of girdle sprinkler irrigation system for coconut in intermediate and dry zone of Sri Lanka. In: Proceedings of the second symposium on plantation crop research, Sri Lanka. October 16–17, 2008, pp 357–367

Loomis RS, Williams WS (1973) Maximum crop productivity-an estimate. Crop Sci 3:67–72

Louis IH, Chandrasekaran NR, Venkateswaran AN, Vijayan KP (1980) Note on the effect of summer irrigation on the dwarf varieties of coconut palm. Indian Cocon J 11(7):5–6

Louis IH, Krishnaswamy V, Venkateswaran AN, Vijayam KP (1981) Effect of spacing and manuring in coconut palms of the east coast of Tamil Nadu. Indian Cocon J 12(1):1–5

Lourduraj CA, Geethalakshmi V, Rajamanickam K, Kennedy FJS (1992) Soybean- a suitable intercrop in coconut gardens. Indian Cocon J 22(11):8–9

Madurapperuma WS, Bleby TM, Burgess SSO (2009a) Evaluation of sap flow methods to determine water use by cultivated palms. Environ Exp Bot 66:372–390

Madurapperuma WS, de Costa WAJM, Sangakkara UR, Jayescara C (2009b) Estimation of water use of mature coconut (*Cocos nucifera* L.) cultivars (CRIC 60 and CRIC 65) grown in the low country intermediate zone using the compensation heat pulse method (CHPM). J Natl Sci Found Sri Lanka 37(3):175–186

Magat SS (1990) Growing conditions and growth habits of coconut in relation to coconut-based farming systems. In: Silva de S (ed) Coconut based farming systems, XXVII COCOTECH meeting, Manila, Philippines, pp 17–40

Magat SS (1993) Coconut production and productivity in the Philippines: realities and opportunities in the early 90s. BUROTROP bulletin no. 6 of December 1993, pp 11–12

Magat SS (1999) Coconut-based farming system: technology notes for practitioners. PCA-ARDB, Diliman, Quezon City, Metro Manila, p 97

Magat SS (2004a) Growing of Intercrops in coconut lands to generate more food and agricultural products, jobs and enhancing farm income. Coconut Intercropping Primer. Published by PCA-Diliman, Quezon City, Dec 2004, p 7

Magat SS (2004b) Coconut-fruit crop (Banana) cropping model. Coconut intercropping guide no. 2. Department of Agriculture, Philippine Coconut Authority pp 1–10

Magat SS, Prudente RL (1974) Coconut nutrition study from polybag nursery to full-bearing. Ann Rep 1974-1975. Agricultural Research Department, Philippine Coconut Authority

Magat SS, Secretaria MI (2007) Coconut-Cacao cropping model. Coconut inter cropping guide no. 7. Department of Agriculture. Philippine Coconut Authority, pp 1–12

Magat SS, Margate RZ, Prudente RL (1977) Utilization of common salt (sodium chloride) as a fertilizer and for the control of leaf spot disease of coconut seedlings. Philippine J Cocon Stud 2(3):39–45

Magat SS, Canja LH, Margate RZ (1994) Response of coconut to increasing levels of leaf pruning and its implications on farm productivity. Cocon Res Dev J 10(2):16–32

Magat SS, Canja LH, Secretaria MI, Margate RZ (2002) Long-term response to coconut leaf pruning (CLP) of selected coconut – annual intercrops cropping systems. Philippine J Cocon Stud 27(1–2):17–28

Maheshwarappa HP, Hegde MR, Dhanapal R, Sairam CV, Vidhan Singh T (2001) Impact of integrated mixed farming system on coconut yield and economic analysis. Indian J Agro 46(1):56–63

Maheswarappa HP (2008) *In-situ* waste management in integrated nutrient management system under coconut (*Cocos nucifera*) based high density multi-species cropping system in tropical soils of India. Indian J Agric Sci 78(11):924–928

Maheswarappa HP, Anithakumari P (2002) Nutmeg – a suitable mixed crop for coconut garden. Indian J Arecanut Spices Med Plants 4(2):82–83

Maheswarappa HP, Hegde MR (1995) Cultivation of fodder crops in coconut garden. Kisan World 22(5):22–23

Maheswarappa HP, Nanjappa HV (2000) Dry matter production and accumulation in arrow root (*Maranta arundinacea* L.) when intercropped with coconut. Trop Agric 77:76–82

Maheswarappa HP, Hegde MR, Dhanapal R, Biddappa CC (1998a) Mixed farming in coconut gardens; its impact on soil, physical and chemical properties, coconut nutrition and yield. J Plantn Crops 26:139–143
Maheswarappa HP, Subramanian P, Dhanapal R, Gopalasundaram P, Hegde MR (1998b) Influence of irrigation and mulching on soil moisture and soil temperature under coconut in littoral sandy soil. J Plantn Crops 26(1):93–97
Maheswarappa HP, Sairam CV, Dhanapal R, Vidhan Singh T, Hegde MR (2000) Economics of coconut based farming system. CORD 2:37–49
Maheswarappa HP, Anithakumari P, Sairam CV (2003) High density multispecies cropping system for root (wilt) affected coconut gardens-its impact on productivity and economic viability. J Plantn Crops 31(1):23–27
Maheswarappa HP, Anithakumari P, Kamalakshi Amma PG, Shanavas M (2005) Influence of integrated nutrient management and high density multispecies cropping system on soil properties, plant nutrition and yield in root (wilt) affected coconut palms. CORD 21(2):18–29
Maheswarappa HP, Srinivasa Reddy DV, Krishnakumar V, Subramanian P, Dhanapal R (2008) Organic cultivation of medicinal plants as inter/mixed crops in coconut gardens. In: National conference on organic farming in horticultural crops with special reference to plantation crops, programme and abstracts. 15–18 October, 2008, CPCRI, Kasaragod, p 65
Maheswarappa HP, Krishnakumar V, Srinivasa Reddy DV, Dhanapal R, John Zachariah T (2012) Performance of different varieties/hybrids of black pepper (*Piper nigrum* L.) as mixed crop in coconut garden. J Plantn Crops 40(2):82–87
Maheswarappa HP, Dhanapal R, Subramanian P, Palaniswami C (2013) Evaluation of coconut based high density multi-species cropping system under organic and integrated nutrient management. J Plantn Crops 41(2):130–135
Maheswarappa HP, Krishnakumar V, Alka Gupta, Geetha Kumari A (2016) Influence of organic sources of nutrients on vanilla (*Vanilla planifolia*) as an intercrop in coconut garden. J Plantn Crops 44(2):85–89
Manjunath BL, Balbatti MS, Dhandar DG (1998) Intercropping in coconut for higher monetary returns. Developments in plantation crops research. In: Proceedings of PLACROSYM XII. Indian Society for Plantation Crops, Kasaragod, India, pp 199–202
Manjunath BL, Ramesh R, Korikanthimath VS (2010) Feasibility of ramie intercropping in coconut. J Plantn Crops 38(3):207–210
Manthriratna MA, Abeywardena V (1979) Planting densities and planting systems for coconut, *Cocos nucifera* L. 2. Study of yield characters and the economics of planting at different densities. Ceylon Cocon Qtly 30:107–115
Mantiquilla J, Gabunada F Jr, Buac R, Laguardia R, Magat S, Margate R (2000) Forages for growing under coconuts in Mindanao, the Philippines. Working with farmers: the key to adoption of forage technologies. In: Stür WW, Horne PM, Hacker JB, Kerridge PC (eds) Procedings of International workshop held in Cagayan de Oro City, Mindanao, Philippines, pp 152–157
Marar KMM, Padmanabhan V (1970) Studies on the effect of cutting leaves on the yield of coconut palms. Cocon Bull 1:7–11
Maravilla JN (1986) Nutrition and fertilization of coconut seedlings in polybags. Philippine J Cocon Stud 11(1):44–56
Maravilla JN, Prudente RL, Magat SS (1978) Fertilizer requirement of coconut seedlings grown on three major coconut soils of Davao. In: Proceedings of 9th annual meeting of the Crop Science Society of the Philippines, May 11–13, 1978
Margate RZ, Magat SS (1983) Coconut based multi-storey cropping. Philippine J Crop Sci 8(2):81–86
Margate RZ, Maravilla JN, Ebunar RM, Eroy MN (1993) Response of coconut and coffee to fertilizers applied on either or both crops in an intercropping system. Philippine J Cocon Stud 18:1–6
Marimuthu R, Natarajan C (2005) Sand curing is essential for obtaining more recovery of quality seedlings in coconut. Indian Cocon J 35(12):10–11

Mathai G (1979) Effect of cultural practices on the yield of coconut palm. Indian Cocon J 10(5):1–2
Mathes DT (1986) Economics of intercropping coffee, cacao and pepper under coconut. Cocon Bull 3(1):9–11
Mathew C, Ramadasan A (1964) Effect of N, P, K and nutrients on the growth of coconut seedlings. Indian Cocon J 17(4):114–117
Mathew J, Pillai GR, Santhakumari G, Varghese K, Varghese K (1993a) Influence of supplemental irrigation on the productivity and water use of adult coconut palms. J Plantn Crops 21(2):81–87
Mathew PA, Nagabhushanam S, Dhandar DG (1993b) Multicropping trials in coconut in Goa. In: Nair MK et al (eds) Advances in coconut research and development. Oxford and IBH Publishing, New Delhi, pp 437–439
Menon KS, Nayar TVR (1978) Effect of intercropping of tuber crops in root wilt affected coconut gardens. In: Nelliat EV, Ranganathan V, Vishveshwara S, Potti SN, Krishna Murthy Rao W, Zachariah PK (eds) Proceedings of PLACROSYM-1. Indian Society for Plantation Crops, Kasaragod, pp 416–424
Menon KPV, Pandalai KM (1960) The coconut palm- a monograph. Indian Central Coconut Committee, Ernakulam, p 384
Mensah AE, Ofosu-Budu GK (2012) Evaluation of coconut-citrus intercropping systems in the context of lethal yellowing disease of coconut in Ghana. African J Food Agric Nutri Dev 12(7):6945–6962
Merilyn VJ, Thomas GV (1992) Distribution of nitrogen fixing Beijerinckia in the rhizosphere of coconut. J Plantn Crops 20(1):149–155
Mialet-Serra I (1998) Preliminary study of aerial development in *Theobroma cacao* L. in a monoculture and intercropped with coconut. In: STD3-TS3-CT92-0132, CP no. 944/98, appendix 8, CIRAD, Montpellier, France, p 95
Mialet-Serra I, Bonneau X, Mouchet S, Kitu WT (2001a) Growth and yield of coconut–cacao intercrops. Exp Agric 37:195–210
Mialet-Serra I, Dauzat J, Auclair D (2001b) Using plant architectural models for estimation of radiation transfer in a coconut-based agroforestry system. Agrofor Syst 53:141–149
Michon MG, Mary F, Bompard J (1986) Multistoreyed agroforestry system in West Sumatra, Indonesia. Agrofor Syst 4(4):315–338
Miguel FN, de Holanda JS, Dias N d S, Gheyi HR, Folegatti MV (2011) Growth and yield of Anão Verde coconut under fertigation with nitrogen and potassium. Revista Brasileira de Engenharia Agrícola e Ambiental 15(7):657–664
Miranda FR, de Oliveira VH, de Montenegro AAT, de Miranda FR, de-Oliveira V (1999) Development and precocity of dwarf coconut palms (*Cocos nucifera* L.) production under different irrigation frequencies. Agrotropica 11(2):71–76
Miranda FR, de Gomes ARM, Oliveira CHC, de Montenegro AAT, Bezerra FML (2007) Evapotranspiration and crop coefficients for green-dwarf coconut in the coastal area of Ceara State, Brazil. Rev Cienc Agron 38:129–135
Mohandas S (2011) Prospects of intercropping medicinal and aromatic plants in coconut garden. Madras Agric J 98(1-3):82–83
Moog FA, Faylon FS (1991) Integrated forage-livestock systems under coconuts in the Philippines. In: Shelton HM, Stur WW (eds) Forages for plantation crops. Proceedings of a workshop, Sanur Beach, Bali, Indonesia, ACIAR proceedings no. 32, pp 144–146
Muliyar MK, Pillai RV (1989) Nursery manual for coconut. Pamphlet No. 6. CPCRI, Kasaragod
Murali Gopal, Alka Gupta, Thomas GV (2006) Prospects of using *Metarhizium anisopliae* to check the breeding of insect pest *Oryctes rhinoceros* in coconut leaf vermicomposting sites. Bioresour Technol 97:1801–1806
Murali Gopal, Alka Gupta, Sunil E, Thomas GV (2009) Amplification of plant beneficial microbial communities during conversion of coconut leaf substrate to vermicompost by Eudrilus sp. Curr Microbiol 59:15–20

Murali Gopal, Alka Gupta, Palaniswami C, Dhanapal R, Thomas GV (2010) Coconut leaf vermiwash: a bio-liquid from coconut leaf vermicompost for improving the crop production capacities of soil. Curr Sci 98:1202–1210
Murali Gopal, Alka Gupta, Thomas GV (2012) Vermicompost and vermiwash add beneficial micro flora that enhance soil quality and sustain crop growth. Int J Innov Hortic 1(2):93–100
Mwinjaka SR, Hocck C, Ngereza A (1994) A socio-economic survey in the priority areas of NCDP, National Coconut Development Programme, Dar es Salam, Tanzania
Mwinjaka S, Chiduza C, Temu AE, Sukume C, Diehl L (1999) Coconut palm replacement model for Tanzanian farming systems. J Agric Econ Dev 3:61–70
Nagarajan M, Sankaranarayanan MP, Nelliat EV (1975) A note on saline water irrigation to coconut palm *Cocos nucifera* Linn. in sandy loam soil. Sci Cult 41:381–383
Nagarajan R, Manickam TS, Kothandaraman GV, Ramaswamy K, Palaniswamy G (1985) Manurial value of coir pith. Madras Agric J 72:533–535
Nagwekar DD, Haldankar PM, Desai AG, Gunjate RT, Rajput JC (1997) Performance of rainfed vegetables as intercrops in coconut plantations. J Plantn Crops 25(2):209–211
Nagwekar DD, Desai AG, Joshi GD, Magdum MB, Khan HH (2002) Performance of spice crops as intercrops in coconut plantation under Konkan condition. In: Rethinam P, Khan HH, Reddy VM, Mandal PK, Suresh K (eds) PLACROSYM XIV. Coconut Development Board, Kochi, pp 333–335
Nagwekar DD, Desai VS, Sawant VS, Haldankar PM, Joshi GD, More TA (2006) Effect of drip irrigation on yield of coconut (*Cocos nucifera* L.) in sandy soil of Konkan region of Maharashtra (India). J Plantn Crops 34:344–346
Nagwekar DD, Sawant VS, Haldankar PM, Jadhav BB, Arulraj S, Maheshwarappa HP (2013) Performance of medicinal and aromatic plants as intercrops in coconut plantations in Konkan region of Maharashtra. J Plantn Crops 41(3):384–388
Nainanayake A, Ranasinghe CS, Tennakoon NA (2008) Effects of drip irrigation on canopy and soil temperature, leaf gas exchange, flowering and nut setting of mature coconut (*Cocos nucifera* L.). J Nat Sci Found Sri Lanka 38:43–50
Nair MA, Sreedharan C (1986) Agroforestry farming systems in the homesteads of South Kerala, South India. Agrofor Syst 4:339–363
Nair MGK, Hegde MR, Yusuf M, Das PK (1991) Mixed cropping. Tech Bull 24. CPCRI, Kasaragod, p 8
Nair PKR (1976) Intensive cropping for stabilized income of enhanced rates from coconut plantations. Financing Agri Oct-Dec 1976
Nair PKR (1977) Multispecies crop combinations with tree crops for increased productivity in the tropics. Gartenbauwissenschqft 42:145–150
Nair PKR (1979) Intensive multiple cropping with coconuts in India: principles, programmes, and prospects. Adv Agron Crop Sci 6. Verlag Paul Parey, West Berlin, p 147
Nair PKR, Balakrishnan TK (1976) Pattern of light interception by canopies in coconut + cocoa crop combination. Indian J Agric Sci 46(10):453–462
Nair PKR, Bavappa KVA (1975) Tubers as intercrops in plantations. J Root Crop 1(1–2):93–94
Nair PKR, Varghese PT (1976) Crop diversification in coconut plantations. Indian Fmg 26(11):17–21
Nair PKR, Varma R, Nelliat EV (1974) Intercropping for enhanced profits from coconut plantations. Indian Fmg 24(4):11–13
Nair PKR, Varma R, Nelliat EV, Bavappa KVA (1975) Beneficial effects of crop combinations of coconut and cocoa. Indian J Agric Sci 45(3):165–171
Nair RR, Pillai AG, Rajagopan A (1988) Water management in coconut. In: Aravindakshan M, Nair RR, Wahid PA (eds) Six decades of coconut research. Kerala Agricultural University, Thrissur, p 151
Nair SK, Subba Rao NS (1977a) Microbiology of the root region of coconut and cocoa under mixed cropping. Plant Soil 46:511–519
Nair SK, Subba Rao NS (1977b) Distribution and activity of phosphate solubilising microorganisms in the rhizosphere of coconut and cacao under mixed cropping. J Plantn Crops 5:67–70

Nambiar MC (1978) Multiple Cropping in littoral sand. In: Nelliat EV, Ranganathan V, Vishveshwara S, Potti SN, Krishna Murthy Rao W, Zachariah PK (eds) Proceedings of PLACROSYM-1. Indian Society for Plantation Crops, Kasaragod, pp 444–445

Nampoothiri KUK, Mathew J, Sukumaran CK (1973) Variation in germination pattern of coconut cultivars and hybrids. J Plantn Crops 1(Suppl):24–27

Nampoothiri KUK, Singh HP, Arulraj S, Thamban C (2000) Coconut cultivation technology. Coconut Devt Board, Govt. of India, p 129

Naresh Kumar S, Kasturi Bai KV (2009) Photosynthetic characters in different shapes of coconut canopy under irrigated and rainfed conditions. Indian J Plant Physiol 14:215–223

Nath JC, Deka KK (2010) Second generation coconut based high density multi species cropping system in Assam. In: Thomas GV, Krishnakumar V, Jerard BA (eds) Improving productivity and profitability in coconut farming. Central Plantation Crops Research Institute, Kasaragod, pp 167–173

Nath JC, Saud BK, Chowdhury D, Deka KK, Sarma UJ (2008) Coconut based high density multi-species cropping system in Assam. J Plantn Crops 36(2):98–102

Nath JC, Deka KK, Saud BK, Maheswarappa HP (2015) Intercropping of medicinal and aromatic crops in adult coconut garden under Brahmaputra valley region of Assam. J Plantn Crops 43(1):17–22

Nayar TVR, Sahasranaman KN (1978) Performance of four fodder crops in coconut garden under rainfed and irrigated conditions. In: Nelliat EV, Ranganathan V, Vishveshwara S, Potti SN, Krishna Murthy Rao W, Zachariah PK (eds) Proceedings of PLACROSYM-1. Indian Society for Plantation Crops, Kasaragod, pp 425–430

Nelliat EV (1968) Effect of frequency of irrigation on newly planted young coconut palms in sandy soil. Indian J Agric Sci 38:737–746

Nelliat EV (1972) NPK nutrition of coconut palm-a review. J Plantn Crops 1(Suppl):70–80

Nelliat EV, Krishnaji N (1976) Intensive cropping in coconut gardens. Indian Fmg 27(9):9–12

Nelliat EV, Padmaja PK (1978) Irrigation requirement of coconut and response to levels of fertilizer under irrigated condition during the early bearing stage. In: Nelliat EV, Ranganathan V, Vishveshwara S, Potti SN, Krishna Murthy Rao W, Zachariah PK (eds) Proceedings of PLACROSYM-1. Indian Society for Plantation Crops, Kasaragod, pp 187–198

Nelliat EV, Bavappa KVA, Nair PKR (1974) Multi-storeyed cropping – a new dimension in multiple cropping for coconut plantations. World Crops 26(6):262–266

Nelliat EV, Nair PKR, Kushwaha BL (1976) Effect of month of harvest of seed nuts and fertilizer application to nursery on vigour and quality of coconut seedlings. J Plantn Crops 4(1):16–20

Nelliat EV, Gopalasundaram P, Thomas Varghese P, Sivaraman K (1979) Mixed cropping in coconut. In: Nelliat EV, Shama Bhat K (eds) Multiple cropping in coconut and arecanut gardens. Tech Bull CPCRI, Kasaragod, Kerala, India, pp 28–34

Nihad K, Krishnakumar V (2015) Light use efficiency and flower quality of *Heliconia stricta* in coconut based cropping system. Acta Hortic 1104:497–504

Nihad K, Jessykutty, Sivaprasad P (2010) Effective utilization of bioresources for yield improvement in Plumbago rosea intercropped in coconut garden. In: Thomas GV, Krishnakumar V, Jerard BA (eds) Improving productivity and profitability in coconut farming. Central Plantation Crops Research Institute, Kasaragod, pp 207–210

Nihad K, Krishnakumar V, Sheela VL (2013) *Heliconia stricta* as intercrop in coconut-impact of cropping system and nutrition on floral traits. J Plantn Crops 41(2):251–254

Nimal PAH (1989) Place of coconut in home gardens. Cocon Bull 6(1):16–19

Norton BW, Wilson JR, Shelton HM, Hill KD (1991) The effect of shade on forage quality. In: Shelton HM, Stur WW (eds) Proceedings of workshop forages for plantation crops, Bali, Indonesia, ACIAR proceedings no. 32, pp 83–88

Novarianto H, Warokka J (2006) Past, present and future coconut research in Indonesia. In: Adkins SW, Foale M, Samosir YMS (eds) Coconut revival – new possibilities for the 'tree of life'. Proc. of the International Coconut Forum, held in Cairns, Australia, ACIAR proceedings no. 125

Oguis LG, Magat SS, Margate RZ (1979) The effect of chlorine sources and ammonium sulphate on the growth of coconut seedlings. Philippine J Cocon Stud 4(3):25–39
Ohler JG (1992) Coconut-based farming systems with perennials. Modern coconut management; palm cultivation and products. EcoPort version by Peter Griffee. FAO, Rome
Ohler JG (1999) Climate and soils. In: Modern coconut management. Intermediate Technology Publications, FAO, Universiteit Leiden, pp 35–46
Ollivier J, Akus W, Ollivier LB, Bonneau X, Kakul T (2001) Replanting/underplanting strategy for old coconut plantations in Papua New Guinea. Oléagineux 8(6):659–665
Ontolan DG (1988) Status report on intercropping and intergrazing in the Philippines. In: Proceedings of workshop on intercropping and intergrazing in coconut areas, 7–11 Sept. 1988, Colombo, Sri Lanka
Opio FA (1987) Intercropping/Intergrazing possibilities in the South Pacific. A paper presented at the 6th annual conference on coconut research and development in the South Pacific, 14–18 June, 1987. Honiara, Solomon Islands
Opio FA (1990) The need for coconut based systems. In: Silva de S (ed) Coconut based farming systems. Proceedings of the XXVII COCOTECH meeting, Manila, Philippines, pp 1–15
Osei-Bonsu K, Opoku-Ameyaw K, Amoah FM, Oppong FK (2002) Cacao-coconut intercropping in Ghana: agronomic and economic perspectives. Agrofor Syst 55:1–8
Ovasuru T (1988) Intercropping and intergrazing in Papua New Guinea. In: Proceedings of workshop on intercropping and intergrazing in coconut areas, 7–11 September, 1988, Colombo, Sri Lanka
Padrones GD, Secretaria MI, Magat SS (2000) Leaf pruning technique in bearing coconut palms: its effects on underplanting young coconuts for pit or ubod production. Philippine J Cocon Stud 25(1–2):1–11
Palaniswami C, George V Thomas, Dhanapal R, Subramanian P, Maheswarappa HP, Upadhyay AK (2007) Integrated nutrient management in coconut based cropping system. Tech bull no.49. ICAR-Central Plantation Crops Research Institute, Kasaragod, India, p 24
Pandey CB, Begum M (2010) The effect of a perennial cover crop on net soil N mineralization and microbial biomass carbon in coconut plantations in the humid tropics. Soil Use Manage 26(2):158–166
Pandey CB, Begum M, Singh SK, Saha D (2014) Coconut-based home gardens: mechanisms of complementarity in sharing of growth resources among home garden trees in the South Andaman Islands of India. Trop Ecol 55(3):339–348
Patel JS (1938) The coconut-a monograph. Govt Press, Madras, p 313
Patil JL, Haldankar PM, Jamdagni BM, Salvi MJ (1991) Influence of inter planting with tree spices on yield of coconut. Indian Cocon J 22(1):15–18
Patil JL, Haldankar PM, Jamadagni BM, Salvi MI (1992) Intercropping of vegetables in coconut. J Plantn Crops 20(2):167–169
Patil JL, Haldankar PM, Rangwala AD, Jamadagni BM (1998) Effect of different irrigation with sea water on survival and growth of coconut. (*Cocos nucifera* L.). South Indian Hortic 46(3–6):361–363
Paull J (2010) From France to the world: the Inteternational Federation of Organic Agriculture Movement (IFOAM). J Soc Res Policy 1(2):93–102
Peiris WBK, Fernando MTN, Hitinayake HMGSB, Dassanayake CKB, Gunathilake HAJ, Subasinghe SDJN (2003) Economic feasibility and biological productivity of coconut-based agroforestry models in Sri Lanka. COCOS 15:38–52
Perera L, Fernando WMU (2002) Decision making in replanting of coconut. COCOS 12(1–2):1–3
Perera L, Peries RRA, Jayatileke R (1996) Improvement of seedling quality in polybags through manipulation of potting media. COCOS 11:69–78
Peries RRA (1984) Some observations on the pre-nursery system for raising coconut seedlings. COCOS 2:10–17
Peries RRA (1995) A comparison of two coconut cultivars for high productivity under irrigation. COCOS 10:1–9

Peries RRA, Everard JMDT (1991) River sand as an alternative to top soil for raising coconut seedlings in polybags. COCOS 9:40–46

Peries RRA, Everard JMDT (1995) Is there ecological adaptation in coconut nursery seedling stage? A preliminary investigation. COCOS 10:10–19

Persley GJ (1992) Replanting the tree of life. Towards an international agenda for coconut palm research. CAB, Oxford, p 156

Pillai RV (1994) Coconut seed production. In: Chadha KL, Rethinam P (eds) Advances in horticulture- plantation and spice crops. Part 1, vol 10. Malhotra Publishing House, New Delhi, pp 655–670

Pomier M, Brunin C (1974) Irrigation des cocotiers à l'eau salée. Oléagineux 29(4):183–186

Pordesimo LO, Noble DH (1989) Evaluation of alternative replanting strategies for small coconut farms in the Philippines using a simulation model. Agric Syst 32(1):27–39

Potty VP, Jayasankar NP (1976) Influence of crop mixing of hybrid Napier on the root zone microflora of the coconut palm. In: Nayar NM (ed) Coconut research and development. Wiley Eastern, New Delhi, pp 300–303

Potty VP, George M, Jayasankar NP (1977) Effect of crop mixing on coconut rhizosphere. Indian Cocon J 8:1–2

Prabhu SR, Subramanian P, Biddappa CC, Bopaiah BM (1998) Prospects and improving coconut productivity through vermiculture technology. Indian Cocon J 29:79–84

Prabhu SR, Subramanian P, George V Thomas (2000) Opportunities for organic cultivation of coconut palms. In: 37th Cocotech meeting–International coconut conference 2000 Chennai, India, pp 37-45

Prasada Rao GS (1986) Effect of drought on coconut production. Indian Cocon J 17(2):11

Premaratne WHE, de Silva MP (1991) Evaluation of a high intensity mixed cropping model in the mid country of Sri Lanka. First international symposium on horticultural economics in developing countries, Alemaya, Ethiopia, 16-23 July 1989. Acta Hortic 270:381–387

Premaratne S, Somasiri SC (2011) Integration of livestock in coconut based agro forestry systems. In: Pushpakumara DKNG, Gunasena HPM, Gunathilake HAJ, Singh VP (eds) Increasing coconut land productivity through agro forestry interventions. Proceedings of symposium held at the Coconut Research Institute of Sri Lanka, pp 67–76

Priya George, Alka Gupta, Murali Gopal, Chandramohanan R, Thomas L, Thomas GV (2012a) *In vitro* antagonism of rhizopheric flurescent pseudomonads of coconut against *Ganoderma applanatum* and *Thielaviopsis paradoxa*, fungal pathogens of coconut. J Plantn Crops 40(2):75–81

Priya George, Alka Gupta, Murali Gopal, Thomas GV (2012b) Screening and *in vitro* evaluation of phosphate solubilising bacteria from rhizosphere and roots of coconut palms (*Cocos nucifera* L.) growing in different states of India. J Plantn Crops 40(1):61–64

Priya George, Alka Gupta, Murali Gopal, Chandramohanan R, Thomas L, Thomas GV (2012c) *In vitro* screening of fluorescent *Pseudomonas* spp. for efficient biocontrol agents against *Ganoderma applanatum* and *Thielaviopsis paradoxa,* fungal pathogens of coconut. J Plantn Crops 39(2):278–284

Proud KRS (2005) A guide to intercropping coconuts. Principles for developing productive upland agriculture in the humid tropics. Prepared for the Upland Development Programme in Southern Mindanao (UDP). Project ALA – 97/68 www.udpmindanao.org

Rajagopalan A, Nair AR, Prasada Rao GSLHV (1988) Water requirement of young T x G coconut hybrids. In: Agrometeorology of plantation crops. Kerala Agricultural University, Thrissur

Rajagopalan A, Viswanathan TV, Nirmala Devi S (1992) Medicinal plants as intercrops in coconut gardens – a preliminary study. J Plantn Crops 20(Suppl):50–51

Rajeshkumar PP, Thomas GV, Gopal M, Alka Gupta (2015) Diversity, richness and degree of colonization of arbuscular mycorrhizal fungi in coconut cultivated along with intercrops in high productive zone of Kerala, India. Symbiosis 65(3):125–141

Ramadasan K, Abdullah I, Teoh KC, Vanialingam T, Chan E (1978) Intercropping of coconuts with cocoa in Malaysia. Planter 54:329–342

Ramadasan A, Kasturi Bai KV, Shivasankar S (1993) Selection of coconut seedlings through physiological and biochemical criteria. In: Nair MK, Khan HH, Gopalasundaram P, Rao EVVB (eds) Advances coconut research and development. Oxford and IBH Publishing, New Delhi, pp 201–208

Ramakrishnan Nayar TV, Sahasranaman KN (1978) Performance of four fodder crops in coconut garden under rainfed and irrigated conditions. In: Nelliat EV, Ranganathan V, Vishveshwara S, Potti SN, Krishna Murthy Rao W, Zachariah PK (eds) Proceedings of PLACROSYM-1. Indian Society for Plantation Crops, Kasaragod, pp 425–430

Ramamoorthy M, Muthusamy M, Meena B (1999) Biodegradation of coir pith using fungal biocontrol agents. Indian Cocon J 29(11):1–3

Ranatunga AS, Liyanage LVK, Perera RAJR (1988) Coconut-based cropping systems in the wet and wet intermediate zones: present constraints and prospects. Cocon Bull 5(2):8–12

Ranaweera SRM, Nanayakkara CM, Tennakoon NA (2010) Comparison of the effects of organic fertilizers with inorganic fertilizers on the growth of eight months old coconut seedlings and the nutrient availability and soil microbial activity of soils. In: Proceedings of 15th International forestry and environment symposium, 26–27 November 2010, Sri Lanka

Rao AS (1989) Water requirements of young coconut palms in a humid tropical climate. Irrigation Sci 10(3):245–249

Rao GSLHVP (1994) Water management in coconut. In: Satheesan KV (ed) Water management for plantation crops. Problems and prospects. CWRDM, Kozhikode, pp 65–70

Rao AS, Vamadevan VK (1982) A climatic study of water availability periods in Kerala. Ann Natl Assoc Geogr 11:1–7

Ratnambal MJ (1995) Deputation report (COGENT/LPGRI Trainers course on coconut breeding research techniques, September 19–29 1995, Monado, Indonesia). CPCRI, Kasaragod, India

Ravi Bhat, Krishnakumar V (2011) Influence of planting material and nutrients on herbage and oil yield of patchouli grown under coconut. J Plantn Crops 39(1):57–61

Remison SU (1987) Coconut research in Nigeria – advances. Agric Int 39:290–292

Remison SU, Iremiren GO (1990) Effect of salinity on the performance of coconut seedlings in two contrasting soils. COCOS 8:33–39

Remison SU, Mgbeze G (1987) Comparative effects of weed and beneficial effect of mulching on coconut seedlings. COCOS 5:19–28

Remison SU, Mgbeze CC (1988) Effects of storage and planting methods on the germination of coconut. Niger J Palms Oil Seeds 9:59–70

Rethinam P (1989) Research highlights of AICRP on Palms. 1972-1979. Central Plantation Crops Research Institute, Kasaragod, pp 11–12

Reynolds SG (1988) Pastures and cattle under coconuts. Rome, FAO plant production and protection paper no. 91, p 321

Reynolds SG (1993) Pastures and livestock under tree crops: present situation and possible future developments. In: Evans TR, Macfarlane DC, Mullen BF (eds) Sustainable beef production from small holder and plantation farming systems in the South Pacific. Proceedings of a workshop. AIDAB, pp 77–117

Reynolds SG (1995) Pasture-cattle-coconut systems. Food and Agriculture Organisation of United Nations, Bangkok, p 668

Rico Gray V, Garcia-Franco JG, Alexandra C, Armando P, Paulino S (1990) Species composition, similarity and structure of mayan homegardens in Tixpeual and Tixcacaltuyub, Yucatan, Mexico. Econ Bot 44:470–482

Rohini Iyer (1983) Rhizosphere microflora in crop mixed coconut soil. Philippine J Cocon Stud 6(2):42–45

Romney DH (1987) Competition from mature coconut trees on under-planted coconut varieties and hybrids. Oléagineux 42:191–196

Romney DH, Whitehead RA, Smith RW, Shaw FD (1968) How to grow better coconuts. A handbook for coconut growers in the Carribean and Central America. FAO, UNPL.IC/5, p 48

Rosenfeld E (2009) Effects of pruning on the health of palms. Arboricult Urban For 35(6):294–299

Roupsard O, Bonnefond J-M, Irvine M, Berbigier P, Nouvellon Y, Dauzat J, Taga S, Hemel O, Jourdan C, Saint-André L, Mialet-Serra I, Labouisse J-P, Epron D, Joffre R, Braconnier S, Rouzière A, Navarro M, Bouillet J-P (2006) Partitioning energy and evapo-transpiration above and below a tropical palm canopy. Agric For Meteorol 139:252–268

Sahasranaman KN (1961) Chillies-an intercrop for coconut plantation. Cocon Bull 15(2):49–52

Sahasranaman KN (1964) It pays to grow groundnut in coconut plantations. Cocon Bull 18(4):128–130

Sahasranaman KN, Pillai NG (1976) Mixed farming in coconut garden. Intensive Agric 14(9):6–10

Sahasranaman KN, Pillai NG, Jayasankar NP, Potty VP, Varkey T, Amma PGK, Radha K (1983) Mixed farming in coconut gardens-Economics and its effect on root (wilt) disease. In: Nayar NM (ed) Coconut research and development. Wiley Eastern, New Delhi, pp 160–165

Sairam CV, Gopalasundaram P, Reddy DVS, Subramanian P, Umarnaheswari L, Hegde MR (1999) Cash flow analysis of coconut based high density multi-species cropping system – a case study. J Plantn Crops 27(1):37–44

Salam MA, Mammen MK (1990) Soil and climate based irrigation scheduling for adult coconut. Indian Cocon J 21(3):2–6

Salam MA, Sreekumar D, Mohanakumaran N, Sathees Babu K (1991) Coconut based homestead farming: a model. Indian Cocon J 22(4):4–12

Sampson HC (1923) The coconut palm: the science and practice of coconut cultivation. John Bale Sons and Danielsson, London, p 262

San Juan NC, Abad RG (1981) Preliminary study on the adaptability of winged bean (*Psophocarpus tetragonolobus*) (L.) as a cover crop under coconut (*Cocos nucifera*). Philippine J Weed Sci 8:1–4

Sanico RLT, Laguna LM (1989) Comparative effects of weeding methods on coconut growth. Philippine J Cocon Stud 19(1):5–7

Santiago RM (1978) Growth of coconut seedlings as influenced by different fertilizer levels and three soil types. Philippine J Cocon Stud 3:15–27

Saseendran SA, Jayakumar M (1988) Consumptive use and irrigation requirement of coconut plantations in Kerala. J Plantn Crops 16(2):119–125

Satyabalan K (1976) Coconut growing in Fiji Islands. Cocon Bull 7:1–4

Satyabalan K (1983) Age of seedling at transplanting in coconut. J Plantn Crops 11:13–20

Satyabalan K (1984) Nursery studies in West Coast Tall coconut.2. Seedling growth and output of quality seedlings in seed nuts germinated during different months. Indian Cocon J 14(12):3–11

Satyabalan K, Mathew J (1983) Identification of pre potent palms in West Coast Tall coconut based on growth of the progeny in the nursery. In: Nayar NM (ed) Coconut research and development. Wiley Eastern, New Delhi, pp 15–22

Scheewe W (2003) Coconut farm diversification- ideas and considerations. Group of Advocates for Sustainable Agriculture, Inc, p 31

Secretaria MI, Magat SS (2004a) PCA generates technology for a productive coconut-gmelina tree cropping system for farmers to increase land productivity and farm income. Cocon Today 19:33–37

Secretaria MI, Magat SS (2004b) Long-term effects of CLP on CBFS: Coconut + Gmelina agroforestry ecosystem. Paper presented during the 17th regional symposium on RDE highlights, Aug. 4–5, 2004. USEP, Davao City

Secretaria MI, Magat SS (2006) Coconut-root crops cropping model. Coconut intercropping guide no. 5. Department of Agriculture, Philippine Coconut Authority, p 22

Secretaria MI, Magat SS, Canja LH (2003) Predicting yield of coconut at two leaf pruning levels under four coconut-based crop combinations. Philippine Agric Sci 86(3):290–302

Senarathne SHS, Gunathilake HAJ (2010) Weed management in mature coconut plantations in Sri Lanka. COCOS 19(2):93–100

Senarathne SHS, Perera KCP (2011) Effect of several weed control methods in tropical coconut plantation on weed abundance, coconut yield and economical value. Int Res J Plant Sci 2(2):25–31

Senarathne SHS, Samarajeewa AD, Perera KCP (2003) Comparison of different weed management systems and their effects on yield of coconut plantations in Sri Lanka. Weed Bio Manage 3:158–161

Sethumadhava Menon K, Ramakrishnan Nayar TV (1978) Effect of intercropping with tuber crops in root (wilt) affected coconut garden. In: Nelliat EV, Ranganathan V, Vishveshwara S, Potti SN, Krishna Murthy Rao W, Zachariah PK (eds) Proceedings of PLACROSYM-1. Indian Society for Plantation Crops, Kasaragod, pp 416–424

Shanthamallaiah NR, Venkatesha Gowda TN, Krishnamonohar R, Shivayogishwar B (1982b) Studies on intercropping coconut with different field crops in Maidan tract of Karnataka. In: Proceedings of PLACROSYM–V. Indian Society of Plantation Crops, Kasaragod, pp 402–404

Shanthamalliah NR, Murugendrakumar C, Krishnamanohar R (1978) Effect of coir dust mulching on water requirement and on the performance of coconut seedlings in the nursery. In: Nelliat EV, Ranganathan V, Vishveshwara S, Potti SN, Krishna Murthy Rao W, Zachariah PK (eds) Proceedings of PLACROSYM-1. Indian Society for Plantation Crops, Kasaragod, pp 261–266

Shanthamalliah NR, Venkatesha Gowda TN, Krishnamanohar R, Shivayogishwara B, Balakrishna P (1982a) Note on the mixed cropping of mulberry in coconut plantation. In: Proceedings of PLACROSYM-V. Indian Society for Plantation Crops, Kasaragod, pp 465–467

Sharma R, Prasad S, Mohan NK, Medhi G (1996) Economic feasibility of growing some root and tuber crops under intercropping system in coconut garden. Hortic J 9(2):167–170

Shelton HM, Humphreys LR, Batello C (1987) Pastures in the plantations of Asia and the Pacific: performance and prospect. Trop Grasslands 21(4):159–168

Shelton HM (1991) Productivity of cattle under coconuts. In: Shelton HM and Stur WW (eds) Forages for plantation crops. Proceedings of a workshop, Bali, ACIAR Proceedings no. 32 pp 92–96

Smith MA, Whiteman PC (1983) Evaluation of tropical grasses in increasing shade under coconut canopies. Exp Agric 19:153–161

Solangi AH, Iqbal MZ (2012) Preliminary studies on morphological diversity of coconut (Cocos Nucifera L.) seedlings by organic and inorganic fertilizer amendments at Karachi, Pakistan. Pak J Bot 44(1):161–164

Solangi AH, Mal B, Kazmi AR, Iqbal MZ (2010) Preliminary studies on the major characteristic, agronomic feature and nutrient value of *Gliricidia sepium* in coconut plantations of Pakistan. Pak J Bot 42(2):825–832

Srinivasa Reddy DV (1998) Comparative performance of coconut nursery raised in polybags and cement tanks on different media. In: Menon ARS, Iyer RD (eds) Proceedings of 7th Swadeshi Science Congress, Swadeshi Science Movement, Kochi, India, pp 148–150

Srinivasa Reddy DV (2000) Agronomic techniques for production of vigorous seedlings of WCT coconut. In: Muralidhharan N, Rajkumar R (eds) Recent advances in plantation crops research. Allied Publisheres, Mumbai, pp 160–163

Srinivasa Reddy DV, Arunachalam V (2002) Patchouli (*Pogostemon cablin*) – a potential aromatic crop for intercropping in coconut gardens. Indian Cocon J 32(10):1–3

Srinivasa Reddy DV, Biddappa CC (2000) Coconut based cropping/farming system in India. J Plantn Crops 28(1):1–18

Srinivasa Reddy DV, Thomas GV (2001) Mixed cropping of black pepper in coconut and arecanut gardens. Technical bulletin no. 42, Central Plantation Crops Research Institute, Kasaragod, India, p 20

Srinivasa Reddy DV, Ratnambal MJ, Gopalasundaram P (1996) Poly bag nursery for production of vigorous seedlings in coconut. Indian Cocon J 27(3):5–7

Srinivasa Reddy DV, Dhanapal R, Palaniswami C, Maheswarappa HP (1998a) Effect of levels and methods of fertilization on growth and recovery of vigorous seedlings in WCT coconut nursery. J Plantn Crops 26(1):58–62

Srinivasa Reddy DV, Bidappa CC, Gopalasundaram P (1998b) Clove (*Syzygium aromaticum* L. Merr and Perry.) mixed cropping in coconut garden. Indian Cocon J 29(6):6–8

Srinivasa Reddy DV, Gopalasundaram P, Biddappa CC (2000) Cultivating clove in coconut garden. Indian Hortic 45(2):29–30

Srinivasa Reddy DV, Naresh Kumar S, Prabhu SR (2001) Evaluation of alternative media to potting mixture for raising coconut seedlings in poly bags. J Plantn Crops 29(1):62–65

Steel RJH, Humphreys LR (1974) Growth and phosphorous response of some pasture legumes grown under coconuts in Bali. Trop Grasslands 8(3):171–178

Stur WW, Shelton HM (1991) Compatibility of forages and livestock with plantation crops. In: Shelton HM, Stur WW (eds) Forages for plantation crops. Proceedings of a workshop, Bali, Indonesia, ACIAR proceedings no. 32, pp 112–116

Stur WW, Reynolds SG, Macfarlane DC (1994) Cattle production under coconuts In: Copeland, JW, Djajanegra A, Sabrani M (eds) International symposium AAAP Animal Science Congress, Bali, Indonesia. Agroforestry and animal production for human welfare. ACIAR proceedings no. 55, Canberra, Australia, pp 106–114

Subramaniam PS, Mohandas S, Marimuthu R, Gurusamy A, Natarajan AS (2010) Multispecies cropping system in coconut for east coast region of Tamil Nadu. In: Thomas GV, Krishnakumar V, Jerard BA (eds) Improving productivity and profitability in coconut farming. Central Plantation Crops Research Institute, Kasaragod, pp 162–168

Subramanian P, Krishnasamy S, Devasagayam MM (1997) Studies on the evaluation of drip irrigation in comparison with surface irrigation (basin) in coconut. South Indian Hortic 45(5/6):255–258

Subramanian P, Biddappa CC, Maheswarappa HP, Dhanapal R, Palaniswani C (2000) Growing of Gliricidia as green manure crop in coconut garden under littoral sandy soil. J Plantn Crops 28:212–217

Subramanian P, Dhanapal R, Palaniswami C, Maheswarappa HP (2006) Coastal sandy soil management through intercropping/alley cropping of Glyricidia in the coconut garden. Cocon Info Int 13(1):23–24

Subramanian P, Dhanapal R, Palaniswami C, Joseph Sebastián (2007) Feasibility studies on growing hybrid Bajra Napier fodder grass as intercrop in coconut under coastal littoral sandy soil. J Plantn Crops 35(1):19–22

Subramanian P, Dhanapal R, Palaniswami C, Alka Gupta, Vidhan Singh T (2008) Comparison of biomass yield of hybrid bajra Napier grass in organic and integrated nutrient management under coconut based mixed farming system In: National conference on organic farming in horticultural crops with special reference to plantation crops, Programme and abstracts. 15–18 October, 2008, Central Plantation Crops Research Institute, Kasaragod, India

Subramanian P, Dhanapal R, Palaniswami C, Maheswarappa HP, Alka Gupta, George V Thomas (2009) Coastal sandy soil management for higher coconut productivity. Technical bulletin no. 50. Central Plantation Crops Research Institute, Kasaragod, India, p 20

Subramanian P, Dhanapal R, Mathew AC, Palaniswami C, Upadhyaya AK, Nareshkumar S, Reddy DVS (2012a) Effect of fertilizer application through micro irrigation technique on nutrient availability and coconut productivity. J Plantn Crops 40:168–173

Subramanian P, Dhanapal R, Palaniswami C, Alka Gupta, Maheswarappa HP (2012b) Agro-techniques for higher coconut productivity under coastal sandy soil. Cocon Info Intl 19(1):54–61

Subramanian P, Dhanapal R, Thomas GV, Maheswarappa HP, Thamban C (2014) Coconut based integrated farming system. Tech bull no. 78, ICAR-Central Plantation Crops Research Institute, Kasaragod, India, p 32

Subramanian P, Krishnakumar V, Alka Gupta, Murali Gopal, Surekha R, Selvamani V (2016) Impact of long term organic cultivation practices on soil health and productivity of coconut. In: Chowdappa P, Muralidharan K, Samsudeen K, Rajesh MK (eds) Abstracts. Third International symposium on coconut research and development, ICAR-CPCRI, Kasaragod, India, p 54

Sudha B, Annamma G (2011) Tillage and residue management for organic carbon sequestration in coconut (*Cocos nucifera*)-based cropping systems. Indian J Agro 56:223–227

Sudhakara K, Kannan K, Nambiar IPS (1989) Effect of defoliation on the productivity of coconut palm. J Plantn Crops 16(suppl):457–461

Suja G, Nair VM, Saraswathy P, Pushpakumari R (2003a) Plant population and sett size effects on white yam (*Dioscorea rotundata* Poir.) intercropped in coconut garden. Trop Agric 80(3):157–162

Suja G, Nair VM, Sreekumar J (2003b) Influence of organic manures, nitrogen and potassium on nutrient uptake and nutrient use efficiency of white yam (*Dioscorea rotundata* Poir) intercropped in coconut garden (*Cocos nucifera*). Indian J Agro 48:168–171

Suja G, Nair VM, Saraswathy P, Sreekumar J (2004a) Implications of organic manure, nitrogen and potassium application in white yam (*Dioscorea rotundata* Poir.) intercropped in coconut garden on nutrient balance of soil. Ann Agric Res 25(1):76–82

Suja G, Nair VM, Saraswathy P, Nayar TVR (2004b) Response of white yam intercropped in coconut garden to organic manures, nitrogen and potassium. J Plantn Crops 32(1):21–27

Sumathykutty Amma B (1964) Preliminary studies on the effect of micronutrients on the germination of coconut seed nuts. Curr Sci 33:49–50

Sumbak JH (1968) Coconut seedling establishment as affected by seedling development at transplanting as well as agronomic practices. Progress Report (1) Oleagineux 23:579–582

Sumbak JH (1970) Coconut seedling establishment as affected by seedling development at transplanting as well as agricultural practices. Papua New Guinea Agric J 22:6–25

Sundaresan N, Rangaswamy M, Mohammed Ibrahim PA (1974) Suitable season for the collection of seed nut in the East coast of Tamil Nadu. Madras Agric J 61:525–530

Suneetha MS, Chandrakanth MG (2003) Intercropping medicinal plants in coconut gardens: a feasible enterprise. Indian Cocon J 34:12–14

Sutherland R, Ziegler A (2007) Effectiveness of coir-based rolled erosion control systems in reducing sediment transport. Appl Geogr 27:150–164

Taffin GD, Zakra N, Pomier M, Braconnier S, Weaver RW, De-Taflin G (1991) Search for a stable cropping system combining coconut and nitrogen-fixing trees. Oleagineux 46(12):489–500

Tennakoon NA (2011) Soil fertility and water management through coconut based agro forestry systems. In: Pushpakumara DKNG, Gunasena HPM, Gunathilake HAJ, Singh VP (eds) Increasing coconut land productivity through agro forestry interventions. Proceedings of a symposium held at the Coconut Research Institute of Sri Lanka, pp 45–56

Thamban C, Sairam CV, Muralidharan K, Krishnakumar V, Arulraj S (2006) Multidimensional analysis of coconut based farming systems adopted by farmers. J Plantn Crops 34(3):720–725

Thamban C, Mathew AC, Subramanian P, Muralidharan KK (2014) Soil and water conservation measures for sustainable coconut production: impact of interventions under Farmers Participatory Action Research Programme. J Plantn Crops 42(1):94–100

Thampan PK (1981) Handbook on coconut palm. Oxford & IBH Publishing, New Delhi, pp 59–101

Thiruvarassan S, Maheswarappa HP (2014) Performance of medicinal and aromatic plants as intercrops in coconut garden. J Plantn Crops 42(2):238–240

Thiruvarassan S, Maheswarappa HP, Subramani T (2014) Evaluation of coconut based multispecies cropping systems for east coast region of Tamil Nadu. J Andaman Sci Assoc 19(1):59–64

Thiruvarassan S, Mohandas S, Ganesan K, Sanbagavalli S (2016) Response of Coconut (*Cocos nucifera* L.) to Fertigation in East Coast Region of Tamil Nadu. Madras Agric J 103(4–6):166–168

Thomas KM (1974) Influence of certain physical and chemical treatments on the germination and subsequent growth of coconut (*Cocos nucifera* L.) seedlings: a preliminary study. East African Agric For J 40:152–156

Thomas GV (1987) Microbial population, enzyme activity and VA - mycorrhiza in the root region of coconut in relation to in situ green manuring. In: Sethuraj MR, Potti SN, Bavappa KVA, Jayarathnam K, Ramaiah PK, Krishnamurthy KK, Das PK (eds) Proceedings PLACROSYM VI. Oxford and IBH Publishing, New Delhi, pp 267–274

Thomas GV (1988) Vesicular-arbuscular mycorrizhal symbiosis in coconut *(Cocos nucifera)* in relation to the root (wilt) disease and intercropping or mixed cropping. Indian J Agric Sci 57(2):145–147

Thomas GV (2010) Technological advances in organic farming of plantation crops. In: Singh HP, Thomas GV (eds) Organic horticulture, principles, practices and technologies. Westville Publishing House, New Delhi, p 440

Thomas GV, George M (1990) Basin management using green manure legumes for root (wilt) affected coconut palms. Indian Cocon J 20:7–8

Thomas GV, Ghai SK (1987) Genotype dependent variation in vesicular- arbuscular mycorrhizal colonization of coconut roots. Proc Indian Acad Plant Sci 97:289–294

Thomas GV, Ghai SK (1991) Influence of *Rhizobium* inoculation and seed pelleting on nodulation of green manure legumes in an acidic coconut soil. J Plantn Crops (Suppl) 18:72–77

Thomas GV, Prabhu SR (2003) Association of diazotrophic and plant growth promoting rhizobacteria with coconut palm (*Cocos nucifera* L.). In: Proceedings of 6th International PGPR workshop, 5–10 October 2003, Calicut, India, pp 20–26

Thomas GV, Shantaram MV (1984) *In situ* cultivation and incorporation of green manure legumes in coconut basins. An approach to improve soil fertility and microbial activity. Plant Soil 80:373–380

Thomas GV, Shantaram MV (1986) Solubilization of inorganic phosphates by bacteria from coconut soils. J Plantn Crops 14:42–48

Thomas GV, Shantaram MV, Saraswathy N (1985) Occurrence and activity of phosphate solubilizing fungi from coconut plantation soils. Plant Soil 87:357–364

Thomas GV, Rohini Iyer, Bopaiah BM (1991) Beneficial microbes in the nutrition of coconut. J Plantn Crops 19:127–138

Thomas GV, Rajagopal V, Bopaiah BM (1993) VA-mycorrhizal association in relation to drought tolerance in coconut. J Plantn Crops (Suppl) 21:98–103

Thomas GV, Prabhu SR, Reeny MZ, Bopaiah BM (1998a) Evaluation of lignocellulosic biomass from coconut palm as substrate for cultivation of *Pleurotus sajor caju*(Fr.) Singer. World J Microbiol Biotechnol 14:879–882

Thomas GV, Reeny MZ, Prabhu SR (1998b) Variation in oyster mushroom production in coconut by-products as influenced by mixing of substrates. In: Proceedings of Swadeshi Science Congress, Swadeshi Science Movement, Kochi, India, pp 40–46

Thomas GV, Biddappa CC, Prabhu SR (2001a) Evaluation of N_2-fixing cover legumes as green manures for N substitution in coconut palm (*Cocos nucifera* L.). Trop Agric (Trinidad) 78:13–18

Thomas GV, Prabhu SR, Subramanian P, Iyer R (2001b) Organic farming technologies in coconut. ATIC series publication no. 4, Central Plantation Crops Research Institute, Kasaragod, Kerala, India, p 14

Thomas GV, Subramanian P, Krishnakumar V, Alka Gupta, Chandramohanan R (2010a) Package of practices for organic farming in coconut. Tech bull no. 64. CPCRI, Kasaragod, India, p 25

Thomas GV, Krishnakumar V, Subramanian P, Alka Gupta (2010b) Organic farming as a viable nutrient management strategy for sustainability in organic farming. In: Thomas GV, Krishnakumar V, Jerard BA (eds) Improving productivity and profitability in coconut farming. Central Plantation Crops Research Institute, Kerala, India, pp 230–237

Thomas GV, Krishnakumar V, Subramanian P, Murali Gopal, Alka Gupta (2012a) Organic farming in coconut – feasibility, technological advancements and prospects. Indian Cocon J 55(7):22–31

Thomas GV, Palaniswami C, Gopal M, Alka Gupta (2012b) Recycling coconut leaf-agro wastes mixture using *Eudrilus* sp. and growth promotion properties of coconut leaf vermicompost. Int J Innov Hortic 1(2):113–118

Thomas GV, Palaniswami C, Prabhu SR, Murali Gopal, Alka Gupta (2013) Co-composting of coconut coir pith with solid poultry manure. Curr Sci 104(2):245–250

Thomas GV, Murali Gopal, Alka Gupta, Subramanian P, Prabhu SR (2016) Bio- resources based biological soil fertility management for sustainable coconut production. In: Chowdappa P et al (eds) Abstracts of 3rd International symposium on coconut research and development, ICAR – Central Plantation Crops Research Institute, Kasaragod, India, 15–17 December 2016

Thompson P (1988) Fodder production and fodder conservation in the Kingdom of Tonga. Mission report to FAO, Project RAS/86/039, p 34

Tripetchkul S, Pundee K, Koonsrisuk S, Akeprathumchai S (2012) Co-composting of coir pith and cow manure: initial C/N ratio vs physico-chemical changes. Int J Recycling Org Waste Agric 1:15

Ubaldo MBA, Magat SS, Rillio EP, Secretaria MI (2006) Crop nutrition and fertilizer management of embryo-cultured seedlings of Makapuno and Laguna Tall coconuts with special reference to their response to different levels of NaCl and chicken manure. CORD 22(2):1–14

Ugbah MM, Akpan EEJ (2003) Effect of nut positioning and husk slashing nut germination and growth of two coconut varieties. Niger J Palms Oil Seeds 15:11–21

Uthaiah BC, Indiresh KM, Shetty TKP (1993) Preliminary studies on the effect of mulches and irrigation on growth of young coconut plants in coastal Karnataka. Indian Cocon J 24(6):5–9

Utomo B (2013) Environmental performance of cocoa production from monoculture system and agroforestry system in Indonesia. Thesis submitted as a part of the requirements for the degree of Master of Science in Environmental Technology and Management the Joint Graduate School of Energy and Environment at King Mongkut's University of Technology, Thonburi, p 105

Van Holm L Jr (1993) A report on Coir dust as a growing medium presented to the 7th floricultural symposium. Colombo Biological nitrogen fixation project. Institute of Fundamental Studies, Hantana, Kandy, Sri Lanka

Varadan KM, Chandran KM (1991) Irrigation requirement of coconut under drip method. In: Silas EG, Aravindakshan M, Jose AI (eds) Coconut breeding and management. Kerala Agricultural University, Thrissur, pp 281–282

Varghese PT, Nelliat EV, Balakrishnan TK (1978a) Beneficial interactions of coconut-cacao combination. In: Nelliat EV, Ranganathan V, Vishveshwara S, Potti SN, Krishna Murthy Rao W, Zachariah PK (eds) Proceedings of PLACROSYM-1. Indian Society for Plantation Crops, Kasaragod, pp 383–392

Varghese PT, Nair PKR, Nelliat EV, Rama Varma, Gopalasundaram P (1978b) Intercropping with tuber crops in coconut garden. In: Nelliat EV, Ranganathan V, Vishveshwara S, Potti SN, Krishna Murthy Rao W, Zachariah PK (eds) Proceedings of PLACROSYM-1. Indian Society for Plantation Crops, Kasaragod, pp 399–415

Veloso AV, Ly T (1982) Effect of chlorine sources and ammonium sulfate and the growth of coconut seedlings grown on four soil types. Ann Trop Res 4(3):224–230

Venkitaswamy R, Christopher LA, Devasenapathy P, Sridharan CS, Prabhakaran NK, Madhiyazhagan R (1997) Effect of drip irrigation on hybrid coconut. Madras Agric J 84:591–593

Vergara NT, Nair PKR (1985) Agroforestry in the South Pacific Region: an overview. Agrofor Syst 3:363–379

Verghese EJ, Krishna Marar MM, Thomas PK (1953) Preliminary studies on the variation in moisture in coconut soils. Indian Cocon J 6:49–56

Vidhana Arachchi LP (1996a) Characterization of physical properties of soils and studies on the development of coconut roots. The project report (12/175/1490 submitted to the Council for Agricultural Research Policy (CARP), Sri Lanka

Vidhana Arachchi LP (1996b) Effect of soil physical properties on growth and activity of coconut (*Cocos nucifera L.*) roots. Thesis submitted in partial fulfillment of the requirements for the degree of Doctor of Philosophy of the Faculty of Applied Science, University of Sri Jayawardenepura, Nugegoda, Sri Lanka

Vidhana Arachchi LP (1998a) An evaluation of different mulches used in soil moisture conservation of coconut lands. CORD 14(2):12–27

Vidhana Arachchi LP (1998b) Preliminary requirements to design a suitable drip irrigation system for coconut (*Cocos nucifera* L.) in gravelly soils. Agric Water Manage 38:169–180

Vidhana Arachchi LP, Jayasekara KS (1988) Coir dust to improve coconut production. Annu Sessions Sri Lanka Assoc Adv Sci B 10:32

Vidhana Arachchi LP, Somasiri LLW (1993) Efficient use of coir dust on Sudu series soil. Annu Sessions Sri Lanka Assoc Adv Sci B 45:73

Vidhana Arachchi LP, Somasiri LLW (1997) Assessment of oxygen availability in soils of different drainage classes and its effect on coconut root growth. Proceedings of Sri Lanka Assoc Adv Sci 53(2):159–160

Vidhana Arachchi LP, Yapa PAJ, Mapa RB, Somapala H (2000) Soil physical constraints and their effect on morphological characters of coconut (*Cocos nucifera* L.) roots. CORD 16(1):14–33

Vijayakumar KR, Mammen G, Pillai GG, Vamadevan VK (1986) Alley cropping of *Leucaena* in coconut gardens in Western Ghats of India: yields of dry matter and organic nitrogen. Leucaena Res Rep 7:72–74

Vijayaraghavan H, Ramachandran TK (1989) Effect of *in situ* cultivation and incorporation of green manure crops on the yield of coconut. COCOS 7:26–29

Viswanathan TV, Radhakrishnan VV, Reghunath BR, Prsannakumari KT (1992) Patchouli as an intercrop in young coconut garden. Indian J Agric Sci 63(9):616–617

Viswanathan TV, Reghunath BR, Radhakrishnan VV, Prasannakumari KT, Sosamma C (1993) Patchouli – a potential essential oil yielding crop for Coconut gardens. Indian Perfumer 37:12–13

Whitehead RA, Smith RW (1968) Results of a coconut spacing trial in Jamaica. Trop Agric (Trinidad) 45:127–132

Wickramaratne MRT (1987) More palms, more nuts, more copra. Cocon Bull 4(1):1–5

Wickramaratne MRT, Padmasiri MHL (1986) Some observations on the position of the soft eye in *Cocos nucifera*. COCOS 4:35–37

Wickramaratne MRT, Coe R, Fernando S (1987) Evaluation of criteria for selection of seed coconuts (*Cocos nucifera* L.). COCOS 5:1–7

Wijebandara DMDI (2010) Phosphorus status of major coconut growing soils of Sri Lanka on the performance of *Pueraria phaseoloides*. COCOS 19:27–40

Wong CC (1991) Shade tolerance of tropical forages: a review. In: Shelton HM, Stur WW (eds) Proceedings of workshop forages for plantation crops. Bali, Indonesia, ACIAR proceedings no. 32, pp 64–69

Wuidart W (1981a) Production of coconut planting material-selection of hybrids in the seed-bed. Oleagineux 36(10):497–499

Wuidart W (1981b) Production of coconut planting material. Conduct of a seed-bed. Oléagineux 36(6):305–309

Wuidart W (1981c) Production of coconut planting material. The polybag nursery. Oleagineux 36(7):367–376

Wuidart W, Nuce de Lamothe M d (1981a) Ripeness of coconut seeds and germination. Oleagineux 36(11):549–554

Wuidart W, Nuce de Lamothe M d (1981b) Seed germination and development of coconut plants in function of nut position. Oleagineux 36(12):600–603

Yusuf M, Varadan KM (1993) Water management studies on coconut in India. In: Nair MK, Khan HH, Gopalasundaram P, Rao EVVB (eds) Advances in coconut research and development. Oxford and IBH Publishing, New Delhi, pp 337–346

Ziegler AD, Sutherland RA (1997) Reduction in interrill sediment transport by rolled erosion control systems. Soil Tillage Res 45:265–278

Ziller R, Fremond Y (1961) Novos resultados com a adubacao do coquiro na Costa do Marfim. Fertilite 13:13–25

Dr. George V. Thomas Former Director, ICAR-Central Plantation Crops Research Institute, Kasaragod, India, has 36 years of research experience on coconut, especially on bio-resources management. He was Chairman, International Coconut Genetic Resources Network; President, Indian Society for Plantation Crops; and Editor, Journal of Plantation Crops. He is a recipient of ICAR Outstanding Team Research Award. Dr. Thomas has 150 research publications and has edited 10 books. Presently, he is Emeritus Scientist, Council for Food Research and Development, Konni, Kerala, India. georgevthomas@yahoo.com

Dr. V. Krishnakumar is currently a Principal Scientist (Agronomy) and Head, ICAR-Central Plantation Crops Research Institute (Regional Station), Kayamkulam, Kerala. He served as Editor of the Journal of Plantation Crops. As an agronomist, he has edited 10 books and published more than 100 research and technical articles as well as contributing chapters in various books and technical bulletins. dr.krishnavkumar@gmail.com

Dr. R. Dhanapal is an agronomist, retired as a Principal Scientist from the ICAR-Sugarcane Breeding Institute, Coimbatore, India. He has 28 years of experience in coconut, mainly in the area of water management. He has 60 publications to his credit. ramaswamydhanapal@yahoo.co.uk

Dr. D. V. Srinivasa Reddy is currently working as Principal Scientist (Agronomy) in ICAR-Agricultural Technology Application Research Institute, Bengaluru, India. He has served the Central Plantation Crops Research Institute (ICAR). He has also worked with the Department of Agriculture and Cooperation, Ministry of Agriculture (Government of India), as Deputy Commissioner (Crops). Dr. Reddy has 98 publications apart from chapters in 5 books. reddydvs@yahoo.com

Chapter 8
Soil Productivity and Nutrition

H. Hameed Khan and V. Krishnakumar

Abstract The natural habitat of coconut is the coastal belt of the tropics where sandy and red sandy loam soils predominate. This chapter describes the major coconut growing soils around the world, their characteristics and management. An understanding of the role of major, secondary and micronutrients and their interactions is discussed in relation to the yielding ability of palms. As a general rule, it is advised that at least a quantity of nutrients exported through harvested produce and above-ground parts are replenished to maintain a balance between export and import of nutrients. For a perennial palm like coconut, a fertilizer management programme should consider soil nutrient reserves, plant nutrient status and nutrients stored in the plant system and evolve a system in preference to a blanket recommendation. Productivity and sustainability of yields in the light of integrated nutrient management are emphasized. A balanced approach evolved through the physiology-related growth parameters and their significance is presented in an integrated approach in the nutrient management of coconut. A review of fertilizer practices in coconut growing countries is provided, and a basis for fertilizer formulations to coconut palms is indicated in the wake of knowledge of DRIS, DFR and other models tested in coconut nutrition. Research results on leaf analysis and diagnostic techniques for site-specific nutrient recommendation to coconut gardens are indicated. A refinement of these models is suggested with respect to different agroclimatic regions. Soil and nutritional aspects associated with certain diseases and disorders across coconut growing regions are also presented.

H. Hameed Khan (✉)
160/14, DB Road, Optic Craft, RS Puram, Coimbatore, Tamil Nadu, India
e-mail: khanhh.in@gmail.com

V. Krishnakumar
ICAR-Central Plantation Crops Research Institute, Regional Station,
Kayamkulam, Kerala, India
e-mail: dr.krishnavkumar@gmail.com

K. U. K. Nampoothiri et al. (eds.), *The Coconut Palm (*Cocos nucifera *L.) - Research and Development Perspectives*,
https://doi.org/10.1007/978-981-13-2754-4_8

8.1 Introduction

The coconut palm grows on different soil types provided; they are reasonably deep and free draining to allow for unrestricted root development and aeration. Examples of major world producing areas indicate that the soils range from coarse sand, which contains up to 97% sand (0.2–2.0 mm), to heavy soils with about 70% clay (Child 1974). It is also observed that coconut grows very well in lower slopes, piedmont slopes of hills and, most importantly, coastal areas, where subsurface moisture drains down slope, which makes them to retain moisture for appreciably longer periods. Since coconut is grown in the tropical condition, most of the soils under coconut are deficient in organic matter except the soils of the humid tropics like lateritic soil of the hilly region and the alluvial soils. Manciot et al. (1979b) suggested a threshold value of 1% organic carbon for coconut soils. The soil nitrogen supply is directly related to the soil organic fractions. However, it varies with soil type.

On the west coast of India, coconut is grown on laterites, coastal sand and red sandy loam, while on the interior areas of the east coast, coconut is grown on deltaic alluvium, coastal sandy areas and localized patches of red and black soils. The morphological and some of the physico-chemical characteristics of typical coconut growing soils from different agroclimatic regions of India have been discussed by Khan et al. (1978). They classified soils under alfisols, entisols, ultisols and vertisols, and soil taxonomy up to subgroup level has been provided. Sankaranarayanan and Velayutham (1976a, b) have described in detail about soils that support coconut in the west coast and east coast of India.

It is commonly observed that coconuts along coastal areas are more productive than those growing inland. Strong reasons for this are more humid conditions and narrow temperature fluctuation and better subsoil water supply kept in constant movement by the ebb and flow of the tide of the sea (Menon and Pandalai 1960). It is claimed that sodium abundant in such coastal areas substitutes partially for the potassium needs of the coconut (Nartea and Reyes 1973).

8.2 Major Coconut Growing Soils and Their Management

The principal soil types that support coconut cultivation in the world are red and laterite, sandy, alluvium, coral, volcanic, clay and peat soils.

8.2.1 Red and Laterite Soils

In India, Sri Lanka and many other countries, coconut farming is seen on the red and laterite soils to a large extent. These soils are highly leached with varying depth of solum and are deficient in bases. Physically, red and laterite soils are loose, porous and well drained, but the water retention power is rather poor. Chemically, these soils are acidic and contain excess amounts of aluminium, iron and manganese. They are also deficient in many of the essential plant nutrients, particularly potassium and phosphorus. Since the nature of soil colloids in these groups of soils is dominantly 1:1 lattice silicate minerals intermixed with non-expanding 2:1 types, they have low cation exchange capacity and poor nutrient retention power. The dominant cations in these types of soils are Al and Fe which would convert the soluble phosphates into insoluble forms. The organic matter content is also low. Consequently, these groups of soils exhibit many production problems for coconut in the absence of appropriate management practices. Nevertheless, they respond well to manuring and support a good stand of productive coconut plantations.

8.2.2 Sandy Soils

The second major soil group on which the coconut is cultivated is the coastal sands. It is commonly found in the coastal belt of almost all coconut growing countries in the world. In India, it is found along the coastal belt both in the east and west coast of the peninsular India. In Sri Lanka, the sandy soils occur all along the west coast from Kalpitiya to Dondra head and are known as "cinnamon" sands. However, on the east coast of Sri Lanka, coarse sandy deposits overlie coral hard pans and are often shallow in depths. Practically, all coconut plantations in West Africa and Madagascar are observed on coastal sandy soils. Similar type of soil is found in Mozambique, in the Philippines and in the east coast of Malaysia. Consequently, this group of soils is deficient in almost all essential nutrients. The soil in general has a single-grained structure, poor silt and clay contents, poor in organic matter and deficient in almost all nutrients. The leaching loss of applied nutrients is heavy in high rainfall areas. Management of these soils requires liberal additions of almost all sources of organic matter, and that generated within the system for successful cultivation of coconut palms.

8.2.3 Alluvium

The third major soil group is the alluvium. Among all the soils, the river or estuarine alluvial deposits with sufficient amount of coarse and fine sand are considered to be the best for coconut cultivation. In India, coconuts are cultivated extensively in the deltaic regions of Cauvery,Godavari, Krishna, Mahanadi, Brahmaputra and Ganga where the soils are deep and loamy. In Sri Lanka, the best estates are located on the tidal loams of the alluvia of the Ma Oya, Deduru Oya and Batholu Oya rivers and their tributaries in the North-West Province and the estuarine deposits of the silted up Negombo, Madampe, Mundel and Puttalam lagoons on the west coast (Child 1974). In the Philippines, the alluvial limestone-derived soils are exceptionally fertile (Cooke 1936). In Indonesia excellent plantations are found on marine alluvial soils. Almost a similar situation is met with in other coconut growing countries of the world. These soils are well-drained, deep and highly fertile. The supply of organic matter and essential plant nutrients is usually more than adequate. As a result, alluvial soils support highly productive coconut groves.

8.2.4 Coral Soils

Coral soil is yet another group of soils extensively cultivated to coconut. This group covers most of the atolls (Polynesia). It is a mixture of coral, sand and rock, almost entirely calcareous, which poses unique problems of nutrient assimilation and deficiencies (Newton 1967). It is considered to be an extreme soil condition where the coral fragments are mixed with sand and pH reaching as high as 8.5–9.0 rendering some critical nutrients to be unavailable.

Their fertility depends on the amount of organic matter content and the degree of weathering. Many studies on these soils have been carried out under the auspices of the South Pacific Commission and the *Institut de Recherches Pour les Huiles et Oleagineux* (IRHO), particularly at the research station at Rangiroa in the Tuamotu Archipelago (Pomier 1964). Deficiencies of iron and manganese as well as potassium and nitrogen have been commonly noted in these soils. In the Solomon Islands and in New Hebrides, these soils are overlaid by alluvial deposits of varying depths, which makes them very fertile, and excellent coconut groves are found there. In India, the soils of Lakshadweep, some parts of Andaman and Nicobar and soils of Maldives Islands also belong to this group. Although these soils are poor in organic matter and major plant nutrients barring phosphorus, calcium and magnesium, the coconut production is excellent due to the supply of almost all plant nutrients through partially saline underground water. Potassium is the most limiting nutrient in such soils. The management of these soils is very difficult due to the highly sensitive ecosystem prevailing in the island habitats.

In Lakshadweep, India, 80% of the geographical area of 3270 ha of the Islands is cultivated to coconut. Krishnan et al. (1997, 2004) observed the soils are entirely derived from coral limestone and are predominantly sandy and loamy sand and majority of soil is classified *as carbonic isohyperthermic family* of *Typic ustipsamments* and in some island locations as *orthants* under soil taxonomy (Soil Survey Staff 1999). A chemical analysis of soil profile is presented in Table 8.1.

8.2.5 Volcanic Soils

The volcanic soils cultivated with coconut are generally fertile. They are found mainly in Indonesia and the Philippines. In Indonesia, coconuts grown on the so-called plains of Anggio on Sangihe Island which were built up from the dejects of

Table 8.1 A model profile of coral soil series with its characteristics (series A)

(a) Mechanical composition

Depth (cm)	Horizon	Sand (%)	Silt (%)	Clay (%)	Texture	E C dS/m	pH (1:2.5)
0–16	Ap	90.6	5.3	4.1	S	0.23	8.5
16–43	C1	92.6	2.4	5.0	S	0.17	8.5
43–84	C2	94.3	2.8	2.9	S	0.16	8.9
84–122	C3	97.1	2.8	1.5	S	0.13	9.1
122–164	C4	97.1	2.9	0.0	S	0.12	9.1

(b) Chemical constituents

Depth (cm)	O.C (%)	Ca CO_3 (%)	Exchangeable cations C (mol) kg^{-1} soil				CEC c (mol) kg^{-1} soil	Base saturation (%)
			Ca	Mg	Na	K		
0–16	0.78	93.1	5.88	0.47	0.32	0.1	1.0	100
16–43	0.59	95.5	5.37	0.48	0.29	0.6	0.6	100
43–84	0.19	96.0	4.47	0.33	0.33	0.5	0.5	100
84–122	0.11	9.8	4.34	0.39	0.29	0.4	0.4	100
122–164	0.13	96.0	4.67	0.37	0.31	0.4	0.4	100

(c) Available nutrients

Depth (cm)	O.C (%)	Available nutrients (kg ha^{-1}) micronutrients (ppm)					
		P_2O_5	K_2O	Fe	Mn	Zn	Cu
0–16	0.78	125	105	1.26	1.11	0.62	1.20

two neighbouring craters of Mount Awu are found to produce about 3 tonnes of copra ha^{-1}. In the Philippines, good coconut plantations are found in Southern Luzon surrounding Mount Banahaw in Tayabas and Mount Mayon in Albay which are volcanic in origin (Child 1974).

8.2.6 *Clay and Peat Soils*

The clays are generally regarded as the least suitable for coconut cultivation, and those with stiff clay subsoil should, as a rule, be avoided. In Sri Lanka, estuarine clays formed by the silting up of estuaries and lagoons are considered to be unsuitable for coconut. These soils become waterlogged during monsoon season and baked and cracked during dry weather (Child 1974). Similar situations are also found to exist in Mozambique. In contrast to other coconut growing countries, large coconut areas in Malaysia come under heavy clays and peat soils. Cooke (1936) reported a type of clay soil in the Philippine Island, Basilan, with a striking resemblance to the clay soils of the west coast of Malaya. Deep ploughing, husk burying, draining and raising leguminous cover crops are some of the practices suggested for improving such areas (Child 1974). In the southern state of Tamil Nadu, large areas of heavy soils are cultivated with coconut.

Peat soils in Malaysia are studied by Cooke (1930), and he concluded that soils containing more than 80% of organic matter are not really suitable for coconut cultivation. In India, peat or Kari soils are found in some parts of Alappuzha and Kottayam districts of Kerala. These soils are characterized by strong acidity (pH 3.0–4.5) and high content of organic matter. They are very rich in total nitrogen but often deficient in phosphorus and calcium. The level of potassium is generally satisfactory. Coconuts are usually grown on raised bunds(mounds). With regular soil reclamation and management practices, the crop performs well on such soils.

8.3 Soil Suitability for Coconut Plantation

Most of the plant species need well-drained, moderately fine to medium textured soils having optimum physical and chemical properties. Soil resource maps, based on such parameters, can aid in predicting the behaviour and suitability of soils for growing different crops and forest species, once the suitability criteria for each crop is established. The National Bureau of Soil Survey and Land Use Planning, India, has proposed land suitability criteria for selecting soils for coconut cultivation (Naidu et al. 2006). Climatic parameters and land quality characteristics as required for a perennial crop like coconut are considered, and the suitability of the site has been rated as S1, S2, S3 and N (Table 8.2).

8.4 Soil Fertility Management

The performance of coconut palms under cultivation is closely linked to the health of the soil, and high yield is tied up with the fertility status of the soil. According to Pandalai (1953), at least six principal soil factors, viz. soil moisture, soil nutrients, soil air, soil temperature, root space and soil toxins, govern the productivity of coconut of which the soil water and nutrient supply regulate the productivity of more than 80%. Soil physical properties are more important as it is more difficult to modify them than to correct mineral deficiencies.

Sharing his experiences from Brazil, Sobral et al. (2007) observed a base saturation of the soil of 60–70% as ideal for coconut. At this level, it is probable that exchangeable Al^{+++} becomes insoluble and exchangeable Ca^{++} and Mg^{++} should exceed 20 mmol/dm^3. It has been observed that although these values have been reached, the levels of Ca and Mg in the leaves remain below the critical level and such lands require application of calcium sulphate and a magensium source to correct the deficiency.

Where the Al, Ca and Mg levels are low, Ca and Mg should be applied to correct the deficiency. In this case, lime should be applied in a circular area, with the trunk at the centre and the edges at the crown projection as the outer limit (Sobral 1998). The quantity of lime to be applied per plant is obtained by the proportion between the quantity hectare^{-1} and the calculated area of the crown projection. In all cases, incorporation of lime is important to correct acidity and place Ca and Mg near the

Table 8.2 Soil-site suitability criteria for coconut cultivation

Soil-site characteristics			Ratings			
Parameters		Units	Highly suitable	Moderately suitable	Marginally suitable	Not suitable
			S1	**S2**	**S3**	**N**
Climatic regime	Mean temperature in growing season	°C	26–29	23–25, 30–32	20–22, 33–34	
	Total rainfall	mm	1500–2500	1000–1500	500–1000	<500
	Dry months (months with <50 mm rainfall)	Months	<3	4–5	6–7	–
Land quality		**Land characteristics**				
Oxygen availability to roots	Soil drainage	Class	Well drained	Moderately well drained	Imperfectly drained, excessively drained	Poorly drained
	Depth of water table	M	2–3	1–2	0.5–1	–
Nutrient availability	Texture	Class	cl, scl, sc, sicl, sil	sl, c (non-swelling), sic	c (swelling), ls, s	
	pH	1:2.5	5.1–6.5	6.6–7.5, 4.5–5.0	7.6–8.5, 4.0–4.4	
Rooting condition	Effective soil depth	Cm	>100	75–100	50–75	<50
	Coarse fragments	Vol %	<15	15–35	35–50	>50
Erosion hazard	Slope	%	<8	8–15	15–30	

Note: *c* clay, *cl* clay loam, *sicl* silty clay loam, *sic* silty clay, *sil* silt, *ls* loamy sand, *sl* sandy loam, *s* sand

roots. Where fertigation is not practised, the interval between liming and fertilization should be at least 60 days.

Very pronounced effect of drainage, soil depth and soil series on coconut yield was noticed by Cosico and Fernandez (1983) in the Philippines. Yield was the highest in soil that is well-drained, with at least 80 cm deep, and whose texture is coarse to medium. Similarly, yield was highest in relatively flat areas and where the climate has only a short dry period. Compaction of the subsoil rather than the topsoil adversely affected yields.

Issaka et al. (2012) reported that soil physical properties did not cause any major limitation for good coconut growth and yield in Western and Central regions of Ghana. However, the soils suffer from multi-nutrient deficiency with nutrient levels of the soils being low to very low and will not support good coconut growth and yield. Except for soil pH, nutrient levels generally showed a decreasing trend in the order top soil > subsoil and below, a feature commonly observed in soil profiles. For growth and productivity, liming to improve the exchangeable basic cations and pH of the soils is recommended. Use of rock phosphate is also recommended for raising the levels of both phosphorus and some basic cations.

8.4.1 Depth of Soil

The depth of the soil is one of the important physical criteria to promote sustained productivity of the palm. Around 73% of roots of coconut are found within a 2 m radius, and most of them were confined to 30–120 cm in depth (Kushwah et al. 1973). Root distribution pattern was also studied by Dhanapal et al. (2000). The presence of hard soil pan, bed rock, a plinthite layer or permanent water table within 1 m depth should be considered as unsuitable soil physical conditions for satisfactory growth of coconut palms. Under such situations, the initial growth would be satisfactory, probably only up to 10–15 years, beyond which the palms show quick decline both in vegetative growth and yield-related parameters unless intensive fertilization is resorted to. Hence, sufficient depth of soil is a prerequisite for planting coconut which gives a good physical support and better anchorage against gusts of winds. Ganarajah (1953) emphasized the necessity of at least 1 m of fertile soil for healthy coconut growth. A minimum depth of 80–100 cm was observed as ideal by Fremond (1964).

8.4.2 Soil pH and Management

The optimum soil pH range is 6.4–7.0, though the palms can also thrive under soil pH range of 5.5–6.3 and 7.1–7.5. Soil pH values below 5.4 or above 7.5 provide marginal condition for the growth of the palms (Child 1974). The acidity of samples of Philippine coconut soils showed a pH range between 6.2 for poor yielding clay and 7.0 for high-yielding volcanic alluvium to 8.3 for high-yielding coral sand (Cooke 1936). The pH of representative Indian soils is reported to be in the range of 5.2–8.0 (Menon and Nair 1952). Manciot et al. (1979b) suggested a pH range of 5.0–8.0 as normal for coconut. According to Fremond (1964), the ideal soil pH for coconut growing soils ranges between 5 and near neutral. However, soil acidity frequently changes depending up on management practices. Continuous application of acid-forming chemical fertilizers reduces the soil pH due to the acid residues. On the other hand, continuous application of organics such as forest leaves, farmyard manure, compost and green manures tends to buffer the soil pH. The soil acidity varies depending on the parent materials from which the soils are formed. Besides the basic pedogenic processes, the presence of dominant cation in the exchange complex determines its acidity. Almost all red and lateritic soils have very low pH, while it is towards neutral for alluvial soils. In the case of sandy soils, the pH is near neutral, whereas the coral soil (Krishnan et al. 1997), which contains calcium as the dominant cation followed by Mg, is in the pH range of 7.8–8.5. The soils of volcanic origin are generally acidic and have a low pH. The measure of soil acidity generally gives an indication about the status of nutrient availability and the possible production-limiting factors prevailing in such soils.

Soil Acidity Most of the red and lateritic soils, where coconut is predominantly cultivated and is formed under humid tropical conditions, have Al^{+++} as the dominant cation, and consequently their pH is very low (4.2–5.8). These highly leached soils are deficient in many of the essential plant nutrients, particularly potassium and magnesium. The presence of high Al and Fe reverts the soluble phosphates into insoluble forms.

Soil Salinity In low-lying areas where the soils are frequently inundated by seawater or submerged with brackish water during certain seasons of the year, excess salinity may create some production problems. Drainage facilities and check bunds can reduce the inundation. Coconuts cultivated in nontraditional saline/alkaline areas face considerable limitation due to the dominance of sodium, carbonates and bicarbonates.

The management practices that are appropriate for problem soils are liming to correct acidity and gypsum application to correct alkalinity as per the lime or gypsum requirement of the soil.

Soil acidity can be managed by applying liming material based on the lime requirement of the soil. Often, the acid-forming cations such as Al^{+++} could be inactivated by using excess of phosphate or organic manures. But in the case of acid sulphate soils, the correction of acidity and suppression of aluminium could be achieved only by using magnesium silicate. This facilitates inactivation of Al as aluminium silicate, and the magnesium sulphate thus formed can be leached out of the system. Thus, the problem of Al^{+++} and sulphate can be tackled by the single application of magnesium silicate. According to Nambiar et al. (1975), the application of seaweed manure reduced the acidity to a certain extent. In acid sulphate soils of Malaysia where coconut is intercropped with cacao, Chew et al. (1984) reported that liming from pH 3.9–4.5 in the top soil increased copra content and copra yields by 10% and nearly doubled the cacao yields. The application of acid-forming chemical amendments such as sulphur or gypsum helps in reducing the salt problem. Application of more organic matter in the planting pits along with soil-sand mixture in equal amounts was also observed to reduce the salt problems. Similarly coir pith has been found to be a good amendment to correct soil alkalinity.

8.4.3 Electrical Conductivity

The total soluble salts in the soil are not a serious problem as far as coconut cultivation is concerned. This is mainly because coconut is a semi-halophytic palm and can tolerate salinity to a greater extent. According to Sankaranarayanan et al. (1958), the palm could tolerate a salinity up to 0.6% which is beyond the tolerance limits of many other crops. Majority of the soils where coconut is cultivated are acidic and highly leached, and as such, the salt problem is not a matter of concern. However, in low-lying areas where the soils are frequently inundated by seawater or submerged with brackish water during certain seasons of the year, excess salinity may create some production problems. Excess salt concentration also enhances the osmotic potential of soil matrix which restricts the water uptake by plants. Na, being a toxic ion under such conditions, interferes with K uptake disturbing stomatal regulation which ultimately causes water loss and necrosis. The application of acid-forming chemical amendments such as sulphur or gypsum would also help in reducing the salt problem. Application of organic matter in the planting pits along with soil-sand mixture will have an ameliorative effect under conditions of excess salinity.

8.4.4 *Organic Matter*

Soil organic matter is the source and sinks for plant nutrients. Coconut growing soils, in general, are deficient in this component. Among the major soil groups, the lateritic soil on the hilltops and slopes and the alluvial soils tend to contain fairly higher amount of organic matter than the coastal sandy soils. The build-up of soil organic matter under tropical humid condition is rather difficult due to faster disintegration and degradation/oxidation. In the case of soils of volcanic origin and marshy and acid sulphate soils, the total soil organic matter content is high, but the humic and fulvic acid fractions which are the active components contributing to soil fertility are low. Manciot et al. (1979b) suggested a threshold value of 1% organic carbon for coconut growing soils.

The underlying productive capacity of the soil is determined by the organic matter status through its beneficial influence on soil physical, chemical and biological environments. This is particularly true for sandy or coral soils. One such beneficial effect has been reported by Nambiar et al. (1983) in the case of littoral sandy soils of west coast of India where it was successfully demonstrated that coconut seedlings can be well established by the application of different organic sources. Their studies also revealed favourable changes in the soil physical and chemical environments which in turn improved the health of the palm. In cases where organic manures are in short supply, on farm availability of organic sources of nutrients could be ensured, through green manuring, cover cropping and conservation of farm biomass. Wherever feasible it is advantageous to generate sufficient organic matter within the plantation.

8.5 Mineral Nutrients on Growth and Productivity of the Palm

The vital aspect of mineral nutrition is to ensure the availability of the essential mineral elements in the soil in the required levels and in the right proportion to the palm, for its sustained growth/productivity throughout its life. The two important approaches for achieving this are (1) to assess the nutrient demand of the palm for expressing its full production potential and (2) to assess the capacity of the soil medium to supply the nutrient.

8.5.1 Soil Nitrogen

The soil nitrogen varies with soil type. Generally, the sandy soils contain very low amount of soil nitrogen, while it is high in the case of swampy, Kari (peat) and alluvial soils. Heavy soils with high amount of clay such as alluvial, black and heavy red soil, tend to retain N in the ammoniacal form dominantly in their exchange complex compared to sandy and red sandy loam soils depending upon the pH of the soil. Swampy and marshy soils contain high amount of organic nitrogen.

The nitrate concentration in arable soils, in general, does not exceed 20–30 ppm. It is highly mobile and susceptible to loss by leaching. The initial showers of the monsoon rain tend to leach most of the nitrate accumulated in the coconut gardens. Such a loss is very high in sandy soil followed by sandy loam and red soils. The nitrate concentration is generally high in the alluvial, forest and black soils. It undergoes reduction process in soils submerged in water at least for a period of 1 or 2 weeks. Thus, all the coastal soils subject to inundation by seawater tend to lose the nitrate by this process. Even some of the upland soils, which receive torrential rain continuously for a week, are susceptible for nitrate reduction. This is especially true for situations like West Coast and North Eastern parts of India.

The soil N is also subjected to loss by volatilization process. Although most of the coconut growing soils are acidic in nature, the N loss through volatilization would take place due to localized changes in soil pH caused by the hydrolysis of urea. This loss varies among soils depending on the H^+ ion buffering capacity, the presence of easily decomposable organic matter content, etc. The volatilization loss from applied urea can be reduced to a large extent by covering with 5 cm soil after fertilizer application. Upadhyay et al. (2007) evaluated N mineralization potential of predominant coconut growing soils, viz. sandy (Oxic quartzipsamments), red sandy (Arenic Paleustults), laterite (Oxic haplustults) and Kari (Tropic fluvaquents) type of soils of Kerala, India, and found it to be more in Kari soils (0.543 μg N day^{-1}) followed by red sandy loam (0.042 μg N day^{-1}), sandy (0.027 μg N day^{-1}) and laterite (0.024 μg N day^{-1}) soils.

Bopaiah et al. (1998) studied the effect of slow-release N and P fertilizers (different combinations of urea, urea formaldehyde, neem-coated urea, lac-coated urea blended along with coir dust, tar, single superphosphate, Mussoorie rock phosphate and muriate of potash). Among the slow-release N fertilizers, urea form (urea formaldehyde), neem cake-coated urea and coir dust mixed with urea have been found to remain for a long period in the sandy soil, thus facilitating availability of N in the more permeable soil, and among the P carriers, Mussoorie phosphate was equally efficient as superphosphate. Dissanayake and Rajapaksha (2016) observed that in the sandy regosols of Sri Lanka, urea blended with neem cake is more suitable to reduce nitrate losses than use of neem extract.

It is inevitable that soils which receive high doses of nitrogenous fertilizers suffer, in the long run, from decreasing base saturation, acidification and a significant drop in soil pH. Organic manures such as green manure crop residues, compost, cow dung, etc. are known to alleviate the negative effects of inorganic fertilizers. Among the different management practices, mixed cropping with cacao and banana, intercropping including pineapple, etc. and mixed farming in coconut gardens as well facilitate recycling of nutrients and more so of nitrogen.

8.5.2 Soil Phosphorus

The P distribution in the soil varies with geomorphology and parent material of the soil as well as soil chemical constituents. In sandy soil, the total as well as available soil P is very low, whereas it is high in alluvial, black and swampy soils. The P status of laterite, lateritic and red soils is medium. However, the P retained in the soil is made available to the crop in course of time. Soils formed under the process of laterization contain high amounts of sesquioxides that revert the soluble P into insoluble complexes. In these soils, the Al^{+} Fe labile P forms are relatively higher than that of saloid-bound and calcium phosphates, whereas the soils of semiarid regions contain dominantly the Ca-P. In the case of alluvial and swampy soils, the organically bound P is higher than that of inorganic fractions. Organic P content was found to increase due to introduction of mixed farming and multiple cropping in coconut gardens. Nambiar et al. (1989) recorded considerable change in the P fractions of red sandy loam soil of Kasaragod, India, due to continuous P fertilization (Table 8.3), and Khan et al. (1985b) reported distribution of P fractions as influenced by P carriers, viz. single superphosphate, nitro-phosphate, ammonium phosphate and rock phosphate on their application to laterite soil.

In a discussion on P requirement of coconuts cultivated in New Zealand in soils derived from limestone, Baseden and Southern (1959) opined that though the soils are low in available P, it is not likely to be a serious factor for coconuts.

The coconut growing soils in Sri Lanka are generally deficient in total as well as the active and available forms of P. Although soils of the ultisols had marginal to moderate amounts of total P, the active and available fractions were extremely low (Loganathan and Balakrishnamurti 1983). Leaf analytical data from the joint FAO/CRI/CCB study revealed that about 85–90% of the holdings have adequate levels or an excess of P. Accordingly, P is placed fourth in the order of priority of nutrients

Table 8.3 Effect of P fertilization on phosphorus fractions (ppm) in red sandy loam soil

Year	Depth (cm)	Saloid-P	Al-P	Fe-P	Ca-P
1983	0–25	2.67	38.9	92.6	3.5
	25–50	2.03	15.1	52.4	3.9
1986	0–25	47.50	217.6	97.7	40.3
	25–50	34.40	277.7	85.8	27.0

for adult coconut in Sri Lanka (Anon 1989). Available P in the Boralu and Pallama soil series of Gampaha District of Sri Lanka was at a medium level of sufficiency, with a mean of 16 mg kg^{-1}, even in unfertilized coconut plantations (Jayakody et al. 2007).

The mobility and diffusion of soil P is very much limited and also its loss from soil through leaching. The loss of P through surface run-off is quite high, particularly during heavy rains. Nevertheless, P build-up would take place due to continuous application of fertilizers containing phosphates (Muliyar and Wahid 1973). On a sandy loam soil in the dry zone of Sri Lanka, the downward movement of P from concentrated superphosphate was greater than from rock phosphate (saphos) (Loganathan and Nalliah 1977). The surface layers (0–15 cm) of soil when applied with concentrated superphosphate had higher P values (60 and 89 mg kg^{-1} for the 8th and 9th year, respectively) than those given rock phosphate (3 and 16·5 mg kg^{-1}). At 40 cm depth, the concentrated superphosphate treatment had 6 and 30 mg P kg^{-1}, but the rock phosphate treatment had almost zero P at and below 40 cm.

There have been many evidences of P build-up in the soil due to continuous fertilization. The studies of Khan et al. (1992) showed that the available soil P (0–30 cm) increased from 84 ppm to 121 ppm when annual fertilization of P was done at 320 g P_2O_5 $palm^{-1}$ for 14 years, whereas in the treatment where P was not applied, it decreased from 84 ppm to 21 ppm during the same period. Khan et al. (1985b) further reported a general build-up of available soil P irrespective of sources of phosphatic fertilizer applied to a typical laterite soil of Malabar area of Kerala, India (Table 8.4). They recommended rock phosphate as the ideal carrier of P in coconut mineral nutrition, as soil contents receiving rock phosphate gave a better reflect on plant P contents and enriched all P fractions in the soil and influenced the yield. The study further revealed that plant P content in coconut can be built up only gradually over a period of time compared to N and K. They also observed mobility of applied phosphorus to lower depths as a result of cultural practices associated with fertilizer application and mobility of P increased with increase in the level of its application. As Kushwah et al. (1973) demonstrated that > 80% of coconut roots were seen below 30 cm depth, and P moved to the roots mainly by diffusion (Barber 1962), they suggested deeper placement of P for better utilization as a rational approach to tree crops such as coconut.

Table 8.4 Influence of P carriers on the available P (Bray I) status in the coconut basins

	Available P status (ppm)					
	1975–1976 (pretreatment)			1980–1981 (posttreatment)		
P-carriers (to supply 320 g P_2O_5 $palm^{-1}$ $year^{-1}$)	0–30 cm	30–60 cm	60–90 cm	0–30 cm	30–60 cm	60–90 cm
Single superphosphate	34.7	38	1.3	132.8	3.7	3.1
Ammonium phosphate	36.0	Traces	Traces	152.7	3.5	3.1
Nitro-phosphate	29.3	Traces	Traces	148.1	5.8	1.3
Rock phosphate	40.7	Traces	Traces	91.1	3.5	2.7

Table 8.5 P budget and balance (kg ha^{-1}) in coconut-based high-density multispecies cropping system

	1983–1984			1985–1986		
Fertilizer levels	Budget	Removal	Balance	Budget	Removal	Balance
*F	198	8	190	395	20	375
2/3 F	170	8	162	315	18	297
1/3 F	148	8	140	243	20	223

*F = Recommended dose of N, P_2O_5, K_2O (500:320:1200 g $palm^{-1}$ $year^{-1}$); 2/3 F = 2/3 of the recommended dose; 1/3 F = l/3 of the recommended dose

Bavappa et al. (1986) have recorded a phenomenal build-up of P (Table 8.5) in the high-density multispecies cropping system involving 18 crops over a period of 3 years. They have also observed similar trend in mixed farming and coconut-cacao mixed cropping systems at the experiments at ICAR-CPCRI.

A review of P fertilization trials in coconut in many Asian and African countries (Wahid et al. 1977) indicated that P application had no desired effect on yield and growth parameters. Where responses were observed, they occurred after many years of continuous P application (Muliyar and Nelliat 1971; Ouvrier and Ochs 1978). Nevertheless, Fremond (1966) and Muliyar and Nelliat (1971) found P response in the presence of K. Chew (1978) indicated that response to P fertilizers could have been due to the generally low levels of fertilizers. With continued use of P fertilizer and consequent build-up in soil, the need for it could be lower. Furthermore, in soils high in P, there would be no P response (von Uexkull 1972); the leaves of coconut palms on these soils reflected sufficient P levels (0.15%). For coconut, Cordova (1965) considered 10 ppm available P in soil as the critical level, and Khan et al. (1985a) observed that an available P status (Bray 1) of 15 ppm might be sufficient in maintaining the palms at sufficiency levels (0.110%) of phosphorus. Limbaga (1986) considered 15–20 ppm as the critical level and 30–35 ppm Olsen P as the optimal P level in surface soil.

The P fixation measurements are very useful for determining the P-deficient and P-sufficient conditions in soil and finding out the amount of P fertilizer to be added to a particular soil in order to supply adequate amount of P nutrient to coconut palm. Asara et al. (2011) studied the P fixation capacities of coconut growing soils from 20 coconut estates in the low country intermediate zone of Sri Lanka. They observed that 14 series showed higher fixation capacities which can be categorized as P-sufficient soils as the soils recorded higher available P and other 6 series showed lower fixation capacities which can be categorized as P-deficient soils as the soils recorded lower available P compared with critical value. The amount of P fertilizer to be added to P-deficient soils to supply adequate amount of P for coconut was 170, 80, 60, 130, 35 and 120 mg kg^{-1} for Kuliyapitiya, Tambarawa, Welipelessa, Ambakele, Sudu and Kiruwana soil series, respectively.

The analysis of Eppawala rock phosphate (a phosphate deposit in Sri Lanka) indicated that the total P content was about 30% and the water-soluble P content was about 0.02% (Perera 2007). Compared with other soluble phosphates, it is low in solubility level, but its solubility increases in acidic media. The low solubility of locally available Eppawala rock phosphate (ERP) makes it unsuitable for direct application for coconut palms in dry zone. In order to increase the solubility, incorporation of goat manure was suggested in the fertilizer schedule (Kulasingh et al. 2013). The Coconut Research Institute, Sri Lanka, recommends 100% of the P requirement of adult and young coconut palms in wet and intermediate zones with locally available ERP.

Phosphate-solubilizing microorganisms were found to be widely distributed in coconut growing soils (George Thomas et al. 1991), and the predominant bacteria solubilizing phosphate were *Pseudomonas* spp. and *Bacillus* spp. (Nair and Subba Rao 1977). Priya George et al. (2012) evaluated phosphate-solubilizing potential of bacterial isolates from rhizosphere soil and roots of coconut palms growing in Kerala, Karnataka, Tamil Nadu, Andhra Pradesh and Maharashtra states of India and, based on their phosphate-solubilizing potential and the ability to produce IAA, ACC-deaminase and siderophores, reported *E. cloacae* 267, *P. putida* Biotype B HSF 132 and *P. plecoglossicida* KnSF 227 as prospective candidates for use as bio-inoculants for organic farming of coconut.

Dwivedi et al. (1981) examined the efficiency of different methods of plant injection and soil placement techniques using carrier-free ^{32}P tagged to single superphosphate and ^{86}Rb, with coconut as test crop. By plant injection technique, radioactivity was detected in a 10-m-tall coconut palm after 4,8,12 and 18 hours of application through cut end of roots, stem, leaf axils and growing root tips, respectively. Out of 4 soil placement methods, the quickest recovery of ^{32}P in palms was detected after 7 days of placement when applied through hole method, 8 days of application by trench method and strip methods and after 11 days in basin method. They observed that the quickest recovery of radioactivity was noted through hole method and the average highest accumulation of activity (CPM/g dry matter), among all the methods, soil placement techniques, was recorded in trench method, where roots were exposed to root activity at 20 cm and 40 cm depths around the bole at a distance of 0.5 m and 1.0 m, respectively. Based on the above findings, circular trench method of nutrient application containing P can be recommended in the fertilizer application programmes (Khan and Bavappa 1986). Nethsinghe (1976) reported that roots of coconut were most active up to a depth of 45 cm and nutrients placed around the bole to a distance of 1.2 m are absorbed quickly.

8.5.3 Soil Potassium

The K content is relatively low in almost all acid soils due to leaching and surface run-off as a consequence of heavy rainfall. The soil K content depends on the nature and composition of the parent material (Graham and Fox 1971). It is low to medium in coconut growing soils of humid tropics but high in alluvial soils and those formed under semiarid conditions such as black soils. The swampy soils also contain moderately high amount of soil K. The potential reserve of soil K in the fixed or non-exchangeable form is an important source of K to coconut. The physico-chemical dynamics of soil K determines its fertility status and supplying power. The quantity-intensity relationship and the potential buffering capacity of soil K virtually regulate its availability to coconut.

Vegetative production appears to have priority over dry matter storage in the inflorescence and nuts when resources are limiting. A shortage in K availability will, therefore, first affect nut size, number of nuts and finally vegetative growth. With severe K deficiency, the rate of new leaf production is reduced, leaves are smaller (fewer and shorter leaflets) and duration of leaf production is less. Young palms supplied with low amount of K take much longer to develop a trunk and to come into bearing. Ouvrier (1984) noticed perfect correlation between K levels in the leaf and husk. Therefore, leaving the husks in the palm basins will help to restore the soil excess K. Fremond and Ouvrier (1971) in the Côte d'Ivoire tested the effect of withholding K on the development of young palms (Table 8.6).

Seedlings which received KCl fertilizer from field planting stage showed progressive improvement in all growth parameters and better partitioning of nutrients over the years, resulting in a consequent increase in the copra production of adult palms significantly, compared to those palms which received KCl fertilizer from bearing stage only.

Compared to the soil threshold value of 179 mg kg^{-1} for exchangeable K for the coconut palms, the status of exchangeable K of most of coconut growing soils is not sufficient for coconut cultivation, and, therefore, potash fertilizer application is necessary.

Pillai (1975), based on the ratings of Muhr et al. (1963), reported that all the soil groups of Kerala, India, under coconut are generally deficient in available K, and no soil group followed high ratings. The deficiencies of N and K are the most common in coconut plantations (Sobral 1998), and the quantities of nutrients removed by the coconut palm are high, since nutrient mining by coconut is a continuous process linked to growth. The nutrient removal ratio of N: K of 1:1.2–1.75 is indicative of

Table 8.6 Effects of periods of first potash fertilizer application on the performance of young coconut palms

Year	Characteristics observed	Time of KCI application: **A** – from field planting	**B** – from bearing stage only	**B** as % of **A**
1956	Number of fronds	8.89	7.69	86.5
1958	Length of frond (cm)	256.00	223.00	87.1
1959	Circumference of trunk	124.10	105.40	69.9
1960	Number of fronds in 1 year	11.70	10.70	91.4
1962	kg of copra ha^{-1}	944.00	272.00	18.7
1966	kg of copra ha^{-1}	2560.00	2272.00	88.8
1970	kg of copra ha^{-1}	2480.00	2096.00	84.5
1961–1970	Cumulative yield, kg ha^{-1}	17344.00	12704.00	73.2

its influence being exerted on the crop (Khan et al. 2000). Silva et al. (2005) observed that, in the nonirrigated area, the N and K for nuts harvested at 7 months were 23 and 76 kg ha $^{-1}$ year $^{-1}$, respectively, whereas in irrigated area, the removal was 74 and 243 kg ha^{-1} year^{-1}.

The nutritional limitations of two coconut growing soils, viz. Wariyapola and Maho series in Sri Lanka for the production of coconut, were inadequate supply of K and plant availability of Ca (Herath et al. 2008). Low soil N, Mg and Ca levels indicated their potential to be limited, if they are not replenished adequately. However, Ekanayaka et al. (2011) observed that high potential buffering capacities of potassium (PBCk) was in coconut growing soil series of Kuliyapitiya, Boralu, Madampe and Melsiripura, belonging to S1, S2, and S4 land suitability classes, indicating that they can buffer K and release it slowly to the soil solution. The Coconut Research Institute, Lunuwila, Sri Lanka (Anon 2008), recommends muriate of potash at 1600 g and 1800 g palm^{-1} year^{-1} for local and improved varieties, respectively, without split application for coconut palms grown in such soil series.

Khan et al. (1982) found that K adsorption was comparatively more and uniform in laterite soils than in red sandy loam, river alluvium and coastal sands cultivated to coconut. The magnitude of the constants K and 1/n and the difference in the values of Freundlich adsorption isotherm were attributed to the contents and nature of clay minerals in these soils. Potassium is likely to get depleted more easily from the root zone around the palm, and the K pool also cannot restore much K needed. As the coconut palm has a heavy demand for K, there is a need for evolving a more suitable way of K management.

The influence of clay minerals in K supply to the nutrient pool was also indicated by Ramanathan and Krishnamoorthy (1976). In the incubation experiments with different coconut growing soils, the highest water-soluble K fraction was obtained in sandy soil followed by laterite, red sandy loam and alluvial soils. The exchangeable K fraction was the highest in red sandy loam than in laterite, alluvial and sandy soil. The non-exchangeable K fraction was the highest in alluvial followed by laterite, red sandy loam and sandy soil (Anon 1967). This variation in the fractional distribution of K is on account of the variation in the mineralogical constituent of the soil and the initial soil K status. The presence of plenty of K-bearing minerals like mica would tend to regulate the release of K into the soil solution. The availability of K in soils is regulated by the levels of other cations like Ca, Mg and Na. Bastin and Venugopal (1986) indicated that the alfisols, which are intensively cultivated with coconut, are generally low to medium in potash status.

Many researchers (Shanmuganathan and Loganathan 1976; Somasiri and Liyanage 1996; Fernando 1999; Giritharan et al. 2000, 2002) studied on K status of the coconut growing soils, and they concluded that exchangeable K appeared to be a better measure of K availability as this form correlated better with leaf K values. However, they further reported that total K was not related to leaf K status because K minerals do not become available for a long-lived crop like coconut. The studies showed that the exchangeable K levels are different with the soil types irrespective of the agroclimatic zones and most of soils have exchangeable K levels below the threshold value. Upadhyay et al. (2005) observed that the optimum K application to maximize the soil solution K for optimum plant nutrition ranged from 662 ppm (full dose) to 692 ppm (No fertilizer). The quantity of K fertilizer required to optimize the soil solution K concentration in red sandy soil was 1150 g K_2O palm^{-1} year^{-1}, which very well matches with the general K recommendation (1200 g K_2O palm^{-1} year^{-1}) for coconut in India.

As K is one of the most essential nutrients to coconut, any damage caused in the early stages of its growth due to its deficiency would be difficult to be corrected in the later stages of growth. Hence, adequate K supply is essential right from the planting of coconut in the main field. The K supply can be regulated by suitable management practices. For example, Nambiar et al. (1983) observed higher build-up of available K in littoral beach sand by blending inorganic fertilizers with different organics. The maximum build-up was obtained when K was applied along with cattle manure followed by coir dust (Table 8.7).

Cultivation of grasses, cacao, banana, pineapple and other compatible crops in coconut gardens tend to recycle large amount of K to the soil (Bavappa et al. 1986). The results of studies conducted at ICAR-CPCRI, Kasaragod, India, showed that the potassium balance and budget considerably increased in the high-density multispecies cropping system involving 18 crops over a period of 3 years (Table 8.8). The management of soil K is not as difficult as that of nitrogen. Regular fertilizer application combined with possible recycling of organic residues would help to build up the soil K sufficient enough to meet the continuous requirement of coconut.

Table 8.7 Effect of blending organic sources with inorganic fertilizers on available potassium (0–50 cm depth)

Organic sources	Available K (ppm)			
	1971	1972	1976	1980
NPK* alone	7.6	9.1	11.6	44.0
NPK+ Coir dust	6.3	12.2	30.8	108.0
NPK. + Coconut *sheddings*	8.2	10.8	39.6	73.0
NPK+ Forest leaves	5.7	10.0	32.4	44.0
NPK+ Cattle manure	9.0	13.0	52.0	95.2

*K was supplied as muriate of potash at 2.0 kg palm^{-1} year^{-1}.

Table 8.8 K budget and balance (kg ha^{-1}) in coconut-based high-density multispecies cropping system

	1983–1984			1985–1986		
Fertilizer levels	Budget	Removal	Balance	Budget	Removal	Balance
*F	610	110	500	1403	488	915
2/3 F	455	100	355	1025	335	690
1/3 F	300	80	220	740	428	312

*F = Recommended dose of N, P, K (500:320:1200 g palm^{-1} year^{-1}); 2/3 F = 2/3 of the recommended dose; 1/3 F = 1/3 of the recommended dose.

8.5.4 Secondary Nutrients in the Soil

Among the secondary nutrients, Mg and S are very important as far as coconut productivity is concerned. Both the nutrients are generally in short supply in many of the coconut growing areas in the humid tropical soils. They are either deficient or tending towards deficiency in most of the soils in Kerala, Karnataka, Maharashtra and North Eastern States of India. The deficiency of S has been reported from Papua and New Guinea (Southern 1969) and some of the East African and South East Asian countries (Ollagnier and Ochs 1972). In the Philippines, even coconut grown on soils derived from volcanic origin could suffer from S deficiency under high rainfall condition throughout the year. The contents of Mg and S are fairly high in black and alluvial soils. Both magnesium and sulphate ions are highly mobile, and, hence, they are easily lost from the soil by leaching.

Ollagnier et al. (1983b) considered an exchangeable Mg level of 0.46 meq % as high. The critical values for exchangeable K and Mg in the Philippines are 0.45 and 2.9 meq %, respectively (Santiago 1978). Margate et al. (1979b) did not observe a K-Mg antagonism in a long-term KCl fertilizer study on a clay loam soil in the Philippines in spite of high application of KCl (8 kg palm^{-1} year^{-1}), with soil levels of 0.45 meq % K and 5.3 meq % Mg.

Bavappa et al. (1986) have reported that the Mg budget and balance in coconut-based high-density multispecies cropping system declined very fast to the extent of 50% over a period of 3 years (Table 8.9). Further, the systems involving coconut-grass and coconut-cacao also showed a similar pattern of decline with respect to Mg.

Nethsinghe (1963) suggested that the exchangeable Mg content of soils had no diagnostic value for predicting Mg availability to coconut and according to him the 0.01 M $CaCl_2$ extractable Mg would give a more reliable estimate of Mg availability. Manciot et al. (1979b) suggested a threshold level of 0.2–0.5 m.e 100 g^{-1} of exchangeable Mg out of 1.0 m.e. 100 g^{-1} of total exchangeable cations. They have further suggested an exchangeable Mg/K ratio of 2.5 as ideal for coconut soils, while Cecil (1981) found a minimum value of 2.0 for the satisfactory supply of Mg to the palm. Field trials conducted in Sri Lanka (de Silva 1966), Côte d'Ivoire (Brunin 1970; Coomans1977), India (Cecil 1981) and other coconut growing countries (Manciot et al. 1979c) have clearly demonstrated the correction of Mg

Table 8.9 Mg budget and balance (kg ha^{-1}) in coconut-based high-density multispecies cropping system

Fertilizer level	1983–1984			1985–1986		
	Budget	Removal	Balance	Budget	Removal	Balance
*F	95	10	85	73	28	45
2/3 F	93	10	83	75	28	47
1/3 F	93	10	83	78	30	48

*F = Recommended dose of N, P, K (500:320:1200 g palm^{-1} year^{-1}); 2/3 F = 2/3 of the recommended dose; 1/3 F = 1/3 of the recommended dose.

deficiency by the application of magnesium salts/minerals like Kieserite/Epsom salt or magnesite/dolomite. Cecil (1981) has recommended the application of 500 g MgO $palm^{-1}$ $year^{-1}$ as Epsom salt ($MgSO_4.7H_2O$) for correcting Mg deficiency in the coastal sandy soils of the Onattukara series of Kerala, India.

Although calcium is one of the essential secondary nutrients, its concentration beyond a certain limit would decrease the coconut yield probably due to its interaction effects with other nutrients. The Ca content in most of the coconut growing soils of the world is satisfactory, and no improvement in growth or yield was obtained by Ca treatments. Besides, the indirect addition of Ca through phosphatic fertilizers is considered to be quite adequate for sustaining the Ca demand of the palm. However, for acid soils, regulated additions of liming materials are needed for correcting the soil acidity and other related problems. The availability of Ca in coral soils is very high and often reaches a concentration where it would interfere with the uptake of K, Mg and Na by coconut. Liberal organic amendments are generally recommended for improving the fertility of such calcareous soils. As fairly high amount of Ca is being recycled through litter fall and prunings of cacao and other intercrops grown in coconut gardens, the management of Ca is not a serious problem in coconut growing soils.

Analysing the nutrient contents of coconut leaves in the gardens of Tanzania, Thomas (1973) indicated that the yield was related to the ratios such as N /P, N /K and Ca/ Mg, but the level of K had to be interpreted in terms of a balance between K and Ca. When the individual nutrients were considered, the levels of N and Ca alone had a positive relation with the yield. The study also revealed that the nutrient composition of the leaves reflected to some extent the nutrient status of the soil. Hence, application of fertilizers having N and Ca might possibly correct the nutrient ratios and thereby improve the productivity of low-yielding coconut palms.

Potassium is not always necessary and responses to KCl often result from the effect of chlorine. It was shown from numerous experiments that leaf Cl levels are nearly always closely linked to those of Ca, K, Mg or even N cations. Thus, in soils with a high exchangeable Ca content, KCl is frequently inadequate to correct a K deficiency, whereas, where there is a low exchangeable Ca level, on the contrary, KCl appears to be one of the best forms for the correction of a Cl deficiency as well as the K deficiency which may be associated with it. Where there is no chlorine deficiency, chlorinated fertilizers, particularly KCl, can be used without problem.

Sulphur is either deficient or approaching to deficiency in most of the coconut growing soils barring acid sulphate soils, alluvial and black soils where the soil S content is medium to high. The deficiency is mainly due to the leaching of soluble sulphates from the soil solution. The anion adsorption capacity of most of the acid soils is also low. Soils very low in organic matter and continuously cropped without S fertilization are likely to suffer from deficiency of S. The use of high analysis fertilizers in modern agriculture restricts the addition of S as an incidental component to the soil. However, the S management can be effectively achieved by suitably selecting some of the S-containing fertilizers like superphosphate, ammonium sulphate, diammonium phosphate, magnesium sulphate, etc. while formulating the fertilizer schedule for coconut. The organic matter is one of the potential sources of S, its management and release pattern is almost similar to that of N from organic matter. As the major fraction of S is held in the organic form, the S management in the soil can be best achieved through proper organic farming practices.

8.5.5 Micronutrients

The tropical acid soils are in an advantageous position with respect to most of the micronutrients especially iron, manganese, copper and zinc. These cations are easily soluble and readily available under acid conditions. The availability of Fe and Mn is generally high in acid laterite and red soils; moderate in alluvial, volcanic, peat and clay soils; low in coastal sandy soils; and very low in coral soils.

The high calcium carbonate content of the coral soils blocks the assimilation of Fe and Mn by the palm. Severe deficiencies of Fe and Mn were reported in the coral tolls of Oceania which were not corrected by soil application of Fe and Mn salts (Pomier 1964, 1969). In acid laterite and red soils, their contents may often reach to toxic levels, particularly under anaerobic conditions.

The concentrations of Cu and Zn are moderate in acid laterite and red soils, high in the case of alluvial soils and low in black and coral soils. However, these nutrients are not generally found limiting in the nutrition of the palm. The requirement of Mo for coconut is very small, and its deficiency has not been felt in coconut farming anywhere in the world. On the other hand, the problem of B deficiency is becoming more and more conspicuous in most of the acid laterite, red, alluvial and coastal sandy soils. The “crown choke disorder” of coconut is considered to be due to boron deficiency which can be corrected by judicious application of borax depending on the severity of the symptoms as well as the age of the palm. Baranwal et al. (1989) recommended soil application of borax decahydrate at 50 g palm^{-1} just after the appearance of crown choke symptom. In slightly advanced stages, two applications of borax, 50 g each, at an interval of 3–4 months were found necessary for the redemption of the disorder.

8.5.5.1 Sodium

Though coconut consumes fairly high amount of Na, its specific function is yet to be established. Sodium is not really indispensable, but it is known that it can replace K to a certain extent when the latter is in short supply. Almost all coconut growing soils of humid tropics are fairly well supplied with Na. The soils originated under semiarid condition like the black and red soils are rich in Na. The soils near coastal belt are also well supplied with Na due to continuous salt spray from the sea. Under no circumstances, Na becomes a limiting element for coconut production. However, under brackish water ecosystems, the marshy soils are frequently inundated with seawater, and they tend to contain very high concentration of Na which might cause its toxicity. Since coconut is a semi-halophytic plant, it can tolerate fairly high amount of Na concentration in the soil solution.

There is not much information available on the soil Na in literature except some sporadic studies. The management of Na as a plant nutrient has not been attempted in greater detail. Nevertheless, it is being used as a soil ameliorant in the hard lateritic soils as common salt. The common salt enhances the crumbling and weathering of hard laterite and, in due course, softens the soils favouring the coconut roots to penetrate deep into the soil easily.

8.5.5.2 Chlorine

Chlorine is generally considered as an essential micronutrient for cultivated crops. Recent research has shown that coconut has a high requirement of Cl and that its deficiency causes detrimental effects on growth, yield and disease resistance in the palm. However, the specific functions of Cl in coconut have not been well established. The Cl content in most of the soils is medium to high. It is being added indirectly to the soil through muriate of potash which contains about 47% of Cl. The soils near coastal belt are well supplied with Cl due to continuous salt spray from sea through aerosol which would be deposited on the soil surface. In semiarid regions, the Cl content in the soil is high due to the chloride-bearing minerals from which the soil has been formed. The amount of Cl in the soil also depends on the amount of chloride brought down by the rain. The highly leached soils located far away from the sea coast usually show deficiency of Cl.

Ollagnier and Ochs (1971a, b), von Uexkull (1972) and Mendoza and Prudente (1972) strongly suggested that chloride must be considered as an essential macronutrient for coconuts. Chlorine deficiency appears to be widespread in many inland areas, especially on well-drained soils (volcanic soils). A survey conducted by the Philippine Coconut Authority revealed that 24 out of 54 coconut-producing provinces have widespread Cl deficiency (Magat 1988). In many cases, responses to applied KCl may be due to chlorine than potassium. Studies conducted in the Philippines (Magat et al. 1975; Magat and Pardones 1984), in the New Hebrides (Daniel and Manciot 1973) and in Indonesia (Manciot et al. 1979c) have clearly demonstrated the correction of Cl deficiency by the addition of Cl-bearing salts like muriate of potash, common salt and ammonium chloride.

Deficiency of Cl affects nut size, number of fruits palm^{-1}, copra yield, nitrogen uptake and the water economy of the plant. Combined application of N, Cl and S was found to increase nut yield, copra nut^{-1} as well as copra yield palm^{-1} year^{-1} by 32–34 nuts palm^{-1} year^{-1}, 27–34% and 10–11 kg palm^{-1}, respectively (Magat et al. 1979). They also observed that Cl and S are both strongly and positively correlated with nut production and copra yields, while only leaf Cl and Ca were highly correlated with copra nut^{-1}. They suggested application of (a) 1.8 kg ammonium sulphate (21–0-0) + 2 kg common salt (sodium chloride) or (b) 1.8 kg ammonium sulphate +2 kg potassium chloride (0–0-60) palm^{-1} year^{-1}.

Coconut palms grown far away from the sea at *Côte d'Ivoire* were found to exhibit Cl deficiency with reduced growth, yield and resistance to drought and increased susceptibility to fungal diseases (Ollagnier et al. 1983a). The manifestations of Cl deficiency seen on coconuts in the Côte d'Ivoire and Indonesia were interpreted as the inability of the plant to maintain its water potential at sufficiently low values because of the deficiency in monovalent anions (Ollagnier et al. 1983b).

The Cl critical level is 0.42–0.45% for young PB 121 hybrids (leaf 4) and 0.5% for adults (leaf 14). Though NaCl fertilizer is very effective and much cheaper than KCl, but it cannot be used alone because of K-Na antagonism (Bonneaux et al. 1993). In order to maintain good K and CI contents at the same time, it is necessary to use a combination of KCl and NaCl.

Prema et al. (1987) studied the effect of NaCl on growth and yield of 24-year-old coconut palms in a laterite soil of Kerala and reported the possibility of substituting 50% of KCl by NaCl as the palms applied with 100% of the recommended dose of KCl and that receiving 50% substitution of KCl by NaCl gave similar nut yields, which were higher than other treatments. Coconut palms were found to display a considerable response to NaCl applications as regards all the yield variables (growth, flowering, number of nuts, nut size, phytosanitary condition, resistance to water stress), at all stages (from the nursery to a bearing plantation) and under a wide variety of soil and climatic conditions (Bonneaux et al. 1997). The cost-effectiveness of NaCl applications, which exists virtually everywhere in Indonesia, increases in zones with a water deficit, since NaCl attenuates the effect of water stress, thereby regularizing production in coconut plantations. NaCl, which is much cheaper and readily availabe, can be used to largely replace KCl. NaCl is, therefore, set to become the main component in fertilization for coconut plantations in Indonesia.

Attention was first paid to the K ion, and its effect on coconut palm growth, right from an early age, was very quickly recognized. Later on, agronomists realized its presence alone could not suffice under water stress conditions and found that when Cl was combined with K, it played a synergistic role in growth and yields of coconuts and had an exclusive role with an increase of resistance to water stress (Ollagnier et al. 1983a). In trials in the Côte d'Ivoire and Indonesia, Ollagnier (1985) found that Cl had a specific effect on drought resistance in coconut as the stomata of Cl-deficient palms do not function normally. A close correlation occurs between potassium and chloride fluxes during stomatal opening. In chlorine-deficient plants, stomatal opening is delayed by about 3 hrs (Marschner 1995). Impairment of stomatal regulation in palms is considered to be a major factor responsible for growth depression and wilting symptoms in chlorine-deficient plants (von Uexkull 1985). Braconnier (1988) studied the physiology of Cl nutrition in coconut palms and provided a clearer picture of the physiological role of the Cl showing that this ion was involved at two levels: in stomata control mechanisms and osmotic phenomena. Osmotic adjustment in response to water stress occurs well before stomata closure and helps to maintain cell turgor, improving cell growth and stomata opening, among other things. Mialet-Serra and Bonneau (2004) observed that (i) Cl and K had an exclusive effect on leaflet size and are increasing their area of 6-year-old PB 121 hybrid coconut palms in Southern Lampung, South Sumatra, Indonesia. Potassium primarily affected petiole length, thereby the total frond length; (ii) synergy between these two ions was found to occur only for the number of leaflets; (iii) Cl and K acted independently on stem diameter, and synergistically on height of coconut palms receiving K or Cl had significantly more voluminous leaf crowns; and (iv) when K and Cl were supplied in a balanced manner, the crown area and leaf area index of coconuts increased leading to early bearing and higher nut yields.

Intensive basic and applied field studies on the usage of NaCl on tall variety and hybrids at PCA-Davao Research Centres gave consistent results that NaCl is a general and ideal fertilizer for coconut, both for corrective and maintenance fertilization. On local tall varieties like the Laguna tall, for a 5-year period, the average annual response to NaCl application at 2 kg palm^{-1} year^{-1} on bearing palms was about 31%, 30% and 69%, respectively, for nut yield, copra weight nut^{-1} and copra yield palm^{-1} (Magat and Margate 1990), while on coconut hybrids as CATD × LAGT, the respective average annual response was 65%, 31% and 65% (Secretaria et al. 2006). Magat (2009) reported that application of NaCl increased coconut yield by 25% and 50% over the unfertilized conditions during the first and subsequent years (2–5), respectively. In terms of copra, average annual yield of 1.78 tonnes ha^{-1} is achievable by application of NaCl over the unfertilized palms. Increasing chloride levels of coconut growing sandy regosol soil (Aquic Quartzipsamments) was found to reduce the leaching losses of nitrate N (Dissanayake et al. 2016) indicating the inhibitory effect of NaCl on soil nitrification.

8.6 Removal of Nutrients by the Palm

The coconut palm is an extreme type of perennial and is unique in that once it starts flowering, the productive phase lasts throughout the year and all through its economic life, which may average 70 years. The palm has many distinct developmental stages in its vegetative and reproductive cycles of growth. Once these are carefully identified and defined, then it should be possible to plan systematic studies covering every stage in its ontogeny. An evaluation and integration of these studies could then be expected to provide a reliable foundation for taking a complete nutrient inventory of the crop. To ensure a full appraisal of the nutrient status of the crop in all its bearings, and a thorough knowledge of all its growth manifestations, a three-pronged approach is in fact envisaged (Nathanael 1961). The *first phase* would constitute the establishment of norms for selected palms of comparable age and optimum performance grown under standard or typical conditions of cultural management and fertilizer regime. The *second phase* would be the study of reactions to known changes in fertilizer regime and levels of nutrient supply on the basis of long-range field experiments on differential manuring, involving single and multiple elements. The *third phase* would be the institution of plant nutritional surveys for the diagnosis of causative factors of deviations from established norms. The three phases of the proposed investigations would combine fundamental long-range work with short-term studies covering seedlings in the nursery stage and adult palms in their vegetative and productive stages of the life cycle. In addition, collateral sand culture experiments would be carried out on seedlings and young palms less than 5 years of age.

Nathanael (1958) suggested three approaches to study the mineral nutrition of the palm, viz. (i) to assess the mineral needs of the palm with the help of field fertilizer trials and by successive approximations, (ii) to analyse the soil for its nutrient-supplying capacity and (iii) to analyse the coconut water and leaf for understanding the level of nutrition of the palm in relation to its productivity and also the available nutrient status of the soil. Nathanael (1969) further elaborated the conceptual basis of assessing the nutrient requirement of the palm by the equation: F = R-S + L, where F is the quantity of the fertilizer nutrient, R is the quantity of the nutrient required by the crop for unrestricted growth, S is the quantity of the nutrient supplied by the soil and L is the portion of the fertilizer nutrient not utilized by the crop. He calculated the annual removal of the major nutrients by a middle-aged palm of the ordinary tall variety as 0.59 kg N, 0.26 kg P_2O_5 and 0.86 kg K_2O for palm yielding 40 nuts $year^{-1}$. When the yield level increases to 60 nuts, the calculated removal would be 0.72, 0.33 and 1.08 kg N, P_2O_5 and K_2O, respectively.

Conventional soil analysis is difficult to be adopted because a large volume of soil and subsoil is used by the coconut roots. Very often the composition of the leaves is not related to soil contents as the depth of effective soil and its structure are found more important than the nutrient concentrations. The sustained argument against soil testing is its inability to provide information on the nutrient absorption capacity of the palm. The soil analysis indicates only the potential capacity of a soil to supply nutrients to the palm, but it neither characterizes sufficiently the mobility of nutrients in the soil nor it provides any information on the plant factors, such as the extent of root growth and root functions which are of decisive importance for nutrient uptake under field conditions.

Visual deficiency symptoms usually appear when the deficiency is acute, possibly after passing through stages of hidden hunger, and the growth rate and yield are severely depressed. Moreover, the occurrence of multiple deficiencies, pest and disease attack, unfavourable environmental conditions, etc., makes the diagnosis complicated. The most widely adopted method is foliar analysis and recommending fertilizer applications based on established critical levels. This method has limitations as it fails to differentiate between metabolically active and inactive fractions of the elements in the tissue samples. The synergistic/antagonistic interaction between elements may lead to misinterpretation of foliar values. Environmental factors, type of cultivars/hybrids and the yield levels of the palm alter the critical levels of the nutrients. Therefore, an integrated approach giving a meaningful interpretation of different methods based on practical experience with respect to each situation is very much needed for making a realistic assessment on the nutrient requirement of the palm. The results of various approaches made to understand the mineral requirement of the palm are described in the following sections.

For the continued growth and productivity of the palm, an understanding of the nutrient demand and nutrient supply in the soil-plant system is essential. The major nutrient demand in the coconut plantation arises due to (i) the nutrients removed by the palm for its growth and yield, (ii) the nutrients lost by leaching and volatilization and weed growth, as well as (iii) the nutrients immobilized by the soil.

Table 8.10 Annual mineral export by the coconut palm

Sl. No	Basis	Nutrients (kg ha^{-1}) N	P	K	Ca	Mg	References
1	Annual removal ha^{-1}	63.9	29.10	95.3	–	–	Jacob and Coyle (1927)
2	6900 nuts ha^{-1} $year^{-1}$	91.9	41.50	136.7			Copeland (1931)
3	Annual removal ha^{-1} (tall type, 7400 nuts)	63–90	10.50–19.00	81–250	3.5–22	5.4–32	Georgi and Teik (1932)
4	Annual removal ha^{-1}	90.8	40.40	131.1	–	–	Eckstein et al. (1937)
5	156 mature palms ha^{-1}	116.6	40.40	141.2	–	–	Carvalho (1947)
6	150 palms ha^{-1} with 50 nuts $palm^{-1}$ $year^{-1}$	58.3	17. 90	53.8	29.1	44.8	Cooke (1950)
7	70 palms ha^{-1}	10.2	2.40	12.9	–	–	Nethsinghe (1960)
8	173 palms ha^{-1} with 40 nuts $palm^{-1}$	56.0	11.90	70.0	33.9	12.5	Pillai and Davis (1963)
9	173 palms ha^{-1} with 40 nuts $palm^{-1}$	96.0	20.80	120.0	61.8	21.9	Ramadasan and Lal (1966)
10	60 nuts $palm^{-1}$	72.0	39.00	108.0	–	–	Nathanael (1969)
11	100 nuts $palm^{-1}$ (nuts-whole palm)	120.0	18.00	85.0	–	–	Khanna and Nair (1977)
		157.0	28.00	346.0			
12	Annual removal ha^{-1} (1.5 t copra ha^{-1})	95.0	9.00	117.0	65.0	–	Ouvrier and Ochs (1978)
13	Annual removal ha^{-1} (6.7 t copra ha^{-1})	174.0	20.00	249.0	70.0	39.0	Manciot et al. (1979b)
14	Annual removal ha^{-1} (200 palms ha^{-1})	142.0	17.00	202.0	82.0	28.0	Omoti et al. (1986a)
15	Dwarf coconut (200 nuts $palm^{-1}$ $year^{-1}$	87.71	12.44	169.8	6.02	9.48	Sobral (1998)
16	15,000 nuts ha^{-1} $year^{-1}$	50.0	6.00	106.0	–	14.0	Somasiri et al. (2003)
17	Coconut yield 7500 nuts ha^{-1}	32.6	4.2	80.1	–	4.4	Gunathilake and Manjula (2006)

At a given soil-plant environment, the export of nutrients by the palm is an indicator of its nutrient requirement in addition to other related soil environmental parameters. In perennial plantation crops such as coconut, it is often difficult to measure nutrient removal since all parts of the plant cannot be analysed for their nutrient contents periodically. However, estimates of annual removal of nutrients by the palm reported by workers from different coconut growing countries of the world are presented in Table 8.10.

An insight into the nutrient exhaust data shows that though there are variations in the absolute values reported by different workers, the results generally indicate that the quantitative sequential order of importance of the major mineral nutrients for the adult coconut palm is $K > N > P \geq Mg > Ca$ (Nathanael 1967). Reports on chemical analysis of the various parts of the palms in Nigeria showed the predominance of $K > N > Ca > Mg > P$ in the export of nutrients (Omoti et al. 1986a).

The sequence of nutrient export by hybrid PB 121 as K > Cl > N > Ca > Na > M g > S > P shows the predominant requirement of chlorine (Ouvrier and Ochs 1978). Quoting the above data and those of Copeland (1931), Manciot et al. (1979b) reported that for the same yield of copra, there is no difference in the quantity of uptake between hybrid and tall coconuts. The uptake of different nutrients by coconut seedlings is also more or less in the same order (K > N > Cl > Ca > Mg > P) with K as the greatest and P the least absorbed (Santiago 1978). It is obvious that the dominant requirement in the nutrition of the palm is for K, while the least required nutrient is P.

The importance of a soil nutrient for the palm changes as the palms grows as demanded by its physiological requirements. The order of nutrient requirement for young palms is N > P > K > Mg, while that for adult palms (fruit-bearing trees) is K > Mg > N > P (Tennakoon 2004). In the coconut belt of south-western Ghana, the order of nutrient requirement for restoration of nut yield potential in mature coconut is K > P > N > Mg (Andoh-Mensah et al. 2003). According to Khan et al. (1986b), Tennakoon (2004) and Okeri et al. (2007), coconut palm responds positively to fertilizer, and the response is normally higher when Mg is included.

Detailed studies were conducted by Pillai and Davis (1963) on the removal of nutrients by West Coast Tall palms in India and that for the hybrid PB 121 in West Africa by Ouvrier and Ochs (1978). Even though the differences in the absolute values reported by these authors are large, there is agreement in the pattern of removal. The percentage removal of nutrients through different parts of the tall palm and the annual export of nutrients by the hybrid PB 121 are presented in Tables 8.11 and 8.12.

Table 8.11 Percentage removal of nutrients by different parts of WCT palm

	Nutrient removal (%)				
Parts of the palm	N	P	K	Ca	Mg
Nuts	43.0	40.0	63.0	15.3	25.0
Peduncles	4.2	7.0	12.1	3.3	11.4
Spathes	3.5	2.9	2.7	4.5	4.9
Leaf with stipules	41.2	45.1	12.4	73.8	56.5
Stem	8.1	5.0	9.8	3.1	2.2
Total	100.0	100.0	100.0	100.0	100.0

Table 8.12 Annual exhaust of major nutrients (kg ha^{-1}) by the hybrid PB 121

Particulars	N	P	K	Ca	Mg	Na	Cl	S
Total uptake for yield (6.7 t copra ha^{-1}) and growth of stem and leaves	174	20.0	249.0	70.0	39.0	54	249	30.0
Uptake for yield alone (6.7 t copra ha^{-1})	108	14.7	193.1	9.3	15.4	20	125	8.5
Difference (uptake for the growth of stem and leaves)	66	5.3	55.9	61.0	23.6	34	124	21.5

Table 8.13 Partitioning of nutrients; percentage removal of nutrients for the yield of nuts and for the growth and development of stem and leaves

Cultivar	Plant part	N	P	K	Ca	Mg	Na	Cl	S
West Coast Tall	Stem + leaves	49	50	22	77	59	–	–	–
	Nuts	51	50	78	23	41	–	–	–
Hybrid PB 121	Stem + leaves	38	25	22	87	62	63	50	70
	Nuts	62	75	78	13	38	37	50	30

For a comparative assessment, the percentage removal of nutrients for the development of bunches with nuts and for the growth of stem and leaves is computed and presented in Table 8.13. For both the tall and the hybrid, the quantity of K exhausted through the harvest of bunches is 78%, and that used for the growth of stem and leaves is only 22%. Among all the nutrients, the largest removal as well as the highest proportion in the nut harvest is K which suggests that maximum utilization of K is for the productivity of the palm irrespective of the variety.

The hybrid palms (PB 121) require more N and P as compared to tall palms and utilize higher proportion of absorbed N and P for the production of more nuts. In tall palms these nutrients are utilized more or less equally in the production of nuts and for growth. For both cultivars, the K removal through bunches is 78% of K uptake. Calcium is utilized more for the production of leaves and stem and least for nuts. Since Ca is not a very mobile element and its concentration in the leaf tissues increases with maturity, it is quite probable that a higher proportion of this element is exhausted through the shedding of leaves. Unlike K, the proportion of Ca used for the development of bunches is the least, while the one used for the growth and development of stem and leaves is the highest suggesting the importance of Ca for the proper growth and functioning of the leaves and stem. Moreover, Ca plays a major part in the formation of cell walls. Thus, among the nutrients, Ca is the least involved in increasing the nut yield of bearing palms, but it helps in the healthy growth of leaves.

In the case of Mg, the quantity removed for the growth of stem and leaves is about 60%, while that exhausted through the harvest of bunches is about 40%. Magnesium is the only metal constituent of the chlorophyll molecule and is concerned in a variety of enzyme reactions in which it acts as the most effective activator. Thus, Mg is an important element in the nutrition of the palm and is mostly required for the effective functioning of the leaves, and through its photosynthetic function, it very much regulates the growth as well as the productivity of the palm.

Though the total quantity of Cl export by the hybrid is equal to that of K, the quantity used for the development of bunches ranks second, next to K, which makes it the second most required nutrient for productivity as reported for PB 121 hybrid. Its requirements for vegetative and reproductive growth are equal, and hence Cl-bearing fertilizers assume significance in coconut nutrition as opined by Oguis et al. (1979). However, its role in the physiology of the palm is not clearly understood in spite of its high requirement. Sodium and sulphur are more utilized for the growth of stem and leaves than for the development of nuts. Pillai and Davis (1963) from India and Ouvrier and Ochs (1978) from Côte d'Ivoire suggested that the dominant requirement of the palm is for potassium and probably chlorine.

The high productivity of the hybrid is invariably accompanied by a proportionate increase in the uptake of nutrients, particularly N and K, from the soil. The above data further suggest the importance of adequate nutrition to hybrids for higher yields and that regular fertilizer application is necessary to maintain a balanced supply of nutrients for sustained productivity (Khan et al. 1986a; Khan 1993).

According to von Uexkull (1978), around 10.5 g K is lost and immobilized for every nut harvested. Low yields (below 40 nuts palm^{-1} year^{-1}) can usually be sustained by the soil reserves of nutrient without replacing the K removed. At about 150 nuts palm^{-1}, yields can usually only be maintained if the total amount removed is replaced by soil application. At the highest yield levels, application should exceed actual removal because applied K is used less efficiently at that level and K concentration in the soil solution should be raised.

The PB 121 is capable of producing up to 6.0 t of copra ha^{-1} year^{-1}. The determination of annual exports of mineral elements (Ouvrier and Ochs 1980) revealed that a yield of 6 t of copra removed a fertilizer equivalent of 1450 g of urea (46% N), 600 g of tricalcium phosphate (38% P_2O_5), 2500 g of KCl (60% K_2O) and 500 g of Kieserite (33% MgO). These exports, however, represent only a part of the total requirements of the palm.

The removal of K from the soil by a coconut plantation with a population of 150 palms ha^{-1} at a production level of 7500 nuts ha^{-1} year^{-1} (50 nuts palm^{-1} year^{-1}) given by Mahindapala and Pinto (1991) indicates that husk removes 34% and nut water 27% of K (Table 8.14).

The annual exhaust computed from 1 hectare of 173 palms in sandy loam soil was 65.6, 29.7, 84.5, 47.4 and 20.3 kg of N, P_2O_5, K_2O, CaO and MgO, respectively, taking into account the nuts, fallen leaves, spathes and the stem growth (Pillai and Davis 1963). The amounts of macronutrients lost through the removal of plant components from the field of Typica × Typica coconut palms yielding an average of 17380 nuts ha^{-1} year^{-1} in Sri Lanka were 116.79 kg N, 14.02 kg P, 245.43 kg K, 40.47 kg Ca and 33.66 kg Mg ha^{-1} year^{-1}. The amounts of micronutrients lost were 1.14 kg Fe, 0.63 kg Mn, 0.13 kg Cu, 0.44 kg Zn and 0.26 kg B ha^{-1} year^{-1} (Somasiri et al. 2003). Potash

Table 8.14 The removal K from a coconut plantation having 150 palms ha^{-1}

Component		Amount of K removed	
		kg ha^{-1}	g palm^{-1}
Flower parts		1	7
Frond	a) Petiole	7	47
	b) Leaflets	32	213
Nut	a) Husk	45	300
	b) Shell and kernel	11	73
	c) Nut water	36	240
Total		132	880

was found to be removed the most, followed by nitrogen, calcium, magnesium and phosphorus. The quantity of nutrients removed varies with the soil type and yield. Palms growing on coastal alluvium removed 70 kg K_2O ha^{-1}, but the average removal from red sandy loam and laterite was around 53 kg K_2O ha^{-1}.

As the trees grow older, the total amounts of P and other nutrients immobilized in the system increase. Chew (1978) and Omoti et al. (1986b) reported that over a 25-year life cycle, about 9–10 kg P ha^{-1} is immobilized in the trunks each year. The removal of P in nuts (husks, shells, albumen and water) varies with yield. Immobilized nutrients are important for palms as they sustain the growth of the palm, and as assimilates increase in the system, the nutrients are partitioned towards vegetative and reproductive needs.

8.7 Long–Term Impact of Fertilization on Nutrient Status

In all coconut growing countries, fertilizer application and its impact on sustainable yields has been demonstrated. Fertilizer application over years has given sufficient lessons to understand its impact on management of groves considering the physical and chemical changes it has brought on soil health, plant nutrient contents and tailor management practices for sustainable yields.

Annual application of NPK fertilizers over an 18-year period to coconut on red sandy loam soil in a high rainfall region of Kerala, India, resulted in a marked increase in soil available P and K but only a minimal increase in mineralizable N (Khan et al. 1986a). In monitoring the healthy coconut groves, they suggested that leaf analysis gives a better reflection of the effect of N fertilizer than soil analysis and slightly lower values even in fertilized plots compared to the critical level of N established elsewhere (Table 8.15). The study also suggested that maintaining 15 ppm available P (Bray I) in soil in the manuring circle is sufficient to maintain optimum P nutrition of palms.

Table 8.15 Effect of different levels of fertilizer on plant (14th leaf) nutrient status of three genotypes

		Percent						ppm			
Genotype	Fertilizer level	N	P	K	Ca	Mg	Na	Zn	Mn	Cu	Fe
WCT	M_0	1.4	0.11	0.7	0.23	0.23	0.25	20	593	6	93
	M_1	1.7	0.11	1.0	0.33	0.17	0.23	18	711	6	86
	M_2	1.7	0.11	1.1	0.27	0.15	0.13	17	648	7	90
CDO × WCT	M_0	1.4	0.11	0.6	0.26	0.22	0.24	19	599	6	99
	M_1	1.6	0.11	1.0	0.26	0.13	0.11	16	585	8	108
	M_2	1.7	0.11	1.1	0.27	0.11	0.10	15	620	9	108
WCT × CDO	M_0	1.4	0.11	0.6	0.25	0.26	0.27	24	521	10	96
	M_1	1.6	0.11	1.0	0.26	0.16	0.14	18	615	8	86
	M_2	1.6	0.11	1.1	0.30	0.15	0.11	16	678	8	73

M_0: No fertilizer, M_1: 500 g N, 220 g P and 830 g K, M_2: 1000 g N, 440 g P and 1660 g K

Table 8.16 Long-term effect of fertilization on available potassium status of red sandy loam soil (Arenic Paleustults) at different soil depths

Fertilizer level N/P/K ($g\ palm^{-1}\ year^{-1}$)	Irrigated condition Available K (ppm)		Rainfed condition Available K (ppm)	
	0–25 cm	25–50 cm	0–25 cm	25–50 cm
M_0 (No fertilizer)	79	38	66	66
M_1 (500:218:833)	110	69	202	153
M_2 (1000:437:1667)	212	129	318	235

Fertilizers applied to coconut for longer periods (1965–1984) also lowered the soil pH (5.17–4.13) probably as a result of increase in Mn content, and it also decreased the plant Zn status, however, insignificantly. The superiority of COD × WCT hybrids over the reciprocal cross and WCT palms was also indicated (Khan et al. 1986a). In an extension of the above study, a modified treatment structure with irrigation to palms was introduced (1984–1985), and the findings were reported by Srinivasa Reddy et al. (2002). They observed that after 32 years of fertilizer application, the available soil K content was 66 ppm in M_0 plot under rainfed condition, which increased to 202 ppm and 318 ppm, respectively, at 0–25 cm soil depth with M_1 (500:220:830 g) and M_2 (1000:440:1660 g N, P, K $palm^{-1}\ year^{-1}$) levels of fertilizer application. Under irrigation, a reduction in soil available K was observed in M_1 and M_2 plots (Table 8.16). Application of K fertilizers raised the leaf K levels to 1.14% (M_1) and 1.25% (M_2) compared to 1.07% in M_0 under rainfed condition probably because of ready availability under the moisture regime. Under irrigation, leaf K content was 1.07% under M_1 and 1.20% under M_2 compared to 0.90% under M_0.

Application of K fertilizer at M_1 level was found to maintain K content of leaves above critical level (0.8–1.0%). This study suggests that doubling the K levels had little effect indicating that rates beyond 830 g K (1000 g K_2O) $year^{-1}$ are probably not needed. Thus, a soil available K (1 N NH_4OAc) content of 50–60 ppm (0.128–0.153 meq 100 g^{-1}) is adequate for maintaining sufficiency levels in coconut palms. Annual application of fertilizers also resulted in a marked increase in available phosphorus and potassium status in soil, but a marginal change in soil available nitrogen status was observed. Foliar contents of N remained below the critical levels of 1.8–2.0%. Phosphorus build-up in the soil due to fertilizers did not reflect in the P contents of diagnostic leaf under both rainfed and irrigated conditions. Application of K fertilizer at M_1 level maintained K content of leaves at 1.07%, i.e. just above the critical level.

Manciot et al. (1979b) reported that 0.15–0.20 meq 100 g^{-1} (59–78 ppm) and Loganathan and Balakrishnamurti (1980) suggested that 0.13 meq 100 g^{-1} (51 ppm) of exchangeable K is sufficient for satisfactory growth of coconut palm.

In another long-term study in littoral sandy soil at Kasaragod, India, the available K status of the soil (0–100 cm depth) increased from 50.2 ppm at K_1 level (750 g K_2O $palm^{-1}\ year^{-1}$) to 95.9 ppm at K_2 level (1250 g K_2O $palm^{-1}\ year^{-1}$) to 105.56 at K_3

level (1750 g K_2O palm^{-1} year^{-1}) (Srinivasa Reddy et al. 1999). These values of soil available K corroborate with the statement by Biddappa et al. (1993) that a soil available K content of 50–60 ppm is adequate for maintaining the sufficiency levels in coconut. Joseph and Wahid (1997) have observed that the application of KCl resulted in a large increase in K reserves in soil to depth of 100 cm. The increase in K content was nearly 200 ppm within this depth. Relatively less accumulation of K was noticed in the 0–50 cm root zone than below it.

Smith (1968) emphasized the importance of K not only for the faster development and vigorous growth but also for reducing the pre-bearing period of palms. The palms, which received adequate nutrition from the beginning, produced more yield than those supplied after maturity. In coconut experiments in the Côte d'Ivoire (Fremond and Ouvrier 1971), the effect of applying K and the time of field planting was compared to withholding K applications until the age of bearing. The latter practice was inferior for all palms. Thus, the yield potential and also the precocity of fruiting of coconut palms depended much on the K supply of the young palms.

In a sandy loam soil at Kasaragod, Kerala, palms which received 1.0 kg N and 1.5 kg K_2O flowered first (Nelliat et al. 1978). Similarly, palms fertilized with KCl and N or NP from transplanting time in an inland-upland area of the Philippines recorded initial flowering in less than 4 years and significantly higher nut and copra production than those palms, which did not receive KCl (Mendoza and Prudente 1972).

In a 5-year study in a high rainfall laterite soil region with different carriers of phosphorus for coconut, viz. single superphosphate, ammonium phosphate, nitrophosphate and rock phosphate, Khan et al. (1985a) suggested rock phosphate as the ideal carrier of P and use of ammonium phosphate, to be discouraged as continued use may lead to extreme acidic conditions in the rhizosphere region though it is a soluble P carrier. They further observed that build-up of P in the plant system could be achieved only over a period of time compared to N and K.

Studying the residual effect of applied phosphorus to coconut palms after 14 years, Khan et al. (1992) observed that though the available P levels reduced to a mean value of 12.72 ppm (0–90 cm depth) from 43.88 ppm, it maintained the leaf P levels (frond 14) as 0.119% without bringing any reduction in yield. In the P-skipped plot, the mycorrhizal infection might have played a significant role in P nutrition (mobilization) of palms (Table 8.17).

Table 8.17 VA mycorrhizal association in different P treatments

Treatment	Per cent colonization	Intensity of infection (%)	Spore count (g^{-10} soil)
P_0	79.29	38.14	128.86
P_1	52.14	20.14	94.00
P2	47.86	17.43	81.86

8.8 Role of Mineral Nutrients

The perennial nature of the coconut palm as well as its extensive root system poses considerable challenges in carrying out plant nutritional studies. However, various studies have been carried out in the different coconut growing countries of the world, and the present state of knowledge on the role of individual nutrients on growth and productivity of the palm is presented in this section.

8.8.1 Nitrogen

Nitrogen is one of the primary nutrients which is universally limiting the growth and yield of crops including that of coconut. It is an essential constituent of amino acids, proteins and nucleic acids and also of the green colouring matter chlorophyll. The shortage of nitrogen results in retarded growth and chlorosis.

8.8.1.1 Behaviour in the Palm

A basic understanding of nitrogen in the coconut plant system was presented by Ohler (1999). Braconnier et al. (1992) analysing 4-year-old coconut trees, 3.5 months after applying isotopically labelled nitrogen (^{15}N), found that total nitrogen distribution in the plant is related to the distribution of dry matter. Thus, leaves contained 66% of total nitrogen of aerial parts. Differences between these findings are probably due to age differences between groups. The highest labelled N percentages were recorded for the spear, bud and green spathes, indicating that developing organs are considerable sinks of nitrogen. By contrast, stipules and dry spathes, which did not grow and had low physiological activity, had low-labelled N percentages. However, the percentage in the stem was surprisingly high, although growth of this organ was nil. The same situation was found in older leaves. Apparently N fertilizer was distributed in all leaves, indicating that labelled N in different organs has rather rapid turnover. These observations confirm the existence of an influx and afflux of nitrogen in each leaf, whatever its rank. The proportion of nitrogen in coconut seems to be in continuous flux. Labelled N was distributed in all parts of coconut, except in stipules and dry spathes, indicating N distribution throughout the tree. Little nitrogen fertilizer was distributed into mature bunches, indicating that N nutrition of albumen comes mainly from the husk and shell. Thus, N fertilization would influence production only 8–10 months after fertilizer application.

The N concentration in the leaf varies with its position on the crown. It progressively increases up to the sixth or seventh leaf and then decreases with maturity and drops down to a low level when senescence advances. It also shows marked seasonal and diurnal variations. Its content is invariably high during cooler and wet periods than hot summer months. The N flux in the crown is minimum during summer months due to the low water content in the soil. Diurnally the N content increases up to 10–11 AM, and then it decreases as the day progresses.

The critical level of 1.8–2.0% N in frond 14 is widely accepted as a guide for regulating the N nutrition of the palm. Thus, leaf analysis is found to be the best diagnostic method for predicting the N demand as well as the possible N deficiency in the palm. The interaction between N and P, N and K and N and S are generally positive on growth and productivity of the palm. In case of nitrogen deficiency, heavy doses of nitrogen fertilizer may result in rapid improvement of leaf colour and photosynthetic capacity, giving yield responses compared with other elements. Ohler (1999) observed that both in the heavy soil (Port Bello) and on a minerally poor sandy soil (Barra) in Mozambique, palms reacted strongly within 1 year for fertilizer application, the canopy colour changing from light yellowish to dark green. The reactions to other elements, when applied without N, were rather insignificant. Manciot et al. (1980) reported that only after N levels in the palms had normalized, potassium deficiency became apparent. Khan et al. (1986a) observed that N content of leaf gives a better indication of fertilizer application with nitrogenous fertilizers than soil analysis.

8.8.1.2 Deficiency Symptoms, Causes and Correction

Visible symptoms of N deficiency on coconut were described by different workers (Fremond et al. 1966; Manciot et al. 1979b). The specific symptoms are yellowing of foliage in varying degrees and stunted growth. In the initial stages, the palm loses the healthy green colour, and the whole foliage exhibits a slight and continuous yellowing. When the deficiency advances, the older leaves are affected the most and may develop uniform golden yellow colour, while the younger leaves turn pale green giving the leaflets a dull appearance with a diffuse underlying yellowing. This is accompanied by the abortion of many of the inflorescences, and the number of female flowers per inflorescence becomes less. N deficiency limited female flower production, rate of bunch production and yield (Smith 1969).

Nitrogen deficiency is commonly observed, under dry climatic conditions which inhibit nitrification and N absorption, on calcareous soils where alkalinity impedes mineralization of organic matter and on sandy soils that are deficient in organic matter and also in waterlogged situations. Deficiency can also result from exhaustion of the soil by continuous mining of nutrients for longer periods without adequate replenishment. The size of the leaves gradually gets reduced, and the number of functioning leaves becomes less. In the advanced stages, the stem below the crown narrows to a "pencil point"-like appearance with few short leaves on the crown. Inflorescences fail to emerge, if any emerges will have very few or no female flowers and ultimately the palm becomes unproductive (Fig. 8.1). In potted seedlings under N deficient condition, the newly emerged leaves appear pale, abnormally short and non-succulent. Nitrogen deficiency is more common in young palms.

Fig. 8.1 Symptoms of nitrogen deficiency. (Photo: Jacob John)

Coconut responds well to the application of nitrogenous fertilizers and organic manures in almost all soil types, and the response is maximum in light-textured sandy soils and lateritic soils. Field experiments conducted in different parts of the world have demonstrated a response to manuring with nitrogenous fertilizers. It increases all the growth parameters of young palms, reduces the pre-bearing age and increases the rate of frond and bunch production. The female flower production and the number of nuts palm^{-1} are increased significantly. The full expression of the beneficial effect of N occurs in the presence of adequate levels of P, K and S in the soil. In New Guinea, the application of S and urea together produced spectacular response on vegetative growth and foliage colour of young palms (Charles 1968). In Jamaica, Smith (1968) found significant increase in growth characters of young palms due to application of nitrogenous fertilizers. A further study in Jamaica showed that N increased the production of bunches (Anon 1967). In India, Muliyar and Nelliat (1971) observed about 17% increase in yield of nuts, but the copra weight nut^{-1} was depressed by N (Table 8.18). It is generally experienced that when nitrogen considerably increased the number of nuts produced, it very significantly lowered the copra content of the nuts (Manciot et al. 1980).

Similar responses to nitrogen applications were reported in different coconut growing countries (Manciot et al. 1979b). In an experiment on quartz sand soil from 1987 to 1992 in north-eastern Brazil, Sobral and Leal (1999) observed that N influ-

Table 8.18 Effect of different levels of N on yield of nuts and copra content

N levels (kg palm^{-1} year^{-1})	Mean annual yield (nuts palm^{-1})	Copra weight (g nut^{-1})
N_O - (0.0)	47.8	179
N_1 - (0.34)	55.9	165
N_2 - (0.68)	54.7	162

enced the number of nuts and considered 1.718% as the critical N level. Sobral (2004) observed an increase in the number of nuts in dwarf green coconut palms when the plant was receiving nitrogen in the form of urea in the fertilizer programme.

8.8.2 *Phosphorus*

Among all the major nutrients, the quantitative requirement of P is the least. According to Manciot et al. (1979b), P uptake is nearly one tenth of total uptake of potassium and that of chlorine as well. It is not usually a limiting factor in most of the coconut growing areas of the world. Although generally known as a macroelement, phosphorus is taken by coconut in modest quantities. Phosphorus is an essential constituent of many vital cellular compounds like ATP, ADP, AMP, RNA, DNA and other phosphorylated sugars and fats. It is mostly concentrated in the growing points like tender leaves and root tips of the palm. The leaf P content decreases with maturity as evidenced by the leaf content across the crown. It is high during cooler and wet periods and low during dry summer months. Both seasonal and diurnal variability influence the leaf P concentration.

8.8.2.1 Behaviour in the Palm

The beneficial role of P on adult coconut palms has always been a debating question as far as its direct role in the yield of palms. Eden et al. (1963) in Sri Lanka, Pandalai and Marar (1964) in India, Smith (1968) in Jamaica, von Uexkull (1972) and Barile and Azuceana Jr (1972) in the Philippines did not realize much improvement in the yield of adult coconut palms due to the application of P-containing fertilizers. Wahid et al. (1975) observed no reduction in either soil or foliar levels of P when fertilizer with phosphorus was discontinued for 1 year. Application of P was found to increase leaf production, girth at collar and root density of coconut seedlings, lower the age of flowering and reduce the incidence of leaf disease caused by the fungus *Helminthosporium incurvatum* in young palms (Loganathan et al. 1984). The growth as indicated by leaf production and age of flowering was largely influenced by application of P. Discussing the influence of Mg on assimilation of phosphorus, Nethsinghe (1963) observed that when Mg was deficient, the movement of P within the plant is hampered and under such conditions the palm may experience double deficiencies of P and Mg and according to de Silva (1973), this could occur even if the palm is supplied with an available source of phosphorus.

8.8.2.2 Deficiency Symptoms, Causes and Correction

Specific instances of P deficiency were rarely reported. There was no characteristic visual symptom apart from slowing down of growth and shortening of fronds (Manciot et al. 1979b). Phosphorus deficiency in young palms in sand culture manifested stunted growth and rosette appearance. The leaves were dark green and they could not come out fully from the stipules. The older leaves showed severe drying (Velasco et al. 1960). The delayed flowering in certain palms in Bandirippuwa (Sri Lanka) was attributed to an induced lowering of P in fronds caused by a lower uptake of Mg by the roots (de Silva et al. 1973). In a global review on mineral nutrition and fertilization of coconut, Manciot et al. (1979b) reported that there are only a few instances in Madagascar, India, Sri Lanka and Côte d'Ivoire in which the favourable effect of P manuring on coconut yield was noticed and that too after several years of continuous P application to palms. Commenting on P manuring of coconuts in Sri Lanka, Halliday and Sylvester (1954) stated that there was no response to P application at Bandiruppuwa where the soil contained adequate reserves of P, but experiments on poor lateritic soils of Veyanagoda and Ahangama with low P resources in the soil gave response to P application at 275 g P_2O_5 palm^{-1} once in 2 years. Summarizing the contribution of. IRHO on the study of mineral nutrition, Fremond (1964) reported that P did not show much beneficial effect in increasing either the yield of nuts or copra content. But in the presence of K, P was found to have beneficial effects. However, nuts were larger where P was applied (Smith 1969).

Being the least mobile element in the soil, the loss of P by leaching is the minimum among all major nutrients. Various studies have shown that the continued use of P fertilizers can substantially increase the available soil P, and the residual effect was observed a number of years later. In Sri Lanka good responses were obtained on very poor latosols containing only traces of available P, when P was applied at 0.12 kg palm^{-1}. The soil P potential was then built up to a level, at which discontinuance of P application did not lead to any reduction in yield for at least 5 years (Child 1974).

Reduction of soil available P from 84 to 59 ppm at 0–30 cm depth and 24 to 13 ppm at 30–60 cm depth in the coconut basins was observed over years by Khan et al. (1983) in India when application of P fertilizer to adult coconut palms was resorted to. Neither a reduction in soil available P nor an increase consequent to fertilizer application at two levels for about 6 years had significantly influenced foliar P levels and yield. The palms, however, were receiving regularly fertilizers containing N and K as per schedule. The possibility of withholding application of phosphate fertilizers to adult coconut groves in soil where available P is around 24 ppm at 30–60 cm depth was suggested. Subsequent studies by Khan et al. (1990) showed that skipping P application for 14 years did not show any adverse effect

Table 8.19 Foliar P content (frond 14) and yield of nuts as influenced by levels of P application

P fertilizer as levels ($palm^{-1}$ $year^{-1}$)	Foliar P (%) (1989)	Yield (nuts $palm^{-1}$) 1985–1986	1986–1987	1987–1988	1988–1989
P_O (Zero P)	0.12	127	97	89	109
P_1 (160gP_2O_5)	0.12	130	85	98	102
P_2 (320 g P_2O_5)	0.11	138	93	96	109

either on yield or leaf P levels which suggest that utilization of built-up reserves in soil is the most ideal and economic way of management of coconut groves. Addition of green manures may further assist in dissolution and availability of P to palms. It also suggests the importance of soil testing in the P nutrition of palms (Table 8.19).

Based on the above study, it has been recommended that when the available soil P (Bray I) in the 0–90 cm soil is less than 10 ppm, apply the full recommended dose of 320 g P_2O_5, and if it is between 10 and 20 ppm, a maintenance dose of 160 g P_2O_5 $palm^{-1}$ $year^{-1}$ may be applied. If the available P is more than 20 ppm, P application can be skipped for a few years and monitored through soil analysis (Khan et al. 1992). A foliar content of 0.11–0.12% P (frond 14) can be regarded as sufficiency level for coconut under Indian conditions. The interaction between N and P and P and K is generally positive on growth and productivity of the palm, while that between P and Mg is positive on foliar P levels. Though the requirement of P is low compared to other nutrients, its role in interaction with other elements and its physiological functions cannot be undermined.

8.8.3 Potassium

Potassium is usually the least needed major nutrient in low-yield agriculture but rises to a dominant position when yields are maximized (von Uexkull 1985). The severity/frequency of K deficiency has been found to be one of the most limiting factors in the economic production of coconut all over the world, and conspicuous increases in yield have been obtained by its correction through potassium fertilization. The coconut palm is a K-demanding crop as large quantities of K are removed by the nuts of the palm (Fremond et al. 1966). A low-yield level of 40 nuts $palm^{-1}$ $year^{-1}$ in case of coconut can be sustained without replenishing the K to the soil; however, at yield levels of 150 nuts, all the K removed must be replenished. von Uexkull (1985) opined that the higher K demand may be due to the coarser root system and midday heat and moisture stress suffered by fully exposed leaves.

Potassium improved all the nut characters, viz. weight of whole nut, weight of husked nut, volume of husked nut and copra weight nut^{-1}, whereas nitrogen had an adverse effect. For palms yielding less than 60 nuts annually, the optimum dose of N ranged between 400 and 650 g and that of potash between 890 and 1210 g $palm^{-1}$ $year^{-1}$ (Muliyar and Nelliat 1971). In a long-term fertilizer experiment in red sandy loam soil in Kerala, significantly higher nut yield besides early bearing was achieved with increased levels of K application. The yield was 7, 68 and 77 nuts palm^{-1} year^{-1} in the 21st year after planting under no fertilizer, 450 g K_2O and 900 g K_2O palm^{-1} year^{-1}, respectively (Wahid et al. 1988).

8.8.3.1 Behaviour in the Palm

Potassium positively influences the number of inflorescences, increasing the production by increasing the number of fruits and the quantity of copra nut^{-1}. Coomans (1974) demonstrated that the K content in the leaf is influenced by production in view of the large quantities exported. Thus, plants of low productivity may have a high K content in the leaf, giving the impression of good nutrition. It is also observed that palms with higher yield load may have a slightly lower K content. Anilkumar and Wahid (1989) found that application of K caused a decrease in Ca and Mg in the leaf. Manciot et al. (1980) proposed as critical level for K on leaf 14 values between 0.8 and 1.0%.

The coconut palm is highly responsive to K which increases resistance to drought and disease, hastens maturity and increases fruit set and the number of harvested nuts (von Uexkull 1985). An adult palm (more than 10 years) yielding 1.8 t ha^{-1} of copra removes between 90 and 130 kg K ha^{-1} in 1 year. The number of coconuts required to produce 1 kg of copra was 4.5 with adequate K nutrition and 8 where palms were under K stress.

Potassium chloride is the most widely used fertilizer for both coconut and oil palms. It increases the size (weight) of the nuts and the copra yield, as well as the Cl and Ca concentration, but slightly decreases K in the leaves (von Uexkull and Sanders 1986).

Potassium is taken up by the palm as K^+ ions, and its absorption is antagonized by the presence of high concentrations of Ca, Mg and Na. The annual removal of K by the palm varies with the cultivars as well as the soil conditions. Total removal of K by the high-yielding varieties and hybrids is generally higher than that of low-yielding types. As mentioned earlier, 78% of the K exhaust is through the harvest of coconut bunches, out of which 60% is present in the husk (Ouvrier and Ochs 1978). That is, about 47% of the total K removed by the high-yielding hybrid is present in the husk.

The K concentration in the leaf varies with its position on the crown. It is high in young leaves which progressively decrease with maturity, indicating mobility of K to the younger leaves. The middle leaves do not show much seasonal variations in their K contents. However, in the young leaves, the K content is more during dry seasons and less during wet seasons, while in the older leaves, this trend is found reversed.

The widely accepted critical levels of K for traditional varieties and high-yielding hybrids are 0.8 to 1.0 and 1.4%, respectively (Manciot et al. 1979c). The K content in the index leaf often shows good correlation with nut yield. As the foliar K levels or the yield of nuts does not generally show significant correlation with soil K levels, leaf analysis is found to be the best diagnostic method for predicting/regulating the K nutrition of the palm. The critical level of K is about 1 g kg^{-1} DM (von Uexkull and Sanders 1986).

Manciot et al. (1979b) reported that there exist strong antagonisms between K-Ca, K-Mg and K-Na in coconut. Often, a higher quantity of fertilizer application was found to decrease Mg level in the tissue. Application of K led to a significant drop in the content of Ca, Mg and Na in the leaf. Further, Mg fertilization is beneficial only when the K is adequate in supply or the K deficiency is corrected. The results showed that Mg application had a beneficial effect on the copra yield only if K fertilizers were also applied. Similarly higher levels of K manuring increased the yield only in the presence of Mg. In fact, higher levels of K application had a depressive effect on copra yield in the absence of Mg fertilization. Wahid et al. (1974) found that the foliar content of K + Na decreased with increase in root CEC while Ca + Mg increased with increase in root CEC. The uptake of cations by the palm was found to be governed by their ratios in soil. Highly significant correlations were obtained between K/Na, K/(Ca + Mg) and K/Mg in soil and their corresponding ratios in the leaf. The negative correlation of root CEC and positive correlation of both soil and leaf potassium with yield indicated the role of potassium in improving the yield of coconut. Coomans (1977) observed that application of K had induced the Mg deficiency in coconut hybrid palms, but Mg application had no effect on leaf K level. However, Brunin (1970) reported that in tall cultivars when the leaf K levels were between 0.7 and 1.2 g kg^{-1}, application of high rates of Mg significantly reduced K contents.

Nutrient interaction studies have clearly demonstrated the antagonism between K-Ca, K-Mg and K-Na in the palm, among which the K-Mg antagonism is more severe and has been well documented (Manciot et al. 1979c; Khan et al. 1986a; Wahid et al. 1988). Heavy rates of K application induce Mg deficiency wherein both the absorption and functional antagonisms are operating in the palm. On the other hand, higher rates of Mg application in poor soils induce K deficiency conditions. The effect of higher application of potassium chloride on foliar nutrient levels is shown in Table 8.20.

Though the foliar contents of all the three cations, viz. Ca, Mg and Na, are depressed by higher doses of K (Khan et al. 1986b), the effect is highly pronounced on Mg ($r = -0.68^{**}$). While formulating higher fertilizer doses, particularly of K, for hybrids and other high-yielding genotypes, it is necessary to include proportionate quantities of Mg fertilizer salts to maintain a proper balance between K and Mg in the soil.

Table 8.20 Effect of higher levels of KCl application on foliar nutrient levels

	Foliar nutrient level (%)					
KCl (kg $palm^{-1}$ $year^{-1}$)	N	P	K	Ca	Mg	Na
Control	1.80	0.091	0.20	0.50	0.57	0.17
5	1.75	0.097	0.98	0.51	0.19	0.29
10	1.74	0.094	1.38	0.40	0.16	0.23
15	1.74	0.097	1.55	0.39	0.13	0.18

8.8.3.2 Deficiency Symptoms, Causes and Correction

The deficiency symptoms of K are usually perceptible only when the soil K levels are below 0.15 meq 100 g^{-1} and the leaf K level (frond 14) falls below 0.4%. The first visible symptom is the development of rusty spots in two longitudinal bands on either side of the midrib with their diameter ranging from 0.5 to 4.0 mm, which is accompanied by slight yellowing of the lamina. The yellowing is more marked towards the tip of the leaflets. When the yellowing intensifies, the older leaves assume an orange-red tinge, while the younger leaves remain green. The yellowing is never uniform. It is more intense along the edges of a leaflet leaving a central band along the midrib green. The individual leaflets are also greener at the base than towards the distal ends where necrosis sets in and the rusty spots coalesce into numerous irregular brown blotches. As the deficiency advances, the yellowed surface becomes necrotic resulting in more of a necrotic appearance than of yellowing (Fig. 8.2).

K deficiency leads to development of poor crown with short fronds. The growth is reduced, the trunk becomes slender, leaflets become short and the number of inflorescences, nut set and nuts $bunch^{-1}$ gets reduced (Salgado 1953; Menon and Pandalai 1960; Fremond et al. 1966; Smith 1969; Anon 1970; Manciot et al. 1979c, 1980; Liyanage 1999). An increase in the level of K in the leaves improves the precocity of flowering and increases number of female flowers, setting percentage, thereby increasing the number of bunches $palm^{-1}$, average copra nut^{-1} and total copra production $palm^{-1}$ (Anon 2010). Severe K deficiency in coconut has been noted on tertiary and quaternary sands of West Africa, on the coastal sands of Sambava (Madagascar), on the coral soils of the Oceania atolls, on the exhausted lateritic soils of India and on the sandy soils of the east coast of Sri Lanka (Manciot et al. 1979c).

Baseden and Southern (1959) in New Ireland observed that K was the main factor limiting coconut production. They reported K deficiency in coconuts on soils

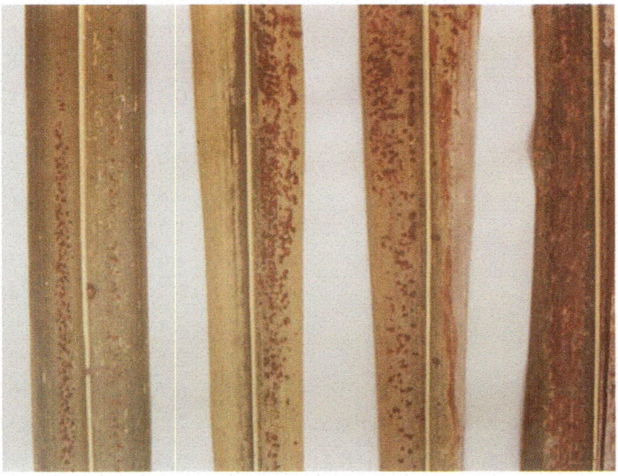

Fig. 8.2 K deficiency symptoms. (Photo: Jacob John)

developed over coral limestone, because of an imbalance in cations and the considerable amount of K removed with the husks and nuts. The wide range of productivity (from <5 to >60 nuts palm^{-1}), occurring among coconut palms on the coral-derived soils of eastern New Ireland, and the intensity of symptoms, such as chlorosis, smallness and sparseness of fronds, is closely related to the K levels found in nut waters, husks fronds and the soils. The better palms, with higher K status, occur on a narrow coastal strip of shallow, neutral to slightly acid, red-brown clay loams, and the poorer palms, with low K status, on much deeper, acid, yellow-brown clay, present on the inland side of the strip. On these soils, coconuts showed a dramatic response to K applications (Charles and Douglas 1965) which was also found in soils with high Mg/K ratios (Sumbak and Best 1976). Dootson et al. (1986) reported significant reduction in all growth parameters of MYD × WAT hybrid coconut in Thailand, except trunk length, due to non-application of K. Nitrogen and K were mutually antagonistic, and antagonism among cations was also observed.

Experiments in Sri Lanka by Salgado (1952) have shown that K has an effect on the earliness of bearing in the palm. Palms receiving K flowered in the fifth year of planting while those without K application flowered in the eighth year. When N application caused an 8% increase in copra yield, K application gave an increase of 25–39% (Salgado 1947).

The spectacular response to K fertilizer application on the yield of nuts at the Coconut Research Station, Balaramapuram, Kerala, on red loam soil (Wahid et al. 1988), showed the importance of K in augmenting coconut yield (Table 8.21). More than tenfold increase in yield was obtained with 450 g K_2O palm^{-1} year^{-1}, while it was about 13-fold increase with 900 g K_2O palm^{-1}year^{-1}.

Responses to K fertilization were reported from different coconut growing countries of the world (Prevot and Fremond 1961; Eden et al. 1963; Foale 1965). Fremond (1964) and Manciot et al. (1979c) observed that the application of K resulted in the improvement of all production factors such as the number of bunches

Table 8.21 Effect of K fertilization on the nut yield of West Coast Tall palms planted in 1964

	Mean yield (nuts palm^{-1})		
Year	K_0	K_1	K_2
1976	0.3	12.2	18.7
1977	1.0	21.3	29.1
1978	0.9	21.4	27.5
1979	0.8	28.1	38.2
1980	1.5	27.9	33.3
1981	4.6	36.1	41.8
1982	6.3	46.3	52.6
1983	3.4	33.5	40.0
1984	2.1	23.9	33.6
1985	6.7	67.7	77.0
Cumulative yield (1976–1985)	27.6	318.4	391.2

K_0 No potassium, K_1 450 g K_2O palm^{-1}year^{-1}, K_2 900 g K_2O palm^{-1} year^{-1}

palm^{-1}, number of female flowers bunch^{-1}, fruit setting, number of nuts bunch^{-1}, copra nut^{-1} and ultimately the total copra outturn palm^{-1}. Potassium application also improved all nut characters studied, viz. weight of whole nut, husked nut volume and copra weight nut^{-1} (Muliyar and Nelliat 1971).

The foliar K levels were also increased simultaneously to the sufficiency level. Fremond and Ouvrier (1971) found that the damage caused by K deficiency in the early stages was not fully corrected by subsequent K additions. Although later additions of K enabled the re-establishment of the palm in good physiological functioning, the palms which suffered K deficiency during the pre-bearing stage remained, on an average, 15% less productive than those never suffered. Therefore, it is necessary that, for maximum productivity, adequate K nutrition should be ensured early from the time of field planting.

8.8.4 Calcium

Calcium is a less mobile element in the plant, and it functions mainly outside the cytoplasm in the apoplast. In contrast to other macronutrients, a high proportion of Ca in plant tissues is present in the cell walls. It is very much concerned in membrane stability and cell integrity and helps the maintenance of acid-base equilibrium in the sap. In contrast to K or Mg, Ca activates only a limited number of enzymes in plants. In coconut palm, Ca is mainly concerned for the proper growth and functioning of stem and leaves rather than the productivity of nuts. Cecil (1981) observed that when all the major nutrients had been regularly applied since field planting, the length and width of the leaflets were significantly increased only with Ca treatment. Its concentration is low in young leaves, which progressively increases with maturity of the fronds. The remobilization of Ca is low in the coconut crown because of the relatively immobile nature of the element. Cellular distribution studies indicated that Ca was found to be high in polar compounds followed by cellulose and lignin fractions of leaf tissues of the palm.

The Ca content in most of the coconut growing soils of the world is satisfactory. The calcium added through phosphatic fertilizer is considered to be quite adequate for sustaining the Ca requirement of the palm. In very acid soils, regulated additions of liming materials are needed for ameliorating the soil acidity and other related problems. Concentration beyond certain limit would decrease the coconut yield probably due to its interaction effects with other nutrients.

8.8.4.1 Behaviour in the Palm

Only limited reports are available on the responses of coconut to Ca treatment. Manciot et al. (1979c) remarked that application of calcium carbonate to tall coconut in Côte d'Ivoire for 4 consecutive years had no beneficial influence on yield. On the other hand, lime dressing alone (Wilshaw 1941) and lime with fertilizers (Vertueil 1934) were reported to have beneficial effects on nut yield in Malaya and Trinidad, respectively, but it is not clear whether the beneficial effects reported by them were the direct influence of Ca in the nutrition of the palm. Cecil (1984) observed some favourable effects of liming on growth as well as foliar Ca levels of young palms, but the foliar levels were not correlated with growth/flowering or initial yields. The foliar Ca content even without Ca treatment was above 0.3% (frond 14) which is considered to be the critical level of Ca by Magat (1979).

The critical level of Ca initially suggested by IRHO was 0.5% for the tall variety (Fremond 1964). But values lower than this have been very widely reported on healthy plantations without any adverse effect on yield or foliar conditions. Manciot et al. (1979c) suggested that 0.3–0.4% Ca content in frond 14 was satisfactory and no improvement on growth or yield could be expected from application of fertilizers containing calcium. In Malaya, Kanapathy (1971) proposed an optimum level of 0.15–0.30% Ca for talls, semi-talls and dwarfs. Magat (1979) opined that the critical level of Ca initially suggested by IRHO appeared to be too high for the Philippine conditions. According to him, the critical level of Ca followed in the Philippines is 0.3%. Cecil (1984) also suggested 0.3% Ca in frond 14 for regulating the Ca requirement of the palm under west coast conditions of India. The leaf Ca contents are generally increased by nitrogen and Ca-bearing phosphatic fertilization, while it is depressed by higher levels of K and Mg fertilizers.

8.8.4.2 Deficiency Symptoms, Causes and Correction

Calcium deficiency in coconut was rarely reported. Visual symptoms of Ca deficiency were recently reported in Côte d'Ivoire on Malayan Yellow Dwarf with leaf Ca levels below 0.1% (Manciot et al. 1979c). The symptoms are yellowing of leaflet tips with yellow to orange ring-shaped spots, spread on the leaflets which later become necrotic and brown, and the leaf dries up. Middle leaves are affected before the oldest. The first symptoms of Ca deficiency appear on leaf numbers 1, 2 and 3, and they become yellow and rounded, turning brown at the centre. The spots are isolated in the early stages, joining and drying later on. In young leaves, the spots are uniformly distributed; however, starting from leaf number 4, the spots are concentrated at the base of the leaf. Plants with such symptoms contained only 0.85 g kg^{-1} Ca in leaf number 4 (Dufour et al. 1984).

Salgado (1947) reported that the lime requirement of coconut in Sri Lanka could be adequately met by the Ca present in Ca-bearing fertilizers. It is suggested that heavy liming is not needed for the management of the palm which is known to grow well under a wide range of soil pH ranging from 3.5 (peat soil) to 8.5 (coral soil). Nevertheless, regulated additions of Ca through Ca-bearing fertilizers or light addition of liming materials may be followed for supplying the Ca requirement of the palm (Cecil 1981). This is all the more important, in view of the heavy loss of Ca by crop removal and also by excessive leaching under high rainfall conditions in the tropics.

8.8.5 *Magnesium*

Magnesium is considered as the fourth important major nutrient in coconut nutrition. Its major function is as the central atom of the chlorophyll molecule. It is actively involved in Mg-dependent ATPase activity in the plant membrane where proton pumping operates. It also has an essential function as a bridging element for the aggregation of ribosomal subunits which is necessary for protein synthesis. When the level of Mg is deficient, in the presence of excessive levels of K, protein synthesis is impaired. Chlorosis of matured leaves is the most obvious visible symptom of Mg deficiency.

8.8.5.1 Behaviour in the Palm

The concentration of Mg is low in younger leaves which gradually increases reaching the maximum in the fully expanded leaves and then decreases with the progress of senescence of the leaves. This shows a high order of Mg fluxes in the coconut crown. Under normal growing conditions, the leaf number 14 or 15 is found to be the buffered leaf where the influx and efflux of Mg are equal. Leaves younger to the buffer leaf import considerable quantity of Mg, while older leaves export it. The Mg content in the leaf is highly influenced by diurnal as well as seasonal variations.

8.8.5.2 Deficiency Symptoms, Causes and Correction

Symptoms of Mg deficiency in coconut were reported by different workers (Coomans 1977; de Silva 1966; Fremond et al. 1966; Manciot et al. 1979c; Jeganathan 1990). Mg deficiency symptoms appear first in the old leaves. When the deficiency becomes severe, there is a necrosis on the extremities of the leaflets, which become dark yellow. At this stage translucent spots become visible. The visual symptoms are characterized by yellowing of the leaflets on the oldest leaves, going from the tips towards the rachis of the leaf. When the deficiency is fairly severe, the leaflet is almost devoid of pigmentation, but the parts nearest the rachis remain green. As the deficiency advances, yellowing becomes intense near the periphery of the leaf blade leaving a narrow longitudinal green band parallel to the midrib on either side of the leaflet. When the deficiency gets worse, yellowing further intensifies, the number of green leaves become less and necrosis sets in at the tips of the leaflets. Magnesium-deficient leaves are more sensitive to sunlight as the part exposed to sunlight shows intense yellowing, while shaded part of the same leaflet remains green (Fig. 8.3).

When the deficiency becomes severe, intense yellowing accompanied by severe necrosis and browning develop, and the mature leaves wither away prematurely leading to a lesser number of functioning leaves on the crown. The frond production rate is reduced, onset of bearing is delayed in young palms and the productivity is adversely affected.

Fig. 8.3 Magnesium deficiency symptoms. (Photo: Jacob John)

Magnesium is observed to be one of the important elements in the nutrition of seedlings and young palms especially when the soil supply is low (Khan et al. 1994). Magnesium deficiency is more common in acid sandy soils in view of their low Mg status and imbalance of K, Mg and Ca. Specific instances of absolute Mg deficiency situations are found in West Africa (Brunin 1969), Sri Lanka (De Silva 1966) and India (Cecil 1969). Nethsinghe (1959) reported Mg deficiency in lateritic soils of high rainfall areas where palms responded to foliar spray with magnesium sulphate. Prolonged use of K fertilizers, especially at higher rates, has been found to depress foliar Mg content and induce Mg deficiency conditions in the palm.

Application of Mg-containing fertilizers (Kieserite ($Mg\ SO_4\ H_2O$)/Epsom salt ($Mg\ SO_4$)/dolomite(Ca $Mg(CO_{3)2}$ /magnesite ($Mg\ CO_3$)) corrects the deficiency very well resulting in the regreening of chlorotic foliage/prevention of chlorosis accompanied by increase in foliar Mg levels and improvement in growth and yield of nuts. Child (1974) suggested that the application of Mg was necessary for the successful nursery culture of coconut in Sri Lanka. Fremond et al. (1966) recommended 60 g $Mg\ SO_4$ seedling^{-1} in the nursery along with similar quantities of double superphosphate and muriate of potash. Balanced application of N, K and Mg from the time of field planting was indispensable for the successful cultivation of hybrid coconut in Côte d'Ivoire (Coomans (1977), and he obtained highly significant response with 600 g Kieserite palm^{-1} year^{-1}. Varkey et al. (1979) obtained 87% decrease in foliar yellowing and 95% increase in yield of affected palms by the application of 500 g $Mg\ SO_4$ palm^{-1} year^{-1}. Kamalakshiamma et al. (1982) applied two levels of Mg, viz. 500 and 1000 g MgO palm^{-1} year^{-1} as $Mg\ SO_4$ along with three levels of NPK to D × T hybrids since planting. The second level of Mg did not generally show any significant increase over the first. Cecil (1981) observed highly significant response on growth, flowering and initial yields of young palms with 500 g MgO palm^{-1} year^{-1}. Nethsinghe (1962) obtained complete recovery of yellowing in 3–5 months by fortnightly spraying with 1–2% solution of Mg SO_4. He further suggested that soil application of Mg SO_4 was more effective than dolomite, while the latter could be used for long-term remedy. It was observed that the main effect of Mg on yield could be as much as 40% when K was in the sufficiency level, but unlike K, magnesium influenced only the number of nuts palm $^{-1}$ and had no effect on the copra nut^{-1} (Manciot et al. 1979c; Cecil 1981). A review of nutrition experiments in Sri Lanka involving magnesium was discussed by Jeganathan (1993).

As a preventive measure against occurrence of Mg deficiency of coconut palms, it is recommended to apply 1 kg of dolomite palm^{-1} year^{-1} along with NPK fertilizers. But when Mg deficiency symptoms appear in palms, 1 kg of Kieserite palm^{-1} has to be applied biannually until the disorder is corrected (Mahindapala and Pinto 1991). Nevertheless, the planters' experience is that it takes a long time to correct Mg deficiency even by application of Kieserite along with NPK fertilizer mixtures, particularly in palms on red-yellow podzolic soils with laterite in the wet zone which are highly leached.

The critical level of Mg initially suggested by IRHO was 0.3% (frond 14) for the tall variety (Fremond 1964). However, lower values have been found in healthy palms without any foliar symptoms of deficiency. In Côte d'Ivoire, the application of Mg showed a highly significant effect in increasing the yield of nuts palm^{-1} and copra palm^{-1}, and the foliar Mg level was also simultaneously increased at a highly significant level from 0.098 to 0.23%. In Sri Lanka, Nethsinghe (1963) suggested that deficiency symptoms could be expected when the Mg content (frond 6) was less than 0.2%.

There has been a general decline in Mg levels, probably due to the potassium chloride applications necessitated by the severe potassium deficiency in the Côte d'Ivoire coconut plantations. The critical levels proposed by the IRHO, 0.3% for Mg, are confirmed by Brunin (1970).

In India, Kamala Devi et al. (1973) observed a mean foliar content of 0.18% Mg in three high-yielding genotypes, viz. high-yielding talls, dwarf × tall and tall × dwarf hybrids. Cecil (1975) recorded 0.08% Mg in palms showing severe Mg deficiency symptoms and 0.18% Mg in apparently healthy palms without any visual symptom of deficiency. Cecil (1981) and Dufour et al. (1984) proposed an optimum level of 0.3% Mg for talls, semi-talls and dwarfs, while 0.2% Mg was reported to be critical under the Philippine conditions (Magat 1976). Manciot et al. (1979c) suggested 0.24% Mg for talls and 0.2% Mg for hybrids during the initial bearing periods. Cecil (1981, 1988) suggested that a critical level of 0.2% (frond 14) may be adopted as a diagnostic aid for regulating Mg nutrition of the palm under west coast condition of India until specific critical levels for each variety/type are established.

8.8.5.3 Interactions

Nitrogen fertilization often depresses the leaf Mg levels, possibly due to the antagonistic effect of NH_4^+ ions on Mg absorption by the palm. On the other hand, phosphatic fertilization generally increased foliar Mg levels and vice versa, indicating a close synergistic relation between P and Mg in the nutrition of the palm. Potassium fertilization generally depresses Mg uptake by the palm. Cecil (1981) observed a depressive effect of K on leaf Mg content when the soil and the leaf Mg contents were low and an effect when the Mg levels were improved due to regular Mg additions. On the other hand, Smith (1967), von Uexkull (1972), Barrent (1977) and Rosenquist (1980) reported an increase in foliar Mg levels by K fertilization. The results of Margate et al. (1979b) showed that even up to a level of 8.0 kg KCl palm^{-1} year^{-1}, the leaf Mg content (frond 14) remained more or less stationary (0.216%). Cecil (1988) suggested that the action of K fertilizers on leaf Mg content largely depends on the balance between K and Mg in the soil. The depressive action is severe when the exchangeable Mg/K ratio in the soil is less than 2.0–2.5. At higher Mg/K ratios, the action is not significant. A negative linear relationship between K and Mg in the soil was found (Limbaga 1986; Giritharan et al. 2000). Khan et al. (1986a) and Goh and Sahak (1988) reported that Mg leaf levels decreased significantly and linearly to increasing rates of K. It has been observed that application of high levels of K fertilizer induces Mg deficiency in coconut palms particularly on red-yellow podzolic soils with laterite in the subsoil (Jeganathan 1990). It has been a general observation that application of NPK coconut fertilizer mixture without application of dolomite induces Mg deficiency in palms on most of the soils. Brunin (1970) and Coomans (1977) reported that it was only after K deficiency has been corrected that Mg manuring was found to have a positive effect on production.

Application of potassium fertilizer decreased the quantity of both exchangeable and water-extractable Mg, and magnesium fertilizer decreased the quantity of exchangeable K in soils (Somasiri 1997). The mutual decreasing effect on the exchangeable fraction of each nutrient is attributed to low cation exchange capacity and base saturation of the soils. The close association of the coconut leaf nutrient contents with soil nutrient status implies that poor chemical characteristics of red-yellow podzolic soils bring about imbalance of K and Mg nutrition in coconut palms. The results showed that application of potassium fertilizer to the coconut palm would drastically affect the magnesium status of the palm on lateritic soils despite Mg fertilizer application. In the presence of K fertilizer, even the application of Kieserite at 1.2 kg palm^{-1}was not sufficient to raise the Mg status of the palm to the sufficiency range. It is very difficult to balance K and Mg nutrition of coconut palms grown on highly leached red-yellow podzolic soils with laterite, just by application of inorganic fertilizers due to their poor soil characteristics such as low cation exchange capacity and base saturation. This problem could be overcome by improving cation exchange capacity of such soils by increasing the humus content of the soils by organic matter incorporation. According to Somasiri et al. (2003), the dolomite supplied in fertilizer schedule does not necessarily meet the Mg lost from a coconut land, and Tennakoon and Bandara (2003) opined that organic manures only have the ability to partially supply the requirement of Mg demand of coconut palm.

8.8.6 Sulphur

Sulphur is an essential component of organic structure like sulpholipids in cell membranes and is a constituent of amino acids cysteine and methionine. It is also a constituent of several coenzymes and prosthetic groups such as ferredoxin, biotin and thiamine pyrophosphate. It also plays a key role in the redox systems in plants. Sulphur deficiency inhibits protein synthesis and causes a reduction in the chlorophyll content.

8.8.6.1 Behaviour in the Palm

S is an important secondary nutrient for coconut. It is involved in oil synthesis, copra quality as well as chlorophyll synthesis; coconut takes up considerable amount of S from the soil for its annual growth and productivity. The S distribution in the coconut crown follows a similar pattern as that of nitrogen. This is primarily because of the close interrelationship between these elements in the biosynthetic processes in coconut. However, unlike nitrogen, S is more uniformly distributed between old and young leaves. Young leaves generally have low concentration of S which is gradually increased up to middle-aged leaves beyond which the concentration is decreased.

8.8.6.2 Deficiency Symptoms, Causes and Correction

In coconut, S deficiency exhibits some similarity to that of N deficiency. Major observations on the S requirement of the palm and its deficiency symptoms were reported by Southern (1969) and Ollagnier and Ochs (1972). Sulphur deficiency symptoms are easily differentiated from those of N deficiency since the latter does not affect the young leaves unless the deficiency is very severe. The colour of affected leaves is yellowish orange to orange in S-deficient palm, while under N deficiency the colour of the foliage is pale green to yellow. The nut size of N-deficient palms may become small, but it produces normal copra on drying. In the case of palms with S deficiency, the kernel becomes rubbery (rubbery copra). S improves oil percentage, protein content and marketability of copra.

The most important source of S is the sulphate present in the soil which is taken up by roots even though atmospheric SO_2 is taken up and utilized by the aerial parts. Soils very low in organic matter and continuously cropped without adequate S fertilization are also likely to suffer from S deficiency. In the Philippines, coconuts grown even on soils derived from volcanic materials could suffer from S deficiency under high rainfall conditions. In India, up to the 1970s, S was added to coconut indirectly through S-containing fertilizers such as superphosphate and ammonium sulphate. Thereafter, the use of sulphur-free fertilizers like urea and rock phosphate was found to cause shortage in supply of sulphur to the palm. Regular use of S-containing fertilizers could take care of the sulphur needs of the palm. Organic farming is an alternate approach for S management. In situ organic recycling and/or additions of organic manures in the coconut basin would improve the organic S status of the soil which is further released into the inorganic form on mineralization. This process also checks the possible loss of S through leaching.

Coconut palms respond well to S fertilization. Maximum response to S is obtained when other nutrient anions, particularly NO_3 and Cl, are also applied at the required levels. Sulphur application increased the total yield of fruits and weight of copra but decreased the weight of kernel nut^{-1} (de Silva et al. 1985). The combined action of S and NO_3 and that of S and Cl is positive on growth as well as on yield of nuts and copra outturn. Ammonium sulphate is found to be a good fertilizer for correcting field conditions of S deficiency which is generally associated with foliar S content below 0.13%. The most frequently recommended fertilizer is ammonium sulphate which raises N and S levels at the same time (Wuidart 1994). Magat et al. (1991) using the data on yield and leaf nutrient contents from a gypsum-fertilization experiment derived a quadratic equation where 74.6% of the variation in yield of nuts $palm^{-1}$ was accountable to the level of sulphur in the leaves. The critical and optimum levels of leaf S in the reference leaf were 0.12 and 0.19%, respectively. They opined that the leaf S deficiency level of 0.13% pointed out by Manciot et al. (1979c) is actually the critical level. It was also indicated that the value of 0.15% S suggested by Southern (1969) in foliar diagnosis is higher by 0.02%, while the range mentioned by Manciot et al. (1979c) as critical level (0.15–0.20% S) covers the optimum level of leaf S (0.19%) obtained in their study.

8.8.7 Sodium

The specific role of Na in plant is not yet established. However, it participates in non-specific functions in plants such as maintenance of cell turgor, electrical neutrality, osmoregulation and detoxification of excess of phenols. In recent years, it has been postulated that Na could release the locked-up K in cell vacuoles under K stress conditions and, thus, make available limited amounts of K for vital metabolic functions.

Briones (1931) made a study on the salt requirement of coconut seedlings in pots and found that moderate quantities of sodium chloride were invigorating, while higher doses were harmful. Addition of NaCl at 0.5 kg young $plant^{-1}$ $month^{-1}$ on a rocky laterite soil gave a distinct difference in vigour, size and colour after 15 months compared to untreated palms (Salgado 1951). The coconut grows well on soils rich in Na although there is no direct relationship between the Na content of soil and that present in the leaves (Fremond 1964). Harmer, quoted by Manciot et al. (1979c),

Table 8.22 Effect of NaCI on growth and flowering of young D × T hybrid and on yield of bearing West Coast Tall palms

Treatments (g adult palm^{-1} year^{-1})	D × T hybrid		West Coast Tall
$K_2O + Na_2O$	Fronds produced 1976 to 1985	Early flowering index	Mean posttreatment yield (nuts palm^{-1} year^{-1})adjusted
0 + 0	73.9	1.0	69.7
1000 + 0	73.3	1.9	86.8
750 + 250	78.3	2.6	73.8
500 + 500	79.3	2.4	84.8
250 + 750	72.7	2.6	83.0
0+ 1000	77.5	2.4	75.0

grouped coconut among the plants which give a moderate response to Na even when there is plenty of potassium, while Jacques (1932) suggested that NaCl was also needed in the nutrition of the palm. Results obtained from young coconuts in the early stages of production showed that application of NaCl could significantly increase the number of inflorescences, the number of female flowers and the number of nuts palm^{-1}. Prema et al. (1987) in their studies in laterite soil region of Kerala observed that performance of palms in terms of yield was at par with palms receiving full dose of K_2O and 50% K_2O substituted with Na_2O. Preliminary studies indicated that Na to a certain extent substitutes the role of K when its supplies are inadequate. The copra content was also increased. Field studies conducted in India showed that addition of 50% of K requirement as NaCl and the rest as KCl did not show any difference in the growth of young palms or the productivity of bearing palms (Table 8.22). The early flowering index of NaCl-treated palms was higher than that of KCl-treated palms (Wahid et al. 1988). Irrigation with seawater and sweet water did not show any difference in the performance of the palms (Menon and Pandalai 1960).

Either as direct manure or as an indirect soil ameliorant, the addition of NaCl in coconut gardens has been a very old and popular practice among coconut growing farmers in Kerala, Java and Colombia (Child 1974; Manciot et al. 1979a). In Kerala, farmers apply it to the soil as well as in the crown of the palm, often admixed with wood ash, on the belief that it could increase the productivity of the palm. However, there is not much evidence on the direct effect of Na in increasing the yield of coconut. Since Na was applied in all the related studies in the form of NaCl, it is quite probable that the improvements in growth and yield obtained might be the effect of Cl which has been considered recently as an important nutrient in the nutrition of the palm for its effect on enhanced growth and productivity.

Fremond et al. (1966) suggested a maximum level of 0.40% Na (frond 14) beyond which adverse effects would be expected. The foliar Na contents reported by different workers normally range from 0.1% to 0.5%. The foliar level of 0.4% Na may be taken only as a rough guide as many coconut groves giving excellent yields have low Na levels around 0.1%. The leaf Na concentrations are generally depressed by K fertilization. Khan et al. (1986a) reported the relationship between K and Na as $r = -0.87^{**}$ in their studies involving tall and hybrid palms.

8.8.8 Chlorine

The specific functions of Cl in plants are not well understood. It is readily taken up by plants as chloride ion. It functions mainly as a highly mobile inorganic anion in processes related to charge compensation and osmoregulation. There is evidence to assume that Cl is required for the *photosystem II* in photosynthesis and it stimulates the membrane-bound proton-pumping ATPase activity. Chlorine can also influence photosynthesis and plant growth indirectly through stomatal regulation. Wilting of leaves, especially the leaf margins, is a typical symptom of Cl deficiency.

The essentiality of chlorine for higher plants was first established by Broyer et al. (1954), but its importance in the nutrition of two tropical oil-yielding crops, oil palm and coconut, was brought to light by Ollagnier and Ochs (1971a, b). They reported that oil palm and coconut gave significant yield increase due to Cl. They further emphasized the high requirement of this element and suggested to rank Cl as a major nutrient for coconut and oil palm. The high requirement of Cl for coconut was established later by Ouvrier and Ochs (1978), and they reported that for the high-yielding hybrid PB 121, the removal of Cl was equal to that of K, and Manciot et al. (1979c) ventured to rank Cl as the second most important nutrient, next to K, for coconut.

The response of coconut to other Cl-containing salts like KCl has also been noted, but the importance of Cl in coconut nutrition in certain countries was reported only in the 1970s (Ollagnier and Ochs 1971a, b; von Uexkull 1972; Magat et al. 1975). Chlorine in the form of either KCl or NaCl is easily absorbed by palms (Magat et al. 1975), and the amount of Cl in the soil depends on rain and seawater (Manciot et al. 1979c).

Chlorine increases the thickness of kernel and copra weight nut^{-1} as a result of a bigger cell volume. It also accelerates plant development in terms of girth and frond production rate. It enhances better absorption of other nutrients like K, Ca and Mg which contributes to accelerated growth and early flowering in coconut. von Uexkull (1972) found that Cl was the only foliar nutrient that was significantly correlated with growth of young palms and yield of bearing palms. He also demonstrated the importance of Cl for coconut when the application of KCl increased the weight of kernel from 117 to 216 g and the composition of Cl, from 0.40 to 2.33 g kg^{-1} in 14th leaf. Multiple regression analysis showed that 55.5% of increased nut production, 74.3% of increased copra yield nut^{-1}and 80.3% of increased copra palm^{-1} were due to chlorine (Margate et al. 1979b).

Similar highly significant correlations between Cl content in the leaves and growth/production of nuts of young/bearing palms were also reported by Magat and Prudente (1975). According to Prudente and Mendoza (1976), Cl is the most likely factor that limits production of inland coconut areas, especially in Davao, followed by N. The addition of 1.6 kg ammonium sulphate plus 1.8 kg muriate of potash tremendously increased production by 191%, 58% and 314% in terms of nut yield, copra weight nut^{-1} and copra production, respectively, over the palms with N fertilizer only. Nitrogen levels in the leaves are highly correlated with nut and copra production, while Cl levels with copra weight nut^{-1}, copra and nuts palm^{-1}.

Coconuts growing near seashore (where Cl is sufficient) are generally more productive than those growing on inland, supposed to be low-Cl areas. Studying the effect of long-term fertilization with different levels of KCl on bearing coconuts in the Philippines, Margate et al. (1979b) concluded that application of KCl at 2.0 kg palm^{-1} year^{-1} gave the highest production of nuts. Though copra weight nut^{-1} correspondingly increased with increasing levels, copra production did not follow a definite trend of response beyond 2.0 kg level of application. The palm receiving with 2.0 kg KCl palm^{-1} year^{-1} showed an increase in leaf Cl status from 0.04 to 0.55% and yielded 128 nuts against 87 nuts by the control palms, an increase of about 47%. The copra weight nut^{-1} was significantly increased with increased rate of KCI, and the response followed a linear pattern up to the highest level of 8.0 kg KCl palm^{-1} year^{-1}. It is very interesting to observe that even up to a treatment level of 8.0 kg KCl palm^{-1} year^{-1}, no significant correlation was obtained between foliar K levels and nuts palm^{-1}, copra nut^{-1} or copra palm^{-1}. The response of coconut in the study was due to the chlorine component and not to K. Mineral nutrition studies in the Philippines indicate that as a source of chlorine, NaCl can effectively replace KCl which is an expensive fertilizer.

The use of NaCl and seawater is an ancient and very common practice among coconut growers in many parts of the world (Bonneaux et al. 1997), and sufficient evidence supports the contribution of Cl to oil palm production. Palms contain little starch that can produce malate as the accompanying anion for K in their guard cells (von Uexkull and Sanders 1986), and, therefore, Cl plays a vital role in movement of stomata. Healthy coconut palms along the seashore usually contain Cl at a concentration of 7–10 g kg^{-1} DM in their foliage. The optimal Cl concentration is usually in the range of 4.5–5.5 g kg^{-1}. At Cl concentrations lower than 2.5 g kg^{-1}, coconut palms may exhibit some visual symptoms of yellowing and/or orange mottling of the older leaves and the leaf tips and edges (von Uexkull and Sanders 1986). For example, the threshold value for EC in a soil extract is 4.5 dS m^{-1}. Above this level, growth and copra yield begin to decline (Hassan and El-Samnoudi 1993). No copra yield was obtained when the EC value of the soil extract exceeded 23.2 dS m^{-1}, and salinity symptoms appeared on the leaves, but the trees survived. Soil salinity leads to an accumulation of Cl, Na and K in the leaves that of Cl was larger than that of Na and was highly correlated with salinity symptoms (Hassan and El-Samnoudi 1993).

Reports on Cl deficiency symptoms in coconut are limited. The visible symptoms are yellowing and/or orange mottling of the older leaves and drying up of the outer edges and tips of leaflets which are very similar to K deficiency. The nuts from Cl-deficient palms are smaller compared to nuts from palms well supplied with Cl. The number of leaves would be less with narrow leaflets, and the leaves are susceptible to leaf spot/blight diseases. The deficiency symptoms are associated with leaf Cl contents (frond 14 less than 0.3%) in adult palms.

Magat et al. (1977) suggested a critical level of 0.7–0.8% Cl in coconut seedlings (frond 3). Ollagnier and Ochs (1971b) proposed that for high yields of coconut, leaf level of 0.5–0.6% Cl was necessary. Magat (1979) suggested the critical level of Cl (frond 14) at 0.3–0.4% and optimum level of 0.5–0.6%. Chlorine is not generally found limiting in coconuts in Kerala, particularly when K is applied in the form of muriate of potash, in most cases leaf Cl levels range from 0.5% to 0.7%, which is well above the critical level.

Magat et al. (1986) noticed improvement in nut production, copra weight nut^{-1} and copra yield palm^{-1} over a 3-year period with the application of NaCl. They also found that only leaf Cl was highly correlated with the yield parameters studied, indicating that the increase in yield was mainly due to the correction of Cl deficiency by NaCl application. From their subsequent studies, Magat et al. (1988) suggested that NaCl could be used to replace KCl, which is an expensive fertilizer, the optimum economic NaCl rate for coconuts grown in the Tugbok clay loam and those grown in similar conditions being 3.8 kg palm^{-1} year^{-1} (yielding 25.9 kg copra), although the rate of 1 kg palm^{-1} year^{-1} already made it possible to obtain more than half of the positive effects corresponding to the optimum economic rate.

Findings at the Davao Research Centre of the Philippine Coconut Authority revealed that the annual application of fertilizer $(NH_4)_2SO_4$ and KCI significantly increased nut and copra yield of bearing palms (Mendoza and Prudente 1972). With the same fertilizer combination, genetically tall coconuts of the "Laguna" form were induced to flower in less than 4 years instead of the normal 6–7 years from field planting. von Uexkull (1972) reported that the response observed from KCl application is due to Cl and not to K, since Cl levels in the leaf were correlated with the growth of young palms and the production of bearing ones. On the other hand, separate studies on the use of NaCl, another Cl source, showed that its application improved the development and yield of the palms. However, the response was not attributed to the effect of Cl (Roperos and Bangoy 1967; Del Rosario 1972; Ramanandan 1973), whereas the positive response to KCI applications observed in earlier studies (Magat et al. 1975; Margate et al. 1979b; Magat et al. 1981b) is attributed to the Cl component of the fertilizer.

Practical recommendations for the application of chloride to coconut are available (IFA 1992). For coconuts under Malaysian conditions, the rate of application ranges from 0.11 kg Cl palm^{-1} at an age of 6 months and increasing progressively to 0.9 kg Cl palm^{-1}.

Positive residual effects of Cl-bearing fertilizers (KCl, NaCl and NH_4Cl) on the yield indices of coconuts were reported by Magat et al. (1991, 1993). They suggested that for every 5 years of regular Cl fertilization (0.80 kg Cl), the application of Cl fertilizers (and even N and S fertilizers) is not required for at least the next 3 years as production of nuts and copra is maintained at high levels. This was mainly attributed to the residual effects as a result of the concentration of Cl in the crop at optimum levels (0.50–0.60%) or above critical levels (0.30%) of leaf Cl. With NaCl application, the positive residual effects on copra (weight nut^{-1} and yield palm^{-1} year^{-1}) could last for a longer period (4–5 years) at fertilization rates of >= 1.76 kg NaCl palm^{-1} (0.97–3.87 kg Cl palm^{-1} year^{-1}).

Drought and chlorine deficiency in Lampung province, Indonesia, were reported by von Uexkull (1992). Palms deficient in chlorine lose water at a much faster rate than those well supplied with Cl. Similarly palms deficient in chlorine require a much longer period to recover. On soils rich in K, NaCl or NH_4Cl can be used as a source of chlorine to palms.

8.8.9 *Micronutrients*

Micronutrients are mostly constituents of important enzyme reactions and involve themselves in the effective functioning of the nutrient system in the plants. The most important observations indicate that mineral nutrient deficiencies, mainly of micronutrients, cause reductions in the number of female flowers spathe^{-1} and the fruits, which eventually succeed easily drop off the plant, a condition generally referred to as "abortion of immature fruits" (Siqueira et al. 1997; Holanda et al. 2007).

8.8.9.1 Iron

Iron is an essential constituent of certain enzymes, especially the cytochromes, which participate in the electron transfer system. Iron is taken up by plants as $Fe2^{+}$ ion and also as chelated iron molecules like Fe-EDTA. The distribution of iron in the coconut crown shows that its concentration is low in young leaves and increases gradually with maturity. The internal recycling of this nutrient is very much restricted. The subcellular distribution of Fe shows that the major fraction is in the immobile stage followed by proteins and polysaccharide fractions. Generally hybrids utilize iron into various functional sites better than tall variety of palms. The critical concentration of Fe is considered at 50 ppm in frond 14. However, lower values are found in palms growing on rich soils without exhibiting any visual symptoms of iron deficiency. Lime-induced chlorosis is the characteristic disorder in coconut on coral and limestone soils that are rich in calcium carbonate. Iron deficiency is also found on peat soils.

The characteristic symptom of iron deficiency in coconut is the gradual development of yellowing in all the leaves. The entire leaflet becomes yellow in longitudinal strips parallel to the veins, and the leaf becomes completely yellow in the advance stage. Necrosis is generally absent in any part of the leaflet. The rachis and leaflets become shorter. Symptoms of iron deficiency may appear similar to that of nitrogen deficiency. However, in the latter case, yellowing is uniform, while in the case of former, strip discolouration is noticed in the initial stages with yellowing becoming general in the advanced stage. The symptoms of Fe deficiency were described by Pomier (1969) on the Pacific Coralline Islands, where the high levels of calcium carbonate render the iron unavailable. Trewren (1987) reported that on coral line soils of Tuvalu (Pacific Ocean Island), stem injection with 30 g ferrous sulphate to palms is considered safe remedy against lime-induced chlorosis. Generally, in tropical soils the presence of iron oxides is adequate to satisfy the crop demand.

8.8.9.2 Manganese

Manganese acts as a cofactor in some of the oxidative enzymes in the plant. It also participates in the oxidation-reduction reactions in the plant. It is absorbed mainly as Mn^{2+} ion and is translocated predominantly as the free divalent cation from the roots to the aerial parts. Mn deficiency is characterized by generalized chlorosis. For coconuts grown in north-eastern Brazil, analysis of leaf number 14 showed great variability in the Mn composition. Sobral (1989), studying the nutritional state of coconuts in Sergipe, showed no direct relationship between Mn in leaves and the burned leaf symptom. It was observed, however, that there is a significant relationship between the composition of Mn in the soil and the leaf.

Davis and Pillai (1966) reported that Mn could increase the yield of nuts in coconut. According to Pomier (1967), the application of K is effective only when the needs of Fe, Mn, and N are satisfied. Kamala Devi et al. (1975) noticed enhancement in the availability of Mn, Fe and Al in soil with lowering of pH by NPK treatments. However, only Mn was taken up by the palm in larger amounts. Water-soluble Mn in soil was highly correlated with plant uptake. Mn has indirect effect in the chlorophyll formation (Mandal 2000), and, therefore, deficiency of Mn can result in reduction of photosynthetic material such as carbohydrates.

The distribution of Mn in the coconut crown follows a pattern similar to that of Fe. Its concentration is low in young leaves and increases with maturity. The critical level of Mn is considered as 60 ppm in frond 14. It is not found to be a limiting nutrient in acid tropical soils. On the other hand, concentrations as high as 700–800 ppm Mn in frond 14 have been reported on lateritic and red sandy loam soils. However, Mn deficiency is found on coral soils, which is corrected by the application of Mn salts as foliar sprays and through trunk injection. Addition of organics charged with Fe and Mn salts is yet another method of supplementing these nutrients in calcareous soils.

8.8.9.3 Boron

Boron is a relatively immobile element in phloem but highly mobile in xylem accounting for accumulation of B in older tissues and also in margins. It is very closely related to the activity of moisture stress especially apical-meristem. When in short supply, normal cell division does not proceed satisfactorily to complete separation of cells whose longitudinal walls remain short. This results in incomplete and irregular leaf expansion, the development of distorted leaves and lack of elongation of internodes. B deficiency occurs due to accumulation of super-optima levels of endogenous IAA, probably due to reduced activity of IAA oxidase.

It improves water relations and translocation of sugars in plants, enhances tissue respiration, and influences N metabolism and the oxidation-reduction equilibria in cells. Boron deficiency affects the cells of the growing regions, and the effects are observed on the differentiating cells leading to the death of the apical growing point preceded by abnormal/deformed growth of young leaves.

In the coconut palms, B deficiency decreases the photosynthetic capacity, since it reduces the electron transportation of *photosystem II* (−12.5%), photosynthesis (35.7%), sweating (−32.2%) and stomatal conductance (−45.6%) (Pinho et al. 2010). Boron deficiency also compromises the coconut palm's radicular system, decreasing the percentage of fine roots and increasing the percentage of thick roots, causing over sprouting, necrosis, darkening and thickening of roots (Power and Woods 1997). With the compromising of the radicular system, the plant can possibly present secondary deficiencies and be more susceptive to hydric deficit, blight and diseases.

In nature, B is moderately rare and occurs principally as borates of calcium and sodium. It occurs in soils in the form of tourmaline (*crystalline borosilicate mineral*). Its availability is maximum within the pH range of 5 and 7. Boron is less available above the pH of 7.5. Excessive liming accentuates boron deficiency.

Symptoms of B deficiency become visible as soon as the leaves emerge, which occurs 2 or more months after the occurrence of the deficiency. Boron deficiency can also be chronic, affecting a series of successive leaves as they develop. As an immobile element, B deficiency causes leaflet fusion and malformation, truncation and reduction in the size of newly emerging leaves (Broschat 2007a; Corrado et al. 1992; Kamalakshiamma and Shanavas 2002). These symptoms could be confused with deficiencies of other micronutrients, such as manganese, zinc or copper, herbicide toxicities or even bud rot diseases (Broschat 2007b; Elliott et al. 2004). Where visual deficiency symptoms are insufficient to diagnose chronic B deficiency, leaf nutrient analysis can be useful (Mills and Jones Jr 1996).

Boron deficiency can be extremely transient, affecting developing leaves for as little as a day or two before normal growth resumes. Under these conditions, the effects of a temporary deficiency only become visible when the affected developing leaf emerges 4 or more months after the deficiency occurred. Palms may experience multiple alternating periods of B sufficiency and deficiency during the time that it takes for the first affected leaf to emerge. Thus, visual deficiency symptoms are an indication that a temporary B deficiency has occurred before leaf emergence but provides no clues as to the current B status of the palm (Rajaratnam 1973).

B deficiency is manifested in the leaflets, which are joined at the extremities (Fig. 8.4). In severe cases the leaflets at the base of the stem are smaller and crest and may even disappear. When B deficiency is very severe, the point of growth completely deforms, preventing the development of the palm (Corrado et al. 1992; Kamalakshiamma and Shanavas 2002; Santos et al. 2003; Broschat 2005; Broschat 2007a). Boron deficiency, in general, reduces root growth (Lima Filho and Malavolta 1997; Viégas et al. 2004), and in the coconut palm, production of total roots is

Fig. 8.4 Boron deficiency symptoms in adult coconut palm (Photo: V. Krishnakumar)

reduced by 30% and of thin roots by 48% (Pinho et al. 2008a). According to Broschat (2009), B deficiency can be transient to chronic and mild to lethal in palms. If this temporary deficiency is severe enough, the entire tip of the leaf beyond the necrotic point will often fall off. This very temporary deficiency is believed to be caused by a single heavy leaching event lasting as little as one day. In rainy climates, this pattern may be repeated every time a heavy leaching rainfall occurs, and as many as 3, such events have been documented during the development of a single leaf of a coconut palm (Broschat 2007a).

Santos et al. (2004), in an evaluation of the nutritional status of a coconut palm, reported that the most important elements for coconut production in decreasing importance were K > Ca > B. Manciot et al. (1980); Rognon (1984) and Sobral (1998) observed that a critical boron level in plant leaves was 10 mg kg^{-1}. Boron deficiency may be limiting yield at high rates of application of sulphate of ammonia and muriate of potash (Rosenquist 1980). B deficiency has become a common and widespread disorder of palms throughout the world (Corrado et al. 1992; Elliott et al. 2004; Broschat 2007a). A deficiency symptom varies from palm to palm, and all the symptoms may not be expressed on a single palm (Kamalakshiamma et al. 2005).

General recommendations for the application of boron to coconut plants are that young palms should receive 30 g of borax applied to the fourth leaf axilla. For yielding palms, it is recommended that borax should be applied directly into the soil at 2 kg ha^{-1} of B as borax (Sobral 1998) when analysis indicates levels lower than 0.2 mg dm^{-3} (hot water soluble) (Teixeira et al. 2005b). The application of B directly into the soil is more efficient than foliar techniques due to the low mobility it shows in plant tissues. Boron applied to the soil has a more persistent effect than when it is deposited in leaf axillae (Pinho et al. 2008b; Broschat 2011). Recommendations about B dosages to be applied to coconut palms are limited in the literature (Teixeira and da Silva 2003; Santos et al. 2003).

Leaf B concentrations did not vary significantly among leaves within the canopy or among leaflets within a single leaf for coconut palm (Broschat 2011). Boron concentrations were significantly higher towards the tips of individual leaflets. Application of Solubor to the soil significantly increased leaf B concentrations in all leaves of coconut palm after 2 months as well as in new leaves produced up to 6 months later. Application of Solubor as a leaf axil drench was much less effective in increasing foliar B concentrations than soil treatment. Thus, there appears to be no advantage to applying B to the axils of palm leaves when soil application is both efficient and effective. In young coconuts, deficiency may be corrected by applying 30 g of borax at the axil of leaf number 4. In adult coconuts, B can be added as borax mixed with other fertilizers and added to the soil. Because the limits of deficiency and toxicity are very close, elevated doses of B may cause toxicity in the plant.

Soil applications of 40–700 g of borax palm^{-1} year^{-1} have been suggested by Dickey (1977), von Uexkull and Fairhurst (1991) and Kamalakshiamma and Shanavas (2002). Some of these workers also suggested applying borates in the leaf axils, but they provided no data on the relative effectiveness of these two methods. Healthy palms had higher B content (6.9–7.9 ppm) compared to (4.7–6.3 ppm) in diseased palms, and Ca/B ratio was found to be lower in healthy palms (Baranwal et al. 1989). They recommended soil application of Borax at 50 g palm^{-1} in Assam and West Bengal where boron deficiency was noticed on young palms.

A fertilizer dosage of N, P and K at 0, 1 and 2 kg ha^{-1} resulted in leaf boron concentration levels lower than the considered critical level (10 mg kg^{-1}) as indicated by Manciot et al. (1980), Rognon (1984) and Sobral (1998). Teixeira and da Silva (2003), in a study of 2-year-old coconut palm of 7 genotypes growing in Bebedouro, State of São Paulo, Brazil, found foliar B levels higher than 10 mg kg^{-1} (43.9–47.9 mg kg^{-1} of B in leaf 9).

An understanding of natural B distribution patterns within palm canopies and leaves is necessary to determine which leaves should be sampled for analysis. In the same context, Rajaratnam (1973) indicated that a temporary B deficiency has occurred before leaf emergence but provides no clues as to the current B status of the palm. Considering this, Oertli (1994) opined that leaf analysis for B content may not always be a good indicator of current B status. Nevertheless, boron contents indicated by leaf analysis are taken as a guide for understanding and boron application to palms (Broschat 1997).

Jayasekara and Loganathan (1988) noticed the symptoms as unsplit, crinkled nature of leaflets, stunted and withered apical leaves and lack of leaflets in some fronds of young palms of 1–3 years. The third leaf of the affected palms had 3.4–7.5 ppm B as compared to 7.6–10.0 ppm B for healthy palms in the same vicinity. Soil application of sodium tetraborate ($Na_2\ BO_4.2H_2O$) at 28 and 56 g palm^{-1} to the affected palms at incipient stages improved the condition of palms within 6 months, and complete recovery was achieved at the end of 8 months. However, the symptoms in the untreated affected palms gradually became acute, and the palms died after 6–8 months. Results suggest that the critical nutrient concentration range for B in the third leaf is 8–10 ppm and the deficiency could be corrected only at the incipient stages by soil application of sodium tetraborate.

Table 8.23 Recommendations for correcting boron deficiency in coconut palms

Sl no	Area	Recommendation
1	West Bengal (India)	20 g palm^{-1} borax decahydrate (11.5%) *
2	Kerala (India)	200 g palm^{-1} borax decahydrate followed by irrigation*
3	Assam (India)	50 g palm^{-1} borax repeat twice in severe cases at an interval of 3–4 months*
4	Côte d'Ivoire (W Africa)	Borax pentahydrate (14.8% B) 15 g palm^{-1} once in 6 months*
5	Sri Lanka	28–56 g sodium tetraborate to incipient stages **

* Compiled by Khan (1993); ** Jayasekara and Loganathan (1988)

Recommendations for correcting boron deficiency in coconut palms is given in Table 8.23.

Some academic studies describe the distribution of B in the leaves of coconut palm. In palms not fertilized with B, the tendency is to have similar contents in young and in old leaves; in those fertilized, however, the content of B increases, but it does not follow a consistent pattern of distribution between young and older leaves (Pinho et al. 2008b; Broschat 2011). The correction of the deficiency in coconut palms is possible after application of B (Santos et al. 2003). Doses of 30 g and 60 g of boric acid palm^{-1}, applied without dilution at the leaf axils or in the soil, respectively, adequately nourish the palm and do not cause toxicity (Pinho et al. 2008b).

Santos et al. (2003) observed that applying boron in young coconut palms promoted the emission of normal leaves, but did not correct the symptom of leaves already affected by B deficiency. Thus, to reduce potential decline in productivity, management of fertilization with B should be done to prevent disability and prevent the formation of abnormal leaves.

Boron deficiency in coconut in the northern region of the State of Rio de Janeiro was found by Mirisola Son (1997) and Santos et al. (2003). For the correction of this deficiency, it was recommended the use of 30 g of borax in the axilla of fourth leaf of young plants. In adult plants, borax could be mixed with other fertilizers and soil (Sobral 1998). Santos et al. (2003) found that providing 30 g of borax, divided into 2 applications of 15 g in the axils of leaves 2, 3 and 4, promoted the emission of normal leaves of coconut dwarf green plants showing epinasty and deflections of the new leaves, symptoms attributed to B deficiency (Sobral 1998; Macêdo et al. 1999; Broschat 2005). The higher palm yield was associated to levels of 0.6 mg dm^{-3} of B in the soil and 23.5 mg kg^{-1} in leaves. The maximum production was obtained in 95% of palms with the use of a boron dosage of 2.1 kg ha^{-1} (Moura et al. 2013). The leaf and leaflets become deformed due to deficiency of B (Pinho et al. 2015). The deficient and sufficient contents of B varied significantly in the canopy but did not vary in leaves. To study the level of B in the coconut palm, they suggested to use samples taken from the youngest leaf and to calculate the relation between the B content in the apex of the leaflet and the content in the centre or in the bottom leaflets. Little is known about the trace element contents of Papua New Guinea soils, but some studies were carried out by Southern and Dick (1969) to diagnose characteristic symptoms of B deficiency through leaf analysis and critical level varied between 5 and 10 ppm, depending on the site.

8.8.9.4 Copper and Zinc

Both copper and zinc are associated in certain enzyme systems in plants. They are taken up by plants as Cu^{++} and Zn^{++} ions and also as chelated molecules. Their deficiencies are not commonly found in coconut.

Copper deficiency has been observed in coconuts, in the Philippines, in new plantings, especially with improved planting materials or hybrids on peat soils (Magat 1991). The characteristic symptoms include severe bending of rachis of the young leaves accompanied by yellowing and drying of leaf tip which appears rimmed with brown and yellow, while the middle portion remains normal green. When the deficiency is severe, new leaves are deformed and are abnormally short giving the palm a runty sagging appearance. Deficiency of Cu in coconuts was described by Ochs et al. (1993) when the palms were grown in peat soils in Indonesia. Firstly, the stems of the new leaves become flaccid and later bend. Almost simultaneously, the extremities of the leaflets start to dry, colour changing from green to yellow and, finally, to brown – appearing burnt. When the deficiency is serious, new leaves become small and chlorotic, and the plant may dry completely. Earlier Ochs and Bonneau (1998) studied Cu and Fe deficiency symptoms in peat soils of Indonesia and opined that leaf contents in commercial plantations gradually increase with age and deficiency is seen in young palms rather than in the older ones. In Brazil, Cu deficiency was found in coconuts planted in Quartz Neosols (SiBCS) (Sandy Quartz) (Sobral et al. 2007).

Zn deficiency causes abnormal growth of the young leaves (Mandal 2000). The abnormal growth of the leaves was observed on the palms under the Kish Island environmental conditions of south Iran (Arzani et al. 2005). These abnormal leaves become rough in the surface and not fully expanded might be due to reduction in the photosynthetic activity of such leaves and further reduction of carbohydrates and lower female flower production.

The optimum levels of copper and zinc in coconut nutrition are not known, and hence, their critical levels are not established. However, levels of 5–7 ppm Cu and 10–15 ppm Zn in frond 14 are found to be adequate for the normal growth of the palm. The foliar contents reported by Southern and Dick (1968) were 2–5 ppm Cu and 19–36 ppm Zn. Pillai et al. (1975) reported mean values of 6.5–7.7 ppm Cu and 8.8–11.9 ppm Zn.

The distribution of zinc in the coconut crown shows that in young leaves the concentration is relatively low which gradually increases in the middle-aged leaves and then it declines. In the case of copper, the concentration does not show much variation among the different leaves. However, the copper concentration is relatively higher in older leaves.

Table 8.24 Fertilizer recommendations for B, Cu, Mn and Zn based on soil and leaf analysis

Nutrient/Analysis method	Soil (mg dm^{-3})	Leaf number and nutrient content (mg kg^{-1}) 9	14	Fertilizer (g $plant^{-1}$)
Boron (hot water)	0–0.6	<17	<20	Borax 50
	> 0.6	>17	>20	–
Copper (DTPA)	0–0.8	<5	<5	Copper sulphate 100
	>0.8	>5	>5	–
Manganese (DTPA)	0–5	<60	<65	Manganese sulphate 100
	>5.0	>60	>65	–
Zinc (DTPA)	0–1.2	<14	<15	Zinc sulphate 120
	>1.2	>14	>15	–

Source: Sobral et al. (2007).

8.8.9.5 Molybdenum

The requirement of molybdenum for coconut is very small, and its role has not been determined in coconut farming anywhere in the world. The optimum requirement and critical level are not known. The recommendation of fertilizers for B, Cu, Mn and Zn based on soil and leaf analysis is given in Table 8.24.

The importance of major, secondary and micronutrients has been discussed with respect to their role in the mineral nutrition of palms. The studies were specific to certain regions of the coconut world and focussed on the soil and plant health in relation to yield and sustainability. The studies have given meaningful interventions to be adopted as a policy to improve soil health and yield. Consolidating the information of all these interventions for each region is necessary and to be revisited and practiced in larger areas through site-specific recommendations. Studies have also given information on economizing nutrient applications. The importance of magnesium and chlorine is largely recognized and that of potassium, nitrogen and phosphorus in their order, besides role of boron in specific sites. This information is to be studied in relation to the mineralogy of soils of different regions and basic studies on the release mechanism of nutrients.

8.9 Diagnostic Techniques for Evaluating Nutrient Requirement

Fertilizer use for a crop could be recommended based on soil testing and leaf/plant analysis. These guidelines restrict overuse of fertilizer, imbalance in the nutrition of crops and contamination water resources with nitrates and other residues.

Furthermore, overuse of fertilizers means higher costs for farmers and waste of resources.

Four diagnostic methods can be used for coconut, viz. soil and plant tissue analysis, deficiency symptoms and field fertilizer trials. Soil testing and plant diagnosis make it possible for farmers to assess the nutrient status of the soil and crop, apply fertilizers for an expected yield and when there is a deficiency to be corrected. In countries where fertilizer applications are low because of economic constraints, soil and plant testing can guide the farmers on fertilizer applications and realize maximum benefit in terms of crop yield.

The diagnosis and recommendation of nutrients for a given stand of a perennial like coconut palm require considerable knowledge as one has to understand the complex nature of the soil at a given location, its nutrients, the storage of the nutrients in plant and the dilution that takes place in the system once it is absorbed and stabilized. Normally the absorbed nutrient becomes highly mobile in the system satisfying the physiological requirements of the sink and stabilizes. At a given day of the hour, leaf samples are collected from designated leaves of the plant following standard procedures.

Plant or soil analysis may be complementary, not competitive, and most of the modern laboratories are equipped to carry out both types of analysis. However, it is recommended to start soil analysis prior to plant testing to gather sufficient knowledge of the characteristics of the soils that support the crop which will receive the fertilizers, and then plant testing may be carried out for observing the effect of fertilizer application as for determining nutrient requirement of crops.

Modern laboratories are equipped with sophisticated instruments for analysis of major nutrients, secondary nutrients and micronutrients following the principles but setting aside the conventional procedures which are time consuming. The soil as a first step is analysed for pH, electrical conductivity, organic matter, moisture and in special cases for bulk density and carries out water analysis as per the need. Sufficient information is in the knowledge of the agronomists for soil sample collection from defined zones and the diagnostic leaf to sample which will reflect the best nutritional status of the crops. The laboratoires can also carry out mechanical analysis of the soil.

In foliar diagnosis, a composite sample of palms grown under similar conditions is collected at defined intervals. For a particular stage or age of coconuts, leaf sampling is done on the selected leaf rank (number) of the palm based on its phyllotaxy (Magat 1979) (Table 8.25), and maturity of the leaf will give best indication of the nutritional status.

All the studies on soil fertility and plant nutrition are mainly concerned in predicting the nutrient requirement of crops. Although some of the conventional methods have been employed in coconut, none of them in isolation has been found successful in predicting the specific needs under field conditions for achieving the desired yields. New approaches like Diagnosis and Recommendation Integrated System (DRIS) (Beaufils 1973), which was successful for tree crops (Sumner 1982) like rubber, were tested for coconut, oil palm, grapes, etc. Differential Fertilizer Requirement (DFR) (Jayasekara 1993) for coconut in Sri Lanka was useful in prescribing nutrient

Table 8.25 Leaf rank for analysis depending on growth stage of coconut palms

Living leaves average count (5–10 leaves)	Stage	Leaf rank to sample
4–6	Nursery	1
7–12	Nursery/field	3 or 4
13–18	Pre-bearing	9
19 or more	Bearing	14

schedules. Saldanha et al. (2017) evaluated the nutritional status and established nutritional standards for the cultivation of hybrid coconut in Moju, Pará, in northern Brazil using DRIS and presented order of nutrient deficiencies. Their study compared with the critical levels established for reference suggested the following critical levels for dwarf coconut as percentage of N 2.02, P 0.15, K 1.72, Ca 0.41, Mg 0.12 and S 0.13 and mg kg^{-1} for B 15, Cu 4, Fe 115, Mn 101 and Zn 21.

Rosa et al. (2011) developed a Lime and Fertilizer Recommendation System for Coconut Crop based on the nutritional balance. The system considers the expected productivity, plant nutrient use efficiency, effective rooting layer, soil nutrient availability as well as other nutrient inputs to estimate the nutrient supply. Comparing the nutrient demand with the nutrient supply, the system defines the nutrient balance. These new approaches, if tested further and refined, will benefit in evolving a meaningful fertilizer recommendation based on soil and leaf analysis.

8.9.1 The Soil Approach

In coconut, the assessment of nutrient requirement through field experimentation would take a longer period. Therefore for a quick recommendation, soil testing is comparatively rapid and less expensive. However, the effectiveness of the procedure is closely related to the extent to which the data could be calibrated with field trails or that could be correlated with productivity. For making soil analysis more effective for coconut, some fundamental aspects like the time (period) of sampling, intensity, nutrient extraction procedures, calibration, etc. have to be standardized. Besides, the effective soil depth and total volume of soil and ground water characteristics of the area are yet other factors which regulate the nutrient reserves left at the disposal of the palm. Variations in soil type also need critical consideration. Thus, the problem of establishing soil critical values is rather a difficult task. However, attempts have been made to find out certain threshold soil values for making fertilizer recommendation to the palm. For example, Nethsinghe (1965) suggested a critical value of 9 ppm available soil P in Sri Lanka, while in India, Khan et al. (1990) suggested that 10 ppm Bray I P (0–90 cm) could sustain adequate P nutrition of the palm. If such soil fertility rating (of nutrients) for coconut under different pedoecological situations is established, soil testing could be profitably used for predicting the nutrient requirement of the palm.

Table 8.26 Suggested P application rate for coconut based on soil available P level

Olsen P (ppm)	P application (g palm^{-1} year^{-1})
Up to 6	180
7–1 0	130
11–15	90
16–20	40
>20	0

For coconut, Felizardo (1982) suggested rates of P application based on soil P levels (Table 8.26). Rates applied in other countries vary, possibly because of differences in soil P, soil type and yield level.

Studies conducted at ICAR-CPCRI, Kasaragod, India (Anon 1985; Khan et al. 1986b), have shown that the P and K requirement in laterite and red sandy loam soils could be predicted with the help of desorption equilibrium models which were constructed taking into account the response or reaction of the soils to added fertilizers. These models were developed with the assumption that 20 ppm P_2O_5 (Bray I) and 80 ppm K_2O (1 N NH_4OAc) in 0–50 cm soil would be optimum for sustaining adequate P and K nutrition of the palm, respectively. The models can be prepared for a given region or province based on soil series level to guide fertilizer application in the absence of leaf analysis data than arbitrary blanket recommendation. The validity of the models has been tested, and the predictability has been satisfactory. However, it requires refinement for different ecosystems.

8.9.2 The Plant Approach

In tree crops like coconut, the plant diagnostic methods give more reliable information on the nutritional status of plants in relation to soil fertility potential. The methods involve both qualitative and quantitative procedures which are designed to correlate plant status with productivity. Among them, one of the earliest approaches is the nutrient deficiency diagnosis in which the plant exhibits characteristic visible symptoms under nutrient stress conditions, and such deficiency could be corrected through judicious application of the concerned nutrient. One such example in coconut is the crown choke disorder which is primarily due to boron deficiency, and the malady could be corrected by the application of boron to the soil. The deficiencies of nitrogen, potassium, magnesium and sulphur in coconut are easily identifiable under field conditions and correction of such deficiencies can be attempted. However, its application is limited because some of the deficiencies do not show clear cut diagnostic symptoms. Diagnosis becomes difficult and inaccurate when more than one element is deficient. Further, the symptoms become clearly visible only when the deficiency is acute and the growth rate and the yield are severely depressed.

During the 1950s, the interest in the use of leaf analysis as an index of crop nutrient status grew rapidly. The concept of foliar diagnosis (Prevot and Ollagnier 1957) led to great expectations that it could be widely used as a means to predict fertilizer needs of tropical oil crops (oil palm and coconut). The coconut lends itself particularly well to nutrient investigation based on tissue analysis because of regular production of foliage and fruit throughout the year. It was strongly felt that plant analysis must have the advantage over soil analysis as the former reveals amounts of plant nutrients absorbed by the crop and interpretation is not dependent on soil nutrient “availability” or “exchange”. Hence, on the assumption that nutrient level or concentration within the leaves of the plant is related to crop growth or yield, leaf analysis was developed as a diagnostic tool. In the Philippines, the serious use of leaf analysis was pioneered in the early 1970s by the Philippine Coconut Research Institute, which has led to the finding of the essentiality of chlorine for coconut production (von Uexkull 1971; Mendoza and Prudente 1972). Now leaf analysis is widely used in many parts of the world to determine the nutritional development or effect of fertilizer application on the performance of coconut palms.

The method depends on the determination of critical nutrient levels and the relationship between nutrient concentrations in standard plant parts and the corresponding growth/yield response curves derived mainly from field fertilizer experiments using different nutrient levels. The foliar levels are then compared with the critical levels for assessing the nutrient needs of the palm. The critical level is defined as the concentration of the nutrient in the standard plant part below which the addition of that particular nutrient has every chance of giving an economic increase in yield. The critical levels of nutrients have been proposed providing a very useful guide for further detailed investigations of the palm that are evidently falling short of potential yield (Table 8.27). The following critical nutrient levels (per cent) proposed by Fremond et al. (1966) in Côte d’Ivoire and Magat (1978) in the Philippines, Cecil (1981) in India and revised sufficiency ranges by Loganathan and Atputharajah (1986) in Sri Lanka are presented.

Plant analysis, in association with soil testing, is a highly useful tool not only in diagnosing the nutritional status but also helps in formulating management decisions for improving the crop nutrition. Plant analysis is the quantitative analysis of the total nutrient content in a plant tissue, based on the principle that the amount of a nutrient in diagnostic plant parts indicates the soil’s ability to supply that nutrient and is directly related to the available nutrient status in the soil. As leaf is the primary centre where the major synthetic processes of the plant take place, variations in the nutritional condition of the leaf can be related to the soil nutrient status and the level of fertilizers to be applied. Leaf analysis is used most commonly as material of a similar stage of maturity, such as 14th frond, which can be used as a standard source for collection of samples for mature coconut.

Even though there are certain limitations, the studies conducted by IRHO, Paris reported by Magat (1979) have illustrated that leaf analysis is a very useful tool for predicting the fertilizer requirement of the palm. The 14th leaf of an adult palm (8 years and above) has been accepted as the index or reference leaf for foliar diagnosis in coconut. In sampling the diagnostic leaf, the first fully opened leaf on the

Table 8.27 Coconut diagnostic leaf (frond 14) critical levels (% dry matter) tall cultivar

	IRHO*	Unilever Friend (1975)	Malaysia Kanapathy (1971)	P C A Magat (1979)	India Cecil (1981)	CRISL**	CRISL**
Nutrient	1	2	3	4	5	6	7
Nitrogen (%)	1.8–2.0	2.00	1.80	1.80	1.8–2.0	1.8–2.1	1.9–2.1
Phosphorus (%)	0.12	0.14	0.12	0.12	0.12	0.11–0.12	0.11–0.13
Potassium (%)	0.8–1.00	1.00	0.80	0.80	0.8–1.0	1.20–1.40	1.2–1.5
Calcium (%)	0.50	0.55	0.15–0.30	0.30	0.30	0.25–0.35	0.35–0.55
Magnesium (%)	0.24–0.28	0.26	0.30	0.20	0.20	0.50	0.25–0.30
Sodium (%)	0.40	0.20	–	–	0.40	0.4	–
Chloride (%)	0.50	–	–	0.30–0.40	0.15	0.3–0.6	0.3–0.40
Sulphur (%)	0.15–0.20	–	–	–	0.20	0.15–0.20	–
Boron (ppm)	–	14	–	9.0–11.0	–	–	–
Manganese (ppm)	60	185	60	–	–	–	–
Iron (ppm)	50	115	50	–	–	–	–
Zinc (ppm)	60	15	60	–	–	–	–
Copper (ppm)	–	12.5	–	–	–		–

*IRHO *Côte d'Ivoire*, Fremond 1966; **CRISL, Loganathan and Atputharajah (1986).

crown is considered as 1, and subsequent leaves are counted to reach the 14th leaf considered as diagnostic leaf. This leaf is regarded as the one which has reached physiological maturity but has not entered the phase of senescence. For young palms up to 4 years of age, the fourth leaf and, for 5–7 years, the 9th leaf are taken for the purpose. For the details on sampling and sample preparation, the reader is directed to refer Prevot and Bachy (1962), Ziller and Prevot (1962) and Taffin and Rognon (1991).

Manuring recommendations should be aimed at the maintenance of the tree's mineral status at an optimum or its correction to that level, so that it is able to cover its needs for its production in the following years. Precisely, leaf analysis enables this status to be appraised and is therefore a good approach to the problem (Ouvrier 1982). The optimum range of the nutrients and fertilizer rates proposed vary with the variety (Table 8.28).

A critical level of 0.12% P in leaf 14 as suggested by Fremond (1966) and Magat (1978) was considered satisfactory for a long time. Magat et al. (1981a) showed a relationship between leaf P and coconut yield. However, Limbaga (1986) showed that yield increases were obtained by increasing frond P up to 0.15% indicating this

Table 8.28 Optimum range of concentration of nutrient elements in frond 14 in mature bearing palms

Nutrient element (as per cent dry matter)	Variety		
	Local Tall	Dwarf	Hybrid
N	1.80–2.00	1.80–2.00	1.80–2.00
P	0.12–0.14	0.12	0.12–0.30
K	0.80–1.00	0.60–0.80	0.90–1.40
Mg	0.20–0.30	0.25	0.25–0.35
Ca	0.30–0.50	0.2–0.5.0	0.32–0.35
S	0.13–0.17	–	0.15–0.16
Na	0.10–0.20	–	0.15–0.17
Cl	0.30–0.60	–	0.45–0.50
Mn (ppm)	60–115	60	–
B(ppm)	9–11	8	–

Source: Tall palms: Magat (1979); Magat (1992); Dwarf palms: Chew (1982); Hybrid varieties: Magat (1988)

to be the optimal level. It was suggested that yield is related to the ratios between foliar N and K, but that the K level should in turn be interpreted in the light of a balance between the monovalent and divalent cations (Smith 1969).The critical K levels are higher for young than for old palms. It should be higher if the pool of available K is low or vice versa.

According to Fairhurst (2003), the optimum requirement for individual nutrients can vary over a considerable range, depending upon factors such as the age of the palms, soil moisture regime, ratio to other nutrient concentrations, type of planting material, etc. Hence, the optimum leaf nutrient concentration must be determined for each agroecological environment taking local soil and climate conditions into consideration.

Discrepancies in the critical nutrient levels are expected as the yield is modified by several other extraneous factors and critical levels established for one region may not hold good for another. Kamala Devi et al. (1983) from a fertilizer experiment also opined that the critical level of nitrogen under the coastal condition of India must be less than the value recommended by IRHO. The critical level obtained for Ca and Mg (0.36 and 0.24–0.25%, respectively) was in the range more or less similar to the concentration reported by Chew (1982) and Ravi Savery et al. (1994).The critical level for S was almost similar to the figure of 0.15–0.20% suggested by Manciot et al. (1980). Foale and Ashburner (2005) summarized the critical nutrient concentrations proposed in the literature with their comments on behaviour of nutrients in the plant system (Table 8.29).

Manciot et al. (1980) working on the critical level of trace elements opined that it was not possible to define critical levels for Fe, Mn and Cu in coconut. The critical values of Cu-5-7 ppm, Zn-15 ppm, Fe-50 ppm and Mn-60 ppm in the 14th leaf suggested by Manciot et al. (1979c) are only proposed values, and critical levels have so far not been established experimentally (Manikandan et al. 1986).

Table 8.29 Critical values of concentration of mineral elements in the leaf tissue (14th frond) of adult tall coconut

Major elements	% Dry matter	Comments
N	1.8–2.0	In tall × dwarf hybrid –2.2
P	0.12	–
K	0.8–1.0	In tall × dwarf hybrid –1.4
Mg	0.20–0.24	Strong inverse sensitivity to extremes of K
Ca	0.30–0.40	Strong inverse sensitivity to extremes of K
Na	Not essential	Substitutes K in case of deficiency
Cl	0.5–0.6	–
S	0.15–0.20	–
Trace elements	**Parts per million**	**Comments**
B	10	–
Mn	>30	Difficult to fix value as very interactive with Fe in strongly alkaline soil; potentially toxic in extreme acid soils
Fe	50	Deficient only in strong alkaline soils
Cu	5–7	Deficiency very rare, not very certain
Mo	0.15	Common value-no response observed yet
Zn	20	Common value-no response observed yet
Al	>38	Nonessential element but always present; potentially toxic at values well in excess of this common level

Source: Manciot et al. (1979b, c); Manciot et al. (1980); de Taffin (1993); Foale and Ashburner (2005)

Acharya and Dash (2006) employing Cate Jr and Nelson (1965) technique and relative yield on random selection of palms observed critical level of major, secondary and micronutrients for palms growing on coastal sandy tract of Odisha (India) which can be considered as sufficiency levels only.

8.9.3 The Soil–Plant Integrated Approach

The third important approach is the soil-plant integration models in which the soil nutrients, the plant nutrient composition and the yield would be viewed as a relationship where soil nutrient contents influence plant (leaf) nutrients and yield, leaf nutrient content is related to yield. Among them, the Mitscherlich-Bray model is being popularly used in most of the field crops. This principle was employed to evaluate the nutrient requirement of three high-yielding genotypes of coconut by Khan et al. (1986b) (Table 8.30). They indicated that the nutrient requirement of D × T hybrid was relatively lower than that of talls to produce the same quantity of yield. Tall × dwarf hybrid was intermediate between D × T and WCT. They further evaluated the efficiency of soil and fertilizer nutrients with respect to productivity. It is also possible to forecast coconut productivity for a given quantity of nutrient inputs.

Table 8.30 Fertilizer recommendation based on Mitscherlich-Bray equation

		Baule units fertilizer to be applied palm^{-1} year^{-1} (g)			
Cultivar	Nutrients	1 (50%)	2 (75%)	3 (87.5%)	4 (93.75%)
WCT	N	370	851	1332	1813
	P_2O_5	370	851	1332	1813
	K_2O	741	1703	2665	3626
COD × WCT	N	75	214	352	490
	P_2O_5	75	214	352	490
	K_2O	150	427	704	980
WCT × COD	N	144	333	521	710
	P_2O_5	144	333	521	710
	K_2O	288	666	1043	1420

WCT: West Coast Tall, *COD:* Chowghat Orange Dwarf

Although leaf analysis may indicate that one or more elements are abnormally low or high in the leaves, it will not necessarily tell which of the yield-limiting factor (e.g. inadequate or excess water, pests and diseases, improper nutrition, improper fertilizer placement, very low or very high pH, over grazing) is the cause. Magat (1976) opined that when leaf analysis is supplemented by soil analysis, a much better insight on the nutritional status of the crop may be obtained. It appears that foliar diagnosis is an efficient method of detecting nutrient deficiencies if they occur singly. If, however, several deficiencies occur simultaneously, greater interpretative caution is required, and other agricultural factors have to be considered. In most cases, field trial or experiment is advisable, especially if the results of soil and leaf analyses, respectively, are inconsistent.

Magat (2000) provided a complete guide for fertilizer recommendation for coconut in the Philippines based on soil and leaf analysis. The guide classified the nutrient content in soils as low/deficient, medium/adequate and high/excessive and provided information on the quantity of nutrients to be applied for a given set of soil test values. A similar guidance was also given for nutrient application based on leaf analysis. He also suggested the fertilizers N, P, K, Cl, Mg and S to be applied to different age group of palms.

For coconut, the advantage of leaf analysis over soil analysis as a basis of fertilizer recommendation is now largely accepted. This has been so, as leaf analysis reveals amounts (concentrations) of nutrients absorbed by the crop and diagnosis is not highly dependent on varying concepts of soil nutrient “availability” or exchange reactions (highly dynamic in behaviour). Thus, with the strong consideration that nutrient levels or concentration in a plant part like the leaf is directly related to its (coconut) growth and / or yield, leaf analysis has been successfully developed as a diagnostic tool to predict the fertilizer needs of the crop.

Magat (1978) revealed the usefulness of leaf analysis as an effective and rapid tool in determining the qualitative needs and estimating the fertilizer rates of coconut based on results of several fertilizer trials. Results obtained by Magat (1978) were reviewed by Manciot et al. (1979b, c) with the conclusion that leaf analysis or foliar diagnosis is undoubtedly a very effective tool in predicting fertilizer needs of existing stands. Moreover, further analysis done on soil, leaf and yield data of the Philippine Coconut Authority (PCA) survey (1975–1980) collected from 1131 sampling areas (57 coconut provinces) with diverse agroclimatic conditions showed that leaf analysis could give a better predictive value of the nutrient needs of the coconut and that leaf nutrients (N, P, K, Ca, Cl and S) are more closely associated with coconut yields (nuts, copra) compared to soil properties and yields (Cosico and Fernandez 1983; Limbaga 1986). It is now well recognized that plant and soil analysis techniques used in complementary roles are indispensable tools for assessing the nutrient status of soils and determining correct fertilizer practice (Nathanael 1967).

Magat (1991) summarized the experience gained on use of leaf analysis in coconut nutrition as follows:

1. Both the soil analysis and leaf analysis can be used as basis of fertilizer recommendations for coconut and these complement each other.
2. As leaf analysis is not affected by soil nutrient "availability" and exchange reactions, it is a more reliable method.
3. For macronutrients as Cl, S and N, leaf analysis is a more accurate tool as these elements are highly mobile in the soil.
4. In both the methods, availability of guides on critical or satisfactory values (soil analysis), critical and optimum levels (leaf analysis) and average nutrient and fertilizer needs, under different stages/ ages of the coconut, should improve reliability of recommendations.
5. Reference critical levels (soil and leaf) should be evaluated under local conditions before wider use.

8.9.4 *Use of Nut Water for Nutrient Analysis*

In Sri Lanka, Salgado (1951, 1954) suggested usefulness of nut water analysis in the diagnosis of nutrient deficiency of palms, and found it to be a better guide in interpreting the response of K fertilizers in terms of nut yield besides its usefulness as an additional tool in the interpretation of experimental yield data on K, Na, Mg and Cl.

According to Salgado (1955), while soil analysis attempts to measure the nutrient supply of a limited stratum of the soil ("intensity factor"), the nut water technique measures the "capacity factor" and takes into account not only the nutrient status of the soil but also the volume of soil from which the palm is drawing its nutrients. It was generally concluded that coconut water is analogous to plant sap and accordingly would indicate the physiological status of the palm and also the soil conditions in which it grows. The dominant requirement of the coconut is K, which is concentrated in the pericarp and water of the nut. K_2O in the nut water can be determined by the gravimetric cobaltinitrite method. It was shown that drought

Table 8.31 Changes in nut water composition during development of coconut (West Coast Tall)

Age of nut (month)	Volume of water (ml)	pH	Total sugars (%)	N	P	K
				mg^{-1}		
4	75	3.5	0.8	32	48	1113
6	310	4.7	3.3	195	118	5320
8	230	5.5	5.6	432	186	7300
10	145	5.9	3.4	336	140	3260
12	100	6.1	1.8	299	108	3181

markedly affected nut size and the volume of nut water, as well as its potash content. The K_2O content of nut water increased with potash applications and can provide an index of the K_2O status of the soil and of the expected yields.

In Sri Lanka, chemical analysis of nut water has been successfully used in the study of P and K nutrition of coconut (Salgado and Abeyawardena 1964). In other countries this approach has also been used with some measure of success (Southern 1956; Lockard et al. 1969). Jeganathan (1990) suggested the possibility of using nut water analysis as an additional tool in the interpretation of field experiment data for Na, K, Mg and CI. Both Ca and Mg have functionally limited roles to perform, both in the liquid and in the solid endosperm, and, therefore, their concentration will be low and so too the changes.

Nagarajan and Pandalai (1965) opined that coconut water furnishes, by and large, a good material to study the nutrient needs of the palms, based on the studies on enzymatic activity during various stages of development of the nut. They found that the enzymes such as catalase, peroxidase and polyphenol oxidase are to be correlated with potash content of nut water. According to them, a higher enzymatic activity of the nut water indicates that the palms required to be supplied with potash.

Coconut water and its relationship with potassium were reported by many workers. An account of increase in the K content of nut water with K application to palms at different rates was given by Jeganathan (1990) in Sri Lanka and Silva et al. (2006) and Riberio et al. (2011) in Brazil. The difference in composition with age of the palms and among cultivars in India has been reported by Kamala Devi and Velayutham (1978). The quality of tender nut water of *anão verde* coconut, grown in Brazil, in relation to doses of N and K through fertigation was evaluated by Neto et al. (2007). An increase in dose of K (258–4872 g plant^{-1} year^{-1}) decreased the salinity and increased °Brix of coconut water.

An account of nutrient content of tender coconut water at different stages of the development of the nuts indicated abundance of K at maturity stage of 8th month (Table 8.31).

Total sugars and potash content indicate that tender coconut water can be considered as a health drink in the 8th month. Kamala Devi and Velayutham (1978) observed least difference among cultivars. Nitrogen and potassium affected the volume of coconut water of a dwarf variety with maximum volume of 417.75 ml found when 818 g of N and 1487 g of K palm^{-1} year^{-1} were applied (Silva et al. 2006).

Holanda et al. (2007) observed that in green dwarf coconut, the critical levels of N and K on frond 14 are between 18.7 and 19.3 g kg^{-1} and between 9 and 10 g kg^{-1}, respectively. The critical level of K is larger than the range of 6 to 8 g kg^{-1} proposed by Magat (2005) for leaf 14 of the dwarf coconut without irrigation.

Nitrogen and potassium levels had a linear effect on the soluble solids content of coconut water, where N had a negative and K exhibited a positive effect. In accordance with published reports elsewhere, Riberio et al. (2011) in Brazil observed that application of KCl increased the concentration and content of K in nut water and kernel of green dwarf coconut and there was no influence of K fertilization on the mass, the volume, the pH, the C.E. and the TSS of the coconut water. Irrigated plantations of dwarf coconut for the production of tender coconut water have expanded considerably in Brazil. Higher yields obtained under irrigation affect the amount of N and K required by dwarf coconut, influencing the relationships between yield and soil and leaf contents of these nutrients. The highest fruit weight and coconut water volume were obtained with the lowest N level (Sobral and Nogueira 2008). Nitrogen and K combinations did not influence the coconut water pH but increased coconut water brix. The K content in the coconut water increased along the K doses. Teixeira et al. (2005a) found that application of N decreased the volume of the coconut water and the fruit weight, while K had the opposite effect. Silva et al. (2006) and Ferreira et al. (2007) observed that N decreased solid content and K increased the same.

Though experimental evidences show the possibility of use of nut water as a diagnostic tool, this technique did not gain wider acceptability in nutrient diagnostic studies on coconut.

8.10 Nutrient Management

8.10.1 Nutrient Management in the Nursery

The haustorium absorbs food materials from the nut water as well as kernel and supplies to the growing plant. The role of haustorium in seedling growth has been discussed by Child (1974). Foale (1968) reported that haustorium decreases from the fourth month after germination suggesting that the young seedlings are in short supply of nutrients for a major part of their 1-year growth. Harries (1970) opined that though food reserves were adequate as far as carbon compounds and nitrogen are concerned, potash application is needed considering its uptake. Based on the studies in India, Nelliat (1973) suggested application of fertilizers to the nursery in December, February and April to supply 40 kg N, 20 kg P_2O_5 and 40 kg K_2O ha^{-1} under west coast conditions to produce good-quality seedlings. For nursery seedlings fertilizer recommendation with emphasis on chloride nutrition as needed for the Philippines' conditions is prescribed by Magat (2000).

8.10.2 Nutrient Management of Young Palms

Young transplanted seedlings require adequate nutrients for better growth on all soils. Fertilizer application to the seedling is very important to guarantee good development of bole, which is important for the productivity of the tree, as it increases the rooting surface. Given a good nutrition, the stem will also attain its maximum width (Ohler 1999). With a very active root system, the young plants respond well to manuring, grow better and start bearing early. An enhanced rate of leaf production with larger number of leaves on the crown results in larger total leaf area leading to increased/required photosynthetic activity which may probably increase building up of adequate carbohydrate reserves in the system. There is a correlation between chlorophyll content in the leaves, rate of apparent photosynthesis and annual yield (Narayanan Kutty and Gopalakrishnan 1991). Ramadasan and Mathew (1977) observed that adequate nutrition to the palms in the juvenile phase leads to build-up of required carbohydrate reserves in the trunk with commencement of flowering, a question of partition of assimilates towards reproductive phase once the vegetative phase is satisfied. The above statement implies that in an intelligent nutrition management programme, importance of fertilizing young palms with adult palm dosage should be resorted to well in advance before they come to flowering. Studies indicate that the recommended adult palm dosage is given from the fourth year onwards. The damage caused by K deficiency in the early stages cannot be fully repaired by later K dressings (Fremond and Ouvrier 1971). Although later applications of K enabled re-establishment of good physiological functioning, the palms which suffered from K deficiency during pre-bearing age remained on an average 15% less productive than those which never suffered from K deficiency. In Sri Lanka, fertilizer recommendation for young palms is based on a 3-year fertilizer application conducted earlier (Jeganathan 1993). The results gave good indications on the importance of fertilizing the palms and unequivocally established that neglect at the seedling stage can have very damaging effect on future production (Loganathan 1977).

8.10.3 Nutrient Management of Adult Palms

The economic importance of nutrition of adult palm has drawn attention of many researchers in all the coconut growing countries, as a sustainable yield only will satisfy the requirement of the grower and increasing need of the industry. A basic understanding of management of the plantation is discussed here. Annually, the palm removes large quantities of nutrients from the soil (Nathanael 1961; von Uexkull 1971; Ouvrier and Ochs 1978). The most rapid growth occurs between the second and fifth year in the life of the coconut palm. The crown grows 30–50 cm year^{-1}, up to about 50–60 years. Dry matter production is around 50–80 kg year^{-1}. In its prime, a coconut palm normally produces 12–15 leaves and about 80–100 nuts year^{-1} (Chan and Elevitch 2006). A balanced application of nutrients is essential to obtain high and sustainable yield, and it is the key to increased plant use efficiency of applied nutrients. It replaces the amount of nutrients removed by the crop besides ensuring that fertilizers are applied in adequate quantities and correct ratios for optimum growth and ensures sustenance of soil and crop productivity. Several field experiments to assess the nutrient need have been carried out under different environmental conditions on adult palms, and some of the earlier reports are available from India (Thampan 1970; Muliyar and Nelliat 1971), in Sri Lanka (Balakrishna 1975), Jamaica (Smith 1964, 1969), the Philippines (Prudente and Mendoza 1976) and Malaysia (Soon and Wat 1972) and on young palms in Côte d"Ivoire (Fremond and Ouvrier 1971). Preliminary results of fertilizer experiment carried out at Tanganyika (east Africa) indicated the necessity of fertilization for increasing the yields (Anderson 1967).

Over years of gathering research information, Magat (2000) published a guide on soil fertility levels in coconut production in which he categorized the soils of the Philippine coconut growing areas into low/deficient, medium/adequate and high/excessive classes for major, secondary and micronutrients assigning values for each nutrient. Corresponding to this, fertilizer is prescribed for different age groups from field planting to palms of 5 years and above. The rates of fertilizers are the ones recommended under moderate level of nutrients, that is, those above the critical levels. When the level of nutrients are above equal or lower than the critical level, rates of nutrient and fertilizer application should be increased to at least 20–50% of values in the reference tables, while when levels of nutrients are higher than critical, the rates of nutrients and corresponding fertilizers should be lower by at least 20–50% of reference values. Under high levels of soil fertility, fertilization is not usually required except for periodic maintenance. Similar recommendation is made taking into account plant nutrient levels. Straight fertilizers, fertilizer mixtures and dolomite- and chloride-bearing fertilizers are recommended for different groups of palms.

In the Philippines, studies have shown the beneficial effects of fertilization (Mendoza and Prudente 1972; Magat et al. 1975; Prudente and Mendoza 1976; Magat et al. 1981a) in increasing copra yields as high as 3 tons ha^{-1} year^{-1}.

Menon and Pandalai (1960), in a review of nutritional studies on coconut palm in India, observed that a minimum of 3 years is required to obtain the full response to fertilizer application to coconut. Muliyar and Nelliat (1971) registered that response to N was obtained from third year onwards and for phosphorus from ninth year onwards at Kasaragod, India. Their studies indicated that for palms yielding less than 60 nuts annually, optimum nitrogen dose ranged between 400 and 650 g palm^{-1} year^{-1}. Nitrogen affected all the nut characteristics studied, viz. weight of whole nut, and that of husked nut, volume of husked nut and copra weight nut^{-1}. These characters were much improved by potassium manuring, while phosphorus had negligible effect. Although nitrogen application increased the yield by 16.9%, copra yield was increased only 6%. With potassium, increase in nut production was 12%, while copra yield was 22%. Large-scale fertilizer demonstration trials conducted all over the west coast of India (John and Jacob 1959) showed that application of 340 g N, 340 g P_2O_5 and 680 g K_2O palm^{-1} year^{-1} had resulted in an increase of 35% in yield of nuts and 44% in copra outturn over the farmer's practice. In certain locations where the above fertilizer dose is not producing expected response, significant increase in yield was obtained when the K_2O level was raised to 900 g palm^{-1} year^{-1}. These trials demonstrated the need for fertilizer inputs to increase yields and the importance of a specific nutrient for correcting the deficiency to further increase the yield.

However, the general fertilizer recommendation for palms in India by ICAR-CPCRI is 500 g N, 320 g P_2O_5 and 1200 g K_2O palm^{-1} year^{-1}, which was arrived based on the agronomy trials. Nelliat (1973) recommended an increased quantity of 1000 g N, 500 g P_2O_5 and 2000 g K_2O palm^{-1} year^{-1} for palms with higher yield potential. Based on long term multilocation trials in different soil and agroclimatic conditions, fertilizer recommendation has been prescribed for different coconut growing regions in India (Table 8.32).

In addition to the recommended fertilizer application for Kerala, 50–60 kg organic manure is also usually applied. Mostly fertilizers are applied in two or three splits according to the rainfall pattern and soil type. Fertigation is also recommended for economizing the quantity of fertilizers being applied. Please refer to Chap. 7 for details.

Table 8.32 Fertilizer recommendation for coconut in different regions of India

	Fertilizer recommendation for adult palms (g palm^{-1} year^{-1})					
State	N	P_2O_5	K_2O	Variety	Soil type	References
Tamil Nadu	560	320	1200	Tall	Red sandy loam/ alluvial soil	
	1000	250	2000	Hybrid	Red sandy loam/ alluvial soil	Venkitasamy (2004)
Karnataka	560	320	1200	Tall	Red sandy loam	Khan et al. (1986a, b)
	1000	250	1000	Hybrid	Red sandy loam/light black soil	
Andhra Pradesh	500	250	1000	Tall	Coastal alluvial type	
Odisha	560	320	1200	Tall	Coastal alluvial type	
	1500	750	1250	Tall	Coastal littoral sand (150 g each Ca and Mg)	
Maharashtra	1000	500	1000	Tall and hybrid	Konkan coastal area	Nagewekar et al. (2004)
Assam	500	500	2000	Tall and hybrid	Alluvial clay loam soil	Nath et al. (2012)
West Bengal	500	250	750	Tall	Alluvial sandy loam soil	
	1000	500	1000	Hybrid	Alluvial sandy loam soil	Ghosh and Maheswarappa (2016)
Kerala	500	320	1200	Tall/ hybrids	Sandy loam, laterite and littoral sandy soil	Khan et al. (1986a, b)

While application of chemical fertilizers over years alone makes the soil fertile and increases productivity of crops, it brings adverse effects on soil and environment. It is essential that fertility and productivity of the soil be restored through integrated nutrient management (INM) approach (Khan et al. 2000). A review of the results of fertilizer experiments carried out in different agroclimatic regions of Sri Lanka (Loganathan 1978) has shown that coconut responds to fertilizer application. Striking responses have been obtained in the poorer soils of the wet zone compared to the relatively richer soils of the intermediate zone. Though the per cent increase in yield was very much higher for the poorer soils, the absolute increase in yield was nearly the same for all soils (about 3–4 kg palm^{-1} year^{-1}). Chew (1978) observed that fertilization increased the number of female flowers by bunch. However, continued applications of fertilizers decreased the size of the fruit to a certain extent, and the amount of copra per fruit though increased the number of nuts.

From the long-term experiment on a lateritic gravelly soil in Sri Lanka, where the rates of fertilizers were progressively increased up to the 16th year, Loganathan and Balakrishnamurti (1975) obtained the optimum yield by applying 1.818 kg sulphate of ammonia, 1.136 kg *saphos* phosphate and 2.043 kg muriate of potash palm^{-1} year^{-1} from the 16th year onwards. Application of 1.362 kg each of sulphate of ammonia, *saphos* phosphate and muriate of potash between the 9th and the 16th years produced a yield of 20.7 kg copra palm^{-1} year^{-1} from the 13th to the 16th years, which is about 150% higher than the plots which received no fertilizer from the seedling stage. Based on the results, they suggested that in the current fertilizer recommendations, for both young and adult palms, the rate of N could be reduced and that of K be increased. Their subsequent studies (Loganathan and Balakrishnamurti 1975) highlighted the importance of balanced fertilizer application. The combination of the highest dosage of N and without P gave the lowest yield, the yield being even lower than the control plots. This has shown that an increase of N without a corresponding increase of P would be detrimental. The positive NP interaction indicates that the benefit from an increase of N could be obtained only if P also is increased and vice versa.

Results of an experiment with adult coconut on a lateritic gravelly soil showed that application of muriate of potash up to 1.8 kg palm^{-1} year^{-1} linearly increased nut and copra yield and copra weight nut^{-1} (Loganathan and Balakrishnamurti 1979), while sulphate of ammonia, up to 4.4 kg palm^{-1} year^{-1}, decreased copra weight nut^{-1}. The optimum rates of fertilizers were 1.1 kg sulphate of ammonia, 0–0.83 kg *saphos* phosphate and 1.8 kg muriate of potash palm^{-1} year^{-1} giving a yield of 12 kg copra palm^{-1} year^{-1}. Prudente and Mendoza (1976), based on the first 25 months yield data of young coconut in the Philippines, also reported that application of N without P would give yield even less than the plots which received no fertilizer.

Use of fertilizer increased yield in the range of 30–200% for palms grown in moderate to virtually poor soil types in Sri Lanka. Palms grown in poor soils or those receiving little or no field care have shown 100% improvement in the yield over a period of 3 to 5 years when they received fertilizer annually at the rate of 1.58–2.26 kg palm^{-1} (de Silva 1973). In the coastal quaternary sands of the Côte d'Ivoire, Pomier and de Taffini (1982) noticed a drop in fertility especially that of N, where the soil has become very poor due to prolonged monocropping of coconut. They suggested adopting regular application of nitrogen fertilizers or raising leguminous cover crops. At the adult stage, an annual manuring of 1.5 kg of sulphate of ammonia tree^{-1} (besides the usual application of 2–3 kg of KCI and 1 kg of Kieserite) was also recommended. de Silva (1981) estimated the aggregate response of coconut production to quantity of fertilizer applied in each year and the weather (rainfall) by analysing aggregate data for 26-year period from 1956 to 1981 and suggested that the aggregative approach appears to be a useful alternative in modelling fertilizer response under non-experimental conditions.

Results of leaf analysis revealed significant widespread N deficiencies at most of the sampled sites and geographic variations in K deficiency. Chlorine deficiency varied with geographic sites and was closely related to the prevailing wind pattern. Preliminary results on nut set and flowering in the trial at Stewart Research Station, Papua New Guinea (Ollivier et al. 1999), revealed a positive response to N- and Cl-based fertilizer applications. This suggests that appropriate fertilizer applications would be beneficial to future coconut production on this particular site. Potassium-chlorine interaction was found significant in the result and most probably related to the K-Na antagonism. This was also observed at Gunung Batin in Indonesia (Bonneaux et al. 1997). Chlorine is the dominant element, and potassium only reveals its effect if the Cl effect is resolved.

The nutrients exported from the soil at highest quantities by palms are mostly N and K (Pillai and Davis 1963; Ouvrier and Ochs 1978). In the commercial coconut plantations in Papua New Guinea, N and K are the most yield-limiting elements (Ollivier et al. 1999). In coconut plantations, harvested nuts, as well as fronds and much of the other residues, are removed, resulting in gradual depletion of plant nutrients from the soil (Somasiri 1987). Furthermore, the nutrients stored in the trunk are not returned to the soil system. Application of fertilizers containing N, P, K and Mg at the recommended rates only partially compensates for this depletion. Studies indicated that nuts remove a considerable quantity of major nutrients (Jeganathan et al. 1977; Ohler 1984; Jayasekara et al. 1991). Though increasing N, P and K rates were tested, there was no yield response to the increased levels of P application (Sobral and Leal 1999).

In Sri Lanka, adult palm mixture containing 800 g of urea, 600 g of rock phosphate and 1600 g of muriate of potash with 1 kg of dolomite palm^{-1} year^{-1} is recommended to meet the demand of nutrients of plantations yielding 7500 nuts ha^{-1} year^{-1}. The recommendation for high-yielding plantations is 1.5 times the above rates (Mahindapala and Pinto 1991). It would be sufficient for plantations yielding up to 11,250 nuts ha^{-1} year^{-1}. Nevertheless, since plantations yielding 12,500–19,000 nuts ha^{-1} year^{-1} (shown by land suitability studies) remove higher quantities of nutrients than the above inputs, they require additional fertilizer nutrients (Somasiri et al. 1994, 2000, 2001). To sustain productivity of such high-yielding plantations, the nutrient-supplying power of the soil is to be maintained with large supplements of organic manures. Details of inorganic fertilizer recommendation for coconut in Sri Lanka (Anon 2016) are given in Table 8.33.

Table 8.33 Inorganic fertilizer recommendation for coconut in Sri Lanka

Fertilizer (g)	Age of palms							
	6 months	1 year	1.5 years	2 years	2.5 years	3 years	3.5 years	4 years up to bearing
Wet and intermediate zone								
Urea	190	235	235	305	305	375	375	470
Eppawela rock phosphate	420	530	530	690	690	850	850	1060
Muriate of potash	190	235	235	305	305	375	375	470
Dolomite	500	500	500	500	500	500	500	500
Dry zone								
Urea	190	235	235	305	305	375	375	470
Eppawela rock phosphate	270	330	330	490	490	600	600	660
Muriate of potash	190	235	235	305	305	375	375	470
Dolomite	500	500	500	500	500	500	500	500

For adult palms

Wet and intermediate zones		Dry zone	
Fertilizer	Amount (g)	Fertilizer	Amount (g)
Urea	800	Urea	800
Eppawela rock phosphate	900	Eppawela rock phosphate	–
Imported rock phosphate	–		600
Muriate of potash	1600	Muriate of potash	1600
Dolomite	1000	Dolomite	1000

Adult palm mixture for the wet and the intermediate zones-APM-W	
Urea	8 parts by weight
Eppawela rock phosphate	9 parts by weight
Muriate of potash	16 parts by weight
NPK composition of the mixture	
11% N, 8% P_2O_5, 29% K_2O (11-8-29)	
Adult palm mixture for the dry zone-APM-D	
Urea	8 parts by weight
Muriate of potash	16 parts by weight
NPK composition of the mixture	
12% N, 6% P_2O_5, 32% K_2O (12-6-32)	

Broadcast method of fertilizer application is recommended best for flat lands. However, on sloping lands where soil erosion or run-off is possible, fertilizer may be applied in full circle trenches cut around the palm or half circle trench on the upper side of the palm. The trench should be 0.9 m away from the base, 0.9 m wide and 10 cm deep.

Based on studies on mineral nutrition of coconut palms, Gunathilake et al. (2008) emphasized that coconut palms responded well to mineral fertilizers. The highest average response to mineral fertilizers was observed in the wet and intermediate zones of Sri Lanka (41%) followed by the dry zone (32%). The fertilizer response was associated with favourable climate (sunshine hours and rainfall). The contents of micronutrients in fronds, nuts and inflorescences are governed by the soil characteristics, and accordingly application of micronutrients to specific soil sites is suggested to improve the soil reserves of various nutrients (Nadheesha and Tennakoon 2008).

Several studies suggest that the nutrients exported through harvested nuts and other usufructs be returned to soil in chemical and organic forms to maintain the productivity of soil. According to Somasiri et al. (2003), nutrients exported by the coconut plant components were 116.79 kg N, 14.02 kg P, 245.43 kg K, 40.47 kg Ca and 33.66 kg Mg and additionally 55.79 kg Na ha^{-1}. Inputs at the recommended rate (4.5 kg of APM and 1.5 kg of dolomite $palm^{-1}$ $year^{-1}$) supplied 87.21 kg N, 17.07 kg P, 188.90 kg K, 104.80 kg Ca and 28.59 kg Mg ha^{-1} $year^{-1}$ indicating a negative budget for exchangeable K (363 kg ha^{-1}), whereas the other macronutrients were present in reasonably high quantities in the experimental site. As the K reserves in the soil were low and fertilizer inputs supplied only about 77% of the requirement of high-yielding coconut, soil K will deplete rapidly, and it is necessary to compensate for either by increasing the quantity of chemical fertilizer or recycling organic products of the palm itself or both.

Studying the nutrient removal by all plant parts of coconut through leaf and soil analysis (164, 53, 37, 22, 19 and 9.85 kg ha^{-1} $year^{-1}$ for K, N, Na, Ca, Mg and P, respectively) and comparing with that of the amount of N, P, K, Ca and Mg applied annually as fertilizers (58, 11.5, 125.9, 69.9 and 19 kg ha^{-1}). Wijebandara et al. (2015) observed that K and Mg, input by application of fertilizer, slightly exceed the nutrient removal. The N, P, Ca and Mg reserves of soil were high and would not deplete rapidly. The K input by application of 3.3 kg of APM fertilizer mixture was about 76.7% of the K removal. It is inferred that if the harvested nuts and fallen plant components are all removed from the plantation, depletion of exchangeable K pool will occur rapidly and should be compensated for either by adding extra 80 kg ha^{-1} of muriate of potash or recycling of fallen fronds and residues of the inflorescences or mulching the manure circle using fresh coconut husks or addition of organic manure. Their specific studies in Boralu soil series also indicated high reserves of available Fe, Mn, Cu and Zn, and micronutrient application would not be required.

The apparent result of reduction in N content with higher yields has to be viewed in the context of N/Mg and N/S ratios (Mathewkutty et al. 1995, 1997). Higher productivity calls for an increase of Mg and S in relation to N suggesting that yield-limiting influences come from a real and apparent deficiency of Mg and S. This indicates that Mg and S have to be applied even at the expense of N. This calls for recommendation of Mg and S for coconut cultivation in Kerala, India. The yield-limiting influences of coconut are not the limitation in major nutrients that are regularly applied but the deficiency of some non-applied elements like Mg and S and excesses of Ca, Fe, Mn, etc.

Micronutrient status of coconut palms receiving 500:250:1200 g $N/P_2O_5/K_2O$ palm^{-1} year^{-1} for 10 years was studied by soil and leaf analysis (Venkitaswamy et al. 2006). They observed that leaf analysis showed sufficiency levels of Fe, Mn, Zn, Cu, whereas soil critical levels established in Tamil Nadu, India (Krishnasamy et al. 1994), indicated deficiency for Fe and Cu suggesting a relook into the soil critical levels for coconut growing regions.

In COD × WCT palms, significant difference in yield was observed with N and K, influencing bunches harvested, female flower-produced inflorescence^{-1}, while P did not influence any of the characters. When N was a major limiting nutrient, influence of P and K was not much expressed (Venkitaswamy et al. 2011). A trivariate extension of the quadratic model without higher-order interaction terms fitted to the yield accounted for 98.3% of the variation. The physical optimum rate of fertilizer requirement worked out to 818, 130 and 1362 g palm^{-1} year^{-1} of N, P and K, respectively, with a yield of 159 nuts palm^{-1} year^{-1}. Application of NPK at 500:108:830 g palm^{-1} year^{-1} recorded 178 g copra nut^{-1} and 26.0 kg copra palm^{-1}.

Upadhyay et al. (2002) worked out response function for coconut taking into account recommended dose as one (500 N: 320, P_2O_5 and 1200 K_2O g palm^{-1} year^{-1}) and transforming other treatments accordingly (Fig. 8.5).

The quadratic response led to optimum fertilizer requirement as 359 N: 229 P_2O_5 and 860 K_2O which gave a nut yield of 151.77 nuts palm^{-1}.

Magat (2009) in the Philippines compared application of multi-nutrient fertilizer 14:0:20:15:4.5:0.2 ($N/P_2O_5/K_2O/Cl/S/B$) and NaCl and found that the former increased coconut yield during the first, second and ensuing years by 50%, 100% and 150%, respectively. The increase in copra yield was 20%, 33% and 66%, respectively, over the latter during the same period. Magat et al. (2009) further reported that the multi-nutrient coconut-specific fertilizer formulation (14–5-20 NPK) with adequate content of other nutrients at 1 kg palm^{-1} is highly suitable, nutritionally balanced and practical for long-term productive coconut farming.

A dosage of 120 kg N ha^{-1} year^{-1} was enough to maintain sufficiency range on leaves, for "dwarf green" coconut cultivated in the State of São Paulo, Brazil (Teixeira et al. 2005a). Application of muriate of potash at 120 kg ha^{-1} was sufficient to maintain soil exchangeable K at the same concentration as found prior to the experiment and influence the leaf K levels above sufficiency range. Foliar P contents were always higher than sufficiency range despite of P fertilization, indi-

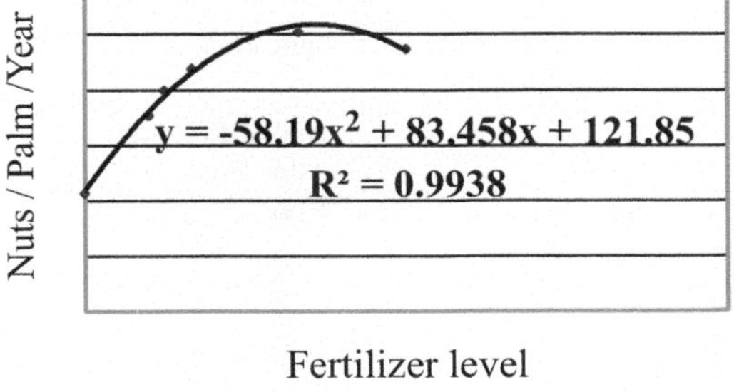

Fig. 8.5 Fertilizer response function for coconut

cating that soil of the experiment site has adequate reserve of phosphorus to support the plantation. In Pakistan, maximum nut yield was obtained with the application of 700 g urea +270 g diammonium phosphate +1600 g muriate of potash palm^{-1} year^{-1} for tall variety of coconut (Baloch et al. 2004).

The fertilizer recommendation for coconut in Malaysia (Table 8.34) and Indonesia (Table 8.35) is given below.

In Trinidad and Tobago, data obtained from soil tests in coconut growing areas, along with the nutrients required to produce 100 nuts year^{-1}, formed the basis to formulate a fertilizer recommendation (Ramkhelawan 2013). The fertilizer formulation recommended was 15-5-20 NPK, and the fertilizer rates palm^{-1} at different age group of palms are shown in Table 8.36.

Alternatively to 15-5-20 coconut palm mixture, straight fertilizers, viz. urea, triple superphosphate (TSP) and MOP, could be procured and blended by weight in the ratio of 3:1:3 (urea/TSP/MOP) and applied at the rate shown in the above table. In addition to the above recommendations, where soils are acidic, 2 kg of finely ground dolomite limestone application is suggested for each adult palm^{-1} year^{-1}.

A programme was developed for establishing "Malayan dwarf" coconut palms in Florida using foliar and granular fertilizer (Malayan dwarf coconut palm granular fertilizer) for the initial 6 months and subsequently with granular fertilizer alone at

Table 8.34 Recommended fertilizer rates for coconut on mineral soils in Malaysia

Fertilizer rate (kg ha^{-1})				
N	P_2O_5	K_2O	Plant population (ha^{-1})	Recommending agency
71.7	93.21	107.55	239	Peninsular Malaysia's Department of Agriculture

Anon (2004)

Table 8.35 Fertilizer rates recommended for coconut in Indonesia

		Fertilizer rate (kg ha^{-1})			
Crop	Growth stage	N	P_2O_5	K_2O	Kieserite
Hybrid coconut	Mature	150	100	105	190
	Immature	65	70	155	125
Tall variety	Mature	75	50	00	95

Anon (2005)

Table 8.36 Rate of fertilizer application of 15-5-20 NPK formulation palm^{-1} year^{-1} followed in Trinidad and Tobago

Age of palm	Application (kg palm^{-1} year^{-1})	Period of application
Adult >4 years	2.00	June and December
1 year and less	0.20	At planting and 6 months later
2 years	0.60	June and December
3 years	1.20	June and December

3 to 4 months intervals to maintain soil fertility (Donselman 1980). The granular fertilizer contained primary plant nutrient sources such as potassium magnesium sulphate, ammonium sulphate, granular sludge, urea form, potassium sulphate and diammonium phosphate and secondary plant nutrient sources such as magnesium sulphate, manganese oxide, borate, iron sulphate and zinc sulphate. The micronutrient foliar spray FER-A-GRO consisted of 17.46 g l^{-1}, tribasic copper of 1.19 g l^{-1}, urea of 5.82 g l^{-1} and a spreader-sticker.

Rosa et al. (2011) developed a new concept for recommendation of fertilizers for coconut in Brazil by considering the expected productivity and plant nutrient use efficiency to estimate nutrient demand and effective rooting layer, soil nutrient availability, as well as any other nutrient input to estimate the nutrient supply and developed a "lime and fertilizer recommendation system" for coconut crop based on the nutritional balance.

Diagnosis and Recommendation Integrated System (DRIS) provides a means of simultaneous identifying imbalances, deficiencies and excesses in crop nutrients and ranking them in order of importance in which leaf analysis values are interpreted on the basis of interrelationship among nutrients, rather than nutrient concentration themselves (Beaufils 1973). The DRIS is based on the comparison of crop nutrient ratios with optimum values from a high–yielding group (DRIS norms). The major advantage of this approach lies in its ability to minimize the effect of tissue age on diagnosis, thus enabling one to sample over a wider range of tissue age than permissible under the conventional critical value approach.

Khan et al. (1988) proposed preliminary N, P and K foliar DRIS norms for coconuts growing in sandy loam soil following the general procedure outlined by Sumner (1982). They indicated that based on the norms developed, nutrient application can be tailored to the optimum needs of production. Mathewkutty et al. (1998) developed DRIS norms for N, P, K, Ca, Mg, S, Cl, Fe, Mn and Zn in middle–aged (30–40 years old) coconut West Coast Tall palms in Kerala based on their ratios in the diagnostic 14th leaf. DRIS successfully diagnosed the deficiency of K in the palms, and accordingly, response in terms of nut yield was obtained when palms with low K index were fertilized with K. But the method failed to diagnose N and P deficiencies. There is ample scope to use DRIS as a diagnostic tool for coconut fertilizer application and refine and adopt Differential Fertilizer Recommendation (DFR).

In Sri Lanka, Jayasekara (1993) developed DFR programme. DFR is based on the leaf and soil analysis nutrient levels, present/potential/target yield and resources of the plantation. The DFR computer model is capable to adjust "sufficiency" ranges and nutrient removal factors for a range of plant, soil, agroclimate and management conditions. The model provides with the required fertilizer for individual field/estate to achieve sustainable high productivity with increased profits.

Biddappa et al. (1984) have developed system models for the integrated nutrient management in coconut-based cropping systems. The basic principle in employing such a model is that the systems are being enriched and depleted of nutrients simultaneously through different processes. This model has been used to evaluate the nutrient budget and balance in different systems, viz. the coconut-based high-density multispecies cropping system, the coconut-cacao mixed cropping system and the coconut-grass mixed farming system. The studies indicated that the nutrient budget and balance of nitrogen and magnesium progressively decreased, while those of P and K increased, indicating the build-up of P and K in the system, while N and Mg got depleted. Such studies help to rationalize the application of fertilizers to coconut and coconut-based cropping systems.

Saldanha et al. (2017) evaluated the nutritional status and established nutritional standards for cultivation of hybrid coconut in Brazil with 134 observations which formed the basis and analysed the nutritional status of the palms. They found most common deficiency as K and possibly excess of Mg. The order of limitations was K > P > Ca > Fe > N > B > Zn > Cu > Mn and Mg. Interpretation of the data with DRIS indicated large divergence for nutrients, viz. Ca, Mg, Cu, Fe, Mn and Zn, when compared to those adopted as regional reference by means of critical levels showing the need for revalidation through experimental work.

8.10.4 Period of Fertilizer Application

Chemical fertilizer should be applied either after heavy rains have passed, or early in the rainy season, preferably some months before the heavy rains will fall (Ohler 1999). Studies conducted in the coconut growing countries under varying soil and climate proposed time/period and frequency of application of fertilizers taking into account better utilization of nutrients by the crop, prevent loss of applied fertilizer to a great extent by leaching and gaseous means. The growth demand for the nutrient and water exists continuously especially with adult palms, and nutrient availability should be ensured adequately for desirable productivity. By virtue of the inherent characteristics of fertilizer and its interaction in the soil environment, availability in the nutrient pool varies with time. Period of fertilizer application in India is linked with the south west monsoon in the west coast and north east monsoon and local irrigation facilities on the east coast and inland areas.

Studies carried out in India and Sri Lanka reported that the increased availability of N and K due to fertilizer application to coconut do not last long under the prevailing agroclimatic conditions and to ensure continuous availability of these nutrients and their use efficiency, N and K fertilizers should be applied in small doses frequently rather than in larger doses (Markose and Nelliat 1975). The inherent characteristics of some of the coconut growing soils in Sri Lanka suggest that the most efficient method of fertilizer application would be to apply small quantities frequently (Sathirasegaram et al. 1966; Kamala Devi et al. 1973). It was also indicated the need to apply slow release nitrogenous fertilizers for efficient utilization by the crop under the heavy rainfall conditions of Kerala, India.

Better growth as well as greener foliage with split application of N and Mg was observed for coconut palms growing in sandy soils, during the first 2 years (Coomans 1977). At an early age of palms, such practice makes it possible to limit losses through leaching by rain in soils with a low power of fixation and in the presence of a root system yet undeveloped. In the conditions of the lower Côte d'Ivoire, the

Table 8.37 Fertilizer recommendation for coconut palms in India (g palm^{-1})

	May–June			September –October		
Age of palm	N	P_2O_5	K_2O	N	P_2O_5	K_2O
1st year	Planting in May–June			50	40	120
2nd year	50	40	125	110	80	275
3rd year	110	80	250	220	160	550
4th year onwards	170	120	400	330	200	800

period from July to September is the most appropriate period as it makes it possible to avoid the excessive leaching in the long duration of the rainy season (May to June), at the same time profiting from the short rainy season for better availability in the nutrient pool. For effective assimilation, fertilizer application is advised during a humid period and avoiding heavy rain, to limit element losses through leaching. When the young palms are in their first and second years, the annual manuring may be given in 2 instalments; and afterwards, there is only one time application, every year. In India and Sri Lanka, the adult palm dosage is partitioned and applied according to the age and growth demand of seedlings as discussed earlier in this chapter (Table 8.37).

Frequency of fertilizer application recommended in Sri Lanka is as follows:

(a) *Young palms*: At least half yearly application is recommended. Wherever possible, split application may be adopted particularly in areas with sandy soils to minimize leaching of fertilizer due to heavy rainfall during monsoon months.
(b) *Bearing palms*: Fertilizer application should be made annually. In areas where heavy rainfall is likely during both the monsoons, and for sandy soils, half yearly application with each monsoon may be adopted to minimize leaching losses.

In West Africa, where there are two rainy seasons (April and August), young coconuts are fertilized just before or at the very beginning of each rainy season. Older palms are only manured in August before the short rainy season (Ouvrier and de Taffin 1985).

The application of K fertilizer decreased the quantity of exchangeable Mg and vice versa (Giritharan et al. 2002). Instead of using 15-5-20 fertilizer formulation, urea, triple superphosphate (TSP) and muriate of potash can be blended in the ratio (by weight) 3:1:3 (urea/TSP/MOP). The Jamaica Coconut Industry Board (Anon 2017) recommends that fertilizer should not be placed in the planting hole because the planted seedling is still receiving its nutrition from the endosperm. The first application of fertilizer should be made 3–6 months after planting by which time active feeding roots will have developed. Subsequent applications should be made at 6-monthly intervals. The fertilizer must be spread over the zone where the active root tips occur. This zone stretches from the base of the palm to the limit of the leaf spread. During the first year, fertilizer should be applied in a circle 15–20 cm from the seedling stem. Subsequent applications should be made in wider circles as the roots grow outwards. For bearing trees, fertilizer should be spread about 2 m from the base of the trunk. On sloping lands, the fertilizer should be incorporated to a depth of 15 cm at various points at the same radius as described previously.

The frequent application of small amounts of N and K, keeping the nutrients within reach of the roots, is one of the benefits of fertigation. Teixeira et al. (2005b) concluded that higher doses 240 kg ha^{-1} of N as ammonium nitrate caused acidification of the soil, reducing saturation.

Ohler (1999) summarized the effect of fertilizer application and the transformation that takes place in the physiology of palms leading improvement in yield. The time required for a yield response to fertilizer application is considerable. Fertilizer application results in a slight increase in yield in the second year followed by a higher yield in the third year, and full response to fertilizer will be obtained from the fourth year onwards.

8.10.5 *Organic Manuring in Coconut*

Organic manures serve not only as a source of plant nutrients but also in restoring soil fertility by increasing chemical, physical and biological properties of soil through soil aggregation. The mineral composition depends on the type of organic manure and has been discussed in detail (Mantiquilla et al. 1994). Where the usufructs of coconut palms are added to the soil as organic manure, considerable amounts of nutrients of the palms are returned to the soil. In newly planted coconut gardens, the fertility status of interspaces can be improved by sowing green manures like *Crotalaria juncea* (sun hemp) and ploughing in situ at its flowering and developing a base in organic farming system.

According to the 2009 statistics report by IFOAM on world organic agriculture, coconut has been listed among the key crops organically cultivated worldwide and suggests that the importance of understanding the prospects of organic coconut cultivation. At least 28,000 hectares of coconuts are grown organically (including conversion areas). However, for some of the world's leading coconut producers (Indonesia, India and Sri Lanka), no details are available, and it can therefore be assumed that the actual coconut area is higher (Willer and Kilcher 2009).

Nirukshan et al. (2016a) compared the soil and plant nutritional status between organically and conventionally cultivated coconut in Sri Lanka. The organically cultivated system had a better soil nutrient status than those receiving inorganic nutrition. Information on the yield of palms was not indicated.

8.10.5.1 Effect on Soil Physico–Chemical and Microbial Properties

The physical and chemical properties of soils depend upon the nature of their formation, and the biological properties are largely governed by the environment which contributes to organic matter status of soil. Coconut is also governed by the physical and chemical properties and the biological environment which influence the mineral nutrition. Generally the organic matter is subject to mineralization and immobilization turnover (MIT) under the influence of microbial population of the soil before release of the nutrients and mostly governed by the complexity of the organic matter available/added to the soil.

Large areas of sandy soils are cultivated to coconut that are poor in basic nutrients, and the nutrient-supplying capacity of these soils is improved by addition of tank silt once a year to improve the soil texture and CEC and enhance water-holding capacity. The organic matter content of these soils can be further improved by addition of around 50 kg green leaf manure palm^{-1}.

Coconut husk piled at the base of the coconut palm as mulch or chopped and composted is found to improve the water-holding capacity of soil by its physical presence and improve the organic matter content on degradation, besides being a rich source of chlorine (40.7%) and potassium (66.9%) (Eroy 1991). Joshi et al. (1982) recorded increase in water-holding capacity of sandy soil and decrease in the bulk density due to continuous application of organic sources mostly coconut usufructs blended with inorganic fertilizers.

Considering the characteristics of husk, it is a practice to bury the husk with convex face upwards around 2 m away from the base to improve water retention characteristics of soil and reduce weed growth. Such a practice in the sandy loam soil improved the yield of East Coast Tall (ECT) palms by 49% over control treatment. As moisture conservation measures for the palm roots to retain and utilize the moisture, it can also be buried in trenches of 3 m × 1.2 m and 50 cm deep and covered with soil (Grimwood 1975).

Burying husk in the surface soil is also found to lower the bulk density of soil (Cadigal and Magat 1977). Husk was found to be beneficial to increase the yield of ECT palms maintained under inorganic fertilizer. Mulching coconut husk around the base of the palm at 2 m distance improved the yield by 49.6% over control. Coconut husk can conserve moisture and store water nearly three times its weight, and a thousand husks can yield 7–8 kg potash, and this practice raised yield by 44.6% over the control (Balasubramanian et al. (1985).

The use of green manures like *Eucheuma spinosum* (a seaweed) and *Leucaena leucocephala* contributed soil organic matter and improved physical properties, besides functioning as source of K and N, and improved the growth and functioning of seedlings especially that of MAWA hybrid (Cadigal and Prudente 1983; Cadigal et al. 1983).

Mahindapala (1989) obtained 100% survival rate of 2-month-old poly bag seedlings transplanted in sandy soil amended with *Gliricidia sepium.* A sandy soil environment responded well on receiving goat dung and inorganic fertilizers when applied together, supporting adult coconut palm with 42% increase in yield and 45% in copra outturn when continuously applied for 5 years.

In coralline soils of Lakshadweep Islands, fertility status of coconut groves was improved with the incorporation of coir dust, coconut sheddings, available forest leaves and cattle manure along with inorganic fertilizers carrying N, P and K continuously over 10 years, reduced the mortality of transplanted seedlings from 50 to 17% and improved the organic matter status from 0.06 to 0.17% (Bavappa 1986).

Table 8.38 Organic manuring schedule for bearing adult coconut palms in India

Mixture	Manures	Quantity (kg)
Mixture 1	F Y M (or) compost	35–70
Mixture 2	Fish guano	5–7
	Wood ash (or)	18–20
	Coconut husk ash	2.5
Mixture 3	Prawn dust	5–7
	Wood ash (or)	18–20
	Coconut husk ash	2.5
Mixture 4	Ground oil cake	5–7
	Wood ash (or)	18–20
	Coconut husk ash	2.5
	Bone meal	1–2

For these coral soils, Krishnamoorthy (1985) advised incorporation of green manures or compost 1.2 m away around the base of the adult palms for conservation of moisture and favourable utilization by the palm roots.

In India, organic manure mixture which is exclusively made up of organic sources is recommended for tall variety palms in coarse-textured soils which are deficient in organic matter (Thampan 1982) (Table 8.38).

Long-term applications of inorganic fertilizers were found to be detrimental to the beneficial soil microorganisms such as *Pseudomonas*, *Azotobacter* and *Bacillus* as well as soil physical parameters (Pushpakumari et al. 2008). They, however, observed significant improvement in parameters like available nutrients in soil and improved nutrient levels in the palm with the application of either goat dung (25 kg) and MOP (800 g) palm^{-1} or poultry manure (30 kg) and MOP (250 g) palm^{-1}. The microbial biomass was much influenced by application of poultry manure (Kondagama et al. 2009). Silva et al. (2008) also reported higher levels of macronutrients, favourable soil physical properties and higher microbial biomass with the addition of soil amendments than the inorganic fertilizer application.

Intercropping coconut with *Gliricidia sepium* is an effective strategy to improve soil chemical, physical and biological properties, viz. soil microbial activity, bulk density, organic matter, total nitrogen, available phosphorus and exchangeable potassium and magnesium dynamics of coconut growing, reaffirming the quality of *G. sepium* for replenishing soil fertility of degraded coconut growing soils in intermediate and dry zone of Sri Lanka (Secretaria and Maravilla 1997).

Long-term nutrient application studies in coconut palms (red-yellow podzolic soil classified as Andigama series; moderately acidic, rich in organic carbon) with cattle manure were found to increase the Fe and Mn levels, while poultry manure increased the Cu and Zn level in soil as evidenced in the contents of coconut diagnostic leaf (Chathurangani et al. 2010). The content of micronutrients in leaf was similar to the reports of the sufficiency for Mn, Cu and B. However, Zn levels in 14th leaf in all the treatments were below the sufficiency levels reported elsewhere.

Udayangani et al. (2013) reported that only goat manure and poultry manure contributed to sufficient level of P, while sufficient level of K residuals was recorded in cattle manure treatment. Addition of organic manure increased soil organic carbon and improved the microbial population and activity. However, the sulphate content added with sulphate of potash (SOP) had an inhibitory effect more on microbial activity than on microbial population (Nirukshan et al. 2016b).

In a *Trichoderma*-activated compost study, the application of city garbage plus swine manure produced significantly taller seedlings with wider girth, higher leaf count and dry matter accumulation and significantly higher N uptake at the rate of 606 g seedling^{-1}. The compost could also be mixed with ammonium sulphate at 75:25 ratio and capable of substituting the latter wholly for coconut seedlings (Ebuna and Cagmat 1992).

Green manure like *G. sepium* helped the establishment of 2-month-old poly bag seedlings at 100% survival rate. Palms grown in sandy soils with application of inorganic fertilizer and extra goat dung increased production of nuts by 42%, while copra production increased by 45% at the end of the fifth year (Mahindapala 1989).

Application of organic manure (either 18 or 24 kg goat dung) supplemented with inorganic fertilizers (NPK or K alone) in the gravel soils commonly found in coconut lands of the low country wet zone of Sri Lanka, increased the microbial counts and microbiologically mediated processes in the soil compared with the control resulting in increased yield and yield attributes (Tennakoon et al. 1995). Mantiquilla et al. (1994) reported consistent positive response in terms of growth and yield of coconut when organic fertilizers were combined with inorganic fertilizer high in chloride.

8.11 Soil and Nutritional Aspects Associated with Certain Disorders

The soil and other environmental factors exert considerable influence in the development, spread as well as intensity of plant diseases. The coconut palm is a very adaptable crop and has been grown under extreme conditions of soils. It is often difficult to specify the influence of a specific soil or nutritional factor associated with different diseases of the palm, mainly because of its perennial nature, the high heterogeneity among the field populations and also the highly heterogeneous soil environment under which it is distributed, particularly the subsoil environment. The influence of the extensive root system is still another factor as it is beyond the control of the experimenter to explore the root functions in the subsoil layers and also in places far beyond the basal region.

The different diseases of the palm, their causes and control have been described in greater detail in Chap. 10. Under ideal soil/ nutritional and other environmental conditions, the palm may be in a better position to offer resistance to pathogenic infections.

Most of the important diseases of the palm are generally found to occur on all soil types under which it is grown (Menon and Nair 1951). However, the diseases appear in a more acute form in areas with unfavourable soil such as poor aeration, waterlogging, poor moisture retentive capacity, high water table/shallow soil depth, poor drainage, etc. Menon (1961) suggested that many of the diseases cannot be primarily caused by unfavourable soil conditions or nutrient deficiencies, but they provide an environment conducive to infection by biological factors. If deficiency of one or more elements in the soil is responsible for the diseases, the incorporation of the deficient element(s) in an available form shall help to restore the palm into normal health or prevent fresh incidence of the disease. Menon and Pandalai (1960) have stated that certain diseases like bud rot and leaf rot are known to be purely parasitic; some like the bronze leaf wilt and "tapering stem" or "pencil point" condition are due to unsuitable soil conditions, while others like stem bleeding and fungal root rot are known to be associated with parasitic infection predisposed by soil conditions. Yet another set of disorders are caused by nutritional deficiencies/imbalances. Some of these problems are enumerated in the following sections.

8.11.1 Bronze Leaf Wilt

The term "bronze leaf wilt" described by Briton-Jones (1940) covered more than one diseased condition including lethal yellowing. However, when critically examined, a state of unhealthy condition of the palm without any abnormality in inflorescence, nuts and roots indicated to be purely of physiological origin (Child 1974). Potassium deficiency is reported to be associated with the disease. Briton-Jones (1940) from observations in Trinidad and St. Lucia suggested very strongly that the disease to be caused by a combination of adverse soil conditions, which include waterlogging, drought or/and impermeable soil strata, leads to shallow root system resulting into a condition of physiological drought. Analysis of the various physical and chemical factors of the soil throughout Trinidad suggested that unbalanced nutrition coupled with the physical conditions of the soil and its water relationship constitutes the primary cause of the disease (Bain 1937). Maramorosch (1964) concluded that the cause of bronze leaf wilt was purely physiological and related to soil and water conditions. Correction of deficiencies and adequate soil and nutritional management restore the palm to normal health (Briton-Jones 1940).

8.11.2 Crown Choke Disease

The occurrence of "crown choke" (crown rot) disease of coconut was reported in 1964 from Kahikuchi (Assam-India) by Chakrabarthy et al. (1970), and later studies were conducted by Brunin and Coomans (1973) in Côte d'Ivoire. Cecil and Pillai (1978) and Baranwal et al. (1989) in India have confirmed the possible role of boron deficiency in the development of the disorder. The "little leaf disease of coconut" described by Ashby (1917) and the "frond choke" disease reported by Dwyer (1937) seem to be similar to the "crown rot" disease reported. Fremond (1965) reported that a form of bud rot of coconut palms in the New Hebrides might be due to boron deficiency. A survey conducted in Assam, India (Anon 1990), showed that about 10.8% of the palms were affected by the disease and the total annual loss in yield due to the malady was then estimated at 6.38 million nuts.

Dufour and Quencez (1979) observed boron deficiency symptoms in 1-year-old coconut seedlings in solution culture. Only the first sign of deficiency was noticed which was the development of small chlorotic spots symmetrically oriented in relation to the main veins of the young leaves. The symptoms of the disease in 3–10-year-old palms have been described by Cecil and Pillai (1978) and Baranwal et al. (1989) and 18–20-month-old palms in Côte d'Ivoire by Brunin and Coomans (1973) which are more or less identical. The conspicuous symptoms of the disease are the fusion of terminal pinnae of young fronds, emergence of shorter fronds that crowd around the apex; development of deformed and crinkled pinnae; development of "hook" at the frond tips and also other parts of the frond; development of fronds with very short unfolded pinnae either on one or both the sides of the rachis with zigzag folding, necrosis on rachis and frond tips; and development of black necrotic stumpy frond without any pinnae in the advanced stage. Finally the growth of the bud is arrested, and the palm succumbs to deformity. The unaffected outer whorls of leaves remain normal throughout and even quite some time after the death of the growing point. Laminal expansion is very much restricted, and the affected pinnae become brittle and thicker than normal. The crowding of young abnormal fronds around the bud gives a choked appearance to the palm which might be the reason for the terming of the disorder as "frond choke". In some cases, the young affected fronds show "witch's broom" appearance. In other cases, the petiole of the new frond becomes very thick and forms a tubular structure enclosing the entire space of the apex.

Brunin and Coomans (1973) could prevent the occurrence of the symptoms on young palms in Côte d'Ivoire by application of borax pentahydrate ($Na_2B_4O_7{\cdot}5H_2O$) at 15 g palm^{-1} at planting and again once after 6 months. For the affected palms, IRHO has recommended the above dose of borax pentahydrate in April and October for 1–3-year-old palms and application of 30–80 g borax pentahydrate as a single dose for 4-year-old palms depending on the symptoms.

Chakrabarthy et al. (1970) made 9 applications of borax (sodium tetraborate), 20 g palm^{-1} in each application through soil. In each set, 3 consecutive applications of boron were given at an interval of 15 days, followed by the rest in 3 months. The field was irrigated after each application of boron. After the first set of application was over, treated plants showed signs of recovery, whereas in the control plants, the conditions remained unchanged. In all the treated plants, healthy leaves emerged after the application of third set of boron. Cecil and Pillai (1978) reported that soil application of borax as decahydrate at 250 g and 500 g palm^{-1} as a single dose followed by irrigation to 5-year-old palms was effective in curing the disease and the recovery was faster in palms treated with 500 g borax. The newly emerged leaves after treatment were normal in appearance.

In West Bengal, India, where the soil pH ranged from 4.9 to 5.0, highly deficient gardens showing symptoms of crown choking recorded 0.19 ppm B, and in healthy gardens the boron level was 0.22 ppm (Baranwal et al. 1989). They recommended soil application of borax at 50 g palm^{-1} just after the appearance of the symptoms. In slightly advanced stages, two applications of borax, 50 g each, at an interval of 3–4 months were found necessary for the redemption of the disorder. They reported a lower content of B in the leaf tissues of diseased palms compared to disease-free palms, both in Assam and West Bengal. Comparison of leaf B levels of samples collected from diseased areas with those of healthy area indicated that leaf B concentration is much less in diseased areas than in palms from healthy area. The Ca/B ratio (expressed in equivalent basis) is significantly lower for healthy palms when compared with diseased palms. It is interesting to note that the Ca/B ratio of palms in healthy area is much lower as compared to that of healthy palms in the affected area.

Cecil and Pillai (1978) reported 5.7 ppm B in affected palms and 9.2 ppm B in healthy palms. Margate et al. (1979a) suggested that the critical level of boron in hybrid coconut seedlings (frond 3) was likely to fall within the range of 13 to 14 ppm. They recommended the application of 1–1.5 g of borax seedling^{-1} in the nursery to prevent the occurrence of B deficiency symptoms. The dose of borax, however, suggested by them is extremely low. Rosenquist (1980) reported a mean foliar content of 10.5 ppm B, and he suggested that 9.5 ppm B in frond 14 was critical. The boron contents of coconut palms (frond 14) growing on different soil types of Kerala reported by Pillai et al. (1975) ranged from 10.9 to 14.6 ppm. They also reported that the available soil boron levels were low. Studies carried out in Kerala showed that boron deficiency causes reduction in coconut productivity and the deficiency could be cured by the application of 300 g and 500 g borax palm^{-1} year^{-1} in 2 split doses for seedlings and adult palms, respectively. The deficiency symptoms associated with B were completely recovered by borax application (Kamalakshiamma and Shanavas 2002).

Apart from N, P, K and Mg deficiencies in coconut so far observed in Sri Lanka, boron (B) deficiency was observed in 12 young coconut palms of age 1–3 years at Poojapitiya in Kandy District. The third leaf of the affected palms had 3.4–7.5 ppm B as compared to 7.6–10.0 ppm B for healthy palms in the same vicinity (Jayasekara and Loganathan 1988). The presence of healthy palms in the same vicinity is not uncommon. Ohler (1999) suggested that this may be due to local differences in soil, individual differences in sensitivity to boron deficiency or individual differences in the root system development, the palms with a more extensive root system being able to obtain a sufficient supply, while those with a smaller root system suffer from the deficiency.

8.11.3 Stem Tapering Disease

The "pencil point" or "tapering stem" disease is generally considered to be due to deficiency of nutrients/water and unfavourable soil conditions, and the malady has been referred to as "starvation of palm" (Briton-Jones 1940; Menon and Pandalai

1960). Ohler (1984) stated that stem tapering is not a disease by itself, but it is the result of various unfavourable growing conditions. The leaves become pale and diminished in size, photosynthetically inefficient, the trunk diameter becomes reduced, the inflorescence production is very much reduced and the inflorescences have very few female flowers which rarely develop into nuts. In course of time, the palm becomes barren, the crown size is progressively reduced and the trunk diameter is gradually reduced to a condition of pencil point until the palm finally dies. The malady is more prevalent in waterlogged and other infertile areas and also in areas with very hard substratum of laterite, coral or other formations which are basically unsuitable for coconut culture. Intensive manuring and management may provide reasonable growth and productivity in the early period which eventually exhausts the limited soil of its nutrient reserves. Multiple deficiencies including those of secondary and micronutrients are often associated with the malady (Menon and Pandalai 1960). In Kerala, the disease is commonly found in swampy or shallow soil situations which are ordinarily not suitable for coconut cultivation. In such areas planting is usually done on mounds, and the land is periodically raised with transported sand or soil which is an expensive practice in most of the areas. As long as the land reclamation process is continued along with generous manuring, the palms grow normal and yield satisfactorily. Subsequent neglect leads to tapering stem problems. Park and Fernando (1941) and Menon and Pandalai (1960) have suggested factors leading to tapering stem disease. The palms do recover when deficiencies are corrected through intensive manuring and soil conditions are improved. Savy (1962) reported that in severe cases the palms do not bear to their full capacity even after recovery. Recovery is possible with intensive soil and nutritional management for sustaining the palms to normal bearing.

8.11.4 Rubbery Copra

The problem of "rubbery copra" in coconut was reported to be due to acute deficiency of sulphur (Southern 1969). Extensive occurrence of sulphur deficiency was reported from Papua New Guinea (Southern 1969) and Madagascar (Ollagnier and Ochs 1972). Sumbak (1975, 1976) reported sulphur deficiency in Madang, which according to him was associated with poor drainage conditions. The main sulphur deficiency symptoms reported are orange-yellow leaves (both young and old) and weakening and arching of the rachis followed by necrosis of leaflets. Chlorosis and necrosis increase with the age of the fronds, and in severe cases, second or even the first leaf may show yellowing. Colouration may vary from greenish yellow to bright yellow and in some cases vivid orange. The number of live fronds becomes fewer. In the advanced stages, the crown loses most of the leaves, and severe necrosis is found on the older leaves. A sulphur-deficient palm may be distinguished from nitrogen-deficient palm in that the young leaves as well as old leaves are discoloured (Southern 1967).

The nuts are usually small with normal kernel thickness, but on drying, the kernel collapses into soft, flexible and leathery copra, often brown in colour, which is usually referred to as "rubbery copra" possessing very poor physical and chemical characteristics. The rubbery copra is characterized by low oil content (as low as 36% on dry basis) and high nitrogen, ash and sugar contents. The oil from rubbery copra has high iodine value and low saponification value compared to the oil from normal copra.

According to Southern (1969), deficiency symptoms on the foliage and incidence of rubbery copra were more associated with the sulphur content of nut water. The nut water of sulphur-deficient palms contained less than 10 ppm sulphate sulphur, while 20 ppm was suggested as critical even though values up to 100 ppm were observed. According to him, with 20 ppm sulphate sulphur in nut water and 150 ppm sulphate, sulphur deficiency is likely to occur in the palm. The deficiency condition can be corrected by the application of sulphur or sulphur-bearing fertilizers. He recommended sulphur-containing fertilizers equivalent to 900 g sulphur palm^{-1} once in 2 years. There was spectacular response on the foliage within 6 months, and the copra quality had become normal in 18 months after sulphur application.

The total sulphur content in frond 14 reported by Pillai et al. (1975) ranged from 0.113 to 0.152%, while Margate et al. (1979b) reported a mean value of 0.162%. The results presented by Manciot et al. (1980) showed that the values varied from 0.163 to 0.238% for talls and 0.175–0.445% for hybrids. They suggested a critical level of 0.15–0.20% sulphur in frond 14, while Magat (1979) suggested a critical level of 0.15%. While discussing the sulphur nutrition of coconut, Cecil and Pillai (1976) reported that S deficiency did not seem to be an immediate problem for coconut in the west coast of India, but continued application of sulphate-free fertilizers could eventually lead to S deficiency conditions. They recommended the inclusion of any one of the S-bearing fertilizers like ammonium sulphate, single superphosphate, ammophos, magnesium sulphate or sulphate of potash in the fertilizer schedule for coconut.

8.11.5 The Coconut Root (Wilt) Disease

The coconut root (wilt) had been considered as a disease of uncertain aetiology until Solomon et al. (1983) observed constant occurrence of phytoplasma in sieve tubes of roots, tender stem, petiole and developing leaf bases of root (wilt)-diseased palm as against their total absence in healthy palms from disease-free areas. The survey carried out by Pillai et al. (1973) indicated that the disease has been found to occur on all soil types of Kerala. However, the spread of the disease has been faster on light-textured sandy, sandy loam and alluvial soils, particularly in low-lying areas and also on heavy textured clays compared to laterite soils. The incidence has been higher in waterlogged low-lying areas adjacent to rivers and canals. Investigations on soil conditions and nutritional factors associated with the disease were initiated in 1939 by Menon and Nair (1949) and Menon et al. (1950,1952) suggested that in addition to biotic agents, the disease might also be associated with nutritional deficiencies. They reported that the soils of disease-affected areas were generally deficient in major nutrients, particularly K, and had a lower content of exchangeable cations and a low pH and cation exchange capacity. The silica/sesquioxide ratio was higher.

Soil sickness characterized by low pH, inadequate drainage, poor aeration, low microbial activity in the rhizosphere and nutrient imbalances together with mineral deficiencies, probably those of K, Ca and Mg were reported to have a predisposing decisive role on the incidence of the disease (Menon et al. 1952; Pandalai et al. 1958a, b; Menon 1961; Verghese 1961; Lal 1964; Cecil 1969). An intensive study of the major soil groups of erstwhile Travancore-Cochin State, India, representing healthy and diseased pockets was conducted by Sankarasubramoney et al. (1954, 1955, 1956) and Pandalai et al. (1958a, b, 1959a, 1959b). Their studies showed, in general, that soils in disease-affected areas were low in available K, total Ca and Fe, exchangeable Ca and Mg, total exchangeable cations, CEC, pH and percentage base saturation. Waterlogging was found to favour disease incidence in the tract, and majority of the diseased areas had a high water table. Cecil and Verghese (1959) observed that the reduction products formed in soils under waterlogged conditions were not responsible for disease incidence. Verghese (1966) indicated the association of faulty nutrient ratios in soils, particularly K/Mg, K/Ca and N/K with disease incidence. Pillai and Pushpadas (1965) observed that coconuts growing on *Kari* tracts (peat soil) having high acidity, often in the pH range of 3–4, had less incidence of disease.

Menon and Nair (1952) were the first to examine the major nutrient status of leaves in relation to the disease. Subsequent studies by Sankarasubramoney et al. (1952), Verghese et al. (1959a), Pillai (1959) and Pandalai (1959) showed that there was a tendency for N, P and K to get accumulated in the leaf tissues of diseased palms and the accumulation increased with the advancement of the disease. Verghese (1959) suggested that probably the mineralogical composition of rocks and some toxic products of weathering could be responsible for the disease and the possibility of water acting as their carrier. Compared to healthy, the diseased palms contained more of N, P, K and silica to the extent of 5.0–13.0, 0.0–13.0, 5.0–39.0 and 59.0–134.0%, respectively (Verghese et al. 1959b). Similar accumulation of nutrients in the leaves of Cadang-Cadang-affected palms was reported by Yualves et al. (1958). Biddappa and Khan (1985) studied the heavy metal status of coconut growing soils of Kerala and found that the contents of DTPA extractable barium, chromium, cadmium, lead, sulphur and vanadium were significantly high in diseased soils compared to healthy.

Further studies showed that there was no significant accumulation of nutrients in the palms in the early stage, while the concentration of N, P and K was significantly higher in the middle and advanced stages of the disease (Cecil 1981). The nutrient exhaust values reported by Pillai and Davis (1963) showed that about 50% of N and P and 78% of K are exhausted through the harvest of bunches with nuts which indicates that when the yield of nuts is restricted basically due to the disease, the excess nutrients are liable to get concentrated in the foliage of the diseased palms. So the accumulation of nutrients, particularly in the advanced stage of the disease, is partly due to the reduced rate of dry matter content of the foliage and partly due to the reduced rate of nut yield with increased rate of disease intensity.

Verghese et al. (1957, 1959b) ruled out the possibility of Cd and Sr toxicity in the disease complex. Cecil (1975) found that the N, P and K contents did not differ between healthy and diseased palms in the early stage of infection; but Ca and Mg contents of healthy palms in disease-free areas were significantly higher than those of apparently healthy or diseased palms in the diseased tracts. He also reported that the palms in the diseased area were in a state of imbalanced nutrition with wide ratios of N/Mg, P/Mg, K/Mg and Ca/Mg indicating a lower content of Mg in proportion to other major nutrients. Imbalance in cationic ratios like K/Na, K/Mg, K/(Ca + Mg) and K/(Ca + Mg + Na) and anionic ratios like P/S and N/S were also reported to be associated with the diseased conditions of the palm (Pillai et al. 1975). A critical evaluation of the earlier studies on the quality of nutrition in relation to the disease suggested that the palms in the disease-affected areas, whether apparently healthy or visibly diseased, are in a state of an unbalanced nutrition, possibly the result of a relatively higher content of N, P and K on the one hand and a lower content of Ca, Mg and S on the other.

Biddappa and Cecil (1984) and Biddappa (1985) studied the deposition of heavy metals in the root and cabbage tissues, respectively, of diseased palms by employing scanning electron X-ray microprobe analyser. High deposition of Al, Mn, Cu and Co in the diseased roots and Cr, Ti, Pb, Bi and Ga in the cabbage tissues of diseased palms was also observed compared to healthy tissues. This was also confirmed by the chemical analysis of a large number of soil and tissue samples under identical conditions. Wahid et al. (1983) studied the non-nutrient elemental composition of soil (0–30 cm) and plant tissues of healthy and root (wilt)-diseased palms from a few selected locations employing energy dispersive X-ray fluorescence technique and found that Ni and Sr were present at a higher concentration in the root of diseased palms compared to healthy.

The soil and nutritional aspects of the disease were reviewed by Cecil and Kamalakshi Amma (1991) who reported that neither the major nor micronutrients had any direct role on the incidence of the disease. While discussing the nutritional disturbances in relation to root (wilt) disease, Pandalai (1959) suggested that lack or unavailability of nutrients was not the cause of tissue abnormalities but was actually the inability of the palm to transact the normal processes at the appropriate site.

Reviewing the fertilizer demonstrations on coconut in the west coast of India, John and Jacob (1959) reported that in root (wilt)-affected areas, NPK applications along with the use of fungicides and insecticides markedly improved the health of the palms and increased the yield, possibly due to the improvement in the fertility levels of the neglected gardens. Sahasranaman et al. (1964) found that application of NPK at 227 g N and P_2O_5 and 454 g K_2O palm^{-1} year^{-1} gave economic yield and maintained general health of the palms, but higher doses of NPK, viz. 681–1362 N and P_2O_5 and 1362–2724 g K_2O palm^{-1} year^{-1}, aggravated the disease and reduced the yield. Cecil (1981) observed that increased levels of N, P and K had an adverse effect on the growth of young palms and on the yield of diseased palms. Application of lime and ash (Chettiar et al. 1959) and Chilean nitrate (John et al. 1959) showed no positive effect on the disease.

Davis and Pillai (1966) reported that the application of micronutrients and Mg did not prevent fresh incidence of disease. They observed that Mg had no significant influence on the yield of healthy palms, but it had decidedly a favourable response on moderately affected palms, while on severely affected palms, the effect was highly significant. Similar differential response of Mg on diseased palms was also reported by Varkey et al. (1979), Cecil (1981) and Anon (1981).

Khan et al. (1985b) did not observe any relationship between the micronutrient composition of diseased palms and the disease index compared to healthy palms. Zinc and molybdenum, both as soil application and foliar spray, had no effect on incidence or intensity of the disease, even though the tissue levels of Zn and Mo increased considerably (Mathew et al. 1986). A systematic micronutrient manurial experiment consisting of all combinations of two levels each of Fe, Mn, Cu, Zn, B and Mo since field planting had shown that the disease was not related to micronutrient nutrition of the palm (Anon 1986). Lal (1968) reported that the foliar yellowing associated with the disease might be largely due to Mg deficiency and the intensity of yellowing decreased markedly when diseased palms were sprayed with 2.0% solution of magnesium sulphate at quarterly intervals (Varkey et al. 1979; Anon 1966). Application of sulphur, calcium sulphate and magnesium sulphate along with NPK was found to increase the yield of affected palms (Lal 1964; Nair and Radha 1959). Lal (1964) reported reduction in foliar yellowing and increase in yield of diseased palms by applying NPK, lime and farmyard manure and spraying with Bordeaux mixture, micronutrients and magnesium sulphate. The results of a field fertility trial with three levels of NPK and two levels each of Ca and Mg on diseased palms showed that the lowest level of NPK tried, viz. 350 g N, 300 g P_2O_5 and 600 g K_2O along with 500 g MgO palm^{-1} year^{-1}, could be the economic dose for the management of the diseased palms (Anon 1981). The above observations suggest that the addition of NPK fertilizers without having a balance with the availability of secondary nutrients, particularly Mg and S, had an adverse effect on the diseased palms, while the inclusion of secondary nutrients showed beneficial effects. Based on the earlier observations, Cecil (1981) conducted a field study on the role of major nutrients in relation to disease incidence for a period of 12 years right from field planting in 1970. The following are the results emerged out of the study in relation to the disease.

1. The incidence of root (wilt) disease was not related to the major element nutrition of the palm. However, an imbalance between N and Ca, K and Ca and Mg was found to be associated with disease incidence. Heavy doses of NPK fertilizers had an adverse effect on growth as well as yield of diseased palms. A balanced supply of nutrients, particularly K, Ca and Mg, was found to be ideal for the management of root (wilt)-affected gardens.
2. The first level of N, P and K, viz. 500 g N, 300 g P_2O_5 and 1000 g K_2O palm^{-1} year^{-1}, was found to be adequate for the normal growth and productivity of WCT palms in the disease-affected areas. As the build-up of P and K in the

soils as well as in coconut palm tissues was high due to continued fertilization, their application has to be regulated based on foliar levels. The critical levels of 0.12% P and 0.8–1.0% K in frond 14 may be followed for regulating the P and K requirements of the palm.

3. Heavy liming is not necessary for the management of disease-affected plantations. Nevertheless, regulated additions of Ca through Ca-bearing fertilizers like rock phosphate/superphosphate or light additions of liming materials may be followed for supplying the Ca requirement of the palm. The concentration of 0.3% Ca in frond 14 may be considered as critical level for regulating the Ca requirement of the palm.
4. Regular Mg treatment since planting gave highly significant response on growth, onset of bearing and yield of young palms with simultaneous improvement in soil and foliar Mg levels. The pre-bearing age was reduced by 9.1 months, and the earliness of bearing was significantly correlated only with foliar Mg levels. Development of foliar yellowing was prevented in young palms, which was aggravated by higher levels of K, particularly in the absence of Mg. It increased the frond production rate and the number of functioning leaves of root (wilt)-affected palms at a highly significant level and improved the yield attributes and yield of affected palms more than those of healthy palms. When the increase in yield of nuts in healthy palms was 37%, the corresponding increase in root (wilt)-affected palms was 60%. Judicious manuring with Mg salts like magnesium sulphate from the time of field planting was highly essential for the successful growth and increased productivity of the palm in diseased area. An annual dose of 500 g MgO adult palm^{-1} was found to correct the Mg deficiency problems in the field. However, the dose may be regulated based on the foliar Mg levels as well as the Mg/K balance in the soil. The foliar content of 0.2% Mg in frond 14 and an exchangeable Mg/K ratio value of 2.0 in the soil may be considered as critical for regulating the Mg requirement of the palm.
5. The disease caused a heavy loss in the yield of nuts as well as copra outturn in the early bearing period. The general reduction in the yield of nuts was 60% and in the yield of copra was 64%. However, the reduction in the oil content of copra was not considerable. The mean values for the oil content of copra from healthy and root (wilt)-affected palms were 68.4 and 67.4%, respectively.
6. The application of P as ammophos (15% P) or that of magnesium sulphate (13% S) could effectively take care of the sulphur requirement of the palm.

The nutritional requirement of COD × WCT hybrids in the root (wilt)-affected area was investigated by Kamalakshiamma et al. (1982) who observed that the hybrid responded favourably to higher levels of NPK for its growth and early flowering. However, the dose of 500 g N, 300 g P_2O_5 and 1000 g K_2O along with 500 g MgO palm^{-1} year^{-1} was found to be ideal for optimum productivity under rainfed condition. A comparative study on the performance of WCT (Cecil 1981) and COD × WCT (Kamalakshiamma et al. 1982) under rainfed conditions and regular fertilization with N, P, K, Ca and Mg since field planting in the same locality showed that the hybrid was superior to WCT with respect to reduced disease incidence and increased nut yield.

Valiathan et al. (1992) reported the presence of high Ce and low Mg levels in the leaf may be responsible for the incidence of root (wilt) disease. They also reported a common geochemical basis for endomyocardial fibrosis in human beings and root (wilt) disease of coconut palms. Taking lead from these results, Wahid et al. (1998) studied the concentrations of major nutrients and micronutrients, rare earth elements (REEs) and Th and nutrient/REE ratios in the leaves of diseased, and apparently healthy coconut palms of the root (wilt) disease-affected tract and healthy palms of the disease-free tract, covering three major soil types of Kerala, viz. alluvial (entisols), laterite (ultisols), and sandy soils (entisols), were examined in relation to the disease. Accumulation of major nutrients, especially K, was generally observed in the leaves of diseased palms. Mg content of leaves of palms growing on laterite soil in disease-affected tract was lower than of palms in the disease-free tract. The leaf concentrations of La (lanthanum), Ce (cerium), Pr (praseodymium), Nd (neodymium), Sm (samarium), Gd (gadolinium) and Th (thorium) did not show significant differences between healthy and diseased palms. The only exception to this trend was Gd whose concentration was less in the diseased and apparently healthy palms growing on laterite soil of the disease endemic area than that in the healthy palms of the disease-free area. Some of the essential plant nutrients (EPN)/La and EPN/Ce ratios were significantly different in palms of the disease-affected tract compared to that in palms of the disease-free tract indicating imbalances in the relative concentrations of EPNs and REEs. These results call for more detailed study of the geochemical differences between the disease-affected and disease-free tracts for identification of the soil chemical factors associated with the incidence of the disease. Wahid et al. (2003) studied the pattern of uptake of 6 REEs from 3 types of soils varying widely in their chemical characteristics and monazite content through soil and tissue analysis and stated that Sm was the least accumulated lanthanide in coconut palm tissues and it was the most favoured element by palm on a comparable scale of substrate concentration.

Critical studies are needed to elaborate the beneficial role of Mg in the management of the disease, as the interaction between disease incidence and Mg deficiency on the productivity of the palm is negative. The root (wilt) disease is now known to be caused by phytoplasma (Solomon et al. 1983). However, a balanced and regulated supply of primary (NPK) and secondary nutrients (Ca, Mg and S) is necessary for the successful management of the disease. Further, the correction of unfavourable soil conditions like waterlogging, inadequate drainage and poor aeration shall help to check the deterioration of the palms.

8.12 Future Strategy

The coconut palm has emerged from the status of a back yard crop to that of a crop of commercial significance stressing the importance of mineral nutrition and sustainable productivity of coconut plantations. There is a need to revisit the agronomical interventions and modify them considering the site-soil-system as an entity. The following lines of investigations are suggested to ensure maximum coconut productivity on a sustainable basis.

1. A differential fertilizer system for fertiliser recommendations in an integrated manner involving plant and soil nutrient status.
2. Critical studies on boron and magnesium for inclusion as components of fertilizer recommendation to alleviate the ill effects of maladies associated with the deficiency of these nutrients.
3. Intensified research on INM involving organic recycling using the usufructs of coconut and component crops. Biochar initiatives in improving the carbon stock and reducing oxidation of organic matter for better nutrient use efficiency in coconut farming systems where sufficient biomass is generated from the plantations.
4. Developing an INM package for reducing the problems related to phytoplasmal diseases, in view of the absence of other control measures.
5. Developing desorption models for major coconut growing areas, to form a basis to prescribe potassium and phosphorus recommendation at a given soil test value.
6. A deeper understanding of the physiology of palms, especially on dry matter partitioning to design life-saving irrigation/fertigation in drought-affected areas.
7. Mandate to the national institutes for coconut palm nutrition survey, to prepare a status map in the major coconut growing tracts of respective countries to design a strategy for improving the yield levels and formulate yield prediction models.
8. Identification and study of the nutritional status of high-yielding palms in relation to soil nutritional status for designation of index gardens.
9. Further critical studies in nutrient budget and energy input-output analysis in high-density multiple cropping systems.
10. Refinement of precision farming technology in relation to coconut palms to ensure sustainable production.

References

Acharya GC, Dash DK (2006) Critical levels of micro-nutrients in coconut leaves in littoral sandy soil. J Plantn Crops 34(1):68–70

Anderson GD (1967) Increasing coconut yields and income on the sandy soils of the Tanganyika Coast. East African Agric For J 32:310–314

Andoh-Mensah E, Bonneau X, Nuertey BN, Dery SK (2003) Effect of mineral nutrition on nut yield and fruit composition of mature coconut palms in the coastal belt of Western Region of Ghana: preliminary studies. CORD 19(2):11–19

Anilkumar KS, Wahid PA (1989) Impact of long-term inorganic fertilization on soil nutrient availability and nutrition of coconut palm. Oleagineux 44:281–284
Anonymous (1966) Annual report for 1965. Central Coconut Research Station, Kayangulam, p 64
Anonymous (1967) Seventh Report of the Research Dept. 1966–67. The Coconut Industry Board, Jamaica
Anonymous (1970) Symptoms of nitrogen, iron and potassium deficiencies in the coconut palm. Fertilite 36:4–66
Anonymous (1981) Annual report for 1978. Central Plantation Crops Research Institute, Kasaragod, p 242
Anonymous(1985) Annual report for 1984. Central Plantation Crops Research Institute, Kasaragod, p211
Anonymous (1986) Annual report for 1985. Central Plantation Crops Research Institute, Kasaragod, p 212
Anonymous(1989) Annual report of the Coconut Research Institute, Lunuwila, Sri Lanka
Anonymous (1990) Coconut crown choke disease – a survey report. Central Plantation Crops Research Institute, Kasaragod, p 14
Anonymous (2004) Fertilizer use by crop in Malaysia. Food and Agriculture Organization of the United Nations, Rome
Anonymous (2005) Fertilizer use by crop in Indonesia. Food and Agriculture Organization of the United Nations, Rome, pp 15–39
Anonymous (2008) Fertilizer usage and land suitability for coconut cultivation. CRI, Lunuwila, pp 12–13
Anonymous(2010) Coconut nutrient management. http://www.Ikisan.com
Anonymous(2016) Coconut Research Institute of Sri Lanka. Planting and Management of Coconut Plantations. www.cri.gov.in. Accessed on 10 Mar 2017
Anonymous(2017) Field maintenance is crucial for healthy coconut production. www.guyana chronicle.com>Pepperpot>Lead stories. Accessed on 15 Jan 2018
Arzani K, Nahandi F, Amiri ME (2005) Performance of flowering and fruit set of coconut *(Cocos nucifera)* palms under the Kish Island environmental conditions of Iran. Indian J Agri Sci 75(11):731–734
Asara AF, Wijebandara DMDI, Gunathilake HAWS (2011) Phosphorus fixation capacities of some coconut growing soils of Sri Lanka. In: Proceedings of 11th agricultural research symposium, pp. 275–279
Ashby SF (1917) Leaf bitten disease of coconuts. Department of Agriculture (Jamaica) Leaflet
Bain FM (1937) Bronze leaf wilt disease of the coconut palm. Govt. Printing Office, I Trinidad
Balakrishna MTS (1975) Inorganic and organic sources of nitrogen and phosphorus fertilizers for coconut. Ceylon Cocon Qtly 26:104–107
Balasubramanian R, Ramanathan T, Vijayaraghavan H (1985) Certain aspects of moisture conservation in coconut gardens. Indian Cocon J 16(2):13–15
Baloch PA, Moizuddin M, Imam M, Abro BA, Lund JA, Solangi AH (2004) Effect of NPK fertilizers and farmyard manure on nut production of coconut (*Cocos nucifera* L.). Asian J Plant Sci 3(1):91–93
Baranwal VK, Manikandan P, Ray AK (1989) Crown choking disorder of coconut. A case of boron deficiency. J Plantn Crops 17(2):114–120
Barber SA (1962) A diffusion and mass flow concept of soil nutrient availability. Soil Sci 93:39–49
Barile CL, Azuceana Jr BA (1972) Trials with fertilizer. In: Emata RC (ed) Coconut Production. United Coconut Association of Philippines Weekly Bulletin, pp. 36–38
Barrent CI (1977) 17th Report of the Research Department. The Coconut Industry Board, Jamaica, pp 15–37
Baseden C, Southern PJ (1959) Evidence of potassium deficiency in coconut palms on coral-derived soils in New Ireland from analysis of nut waters, husks, fronds and soils. Papua New Guinea Agr J 11:101–115
Bastin B, Venugopal VK (1986) Available nutrient status of some red soils (alfisols) from different regions of Kerala. Agri Res J Kerala 24(1):22–29
Bavappa KVA (1986) Research at CPCRI, Kerala, India

Bavappa KVA, Kailasam C, Khader KBA, Biddappa CC, Khan HH, Kasturi Bai KV, Ramadasan A, Sundararaju P, Bopiah BM, Thomas GV, Misra LP, Balasimha D, Bhat NT, Shama Bhat K (1986) Coconut and arecanut based high density multispecies cropping systems. J Plantn Crops 14(2):74–87

Beaufils ER (1973) Diagnosis and Recommendation Integrated System (DRIS). Soil Sci. Bull. No. 1. University of Natal, South Africa, p 132

Biddappa CC (1985) Identification of heavy metal deposit in the cabbage tissue, of diseased coconut palms (*Cocos nucifera* L.) by using electron microprobe X-ray micro analyser. Curr Sci 54:679–682

Biddappa CC, Cecil SR (1984) Electron microprobe X-ray micro analysis of diseased coconut (*Cocos nucifera* L.). Plant Soil 79:445–447

Biddappa CC, Khan HH (1985) A note on the heavy metal status of coconut growing soils of Kerala. J Indian Soc Soil Sci 33:915–916

Biddappa CC, Khan HH, Joshi OP, Manikandan P, Bavappa KVA (1984) Integrated nutrient management in perennial based high density cropping through system approach. A theoretical consideration. In: Sethuraj et al. (eds) Proceedings of PLACROSYM VI, Oxford and IBH Publishing Co. Ltd. New Delhi, India, pp. 299–309

Biddappa CC, Khan HH, Cecil SR (1993) Potassium in coconut growing soils. In: Singh M et al (eds) Proceedings of symposium on potassium for plantation crops. Potash Research Institute of India, Gurgaon, Bangalore, pp 111–120

Bonneaux X, Ohs R, Kitu WT, Yuswohadi (1993) Chlorine, an essential element in the mineral nutrition of hybrid coconuts in Lampung (Indonesia). Oleagineux 48(4):179–189

Bonneaux X, Boutin D, Bourgoing R, Sugarianto J (1997) Sodium chloride, an ideal fertilizer for coconut palms in Indonesia. Plantations Recherche Develop 4(5):336–346

Bopaiah MG, Nambiar CKG, Biddappa CC (1998) Effect of slow release nitrogen and phosphorus fertilizer sources in coconut growing coastal sandy soil. In: Mathew NM, Jacob K (eds) Proceedings of PLACROYSM XII, developments in plantation crops research. Indian Society for Plantation Crops, Kasaragod, pp 182–187

Braconnier S (1988) Physiologie de la nutrition en chlorure chez Ie palmier it huile et Ie cocotier. Thesis. Universite des sciences et techniques du languedoc, Montpellier, p 207

Braconnier S, Zakra N, Weaver R, Ouvrier M (1992) Fertilizer nitrogen distribution in coconut hybrid PB 121. using ^{15}N. Oleagineux 47(2):63–69

Briones GR (1931) A study on the salt requirement of coconut seedlings grown in pots. Philippine Agric 20:352–361

Briton-Jones HR (1940) The diseases of the coconut palm. Bailliere, Tindall and Cox, London, p 176

Broschat TK (1997) Nutrient distribution, dynamics, and sampling in coconut and Canary Island date palms. J American Soc Hort Sci 122:884–890

Broschat TK (2005) Boron deficiency in palms. University of Florida, IFAS Extension

Broschat TK (2007a) Boron deficiency symptoms in palms. Palms 51:115–126

Broschat TK (2007b) Boron deficiency, phenoxy herbicides, stem bending, and branching in palms—is there a connection? Palms 51:161–163

Broschat TK (2009) Palm nutrition and fertilization. Hort Technol 19(4):690–694

Broschat TK (2011) Uptake and distribution of boron in coconut and paurotis palms. Hortic Sci 46(12):1683–1686

Broyer TC, Carlton AB, Johnson CM, Stout PR (1954) Chlorine: A micronutrient element for higher plants. Plant Physiol 29:526–532

Brunin C (1969) Les symptomes de carence en magnesium chez le cocotier. Oleagineux 24:679–680

Brunin C (1970) The magnesium nutrition of coconut plantations in Cote d'Ivoire. Oleagineux 25:269–274

Brunin C, Coomans P (1973) Boron deficiency in young coconuts in the Ivory Coast. Oleagineux 28(5):229–234

Cadigal VL, Magat SS (1977) Agronomic utilization of coconut husk at the DRC In: ARB-PCA Annual Report 1977. PCA, Diliman

Cadigal VL, Prudente RL (1983) Utilization of *Eucheuma spinosum* as a fertilizer for coconut seedlings. In: Proceedings of 14th Annual Scientific Meeting, Crop Science Society of the Philippines, IRRI, Los Baños, Laguna

Cadigal VL, Escoton AD, Prudente RL (1983) The potential of giant ipil-ipil and agrispon as fertilizers for coconut seedlings. In: Proceedings of 14th Annual Scientific Meeting, Crop Science Society of the Philippines, IRRI, Los Baños, Laguna

Carvalho RF d (1947) The Fertilizing of Coconut Palms. Bol Sec Agric Pernambuco 14:275–278

Cate RB Jr, LA Nelson (1965) A rapid method for correlation of soil test analysis with plant response data. North Carolina Agric Exptl Stn, International Soil Testing Series Bulletin No. 1

Cecil SR (1969) Nutritional aspects of the coconut palm in health and disease. M.Sc. Thesis, University of Kerala, Trivandrum, India

Cecil SR (1975) Mineral composition of coconut leaves in relation to root (wilt) disease. J Plantn Crops 3:34–37

Cecil SR (1981) Mineral nutrition of the coconut palm (*Cocos nucifera* L.) in health and disease with special emphasis on calcium and magnesium. Ph.D. Thesis, University of Kerala, Trivandrum, India

Cecil SR (1984) Effect of N, P, K, Ca and Mg on leaf nutrient content and yield of young coconut. In: Sethuraj MR et al (eds) Proceedings of PLACROSYM VI. Indian Society for Plantation Crops, Kasaragod, pp 331–351

Cecil SR (1988) Calcium and magnesium nutrition of coconut palm. In: Silas EG, Aravindakshan M, Jose AI (eds) Coconut breeding and management. Kerala Agricultural University, Thrissur, pp 34–35

Cecil SR, Verghese EJ (1959) Chemical studies on the leaf rot and root (Wilt) diseases of coconuts in Travancore- Cochin. III. Reduction products formed under waterlogged conditions-Preliminary investigations. Proceedings of first conference on coconut research workers in India, Trivandrum, India, pp 320–333

Cecil SR, Pillai NG (1976) Role of sulphur in coconut nutrition. Cocon Bull 7(3):1–3

Cecil SR, Pillai NG (1978) Role of boron in coconut nutrition. Indian Cocon J 9:1–3

Cecil SR, Kamalakshi Amma PG (1991) Soils and Nutrition. In: Coconut Root (wilt) Disease Monograph Series No. 3:45–52. Central Plantation Crops Research Institute, Kasaragod, India

Chakrabarthy BK, Nath BK, Dhar BK, Goswami RN (1970) Note on crown rot disease of coconut. Indian J Agric Sci 40:502–594

Chan E, Elevitch CR (2006) *Cocos nucifera L.* ver. 2.1. In: Rlevitch CR (ed) Profiles for Pacific Island agroforestry. Permanent Agricultural Resources, Hawaii

Charles AE (1968) Report on preliminary observations from a coconut variety trial. In: Proceedings of Third Session, FAO Technical Working Party. Coconut Production, Protection and Processing, Jakarta, Indonesia

Charles AE, Douglas LA (1965) Coconut experiment work in New Ireland. III. Progress of fertilizer trials 1958–1964. Papua New Guinea Agri J 17:76–86

Chathurangani MD, Wijebandara DMDI, Tennakoon NA, Weerakody WJSK (2010) Effect of organic manure and inorganic fertilizer on micronutrient status of coconut lands.In: Proceedings of 10th agricultural research symposium, pp. 175–179

Chettiar MR, Verghese El, Sankaranarayanan MP (1959) Chemical studies on the leaf and root (Wilt) diseases of coconuts in Travancore-Cochin. V. Effect of lime and ash on disease conditions. Proceedings of First Conference of Coconut Research Workers, Trivandrum, India

Chew PS (1978) Nutrition of coconuts-A review for formulating guidelines on fertilizer recommendations in Malaysia. Planter 54:115–141

Chew PS (1982) A fertilizer policy for coconuts on coastal clay soils in Peninsular Malaysia Proceedings of MARDI National Coconut Conference,1982

Chew PS, Kee KK, Ooi LH (1984) Management of coconuts and cocoa on acid sulphate soils. The Planter, Incorporated Society of Planters, Kuala Lumpur, 60(704): 483–498

Child R (1974) Coconuts, 2nd edn. Longman Group Ltd., London, p 335

Cooke FC (1930) Coconuts on peat. Malayan Agric J 18(2):587–595

Cooke FC (1936) The coconut industry of the Philippine Islands, Dept. Agric. S.S. and F.M.S. Bull. (Gen. Series) No. 23.1
Cooke FC (1950) The tapering disease of coconuts. Ceylon Cocon Qtly 1(1):17–21
Coomans P (1974) Variations saisonaires of the mineral composition Des feuilles de cocotier. Oleagineux 29:297–304
Coomans P (1977) First experimental results of hybrid coconut fertilization in the Ivory Coast. Oleagineux 32(4):164–166
Copeland EB (1931) The coconut. Macmillan & Co. Ltd., London, p 233
Cordova AS (1965) The soil chemical status and foliar nutrient composition of coconut tree in some Philippine coconut plantation. M.S. Thesis, University of the Philippines at Los Baños, Laguna, Philippines
Corrado F, Quencez P, Talliez B (1992) Boron deficiency in oil palm: symptoms and corrections. Oleagineux 47:719–725
Cosico WE, Fernandez NC (1983) Effect of some land qualities and soil properties on productivity of coconut in the Philippines. Philippine J Cocon Studies 8(1&2):25–30
Daniel C, Manciot R (1973) La nutrition en chiore desjeunes cocotiers aux Nouvefles-Hebrides. Oleagineux 28(2):71–72
Davis TA, Pillai NG (1966) Effect of magnesium and certain micronutrients on root (wilt) affected and healthy coconut palms in India. Oleagineux 21:669–674
Del Rosario MAS (1972) Sea water and coconuts. Mindanao Trib 3(1)
de Silva MAT (1966) Magnesium deficiency and nutrient imbalance in coconut. Ceylon Cocon Qtly 17:125–128
de Silva MAT (1973) Fertilizer Experiments and Coconut Yield. Ceylon Cocon Planters Rev 7:92–94
de Silva MAT, Anthony Pillai GM, Mathes DT (1985) The sulphur nutrition of coconut. COCOS 3:22–28
de Silva MAT, Abeywardena V, George GD, Balasubramanian K (1973) Nutritional studies on initial flowering of coconut (var. Typica). 1. Effect of Mg-deficiency and Mg- P relationship. Ceylon Cocon Qtly 24:107–113
de Silva NTMH (1981) Response of coconuts to fertilizer and advice to Sri Lankan growers: an aggregative approach. Ceylon Cocon Qtly 32:72–79
de Taffin G (1993) Le Cocotier (The coconut). Maisonneuve et Larose, Paris, p 199
Dhanapal R, Maheswarappa HP, Subramanian P (2000) Response of coconut roots to the methods of irrigation in littoral sandy soil. J Plantn Crops 28(3):208–211
Dickey RD (1977) Nutritional deficiencies of woody ornamental plants used in Florida landscapes. University of Florida Agric Expt Stn Bull 791
Dissanayake DMPD, Rajapaksha RMCP (2016) Controlling nitrate leaching losses from fertilized sandy Regosol using neem (*Azadirachta indica*). In: Fifth young scientists forum symposium national science and technology commission, Colombo, Sri Lanka, pp 1–4
Dissanayake DMPD, Kumari LMSS, Herath HMIK, Herath HMSK, Wijebandara DMDI, Nirukshan GS (2016) Effect of sodium chloride application on nitrification rates in coconut growing sandy regosol. In: Proceedings of the sixth symposium on plantation crop research "Plantation Agriculture towards National Prosperity", Volume 1. BMICH, Colombo, Sri Lanka, pp 185–194
Donselman HM (1980) Malayan dwarf coconut palm fertilization tests and recommendations. Proc Florida State Hort Soc 93:74–75
Dootson J, Klopeng K, Ngankoranatigarn P, Rattanapruk M (1986) Effects of fertilizer application on early vegetative growth and precocity of MYD x WAT hybrid coconut in Thailand. In: Proceedings of the international conference on cocoa and coconuts: progress and outlook. The Incorporated Society of Planters, Kuala Lumpur, Malaysia, pp 869–880
Dufour FO, Quencez P (1979) Study of trace bent nutrition of oil palm and coconut grown on nutrient solutions. Oleagineux 34(7):324–330
Dufour FO, Quyillec G, Olivin J, Benard JL (1984) Revelation of a calcium deficiency in coconut. Oleagineux 39(3):133–142
Dwivedi RS, Ray PK, Ninan S (1981) Studies on methods of inorganic nutrient application in coconut. Plant Soil 63:449–456

Dwyer REP (1937) The diseases of coconuts (*Cocos nucifera*) in New Guinea. New Guinea Agric Gaz 3:28–93

Ebuna RM, Cagmat RB (1992) Generation of *Trichoderma*-activated compost fertilizer: its decomposition and fertilizer efficiency on coconut seedlings. ASD Annual Report 1992. PCA-DRC, Davao City

Eckstein O, Bruno A, Turrentine JW (1937) Potash deficiency symptoms. Ackerbau. M. B. H, Berlin

Eden T, Gowert C, Salgado MLM (1963) A factorial fertiliser experiment on coconuts. Aust J Exp Agric 31:283–295

Elliott ML, Broschat TK, Uchida JY, Simone GW (2004) Compendium of ornamental palm diseases and disorders. American Phytopathol Soc Press, St. Paul

Ekanayaka EMRA, Nadheesha MKF, Weerakkody WJSK (2011) Potential buffering capacities of some coconut (*Cocos nucifera*) growing soils in relation to potassium in Sri Lanka. In: Proceedings of 11th agricultural research symposium, pp 337–341

Eroy MN (1991) Highlights 91. PCARRD-DOST, Los Baños, Laguna

Fairhurst T (2003) The oil palm leaf sampling. Review article published by Potash and Phosphate Institute, East and South East Asia Programs, 126 Watten Estate Road, Singapore

Felizardo BC (1982) Intercropping coconut with sweet corn and sweet potato. In: 4th PCRDF Professorial Chair Lecture, Department of Soil Science, University of the Philippines at Los Baños, Laguna, Philippines

Fernando KSKS (1999) Study of the microbiological and related physical and chemical properties of different coconut soils after amendment with organic manures. M.Phil Thesis, University of Kelaniya, Kelaniya, Sri Lanka

Ferreira NM, Holanda JS, Folegatti MV, Gheyi HR, Pereira WE, Cavalcante LF (2007) Qualidade de fruto de coqueiro-anão verde em função de nitrogênio e potássio na fertirrigação. R Bras Eng Agríc Amb 11:453–458

Foale MA (1965) Nutrition du jeune coctier dans les Ties Russell. Oleagineux 20(10):585–588

Foale MA (1968) Seedling growth pattern. Australian J Agric Res 19(5):781–789

Foale MA, Ashburner GR (2005) The coconut palm. In: Chopra VL, Peter KV (eds) Hand book of industrial crops. The Hayworth Press Inc, Binghamton, NY, pp 235–281

Fremond Y (1964) The contribution of IRHO to the study of mineral nutrition of the coconut palm. In: Proceedings of Second Session FAO Technical Working Party, Coconut Production, Protection, and Processing, Colombo, Sri Lanka

Fremond Y (1965) Nutrition minerale du cocotier. Oleagineux 20:89–95

Fremond Y (1966) Contribution of IRHO to the study of mineral nutrition of the coconut palm. In: FAO Technical Working Party on Coconut Production, Protection and Processing, Colombo, pp 97–107

Fremond Y, Ouvrier M (1971) Importance to the young coconut palm of suitable mineral nutrition from the time of field planting on a beach sand. Oleagineux 26:609–618

Fremond Y, Ziller R, De Lamothe MN (1966) The coconut palm. In: I.R.H.O. and International Potash Institute, Bern, Switzerland, pp 44–50

Friend D (1975) The Solomon Islands and Lever Pacific Plantations property Limited, Joint Coconut Research Scheme, In: Fertilizer recommendation based on soil and leaf analysis. Cocoinfo Intl 7(1):1971–74

Ganarajah T (1953) Planting coconut on old rubber lands. Ceylon Cocon Qtly 4(1):16–18

George V Thomas, Iyer R, Bopaiah BM (1991) Beneficial microbes in the nutrition of coconut. J Plantn Crops 19(2):127–138

Georgi CDV, Teik GL (1932) The removal of plant nutrients in coconut cultivation. Malayan Agric J 20:358–364

Ghosh DK, Maheswarappa HP (2016) Integrated nutrient management for higher productivity of hybrid coconut in West Bengal. SATSA Mukhapatra Ann Tech 20:145–153

Giritharan D, Somasiri LLW, Perera BSVJ (2000) Studies on seasonal fluctuation of exchangeable cations in a loamy sand soil of a coconut plantation. AGRIEAST 2:49–55

Giritharan D, Somasiri LLW, Perera BSVJ (2002) Effect of the methods of applying potassium and magnesium fertilizers on their interactions in the soil and uptake by coconut (*Cocos nucifera* L). COCOS 14:69–86

Goh HS, Sahak D (1988) The effects of nitrogen and potassium on the growth and foliar nutrient levels of immature MAWA coconut palms grown on an ultisol. MARDI Res J 16(1):63–72

Graham ER, Fox RL (1971) Tropical soil potassium as related to labile pool and calcium exchange equilibrium. Soil Sci 111:318–322

Grimwood BE (1975) Coconut palm products: their processing in developing countries. FAO-UN, Rome

Gunathilake HAJ, Manjula BAS (2006) Balanced fertilization for sustainable coconut and coconut-intercrop systems in Sri Lanka. In: Benbi DK, Brar MS, Bansal SK (eds) Balanced fertilization for sustaining crop productivity. IPI Proceedings. pp 415–425

Gunathilake HAJ, Herath HMSK, Nosov V (2008) Site-specific nutrient management in coconut plantations. e-ifc 17:1–4

Halliday DW, Sylvester JB (1954) Phosphorus fertilizer for plantation crops. Ceylon Cocon Qtly 5:141–163

Harries HC (1970) The Malayan dwarf supercedes the Jamaican Tall coconut. 1. Reputation and performance. Oleagineux 25:527–531

Hassan MM, El-Samnoudi IM (1993) Effect of soil salinity on yield and leaf mineral content of date palm trees. Egypt J Hortic 20:315–322

Herath HMIK, Ranasinghe K L, Tennakoon NA (2008) Use of soil and foliar nutrient levels to identify the nutritional limitations of two coconut growing soils.In: Nainanayake NPAD, Everard JMDT (eds) Proceedings of the 2nd Symposium on Plantation Crop Research - Export Competitiveness Through Quality Improvements, Coconut Research Institute, Lunuwila, Sri Lanka held in BMICH, Colombo, pp 350–356

Holanda JS, Neto MF, da Silva RA, Chagas MCM, Sobral LF, Gheyi HR (2007) Technologies for the intensive production of green dwarf coconut. Res Bull 34

IFA (International Fertilizer Industry Association) (1992) World fertilizer use manual. Paris, p 632

Issaka RN, Senayah JK, Andoh-Mensah E, Ennin SA (2012) Assessment of fertility status of soils supporting coconut (Cocus nucifera) cultivation in Western and Central Regions of Ghana. West African J Appl Eco 20(1):47–56

Jacob A, Coyle V (1927) Fertilizer requirements of tropical plants and soils. Booklet printed in Germany

Jacques C (1932) Manuring of coconut palms in New Caledonia. Rev Agric Nouvelle Caledonia, Nov–Dec 1932

Jayakody JADSS, Tennakoon NA, Fernandopulle MND (2007) Available soil phosphorus in coconut plantations in Boralu and Pallama soil series, under different fertilizer regimes in Gampaha district of Sri Lanka. COCOS 18:31–44

Jayasekara KS (1993) Differentia fertilizer recommendation for coconut based on nutrient and productivity level. In: Nair MK, Khan HH, Gopalasundaram P, Rao EVVB (eds) Advances in coconut research and development. Oxford & IBH, New Delhi, pp 395–405

Jayasekara KS, Loganathan P (1988) Boron deficiency in young coconut (*Cocos nucifera* L.) in Sri Lanka: symptoms and corrective measures. COCOS 6:31–37

Jayasekara KS, Jayasekara C, Periyathambi S (1991) Report of the soils and plant nutrition division. In: Annual Report of Coconut Research Institute, Lunuwila, Sri Lanka, pp 87–110

Jeganathan M (1990) Studies on potassium-magnesium interaction in coconut (*Cocos nucifera* L.). Plant Soil 124:265–271

Jeganathan M (1993) Coconut nutrition: experiences in Sri Lanka. In: Nair MK, Khan HH, Gopalasundaram P, Rao EVVB (eds) Advances in coconut research and development. Oxford & IBH Publishing Co. Pvt. Ltd., New Delhi, pp 291–311

Jeganathan M, Appuhamy PADGA, Mendis BJAF, Geoge GD (1977) Study on the annual exhaust of soil nutrients by the Typica x Pumila hybrid palm. Proc Sri Lanka Assoc Advan Sci 32(1):43

John CM, Jacob K (1959) Fertilizer demonstration on coconut in west Coast – a review. Proceedings of First Conference of Coconut Research Workers in India, Trivandrum, India

John PC, Chandy PV, Verghese U (1959) Chemical studies on the leaf and root (wilt) disease of coconuts in Travancore Cochin. VI. Chilean nitrate as a curative for diseased palms and as a fertility for coconut seedlings. In: Proceedings of First Conference of Coconut Research Workers in India, Trivandrum, India, 358–365

Joseph J, Wahid PA (1997) Dynamics of soil nutrient reserves in coconut rhizosphere as influenced by ling term inorganic fertilization. J Plantn Crops 25(1):44–51
Joshi OP, Nambiar CKB, Hameed Khan H (1982) Effect of organic manuring on some physical properties and water retention of coastal sand. Philippine J Cocon Studies 7(1 & 2):42–46
Kamala Devi CB, Velayutham M (1978) Changes in the chemical composition of nut water and kernel during development of coconut. In: Nelliat EV et al (eds) Proceedings of PLACROSYM I. Indian Society for Plantation Crops, Kasaragod, pp 340–346
Kamala Devi CB, Nelliat EV, Pillai NG (1973) Nutritional studies on high yielding coconut genotypes. J Plantn Crops 1:67–69
Kamala Devi CB, Wahid PA, Pillai NG, Sankaranarayanan MP (1975) Effect of NPK fertilizers on trace element nutrition of three coconut genotypes. J Plantn Crops 3(1):11–13
Kamala Devi CB, Velayutham M, Haridasan M (1983) Soil and leaf analysis in relation to the nutrition of high yielding coconut genotypes. Wiley Eastern Publication, New Delhi, pp 115–119
Kamalakshiamma PG, Shanavas M (2002) Boron deficiency in coconut-Symptoms and corrections. Indian Cocon J 32:1–5
Kamalakshiamma PG, Cecil SR, Pillai NG, Mathew AS, Nambiar PTN (1982) Performance of Dwarf x Tall hybrid coconut in root (wilt) affected areas of Kerala under different fertilizer schedules. In: Bavappa KVA et al (eds) Proceedings of PLACROSYM V. Indian Society for Plantation Crops, Kasaragod, pp 405–410
Kamalakshiamma PG, Shanavas M, Prema D, Mathew J (2005) Effect of borax application on young coconut palms. CORD 21(2):74–85
Kanapathy K (1971) Preliminary work on foliar analysis as a guide to the manuring of coconut Malaysia. In: Westre R, Earp E (eds) Cocoa and Coconuts in Malaysia. Incorporated Society of Planters, Kuala Lumpur, pp 357–366
Khan HH (1993) Fertilizer management in Coconut. In: Tandon HLS (ed) Fertilizer Management in Commercial Crops. Fertilizer Development and Consultation Organisation, New Delhi, pp 96–118
Khan HH, Bavappa KVA (1986) Management of P in plantation crops. In: Proceedings of World Phosphate Institute Second Regional Seminar- Crop Production Techniques and Fertilizer Management in Rainfed Agriculture in Southern Asia. Jan 22–25, New Delhi, India, pp 513–528
Khan HH, Sankaranarayanan MP, Narayanan BB (1978) Characteristics of coconut soils of India 1. Morphology, some physico-chemical characteristics and taxonomy. In: Nelliat EV et al (eds) Proceedings of PLACROSYM I. Indian Society for Plantation Crops, Kasaragod, pp 54–79
Khan HH, Joshi OP, Biddappa CC (1982) Adsorption-desorption and selective distribution of phosphorus and potassium in relation to soil fertility potential in major coconut growing soils of India. In: Ahamed Bavappa KV et al. (eds) Proceedings of PLACROSYM-V. Indian Society for Plantation Crops, Kasaragod, India, pp 411–424
Khan HH, Sankaranarayanan MP, George MV, Narayana KB (1983) Effect of phosphorus skipping on the yield and nutrition of coconut palm (*Cocos nucifera* L.). J Plantn Crops 11(2):129–134
Khan HH, Biddappa CC, Joshi OP, Cecil SR (1985a) Micronutrient distribution in the crown of healthy and root (wilt) diseased palms. J Plantn Crops 13:65–69
Khan HH, Sankaranarayanan MP, Joshi OP, George VM, Narayana KB (1985b) Comparative efficiency of selected phosphates as P-carriers for coconut. Trop Agric 62:57–61
Khan HH, Gopalasundaram P, Joshi OP, Nelliat EV (1986a) Effect of NPK fertilization on the mineral nutrition of coconut genotypes. Ferti Res 10:185–190
Khan HH, Gopalasundaram P, Joshi OP, Manikandan P, Gopalasundaram P (1986b) Fertilizer recommendation for coconut based on Mitscherlich-Bray equation. J Plantn Crops 14:65–73
Khan HH, Suresh KK, Biddappa CC (1988) Preliminary foliar DRIS norms for coconut. Madras Agric J 75:105–108
Khan HH, Biddappa CC, Nagarajan M, Thomas GV, Cecil SR (1990) Residual effect of long term phosphorous application on soil and plant nutrient contents, microbial association and yield of coconut. In: Nambiar KKN, Iyer RD, Rao EVVB, Krishnamurthy Rao W, Premkumar T, Hegde MR (eds) Proceedings of PLACROSYM IX. Indian Society for Plantation Crops, Kasaragod, pp 14–18

Khan HH, Biddappa CC, Nagarajan M, George T, Cecil SR (1992) Residual effect of long-term phosphorous reserves in soil and plant nutrient contents, microbial association and yield of coconut. J Plantn Crops 20(Suppl):14–18

Khan HH, Biddappa CC, Cecil SR (1994) Nutrition of coconut. In: Chadha KL, Rethinam P (eds) Advances in horticulture, vol 9. Malhotra Publishing House, New Delhi, pp 377–394

Khan HH, Upadhay AK, Palaniswami C (2000) Integrated nutrient management of plantation crops. In: Rethinam P et al (eds) Plantation crop research and development in the new millennium. Proceedings of PLACROSYM XIV. Coconut Development Board, Kochi, pp 9–22

Khanna PK, Nair PKR (1977) Evaluation of fertilizer practices for coconuts under pure and mixed cropping systems in the West Coast of India. Conference on classification and management of tropical soils. Malaysian Soc Soil Sci, Kuala Lumpur, Paper No 50

Kondagama MCK, Nadheesha MKF, Weerakkody WJSK, Tennakoon NA (2009) Long term effect of organic manure application to coconut measured by nutrients and microbial parameters. In: Proceedings of 9th agriculture research symposium, 11–12 August 2009. Wayamba University of Sri Lanka, Makandura, Gonawila, pp 166–171

Krishnan P, Naidu LGK, Shamsuddin VM, Raghu Mohan NG, Sehgal J, Velayutham M (1997) Soils of Lakshadweep: their kinds, distribution, characterization and interpretations for optimising land use. National Bureau of Soil Survey and Land Use Planning Regional Centre, Bangalore

Krishnan P, Nair KM, Naidu LGK, Srinivas S, Arti K, Nasre RA, Ramesh M, Gajbhiye KS (2004) Land, soil and land use of Lakshadweep Coral Islands. J Indian Soc Soil Sci 52(3):226–231

Krishnamoorthy B (1985) How to increase yield of coconuts in Lakshadweep. Indian Cocon J 16(6):14–15

Krishnasamy R, Shanmugham M, Poongothai S, Valliappan K, Mani S, Malliga V, Savithri P, Devarajan R (1994) Twenty five years of micronutrient research in soils and crops of Tamil Nadu.(1967–1992). Res Bull, Dept Soil Sci Agri Chem, Tamil Nadu Agricultural University, Coimbatore

Kulasingh KPN, Wijebandara DMDI, Nugawela A, Fernandopulle MND (2013) Performance of Eppawala Rock Phosphate with organic manure as a phosphate source for adult coconut palms in Dry Zone (DL3). In: Proceedings of 12th agricultural research symposium, pp 180–184

Kushwah BL, Nelliat EV, Markose VT, Sunny AF (1973) Rooting pattern of coconut. Indian J Agron 18:71–74

Lal SB (1964) Advances in research on the root (wilt) disease problems of Kerala. Proceedings of Second Session. FAO Technical Working Party, Coconut Production, Protection and Processing. Colombo, Sri Lanka

Lal SB (1968) Root (wilt) disease- resume of work done since 1964. In: Proceedings of Third Session FAO Technical Working Party, Coconut Production, Protection and Processing, Jakarta, Indonesia

Lima Filho OF, Malavolta E (1997) Sintomas de desordens nutricionais em estévia [*Stevia rebaudiana* (Bert.) Bertoni]. Sci Agric 54:53–61

Limbaga CM (1986) Correlation studies among soil and leaf nutrient levels and coconut yield. M.S thesis, University of the Philippines at Los Baños, Laguna, Philippines, p 289

Liyanage M de S (1999) A guide to scientific cultivation and management of coconut, Sri Lanka, Hitech Printers, Nugegoda, p 174

Lockard RG, Ballaux JC, Azucena B (1969) The results of pretreatment leaf and nut water analysis from ten coconut fertilizer experiments. Philippine Agric 53:276–288

Loganathan P (1977) Profitability of fertilizing young coconuts: results of three long-term experiments. Ceylon Cocon Qtly 28:68–72

Loganathan P (1978) Some recent developments in fertilizing coconut. Ceylon Cocon Qtly 29:87–96

Loganathan P, Atputharajah PP (1986) Effect of fertilizers on yield and leaf nutrient concentration in coconut. Trop Agric 63:143–148

Loganathan P, Balakrishnamurti TS (1975) Response of *Cocos nucifera* to N, P and K fertilizer application from the time of field planting on a lateritic gravel soil in Sri Lanka. Ceylon Cocon Qtly 26:89–98

Loganathan P, Balakrishnamurti TS (1979) Effect of NPK fertilizers on the yield and leaf nutrient concentration of adult coconut on a lateritic gravelly soil in Sri Lanka. Ceylon Cocon Qtly 30:81–90

Loganathan P, Balakrishnamurti TS (1980) Effect of NPK fertilizers on young coconut (*Cocos nucifera*) on a sandy soil in Srilanka. Expt Agri 1:41–48

Loganathan P, Balakrishnamurti TS (1983) Fertilizer requirements of adult coconut palms in Sri Lanka. In: Nayar NM (ed) Coconut research and development. Wiley Eastern Limited, New Delhi, pp 127–135

Loganathan P, Nalliah V (1977) Downward movement and transformation of phosphorus in soils after long-continued fertilizer application to coconut (*Cocos nucifera*). J Agri Sci Cambridge 89:279–284

Loganathan P, Dayaratne PMN, Shanmuganathan RT (1984) Evaluation of the Phosphorus status of some coconut growing soils of Sri Lanka. COCOS 2:29–43

Macêdo JLV, Gasparoto L, Cunha RNV (1999) Boron deficiency in coconut: symptoms and correction. Manaus EMBRAPA-CPAA 1999:1–3

Magat SS (1976) Soil and leaf analyses in relation to coconut yield. Philippine J Cocon Studies 1(2):1–9

Magat SS (1978) A review of fertilizer studies on coconut. Philippine J Cocon Studies 3(2):51–72

Magat SS (1979) The use of leaf analysis in the coconut field fertilizer trials in the Philippines. Philippine J Cocon Studies 4(1):32–39

Magat SS (1988) Fertilizer recommendation for coconut based on soil and leaf analyses. Philippine J Cocon Studies 16(2):25–30

Magat SS (1991) Coconut nutrition-experiences in the Philippines. In: Nair MK et al (eds) Advances in coconut research and development. Oxford and IBH, New Delhi, pp 277–298

Magat SS (1992) Coconut In: IFA World Fertilizer Use Manual. International Fertilizer Industry Association, Paris, pp 234–244

Magat SS (2000) The use of leaf analysis in the conduct of field trials in the Philippines. Philippine J Cocon Studies 4(1):32–39

Magat SS (2005) Coconut. In: World Fertilizer Use manual. Intl Fertilizer Industry Association, 2005. http://www.fertilizer.org. Accessed on 10 Mar 2017

Magat SS (2009) Coconut yield and profitability under two practical fertilizer options: common salt (sodium chloride) and multinutrient 14N-0P2O5-20K2O-15Cl-4.5S-0.02B applications on bearing trees *(technology notes)* Philippine J Cocon Studies 35(1&2): 18–32

Magat SS, Margate RZ (1990) The chlorine needs of coconuts. Cocon Res Dev J 7(1):30–51

Magat SS, Pardones GD (1984) Initial response of young bearing palms to various sources of chlorine. Philippine J Cocon Studies 12:21–29

Magat SS, Prudente RL (1975) Progress of mineral nutrition studies on coconut in the Philippines. FAO Technical Working Party on Coconut, Kingston, Jamaica

Magat SS, Cadigal VL, Habana JA (1975) Yield improvement of coconut in elevated inland area of Davao by KCI fertilization. Oleagineux 30(10):413–418

Magat SS, Margate RZ, Prudente RL (1977) Utilization of common salt (sodium chloride) as a fertilizer and for the control of leaf spot disease of coconut seedlings. Philippine J Cocon Studies 2(3):39–45

Magat SS, Maravilla JN, Padrones GD (1979) Increasing productivity of Inland coconuts by chlorine and sulfur fertilization. Philippine J Crop Sci 4(4):177–182

Magat SS, Habana JA, Escoton AG, Labarcon AA (1981a) Studies on the nutritional status of existing coconut stands in the Philippines. I. Leaf survey. In: Agricultural research report 1981. Philippine Coconut Authority Manila, pp 16–22

Magat SS, Maravilla JN, Padrones CD (1981b) Increasing productivity of inland coconuts by nitrogen, chlorine, and sulfur fertilization. Philippine J Cocon Studies 6(1):46–53

Magat SS, Margate RZ, Habana JA (1986) Sodium chloride (common salt) fertilization of bearing coconuts. 1: early yield response. Philippine J Cocon Studies 11(1):37–45

Magat SS, Margate RZ, Habana JA (1988) Effects of increasing rates of sodium chloride (common salt) fertilization on coconut palms grown under an inland soil (Tropudalfs), of Mindanao, Philippines. Oleagineux 43(1):13–19

Magat SS, Padrones GD, Habana JA, Alforja LM (1991) Residual effects of chloride fertilizers on yield of coconuts grown on an inland soil of Davao (Mindanao, Philippines). Philippine J Cocon Studies 16(1):1–8

Magat SS, Padrones GD, Alforja LM (1993) Residual effects of three chloride fertilizers on yield and leaf nutrient levels of coconuts grown on an inland soil of Davao (Mindanao, Philippines). Oleagineux 48:237–242

Magat SS, Secretaria MI, Mantiquilla JA, Margate RZ (2009) Integrated Soil Fertility Management (ISFM) on Coconut + Lanzones *(Lansium domesticum* Carr) Agro-ecosystem in Southern Mindanao, Philippines (1993–2007): with Emphasis on the Multi-nutrient Coconut Specific Mineral Fertilizer. Part I. Crop Productivity, Quality and Profitability. CORD 25(1):1–23

Mahindapala R (1989) Coconut Research Institute of Sri Lanka Report. CRB, Sri Lanka

Mahindapala R, Pinto JLJG (1991) Coconut cultivation. Coconut Research Institute, Lunuwila, p 162

Manciot R, Ollagnier M, Ochs R (1979a) Mineral nutrition and fertilization of the coconut around the world. Oleagineux 34:499–510

Manciot R, Ollagnier M, Ochs R (1979b) Mineral nutrition and fertilization of coconuts around the world. Oleagineux 34(11):511–515

Manciot R, Ollagnier M, Ochs R (1979c) Mineral nutrition and fertilization of the coconut around the world II – study of the different elements. Oleagineux 34(12):576–580

Manciot R, Ollagnier M, Ochs R (1980) Mineral nutrition and fertilization of the coconut around the world. Part III Study of the different elements. Oleagineux 35(1):23–27

Mandal RC (2000) Coconut production and protection technology. Published by Agrobios (India) p 168

Manikandan P, Joshi OP, Khan HH, Biddappa CC (1986) Available micro-nutrient status of the coconut- tract of Kasaragod district of Kerala. J Plantn Crops 14(2):134–146

Mantiquilla JA, Canja LH, Margate RZ, Magat SS (1994) The use of organic fertilizer in coconut (A research note). Philippine J Cocon Studies 19(1):8–13

Maramorosch K (1964) A survey of coconut diseases of unknown etiology. FAO, Rome, p 38

Margate RZ, Magat SS, Abad RG (1979a) Boron requirement of hybrid coconut seedlings grown in an inland coconut soil of Davao. Philippine J Cocon Studies 4(3):6–14

Margate RZ, Magat SS, Alforja IM, Habana JA (1979b) A long term KCI fertilization study of bearing coconuts in an in and-upland area of Davao (Philippines). Oleagineux 34(5):235–242

Marschner H (1995) Mineral nutrition of higher plants. Academic Press, New York

Markose VT, Nelliat EV (1975) Frequency of fertilizer application to bearing coconut palms-effect on yield and yield components. J Plantn Crops 3(1):16–19

Mathew AS, Jose AI, Abraham Jose, Cecil SR. Amma PGK (1986) Effect of zinc and molybdenum on root (Wilt) affected coconut palms. Agric Res J Kerala 24:169–174

Mathewkutty TI, Potty NN, Tajuddin E, Markose VT (1995) Intriguing influence of Mg and S in coconut productivity. In: Iyengar PK (ed) Proceedings of the VIIth Kerala science congress. Kerala State Committee on Science, Technology and Environment, Thiruvananthapuram

Mathewkutty TI, Potty NN, Tajuddin E (1997) Significance of non-applied elements in coconut productivity. COCOS 12:72–78

Mathewkutty TI, Wahid PA, Tajuddin E (1998) Diagnosis and Recommendation Integrated System (DRIS) in coconut palm (*Cocos nucifera* L.). 1: Development of the norms. J Plantn Crops 26:31–40

Mendoza AMR, Prudente RL (1972) Fertilization studies on coconut in the Philippines Coconut Research Institute. Proceedings of National Science and technology Week, Manila

Menon KPV (1961) Diseases of undetermined causes with special reference to root (Wilt) disease of South India. In: Proceedings of First Session FAO Technical Working Party Coconut Production, Protection and Processing. Trivandrum, India, pp 58–84

Menon KPV, Nair UK (1949) The root (wilt) disease of coconut in Travancore–Cochin. Indian Cocon J 3(5–10):40–44

Menon KPV, Nair UK (1951) Scheme for the investigation of the root and leaf disease of the coconut palm in South India. Consolidated final report of the work done from 8th March 1937 to 31st March 1948. Indian Cocon J 5:5–19

Menon KPV, Nair UK (1952) Scheme for the investigation of the root and leaf diseases of the coconut palm in South India. Consolidated final report of the work done. Indian Cocon J 5:81–100
Menon KPV, Pandalai KM (1960) The coconut palm. A Monograph. Indian Central Coconut Committee, Ernakulam
Menon KPV, Sankarasubramoney H, Pandalai KM (1950) Investigation on the diseases of the coconut palm in Travancore-Cochin States: studies on soil conditions in relation to disease incidence. Indian Cocon J 3(81–87):99–103
Menon KPV, Nair UK, Pandalai KM (1952) Influence of waterlogged soil conditions in some fungal parasites on the roots of the coconut palm. Indian Cocon J 3:71–79
Mialet-Serra I, Bonneau X (2004) Effects of chlorine and potassium mineral nutrition on the leaf morphology of the PB 121 hybrid coconut palm (Malayan yellow dwarf x West African tall). CORD 20(2):1–11
Mills HA, Jones JB Jr (1996) Plant analysis handbook II. A practical sampling, preparation, analysis, and interpretation guide. Macro-Micro Publishing, Athens, GA
Mirisola Son LA (1997) Nutritional status of dwarf coconut palm (*Cocos nucifera* L.) in Northern Fluminense. Thesis (MA) - State University of Norte Fluminense, Campos dos Goytacazes p 57
Moura JZ, Prado RM, Benvinod RN, Alencar CL (2013) Applying boron to coconut palm plants: effects on the soil, on the plant nutritional status and on productivity boron to coconut palm trees. J Soil Sci Plant Nutr 13(1):79–85
Muhr GR, Datta NP, Sankarasubramoney H, Leley VK, Donahue RL (1963) Soil testing in India. 2. United States Agency for International Development Mission to India, New Delhi, p118
Muliyar MK, Nelliat EV (1971) Response of coconut palms (*Cocos nucifera* L.) to N, P and K fertilizer application on the West Coast of India. Oleagineux 26(11):687–691
Muliyar MK, Wahid PA (1973) Note on the movement and availability of phosphorus in the soil as influenced by heavy phosphorus with special reference to coconut. Indian J Agric Sci 43:527–528
Nadheesha MKF, Tennakoon A (2008) Removal of micronutrients from high and moderate yielding coconut plantations in Sri Lanka. In: Proceedings of Second Symposium on Plantation Crop Research, Colombo, Sri Lanka on 16 &17 October 2008. pp 164–169
Nagarajan M, Pandalai KM (1965) The influence of some agricultural practices on the enzyme activity of coconut water. Curr Sci 34(24):685–686
Nagewekar DD, Sawant V, Magdum MB, Joshi GD, Khan HH, Jambhale ND (2004) Effect of NPK fertilizer on yield of COD x WCT hybrid of coconut under Konkan condition of Maharashtra (India). J Plantn Crops 32(suppl):186–189
Naidu LGK, Ramamurthy V, Challa O, Hegde R, Krishnan P (2006) In: Manual soil site suitability criteria for major crops. National Bureau of Soil Survey & Land Use Planning, Publ No 129(7):59–61
Nair UK, Radha K (1959) Manuring-cum-spraying for the control of coconut palm diseases. In: Proceedings of First Conference Coconut Research Workers, Trivandrum, India
Nair SK, Subba Rao NS (1977) Distribution and activity of phosphate-solubilizing microorganisms in the rhizosphere of coconut and cocao under mixed cropping. J Plantn Crops 5:67–70
Nambiar CKB, Wahid PA, Markose VT (1975) Evaluation of seaweed as manure to coconut. J Plantn Crops 3(1):6–8
Nambiar CKB, Khan HH, Joshi OP, Pillai NG (1983) A rational approach to the management of coastal sand for establishment and production of coconuts. J Plantn Crops 11:24–32
Nambiar CKB, Bidappa CC, Bopaiah MG, Kailasam C (1989) Influence of high density multispecies cropping on P and K forms in soil. J Plantn Crops 17(1):36–43
Narayanan Kutty MG, Gopalakrishnan PK (1991) Yield components in coconut palms. In: Silas EG, Aravindakshan M, Jose AI (eds) Coconut breeding and management. Kerala Agricultural University, Thrissur, pp 94–98
Nartea RN, Reyes ED (1973) Potassium release from some Philippine soils by crop uptake and chemical procedures Proceedings of the Crop Sci Soc of the Philippines 1973
Nath JC, Arulraj S, Maheswarappa HP (2012) Integrated nutrient management of COD x WCT hybrid coconut under alluvial clay-loam soil of Assam. J Plantn Crops 40(2):105–110

Nathanael WRN (1958) Diagnostic approaches to problems in crop nutrition. Ceylon Cocon Qtly 9(3–4):11–29
Nathanael WRN (1961) Coconut nutrition and fertilizer requirements—The Plant Approach. Ceylon Cocon Qtly 12:101–120
Nathanael WRN (1967) The application of fertilizer to adult coconut palm in relation to theoretical concepts. Ceylon Cocon Qtly 18:5–30
Nathanael WRN (1969) Theory of fertilization and application of fertilizers in adult coconut. Fertilite 35:11–27
Nelliat EV (1973) N, P, K nutrition of coconut palm-A review. J Plantn Crops 1.(Suppl)1:79–80
Nelliat EV, Nair RV, Thomas Varghese P (1978) Response of high yielding coconut genotypes to fertilizer levels under rainfed conditions. In: Nelliat et al. (eds) Proceedings of PLACROSYM-I. Indian Soiety for Plantation Crops, Kasaragod, pp 87–100
Nethsinghe DA (1959) Annual report of the soil chemist for 1959. Ceylon Cocon Qtly 10:17
Nethsinghe DA (1960) The manuring of adult coconut palms. Ceylon Cocon Planters Rev 1:5–12
Nethsinghe DA (1962) Magnesium deficiency in coconut palm. Ceylon Cocon Planters Rev 2:3–6
Nethsinghe DA (1963) Report of the soil chemist. Ceylon Cocon Qtly 14:8–20
Nethsinghe DA (1965) Annual report of the soil chemist. Ceylon Cocon Qtly 16:9–24
Nethsinghe DA (1976) Placement of fertilizers. Ceylon Cocon Planters Rev 5:10–14
Neto MF, Holanda JS, Folegatti MV, Gheyi HR, Pereira WE, Cavalcante LF (2007) Quality of fruit of green dwarf coconut in function of Nitrogen and potassium in fertigation. Revista Brasileira de Engenharia Agrícola e Ambiental 11(5):453–458
Newton K (1967) In: Coconut Research at Rangiroa by Pomier M, South Pacific Commission Tech Paper No. 153
Nirukshan GS, Dissanayake DMPD, Herath I, Tennakoon NA (2016a) Comparison of plant and soil nutritional status between organically and conventionally cultivated two coconut lands in the intermediate zone of Sri Lanka. In: 5th Young Scientist Forum Symposium National Science and Technology Commission, Colombo
Nirukshan GS, Herath HMIK, Wijebandara DMDI, Dissanayake, DMPD (2016b) Soil Microbial Population and Activity Affected by Fertilizer and Manure Addition in a Coconut Growing Sandy Regosol. In: Proceedings of the sixth symposium on plantation crop research "Plantation Agriculture towards National Prosperity" 2–4 November, 2016 BMICH Colombo, Sri Lanka, pp 163–172
Ochs R, Bonneau X (1998) Copper and Iron deficiency symptoms in coconut on peat soils in Indonesia. Oleagineux 48:65–76
Ochs R, Bonneau X, Qusairi L (1993) Copper mineral nutrition in hybrid coconuts on peat soils. Oleagineux 48(2):65–76
Oertli JJ (1994) Non-homogeneity of boron distribution in plants and consequences for foliar diagnosis. Commun Soil Sci Plant Anal 25:1133–1147
Oguis LG, Magat SS, Margate RZ (1979) The effect of chlorine sources and ammonium sulphate on the growth of coconut seedlings. Philippine J Cocon Studies 4(3):25–38
Ohler JG (1984) Coconut tree of life. Food and Agriculture organization of the United Nations, Rome, p 446
Ohler JG (1999) Modern Coconut Management: Palm Cultivation and Products. Intermediate Technology Publications p 458
Okeri HA, Alonge PO, Udoh JJ (2007) Assessment of nutritional status of soil supporting coconut (*Cocus nucifera*) cultivation in some localities of Edo State of Nigeria. African J Biotech 6(3):258–262
Ollagnier M (1985) Ionic reactions and fertilizer management in relation to drought resistance of perennial oil crops (oil palm and coconut). Oleagineux 40(1):1–10
Ollagnier M, Ochs R (1971a) Chlorine- a new essential element in oil palm nutrition. Oleagineux 26(1):1–15
Ollagnier M, Ochs R (1971b) The chlorine nutrition of oil palm and coconut. Oleagineux 26(6):367–372
Ollagnier M, Ochs R (1972) Sulphur deficiencies in the oil palm and coconut. Oleagineux 27:193–198

Ollagnier M, Ochs R, Pomier M, Tafffin G d (1983a) The action of chlorine on the hybrid coconut PB-121 in the Ivory Coast and in Indonesia. Oleagineux 38(5):309–317

Ollagnier M, Ochs R, Pomier M, Tafffin G d (1983b) Effect of chlorine on the hybrid coconut PB-121 in the Ivory Coast and Indonesia: growth, tolerance to drought, yield. Oleagineux 38(5):318–321

Ollivier J, Akus W, Bonneau X (1999) Coconut nutrition in Papua New Guinea. CORD 15(2):76–105

Omoti U, Ataga DO, Amalu UC (1986a) Dry matter production and macro-nutrient content of Nigerian tall coconut. In: Pushparajah E, Soon CP (eds) Cocoa and coconuts: progress and outlook. Incorporated Society of Planters, Kuala Lumpur, pp 803–817

Omoti U, Amalu UC, Ataga DO (1986b) Distribution and morphology of roots of the Nigerian tall coconut in relation to nutrient absorption and fertilizer placement. In: Proceedings of the international conference on cocoa and coconuts: progress and outlook. The Incorporated Society of Planters, Kuala Lumpur, pp 753–770

Ouvrier M (1982) Mineralisation of the bunch in the hybrid coconut PB-121, from the flower to maturity. Oleagineux 37(5):235–236

Ouvrier M (1984) Nutrient removal in the harvest of the hybrid coconut, PB-121 depending on potassium and magnesium manuring. Oleagineux 39(5):269–271

Ouvrier M, Ochs R (1978) Mineral exportation of the hybrid coconut. PB 121. Oleagineux 33:437–443

Ouvrier M, Ochs R (1980) Nutrient removal by hybrid coconut port - Bouet 121 (MAWA). In: Proceedings of the international conference on cocoa and coconuts. Incorporated Society of Planters, Kuala Lumpur, pp 595–605

Ouvrier M, de Taffin G (1985) How to fertilize coconut groves 1 Smallholdings. Oleagineux 40:275–278

Pandalai KM, Marar MMK (1964) Fertilizer and cultivation investigation on coconut in India, response trends. In: Proceedings of Symposium on Fertile Indian Soils. 1962. Bull Natn Inst Sci, India 26:307–321

Pandalai KM (1953) Certain aspects of soil suitability in relation to coconut cultivation. Indian Cocon J 6:89–93

Pandalai KM (1959) Some aspects of nutritional disturbances in relation to certain disorders in the coconut palm. In: Proceedings of First Conference of Coconut Research Workers. Trivandrum, India, pp 118–131

Pandalai KM, Sankarasubramoney H, Menon KPV (1958a) Studies on soil conditions in relation to the root and leaf diseases of the coconut palm in Travancore Cochin. Part IV. Total and exchangeable calcium and magnesium contents of coconut soils. Indian Cocon J 11:49–66

Pandalai KM, Sankarasubramoney H, Menon KPV (1958b) Studies on soil conditions in relation to the root and leaf diseases of the coconut palm in Travancore Cochin. Part V. Exchangeable cations, cation exchange capacity, and pH of coconut soils. Indian Cocon J 11:87–101

Pandalai KM, Sankarasubramoney H, Menon KPV (1959a) Studies on soil conditions in relation to the root and leaf diseases of the coconut palm in Travancore Cochin. Part VI. The combined hygroscopic water, loss on ignition and water table aspects of coconut palm. Indian Cocon J 12:87–100

Pandalai KM, Sankarasubramoney H, Menon KPV (1959b) Studies on soil conditions in relation to the root and leaf diseases of the coconut palm in Travancore Cochin. Part VII. The mechanical composition of coconut soils of healthy and diseased areas. Indian Cocon J 12:101–113

Park M, Fernando M (1941) Diseases of village crops in Ceylon, Colombo, 1941

Perera MLU (2007) Study on the solubility of Eppawala Rock Phosphate. University of Sri Lanka. Dissertation submitted in partial fulfillment of the requirements form the degree of Master of Science in Analytical Chemistry of the University of Colombo, Sri Lanka

Pillai NG (1959) Physiological Aspects of Nutrition in the Coconut Palm. M.Sc. Thesis, University of Kerala, Trivandrum, India

Pillai NG (1975) Studies on mineral nutrition of coconut palm with special reference to the root (wilt) disease. Ph.D. Thesis University of Kerala, Trivandrum

Pillai NG, Davis TA (1963) Exhaust of macronutrients by the coconut palm: a preliminary study. Indian Cocon J 16:81–87

Pillai NG, Pushpadas MV (1965) Kari soils are out of bounds for root (wilt). Cocon Bull 19:33–39

Pillai NG, Lal SB, Shanta P (1973) Distribution and intensity of root (wilt) disease of coconut in Kerala. J Plantn Crops 1(Suppl):107–112

Pillai NG, Wahid PA, Kamala Devi CB, Ramanandan PL, Cecil SR, Amma, PGK, Mathew AS, Nambiar CKB (1975) Mineral nutrition of root (wilt) affected coconut palm. In: Proceedings of Fourth Session FAO Tech Working Party Coconut Production, Protection and Processing. Kingston, Jamaica

Pinho LGR, Monnerat PH, Pires AA, Marciano CR, Soares YJB (2008a) Distribuição de nutrientes e sintomas visuais de deficiência de boro em raízes de coqueiro-anão verde. Revista Brasileira de Ciências do Solo 32:2581–2590

Pinho LGR, Monnerat PH, Pires AA, Santos ALA (2008b) Boron Absorption and Redistribution in Green Dwarf Coconut Tree. Brazilian Agri Res 43(12):1769–1775

Pinho LGR, Campostrini E, Monnerat PH, PiresNetto AT, Marciano CR, Soares YJB (2010) Boron Deficiency Affects Gas Exchange and Photochemical Efficiency (JPI Test Parameters) in Green Dwarf Coconut. J Plant Nutr 33:439–451

Pinho LGR, Monnerat PH, Pires AA, Freitas MSM, Marciano CR (2015) Diagnosis of boron deficiency in green dwarf coconut palm. Agric Sci 6:164–174

Pomier M (1964) Restoration and nutrition of established coconut plantations on coral soils. Oleagineux 19:615–620

Pomier M (1967) Coconut Research at Rangiora. Tech Paper No 153 South Pacific Commission, Noumea (from Hort Abstracts), 38 p 551

Pomier M (1969) Mineral nutrition of young coconut palm in coral soils. Oleagineux 24(1):13–19

Pomier M, de Taffini G (1982) Study of fertilization and regeneration of soils in the replanting of coconut palm. Oleagineux 37(10):460–461

Power PP, Woods WG (1997) The chemistry of boron and its speciation in plants. Plant Soil 193:1–13

Prema D, Jose AI, Nambiar PKN (1987) Effect of sodium chloride on growth and yield of coconut palms in a laterite soil. Agri Res J Kerala 25(1):66–73

Prevot P, Bachy A (1962) Foliar diagnosis in coconut - The effect of leaf position and vegetative growth on the content of various elements. Oleagineux 17:451–458

Prevot P, Fremond Y (1961) Manuring of the coconut palm in Dahomy. Fertilite 13:3–12

Prevot P, Ollagnier M (1957) Methods d'utilization du diagnostic foliare. IRHO, Paris

Priya George, Alka Gupta, Murali Gopal, Litty Thomas, Thomas GV (2012) Screening and *in vitro* evaluation of phosphate solubilizing bacteria from rhizosphere and roots of coconut palms (*Cocos nucifera* L.) growing in different states of India. J Plantn Crops 40(1):61–64

Prudente RL, Mendoza AMR (1976) Response of inland coconuts to inorganic fertilization from field-planting. Philippine J Cocon Studies 1(1):27–35

Pushpakumari HGW, Nadheesha MKF, Tennkoon NA, Ranaweera B (2008) Effect of some commercial bio fertilizers on the growth of coconut seedling and on the availability of soil nutrients. In: Proceedings of 8th Agricultural symposium, Faculty of Agriculture Plantation Management, Wayamba University of Sri Lanka, Makandura, Gonawila (NWP) pp 274–278

Rajaratnam JA (1973) Application, absorption, and translocation of boron in oil palm. III. Leaf analysis for diagnosing boron requirements. Expt Agri 9:257–262

Ramadasan A, Lal SB (1966) Exhaust of nutrients from coconut garden – Factors affecting production. Cocon Bull 20:173–175

Ramadasan A, Mathew C (1977) Relationship of the carbohydrate reserve in the trunk with commencement of flowering in young west coast tall coconut palms. J Plantn Crops 5:125–126

Ramanandan PL (1973) Use of common salt in coconut gardens. Cocon Bull 4(6):2–3

Ramanathan KM, Krishnamoorthy KK (1976) A study on the relationship between certain soii characters and K fixation. Bull Indian Soc Soil Sci 10:42–45

Ramkhelawan E (2013) Growers' guide for coconut. Ministry of Food production. Caroni South Bank Road, Trinidad. Central Experiment Station, Centeno

Ravi Savery MAJ, Madhiazhagan K, Rettinassababathy C (1994) Leaf sampling and critical nutrient levels in coconut. Indian Cocon J 25(8):6

Riberio G, Monneart PH, Campanharo M, Rabello WS (2011) Fruit quality of green dwarf coconut tree in response to potassium fertilization. Revista Caatinga 24(2):187–191

Rognon FC (1984) In: Martin-Prevel P, Gagnard J, Gautier P (1984) L'analyse végétale dans le contrôle de l'alimentation des plantes tempérées et tropicales. Tec & Doc, Paris, p 447–457

Roperos NI, Bangoy VO (1967) The effect of fertilizer on the yellowing of coconuts in Davao. Philippine J Plant Ind 32:97–107

Rosa GNGP, de Novais RF, Alvarez VHH, de Barros NF, de Albuquerque Villani EM (2011) Lime and fertilizer recommendation system for coconut trees. Rev Ceres (Impr) 58(1):90–99

Rosenquist EA (1980) A coconut fertilizer trial on the podsolic soils of North Sumatra. Oleagineux 35(5):241–244

Sahasranaman KN, Radha K, Pandalai KM (1964) Effect of manuring and intercultivation on the yield of coconut in relation to leaf rot and root (wilt) diseases. Indian Cocon J 18:3–11

Saldanha ECM, De Silva ML Jr, Lins PMP, Farias SCC, Wadat Salvador PG (2017) Nutritional diagnosis in hybrid coconut cultivated in North Eastern Brazil through Diagnosis Recommendation and Integrated System (DRIS). Rev Bras Frutic 39(1):e-728

Salgado MIM (1947) The use of lime and salt for manuring coconut palm. Trop Agric 103:121–123

Salgado MIM (1951) The manuring of under planted seedlings. Ceylon Cocon Qtly 2:161–164

Salgado MIM (1952) Potash and its importance in coconut manuring. Ceylon Cocon Qtly 3:191

Salgado MIM (1953) Potash deficiency in coconut. Oleagineux 8:297–298

Salgado MIM (1954) The nutrient content of nut water in relation to available soil nutrients as a guide to the manuring of coconut palm (*Cocos nuçifera* L.)- A new diagnostic method. In: Plant analysis and fertilizer problems. Eighth international congress of botany. WHO, Paris, pp 217–237

Salgado MIM (1955) The nutrient content of nut water in relation to available soil nutrients as a guide to the manuring of the coconut palm-a new diagnostic method. In: Prevot P (ed) Plant analysis and fertilizer problems. IRHO, Paris, pp 217–238

Salgado MLM, Abeyawardena V (1964) Nutritional and physiological studies on coconut water. Part 1: Further aspects of the nutrient content of nut water in relation to soil nutrients. Ceylon Cocon Qtly 15:95–108

Sankaranarayanan MP, Velayutham M (1976a) Soils that support coconut in the west coast. Indian Fmg 27:21–23

Sankaranarayanan MP, Velayutham M (1976b) Coconut soils on the east coast of India. Cocon Bull 7:13

Sankaranarayanan MP, Verghese EJ, Menon KPV (1958) A note on the tolerance of salinity by coconut palm. Indian Cocon J 11:133–139

Sankarasubramoney H, Pandalai KM, Menon KPV (1952) On the nutritive content of soil and plant tissues in relation to the root and leaf diseases of the coconut palm. Indian Cocon J 6:7–18

Sankarasubramoney H, Pandalai KM, Menon KPV (1954) Studies on soil conditions in relation to the leaf and root diseases of coconuts in Travancore-Cochin. Part I. Nitrogen, organic matter content and C:N ratio of coconut soils. Indian Cocon J 8:5–25

Sankarasubramoney H, Pandalai KM, Menon KPV (1955) Studies on soil conditions in relation to the leaf and root diseases of coconuts in Travancore-Cochin. Part II. Total phosphorus available phosphorus and iron content of coconut soils. Indian Cocon J 9:20–29

Sankarasubramoney H, Pandalai KM, Menon KPV (1956) Studies on soil conditions in relation to the leaf and root diseases of coconut in Travancore-Cochin, Part III. Total, available and exchangeable potassium contents of coconut soils. Indian Cocon J 9:90–100

Santiago RM (1978) Growth of coconut seedlings as influenced by different fertility levels in three soil types. Philippine J Cocon Studies 3(4):15–27

Santos AL, Monnerat PH, Alves EAB (2003) Foliar Boron content depending on the application of Borax in Leaf Axil of Green Dwarf Coconut in Norte Fluminense. Pesquisa Agropecuária e Desenvolvimento Sustentável 1:165–171

Santos AL, Monnerat PH, Carvalho AJCD (2004) Estabelecimento de normas DRIS para o diagnóstico nutricional do coqueiro-anão verde na região norte fluminense. Rev Bras Frutic 26:330–334

Sathirasegaram K, Rajaratnam N, Ferdinandez DEF (1966) Studies on the nutrient status of some coconut soils in Ceylon. Ceylon Cocon Qtly 17:42–53

Savy SM (1962) Coconut disease situation in Seychelles. In: Proceedingsof First Session FAO Tech, Working Party, Coconut Production, Protection, and Processing. Trivandrum, India, pp 28–29

Secretaria MI, Maravilla JN (1997) Response of hybrid coconut palms to application of manures and fertilizers from field-planting to full-bearing stage. Plant Recherche Develop 4(2):126–138

Secretaria MI, Magat SS, Aterado ED, Margate RZ (2006) Integrated crop management on local coconut hybrid, Catigan dwarf x Laguna tall (CATD x LAGT): early crop development years. Cocon Res Dev J 22(1):16–26

Shanmuganathan RT, Loganathan P (1976) Potassium status of some coconut growing soils of Sri Lanka. Ceylon Cocon Qtly 27:34–46

Silva JQ, Ramos VMY, Reis HS, Lima E (2005) Export of nutrients by the fruit of the green dwarf coconut in an irrigated area on the southern coast of Rio de Janeiro. In: Brazilian Congress on Science of Solo, 30. Recife, 2005. Annals. Recife, Brazilian Society of Soil Science

Silva RA, Cavalcante LF, Holanda JS, Pereira WE, Moura MF, Neto MF (2006) Fruit quality of the green dwarf coconut fertilized with nitrogen and potassium. R Bras Frutic 28(2):310–313

Silva MMUD, Nadheesha MKF, Tennakoon NA (2008) Availability of major nutrients in the soil after long term application of organic manure for coconut. Fulfillment of six months in-plant training program, HNDT (Agriculture). Thesis, Labuduwa, Advanced Technological Institute

Siqueira ER, Ribeiro FL, Aragão WM (1997) Melhoramento genético do coqueiro. In: JMS F, DRN W, LAA S (eds) Cultura do coqueiro no Brasil. EMBRAPASPI, Aracaju, pp 73–98

Smith RW (1964) Mineral nutrition of coconut in Jamaica. A progress report. In: Proceedings of second session FAO technical working party coconut production protection & processing. Colombo, Sri Lanka

Smith RW (1967) Seventh report of the Research Department from June 1966 to June 1967. The Coconut Industry Board, Jamaica, West Indies, pp 25–65

Smith RW (1968) Effect of fertilizer on the early growth and yield of Malayan Dwarf coconuts in Jamaica. Proceedings of Third Session FAO Technical Working Party, Coconut Production, Protection, and Processing, Jakarta, Indonesia

Smith RW (1969) Fertilizer responses by coconuts *(Cocos nucifera)* on two contrasting Jamaican soils. Exp Agric 5:133–145

Sobral, LF(1989) Nutritional status of the coconut groves of Sergipe Aracaju. EMBRAPA-CNPCo p 19

Sobral LF (1998) Coconut nutrition and fertilization. In: Ferreira JMS, Warwick DRN, Siqueira LA (eds) The coconut culture in Brazil. Aracaju, Embrapa, pp 129–157

Sobral LF (2004) Green Dwarf coconut ferti-irrigation with N and K on the Neopolis Plateau. In: FERTIBIO 26 Lages. Annals. Lages: Udesc

Sobral LF, Leal MLS (1999) Response from coke to fertilization with uréa, simple superphosphate and potassium chloride in two soils in the northeast of Brazil. R Bras Ci Solo 23:85–89

Sobral LF, Nogueira LC (2008) Influence of N and K applied by fertirrigation on soil properties, critical leaf levels and yield of dwarf coconut. R Bras Ci Solo 32:1675–1682

Sobral LF, de Freitas JAD, de Holanda JS, Fontes HR, Cuenca MAJ, Resende RS (2007) 5. Coconut – green Dwarf variety. In: Crisóstomo LA, Naumov A (eds) Fertilizing for high yield and quality- tropical fruits of Brazil. IPI Bulletin No.18, pp 89–102

Soil Survey Staff (1999) Soil Taxonomy, 2nd Agricultural Handbook 436, Natural Resources Conservation Service, USDA, Washington DC, USA, p 869

Solomon JJ, Govindankutty MP, Njienhaus F (1983) Association of mycoplasma like organisms with the coconut root (wilt) disease in India. Z Pflkrankh Pflschutz 90:295–297

Somasiri LLW (1987) Fertilizer increases your coconut yield. Cocon Bull 4:15–22

Somasiri LLW (1997) The interaction between potassium and magnesium in red yellow podzolic soils with laterite and its effect on nutrition of coconut palms (*Cocos nucifera* L). COCOS 12:18–32

Somasiri LLW, Liyanage DDKS (1996) Evaluation of the suitability and calibration of methods for the determination of available potassium of coconut growing soils. In: Annual Report of Coconut Research Institute, Lunuwila, p 118

Somasiri LLW, Nadarajah N, Amarasinghe L, Gunathilake HAJ (1994) Land suitability assessment of coconut growing areas in the coconut triangle. Occasional Publication Series No.3, Coconut Research Institute of Sri Lanka, pp 1–39

Somasiri LLW, Wijebandara DMDI, Sabaratnam S, Panditharatna BDP, Perera USS (2000) Removal of plant nutrients of soils by nuts of coconut. Proc Sri Lanka Assoc AdvanSci 56(1):77

Somasiri LLW, Wijebandara DMDI, Panditharatna BDP, Sabaratnam S, Kurundukumbura CPA, Pathirana KPA (2001) Study on nutrient removal by Typica x Typica coconut palms in high potential lands. Proc Sri Lanka Assoc Advan Sci 57(1):40

Somasiri LLW, Wijebandara DMDI, Panditharathna BDP, Sabaratnam S, Kurundukumbura CPA (2003) Loss of nutrient in a high yielding coconut plantation through removal of plant materials from field. COCOS 15:12–22

Soon CP, Wat LK (1972) Preliminary results from a coconut manuring trial on coastal clay.In: Cocoa and Coconuts in Malaysia, Kuala-Lumpur. Incorporated Society of Planters, pp 376–384

Southern PJ (1956) The flame spectrophotometry: determination of potassium, sodium, calcium and magnesium in nut water. Papua New Guinea Agric J 11:69–76

Southern PJ (1967) Sulfur deficiency in Coconuts, a widespread Field Condition in Papua and New Guinea. Part 1. The field and Chemical Diagnosis of Sulfur Deficiency. Papua New Guinea Agric J 19(1):18–44

Southern PJ (1969) Sulphur deficiency in coconuts. Oleagineux 24(4):211–220

Southern PJ, Dick K (1968) The distribution of trace elements in the leaves of the coconut palm and the effects of trace element injection. Oleagineux 23:521–527

Southern PJ, Dick K (1969) Trace element deficiencies in tropical tree crops in Papua and New Guinea, Res Bull No 7, Dept of Agri, Stock and Fisheries, Port Moresby

Srinivasa Reddy DV, Biddappa CC, Gopalasundaram P, Upadhyay AK, Zavier TV (1999) Long term effect of graded fertilizer levels on the productivity of WCT coconut in littoral sandy soil of Kasaragod. J Plantn Crops 27(3):201–206

Srinivasa Reddy DV, Upadhyay AK, Gopalasundaram P, Hammed Khan H (2002) Response of high yielding coconut variety and hybrids to fertilization under rainfed and irrigated conditions. Nutr Cycl Agroecosyst 62:131–138

Sumbak JH (1975) Progress of two fertilizer trials in Papua New Guinea. In: Proc Fourth Session FAO Tech, Working Party, Coconut Production, Protection, and Processing Kingston, Jamaica

Sumbak JH (1976) Progress of two Coconut Fertilizer trials in Papua New Guinea. Oleagineux 31(10):427–434

Sumbak JH, Best E (1976) Fertilizer response with coconuts in coastal Papua. Papua New Guinea Agri J 27(4):93–102

Sumner ME (1982) The Diagnosis and Recommendation Integrated System (DRIS). Paper presented in Soil Plant Analysts' Seminar on Soil Testing and Plant Analysis, Anaheim, CA, December 1, 1982

Taffin G, Rognon F (1991) Coconut leaf analysis. Oleagineux 5(4):169–173

Teixeira LAJ, da Silva JNA (2003) Mineral nutrition of populations and hybrids of coconut (*Cocos nucifera* L.) grown in Bebedouro (SP). Brazil Rev Bras Frutic. Jaboticabal SP 25(2):371–374

Teixeira LAJ, Bataglia OC, Buzetti S, Furlani E Jr, Isepon JS (2005a) NPK fertilization on Dwarf Green coconut (*Cocos nucifera* L.) - yield and fruit quality. Brazil J Fruticult 27(1):120–123

Teixeira LAJ, Bataglia OC, Buzetti S, Furlani E Jr (2005b) Fertilization with NPK on dwarf coconut (*Cocos nucifera* L.) Soil chemical attributes and plant nutrition. Brazil J Fruticult 27(1):115–119

Tennakoon A (2004) Soil fertility status and fertilizer recommendation for coconut in Sri Lanka. http://www.ippipotash.org/udoc's/soil

Tennakoon NA, Bandara SDS (2003) Nutrient content of some locally available organic materials and their potential as alternative sources of nutrients for coconut. COCOS 15:23–30
Tennakoon NA, Mahindapala R, Widanapathirana S (1995) Effects of organic manure on the quality of coconut soils. J Nat Sci Council Sri Lanka 23:171–182
Thampan PK (1970) Manuring the coconut palm. Cocon Bull 1:3–7
Thampan PK (1982) Handbook on Coconut Palm. Oxford and IBH Publishing Co. Kolkotta, pp 81–157
Thomas KM (1973) Nutrient composition of coconut leaves and its relationship to nut yields in Tanzania. East African Agri For J 39:170–175
Trewren K (1987) Tuvalu technical report on tree crops research 1986. Ministry of Commerce and Natural Resources, Dept of Agri, Vaiaku
Udayangani WN, Tennakoon NA, Weerakkody WJSK (2013) Residual effect of long-term organic and inorganic fertilizer application to coconut measured by major nutrients and some soil physical parameters. In: Proceedings of 12th agricultural research symposium, pp 226–230
Upadhyay AK, Palanisamy C, Srinivasa Reddy DV, Hameed Khan H, Subramanian P (2002) Effect of fertilizer levels on the nutrition and productivity of coconut based cropping system. Paper presented in 17th World Congress Soil Science, 2002, Thailand, pp 371
Upadhyay AK, Palaniswarni C, Hameed Khan H, Balamurugan J (2005) Influence of fertilizer regime on potassium dynamics in a red sandy loam soil. J Plantn Crops 33(2):95–98
Upadhyay AK, Palaniswami C, Biddappa CC (2007) Mineralization of nitrogen in some soils of Kerala. J Plantn Crops 35(2):73–75
Valiathan MS, Eapen JT, Mathews CK (1992) Reciprocal relationship between magnesium and cerium as a common basis for coconut root (wilt) and humen cardiomyopathy. Curr Sci 63(9–10):565–567
Varkey T, Amma PGK, Ramanandan PL, Nambiar PTN (1979) Foliar yellowing in coconut palms. In healthy and root (wilt) affected areas. J Plantn Crops 7:117–120
Velasco JR, Dc Cuzman EV, Mercado BT (1960) Growth of young coconut plants in sand culture. Philippine Agric 43:548–576
Venkitasamy R (2004) Effect of NPK levels on dwarf x tall (COD x WCT) hybrid coconut in Tamil Nadu. J Plantn Crops 32(suppl):169–172
Venkitaswamy R, Hameed Khan H, Poongothai S (2006) Effect of long term application of graded levels of NPK fertilizers on soil and leaf micronutrient status of hybrid coconut. J Plantn Crops 34(1):63–65
Venkitaswamy R, Hameed Khan H, Palanisamy V (2011) Effect of graded levels of N,P,K on the reproductive characters and yield of hybrid coconut COD x WCT. Trop Agric 88(2):79–86
Verghese EJ (1959) Chemical studies on the leaf rot and root (wilt) diseases of coconuts in Travancore-Cochin. I. Some interesting observations on the incidence of the disease. In: Proceedings of First Conference Coconut Research Workers, Trivandrum, India, pp 306–309
Verghese EJ (1961) Statement of work done on the chemical aspects of root (wilt) disease of coconut in Travancore-Cochin. In:Proceedings of First Sessn. FAO Technical Working Party Coconut Production, Protection & Processing. Trivandrum, India, pp 92–103
Verghese EJ (1966) Fertility status of coconut soils with special reference to the leaf and root (wilt) diseases of the coconut palms in Kerala. Agri Res J Kerala 4(2):49–60
Verghese EJ, Sankaranarayanan MP, Menon KPV (1957) A note on the strontium content of coconut leaves and soils in relation to leaf and root (wilt) diseases of Coconut in Travancore-Cochin. Indian Cocon J 10(2):25–31
Verghese EJ, Sankaranarayanan MP, Menon KPV (1959a) Chemical studies of the leaf and root (wilt) diseases of Coconut in Travancore-Cochin. II. Nutrient content of leaves of healthy and diseased palms. In: Proceedings of first conference coconut research workers. Trivandrum, India, pp 310–319
Verghese EJ, Sankaranarayanan MP, Menon KPV (1959b) Chemical studies of the leaf and root (wilt) diseases of Coconuts in Travancore-Cochin. III. Cadmium content of coconut soils

and leaves. In: Proceedings of first conference coconut research workers, Trivandrum, India, pp 366–370

Vertueil J d (1934) Manurial experiments with coconuts on Perseverance Estate. Trop Agri (Trinidad) 11:313–315

Viégas IJM, Thomaz MAA, Silva JF d, Conceição HEO, Naiff APM (2004) Efeito da omissão de macronutrientes e boro no crescimento, nos sintomas de deficiências nutricionais e na composição mineral de plantas de camucamuzeiro. Rev Bras Frutic 26:315–319

von Uexkull, HR (1971) Manuring of coconut. In: Proceedings of the Conference on cocoa and coconut. Kuala Lumpur, pp 386–399

von Uexkull HR (1972) Response of coconut to potassium chloride in the Philippines. Oleagineux 27(1):13–19

von Uexkull HR (1978) Potassium requirements of some tropical tree crops (oil palm, coconut palm, rubber, coffee, cocoa. In: Proceedings of 11th congress international Potash Institute, pp 71–104

von Uexkull HR (1985) Potassium nutrition of some tropical plantation crops. In: Munson RD (ed) Potassium in agriculture. ASA/CSSA/SSSA, Madison, pp 929–954

von Uexkull HR (1992) Drought and Chloride- a deadly combination for hybrid coconuts. Better Crops Int 1992 pp 89

von Uexkull HR, Sanders JL (1986) Chloride in the nutrition of palm trees.In: Jackson TL (ed) Special Bulletin on Chloride and Crop Production No. 2, Potash & Phosphate Institute, Atlanta, Georgia, pp 84–99

von Uexkull HR, Fairhurst TH (1991) The oil palm: Fertilizing for high quality and yield. Int Potash Inst Bull No. 12

Wahid PA, Kamala Devi CB, Pillai NG (1974) Interrelationships amongst root CEC, yield and mono- and divalent cations in coconut (*Cocos nucifera* L.). Plant Soil 40(3):607–617

Wahid PA, Kamala Devi CB, Philip G, Pillai NG (1975) Effect of discontinuation of fertilizer application on NPK nutrition of coconut palm. J Plantn Crops 3(2):58–60

Wahid PA, Kamala Devi CB, Haridasan M (1977) A critical review of phosphate fertilization to coconut. Philippine J Cocon Studies 2(4):1–8

Wahid PA, Kamalan NV, Venugopal VK, Kataria SK, Govil R, Kapoor SS, Lal M (1983) X-ray fluorescence spectra of root (wilt) affected coconut palm. J Plantn Crops 11:91–100

Wahid PA, Nambiar PKN, Jose AI, Rajaram KP (1988) Soils and nutrition. In: Silas EG, Aravindakshan M, Jose AI (eds) Six decades of coconut research. Kerala Agricultural University, Thrissur, pp 46–80

Wahid PA, Kamalam NV, Vijayalakshmi R, Krishna Prabhu R, Mahalingam TR, Eapen JT, Dong HS, Jaiswal DD, Balakrishnan PV, Venugopal VK, Suresh RR, Unnithan VKG, Thomas J (1998) Foliar levels of Rare Earth Elements and Thorium in coconut palm in relation to root (wilt) disease. Curr Sci 75(11):1180–1183

Wahid PA, Kamalam NV, Krishna Prabhu R, Sekhar JK, Vijayalakshmi S, Mahalingam TR, Ajith Kumar CE (2003) Rare Earth element fluxes in diverse soils and their absorption by coconut palm. J Plant Nutr 26(3):1427–1438

Wijebandara DMDI, Kurudukubura CPA, Panditharathna DP (2015) Nutrient removal through plant components of low yielding coconut plantation. COCOS 21:1–13

Willer H, Kilcher L (2009) The World of Organic Agriculture. Statistics and Emerging Trends 2009, IFOAM, Bonn, and FiBL, Frick

Wilshaw RGH (1941) Results of manorial and cultivation experiments in coconut. Malayan Agri J 29:145/267–151/272

Wuidart W (1994) Sulphur deficiency symptoms in coconut. Oleagineux 49(1):31–34

Yualves LA, Covaar RR, Ocfemia GO (1958) Chemical analysis of cadang-cadang infected coconuts. Philippine Agric Sci 41:491–494

Ziller R, Prevot P (1962) Foliar diagnosis: a method of studying mineral nutrition- its application to the coconut palm. Indian Cocon J 15:156–159

Dr. H. Hameed Khan retired as Project Coordinator, All India Coordinated Research Project on Palms. He is basically a Soil Scientist, specialized in the mineral nutrition of coconut palm and coconut-based cropping and farming system. He has received Indian Society of Soil Science Award. Dr. Khan served as Editor, Journal of Plantation Crops and Journal of Agricultural Resource Management. He has to his credit 110 publications and has edited 2 international conference proceedings. khanhh.in@gmail.com

Dr. V. Krishnakumar is currently a Principal Scientist (Agronomy) and Head, ICAR-Central Plantation Crops Research Institute (Regional Station), Kayamkulam, Kerala. He served as Editor of the Journal of Plantation Crops. As an agronomist he has edited 10 books and published more than 100 research and technical articles as well as contributing chapters in various books and technical bulletins. dr.krishnavkumar@gmail.com

Chapter 9
Physiology and Biochemistry

S. Naresh Kumar, K. B. Hebbar, K. V. Kasturi Bai, and V. Rajagopal

Abstract Coconut palms are grown in diverse environments in different tropical regions of the world. Palms experience several cycles of different stresses during their life period and respond to external factors at morphological, anatomical, physiological, biochemical and molecular levels. Significant developments are made in unravelling the physiological and biochemical mechanisms underlying the productivity performance of coconut in various agroclimatic conditions. Delineating the seedling and adult palm growth, dry matter partitioning, compositional changes in developing nuts, oil and fatty acid profile in germplasm, drought tolerance mechanism, in situ drought-tolerant palms, responses to root (wilt) disease, drought management strategies, response to climatic parameters and climate change impact and adaptations are important milestones in this area of research. Further efforts eventually led to the development of InfoCrop-COCONUT model for simulating growth, development and yield, opening new vistas in research programmes for improving coconut management for higher yields. Since climate change is projected to have positive and negative impacts on coconut yield depending on the region, it is important to use advanced technologies for harnessing positive impacts while countering measures for overcoming the negative effects. This chapter summarizes the major research understandings in physiology and biochemistry of coconut palms.

S. Naresh Kumar (✉)
Centre for Environment Science and Climate Resilient Agriculture,
ICAR-Indian Agricultural Research Institute, New Delhi, India
e-mail: snareshkumar.iari@gmail.com

K. B. Hebbar · K. V. Kasturi Bai
ICAR-CPCRI, Kasaragod, Kerala, India
e-mail: balakbh64@gmail.com; kasturikv@yahoo.com

V. Rajagopal
ICAR-CPCRI, Kasaragod, Kerala, India

Society for Hunger Elimination, Tirupati, Andhra Pradesh, India
e-mail: rajvel44@gmail.com

K. U. K. Nampoothiri et al. (eds.), *The Coconut Palm (*Cocos nucifera *L.) - Research and Development Perspectives*,
https://doi.org/10.1007/978-981-13-2754-4_9

9.1 Introduction

Physiological and biochemical studies in coconut deal with the plant physiological and biochemical aspects that regulate the production potential of coconut palms. The productivity of coconut depends on the biological efficiency and is mainly dependent on dry matter production and partitioning towards the economically important parts. This in turn is regulated by genetic makeup, agroclimatic conditions, management as well as soil characteristics. The literature available on biological efficiency of coconut is comprehensive, generated through exhaustive experimentation. There exists substantial gap between potential nut yield and actual realized yield which is mainly due to low inputs, limited management, poor soil conditions, climatic stresses, nutrient deficiencies and biotic factors. Further, coconut plantations face vagaries of nature, quite often due to their perennial nature, flowering behaviour and long life span. Stable dry matter production, efficient use of inputs, tolerance to abiotic stresses, resilience to climatic stresses and improved harvest index (HI) are the key factors for sustained growth and yield. For achieving these, an in depth understanding of physiology and biochemistry of the crop is essential. In this chapter, the physiological and biochemical processes that regulate the crop growth and development starting from seed germination through early seedling growth, flowering and post-fertilization stages culminating in the nut yield are dealt with.

9.2 Growth and Development

9.2.1 Physiological and Biochemical Changes During the Early Phase of Germination

The coconut palm is a monocotyledonous palm with the fruit (drupe) having a thick epicarp, fibrous mesocarp and a hard endocarp (shell) lined by solid endosperm. Both the epicarp and mesocarp together constitute the husk (Juliano 1926). The fruit has a single embryo enveloped by the liquid endosperm (nut water). The embryo is embedded in the solid endosperm at one end of the endocarp. The embryo starts germinating when the mature nut comes into contact with moisture for sufficient duration. The nut contains all essential nutrients, hormones and carbon sources required for the germination and growth of the embryo.

9.2.1.1 Germination

Mature nuts (11–12 months old) are suitable for sowing (Nelliat et al. 1976), and storage of seed nuts in the shade for 1 month (seasoning) is essential for breaking the dormancy (Fremond and Lamothe 1966). Soaking the nuts in water or in a solution of 0.01 M potassium nitrate and 0.02 M sodium carbonate for 48 h (Thomas 1974), chopping the husk from both ends of the nuts, injection of different hormones (Liyanage 1952; Deshpande and Kulkarni 1962) as well as major and minor nutrients (Menon and Pandalai 1960; Sumathykutty Amma 1964) hasten the germination process. Nuts sown horizontally give rise to better seedlings than those planted vertically (Espino 1923; Lumige 1969) due to constant contact of nut water with the embryo. Please refer to Chap. 7 for details on nursery management.

The first morphological sign of germination is the enlargement of the embryo and protrusion of the apical mass around the shell (Kartha 1981). For this, the dwarfs require 10 days, hybrids 15 days, while the slowly germinating talls require an average of 20 days (Wuidart 1981). In the agronomic context, emergence of plumule out of the husk is considered as germination. As per this, the nuts of tall variety take 11–12 weeks to germinate, and the percentage of germination reaches the maximum between the 17th and 18th week.

The embryo commences to grow in two directions. The plumule moves towards the soft eye to develop as shoot, and the other end of embryo develops into an absorbent spongy growth known as the haustorium, which consists of loosely connected thin-walled cells with interspaces (Selvaratnam 1952; Child 1964). The haustorium absorbs food materials from the nut water as well as kernel and supplies to the growing plant. Physiological and biochemical changes during the early phase of germination include the solubilization of stored food material and its utilization, which is influenced by various factors like absorption of water, activity of enzymes and hormones.

Absorption of Water During the early phase of germination, there is a gradual increase in the amount of water absorbed by the nut which is higher in talls as compared to early germinating dwarfs. Seed reserves support the growth and development of haustorium and emergence of root and shoot. Dry matter of haustorium is more than that of shoot and root, implying the essential role of haustorium in the early phase (Manjula 1990). The volume of haustorium differs among the talls (~315 cm^3), dwarfs (~155 cm^3) and hybrids (~205 cm^3).

Solubilization of Seed Reserves During Germination Coconut endosperm comprises mainly of lipids (68–70%), sugars (6–7%) and protein (6–9%) (Nathanael 1967). The major reserve food material is, thus, lipid which gets solubilized faster, during the very early stage, i.e. 28 days and 42 days after sowing, respectively, in the case of Dwarfs and Talls. This is evident from a decrease in the oil content in the kernel, with a concomitant increase in the major lipid solubilizing enzymes, i.e. lipase. There is a threefold increase in the activity of lipases in dwarfs with a comparatively higher reduction in the oil content (4.1–4.8%) than the talls and hybrids (2.3–3.2%) which explains the early germination in dwarfs.

Kernel sugars are composed mainly of mannans (Balasubramaniam 1983) which are broken down by β-mannosidases. During the development of haustorium, total soluble sugars and reducing sugars decline in kernel, along with decreased activity of hydrolytic enzymes, viz., amylase and invertase. These products are absorbed by haustorium up to 13 weeks. Thus, there exists a pool of soluble sugars in the haustorium, formed by the activities of β-mannosidases and β-amylases. From this pool, sugars are supplied to the growing embryo. Seedlings utilize the soluble carbohydrates, and excess are stored as starch to be utilized for growth (Balasubramaniam et al. 1973). Up to the formation of haustorium, protein content increases gradually and then declines rapidly thereafter. The protease activity, in general, remains parallel with the protein levels (Manjula et al. 1993). In coconut, the liquid endosperm, i.e. nut water, serves as a rich source of nutrients between kernel and developing embryo. Up to the time of haustorium development, the sugars and amino acids increase in the nut water due to the dissolution of material from the kernel. These compounds decline with the corresponding development of embryo. Breakdown products of lipids, carbohydrates and proteins from kernel are dissolved in liquid endosperm, which then act as a via media for transferring the components to the growing embryo (Manjula et al. 1995). By about 22 weeks after germination, with the appearance of green leaves, the seedling gradually becomes autotrophic.

9.2.2 Seedling Growth

Growth Characters The development of the seedling occurs in two phases. In the first phase (up to 4 months), the relative growth rate (RGR) falls from 72 g $100g^{-1}$ $week^{-1}$ to 7.5 g 100 g^{-1} $week^{-1}$and thereafter remains constant during the second phase (5–15 months). Till the 4th month, the plant is entirely dependent upon the endosperm for its growth, while at 15th month, it becomes fully autotrophic (Foale 1968). The first leaf unfolds at 2 months after germination. Leaves produced up to 4th month are quite small with total leaf area (LA) of 8 dm^2. As the seedlings grow from 6th month to 12th month, there is a steady increase in LA, typically, from 19 dm^2 to 51 dm^2 and shoot dry mass (SDM) from 41g to 133 g. Leaf area index (LAI) reaches 5.6 at the 12-month stage (Kasturi Bai and Ramadasan 1990). The rate of leaf production remains fairly constant and is an important characteristic feature showing positive correlation with earliness of flowering.

Measurement of Seedling Growth and Vigour Seedling selection for high vigour is of paramount importance in establishing a stand of superior yielders (Table 9.1). Conventionally, selection of vigorous seedlings is done on the basis of a few easily measurable morphological characteristics such as girth at collar, total number of leaves, plant height, length and breadth of leaves or leaflets and early splitting of leaves.

Since growth is a function of LA development and dry matter (DM) production, they are used as criteria for selection of superior seedlings (Liyanage 1953; Menon and Pandalai 1960). Several regression equations have been developed for non-destructive estimation of LA and DM production (Foale 1968; Ramadasan et al. 1980; Satheesan et al. 1983; Shivashankar et al. 1986). Since these equations differ

Table 9.1 Correlation of seedling characters with nut yield of palm

Character (1-year-old seedling)	6-year-old palms	
	Cumulative yield	Female flower production
Girth	0.65**	0.69**
No. of leaves	0.55**	0.61**
Leaf area	0.73**	0.83**
SDM	0.62**	0.72**

**Significant at p = 0.01

based on the morphology of seedling and cultivar variations, it is necessary to define growth conditions of seedlings to ensure accuracy of estimations.

$$\text{Seedlings' dry weight (g)} = -112.5 + 12.6\text{G} + 0.23\text{H} + 5.6\text{N} + 0.02\text{T}$$

where G = Girth at collar (cm), H = height (cm), N = no. of leaves, and T = Total leaf area = 27.4 + 0.61 X, where X = length (cm) × breadth (cm) of individual leaves

Among the various seedling characteristics, though height of seedlings and number of leaves are correlated with dry weight of shoot in WCT, their direct bearing on the dry weight of the shoot is negligible. On the other hand, LA and girth at collar significantly influence the shoot dry matter (SDM). Thus, the important determinants of the vigour of seedling are LA, girth at collar and SDM (Ramadasan et al. 1980). The LA of 6-month-old seedlings is correlated with SDM of 12-month-old seedlings ($r^2 = 0.55$) (Kasturi Bai and Ramadasan 1990), implying that the vigour of the seedlings can be determined before the attainment of complete autotrophy. Coconut seedlings show a gradual change in the leaf morphology from pinnate to bifurcate leaves with advancing age and as such the applicability of regression equations is limited to the age of the plant, especially in the first few years of growth.

In many crop plants, the hybrid vigour and yield are correlated with the activity of nitrate reductase (NR). Since NR plays a key role in the utilization of nitrate and consequently DM production, NR activity in coconut seedlings is significantly correlated with the leaf area development and DM production (Shivashanker et al. 1985). Heterotic hybrid seedlings of COD x WCT had higher NR activity than the intermediate and dwarfs, and it was significantly correlated with dry matter accumulation ($r^2 = 0.55$).

9.2.3 *Leaf and Crown*

Young coconut seedlings have pinnate type of leaves, which are not split into leaflets. These initial leaves act as the source of photosynthates to the developing seedling and subsequently emerging leaves start splitting (called as frond). As palms grow, the size of newly emerging fronds also becomes bigger. Every palm has a crown of leaves orientated at the top of the trunk consisting of opened leaves and those in the bud in various stages of development. The number of leaves in the crown varies depending on the age and the conditions prevailing, viz., climatic conditions, methods of cultivation, cultivar and management. The leaves are produced in spiral succession, generally at one leaf month^{-1}. In highly productive well-grown palms, the number of leaves is about 30–42, and one inflorescence emerges in the axil of each leaf. At any given time, generally 12–15 spadices can be seen. There are generally 3 sets of leaves in a coconut canopy, (i) above inflorescence bearing leaves, (ii) inflorescence bearing leaves and (iii) leaves below inflorescence bearing leaves (Naresh Kumar and Kasturi Bai 2009a). Leaves in coconut canopy are arranged in 3 shapes, viz. oval, X shape and semicircle shape. Leaf orientation is most suited for proper light interception by all leaves in oval-shaped canopy (Naresh Kumar and Kasturi Bai 2009a).

Development and Structure of the Leaf The primordium of the leaf is first differentiated about 30 months prior to its emergence from the leaf sheath. In the beginning, the leaf primordium is a small indistinct protuberance at the base and on the side of the growing point. By periclinal and anticlinal divisions of the cells, the protuberance, in about a month's time, increases in size and assumes a finger-like shape with 7 to 10 layers of cells. At this very early stage, the 3 primary meristems, viz. the dermatogens, the periblem and the plerome, can be distinguished in the growing point. The outermost layer of cells forms the dermatogen which is continuous with the growing point. The 3 layers of cells below the dermatogen form the periblem, and the plerome is innermost in the growing point. During the succeeding month, the size of the protuberance considerably increases and assumes the shape of a hollow cone enclosing the next younger leaf and the growing point. At this stage, the initial cells of the procambial strands, which will ultimately become the vascular bundles, first make their appearance in the central tissue or plerome of the base of the second leaf from the growing point of the stem. The procambial strands consist of elongated cells actively dividing along their length lying in the middle of the leaf, where they are in 5 to 6 rows, while along the margins, they lie only in one central row. The number of strands later increases as the leaf develops in size. The young leaves are without laminae, and they consist of only the petiole portion at the apex of which is a growing point with the primary meristems. The leaflets are first differentiated at the tip of the seventh leaf from the growing point in about 8 months after it originated. The differentiation of the leaflets occurs first below the top of the young petiole on either side of it.

Leaf area and Dry Weight Non-destructive methods to estimate the LA and leaf dry weight were developed (Ramadasan et al. 1985; Ramadasan and Mathew 1987). Regression equations were developed taking into account all the absolute values derived from the base, middle and top 3 pairs of leaflets from a single leaf. Total area and dry weight of the leaf can be determined with ease by estimating the dry weight of middle 6 leaflets and by leaflet count of the same leaf. Once the leaf is completely unfolded, there is no perceptible increase in LA and dry weight.

The length of the leaf (including the rachis) normally ranges from 4 to 6 m, while the rachis length varies from 1 to 1.5 m (Kasturi Bai 1993). The LA of individual leaf ranges between 3.8 and 5.1 m^2, whereas the dry weight ranges from 0.97 to 1.36 kg. Differences in area and dry weight between leaves of the same age in a palm are low. Thus, the area and dry weight of leaves produced per unit time can be estimated by multiplying the estimated LA or dry weight of a single leaf by the number of leaves produced in a year. Variations exist in the LA and dry weights among the cultivars. Equations for non-destructive estimation of the total LA based on the area of 12 leaflets and the total number of leaflets per leaf (Jayasekara and Mathes 1992) and for the determination of the dry weight of the leaf by measuring the width and length of the petiole (Friend and Corley 1994) were developed.

$$\text{Leaf dry weight} = -3.14 + 0.0197X_1 + 0.0202X_2$$

where X_1 = dry weight of six leaflets (g) and X_2 = no. of leaflets

Leaf Anatomy Besides leaf area and dry weight, leaf anatomy has also been recognized as an important component of productivity. Cultivar differences have been observed in leaf thickness and tissue density (Ramadasan and Satheesan 1980). Hybrids have been found to possess higher tissue density than the WCT palms. However, in coconut, the leaf tissue density, in general, is low as it contains more air space volume which is characteristic of all C_3 species as against high tissue density and low air space volume in C_4 species.

Leaflet mean thickness is 341 μm with bottom portion of the leaflet being thicker and tapered towards tip portion. Epidermal cells are closely attached to form a compact layer devoid of intercellular spaces. The upper epidermis is thicker with large-sized cells than lower epidermis. Cuticle on upper epidermis is twofold thicker than the cuticle on lower epidermis (2.49 μm). Cuticle is even thicker at midrib and at edges of leaflet. Coconut leaflets are hypostomatus. The guard cells have hook-like protuberances at both ends, a characteristic typical to Palmae. Elongated epidermal cells surround the guard cells along their entire length. These cells are larger than guard cells. On the lower epidermis, multicellular, shortly stalked scales occur at regular intervals in short depressions. These scales contain tannins (Menon and Pandalai 1960; Naresh Kumar et al. 2000a). This distribution of palisade and spongy parenchyma makes the leaflet dorsiventral (Naresh Kumar et al. 2000a).

Anatomical features of leaflet vary among the cultivars. Leaflets are significantly thicker in WCT, FMS (405 and 363 μm, respectively) with thick cuticle on both adaxial and abaxial side of leaflet compared to other cultivars in which the leaflet thickness ranges between 308 and 339 μm. However, GBGD and MYD have very thin cuticle on both sides. Upper epidermal cell size, guard cell size, xylem tracheid lignification, sub-stomatal cavity size and hypodermal cell size also vary among the cultivars (Naresh Kumar et al. 2000a). Further, they reported that the parenchyma cells (spongy and palisade) are significantly bigger in WCT and FMS compared to other cultivars. Palisade parenchyma cells are the smallest in WCT x COD, while PHOT has smaller spongy parenchyma cells than other cultivars. Xylem tracheids in vascular bundles have scalariform thickenings ranging from 2.85 μm in MYD to 3.38 μm in WCT. Water cells also are significantly bigger in WCT and FMS. The indigenous cultivars had extreme values for all the traits related to leaflet anatomy (WCT had maximum values, and GBGD and COD x WCT had minimum values). However, the exotic cultivars had medium values thus indicating the possibility of using some of these parameters to identify ecotypes in coconut.

9.2.4 Stem

The trunk of coconut grows erect to a height of 10–24 m. The annual growth of stem is recognizable by 'scars' (mark of leaf base). As the new leaves are added, the old ones are shed leaving the scars of leaf base which correspond to a year of growth of the palm. Since the trunk just below the crown mainly contributes to the growth of the stem, dry matter production of this apical portion is used for estimation.

In general, coconut palm produces 12–14 leaves annually, and the data on these segments is sufficient to estimate the quantity of dry matter produced by way of stem growth annually. Besides, any differences in the stem growth rate due to variation in environment variables will be reflected in this portion of stem. Increment in stem height can be easily determined by marking the stem portion just below the crown and taking the measurement after a period of time, typically 1 year. Coconut stem is the main reservoir of carbohydrates which are mainly stored in soluble form, largely as sucrose, while roots have no storage function (Mialet-Serra et al. 2005). Canopy pruning did not influence stem reserves showing a very low phenotypic plasticity of coconut (Mialet-Serra et al. 2008). The increment in stem height year^{-1}, ranged between 23 and 37 cm, varies significantly between cultivars and hybrids. Talls, Dwarfs and hybrids differed for stem DM production between 1.7 and 7.3 kg palm^{-1} year^{-1} (Kasturi Bai et al. 1996a). The density of stem decreased gradually from just below canopy (0.32 g cm^{-3}) to base (0.8 g cm^{-3}) with a mean density of 0.517 g cm^{-3} (Naresh Kumar et al. 2008a).

A non-destructive method was developed to estimate the standing stem biomass (Naresh Kumar et al. 2008a).

Standing stem dry weight (kg) = Length (in m) × girth^2 (in m, at 1.5 m a bove ground level) × 41.14142.

For annual stem dry matter estimation (Ramadasan et al. 1985), the equation is

$$Y = -113.44 + 93.67 \times \text{Length of three leaf scar segments}$$

9.2.5 Flowering

Commencement of flowering in coconut is noticed by the appearance of first inflorescence (spadix) in the leaf axil. The initiation of inflorescence primordium occurs in the 10th to 14th leaf axil, and the flowering in the Tall variety takes place around the age of 5 years or at the 45th leaf stage of growth or beyond (Patel 1938; Pillai et al. 1973). Genotype x environment influence on spadix initiation is evident from their higher production during March to September when average day length is more (Wickramasurya 1968). WCT seedlings responded to long-day treatment in terms of chlorophyll fractions, vigour and early initiation of inflorescence primordium. The primordial initiation took place in the 10th leaf axil in the long-day treatment palms and in the 14th leaf axil, in the controls. Although inflorescence primordial growth is initiated in the 10th to 14th leaf axil, all these in most instances are aborted until the 45th leaf stage after transplantation (Pillai et al. 1973).

Factors Influencing the Onset of Flowering In fruit trees, a high carbohydrate reserve in the stem is an essential prerequisite for early initiation of flowering. In coconut, seasonal and cultivar variation exists in the carbohydrate fractions. Hybrids are superior to Tall (WCT) in their efficiency of mobilization of carbohydrate fractions to inflorescence primordium (Kasturi Bai and Ramadasan 1983). The inflorescence from the primordial stage takes about 26 months to emerge out of the leaf axil (Patel 1938). In juvenile palms, the emergence of first inflorescence is noticed during the months of August to October. This period coincides with the decrease in the insoluble carbohydrate fraction (starch) as against an increase in the non-reducing sugar fraction (Kasturi Bai and Ramadasan 1978). Increased productivity in irrigated condition can be due to the assured availability of soluble carbohydrate fractions for the initiation and the development of inflorescence.

Stem carbohydrate reserves significantly differ between the flowered and non-flowered palms. In 8-year-old WCT palms, a higher ratio of carbohydrates to nitrogen (C/N ratio) as well as higher leaf number is observed in palms that have commenced flowering over those which have not (Ramadasan and Mathew 1977). Not only the development of inflorescence but also the female flower production is regulated by the carbohydrate metabolism. In Kerala, India, maximum female flower production occurs during the period of March to May (Menon and Pandalai 1960), whereas in Sri Lanka it is up to September (Abeywardena 1968). Concomitant with the increase in female flower production from March to May, there is an increase in the insoluble carbohydrate fraction in stem and leaf and a decrease in soluble fraction. Insoluble fraction is positively correlated with female flower production, whereas the sugar content is negatively correlated (Kasturi Bai and Ramadasan 1982). The derangement of the carbohydrate metabolism due to Mg and P deficiencies leads to an impaired nitrogen metabolism and delays the initiation of flowering (De Silva et al. 1973). Glucose, free amino acid, total nitrogen, protein nitrogen and nonprotein nitrogen were lower in the laminae of nonbearing coconut palms than in the bearing palms of similar age (Balasubramaniam et al. 1974).

Hastening the age of flower production is of particular interest in palms which do not flower even after 5 or 6 years. Though exogenous application of gibberellin and kinetin (500 ppm and 1000 ppm) enhanced only the elongation of the petiole of the youngest leaf, injecting 1000 ppm gibberellin into the trunk of 2-year-old seedlings could induce flowering in the 36th leaf axil instead of 45th leaf axil. Thinning of immature bunches at 4 months of maturity and extraction of the inflorescence sap increased female flower production (Mathes 1984).

Coconut Inflorescence Sap Coconut inflorescence sap is the phloem sap collected from the unopened spadix. The sap collected by traditional method (called *neera*) is prone to microbial fermentation. A simple technology developed at ICAR-CPCRI 'coco-sap chiller' for the collection of sap under low temperature keeps it fresh and unfermented during collection (Hebbar et al. 2013, 2015a, b). The sap, thus, collected can be stored for prolonged time under refrigerated condition (−1 to −3 °C). It tastes sweet and is delicious with no alcohol and devoid of contamination, as it is collected in an aseptic condition. This fresh, hygienic and unfermented sap (called Kalparasa) is slightly alkaline (pH 7–8) and golden brown in colour, while neera is oyster white in colour, with a pH 6 or below and an astringent smell. Kalparasa also contains amino acids, total phenols, flavonoids and antioxidants 2.5, 1.5, 4.6 and 1.8 times higher than neera, respectively (Hebbar et al. 2015a). It is also rich in vitamins C, E and Niacin. Further, the products of Kalparasa such as sap concentrate and sugar were also found to be rich in amino acids, polyphenols, flavonoids, vitamins and antioxidants (Hebbar and Chowdappa 2016). Extracting the sap before the nut production is advantageous in terms of energy production (Hebbar et al. 2015a; Hebbar and Chowdappa 2016). However, a balance between sap production and nut production must be maintained for yield sustainability of the palm.

9.2.6 Nut Development

The ovary development in coconut from the time of initiation of inflorescence primordium to full maturity of the nut can be divided into two major phases, i.e. prefertilization phase taking about 32 months and the post-fertilization phase continuing for another 12 months. The growth of the fruit begins immediately following fertilization, with a rapid development of the pericarp at the basal region which remains soft and white until the fruit is nearly mature. The endocarp is already differentiated as a soft, creamy white structure long before the time of fertilization. During the development of fruit to maturity, the embryo sac increases in size, leaving a large cavity at the centre.

The coconut water (liquid endosperm) is formed in small quantities in the third month of development of the nut and reaches maximum in the eighth month and declines thereafter as the nut ripens. The endosperm of the coconut develops as a coenocytic liquid, containing many free nuclei and some cells (Cutter et al. 1955; Bhatnagar and Johri 1972). The cells coalesce towards the periphery of the embryo sac. Additional cells are formed when free nuclei adhere resulting in the formation of cellular endosperm. In the mature coconut, the liquid, which is of cytoplasmic origin, does not contain free nuclei or free cells. The shell begins to form during the fourth month of nut development and continues to grow up to the twelfth month. The kernel is the last component to form in the seventh month and its growth continues up to the eleventh month when the maximum value of dry weight is reached. The reddish testa, which assumes a brownish tint when matures, is laid down before the formation of the kernel. Total nut production palm^{-1} year^{-1} shows great variation among cultivars/hybrids, but bunch production or spikelet bunch year^{-1} does not vary significantly among them. However, female flower production varies significantly among cultivars/hybrids (100–400) with higher production observed in hybrids than in the cultivars (Kasturi Bai et al. 1996a).

9.2.6.1 Biochemistry of the Developing Nut

The liquid endosperm plays a vital role in fruit development by acting as a reservoir of precursors for the synthesis of fruit constituents. The major constituents of the liquid endosperm are sugars and minerals, while fat and nitrogenous substances form a minor fraction (Kamala Devi and Velayudham 1978).

Changes in Carbohydrates Reducing sugars form the major part of immature nuts. Sugars in liquid endosperm of immature nuts (3–4 months) are almost entirely reducing sugars, which increases to a concentration of about 5% by seventh month. During this period of fruit development, sugar (sucrose) in inflorescence phloem sap entering the fruit is completely converted to glucose and fructose by the action of invertase present in the stalk (Balasubramaniam 1983; Balasubramaniam and Alles 1989). With the formation of solid endosperm (kernel) at about seventh month, non-reducing sugars appear and increase in concentration both in kernel and in nut water, while the levels of reducing sugars and total sugars decrease (Balasubramaniam 1983). Simultaneously, polysaccharides in kernel increase following a sigmoid pattern. The initial lag phase of polysaccharide synthesis continues until the establishment of the jelly-like endosperm on the entire surface area of nucellus. This is followed by a logarithmic phase in which the endosperm gradually thickens and turns hard. Stationary phase is characterized by the formation of a hard solid endosperm in which almost all cell wall polysaccharides are laid down. Fat content of the kernel increases during the same period indicating that the glucose of the liquid endosperm is converted to sucrose and to the polysaccharides of the kernel, besides being utilized as a precursor for the synthesis of fats (Naresh Kumar and Balakrishnan 2009). The total carbohydrate content of the kernel increases from 31.9% in the 6 month to 44.9% at maturity, during which time free sugars increase from 11.3% to 23.6% (Sierra and Balleza 1972). Sucrose contributes to the synthesis of nearly 20% of the lipids stored in the endosperm and a large number of polysaccharides including cellulose stored in its pericarp (Van die 1974).

Composition of Carbohydrates in Nut Water Coconut water contains mainly monosaccharides, viz. glucose, fructose and sucrose, and traces of trisaccharide raffinose (galactose-glucose-fructose) and tetrasaccharide stachyose (glucose-galactose-glucose-fructose). A polysaccharide (an arabinogalactan) was isolated from the liquid endosperm of mature coconut (White et al. 1989). It has a molecular weight exceeding 500 kDa and is composed of galactose and arabinose with minor amounts of mannose and glucose. Two smaller polysaccharides containing significant amounts of xylose or mannose and lesser amounts of arabinose and another polysaccharide, composed exclusively of uronic acid residues, have also been reported.

Carbohydrate Composition of Kernel Galactomannans and galactoglucomannans are the major polysaccharides of mature coconut kernel and copra meal, respectively (Rao et al. 1961; Kooiman 1971; Balasubramaniam 1976). The copra meal contains substantial amounts of cellulose, minor amounts of arabinoxylogalactan, galactomannan and traces of arabinomannogalactan and galactoglucomannan (Saitagaroon et al. 1983). Primary cell wall of immature cells is formed of cellulose, polysaccharides and galactomannans. The mannans are absent in the kernel of 5- to 6-month-old nuts, while the mature nut consists chiefly of mannans, containing at most a few per cent of galactose residues (Balasubramaniam 1976), and during fruit maturation, the galactomannans (galactose to mannose 1:7) increase (Balasubramaniam et al. 1974). During the transition of the endosperm from the hydrated gelatinous phase to the dehydrated solid mature state, most of the galactose groups are removed from the cell wall (Kooiman 1971). This process is catalysed by the enzyme D-galactosidase present in the endosperm (Mujer et al. 1984a, b).

Changes in Amino Acids The developing fruit contains a variety of nitrogenous substances, of which free amino acids constitute a major part. During nut maturity, the total nitrogen and nonprotein nitrogen (NPN) progressively increase although percentage levels decrease marginally on whole nut basis. The NPN content remains above 60% at all stages of maturity (Jayalekshmy et al. 1988). In the maturing nut, the free amino acid content in the coconut water increases from 4 mg to 16 mg 100 ml^{-1}, whereas concentration of bound amino acids does not show any marked change. The first traces of α-amino butyric acid appear in liquid endosperm with the formation of kernel at about 5–6 months after the fruit set and subsequently increases with maturity. The α-alanine and β-aminobutyric acid are present in large quantities in the endosperm of the immature nut. Proline content is the highest in the immature nut and decreases with maturity leaving only a trace in the water of the mature nut. In the immature nut water, about 70% of the free amino acids are made up of glutamine, arginine, asparagine, alanine and aspartic acid, while alanine, a-amino butyric acid and glutamic acid constitute about 75% of the free amino acids of mature nut water. With increasing age of the fruit, aspartic acid and glutamic acid increase slightly, while alanine increases markedly, whereas other amino acids do not show much change (Baptist 1956). In coconut kernel, amino acids such as methionine, tyrosine, tryptophan and phenyl alanine are present in high levels (Hagenmaier et al. 1972).

Changes in Protein The protein content of nut water increases from 0.13% to 0.29%, while it decreases in the kernel from 8.3% in the fifth month to 6.2% at maturity. The coconut proteins have been classified according to their solubility behaviour, amino acid content and heat denaturation. About 84% of the coconut proteins are of high molecular weight (150 kDa), and the rest belong to low molecular weight class (24 kDa). Based on solubility studies, it was identified that 90% protein could be classified as albumins and globulins.

Changes in Lipids Fatty acid biosynthesis starts at about fifth month, when the kernel begins to form and increases up to 12th month. The fat content of the nut water increases up to 10th month and rises gradually thereafter reaching maximum in the 12th month. At the same time, the fat content of the kernel also increases, and the oil content per nut is at its maximum in the 12-month-old nuts. In coconut dry kernel, oil is present in the range of 57–70% depending on cultivar/hybrid. In coconut oil neutral lipids form major fraction (~94%) followed by the glycolipids (3.5%) and phospholipids (2.5%) (Naresh Kumar et al. 2000b). Lysophosphatidylethanolamine comprises about 23% of all phospholipids in coconut kernel. The rest of the phospholipids are composed of phosphatidylinositol, phosphatidylserine, phosphatidylcholine, phosphatidylethanolamine and other unidentified phospholipids in equal abundance (Monera and del Rosario 1982).

Changes in Fatty Acid Composition of Endosperm In the developing fruits, the pattern of variation of fatty acids in both kernel and nut water appears to be similar, although significant deviations in the relative abundance of component fatty acids have been found (Oo and Stumpf 1979; Padua - Resurreccion and Banzon 1979; Jayalekshmy et al. 1988; Naresh Kumar et al. 2004; Naresh Kumar 2007; Naresh Kumar and Balakrishnan 2009). In general, the relative proportions of fatty acids increase up to C 14:0 during maturation, while a corresponding decrease in the high unsaturated fatty acids occurs (Fig. 9.1).

The most characteristic feature is that the content of lauric acid (C 12:0) in kernel rapidly increases with maturity up to 42.4–52.5% and that of myristic acid is maintained at about 18–23% of the total in the mature nut, while the content of most other fatty acids remains far below 20% (Naresh Kumar and Balakrishnan 2009; Naresh Kumar 2011a). Palmitic acid decreases from 14% to 9% and oleic acid from 23% to 6% from 6-month-old tender nuts to mature nuts. Linoleic acid gets reduced from 5% to 1% during nut development. Palmitoleic and linolenic acids are found in low quantities in 6-month-old nuts, and their percentage decreases with nut maturity. Four of the fatty acids in nut water, C14:1, C15:0, C16:1 and C17:0, which are present in the early stages, disappear as the nut matures, whereas the content of the short-chain fatty acids, C6:0, C8:0 and C10:0, which are present in the negligible levels initially, rises with maturity both in water and kernel. The long-chain fatty acids C18:3 and C22:0 have been found at all stages of fruit growth in nut water but not in kernel. The concentration of long-chain fatty acids is more in 5-month-old nuts as compared to the mature ones. In the mature nut, saturated fatty acids, mostly of short- and medium-chain length, like caprylic (C 8:0), lauric (C 12:0) and myristic (14:0) acids, comprise nearly 83–89%. Concentration of long-chain unsaturated fatty acids in developing coconut kernel is high at 5 and 6 months after fertilization and decreases towards maturity. Concentration of fatty acids also varies during growth (Naresh Kumar and Balakrishnan 2009). Fatty acids degrade in stored oil due to photooxidation, chemical reactions and microbial activity. During storage, the small- and medium-chain fatty acids are degraded to form the free fatty acids (FFAs). Additives such as citric acid (100 ppm) or common salt (NaCl at 1%) prolonged shelf life with least increase in FFAs (Naresh Kumar and Balakrishnan 2012).

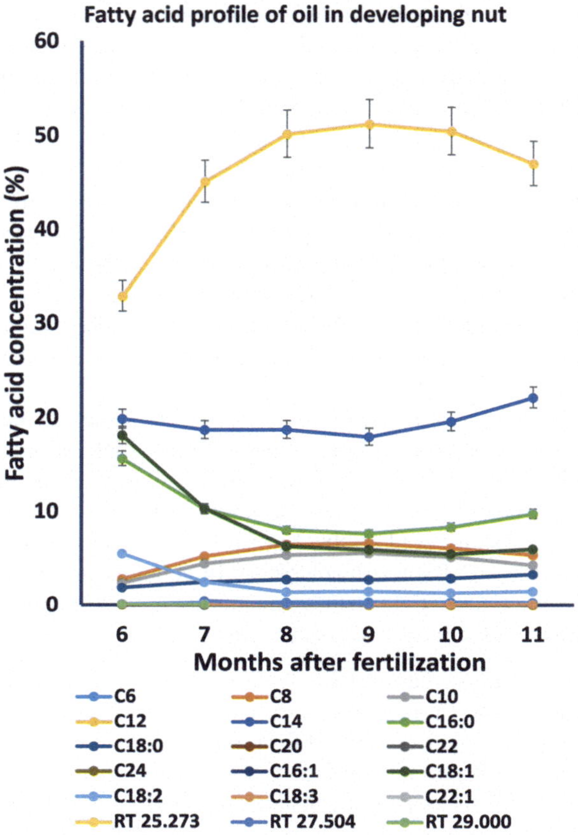

Fig. 9.1 Fatty acid profile of a developing coconut. (Modified from Naresh Kumar and Balakrishnan 2009)

Growth Regulators in Developing Fruits Coconut contains numerous growth promoting compounds, and to date several growth promoters have been isolated in pure form from various parts of the coconut fruits. 1,3-diphenylurea, gibberellin, myo-inosital, scyllo-inositol, sorbitol, phyllococosine and auxin-gibberellin-like substances were identified from liquid endosperm. Zeatin and zeatin ribosides were identified both in mature fruit and liquid endosperm. Purine-like substances and 9-D-ribofuranosyl zeatin were identified only in mature endosperm (Van Staden and Drewes 1975; Dix and Van Staden 1982).

Other Constituents The developing coconut contains many other compounds having specific biochemical roles. Among these, shikimic acid and quinic acid found at all stages of fruit growth are possibly involved in aromatic biosynthesis. Buttons and green variety contain leucocyanidin and leucopelargonidin, while the red dwarf contains small amount of flavonoid pigments in addition to leucoanthocyanidins. Mature coconut contains a number of vitamins in which ascorbic acid is the major component. Coconut water contains an RNA in the soluble form and is found to be not associated with any cell organelles. Phytin is reported to be present in coconut water and plays a role in supplying phosphate for the synthesis of nucleic acid.

Activity of Enzymes The endosperm, being a site of active metabolism, contains a large number of enzymes. The onset of cellular differentiation in the developing coconut coincides with a rapid rise in acid phosphatase activity. Acid phosphatase activity in the kernel reaches a minimum when the maturation is complete and remains so even during germination. The activities of pyrophosphatase 3′nucleotidase, ribonuclease and deoxyribonuclease also remain similar in the kernel of mature and germinating coconuts. Isolation of RNA polymerases from coconut nuclei (Mondal et al. 1970) of immature coconuts suggests the existence of transcriptional process in them. The enzymes present in coconut include acid phosphatase (immature kernel, nut water), decarboxylases, aspartate amino transferase, RNA polymerase (in immature kernel), pyrophosphatase 3′-nucleotidase, ribonuclease, glycerol dehydrogenase, amylase, lipase, phospholipase and deoxyribonuclease, mannan synthetase, GDP mannose, pyrophosphorylase, esterase (in mature kernel), carbonic anhydrase (1- to 2-month-old fruit) and CAMP-dependent protein kinase (nut water) and invertase (6- to 7-month-old fruit stalk and mesocarp) (Padmaja et al. 1980; Balasubramaniam 1983; Balasubramaniam and Alles 1989; Manjula 1990). Peroxidase is present in 7- to 12-month-old normal and makapuno nuts, while a D-galactosidase and tryptophan aminotransferase are present in 12-month-old nut and kernel (Mujer et al. 1984b).

9.2.7 Variability in Coconut Germplasm for Oil and Fatty Acid Profile

In general, oil content is high in talls than in dwarfs. Among the Talls, Laccadive Micro Tall has the highest oil content, while Green Dwarf has the highest oil content among dwarfs. Lipid fraction content also varies with cultivar. The neutral fraction is maximum in WCT, TCT, MYD x WCT, JVT, FIJT and PHOT. Hybrids have lower concentrations of neutral lipids (90–94%) (Naresh Kumar et al. 2004). Analysis of fatty acid profile of oil from 60 Talls, 14 Dwarfs and 34 hybrids, which included collections from coconut-growing areas of 13 countries as well as from the indigenous ones, indicated wide variation (Naresh Kumar 2007, 2011a). Variability in the saturated to unsaturated fatty acid ratio (6.32–17.6) and contents of lauric acid (42.4–52.5%) and other fatty acids are reported. Further, coconut germplasm also significantly varied for concentration of long-chain unsaturated fatty acids followed by that for long-chain saturated fatty acid and small- and medium-chain fatty acids (Naresh Kumar 2011a). Seasonal variations exist for fatty acid profile except myristic acid which did not vary significantly. The concentration of linolenic acid was the highest in nuts harvested during October. Small- and medium-chain fatty acids were high in nuts collected during January which gradually decreased up to October

(Naresh Kumar and Balakrishna 2009). Eriophyid mite infested nuts had significantly low concentration of unsaturated fatty acids and further decreased with the intensity of infestation (Sujatha et al. 2006).

9.2.8 Total Dry Matter Production (TDM)

Coconut palm has indeterminate growth pattern, producing an inflorescence at each leaf axil at intervals varying from 25 to 30 days, depending on the environmental conditions and age of the palm. Consequently, at any time, a healthy coconut palm carries 12–14 inflorescences or bunches with varying numbers of fruits at different developing stages. TDM constitute the above-ground and below-ground (root) dry biomass, but generally only above-ground biomass is estimated. Total above-ground dry matter constitutes both vegetative dry mater (VDM) and reproductive dry matter (RDM). The dry weight of stem and leaf together constitutes the VDM of the palms and is a useful character for selection for vigorous palms. Low VDM is associated with high harvest index. Cultivar, crop management and agroclimatic conditions cause variation in annual dry matter partitioning to stem (2–6%), leaf (18–33%), inflorescence (4–6%) and nut (61–70%) (Rajagopal et al. 1989b; Kasturi Bai et al. 1996a, 1997; Siju Thomas et al. 2005; Naresh Kumar 2009) as indicated in Fig. 9.2. Significant variation exists in the VDM between the Tall, Dwarf, and hybrid (Kasturi Bai 1993; Kasturi Bai et al. 1996a). Dwarfs produce less TDM mainly due to the low number and small size of leaves. This leads to small and thin stem with narrow leaf

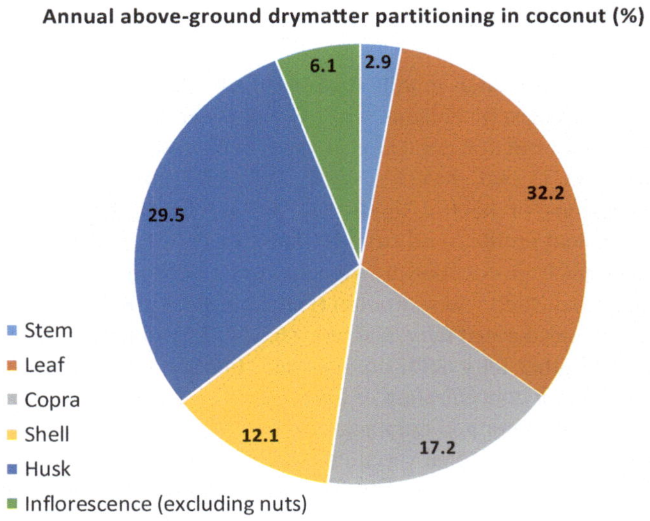

Fig. 9.2 Annual above-ground dry matter partitioning (per cent) in coconut. Values are mean of three talls, one dwarf cultivar and five hybrids

scars. On the other hand, due to big and long leaves as well as more production of leaves, the Talls produce higher TDM, leading to long, sturdy and thicker stem. The increment in stem height year^{-1} is also higher in Talls as compared to the Dwarfs.

The total dry weight of spathes, bunches and nuts constitute the RDM. Dry weights of the spadices and the spathes constitute about 5% of the total RDM production. Thus, RDM production depends mainly on the nut production and partitioning of nut dry matter towards its components, viz. husk, shell and copra. Based on the dry matter accumulation in the vegetative and reproductive parts, TDM production also greatly varies between the cultivars and hybrids. Kasturi Bai et al. (1996a) reported a TDM production of 17 t ha^{-1} year^{-1} in WCT x COD hybrids. However, the highest value reported is 30 t ha^{-1} year^{-1} in Dwarf x West African Tall hybrid in Cote d'Ivoire (Corley 1983). Stability of dry matter and yield also varies across locations (Siju Thomas et al. 2005). The potential dry matter production is estimated at 51 Mg ha^{-1} year^{-1} (Corley 1983; Foale 1993), while in India, the simulated TDM potential production ranged between 52 and 62 Mg ha^{-1} year^{-1}(Naresh Kumar et al. 2008b). Thus, a huge gap exists in the realization of yield and the production potential of coconut palms.

Harvest Index The harvest index (HI) has been considered as an important criterion in biological and economic yield. Because of the limitation in estimating the total biomass including the roots, Ramadasan and Mathew (1987) coined the term 'Annual Productivity Index (API)'. The HI in coconut was estimated by taking into account the annual increment in DM production and expressed as the ratio between dry weight of economic product to TDM production. Being a crop of continuous productivity, API is an appropriate criterion comparable to the HI of annual crops. In coconut, since all the parts are economically important, several values of HI could be calculated. The values of API estimated ranged from 0.4 to 0.5 in a group of palms in which the annual yield ranged from 45 to 91 nuts. Harvest indices are calculated based on the total DM production and its partitioning towards the annual copra out turn (Kasturi Bai et al. 1996a). Hybrids have HI indicating better nut composition than Talls and Dwarfs. The HI based on the copra out turn ranged from 0.13 to 0.23.

9.3 Physiological Basis of Yield Potential

The coconut palm exhibits wide variability in productivity ranging from 30 nuts to 400 nuts palm^{-1} year^{-1}. This is mainly due to the efficiency of the palms in the dry matter production and partitioning towards yield. The relationship between height of the palm, number of leaves on the crown and the annual yield indicated that leaf area, photosynthetic efficiency and dry matter production are important parameters for high production potential of coconut palms. In addition to management, agroclimatic conditions influence the photosynthetic efficiency, dry matter production and economic yield, thus determining the production potential of palms (Siju Thomas et al. 2005). Significant differences in the number of leaves on crown and chlorophyll content have been observed between yield groups, and high correlation between these parameters and annual yield of nuts have been reported. Cultivars with high rate of photosynthesis and lower respiration produce higher yields than the cultivars with high respiratory rates. The net assimilation rate, shoot dry weight and chlorophyll concentration have heritability values of 0.64, 0.74 and 0.81, respectively (Ramadasan et al. 1985).

Leaf Photosynthesis Under favourable environmental conditions, coconut develops a profuse canopy with an estimated potential dry matter production of 51–62 Mg ha^{-1} $year^{-1}$ (Foale 1993; Naresh Kumar et al. 2008b). In situ measurements of coconut seedling photosynthetic rates indicated saturated photosynthetic photon flux density of about 1400 μ moles photons m^{-2} s^{-1}, typical of C3 plants (Naresh Kumar and Kasturi Bai 2009a). Net photosynthetic rates vary between 2 and 15 μ mol CO_2 m^{-2} s^{-1} depending on the age of palm, moisture and nutrient status as well as weather conditions. On the other hand, rate of leaf respiration was relatively similar in young and old leaves −2 to −1 μ mol CO_2 m^{-2} S^{-1}. Among C3 plants, coconut has been shown to be relatively less efficient in conversion of photosynthetically active radiation (PAR) energy to biomass. The highest energy conversion of PAR into dry matter in coconut has been estimated to be 1.2–1.4 g MJ^{-1} (Corley 1983). Within coconut canopy, net photosynthetic (Pn) rates were higher in second to tenth leaves from the top and decline considerably making it a source-limited crop (Naresh Kumar and Kasturi Bai 2009b). A decline in photosynthetic activity of fronds has been observed from 12 to 14 months which is accompanied with a decrease in leaf nitrogen content and decline in incident PAR due to self-shading effect of upper canopy leaves. Canopy shape plays a role in the overall performance of photosynthesis and water use efficiencies and productivity in coconut. Palms with oval-shaped canopy have higher photosynthesis efficiency, water use efficiency and productivity than those with X-shaped and semicircle-shaped canopies (Naresh Kumar and Kasturi Bai 2009b). The photosynthetic nitrogen use efficiency (PNUE) declined with age of leaf (Jeyasekara et al. 1996). Manifestations of water-deficit stress in coconut cause varied responses including low stomatal conductance and water potential that often impair Pn and transpiration (Repellin et al. 1994, 1997; Rajagopal and Kasturi Bai 2002; Naresh Kumar and Kasturi Bai 2009a, b). As a consequence, carbon assimilation rate is impaired in both Tall (Reppellin et al. 1997; Prado et al. 2001) and Dwarf cultivars (Gomes et al. 2007) in response to atmospheric and soil water deficit (Naresh Kumar et al. 2002b, 2006a, b; Hebbar et al. 2016b, c). Limited diffusion of atmospheric CO_2 to intercellular spaces due to stomatal closure has been attributed with photosynthetic reductions (Reppellin et al. 1994, 1997). Later physiological studies have delineated that non-stomatal factors also contribute to the reduction in Pn both during a period of severe water deficit and during the recovery phase after resuming irrigation (Rajagopal et al. 2000a, b; Gomes and Prado 2007; Gomes et al. 2007). In addition, the photochemical efficiency (FV/Fm) reduced with decreasing water potential suggesting damage to photosynthetic apparatus under stress (Kasturi Bai et al. 2006b). Prolonged exposure to high light intensities caused photo-oxidative stress in coconut seedlings damaging cell membrane, caused leaf senescence and in severe cases resulted in seedling death (Naresh Kumar and Kasturi Bai 2009a). Similarly, zygotic embryo cultured plantlets undergo photosynthetic acclimatization during their growth in in vitro condition and during subsequent transfer to acclimation chambers and shade nets (Naresh Kumar et al. 2001).

Gas exchange measurements are linked to water and gas economy of the plants. Coconut palms under the influence of water depletion stress respond with decreased stomatal conductance (g_s, mol m^{-2} s^{-1}) (Rajagopal et al. 1988, 1989b). Stomatal resistance (*rs*) differs during the period of adequate soil water availability and during soil water deficit among the cultivars and hybrids of coconut. During non-stress, *rs* shows the least variation among the cultivars, except in WCT x WCT, which exhibited relatively more *rs*. However, under stress conditions, where high evaporative demand in the atmosphere prevails, cultivars exhibit differential adaptability through stomatal regulation which is high in Talls and hybrids, whereas, in Dwarfs, it is almost 50% less than that in hybrids (Rajagopal et al. 1990; Naresh Kumar et al. 2002a; Siju Thomas et al. 2008; Hebbar et al. 2016a, b). This indicates the higher transpiration loss of water in Dwarfs than in Talls and hybrids. Among the hybrids studied, COD x WCT had significantly low *rs*. Rainfed palms had higher LAVPD (leaf to ambient vapour pressure deficit), leaf temperature (T_{leaf}) and leaf to ambient temperature difference (ΔT), whereas the irrigated palms had higher Pn, $\Psi_{leaf,}$ and transpiration rates (*E*). The *LAVPD* and ΔT influenced the *gs* and water relations of coconut. Strong stomatal control of plant water status has been demonstrated in mild to moderate water stress (Naresh Kumar et al. 2002a).

9.3.1 Leaf Water Potential (ψ_{leaf})

Leaf water potential (ψ_{leaf}), an indicator of plant water status, has a vertical gradation from middle leaf upwards, the magnitude being higher under rainfed condition. Spindle leaf had significantly higher ψ_{leaf} throughout the day irrespective of rainfed or irrigated conditions (Voleti et al. 1993a, b). Under rainfed conditions, ψ_{leaf} showed a reduction from the spindle to the first leaf. Characteristic midday depression in ψ_{leaf} was evident in both the spindle and the first leaf. The midday depression in ψ_{leaf} was more in rainfed palms than in irrigated ones (Rajagopal et al. 2000a). Variation among the cultivars for ψ_{leaf} also was noted. Seasonal variations in the ψ_{leaf} occur, depending on the weather, type of soil and soil water availability. For example, in rainfed palms, ψ_{leaf} was high in December (−1.10 MPa) and decreased significantly in May (−1.75 Mpa) (Shivashankar et al. 1991; Voleti et al. 1993a, b). In general, ψ_{leaf} is lower in palms grown in red sandy loam than those in laterite soil. In irrigated condition, the ψ_{leaf} is maintained at relatively high level corresponding with soil moisture availability even during the non-rainy period (March to May) (Rajagopal et al. 1989b). A rapid screening method was developed based on ψ_{leaf} in excised leaflets (Rajagopal et al. 1988) for easy handling of a large number of cultivars. The

ψ_{leaf} declined with time to different degrees among the varieties, indicating the degree of tolerance. Hence, water requirement of coconut palms depends on many factors such as age, height, stomatal frequency, wax content and stomatal control (Passos and Silva 1990; Rajagopal et al. 1990; Nogueira et al. 1998).

9.3.2 Chlorophyll Index

Foliar chlorophyll content is a good indicator of plant stress and plant health because of its effects on photosynthesis and growth (Datt 1999). Environmental (drought and high temperatures) and nutrient (particularly N) stresses commonly cause loss of leaf chlorophyll content leading to poor photosynthesis, growth, biomass and economic yield. Because chlorophyll is mostly made up of N-containing enzymes and other organic compounds, stress-restricted uptake of N causes early senescence, which is commonly expressed as loss of chlorophyll content and loss of green leaf area. The ability of a plant to maintain chlorophyll content and green leaf area for a longer duration under stress will allow plants to remain photosynthetically active for a longer period of time. Heterotic hybrid seedlings of COD x WCT had higher NR activity as well as chlorophyll and carotenoid concentrations than the intermediate and dwarf types (Shivashankar et al. 1985). Non-destructive method for estimation of chlorophyll in coconut leaves is standardized (Hebbar et al. 2016c).

Chlorophyll Fluorescence Chlorophyll fluorescence is used to evaluate the plant health status, and photochemical efficiency of photosystem II (PS II; Fv/Fm) is routinely used as an indicator of the degree of stress in plants. Estimation of chlorophyll fluorescence as a measure of photosynthetic efficiency in coconut seedlings is a promising indicator to identify potential genotypes that show resistance to drought. The chlorophyll fluorescence PS II efficiency parameter (Fv/Fm), an indicator of extent of physiological stress in leaf, has been found to be higher in irrigated palms compared to the rainfed palms (Kasturi Bai et al. 2006a, b, 2008). Based on PS II efficiency (Fv/Fm) screening, coconut accessions have been classified into high and low groups. Hence, screening of drought-tolerant palms based on yield-related photosynthetic parameters could be achieved effectively by measuring chlorophyll fluorescence (Nainanayake 2007). Coconut seedlings exposed to high light intensities have low Fv/Fm and photochemical quenching than those in shade condition indicating that high light condition causes stress to coconut seedlings implying the need for protection from photo-oxidative stress (Naresh Kumar and Kasturi Bai 2009a).

9.3.3 Osmotic Adjustment

Osmotic adjustment is an important physiological mechanism to combat stress conditions, mainly under water deficit condition. Coconut palms accumulate organic solutes such as sugars and amino acids during stress period as compared to non-stress period. Cultivar differences were not found significant in sugar accumulation during non-stress and stress periods. Generally, 6.2–16.3% increase in sugar accumulation was observed during stress period compared to non-stress period. Similarly free amino acid accumulation increased by 38% during stress period as compared to non-stress period. Drought-tolerant types accumulated more of these solutes than the susceptible types during severe stress condition (Kasturi Bai and Rajagopal 2000; Kasturi Bai et al. 2006a). Similarly, accumulation of proline, starch and soluble sugars in the leaves of coconut palms subjected to drought stress has been demonstrated. A gradual increase in proline content was observed in coconut leaves under drought though no significant differences could be observed between cultivars. Furthermore, total soluble sugar in coconut cultivars have been shown to cause osmotic adjustment during drought stress (Lakmini et al. 2006).

9.3.4 Epicuticular Wax

Epicuticular wax (ECW) from the leaf surface of coconut has served as an important parameter for evaluating genotypes for drought tolerance. Generally, chloroform is being used as an organic solvent for the extraction of ECW from coconut leaf surface, though benzene gives better separation of wax components (Voleti and Rajagopal 1991). A three- to fourfold increase in ECW was observed in coconut forms subjected to drought stress (Voleti and Rajagopal 1991; Kurup et al. 1993). The wax formed in the leaf surface is greatly influenced by the physiological age of palms and leaves. Seedling leaves exhibited low (almost 50% less) ECW than that on the leaves of adult palms subjected to same intensity of moisture stress. Composition of wax content during different stages of palm's growth using TLC showed qualitative differences. Hydrocarbons and esters were formed as major components during non-stress period, whereas alcohols were identified during stress period, and fatty acids were detected during post-stress.

9.3.5 Water Consumption and Water Use Efficiency

Evidences of efficient stomatal regulation to maintain plant water status under a mild to relatively high water deficit stress are reported (Naresh Kumar et al. 2002b; Prado et al. 2001; Passos et al. 2003, 2005; Gomes et al. 2007; Gomes and Prado 2007). It has also been documented that water requirement of a palm depends on various factors such as its height, leaf area, soil type, growth conditions, age (Nogueira et al. 1998) and weather factors (Kasturi Bai et al. 1988). Even within the coconut canopy, the lower leaves have low Pn and WUE than others (Naresh Kumar and Kasturi Bai 2009b). Furthermore, Dwarf varieties of coconut consume more water than Tall varieties because of high transpiration rate, stomatal frequency, and limited epicuticular wax on the leaf surface (Rajagopal et al. 1990), as well as a poorer stomatal control of water loss (Passos and Silva 1990). However, some Dwarf varieties displayed better adaptation strategies to water deficit stress than Talls and hybrids owing to effective control on stomatal transpiration. The WUE has been shown to vary among varieties of Talls and Dwarfs and also among ecotypes of the same variety (Prado et al. 2001; Gomes et al. 2002). The tall genotypes, viz. Kalpadhenu and FMST, had high WUE under 100% field capacity due to their higher root biomass. On the other hand, under water deficit stress, Dwarf maintained higher WUE despite higher stomatal conductance (Hebbar and Chaturvedi 2015).

9.3.6 *Biochemical Traits*

Biochemical response of coconut palms to water deficit stress and information concerning the protoplasmic tolerance to drought stress has led to the conclusion that coconut leaves have efficient biochemical systems that protect cell membranes and their intracellular components. Lipid composition, lipid peroxidation level and the activities of enzymes related to oxidative stress are good indicators of dehydration tolerance in leaves of coconut. Water deficit induced a reduction in total leaf lipid content, mainly that of the chloroplast membranes, an effect particularly expressive in the less drought-tolerant genotypes (Repellin et al. 1994). In addition, an increase in the degree of lipid unsaturation in response to severe drought was also observed, which seems to be related to the maintenance of membrane fluidity, mainly in the chloroplasts (Repellin et al. 1997). Drought-tolerant coconut cultivars showed a lower level of lipid peroxidation and higher activity of catalase, superoxide dismutase and peroxidase than cultivars empirically classified as drought susceptible. Indeed, peroxidation level was negatively correlated ($r^2 > 0.73$) with activity of antioxidant enzymes (Shivashankar et al. 1991; Chempakam et al. 1993).

Hybrids, viz. LCT x GBGD, LCT x COD and WCT x COD, and the Talls, viz. JVGT, FMST, PHOT and CCNT, show higher ECW content than the other cultivars including Dwarfs. The transpiration (*E*) is inversely proportional to the content of ECW on the leaf surface (Rajagopal et al. 1990). Coconut palms subjected to drought stress have been demonstrated to have efficient enzymatic removal of ROS due to the increased activities of enzymes such as superoxide dismutase (SOD), catalase (CAT), peroxidase (POX) and polyphenol oxidase (PPO) (Kasturi Bai et al. 1996b; Mukesh Kumar et al. 2016) as well as acid phosphatase (APH) and L-aspartate:2-oxoglutarate amino transferase (AAT) in adult WCT palms, while activities of malic dehydrogenase (MDH) and nitrate reductase (NR) decreased (Shivashankar 1990; Shivashankar et al. 1991). Similarly, increased enzymatic activities were observed with PEG induced osmotic stress (Shivashankar 1990). However, the intensity of change was high because of rapid stress induction. Increase of APH and AAT activities in coconut leaves subjected to osmotic stress is correlated with decline in ψ_{leaf} (Rajagopal et al. 1988). Furthermore, enzymatic activity of NR decreases with a little change in relative water content (RWC) of leaves, and the activity loss was found to be high in susceptible cultivars. Besides, a correlation has been established between thermal stability of NR in vivo and drought tolerance in several coconut cultivars (Shivashankar 1992). Membrane integrity studies on coconut subjected to drought stress showed negative correlation between leaf water potential and MDA content (Kasturi Bai et al. 2011). Besides drought, influence of atmospheric concentration of CO_2 and temperature are considered to be major climatic factors that influence the growth, development and productivity of coconuts (Naresh Kumar et al. 2008b). Studies on antioxidant enzymes of coconut seedlings under elevated CO_2 and temperature revealed that specific activities of the enzymes were greatly altered. Elevated CO_2 caused significant increase in the activities of enzymes such as SOD, POX and CAT, whereas elevated temperature levels reduced the POX activity even though increased activities of SOD and CAT were observed. However, the activities of PPO decreased under both elevated CO_2 and temperature conditions. Increased activities of SOD, CAT and POX are pertinent to maintain integrity of the cell. It was also proposed that increased activity of SOD and CAT compensated for the decreased activity of POX in coconut seedlings under high ambient temperature stress (Sunoj et al. 2014). Moreover, low water potential of coconut seedlings under elevated temperature is compensated due to improved antioxidant potential and hence maintained a low growth. This study also identified WCT and COD x WCT as tolerant due to greater membrane stability and lower MDA content, under changing climatic conditions.

9.4 Physiological and Environmental Constraints for Higher Productivity

Weather variables like rainfall, day/night temperature regimes, relative humidity, sun shine duration and vapour pressure deficit play pivotal roles in crop growth, development and yield. Short-term responses of coconut to water stress such as low stomatal conductance, Ψleaf, net photosynthetic rate and transpiration rate have been demonstrated as mentioned above. As coconut is a source-limited crop (Naresh Kumar et al. 2008b), low Pn rates and mutual shading of leaves in canopy contribute to yield loss, apart from external stresses. The influence of weather on nut yield in coconut starts from inflorescence initiation and lasts till nut maturity (about a 44 month period). Coincidence of dry spell with critical stages, viz. initiation of inflorescence primordium, ovary development and button size nut, significantly reduces nut yield (Rajagopal et al. 1996). See Fig. 9.3 for details.

Earlier studies also indicated relationship between rain fall, other weather variables and nut yield in coconut (Peiris et al. 1995; Peiris and Thattil 1998; Naresh Kumar et al. 2008b, 2009a, b). Longer dry spell affects the nut yield for the next 4 years to follow with stronger impact in fourth year, irrespective of the total rainfall as depicted in Fig. 9.4 (Naresh Kumar et al. 2007b, 2009a).

Climatic extreme events such as tropical cyclones affect coconut palms, and severely affected palms take about 6 years to recover to give satisfactory nut yield

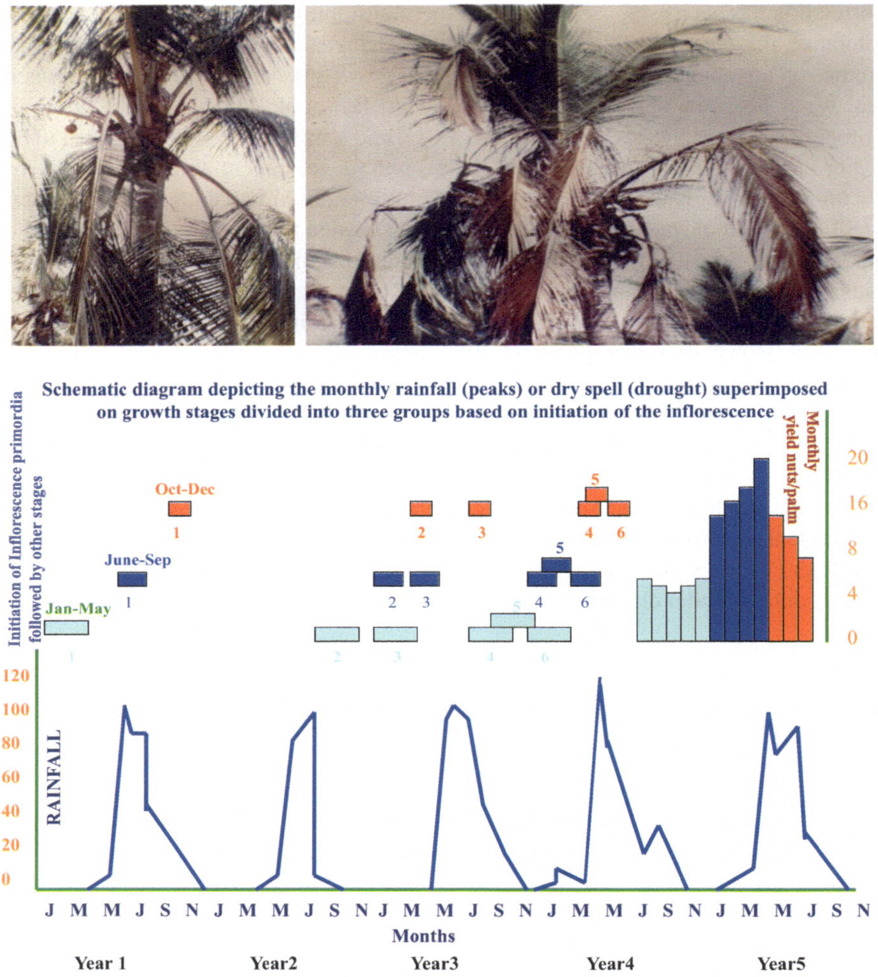

Fig. 9.3 Symptoms of drought stress in coconut palm. Coincidence of dry spell with sensitive stages of inflorescence and nut development and resultant nut yield. (redrawn from Rajagopal et al. 1996)

(Naresh Kumar et al. 2007b, 2009a; Naresh Kumar 2011a). Further, temperatures below 10 °C and above 40 °C affect effective leaf area, nut set and yield (Naresh Kumar et al. 2008b). Thus, unfavourable climatic conditions contribute to yield reduction. Simulation analysis indicated spatial variation and substantial yield gap varying from 34% to 85% depending on management in major coconut-growing regions in India (Naresh Kumar et al. 2008b). In addition to these, the biotic stresses further reduce nut yield. Therefore, genetic improvement for tolerance to biotic and abiotic stresses, as well as development of improved agronomic management to mitigate these stresses are needed for reducing yield gap.

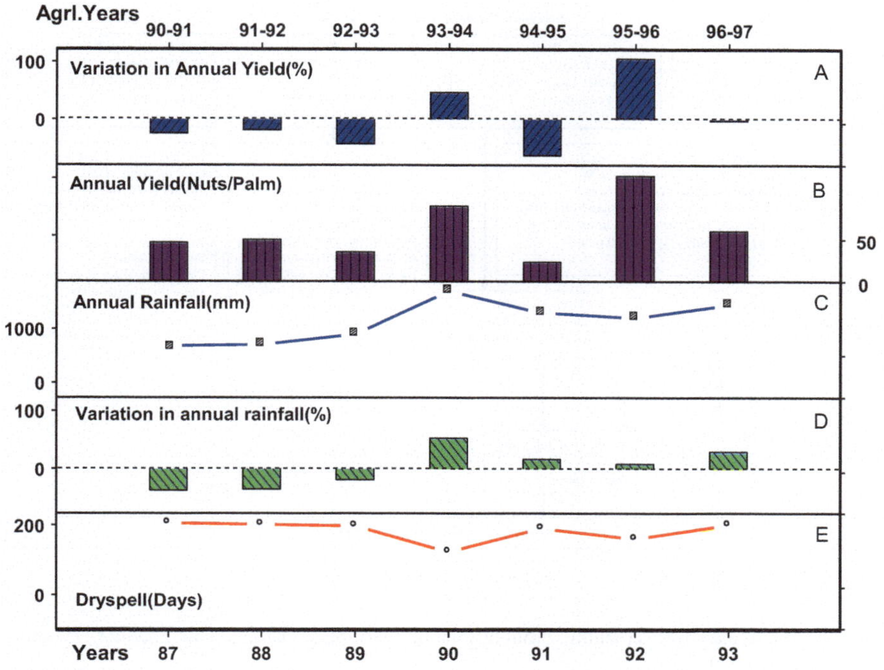

Fig. 9.4 Influence of dry spell and variation on annual rainfall on annual nut yield. Note the year lag (4 years) between lower and upper X axes. Dry spell, variation in annual rainfall and annual rainfall are represented by lower X1 axis. Annual yield and variation in annual yield are represented by upper X2 axis. Please note the difference in year in X1 and X2 axis

9.4.1 Selection Criteria for Drought Tolerance

Analysis of a range of cultivars over several years and with varying agronomic management showed drought as one of the major causes of yield loss in coconut (Rajagopal et al. 1996). The length of dry spell is positively correlated with nut yield reduction in different agroclimatic regions of India (Naresh Kumar et al. 2007b). Extensive research work carried out on coconut led to the delineation of drought tolerance mechanism (Fig. 9.5) and development of screening methods for identification of drought-tolerant genotypes (Fig. 9.6). Some of the essential anatomical, biochemical and physiological parameters such as cell size and number, sub-stomatal cavity size, stomatal frequency, epicuticular wax content, leaf thickness, stomatal resistance, water potential components, cell membrane stability and scavenging enzyme activities have been stipulated for assessing moisture stress in plants (Rajagopal et al. 1991, 2005; Kasturi Bai 1993; Champakam et al. 1993; Rajagopal and Kasturi Bai 2002; Naresh Kumar et al. 2000a, 2006a, 2016b). Some or all of these parameters were extensively used to screen coconut germplasm for drought tolerance (Pomier and de Taffin 1982; Rajagopal et al. 1990; Kasturi Bai

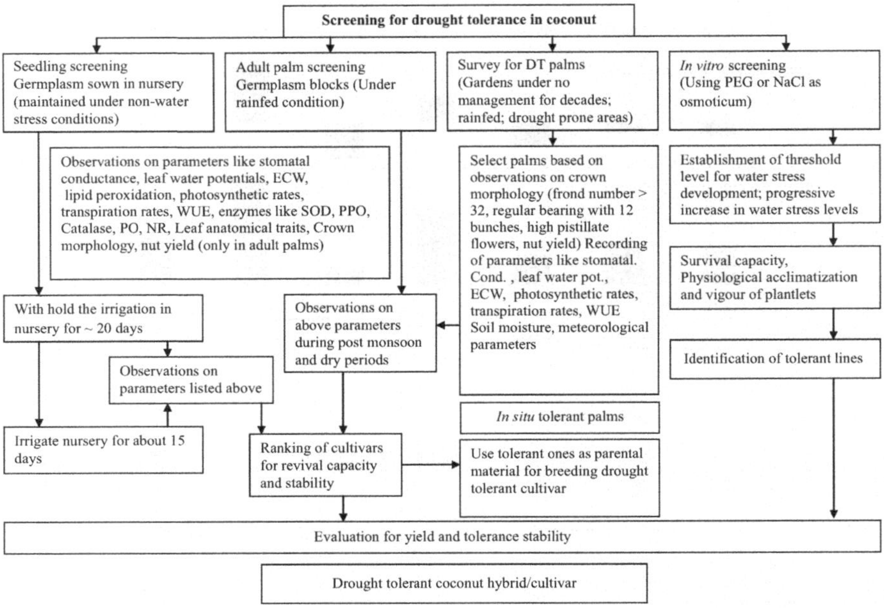

Fig. 9.5 Strategies for screening drought-tolerant germplasm/varieties in coconut. (Source: Naresh Kumar et al. 2016a, b)

1993; Repellin et al. 1994; Naresh Kumar et al. 2000a). Screening of parental lines and hybrids at the nursery stage indicated variations in tolerance to drought stress. The drought-tolerant cultivars/hybrids included were WCT, LCT, FMS, WCT x COD, LCT x GBGD and LCT x COD. Screening at the adult palm stage indicated WCT x WCT, FMST, LCT, WCT x COD, LCT x GBGD and LCT x COD as drought tolerant and MYD as drought susceptible (Rajagopal et al. 2000a, b).

9.4.2 *Other Abiotic Stresses*

Abiotic stresses that affect the growth, development and yield of coconut include flood, high light intensity and temperature apart from drought stress. Tropical cyclones uproot the palms, while flooding for long duration causes anoxia stress to roots affecting the water and nutrient uptake. The heat-stable protein fraction (HSPF) increased in leaf tissue of seedlings subjected to stresses like water scarcity, high temperature and flooding (Naresh Kumar et al. 2007a). The HSPF increased despite a decrease in total protein concentration. The HSPs that expressed during stress included ~66 KDa and ~76 KDa (water stress), ~66 KDa and ~76 KDa and LMW HSP of 14.4 KDa (high light intensity of ~1500 mmol/m^2/s) ~30 KDa and ~17 KDa (temperature and flooding). The LMW protein (~20.1 KDa) disappeared during water stress, while a ~30 KDa protein in root tissue was observed under temperature stress (Naresh Kumar et al. 2007a). High light intensity, as mentioned

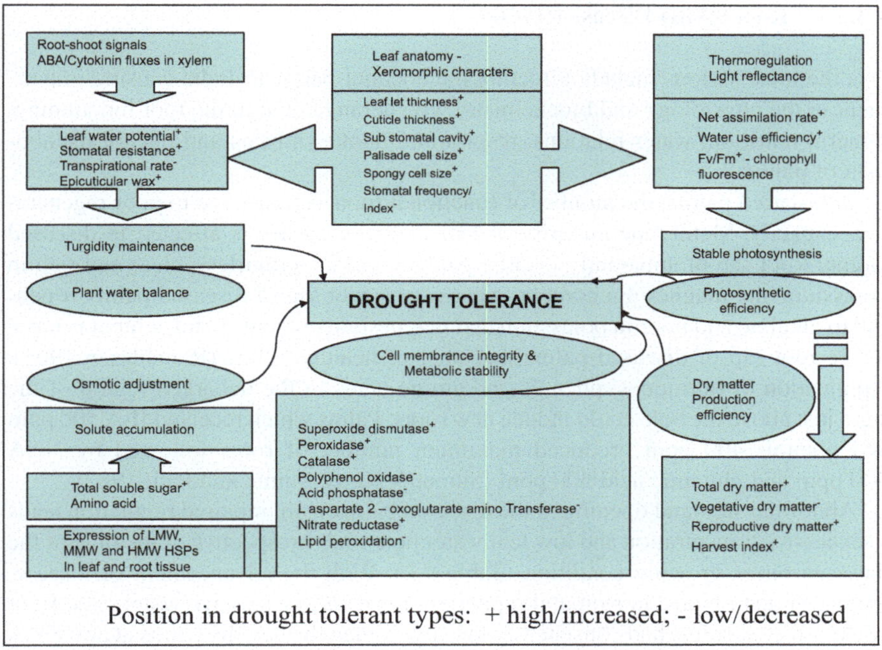

Functional relationship in mechanism of drought tolerance in coconut

Fig. 9.6 Adaptive strategies of coconut palm under stressful conditions. (Source: Naresh Kumar et al. 2016a, b)

earlier, causes oxidative stress with cascading effects such as membrane lipid peroxidation, reduced photochemical quenching, reduced water potential and chlorophyll bleaching leading to leaf scorching and seedling death (Naresh Kumar and Kasturi Bai 2009a). More details on the effect of abiotic stresses on coconut are presented in several publications (Naresh Kumar et al. 2016b).

Coconut is capable of using Na effectively, and application of NaCl is recommended. Coconuts grown on beaches and near brackish waters indicated salt tolerance. For instance, dwarf coconut seedlings are tolerant to a salinity level of 5.2 dS m^{-1} and moderately tolerant to salinity of 10.1 dS m^{-1}. This indicates the possibility of using saline water for raising quality seedlings. Moreover, studies on organic solute retention revealed that root system of dwarf coconut plays a role in tolerance of seedlings to salinity (Lima et al. 2017).

9.4.3 Biotic Stresses

The physiological responses of palms to pathological diseases like root (wilt), lethal yellowing and basal stem rot were used to develop diagnostic techniques for early detection of diseases.

9.4.3.1 Root (Wilt) Disease (RWD)

It is the most serious malady affecting the coconut palms in India causing impairment in the physiology and biochemistry, viz. derangement in the root functioning, mineral nutrition, water relations, respiration, photosynthesis and phenol metabolism of palms.

In diseased palms, the number of functional roots reduces due to poor regeneration capacity. Membrane integrity of leaf and root tissues is affected in diseased palms. Root sap of apparently healthy palms is acidic, odourless, clear and rich in potassium and magnesium content, whereas the root sap of diseased palms is neutral to alkaline and poor in potassium and magnesium content. Solid content is more in the root sap of diseased palms than that of healthy palms (Ramadasan 1964). Application of hormones, phenols and amino acids in the debarked region of the stem just above the bole could induce new roots. Palms which received IBA 500 ppm and phenols 400 ppm produced maximum number of roots followed by NAA 500 ppm and glutamic acid 500 ppm (Sumathy Kutty Amma and Patil 1982).

Abnormal stomatal opening in the diseased palms with impaired regulation leads to excessive transpiration and low leaf water potential, irrespective of the time of the day, season or growing condition (Table 9.2). With the advancement of disease, increased disturbance in stomatal regulation resulted in excessive water loss. Root (wilt) affected palms had consistently lower ψ_{leaf} than the healthy palms at any given time (Rajagopal et al. 1986, 1987).

The flow rate of phloem sap from the inflorescence of coconut depends on the age and disease severity (Rajagopal et al. 1989a). In general, in a palm, 4 to 5 inflorescences yield the phloem sap. However, rate of sap flow decreases when the successive inflorescences are tapped (Fig. 9.7).

The rate of sap flow is higher during the day than at night, and the sap quality also varies in pH, osmotic concentration and total and reducing sugars which are generally higher during the day (Rajagopal et al. 1989a). The concentration of biochemical constituents was low in the phloem sap collected from inflorescences of diseased palms. Concentrations of arginine, aspartic acid and tyrosine as well as

Table 9.2 Changes ψ_{leaf} (Mpa) with age of leaf and leaflet condition in root (wilt) affected palms

Leaf position in canopy	Apparently healthy palm	Ψ leaf (Mpa)	Diseased palm	Ψleaf (Mpa)
Spindle	Yellow to light green thick, stiff	−0.37	White to dull cream, thin papery, brown spots	0.72
First whorl	Green, normal, erect	−0.68	Light green, slight flattening and bending at the tip	−1.24
Middle	Dark green normal, erect	−0.79	Flaccidity, yellowing, necrosis	−1.28
Outer whorl	Dull green, senescent, but normal	−0.89	Ribbing, necrosis, abnormal, senescence	−1.26

Rajagopal et al. 1986

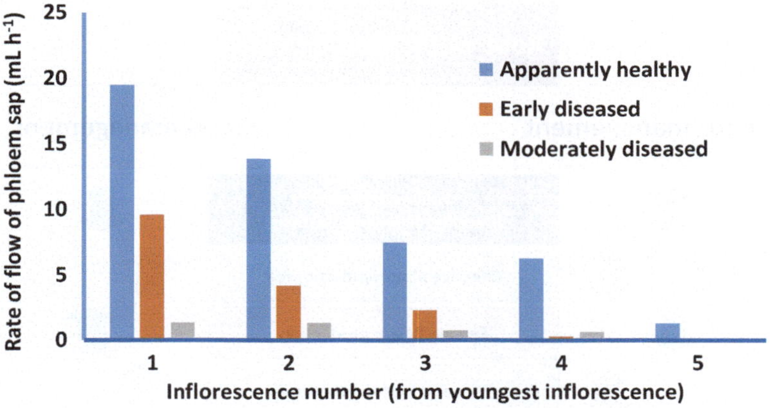

Fig. 9.7 Rate of flow of phloem sap (mL h^{-1}) from successive inflorescences

glucose and galactose were higher in sap from the diseased palms than from apparently healthy palms (Chempakam et al. 1991).

The root (wilt) diseased palms show higher respiratory rates in leaf and roots compared to apparently healthy palms (Michael 1978). On the other hand, chlorophyll concentration and photosynthetic rates are high in apparently healthy palms compared to those in root (wilt) affected palms. An impairment of sugar metabolism, translocation and distribution and a reduction in the C/N ratio were noticed in the roots and leaves of diseased palms (Varkey et al. 1969). Derangement in the nitrogen metabolism resulted in increase in the nonprotein nitrogen content, concomitant with a sharp decrease in the water-soluble nitrogen and protein nitrogen fractions in the diseased tissue (Padmaja et al. 1981). Free amino acids in the leaves, particularly arginine, increases with the incidence and intensity of the disease. The activity of carbonic anhydrase (CA), the enzyme-limiting carbon metabolism, is low in leaves of diseased palms as compared to that in healthy palms. On the other hand, higher activity of cellulase and pectin lyase activity was noticed in roots of diseased palms as compared to the healthy ones (Padmaja and Sumathykutty Amma 1979). Isotopic (^{32}P) studies revealed delayed uptake of phosphorus in diseased palms compared to healthy palms (Dwivedi et al. 1979). This indicates less utilization of absorbed phosphorus in the synthesis of phosphorus constituted organic substances in diseased palms.

Physiological Trait-Based Diagnosis for Early Detection Based on the abnormal stomatal opening phenomena in diseased palms, an early diagnostic tool for root wilt has been developed. In this method, stomatal resistance is used as a criterion for the pre-visual detection of root (wilt) disease in coconut (Rajagopal et al. 1988). This is comparable to the serological test, using the cross-absorption technique, to detect the disease before the manifestation of visual symptoms. Both the tests could detect disease-affected palms 6–20 months earlier than the actual manifestation of flaccidity symptom. Stomatal resistance determinations are also employed for diag-

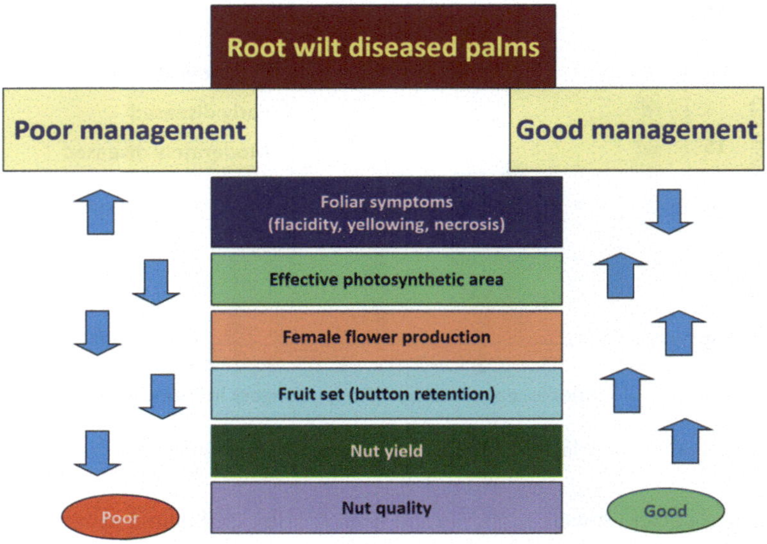

Fig. 9.8 Influence of good management on source and sink parameters and yield of root (wilt)-affected coconut gardens

nosing the lethal yellowing affected coconut palms (Eskafi et al. 1986). The leaf water potential measurements are also effective for the early diagnosis of palms affected by phytoplasmas as shown both in LYD (Eskafi et al. 1986) and RWD (Rajagopal et al. 1986).

Physiological Basis for Management of Root (Wilt) Disease The CGD (Kalpasree) is identified as a relatively tolerant variety to root (wilt) disease. Kalpa Sankara (CGD x WCT) has been released for cultivation in the RWD affected tracts, considering its low disease incidence and high yield potential (Shareefa and Regi Thomas 2016). Management of coconut root (wilt)-affected gardens and managing leaf rot are viable options to maintain nut yield in affected gardens (Fig. 9.8). Increase in photosynthetically active LA resulted in improved TDM production resulting in higher nut yield.

9.4.3.2 Lethal Yellowing Disease (LYD)

Physiological and biochemical disorders in LYD-affected palms have been reported by Eskafi et al. (1986). Extensive root necrosis observed in diseased palms (Eskafi et al. 1986) has been correlated with foliar yellowing (Eden Green and Waters 1982). Lower stomatal conductance during noon, concomitant with greater stomatal closure, has been observed in LYD-affected coconut palms than in healthy palms (Eskafi et al. 1986). A permanent stomatal closure was observed with the development of LY symptoms. Ψ leaf was in the range of -0.1 to -1.0 MPa between night and midday in healthy palms, whereas diseased palms rarely exhibited Ψ leaf lower than -0.4 Mpa. This implies that stomatal closure is exhibited in the diseased palms (Mc Donough and Zimmermann 1979). Decrease in turgor, osmotic and water potentials with disease intensity were also observed (Leon et al. 1995). Due to the permanent closure of stomata, water transport through xylem was affected in LYD-affected palms. When ^{32}P was applied at the base of the petiole of diseased palms, xylem transport was reduced by 75%, while the reduction was 100% when applied through either stem or roots (Eskafi et al. 1986). The photosynthetic rate and protein content were maintained at higher level during the early stages of the disease, whereas during the advanced stage, a sharp decline in these traits was noticed. Free amino acids and disease susceptibility are related (McCoy et al. 1983). Arginine content in the leaf is an important diagnostic tool for the early detection of LY disease. Sharp decline in respiration, total sugars and reducing sugars has been observed in both the primary and secondary roots of LYD-affected palms (Oropeza et al. 1995).

A comparison made between the LY disease of Jamaica and RWD of India (Rajagopal 1991) showed certain similarities and distinct differences in some of the characteristics (Table 9.3). Although both are caused by phytoplasmas, LYD-affected palms differ from the RWD not only in the primary foliar symptoms but also in the water relations. The water transport is affected to different degrees in both the diseases, but it is the stomatal regulation which seems to play a key role in the ultimate expression of foliar symptoms. Rapid closure of stomata due to infection led to a

Table 9.3 Comparison between the lethal yellowing disease and root (wilt) disease

Description	LYD	RWD
Earliest record/location	1904, Cayman Islands	1874–1884, Kottayam, Kerala, India
Symptoms, primary associated	Yellowing, inflorescence necrosis, immature fruit drop	Flaccidity, yellowing, necrosis
Nature of disease	Lethal	Debilitating
Causal organism	Phytoplasma-like organisms	Phytoplasma-like organisms
Insects associated with transmission	Plant hopper (*Haplaxius crudus* (van Duzee))	Lace bug (*Stephanitis typica*) and *Proutista moesta (Westwood)*
Transmission through	Not done	*Cassytha* filiformis
Water absorption/ transport	Decreased	Decreased
Stomata	Closed	Abnormal opening
Transpiration rate	Normal	Excessive
Leaf water potential	Low	Low
Early diagnosis	Stomatal resistance and water potential	Serological reaction, Stomatal resistance, water potential
Flow of phloem sap	Nil or very low	Low
Arginine content	High in leaves	High in leaves and phloem sap
Antibiotic therapy	Tetracycline	Tetracycline

Modified from Rajagopal (1991)

typical symptom expression, viz. yellowing in LY diseased palms, whereas impaired stomatal regulation (abnormal opening) led to excessive transpirational loss of water and resulted in wilt (flaccidity) symptom in RWD palms. Interestingly, although both the water absorption and transpiration rates are affected, the RWD remained as a debilitating disease in contrast to the lethal nature of LY disease. This shows that the metabolic status of these palms differed in some respects. The insects involved in the transmission of phytoplasmas are also different for the two diseases.

9.4.3.3 Basal Stem Rot

Basal stem rot (BSR) disease of coconut is caused by two species of Ganoderma, viz. *Ganoderma applanatum* and *Ganoderma lucidum*. The disease is referred to as Ganoderma Wilt or Thanjavur Wilt or Anabe roga in different parts of India. In Sri Lanka, the disease is caused by *G. boninense*. The diseased palms are characterized by low stomatal resistance and high transpiration rate, as compared to healthy palms. The transpiration rate increases with the severity of the disease, and higher transpiration rates are observed at least 6 months before the expression of disease symptoms. In diseased palms, the contents of mineral nutrients, viz. N, P, K, Ca and Mg in the leaf, stem, bole and root tissue, decreased (Anbalagan et al. 1987). An increase in total phenols (20–35%) and orthodihydroxy phenols (40–48%) and a reduction in total and reducing sugar levels have been noted in the diseased palms. Impairment in membrane stability has also been observed in the diseased palms as expressed by high electrolyte leakage in root and leaf tissues of diseased palms than in healthy palms. Similarly, the leaves of diseased palms show a reduction in the relative water content as compared with healthy leaves. For the early detection of disease, EDTA colour test was employed (Natarajan et al. 1986).

9.5 Conditions for Optimum Growth and Yield

The coconut palm (*Cocos nucifera* L.), mainly a crop of humid tropics, grows mostly between 26°N and 27°S and up to the altitudes of 600 m above mean sea level. A well-distributed rainfall (annual rainfall between 130 and 230 cm), mean annual temperature of 27 °C with diurnal variation of 5 °C and abundant sunlight ranging from 250 to 350 W m^{-2} with annual sunshine of 2000 h (at least 120 h per month) are optimum conditions for good growth and nut yield in coconut.

9.5.1 Characterization of the Growing Environment

Inadequate or uneven distribution of rainfall and summer dry spells cause water stress situation in rainfed plantations. Minimum temperature of above 10 °C is required for flowering, whereas a temperature below 10 °C for a month causes nut fall, and similarly, temperature above 40 °C during April–July leads to decline in effective functional leaf area index, dry matter production and consequently nut yield (Naresh Kumar et al. 2008b). Prevalence of high ambient leaf temperatures and high VPD (Escbach et al. 1982; Rajagopal et al. 2000a), low water potentials (Repellin et al. 1997; Rajagopal et al. 2000a) and stomatal and non-stomatal limitations (Gomes et al. 2008; SijuThomas et al. 2008) lead to impaired photosynthetic rates and lesser dry matter accumulation resulting in low yield. Coconut being a perennial crop, previous years' rainfall influence total annual yield (Peiris et al. 1995; Peiris and Thattil 1998). The inflorescence and nut development stages are more sensitive to the coinciding dry spell (Rajagopal et al. 1996). Coconut adapts to different weather conditions through accumulation of biochemicals (SijuThomas et al. 2006). Besides, the quantum of rainfall and length of dry spells experienced over the preceding 4 years heavily hamper the productivity of coconut in different agroclimatic zones of India (Naresh Kumar et al. 2007b). Weather change in the past has been influencing the coconut yields in India (Naresh Kumar et al. 2009b). In coastal Andhra Pradesh, 1995 cyclone severely affected coconut palms and it took seven years (2003–04) to recover to pre-cylone period yield levels (Naresh Kumar 2011b).

9.5.2 Physiological Optimization for Growth and Development in Resource Constraint Environments

Physiological-Based Agronomic Management Options Coconut palms exposed to moisture stress for 16–24 days show a reduction of 15–18% in vegetative dry matter (VDM) and 20–22% reduction of reproductive dry matter (RDM) production compared to control palms kept under well-watered conditions. Similarly, palms grown under rainfed conditions showed a reduction of 29% in RDM and 19% in VDM (Rajagopal et al. 1989b). Drought management strategies mainly include the conservation of available soil moisture and efficient use of available water resources for high production. Improvement of soil moisture status through conservation measures such as burial of husks or coir pith in basins, mulching of dried leaves including gliricidia, farm waste and compost application in the basins resulted in increased nut yield (Rajagopal and Naresh Kumar 2001; Naresh Kumar et al. 2006b). Application of water greatly increased photosynthetic rates and improved stomatal conductance. Moreover, irrigation significantly increased the Pn rates and stomatal conductance. Among the various methods of irrigation, drip irrigation resulted in conducive physiological efficiency of source and sink for high yield and water use efficiency (Table 9.4) Naresh Kumar et al. 2002a). Palms grown under drip irrigation have been demonstrated to have high leaf water potential that could be attributed to better regulation of stomatal conductance leading to improved water use efficiency. Efficient irrigation conditions such as drip irrigation improved production of female flowers and nut set. Other agronomic practices for soil management

Table 9.4 Source-sink response of coconut to different water availability conditions

Source	Sink	Condition	Yield and WUE
Low $\psi_{leaflet}$, *E*, *gs* and *Pn*	*Less FFP and nut retention*	Rainfed	Low/low
Low $\psi_{leaflet}$, high *E*, *gs* and *Pn*	*More FFP and nut retention*	Basin irrigated	High/low
High $\psi_{leaflet}$, medium *E*, *gs* and high *Pn*	*More FFP and nut retention*	Drip	High/high

to conserve soil water include organic farming and tillage practices like summer ploughing, soil mulching and addition of soil stabilizers.

Among the nutrients, potassium chloride or sodium chloride nutrition imparts drought tolerance to coconut through stomatal regulation (Braconnier and Bonneau 1998). Absence of chloroplasts in the coconut leaf guard cells causes non-availability of malate. The Cl ion possibly replaces malate as an osmoticum to maintain water potential. Hence Cl ion (from KCl or NaCl) is essential for coconut growth in dry conditions. By maintaining cell osmoregulation, the Cl ion increased water absorption and reduced transpiration from leaf (Escbach et al. 1982; Ollagnier et al. 1983). The critical level of Cl in 14th leaf is about 0.7%. Application of KCl increased the drought tolerance of palms under dry conditions as potassium nutrition also plays an important role in drought tolerance in coconut (Ollagnier et al. 1983; Rajagopal et al. 2000b; Rajagopal and Naresh Kumar 2001). Deficiency of K and Cl results in yellowing and drying of leaves due to impaired water potential. Palms fertilized with higher levels K_2O have shown higher stomatal regulation under rainfed condition. For water logging situations, planting on raised bunds is a good practice. The seedlings are to be provided with shade to protect from high light intensity stress.

Improving Genotypic Adaptation Genetic analysis of drought-responsive physiological traits such as water potential, Pn and transpiration indicated that transpiration rate and leaf water potential showed higher specific combining ability effects due to predominant nonadditive gene action (Rajagopal et al. 2007). Using morphological criteria (number of leaves, nut yield), physiological parameters (Pn, gs, WUE), stable isotope discrimination (C^{13}) and biochemical constituents, some in situ drought-tolerant palms were identified for use in population improvement programme (Naresh Kumar et al. 2002b). Strategies for improving drought tolerance in coconut (Fig. 9.5) need to be implemented for improving coconut yield especially in rainfed conditions and water-limited conditions (Naresh Kumar et al. 2016a).

9.6 New Tools for Complex Tasks

From early 2000, plant physiology research in plantation crops has been integrated with information processing and molecular biology to enhance the potential for exploiting physiological understanding in crop improvement. Spectral reflectance signature based on the chlorophyll character can be used to identify the vegetation or material, if the sensing system has sufficient spectral resolution to distinguish particular spectrum of the targeted vegetation or material. Though as early as 1968 the first application of remote sensing technology in agriculture was on mapping coconut root (wilt) affected area, it was not taken forward. Limited effort has been made to obtain biomass, vigour and canopy cover on a temporal and spatial scale using Remote Sensing and GIS. It has huge potential in determining the coconut area and crop condition. In combination with simulation model, the technology can be used as decision support system for improved management for higher productivity. Simulation models enhance the ability to handle large amount of data and to explore the dynamics of physiological processes and their complex interactions with the environment. Development of simulation model (InfoCrop-COCONUT) opened up several possibilities for spatiotemporal studies (Naresh Kumar et al. 2008b). Model-based assessment of coconut productivity under the current scenario and changing climatic scenarios in the future has been worked out. This study also warned that where productivity levels are expected to be gaining, current poor management practices would be limiting factors to reap the benefits of elevated CO_2 levels in future climatic scenario.

Though application of genetic engineering and genome editing is limited in plantation crops, the molecular biology studies are improving the understanding on stress responses of coconut. Expression of low, medium and high molecular weight stress proteins in seedlings exposed to drought stress is reported (Naresh Kumar et al. 2007a). The RAPD and ISSR markers which correlated with leaf water potential in coconut cultivars are identified (Manimekalai et al. 2004, 2005). Water stress-related MAPK genes were sequenced in coconut (Bobby et al. 2012). Though a viable regeneration technique is yet to be achieved, efforts are going on to develop molecular markers for drought tolerance for marker-assisted selection (MAS). From the RFLP analysis (Lebrun et al. 1998, 1999) to SCoT analysis (Rajesh et al. 2017), several studies have quantified the genetic relationship and diversity among coconut accessions. A cDNA clone (467 bp) encoding MAPK, significantly homologous to that of maize, rice and wheat, was isolated from leaves of water-stressed coconut plantlets. Coconut MAPK belongs to the serine/threonine kinases plant TEY MAPK subfamily group A (Bobby et al. 2012). The water stress-responsive candidate genes such as AP2, CBF, MAPK, NAC and 14-3-3 were upregulated during stress. These, as well as new molecular markers and techniques, will be extremely useful to identify the lines with desirable traits for crop improvement.

9.6.1 Simulation Modelling of Growth and Development

Simulation models are being increasingly used as decision support systems in agricultural management. Crop simulation models are effective tools for the assessment of growth and yield of crops in diverse environmental and management conditions. These are being used for identifying optimal resource management options, desired cultivar characteristics, performance evaluation of weather forecasters, regional yield prediction, yield gap analysis, crop zonation, research need identification and climate change studies. They become even more important for perennial crop management and for research and development decisions. A model for simulating growth, development and yield of coconut, InfoCrop-COCONUT, was developed, calibrated and validated using data on palm, soil and weather over the years (Naresh Kumar et al. 2008b). The InfoCrop model developed not only could simulate dry matter production of coconut, DM partitioning and nut yield but also was useful for ascertaining potential coconut yields in various agroclimatic zones of the country (Naresh Kumar et al. 2008b). Furthermore, the model obviates the need for agronomic or plant breeding experiments as it can help simulate multilocation trials and forecasts yield levels of coconut. Using the model, climate change impacts on coconut plantations in India, adaptation options and adaptation gains were quantified at spatial scale (Naresh Kumar and Aggarwal 2013). The assessment indicated positive impacts of climate change on coconut in the western coastal region, Kerala, parts of Tamil Nadu, Karnataka and Maharashtra and also in North-Eastern states, islands of Andaman and Nicobar and Lakshadweep, while negative impacts are projected for Andhra Pradesh, Odisha, West Bengal, Gujarat and parts of Karnataka and Tamil Nadu in India. On all India basis, even with current management, climate change is projected to increase coconut productivity by about 1.9–6.8% towards the end of the century. A study indicated the current productivity can be improved by 20% to almost double if all plantations in India are under improved management.

9.7 Future Strategy

Classic plant physiological techniques have been effectively employed to study the physiology of coconut which helped identification of cultivars or forms suitable for drought tolerance, improved water use efficiency, photosynthetic efficiency and with good root characteristics. Further, suitable climatic parameters have been worked out to identify target environment, and operating future environment has also been identified using simulation model and growth gains under elevated CO_2 and temperature. Future thrust may involve identification of natural sources of resistance to biotic and abiotic stresses using advanced phenomics, metabolomics, proteomics and genomic analysis approaches complementary to physiological and biochemical studies. Evaluating identified drought-tolerant genotypes in different agroclimatic zones, and using identified in situ drought-tolerant palms in population improvement programmes should be strengthened. Detailed physiological studies are required on developing physiological thresholds for sensor-based precision management practices. There is also a need to develop DSS for coconut management using simulation model and remote sensing technology. The simulation model can be used to optimize the plantation management, identify research and development priorities and optimize the multilocation experimentation, among many other potential applications such as precision management of plantations.

References

Abeywardena (1968) Forecasting coconut crops using rain fall data. III Session of FAO Tech working party on coconut production, protection and processing. Jakarta, Indonesia. LCNP 68(29):49

Anbalagan R, Shanmugam N, Bhaskaran R, Vijayaraghavan H (1987) Biochemical constituents of coconut palms as influenced by Thanjavur Wilt disease. Abstracts of international symposium on Ganoderma wilt disease on palms and other perennial crops. Tamil Nadu Agricultural University, Coimbatore, pp 17–18

Balasubramaniam K (1976) Polysaccharides of the maturing and matured coconuts. J Food Sci 47:1307–1337

Balasubramaniam K (1983) Biochemical changes in coconut during maturation and germination. In: Nayar NM (ed) Coconut research & development. Wiley Eastern Ltd., New Delhi, pp 221–228

Balasubramaniam K, Alles NH (1989) A preliminary study of the invertase activity in coconut. Ann Bot 64:253–257

Balasubramaniam K, Atukorala TMS, Wijesundara S, Hoover AA, De Silva MAT (1973) Biochemical changes during germination of the coconut. Ann Bot 37:439–445

Balasubramaniam K, Sothary RD, De Silva MAT (1974) Nutritional studies on initial flowering of coconut (Var. Typica) III. Free amino acids in leaves of bearing and non- bearing palms. Ceylon Cocon Qtly 25:149–152

Baptist NG (1956) L-amino acid and other free amino acids in the coconut. Nature 178:1403–1404

Bhatnagar SP, Johri BM (1972) In: Kozlowski TT (ed) Seed biology, vol 1. Academic Press, New York, p 78

Bobby P, Rajesh MK, Naresh Kumar S (2012) Molecular cloning and characterization of a partial mitogen-activated protein kinase cDNA, CnMAPK1 induced in response to water stress in coconut. Indian J Hortic 69(2):145–150

Braconnier S, Bonneau X (1998) Effects of chlorine deficiency in the field on leaf gas exchanges in the PB121 coconut hybrid. Agronomie 18(8–9):563–572

Chempakam B, Rajagopal V, Gopalam A (1991) Identification of biochemical constituents in die phloem sap from root(wilt) diseased and apparently healthy coconut palms. Plant Physiol Biochem 18:21–25

Chempakam B, Kasturi Bai KV, Rajagopal V (1993) Lipid peroxidation and associated enzyme activities in relation to screening for drought tolerance in coconut (*Cocos nucifera* L.). Plant Physiol Biochem 20:5–10

Child R (1964) Coconuts. Longmans, London, p 216

Corley RHV (1983) Potential productivity of tropical perennial crops. Exp Agric 19:217–237

Cutter VM Jr, Wilson KS, Freeman B (1955) Nuclear behaviour and cell formation in the developing endosperm of *Cocos nucifera*. American J Bot 42(2):109–115

Datt B (1999) A new reflectance index for remote sensing of chlorophyll content in higher plants: tests using *Eucalyptus* leaves. J Plant Physiol 154:30–36

De Silva MAT, Abeywardena V, George GD, Balasubramaniam K (1973) Nutritional studies on initial flowering of coconut (var. *typica)*. Effect of Mg deficiency and Mg-P relationship. Ceylon Cocon Qtly 24:107–113

Deshpande BR, Kulkarni VG (1962) Studies on germination of coconut. Part 1. Cocon Bull 16:336–338

Dix L, Van Staden J (1982) Auxin and Gibberellin -like substances in coconut milk and malt extract. Plant Cell Tissue Organ Cult 1:239–246

Dwivedi RS, Ray PK, Ninan S (1979) Absorption, distribution and utilisation of radioactive phosphorus in unhealthy and root (wilt) diseased coconut palms. J Nucl Agric Biol 8:33–35

Eden G, Waters H (1982) Collection and properties of phloem sap from healthy and lethal yellowing diseased coconut palms in Jamaica. Phytopathol 72:667–672

Escbach JM, Massimino D, Mendoza MR (1982) Effect of chlorine deficiency on germination, growth and photosynthesis of coconut. Oleagineux 37:115–125

Eskafi F, Basham H, Mc Coy R (1986) Decreased water transport in lethal yellowing diseased coconut palms. Trop Agric 63:225–228

Espino RB (1923) On the germination of coconut seedlings. Philippine Agric 11:191–200

Foale MA (1968) The growth of the young coconut palm (*Cocos nucifera* L.). 1. The role of the seed and photosynthesis in seedling growth up to 17 months of age. J Agric Res 19:781–789

Foale MA (1993) Physiological basis for yield in coconut. In: Nair MK, Khan HH, Gopalasundaram P, Rao EVVB (eds) Advances in coconut research and development. Oxford & IBH Publishing Co., New Delhi, pp 181–189

Fremond Y, De Lamothe MDN (1966) Dechracteristiaues et production de cocotier hybrids Nain jaune “Malaisie x Grand ouest African”. Oleagineux 26:459–464

Friend D, Corley RHV (1994) Measuring coconut palm dry matter production. Exp Agric 30:223–235

Gomes FP, Prado CHBA (2007) Ecophysiology of coconut palm under water stress. Brazilian J Plant Physiol 19:377–391

Gomes FP, Mielke MS, Almeida A-AF (2002) Leaf gas exchange of green dwarf coconut (*Cocos nucifera* L. var. nana) in two contrasting environments of the Brazilian north-east region. J Hortic Sci Biotechnol 77:766–772

Gomes FP, Oliva MA, Mielke MS, Almeida A-AF, Leite HG, Aquino LA (2007) Photosynthetic limitations in leaves of young Brazilian Green Dwarf coconut (*Cocos nucifera* L. ‘*nana*’) palm under well-watered conditions or recovering from drought stress. Environ Exp Bot 62(3):195–204. https://doi.org/10.1016/j.envexpbot.2007.08.006

Gomes FP, Marco MAO, Mielke S, de Almeida AAF, Leite HG, Aquino LA (2008) Photosynthetic limitations in leaves of young Brazilian Green Dwarf coconut (*Cocos nucifera* L. ‘nana’) palm under well-watered conditions or recovering from drought stress. Environ Exp Bot 62:195–204

Hagenmaier R, Cater CM, Mattil KF (1972) A characterization of two chromatographically separated fractions of coconut protein. J Food Sci 37:4–7

Hebbar KB, Chaturvedi VK (2015) Impact and adaptation strategies of coconut to climate change. In Proceedings of Kerala Environment Congress-2015, Centre for Environment and Development, Thiruvananthapuram, India, May 06–08, 2015, pp 73–78

Hebbar KB, Chowdappa P (2016) Transforming coconut farmer into entrepreneur. In ICAR-CPCRI 2016- Centenary Souvenir-1916 to 2016, pp 75–83

Hebbar KB, Mathew AC, Arivalagan M, Samsudeen K, Thomas GV (2013) Value added products from Neera. Indian Cocon J LVI(4):28–33

Hebbar KB, Arivalagan M, Manikantan MR, Mathew AC, Thamban C, Thomas GV, Chowdappa P (2015a) Coconut inflorescence sap and its value addition as sugar – collection techniques, yield, properties and market perspective. Curr Sci 109. https://doi.org/10.18520/v109/i8/1411-1417

Hebbar K B, Arivalagan M, Manikantan MR, Mathew AC, Chowdappa P (2015b) Kalparasa collection and value addition. Technical Bulletin No. 92, ICAR-CPCRI, Kasaragod, Kerala, India, p 28

Hebbar KB, Mukesh Kumar B, Chaturvedi VK (2016a) Plantation crops: climatic risks and adaptation strategies. Indian J Plant Physiol. https://doi.org/10.1007/s40502-016-0265-9

Hebbar KB, Mukesh Kumar B, Arivalagan M, Chaturvedi VK (2016b) Physiological and biochemical response of coconut to climate change variables. In: Hebbar KB, Naresh Kumar S, Chowdappa P (eds) Impact of climate change on plantation crops. Daya Publishing House, New Delhi, pp 45–60

Hebbar KB, Subramanian P, Sheena TL, Shweta K, Sugatha P, Arivalagan M, Prasad PVV (2016c) Chlorophyll and nitrogen determination in coconut using a non-destructive method. J Plant Nutr 39(11):1610–1619. https://doi.org/10.1080/01904167.2016.1161781

Jayalakshmy A, Arumugham CS, Narayanan C, Mathew AG (1988) Changes in chemical composition of coconut water during maturation. Oleagineux 43:409–414

Jayasekara C, Mathes DT (1992) A method to determine leaf area of a frond and the whole canopy of an adult palm. Indian Cocon J 23:7–13

Jayasekara C, Nainanayake NPAD, Jayasekara KS (1996) Photosynthetic characteristics and productivity of the coconut palm. COCOS 11:7–20

Juliano JB (1926) Origin, development and nature of the stony layer of the coconut (*Cocos nucifera* L.) Philippine. J Sci 30:187–200

Kamala Devi CB, Velayudham M (1978) Changes in the chemical composition of nut water and kernel during development of coconut. In: Nelliat EV (ed) Proceedings of PLACROSYM I. Indian Society for Plantation Crops, Kasaragod, pp 340–346

Kartha ARS (1981) Embryo of coconut and its germination – a short scientific note. J Plantn Crops 9(2):125–127

Kasturi Bai KV (1993) Evaluation of coconut germplasm for drought tolerance. Ph.D. Thesis, Mangalore University, Mangalore, India, pp 257

Kasturi Bai KV, Rajagopal V (2000) Osmotic adjustment as a mechanism for drought tolerance in coconut (*Cocos nucifera* L.). Indian J Plant Physiol 5(4):320–323

Kasturi Bai KV, Ramadasan A (1978) Changes in leaf carbohydrates in relation to commencement of flowering in coconut palm. In: Nelliat EV (ed) Proceedings of PLACROSYM I. Indian Society for Plantation Crops, Kasaragod, pp 321–330

Kasturi Bai KV, Ramadasan A (1982) Changes in the carbohydrate fractions in relation to female flower production in coconut. J Plantn Crops 10(2):124–126

Kasturi Bai KV, Ramadasan A (1983) Changes in the levels of carbohydrate as a function of environmental variables in hybrids and tall coconut palms. In: Nayar NM (ed) Coconut research and development. Wiley Eastern Limited, New Delhi, pp 203–209

Kasturi Bai KV, Ramadasan A (1990) Growth studies in coconut seedlings. J Plantn Crops 18(2):130–133

Kasturi Bai KV, Voleti SR, Rajagopal V (1988) Water relations of coconut palms as influenced by environmental variables. Agric For Meteorol 43:193–199

Kasturi Bai KV, Rajagopal V, Prabha CD, Ratnambal MJ, George MV (1996a) Evaluation of coconut cultivars and hybrids for dry matter production. J Plantn Crops 24:23–28

Kasturi Bai KV, Rajagopal V, Chempakam B, Prabha C (1996b) Assay of enzymes in coconut cultivars and hybrids under non-stress and stress condition. J Plantn Crops 24:548–554

Kasturi Bai KV, Rajagopal V, Balasimha D, Gopalasundaram P (1997) Water relation, gas exchange and dry matter production in coconut (*Cocos nucifera* L.) under un-irrigated and irrigated conditions. CORD XIII(2):45–58

Kasturi Bai KV, Rajagopal V, Arunachalam V (2006a) Assessment of diversity in coconut varieties for drought responsive physiological traits. J Plantn Crops 34(2):118–120

Kasturi Bai KV, Rajagopal V, Naresh Kumar S (2006b) Chlorophyll fluorescence transients with response to leaf water status in coconut. Indian J Plant Physiol 11:410–414

Kasturi Bai KV, Naresh Kumar S, Rajagopal V, Vijayakumar K (2008) Principal component analysis of chlorophyll fluorescence transients for tolerance to drought stress in coconut seedlings. Indian J Hortic 65(4):471–476

Kasturi Bai KV, Sugatha P, Muralikrishna KS, John Sunoj VS (2011) Physio chemical changes in coconut leaf at different maturity stages. J Plantn Crops 39:319–321

Kooiman P (1971) Structure of the galactomannans from seeds of *Annona muricata, Arenga saccharifera. Cocos nucifera, Convolvurus tricolor* and *Sophora japonica.* Carbohydr Res 20:329–337

Kurup VVGK, Voleti SR, Rajagopal V (1993) Influence of weather variables on the content and composition of leaf surface wax in coconut. J Plantn Crops 2:71–80

Lakmini WGD, Nainanayake NPAD, De Costa WAJM (2006) Biochemical changes of four different coconut (*Cocos nucifera* L.) forms under moisture stress conditions. J Agric Sci 2(3):1–7

Lebrun P, N'cho YP, Seguin M, Grivet L, Baudouin L (1998) Genetic diversity in coconut (*Cocos nucifera* L.) revealed by restriction fragment length polymorphism (RFLP) markers. Euphytica 101:103-1081-6 Available from: https://www.researchgate.net/publication/313476956_Biotechnology_of_Plantation_Crops_Coconut

Lebrun P, Grivet L, Baudouin L (1999) Use of RFLP markers to study the diversity of the coconut palm. In: Santamaria JM (ed) Current plant science and biotechnology in agriculture-current advances in coconut biotechnology. Kluwer Academic Publishers, Dordrecht, pp 73–87

Leon R, Alpizar L, Escamilla A, Santamariaj J, Oropeza C (1995) Physiological and biochemical changes in shoots of lethal yellowing-affected coconut palms. New Phytol 134:227–234. http://onlinelibrary.wiley.com/doi/10.1111/j.1469-8137.1996.tb04627.x/pdf

Lima BL, de Lacerda CF, Miguel Ferreira Neto, da Silva Ferreira JF, Bezerra AME, Marques EC (2017) Physiological and ionic changes in dwarf coconut seedlings irrigated with saline water. Rev Brasilian eng agríc ambient. [online]. vol. 21, n. 2, pp 122–127 ISSN 1415-4366. https://doi.org/10.1590/1807-1929/agriambi.v21n2p122-127

Liyanage DV (1952) Report of the acting botanist, annual report of coconut research scheme for 1950. Ceylon Government sessional paper. 14:22–30

Liyanage DV (1953) Selection of coconut seed nuts and seedlings. Ceylon Cocon Qtly 4:127–129

Lumige LC (1969) A note on the effects of coconut water transfusion on the germination and growth of water less coconuts. Philippine J Plant Ind 34:34–39

Manimekalai R, Nagarajan P, Bharathi M, Naresh Kumar S (2004) DNA polymorphism among coconut (*Cocos nucifera*) cultivars and reciprocal cross derivatives. J Plantn Crops 32(suppl):17–122

Manimekalai R, Nagarajan P, Bharathi M, Karun A, Naresh Kumar S, Kumaran PM (2005) Genetic variation of selected progeny lines of coconut (*Cocos nucifera* L.) based on simple sequence repeat markers. Trop Agri Res (Sri Lanka) 17:58–66

Manjula C (1990) Physiological and biochemical changes during the germination of coconut seeds. M.Phil Dissertation, Mangalore University, India, pp 79

Manjula C, Champakam B, Rajagopal V (1993) Solubilization and utilization of seed reserve during germination of coconut. J Plantn Crops 21(suppl):313–321

Manjula C, Champakam B, Rajagopal V (1995) Changes in nut water constituents during seed germination in coconut. Plant Physiol Biochem 22(2):169–172

Mathes DT (1984) Effect of extraction of inflorescence sap and the removal of immature fruit bunches on the production of female flowers in coconut. COCOS 2:44–47

Mc Coy RE, Howard FW, Tsai JH, Donselman HM, Thomas DL, Basham HG, Atilano RA, Eskafi FM, Britt L, Collins ME (1983) Lethal yellowing of palms. Univ Florida Agric Exp Stn Bull No. 834

McDonough J, Zimmermann J (1979) Effect of lethal yellowing on xylem pressure in coconut palms. Principes 66:1038–1042

Menon KPV, Pandalai KM (1960) The coconut palm. A monograph. Indian Central Coconut Committee, Kochi, Kerala, India, p 303

Mialet-Serra I, Clément A, Sonderegger N, Roupsard O, Jourdan C, Labouisse JP, Dingkuhn M (2005) Assimilate storage in vegetative organs of coconut (*Cocos nucifera* L.). Exp Agric 41:1–14

Mialet-Serra A, Clement-Vidal A, Roupsard O, Jourdan C, Dingkuhn M (2008) Whole-plant adjustments in coconut (*Cocos nucifera*) in response to sink–source imbalance. Tree Physiol 28(8):1199–1209

Michael KJ (1978) Respiratory rate and nut yield in root (wilt) diseased coconut palms. J Plantn Crops 6:1–3

Mondal H, Mandal RK, Biswas BB (1970) Absence of free nuclei in coconut water as revealed from biochemical studies. Nucleus 13:10–18

Monera OD, Del Rosario EJ (1982) Physiochemical evaluation of the natural stability of coconut milk emulsion. Ann Trop Res 4:47–54

Mujer CV, Ramirez DA, Mendoza EMT (1984a) L-D- Galactosidase deficiency in coconut endosperm. Its possible pleiotropic effects in makapuno. Phytochemistry 23(4):893–894

Mujer CV, Ramirez DA, Mendoza EMT (1984b) Coconut L-D Galactosidase isoenzymes isolation, purification and characterisation. Phytochemistry 23(6):251–1254

Mukesh Kumar B, Sugatha P, Niral V, Hebbar KB (2016) Variability in superoxide dismutase isoforms in tall and dwarf cultivars of coconut (*Cocos nucifera* L.) leaves. Indian J Agric Biochem 29(2):184–188. https://doi.org/10.5958/0974-4479.2016.00029.0

Nainanayake AD (2007) Use of chlorophyll fluorescence parameters to assess drought tolerance of coconut varieties. COCOS 18:77–105

Naresh Kumar S (2007) Capillary gas chromatography method for fatty acid analysis of coconut oil. J Plantn Crops 35(1):23–27

Naresh Kumar S (2009) Carbon sequestration in coconut plantations. In: Aggarwal PK (ed) Global climate change and Indian agriculture-case studies from ICAR network project. ICAR, New Delhi, pp 129–134

Naresh Kumar S (2011a) Variability in coconut (*Cocos nucifera* L.) germplasm and hybrids for fatty acid profile of oil. J Agric Food Chem 59(24):3050–13058

Naresh Kumar S (2011b) Climate change and Indian agriculture: current understanding on impacts, adaptation, vulnerability and mitigation. J Plant Biol 37(2):1–16

Naresh Kumar S, Aggarwal PK (2013) Climate change and coconut plantations in India: impacts and potential adaptation gains. Agric Syst 117:45–54

Naresh Kumar S, Balakrishnan A (2009) Seasonal variations in fatty acid composition of oil in developing coconut. J Food Qual (USA) 32:158–176

Naresh Kumar S, Balakrishnan A (2012) Quality and fatty acids of coconut oil under different conditions and duration of storage. Indian J Plant Physiol 17(1) NS:79–83

Naresh Kumar S, Kasturi Bai KV (2009a) Photo-oxidative stress in coconut seedlings: early events to leaf scorching and seedling death. Brazilian J Plant Physiol 21(3):223–232

Naresh Kumar S, Kasturi Bai KV (2009b) Photosynthetic characters in different shapes of coconut canopy under irrigated and rainfed conditions. Indian J Plant Physiol 14(3) NS:215-223

Naresh Kumar S, Rajagopal V, Karun A (2000a) Leaflet anatomical adaptations in coconut cultivars for drought tolerance, Recent advances in plantation crops research. CPCRI, Kasaragod, pp 225–229

Naresh Kumar S, Champakam B, Rajagopal V (2000b) Fatty acid composition of coconut oil among the cultivars – an insight into industrial application. Indian Cocon J 31(3):25–28

Naresh Kumar, Rajagopal V, Karun A (2001) Photosynthetic acclimatization in zygotic embryo cultured plantlets of coconut (*Cocos nucifera* L.). CORD 18(2):1–12

Naresh Kumar S, Rajagopal V, Laxman RH, Dhanapal R, Maheswarappa HP (2002a) Photosynthetic characteristics and water relations in coconut palms under drip irrigation levels. In: Rathinam P, Khan HH, Reddy VM, Mandal PK, Suresh K (eds) Proceedings of PLACROSYM XIV. 'Plantation Crops Research and development in new Millennium' Coconut Development Board, Kochi, pp 116–120

Naresh Kumar S, Rajagopal V, Siju Thomas T, Cherian VK, Hanumanthappa M, Kumar A, Sreenivasulu B, Nagvekar D (2002b) Identification and characterization of in situ drought tolerant coconut palms in farmer's fields under different agro-climatic zones. In: Sreedharan

K, Vinod Kumar PK, Jayarama (eds) Proceedings of PLACROSYM XV. Indian Society for Plantation Crops, Kasaragod, pp 335–339

Naresh Kumar S, Champakam B, Rajagopal V (2004) Variability in coconut cultivars for lipid and fatty acid composition of oil. Trop Agric (Trinidad and Tobago) 81(1):34–40

Naresh Kumar S, Rajagopal V, Kasturi Bai KV (2006a) Adaptive strategies of coconut palm under stressful conditions. In: Hemantharanjan A (ed) Advances in plant physiology, vol 9. Scientific Publishers, Jodhpur, pp 155–167

Naresh Kumar S, Rajagopal V, Siju Thomas T, Cherian VK (2006b) Effect of conserved soil moisture on the source-sink relationship in coconut (*Cocos nucifera* L.) under different agro-climatic conditions in India. Indian J Agric Sci 76(5):277–281

Naresh Kumar S, Kasturi Bai KV, George J, Balakrishnan A, Thomas ST (2007a) Stress responsive proteins in coconut seedlings subjected to water, high- Light, and flooding and high- temperature stresses. Indian J Hortic 64(4):373–380

Naresh Kumar S, Rajagopal V, Siju Thomas T, Cherian VK, Ratheesh Narayanan MK, Ananda KS, Nagawekar DD, Hanumanthappa M, Vincent S, Srinivasulu B (2007b) Variations in nut yield of coconut (*Cocos nucifera* L.) and dry spell in different agroclimatic zones of India. Indian J Hortic 64:309–313

Naresh Kumar S, Kasturi Bai KV, George J (2008a) A method for non-destructive estimation of dry weight of coconut stem. J Plantn Crops 36(3):296–299

Naresh Kumar S, Kasturi Bai KV, Rajagopal V, Aggarwal PK (2008b) Simulating coconut growth, development and yield using Info Crop-coconut model. Tree Physiol (Canada) 28:1049–1058

Naresh Kumar S, Rajagopal V, Cherian VK, Siju Thomas T, Sreenivasulu B, Nagvekar DD, Hanumanthappa M, Bhaskaran R, Vijaya Kumar K, Ratheesh Narayanan MK, Amarnath CH (2009a) Weather data based descriptive models for prediction of coconut yield in different agro-climatic zones of India. Indian J Hortic 66(1):88–94

Naresh Kumar S, Rajeev MS, Vinayan, Nagwekar DD, Venkitaswmy R, Raghava Rao DV, Boraiah B, Gawankar MS, Dhanapal R, Patil DV, Kasturi Bai KV (2009b) Trends in weather and yield changes in past in coconut growing areas in India. J Agrometeorol 11(1):15–18

Naresh Kumar S, John Sunoj, Muralikrishna KS (2016a) Impacts of climate change. In: Chowdappa P, Niral V, Jerard BA, Samsudeen K (eds) Coconut. Daya Publishing House, New Delhi, pp 355–374

Naresh Kumar S, Rajagopal V, Kasturi Bai KV (2016b) Abiotic stress tolerance in horticultural crops: coconut and arecanut. In: Srinivasa Rao NK, Shivashankara KS, Laxman RH (eds) Abiotic stress physiology of horticultural crops. Springer, New Delhi, pp 269–306

Natara Ian S, Bhaskaran R, Shanmugam N (1986) Preliminary studies to develop techniques for early detection of Thanjavur Wilt in coconut. Indian Cocon J 17(3):3–6

Nathanael WRN (1967) Report of the chemist. Ceylon Cocon Qtly 18:15–35

Nelliat EV, Nair PKR, Kushwaha BL (1976) Effect of month of harvest of seed nuts and fertilizer application to nursery on vigour and quality of coconut seedlings. J Plantn Crops 4:16–20

Nogueira LC, Nogueira LRQ, Miranda FR (1998) Irrigação do coqueiro. In: JMS F, DRN W, Siqueira LA (eds) A cultura do Coqueiro no Brasil. EMBRAPA, Brasília, pp 159–187

Ollagnier M, Ochs R, Pomier M, De Taffin G (1983) Effect of chlorine on the hybrid coconut P.B. 121 – in the Ivory Coast and Indonesia – Growth, tolerance to drought, yield. Oleagineux 38(5):309–321

Oo KC, Stumpf PK (1979) Fatty acid biosynthesis in the developing endosperm of *Cocos nucifera*. Lipids 14(2):132–143

Oropeza C, Alpizar L, Islas-Flores I, Escamilla A, Santa-Maria J (1995) Physiology and biochemistry of lethal yellowing affected *Cocos nucifera* palms. In: Oropeza et al (eds) Lethal yellowing: research and practical aspects. Kluwer Academic Publishers, Dordrecht, pp 65–67

Padmaja G, Sumathykutty Amma B (1979) Cellulose activity in the roots of coconut affected by root (wilt) disease. J Plantn Crops 7:101–104

Padmaja G, Sumathy Kutty Amma B, Mathew C, Ray PK, Dwivedi RS (1980) Carbonic anhydrase activity versus acidity in low yielding and high yielding coconut palms. Indian J Exp Biol 18(10):1209–1211

Padmaja G, Sumathykutty Amma B, Mathew C, Nambiar PTN, Dwivedi RS (1981) Alterations in the leaf protein content of coconut affected by root (wilt) disease. Indian J Plant Physiol 24:42–46

Padua - Resurreccion AN, Banzon JA (1979) Fatty acid composition of the oil from progressively maturing bunches of coconut. Philippine J Cocon Stud 4(3):1–16

Passos EEM, Silva JV (1990) Fonctionnement des stomates de cocotier (*Cocos nucifera*) au champ. Can J Bot 68:458–460

Passos EEM, Passos CD, Aragão WM (2003) Trocas gasosas em genótipos de coqueiro anão sob condições de deficiência hídrica. Aracaju, Embrapa Tabuleiro Costeiros (Comunicado Técnico 11), p 20

Passos CD, Passos EEM, Prado CHBA (2005) Comportamento sazonal do potencial hídrico e das trocas gasosas de quatro variedades de coqueiroanão. Rev Bras Frutic 27:248–254

Patel JS (1938) The coconut. A monograph. Govt Press, Madras, p 102

Peiris TGS, Thattil RO (1998) The study of climate effects on nut yield of coconut using parsimonious models. Exp Agric 34:189–206

Peiris TGS, Thattil RO, Mahidapala R (1995) An analysis of the effect of climate and weather on coconut. Exp Agric 31:451–460

Pillai RV, Nair RB, Mathew C, Bavappa KVA, Ramadasan A (1973) Studies on photoperiodic responsive reaction in coconut. J Plantn Crops l(Suppl):89–92

Pomier M, De Taffin G (1982) The tolerance to drought of some coconut hybrids. Oleagineux 37(2):55–62

Prado CHBA, Passos EEM, Moraes JAPV (2001) Photosynthesis and water relations of six tall genotypes of *Cocos nucifera* in wet and dry seasons. S African J Bot 67:169–176

Rajagopal V (1991) Is water stress associated with root (wilt) disease of coconut? J Plantn Crops 19:91–101

Rajagopal V, Kasturi Bai KV (2002) Drought tolerance mechanism in coconut. Bur Bull 17:21–22

Rajagopal V, Naresh Kumar S (2001) Avenues to improve productivity potential under drought condition-A case study on coconut. In: Abstracts of National Seminar on role of plant physiology for sustaining quality and quantity of food production in relation to environment. 5–7 Dec 2001, UAS, Dharwad

Rajagopal V, Patil KD, Sumathykutty Amma B (1986) Abnormal stomatal opening in coconut palms affected with root (wilt) disease. J Exp Bot 37:398–1405

Rajagopal V, Sumathykutty Amma B, Patil KD (1987) Water relations of coconut palms affected with root (wilt) disease. New Phytol 105:289–293

Rajagopal V, Shivashankar S, Kasturi Bai KV, Voleti SR (1988) Leaf water potential as an index of drought tolerance in coconut. Plant Physiol Biochem 15:80–86

Rajagopal V, Chempakam B, Robert Cecil S, Kamalakshy Amma PG (1989a) Studies on phloem sap collection from healthy and root (wilt) diseased coconut palms (*Cocos nucifera*. L.). Plant Physiol Biochem 16:52–56

Rajagopal V, Ramadasan A, Kasturi Bai KV, Balasimha D (1989b) Influence of irrigation on leaf water relations and dry matter production in coconut palms. Irrig Sci 10(1):73–81

Rajagopal V, Kasthuri Bai KV, Voleti SR (1990) Screening of coconut genotypes for drought tolerance. Oleagineux 45:215–223

Rajagopal V, Voleti SR, Kasturi Bai KV, Shivashankar S (1991) Physiological and biochemical criteria for breeding for drought tolerance in coconut. In: Silas EG, Aravindakshan M, Jose AI (eds) Coconut breeding and management. Kerala Agricultural University, Thrissur, pp 136–143

Rajagopal V, Shivashankar S, Jacob M (1996) Impact of dry spells on ontogeny of coconut fruits and it's relation to yield. Plant Res Dev 3:251–255

Rajagopal V, Naresh Kumar S, Kasturi Bai KV, Laxman RH (2000a) Day time fluctuations in photosynthetic parameters and water relations in juvenile palms of coconut grown under rainfed and irrigated conditions. J Plant Biol 27:27–32

Rajagopal V, Kasturi Bai KV, Naresh Kumar S (2000b) Adaptive mechanism of coconut palms in the changing environment conditions for higher production. In: Extended summaries Vol 2. Natural Resources- Agrobiodiversity, International conference on managing natural resources for sustainable agricultural production in the 21st century, New Delhi, pp 778–779

Rajagopal V, Kasturi Bai KV, Naresh Kumar S (2005) Breeding for drought tolerance in coconut: status and potentials. In: Batugal P, Ramanatha Rao V, Oliver J (eds) Coconut genetic resources. IPGRI publication, IPGRI-Regional Office for Asia, The Pacific and Oceania, Serdang, pp 282–301

Rajagopal V, Kasturi Bai KV, Naresh Kumar S, Niral V (2007) Genetic analysis of drought responsive physiological characters in coconut (*Cocos nucifera* L.). Indian J Hortic 64(2):181–189

Rajesh MK, Sabana AA, Rachana KE, Rahman S, Jerard BA, Karun A (2017) Genetic relationship and diversity among coconut (*Cocos nucifera* L.) accessions revealed through SCoT analysis. Biotechnology 5(6):999–1006. https://doi.org/10.1007/s13205-015-0304-7

Ramadasan A (1964) Physiology of Wilt disease in coconut palms. In: Proceedings of 2nd Sessn. FAO Working Party. Coconut Production Protection and Processing. Colombo, pp 257–272

Ramadasan A, Mathew C (1977) Relationship of the carbohydrate reserve in the trunk with commencement of flowering in young West Coast Tall coconut palms. J Crop 5:125–126

Ramadasan A, Mathew J (1987) Leaf area and dry matter production in adult coconut E.J., palms. J Plantn Crops 15(1):59–63

Ramadasan A, Satheesan KV (1980) Certain leaf anatomical characteristics of two coconut cultivars and hybrids. J Plantn Crops 8:55–57

Ramadasan A, Satheesan KV, Balakrishnan R (1980) Leaf area and shoot dry weight in coconut seedling selection. Indian J Agric Sci 50:553–554

Ramadasan A, Kasturi Bai KV, Shivashankar S, Vijaya Kumar K (1985) Heritability of seedling vigour in coconut palms. J Plantn Crops 13:136–138

Rao CVN, Choudhury D, Bagchi P (1961) A water soluble galactomannan from coconut *(Cocos nucifera). Pan I* Methylation studies. Canadain J Chem 39:375–381

Repellin A, Daniel C, Zuily-Fodil Y (1994) Merits of physiological tests for characterizing the performance of different coconut varieties subjected to drought. Oleagineux 49:155–168

Repellin A, Pham Thi AT, Tashakorie A, Sahsah Y, Daniel C, Zuily-Fodil Y (1997) Leaf membrane lipids and drought tolerance in young coconut palms (*Cocos nucifera* L.). Eur J Agron 6:25–33

Saitagaroon S, Kawakishi S, Naniki M (1983) Characterization of polysaccharides of copra meal. J Sci Food Agric 34:855–860

Satheesan KV, Narasimhayya G, Ramadasan A (1983) A rapid method for estimation of leaf area of one year old seedlings of tall variety of coconut palms. J Plantn Crops 11:47–49

Selvaratnam EM (1952) Embryo of the coconut. Nature 69:714–715

Shareefa M, Thomas RJ (2016) Planting material production strategies for the root (wilt) disease prevalent tract. In: Chowdappa P, Samsudeen K, Thamban C, Rajesh MK (eds) Planting material production in coconut. Today and Tomorrow Printers and Publishers, New Delhi, p 130

Shivashankar S (1990) Studies on soluble enzyme systems of coconut (*Cocos nucifera,* L.) cultivars. Ph.D. thesis, Biochemistry, Mysore University, Mysore, India

Shivashankar S (1992) Thermal stability of nitrate reductase in relation to drought tolerance in coconut. J Plantn Crops 20(Suppl):267–273

Shivashankar S, Rajagopal K, Ramadasan A (1985) Chlorophylls and nitrate reductase activity in relation to heterosis in coconut seedlings. Ann Bot 55:755–758

Shivashankar S, Kasturi Bai KV, Vijayakumar K (1986) A non-destructive method for the estimation of leaf area and shoot dry mass in the seedlings of coconut (*Cocos nucifera,* L) hybrids. Philippine J Cocon Stud 11(2):9–12

Shivashankar S, Kasturi Bai KV, Rajagopal V (1991) Leaf water potential, stomatal resistance and activity of enzymes during the development of moisture stress in coconut palm. Trop Agric 68:106–110

Sierra ZN, Balleza CF (1972) Proximate analysis of the coconut endosperm at progressive stages of development. National Institute of Science and Technology Annual Report 1971–1972, p 3

Siju Thomas T, Rajagopal V, Naresh Kumar S, Arunachalam V, Vinu K Cherian (2005) Stability analysis for dry matter production and yield components of coconut in two agro-climatic regions of India. Indian J Plant Physiol 10(1) (NS):1-8

Siju Thomas T, Naresh Kumar S, Cherian VK, Rajagopal V (2006) Role of certain biochemical compounds in adaptation of coconut to different weather conditions-a study at two agro-climatic regions of India. Indian J Hortic 63(1):1–7

Siju Thomas T, Naresh Kumar S, Cherian VK, Rajagopal V (2008) Gas exchange parameters and canopy area in relation to coconut productivity in two-agro climatic regions of India. Trop Agric 85:24–35

Sujatha A, Naresh Kumar S, Chalapathi Rao NBV (2006) Studies on impact of mite infestation on nut characters and oil content in coconut variety East Coast Tall (ECT). J Plantn Crops 34(1):32–35

Sumathykutty Amma B (1964) Preliminary studies on the effects of micro nutrients on the germination of coconut seednuts. Curr Sci 33:49–50

Sumathykutty Amma B, Patil KD (1982) An attempt to generate roots in root (wilt) diseased coconut palms with hormones, Phenols and amino acids. In: Proceedings of International Workshop on special problems in physiological investigations of Tree crops. RRII, Kottayam, pp 55–58

Sunoj VJ, Kumar SN, Muralikrishna KS (2014) Effect of elevated CO2 and temperature on oxidative stress and antioxidant enzymes activity in coconut (*Cocos nucifera* L.) seedlings. Indian J Plant Physiol 19(4):382–387

Thomas KM (1974) Influence of certain physical and chemical treatments on the germination and subsequent growth of coconut (*Cocos nucifera* L) seedlings, preliminary study. East African Agric J 40:152–156

Van die J (1974) The developing fruits of *Cocos nucifera* and *Phoenix dactylifera* as physiological sinks importing and assimilating the mobile aqueous phase of the sieve tube system. Acta Bot Neerl 23(4):521–540

Van Staden J, Drewes SE (1975) Identification of zeatin riboside in coconut milk. Physiol Plant 34:106–109

Varkey T, Michael KJ, Ramadasan A (1969) Note on the investigation on the nitrogen and phosphorus metabolism of coconut palm affected by root (wilt) disease. Indian J Agric Sci 39:25–26

Voleti SR, Rajagopal V (1991) Extraction and identification of epicuticular wax in coconut. Plant Physiol Biochem l8(2):88–90

Voleti SR, Kasturi Bai KV, Nambiar CKB, Rajagopal V (1993a) Influence of soil type on the development of moisture stress in coconut. Oleagineux 46:505–509

Voleti SR, Kasturi Bai KV, Rajagopal V (1993b) Water potential in the leaves of coconut (*Cocos nucifera* L.) under rain fed and irrigated conditions. In: Nair MK, Khan HH, Gopalasundaran P, Rao EVVB (eds) Advances in coconut research and development. Oxford & IBH Publishing, New Delhi, pp 243–245

White AR, Elmore HW, Watson MB, Gill JP (1989) Purification and partial characterization of polysaccharides from coconut milk. Ann Bot 64:205–209

Wikremasurya CA (1968) Some observations on the effect of photo period on the flowering behaviour of the coconut palm, *Cocos nucifera,* L. Ceylon Cocon Qtly 19:152–160

Wuidart W (1981) Production of coconut planting material. The polybag nursery. Oleagineux 36(7):367–376

Dr. S. Naresh Kumar is presently Professor and Principal Scientist at the Centre for Environment Science and Climate Resilient Agriculture, Indian Agricultural Research Institute, New Delhi, India. His contributions are mainly in climate change and simulation modelling research. He is a Government of India Expert for UNFCCC IPCC reports, Lead Scientist of Info Crop models and Member of the Global Steering Council, AgMIP. He has published 100 research papers and authored/edited 6 books. snareshkumar.iari@gmail.com

Dr. K. B. Hebbar is currently Head of Plant Physiology, Biochemistry and Post Harvest Technology at ICAR-CPCRI, Kasaragod. He has made significant contribution in the field of climate change and abiotic stress tolerance of various field and plantation crops. He was awarded Borlaug Fellow under Global Research Alliance programme on climate change for his work in unravelling the mechanism of the sudden wilt of cotton under changing climate conditions. balakbh64@gmail.com

Dr. K. V. Kasturi Bai retired as Head of the Division of Plant Physiology, Biochemistry and Post Harvest Technology at ICAR-CPCRI, Kasaragod. Her areas of interest are flowering, water management and stress and production physiology aspects of coconut. She has published 200 research articles. kasturikv@yahoo.com

Dr. V. Rajagopal former Director, ICAR-CPCRI, is a Plant Physiologist. He has 140 research papers and 18 books and book chapters to his credit. He is a recipient of Dr. RD Asana Medal and Dr. CS Venkataram Memorial Award for the distinguished plantation crop scientist. He is the Founder President of the NGO, Society for Hunger Elimination. rajvel44@gmail.com

Chapter 10
Coconut: Maladies and Remedies

Rohini Iyer, M. Gunasekharan, and Vinayaka Hegde

Abstract Coconut, a perennial palm with a life span of many decades, suffers from a number of biotic and abiotic stresses during various stages of its growth. Microbial pathogens including fungi, bacteria, viruses, viroids and phytoplasmas are known to be the causative organisms of these diseases. Among the 173 fungal species reported on coconut, only a few are lethal, while others cause economic losses of varying degrees. The causal agents of various diseases, formerly regarded as *diseases of unknown etiology*, have now been determined. While curative measures are available for some maladies like stem bleeding, prophylaxis is the best option for some of the lethal diseases like bud rot. Some of the debilitating non-lethal diseases have been found to respond favourably to nutritional management measures. Hence, adopting an integrated approach involving tolerant varieties, cultural, nutritional, prophylactic and phytosanitary measures is the best option for managing the diseases of coconut at present. In view of difficulties due to the tall stature of the palm, modern technologies like computer-aided, remote-controlled drones should be developed for plant protection operations. Remote-sensing methods using advanced technologies can help in regular surveillance for disease incidence, intensity and spread. Molecular techniques can also help in early diagnosis and in resistance breeding. Region-specific integrated disease management strategies for specific maladies have to be developed for attaining maximum efficacy in disease control. The distribution, symptoms, etiology, epidemiology and possible control measures of fungal, bacterial, virus, viroid and protozoan diseases affecting coconut are detailed in this chapter.

R. Iyer (✉)
Navasakti Trust, Karunagappally, Kollam, Kerala, India
e-mail: navasakti@gmail.com

M. Gunasekharan
Bharathiyar University, Coimbatore, Tamil Nadu, India
e-mail: mgsekar@gmail.com

V. Hegde
ICAR-CPCRI, Kasaragod, Kerala, India
e-mail: hegdev64@gmail.com; vinayaka.hegde@icar.gov.in

K. U. K. Nampoothiri et al. (eds.), *The Coconut Palm (Cocos nucifera L.) - Research and Development Perspectives*,
https://doi.org/10.1007/978-981-13-2754-4_10

10.1 Introduction

Various fungal, virus, viroid and protozoan diseases reported so far are listed in Table 10.1. Please refer to Chap. 11 for 'Phytoplasmal Diseases'.

10.2 Fungal Diseases

10.2.1 Bud Rot

10.2.1.1 Geographic Distribution

The first report of bud-rot incidence was from Grand Cayman, an Island in the West Indies, in 1834 (Tucker 1926). Since then, Quillec et al. (1984) reported the incidence from Sri Lanka, Indonesia, the Philippines, Colombia, Papua New Guinea, Vanuatu, Fiji, French Polynesia, the Dominican Republic and Côte d'Ivoire. Butler (1906) had first reported this disease from India.

10.2.1.2 Symptoms

Unopened tender leaf or spindle is affected leading to rotting of the terminal bud and death of palms. The first visible symptom is withering of the spindle marked by its pale colour. The spear leaf or spindle turns brown and bends down. Light brown specks are present on the petiole bases of the youngest leaf, and on those of the older leaves, large yellowish to brownish necrotic areas may be observed. The affected spindle can easily be pulled out as the basal portion of the spindle including the terminal bud is completely rotten, emitting a foul smell (Fig. 10.1).

10.2.1.3 Aetiology

Though primary infection is a dry rot caused by *Phytophthora palmivora*, Butl., subsequent colonisation by secondary invaders such as *Fusarium* sp. and bacteria like *Xanthomonas, Pseudomonas* and *Erwinia* results in wet rotting (Radha and Joseph 1974), with the affected tissues emitting a foul smell. The older leaves remain green and retain their position for several months, which is very characteristic of this disease (Ohler 1999). Even though the palm may not die immediately, it succumbs finally due to loss of the single apical bud. The older nuts persist on the crown for some time, while the younger ones may fall off (Radha and Joseph 1974). Later, the inner leaves also fall off one by one leaving only the outer whorl of mature leaves on the crown.

Table 10.1 List of diseases reported on coconut

Sl. no.	Disease	Pathogen/causal agent	Distribution	References
I. *Major fungal diseases*				
1.	Bud rot	*Phytophthora palmivora, P. heveae, P. katsurae, P. nicotianae, Fusarium moniliforme, F.solani, Graphium* sp.	West Indies, India, Côte d'Ivoire, Indonesia, Jamaica, Puerto Rico, Africa, Peninsular Malaysia and Philippines	Butler (1906), Tucker (1926), Menon and Pandalai (1960), Quillec et al. (1984), and Uchida et al.(1992)
2.	Fruit rot or mahali	*P. arecae, P. katsurae*	India, Sri Lanka and Côte d'Ivoire	Erwin and Ribiero (1996) and Quillec et al. (1984)
3.	Basal stem rot	*Ganoderma lucidum, G. applanatum, G. zonatum, G. boninense*	India, Florida, USA, South America, Java, Tropical Africa, Australia, Japan, Indonesia, Malaysia, Philippines, Samoa, Sri Lanka and Tasmania	Pieries (1974), Bhaskaran and Ramanathan (1984), and Satyanarayana et al. (1985)
4.	Stem bleeding	*Thielaviopsis paradoxa/ Chalara paradoxa*	Sri Lanka, India, Indonesia, Malaysia, Philippines, Fiji, Ghana, Trinidad	Petch (1906), Sundararaman (1922), and Britton-Jones (1940)
5.	Leaf rot	*Exserohilum rostratum/Colletotrichum gloeosporioides/Fusarium solani F. moniliforme*	India	Varghese (1934), Menon and Pandalai (1960), Radha and Lal (1968), and Srinivasan and Gunasekaran (1999)
6.	Grey leaf blight	*Pestalotiopsis palmarum*	Guyana, India, Malaysia, New Hebrides, Sri Lanka, Trinidad, Nigeria	Copeland (1931), Cook (1971), and Holliday (1980)
7.	Leaf blight	*Lasiodiplodia theobromae*	India	Johnson et al. (2014)
II. *Virus disease*				
8.	Coconut foliar decay or Vanuatu wilt	*Coconut foliar decay virus (CFDV)*	Vanuatu	Calvez et al. (1980) and Randles et al. (1986)
III. *Viroid diseases*				
9.	Coconut cadang-cadang disease	*Coconut cadang-cadang viroid (CCCVd)*	Philippines	Randles (1975)
10.	Coconut tinangaja	*Coconut tinangaja viroid (CtiVd)*	Guam on Mariana Islands	Boccardo (1985)

(continued)

Table 10.1 (continued)

Sl. no.	Disease	Pathogen/causal agent	Distribution	References
IV. *Protozoan disease*				
11.	Fatal wilt or heart rot	*Phytomonas staheli*	Central America (Costa Rica, Honduras and Nicaragua), South America (Brazil, Colombia, Ecuador, Guyana, Peru, Surinam and Venezuela) and the West Indies (Grenada, Trinidad and Tobago)	Waters (1978)
V. *Other minor diseases*				
12.	Lethal bole rot	*Marasmiellus cocophilus*	Kenya and Tanzania	Jackson and Firman (1982) and Jackson and McKenzie (1988)
13.	Anthracnose	*Colletotrichum gloeosporioides*	Brazil, India	Almeida et al. (1978)
14.	Leaf spots	*Bipolaris incurvata*	Hawaii, Florida, Jamaica, Asia, Australia, Oceania (French Polynesia, Fiji), Philippines, Seychelles	Uchida and Aragaki (1991)
15.	Algal leaf spot	*Cephaleuros virescens, Cephaleuros parasiticus*	Hawaii	Ploetz et al. (1999) and Harrison and Jones (2003)
16.	Thread blight	*Pellicularia filamentosa, Corticium penicillatum*	Sri Lanka, Fiji, Papua New Guinea, Samoa and Solomon Islands in Oceania	Kohler et al. (1997)

Sometimes the same bud-rot pathogen was found to infect coconut fruits and cause fruit rot and immature nut fall. Water-soaked lesions appear on the surface of the nuts, especially near the perianth, and the lesions turn brown and the nut gets detached from the bunch. Nut drop and rotting are commonly seen during rainy season.

10.2.1.4 Epidemiology

Bud-rot disease is favoured by conditions of high humidity such as those found in low-lying badly drained lands, in plantations with a very dense stand and under extensive rainfall. Disease development is related to relative humidity (RH) (Menon

Fig. 10.1 Bud-rot disease. (Photo: Jacob John)

and Pandalai 1960; Darwis 1992). Rainfall aggravates the infection and young palms in the low-lying and moist conditions are more susceptible (Thevenin et al. 1992; Brahmana et al. 1992; Pohe 1992; Steer 1992). In the Philippines, Rillo and Paloma (1989) found higher incidence of *Phytophthora,* which was always preceded by high rainfall during previous months. Even though palms of all ages are susceptible to the disease, it was more rampant in palms which are below 20 years. The period elapsing between the first infection and withering of heart-leaf depends on a number of factors, the more important of which are RH and the point of infection. This period might vary from 3 to 9 months or even longer. With the onset of dry weather, the infection becomes less severe and pathogen remains dormant. The fungus was found to survive in the frond base or basal part of the crown (Menon and Pandalai 1960; Radha and Joseph 1974) or in roots (Harris et al. 1984). Seasonal factors are found to be associated with the incidence and spread of the disease. Disease incidence is found to be severe during monsoon when the RH is high. Studies conducted at ICAR-CPCRI, Kayamkulam, India (Radha and Joseph 1976), revealed that the favourable period of infection is when the RH is above 94% and temperature is below 24 °C. The rate of disease development was determined by the number of preceding favourable days. In young palms, between the ages of 5 and 10 years, the occurrence of favourable days is more frequent, and hence, more disease incidence was noticed in such young palms (Radha and Joseph 1974). An infection cycle of *P. palmivora* in coconut is completed within 6 days under favourable conditions of humidity (98–100%) and temperature (22–24 °C). Inoculation of young seedlings (1–2 years old) grown in pots and provided with these conditions resulted in infection and death of seedlings. It took 1 week for the fungus to complete its life cycle from sporangia to manifestation of the disease symptoms (Radha and Joseph 1974). Coconut is a sun-loving crop, requiring high solar radiation and high humidity. However, high rainfall intensity in certain areas causes low solar radiation and high air humidity.

Bud-rot disease incidence was high in the hilly tracts where cooler temperatures coupled with very high RH prevailed for extended periods in the coconut crown because of the altitude of the location. The cooler post-monsoon weather also favours the formation of dew for extended periods, and this factor is conducive to the development of fresh disease incidence in the hilly tracts compared to the plains. In the plains, disease incidence was recorded only up to September. Initial incidence of bud-rot disease is always dependent upon the monsoon showers. However, the occurrence of bud rot in subsequent months, i.e. from October to January, could be attributed to the favourable microclimatic conditions inside the crown with consistently high humidity, low night temperatures and the presence of dew-water droplets (Rasmi et al. 2004).

10.2.1.5 Dissemination of the Pathogen

Epidemiological models applied to plots affected by *Phytophthora* in Côte d'Ivoire suggested the existence of two propagation phases, an aggressive phase during which transmission occurred from palm to palm and a regular phase during which new cases appeared some distance away from initial foci (Renard and Darwis 1993). The disease is primarily disseminated by wind and windblown raindrops and to a lesser extent by insects, birds and climbers. The disease spreads over large areas due to the influence of environmental condition in the plots. The wind pattern and direction are required to correctly interpret the disease spread. Dauzat and Lecoustre (1992) reported that there are two phases for bud-rot disease propagation. One is the 'cluster' phase during which contamination would be seen primarily to spread around contamination foci, and another is a regular phase during which new contamination occurs from foci. Rainwater acts as a carrier for the infectious propagules and plays an important role in the spread of the disease (Brahmana et al. 1992; Thevenin et al. 1992). Insects also spread the pathogen over large areas by moving from one place to another; their legs and mandibles seem to be helpful in transferring the propagules (Pohe 1992). Radioactive tracer studies have shown that certain flying insects, beetles, caterpillars and even snail species are involved in the spread of the pathogen. Evans (1973) reported two species of *Sogatella* as vectors of dry bud rot of young coconuts in Côte d'Ivoire.

10.2.1.6 Disease Management

Control measures are effective only when they are adopted in the initial stages of the disease. If the disease is detected when the central shoot just withers, chemical control by the application of 10% Bordeaux paste on the affected portion can check the disease. Bordeaux paste has to be applied after thorough cleaning and removal of infected material. The treated portion should be given a protective covering of plastic sheet to prevent washing off of the paste by rain. As a prophylactic measure, adjacent healthy palms should also be sprayed with 1% Bordeaux mixture. Other fungicides advocated earlier against bud-rot disease include phenylmercuric urea applied in the form of powder or pellets in the leaf axils (Pieris 1962) or stem injection or root infusion of Aliette or Ridomil or Akomin (phosphonic acid). According to Schutt (1975) and Nambiar and Rawther (1993), regular spraying with copper fungicides at 40-day intervals, especially before and after monsoon, was found to be an effective preventive measure. In copper-sensitive palms, keeping perforated sachets containing Dithane M-45 in the leaf axils during rainy season is useful. Radha and Joseph (1974) found that Demosan (1200 ppm) effectively checked infection in laboratory tests. Renard and Quillec (1984) and Brahmana et al. (1992) reported that injection of the coconut trunk with systemic fungicides like Aliette (fosetyl-Al) and Ridomil (metalaxyl) at 3 g a.i. palm^{-1} was effective in protecting the palms from bud rot caused by *P. heveae*. Rohini Iyer (1997) found that stem injection/root feeding with systemic fungicides like aureofungin-sol (46.5%) 36.4 g palm^{-1}, Calixin (tridemorph 80%) 21 ml palm^{-1}, Aliette (fosetyl-Al 80% wp) 21 g palm^{-1} and Akomin (phosphonic acid) 16.8 ml palm^{-1} can protect the crown from *Phytophthora* attack for a period of 8 weeks.

Another long-term method of control is through the identification of resistant genotypes and breeding for resistance/tolerance. Characterisation of resistant palms can be made by assessing the field performance of planted varieties. Quillec et al. (1984) observed that MAWA hybrids were less sensitive than West African Tall (WAT). Brahmana and Kelana (1988) found that in Indonesia, dwarf palms are more sensitive while others are more tolerant. Among dwarf varieties, Nias Yellow Dwarf (NYD) seems to be the most susceptible one (Bennett et al. 1986). Bud rot is also observed predominantly in areas planted with PB-121 (MYD × WAT) coconut hybrids. Exotic talls such as Rennel Tall and local Indonesian talls are more tolerant (Renard and Darwis 1993). Malayan Yellow Dwarf × West African Tall hybrids are highly susceptible to bud rot in Indonesia (Mangindaan et al. 1992). Hybrids of Malaysian Yellow Dwarf × Rennel Tall were found to be less affected than PB-121. The NIWA hybrids obtained by crossing NYD × WAT are susceptible to bud rot. Rillo and Paloma (1989) found that *Phytophthora* incidence was higher in Cameroon Red Dwarf and MYD compared to Catigan. Coffey (1990) reported that in Indonesia, dwarf selections such as *Jombang* and *Raji* appeared to be more resistant to nut fall in inoculation trials.

In Asia, local ecotypes are generally more tolerant of *P. palmivora* than introduced ones, although the Polynesian Tall and Rennel Tall are less severely affected than the Bali Tall in North Sumatra. Malayan Red Dwarf (MRD) is susceptible to bud rot caused by *P. katsurae* in Côte d'Ivoire and is also susceptible to the same damage in Jamaica, whereas Red Dwarf × West African Tall coconut hybrids are more tolerant to bud rot than the Red Dwarf parent. Both Malayan Red Dwarf and West African Tall were susceptible to bud rot, whereas the Malayan Yellow Dwarf (MYD) and the Polynesian, Rennel and Malaysian Talls are highly tolerant in Jamaica. In the Philippines, MAWA hybrid (Malayan Yellow Dwarf × West African Tall was found to be susceptible to bud-rot infection. For reducing economic loss, it is recommended that the choice of planting material should be done by taking into account the environmental conditions also (Mangindaan et al. 1992).

Proper spacing among the palms is important for the management of the disease. Too close planting encourages rapid spread of the disease. A good spacing between palms favours air movement and dissipation of the excess humidity that can build up in the crowded gardens (Ohler 1984). Lowlands with generally high humidity are very favourable for the development of the disease, especially when the drainage is poor.

Organic matter application favours the growth of a variety of microbes including antagonists such as *Trichoderma* and *Gliocladium* spp. which multiply on them and help in reducing the population of soilborne pathogens like *Phytophthora*. Biological control is geared towards identifying microorganisms effective against *Phytophthora*. Regular use of site-specific fungicide is not recommended because fungicide-resistant isolates or strains of *Phytophthora* spp. may develop due to their continued use (Cohen and Coffey 1986). Copper injury, especially to certain dwarf varieties (Schutt 1975), high cost of fungicides and lack of trained labourers for spraying are constraints in advocating fungicidal control for bud rot. Though many bacteria, actinomycetes and fungi are antagonistic to *Phytophthora*, their activity in the field is limited. The use of microbial antagonism against *Phytophthora* is an important component of disease control (Malayczuk 1983; Shea and Broadbent 1983). However, the prospect of using biological antagonism is still uncertain, especially on a practical short-term basis (Shea and Broadbent 1983). The development of effective biological control of *Phytophthora* species has been fraught with difficulties because of their ability to produce several forms of inoculum (zoospores, sporangia, chlamydospores and mycelium) rapidly and repeatedly, wide host range, ability to penetrate and infect a host plant within a few hours and to exist in soil at depths of even 1 metre allowing them to escape most of the antagonists. Several genera of bacteria, actinomycetes and fungi have been shown to parasitise and lyse *Phytophthora* propagules in soil (Sutherland et al. 1984). These antagonists exert a lytic effect on mycelium, chlamydospores and oospores. Among the different fungi isolated from endemic plots, *Trichoderma harzianum* and *Trichoderma viride* were identified as the most effective antagonistic fungi of the bud-rot pathogen, *P. palmivorain vitro*. However, the result of pot experiments revealed that *T. harzianum* has a higher competitive saprophytic ability in soil compared to that of *T. viride*. Moosa et al. (1998) reported the occurrence of endophytic *Bacillus amyloliquefaciens* antagonistic to *Phytophthora palmivora* in coconut seedlings. The major problem for the use of fungicides is that crops like black pepper and cardamom are grown as intercrops in coconut gardens, whose export is highly sensitive to pesticide residues. Chemical control, though effective, is undesirable as it pollutes the environment, and the residues left in the products are hazardous to human health. To minimise the use of pesticides, biological control becomes imperative in integrated management of the disease. At present, there is a need for an effective, broad-based integrated control of *Phtyophthora* disease, involving the use of resistant varieties, improved cultural practices, new translocation fungicides and effective biological control methods (Tuset et al. 1984, 1992).

10.2.2 Basal Stem Rot

Basal stem rot is one of the most destructive diseases of coconut occurring in various coconut-growing regions of the world. The disease is known in India by various names such as '*Ganoderma* wilt' in Andhra Pradesh, 'Anabe' in Karnataka (Venkatarayan 1936; Rao et al. 1966) and as 'Thanjavur wilt' in Tamil Nadu (Vijayan and Natarajan 1972).

The disease has been reported from various places all over the tropical world, viz. India, Sri Lanka, West Indies, Seychelles and Guam. The disease was first reported by Coleman in 1911 on areca nut palms in Mysore, India. Pieries (1974) reported the disease as 'basal stem rot' from Sri Lanka. A severe outbreak occurred in 1952 in Thanjavur district of Tamil Nadu, India, and hence, the name 'Thanjavur wilt', although the disease is noticed in all districts of Tamil Nadu. Bhaskaran and Ramanathan (1984) found that the disease incidence ranged from 0.6% to 4.9% in Tamil Nadu. In the severely affected gardens, the disease incidence was as high as 31.4% (Bhaskaran et al. 1984). Apart from Tamil Nadu, the disease is also reported from Andhra Pradesh, Karnataka and Kerala in India. A disease with similar symptoms has been noticed in some parts in the Indian states of Maharashtra and Gujarat also.

10.2.2.1 Symptoms

Though the root system is affected, visible symptoms are seen in the crown as wilting of leaflets, similar to those of severe drought. The outer whorl of leaves turns yellowish, then gradually becomes brown and droops down from their point of attachment and hangs vertically downwards to form a skirt around the trunk apex. The apex of the trunk shows tapering with the advancement of the disease, and bleeding symptoms may appear on the bole region. The drooping leaves fall off one by one leaving only a few leaves at the apex. The crown is easily blown off by wind (Bhaskaran et al. 1984), leaving only the decapitated trunk. Male flowers become necrotic starting from the tip and spreading to the base of the spikelets. The few female flowers are also poorly developed. In the early stages, there is no button shedding seen. As the disease progresses, normal development of flowers and bunches is fully arrested. Most of the palms bear profusely just at the time of initiation of visible symptoms. As the leaves droop down, the subtended bunches also hang down. The nuts become barren and gradually the production stops. The roots are first affected and destroyed. The cortical region of the affected roots turns brown first, then the stele and the roots become friable and disintegrate. As the roots in contact with the soil die back, the palm puts forth new roots from higher up the trunk and sometimes new roots may be seen coming from healthy tissue piercing through the affected tissue. At the base of the stem, a characteristic reddish brown discolouration develops, accompanied by the exudation of a brown viscous gummy substance. Initially these bleeding patches appear on several places as parallel vertical streaks. They soon coalesce, forming a discoloured band around the trunk. These brownish patches may extend up to 1 m from ground level. Occasionally, infected palms do not show any bleeding patch.

Symptoms of dry rot of internal tissue at the base of the stem are characteristic. Transverse and longitudinal sections in these areas show a light brownish rotting tissue marked by darker bands, often with an irregular outline. The edge of the lesion is marked with a distinct yellow margin 0.5–1.0 cm wide. The bole decays rapidly resulting in the formation of large cavities. In some palms, the bark from the base of the palm peels off. The sporophores (fruiting bodies) of *Ganoderma* appear as 'brackets' at the base of the trunk, generally after the death of the palm (Fig. 10.2) (Bhaskaran et al. 1989), and in some palms, just above the soil level. Usually it takes from 6 to 24 months for the affected palms to die.

10.2.2.2 Aetiology

In India, *Ganoderma lucidum* (Leyss.) Karst. was first reported to be the causal agent of the disease (Rao et al. 1966). However, later reports (Bhaskaran et al. 1989) showed that *G. applanatum* (Pers.) Pat. is also involved. Pieries et al. (1975) reported *G.boninense* Pat. as the inciter of basal stem rot in Sri Lanka.

The fungus is a soil dweller inhabiting dead as well as living plant material in the soil. It is a root parasite entering the host through wounds. The spread of the disease takes place mainly through soil and through root to root contact.

The fruit body (basidium) is bracket-shaped, perennial, stipitate, lateral and sometimes sessile. Its size varies from less than 2 to more than 50 cm in diameter.

Fig. 10.2 Basal stem rot disease. Note the fruiting bodies. (Photo: Jacob John)

The thickness of the fruiting body likewise varies up to 13 cm. The upper surface of the bracket is shiny, oxblood in colour, with solid thick-walled covering and concentrically furrowed. The shiny surface of the cap and stalk is most characteristic. When examined closely with naked eye or hand lens, numerous minute holes or pits will be seen all over the undersurface. It is in these tiny pores that the fungus produces its 'basidiospores'. The fungus is heterothallic and tetrapolar. The hyphae are hyaline, 1–2 μ in diameter covered with a deposit of calcium oxalate crystals. Clamp connections occur profusely in older hyphae. The fungus has a wide host range and attacks a variety of palms and several forest, avenue and fruit trees as well. According to Naidu et al. (1966), hosts belonging to 19 families, 36 genera and 48 species have been recorded. Some of these are *Areca catechu*, *Cocos nucifera*, *Cassia siamea*, *C. javanica*, *Pongamia glabra*, *Eucalyptus* spp., *Azadirachta indica*, *Morus alba*, *Artocarpus fraxinifolius*, *Acacia arabica*, *Casuarina equisetifolia* and *Dalbergia sissoo*. Papa Rao and Rao (1966) reported low frequency of incidence of the disease in heavy soils of Andhra Pradesh, probably due to their high moisture retention capacity. Pieries et al. (1975) also found that the disease progressed rapidly in dry areas, the destruction of the palms being faster (6–30 months) in areas receiving less rainfall (up to 1000 mm $year^{-1}$), as compared to areas receiving 2000 mm rainfall $year^{-1}$. Lewin et al. (1983) reported that soil moisture stress during the summer predisposed the palms to infection. A positive correlation between mean, maximum temperature and the number of bleeding patches has been observed by Jagannathan and Ramaswamy (1975).

Though the disease is prevalent in palms of all ages, Vijayan and Natarajan (1972) reported that palms in the age group of 10–30 years were more susceptible. Linear spread was found to be influenced by low rainfall and RH (Ramapandu et al. 1981). Since diagnosis of the disease at a very early stage is essential for taking up effective control measures, work has been done using colorimetric methods (Natarajan et al. 1986) or by methods employing physiological parameters like transpiration rate and stomatal resistance (Vijayaraghavan et al. 1987), which have given encouraging results. A diagnostic method using fluorescent antibody technique developed against *Ganoderma* disease in areca nut by Koti Reddy et al. (1984) is being employed in coconut also (Sampath Kumar and Nambiar 1993).

10.2.2.3 Disease Management.

On-farm trials conducted in Tamil Nadu and Andhra Pradesh states of India showed that the following integrated approach was very effective for containing the disease (Bhaskaran et al. 1989):

1. Removal of dead palms and palms in advanced stages of the disease and destruction of the boles and root bits of the diseased palms.
2. Isolation of neighbouring healthy palms by digging isolation trenches of 50 cm × 50 cm around the diseased palm.
3. Soil drenching with 40 l of 1% Bordeaux mixture thrice a year for 1 year.

Table 10.2 Performance of coconut cultivars and hybrids in *Ganoderma*-sick soils of Tamil Nadu, India

Sl. no.	Cultivars	No. of palms field planted	Percent survival	Mean nut yield (palm year^{-1})
1.	San Ramon	15	33.3	36
2.	Laccadive Ordinary	15	6.7	36
3.	British Solomon Islands	12	8.3	98
4.	Java Giant	15	26.7	80
5.	Straits Settlement Green	15	46.7	49
6.	WCT × COD	15	40.0	86
7.	COD × WCT	15	26.7	121
8.	VHC-1	10	10.0	66
9.	East Coast Tall (ECT)	10	40.0	86
10.	ECT × wilt-tolerant ECT	18	66.7	122

4. Addition of 50 kg farm yard manure or green leaf manure or 200 kg tank silt palm^{-1} year^{-1}.
5. Chiselling off the affected tissues and applying aureofungin-sol or Calixin (tridemorph) 5%.
6. Raising *Ganoderma*-resistant crops like banana as intercrop wherever irrigation is possible.
7. Root feeding of 2 g of aureofungin-sol and 1 g of copper sulphate in 100 ml of water thrice a year at quarterly intervals. Alternatively, use tridemorph (Calixin 5 ml in 100 ml of water). Fungicide treatments will be effective only for palms in early stages of the disease.
8. Ploughing and flood irrigation are to be avoided to prevent the spread of infective propagules. Irrigation through drip or channel is recommended.
9. Application of neem cake at 5 kg palm^{-1} year^{-1}.
10. Application of 500 g of *Trichoderma harzianum* multiplied in 50 kg farm yard manure as biocontrol.

Screening of 6 cultivars and 4 hybrids for their reaction to basal stem rot revealed that the hybrid East Coast Tall × wilt-tolerant ECT is more tolerant to basal stem rot, with 66.7% survival, compared to other cultivars (Bhaskaran et al. 1984). The details are given in Table 10.2.

10.2.3 Stem Bleeding

It is a debilitating disease prevalent in India, the Philippines, Malaysia, Trinidad, Fiji, Ghana, Papua New Guinea, Indonesia, Bangladesh, Sri Lanka, Mexico and many other coconut-growing countries. The disease was first reported from Sri

Lanka (Petch 1906) and later from India (Sundararaman 1922). Subsequently, its occurrence has been reported from Brazil (Warwick et al. 2009) and Hainan, China (Yu et al. 2012), although there is no report on the exact extent of loss due to this disease. The typical symptom is the oozing out of reddish brown liquid through the growth cracks on the trunk which later turns black on drying forming encrustation with brownish orange margin (Fig. 10.3). Bleeding patches progress both upwards and downwards and cover major portion/part of the trunk.

10.2.3.1 Aetiology

The aetiology of the disease was not established for a long time. Lilly (1984) isolated *Phomopsis coconina* Cke. (Punith) and *Schizophyllum commune* Fr. from the stem-bleeding-affected palms. However, their pathogenicity could not be proved in spite of repeated attempts. Although Menon and Pandalai (1960) suspected *Thielaviopsis (Ceratocystis) paradoxa* (de Synes) as the incitant of stem bleeding, it was Nambiar et al. (1986) who established *T. paradoxa* as the aetiological agent of the disease, through artificial inoculations. Later, the perithecial stage, *Ceratocystis paradoxa* (Dade) Moreau, of the causal agent was isolated from infected palms (Ramanujam et al. 1993). *T. paradoxa* produces pale brown to brown hyphae. Conidiophores are slender, arising laterally from the hyphae and producing cylindrical to oval endoconidia. When mature, they are hyaline to pale brown and smooth-walled (6–24 × 2–5.5 μ). Chlamydospores are also formed terminally in chains and are obovate, thick-walled, brown and 10–25 × 7.5–20 μ in size. The perithecial stage is *Ceratostomella* (*Ceratocystis*). Perithecia are partly immersed,

Fig. 10.3 Stem-bleeding disease of coconut. Note the oozing out of reddish brown liquid in the trunk. (Photo: Jacob John)

light brown and 190–350 μ in diameter with numerous appendages; long and black necked, tapering up to 1400 μ, osteolar and hyaline; and ascospores ellipsoid, often with unequally curved sides, hyaline, nonseptate, smooth and 7–10 × 2.5–4 μ. The optimum temperature for mycelial growth was reported to be 30 °C (Nishita Naik 1990). As many workers failed to isolate the fungus from infected trees, Anil Kumar and Nambiar (1991) developed a simple and highly reproducible technique for isolating *T. paradoxa* from the diseased tissues. A baiting technique was also developed for the isolation of *T. paradoxa* from the infected soils using sterile frond bits as baits. Gowda (1987) recorded variability among the 5 isolates derived from 5 localities in Karnataka and Kerala, India, with regard to colour, nature of colony, growth rate and conidial and chlamydospore production on various media. Nisitha Naik (1990) has demonstrated the variability among 7 isolates based on their optimum temperature, pH, carbon, nitrogen and vitamin requirements. Some of the isolates exhibited a characteristic fruity (pineapple) smell. Ramanujam et al. (1996) distinguished 2 sub-groups among the 12 isolates of *T. paradoxa* based on their pathogenic reaction to detached coconut petioles and also corresponding to their electrophoretic reaction.

All the 26 coconut cultivars (16 Talls, 6 Dwarfs and 4 hybrids), which were tested for susceptibility to *T. paradoxa* using the detached petiole technique, were found to be susceptible with varying degrees of disease intensities (Ramanujam et al. 1996). Banawali Green Round, Banawali Brown Round and Malayan Orange Dwarf were less susceptible, while Malayan Green Dwarf, Chowghat Orange Dwarf and Philippines Ordinary cultivars were more susceptible.

10.2.3.2 Epidemiology

Nambiar and Sastry (1988) reported that development of growth cracks, poor drainage, soil moisture stress, hard pan formation in soil, imbalanced nutrition, excessive soil salinity, stem injury, lightning and insect attack by *Diocalandra* and *Xyleborus* are the predisposing factors responsible for disease development as this fungus is a weak soilborne pathogen. It enters the coconut stem tissue through growth cracks and wounds and multiplies in the host tissue, producing endoconidia and chlamydospores, which survive in the soil during unfavourable weather conditions. When the conditions are favourable, the chlamydospores in the soil germinate and become capable of infecting coconut. Radha (1962) showed that fluctuations in soil moisture or ill-drained soil conditions could cause severe stem-bleeding disease. Mathew and Ramanandan (1980) could not find any significant differences in major nutrient contents between healthy and diseased palms. They also did not find any relation between disease incidence and soil pH and electrical conductivity. Nambiar and Ramanujam (1993) reported that chlorine deficiency does not seem to be a contributing factor for stem-bleeding disease in India, especially in Kerala, India, where on the banks of backwaters, this disease is noticed. Nambiar et al. (1989) studied the conditions required for infection through artificial inoculation and found that the disease development was faster during July–November when high humidity and optimum temperature prevailed. Usman (1988) reported that maximum survival of *T. paradoxa* propagules was noticed in red loam soil followed by laterite soil and the least was noticed in sandy soil. The fungus attacks a wide variety of hosts like areca nut, palmyrah palm, date palm, banana, pineapple, sugarcane and papaya.

10.2.3.3 Disease Management

Prior to confirmation of the aetiological agents of the disease, control measures recommended mainly consisted of phytosanitary practices involving removal of affected bark tissues with a chisel and application of hot coal tar or Bordeaux mixture to protect the wound. Nambiar and Sastry (1988) reported improvement of palm conditions when carbendazim or tridemorph was root-fed. Further studies also indicated effectiveness of these chemicals in reducing the disease intensity and increasing the yields (Anil Kumar et al. 1992; Ramanujam et al. 1993). Their results also helped in the detection of residues of carbendazim in the stem along the feeding site only, while tridemorph was detected on the feeding side as well as on the opposite side of coconut trunk. No residues of carbendazim and tridemorph were detected in the nut water, in the palms, which received 5 g and 8 ml of chemicals, respectively (Ramanujam et al. 1993).

Since wounds on the trunks predispose the palms to infection, care should be taken not to injure the palms while doing cultural operations in the coconut garden. Providing summer irrigation and conserving the soil moisture by adopting suitable conservation practices are beneficial in reducing growth cracks. Application of neem cake (5 kg palm^{-1} year^{-1}) was found to increase soil microflora including *Trichoderma* population, which was found inhibitory to the pathogen in vitro (Gowda 1987) and on detached coconut leaf petiole (Usman 1988). They identified *Trichoderma viride*, *T. harzianum* and *Aspergillus niger* as potential antagonists to the pathogen. Later, Ramanujam (1997) identified *Gliocladium virens* as the most effective antagonist against *T. paradoxa* and also recorded rice bran neem cake (1:1) as the best substrate for mass production of this biological control agent.

Ramanujam (1997) also developed integrated management practices for effective management of stem bleeding of coconut involving root feeding of 5% tridemorph (100 ml at quarterly intervals) and wound dressing (50–200 ml) of tridemorph (4%) followed by coal tar sealing (100–300 g) and soil application of *Gliocladium virens* (1 kg), neem cake (5 kg), FYM (50 kg) and NPK fertiliser (500:320:1200 g palm^{-1} year^{-1}). Root feeding with hexaconazole (2 ml in 100 ml water at quarterly intervals) and wound dressing with the same fungicidal solution have been advocated. Application of paste of *Trichoderma harzianum* talc formulation on the bleeding patches on the trunk was also effective in preventing the spread of stem bleeding.

10.2.4 Leaf Rot

Radha (1961) first coined the name ‘leaf rot’ for foliar necrosis of coconut frond in the root (wilt)-affected tract of southern Kerala, India. Since the beginning of the last century, it was well established that the palms affected by root (wilt) disease are generally superimposed by leaf rot disease (Sundararaman 1925; Varghese 1934; Nagaraj and Menon 1956; Srinivasan 1991). In the palms weakened by the root (wilt), *Phytoplasma* might result in the breakdown of their defence mechanism leading to susceptibility to leaf rot disease. Crop loss due to leaf rot alone was not available earlier, as it is superimposed on root (wilt). Menon and Nair (1948) estimated the loss due to leaf rot as 5.6 million nuts annually. This is besides the loss in quality of the leaves rendering them unfit for thatching and other purposes. The loss due to leaf rot has been computed at 461 million nuts in Kerala, India, as it is prevalent in 0.41 million ha (Srinivasan and Gunasekaran 1999).

10.2.4.1 Symptoms

Leaf rot starts as minute, water-soaked lesions on the emerging spindle with different shades of colour. These lesions enlarge, coalesce freely leading to extensive rotting. The rotted portions dry up, turn black and fall off. Tips of leaflets and midribs often become black and shrivelled. The inner whorls of leaves are more vulnerable to the disease (Fig. 10.4). Continuous attack of newly emerging spindle leaves results in the gradual exhibition of similar symptoms in all the leaves in the crown (Srinivasan and Gunasekaran 1992). Sometimes the decayed leaflets are glued together so that spindle does not open out. Though the disease does not kill the palm outright, its slow progress in the crown causes steady decline in nut yield. Palms of all ages are susceptible to the infection (Radha and Lal 1968; Srinivasan and Gunasekaran 1992).

Fig. 10.4 Leaf rot disease of coconut. (Photo: ER Asokan)

10.2.4.2 Aetiology

Radha and Lal (1968) confirmed the association of a number of fungi with leaf rot disease. They were identified as *Colletotrichum gloeosporioides* (Penzig) Penzig and Sacc., *Exserohilum rostratum* (Drechler) Leonard and Suggs., *Gliocladium vermoeseni* (Biourge) Thom., *Cylindrocladium scoparium* Morgan, *Fusarium solani.* (Mart.) Sacc., *Thielaviopsis paradoxa* (Date), *Rhizoctonia solani* J.G. Kühn and *Curvularia* spp. Boedijn. Of these, *C. gloeosporioides* (Penzig) Penzig and Sacc. and *E. rostratum* (Drechler) are considered as major pathogens of leaf rot disease based on their frequency of occurrence and pathogenicity (Srinivasan and Gunasekaran 1999).

10.2.4.3 Epidemiology

The tender leaf is the most susceptible to the disease (Lilly 1963). The susceptibility of the seedlings decreases with age. Seedlings up to 19 months may get severe infection. Leaf rot infection is more severe during the monsoon season when the conditions of high humidity are prevalent (Menon and Nair 1951). Severity of leaf infection by *Helminthosporium halodes* was found to be correlated with high temperature and low humidity present during monsoon period (Radha et al. 1961; Radha and Lal 1968). The population dynamics of leaf rot disease-causing pathogens in relation to environmental variables was studied by monthly isolations from the spindle leaves of diseased palms. The incidence of *C. gloeosporioides* was higher in frequency, and its population is high during monsoon with a peak in June–July. Its incidence was positively correlated with rainfall and RH and negatively correlated with maximum temperature and sunshine hours. Thus, *C. gloeosporioides* was implicated as the principal pathogen of leaf rot during monsoon. Incidence of *E. rostratum* was less frequent and not well-correlated with weather. *Fusarium* spp. and *R. solani* were isolated most commonly during the dry season of January–May (Srinivasan and Gunasekaran 1996). There was no significant difference in the amino nitrogen levels, ascorbic acid, total phenols or sugars between leaves of healthy and leaf rot-affected palms. However, higher moisture levels, total and non-protein nitrogen, P, K, Ca and Mg were observed in tender leaves (Lilly 1963). Lilly and Ramadasan (1979) found that as a result of infection, the total phenols increased in the leaves.

10.2.4.4 Disease Management

Nair and Radha (1959) and Radha (1961) reported that regular manuring and spraying with copper fungicides checked the disease. By regular spraying, the intensity of leaf rot could be brought down from 40% to 7.8%. Spraying the leaves sequentially with Bordeaux mixture (1%), mancozeb (0.3%) and copper oxychloride (0.5%) at quarterly intervals after removing severely affected leaves was found to reduce further incidence of the disease. Subsequently, in vitro assay of contact fungicides [Mancozeb (Indofil M-45), copper oxychloride (Fytolan), Captaf (Captan) and Thiram] and systemic fungicides [hexaconazole (Contaf), tridemorph (Calixin) and aureofungin-sol)] against *C. gloeosporioides* and *E. rostratum* indicated that hexaconazole (Contaf) exhibited a broad-spectrum activity inhibiting all the pathogens of leaf rot disease (Srinivasan and Gunasekaran 1999). A field trial conducted for 3 years on 20-year-old palms revealed that pouring of tridemorph (Calixin) (1%) into leaf axils and spraying of Mancozeb (Indofil) and Dithane M-45 (0.3%) along with phytosanitary practices reduced the disease intensity (Srinivasan and Gunasekaran 1999). The bacterial antagonist *Pseudomonas fluorescens* inhibited the growth of *C. gloeosporioides* and *E. rostratum* under in vivo conditions and reduced the leaf rot onset. Of the 96 phylloplane and 21 rhizosphere isolates from coconut, 2 isolates each from phylloplane and rhizosphere have been identified as effective native antagonists against pathogens, *C. gloeosporioides* and *E. rostratum* (Srinivasan and Gunasekaran 1999; Gunasekaran et al. 2001).

Radha (1961) had observed that Andaman Ordinary and Papua New Guinea cultivars were more resistant to leaf rot. The control of leaf rot gained significance because of vulnerability of root (wilt)-affected palms to leaf rot. An integrated management system involving need-based pruning of infected spindle leaf and a few leaves close to spindle and the use of hexaconazole (Contaf) 2 ml in 300 ml of water poured onto the spindle are the most important measures for controlling leaf rot (Srinivasan and Gunasekaran 2000).

10.2.5 Leaf Blight

This disease is widespread over all coconut-producing countries, but is of little importance in well-managed plantations. The disease causes serious damage in seedlings as well as in adult palms. Grey leaf blight incidence reduces coconut yield to the extent of 10–24% (Karthikeyan et al. 1997). Coconut palms severely affected with grey leaf blight flowered relatively later than the less affected ones (Abad and Blancaver 1975).

10.2.5.1 Symptoms

Initially, symptoms develop only on the outer whorl of leaves, especially on older leaves. Symptoms appear as minute, yellow spots, each surrounded by a greyish margin on the leaflets of older leaves. They are oval in shape and are about 1–5 cm long. The centre of these spots later becomes greyish, and spots may coalesce, giving the leaves a burnt appearance. Complete drying and shrivelling of the leaf blades occur in the later stages. Such a condition is referred to as 'blight'. On the upper leaf surface, globose, spherical, rectangular or ovoid black pycnidia of the fungus are formed. Some varieties are more susceptible than others.

10.2.5.2 Aetiology

The fungus *Pestalotiopsis palmarum* (Cooke) Steyaert is the causal agent of the disease. *Conidiomata acervulii* are globose or ellipsoidal, subepidermal in origin. Conidiophores are indistinct. Conidiogenous cells are discrete, simple, short and filiform. Conidia measure 17–25 × 4.5–7.5 μm; are fusiform to ellipsoid, mainly straight and four-septate; have 3 median cells; and are concolorous and olivaceous, with lower cell sometimes paler, together measuring 11.5–16.5 μm long; with hyaline apical and basal cells; with 3 appendages, 5–25 μm long, arising from the apex of the apical cell; and with filiform basal appendage, 2–6 μm long. Over the years, there has been some confusion with the names *Pestalotia* and *Pestalotiopsis*. Guba (1961) accepted over 200 names in *Pestalotia*, but Sutton reviewed the genera and placed those species with five-celled conidia into *Pestalotiopsis*, while retaining *Pestalotia* for those species with six-celled conidia. Thus, the older literature uses the name *Pestalotia*, while the modern literature usually refers to *Pestalotiopsis* as the most common genus encountered on coconut. Maharachchikumbura et al. (2012) described several species based on molecular studies and epitypification of species. Brown (1973) found that this fungus was the cause of the most common leaf spot of coconuts in Solomon Islands and noted distinct differences between lesions associated with *P. palmarum* and with 3 other *Pestalotiopsis* species on coconut.

10.2.5.3 Disease Management

The disease incidence indicates the poor nutritional status of the affected palms. Imbalanced nutrition such as potash deficiency or too much inorganic nitrogen causes the seedling to be more susceptible (Karthikeyan et al. 1997). Leaf damage by insects may also provide an entry point to the fungus. The best way to avoid the disease is by improving the growing condition of the affected palms. Diseased palms may be treated with fortnightly sprayings of Bordeaux mixture, copper fungicides or carbamates containing zinc or manganese. Regular application of potassium chloride was reported to reduce the disease incidence. Cutting and removal of severely affected lower leaves and spraying of fungicides like carbendazim (0.1%) or Bordeaux mixture (1%) to affected palms immediately after the appearance of symptoms are advocated.

10.2.6 Lasiodiplodia Leaf Blight of Coconut

Reports of this disease have come from various parts of the world, such as Trinidad, Brazil, Malaysia, Sri Lanka and India (Ram 1993; Bhaskaran et al. 2007; Monteiro et al. 2013). The fungus accelerates the death of palms having already been weakened by other causes such as lack of drainage, moisture stress and malnutrition.

Leaf blight is an emerging serious problem in certain districts of Tamil Nadu, India (Johnson et al. 2014). The same fungus also infects seed nuts (Raju 1984). Though leaf blight is present in coconut-growing areas of other states of India, the disease is not a serious problem.

10.2.6.1 Symptoms

The pathogen causes damage to both leaf and nuts. Affected leaflets start drying from the tip downwards and exhibit a charred or burnt appearance (Fig. 10.5). The leaves in lower 3 to 4 whorls are affected. Leaf blight induces apical necrosis of lower leaves with an inverted 'V' shape, and symptoms are similar to those induced by drought (water deficit) and other stresses. The leaflets have extensive necrotic lesions with defined edges and without transition areas between the necrotic and healthy tissues. The pathogen can internally colonise the rachis, inducing internal necrosis that moves upward towards the stem (systemic invasion). The necrotic tissues develop exposed cracks that release gums under the leaf rachis and at petiole insertion (Souza-Filho et al. 1979). On coconuts, small black sunken region appears near the perianth of immature nuts. The eriophyid mite-attacked nuts are infected by the pathogen causing immature nut fall and rotting. When nearly mature or fully mature nuts were infected, the infection spreads internally into mesocarp without any external symptoms. The affected nuts are desiccated, shrunk, deformed and drop prematurely causing 10–25% loss in nut yield (Venugopal and Chandramohanan 2006).

Fig. 10.5 ***Lasiodiplodia*** leaf blight disease of coconut. (Photo: V Hegde)

10.2.6.2 Aetiology

The disease is caused by the fungus *Lasiodiplodia theobromae* (Pat.) The fungus is geographically widespread but is most common in the tropics and sub-tropics. It is plurivorous and has been associated with approximately 500 hosts. The pathogen has been reported to cause numerous diseases, including dieback, root rot, fruit rots, leaf spot and witches' broom, among many others (Punithalingam 1980). The main feature of the fungus is the presence of pycnidial paraphyses and longitudinal striations on mature conidia. So far, 20 species have been described and are differentiated on the basis of conidial and paraphyses morphology. Further reports indicate that *L. theobromae* is a complex of different cryptic species (Alves et al. 2008). Large number of isolates from coconut leaf blight needs to be collected and characterised to determine the exact species status.

10.3 Virus Disease

10.3.1 Coconut Foliar Decay or Vanuatu Wilt

The coconut foliar decay is a virus disease of introduced coconut palms in Vanuatu. It is also known as foliar decay *Myndus taffini* or New Hebrides coconut disease. The name '*Myndus taffini*' comes from the plant hopper insect that transmits the disease. The disease is economically important because of its influence on regional coconut industry and internationally on quarantine considerations.

10.3.1.1 Symptoms

The first symptom on palms in the field is yellowing of a few leaflets on any of the fronds between positions 7 and 11 from the spear leaf. The yellowing spreads along the fronds which break near the base so that they hang down through the still green lower leaves. As the younger leaves age, reaching positions 7–11, they, too, turn yellow, break and hang down. In due course of time, diseased palms will have the top and midsection fronds broken and hanging through the still green fronds below. As the disease progresses, the trunk narrows towards the top, and the palm dies after 1–2 years, except for those that are tolerant to the disease and show remission of symptoms. Foliar decay is more serious on Malayan Red Dwarf coconut introduced to Vanuatu. Coconut cultivars, viz. Vanuatu Tall and Vanuatu Dwarf, are usually not affected by the disease, which, though are hosts for the foliar decay virus, act as symptomless carriers (Randles et al. 1992; Hanold et al. 2003).

10.3.1.2 Aetiology

The disease is caused by a very small circular single-stranded DNA virus, which is named as *Coconut foliar decay virus* (CFDV, an unassigned species under Family *Nanoviridae*) (Randles et al. 1986). The virus is found at very low concentrations in coconut palms. It is difficult to see the virus particle in the sap viewed by electron microscopy. The virus occurs in the phloem of the palm. Coconut is the only known host for the virus, which occurs in leaves, roots, embryo, trunks and even on the husk of the nut. Seed transmission is not yet established.

10.3.1.3 Transmission

The disease is transmitted by *Myndus taffini* Bonfils (*Cixiidae*), a plant hopper which breeds on the roots of *Hibiscus tiliaceus,* a tree commonly found in Pacific seashores. The adults of this insect feed on coconut leaves. Wefels et al. (2015) reported the molecular evidence for a persistent circulative association between CFDV and its vector. Both the vector and virus are apparently limited in distribution in Vanuatu archipelago.

10.3.1.4 Disease Management

The disease is best controlled by either planting selected Vanuatu Tall or the hybrid, Vanuatu Tall × Vanuatu Red Dwarf, which are tolerant to the disease. The removal of the host tree of the insect that spreads the virus is likely to be beneficial though not practically possible. The '*FAO/IBPGR Technical Guidelines for the Safe Movement of Coconut Germplasm*' have recommended that coconuts should be moved as embryos growing in a sterile tissue culture medium (Anitha Karun et al. 2002). As a special note, the guidelines recommend that embryos, seedlings and palms from which pollen is collected should be tested for viroids and *Coconut foliar decay virus*. If that is not possible, seedlings may be monitored for viroids and *Coconut foliar decay virus* in an intermediate (third country) quarantine centre.

10.4 Viroid Diseases

Two viroid diseases, viz. coconut cadang-cadang and tinangaja, are recorded on coconut, and their distribution is limited to the Philippines and Guam, respectively.

10.4.1 Coconut Cadang-Cadang Disease

In the early 1930s, a devastating epidemic of cadang-cadang, the lethal disease of coconut palm, was reported from southern Luzon in the Philippines (Randles 1987). This disease caused tremendous economic losses in coconut plantations in the Philippines. Cadang-cadang disease is widely distributed on the Bicol Peninsula, Masbate, Catanduanes, Northern Samar and other smaller islands in the Philippines. Outbreaks have been found in and around Infanta, Quezon, in Eastern and Western Samar and Maripipi. Small isolated groups of infected palms have been found northeast of the main boundary at Atimonan. At present, the northernmost boundary of disease occurrence is at General Nakar, Quezon, and the southernmost at Calicoan, Guiuan and Eastern Samar.

10.4.1.1 Symptoms

In the early stage, newly developing nuts become more spherical and have equatorial scarifications. Chlorotic leaf spots begin to appear and inflorescences become stunted. In the medium stage, spathe, inflorescence and nut production decline and then cease. Leaf spots become more numerous. By the late stage, the fronds decline in size and number and the leaflets become brittle. Leaf spots coalesce, giving a general chlorotic look. The crown size is reduced and later the palm dies. This progression of symptoms is remarkably constant with some variation in intensity. The early stage lasts for 2–4 years, the medium stage for about 2 years and the late stage for about 5 years. Usually, palms become infected only after they have reached the age of flowering. In rare cases, where younger palms become infected, they are stunted and fail to produce inflorescences, although they survive well past the age of first flowering.

10.4.1.2 Aetiology

The detection of 2 small disease-associated RNAs in 1975 provided the initial clue to the aetiology of cadang-cadang. Electron microscopy, nucleotide sequencing and transmission experiments that demonstrated the infectivity of these RNAs finally proved that cadang-cadang is caused by a viroid. It is now referred to as the coconut cadang-cadang viroid or CCCVd. Viroids are the smallest known pathogens and have been found only in plants. Unlike viruses, they do not have a protein coat and consist solely of a small circular, single-stranded infectious RNA molecule that can replicate in the host cell and get transmitted independently of any other microorganism (Hanold and Randles 1991).

10.4.1.3 Epidemiology

The mode of natural inoculation in the field is not known. No insect vector has been found (Randles 1987). Positive transmission was obtained through assisted pollination of mother palms with pollen from diseased palms. A small percentage of the progenies, as well as seed nuts collected from cadang-cadang-infected palms, was positive for CCCVd and was also successfully transmitted to palms through contaminated harvesting scythes.

10.4.1.4 Disease Management

At present, there is no direct control measure that can be recommended to manage cadang-cadang, but several possible strategies can be considered (Randles 1987). Strict enforcement of quarantine regulations by concerned government agencies on the safe movement of coconut germplasm from infected areas will prevent further spread of cadang-cadang into disease-free areas. Continued research on cadang-cadang runs parallel to the coconut improvement programme in the Philippines. To minimise the risk of an epidemic occurring in new plantings, attempts have been made to find individuals or populations that are resistant or tolerant to cadang-cadang.

10.5 Future Strategy

Coconut being a perennial crop with a life span of several decades is vulnerable to the vagaries of nature at its various growth stages. Out of the several pathogens infecting coconut, the 3 major pathogens threatening coconut production world over are *Phytoplasma*, *Phytophthora* and *Ganoderma*.

Managing bud rot and immature nut fall caused by *Phytophthora palmivora* and *P. katsurae* is a major challenge owing to the difficulty in taking up curative and prophylactic measures in tall palms. Hence, modern technologies like computer-aided, remote-controlled drones could help in such operations. Remote-sensing technologies can help in regular surveillance for disease incidence, intensity and spread which can aid in planning and executing disease management measures.

Molecular techniques should be increasingly used to help in disease detection and in resistance breeding. Though PCR-based techniques for diagnosis of *Phytoplasma*, viruses and viroids are available, development of field-level, reliable and rapid diagnostic kits is essential.

Region-specific integrated disease management strategies for specific maladies have to be developed for attaining maximum efficacy in disease control. Efficient long- and short-term disease-forecasting measures are to be developed to undertake timely prophylactic measures.

Super palms, both in healthy areas and disease hot spots possessing pre-potency, should be identified on a regular basis to serve as potential donors of disease resistance.

Periodic crop loss surveys and constant monitoring and surveillance are essential to compute the economic loss caused by diseases to guide in policy decisions to combat serious diseases and to check the re-emergence of minor diseases and emergence of new diseases in the prevailing era of climate change. Strict implementation of domestic and international quarantine measures is to be taken for preventing the disease spread. Detailed studies on epidemiology and vectors of transmission are to be given importance.

Integrated management measures are to be popularised and supported by appropriate policy decisions to reduce economic loss to the farmer due to these maladies.

References

Abad RG, Blancaver RC (1975) Coconut leaf spot/blight and their control. PCA-ARD Annual Report 1975–76

Almeida JL, Aquino M, de LN de(1978) Antracnose ou podridao dos fruitos de coqueiro. *Cocos nucifera* no estado de Pernambuco. Pesquisa Agropecuaria Pernabucana 2:37–44. (Cited by Rasmi AR 2005. 'Management of bud-rot in young coconut palms'. Ph.D. thesis, Mangalore University, Mangalore, p 275)

Alves A, Crous PW, Correia A, Phillips AJL (2008) Morphological and molecular data reveal cryptic speciation in *Lasiodiplodia theobromae*. Fungal Divers 28:1–13

Anil Kumar, Nambiar KKN (1991) An improved method of isolation of *Thielaviopsis paradoxa* from stem-bleeding affected coconut palm. CORD 7:49–53

Anil Kumar, Nambiar KKN, Voleti SR (1992) Uptake, translocation and persistence of Carbendazim in coconut in relation to stem bleeding disease. CORD 8:40–47

Anitha Karun, Parthasarathy VA, Kumaran PM, Iyer RD, Sajini KK (2002) *Coconut* embryo-culture protocol for germplasm collection, Tech Bull. ICAR-CPCRI, Kasaragod

Bennet CPA, Roboth O, Sitepu G, Lolong A (1986) Pathogenicity of Phytophthora palmivora (Butler) causing premature nut fall disease of coconut (*Cocos nucifera* L.). Indonesian J Crop Sci 2(2):59–70

Bhaskaran R, Ramanathan T (1984) Occurrence and spread of Thanjavur wilt disease of coconut. Indian Cocon J 15(6):1–3

Bhaskaran R, Ramanathan T, Ramaiah M (1984) Thanjavur wilt of coconut; its occurrence, spread and management. In: Sethuraj et al (eds) Proceedings of PLACROSYM-VI. Oxford and IBH Co. Ltd, New Delhi, pp 107–113

Bhaskaran R, Rethinam P, Nambiar KKN (1989) Thanjavur wilt of coconut. J Plantn Crops 17:69–79

Bhaskaran R, Vaithilingam R, Ramanathan A, Natarajan C, Marimuthu R, Subramanian KV (2007) A new lethal disease of coconut in Tamil Nadu. Indian Cocon J 38(3):2–4

Boccardo G (1985) Viroid etiology of tinangaja and its relationship with cadang-cadang disease of coconut. In: Maramorosch K, JJ MK Jr (eds) Subviral pathogens of plants and animals: viroids and prions. Academic Press, Orlando, pp 75–99

Brahmana J, Kelana A (1988) Bud-rot disease on PB – 121 hybrid Coconut at Bangun Purba PTP VI. North Sumatra CORD 4(2):76–87

Brahmana J, Lubis AU, Desmier De Chenon R (1992) Evolution of coconut bud-rot disease and control strategies (observations in Bangum, Purba and North Sumatra). In: Proceedings of coconut phytophthora workshop, 26–30 October1992. Manado, Indonesia, pp 33–37

Britton-Jones HR (1940) Diseases of coconut palm. Builliere, London

Brown JF (1973) Disease of coconut in the British Solomon Islands. Plant Disease Rep 57:856–860

Butler EJ (1906) Some diseases of palms. Agric J India 1:299–310

Calvez C, Renard JL, Marty G (1980) Tolerance of the hybrid coconut local × Rennell to New Hebrides diseases. Oleagineux 35:443–451

Coffey MD (1990) Integrated control of *Phytophthora* bud-rot and premature nut fall of coconut in Indonesia. Abstract in '3rd international congress on plant protection in the tropics', March 20–23, 1990, p 105

Cohen Y, Coffey MD (1986) Systemic fungicides and the control of oomycetes. Annu Rev Phytopathol 24:311–338

Coleman LC (1911) 'Anabe roga' of 'supari'. Annual Report for 1909–1910. Agric Chemist, Mysore, Department of Agriculture, Bangalore, p 32

Cook AA (1971) Diseases of tropical and sub-tropical fruits and nuts. Hafner Press, New York. Cited by Harrison N.A & Jones, p 2003

Copeland EB (1931) The Coconut, 3rd edn. Macmillan & Co., London, p 233

Darwis SN (1992) *Phytophthora* in relation to climate and coconut cultivar. In: Proceedings of coconut phytophthora workshop 26–30 Oct.1992, Manado, Indonesia, pp 17–20

Dauzat J, Lecoustre RL (1992) Spatio-temporal analysis of disease distribution, description of a few methods and initial results on *Phytophthora*. In: Proceedings of coconut phytophthora workshop 26–30 October 1992, Manado, Indonesia, pp 57–68

Erwin DC, Ribeiro OK (1996) Phytophthora diseases worldwide. American Phytopathol Society Press, St. Paul

Evans HC (1973) New developments in black-pod epidemiology. Cacao Growers Bull No 20:10–16

Gowda PV (1987) In vitro studies on the interaction of antagonistic mycoflora with *Theilaviopsis paradoxa,* the causal agent of stem-bleeding disease of coconut. M.Phil., dissertation, Mangalore University, Mangalore, p 197

Guba EF (1961) Monograph of *Monochaetia* and *Pestalotia*. Harvard University Press, Harvard, p 342

Gunasekaran M, Alka Gupta, Srinivasan N (2001) Biological control of leaf-rot pathogens with *Pseudomonas fluorescens*. In: National symposium on eco-friendly approaches for plant disease management. University of Madras, Chennai

Hanold D, Randles JW (1991) Coconut cadang-cadang disease and its viroid agent. Plant Dis 75(4):330–335

Hanold D, Morin JP, Labouisse JP, Randles JW (2003) Foliar decay disease in Vanuatu. In: Loebenstein G, Thottappilly G (eds) Plant virus diseases on crops in developing countries. Kluwer Academic Publishers, Dordrecht, pp 583–596

Harris DC, Card ON, Justin SHFW, Passey AJ (1984) *Phytophthora palmivora* on the cultured roots of coconut. Trans Brit Mycol Soc 82(2):249–255

Harrison NA, Jones P (2003) Diseases of coconut. In: Ploetz RC (ed) Diseases of tropical fruit crops. CABI Publ, Wallingford, pp 197–225

Holliday P (1980) Fungus diseases of tropical crops. Cambridge University Press, Cambridge

Jackson GVH, Firman ID (1982) Seed-borne marasmioid fungi of coconut. Plant Pathol 31(2):187–188

Jackson GVH, McKenzie EHC (1988) *Marasmiellus cocophilus on* coconut in Solomon Islands. FAO Plant Protection Bull 36(2):91–97

Jagannathan T, Ramasamy R (1975) Annual report 1974–76. Tamil Nadu Agricultural University, Coimbatore, p 173

Johnson B, Meena, Rajamanickam K (2014) Biological management of leaf blight disease of coconut using rhizosphere microbes. J Plantn Crops 42(3):364–369

Karthikeyan A, Bhaskaran R, Giridharan S (1997) Management of leaf blight disease of coconut. The Planter 73(858):495–501

Kohler F, Pellegrin F, Jackson G, McKenzie E (1997) Diseases of cultivated crops in Pacific Island countries. South Pacific Commission. Pirie Printers Pty Limited, Canberra, p 181

Koti Reddy M, Ananthanarayanan TV, Nambiar KKN (1984) Detection of *Ganoderma lucidum* in betel-nut by the fluorescent antibody technique. Trans Brit Mycol Soc 82:559–561

Lewin HD, Sindha Madhar A, Sethurarnan V (1983) Effect of fungicides on the control of Thanjavur wilt disease of coconut. In: Proceedings of national seminar on management of diseases of oilseed crops. Agrl. College and Res Inst, Madurai, pp 99–100

Lilly VG (1963) Host-parasite relationships *of Helminthosporium halodes* on coconut palms. Indian Cocon J 16:149–153
Lilly VG (1984) Record of *Phomopsis coconina* (Cooke) Punith. in stem-bleeding affected coconut palms. Curr Sci 53:704–706
Lilly VG, Ramadasan A (1979) Changes in phenolic content of leaf in relation to the development of leaf-rot. Indian Phytopathol 32:112–113
Maharachchikumbura SSN, Guo LD, Cai L, Chukeatirote E, Wu WP, Sun X, Crous PW, Bhat DJ, McKenzie EHC, Bahkali AH, Hyde KD (2012) A multi-locus backbone tree for *Pestalotiopsis*, with a polyphasic characterization of 14 new species. Fungal Divers 56:95–129
Malayczuk N (1983) Microbial antagonism to *Phytophthora*. In: Erwin DC, Bartmicki–Garcia S, Tsao PH (eds) Phytophthora – its biology, taxonomy, ecology and pathology. American Phytopathol Society, St Paul, pp 197–218
Mangindaan H, Thevenin JM, Knarie S, Motulo HFJ (1992) The susceptibility of coconut varieties to *Phytophthora* in Indonesia: The effect of environmental factors. Proceedings of Coconut *Phytophthora* Workshop. October 1992, Manado, Indonesia, pp 39–46
Mathew AS, Ramanandan PL (1980) Incidence of stem-bleeding disease of coconut palm in relation to pH and electrical conductivity in soils. J Plantn Crops 8:40–42
Menon KPV, Nair UK (1948) The leaf-rot diseases of the coconut in Travancore and Cochin. Indian Cocon J 1:33–39
Menon KPV, Nair UK (1951) Scheme for the investigation of the root and leaf diseases of the coconut palms in South India. Final report of the work done from 8th March 1937 to 31st March, 1938. Indian Cocon J 5:81–100
Menon KPV, Pandalai KM (1960) The coconut palm – a monograph. Indian Central Coconut Committee, Ernakulam, p 384
Monteiro CM, Caron ES, da Silveira SF, Almeida AM, Souza-Filho GR, de Souza AL (2013) Control of foliar diseases by the axillary application of systemic fungicides in Brazilian coconut palms. Crop Prot 52:78–83
Moosa H, Ahmed R, Iyer R, Prabhu SR (1998) 'Occurrence of endophytic *Bacillus* antagonistic to *Phytophthora palmivora* in coconut seedlings. Paper presented at international conference of plant pathology held at Edinburgh, Scotland, from 9 to 16 August, 1998
Nagaraja AN, Menon KPV (1956) Note on the etiology of the root (wilt) disease of coconut palms in Travancore-Cochin. Indian Cocon J 9:161–165
Naidu GB, Sampathkumar SN, Sannamarappa M (1966) 'Anaberoga', *Ganoderma lucidum* (Lays) Karst on arecanut palm- a review and further considerations. Lal Baugh 11:14–20
Naik N (1990) Variability and characteristics of different isolates of *Thielaviopsis paradoxa,* the pathogen of stem bleeding disease. M. Phil. dissertation, Mangalore University, Mangalore, p 76
Nair UK, Radha K (1959) Manuring-crown-spraying for the control of Leaf rot of coconut. In: Nair MK, Khan HH, Gopalasundaram P, Rao EVVB (eds) Advances in coconut research and development. Oxford and IBH Publishing Co. Ltd, New Delhi, pp 546–561
Nambiar KKN, Joshi Y, Venugopal MN, Mohan RC (1986) Stem bleeding disease of coconut: reproduction of symptoms by inoculation with Thielaviopsis paradoxa. J Plantn Crops 14(2):130–133
Nambiar KKN, Kalpana Sastry R (1988) Stem bleeding disease of coconut: current status & approaches for its control. Philippine J Cocon Studies 12:30–32
Nambiar KKN, Ramanujam B (1993) Current studies on etiology and management of stem bleeding disease of coconut. Indian Cocon J 23:2–4
Nambiar KKN, Rawther TSS (1993) Fungal diseases of coconut in the world. In: Nair MK, Khan HH, Gopalasundaram P, Rao EVVB (eds) Advances in coconut research and development. Oxford and IBH Publishing co. Pvt Ltd, New Delhi, pp 545–562
Nambiar KKN, Anil Kumar, Kalpana Sastry R, Joshi Y (1989) Seasonal effect on infection by stem bleeding pathogen, *Thielaviopsis paradoxa*. Curr Sci 58:34–35
Natarajan S, Bhaskaran R, Shanmugham N (1986) Preliminary studies to develop techniques for early detection of Thanjavur wilt in coconut. Indian Cocon J 17:3–6
Ohler JG (1984) Coconut- Tree of life. FAO, Rome, p 446

Ohler JG (1999) Modern coconut management: palm cultivation and products. FAO, Rome, p 474
Papa Rao A, Govinda Rao P (1966) A survey of coconut diseases in Andhra Pradesh. Andhra Agri J 13:208–217
Peiries OS (1974) *Ganoderma* basal stem rot of coconut: a new record of the disease in Sri Lanka. Plant Disease Reporter 58(4):293–229
Peiries OS, Liyanage A de S, Mahindapala R, Subasinghe SMP (1975) The incidence of *Ganoderma* root and bole rot of coconut in Sri Lanka. The Ceylon Cocon Qtly 26:99–10
Peiris JW (1962) Bud-rot of coconuts. Ceylon Cocon Rev 3:37–40
Petch T (1906) Diseases of coconut palm. Trop Agri (Trinidad) 27:489–491
Ploetz R, Harrison N, Jones P (1999) Diseases of coconut palm (*Cocos nucifera* L.). APS Publications, New York
Pohe J (1992) Factors involved in the development of nut fall due to *Phytophthora* in *Coté-dé-Ivoir é*. In: Proceedings of Coconut Phytophthora Workshop, 26–30 October 1992, Manado, Indonesia, pp 47–50
Punithalingam E (1980) Plant diseases attributed to *Botryodiplodia theobromae* Pat. J Cramer, Vaduz Biblio Mycol 71:1–123
Quillec G, Renard JL, Ghesqueire H (1984) *Phytophtora hevea* of coconut: Role in bud rot and nutfall. Oléagineux 39(10):477–484
Radha K (1961) Leaf rot disease of coconut. In: Proceedings of first session, FAO technical working party coconut production, protection and processing, Trivandrum, pp 30–36
Radha K (1962) Stem bleeding disease of coconut. Cocon Bull 16:347–350
Radha K, Joseph T (1974) Investigation on the bud-rot disease (Phytophthora palmivora) of coconut. Final report PL-480 scheme, CPCRI, Kayangulam, p 32
Radha K, Joseph T (1976) A reappraisal of associated factors. In: Nayar NM (ed) Coconut research and development. Wiley Eastern Limited, New Delhi, pp 333–340
Radha K, Lal SB (1968) Some observations on the occurrence of leaf-rot disease of coconut and associated factors. 3rd session, FAO technical working party coconut production protection and processing, Jakarta, pp 1–5
Radha K, Sukumaran CK, Prasanna Kumari TO (1961) Studies on the leaf-rot disease of coconut. Fungal infection in relation to environmental conditions. Indian Cocon J 15:1–11
Raju CA (1984) Effect of methyl bromide fumigation on the fungi associated with seed coconuts. Philippine J Cocon Stud 9:1–2
Ram C (1993) Características culturais, esporulação e virulência do “strain” do *Botryodiplodia theobromae*, agente causal da queima-das-folhas do coqueiro (*Cocos nucifera*). Fitopatologia Brasiliera 18(2):143–146
Ramanujam B (1997) Studies on biocontrol of *Theivaviopsis paradoxa* (De Synes) Von Honel., the pathogen of stem-bleeding disease of coconut using antago-nistic fungi. Ph.D. Thesis, Mangalore University, Mangalore, India, p 228
Ramanujam B, Chandramohanan R, Nambiar KKN (1993) Studies on leaf spot disease of arecanut: symptoms, cause and control. J Plantn Crops 21(Suppl):123–128
Ramanujam B, Nambiar KKN, Ratnambal MJ (1996) Screening of coconut cultivars/hybrids against *Thielaviopsis paradoxa* (de Seynes) Hohnel, using petiole inoculation technique. In: Mathew NM, Kuruvilla Jacob C (eds) Developments in plantation crops research, Proceedings of PLACROSYM XII. Indian Society for Plantation Crops, Kasaragod, pp 284–286
Ramapandu S, Satyanarayana Y, Rajamannar M, Chiranjeevi V (1981) Seasonal variation in the manifestation of Ganoderma wilt disease of coconut. Indian Cocon J 12(3):5–6
Randles JW (1975) Association of two ribonucleic acid species with cadang-cadang diseases of coconut plam. Phytopathol 65:163–167
Randles JW (1987) Coconut cadang-cadang. In: Diener TO (ed) The Viroids. Plenum Press, New York, pp 265–277
Randles JW, Julia JF, Calvez C, Dollet M (1986) Association of single-stranded DNA with foliar-decay disease of coconut palm in Vanuatu. Phytopathol 76:889–894
Randles JW, Miller DC, Morin JP, Rohde W, Hanold D (1992) Localisation of coconut foliar-decay virus in coconut palm. Ann Appl Biol 121:601–617

Rao AP, Laxminarayan C, Pandit SV, Govindarao P (1966) Studies on spread of Tatipaka disease of coconut. Andhra Agri J 13(5):169–173

Rasmi AR, Rohini Iyer, Vijayakumar K (2004) Role of relative humidity and temperature on the survival of *Phytophthora palmivora*, the incitant of coconut bud-rot. J Plantn Crops 32(2):28–34

Renard JL, Darwis SN (1993) Report on the coconut *Phytophthora disease*. Seminar, Manado, Indonesia, October, 1992. Oleagineux 48(6):301–305

Renard JL, Quillec G (1984) *Phytophthora heveae* of coconut. II control methods. Oléagineux 3911:529–534

Rillo EP, Paloma MBF (1989) Reaction of some coconut populations and hybrids to *Phytophthora* disease in the Bicol region. Philippine J Cocon Studies 14(1):2–6

Rohini Iyer (1997) Effect of systemic fungicides on control of bud rot of coconut. In: International conference on 'Integrated Plant Disease Management for Sustainable Agriculture', Nov. 1997 New Delhi, India. (Abstracts) Indian Phytopathological Society Golden Jubilee- p 212

Sampath Kumar SN, Nambiar KKN (1993) Relationship among *Ganoderma* species affecting coconut and arecanut palms. J Plantn Crops 2l(Suppl):119–122

Satyanarayana Y, Ramapandu S, Rajamannar M, Chiranjeevi V (1985) Control of Ganoderma-wilt disease of coconut. Indian Cocon J 14(5):3–5

Schutt B (1975) A possible phytotoxic effect of copper on dwarf coconut palms. Surinamese Landbouw 23(1):26–30

Shea SR, Broadbent P (1983) Developments in cultural and biological control of *Phytophthora* diseases. In: Erwin DC, Bartimcki-Garcia S, Tsao PH (eds) Phytophthora – its biology, taxonomy, ecology and pathology. American Phytopathological Society, St, Paul, pp 335–350

Souza Filho BF, Santos Filho HP, Robbs CE (1979) Etiologia da queima das folhas do coqueiro. Fitopatol Bras 4:5–10

Srinivasan N (1991) Occurrence of coconut leaf-rot in association with root (wilt) disease. Indian Cocon J 21:14–18

Srinivasan N, Gunasekaran M (1992) An appraisal of symptom expression in coconut due to leaf rot disease. Indian Cocon J 23(7):2–6

Srinivasan N, Gunasekaran M (1996) Incidence of fungal species associated with leaf-rot disease of coconut palms in relation to weather and stage of lesion development. Ann Appl Biol 129:433–449

Srinivasan N, Gunasekaran M (1999) *Fusarium solani and Fusarium moniliforme* var. *intermedium* in cocon leaf-rot disease. Indian Phytopathol 52:160–162

Srinivasan N, Gunasekaran M (2000) Leaf-rot disease of coconut, Tech Bull No.38. Central Plantation Crops Research Institute, Kasaragod, p 14

Steer ED (1992) Coconut bud-rot disease in Jamaica. Proceedings of Coconut Phytophthora Workshop 26–30, Oct 1992, Manado, Indonesia, pp 51–56

Sundararaman S (1922) The coconut bleeding disease. Pusa Agric Res Inst Bull 127:8

Sundararaman S (1925) Preliminary note on coconut leaf-rot of Cochin. Year Book, Madras Agric Dept, Madras, pp 6–8

Sutherland ED, Baker KK, Lockwood JL (1984) Ultrastructure of *Phytophthora megasperma f.* sp. *glycinea* oospores parasitized by *Achnoplanes missouriensis* and *Humicola fuscoatra*. Transaction of the British Mycol Soc 82:726–729

Thevenin JM, Motulo HFJ, Kharie S, Mangindann HF, Warokkha JS (1992) Epidemiological studies on *Phytophthora* diseases of coconut in north Sulawezi Province, Indonesia. In: Proceedings of Coconut Phytophthora Workshop 26–30 Oct. 1992, Manado, Indonesia, pp 21–25

Tucker CM (1926) *Phytophthora* bud rot of coconut palms in puerto rico. J Agri Res 32:471–498

Tuset JJ, Ainarejos C, Garcia J (1984) Present status of *Phytophthora* disease of Citrus in Spain. Proceedings of International Soc Citriculture 2:338–343

Tuset JJ, Hinare Jos C, Buj A, Molins A, Cebolla V (1992) *Myrothecium roridum* and *M. verrucana,* as potential antagonists of *Phytophthora* spp. on coconut. In: Proceedings of Coconut Phytophthora Workshop, 26–30 October, 1992, Manado, Indonesia, pp 171–176

Uchida JY, Aragaki M (1991) *Bipolaris* and *Exserohilum* leaf-spots. In: Chase AR, Broschat TK (eds) Diseases and disorders of ornamental palms. APS Press, St Paul, p 55

Uchida JY, Ooka JJ, Nagata M, Kadooka CY (1992) A new *Phytophthora* fruit-and heart-rot of coconut in Hawaii. Plant Dis 76:925–927

Usman NM (1988) Studies on stem bleeding disease of coconut. 1. Survival of *Thielaviopsis paradoxa* in soil and 2. *In vitro* interaction of *Thielaviopsis paradoxa* with antagonistic fungi. M.Phil., dissertation, Mangalore University, Mangalore, p 74

Varghese MK (1934) Diseases of coconut palm. Department of Agriculture and Fisheries, Travancore, p 105

Venkatarayan SV (1936) The biology of *Ganoderma lucidum* on arecanut and coconut palms. Phytopathol 26:153–125

Venugopal S, Chandra Mohanan R (2006) Role of fungi in fruit rot and immature nutfall of coconut. CORD 22(2):33–40

Vijayan KM, Natarajan S (1972) Some observations on the coconut wilt disease of Tamil Nadu. Cocon Bull 2(12):2–4

Vijayaraghavan H, Ramdoss N, Raveendran TS, Rethinam P (1987) Effect of toddy tapping on Thanjavur wilt disease of coconut. Indian Cocon J 17(12):3–5

Warwick Dulce RN, Edson EM, Passos (2009) Outbreak of stem-bleeding in coconuts caused by *Thielaviopsis paradoxa* in Sergipe, Brazil. Trop Plant Pathol 34(3):175–177

Waters H (1978) A wilt disease of coconuts from Trinidad associated with *Phytomonas* sp., a sieve-tube-restricted protozoan flagellate. Ann Appl Biol 90:293–302

Wefels E, Morin JP, Randles JW (2015) Molecular evidence for a persistent circulative association between Coconut foliar decay virus and its vector *Myndus taffini*. Australasian Plant Pathol 44:283–288

Yu FY, Niu XQ, Tang QH, Zhu H, Song WW, Qin WQ, Lin CH (2012) First report of stem bleeding in coconut caused by *Ceratocystis paradoxa* in Hainan, China. Plant Dis 96:290–290. https://doi.org/10.1094/PDIS-10-11-0840

Dr. Rohini Iyer holds a PhD in Mycology and Plant Pathology. She standardized safe storage methods for seed rhizomes of ginger. She has extensively studied the VA-mycorrhizal association in plantation crops and developed disease management strategies for control of bud rot and stem-bleeding diseases of coconut through the application of biocontrol agents. She served as short-term ITEC Coconut Expert in Suriname Republic. She has 114 publications including 2 books and 11 book chapters. Dr. Rohini Iyer retired as Head, Crop Protection Division from ICAR-CPCRI. Currently she is the Honorary Managing Trustee of 'Navasakti Trust', an NGO in Kerala. navasakti@gmail.com

Dr. M. Gunasekharan holds a Ph.D. in Applied Botany. He served as a Scientist at ICAR-CPCRI and Central Institute of Cotton Research, Coimbatore, India. His areas of research include etiology, epidemiology, molecular detection and management of coconut diseases. He has published 67 research papers, 15 science communications, 7 review articles and 3 technical bulletins, and contributed 21 book chapters. He was awarded the Rothamsted International Fellowship, CIRAD International Fellowship and the Glory of India Gold Medal. He is currently working as a Honorary Professor of Plant Pathology in a private Agriculture College after superannuation. mgsekar@gmail.com

Dr. Vinayaka Hegde is currently the Head, Division of Crop Protection, ICAR-CPCRI, Kasaragod, Kerala, India. He is a Plant Pathologist by qualification. He was awarded IARI Merit Medal for his outstanding research work and UAS Gold Medal. He has served in ICAR-CPRI and CTCRI. Dr. Hegde is presently working on diagnostics and management of diseases of coconut, areca nut and cacao. He specialized in molecular diagnostics and management of plant diseases and published 55 research papers, authored 6 technical bulletins, 6 book chapters and 10 popular articles. hegdev64@gmail.com; vinayaka.hegde@icar.gov.in

Chapter 11
Phytoplasmal Diseases

J. J. Solomon, Vinayaka Hegde, Merin Babu, and L. Geetha

Abstract Phytoplasmas are phloem-restricted phytopathogenic mollicutes transmitted by hemipteran insects. Phytoplasmas belonging to different 16S rDNA groups and *Candidatus Phytoplasma* species have been reported to be associated with coconut. Several economically important phytoplasmal diseases affecting coconut have been reported worldwide, which include the lethal yellowing disease in Caribbean and African regions as well as the non-lethal maladies like root (wilt) disease and Weligama leaf wilt in Asia. Application of molecular technologies has enabled the development of phytoplasmal diagnostic techniques. However, molecular detection of coconut phytoplasmas is many a time intriguing owing to its exclusive phloem confinement, non-uniform distribution and subminimal titres in palms. Palm-phytoplasmal interactions as well as transmission mechanism by insects are poorly understood. Since the phytoplasmal diseases cannot be cured, research on incursion management is currently based on strengthening quarantine and breeding for disease resistance. Management practices are also available to obtain satisfactory yield even from disease-affected palms in the case of non-lethal phytoplasmal diseases. Outbreak of phytoplasmal diseases as a victim of climate change is a reality in view of extensive migration and survival superiority of insect vectors even under climate extremities.

J. J. Solomon (✉)
B.T-01, Porkudam Apartments, SBI officers colony, Madurai, Tamil Nadu, India
e-mail: solomon.jeyasingh@gmail.com

V. Hegde
ICAR-CPCRI, Kasaragod, Kerala, India
e-mail: hegdev64@gmail.com; vinayaka.hegde@icar.gov.in

M. Babu
ICAR-CPCRI, Regional Station, Kayamkulam, Kerala, India
e-mail: merinbabu1@gmail.com

L. Geetha
Nellithara House, Oachira, Kollam District, Kerala, India
e-mail: geetha.lekshmi@uaeu.ac.ae

K. U. K. Nampoothiri et al. (eds.), *The Coconut Palm (Cocos nucifera L.) - Research and Development Perspectives*,
https://doi.org/10.1007/978-981-13-2754-4_11

11.1 Introduction

Diseases caused by phytoplasmas pose serious threats to the sustainability of coconut cultivation worldwide. Since the nineteenth century, numerous outbreaks of phytoplasmal diseases leading to the loss of millions of coconut palms have been reported from Asia, Africa and America. Though most of them were recorded in the 1800s, the etiology remained elusive till the discovery of phytoplasma (formerly known as *Mycoplasma*-like organisms, MLOs) by Doi et al. (1967). Most of them were named based on the symptoms or geographic locations. Phytoplasmas are wall-less, pleomorphic, unicellular, nutritionally fastidious, phloem-limited, vector-transmitted phytopathogens with a mean diameter of 200–800 nm and genome size ranging from 530 kb to 1350 kb (Sears and Kirkpatrick 1994; Harrison et al. 2015).

Phytoplasmas constitute a unique monophyletic clade in the class *Mollicutes*. Due to the paucity of distinct phenotypic criteria, currently its classification is based on sequencing and RFLP analysis of the conserved 16S rDNA region (Lee et al. 1998; Martini et al. 2007; Wei et al. 2007). The introduction of molecular tools streamlined the phytoplasma taxonomy and enabled the characterisation and differentiation of these phytopathogens. Phytoplasmas belonging to different taxonomic groups have been reported to be associated with diseases of coconut palms. Some of them are fatal and widespread, whereas others are only debilitating or are limited to certain geographic regions. Phytoplasma diseases of coconut are listed in Table 11.1.

11.2 Lethal Yellowing (LY) and Lethal Yellowing-Type Diseases (LYD)

11.2.1 Occurrence, Distribution and Crop Loss

The most well-known and widespread phytoplasmal disease of coconut is lethal yellowing (LY). The disease was noticed in Cayman Islands in the Caribbean region as early as in 1834, and the first scientific report was from Jamaica in 1891 (Fawcett 1891; Howard 1983; Been 1995). The disease was reported from Cuba in the early 1900s (De La Torre, 1906 quoted by Ntushelo et al. 2013). The presence of the disease in Haiti dates back to the 1920s (Llauger et al. 2002).

From the western end of the island where the disease remained endemic for many years, it moved eastwards across the mountain barrier against prevailing winds to eastern part of the island in the 1960s (Grylls and Hunt 1971). The disease gained epidemic proportions following hurricanes of 1932 and 1933, and by 1982, it had spread over the entire west of the island (Martyn 1949; Nutman and Roberts 1955). LY has been reported from Key West Florida of the USA in 1937 spreading to Miami in 1971 (Martinez and Roberts 1967; Seymour et al. 1972; McCoy 1976), Yucatan Peninsula of Mexico in 1977 (Romney and Harries 1978), Dominican

Table 11.1 Phytoplasma associated with major diseases of coconut, vectors involved and the geographic distribution

Sl. no	Disease	16Sr group	Distribution	Vector	References
1.	Lethal yellowing	16SrIV-A	Florida, Jamaica, Honduras, St. Kitts and Nevis, Cuba	*Haplaxius crudus* (van Duzee)	Harrison et al. (1994a, 2008), Myrie et al. (2006, 2012), Brown et al. (2007) and Howard et al. (1983, 1984)
		16Sr IV-B	Yucatan (Mexico) and Cuba		Tymon et al. (1998)
		16Sr IV- E	Dominican Republic		Martinez et al. (2008)
2.	Lethal yellowing-type diseases				
A	Cape St. Paul wilt disease (CSPWD)	16SrXXII-B	Ghana	*Myndus adiopodoumeensis* Synare and *Diostrombus* spp.[a]	Dabek et al. (1976), Johnson and Harries (1976), Danyo (2011), and Dery et al. (1997b)
B	Kaincope disease (maladie de Kaincopé)	Not characterised	Togo	Not identified	Nienhaus and Steiner (1976) and Dollet and Giannotti (1976)
C	Awka wilt disease (AWD) or bronze leaf wilt	16SrXXII-A	Nigeria	Not identified	Bull (1955), Ekpo and Ojomo (1990) and Osagie et al. (2013, 2016)
D	Kribi disease	Not characterised	Cameroon	Not identified	Dollet et al. (1977)
E	Côte d'Ivoire lethal yellowing (CILY)	16SrXXII-B	Côte d'Ivoire	Not identified	Konan Konan et al. (2013) and Arocha-Rosete et al. (2014, 2015)

(continued)

Table 11.1 (continued)

Sl. no	Disease	16Sr group	Distribution	Vector	References
F	Lethal decline/ Lethal disease (LD)	16SrXXII-A	Mozambique	*Platacantha lutea*[a]	Harrison et al. (2014), Arocha-Rosete et al. (2014) and Dollet et al. (2011)
		16Sr IV-C	Tanzania and Kenya	*Diastrombus mkurangai*, *Meenoplus* spp.[a]	Schuiling and Mpunami (1990), Schuiling et al. (1992a), Nienhaus et al. (1982), Mpunami et al. (1999, 2000) and Bonnot et al. (2010)
G	Bogia coconut syndrome (BCS)	Affinity towards 16Sr IV	Papua New Guinea	*Zophium apupillata*, *Lophopssaccharicida*, Zoraidini species (unidentified), Ricaniidae species (unidentified), *Taparella amata* and *Colgar* sp.[a]	Kelly et al. (2011), Pilotti et al. (2014) and Lu et al. (2016)
3.	Root(wilt) Disease	16Sr XI-B	India	*Proutista moesta* (Westwood), *Stephanitis typica* (Distant)	Solomon et al. (1983), Rajan and Mathen (1985), Mathen et al. (1990) and Manimekalai et al. (2010)
4.	Tatipaka disease	Not characterised	India	Not identified	Rao et al. (1956), Rethinam et al. (1989) and Rajamannar et al. (1993)
5.	Coconut yellow decline	16SrXIV, 16Sr XXXII- B	Malaysia	Not identified	Nejat et al. (2009a)
6.	Kalimantan wilt	16Sr XI and 16Sr XIII	Indonesia	Not identified	Sitepu et al. (1988), Warokka (2005) and Bertaccini et al. (2014)

(continued)

Table 11.1 (continued)

Sl. no	Disease	16Sr group	Distribution	Vector	References
7.	Weligama coconut leaf wilt disease	16SrXI	Sri Lanka	*Goniagnathus (Tropicognathus) punctifer*, *Recilia dorsalis* Motschulsky, *Kolla ceylonica* (Melichar), *Idioscopus clypealis* (Lethierry), *Proutista moesta* (Westwood), *Proutista* sp., *Nisia nervosa* (Motschulsky), *Stephanitis typica* (Distant)[a]	Wijesekara et al. (2008), Perera et al. (2010), Wijesekara et al. (2013) and Kumara et al. (2015)
8.	Blast	Not characterised	Côte d'Ivoire	*Recilia mica* Kramer	Quillec et al. (1978) and Julia (1979)

[a]Not confirmed by vector transmission

Republic (Carter and Suah 1964), New Providence island of Bahamas (Leach 1946; Howard 1983) and Honduras (Ashburner et al. 1996). The disease reached the state of Oaxaca in the Pacific coast of Mexico by 1997 (Harrison et al. 2002a). The disease is currently a matter of concern to coconut cultivation in North and Central America and the Caribbean Basin (Brown et al. 2007).

The name 'lethal yellowing' is often used to refer to yellowing-type diseases of palms in several locations in humid tropics. Though these diseases show similarity in symptoms with the typical LY, they differ in geographical distribution patterns, rates of spread, varietal susceptibility, host range and vector. These diseases are collectively referred to as 'lethal yellowing-type diseases' (LYD). LYD occurring in West Africa include Kaincope disease (maladie de Kaincopé) in Togo (Nienhaus and Steiner 1976; Steiner 1976a), Awka wilt disease (AWD) or bronze leaf wilt in Nigeria (Bull 1955; Ekpo and Ojomo 1990), Kribi disease in Cameroon (Dollet et al. 1977), Cape St. Paul Wilt Disease (CSPWD) in Ghana (Dabek et al. 1976; Johnson and Harries 1976) and lethal yellowing (CILY) in Côte d'Ivoire (Konan Konan et al. 2013; Arocha-Rosete et al. 2014). In East Africa, LYD is referred to as 'lethal decline' (LD) and has been reported from Tanzania (Schuiling and Mpunami 1990 quoted by Mpunami et al. 1999), Mozambique (Mpunami et al. 1999; Bonnot et al. 2010) and Kenya (Nienhaus 1984 quoted by Mpunami et al. 1999). LYD named as Bogia coconut syndrome (BCS) has also been reported from the Oceanian country Papua New Guinea (Kelly et al. 2011).

Devastating losses of coconut plantations due to LY have been recorded since the late nineteenth century. In Jamaica alone, at least 5 million Jamaica Tall palms were killed within 20 years following its spread to the main coconut-growing areas in the east of the island (Romney 1983). The disease caused the death of palms and reduction in coconut production in Haiti. Within a short span of 15 years (1979–1994), the coconut production declined from 60 million to 30 million nuts a year. The disease killed about 80% of palms in the western region of Haiti (Donis 2002). Losses in coconut production in Belize due to LY were estimated to be 25% in cays and 75% in the main land (Quiroz 2002). During a span of 2 decades from the 1970s to 1990s, lethal yellowing epidemics decimated most of the Jamaican Tall variety in Florida and Jamaica and spread to neighbouring regions including the Pacific coast of the Americas (Harrison et al. 2002a). The destructive disease caused the death of about 90% of the palms in the Honduran Atlantic coast and Jamaica (Doyle 2001; Myrie 2002). The disease caused the destruction of 650,000 palms in Yucatan Peninsula and 100,000 coconut palms in Florida (Mora-Aguilera 2002).

LYD has devastated millions of palms across several countries in Africa. CSPWD has almost destroyed the coconut industry in the Volta region and has wiped out 5500 ha of coconut palms in south western Ghana putting about 90% of the country's coconut at risk (Ofori and Nkansah-Poku 1997). Over a period of 30 years, 'lethal decline' destroyed about 56% palms in southern Tanzania (Mpunami et al. 1999), and CSPWD has annihilated about 1 million coconut palms in Ghana (Nipah et al. 2007). In Nigeria, 'Awka wilt' killed over 98% of the West African Tall (WAT) palms in 11 years (Odewale et al. 2010).

11.2.2 Symptoms

The pattern of appearance and sequential progression of disease symptoms usually vary based on the phytoplasma group, geographical location, host species and variety (Dollet et al. 2009; Harrison et al. 2014). The first visual symptom on infected bearing coconut palms is the premature nut fall/fruit drop. Most of the fallen nuts have brown to black water-soaked regions under the calyx. This is followed by blackening/necrosis of newly emerging inflorescences. Most of the male flowers will be dead, and there will not be any fruit set in the affected inflorescences. As the inflorescence necrosis phase progresses, yellowing of the leaves starts from the leaves in the outer whorls, advancing upwards, affecting the younger, middle and finally the upper leaves. The yellow leaves turn brown and dry and die. The dried leaves remain hanging on the crown for a few days before falling. Eventually, the whole crown perishes, leaving a bare trunk or 'telephone pole'. The estimated time lag from initial infection by the pathogen to the appearance of first symptom has been variously reported as follows: in mature bearing palms, 230–450 days (Romney 1972) and 210–450 days (Heinze et al. 1972) and for young nonbearing palms, at least 240–270 days (Dabek 1974). The time between probable initial infection and death of mature palms has been reported as 3–6 months (Grylls and Hunt 1971) or 4–5 months (McCoy et al. 1983). In addition to these symptoms in above ground parts, death of root tips coincides with the foliar yellowing symptom. Roots also show necrosis, which becomes more extensive as the disease progresses (Eden-Green 1979).

Though symptoms of LYD in West Africa and Tanzania are similar to LY in the American continent (Mpunami et al. 1999), some differences are noticed. While in Kaincope and SPWD, the spear is often found to be dead with many fronds still remaining green; in lethal yellowing, the spear leaf appears unaffected except for some necrotic patches (Johnson and Harries 1976). In some LY-affected palms, the spear leaf or a mid-crown leaf occasionally shows yellowing prematurely (McCoy et al. 1983). Sometimes inflorescence necrosis becomes conspicuous only after the appearance of frond yellowing as reported in Guatemala (Mejia et al. 2004). Unlike LYD, inflorescence necrosis is absent in BCS reported from Papua New Guinea (Kelly et al. 2011).

11.2.3 Aetiology

Preliminary aetiological investigations were focussed on the role of bacteria, fungi, virus, viroids and other abiotic factors. Despite the severity of economic losses and intensive research, the cause of LY and LYDs remained as a mystery till the 1960s. The discovery of the association of phytoplasmas with plant diseases during the 1960s sparked the search for phytoplasmas in LY-affected palms. In 1972, three groups independently reported the occurrence of phytoplasma in the phloem of coconut palms showing LY symptoms by transmission electron microscopy (Beakbane et al. 1972; Heinze et al. 1972; Plavsic-Banjac et al. 1972). Electron microscopic observations of LY-diseased coconut palms showed that phytoplasma was present in partly expanded inflorescences and spear leaves, but not found in fully expanded inflorescences, leaves or stems (Parthasarathy 1974).

The physical presence of phytoplasma in tissue from spear leaf and unopened inflorescences of coconut palms affected with Kaincope disease in Togo and Cape St. Paul wilt in Ghana was also confirmed by transmission electron microscopy (Dabek et al. 1976; Dabek 1977). Association of phytoplasma with LD in East African countries has also been established by fluorescence microscopy using the fluorochromes DAPI (Deutsch and Nienhaus 1983).

A cause-effect relationship between phytoplasmas and the disease was established by the differential response of LY- and LYD-affected palms to antibiotics. LY palms treated with penicillin showed no recovery response, whereas symptom remission occurred when they were treated with oxytetracycline (Hunt et al. 1974; McCoy 1972). Remission of symptoms in palms after treatment with tetracycline further supported the phytoplasmal etiology of Kaincope disease in Togo and lethal disease in Tanzania, Kenya and Mozambique (Steiner 1976b; Kaiza 1987).

The advent of molecular techniques has enabled systematic identification and differentiation of different phytoplasmas. The taxonomic classification of phytoplasma is now based on the highly conserved genes coding for the 16S rRNA and the spacer region between 16S and 23S rRNA (Lee et al. 1993; Schneider et al. 1993). Though the similarity in symptoms initially supported a common etiology for LY and LYDs, the differences in epidemiology and coconut ecotype susceptibility led to the speculation that coconut-associated phytoplasmas in Africa are probably distinct from those affecting palms in the Caribbean Basin (Schuiling et al. 1992a, b). Restriction fragment length polymorphism (RFLP) profiles of the PCR amplification product also showed differences in restriction pattern indicating that the phytoplasma of LY in the Caribbean and East and West Africa are of different strains (Harrison et al. 1994b). Studies on phytoplasma associated with LY-diseased palms in the Americas have shown genetic differences between the LY phytoplasmas in Florida, México and Jamaica based on the analysis of rDNA (Harrison et al. 2002b) and between Cuba and Mexico based on the analysis of non-ribosomal DNA (Llauger et al. 2002).

Phytoplasmas associated with LY and LYD of palms are most commonly placed in the 16SrIV group although some have now been reclassified into group 16SrXXII (Lee et al. 1998; Harrison et al. 2014). The 16SrIV group is subdivided into A–F reflecting the genetic variation, host range and vectors (Harrison et al. 2002a, b; Brown et al. 2006; Harrison et al. 2008; Martinez et al. 2008; Vázquez-Euán et al. 2011). The subgroups 16SrIV-A, 16SrIV-B, 16SrIV-C and 16SrIV-E are pathogenic to coconut palms. The phytoplasma-causing LY to coconut palms in Florida, Jamaica, Honduras and Nevis falls in 16SrIV-A (Myrie et al. 2006; Brown et al. 2007; Harrison et al. 2008), whereas those causing LY in Mexican Yucatan and Cuba are placed in 16SrIV-B subgroup (Tymon et al. 1998; Llauger et al. 2002). The phytoplasma associated with lethal disease of coconut palms in Kenya and Tanzania belongs to 16Sr IV-C (Tymon et al. 1998). The subgroup 16Sr IV-E has been reported to be associated with LY of palms in southern coast of Dominican Republic (Martinez et al. 2008). The partial 16S rDNA sequence of phytoplasma associated with BCS in Papua New Guinea matched most closely (96%) with that of group 16SrIV (Kelly et al. 2011). Oropeza et al. (2011) analysed the distribution of phytoplasma in LY-affected palms using PCR, and pathogen titre was found to be high in the stem, young leaves, inflorescences, stem apex and root apex when compared to intermediate leaves and roots without apex.

Studies on the LYD pathogen in East Africa using 16S rRNA gene sequence and RFLP analysis showed that phytoplasma associated with Mozambique LYD is distinct from that of Kenya and Tanzania but closely related to those from West Africa, CSPWD in Ghana and AWD in Nigeria (Mpunami et al. 1999). Virtual RFLP profiles of the F2n/R2 portion of the 16S rRNA gene and pattern similarity coefficients revealed that the phytoplasma strains associated with coconut LYD in Mozambique, Nigeria, Ghana and Côte d'Ivoire represent a distinct species-level lineage and delineated into a novel taxon, *Candidatus Phytoplasma palmicola* (16SrXXII). Phytoplasma associated with LYD in Mozambique and Nigeria belongs to 16SrXXII-A, and those causing diseases in Ghana and Côte d'Ivoire are placed in 16Sr XXII-B subgroup (Arocha-Rosete et al. 2014; Harrison et al. 2014). It is also common that diverse groups of phytoplasmas can occur in a geographic region. The association of 3 different groups, viz. West African subgroup (XXII-B), East African subgroup (IV-C) and a novel unclassified *Candidatus Phytoplasma pini*-related strain with LYD in Mozambique, supports this theory (Bila et al. 2015a, b).

11.2.4 Transmission

With the establishment of phytoplasmal etiology of LY and LYD, the search for insect vectors narrowed down to Auchenorrhyncha insects of the order Hemiptera, i.e. leafhopper and plant hopper. The most definitive method for determining the vector species within a given pathosystem is a transmission test. But this approach for coconut has several logistical constraints, viz. difficulties involved in the caging and maintenance of the perennial and woody palms during the incubation period, collection and release of large number of vectors that have acquired phytoplasma, etc. Though conventional vector transmission trials for LY in palms have been carried out since the 1960s, very few were successful (Tsai 1980; Eden-Green 1995). To date, the only established vector of LY is the plant hopper *Haplaxius crudus* (van Duzee) (formerly *Myndus crudus*) in Florida (Howard et al. 1983).

Several insects have been reported as putative vectors of LY and LYD based on PCR results. However, their vector status has not yet been established by cage transmission trials. There have been a great number of unsuccessful palm phytoplasma vector transmission trials. In Ghana, studies on putative insect vectors of CSPWD were initiated in 1990 (Philippe et al. 2009). Even though *Myndus adiopodoumeensis* Synare (Homoptera: Cixiidae) and *Diostrombus* spp. (Homoptera: Derbidae) tested positive for CSPWD phytoplasma in PCR, these hoppers failed to elicit the disease in subsequent cage transmission trials. In spite of the extensive transmission studies conducted, the vector of CSPWD in Ghana remains unknown (Danyo 2011).

An alternate method that helps to surpass the use of caged palms in the preliminary screening trials is the feeding medium approach (Tanne et al. 2001). The method involves assaying phytoplasma DNA in the sucrose solution, on which the insect feeds, through a parafilm barrier. Even though this technique does not provide a conclusive proof of vector status, it is advantageous as the detection of phytoplasma DNA in the feeding media by PCR indicates the competency of the insect to introduce phytoplasma during feeding activity. Lu et al. (2016) combined single-insect feeding medium tests with loop-mediated isothermal amplification (LAMP) to identify putative vectors of BCS.

The possibility of transmission of phytoplasma through seed nuts from diseased palms has been regarded as highly unlikely due to the phloem-limited nature of the pathogen. However, Harrison and Oropeza (1997) reported the presence of lethal yellowing phytoplasma in at least one seed from nuts of 3 coconut palms manifesting primary stage symptoms of LY using pathogen-specific PCR. DNA of the LY phytoplasma was detected in 18.06% of embryos from fruits of 4 diseased Atlantic Tall coconut palms by PCR assays (Cordova et al. 2003). Nipah et al. (2007) detected CSPWD phytoplasma in 9 out of the 52 embryos from diseased West African Tall (WAT) palms by PCR employing phytoplasma universal primer pair P1/P7 nested with CSPWD-specific primer pair G813f/AwkaSR. Since the embryo lacks direct connection with sieve elements, the mechanism of transmission remains unclear. Oropeza et al. (2017) demonstrated for the first time using nested PCR and TaqMan real-time PCR the occurrence of LY phytoplasma transmission from coconut embryos to plantlets obtained by *in vitro* plumule culture. However, there is no conclusive evidence, so far, for the presence of intact phytoplasma in the embryos and seedlings raised from nuts of LY- and LYD-affected palms.

11.2.5 Detection

LY-diseased palms are traditionally identified by symptoms, electron microscopy or fluorescence microscopy. Accumulation of DNA in extra-nuclear sites indicative of phytoplasmal infection could be demonstrated by staining the tissues with DAPI (4′,6′-diamidino-2-phenylindole), a fluorochrome that binds to DNA and fluoresces under UV radiation (Russel et al. 1975). Although the microscopic techniques are useful for establishment of association of phytoplasma with the disease, it cannot differentiate one phytoplasma from another. On the other hand, procedures of electron microscopy for the phytoplasmal detection are time-consuming. Very low concentration of the pathogen in coconut palm species and uneven distribution within vascular tissues make its detection difficult. The advent of molecular techniques has made identification and differentiation of different phytoplasmas comparatively easier. Harrison et al. (1992) have developed DNA probes for detection of phytoplasma of Caribbean origin. Similarly, oligonucleotide primer for selective amplification of LY phytoplasma by DNA polymerase chain reaction (PCR) has been standardised (Rhode et al. 1993; Harrison et al. 1994a; Mpunami et al. 1997; Tymon and Jones 1997).

Spear leaves from around the apical meristem, which is rich in phloem, are the most reliable source of phytoplasma detection in palms (Harrison et al. 1999). Often the old palms are tall and sample collection from the crown is difficult. Trunk shavings obtained by drilling a hole 10–15 cm into the trunk is a non-destructive and less laborious method of sample extraction for successful phytoplasma detection (Harrison et al. 2002b). Myrie et al. (2011) developed improved methods, viz. a multiplex direct-PCR system, PCR on 16S rRNA and hemolysin genes; a SYBRGreen system based on the *GroEL* gene; and real-time PCR using TaqMan probes based on the 16S rDNA and *GroEL* gene for detection of LY phytoplasma-affected palms in Jamaica. These methods with improved sensitivity and specificity together with quantisation capability enhance the efficiency of detection. Córdova et al. (2014) developed a TaqMan real-time PCR assay for detection and quantification of LY phytoplasmas belonging to 16SrIV A, D and E subgroups in America. For the detection of CSPWD, sec A gene-based PCR assay was developed by Yankey et al. (2014).

11.2.6 Disease Management

The non-availability of technologies for the curative treatment of LY and LYD accentuates the need for periodic surveillance and removal of affected palms to prevent the spread of the disease. Black's approach, pioneered by a Jamaican farmer Michael Black, involves an integration of on-farm quarantine, strict weekly surveillance, cutting down and burning of palms with disease symptoms and replanting with a variety selected for high yield and disease resistance as well as whole-farm weed control, and a good fertilisation regime significantly reduced the disease incidence over years (Myrie et al. 2011; Gurr et al. 2016). The removal of infected palms slowed down the disease spread in Dominican Republic and Ghana (Martinez et al. 2008, 2010; Nkansah et al. 2009).

Though oxytetracycline antibiotics have been shown to reduce phytoplasma infection in palms (McCoy et al. 1976), they have never been considered as a management strategy in view of the prohibitive costs and health hazards. Management of the vectors by employing various techniques like insecticide spraying, replacing the ground cover with non-host grass species, mulching and mass trapping are not considered as sustainable management strategies (Howard and Oropeza 1998).

Use of resistant varieties is the best option for management of a phytoplasmal disease. The Jamaica Tall grown in Jamaica and Caribbean is highly susceptible to LY. In Jamaica, field testing for LY resistance of coconut germplasm was carried out in 6 resistance trials planted from 1961 to 1970. Among the cultivars, the Jamaican Tall/Atlantic Tall (AT) was the most susceptible, with 90% mortality. The Sri Lankan, Indian and Malayan Dwarfs and the King Coconut appeared to be highly resistant to LY with less than 5% mortality. However, in some locations in Florida and Jamaica, the susceptibility of Malayan Dwarf ranged between 42.9–100% and 14–40%, respectively (Howard et al. 1987). Though Malayan Dwarfs possessed high degree of resistance to the LY, its poor productivity under marginal conditions hindered the adoption by farmers. Experiments were conducted in Jamaica during the 1970s for the production of a hybrid with high disease resistance and productivity. The Panama Tall, with an intermediate mortality of 44%, was used as pollen donor in a programme to produce hybrid palms. The 'Maypan' hybrid (Malayan Dwarf × Panama Tall), combining the higher resistance of the Dwarf cultivar with large size and adaptability of the Tall, has not only the advantage of the hybrid vigour but also was found to be only 10% susceptible to LY. The Maypan has formed a good compromise between resistance level, yield and product quality. The faster growth rate coupled with greater adaptability to different habitats and poor soils is the added advantage of Maypan over the Malayan Dwarf. After devastating losses of the local Jamaican Tall variety to LYD, the Malayan Dwarf varieties and Maypan became the primary foci of coconut replanting programmes, which led to a recovery of the coconut industry (Been 1995; Ashburner and Been 1997; Harrison et al.

2002a). High mortality rate of MYD and Maypan planted in LY-infected regions in recent years indicate that these cultivars cannot be considered resistant to LY as previously thought. In a coconut cultivar trial at Fort Lauderdale, Florida, Malayan Dwarfs and Maypan hybrids showed a high degree of susceptibility of 70% and 83%, respectively (Broschat et al. 2002). Massive destruction of Malayan Dwarfs and its hybrid by LYD triggered research exploring the possibility of the genetic contamination of parents, variation or mutation of pathogen/vector that altered the host-vector-pathogen interactions. Research on this line provided evidences on genetic contamination in Panama Tall and MYD in Jamaica. This might have caused some loss of resistance but is insufficient to explain a massive outbreak of the disease (Baudouin et al. 2008; Lebrun et al. 2008). In a trial established in 1991 on the northern coast of Yucatan, Tall and Dwarf populations representing the genetic diversity in Mexico were exposed to LY for 15 years. The result of this experiment indicates that together with MYD, coconut populations from the Pacific coasts of Mexico could be an important source of germplasm to deal with LY (Zizumbo-Villarreal et al. 2008). It is imperative to explore populations, other than Malayan Dwarf, with good resistance to broaden the genetic base.

The Malayan Dwarfs which showed resistance to LY in Jamaica were found to be susceptible to CSPWD in Ghana. The high CSPWD resistance shown by the Sri Lankan Green Dwarf (SGD) and Vanuatu Tall (VTT) varieties in the Dixcove trials (1981–1983) led to the development of the SGD × VTT hybrid which is being used for replanting (Mariau et al. 1996; Dery et al. 1997a; Quaicoe et al. 2009). The good agronomic characteristics of SGD × VTT coupled with its resistance to CSPWD make it a promising hybrid to revamp the coconut industry in Ghana (Dare et al. 2010). Field resistance of Green Dwarfs to AWD in Nigeria was reported by Odewale et al. (2006).

11.3 Blast

Blast disease of coconut was first noticed from Côte d'Ivoire in 1971 (Quillec et al. 1978). The disease generally occurs along with another disease, dry bud rot, noticed on nursery seedlings and in juvenile palms. The symptoms are identical to what is noticed in the blast disease in oil palm in Côte d'Ivoire. The characteristic symptoms are wet rot of spear and roots, yellow brown colouring of the bole and rapid drying of the plant starting with the oldest leaves. The disease occurs mainly in the first year of planting and is rarely noticed in older plants. The disease could be transmitted to periwinkle, which exhibited wilting and yellowing symptoms (Dollet 1979). EM examination of symptomatic periwinkle revealed intraphloemic phytoplasma. Tetracycline experiment further confirmed phytoplasmal etiology.

The disease is transmitted by *Recilia mica* (Jassidae, Deltocephalinae) (Julia 1979). The insect preferably multiplies in grasses. Spraying with Temik reduced blast incidence to the extent of 40%. Shading and phytosanitation such as clean weeding in and around nursery seedlings help to keep the insect population low. Growing cover crop such as *Pueraria* suppresses the growth of the grasses thereby reducing the vector population in the field.

11.4 Root (Wilt) Disease

11.4.1 Occurrence, Distribution and Crop Loss

Root (wilt) is one of the major devastating diseases affecting coconut palms in India. The occurrence of the disease was first noticed as early as in 1874 in the erstwhile state of Travancore (present Kerala state, India). The disease became conspicuous after the great floods of 1882. The disease was initially reported from 3 independent locations each at a distance of about 50 km (Butler 1908; Pillai 1911; Varghese 1934). It is a major production constraint in southern districts of Kerala, creating serious economic distress to the agrarian families. According to a survey made during 1984–1985, the disease occurs in a contiguous manner in an area of 4,10,000 ha in 8 out of the 14 districts of the Kerala state starting from Thiruvananthapuram district in the south extending up to Thrissur district and in isolated pockets in the remaining 6 northern districts. The crop loss due to the disease was estimated to be about 968 million nuts (Anon 1985). The total estimated monetary loss in terms of loss in husk, copra yield and leaf number and quality of leaves on the basis of 1984 price index for coconut was of the order of about Rs.3000 million. As per the sample survey conducted in 1996, the disease intensity in the contiguous diseased tract ranged between 2.1% in Thiruvananthapuram district and 48.0% in Alappuzha district, and there was an overall reduction in disease incidence from 32.37 to 24.05. This reduction is attributed to the removal of about 6 million diseased palms, replanting with quality seedlings, adoption of disease management practices and crop conversion from coconut to rubber (Anon 1996). The disease has also been reported from the bordering districts of Tamil Nadu, Karnataka and Goa (Solomon et al. 1999; Chandramohanan 2010). It is a continuing threat to the coconut palms in India.

The disease is non-lethal but debilitating and palms of all age groups are affected. Disease contraction in the pre-bearing age delays flowering and affects the vitality of the palm (Ramadasan et al. 1971). About 65% of the root (wilt)-diseased palms are affected by a fungal disease known as leaf rot (Srinivasan 1991). The leaf rot disease superimposed on root (wilt) diseased palms leads to rapid decline and reduction in yield (Menon and Nair 1948, 1951). The disease occurs in all major soil types (Menon 1938), but the spread is faster in sandy, sandy loam, alluvial and in heavy textured soils than in laterites. Relatively higher incidence is recorded in waterlogged low-lying areas adjacent to rivers and canals (Pillai et al. 1973). The pattern of spread is erratic, occurs in jumps or leaps, characteristic of insect transmission (Pillai et al. 1980).

11.4.2 Symptoms

The most obvious diagnostic symptom of the disease is the abnormal bending of the leaflets termed 'ribbing' or 'flaccidity'. Foliar yellowing of the outer whorl of leaves and marginal necrosis are the other associated symptoms (Radha and Lal 1972) (Fig. 11.1).

In seedlings and juvenile palms, yellowing of foliage is virtually absent, and flaccidity is the only conspicuous symptom. Instance of diseased palms having green outer leaves, with yellowing of leaves in some of the inner whorls, is also noticed (Menon 1938). With the progress of the disease, extensive rotting of the roots is observed. The main roots and rootlets start drying from the tip backwards (Menon and Nair 1951). Shedding of immature nuts is yet another symptom observed in some cases. Drying up of spathe and necrosis of spikelets from the tip downwards in unopened inflorescence are noticed in certain cases (Maramorosch 1964). A high percentage of pollen produced is either sterile or less viable (Varkey and Davis 1960). Meiotic irregularities are also observed (Nambiar and Prasannakumari 1964). The vitality of the diseased palms is so adversely affected that they produce small spathes with fewer female flowers. In the advanced stage, the crown gets very much reduced in size and ceases to produce inflorescence (Menon 1938). The nuts from diseased palms have thinner husk, and fibres are definitely weaker and less firm (Varghese 1934). The kernel is thinner and never dries up but remains soft and flexible. The tender coconut water of diseased palm is insipid. The oil content is very much reduced and the oil loses its flavour as well (Menon 1938).

Fig. 11.1 Root (wilt) disease of coconut super imposed with leaf rot. Note the symptoms such as flaccidity, foliar yellowing and necrosis. (Photo: V. Krishnakumar)

An indexing method for quantifying the disease intensity giving due weightage to the 3 predominant foliar symptoms has been developed (George and Radha1973). Disease index, $\mathrm{DI} = \Sigma \frac{F + Y + N}{L} \times 10$, where F (0–5), Y (0–3) and N(0–2) are the grade points assigned to flaccidity, yellowing, necrosis and L, is the total number of leaves in the palm. For indexing young palms below 10 years, more weightage is to be given to flaccidity as it is found to be the prominent symptom and disease index $\left(\mathrm{DI} = \Sigma \frac{15F + 5Y + 5N}{L} \times 10 \right)$. Disease index can vary from 0 to 100,where 0 represents the total absence of all the symptoms, indicating that the palm is visually disease-free (apparently healthy) and 100 means the presence of all the symptoms in the acute stage on all the leaves. Based on the disease index, the palms can be categorised into disease early (DI < 20), disease middle (DI 20–50) or disease advanced (DI > 50). The indexing method was further simplified by Nambiar and Pillai (1985) by scoring the symptoms on the leaves present in any of the 5 spirals. Usually, only 4–5 leaves are present in a spiral, and hence, the number of observations per palm to be taken for calculation of disease index is less. The indexing system helps in quantifying disease severity in a simple numerical expression that can be analysed statistically.

Comparative anatomical studies of the tender unopened leaves of healthy and root (wilt) affected palms revealed general stunting of epidermal cells, reduction in thickness of the cuticle on the adaxial side and differential rate of division of the upper epidermis in leaves of diseased palms. Reduction in wall thickness of cells, sclerenchymatous fibres and bundle sheath is also observed. These anatomical changes may be contributing to the downward curling of leaflets. In diseased palms, wall thickness of metaxylem elements is reduced resulting in uneven shape of the xylem component. The uptake and translocation of solutes are also impeded with the formation of tyloses in xylem vessels as well as gummosis and necrotic obliteration of phloem in the roots of diseased palms (Govindankutty 1981). The distribution of stomata per unit area is found to be more in leaflets of diseased palms (Joseph and Shanta 1964). The stomatal regulation is also adversely affected due to the disease. The stomata in diseased palms fail to close in response to soil and atmospheric drought (Rajagopal et al. 1986a, b). The overall disturbance in absorption and transpiration may also be contributing to the flaccidity symptoms.

11.4.3 Aetiology

Ever since its report in the 1880s, the causal agent of RWD remained as an enigma for a century. Preliminary investigations on the aetiology of the disease were primarily focused on fungal pathogens associated with the root rot of coconut.

Physiological and biochemical changes observed in diseased palms such as derangement in uptake and translocation of solutes, higher respiration and transpiration rate, altered nitrogen metabolism and an accelerated phenol metabolism are indicative of a pathogen-mediated aberrant host metabolism than of a physiological disorder (Rajagopal et al. 1998). Soil sickness as a contributing factor of the disease was excluded based on field fertility trials as well as tissue and soil analysis (Cecil and Amma 1991). The sporadic occurrence and spreading nature of the disease implied the involvement of a pathogen as the cause of the disease.

Etiological investigations gained new dimensions with the advancement in plant pathology. The role of fungi, bacteria, virus, viroid and nematodes in inducing the disease was ruled out based on the results of the inoculation studies and pathogenicity experiments (Nagaraj et al. 1954; Shanta and Menon 1959; Summanwar et al. 1969; Maramorosch and Kondo 1977; Randles and Hatta 1980; Joseph and Lilly 1991; Jayasankar and George 1991; Sosamma and Koshy 1991). Electron microscopic examination of juvenile tissues like sub-meristem, petiole of developing leaves, rachilla of unopened inflorescence and root tips of diseased palms revealed the presence of phytoplasma (Solomon et al. 1983). These pleomorphic phloem-bound mollicutes are observed only in sieve tubes. The size of the coccoid forms ranged from 250 nm to 400 nm. Distribution of the phytoplasma within the vascular bundle is rather sparse, and not all sieve elements in a patch contained them. In the older leaves, moribund forms, lacking internal contents, only were observed. Constant association of phytoplasma with the disease has been conclusively established with the finding of the organism in more than 75 diseased palms and their absence in an equal number of healthy palms studied (Solomon et al. 1999).

Light microscopic studies with Diene's staining and fluorescent microscopy with 4,6-diamidino-2 phenylindole 2 HCI (DAPI) or Hoechst 33258 lent support to the association of phytoplasma in the phloem tissues of RWD affected palms (Solomon et al. 1987). The results of the experiments on antibiotic therapy (Pillai and Raju 1985; Chowdappa et al. 1989), dodder transmission (Sasikala et al. 1988) and vector transmission (Mathen et al. 1990) provided conclusive evidences on the association of phytoplasma with RWD. Molecular characterisation of 16S rRNA sequences of the RWD pathogen added further evidence to the phytoplasmal etiology of the disease. The phytoplasma-causing RWD has been characterised as *Candidatus Phytoplasma oryzae*-related strain belonging to 16SrXI-B subgroup (rice yellow dwarf group) (Manimekalai et al. 2010, 2014a).

11.4.4 Transmission

Phytoplasmas are generally transmitted by insects belonging to the Homoptera group. A plant hopper, *Proutista moesta*, and a leaf hopper, *Sophonia greeni,* both belonging to the Homopteran group, have been consistently found in coconut foliage besides lace bug, *Stephanitis typica* (Heteroptera:Tingidae). Acquisition of phytoplasma by these putative insect vectors while feeding on diseased palm was studied. Phytoplasma was observed in the salivary glands of lace bug 18–23 days after feeding on diseased palms (Mathen et al. 1987) and also in plant hoppers fed on diseased palms for over 30 days (Rajan et al. 2002). The vector role of lace bug and plant hopper has been conclusively established in transmission experiments (Mathen et al. 1990; Rajan 2011).

Field experiments were conducted to assess the control of aerial insects on fresh incidence of disease. The results were not encouraging. This may be due to the perennial nature of the crop, the presence of insect vector throughout the year, the persistent mode of transmission of phytoplasma and the reinfestation of sprayed plants within a short period (Solomon et al. 1999). The phytoplasma also could be experimentally transmitted from coconut to periwinkle, an indicator host, through dodder laurel (*Cassytha filiformis*). Periwinkle grown in insect-proof cages bridged to diseased coconut palms through dodder laurel developed chlorotic spots in the interveinal areas and at vein endings of fully opened leaves in 3–4 weeks of haustorial establishment. Detection of phytoplasma in periwinkle, dodder laurel and the source palm confirmed transmission of the disease (Sasikala et al. 1988). Though Manimekalai et al. (2014b) confirmed the presence of DNA of root (wilt) phytoplasma in 16.67% embryos of nuts collected from diseased palms, the potential of seed transmission of phytoplasma was checked by subjecting the germinated seedlings raised from mature nuts collected from diseased palms. But phytoplasma DNA could not be detected in any of the raised seedlings.

11.4.5 Diagnosis

Preliminary investigations on the development of diagnostic techniques were based on the biochemical tests of altered host metabolisms detectable in the form of either accumulation or depletion of substances consequent to differential enzymatic activity in diseased palms (Joseph and Shanta 1963; Lal 1968; Dwivedi et al. 1977). As these changes can also be induced by other biotic and abiotic stresses, more thrust was given on the development of a rapid and reliable diagnostic techniques. Physiological changes like stomatal resistance and transpiration rate in healthy and diseased palms were also used for diagnosis of RWD. High stomatal resistance with a correspondingly low transpiration rate was recorded in healthy palms in contrast to low stomatal resistance and high transpiration rate in diseased palms (Rajagopal et al. 1986a). With the establishment of the phytoplasmal aetiology, protocols were standardised for the purification of RWD phytoplasma (Mayil Vaganan et al. 2001), production of polyclonal antisera specific to coconut RWD phytoplasma and direct antigen coated-enzyme linked immunosorbent assay (DAC-ELISA) for the detection of RWD phytoplasma even 24 months before symptom manifestation (Sasikala et al. 1998, 2001, 2004, 2010).

With the advent of molecular biology, efforts were made to develop nucleic acid-based detection techniques for RWD phytoplasma. Preliminary attempts made to detect the coconut RWD phytoplasma using 16S rRNA sequence-based universal primers failed to give consistent results in direct and nested PCRs. Molecular detection of phytoplasma associated with RWD was achieved by modification of phytoplasma enrichment technique for DNA extraction by addition of 5% polyvinylpyrrolidone, designing 6 highly sensitive primers and semi-nested PCR technique (Manimekalai et al. 2010; Ramaswamy et al. 2016). A real-time PCR protocol was also developed for detection of RWD phytoplasma (Manimekalai et al. 2011). However, it was observed that there was no consistency in detection of phytoplasma by nested PCR, real-time PCR and LAMP when large number of root (wilt) affected coconut samples were tested, and further refinement of these techniques are necessary for reliable and rapid detection of RWD in the early stage of infection (Hegde et al. 2016).

11.4.6 Breeding for Resistance

Since there are no prophylactic or curative measures available for managing the RWD, an enduring solution to the problem lies in breeding varieties resistant to the disease. Systematic evaluation trials at ICAR-CPCRI have led to the release of 2 coconut varieties and 1 hybrid for cultivation in the root (wilt) disease-prevalent areas. Purity of D × T hybrid seedlings is assessed using SSR markers (Krishnakumar et al. 2015). For details please see Chap. 5 on varietal resistance.

11.4.7 Disease Management

A noticeable feature of the disease is that it is not lethal but debilitating. Field experiments were conducted to assess the possibility of preventing the fresh incidence of the disease through control of aerial insects. The results were not encouraging. This may be due to the perennial nature of the crop, the presence of insect vector throughout the year, the persistent mode of transmission of phytoplasma and their infestation on sprayed palms within a short period (Solomon et al. 1999). Though Pillai et al. (1991) reported remission of symptoms by OTC treated palms, it cannot be recommended either as a prophylactic or curative measure as the antibiotic has to be given repetitively and the cure is temporary. Prohibitive cost of the antibiotic and the caution required in its use for treating plant diseases are the other factors which weigh against it being recommended as a treatment.

Since there are no measures available to control the disease to perfection, coconut farmers have to 'live with the disease', and certain integrated disease management practices are advocated to ensure a satisfactory yield even in a disease-prevalent area (Sahasranaman et al. 1983, Amma et al. 1983; Bavappa et al. 1986; Muralidharan et al. 1991; Krishnakumar et al. 2015).

1. Removal of palms: All disease advanced and uneconomic palms with annual yield of less than 10 nuts are to be removed.
2. Replanting: Replanting with disease-resistant varieties or elite seedlings from high-yielding disease-free palms located in heavily disease-affected tracts.
3. Biopriming: Biopriming of seedlings with *Pseudomonas fluorescens* to impart tolerance.
4. Application of organic manures: Application of 25 kg farmyard manure or 10 kg vermicompost enriched with *Trichoderma harzianum* at 100 g.
5. Biomass recycling: Application of leguminous green manure crops and *Gliricidia* leaves.
6. Fertiliser application: Application of recommended dose of fertilisers (500g N, 300 g P_2O_5,1250 K_2O and 250 g $MgSO_4$ palm^{-1}year^{-1}) in two splits.
7. Liming: Application of lime/dolomite supplemented with magnesium sulphate.
8. Irrigation: Irrigation with 250 l of water palm^{-1}week^{-1}, soil moisture conservation and providing adequate drainage wherever necessary.
9. Intercropping and farming system: Raise intercrops in rotation, adopting mixed cropping/mixed farming coupled with recycling of organic matter.
10. Adopting recommended management strategies for leaf rot disease, rhinoceros beetle and red palm weevil.

11.5 Tatipaka Disease

11.5.1 Occurrence, Distribution and Crop Loss

Tatipaka is a slow debilitating phytoplasmal disease of coconut in India. Its distribution is confined to East and West Godavari, Srikakulam, Nellore, Krishna and Guntur districts of Andhra Pradesh state in India. The disease was first noticed by the farmers of Tatipaka village after the cyclone of 1949, and the disease was named after the village from where it was observed (Rao et al. 1956). However, the reports of survey conducted during 1956–1958 in Srikakulam district indicated the presence of the disease 20 years prior to the cyclone (Subbiah and Rao 1963; Pandit et al. 1969; Rethinam et al. 1989). The disease generally occurs in heavy black deltaic soils than in sandy, sandy loam and red loam soils. It is observed in both well-managed and neglected gardens. A survey made during 1985–1990 in the central delta of the river Godavari 'Konaseema' (which accounts for 60% of the area under coconut in the state) revealed that the disease is prevalent in 85 out of 201 villages (Narasimhachari et al. 1991). Altogether, 8179 palms were identified as Tatipaka diseased by field surveys (Rajamannar et al. 1993).

The disease is non-lethal but of a debilitating nature, generally affecting palms in the age groups of 20–60 years. Palms below 20 years are very rarely affected

(Rethinam et al. 1989). The spread of the disease is not contiguous but sporadic at a slow pace of 3.5% over a period of 5 years. Later surveys revealed that the disease incidence in the field is less than 1% (Srinivasulu et al. 2006).

11.5.2 Symptoms

The disease-affected palms generally bear profusely for 2–3 years before the expression of visual symptoms, and more number of dark leaves (often fasciated) appear in the crown. With the onset of disease, there is a reduction in both number and size of leaves. The leaves exhibit characteristic chlorotic water-soaked spots, and the fronds bend abnormally, sometimes twisting in loops. In the advanced stage with the narrowing of leaflets and reduction in size of crown, the affected palm looks like a date palm. The spathes produced are very small with very few rachillae. The bunches carry a mixture of normal and atrophied nuts. The atrophied nuts are barren with thinner spongy mesocarp with or without shell, copra and nut water. The undersized nuts show longitudinal cracks with occasional gumming. In the advanced stage of the disease, the stem tapers and produces smaller spathes and inflorescences, which ultimately do not bear any fruit. Reduction in number and size of the roots and extensive rotting of roots are the characteristic underground symptoms of the disease (Rao 1966; Ramapandu and Rajamannar 1981).

11.5.3 Aetiology

The disease was listed as one of the uncertain aetiologies till the 1990s. The involvement of fungi and bacteria in the incidence of the disease was conclusively ruled out as these microbes could not be consistently isolated from diseased palms. Sap transmission and purification studies have also indicated the non-viral nature of the disease (Ramapandu and Rajamannar 1981; Rajamannar et al. 1984). Polyacrylamide gel electrophoretic analysis of isolated nucleic acid from diseased palms did not confirm the association of any viroid-type pathogen (Randles and Hatta 1980).

Remission of symptoms observed in diseased palms treated with tetracycline hydrochloride indicated the association of a phloem-limited phytoplasma with the disease (Rajamannar et al. 1993). This finding was corroborated by light microscopy using Diene's staining, a fluorescent microscopy with aniline blue as fluorochrome and antibiotic therapy (Rajamannar et al. 1994). Electron microscopic examination of spindle, leaves and rachillae of unopened inflorescences from diseased palms revealed the presence of phytoplasma in sieve tubes further confirming the aetiology of the disease (Srinivasulu et al. 2006).

11.5.4 Disease Management

Attempts to control the disease by foliar and soil application of various chemicals, hormones and nutrients did not yield any encouraging result. Root regeneration of affected palms also could not ameliorate the condition of palms (Rajamannar et al. 1993). Since there are no prophylactic or curative measures available for treating phytoplasmal diseases, the options left are to arrest the spread of disease by systematic surveillance and rouging of diseased palms as and when identified and using disease-resistant planting material. The programme of identifying the diseased palms and eradication is a more practical step as the disease is confined to a limited geographical region, the number of diseased palms is comparatively less and the spread is reported to be very slow. The programme, if systematically implemented, will help in total eradication of the disease. Survey of the heavily diseased area indicated the cultivar, Gangabondam, to be free of the disease, whereas the incidence was high in East Coast Tall (Rajamannar et al. 1994). This observation has to be confirmed so that this cultivar could be used as a parent in the future breeding programmes for evolving disease-resistant varieties.

11.6 Coconut Yellow Decline (CYD)

CYD is a debilitating disease that reduces the productivity of coconut in Malaysia.

11.6.1 Occurrence, Distribution and Crop Loss

Coconut lethal yellowing-like symptoms were first reported from Malaysia by Sharples (1928). The disease was not widespread in Malaysia till the end of the twentieth century. However, Nejat et al. (2009a) observed typical symptoms of phytoplasmal infection, viz. yellowing and drying of fronds, in Malayan Dwarf ecotypes in coconut plantations in Selangor State in Malaysia. On the basis of disease symptoms, this infectious disease has been named Coconut Yellow Decline (CYD).

11.6.2 Symptoms

Yellowing of leaves in the outer whorl, which eventually turn light brown, is the characteristic preliminary symptom of the disease. The yellowing spreads rapidly to the younger leaves, and ultimately the emerging spear leaves become chlorotic. The affected palms also show premature nut fall and inflorescence necrosis. As the disease advances, gradual collapse of fronds and terminal rot of the growing point of immature palms occur. Affected palms generally die within 5 months after the appearance of initial symptoms (Nejat et al. 2009a).

11.6.3 Aetiology

CYD in Malaysia is reported to be caused by 2 different groups of phytoplasma. Based on sequences of 16S rDNA, the CYD isolates found in Malayan Red Dwarf (MRD) and Malayan Tall (MT) palms at Serdang were found to belong to the Bermuda grass white leaf group (16SrXIV, *Candidatus Phytoplasma cynodontis* group), while those from Malayan Yellow Dwarf (MYD) palms at Banting formed a novel group *Candidatus Phytoplasma malaysianum* (16Sr XXXII-B) (Nejat et al. 2009a, b). Both 16SrXIV and 16Sr XXXII-B groups causing CYD exist in Selangor State of Malaysia, reiterating the theory of occurrence of diverse groups of phytoplasma in a geographic region. The whole transcriptome profile of MRD in response to infection by 16SrXIV *Candidatus Phytoplasma cynodontis* group was evaluated by Nejat et al. (2015) using RNA-Seq technique. The transcriptomes' profile indicated that out of the 39,783 differentially expressed unigenes, 21,860 were downregulated and 18,013 were upregulated following phytoplasma infection.

11.6.4 Detection

Nejat et al. (2010) developed a real-time PCR assay using a 16S rDNA-based TaqMan primer–probe set, for sensitive, quantitative and rapid detection of the 16Sr XXXII-B of CYD phytoplasmas. The technique developed could not detect the 16SrXIV phytoplasma associated with CYD, and hence, it can be used to distinguish between infections caused by the two different CYD phytoplasmas present in Malaysia. Though remarkable developments have been achieved in the genomics of the pathogen, the vector involved in the transmission of the disease still remains unknown.

11.7 Kalimantan Wilt

11.7.1 Occurrence, Distribution and Crop Loss

Kalimantan wilt (KW) is a major phytopathological constraint that has caused extensive damage to coconut plantations in Central Kalimantan of Indonesia. The occurrence of the disease was initially noticed by farmers in 1978, and an outbreak of the disease was reported in 1988 in East Kotawaringin and Kapuas in Central Kalimantan (Sitepu et al. 1988). Based on the survey conducted in 1997, it was reported that the disease affects more than 100,000 palms in Kecamatan Mentaya Hilir Selatan and Kecamatan Pulau Hanaut. Coconut cultivation occupies a prime position in the agrarian economy of Indonesia. The loss of palms due to this phytoplasmal disease resulted in the economic upheaval of thousands of farm families in the country.

11.7.2 Symptoms

The first visible diagnostic symptom of the disease is the wilting and drying of older fronds. The petioles of the dried leaves break near the base and hang down, skirting the trunk. Blackening and rotting of the young developing inflorescences are another characteristic symptom. Older inflorescences may be normal or partially rotten. In newly opened inflorescences, rotting or blackening is seen on the middle but not on the tip as seen in LY-affected palms. Another commonly observed symptom is the presence of brown streaks along the petiole and midribs of green fronds. The affected palm dies within a period of 4 months after the appearance of the initial symptom (Warokka 1998).

11.7.3 Aetiology

The pathogenic role of bacteria, fungi, viruses, viroids and nematodes in inducing the KW symptoms could not be established (Warokka 1998). The phytoplasmal aetiology of Kalimantan wilt was confirmed by using nested PCR (Warokka et al. 2006). Based on the 16S rRNA sequences, the phytoplasmas causing KW have been reported to belong to 16Sr XI and 16Sr XIII groups (Warokka 2005 quoted by Bertaccini et al. 2014).

11.8 Weligama Coconut Leaf Wilt Diseases (WCLWD)

11.8.1 Occurrence and Distribution

Weligama coconut leaf wilt disease (WCLWD) was first reported in 2006 from the Weligama Divisional Secretariat Division (DSD) in the Matara district in southern Sri Lanka. The disease is now prevalent in Matara, Galle and Hambantota districts (Wijesekara et al. 2008; Perera et al. 2010; Everard 2013).

11.8.2 Symptoms

The symptoms of WCLWD are akin to that of root (wilt) disease reported from India. The earliest and characteristic diagnostic symptom of the disease is the flattening and downward bending of leaflets giving the frond a ribbed or flaccid appearance. In the initial stage of the disease, the crown of affected palms appears dark green in colour. The degree of flaccidity of leaflets varies among the fronds in a single palm. Intense yellowing of outer whorls of leaves is also a specific symptom of the disease. The yellowing becomes more prominent just after the rainy season. Flaccidity of leaflets is evident on seedlings younger than 3 years, whereas yellowing is seen only in older palms. Occasionally, yellowing of 6 to 8 fronds in the middle whorl is observed in affected palms. Yellowing is followed by drying up of the leaflets that starts from the margins of the affected fronds, and dried fronds hang in the crown for some time before falling off. The fronds also curl downward giving a ragged appearance to the crown. In some palms, the tips of fronds get twisted, break and hang down. Unopened spear leaves lose their rigidity and bend downwards in severely affected palms. Due to reduction in the number of fronds, the crown appears smaller and the trunk begins to taper (Fig. 11.2).

Below ground symptoms include the high degree of necrosis of root tips in moderately affected palms and in palms with high disease severity; no young roots are observed due to root degeneration. The roots of the affected palms show intense branching followed by necrosis of the stellar region of young white roots

Fig. 11.2 Weligama coconut leaf wilt disease. (Photo: GV Thomas)

(Wijesekara et al. 2008). As the disease progresses, female flower production declines and the palm becomes unproductive. Palms of any stage are susceptible to the disease, and palms over 3 years are most commonly affected. Furthermore, like RWD, WCLWD also predisposes the palms to a fungal disease complex called leaf rot disease. Weerakkody (2010) reported 40–60% reduction in yield in the advanced stages of the disease. Weerakoon et al. (2011) evaluated the impact of WCLWD on morphology, physiology and yield of coconut palms and found that the disease causes significant yield reduction. The whole nut weight of the palms diminishes with the severity of the disease. Though the copra yield is reduced by the WLCWD, the oil content is not affected much.

Waidyarathne and Samarasinghe (2014) developed artificial neural networks to assess naturally existing WCLWD severity status using data mining approaches. Attempts were made to develop methods to detect diseased coconut palms with prominent foliar yellowing symptom using multispectral satellite images of 0.5 m resolution. Adult coconut palms in the advanced stages of the disease can be detected by this method with more than 80% accuracy. But this method is not suitable for the precise detection of the disease in seedlings and in early stages of the disease. The overall accuracy of this approach is only 60–70% (Nainanayake et al. 2016a).

11.8.3 Aetiology

Analysis of leaflets from WCLWD-affected palms grown in fertilised lands revealed that the yellowing is not related to nutrient deficiency, particularly, magnesium, potassium and calcium (Wijesekara et al. 2008). The similarity of symptoms to the RWD and the symptom remission in oxytetracycline-treated palms supported the hypothesis of phytoplasmal etiology. Hence, the DNA extracted from midribs of spear leaves was subjected to nested PCR with phytoplasma universal primers R16F2n/R16R2 and R16mF2/R16R2 nested with fU5/rU3, P1/P7 nested with Chrfor/rU3 and direct PCR with Pc399/P1694. PCR products of expected sizes were obtained from diseased but not from healthy palms from a disease-free area. The sequences generated from the PCR products were submitted to similarity search (BlastN) in the NCBI database which confirmed that a phytoplasma belonging to the 16SrXI *Candidatus Phytoplasma oryzae* group is associated with WCLWD. The phytoplasma was found to be 99% similar to sugarcane white leaf phytoplasma, sugarcane grassy shoot phytoplasma and RWD phytoplasma, but not identical (Perera et al. 2010, 2012; Wijesekara et al. 2013).

11.8.4 Transmission

The confirmation of the phytoplasmal etiology instigated the research on vectors transmitting the disease. Kumara et al. (2015) collected 32 homopteran and a few hemipteran species from WCLWD-affected coconut plantations and subjected to the nested PCR using universal phytoplasma-specific primers, P1/P7 and Pc399/P1694. Eight homopteran species, viz. *Goniagnathus (T.) punctifer*, *Recilia dorsalis* Motschulsky, *Kolla ceylonica* (Melichar), *Idioscopus clypealis* (Lethierry), *Proutista moesta* (Westwood), *Proutista* sp., *Nisia nervosa* (Motschulsky) and an unknown Cixiid and a hemipteran species, *Stephanitis typica* (Distant), gave positive bands at 1280 bp which on sequencing showed similarity to WCLWD phytoplasma sequence (Gene Bank: EU635503), suggesting them as putative vectors of WCLWD. The vectoral role of these insects has to be confirmed by transmission experiments.

11.8.5 Detection

Concerted efforts are being made to develop serological techniques for the detection of WCLWD. The antibodies developed for the serological detection by using purified antigen from WLCD-affected spear leaves were found to be not specific to the pathogen, and therefore, production of monoclonal antibodies was initiated (Wijesekara et al. 2013). Detection of WLCWD presently depends on PCR amplification of DNA using phytoplasma-specific primers and characterisation of amplicon by sequencing. As the phytoplasma is present in very low titer in WCLWD-affected coconut palms, phytoplasma enrichment procedure is followed for DNA extraction. Siriwardhana et al. (2012) reported a DNAzol direct DNA extraction protocol followed by LAMP colorimetric assay for the detection of phytoplasma associated with WCLWD. The protocol has to be validated with more number of samples and needs further refinement before field application.

11.8.6 Management

Since the disease is confined to the Southern Province only, the Coconut Research Institute (CRI) of Sri Lanka decided to adopt intensive measures to prevent its possible spread to the rest of the country. Accordingly, a 3-km-wide and 80-km-long buffer zone was identified demarcating the diseased area. This buffer zone is inspected for the occurrence of the disease. The diseased and uneconomic palms are removed to curtail further spread. The spread of the disease was further checked by prohibiting the transportation of any palm species and their live parts out of the demarcated boundary. Findings of preliminary experiments conducted in the affected area have shown that the spread of the disease could be effectively contained by these measures (Nainanayake et al. 2013). However, isolated incidences in some places quite far from the boundary are being reported (Nainanayake et al. 2016b).

Research work on breeding for resistance was also initiated to identify tolerant varieties. Sri Lanka Green Dwarf is identified as a promising cultivar with a very high degree of resistance to WCLWD. The *Nana* coconut also shows a fair degree of resistance, though not as high as that of Sri Lanka Green Dwarf (Perera et al. 2015). SLGD is being used for resistance breeding programmes for the production of planting material for replanting in WCLWD endemic areas.

11.9 Future Strategy

Phytoplasmal diseases continue to be a serious threat to the coconut cultivation as they are non-curable. The difficulty in culturing the phytoplasma under axenic conditions remains a bottleneck in unveiling the mechanisms involved in pathogenesis. Despite the advancements in phytoplasma genomics, the information on the molecular basis of disease development on coconut palms is sparse. Recent developments in molecular biology should be fully utilised for elucidating the complex biological processes involved in the palm-phytoplasma-vector interactions. The knowledge on the effect of climate change on vector dynamics and disease spread is a prerequisite to formulate future strategies to mitigate the impact of phytoplasmal diseases. It is imperative to contain the spread of the disease within the current geographical limits by appropriate quarantine measures. Periodic surveillance in the diseased tract and monitoring for new incidence of disease and prompt removal will go a long way in arresting fresh outbreaks. The best option to control phytoplasma diseases of course is evolving disease-resistant/disease-tolerant planting material and hence deserves priority attention.

References

Amma PGK, Cecil RS, Pillai NG, Mathew AS, Nambiar PTN (1983) Performance of Dwarf × Tall hybrid coconut in root (wilt) affected areas of Kerala under different fertilizer levels. In: Bavappa KVA, Iyer RD, Das PK, Sethuraj MR, Ranganathan V, Nambiar KKN (eds) Proceedings of PLACROSYM V. Indian Society for Plantation Crops, Kasaragod, pp 405–410

Anonymous (1985) Coconut root (wilt) disease. Intensity, production loss and future strategy. CPCRI, Kasaragod, 45

Anonymous (1996) Coconut root (Wilt) survey. Department of Agriculture, Government of Kerala, Thiruvananthapuram, p 22

Arocha-Rosete Y, Konan Konan JL, Diallo AH, Allou K, Scott JA (2014) Identification and molecular characterization of the phytoplasma associated with a lethal yellowing-type disease of coconut in Côte d'Ivoire. Can J Plant Pathol 36:141–150

Arocha-Rosete Y, Konan-Konan JL, Diallo AH, Allou K, Scott J (2015) Analyses based on the 16S rRNA and secA genes identify a new phytoplasma subgroup associated with a lethal yellowing-type disease of coconut in Côte d'Ivoire. Phytopathol Mol 5:S57

Ashburner GR, Been BO (1997) Characterization of resistance to lethal yellowing in *Cocos nucifera* and implications for genetic improvement of this species in the Caribbean region. In: Eden-Green SJ, Ofori F (eds) Proceedings of an international workshop on lethal yellowing-like diseases of coconut. Natural Resources Institute, Chathman, pp 173–183

Ashburner G, Cardova I, Oropeza C, Illingworth R, Harrison N (1996) First report of coconut lethal yellowing disease in Honduras. Plant Dis 80:960

Baudouin L, Lebrun P, Berger A, Myrie W, Been B, Dollet M (2008) The Panama Tall and the Maypan hybrid coconut in Jamaica: did genetic contamination cause a loss of resistance to Lethal Yellowing? Euphytica 161:353–360

Bavappa KVA, Jayasanker NP, Radha K, Rethinam P (1986) Coconut root (wilt) disease, Present status of research and management. Technical bulletin no. 14, CPCRI, Kasaragod, p 10

Beakbane A, Clater C, Posnette A (1972) Mycoplasmas in the phloem of coconut, *Cocos nucifera* L. with lethal yellowing disease. J Hort Sci Biotech 47:265

Been BO (1995) Integrated pest management for the control of lethal yellowing quarantine, cultural practices and optimal use of hybrids. In: Oropeza C, Howard FW, Ashburner GR (eds) Lethal yellowing research and practical aspects. Kluwer Academic Publishers, Dordrecht, pp 101–109

Bertaccini A, Duduk B, Paltrinieri S, Contaldo N (2014) Phytoplasmas and phytoplasma diseases: a severe threat to agriculture. American J Plant Sci 5:1763–1788

Bila J, Mondjana A, Samils B, Högberg N (2015a) High diversity, expanding populations and purifying selection in phytoplasmas causing coconut lethal yellowing in Mozambique. Plant Pathol 64:597–604

Bila J, Mondjana A, Samils B, Högberg N (2015b) Potential novel '*Candidatus* Phytoplasma pini'-related strain associated with coconut lethal yellowing in Mozambique. Phytopathol 5:S59

Bonnot F, de Franqueville H, Lourenço E (2010) Spatial and spatiotemporal pattern analysis of coconut lethal yellowing in Mozambique. Phytopathol 100:300–312

Broschat TK, Harrison NA, Donselman H (2002) Losses to lethal yellowing cast doubt on coconut cultivar resistance. Palms 46:185–189

Brown SE, Been BO, McLaughlin WA (2006) Detection and Variability of the Lethal Yellowing Group (16SrIV) Phytoplasma in the *Cedusa* sp. (Hemiptera: Auchenorrhyncha: Derbidae) in Jamaica. Ann Appl Bio 149:53–62

Brown SE, Been BO, McLaughlin WA (2007) The lethal yellowing (16Sr IV) group of phytoplasmas. Pest Technol 1(1):61–69

Bull RA (1955) Bronze leaf wilt of coconut palms in Nigeria. J West African Inst Palm Res 3:70–72

Butler EJ (1908) Report on coconut palm disease in Travancore. Agric Res Inst Pusa Bull (9):23

Carter W, Suah JRR (1964) Studies on the spread of lethal yellowing disease of the coconut palm. FAO Plant Prot Bull 12:73–78

Cecil SR, Amma PGK (1991) Soils and nutrition. In: Nair MK et al (eds) Coconut root (wilt) disease, Monograph series no.3. CPCRI, Kasaragod, pp 45–52

Chandramohanan R (2010) Symptomatology and etiology. In: Thomas GV, Chandramohanan R, Jacob PM, Krishnakumar V (eds) Coconut root (wilt) disease management. Central Plantation Crops Research Institute, Kasaragod, pp 1–18

Chowdappa P, Gopinathan Pillai N, Gunasekaran M, Solomon JJ (1989) Distribution and retention of Oxytetracycline hydrochloride injected into coconut palms affected by root(wilt) disease. J Plantn Crops 17(2):108–113

Cordova I, Jones P, Harrison NA, Oropeza C (2003) *In situ* PCR detection of phytoplasma DNA in embryos from coconut palms with lethal yellowing disease. Mol Plant Pathol 4(2):99–108

Córdova I, Oropeza C, Puch-Hau C, Harrison N, Collí-Rodríguez A, Narvaez M, Nic-Matos G, Reyes C, Sáenz L, Edizioni ETS (2014) A real-time PCR assay for detection of coconut lethal yellowing phytoplasmas of group 16SrIV subgroups A, D and E found in the Americas. J Plant Pathol 96:343–352

Dabek AJ (1974) Biochemistry of coconut palms affected with the lethal yellowing disease in Jamaica. J Phytopathol 81:346–353

Dabek AJ (1977) Electron microscopy of Kaincope and Cape St. Paul Wilt diseased coconut tissue from West Africa. Phytopathol 38:341–346

Dabek AI, Johnson CG, Harries HC (1976) Mycoplasma-like organisms associated with Kaincope and Cape St. Paul Wilt diseases of coconut palms in West Africa. PANS 22(3):354–358

Danyo G (2011) Review of scientific research into the Cape Saint Paul Wilt Disease (CSPWD) of coconut in Ghana. African J Agric Res 6:4567–4578

Dare D, Andoh-Mensah E, Owusu-Nipah J, Yankey N, Quaicoe RN, Nkansah-Poku J, Dery SK (2010) Evaluation of some basic traits of a promising coconut hybrid: Srilankan Green Dwarf crossed to Vanuatu Tall (SGD × VTT). J Sci Technol 30(3):9–14

Dery SK, N'Cho YP, Sangare A, Arkhurst ED (1997a) Cape St Paul wilt disease: resistance screening and prospects for rehabilitating the coconut industry in Ghana. In: Eden-Green SJ, Ofori F (eds) Proceedings of an international workshop on lethal yellowing-like disease of coconut, Elmina, Ghana. Natural Resources Institute, Chatham, pp 147–151

Dery SK, Philippe R, Mariau D (1997b) Preliminary observations Auchenorrhyncha (Homoptera): suspected vectors of St. Paul wilt disease in Ghana. In: Eden-Green SJ, Ofori F (eds) Proceedings of an international workshop on lethal yellowing-like disease of coconut. Natural Resources Institute, Chatham, pp 269–276

Deutsch E, Nienhaus F (1983) Further studies on distribution of mycoplasma like organisms in different tissues of lethal diseased coconut palms in Tanzania. Z Pflkrankh Pjlschutz 90:278–284

Doi Y, Teranaka M, Yora K, Asuyama H (1967) Mycoplasma or PLT Group like microorganisms found in the phloem elements of plants infected with Mulberry Dwarf, Potato Witches' Broom, Aster Yellows or Paulownia Witches' Broom. Jpn J Phytopathol 33:259–266

Dollet M (1979) Research on the etiology of blast of oil and coconut palms. In: Thomas DL, Howard FW, Donselman HM (eds) Proceedings of the fourth meeting of the international council on Lethal Yellowing. Pub. FL-80-1. University of Florida, Fort Lauderdale, p 19

Dollet M, Giannotti J (1976) Kaincope disease: presence of mycoplasma in the phloem of diseased coconuts. Oleagineux 31:169–171

Dollet M, Giannotti J, Renard JL, Ghosh SK (1977) Study of lethal yellowing of coconuts in Cameroon: Kribi disease. Observations of mycoplasma like organisms. Oleagineux 32:317–322

Dollet M, Quaicoe R, Pilet F (2009) Review of coconut "Lethal Yellowing" type diseases diversity, variability and diagnosis. Oléagineux 16:97–101

Dollet M, Macome F, Vaz A, Fabre S (2011) Phytoplasmas identical to coconut lethal yellowing phytoplasmas from Zambesia (Mozambique) found in a pentatomide bug in Cabo Delgado province. Bull Insectol 64:S139–S140

Donis J (2002) Coconut cultivation and lethal yellowing status in Haiti. In: Proceedings of expert consultation on sustainable coconut production through control of lethal yellowing disease. CFC/FAO/CIB, Kingston, pp 90–92

Doyle MMR (2001) Crisis research: managing lethal yellowing disease. Biotechnol Dev Monit 44/45:12–15

Dwivedi RS, Mathew C, Ray PK, Amma BSK, Ninan S (1977) Rapid biochemical test to detect root (wilt) disease of coconut. Curr Sci 46:611–612

Eden-Green SJ (1979) Root symptoms in coconut palms affected by lethal yellowing disease in Jamaica. FAO Plant Prot Bull 4:119–122

Eden-Green SJ (1995) A brief history of lethal yellowing research. In: Oropeza C, Howard FW, Ashburner GR (eds) Lethal yellowing: research and practical aspects. Kluwer Academic Publishers, Dordrecht, pp 17–33

Ekpo EN, Ojomo EE (1990) The spread of lethal coconut diseases in West Africa: incidence of Awka disease (or bronze leaf wilt) in Ishan area of Bendel State of Nigeria. Principes 34:143–146

Everard JMDT (2013) Integrated management of Weligama coconut leaf wilt disease program in Sri Lanka: strategies, implementation and challenges. In: Gunasena HPM, Gunathilake HAJ, Fernando LCP, Everard MDT, Appuhamy PAHN (eds) Weligama coconut leaf wilt disease six years after. Coconut Research Institute, Lunuwila, pp 13–17

Fawcett W (1891) Report on the coconut disease at Montego Bay. Bull Bot Dept (Jamaica) 23:2

George MV, Radha K (1973) Computation of disease index of root (wilt) disease of coconut. Indian J Agric Sci 43:366–370

Govindankutty MP (1981) Histopathological studies. In: Review of research on coconut root (wilt) disease. Central Plantation Crops Research Institute, Regional Station, Kayamkulam, pp 49–54

Grylls NE, Hunt P (1971) A review of the study of the aetiology of coconut lethal yellowing disease. Oleagineux 26:311–315

Gurr GM, Johnson A, Ash G, Wilson B, Ero M, Pilotti C, Dewhurst C, You M (2016) Coconut lethal yellowing diseases: a phytoplasma threat to palms of global economic and social significance. Front Plant Sci 7:5121

Harrison NA, Oropeza C (1997) Recent studies on detection of lethal yellowing disease phytoplasmas in the Americas. In: Eden-Green SJ, Ofori F (eds) Proceedings of an international workshop on lethal yellowing-like diseases of coconut. Natural Resources Institute, Chatham, pp 221–234

Harrison NA, Bourne CM, Cox RL, Tsai JH, Richardson P (1992) DNA probes for detection of mycoplasma-like organisms associated with lethal yellowing of palms in Florida. Phytopathol 82:216–224

Harrison NA, Richardson PA, Kramer JB, Tsai IH (1994a) Detection of the mycoplasma-like organisms associated with lethal yellowing of palms in Florida by Polymerase Chain Reaction. Plant Pathol 43:998–1008

Harrison NA, Richardson P, Jones P, Tymon A, Eden Green S, Mpunami A (1994b) Comparative investigation of MLO's associated with Caribbean and African coconut lethal decline diseases by DNA Hybridization and PCR Assays. Plant Dis 78:507–511

Harrison N, Cordova I, Richardson P, Dibonito R (1999) Detection and diagnosis of lethal yellowing. In: Oropeza C, Verdeil JL, Ashburner GR, Cardeña R, Santamaría JM (eds) Current advances in coconut biotechnology, vol 35. Springer, Dordrecht, pp 183–196

Harrison NA, Narváez M, Almeyda H, Cordova I, Carpio ML, Oropeza C (2002a) First report of group 16SrIV phytoplasmas infecting coconut palms with leaf yellowing symptoms on the Pacific coast of Mexico. Plant Pathol 51(6):808

Harrison NA, Myrie W, Jones P, Carpio ML, Castillo M, Doyle MM, Oropeza C (2002b) 16S rRNA interoperon sequence heterogeneity distinguishes strain populations of palm lethal yellowing phytoplasma in the Caribbean region. Ann Appl Biol 141:183–193

Harrison NA, Helmick EE, Elliott ML (2008) Lethal yellowing-type diseases of palms associated with phytoplasmas newly identified in Florida, USA. Ann Appl Biol 153:85–94

Harrison NA, Davis RE, Oropeza C, Helmick EE, Narváez M, Eden-Green S, Dollet M, Dickson M (2014) '*Candidatus* Phytoplasma palmicola', associated with a lethal yellowing-type disease of coconut (*Cocos nucifera* L.) in Mozambique. Intl J Syst Evolut Microbiol 64:1890–1899

Harrison NA, Gundersen-Rindal D, Davis RE (2015) "*Candidatus* Phytoplasma". Bergey's Manual of Systematics of Archaea and Bacteria 1–33

Hegde V, Nair S, Gangaraj KP, Merin Babu (2016) PCR and LAMP based detection of phytoplasma associated with root (wilt) disease of coconut and yellow leaf disease of arecanut. In: Paper presented in 6th international conference: plant, pathogens and people, 23–27 February, 2016, New Delhi, Abstract, p 160

Heinze KG, Petzold H, Martwitz R (1972) Beitrug Zur Atiologie der Todichen vergilbung der Kokospalme. Phytopatholgische Zeitschrift 74:230–237

Howard FW (1983) World distribution and possible geographic origin of palm lethal yellowing disease and its vectors. FAO Plant Prot Bull 31:101–113

Howard FW, Oropeza C (1998) Organic mulch as a factor in the nymphal habitat of Myndus crudus (Hemiptera: Auchenorrhyncha: Cixiidae). Florida Entomol 81:92–97

Howard FW, Norris R, Thomas D (1983) Evidence of transmission of palm lethal yellowing agent by a plant hopper, *Myndus crudus* (Homoptera, Cixiidae). Trop Agric 60:168–171

Howard FW, Williams DS, Norris RC (1984) Insect transmission of lethal yellowing to young palms. Intl J Entomol 26:331–338

Howard FW, Atilano R, Barrant CI, Harrison NA, Theobold WF, Williams DS (1987) Unusually high lethal yellowing disease incidence in Malayan Dwarf coconut palms on localized sites in Jamaica and Florida. J Plantn Crops 15:86–100

Hunt P, Dabek AI, Schuiling M (1974) Remission of symptoms following tetracycline treatment of lethal yellowing affected coconut palms. Phytopathol 64:307–312

Jayasankar NP, George M (1991) Etiology – B. Bacteria. In: Coconut root (wilt) Disease, Monograph series no.3. Central Plantation Crops Research Institute, Kasaragod, pp 17-18

Johnson CG, Harries HC (1976) A survey of Cape St. Paul wilt of coconut in West Africa. Ghana J Agric Sci 9:125–129

Joseph T, Lilly VG (1991) Etiology – A. Fungi. In: Coconut root (wilt) disease, Monograph series no.3. Central Plantation Crops Research Institute, Kasaragod, pp 13–15

Joseph T, Shanta P (1963) Studies on a colour test for the root (wilt) disease of coconut. Abstracts of 50th Indian Sci. Cong Pt III, pp 618–619

Joseph T, Shanta P (1964) Anatomical changes in the tissues of the tender leaves of coconut palms affected by the root (wilt) disease. J Indian Bot Soc 42:61–65

Julia JF (1979) Mise en evidence et identification des insectes responsables des maladies juveniles du cocotier et du palmier a huile en Cote d' Ivoire. Oleagineux 34:385–393

Kaiza DA (1987) History and etiology of the lethal disease of coconut palm (*Cocos nucifera L.*) in Tanzania. M.Sc. thesis, Sokoine University of Agriculture, Tanzania

Kelly PL, Reeder R, Kokoa P, Arocha Y, Nixon T, Fox A (2011) First report of a phytoplasma identified in coconut palms (*Cocos nucifera*) with lethal yellowing-like symptoms in Papua New Guinea. New Dis Rep 23:9

Konan Konan JL, Allou K, Atta Diallo H, Saraka Yao D, Koua B, Kouassi N, Benabid R, Michelutti R, Scott J, Arocha-Rosete Y (2013) First report on the molecular identification of the phytoplasma associated with a lethal yellowing-type disease of coconut palms in Cote d' Ivoire. New Dis Rep 28:3

Krishnakumar V, Babu M, Thomas RJ, Josephrajakumar, Anithakumari P, Hegde V, Chowdappa P (2015) Root (wilt) disease of coconut- Bench to Bunch strategies. ICAR-CPCRI Tech Bull 91:28

Kumara ADNT, Perera L, Meegahakumbura MK, Aratchige NS, Fernando LCP (2015) Identification of putative vectors of weligama coconut leaf wilt disease in Sri Lanka. In: Chakravarthy AK (ed) New horizons in insect science: towards sustainable pest management. Springer, Delhi, pp 137–146

Lal SB (1968) Root (wilt) disease. Resume of work done since 1964. In: Proceedings of 3rd session of the FAO Tech Working Party on Coconut Production, Protection and Processing, Jakarta, Indonesia, pp 9–19

Leach R (1946) The unknown disease of coconut palm in Jamaica. Trop Agric (Trinidad) 43:59–68

Lebrun P, Baudouin L, Myrie W, Berger A, Dollet M (2008) Recent lethal yellowing outbreak: why is the Malayan Yellow Dwarf coconut no longer resistant in Jamaica? Tree Genet Genome 4(1):125–131

Lee IM, Hammond RW, Davis RE, Gundersen DE (1993) Universal amplification and analysis of pathogen 16S rDNA for classification and identification of mycoplasma like organisms. Phytopathol 83:834–842

Lee IM, Gundersen-Rindal DE, Bertaccini A (1998) Phytoplasma: ecology and genomic diversity. Phytopathol 88:1359–1366

Llauger R, Becker D, Cueto J, Peralta E, González V, Rodríguez M, Rohde W (2002) Detection and molecular characterization of phytoplasma associated with lethal yellowing disease of coconut palms in Cuba. J Phytopathol 150:390–395

Lu H, Wilson BAL, Ash GJ, Woruba SB, Fletcher MJ, You M, Yang G, Gurr GM (2016) Determining putative vectors of the Bogia Coconut Syndrome phytoplasma using loop-mediated isothermal amplification of single-insect feeding media. Sci Rep 6:35801

Manimekalai R, Soumya VP, Kumar RS, Selvarajan R, Reddy K, Thomas GV, Sasikala M, Rajeev G, Baranwal VK (2010) Molecular detection of 16SrXI group phytoplasma associated with root (wilt) disease of coconut (*Cocos nucifera*) in India. Plant Dis 94:636

Manimekalai R, Nair S, Soumya V, Roshna O, Thomas G (2011) Real-time PCR technique-based detection of coconut root (wilt) phytoplasma. Curr Sci 101(9):1209–1213

Manimekalai R, Soumya VP, Nair S, Thomas GV, Baranwal V (2014a) Molecular characterization identifies 16SrXI-B group phytoplasma (‘*Candidatus Phytoplasma oryzae*’-related strain) associated with root wilt disease of coconut in India. Sci Hort 165:288–294

Manimekalai R, Nair S, Soumya VP (2014b) Evidence of 16SrXI group phytoplasma DNA in embryos of root wilt diseased coconut palms. Aust Plant Pathol 43:93–96

Maramorosch K (1964) A survey of coconut diseases of unknown etiology. FAO, Rome, p 39

Maramorosch K, Kondo F (1977) Electron Microscopy of leaf sections from Kerala wilt diseased coconut palm. J Plantn Crops 5:20–22

Mariau D, Dery SK, Sangare A, N’Cho YP, Philippe R (1996) Coconut lethal yellowing diseaseand planting material tolerance. Plant Rech Dev 3:105–112

Martinez AP, Roberts DA (1967) Lethal yellowing of coconut in Florida. Proc Fla State Hortic Soc 80:432–436

Martinez RT, Narvaez M, Fabre S, Harrison N, Oropeza C, Dollet M (2008) Coconut lethal yellowing on the southern coast of the Dominican Republic is associated with a new 16SrIV group phytoplasma. Plant Pathol 57:366–366

Martinez RT, Baudouin L, Berger A, Dollet M (2010) Characterization of the genetic diversity of the Tall coconut (*Cocos nucifera* L.) in the Dominican republic using microsatellite (SSR) markers. Tree Gene Genome 6:73–81

Martini M, Lee IM, Bottner KD, Zhao Y, Botti S, Bertaccini A, Harrison NA, Carraro L, Marcone C, Khan AJ, Osler R (2007) Ribosomal protein gene-based phylogeny for finer differentiation and classification of phytoplasmas. Intl J Syst Evol Microbiol 57:2037–2051

Martyn EB (1949) Further observation on the unknown disease of coconut. Trop Agric (Trinidad) 26:110–112

Mathen K, Solomon JJ, Rajan P, Geetha L (1987) Electron microscopic evidence on the role of *Stephanitis typica* (Distant) as vector of coconut root (wilt) disease. Curr Sci 56:1339–1341

Mathen K, Rajan P, Radhaknshnan Nair CP, Sasikala M, Gunasekaran M, Govindankutty MP, Solomon JJ (1990) Transmission of root (wilt) disease to coconut seedlings through *Stephanitis typica* (Distant) (Heteroptera; Tingidae). Trop Agric (Trinidad) 67(1):69–73

Mayil Vaganan M, Alka G, Chithra KR (2001) Biochemical evidences for purification of phytoplasma of root (wilt) disease of coconut. J Plant Biol 28:55–60

McCoy RE (1972) Remission of lethal yellowing in coconut palm treated with tetracycline antibiotics. Plant Dis Rep 56:1019–1021

McCoy RE (1976) Comparative epidemiology of the lethal yellowing, Kaincope, and cadang-cadang diseases of coconut palm. Plant Dis Rep 60:498–502

McCoy RE, Carroll VJ, Poucher CP, Gwin GH (1976) Field control of coconut lethal yellowing with oxytetracycline-hydrochloride. Phytopathol 66:1148–1150

McCoy RE, Howard FW, Tsai JH, Donselman HM, Thomas DL, Basham HG, Atilano RA, Eskafi FM, Britt L, Collins ME (1983) Lethal yellowing of palms, Agri Expt Station Bull No. 834. University of Florida, IFAS, Gainesville

Mejia F, Palmieri M, Oropeza C, Doyle M, Harrison N, Aguilar E, Narváez M, Estrada R, Ortiz G (2004) First report of coconut lethal yellowing disease in Guatemala. Plant Pathol 53:800

Menon KPV (1938) Report on the scheme of investigations on the diseases of coconuts in South India, 1937. Government of India Press, New Delhi, p 16

Menon KPV, Nair UK (1948) The leaf rot disease of coconut in Travancore and Cochin. Indian Cocon J 1:33–39

Menon KPV, Nair UK (1951) Scheme for the investigation of the root and leaf diseases of the coconut palm in South India. Consolidated final report of the work done from 8th March 1937 to 31st March, 1948. Indian Cocon J 5(1):3–19

Mora-Aguilera G (2002) Dispersal potential of lethal yellowing of the coconut palm: a quarantine disease in Mexico. In: Proceedings of expert consultation on sustainable coconut production through control of lethal yellowing disease, Kingston, Jamaica. CFC/FAO/CIB, pp 128–133

Mpunami A, Jones P, Dickinson M, Tymon A, Harrison NA (1997) Detection of coconut lethal disease in Tanzania by use of DNA probes and PCR. In: Eden Green SJ, Ofori F (eds)

International workshop on lethal yellowing-like disease of coconut, Elmina, Ghana, November 1995. Natural Resources Institute, Chatham, pp 205–218

Mpunami AA, Tymon A, Jones P, Dickinson MJ (1999) Genetic diversity in the coconut lethal yellowing disease phytoplasmas of East Africa. Plant Pathol 48:109–114

Mpunami A, Tymon A, Jones P, Dickinson MJ (2000) Identification of potential vectors of the coconut lethal disease phytoplasma. Plant Pathol 49:355–361

Muralidharan A, Jayasanker NP, Antony KJ, Rethinam P (1991) Management of coconut root (wilt) disease. In: Nair MK et al (eds) Coconut root (wilt) disease, CPCRI monograph series no.3. CPCRI, New Delhi, pp 73–80

Myrie W (2002) Current outbreaks of lethal yellowing in Jamaica and the use of molecular diagnostic techniques in phytoplasma detection. In: Proceedings of expert consultation on sustainable coconut production through control of lethal yellowing disease, Kingston, Jamaica. CFC/FAO/CIB, pp 107–115

Myrie W, Paulraj L, Dollet M, Wray D, Been B, McLaughlin W (2006) First report of lethal yellowing disease of coconut palms caused by phytoplasma on Nevis Island. Plant Dis 90:834–834

Myrie W, Oropeza C, Saenz L, Harrison NA, Mercedes Roca M, Cordova I (2011) Reliable improved molecular detection of coconut lethal yellowing phytoplasma and reduction of associated disease through field management strategies. Bull Insectol 64:S203–S204

Myrie W, Douglas CJ, Harrison NA, McLaughlin WA, James M (2012) First report of lethal yellowing disease associated with subgroup 16SrIV,a phytoplasma on St. Kitts in the Lesser Antilles. New Dis Rep 26:25

Nagaraj AN, Davis TA, Menon KPV (1954) Sap transfusion, a new device for virus transmission trials in palms. Indian Cocon J 7:91–98

Nainanayake AD, Wijesekara HTR, Gunathilake I (2013) Disease spreading pattern – GIS mapping. In: Gunasena HPM, Gunathilake HAI, Fernando LCP, Everard IMDT, Appuhamy PAHN (eds) Coconut Research Institute, Sri Lanka, pp 48–52

Nainanayake AD, Gunathilake J, Kumarathungai MDP, Gunawardena PM, Wijesekara HTR (2016a) Limitation in the use of spectral analysis to detect Weligama Coconut Leaf Wilt Disease affected palms in Southern Sri Lanka. COCOS 22:13–24

Nainanayake AD, Kumarathungai MDP, de Silva PHPR (2016b) A survey of land for Weligama Coconut Leaf Wilt Disease affected palms outside the declared boundary in the Southern Province. COCOS 22:57–64

Nambiar PTN, Pillai NG (1985) A simplified method for indexing root (wilt) affected coconut palm. J Plantn Crops 13(1):35–37

Nambiar MC, Prasannakumari TO (1964) Effect of root (wilt) disease on microsporogenesis in coconut (*Cocos nucifera* L.). Indian Cocon J 17:93–100

Narasimhachari CL, Prasadji JK, Rajamannar M (1991) Survey and confirmation of Tatipaka diseased palms in central delta of river Godavari in Andhra Pradesh. Indian Cocon J 22(3):7–10

Nejat N, Sijam K, Abdullah SNA, Vadamalai G, Dickinson M (2009a) Molecular characterization of a phytoplasma associated with Coconut Yellow Decline (CYD) in Malaysia. American J Appl Sci 6:1331

Nejat N, Sijam K, Abdullah S, Vadamalai G, Dickinson M (2009b) Phytoplasmas associated with disease of coconut in Malaysia: phylogenetic groups and host plant species. Plant Pathol 58:1152–1160

Nejat N, Sijam K, Abdullah SNA, Vadamalai G, Sidek Z, Dickinson M, Edizioni ETS (2010) Development of a TaqMan real-time PCR for sensitive detection of the novel phytoplasma associated with coconut yellow decline in Malaysia. J Plant Pathol 92(3):769–773

Nejat N, Cahill DM, Vadamalai G, Ziemann M, Rookes J, Naderali N (2015) Transcriptomics-based analysis using RNA-Seq of the coconut (*Cocos nucifera*) leaf in response to yellow decline phytoplasma infection. Springer, Berlin/Heidelberg. Mol Gene Genomics 290(5):1899–1910

Nienhaus F, Steiner KG (1976) Mycoplasma like organisms associated with Kaincope disease of coconut palms in Togo. Plant Dis Rep 60:1000–1002

Nienhaus F, Schuiling M, Gliem G, Schinzer U, Spittel A (1982) Investigations on the etiology of the lethal disease of coconut palm in Tanzania. Z Pjlkrankh Pjlschutz 89:185–193
Nipah JO, Jones P, Dickinson MJ (2007) Detection of lethal yellowing phytoplasma in embryos from coconut palms infected with Cape St. Paul wilt disease in Ghana. Plant Pathol 56:777–784
Nkansah JP, Philippe R, Quaicoe RN, Dery SK, Ransford A (2009) Cape Saint Paul Wilt Disease of coconut in Ghana: surveillance and management of disease spread. Oléagineux 16:111–115
Ntushelo K, Elliott ML, Harrison NA (2013) Palm yellows phytoplasmas and their genetic classification. African J Biotech 12:3376–3382
Nutman FJ, Roberts PM (1955) Lethal Yellowing: the unknown disease of coconut palm in Jamaica. Emp J Exp Agric 23:257–267
Odewale JO, Nair RV, Enaberue L (2006) Field resistance to lethal yellowing disease in some coconut varieties in Nigeria. Trop Agric 83(4):135
Odewale JO, Odionwaya G, Osagie JI, Ahanon JM (2010) Rate of lethal yellowing disease (LYD) spread in coconut (*Cocos nucifera* L) plantation of tall inter planted with Dwarf varieties. In: Proceedings of the 19th conference of Botanical Society of Nigeria, Umaru Musa Yar"addua University, Katsina, Nigeria, p 47
Ofori F, Nkansah-Poku J (1997) Cape St Paul Wilt disease of coconut in Ghana: history of its occurrence and spread. In: Proceedings of international workshop on lethal yellowing-like diseases of coconut. Natural Resources Institute, Chatham, pp 27–32
Oropeza C, Cordova I, Chumba A, Narváez M, Sáenz L, Ashburner R, Harrison N (2011) Phytoplasma distribution in coconut palms affected by lethal yellowing disease. Ann Appl Biol 159:109–117
Oropeza C, Cordova I, Puch-Hau C, Castillo R, Chan JL, Sáenz L (2017) Detection of lethal yellowing phytoplasma in coconut plantlets obtained through *in vitro* germination of zygotic embryos from the seeds of infected palms. Ann Appl Biol 171:28–36
Osagie IJ, Ojomo EE, Pilet F (2013) An update on Awka wilt disease of coconut caused by phytoplasma in Nigeria. Acta Phytopathologica Sinica 43(suppl):204
Osagie IJ, Ojomo EE, Pilet F (2016) Detection of phytoplasma associated with Awka wilt disease of coconut palms in Nigeria by polymerase chain reaction. Intl J Appl Res Tech 5(4):49–55
Pandit SV, Simhachalam G, Lal SB (1969) A note on Tatipaka or decline disease of coconut in Andhra Pradesh. India. In: Proceedings of 3rd session. FAO Technical Working Party on Coconut Production, Protection and Processing, Jakarta, 9–19 September 1968. Paper No.41, p 182
Parthasarathy MV (1974) Mycoplasma-like organisms associated with lethal yellowing disease of palms. Phytopathol 64:667–674
Perera L, Meegahakumbura MK, Wijesekara HTR, Kalani NGA, Munasing CE, Fernando WBS, Dickinson MJ (2010) Detection of Weligama coconut leaf wilt disease in Sri Lanka by polymerase chain reaction. In: George VT, Krishnakumar V, Augustine JB, Niral V, Josephrajkumar A (eds) Proceedings of the international conference on coconut biodiversity for prosperity. Niseema Printers and Publishers, Kochi, p 146
Perera L, Meegahakumbura MK, Wijesekara HRT, Fernando WBS, Dickinson M (2012) A phytoplasma is associated with Weligama Coconut Leaf Wilt Disease in Sri Lanka. J Plant Pathol 94:205–209
Perera SACN, Herath HMNB, Wijesekera HTR, Subhathma WGR, Weerakkody WATL (2015) Evaluation of coconut germplasm in Weligama and Matara area of the Southern Province of Sri Lanka for resistance to Weligama Coconut Leaf Wilt Disease. COCOS 21:15–20
Philippe R, Reignard S, Descamps S, Nkansah-Poku J, Quaicoe R, Pilet F, Fabre S, Dollet M (2009) Study on the transmission of coconut Lethal Yellowing in Ghana. Oléagineux 16:102–106
Pillai NK (1911) Naleekeram. Vidyabhivardhini Press, Quilon, p 112
Pillai NG, Raju CA (1985) A simple pressure injection device for application of antibiotics on coconut palm. J Plantn Crops 13:151–154
Pillai NG, Lal SB, Shanta P (1973) Distribution and intensity of root (wilt) disease of coconut in Kerala. J Plantn Crops 1(Suppl):107–112

Pillai NG, Sasikala M, Mathen K (1980) Epidemiology of coconut root(wilt) disease. J Plantn Crops 8:90–97

Pillai NG, Chowdappa P, Solomon JI, Mathew J (1991) Remission of symptoms of root (wilt) disease of coconut injected with Oxytetracycline-HCI. J Plantn Crops 19(1):14–20

Pilotti CA, Dewhurst CF, Liefting LW, Kuniata L, Kakul T (2014) Putative vectors of a phytoplasma associated with coconut (*Cocos nucifera*) in Madang Province Papua New Guinea. Intl J Agric For 4:365–372

Plavsic-Banjac B, Hunt P, Maramorosch K (1972) Mycoplasma like bodies associated with lethal yellowing disease of coconut palms. Phytopathol 62:298–299

Quaicoe RN, Dery SK, Philippe R, Baudouin L, Nipah JO, Nkansah-Poku J et al (2009) Resistance screening trials on coconut varieties to Cape Saint Paul Wilt Disease in Ghana. Oléagineux 16:132–136

Quillec G, Morin JP, Renard JL, Marian D (1978) Disease of young coconut palms. Oleagineux 33:500–501

Quiroz L (2002) Information on coconut Lethal Yellowing in Belize. In: Proceedings of expert consultation on sustainable coconut production through control of Lethal Yellowing Disease. CFC/FAO/CIB, Kingston, p 87

Radha K, Lal SB (1972) Diagnostic symptoms of root (wilt) disease of coconut. Indian J Agric Sci 42:410–413

Rajagopal V, Patil KD, Mathew C, Amma BSK (1986a) Abnormal stomatal opening in coconut palms affected with root (wilt) disease. J Exp Bot 37:1398–1405

Rajagopal V, Patil KD, Mathew C, Amma BSK, Varkey T (1986b) Water relation studies in root (wilt) affected coconut palms. In: Final report Central Plantation Crops Research Institute, Regional Station, Kayamgulam, Kerala, India, p 24

Rajagopal V, Mathew C, Varkey T, Joseph KV (1998) Physiology and biochemistry. In: Nampoothiri KUK, Koshy PK (eds) Coconut root (wilt) disease. Central Plantation Crops Research Institute, Kasaragod, pp 65–73

Rajamannar M, Rao AP, Ramapandu S, Subbiah J (1984) Etiology and control of Tatipaka disease of coconut – an appraisal. In: Proceedings of all India symposium coconut diseases. Coconut Research Station, Veppankulam, pp 35–39

Rajamannar M, Prasadji JK, Rethinam P (1993) Tatipaka disease of coconut. Current status. In: Nair MK, Khan HH, Gopalasundaram P, Rao EVVB (eds) Advances in coconut research and development. Oxford and IBH Publishing Co. Pvt Ltd, New Delhi, pp 591–595

Rajamannar M, Prasadji JK, Rethinam P (1994) Tatipaka disease of coconut. In: Chadha KL, Rethinam P (eds) Advances in horticulture 10 – plantation and spice crops, Part 2. Malhotra Publishing House, New Delhi

Rajan P (2011) Transmission of coconut root (wilt) disease through plant hopper, *Proutista moesta* Westwood (Homoptera: Derbidae). Pest Manag Hort Ecosyst 17:1–5

Rajan P, Mathen K (1985) *Proutista moesta* (Westwood) and other additions to the insect fauna on coconut palms. J Plantn Crops 13:135–136

Rajan P, Nair CPR, Solomon JJ, Geetha L (2002) Identification of Phytoplasma in the salivary glands of *Proutista moesta* (West wood), a putative vector of coconut root (wilt) disease. J Plantn Crops 30(2):55–57

Ramadasan A, Shanta P, Lal SB (1971) Resistance or susceptibility, age of bearing and yield in young coconut palms in relation to the development of root (wilt) disease. Indian J Agric Sci 41:1107–1109

Ramapandu S, Rajamannar M (1981) Further investigations on the etiology and control of Tatipaka disease of coconut. Paper presented at the third international symposium of plant pathology, New Delhi

Ramaswamy M, Soumya VP, Nair S, Rani RG, Rao GP (2016) PCR detection of phytoplasmas from different tissues of root (wilt) diseased coconut palms. Phytopathogenic Mollicutes 6(1):23–28

Randles JW, Hatta T (1980) Etiological studies on virus like diseases of coconut and cardamom in India. In: Biennial report of the Waite Agric Res Inst 1978–1979, Melbourne, Australia
Rao AP (1966) Identification of Tatipaka disease of Coconut. Andhra Agric J 13:112–113
Rao PG, Reddy GS, Subbiah J (1956) A new disease of coconut in Andhra State. Indian Cocon J 9:215–222
Rethinam P, Rajamannar M, Narasimhachari CL (1989) Tatipaka disease of coconut in Andhra Pradesh. Indian Cocon J 20(1):1–4
Rhode W, Kullaya A, Mpunami A, Becker D (1993) Rapid and sensitive diagnosis of mycoplasma-like organisms associated with lethal disease of coconut palm by a specifically primed Polymerase chain reaction for the amplification of 16 SrDNA. Oleagineux 48:319–322
Romney DH (1972) Past studies on and present status of lethal yellowing disease of coconuts. Intl J Pest Manag 18:386–395
Romney DH (1983) Brief review of coconut lethal yellowing. Indian Cocon J 13:1–8
Romney DH, Harries HC (1978) Distribution and impact of lethal yellowing in the Caribbean. In: Proceedings of the 3rd meeting of the international council of lethal yellowing, University of Florida Agricultural Research Centre, Fort Lauderdale, 78:6–7
Russel WC, Newman C, Williamson DH (1975) A simple cytochemical technique for demonstration of DNA in cells infected with mycoplasmas and viruses. Nature 25:461–462
Sahasranaman KN, Pillai NG, Jayasankar NP, Potti VP, Varkey T, Kamalakshy Amma PG, Radha K (1983) Mixed farming in coconut gardens: economics and its effect on root (wilt) disease. In: Nayar NM (ed) Coconut research and development. Wiley Eastern, New Delhi, pp 160–165
Sasikala M, Mathen K, Govindankutty MP, Solomon JJ, Geetha L (1988) Transmission of a mycoplasma-like organism from *Cocos nucifera* with root(wilt) disease to *Catharanthus roseus* by *Cassytha filiformis*. Netherlands J Plant Pathol 94:191–194
Sasikala M, Amma BSK, Rajagopal V (1998) Diagnosis. In: Nampoothiri KUK, Koshy PK (eds) Coconut root (wilt) disease. CPCRI, Kasaragod, pp 45–51
Sasikala M, Chithra R, Solomon JJ, Rajeev G (2001) Development of DAC – indirect ELISA for the rapid detection of coconut root (wilt) disease. CORD 17:23–35
Sasikala M, Prakash VR, Ajithkumar R, Chandra Mohanan R (2004) Selection of root (wilt) disease free coconut elite mother palms using serological test for production of quality planting materials for disease-prevalent tracts. Indian Cocon J 35:16–19
Sasikala M, Rajeev G, Prakash VR, Amith S (2010) Modified protocol of ELISA for rapid detection of coconut root (wilt) disease. Central Plantation Crops Research Institute (CPCRI), Kasaragod, India. J Plantn Crops 38(1):16–19
Schneider B, Ahrens U, Kirkpatrick BC, Seemuller E (1993) Classification of plant-pathogenic mycoplasma- like organisms using restriction-site analysis of PCR- amplified 16s rDNA. J Gen Microbiol 139:519–527
Schuiling M, Kaiza DA, Mpunami A (1992a) Lethal disease of coconut palm in Tanzania 2- History, distribution, and epidemiology. Oleagineux 47:516–521
Schuiling M, Mpunami A, Kaiza DA, Harries HC (1992b) Lethal disease of coconut palm in Tanzania 3- Low resistance of imported germplasm. Oleagineux 47:693–697
Sears BB, Kirkpatrick BC (1994) Unveiling the evolutionary relationships of plant pathogenic mycoplasma like organisms. ASM News 60:307–312
Seymour CP, Miller JW, Roberts DA (1972) An outbreak of lethal yellowing of coconut palms in Miami, Florida. Phytopathol 62:788
Shanta P, Menon KPV (1959) The virus nature of wilt (root) disease of coconut – experimental transmission of it into cowpea (*Vigna sinensis* Endl.) In: Proceedings of conference, Coconut Research Workers India, pp 268–271
Sharples A (1928) Palm diseases in Malaya. Malay Agric J 16(9–10):313–360
Siriwardhana PHAP, Gunawardena BWA, Steve Millington SJ (2012) Detection of phytoplasma associated with Waligama Coconut Leaf Wilt Disease in Sri Lanka by loop mediated isothermal amplification assay performing alkaline polyethylene glycol based DNA extraction. Microbiol Biotech Res 2(5):712–716

Sitepu D, Mangindaan H, Mawikere J (1988) Outbreak of a coconut wilt disease in Central Kalimantan. In: UNDP/FAO integrated coconut pest control project, annual report 1988. Balai Penelitian Kelapa, Manado, North Sulawesi

Solomon JJ, Govindankutty MP, Nienhaus F (1983) Association of mycoplasma-like organisms with the coconut root (wilt) disease in India. Z Pjlkrankh Pjlschutz 90:295–297

Solomon JJ, Govindankutty MP, Mathen K (1987) Detection of mycoplasma like organism in root (wilt) disease affected coconut and its putative vector. In: Regional workshop on plant mycoplasma, New Delhi, India

Solomon JJ, Nair CPR, Srinivasan N, Gunasekaran M, Sasikala M (1999) Coconut root (wilt) – the malady and remedy. J Plantn Crops 27(2):71–92

Sosamma VK, Koshy PK (1991) Etiology-nematodes. In: Root (wilt) disease, Monograph series no. 3. Central Plantation Crops Research Institute, Kasaragod, pp 19–26

Srinivasan N (1991) Occurrence of coconut leaf rot in relation to root (wilt) disease. Indian Cocon J 21(10):14–18

Srinivasulu B, Nagamalleswari DY, Kumar KVK, Aruna K, Prasadji JK, Rao DVR (2006) Transmission electron microscopy confirms phytoplasmal etiology of Tatipaka disease of coconut. J Plantn Crops 34(3):500–501

Steiner KG (1976a) Epidemiology of Kainkope disease of coconut palms in Togo. Plant Dis Rep 60(7):613–617

Steiner KG (1976b) Remission of symptoms following Tetracycline treatment of coconut palms affected with Kaincope disease. Plant Dis Rep 60:617–620

Subbiah J, Rao PG (1963) Studies in coconut decline in Andhra. Andhra AgriC J 10:58–67

Summanwar AS, Raychaudhuri SP, Jagadish Chandra K, Namaprakash LSB (1969) Virus associated with coconut root (wilt) disease. Curr Sci 38:208–210

Tanne E, Boudon-Padieu E, Clair D, Davidovich M, Melamed S, Klein M (2001) Detection of phytoplasma by polymerase chain reaction of insect feeding medium and its use in determining vectoring ability. Phytopathol 91:741–746

Tsai JH (1980) Lethal yellowing of coconut palms – search for a vector. In: Harris K, Maramorosch K (eds) Vectors of plant pathogens. Academic, New York, pp 177–200

Tymon A, Jones P (1997) Comparative analysis of coconut phytoplasmas from East and West Africa. In: Eden-Green SJ, Ofori F (eds) International workshop on lethal yellowing-like disease of coconut, Elmina, Ghana, November 1995. Natural Resources Institute, Chatham, pp 197–203

Tymon AM, Jones P, Harrison NA (1998) Phylogenetic relationships of coconut phytoplasmas and the development of specific oligonucleotide PCR primers. Ann Appl Biol 132:437–452

Varghese MK (1934) Diseases of coconut palm, p 105

Varkey T, Davis TA (1960) Studies on coconut pollen with reference to the leaf and root (wilt) diseases. Indian Cocon J 14:1–7

Vázquez-Euán R, Harrison N, Narvaez M, Oropeza C (2011) Occurrence of a 16SrIV group phytoplasma not previously associated with palm species in Yucatan, Mexico. Plant Dis 95(3):256–262

Waidyarathne KP, Samarasinghe S (2014) Artificial neural networks to identify naturally existing disease severity status. Neural Comput Appl. https://doi.org/10.1007/s00521-014-1572-6

Warokka JS (1998) The impact and etiology of coconut wilt disease in Indonesia. In: Allorerung D, Harries HC, Jones P, Warokka JS (eds) Proceedings of the workshop on lethal diseases of coconut caused by phytoplasma and their importance in Southeast Asia. APCC, Jakarta, pp 27–35

Warokka JS, Jones P, Dickinson MJ (2006) Detection of phytoplasmas associated with Kalimantan wilt disease of coconut by the polymerase chain reaction. J Littri 12:154–160

Weerakkody WATL (2010) Impact of Weligama Coconut Leaf Wilt Disease (WCLWD) on morphological, physiological and yield aspects of affected coconut palms. A Project Report presented to the University of Peradeniya, Sri Lanka. dlib.pdn.lk/bitstream123456789/3147/Weerakody%202010.pdf

Weerakoon WMSSB, Yalegama LLWC, Fernandopulle MND (2011) Evaluation of copra quality and quantity of Weligama Wilt affected coconuts. In: Proceedings of 11th Agri Res Symp, pp 319–322

Wei W, Davis RE, Lee I-M, Zhao Y (2007) Computer-simulated RFLP analysis of 16S rRNA genes: identification of ten new phytoplasma groups. Intl J Syst Evol Microbiol 57:1855–1867

Wijesekara HTR, Perera L, Wickramananda IW, Herath I, Meegahakumbura MK, Fernando WBS, De Silva PHPR (2008) Preliminary investigation on Weligama coconut leaf Wilt disease: a new disease in Southern Sri Lanka. In: Nainanayake NPAD, Everard JMDT (eds) Proceedings of the second symposium on plantation crop research – Export competitiveness through quality improvements. Coconut Research Institute, Lunuwila

Wijesekara HTR, Perera AAFLK, Meegehekumbura MGMK, Dassanayaka EM, Ranasinghe C (2013) Serological and Molecular Techniques. In: Gunasena HPM, Gunathilaka HAJ, Fernando LCP, Everard JMDT, Appuhamy PAHN (eds) Weligama Coconut Leaf wilt disease six years after. Coconut Research Institute of Sri Lanka, Lunuwila, pp 56–70

Yankey EN, Swarbrick PJ, Nipah JO, Quaicoe RN, Dickinson MJ, Edizioni ETS (2014) Detection of the Cape St. Paul wilt phytoplasma in coconut palms in Ghana through the combination of visual symptoms assessment and molecular diagnosis using a secA gene based assay. J Plant Pathol 96(2):281–285

Zizumbo-Villarreal DP, Colunga-García Marín M, Fernández-Barrera N, Torres-Hernández, Oropeza C (2008) Mortality of Mexican coconut germplasm due to lethal yellowing. PGR Newsletter no, vol 156, pp 23–33

Dr. J. J. Solomon retired as Principal Scientist (Plant Pathology) and acting Head of ICAR-CPRI, Regional Station, Kayamkulam. His field of specialization is phytoplasmology, serology, virology and electron microscopy. He has published 75 research papers and contributed chapters in 6 books. Dr. Solomon is a Member of IBPGR/FAO and a FAO consultant. solomon.jeyasingh@gmail.com

Dr. Vinayaka Hegde is currently the Head, Division of Crop Protection, ICAR-CPCRI, Kasaragod, Kerala, India. He is a Plant Pathologist by qualification. He was awarded IARI Merit Medal for his outstanding research work and UAS Gold Medal. He has served in ICAR-CPRI and CTCRI. Dr. Hegde is presently working on diagnostics and management of diseases of coconut, areca nut and cocoa. He specialized in molecular diagnostics and management of plant diseases and published 55 research papers, authored 6 technical bulletins, 6 book chapters and 10 popular articles. hegdev64@gmail.com, vinayaka.hegde@icar.gov.in

Dr. Merin Babu is currently working as Scientist (Plant Pathology) at ICAR-CPCRI, Regional Station, Kayamkulam, Kerala, India. Her specialization is in Pathology. She has 9 years of research experience and has been working on diagnostics and management of phytoplasma diseases of coconut. She has published 17 research papers apart from 4 book chapters and 10 popular articles. merin.babu1@gmail.com

Dr. L. Geetha is a postgraduate in Botany and worked at ICAR-CPCRI, Regional Station, Kayamkulam, Kerala, India. Currently she holds the position of Research Assistant (Molecular Biology) at Khalifa Center for Genetic Engineering and Biotechnology, UAE University, UAE. Her research focuses on plant molecular techniques and electron microscopy. She has published over 14 research articles in peer-reviewed journals. geetha.lekshmi@uaeu.ac.ae

Chapter 12
Pest Dynamics and Suppression Strategies

A. Josephrajkumar, Chandrika Mohan, P. S. Prathibha, Rajkumar, T. Nalinakumari, and C. P. R. Nair

Abstract Dynamism of pests attacking coconut is ever increasing realizing action threshold many a times diminishing palm productivity and threatening livelihood security. Living with pests is, thus, gaining momentum for sustaining coconut yield and safeguarding environment considering the perennial and homestead nature of palms. Bionomics, geo-maps and bio-intensive area-wide management of key pests of coconut, viz. rhinoceros beetle, red palm weevil, black-headed caterpillar, eriophyid mite, white grubs and rodents, as well as emerging pests such as coreid bug, slug caterpillars, whiteflies including the invasive rugose spiralling whitefly, mealybugs, scale insects, ash weevils and nut borers are lucidly presented with all cutting-edge eco-friendly technologies to counter their menace. Storage pests of copra and techniques to unhinge are drawn around. Palm nematodes and potential invasive pests such as *Brontispa longissima*, *Aspidiotus rigidus* and *Wallacea jarawa* are underscored with strategies on emergency preparedness and incursion management. Biorational approaches and health management of palms are greatly emphasized to suppress pests in the most economical and convincing manner. Good agricultural practices, agroecosystem-based stimulo-deterrent strategies, semiochemical tools and nanoporous delivery mechanism of pheromones, diversity farming methods and farmer-participatory community approaches including farmer field school mode to empower farmers with all innovative pest management actions towards harnessing sustainable palm yield in all coconut-growing countries are accentuated.

A. Josephrajkumar (✉) · Chandrika Mohan · C. P. R. Nair
ICAR-CPCRI, Regional Station, Kayamkulam, Kerala, India
e-mail: joecpcri@gmail.com; cmcpcri@gmail.com; Cprnair_47@yahoo.com

P. S. Prathibha · Rajkumar
ICAR-CPCRI, Kasaragod, Kerala, India
e-mail: prathibhaspillai007@gmail.com

T. Nalinakumari
Kerala Agricultural University, Thiruvananthapuram, Kerala, India
e-mail: achutnk@yahoo.co.uk

K. U. K. Nampoothiri et al. (eds.), *The Coconut Palm (*Cocos nucifera *L.) - Research and Development Perspectives*,
https://doi.org/10.1007/978-981-13-2754-4_12

12.1 Introduction

Coconut is depredated by at least 830 insect and mite species, 173 fungi and 78 species of nematodes drastically affecting the productivity (CPCRI 1979). In the monograph of coconut pests, Lever (1969) listed 110 pests including 2 species of mite. Damage by insect, mite and vertebrate pests induces a crop loss as high as 30% in palms (Gitau et al. 2009). The annual crop loss due to pest complex was estimated as 618.50 million nuts in Kerala, a predominant coconut-growing state in India (Abraham 1994). It is practically impossible to eliminate insect pests from palms and, therefore, must be 'lived with' so as to incur minimum economic damage as far as possible (Howard et al. 2001). The major pests infesting coconut would be classified as (1) borers, (2) defoliators, (3) sap feeders, (4) subterranean pests, (5) mammalian pests, (6) emerging pests, (7) nematodes, (8) potential invasive pests and (9) storage pests.

12.2 Borers

12.2.1 Asiatic Rhinoceros Beetle, Oryctes rhinoceros *Linn. (Coleoptera: Scarabaeidae)*

12.2.1.1 Distribution

O. rhinoceros is a cosmopolitan pest endemic to all coconut-growing countries. It was first reported to be damaging coconut palms in 1889 from Malaysia (Ridley 1919) and subsequently described in India by Pillai (1919).

12.2.1.2 Crop Loss

A loss in yield of 5.5–9.1% due to beetle attack was estimated in Kerala, India (Ramachandran et al. 1963). Nair (1986) reported damage to spathe causing reduction in coconut yield up to 10% in India. During 1969, the crop loss in South Pacific countries alone was estimated as at least $US 1,100,000 due to *O. rhinoceros* damage (Catley 1969). Stride (1977) found leaf and central crown damage levels of 40–60%, 60–80% and 80–100% resulting in nut losses to the tune of 12, 17 and 23%, respectively. Of late, incidence of rhinoceros beetle is very alarming in Samoa and Solomon Islands causing crop loss of about $US 10 million.

12.2.1.3 Host Plants

Besides coconut, the pest attacks palmyrah palm (*Borassus flabellifer*), toddy palm (*Phoenix sylvestris*), oil palm (*Elaeis guineensis*), date palm (*Phoenix dactylifera*), areca palm (*Areca catechu*) and other palms. Occasionally, it was also found feeding on agave, sugarcane, pineapple, tree fern, banana and taro (Menon and Pandalai 1960).

12.2.1.4 Taxonomy

The genus *Oryctes* has about 42 species, out of which 15 species are pests of coconut palm (Gressitt 1953). The predominant species in East Africa are *Oryctes monoceros* Oliv., *Oryctes gigas* Cast., *Oryctes boas*, Fa. and *Oryctes owariensis* P.de B. In Mauritius and Reunion, *Oryctes chevrolati* Geu. and *Oryctes tarandus* Ol. are the pests of coconut. *Oryctes latecavatus* Fair., *Oryctes pyrrhus* Burm., *Oryctes ranavalo* Coq. and *Oryctes simiar* are reported from Madagascar, and *Oryctes trituberculatus* Lamb. causes damage to coconut palm in Malaya. *Oryctes australis* was reported as a pest on juvenile palms in Papua New Guinea (Menon and Pandalai 1960; Lever 1969).

12.2.1.5 Symptoms

The robust adult beetles cause damage to palms of all age groups by boring into the unopened spear leaves and spathes and chewing off the soft internal tissues after imbibing the juice. As the pest bores deeper into the host, it pushes out the chewed up tissues as fibres, which are seen extruding from the entry points. Once these injured spindles open up, the green leaves present a geometric 'V'-shaped cut pattern (Fig. 12.1). The damage to inflorescence is seen as round to oblong holes on the spathes which soon dry up resulting in complete loss of nuts in the affected bunch. Attack on juvenile palms results in stunted growth and delayed flowering (Ghosh 1911; Pillai 1919; Nirula 1955b; Bedford 1980; Howard et al. 2001; Rajan et al. 2009).

Of late, entry of rhinoceros beetle through collar region of the freshly transplanted coconut seedlings and eating away the growing spear leaf resulting in wilting as well as improper establishment of seedlings was observed. In certain cases, the spear leaf gets twisted resulting in elephant tusk-like symptom, and farmers uproot such deformed seedlings/juvenile palms as the growing point remains stunted. Recurrent attack on the same seedling is often noticed due to the emanation of odour in the healing process of the injured portions (Josephrajkumar et al. 2015).

O. rhinoceros infestation causes another serious damage by providing egg-laying sites for the lethal pest, viz. red palm weevil (*Rhynchophorus ferrugineus*), and for

Fig. 12.1 Symptom of rhinoceros beetle attack. (Photo: Josephrajkumar)

entry of fungal pathogens like *Phytophthora* sp. (Nirula 1955b; Menon and Pandalai 1960; Bedford 1980; Rajan et al. 2009).

12.2.1.6 Bioecology

Detailed investigations on bioecology and distribution were conducted by Nirula (1955a, b). The adult is a stout black beetle 35–50 mm in length and 14–21 mm in breadth. It has an erect, slightly tapering cephalic horn, which is longer in males. The pygidium has reddish brown hairs in the female. Antennae are lamellate possessing short and strongly toothed mandible. The beetle breeds in the decaying organic matter like cattle dung, farmyard manure, compost heaps, coconut stumps, dead and decaying coconut logs, etc., where the adults lay eggs and complete the larval and pupal stages. Antony and Kurian (1975) observed that in cattle dung pits, the pest breeds in a temperature range of 10–50 °C and 30–60% moisture. The range of breeding sites utilized by *O. rhinoceros* underlies the versatility in breeding and survival dynamics (Menon and Pandalai 1960; Bedford 2013).

Eggs are white and globular in shape with fairly hard shell. Incubation period prolongs for 10 days in west coast of India (Nirula 1955b). Larvae are creamy white in colour with dorsally arched body (C-shaped) with well-developed head capsule, mandibles and thoracic legs. Fully grown grubs are stout and fleshy and may grow up to 60–100 mm length (Wood 1968; Ooi 1988). Nine pairs of spiracles are present on lateral side of the body. Larval period is about 130 days with three instars. The pupal period varies from 20 to 29 days and uniformly brown in colour measuring 50–70 mm lengthwise and 20–25 mm breadthwise (Menon and Pandalai 1960). Life stages of rhinoceros beetle are given in Fig. 12.2.

Adult longevity is 3–6 months. Adults are active during night and remain hidden during daytime in the feeding sites or breeding sites where mating occurs. Multiple mating may occur, but are not essential as spermatozoa remain viable for about

Fig. 12.2 Life stages of rhinoceros beetle. (Photo: Josephrajkumar)

6 months in the spermatheca (Hurpin and Fresneau 1973). Average fecundity per female is 108 eggs. It has been estimated that a single pair of beetle under optimum condition can reproduce 15 million beetles in a span of 3 years (Nirula 1955b). Pest occurs throughout the year with peak adult emergence during June to September. Population of the beetle was observed to be high in young plantations with multiple breeding sites and high rainfall and humidity.

12.2.1.7 Integrated Pest Management

Since the pest is an active flyer, integrated pest management (IPM) strategies adopted on a community basis are essential to bring an effective control (Nair et al. 1998). The major components of IPM package consist of sanitation, ecological engineering, mechanical, chemical, biological and semiochemical methods.

Sanitation Method The dead and decaying coconut logs and other organic debris in the vicinity of coconut plantations are to be properly disposed off, since this act as prolific breeding grounds of the beetle. Never mulch the seedlings with coconut leaves during initial phase of establishment, and planting two turmeric rhizomes on either side of the seedlings was found to reduce pest attack. Planting at correct spacing (Tall and hybrids at 7.5 m × 7.5 m; Dwarfs 7 m × 7 m) with adequate light is very essential to reduce pest attack (Rajan et al. 2009; Josephrajkumar et al. 2015).

Ecological Engineering Intercropping of palms with nutmeg, rambutan, curry leaf and banana along the interspaces disorients the pest away from the source due to crop-habitat diversification induced pest-repulsion cues. Crop heterogeneity is, therefore, preferred for pest regression, income generation as well as continuous employment of farmers. Installation of bird perch and flowering plants like coral vines (*Antigonon leptopus*) maintains pest defenders and executes ecosystem services. Damage by rhinoceros beetle was reduced in coconut-based cropping garden than in monocropped garden. A greater environmental heterogeneity, high species diversity and less host density favour less rhinoceros beetle attack on palms (Josephrajkumar et al. 2014b; Josephrajkumar and Krishnakumar 2016).

Mechanical Method This method involves periodic and systematic examination of the palm crown and removing the adult beetle by means of a metal hook during peak periods of pest abundance (June to September). In order to suppress *O. rhinoceros*, all breeding places should be rendered innocuous, or the grubs should be collected regularly (Menon and Pandalai 1960; Rajan et al. 2009).

Prophylactic Method Application of powdered oil cakes of neem (*Azadirachta indica* A. Juss.), or marotti (*Hydnocarpus wightiana* Bl.) or pongamia cake (*Pongamia pinnata* Linn.) at 250 g mixed with equal volume of sand, into the topmost three leaf axils around the spindle leaf thrice a year during May, September and December is recommended as a prophylactic measure against rhinoceros beetle and red palm weevil (Chandrika Mohan et al. 2000; Josephrajkumar et al. 2014b). Placing three or four naphthalene balls in the leaf axil at the base of spindle leaf at 12g palm^{-1} and covering them with sand to prevent quick evaporation provide good protection against the pest for 45–60 days (Gurmit Singh 1987; Sadakathulla and Ramachandran 1990).

Placement of two perforated sachets containing chlorantraniliprole (3 g) or fipronil (3 g) was found effective in monsoon phase for successful seedling establishment and warding off rhinoceros beetle attack. During dry period, 100 ml of water may be poured over the sachet after placement to release the molecule. Placement of two botanical cakes made out of extracts from *Clerodendron infortunatum* Linn. and *Chromolaena odorata* Linn. (each tablet weighing 1.9 g, 2.5 cm diameter and 4.0 mm thickness) on the topmost leaf axils reduced 54% leaf damage and was found superior than chlorantraniliprole sachets (34%). A paste based on botanical extracts/oil was developed and swiped over the spindle and adjoining petioles at 10 g to safeguard juvenile palms for about 3 months from rhinoceros beetle attack (CPCRI 2016).

Biological Control This method accomplishes long-term and sustainable rhinoceros management. Pigs, rodents, fowls, frogs and toads are generalist predators.

Nirula (1955b) reported carabid, hysterid and elaterid beetles and their larvae in the breeding sites of *O. rhinoceros*, but none of these have any appreciable effect on the population of the grub. Insect predators are observed in the natural breeding grounds of the beetle, which feed on the eggs and early-instar grubs of the beetle. Most important predators are *Santalus parallelus* Payk., *Pheropsophus occipitalis* Macleay, *Pheropsophus lissoderus*, *Chelisoches morio* (Fab.) and species of *Scarites*, *Harpalus* and *Agrypnus* (Kurian et al. 1983). As these predators bio-suppress the pest population, their conservation is essential.

Two potential microbial agents, viz. green muscardine fungus (*Metarhizium anisopliae*) and *Oryctes rhinoceros* nudivirus (*Or*NV), cause diseases to the immature and adult stages. Their use is advantageous as these are relatively host specific, eco-friendly and compatible with other management tactics.

Green Muscardine Fungus M. anisopliae (Metchnikoff) Sorokin is an entomopathogenic fungus which kills the pest in conditions of low temperature and high humidity (Antony and Kurian 1975). The susceptibility of *O. rhinoceros* to the fungus was first reported in Western Samoa in 1913 and in India by Nirula et al. (1955a, b). *M. anisopliae* var. *major* (spore size 10–14 mm) is a highly virulent strain widely used for the control of the pest. The fungus could be mass multiplied using cheaper substrates in both solid and liquid media, and the spores could be harvested and treated in the breeding site at 5×10^{11} spores m^{-3} (Mohan and Pillai 1982; Danger et al. 1991). The infective potential of a strain may become reduced by culturing on media but is restored considerably following infection of a host. In vermicomposting sites, treatments with *M. anisopliae* spores killed all third-instar larvae, with the highest dose giving the fastest kill, taking 8 days when favoured by high humidity (Gopal et al. 2006). It was observed in Malaysia that without spore applications, 2-year-old shredded rotting coconut trunk debris, with 0–2% *M. anisopliae* infections and cover crop over growth, could contain substantial numbers of larvae (Tey and Ho 1995).

Currently, *M. anisopliae* is multiplied on semi-cooked rice medium for field application in organic manure as well as vermicompost pits at 100 g m^{-3}. This technology through farmer-participatory and women group approach has created a great impact on the long-term bio-suppression of the pest in community mode. Mass production technology of *M. anisopliae* was standardized in semi-cooked rice grains yielding a spore count of 3×10^7 cfu g^{-1} of the culture. For liquid broth culture of *M. anisopliae*, carbon source dextrose could be replaced with less expensive jaggery. Tapioca jaggery broth and coconut water jaggery broth were found highly suitable for the liquid broth culture of *M. anisopliae*.

Oryctes rhinoceros Nudivirus (OrNV) *Or*NV was first reported from Malaysia by Huger (1966) and named as rhabdion virus of *O. rhinoceros*. The virus multiplies in the midgut and fat body of grubs and in the midgut of adults and becomes ‘flying virus factories’ aiding dispersal while defecating. Grubs die from the infection, and their cadavers release emerging virus into the breeding sites, where adults are infected by ingesting it there or during mating. The adult life span is shortened, and females cease oviposition (Bedford 2013). *Or*NV is very effective and kills the grub in 15–20 days time significantly suppressing the longevity and fecundity of adult beetles. Previously known as baculovirus, *Or*NV is presently grouped under nudivirus category with a rod-shaped virion as infective agent (Burand 1998). *Or*NV prepared from the infected *O. rhinoceros* adults and stored in sterile vials remain infective for 2 weeks to 3 months at 28 °C or for 20 weeks to 1 year at 4 °C (Zelazny et al. 1987). In India, *Or*NV-infected grubs after developing the symptoms could be stored in the deep freezer at −40 °C indefinitely retaining its virulence (Rajan et al. 2009).

During 1967, the Malaysian isolate of *Or*NV was introduced into *O. rhinoceros* population in many South Pacific Islands. Due to the natural spread and establishment of the disease, the pest population got reduced tremendously. Pest suppression continued for several years in these regions (Bedford 1980). *Or*NV was used to control *O. monoceros* in the Islands of Seychelles (Lomer 1986) and Tanzania (Purrini 1989). Similarly, release of *Or*NV-infected beetle led to the bio-suppression of *O. rhinoceros* in Papua New Guinea and Mauritius (Monty 1978; Gorick 1980).

Diagnosis of OrNV Giemsa stain smear technique was found very simple and convenient to detect hypertrophied nuclei of infected adult midgut cells in India (Mohan et al. 1983) and Papua New Guinea (Gorick 1980). The Giemsa method was supplemented by ELISA (Enzyme-Linked ImmunoSorbent Assay) techniques (Longworth and Carey 1980) and then followed by PCR techniques for detecting *Or*NV DNA in midgut tissues from adults or larvae (Ramle et al. 2001).

Zelazny (1981) reported the occurrence of *Or*NV in India and elaborate work was undertaken by Mohan et al. (1983). Studies conducted at ICAR-CPCRI indicated that *Or*NV-infected grubs become less active and stop active feeding. As a result of virus multiplication in the midgut epithelium, fat body disintegrates and haemolymph content increases. This causes translucency in the abdominal region which is an important exopathological symptom of the *Or*NV infection and mortality in grubs. The infected beetles disseminate the virus through faecal matter into the surroundings after 3–9 days of inoculation at 0.3 mg virus adult^{-1} day^{-1} (Mohan et al. 1983; Rajan et al. 2009).

Field Dissemination The simplest and best practical method of dissemination of *Or*NV is by releasing the infected adult beetles in the field. Healthy adult beetles are allowed to crawl on the viral inoculums at 1g infected midgut 100 ml^{-1} of buffer, for half an hour. The beetles are then kept under starvation for 12–24 h. The beetles are released preferably at dusk in the infested coconut gardens at 12–15 beetles ha^{-1} (Mohan et al. 1985; Rajan et al. 2009). Horizontal spread of *Or*NV was reported as 1 km month^{-1} (Jacob 1996). Introduction of *Or*NV in Minicoy and Androth Islands of Lakshadweep (Mohan et al. 1989; Mohan and Pillai 1993), Chittilappilly,

Thrissur, Kerala (Biju et al. 1995) and Sipighat, Andaman Island (Jacob 1996) successfully reduced the population of *O. rhinoceros* and its damage on coconut.

Insect Growth Regulatory Effect Restricting and managing the breeding sites could check the proliferation of the pest. The weed plant, *Clerodendron infortunatum* Linn. (Verbenaceae), is very effective in controlling the pest buildup in the breeding sites. Alkaloids of *C. infortunatum* have got a juvenile hormone analogous activity on *O. rhinoceros*. Formation of larval-pupal mosaic, pupal-adult intermediates or adultoids (adults with malformed wings) is some of the common abnormalities elicited. These abnormal adults were unable to fly and survived for only 6–8 days against the longevity of 2–3 months for healthy adults (Chandrika Mohan and Nair 2000a).

Attractants An effective trapping method with rotting castor cake slurry kept in mud pots has been developed to attract rhinoceros beetle (Rajamanickam et al. 1992). Hallett et al. (1995) reported the potential of using ethyl 4-methyl octanoate (E4MO), a male-produced aggregation pheromone, to attract and control rhinoceros beetle. Trapping of adult beetles in a PVC trap using the pheromone 'Oryctalure (E4MO)' is a recent innovative method in the IPM for rhinoceros beetle. The traps are set up in the gardens, and beetles trapped inside are collected periodically. These beetles can be used for field release after inoculation with *Or*NV (Rajan et al. 2009). The effectiveness of traps as management tools may well change with population size and only a portion of the population in the correct physiological condition at that time may go to traps, as reported in trap catches for *O. monoceros* (Allou et al. 2008).

12.2.1.8 Area-Wide Demonstration

In a large-scale demonstration of rhinoceros beetle control by *M. anisopliae* as well as *Or*NV conducted at Kerala, India, in 2400 ha, it was found that *O. rhinoceros* damage on spear leaf and spathe could be reduced by 95.8% and 62.5%, respectively (Nair et al. 2008). Area-wide technology adoption facilitated by ICAR-CPCRI covering 1500 ha in Alappuzha district indicated 76–85% reduction in leaf damage by rhinoceros beetle over the pretreatment period. Palm infestation by rhinoceros beetle was reduced to 54.14% from the initial level of 82.35% in a period of 3 years. Coconut yield increased to 13.1% nuts palm^{-1} year^{-1} due to these interventions. The Farmer Field School (FFS) was found to be an ideal method for technology transfer in coconut health management system (CPCRI 2016), and the average knowledge score on pest management of those farmers attending FFS was 51.69 compared to 32.80 in case of non-FFS farmers.

12.2.2 *Red Palm Weevil,* Rhynchophorus ferrugineus *Olivier (Coleoptera: Curculionidae)*

12.2.2.1 Distribution

Red palm weevil (RPW), *Rhynchophorus ferrugineus* Oliv, is a concealed and lethal pest of coconut palm. First information about the pest in India was published in 1891 in the *Indian Museum Notes*. Lefroy (1906) reported RPW as a local sporadic major pest of coconut and other palms throughout India. RPW was reported as a pest of coconut from Sri Lanka (Green 1906), Indonesia (Leefmans 1920), Burma (Ghosh 1923) and the Philippines (Copeland 1931). Although Buxton (1920) reported this pest on date palm from Mesopotamia, it was during the mid-1980s in the Middle East that the pest attained a major pest status (Abraham et al. 1998). Subsequently, RPW moved from North Africa into Europe, and it was reported in the South of Spain for the first time (Cox 1993). In Malaya, red stripe weevil, *Rhynchophorus schach* Olivier, was reported, and its biology worked out (Corbett 1932). *Rhynchophorus palmarum* is reported from coconut palms in the western tropics.

12.2.2.2 Crop Loss

It is estimated that 0.5% of the palms was attacked by the pest every year in erstwhile Travancore-Cochin, India, alone (Menon and Pandalai 1960). Due to the lethal nature of the pest and high value of the crops involved, the assumed action threshold for RPW in coconut and date palm is very low. In big plantations, 1% infested palm is the assumed action threshold.

12.2.2.3 Host Plants

Besides coconut, RPW attacks 26 species of palms (Menon and Pandalai 1960; Esteban-Duran et al. 1998; Malumphy and Moran 2009).

12.2.2.4 Taxonomy

The genus *Rhynchophorus* contains ten species, of which seven, including *R. ferrugineus* and *R. vulneratus*, are known to attack palms (Booth et al. 1990). A key to related genera and the revision of this species was provided by Wattanapongsiri (1966). Reginald (1973) suggested that *R. ferrugineus* is the typical *Rhynchophorus* species occurring worldwide. The species attacking palms are *Rhynchophorus ferrugineus*, *R. bilineatus*, *R. quadrangulus*, *R. palmarum*, *R. lobatus*, *R. distinctus* and *R. ritcheri*. Through molecular-genetic data, *R. vulneratus* was found synonymous to *R. ferrugineus* (Hallett et al. 2004).

12.2.2.5 Symptoms

RPW is reported to attack juvenile palms mostly aged below 20 years (Nirula 1956a; Abraham et al. 1998). Being an internal feeder, it is difficult to diagnose the pest infestation during early stages. It is mostly noticed when the palm has been fatally infested and is irrecoverable (Ghosh 1911; Menon and Pandalai 1960). Mechanical injury on palms; damage by rhinoceros beetle, as well as infection by leaf rot; and bud rot diseases, inappropriate crop geometry, etc. predispose RPW females for egg laying (Abraham and Kurian 1975; Abraham et al. 1998; Josephrajkumar et al. 2014a). According to Abraham et al. (1998) and Rajan et al. (2009), damaged palms exhibit one of the following symptoms depending on the stage of attack, yellowing and later wilting of the inner and middle whorl of leaves, small circular holes as well as tunnels on the palm trunk with oozing out of a brown viscous fluid, emanation of fermented odour around the infested tunnel, longitudinal splitting of leaf base with solidified gums, gnawing sound of grubs and presence of cocoon or chewed up fibres at leaf axil or palm base, ultimately resulting in the toppling of the crown (Fig. 12.3).

Entry of the pest through the crown is the most common and most fatal type of infestation. The grubs in such cases stay very close to the cabbage portion (growing point) of the palm, and this results in drying of the young heart leaf. In seedlings and younger palms, entry of the pest through the bole region is generally noticed. Shallow planting and injuries to the soft stem due to mechanized cultivation practices are the major reasons for this type of pest entry. In Dwarfs, entry is noticed through leaf axil also, if injured usually due to heavy bearing bunches.

Fig. 12.3 Toppling of crown of coconut palm due to attack of red palm weevil. (Photo: Josephrajkumar)

12.2.2.6 Bioecology

Life history of RPW has been well studied and documented in many countries including India, Indonesia, Myanmar, the Philippines, Iran and Spain (Ghosh 1911; Menon and Pandalai 1960; Esteban-Duran et al. 1998; Murphy and Briscoe 1999). Several overlapping generations comprising of different stages of the insect could be seen in the infested palm. The adult weevils measure 35 mm long and 12 mm wide and is ferruginous brown in colour. The snout is elongated in both sexes, and the dorsal apical half of the rostrum in males is covered with a patch of short tuft of brown hairs. The snout of female is bare, relatively slender and a little longer than the males. Oviposition by female weevils is confined to the softer tissues of the palms in holes made with its rostrum (Nirula 1956a). The weevils are crepuscular and, according to Leefmans (1920), are capable of long flights and can locate their hosts in widely separated areas. The weevils after emergence from the pupal case will remain inside the cocoon for about a week making them sexually mature during the period of inactivity (Hutson 1933). A single female lays 58–131 eggs (creamish white, oval and broader at one end) and emerges out as whitish-yellow grubs that could sustain for a period of 25–105 days (Faleiro 2006). Computer-aided flight mill studies to analyse the flying ability of *R. ferrugineus* revealed that 54% adult weevils are short-distance flyers (covering <100 m), 36% are medium-distance flyers (covering 100–5000 m) and 10% are categorized as long-distance (>5000 m) flyers (Ávalos et al. 2014).

The fully grown apodous grubs are conical shaped, bulged in the middle and pointed towards both the ends. The body is 13 segmented with the brownish head pointed downwards. Mouth parts are well chitinized enabling the grubs to burrow the hard woody portions. Grubs chew the palm tissue and move interior leaving behind the chewed up frass emanating foul odour (Menon and Pandalai 1960; Faleiro 2006). Viado and Bigornia (1949) reported nine larval instars for RPW, while El-Shafie et al. (2013) reported eight larval instars sustaining for a period of 25–105 days. The life stage of RPW is depicted in Fig. 12.4.

The grubs construct an oval fibrous cocoon strong wound and are arranged spirally. Interior of the cocoon is smooth and plastered to house the exarate pupa with prominent antenna and eyes. The pupal period ranges from 11 to 45 days (Menon and Pandalai 1960; Faleiro 2006).

Rahalker et al. (1978) developed an artificial diet for rearing RPW based on sugarcane bagasse, while El-Sebay et al. (2003) developed a diet based on potato and carrot. Recently, a meridic diet consisting of agar, distilled water, commercial yeast as well as laboratory produced amino and fatty acid rich brewer's yeast (*Saccharomyces cerevisiae*), wheat meal, corn flour, benzoic acid, ascorbic acid, sorbic acid, vitamin mix and tetracycline hydrochloride was developed for the successful rearing of red palm weevil (El-Shafie et al. 2013).

Fig. 12.4 Life stages of red palm weevil. (Photo: Josephrajkumar)

12.2.2.7 Integrated Pest Management

Prevention of pest entry into the palm is the major step to be adopted in the IPM package. Maintenance of field sanitation by removal, splitting and burning of dead palms, which harbour various stages of the pest, helps a lot in reducing fresh incidence and hence should be considered as an important step in IPM. A systematic intervention of farm and palm hygiene is very critical for pest avoidance.

Early Detection Davis (1964) first developed an electronic detector which consisted of an amplifier, a probe and an earphone which was subsequently modified as built-in speaker for the detection of RPW. Power is supplied by a 3 volt dry battery cells to the amplifier. The probe lead has to be plugged into the amplifier and the needle pressed on the palm trunk to detect the gnawing sound of the grub. Abraham et al. (1966) studied methods of entry of the weevil into the palm, the important symptoms and the detection of the pest infestation by an electronic amplifier. Ramachandran and team developed a prototype detector which could not specifically pinpoint the presence of grubs due to extraneous noise factor. A prototype acoustic detection was developed with the technical support of CDAC, Thiruvananthapuram for early detection of the pest, which is to be refined (CPCRI 2014).

Cultural Control Maintaining optimum palm density is very important not only for harnessing the highest benefits of light energy but also for reducing the release of volatiles orienting the pest towards the host. Intercrops admix and diminish the volatile cues disorienting RPW away from host (Josephrajkumar et al. 2014a). Shallow planting incites more damage (Rajan et al. 2009). Systematic diagnosis through close monitoring and vigilant scouting is the key for early diagnosis. Avoiding physical injury to palms is very critical to reduce pest incidence. Cutting fronds leaving at least 1.2 m from trunk, evading knife injury on crown region and avoiding injury to bole and frond need to be overemphasized (Abraham et al. 1989; Rajan et al. 2009; Josephrajkumar et al. 2014a).

Sterile Male Release Technique Control of *R. ferrugineus* using radiosterilized males (1.5 Krad for 1–2 days) was undertaken by Rahalkar et al. (1973). A total of over 5000 radiosterilized males were released in a 320 ha young coconut plantation in Kerala, India, with limited success due to the fact that the floating female weevils were already mated inside the infested palm. Ramachandran (1998) studied biotypical variability among four populations of red palm weevil from different parts of India and found populations are genetically different and strainal variability exists between them. The weevil has a very high fitness due to high production potential and the absence of effective parasites, predators and pathogens. In addition, gamma radiation also did not have significant effect on the F_2 generation (Ramachandran 1991).

Biological Suppression Murphy and Briscoe (1999) outlined the prospects for biological suppression as a component of RPW-IPM programme. More than 50 natural enemies have been reported to attack *Rhynchophorus* species (Mazza et al. 2014). The highly potent cytoplasmic polyhedrosis virus (CPV), recorded first in India, infected all stages of RPW and could reduce insect population (Gopinadhan et al. 1990). Dangar and Banerjee (1993) discovered bacterial isolates belonging to *Bacillus* sp., *Serratia* sp. and the coryneform group in larvae and adults in India. Alfazariy (2004) reported successful control of RPW in laboratory conditions by infection with *B. thuringiensis* subspecies *kurstaki* isolated from larvae in Egypt. *Pseudomonas aeruginosa* (Schroeter) isolated from infected larvae collected in Kerala, India, induced mortality in early-instar grubs (Banerjee and Dangar 1995). Salama et al. (2004) isolated three potent spore-forming bacilli, as variants of *B. sphaericus*, *Bacillus megaterium* de Bary and *Bacillus laterosporus* Laubach, which caused larval mortality ranging between 40% and 60%. *B. sphaericus* which produced spherical endospores and crystalline endotoxins was the most virulent pathogen responsible for the larval mortality.

Besides phoretic mites, some ectoparasitic podapolid mites associated with *Rhynchophorus* species, such as *Rhynchopolipus rhynchophori* (Ewing), were used as a successful biocontrol agent against RPW (Abdullah 2009). Natural insect enemies of *Rhynchophorus* include several species belonging to the orders Dermaptera, Heteroptera, Coleoptera, Diptera and Hymenoptera (Murphy and Briscoe 1999). Among earwigs, *Chelisoches morio* (Fabricius) was reported as a common predator of RPW eggs and larvae in the canopy of coconut plantations in India (Abraham and Kurian 1973), and *Euborellia annulipes* (Lucas) was found in RPW-infested palms in Sicily (Massa and Lo Verde 2008). Among Hymenoptera, *Scolia erratica* Smith was found in Malaysia as a parasitoid of RPW (Burkill 1917). Krishnakumar and Sudha (2002) observed Indian tree pie bird, *Dendrocitta vagabunda parvula* as an effective predator on RPW adults.

Beauveria bassiana (Balsamo) Vuillemin and *M. anisopliae* (Metchnikoff) Sorokin are two of the most commonly studied species of entomopathogenic fungi. *B. bassiana* isolated from the United Arab Emirates, Italy and Spain, though found effective against RPW grubs and weevils, a commercial formulation of *B. bassiana*, had little effect on RPW (Abdel-Samad et al. 2011). Francardi et al. (2012) showed that *M. anisopliae* isolated from RPW in Italy had the highest efficacy against RPW larvae and adults with high efficacy on dry spores. Cito et al. (2014) reported the first recovery of *Metarhizium pingshaense* associated to RPW in Vietnam killed RPW mainly due to efficient protease activity and toxin production.

Banu et al. (2003) reported the infectivity of red palm weevil grubs and adults to entomopathogenic nematodes (*Heterorhabditis indica*, *Steinernema glaseri* and a local isolate *Steinernenam* sp.). Encouraging results were obtained using *Steinernema* sp. isolated from RPW pupae and adults in Egypt, when tested against field population (Shamseldean and Atwa 2004). The use of chitosan as adjuvant could protect *S. carpocapsae* from environmental conditions increasing the length of efficacy period (Llacer et al. 2009). Josephrajkumar et al. (2014c) reported higher virulence of local entomopathogenic nematode (EPN) strain of *H. indica* (LC_{50} 355.5 IJ) in the suppression of *R. ferrugineus* grubs as well as greater susceptibility (82.5%) of prepupal stage than that of grubs. Synergistic interaction of *H. indica* (1500 IJ) with imidacloprid (0.002%) against red palm weevil grubs indicated combined application of *H. indica*-infected *Galleria mellonella* cadavers and imidacloprid (0.002%) would be an effective strategy in the field-level management of RPW in coconut.

Attractants Henry (1917) suggested, for the first time, that fermenting kitul palm *Caryota urens* wood could be effective in trapping RPW adults. However, it was found that split coconut logs smeared with fresh toddy were more effective in trapping RPW in India (Abraham and Kurian 1975). Kurian et al. (1984) investigated several attractants for red palm weevil management. Log traps consisting of tender coconut stems, 50 cm long and spilt longitudinally, were treated with macerated fruits, molasses (jaggery from sugarcane), acetic acid, yeast and toddy, singly or in combination, and it was found that logs treated with coconut toddy + yeast + acetic acid attracted more weevils which could therefore be used in the management of RPW.

Hallett et al. (1993) identified and synthesized the ferrugineol (4-methyl-5-nonanol) from male-produced aggregation pheromone, and subsequent experiments confirmed that the addition of a second component (4-methyl-5-nonanone) in small amounts enhanced the capture rate of weevils by 65% (Abozuhairah et al. 1996). For operational ease in servicing food bait as well as trapping efficiency, bucket traps were used for trapping RPW (Oehlschlager et al. 1993). Nair et al. (2000a) found that the plantains and sugarcane were equally effective as food baits in enhancing the catch of weevils in pheromone trap and good bait relies on the high sugar content. Effective trapping of the weevils could be accomplished in the field when traps were set up with pheromone and food bait together (Nair and Nair 2003). Mayilvaganan et al. (2003) had assayed locally synthesized ferrugineol dispensed through capillary vials for trapping *R. ferrugineus* and found enhanced catches of female weevils. RPW was successfully brought down within 2 years in Israel by mass trapping the pest at 10 traps ha^{-1} in infested plantations; however, monitoring its activity could be accomplished at one trap in 3 ha (Soroker et al. 2005). Thomas et al. (2014) reported refinement in aggregation pheromone (4-methyl-5-nonanol and 4-methyl-5-nonanone 9:1) in tandem with food baits to attract RPW. Though effective, lures need to be replaced once in 3- to 4-month interval. Use of nanoporous materials as a novel carrier for loading the pheromone and kairomone of RPW trapped higher number of weevils than the commercial lure and sustained for more than 6 months. Mostly the trapped female weevils in pheromone traps are young, fertile and gravid; however, the infestation rate of palms around the trap is higher in certain agroecosystem which further requires fine-tuning in field delivery of pheromones and the field placement of traps. A pull-push strategy of moving weevils away from the field and orienting towards the trap is the need of the hour.

Trypsin Inhibitor Josephrajkumar et al. (2016a) highlighted suppression of growth and endopeptidases of red palm weevil using proteinase inhibitors. Serine protease inhibitors, viz. aprotinin (50 μg), soybean trypsin inhibitor (50 μg) and phenyl methyl sulphonyl fluoride (1700 μg), inhibited the gut proteinases of *R. ferrugineus* such as trypsin, elastase-like chymotrypsin and leucine aminopeptidase. Digestion and nutrient uptake of the insect are affected leading to impaired growth and development.

Chemical Control Several phased-out chlorinated hydrocarbons have been found effective in the control of RPW which are not described here. Mathen and Kurian (1967) recommended carbaryl 1% for the effective management of the pest. The molecule is relatively cheap and non-phytotoxic. Injection of Procyon E suspension after plugging all holes in the stem suppressed existing infestation. Abraham et al. (1975) reported the effectiveness of trichlorphon (1%) in recovering 92% of infested palms. It was found that 10% cashew apple extract was as effective as 1% carbaryl in killing RPW grubs in the laboratory (Krishnakumar and Maheshwari 2003). Laboratory evaluation showed that carbosulfan (0.05%) was superior against adult weevils and fifth-instar larvae (Beevi et al. 2004). Curative treatment of imidacloprid (0.02%) or spinosad (0.013%) was found effective in the management. As the pest has prolonged larval period including nine moultings, use of insect growth regulator would be an appropriate management strategy. Application of lufenuron (0.01%) leading to defective morphogenetic moults and malformed adults spurs long-term strategy in biorational approach to RPW management (Josephrajkumar et al. 2014c). At a global level, chemical groups, phenylpyrazole- and neonicotinoid-based insecticides, are used as preventive and curative applications to control RPW (Kaakeh 2006; Llácer et al. 2012; Al-Shawaf et al. 2013). Injected imidacloprid was capable to cause more than 90% mortality in young grubs for more than 2 months after treatment (Dembilio et al. 2015). With the limitation of insecticides causing environmental and health hazards, the focus has been currently oriented towards the use of eco-friendly strategies in the management of RPW.

Abraham et al. (1989) tested the IPM strategy for management of RPW in farmer's field in Kerala, India, and found the systematic adoption of IPM practices could bring down the pest population to nil and prevent fresh incidences. Josephrajkumar et al. (2014c) suggested refinement in the schedule for integrated management of red palm weevil. Closeness of farmer and systematic scouting would enhance better understanding about regular physical changes in palm for early diagnosis. Maintaining optimal palm density supplemented with intercrops for diminishing volatile cues through crop-habitat diversification strategy diminished pest attack. Prophylactic leaf axil filling with botanical cakes developed by ICAR-CPCRI and new-generation molecule, chlorantraniliprole at 3 g in perforated polythene sachet, safeguarded palms from invasion by rhinoceros beetle for 4–5 months as palm injury incites RPW attack. Influence of insect growth regulator, lufenuron (0.01%), leading to defective morphogenetic moults and malformed adults spurs long-term strategy in biorational approach to RPW management. Integrated management technologies involving complete destruction of infested palms, close monitoring and sustained surveillance for early diagnosis, leaf axil filling of chlorantraniliprole sachet, curative management with imidacloprid (0.02%) and pheromone trap at 1 trap ha^{-1} were found effective in pest suppression. Community-level technology convergence and large-area adoption of IPM technologies conducted in 2150 ha in Kerala (Bharanikavu, Cheppad), Tamil Nadu (Palladam), Andhra Pradesh (Ambajipet) and Karnataka (Bidramamandi) could reduce the pest incidence to 56.8%. Saving approximately 1% of palms from the pest damage all over the country with complete recovery brings a huge monetary gain to the country.

12.2.3 Nut Borer Cyclodes omma *Van der Hoev (Noctuidae: Lepidoptera)*

Cyclodes omma is an occasional pest that feeds on immature nuts inducing button shedding. Different stages of hairy caterpillars bore into the meristematic tissues surrounding the tepel and make prominent and typical bore holes. Caterpillars construct fibrous cocoons around the crown region. Adult moths are ochre green tinged with bilayered ring spots towards apical margin (Menon and Pandalai 1960).

12.3 Defoliators

12.3.1 Coconut Black-Headed Caterpillar **Opisina arenosella** *Walker (Lepidoptera: Oecophoridae)*

12.3.1.1 Distribution

Coconut black-headed caterpillar, *O. arenosella* Walker (synonym *Nephantis serinopa* Meyr.) (Oecophoridae: Lepidoptera), is a serious defoliator of coconut palm. The oldest record of the pest is by Green (1898) from Batticaloa of Ceylon (Sri Lanka). *O. arenosella* was first described by Meyrick in 1905. In India, *O. arenosella* was noted for the first time during 1907 on palmyrah palms in Tamil Nadu and first record of its presence on coconut leaves was from Andhra Pradesh in 1909 (Rao et al. 1948). During 1917 and 1920, severe outbreaks of *O. arenosella* were observed in Kerala and Karnataka, India (Pillai 1919; Rao 1924; Venkatasubban 1932). Since then, it has been reported causing serious damage periodically in major coconut-growing tracts in coastal and backwater areas and in the vicinity of water bodies in the inlands of peninsular India. Pest appears in several locations in a discontinuous manner; but under favourable conditions, it multiplies very rapidly resulting in sporadic outbreaks. This pest was reported to be a serious menace to coconut cultivation in other countries, viz. Sri Lanka (Jayaratnam 1941; Dharmaraju 1963), Myanmar (Ghosh 1923) and Bangladesh (Alam 1962).

12.3.1.2 Symptoms

The caterpillars construct galleries of silken webs reinforced with excreta and scrapes of leaf bits on the abaxial side of leaflets. By hiding in these galleries, they feed on the chlorophyll-containing parenchymatous tissue leaving the thin upper epidermis (Nirula 1956b). The affected portions get dried and form conspicuous grey patches on the upper surface of the leaflets which is the first noticeable symptom of infestation. Usually the feeding and drying start from the outer whorl of fronds and proceed inwards. Close examination of leaflets shows the presence of larval galleries on the lower side with live or dead stages of the pest. Severe pest damage results in complete drying of middle to inner whorl of fronds also. From a distance, the crown of such palms gives a burned appearance (Fig. 12.5). Since drying of fronds occurs due to many reasons, dried fronds should be cut and examined for the presence of larval galleries to identify the pest problem correctly. The pest infests coconut palms of all age groups round the year to varying intensities.

Fig. 12.5 Symptom of attack of leaf-eating caterpillar. (Photo: Josephrajkumar)

12.3.1.3 Crop Loss

Damage results in drying of the outer and middle whorl of leaves, reduction in the rate of production of spikelets, retardation of growth, reduction in photosynthetic efficiency and decline in yield. In addition, the damage renders the leaves unsuitable for thatching. Young palms often die due to the pest attack. The occurrence of the pest causing over 40% damage is considered to be an outbreak (Perera 1993). Joy and Joseph (1972) reported 63% reduction in yield due to *O. arenosella* attack. Annual loss in the pest-infested fields is estimated as INR 7280 ha^{-1} (Rajagopal and Arulraj 2003). Chandrika Mohan et al. (2010b) highlighted a 45% nut yield loss in the succeeding year of severe pest infestation.

12.3.1.4 Host Plants

Although *O. arenosella* is best known as a pest of coconut palm, palmyrah (*Borassus flabellifer* Linn.) is presumed to be the original host. Several palm species including ornamental palms and other crops such as banana, cashew and jack were reported as potential hosts (Rao et al. 1948; Lever 1969; Baburayanayak 1970; Butani 1975;

Kapadia 1982; Talati and Kapadia 1984; Talati and Butani 1988; Murthy et al. 1995a; Manjunath 1985; AICRP 1997; Chandrika Mohan and Shameer 2002; Sujatha and Singh 2002).

12.3.1.5 Bioecology

Rao (1924) studied the occurrence of the pest along the East and West coasts of India. Detailed account of bionomics, life history and morphology of *O. arenosella* was given by Nirula et al. (1951), Nirula (1956b) and Antony (1962). The moth is small, ash grey in colour, measuring 10–15 mm in length with a wing spread of 20–25 mm. The male moth is smaller than female with a slender abdomen ending in a short brush of scales, while in the female, the abdomen is stouter and pointed at the tip. The moths are nocturnal in habit and during daytime are found resting on the undersurface of the leaves. They are poor fliers, a fact that might have contributed to their relatively slow rate of spread and the localized nature of infestations. The moths are not found attracted to light, and strong wind and rain have little effect upon the resting moths. In the laboratory with honey as food, the male moth lives for 7 days on an average and the female for 5 days. Female moths lay eggs in irregular groups in webbed galleries on the underside of the previously attacked leaves of coconut palm. Nirula (1956b) noted that fecundity varied from 59 to 252 eggs, the capacity being influenced by climatic and other factors. The egg is oval in outline, 0.6–0.8 mm long and 0.3–0.4 mm broad, creamy white in colour when freshly laid and turn pinkish before hatching. The surface of the egg is shiny and is covered with irregular faint reticulations. Incubation period varies from 3 to 8 days. Newly hatched caterpillar is about 1.5 mm long and pink coloured with black head and dark thoracic plates. The abdomen is smooth and shining with long setae on the sides. The caterpillar spins a silken gallery on the abaxial frond surface, remaining in it while feeding. Fully developed caterpillar measures 13–18 mm in length and is greenish in colour with reddish brown stripes on the dorsal side (Nirula 1956b; Menon and Pandalai 1960) (Fig. 12.6).

There is difference of opinion among the workers regarding the number of instars. Nirula et al. (1951), Lever (1969) and Mohamed et al. (1982) observed five instars, whereas Antony (1962), Santhosh Babu and Prabhu (1987) reported six and eight instars, respectively. The caterpillars, after feeding voraciously on coconut leaflets, enter the prepupal stage and spin cocoons with white silken fibre. Prepupa is 12–15 mm long and is light pink in colour. The pupa is reddish brown in colour and measures 9–14 mm in length.

Population buildup in field varies with respect to climate and locations, and pest dynamics vis-à-vis abiotic factors are well documented (Nirula 1956b; Joy and Joseph 1972; Narendran et al. 1979; Mohamed et al. 1982; Sathiamma et al. 1973; Puttarudriah and Shastry 1964; Nadarajan and ChannaBasavanna 1980; Pushpalatha and Veeresh 1995; Sujatha 2001). In Sri Lanka, Perera et al. (1989) correlated the population outbreaks with climatic and biotic factors. Analysis of records of population outbreaks from 1965 to 1985 revealed cycles in the populations of approximately

Fig. 12.6 Life stages of leaf-eating caterpillar. (Photo: Josephrajkumar)

one generation period. Parasitism remained high throughout the outbreaks, and there was some evidence that pupal parasitism increased towards the end of the outbreak.

12.3.1.6 Integrated Pest Management

Low to moderate pest infestations in the field can be effectively managed by the biological control methods. However, an IPM strategy is recommended for management for *O. arenosella* in severe outbreak conditions. In IPM, biological control with releases of stage-specific parasitoids is supplemented with mechanical and chemical control approaches.

Mechanical Method Early to mild stages of infestation can be reduced by cutting and burning the badly affected fronds/leaflets. In case of very severe infestation also, removal and burning of fully dried two to three outer whorl of fronds helps in removing the pupae and other pest stages. Careless disposal of the pest-affected leaves in the vicinity of healthy palms can lead to newer infestations. This aspect has to be well taken care of while transporting pest-infested fronds or leaflets to pest-free areas as such or using pest-infested fronds for wrapping other commodities for transporting to newer areas.

Chemical Method Since a very rich natural enemy fauna is associated with *O. arenosella* in the field, chemicals are generally not being encouraged for the management of this pest. In case of very severe outbreaks, one spray of malathion (0.05%) (which is comparatively safe to natural enemies) is recommended to bring down the active pest stages. Spray solution has to reach underneath of the leaves to drench the larval galleries of the pest. If any chemical spraying is undertaken, a waiting period of 3 weeks is to be observed before the release of parasitoids. Due to the difficulty experienced in recent years in getting skilled labourers for palm climbing, chemical spraying recommendation is at a low profile in IPM.

Biological Method The black-headed caterpillar is attacked by several species of parasitoids, predators and microorganisms in the natural environment, and hence biological control provides the best solution for the management of *O. arenosella* on a perennial crop like coconut (Rao et al. 1948; Nirula 1956b; Nagarkatti 1973; Perera 1984, 1987; Cock and Perera 1987; Pillai and Nair 1993; Singh 1999; Chandrika Mohan and Sujatha 2006).

Natural Enemies of O. arenosella Dharmaraju (1962) provided a checklist and distribution of natural enemies of the pest in India and Sri Lanka. The list was updated by Cock and Perera (1987) and Pillai and Nair (1993). The coconut caterpillar supported about 60 species of natural enemies belonging to Class Insecta (Pillai and Nair 1993).

Parasitoids Among the parasitoids recorded from the pest, the larval parasitoids *Goniozus nephantidis* (Bethylidae) and *Bracon brevicornis* (Braconidae), the prepupal parasitoid *Elasmus nephantidis* (Elasmidae) and the pupal parasitoid *Brachymeria nosatoi* (Chalcididae) are the most promising ones which are used for augmentative releases for pest suppression. The major desirable attributes of these parasitoids are their greater host-searching ability, production of higher proportion of females, occurrence throughout the year and distribution in all pest-infested areas. Techniques have been developed for mass production of the promising parasitoids. Close monitoring of the endemic areas is essential to detect pest buildup at the initial stage itself for release of parasitoids at the appropriate period. Usually initial pest buildup was observed after post-monsoon (October to November) in Kerala and coastal Karnataka, India. The most important aspect in field release of parasitoids is that the release should synchronize with the stage of pest in the field. Presence of larvae in the leaflets should be confirmed before release of parasitoids to avoid wastage of parasitoids. George et al. (1982) developed a technique to estimate population of *O. arenosella* in the field. Sathiamma et al. (1987) worked out the norms for release of larval, prepupal and pupal parasitoids based on the estimated population of the larvae, prepupae and pupae of *O. arenosella* in the field. Releases are to be done at fortnightly intervals. The parasitoid *G. nephantidis* is released if the pest is at third-instar larval stage or above at 20 parasitoids palm^{-1} and *B. brevicornis* at 30 parasitoids palm^{-1}. The prepupal parasitoid *E. nephantidis* and pupal parasitoid *B. nosatoi* are released at 49% and 32%, respectively, for every

100 prepupae and pupae estimated to be present on the palm. In a multistage condition of the pest, a combined release of all the parasitoids at 40% of each of the target pest stage is required.

Before releasing the parasitoids in field, they should be fed with honey, and newly emerged parasitoid can be released in the field after 3 days of emergence. *G. nephantidis* and *B. Brevicornis* could easily be mass multiplied on larvae of the rice moth *Corcyra cephalonica*. The prepupal parasitoid *E. nephantidis* is a highly host- and stage-specific parasitoid and always requires a steady supply of prepupa of *O. arenosella* for mass multiplication. This is the major constraint for mass production of *E. nephantidis* in the parasite breeding laboratories. Other major parasitoids associated with *O. arenosella* in the field include early larval parasitoid *Apanteles taragamae*, larval-pupal parasitoid *Meteroidea hutsoni* and pupal parasitoids *Xanthopimpla* spp. and *Trichospilus pupivora* (Ghosh and Abdurahiman 1984, 1985; Chandrika Mohan 2005; Sujatha and Singh 1999; Pillai and Nair 1982, 1983, 1989; Joy and Joseph 1977). Perera (1993) opined that the performance of different parasitoids of *O. arenosella* was highly variable and is perhaps the reason for the periodic occurrence of pest outbreaks. Superiority of the solitary parasitoids over gregarious species in the biological suppression of the pest is well documented in Kerala, India (Pillai and Nair 1995). There are indications that the relative distribution as well as levels of parasitism of the natural enemies varies considerably in different climatic conditions and localities.

Exotic Parasitoids In many biocontrol successes, the introduction of an exotic parasitoid or predator is involved, and in most cases, a single exotic species has eventually been able to reduce the population to a non-pest status. Systematic efforts to breed the exotic natural enemies of *O. arenosella* and release in pest-infested coconut gardens were not successful. Exotic tachinid parasitoids, such as *Stomatomyia* (*Spoggosia*) *bessiana* Baranov from Sri Lanka and *Bessa remota* Aldrich introduced from Malaysia, released in India were not successful (Rao et al. 1971; Jayanth and Nagarkatti 1984; Cock and Perera 1987).

Predators The efficiency of the predators is neither fully assessed nor widely used in the biological control programme of *O. arenosella* (Pillai and Nair 1993). Insect and spider predators are abundant in the coconut ecosystem. The dominant insect predators are the carabid *Parena nigrolineata* Chaudoir and *Calleida splendidula* Fabricius, anthocorid *Cardiastethus* spp., chrysopid *Ankylopteryx octopunctata candida* Fabricius and coccinellid beetles (Pillai and Nair 1986, 1993). Rao et al. (1979) reported four species of spiders (*Cheiracanthium melanostoma* (Thorell), *Olios lamarcki* (Licrielle), *Heteropoda leprosa* Simon, *Marpissa calcuttaensis* (Tikader)) in association with *O. arenosella* in East Godavari district of Andhra Pradesh. A total of 26 species of spiders are recorded on the pest of which *Rhena*, *Sparassus* and *Cheiracanthium* are the major predators (Sathiamma et al. 1987). Surveillance in Karnataka, India, revealed the occurrence of efficient insect predators, viz. *C. exiguus*, *Buchananiella sodalis* Buchanan-White, *Mallada astur* (Banks), *Jauravia* sp. and *Phytoseiulus* sp., and ants, namely, *Tapinoma* sp.,

Monomorium pharaonis (Linnaeus), *M. latinode* Mayr, *Crematogaster* sp. and *Tetraponera rufonigra* (Jerdon) (Sujatha and Singh 1999). In Sri Lanka, Way et al. (1989) observed that 11 species of ants nested in coconut palm spathes and some of them are predators of eggs of *O. arenosella.*

Mites *Pyemotes ventricosus* Newport belonging to Pymotidae family was a gregarious ectoparasitic mite collected from the field. It parasitizes mostly the larval and pupal stages and occurs during rainy season (Mathen et al. 1968). A predatory mite, *Phytoseiulus* spp., was recorded in both coastal and interior zones of Karnataka, India (Sujatha 2001). Studies on seasonal variation revealed their numerical abundance during summer and post-monsoon in coastal area. Population of the mite ranged from 86.25 to 261.25 per 20% sample leaflets palm^{-1}. *Phytoseiulus* sp. and *C. exiguus* in the ratio 2:1 were associated with low population of *O. arenosella* during summer in interior Karnataka.

Pathogens *Bacillus thuringiensis* Berliner, *Serratia marcescens* Bizio and *Aspergillus flavus* Link are observed to be pathogenic to the pest in the field (Antony and Kurian 1961; Muthukrishnan and Rangarajan 1974; Oblisami et al. 1969; Kanagaratnam et al. 1983; Gopal et al. 2000). Philip et al. (1982) and Narayanan and Veenakumari (2003) reported a nuclear polyhedrosis virus affecting the caterpillars in Kerala and Karnataka, India, respectively. Sujatha (2001) observed 15% larval mortality due to bacterial and fungal pathogens soon after the monsoon rains (October) in coastal Kerala, India.

Attractants Kairomones could be used to enhance the efficiency of potential parasitoids (Bakthavatsalam and Singh 1996). Hexane wash of gallery and body of *O. arenosella* elicited positive response from the efficient parasitoids, namely, *G. nephantidis*, *E. nephantidis* and *B. brevicornis*. Analysis of infochemicals using GCMS revealed the presence of dodecane, pentadecane, hexadecane, heptadecane, eicosane and tricosane in the gallery of *O. arenosella*, and larval wash showed terpenoids (Bakthavatsalam et al. 1999). Subaharan et al. (2005) reported improvement in host-searching efficiency of *G. nephantidis* by exposing the newly emerged parasitoids to the host odours (smell of the volatiles of the injured *O. arenosella* larvae and gallery volatiles). The attraction of males of *O. arenosella* to conspecific virgin females was studied on coconuts at three locations in Sri Lanka during November to December 1996. The number of males caught in traps baited with virgin females was significantly greater than the number caught in unbaited traps (Fernando and Chandrasiri 1997). A female sex pheromone was reported by Murthy et al. (1995b). Four components of the pheromone of *O. arenosella* were identified by the Natural Resources Institute, UK (NRI), using insects from Sri Lanka (Cork and Hall 1998).

12.3.1.7 Field Performance of the Bioagents and Demonstration of IPM

Field performance on biological suppression of coconut leaf-eating caterpillar through release of stage-specific parasitoids was established as early as the 1920s. In 1929, a boat laboratory was functioning to breed and transport parasitoids in the Travancore and Cochin (erstwhile Kerala State, India), belt with very successful results (Nirula 1956b). Using IPM technologies or exclusive release of promising biocontrol agents, many demonstrations were laid out from 2000 by ICAR-CPCRI in coastal Kerala and Karnataka, India.

Studies conducted in an endemic area during 1990–1993 at Kollam District, Kerala, India, with the field release of the three stage-specific parasitoids, viz. *G. nephantidis*, *E. nephantidis* and *B. nosatoi*, at fixed norms and intervals in *O. arenosella*-infested coconut garden (2.8 ha) resulted in highly significant reduction (94%) in *Opisina* population (Sathiamma et al. 1996). Regular monitoring and release of stage-specific parasitoids induced 52.6% and 94.7% reduction in pest population after 1 year and 2 years, respectively, of parasite release in a heavily infested tract in Kollam Dist., Kerala (Chandrika Mohan and Sujatha 2006). A large-area field validation of the bio-suppression technology of coconut black-headed caterpillar with regular monitoring and release of stage-specific parasitoids, viz. *G. nephantidis*, *B. brevicornis*, *E. nephantidis* and *B. nosatoi*, was taken up during 1999–2002 in different geographic locations in coastal Karnataka and coastal Kerala (India) comprising of a total of 1,400 ha which could achieve 93–100% reduction in *O. arenosella* population in a period of 2 years (Chandrika Mohan et al. 2010a). Ghode et al. (1987) and Mohanty et al. (2000) had reported biological pest suppression of *O. arenosella* in coastal districts of Odisha, India, by the release of parasitoids. Successful biocontrol of *O. arenosella* was reported in Andhra Pradesh, India (Sujatha and Chalam 2009), and timely augmentation of *G. nephantidis* at 10 adults palm^{-1} at 15-day interval suppressed the pest population in Karnataka, India, at a cost of Rs. 2100 ha^{-1} (Venkatesan et al. 2006).

The higher frequency of occurrence of G. *nephantidis* in all the locations during post release period indicated the suitability of this larval parasitoid to adapt in various locations in the coastal belts of Kerala and Karnataka, India. Nutritional management of the palm with balanced dose of recommended fertilizers and proper irrigation to rejuvenate the pest-affected palms are essentially required to regain the yield potential of pest-infested palms.

The IPM package should be adopted in a larger area as a community programme with active involvement of farmers and development agencies for achieving the desired goal. It is of utmost importance that the parasitic fauna of a locality is to be studied before initiating the biological control programme. Role of solitary parasitoids in pest suppression is well documented. Clipping off the pest at the very initial stage and releasing appropriate dose of parasitoids synchronizing with the stages of the pest during lean periods in the field could check the pest from attaining a severe infestation level. Monitoring the infested field during successive years is the most important step for regulating further releases and for reducing the cost of parasitoid release. Conservation of promising predators (anthocorids, chrysopids, coccinellids and ants) helps a long way in natural suppression of the pest, and hence, spraying of palms with chemical pesticides has always been harmful to the natural enemies. Palmyrah can serve as an alternate host for the conservation of promising natural enemies of *O. arenosella*.

12.4 Sap Feeders

12.4.1 Coconut Eriophyid Mite **Aceria guerreronis** *Keifer (Acarina: Eriophyidae)*

12.4.1.1 Distribution

Coconut eriophyid mite, *Aceria guerreronis* Keifer (Eriophyidae: Acarina), is the most destructive pest among the various species of eriophyid mites affecting coconut palm in 30 countries of Tropical America, Africa and Asia (Mariau 1969). In India, coconut eriophyid mite was first reported from Amballur Panchayat in Ernakulam district of Kerala during 1998 (Sathiamma et al. 1998). Within a short span of time, the mite had spread rapidly to all major coconut-growing regions of the country, and currently its incidence is seen in the entire coconut-growing states of west and east coasts of India and north-east part of India (Nair et al. 2000a; Ramaraju et al. 2000; Mallik et al. 2003; Khan et al. 2003). The occurrence of the pest was also reported from Lakshadweep Islands (Mullakoya 2003). The history of the occurrence of *A. guerreronis* on coconut starts with the first report from the Guerrero State, Mexico by Keifer in 1965. In the same year, it was reported from Rio de Janeiro, Brazil (Ortega et al. 1967). It was widely noticed in several countries from South America and neighbouring Caribbean Islands by 1968. During the 1970s and early 1980s, severe damage of the pest was reported from Central America and West African countries. Tanzania witnessed an outbreak of the pest during 1980. In Sri Lanka, the pest occurrence almost coincided with that of India when the pest was recorded in the later part of 1997 at Kalpitiya Peninsula in the north-west province (Fernando et al. 2000). The mite has never been reported in the presumed region of coconut origin, namely, between the remaining of Southeast Asia and Papua New Guinea (Chan and Elevitch 2006). The mite might have moved from its original host to coconut after it became extensively cultivated in the Americas or Africa, continents where the mite was first found (Moore and Howard 1996).

12.4.1.2 Hosts

A. guerreronis was primarily recorded on coconut. It was also reported from cocosoid palm, *Lytocaryum weddellianum* (H. Wendl.), in Brazil (Flechtmann 1989); queen palm, *Syagrus romanzoffiana* (Cham.) Glassm., in southern California, USA (Ansaloni and Perring 2002); and palmyrah palm (*Borassus flabellifer*) in India (Ramaraju and Rabindra 2002). In cocosoid and queen palm, it was observed on nursery seedlings.

12.4.1.3 Bionomics and Nature of Damage

Coconut mite is a microscopic creamy white, vermiform organism measuring 200–250 microns in length and 36–52 microns in breadth. The body is elongated, cylindrical and finely ringed and bears two pairs of legs at the anterior end (Keifer 1965). Mites attain sexual maturity within a week's time and start laying eggs. An adult mite lays about 100–150 eggs. The eggs hatch into protonymphs and deutronymphs and finally to adults. The total life cycle is completed in 7–10 days.

In coconut, mites infest the developing young buttons after pollination and are seen in the floral bracts (tepals) and the soft meristematic portions beneath the perianth. Entry of the mite into the developing nuts takes place during the early phase of the development immediately after fertilization. Very young fruits are almost entirely covered by the perianth (Fig. 12.7), which is tightly adherent to the fruit surface, giving maximal protection against mite. However, as the fruits grow, the space underneath the bracts increases, in many cases allowing the mite to first have access to the protected tissues when fruits are about a month old (Moore and Howard 1996). The mites, thus gaining entry into the nuts, multiply and form active colonies containing various stages of development, viz. eggs, nymphs and adults. Usually in a developing nut, the coconut mite colonies are seen as two or three congregations on the meristematic regions of the buttons below the perianth.

Fig. 12.7 Symptom of coconut eriophyid mite infestation. (Photo: Josephrajkumar)

Mite could be found on a small proportion of nuts indicative of its egg laying a few days earlier, and that the level of infestation increased progressively afterwards. Thus, it seems that infested young buttons could also be aborted, reducing the number of infested buttons on the palms to low levels, possibly turning their detection difficult. Mite density was found to be higher in 3- to 7-month-old nuts after fertilization which thereafter gets reduced due to buildup of natural enemies (Moore and Alexander 1987; Varadarajan and David 2002).

Under favourable conditions, the high reproductive potential and shorter life cycle of mite result in the enormous multiplication of the colonies. When colony size becomes substantially increased, mite comes out of the interspaces between the tepals of the developing nut for dispersal. The dispersal of the pest takes place mainly through wind. Honeybees and other insects visiting inflorescence of coconut also act as agents for dispersal. The mite infestation symptoms are observed approximately 1 month after the initial colonization of the mite inside the fertilized buttons. Appearance of elongated white streaks below the perianth is the first external visual symptom on young buttons. In many cases, a yellow halo develops around the perianth. Within a few days, this halo develops into yellow triangular patch pointing towards the distal end of the button. This can be clearly seen in 2- to 3-month-old buttons. In a short time, the yellow patch turns into brown and shows necrotic patches on the periphery of the perianth (Moore and Howard 1996; Haq 2001; Nair 2002). As the nut grows, the injuries form warts and longitudinal fissures on the nut surface. In severe infestation, the husk develops cracks, cuts and gummosis. Shedding of buttons and young nuts and malformation of nuts as a result of retarded growth are the other indications associated with severe attack of the pest (Rajan et al. 2009).

12.4.1.4 Population Dynamics

High mite population is related to high temperature, low humidity and diminished level of precipitation (Mariau 1969; Moore et al. 1989; Nair 2002; Fernando and Aratchige 2010). In India, the pest activity has been observed throughout the year with the population peak during the summer months. Studies undertaken in Kerala coast (India) revealed that a period of high temperature with intermittent rains causing high humidity favoured higher multiplication and rapid spread of the mite (Nair et al. 2003). Investigations on population dynamics in Tamil Nadu, India, revealed that maximum population existed during November and May. Mathew et al. (2000) observed monthly variations in total population of mite with a peak in February to March and a sharp decline in subsequent rainy months indicating a negative relationship between rainfall and mite population. Perhaps the cryptic habitat of mite under the perianth may protect them from direct external abiotic stresses.

12.4.1.5 Crop Loss

Mite infestations cause extensive premature fruit drop (Nair 2002; Wickramananda et al. 2007), significant reduction in coconut fibre length and tensile strength (Naseema Beevi et al. 2003), as well as a reduction in husk availability for the coir industry (Wickramananda et al. 2007). Yield loss at various levels has been reported worldwide as a result of infestation by the pest. In general, pest incidence and extent of loss are comparatively high during the initial few years of pest occurrence in a particular locality. Yield loss depends on the cultivar, health and general maintenance of the crop as well as intensity of infestation. Feeding by a few mites causes only cosmetic damage to the husk without affecting the quality and quantity of copra and coconut water. In India during 1998, the pest outbreak was reported where 70% of nuts were affected with malformation and reduction in nut size (Nair 2002). But observations recorded during subsequent years revealed overall reduction in incidence and intensity of pest in areas of its initial occurrence (Nair et al. 2004). In Kerala, though pest damage has been reported initially ranging from 50% to 70%, later surveys carried out during 2000 have shown significant reduction in crop loss indicating an average loss of 30.94% in terms of copra and 41.74% in husk production mainly due to buildup of natural enemies (Muralidharan et al. 2001).

Similar studies undertaken in Tamil Nadu, India, during 2000 revealed an average loss of copra yield to the tune of 27.5% (Ramaraju et al. 2000). A reduction in copra yield ranging from 18% to 42% was observed in Karnataka, India, when severe infestation symptoms were seen on more than 50% of surface area of infested nuts (Mallik et al. 2003). Mite damage caused significant reduction in quality of fibres in terms of fibre length and tensile strength. Studies undertaken at Kerala Agricultural University during 2003 revealed that fibres from moderately to severely infested nuts suffered 26% to 53% reduction in length (Naseema Beevi et al. 2003). Surveys carried out by ICAR-CPCRI in Kerala, India, during 2004 registered lower levels of pest incidence with comparatively less intensity of infestation. The loss in terms of copra in southern districts of Kerala ranged from 8% to 12% compared to an average loss of 25% in initial years (Rajan et al. 2007).

12.4.1.6 Integrated Pest Management

Chemicals Over five dozen systemic and contact insecticides have been evaluated world over and recommended from time to time for management of coconut mite. In India also, a wide spectrum of pesticides have been tried by various research agencies including both central institutes and state agricultural universities (Nair et al. 2000b; Ramaraju et al. 2000; Saradamma et al. 2000; Mallik et al. 2003). Though these pesticides were effective in the field when given as spray or root feeding or stem injection, none of the chemicals has been recommended for larger adoption due to environmental reasons. The massive crown of the palm, large area to be covered in a short spell of time, need for repeated application, residual toxicity of pesticides, labour-intensive mode of application, etc. were other factors which were unfavourable for the wider use of chemical pesticides.

Currently, botanical pesticides, viz. neem-based biopesticides, are recommended for management of the pest in the field. Spraying of neem oil-garlic soap mixture at 2% or commercial botanical pesticides containing azadirachtin 10,000 ppm at 0.004% or root feeding with neem formulations containing azadirachtin 50,000 ppm (7.5 ml) or azadirachtin 10,000 ppm (10 ml) mixed with equal volume of water is recommended for mite management (Nair et al. 2000b, 2003; Saradamma et al. 2000; Mallik et al. 2003; Rajan et al. 2009).

Biological Control Use of predators and pathogens was not attempted in the initial phase of pest outbreak. More emphasis is being given to bioagents for long-term management of the pest. Among the biocontrol agents, predators and pathogens constitute the major groups of natural enemies. So far no parasitoid has been reported on *A. guerreronis*.

Predators Classical biological control is considered as an appropriate approach to provide a sustainable solution to the mite problem as the pest is invasive in nature. Careful import and introduction of natural enemies from the centre of origin of the pest into the new region is very crucial. In this context, efficient natural enemies of mite should be prospected in the tropical areas in the Americas, considered to be its possible area of origin (Navia et al. 2005).

In India, many predatory mites including *Amblyseius largoensis* (Muma), *Neoseiulus paspalivorus* De Leon and *Bdella distincta* Baker and Balogh were reported as potential predators of the mite (Ramaraju et al. 2000; Haq 2001; Nair et al. 2005). *N. paspalivorus* and *N. baraki* are the most abundant predators on nuts attached to the palms, melicharids and blattisociids are predominant on fallen coconuts, whereas *A. largoensis* is predominant on leaves (Lawson-Balagbo et al. 2008). In Sri Lanka, *N. baraki* was found predominant than *N. paspalivorus* (Fernando and Aratchige 2010). Though interpopulation crosses showed complete reproductive isolation between *N. paspalivorus* and *N. baraki*, molecular diagnosis indicated the existence of cryptic species (Negloh et al. 2008). Two successful methods, viz. tray-type arena and sachet-type method, have been developed in Sri Lanka to mass produce *N. baraki*, and the augmentative field release of which reduced mite incidence (Fernando et al. 2010). Occurrence of flattened idiosoma and short legs would allow *N. paspalivorus* and *N. baraki* enter into the microhabitat of eriophyid mite. It was also observed that *N. paspalivorus* usually enters the fruit about a month after mite entry, leading to highly diverging population curves of both species (Negloh et al. 2008).

The predatory mite population was registering an increasing trend of incidence and better establishment in nature over the years in India. From an initial occurrence recorded in 37.1% samples, predator population increased to 62.3% and 80% in 2001 and 2011, respectively. The activity of the predators was high during June to December and frequently noticed in 4- to 6-month-old nuts. The predatory mites are larger in size compared to the coconut mite, and hence they gain entry only later into the nuts. This is one of the limiting factors for the wider use of the predators. However, conservation of the predatory fauna in the ecosystem is beneficial to regulate the coconut mite in nature (Nair et al. 2005; Rajan et al. 2009).

Pathogens Beevi et al. (1999) reported *Hirsutella thompsonii* var. *synnematosa* Samson, McCoy and OoDonnell, infecting eriophyid mite in India. Cabrera (2002) reported two fungi, *H. thompsonii* Fisher and *Hirsutella nodulosa* Petch, infecting mite. Painstaking efforts by a multidisciplinary team in search of potential entomopathogen infecting mite encountered seven isolates of actinomycetes, four yeast isolates, three fungal isolates and two bacterial isolates from mite-infested nuts during the initial years of mite invasion. Gopal et al. (2002) established moderate pathogenicity against mite in the following eight microbial isolates *Fusarium moniliforme* [*Gibberella fujikuroi*], *Aspergillus niger*, *Penicillium* sp., *Aspergillus flavus*, *Scopulariopsis brevicaulis*, *Cladophialophora* sp., *Pseudomonas* sp., *Bacillus* sp., *S. brevicaulis* and *Cladophialophora* sp. Notwithstanding the higher mite incidence during the initial years of emergence (1999–2000), the percentage incidence of mite diminished in subsequent years (2010–2012) with the population buildup of natural enemies especially the predatory mites (*Neoseiulus baraki*) as well as the acaropathogen, *Hirsutella thompsonii* Fisher, in the system.

Searching for natural enemies, ICAR-CPCRI could collect more than 40 isolates of the acaropathogenic fungus, *Hirsutella thompsonii*, from all over the country which are being maintained at the institute. Based on the bio-efficacy studies, one virulent isolate collected from Kayamkulam (Kerala) was molecularly characterized confirming species identity. Talc preparation of this *H. thompsonii* at 20 g l^{-1} of water palm^{-1} containing 1.6×10^8 cfu with a frequency of three sprayings each year resulted in 63–81% reduction in mite incidence. It was found effective in many locations, though seasonal variation in efficacy existed. Coconut water was found as an ideal medium for mass production of *H. thompsonii*, as evidenced by comparable growth rate (1.91 cm in 20 days), spore production (12.9×10^4 cm^{-3}) and yield of dry mycelium mat (1.017 g 100 ml^{-1}) to that of standard fungal growth media (Chandrika Mohan et al. 2016a). Variability of microclimatic conditions, particularly in relative humidity among sites, and virulence of isolates may be likely explanations, which need to be looked into carefully before considering the application of *H. thompsonii*.

Palm Health Management Damage by eriophyid mite generally increased with increasing levels of nitrogen in coconut leaves, and it was suggested that higher levels of potassium could result in less damage by the mite (Moore et al. 1991). The use of organic fertilizers and potassium was reported to result in reduced mite damage in India (Muthiah et al. 2001).

A decreasing trend in mite incidence and intensity was observed in gardens where balanced NPK application and recycling of organic matter were practised. A unified recommendation was therefore formulated with IPM and INM components for adoption in all coconut-growing tracts during 2003. IPM strategies involved phytosanitary measures in coconut garden including crown cleaning, burning of fallen mite-infested nuts and spraying azadirachtin 0.004% on affected younger bunches thrice a year during December to January, April to June and September to October. Wherever spraying is difficult for adoption, root feeding with azadirachtin 5% (7.5 ml + 7.5 ml) or azadirachtin 1% (10 ml + 10 ml) formulation thrice as in the case of spraying was recommended. In synergy with IPM package, adoption of recommended package of practices was also recommended for effective recovery from mite infestation.

Varietal Susceptibility The tepal traits, colour, shape and size of the nut influence the degree of damage. Among these, shape of the nut (round shape) and tepal traits (tight perianth) are important attributes for mite tolerance. A coconut variety exhibiting resistance to eriophyid mite is not reported from any country. However, varieties like Malayan Yellow Dwarf (MYD), Malayan Red Dwarf, Rennal Tall, Cameroon Red Dwarf, Equatorial Green Dwarf and Hybrid [MYD × West African Tall (WAT)] were reported to show varying degrees of tolerance to mite attack in different countries of the world (Rethinam 2003). In India, work done at ICAR-CPCRI revealed that Chowghat Orange Dwarf (COD) variety shows maximum tolerance to mite infestation in the field. Malayan Green Dwarf (MGD), Laccadive Micro and Spicata also recorded comparatively lower mite incidence. West Coast Tall (WCT) and Laccadive Tall (LCT) recorded maximum incidence in the field. WCT with green colour and oblong nuts recorded higher level of mite incidence as compared to WCT with reddish bronze colour and round nuts. Among the hybrids, DxT with COD as mother parent exhibited high level of tolerance compared to Hybrids with Chowghat Green Dwarf (CGD) as mother parent (Nair et al. 2000b). Ramaraju et al. (2000) reported Kenthali variety to have lower surface damage by mite, while Tiptur Tall is the most susceptible variety. In 2015, a high-yielding, tall selection recorded with lesser mite infestation at field level was selected at ICAR-CPCRI from Kulasekharam coconut population, christened as Kalpa Haritha and released for cultivation (Josephrajkumar et al. 2016b). For details, please refer Chap. 5.

Success Story IPM technologies developed by ICAR-CPCRI involving 2% neem oil-garlic emulsion spray, root feeding of azadirachtin 10000 ppm at 10 ml + 10 ml water and soil and palm health management practices reduced pest incidence to the tune of 71.4%. From an initial pest incidence of 58.6% observed in Kerala, Tamil Nadu, Andhra Pradesh and Karnataka (India), the pest incidence was reduced to 16.3% in a period of 2 years indicating the success of the technology at national level. Natural buildup of predatory mites as well as acaropathogenic fungus, *H. thompsonii*, could sustain in the field.

12.5 Subterranean Pests

12.5.1 Coconut White Grub **Leucopholis coneophora** *Burm. (Coleoptera: Scarabaeidae)*

12.5.1.1 Nature of Damage and Species Complex

White grub, *Leucopholis coneophora* Burm, is a univoltine pest of coconut and intercrops grown in pockets of sandy loam soils in southern parts of peninsular India. It was first reported as a pest of coconut by Nirula et al. (1952). It damages seedlings and adult palms by feeding on roots, boring the bole and collar regions, and severe infestation leads to death of the seedlings. In adult palms, they feed on roots impairing the conduction of water and nutrients and thus lead to yellowing of fronds, gradual shedding and complete yield loss. Survey conducted in Kerala and parts of Karnataka, India, indicated the predominance of *L. coneophora* along the coastal belt, where coconut-based cropping system is practised. It has an annual life cycle and prefers loose sandy soil as noticed by Nirula et al. (1952). Kumar (1997) described it as *L. coneophora* – coastal strain that is occurring at an altitude up to 200 m above MSL. Another species of *Leucopholis* which is morphologically very much similar to *L. coneophora* dominating in coconut gardens of Dakshina Kannada district in Karnataka, India, was described and identified as *Leucopholis burmeisteri* Brenske (Nair and Daniel 1982; Veeresh et al. 1982). According to Kumar (1997), the identified morphological characters by Veeresh et al. (1982) between *L. coneophora* and *L. burmeisteri* were not strong enough to permit the delineation of the two populations up to specific status. Though they exhibited distinct differences in biology, the two populations may be two different etho-species which are difficult to delineate morphologically and can be considered to be two geographical clines. Kumar (1997) designated these two species as *L. coneophora* – coastal strain which occurs at altitude up to 200 m above MSL – and *L. coneophora*, hill strain which occurs at altitude >200 m above MSL. Another species of palm white grub, *L. lepidophora*, is observed to be infesting palms in Western Ghats. *L. lepidophora* larvae and adults are morphologically and biologically distinct from *L. coneophora*. These grubs prefer clayey loam soil (Veeresh et al. 1982; Kumar 1997). Study of phylogenic relation by partial amplification of 16s rRNA and *COI* gene of *L. burmeisteri*, *L. coneophora* and *L. lepidophora* (collected from Dharmasthala, Kasaragod and Sringeri, India, respectively) revealed 98, 83 and 89% similarity, respectively, with *L. burmeisteri sulyareca* isolate, and high resolution melt (HRM) analysis revealed the existence of single nucleotide polymorphism among the three species.

12.5.1.2 Host Range

Leucopholis coneophora is highly polyphagous in nature. Apart from coconut, it feeds on root of arecanut as well as rhizomatous and tuberous intercrops raised in palm garden, viz. banana, colocasia, cassava, elephant foot yam, greater and lesser yams, sweet potato, fodder grasses, etc. It is reported to be feeding on roots of rubber and cocoa also.

12.5.1.3 Bionomics

L. coneophora has annual life cycle and adult emergence coinciding with the setting of south-west monsoon. Emergence occurs daily in the evening hours when luminance fall below 124.37 ± 75.5 lx (around 6.35 pm IST in June) and 1.2 ± 0.4 lx (around 7.10 pm IST), and active swarming is sustained for 3 weeks. On emergence, beetles feed on leaves of weeds, mango, cashew, ficus, etc. Females lay the eggs in interspaces in soil. The eggs have an incubation period of 23 days. Larva is pestiferous and passes through three instars which is prolonged for 260–270 days. First-instar larvae feed on organic matter and roots of grass and are seen at depths of 15–20 cm. They are observed during second half of May to beginning of October. Second-instar grubs are largely distributed at depths of 15–45 cm which could be observed from the first half of July to the first half of November. Late second- and third-instar larvae move towards the root zone and start feeding on palm roots. Third-instar grubs were seen from the first half of the October to the end of July of the succeeding year. With the movement of moisture in soil, larvae move deeper and deeper and subsequently pupate during summer (Abraham 1983). During the next monsoon season aestivating pupae emerge out. During 1976–1978, prolonged adult activity period of 60 days was recorded. Adult emergence initiated during first half of March continued at low level up to second half of May or till early part of June. Scanty emergence continued up to August and September (Abraham 1993). But, more recently a narrow window of adult activity that extended for a maximum period of 3 weeks was noticed in ethological study of *L. coneophora* during 2011–2013 (Prathibha et al. 2013). There has been a huge shift in the emergence pattern of *L. coneophora*. Climate change pertaining to rainfall pattern, distribution and soil temperature could be the major reason for this. A hike in soil temperature (an average increase of 0.22 °C in daily soil temperature from March to September) was noticed during 2011–2013 than those temperature regimes noticed in 1977 and 1978.

12.5.1.4 Integrated Pest Management

An IPM strategy comprising of mechanical, chemical and biological methods is recommended to effectively manage white grubs.

Mechanical: Handpicking and Destruction of Beetle Mechanical capture and destruction of cockchafers between 6.35 pm and 7.15 pm for 2–3 weeks commencing from the first day of monsoon is advisable as a mechanical tool in IPM (Abraham 1983; Prathibha et al. 2013; Prathibha 2015). As the peak swarming period is short and beetle congregate during swarming, this method can be well practised. It is found that capture of beetles by handpicking is significantly higher than light trapping.

Biological An array of natural enemies was reported on *L. coneophora*. A solitary ecto-larval parasitoid, *Campsomeriella collaris* Fab. (Hymenoptera: Scoliidae), and parasitism by *Prosena* spp. nr *siberita* (Tachinidae: Diptera) as well as a solitary endo-larval parasitoid were reported on *L. coneophora* grub for the first time from organically managed coconut garden. Entomopathogenic bacterium *Serratia entomophila* caused 'amber disease' to *L. coneophora* grubs. White muscardine fungus *Beauveria brongniartii* and green muscardine fungus *Metarhizium* spp. were obtained from infected *L. coneophora*. Epizootic due to caterpillar fungus *Cordyceps* spp. was noticed on third-instar *L. coneophora*. Two species of entomopathogenic nematodes *Steinernema carpocapsae* and *Heterorhabditis indica* are being used in the management of palms against root grubs. In coconut ecosystem, drenching aqua suspension of EPNs *Steinernema carpocapsae* in the interspaces (5–10 cm depth) at 1.5 billion IJ ha^{-1} was found effective. Soil application of EPN should be continued based on the white grub population.

Chemical Control Use of chemical insecticide is a vital component in IPM of root grub, and it is successful when applied in the right stage and season. During the early 1950s, organo chlorine compounds as dust formulations were commonly used for the management of root grubs. Application of 5% chlordane at 28 lb acre^{-1} gave good control of *L. coneophora* grubs in coconut garden (Nirula and Menon 1957; Valsala 1958). Similarly, application of 10% HCH at 56 lb acre^{-1} once a year after south-west monsoon was recommended against white grubs in coconut which was superior to DDT dusting (Nirula 1958). During the 1970s chlorinated hydrocarbons were replaced with organophosphates (OP) and carbamates for use in management of root grubs. Granules such as carbaryl, carbofuran, phorate, quinalphos and thiodemeton at 4, 6 and 8 kg a.i. ha^{-1} evinced 36% reduction in *L. coneophora* grub population (Abraham 1979). In the 1980s, use of emulsifiable concentrate (EC) formulations of chlorinated hydrocarbons, (chlordane, aldrin, dieldrin, heptachlor, etc.) and OP compounds like chlorpyrifos and quinalphos became popular. Drenching the root zone with chlorpyrifos (0.04%) is recommended for the management. In the early 2000s, soil application of neonicotinoid insecticide imidacloprid at 120 g a.i. ha^{-1} or fourth-generation synthetic pyrethroid bifenthrin at 2 kg a.i. ha^{-1} was found effective in the management of the palm white grubs.

A refined IPM strategy was formulated for the effective management of the pest. The various strategies include:

- Handpicking and destruction of adult beetles during peak emergence
- Blanket application of bifenthrin at 2 kg a.i. ha^{-1} (Talstar 10 EC at 20 l ha^{-1} in 500 l of water) when first-instar stage of grubs dominate in the field

- Drenching aqua suspension of EPNs *Steinernema carpocapsae* in the interspaces (5–10 cm depth) at 1.5 billion IJ ha^{-1} during September to October as well as during November to December
- Second round need-based root zone application of chlorpyrifos 20 EC at 7 ml $palm^{-1}$ after 45 days of first round insecticide application
- Repeated ploughing to expose the grubs to predators/digging and removal of grubs during October to December.

12.6 Mammalian Pests

*12.6.1 Rat (***Rattus rattus***)*

12.6.1.1 Distribution

In coconut plantations, eight different species of rodents were observed to coexist (Advani 1984; Advani 1985). Among them, *Rattus rattus wroughtoni* was the most predominant one (45%) followed by the field mouse *Mus booduga* (31%). Other rodents found in association with these mammals were the tree mouse, *Vandeleuria oleracea* (12%); the Western Ghats squirrel, *Funambulus tristriatus* (7%); *R.r. rufescens* (4%); and the Indian gerbil, *Tatera indica* (1%). The burrows of the lesser bandicoot *Bandicota bengalensis* and the larger bandicoot *Bandicota indica* were also found in certain gardens. *R.r. wroughtoni* lived mostly on the tree canopy, whereas *M. booduga* remained on the ground, thus minimizing competition for food and shelter.

Rat (*Rattus rattus*) is the major and threatening mammalian pest of coconut in the island ecosystem both in Lakshadweep and Bay Islands. The damage intensity varied from 14.3% during 1988 to 20.4% during 1990 in the mainland. Advani (1984) has reported that the damage intensity to coconut was more in coconut-cocoa mixed cropping systems (28.5%) than in coconut monocrop system (21%). In Lakshadweep Islands, nut loss as high as 50% was recorded. Detailed studies on population structure; movement pattern; breeding behaviour including breeding season, ovulation rate and litter size; post-natal development; juvenile emergence; and adult persistence were studied by Advani (1985).

Colonies of rats are found on the crowns of the coconut palm feeding on nut. In closely planted coconut gardens, rats jump from tree to tree. All palms are not invaded by the rats perhaps selected palms that yield sweet nut water and pulp are highly preferred. All stages of the nuts were found to be fed by the rats in Minicoy Island making a typical circular hole by gnawing and feed on the inner contents. Gnawing sound of rats is quite audible during dusk, and all islanders are well familiarized to the sound. Under severe conditions, even the emerging spathes are very badly eaten by the rats in the island. Rats are also habituated to make breeding nests using leaflets on the crown of the palms.

Some of the major reasons attributed for the increased rat damage in the island are higher density of coconut palms; inadequate crown cleaning and delayed harvest of nuts; heaping fallen fronds and husks in the farm; absence of predators like snakes, owls, etc. in the island; and improper care provided by the farmers.

The bandicoot rat, popularly known as Malabar rat or pig rat, causes serious damage to juvenile coconut palms (1–2 years old). It eats the growing stem resulting in the complete collapse of the young palm. Krishnakumar et al. (2014) reported that Gangabondam was highly preferred by *R. rattus* followed by Laccadive Orange Dwarf, Laccadive Green Dwarf and Laccadive Yellow Dwarf. Varieties/hybrids that are highly preferred by *R. rattus* are, therefore, likely to possess sweet nut water and pulp. Flying fox, robber crabs and pigs (Andaman Islands) were also reported as mammalian pests on coconut (Menon and Pandalai 1960).

12.6.1.2 Management

Nirula (1954) suggested the use of poisonous bait, fumigation of rat burrows, trapping, erection of physical barriers, application of chemical repellents and encouragement of natural enemies for the management of rats.

Timely removal of weeds as part of orchard sanitation and regular crown cleaning operation would expose the rats to predators. Banding of coconut trees with GI sheet 25–30 cm wide at a height of 2 m above-ground level is effective in controlling arboreal rats if coconut palms are raised at recommended spacing. Trapping using live or death traps is the safest but labour-intensive method for controlling rodents (Bhat and Sujatha 1987). Bamboo traps are used for the control of burrowing rodents damaging seedlings.

Control methods by poison baiting with multiple- and single-dose anticoagulants were investigated by Bhat and Sujatha (1987). The damage of *R.r. wroughtoni* could be completely controlled by applying warfarin/fumarin wax blocks on the crown of coconut palms three times at an interval of 3 days at 105 g, 70 g and 35 g baits palm^{-1} on 30 palms ha^{-1}. Among the three promising single-dose anticoagulants, viz. brodifacoum, bromadiolone and flocoumafen, the first one was found to be more toxic than the other two. However, since only bromadiolone is registered in India, more detailed studies were conducted with that poison to control *R.r. wroughtoni* in coconut gardens. The rats did not exhibit any bait shyness when exposed to bromadiolone baits. In the field trial, one-time application of bromadiolone (0.005%) wax blocks on the crown of rat-infested coconut palm (30 palms ha^{-1}) reduced the damage by 79.6%. Application of poison baits twice at an interval of 12 days has completely controlled the damage.

Placement of bromadiolone cake (0.005%) (Roban®, two pieces of one cake palm^{-1}) on the crown of those rat-preferred palms was found to reduce the rat damage to a greater extent. ICAR-CPCRI has developed a rod device for easy placement of Bromadiolone cake on the palm top blockading the route map of rats and forcing for consumption (Krishnakumar et al. 2014). Placement of protection barriers such as tin sheets to prevent climbing of rats had limited success in the island condition. Traditional practices such as hanging a fertilizer bag on the petiole and banding the palms with polythene sheets could reduce rat damage in isolated palms in the main land.

12.7 Emerging Pests

Emerging pests are those pests that have newly appeared in the system or those that have existed but are rapidly increasing in incidence on a geographic range. In the climate change scenario, emerging pests need to be closely monitored to avoid gradient outbreaks. Reports on the incidence of a few insects on coconut plantations in different parts of India attaining the status of emerging pests are presented.

12.7.1 Coreid Bug **Paradasynus rostratus** *Dist. (Coreidae: Hemiptera)*

Paradasynus rostratus was reported as a serious emerging pest on coconut from Southern districts of Kerala, India, incidence ranging from 23.4% to 40.6%. The bug causes heavy crop loss by shedding of developing buttons and immature nuts ranging from 18.2% to 66.4% in endemic gardens. The characteristic symptoms include deep furrows, crinkles and gummosis on nut surface. A prominent spindle-like necrotic lesion could be observed inside the fallen buttons (Kurian et al. 1972). In combination with eriophyid mite, there is tremendous reduction in copra content (Mohan and Faizal 2004). *P. rostratus* was first reported from Kerala on coconut by Kurian et al. (1972).

In addition to coconut, the bugs attack other crops like cashew (tender nuts), guava (tender fruits), custard apple (tender fruits), cocoa (tender pods) and neem (young flushes). The pest is active in coconut gardens from June to February with peak population during October to December (Rajan and Nair 2005). An egg parasitoid, *Chrysochalcisea indica*, parasitizing egg mass; a reduviid predator, *Endochus inornatus*, predating on nymphal stages; and weaver ant, *Oecophylla smaragdina*, predating all stages of pest were recorded as natural enemies.

Nut infestation symptoms by coreid bug as well as eriophyid mite were characteristically distinguished by Chandrika Mohan and Nair (2000b). Physiological modulation of qualitative chemical parameters of coconut oil from coreid infested nuts was well documented (Mayilvaganan and Nair 2002). Visalakshy et al. (1987) reported carbaryl (0.1%) as an effective insecticide against coreid bug. Spraying of neem seed oil-garlic emulsion (2%) and profenophos (0.05%) was found effective for the management of coreid bug (Ambily et al. 2009).

12.7.2 Slug Caterpillars

12.7.2.1 *Conthyla rotunda* Hamp. (Limacodidae: Lepidoptera)

C. rotunda is a sporadic pest causing considerable damage to coconut palms. During severe outbreaks, caterpillars devour leaves and at times feed on spathe and nuts also. Defoliation causes extensive yield loss (Menon and Pandalai 1960). Though reported in 1916, Ayyar (1917) described its damage potential from Kochi, India. Occurrence of *C. rotunda* was confined to west coast of India, and its biology and management including natural enemies were studied by Nirula et al. (1954). On an average, the life cycle is completed in 52 days. *C. rotunda* was also reported on banana, wild arrowroot and tea bushes. During February 2010, outbreak of *C. rotunda* was reported from Kerala causing complete defoliation of the leaves resulting in drastic reduction in nut yield (Rajan et al. 2010a, b). Outbreak of a particular species corresponded to the geography-induced weather factors as well as the intrinsic biotic factors.

12.7.2.2 *Darna (Macroplectra) nararia* Moore (Limacodidae: Lepidoptera)

D. nararia was reported from South India and Sri Lanka (Fletcher 1914) and also from East and West Godavari districts of Andhra Pradesh, India (Nirula 1955a). It is also a pest on tea in Sri Lanka and *Pithecellobium dulce* in Tamil Nadu, India. In Eastern and Western Godavari districts of Andhra Pradesh, the outbreak of *D. nararia* was recorded during March to May, 2009 and the level of infestation ranged from 5% to 85%. Temperature and moisture are critical factors in the growth of herbivorous insects and their invertebrate predators, parasitoids and pathogens. Outbreaks can be of eruptive in nature driven by intrinsic population processes and trophic-level factors. It can also be gradient in nature as a consequence of changing environmental factors favouring population growth. High temperature (>39 °C) coupled with high relative humidity favoured emergence of *M. nararia* in higher population (Rajan et al. 2010a, b). Infested gardens showed a scorched/burnt appearance of the middle and outer whorl of leaves and the crop loss reached as high as 90–95% in severely affected gardens. Feeding damage was exacerbated by grey leaf blight fungus, *Pestalotiopsis palmarum*, which infect leaf tissue breached and damaged by early larval instars. This association has been reported from slug caterpillar species such as *Chalcocelis albiguttatus*, *Darna catenatus*, *D. trima*, *Parasa balitkae*, *Setora nitens* and *Thosea lutea* (Holloway et al. 1987; Howard et al. 2001). In the field, some caterpillars were found infected by entomopathogens. Light trapping is suggested as an effective monitoring tool and a feasible mechanical control strategy of the pest.

12.7.2.3 *Parasa lepida* (Cramer)

P. lepida is one of the slug caterpillar species, found in India, Southeast Asia including the Sunda Islands and the Philippines, China and Okinawa, Kyushu, Shikoku and Honshu of Japan (Hirashima 1989). It is widely prevalent in coconut gardens on the west coast of India, but severe infestation occurs sporadically. The biology of the pest was described by Menon and Pandalai (1960). Juvenile palms are invaded, and caterpillars with poisonous scoli (stinging spines) devour coconut leaflets leaving only the midrib. Besides coconut, this pest attacks castor, mango, banana, rose, pomegranate, palmyrah, camellia, poplar and willow. There is a risk of irritation and inflammation of the skin from inadvertent contact with the insect. Most insecticides evaluated were found effective. However, delivery of chemicals in tall palms poses environmental risk. A predatory caterpillar, *Phycita dentilinella* H., was recorded inside the cocoon of the pest destroying pupae (Ayyar 1940). Early stages of the caterpillars that are gregarious could be mechanically clipped along with the portion of leaf and destroyed. During 2007, a sex pheromone ((Z)-7,9-decadien-1-ol (Z7, 9-10:OH)) has been isolated from virgin females and attracting conspecific males in the field (Wakamura et al. 2007).

12.7.3 Whiteflies

Two species of whiteflies, viz. Arecanut whitefly *Aleurocanthus arecae* David (Aleyrodidae: Hemiptera) and spiralling whitefly *Aleurodicus dispersus* Russell (Aleyrodidae: Hemiptera), were recorded in mild to medium intensities in various tracts of Kerala, Tamil Nadu and Lakshadweep Islands of India. These soft-bodied insects were found in congregation on the undersurface of matured coconut leaflets. These whiteflies were suppressed in nature by the natural enemies preventing a pest outbreak.

12.7.3.1 Arecanut Whitefly *Aleurocanthus arecae* David (Aleyrodidae: Hemiptera)

A. arecae is noticed on young coconut palms of both tall and dwarf cultivars generally during April to May. Black envelope of sooty mould fungus is seen on upper surface of infested leaflets. Adult whiteflies have smoky grey wings, and ants are not commonly associated with the pest congregation. Pupae have blackish setae on the body (Chandrika Mohan et al. 2007).

12.7.3.2 Spiralling Whitefly *Aleurodicus dispersus* Russell (Aleyrodidae: Hemiptera)

As the name suggests, adults of *A. dispersus* have a typical spiralling fashion of egg laying and found in mild to moderate levels during March to May. It is a highly polyphagous pest infesting coconut leaflets and a wide array of crops in coconut plantations. In Minicoy (Lakshadweep Island, India), the spiralling whitefly nymphs were reported more on papaya, banana, cassava and castor and were found parasitized by the aphelinid parasitoids. Adults measure about 2 mm length with white wings (Ramani 2000; Josephrajkumar et al. 2010a).

12.7.3.3 Rugose Spiralling Whitefly *Aleurodicus rugioperculatus* (Aleyrodidae: Hemiptera)

Rugose spiralling whitefly, *Aleurodicus rugioperculatus* Martin (Fig. 12.8), is an invasive pest of coconut reported from Pollachi, Tamil Nadu and Palakkad, Kerala (India) during July to August 2016. The pest was initially presumed as *A. dispersus*, which, however, could not incite to the present level of damage on coconut when first appeared in 1995 (Prathapan 1996; Mani 2010). The extensive damage and specific confinement on coconut reinforced the occurrence of new *Aleurodicus* sp. (Chandrika Mohan et al. 2016b). The identity as *A. rugioperculatus* was subsequently confirmed by Shanas et al. (2016) and Selvaraj et al. (2017), based on puparial features. Occurrence of reticulated cuticle on dorsum, presence of compound pores in abdominal segments VII and VIII, presence of corrugation on the surface of operculum and acute shape of the apex of lingual were reported as unique features of *A. rugioperculatus*.

Distribution RSW adults can be distinguished by their large size and the presence of a pair of irregular light brown bands across the wings (Stocks and Hodges 2012). Rugose spiralling whitefly was first described by Martin in 2004 from samples collected in Belize on coconut palm leaves (Martin 2004) and subsequently in Florida from Miami-Dade County in 2009. The whitefly genus *Aleurodicus* Douglas encompasses 35 species, of which only the spiralling whitefly *Aleurodicus dispersus* Russel was so far known to occur in India (Martin 2008).

The pest, distributed in Central and North America, is limited to Belize, Mexico, Guatemala (Evans 2008) and the United States. In the continental United States, the first established population of RSW was reported from Florida in 2009, and since then, its distribution range has expanded considerably within the state. There have been reports of damage caused by this pest to ornamental plant hosts in at least 17 counties of Florida, with the maximum damage reported from Broward,

Fig. 12.8 Rugose spiralling whitefly. (**a**) Eggs of rugose spiralling whitefly (RSW). (**b**) Mobile crawler. (**c**) Nymphs. (**d**) Adult. (**e**) Damage on leaf. (**f**) Damage on petiole. (**g**) Damage on nut. (**h**) Leaf encrustation by RSW. (**i**) *Encarsia guadeloupae*. (**j**) Parasitized pupa, exit hole of the parasitoid. (**l**) Sooty mould laden leaf. (**m**) Sooty mould scavenger beetle, *Leiochrinus nilgirianus*. (Photo: Josephrajkumar)

Collier, Lee, Martin, Monroe, Miami-Dade, Palm Beach and St. Lucie counties. In India, in a period of 6 months, it could spread in limited pockets of Kerala (Palakkad, Malappuram, Thrissur, Kozhikode, Kannur, Ernakulam, Pathanamthitta, Alappuzha, Kollam and Thiruvananthapuram districts), Tamil Nadu (Pollachi) and isolated parts of Andhra Pradesh. Though the invasive pest, *A. rugioperculatus*, has been reported in different districts of Kerala, India, it has not caused any economic setback so far as it confined its feeding mostly on the older coconut leaves that has completed the bearing stage. Extensive desapping of whitefly would induce stress on the palms due to removal of water and nutrients, but neither colour change nor necrosis of leaves is reported. The economic loss is restricted mainly on the loss of photosynthetic efficiently due to the formation of sooty mould fungus (*Leptoxyphium* sp.).

Host Range Stocks and Hodges (2012) reported about 95 host plants of *A. rugioperculatus* in Florida, USA. Further, Francis et al. (2016) reported a broader host range of 118 species in 43 families. Shanas et al. (2016) reported a total of 17 plant species in 11 families from Kerala. In the recent survey, Chandrika Mohan et al. 2016b) observed at least ten alternate host plants (*Psidium guajava*, *Musa* sp., *Myristica fragrans*, *Colacasia* sp., *Garcinia* sp., *Annona muricata*, *Murraya koenigii*, *Spondias mombin*, *Mangifera indica* and *Artocarpus heterophyllus*) in coconut homesteads, but the pest is relatively more confined to coconut, and the reason for its selective preference indicates its host preference to coconut.

Natural Enemy Despite higher level of whitefly incidence in coconut, there is no absolute economic loss due to the pest, because *A. rugioperculatus* is confined mostly on the lower leaf whorls of coconut which had completed the nut production phase. During the observation made in November 2016 in Kerala, India, the pest population has come down quite significantly. Upon close examination, as high as 60% pupae of *A. rugioperculatus* was found parasitized by aphelind parasitoid, *Encarsia guadeloupae*, and many do have exit hole in situ in the field indicating the emergence of the parasitoid. *E. guadeloupae* is a tiny parasitoid of size <1.0 mm found effective in the bio-suppression of the whitefly under natural condition. On account of its miniature size, mass production strategies are not successful so far; however, reintroduction of parasitized pupae in the emerging zones of whitefly attack is feasible and quite practical.

Management

1. Application of 1% starch solution on leaflets to flake out the sooty moulds.
2. Installation of yellow sticky traps on the palm trunk to trap adult whiteflies.
3. Encourage buildup of parasitoids (*E. guadeloupae*) and reintroduce parasitized pupae to emerging zones of whitefly outbreak.
4. In severe case, spray neem oil 0.5%, and no insecticide is recommended.

12.7.4 Mealybugs

Fifty-seven species of pseudococcids have been recorded on palms. Half of the palmivorous species belong to the genera *Dysmicoccus*, *Planococcus*, *Pseudococcus* and *Rhizoecus*. Among the wide array of mealybug species, *Dysmicoccus* has the most palmivorous species (eight) including three species known only from palms. The most commonly reported mealybug pests of palms are highly polyphagous species, are distributed worldwide and are primarily known as pests of crops other than palms. Classical examples include *Dysmicoccus brevipes*, *Nipaecoccus nipae* and *Pseudococcus longispinus*. A few mealybug species, viz. *Dysmicoccus hambletoni*, *Dysmicoccus cocotis*, *Dysmicoccus finitimus*, *Neosimmondsia hirsuta*, *Palmicultor palmarum*, *Phenacoccus sakai*, *Planococcoides anaboranae*, *Pseudococcus portiludovici*, *Tylococcus malaccensis*, *Crinitococcus palmae* and *Cyperia angolica*, are almost restricted to palms (Nair 1983; Chandrika Mohan et al. 2016c).

In coconut, nine important species of mealybugs are reported from India, viz. *Palmicultor palmarum* Ehron., *Dysmicoccus cocotis* Maskell, *Pseudococcus longispinus* Targ., *Pseudococcus cryptus*, *Planococcus lilacinus*, *Pseudococcus microadonidam*, *Nipaecoccus nipae* Maskell, *Dysmicoccus finitimus* and *Rhizoecus* sp. In Guam, *Coccidohstrix insolita* was observed infesting coconut palm (Aubrey et al. 2014).

12.7.4.1 Species Diversity

P. palmarum infests young seedlings and juvenile palms on spear leaf. Red ants are normally associated with *P. palmarum*. *P. palmarum* was found mainly on coconut in Micronesia, Bangladesh, Hawaii and Bahamas (Williams 1981). In India, *Palmicultor* sp. took 22 days to complete its life cycle. Adult females and males survived for 18 and 3 days, respectively, and one female produced 37–89 offsprings (Jalaluddin and Mohanasundaram 1993). *Dysmicoccus finitimus* is found colonizing the spadix of coconut in southern India, Sri Lanka, Cocos Islands and peninsular Malaysia (Williams 1994). Infestation by *D. finitimus* was also recorded from the spathe of coconut palms from Kerala (CPCRI 2012). In India, *Dysmicoccus brevipes* was reported from the perianth of immature nuts in coconut (Radhakrishnan et al. 2003). *Rhizoecus* sp. infests roots of coconut palms in sandy areas. Infested seedlings turn yellowish and loose vigour (Nair et al. 1980). *P. microadonidam* and *P. lilacinus* are known to infest coconut in India (Mohandas and Remamony 1993). The occurrence of *N. nipae* was reported from West Bengal (Green, 1908). Re-emergence of the pest was reported in India after a time gap of 100 years. It was recorded on tender feeder roots of coconut seedling in Kayamkulam, Kerala, India. *N. nipae* was not located on any other arboreal parts of the palm (Josephrajkumar et al. 2012). The leaf mealybug, *Pseudococcus cryptus*, was found colonizing at moderate level on the leaves. Infested colonies were foraged by ants. *P. cryptus* was reported from Tamil Nadu, India (CPCRI 2012).

12.7.4.2 Management

Regular and systematic monitoring for early detection is very important for suppressing the pest. Collection and destruction of infested plant parts would reduce the pest load in the field. Conservation of natural enemies such as *Pullus* sp., *Scymnus* sp. (Coccinellidae), *Spalgis epius* (Lycaenidae), *Bergineus maindroni* (Mycetophazidae), *Dicrodiplosis* sp. (Cecidomyiidae) and *Homalotylus oculatus* (Encyrtidae) could exert natural pest reduction. Regular monitoring and spot application twice with dimethoate 0.05% at 20-day interval during summer were found effective in the management of the pest (Nair 1983).

12.7.5 Scale Insects

A total of seven species of scale insects were reported infesting coconut palms/ seedlings (Rajan et al. 2010a, b). These include four species of armoured scales and three species of soft scales. Armoured scales were recorded as emerging pests which include the coconut scale, *Aspidiotus destructor* Sign.; oriental scale, *Aonidiella orientalis* Newstead; mussel scale, *Lepidosaphes mcGregori* Banks; and needle scale, *Chionaspis* sp. The three species of soft scales recorded from coconut were wax scale, *Ceroplastes floridensis* Comstock; brown soft scale, *Coccus hesperidum* Linn.; and stellate scale, *Vinsonia stellifera* Westwood.

The coconut scale *A. destructor* was recorded from all states of South India, Lakshadweep Islands as well as Assam and West Bengal. Scale insects inhabit mainly on the undersurface of the leaves, and due to their infestations, the leaves turn yellow to brown. Vidyasagar (1989) elaborated symptomology of various scale insects infesting coconut. Kapadia (1984) reported five species of scale insects, viz. *Chrysomphalus aonidium* (Linn.), *Pseudaulacaspis cockerelli* (Cooley), *Aonidiella orientalis* (Newst.), *Hemiberlesia lataniae* (Sign.) and *Lepidosaphes* sp., infesting on nuts, leaf sheaths and leaflets of coconut. Jalaluddin and Mohanasundaram (1987) studied the biology of *A. destructor* and reported three sprays of 2.5% soap solution at weekly intervals induced complete mortality of *A. destructor* in the nursery. *A. orientalis* was also recorded from all major areas of South India and Lakshadweep Islands and are mostly confined to rachillae and nuts. The adult mussel scale feeds from undersurface of leaves and was mainly observed in Kerala and Minicoy Islands. White needle scale, *Chionaspis* sp., was mainly recorded from coconut seedlings in both Tamil Nadu and Kerala (India) (Josephrajkumar et al. 2010b).

Soft scales were seen as emerging pests in island conditions as well as in a few tracts of Kerala and Tamil Nadu, India. Wax scale has characteristic mound or projections on its body and usually seen singly on coconut leaflets. The brown soft scales are closely associated with black ants and infested leaves show sooty mould. Stellate scales are star-shaped and are seen in few numbers on coconut leaflets. Generally scales are bio-suppressed in nature by their natural enemies, and the main predators of this group include coccinellid beetles and cybocephalids. Aphelinid parasitoids also play a key role in suppression of scales along with predators (Josephrajkumar et al. 2010b).

12.7.6 Aphids

During 2010, the palm aphid, *Cerataphis brasiliensis* Hempel (Aphididae: Hemiptera), was noticed on unfurled spindle leaf of Malayan Green Dwarf variety of coconut seedlings at Kayamkulam, India. These aphids resemble whitefly pupae in many aspects especially the presence of fringe white waxy filaments circumscribing the body. Adult aphids have a mid-dorsal ridge and inconspicuous siphunculi. The aphid colony was found attended by two ant species, viz. *Solenopsis* sp. and *Oecophylla smaragdina* (Josephrajkumar et al. 2011b).

12.7.7 Ash Weevils

During 2010, a new species of Asian grey weevil, *Myllocerus undatus* (Curculionidae: Coleoptera), was recorded as an emerging pest of coconut seedlings at Kayamkulam and Karunagapally, Kerala, India. Mild to medium level of infestation damaging 5–10% leaf lamina of unsplit leaves with typical notching-like symptom along the leaf margins was noticed on majority of the coconut seedlings in these areas. The characteristic feature of this weevil is the presence of three-spined hind femur and is considered as an invasive pest from Sri Lanka (Josephrajkumar et al. 2011a). Coconut ash weevil, *Myllocerus curvicornis* Fab., was also noticed on coconut leaves of young palms especially on outer whorls in different tracts of Kerala (Ponnamma et al. 1982).

12.8 Palm Nematodes

Plant parasitic nematodes are obligate root parasites with well-developed protrusible stylet used for puncturing the plant tissues for food intake and release of the toxin. Among the plant parasitic nematodes reported, burrowing nematode (*Radopholus similis*), lesion nematode (*Pratylenchus coffeae*) and red ring nematode (*Rhadinaphelenchus cocophilus*) causing red ring disease and root-knot nematode (*Meloidogyne* spp.) infest intercrops grown in interspaces of coconut all over the world (Koshy 1986a, b). The characteristic symptoms of nematode infestation are formation of root galls, root rot, necrotic lesions, discolouration of roots, stubby root and root proliferation. Studies relating to the pathogenicity of nematode disease in coconut indicated that higher nematode population cause debilitation in growth and yield of adult palms. Palms with nematode infestation are more susceptible to other diseases caused by fungi and bacteria and further aggravate the disease condition. At seedling stage, higher nematode population causes severe root lesions and death of seedlings. In addition *Scuttellonem brachyurum*, *Rotylenchulus reniformis* and *Helichotylenchus elegans* were also recorded from coconut plantations in West Bengal (Rama and Dasgupta 2000). Soil examination from Kerala collected from coconut-growing locations yielded 29 genera of plant parasitic nematodes, viz. *Aphelenchus* (7%), *Aphelenchoides* (3%), *Atylenchus* (1%), *Caloosia* (6%), *Criconemoides* (11%), *Dolichodorus* (6%), *Ditylenchus* (9%), *Helicotylenchus* (50%), *Hemicriconemoides* (2%), *Hemicycilophora* (3%), *Hirshmanniella* (2%), *Hopololaimus* (25%), etc. (Koshy et al. 1980).

12.8.1 Radopholus similis *(Cobb, 1893) Thorne, 1949*

12.8.1.1 Occurrence and Distribution

The burrowing nematode, *R. similis*, occurs in most tropical and subtropical areas of the world including Florida, Jamaica and Sri Lanka (Van Weerdt et al. 1959a, b; Ekanayake 1964; Latta 1966). In India, *R. similis* was reported from coconut palms in Kerala (Weischer 1967; Koshy et al. 1975). Surveys conducted in coconut plantations in South India recorded 24% incidence of *R. similis* in coconut which co-evolved with black pepper and certain cultivars of banana in the western hills of South India (Koshy 1986a).

12.8.1.2 Nature and Symptoms of Damage

Above-Ground Symptoms Burrowing nematode-infested coconut palms exhibit general decline symptoms which include stunting, yellowing, reduction in leaf number and size, delayed flowering, button shedding and reduced yield.

Below-Ground Symptoms Symptoms on roots are very specific. *R. similis* infestation produces small, elongated, orange-coloured lesions on tender creamy white roots. Consequent to nematode parasitization and multiplication, these lesions enlarge and coalesce to cause extensive rotting of roots. On merging of lesions, cracks develop on the epidermis of the semihard orange-coloured main roots. Lesions and rotting are confined to the tender portion of roots which are not conspicuous on the secondary and tertiary roots. Lesions are not usually seen on the old, hard, dark brown roots. Tender roots of heavily infested coconut seedlings become spongy in texture.

The nematode also attacks the plumule, leaf bases and haustoria of seedlings. The drastic reduction in the number and mass of tertiary feeder roots on parasitization by the nematode limits plant growth (Koshy and Sosamma 1987). Lesions are also not conspicuous on the secondary and tertiary roots since these are narrow and rot quickly upon infestation. In the early stage of infestation, roots develop separate cavities that later merge with each other consequent to feeding and multiplication of nematodes. Multiple cavities and their coalescence destroy the cortex to a great extent, but the stelar tube remains intact. Eggs and all immature stages of nematodes could be located in the cavities in longitudinal sections (Koshy and Sosamma 1982, 1987; Sosamma and Koshy 1991, 1998).

12.8.1.3 Diagnostic Features

Female Vermiform, migratory, endoparasitic, lip region rounded, strong well-developed stylet and oesophagus. Two outstretched ovaries, tail conoid to blunt, rounded terminus

Male Vermiform, migratory and not parasitic. Lip region sub-spheroid, offset, slender stylet, degenerated oesophagus, single testis, spicules paired with bursa extending two thirds length of the tail

Juveniles Vermiform, migratory and parasitic. Lip region rounded, well-developed stylet and oesophagus

12.8.1.4 Biotypes/Pathotypes

Two morphologically indistinguishable races of *R. similis* are known. One is the 'banana race' which parasitizes banana and not citrus and 'citrus race' which parasitizes both banana and citrus (Du Charme and Birchfield 1956). But, the citrus race has been elevated to species rank and named *Radopholus citrophilus* because of enhanced haploid chromosomes coupled with variation in protein profile and pheromone behaviour (Huettel et al. 1984). The coconut isolate of *R. similis* isolated from Kerala, India, has a haploid number of four chromosomes ($n = 4$) and does not infest any of the *Citrus* spp. and *Poncirus trifoliata* (Koshy and Sosamma 1977).

12.8.1.5 Alternate Hosts

Burrowing nematode has a wide host range, and among the 115 plant species tested, 48 are potential hosts which included several crops and weeds in coconut gardens (Koshy and Sosamma 1975; Sosamma and Koshy 1977, 1981). The common alternate hosts are arecanut, banana, black pepper, betel vine and ginger.

12.8.1.6 Survival and Dispersal

Burrowing nematode population survives under field conditions for 6 months in moist soil (27–36 °C) and 1 month in dry soil (29–39 °C), whereas it survives for 15 months in moist soil (25.5–28.5 °C) and 3 months in dry soil (27–31 °C) under greenhouse conditions. The nematode survives in root stumps of felled coconut palms up to 6 months (Sosamma and Koshy 1986). One-year-old coconut seedlings raised in such infested nurseries harbour large populations of the burrowing nematode within roots both internal and external to the husk. Transportation of infested seedlings disseminates nematodes to far-off places. Moreover, in the era of liberalized trade and tariff, plants such as *Anthurium* spp., *Calathea* spp., etc. could disseminate *R. similis* effectively (Koshy 1986a).

12.8.1.7 Nematode Management

Cultural The cultural practices existing in Kerala and Karnataka, India, such as application of oilcakes, farmyard manure and growing of sunhemp in the basins and interspaces and their incorporation as green manure may reduce nematode multiplication. In addition, growing of the crops such as cocoa that enrich the soil with sizeable quantum of shed foliage helps in the buildup of beneficial organisms and antagonistic microorganisms that may hinder nematode multiplication.

The root-knot- and burrowing nematode-susceptible crops such as ginger, turmeric, papaya, solanaceous vegetables and elephant foot yam may be avoided as intercrops, and if grown, planting sites may be changed every year. The practice of taking up vegetable cultivation at the same site for many years consecutively helps in the population buildup of root-knot nematodes. Preference may be given for growing nematode-resistant crops such as nutmeg, cinnamon, cloves, colocasia and tapioca. Maximum care should be taken to avoid crop combinations that are susceptible to the same nematode species. Sanitation methods such as opening pits when the weather is dry to expose the immature stages to sunshine and burning of trash in planting pits and avoiding water run-off from infested to uninfected pits are also advisable in managing nematode problem in coconut.

Mechanical Cut and remove all roots external to the husk of seedlings raised in the field before planting, and nematode-infested intercrops are to be destroyed.

Chemical Control Increased incidence of *R. similis* was noticed in coconut nurseries when banana was used as a shade crop. More than 50% of the seedlings raised in such nurseries failed to establish in the main field. Nematicide treatment was found very effective. With the phasing out of carbofuran and phorate, carbosulphan (0.05%) was found effective.

Biological Coconut seedlings raised in polybags filled with potting mixture enriched with bioagents such as *Paecilomyces lilacinus*, *Pasteuria penetrans* and mycorrhizae suppressed nematode population. A consortium of mycorrhizae consisting of multiple endophytes, i.e. *Acaulospora bireticulata*, *Glomus fasciculatum*, *G. macrocarpum*, *G. mosseae*, *G. versiforme*, *Sclerocystis rubiformis* and *Scutellospora nigra*, was found effective in improving the plant growth and reducing *R. similis* infestation on coconut seedlings (Sosamma 1994).

Host Plant Resistance Use tolerant or less susceptible cultivars or their hybrids in infested areas. Most of the coconut varieties were found susceptible to *R. similis* in India. The dwarf cultivars, viz. Kenthali and Klappawangi, recorded the least nematode multiplication and lesion indices. Similar reactions were noticed in hybrids such as Java Giant × Kulasekharan Dwarf Yellow, Kulasekharam Dwarf Yellow × Java Giant, Java Tall × Malayan Yellow Dwarf and San Ramon × Gangabondam (Sosamma et al. 1980, 1988; Sosamma 1984).

12.8.2 *Red Ring Disease:* Rhadinaphelenchus cocophilus *(Cobb 1919) Goodey 1960*

12.8.2.1 Occurrence and Distribution

Red ring disease of coconut caused by *R. cocophilus* and transmitted by the black palm weevil *R. palmarum* was reported from the West Indies (Trinidad, Tobago, Grenada and St Vincent) and Latin America (Venezuela, Guyana, Surinam, French Guyana, Colombia, Ecuador, Peru, Mexico, Brazil, Panama, Nicaragua, Costa Rica, Honduras, Belize and El Salvador) but is not known to occur in India as well as in Southeast Asia. The disease was first reported in Trinidad by Hart in 1905. Later, it was reported in Grenada (Nowell 1918). *R. cocophilus* was first described by Cobb (1919) as *Aphlenchus cocophilus*, and later Goodey (1960) designated it as *R. cocophilus*. This nematode is also associated with a little leaf disease of coconut and oil palm in Surinam (Van Hoff and Seinhorst 1962). In severely affected coconut gardens, up to 60% crop loss was recorded.

12.8.2.2 Nature and Symptoms of Damage

The most important characteristic symptom is the occurrence of internal lesions. In a cross section of the stem, lesions appear as orange- to brick red-coloured ring (2–4 cm wide). Young palms (2–10 years old) easily succumb to red ring disease. Chlorosis first appears at the tips of the oldest leaves and spreads towards their bases. The brown lower leaves may break across the petiole, or they become partly dislodged at the base and hang down. Nuts are shed prematurely. The crown often topples over in about 4–6 weeks after symptom development. However, the trunk remains standing in the field for several months until it decays.

12.8.2.3 Biology

Diagnostic Features

(a) Female – body very slender, offset head, prominent cephalic framework composed of strong sclerotized arches. Spear with strong basal knobs. Median bulb elongated. Vulva covered by flap of cuticle leading into curved vagina.

(b) Male – body slender. Offset head, prominent cephalic framework. Spear with strong basal knobs, elongated median bulb. The most important characteristic feature is the tail, which on death curves to about four-fifths of a circle. Spicules are slightly arcuate and have prominent rostrum. The tail bears a terminal bursal flap. Four pairs of ventrosubmedian papillae are present.

(c) Juveniles: Pre-adult juvenile has conical rounded head, not set off from the body. The tail is shorter than that of female with a short sharp terminal bearing mucron.

12.8.2.4 Alternate Hosts

Hosts of *R. cocophilus* are confined to the family Arecaceae where the nematode is known to infect over 17 species. Most palm species appear to be susceptible to inoculation by red ring nematode. The most economically important species with red ring disease susceptibility are coconut palm, the African oil palm and the date palm.

12.8.2.5 Nematode Management

There are no simple means of controlling red ring disease, and no effective measures are available as yet for control of the nematodes in living palms. Control is based on prevention rather than cure especially involving the destruction of infested palm material as well as trapping and killing of the weevil vectors before they spread the nematodes.

Cultural Since the nematode did not multiply in the insect nor survived for any appreciable time in the dead tree, the only known reservoir of inoculums was the diseased tree in which the vector palm weevils developed. Thus, the elimination of the diseased tree by burning or poisoning, as soon as red ring symptoms appeared, reduced significantly the available source of the pathogen and also controlled the population of the vectors.

Biological The vector weevil is found to be parasitized by several species of Rhabditidae or Heterorhabditidae throughout Latin America. Since the vector insects can be highly parasitized with the above nematodes, selective pressure can be introduced against the vectors. Such measures are being employed in Trinidad with a species of Rhabditidae (Griffith and Koshy 1989).

Chemical The leaf axils of diseased palms should be sprayed with 0.1% Lannate (Methomyl) for the suppression of weevil. Guard baskets made of 2 cm mesh wire are used to protect frequent outbreaks of the disease. These baskets are filled with fresh infected tissue and sprayed with 0.1% Lannate suspension. The palm weevils are attracted to the tissues in the basket. After 2 weeks, the tissues in the basket are burnt. One guard basket is used per 0.4 ha of palms.

12.8.3 Root-Knot Nematode: Meloidogyne incognita

Meloidogyne incognita infests intercrops in coconut system and not on the main crop. In ginger and turmeric, the root-knot nematode causes galling and rotting of roots and underground rhizomes. The nematode also causes severe injuries by way of gall formation in black pepper and vegetable crops.

12.8.3.1 Nematode Management

Control of root-knot nematode infestation in susceptible crops like black pepper, turmeric, ginger, various vegetables and fruit crops in coconut system could be managed by adopting integrated approaches such as crop rotation, selecting less susceptible crops, changing of planting site every year, fallowing and growing of antagonistic crops like marigold in alternate rows or in patches to reduce the nematode buildup in soil (Rajkumar et al. 2016). Regular application of biological agents such as *Trichoderma*, VAM and *Paecilomyces lilacinus* reduced the damaging effect of nematodes (Sosamma et al. 1990).

12.9 Potential Invasive Pests

Alien invasive species (AIS) is a non-native exotic pest which becomes established in natural or seminatural ecosystems or habitat and threatens native biological diversity. The spread of AIS is now recognized as one of the greatest threats to the ecological and economical well-being. Invasions by alien species imbalance native ecosystems and are likely to breed profusely in the absence of natural enemies in the new environment and cause upsets in biodiversity outcompeting native species.

The introduction of new pests into a locality is brought out in various ways such as (1) through a host as the carrier; (2) inert packing materials carrying the quiescent stages of the pest; (3) insect vectors, birds and air currents; and (4) deliberate, illegal introduction as bioweapons. Though the first two modes of distribution are curtailed by quarantine measures, the latter two are beyond the limitations of pest control by exclusion. This creates a need for biosecurity involving integrated approach that encompasses the policy and regulatory frameworks to analyse and manage the risks in the sectors of food safety and other environmental risks (Shetty et al. 2008).

Biosecurity covers the introduction of plant and animal pests and diseases, introduction of genetically modified organisms and their products and introduction and management of invasive alien species and genotypes. As such it is a holistic concept having a direct relevance to the sustainability of agriculture, food safety and protection of the environment including biodiversity. It is in this context that the likely advent of invasive insect pests like coconut leaf beetle (CLB), *Brontispa longissima* Gestro (Chrysomelidae: Coleoptera), and armoured scale insect, *Aspidiotus rigidus* Reyne (Diaspididae: Hemiptera), to India would be devastating and more likely an issue of biosecurity in our country.

12.9.1 Impending Biosecurity Risks

Coconut leaf beetle, *Brontispa longissima* Gestro, and the armoured scale insect, *Aspidiotus rigidus*, ravaging Maldives and the Philippines, respectively, have not been reported in India till 2016; however, there is an impending danger since it is in the doorsteps already (Rajan et al. 2012; Watson et al. 2014).

12.9.1.1 *Brontispa longissima*

The outbreak of the *B. longissima* in Myanmar and Maldives in recent years poses a great threat and concern to the nearby countries such as India, Sri Lanka and Bangladesh. It is feared that the pest will find its way from Maldives to Sri Lanka and southern parts of India to derail the economy of these important coconut-growing regions of the world. Since invasive pests fail to restrict along political/agroecological boundaries, these countries are ever in red alert zones (Rethinam and Singh 2004). Coconut leaf beetle (CLB) was originally described in 1885 from Aru Islands in Indonesia and from Papua New Guinea. Over a period of 124 years, it has widely spread in over 25 countries in Asia, Australia and Pacific Ocean Islands attacking a number of cultivated and wild ornamental palm species in addition to coconut palms. It is currently distributed in Australia, Pacific Ocean Islands, Malaysia, Singapore, Cambodia, Laos, Thailand, Vietnam, Maldives, the Philippines, Myanmar and China. In Solomon Islands, it is estimated that about 5% of CLB-infested palms die annually. In 1980, coconut palms grown in more than 10,000 ha area in seven provinces in Indonesia were attacked by this beetle. In Maldives, pest outbreaks occurred in several islands of South Attol causing extensive damage to coconut production both in inhabited and uninhabited islands. CLB had caused serious threat to the income generation from tourism industry as well as food security in countries like Maldives, Thailand and Vietnam (Rethinam and Singh 2004; Rajan et al. 2012).

Adult beetles (Fig. 12.9) measure 7.5–10.0 mm long and 1.5–2.0 mm wide, with a conspicuous orange to reddish pronotum. The anterior part of elytra is also orange to reddish in colour. Grubs and adult beetles inhabit the developing unopened still-folded heart leaves of coconut palm and feed on leaf tissues (Fig. 12.10).

The spread of *B. longissima* is mainly through the movement of infested seedlings. Since the flight range of the beetles is low, the natural spread is at a very slow pace. Shipments of ornamental palms from countries having the pest infestation have been the main source of spread within the Asia-Pacific region. Pest management is mainly effected by release of biocontrol agents. Two parasitoids of coconut leaf beetle, viz. *Tetrastichus brontispae* Ferriere (Hymenoptera: Eulophidae), a

Fig. 12.9 Adult beetle of *Brontispa longissima*

Fig. 12.10 Symptom of damage of *Brontispa longissima*

pupal parasitoid, and *Asecodes hisparum* Boucek (Hymenoptera: Eulophidae), a larval parasitoid, have been successfully used in several countries to control the beetle (Rethinam and Singh 2004; Rajan et al. 2012).

A close relative of *B. longissima*, viz. *Plesispa reichei* Chapuis, was reported from Sri Lanka. Belonging to the same family, *P. reichi* is reported only from the island nation feeding on coconut leaflets, and its pronotum is gradually narrowed than *B. longissima*. Presently, the incidence is very sporadic and does not cause any economic damage as per reports from Sri Lanka (Josephrajkumar et al. 2016b).

12.9.1.2 *Wallacea jarawa* (Chrysomelidae: Coleoptera)

A close relative of the chysomelid beetle, *B. longissima*, viz. *Wallacea jarawa*, feeding on the spindle region of coconut seedlings (Fig. 12.11) was recorded during 2014 from South and Little Andaman. The feeding niche of *Wallacea jarawa* confining on coconut spindle is a matter of concern. Though 80–90% of seedlings were infested by the pest damaging about 40% of leaf area, there was no seedling mortality; however, the pest was not observed from any adult palm during the snap survey conducted during October 2014. Invasive nature of *Wallacea jarawa* is under scrutiny, as a close relative, *Wallaceana* sp., was reported from Indonesia.

Adult beetles are brownish with six rows of constrictions on each elytron and measured 4.72 mm long and 0.9 mm wide. They are active fliers, may be for a short distance. Grubs possess short-lateral spines on each body segments and have prominent mandibles for active feeding. They measure 5.75 mm long and 0.8 mm wide. Grubs and adults remain within the folds of the spindle leaves and feed from within. Typical feeding damage can be seen within the leaf folds before unfurling along with faecal matters. In severe cases, the feeding streaks coalesce forming broader lesion with brown margin. Though a few feeding adult beetles were observed in between the leaf folds of emerged leaves, the grubs were mostly confined within the spindle region only. Pupae are located at the point of leaflet attachment to the main petiole. Pupae are exarate with exposed appendages and well-developed wing pads and are mostly located on the point of attachment of leaflet with the main petiole (Prathapan and Shameen 2015; Josephrajkumar et al. 2016b).

Fig. 12.11 Grubs of *Wallacea jarawa* on coconut seedling. (Photo: Josephrajkumar)

12.9.1.3 *Aspidiotus rigidus* (Diaspididae: Hemiptera)

Hard scale, *A. rigidus*, is a close relative of *Aspidiotus destructor*, a minor pest reported from Kerala, Tamil Nadu and other coconut-growing tracts of India. Gradient outbreak of coconut scale insect, *A. destructor*, was observed in Kerala during August to September 2012. Though the pest attack was confined in a limited pocket on coconut leaflets along a homestead farm pond, rise in maximum temperature and reduction in relative humidity and rainfall during June to July 2012 could be the major reasons for the immediate flare-up of the pest which was otherwise not reported as a major pest of the region. Population buildup of the pest was so high that caused severe yellowing as well as drying of coconut leaflets in the region. This could be one of the earlier reports on temperature-induced pest outbreak from India. Comparison of maximum temperature, relative humidity and rainfall data of June 2011 with that of June 2012 revealed increase in 0.8 °C of maximum temperature and reduction in relative humidity and rainfall to the tune of 4.1% and 91.8 mm, respectively. Though *A. destructor* is under check by natural enemies, *A. rigidus* is reported to be a ravaging pest in the Philippines incurring huge loss to coconut growers in that country (Watson et al. 2014). The mobile stage being the crawlers and males are easily drifted away by wind or passively carried through any inert packaging materials, nuts, leaflets, dried spathes, etc. (Josephrajkumar et al. 2016b).

A planned and holistic programme through awareness creation, capacity building on incursion management and strict quarantine are essentially warranted to combat invasions. Creation of an incursion management team comprising of experts from all disciplines as well as an emergency preparedness module would be the need of the hour to tackle accidental introduction of invasive pests.

12.10 Pests of Stored Copra

Copra, a most important commodity in international trade, is being infested by a number of pests, wherever it is stored for more than 5 months at a stretch in godowns and oil mills. The pests cause quantitative and qualitative losses to the commodity. Reports of pests infesting copra are given by various workers – Rutgers (1918) in Sumatra, De Fremery (1929) in Amsterdam, Corbett and Ponnaiah (1937) in Malaya, Peter (1974) in Gilbert and Ellice Islands, Rai and Singh (1977) in Guyana, Laborius et al. (1980) in West Samoa, Mathen (1961) and Nalinakumari (1989) in India and Zipagan and Pecumbaba (1974) in the Philippines. Lever (1969) and Nalinakumari et al. (1992) described the extent and damage caused by various insect pests of copra in detail and suggested management practices.

12.10.1 Ham Beetle: Necrobia rufipes *De Geer (Coleoptera: Cleridae)*

This pest occurs throughout the tropics and is virtually cosmopolitan. This beetle is common and becomes serious in wet season in copra stores in many countries in Asia and the Pacific. The occurrence of this insect is an indicator of the degree of uncleanliness of copra stores – their presence in large numbers indicating inferior management in both preparation of copra and unhygienic condition of stores (Lever 1969). The beetle is blue green, with reddish brown legs and antenna. The life cycle is about 66 days in tropical temperature. This beetle damages copra by extensive feeding by both adults and grubs. The larvae make ramifying tunnels inside the copra and feed within, and pupation also occurs inside the tunnels. The adults do not enter the tunnels. Instead they remain on the surface of copra pieces and eat inwards from the cut edges. The adults have the habit of feeding on their own young ones (Simmons and Ellington 1925). The feeding of young ones along with copra increases the longevity of the adult beetle sixfold when compared with feeding of copra alone (Nalinakumari 1989). In growth indices studies undertaken in nine varieties/cultivars, faster growth and multiplication of this pest were noticed in copra obtained from common cultivars, Laccadive micro and Laccadive ordinary (Nalinakumari and Mammen 1998). The development of immature stages of the beetle is faster, and the quantitative loss caused is up to 12% at the end of 6 months of storage of copra maintained at 8% moisture level (Nalinakumari et al. 1993a).

12.10.2 Sawtoothed Grain Beetle: Oryzaephilus surinamensis *Linn. (Coleoptera: Silvanidae)*

This beetle is slender, dark brown and flat and has a row of sharp teeth like projections on either side of the prothorax. Adults bore in through the cut end of the copra pieces and make extensive galleries between the kernel and testa. The eggs are laid in the galleries, and the emerging larvae feed within the galleries cause severe damage. The galleries within copra are not visible externally by holes or frass. This beetle prefers copra produced from the Dwarf × Tall and Tall × Dwarf hybrids. The damage was less in copra obtained from Chowghat Green Dwarf and WCT × CGD. The faster development and high population of the pest is observed in copra kept at 6% moisture level when compared with 4 or 8% moisture. The loss in weight of copra is 11.7% after 6 months of storage at the 6% moisture level (Nalinakumari et al. 1993a; Nalinakumari and Mammen 1996).

12.10.3 Almond Moth: **Ephestia cautella** *Walk. (Lepidoptera: Phycitidae)*

Almond moth has greyish wings with transverse stripes. It lays eggs on the inner surface of copra which are glued on to it. The egg is pearly white. It hatches in 3–4 days. The caterpillar is pinkish and is 12 mm long when fully grown. The duration of the caterpillar stage varies from 22 to 49 days according to the nature and condition of the food as well as temperature and humidity. The cycle from eggs to emergence of adult averages 45 days under tropical conditions. The emerging larvae make a silk woven mat on the inner surface of copra and remain within. The larvae feed by scraping from the surface and cause significant damage. The pupation occurs within the silken mat, and the emerging adults migrate to fresh copra.

12.10.4 Foreign Grain Beetle: **Ahasverus advena** *Walt. (Coleoptera: Silvanidae)*

This pest occurs in all tropical regions. This small reddish brown oval beetle with smooth elytra and clubbed antennae bores into the copra in between dried endosperm and testa. The damage is caused by the adults and larvae in the same manner as is done by *O. surinamensis*.

12.10.5 Cigarette Beetle: **Lasioderma serricorne** *Fab. (Coleoptera: Anobiidae)*

Cigarette beetle is light brown with the thorax and head bending downwards and presenting a humped appearance. Adults bore into dried slices of copra through the testa, and the entry holes could be seen prominently on the surface. The beetles feed by making galleries and lay eggs, and the emerging larvae feed from within and complete their life cycle inside the produce.

12.10.6 Coffee Berry Borer: **Araecerus fasciculatus** *De Geer (Coleoptera: Anthribidae)*

This greyish beetle has small dark patches on the elytra and prothorax. The adults bore into the copra through the inner side, and entry holes could be prominently seen. The beetles make galleries within the copra, lay their eggs and complete their life cycle inside. The adults and larvae also feed on copra remaining inside the galleries.

12.10.7 Red Flour Beetle: **Tribolium castaneum** *H. (Coleoptera: Tenebrioniidae)*

This beetle is distributed in all regions in India. This is a reddish brown, flat elongate beetle. Its life cycle at 30 °C is 26 days on an average. Adults of this beetle are found in small numbers in copra. They feed and breed on powdered copra and oil cakes.

12.10.8 Short-Winged Beetle: **Carpophilus dimidiatus** *F. (Coleoptera: Nitidulidae)*

This pest is also another cosmopolitan species. This dark brown, flat oval beetle has truncated elytra which does not cover the tip of the abdomen. It develops in about 5 weeks with a range of 26–44 days. The grubs are cream coloured, cylindrical and transparent with rudimentary legs. The adults and grubs are seen in large numbers in the godowns feeding only on decayed kernel but not on good quality copra.

12.10.9 Black Larder Beetle: **Dermestes ater** *De Geer (Coleoptera-Dermestidae)*

This is another widespread beetle that is likely to infect copra. The length of the adult is 8 mm; it is much larger than any other pests mentioned above. The life cycle covers about 6 weeks at 30 °C (Lever 1969).

12.10.10 Management of Stored Pests

Regular chemical treatment of the floor and walls of the storage sheds is to be undertaken to check the insect population and to keep it at a low and virtually harmless level. A malathion spray can be used at 450 g of 25% malathion dispersible powder, in 5 l water for 100 m^2 of floors and walls and applied monthly (Lloyd and Hewitt 1958).

For effective and long-term management of pests, well-dried copra preferably at 5% moisture content should be stored in polythene-/alkathene-lined gunny bags or netted polythene bags. As far as possible, avoid heap storage of copra stock (Nalinakumari 1989). Prophylactic application of malathion at 0.4% on gunny bags before storing copra protects the produce up to 6 months. The residue of the toxicant is below tolerance limit up to 15 days after treatment (Nalinakumari et al. 1993b).

For complete control of the major pests, 4.5 $g^{-1}m^3$ (2.5 g $ai^{-1}m^3$) of aluminium phosphide with 1-day exposure period has been found adequate. The residue of phosphine in copra fumigated as above was below tolerance limit after aeration for 24 h (Nalinakumari et al. 1993b).

To ensure that serious infestation does not develop in copra during storage, the following measures should be adopted (Lever 1969).

- Clean uninfested sacks should be employed. Sacks should be treated with steam to ensure disinfestation.
- The floors, corners and walls should be brushed regularly and thoroughly, and all dust and debris are to be removed.
- Bags should be piled neatly and systematically, with a space left between the piles and walls to increase ventilation and facilitate cleaning.
- The stocks must be removed from the stores for shipment in the order in which it was put, the oldest batch always being removed first.

12.11 Future Strategy

With the impact of climate change and modulation in pest dynamics, a holistic approach is very critical for effective implementation of IPM strategy. Farmer-participatory community mode should be given the topmost priority for successful translation of technologies and field reality. A systematic monitoring and overall palm health management are vital for successful coconut production. Good agricultural practices with emphasis on area-wide approach are very crucial for sustainable pest management. Pest management has now focused towards agroecosystem approach for accomplishing health management. Advancements in nanoporous delivery of pheromones, RNA interference, neuropeptide nanoformulations of entomopathogens and botanicals, enhancing the virulence of *Entomophaga* against biotic and abiotic stresses as well as smart upscaling technologies would go a long way in effective realization in palm health management. Early detection gadget for red palm weevil and effective forewarning strategies would foster IPM farmer-centric and eco-friendly.

Sensitization programmes under the Farmer Field School, information and communication technologies and mobile applications should be effectively clubbed and co-synergized. A planned and holistic programme through awareness creation, capacity building on incursion management and strict quarantine are essentially warranted to combat invasions due to biosecurity threats. Creation of an incursion management team would be the need of the hour to tackle accidental introduction of invasive pests. In an era of genomics, undeciphering signal transduction in pheromone chemistry and molecular identification in areas of taxonomic conflict would make pest management holistic and economically sound.

References

Abdel-Samad SSM, Mahmoud BA, Abbas MST (2011) Evaluation of the fungus, *Beauveria bassiana* (Bals.) Vuill as a bio-control agent against the red palm weevil, *Rhynchophorus ferrugineus* (Oliv.) (Coleoptera: Curculionidae). Egypt J Biol Pest Control 21:125–129

Abdullah MAR (2009) Biological control of the red palm weevil, *Rhynchophorus ferrugineus* (Olivier) (Coleoptera: Curculionidae) by the parasitoid mite, *Rhynchopolipus rhynchophori* (Ewing) (Acarina: Podapolipidae). J Egypt Soc Parasitol 39:679–686

Abozuhairah RA, Vidyasagar PSPV, Abraham VA (1996) Integrated management of red palm weevil, *Rhynchophorus ferrugineus* in date palm plantations of the Kingdom of Saudi Arabia. In: Proceedings of the XX international Congress of Entomology (D029) 25–26 August. Firenze, Italy, p 541

Abraham CC (1994) Pests of coconut and arecanut. In: Chadha KL, Rethinam P (eds) Advances in horticulture-plantation crops and spices crops, Part II. Malhotra Publishing House, New Delhi, pp 709–726

Abraham VA (1979) Investigations on cockchafer beetle *Leucopholis coneophora* Burm. Annual report, Central Plantation Crops Research Institute, Kasaragod, p 276

Abraham VA (1983) Biology, bionomics and control of coconut cock chafer, *Leucopholis coneophora Burm.* Ph.D. thesis, Kerala Agricultural University, Thrissur

Abraham VA (1993) Behaviour of adult beetles of the coconut white grub, *Leucopholis coneophora* Burm. Indian Cocon J 24:2–4

Abraham VA, Kurian C (1973) *Chelisoches morio* F. (Forficulidae: Dermaptera), a predator on eggs and early instar grubs of the red palm weevil *Rhynchophorus ferrugineus* F. (Curculionidae: Coleoptera). J Plantn Crops 1:147–152

Abraham VA, Kurian C (1975) An integrated approach to the control of *Rhynchophorus ferrugineus* F. the red weevil of coconut palm. In: Proceedings of fourth session of the FAO technical workshop party on coconut production, protection and processing, September 14–25, Kingston, Jamaica, pp 1–5

Abraham VA, Mathen K, Kurian C (1966) Aids to detect reed palm weevil infestation in coconut palm. Cocon Bull 20:148–152

Abraham VA, Abdulla Koya KM, Kurian C (1975) Evaluation of seven insecticides for the control of red palm weevil (*Rhynchophorus ferrugineus* Fab.). J Plantn Crops 3:71–72

Abraham VA, Abdulla Koya KM, Kurian C (1989) Integrated management of red palm weevil (*Rhynchophorus ferrugineus* Oliv.) in coconut gardens. J Plantn Crops 16(Suppl):159–162

Abraham VA, Al-Shuaibi M, Faleiro JR, Aozuhairah RA, Vidyasagar PSPV (1998) An integrated approach for the management of red palm weevil, *Rhynchophorus ferrugineus* Olivier-A key pest of date palm in the Middle-East. Sultan Qaboos Univ J Sci Res Agric Sci 3:77–83

Advani R (1984) Ecology, status and post-natal development of the black rat infesting coconut and cacao. In: Bavappa KVA et al. (eds) Proceedings of PLACROSYM V. Indian Society for Plantation Crops, Kasaragod, pp 626–632

Advani R (1984) Rodent pest management in the coconut plantations of India and its islands. Indian Cocon J 15(10):3–9

AICRP (1997) Pests on coconut. AICRP on Palms – Annual report 1997. Central Plantation Crops Research Institute, Kasaragod, p 68

Alam M (1962) A list of insects and mites of Eastern Pakistan report of Agriculture Department. Dhaka, p 107

Alfazariy AA (2004) Notes on the survival capacity of two naturally occurring entomopathogens on the red palm weevil, *Rhynchophorus ferrugineus* (Olivier). Egypt J Biol Pest Control 14:423

Allou K, Morin JP, Kouassi P, Hala NF Rochat D (2008) Sex ratio and female sexual status of the coconut pest, *Oryctes monoceros* (Coleoptera: Dynastidae), differ in feeding galleries and pheromone-baited traps. Bull Entomol Res 98:581–586

Al-Shawaf AM, Al-Shagag A, Al-Bagshi M, Al-Saroj S, Al-Bather S, Al-Dandan AM, Abdallah AB, Faleiro JR (2013) A quarantine protocol against red palm weevil *Rhynchophorus ferrugineus* (Olivier) (Coleoptera: Curculionidae) in date palm. J Plant Protect Res 53:409–415

Ambily P, Nandakumar C, Bai H (2009) Management of coreid bug, *Paradasynus rostratus* Dist. on coconut palm in homesteads having mixed cropping. Entomon 34(2):63–70

Ansaloni T, Perring TM (2002) Biology of *Aceria guerreronis* Keifer on queen palms. In: Morales-Malacara JB, Rivas G (eds) XI international congress of acarology. Program and Abstract Book. Universidad Nacional Autónoma de México, Distrito Federal México, p 145

Antony J (1962) Bionomics, life history and morphology of *Nephantis serinopa* M., a major pest of coconut palm. M.Sc. thesis, Kerala University, Trivandrum, p 105

Antony J, Kurian C (1961) A note on the bacterial disease of *Nephantis serinopa* Meyrick a lepidopterous pest of coconut palm. Indian Cocon J 14:82–84

Antony J, Kurian C (1975) Physical and biotic factors which exert a check on the population density of *Oryctes rhinoceros* Linn. in India. In: Paper presented in at the fourth session of the FAO Technical Working Party on coconut production, protection and processing, Kingston, Jamaica, September, vol 14–25, p 1975

Aubrey M, Gillian W, Watson JB (2014) First record of eggplant mealybug, *Coccidohystrix insolita* (Hemiptera: Pseudococcidae), on Guam: potentially a major pest Biodiversity. Data J 2014(2):e1042. https://doi.org/10.3897/BDJ.2.e1042

Ávalos JA, Martí-Campoy A, Soto A (2014) Study of the flying ability of *Rhynchophorus ferrugineus* (Coleoptera: Dryophthoridae) adults using a computer-monitored flight mill. Bull Entomol Res 104:462–470

Ayyar TVR (1917) A new pest of the coconut palm on the West Coast (*Conthyla rotunda*). Madras Agric Dept Year Book, pp:76–83

Ayyar TVR (1940) A handbook of economic entomology for South India. Government Press, Madras, pp 328–336

Baburayanayak N (1970) Studies on the biology and some aspects of ecology of the coconut black headed caterpillar, *Nephantis serinopa* Meyrick (Lepidoptera: Xylorictidae). M.Sc. thesis, UAS, Bangalore, India, p 229

Bakthavatsalam N, Singh SP (1996) L-Tryptophan as an ovipositional attractant for *Chrysoperla carnea* (Stephens) (Neuroptera: Chrysopidae). J Biol Control 10:21–22

Bakthavatsalam N, Singh SP, Tandon PL, Chaudhary M, Preethi S (1999) Behavioral responses of key parasitoids of *Opisina arenosella* Walker (Lepidoptera: Noctuidae) to the kairomones. J Biol Control 13:7–14

Banerjee A, Dangar TK (1995) *Pseudomonas aeruginosa* a facultative pathogen of red palm weevil (*Rhynchophorus ferrugineus*). World J Microbiol Biotechnol 11:18–620

Banu JG, Rajendran G, Subramanian S (2003) Susceptibility of red palm weevil, *Rhynchophorus ferrugineus* (Olivier) to entomopathogenic nematodes. Ann Plant Protect Sci 11:104–106

Bedford GO (1980) Biology, ecology and control of palm rhinoceros beetle. Annu Rev Entomol 25:309–339

Bedford GO (2013) Biology and management of palm dynastid beetles-Recent advances. Ann Rev Entomol 58:353–372

Beevi SP, Beena S, Lyla KR, Varma AS, Mathew MP, Natarajan IL (1999) *Hirsutella thompsonii* var. *synnematosa* Samson, McCoy and O'Donnell on coconut mite, *Aceria guerreronis* (Keifer), a new report for India. J Trop Agric 37:91–93

Beevi SN, Maheswari P, Mohan P, Paul A (2004) Laboratory screening of newer insecticides against red palm weevil, *Rhynchophorus ferrugineus* (Oliv.). Entomon 29:401–404

Bhat SK, Sujatha A (1987) Relative efficiency of two live traps in capturing small mammals in coconut-cocoa mixed habitat. J Plantn Crops 15:40–142

Biju B, Sudhadevi K, Danger TK, Sathiamma B (1995) Biological suppression of *Oryctes rhinoceros* by re-release of *Baculovirus oryctes* in an infected contiguous area. J Plantn Crops 23:62–63

Booth RG, Cox ML, Madge RB (1990) IIE guides to insects of importance to man. 3. Coleoptera. CAB International, Wallingford

Burand JP (1998) Nudiviruses. In: Miller LK, Bal LA (eds) The insect viruses. Plenum, New York, pp 69–90

Burkill IH (1917) *Scolia erratica* Smith, a parasite of the red coconut weevil, *Rhynchophorus ferrugineus*. Straits Settlements. Garden Bull 1:399–400

Butani DK (1975) Insect pests of fruit crops and their control-date palm. Pesticides 9:40–42

Buxton PA (1920) Insect pests of dates and the date palm of Mesopotamia and elsewhere. Bull Entomol Res 1:287–303

Cabrera RIC (2002) Biological control of the coconut mite *Aceria guerreronis* (Acari: Eriophyidae) with the fungus *Hirsutella thompsonii* and its possible integration with other control methods. In: Fernando LCP, de Moraes GJ, Wickramananda IR (eds) Proceedings of the international workshop on the coconut mite (*Aceria guerreronis*). Coconut Research Institute, Lunuwila, pp 89–103

Catley A (1969) The coconut rhinoceros beetle *Oryctes rhinoceros* (L.). PANS 15:18–30

Chan E, Elevitch CR (2006) *Cocos nucifera* (coconut). Species profiles for Pacific Island Agroforestry. http://www.traditionaltree.org

Chandrika Mohan (2005) Studies on *Apanteles taragamae* Vier. an early instar parasitoid of coconut leaf eating caterpillar *Opisina arenosella* Walker. Ph.D. thesis submitted to Kerala University, Thiruvananthapuram, Kerala, India, p 159

Chandrika Mohan, Nair CPR (2000a) Distinguish the nut infestation by eriophyid mite and coreid bug in coconut. Indian Cocon J 30(3):60–61

Chandrika Mohan, Nair CPR (2000b) Effect of *Clerodendron infortunatum* on grubs of coconut rhinoceros beetle, *Oryctes rhinoceros*. In: Muraleedharan N, Rajkumar R (eds) Recent advances in plantation crops research, 2000. Allied Publishers Ltd, New Delhi, pp 297–299

Chandrika Mohan, Shameer KS (2002) Laboratory screening of intercrops for feeding by *Opisina arenosella* Walk. the leaf eating caterpillar pest of coconut. In: National conference on coastal agricultural research, 6–7 April 2002, Extended Summaries, pp 140–142

Chandrika Mohan, Sujatha A (2006) The Coconut leaf caterpillar, *Opisina arenosella* Walker. CORD 22(Special Issue):25–78

Chandrika Mohan, Nair CPR, Rajan P (2000) Scope of botanical pesticides in the management of *Oryctes rhinoceros* L. and *Rhynchophorus ferrugineus* Oliv. affecting coconut palm. Abstract No. Sl-0-17, Entomocongress, November 5–8, 2000, Trivandrum

Chandrika Mohan, Rajan P, Nair CPR (2007) Incidence of whitefly (*Aleurocanthus arecae* David) on coconut foliage. J Plantn Crops 35:119–121

Chandrika Mohan, Nair CPR, Rajan P (2010a) Large scale field validation of biocontrol strategy for management of coconut leaf eating caterpillar *Opisina arenosella* Walker utilizing indigenous parasitoids. In: Singh HP, Thomas GV (eds) Organic horticulture – principles, practices and technologies. Westville Publishing House, New Delhi, p 440

Chandrika Mohan, Nair CPR, Nampoothiri CK, Rajan P (2010b) Leaf-eating caterpillar (*Opisina arenosella*)-induced yield loss in coconut palm. Int J Trop Insect Sci 30(3):132–137

Chandrika Mohan, Radhika MK, Josephrajkumar A, Rajan P (2016a) Coconut water as a promising culture media for *Hirsutella thompsonii* Fisher, a pathogen of coconut mite. In: Chakravarthy AK, Sridhara S (eds) Arthropod diversity and conservation in the tropics and sub-tropics. Springer, Singapore, pp 173–181

Chandrika Mohan, Josephrajkumar A, Hegde V, Krishnakumar V, Renjith PB, Anjali AS, Chowdappa P (2016b) Gradient outbreak and bio-suppression of spiralling whitefly in coconut gardens in South India. Indian Cocon J 59(8):9–12

Chandrika Mohan, Rajan P, Josephrajkumar A (2016c) Plantation crops. In: Mani M, Shivaraju C (eds) Mealybugs and their Management in Agricultural and Horticultural crops. https://doi.org/10.1007/978-81-322-2677-2_58

Cito A, Mazza G, Strangi A, Benvenuti C, Barzanti GP, Dreassi E, Turchetti T, Francardi V, Roversi PF (2014) Characterization and comparison of *Metarhizium* strains isolated from *Rhynchophorus ferrugineus*. FEMS Microbiol Lett 355:108–115

Cobb NA (1893) Nematodes, mostly Australian and Fijian. Macleary Mem. Vol. Linn. Soc. N South Wales, p. 252–308

Cobb NA (1919) A newly discovered nematode *Aphelenchus cocophilus* n. sp. connected with a serious disease of the coconut palm. West Indian Bull 17:203–210

Cock MJ, Perera PACR (1987) Biological control of *Opisina arenosella* Wlk. (Lepidoptera: Oecophoridae). Biocont News Inform 8(4):283–310

Copeland EB (1931) The coconut. Macmillan and Co. Ltd, London, pp 68–113

Corbett GH (1932) Insects of coconut in Malaya. Malay State Federated Dept Agric Ser Bull 10:1–106

Corbett GH, Ponnaiah (1937) Fungi and bacteria associated with copra in Malaya. Deparment of Agriculture. Strait settlement and Federated Malay States Bull No.28 (Science series)

Cork A, Hall DR (1998) Application of pheromones for crop pest management in the Indian subcontinent. J Asia Pac Entomol 1(1):35–49

Cox ML (1993) Red palm weevil, *Rhynchophorus ferrugineus* in Egypt. FAO Plant Protn Bull 41:30–31

CPCRI (1979) Nematodes, fungi, insect and mites associated with the coconut palm. Tech Bull No. 2. ICAR-CPCRI, Kasaragod, p 236

CPCRI (2012) Two new species of mealy bugs identified. CPCRI News 31(2):2

CPCRI (2014) Annual report 2014-2015. ICAR-Central Plantation Crops Research Institute, Kasaragod, p 139

CPCRI (2016) Annual report 2015-2016. ICAR-Central Plantation Crops Research Institute, Kasaragod, p 210

Dangar TK, Banerjee A (1993) Infection of red palm weevil by microbial pathogens. In: Nair MK, Khan HH, Gopalasundaram P, Rao EVVB (eds) Advances in coconut research and development. Oxford IBM, Publishing Co, New Delhi, pp 531–533

Danger TK, Geetha L, Jayapal SP, Pillai GB (1991) Mass production of entomopathogen, *Metarhizium anisopliae* in coconut water wasted from copra making industry. J Plantn Crops 19(1):54–69

Davis TA (1964) Devising instruments for coconut research. In: Proceedings of 2nd session FAO technical Working Party, Coconut Production, Protection & Processing, Colombo, pp 78–85

De Fremery (1929) Omnivorous insects in copra. Report from De Indischa Mercuur. In: No.20 Amsterdam

Dembilio Ó, Riba JM, Gamón M, Jacas JA (2015) Mobility and efficacy of abamectin and imidacloprid against *Rhynchophorus ferrugineus* in *Phoenix canariensis* by different application methods. Pest Manag Sci 71:1091–1098

Dharmaraju E (1962) A checklist of parasites, hyper parasites, predators and pathogens of the coconut caterpillar, *Nephantis serinopa* Meyrick, recorded in Ceylon and in India and their distribution in these countries. Ceylon Cocon Qtly 13(3–4):102–111

Dharmaraju E (1963) Biological control of coconut leaf caterpillar *Nephantis serinopa M.* in Ceylon. Bull Cocon Res Inst (Ceylon) 21:46

Du Charme P, Birchfield W (1956) Physiological races of the burrowing nematode. Phytopathol 46:615–616

Ekanayake UBM (1964) Annual report of the crop protection officer, Coconut Research Institute for 1963. Ceylon Cocon Qtly 15:50–57

El-Sebay Y, El-Lattef MAK, Makhlouf TM (2003) Laboratory rearing of red palm weevil, *Rhynchophorus ferrugineus* Oliv. (Coleoptera: Curculionidae) in artificial diet. Egyptian. J Agric Res 81:551–554

El-Shafie AAF, Faleiro JR, Abo-El-Saad MM, Aleid SM (2013) A meridic diet for laboratory rearing of red palm weevil, *Rhynchophorus ferrugineus* (Coleoptera: Curculionidae). Sci Res Essay 8(39):1924–1932

Esteban-Duran J, Yela JL, Beitia Crespo F, Alvarez J (1998) Biology of red palm Weevil, *Rhynchophorus ferrugineus* Olivier in the laboratory and field, life cycle, biological characteristics in its zone of introduction in Spain, biological method of detection and possible control. Boletin de Sanidad Vegetal Plagas 24:737–748

Evans GA (2008) The whiteflies (Hemiptera: Aleyrodidae) of the world and their host plants and natural enemies. Available from: http://keys.lucidcentral.org/keys/v3/whitefly/PDF_PwP%20 ETC/world-whitefly-catalog-Evans.pdf

Faleiro JR (2006) A review of the issues and management of the red palm weevil *Rhynchophorus ferrugineus* (Coleoptera: Rhynchophoridae) in coconut and date palm during the last one hundred years. Intl J Trop Insect Sci 26(3):134–154

Fernando LCP, Aratchige NS (2010) Status of coconut mite *Aceria guerreronis* and biological control research in Sri Lanka. In: Sabelis MW, Bruin J (eds) Trends in acarology. Springer, Amsterdam, pp 379–384

Fernando LCP, Chandrasiri KAS (1997) More evidence on the attraction of males to the female sex pheromone of *Opisina arenosella* Walker. COCOS 12:79–81

Fernando LCP, Wickramananda IR, Aratchige NS (2000) Status of coconut mite, *Aceria guerreronis* in Sri Lanka. In: Proceedings of international workshop on coconut eriophyid mite, CRI, Sri Lanka, p 128

Fernando LCP, Waidyarathne KP, Perera KFG, de Silva PHPR (2010) Evidence for suppressing coconut mite *Aceria guerreronis* by inundative release of the predatory mite, Neoseiulus baraki. Biol Control 53:108–111

Flechtmann CHW (1989) *Cocos weddliana* Wendel (Palmae, Arecaceae) a new host plant for *Eriophyes guerreronis* (Keifer 1965) Acari: eriophyidae in Brazil. Int Acarol 15:241

Fletcher TB (1914) Some South Indian insects. Government Press, Madras, p 565

Francardi V, Benvenuti C, Roversi PF, Rumine P, Barzanti G (2012) Entomopathogenicity of *Beauveria bassiana* (Bals.) Vuill. and *Metarhizium anisopliae* (Metsch.) Sorokin isolated from different sources in the control of *Rhynchophorus ferrugineus* (Olivier) (Coleoptera Curculionidae). Redia XVC:49–55

Francis AW, Stocks IC, Smith TR, Boughton AJ, Mannion CM, Osborne LS (2016) Host plants and natural enemies of rugose spiraling whitefly (Hemiptera: Aleyrodidae) in Florida. Florida Entomol 99(1):150–153

George MV, Sathiamma B, Vijaya Kumar K (1982) A technique to estimate field populations of coconut leaf eating caterpillar. In: Bavappa KVA et al (eds) Proceedings of PLACROSYM, vol V. Indian Society for Plantation Crops, Kasaragod, pp 638–643

Ghode MK, Nayak UK, Pawar AD (1987) Biological control of coconut black headed caterpillar *Opisina arenosella* Walker (Lepidoptera: Cryptophasidae) in coastal tracts of Orissa. In: National symposium on integrated pest control-progress and perspective. Orissa, India, pp 21–22

Ghosh CC (1911) Life history of Indian insects. III. The rhinoceros beetle (*Oryctes rhinoceros*) and the red or palm weevil (*Rhynchophorus ferrugineus*). Memoirs of the Department of Agriculture in India. Entomol Ser 2(10):193–215

Ghosh CC (1923) *Oryctes rhinoceros* and other important palm pests in Burma. In: Report of proceedings of the V, ENTOMOLOGICAL meeting. Pusa, pp 99–103

Ghosh SM, Abdurahiman UC (1984) Bioethology of *Meteoridea hutsoni* Nixon (Hymenoptera: Braconidae), a parasite of *Opisina arenosella* Wlk. the black headed caterpillar pest of coconut. Entomon 9(31):31–34

Ghosh SM, Abdurahiman UC (1985) Bioethology and evaluation of comparative efficiency of *Apanteles taragamae*, a larval parasitoid of coconut caterpillar, *Opisina arenosella*. Philippine J Coconut Stud 10(2):4–8

Gitau CW, Gurr GM, Dewhurst CF, Fletcher MJ, Mitchell A (2009) Insect pests and insect-vectored diseases of palms. Australian J Entomol 48(4):328–342

Goodey JB (1960) *Rhadinaphelenchus cocophilus* (Cobb 1919). The nematode associated with red ring disease of coconuts. Nematologica 5:98–10

Gopal M, Alka G, Sathiamma B, Chandrika Mohan, Nair KRC, Soniya VP (2000) A fungal pathogen of lace bug and leaf eating caterpillar, two pests of coconut palm. CORD 16(1):49–60
Gopal M, Saleena A, Mc Coy CW (2002) Study of microflora associated with coconut eriophyid mite as a preliminary step towards pathogen isolation. Curr Sci 82(1):22–24
Gopal M, Gupta A, Thomas GV (2006) Prospects of using *Metarrhizium anisopliae* to check the breeding of insect pest *Oryctes rhinoceros* in coconut leaf vermicomposting sites. Bioresour Technol 97:1801–1806
Gopinadhan PB, Mohandas N, Nair KPV (1990) Cytoplasmic polyhedrosis virus infecting red palm weevil of coconut. Curr Sci 59:577–580
Gorick BD (1980) Release and establishment of the baculovirus disease of *Oryctes rhinoceros* in Papua New Guinea. Bull Entomol Res 70(3):445–453
Green EE (1898) A coconut palm caterpillar. Trop Agricult 26:298–308
Green EE (1906) Rhynchophorus ferrugineus. Trop Agricult, p 27
Green EE (1908) Remarks on Indian scale insects (Coccidae), Part 3. With a catalogue of all species hitherto recorded from the Indian continent. Memoirs of the Department of Agriculture in India. Entomol Ser 2:15–46
Gressitt JL (1953) The coconut rhinoceros beetle. Bernice P. Bishop Mus Bull 212:1–157
Griffith R, Koshy PK (1989) Nematode parasites of coconut and other palms. In: Luc M, Sikora R, Bridge J (eds) Plant Parasitic nematodes in sub tropical and tropical agriculture. CAB International, Wallingford, pp 363–386
Gurmit Singh (1987) Naphthalene balls for the protection of coconut and oil palm against *Oryctes rhinoceros* Linn. Planter 63(2):286–292
Hallett RH, Gries G, Borden JH, Czyzewska E, Oehlschlager AC, Pierce HD Jr, Angerilli NPD, Rauf A (1993) Aggregation pheromone of two Asian Palm weevils, *Rhynchophorus ferrugineus* and *R. vulneratus*. Naturwissenschaften 80:328–331
Hallett RH, Perez AL, Gries G, Gries R, Pierce HD Jr, Yue J, Oehlschlager AC, Gonzalez LM, Borden JH (1995) Aggregation pheromone of coconut rhinoceros beetle, *Oryctes rhinoceros* (L.) (Coleoptera: Scarabaeidae). J Chem Ecol 21(10):1549–1570. https://doi.org/10.1007/BF02035152
Hallett RH, Crespi BJ, Borden JH (2004) Synonymy of *Rhynchophorus ferrugineus* (Olivier) (1790) and *R. vulneratus* (Panzer), 1798 (Coleoptera, Curculionidae, Rhynchophoridae). J Nat Hist 38:2863–2882
Haq MA (2001) Culture and rearing of *Aceria guerreronis* and its predators. Entomon 26:297–302
Hart JH (1905) Coconut disease, Bulletin of Miscellaneous Information. R Bot Gard Trinidad 6:241–243
Henry GM (1917) The coconut red weevil, *Rhynchophorus ferrugineus*. Trop Agric 48:218–219
Hirashima Y (1989) A check list of Japanese insects. Entomological Laboratory, Faculty of Agriculture, Kyushu University, Fukuoka, p 1767
Holloway JD, Cock MJW, Desmier de Chenon R (1987) Systematic account of South-east Asian pest Limacodidae. In: Cock MJW, Godfray HCJ, Holloway JD (eds) Slug and Nettle caterpillars. CAB International, Wallingford, pp 15–117
Howard FW, Moore D, Giblin-Davis RM, Abad RG (2001) Insect on palms. CAB International, London, p 400
Huettel RN, Dickson DW, Kaplan DT (1984) *Radopholus citrophilus* sp. n. (Nematoda), a sibling species of *Radopholus similis*. Proceedings of Helminthol Soc Wash 51:32–35
Huger AM (1966) A virus disease of the Indian rhinoceros beetle, *Oryctes rhinoceros* Linn., caused by a new type of insect virus, Rhabdion virus oryctes. J Invertr Pathol 8:38–41
Hurpin B, Fresneau M (1973) Etude en laboratoire des facteurs de fecondite de *Oryctes monoceros* Ol. et *O. rhinoceros* L. Ann Soc Entomol 9:89–117
Hutson JC (1933) The red weevil of coconut. Department of Agriculture, Ceylon, Leaflet No.22
Jacob TK (1996) Introduction and establishment of baculovirus for the control of rhinoceros beetle, *Oryctes rhinoceros* in the Andaman Islands (India). Bull Entomol Res 86:257–262

Jalaluddin M, Mohanasundaram M (1987) Biological studies in *Aspidiotus destructor* Sign. a pest of coconut in Tamil Nadu. Indian Cocon J 19(11):14–19

Jalaluddin M, Mohanasundaram M (1993) Biological studies of the coconut mealybug *Palmicultor* sp. Indian Cocon J 23(11):9–11

Jayanth KP, Nagarkatti S (1984) Testing *Bessa remota* (Dip: Tachnidae) against *Opisina arenosella* (Lepidoptera: Cryptophasidae) and other lepidopteran hosts in India. Entomophaga 29(4):415–419

Jayaratnam TJ (1941) A study of the control of the coconut caterpillar *Nephantis serinopa* Meyr. in Ceylon with special reference to its eulophid parasite *Trichospilus pupivora* Ferr. Trop Agric (Ceylon) 96:3–21

Josephrajkumar A, Krishnakumar V (2016) Good agricultural practices in coconut cultivation. Indian Cocon J 59(2):13–15

Josephrajkumar A, Rajan P, Chandrika Mohan (2010a) Scale insects, mealybugs and whiteflies of coconut palm and their management. In: Ananda Bose CV (ed) A Nutshell: essays on Coconut Coconut Development Board. Kera Bhavan, Kochi, pp 199–206

Josephrajkumar A, Rajan P, Chandrika Mohan, Jacob PM (2010b) Re-invasion and bio-suppression of spiralling whitefly, *Aleurodicus dispersus* Russell on coconut in Minicoy Island. Indian Cocon J 53(2):20–23

Josephrajkumar A, Rajan P, Chandrika Mohan, Thomas RJ (2011a) First record of Asian grey weevil (*Myllocerus undatus*) on coconut from Kerala, India. Phytoparasitica 39(1):63–65. https://doi.org/10.1007/s12600-010-0137-1

Josephrajkumar A, Rajan P, Chandrika Mohan, Krishnakumar V (2011b) Report on the palm aphid, *Cerataphis brasiliensis* on 'Kalparaksha' coconut cultivar. Phytoparasitica 39(4):389–391. https://doi.org/10.1007/s12600-011-0174-4

Josephrajkumar A, Rajan P, Mohan C, Thomas RJ (2012) New distributional record of buff coconut mealybug (*Nipaecoccus nipae*) in Kerala, India. Phytoparasitica 40:533–535. https://doi.org/10.1007/s12600-012-0260-2

Josephrajkumar A, Thomas S, Shanavas M, Chandrika Mohan (2014a) Managing the hidden villain in coconut garden. Kerala Karshakan e-Journal (September issue) 2(4):21–27

Josephrajkumar A, Chandrika Mohan, Shanavas M, Thomas S, Namboothiri CGN (2014b) Defending rhinoceros beetle attack on coconut through botanicals and ecological engineering. In: Dinesh R, Eapen SJ, Senthil Kumar CM, Ramakrishnan Nair R, Devasahayam S, John Zachariah T, Anandaraj M (eds) Abstracts PLACROSYM XXI. International symposium on plantation crops. ICAR-Indian Institute of Spices Research, Kozhikode, p 129

Josephrajkumar A, Chandrika Mohan, Thomas S, Namboothiri CGN, Shanavas M (2014c) Subduing red palm weevil attack on coconut through fine-tuned management approaches. In: Muralidharan K, Rajesh MK, Muralikrishna KS, Vijayan J, Jeyasekhar S (eds) Book of abstracts national conference on sustainability of coconut, arecanut and cocoa farming technological advances and way forward. CPCRI, Kasaragod, p 86

Josephrajkumar A, Chandrika Mohan, Krishnakumar V (2015) Management of rhinoceros beetle. Indian Cocon J 58(7):21–23

Josephrajkumar A, Chandrika Mohan, Chaturvedi VK (2016a) Suppression of growth and endopeptidases of red palm weevil, *Rhynchophorus ferrugineus* (Olivier) infesting coconut using proteinase inhibitors. Entomon 41(4):283–292

Josephrajkumar A, Chandrika Mohan, Merin Babu, Augustine Jerard B, Krishnakumar V, Hegde V, Chowdappa P (2016b) Invasive pests of coconut. Technical bulletin No. 93, ICAR-CPCRI, Regional Station, Kayamkulam, India, p 28

Joy PJ, Joseph KJ (1972) Incidence of *Nephantis serinopa* M. in Badagara area (Kerala), its past history and future prospects for control. Cocon Bull 2:2–5

Joy PJ, Joseph KJ (1977) Role of pupal parasites in the biological control of the black headed caterpillar. Indian Cocon J 8(1):1–2 &7

Kaakeh W (2006) Toxicity of imidacloprid to developmental stages of *Rhynchophorus ferrugineus* (Curculionidae: Coleoptera): Laboratory and field tests. Crop Protect 25:432–439

Kanagaratnam P, Pethiyagoda U, Velu MS (1983) Effect of four commercial preparations of *Bacillus thuringiensis* on *Opisina arenosella* Walker. COCOS 1:7–10

Kapadia MN (1982) Note of laboratory evaluation of some insecticides against the leaf eating caterpillar, *Nephanitis serinopa* Meyr. Indian Cocon J 13(3):10–20

Kapadia MN (1984) New records of coconut scale insects and their natural enemies in Gujarat. Indian Cocon J 14(11):9–10

Keifer HH (1965) Eriophyid studies B-14. California Departmen of Agriculture, Bureau of Ento, p 20

Khan HH, Sujatha S, Sivapuja PR (2003) Regionally different strategies in managing eriophyid mite. In: Singh HP, Rethinam P (eds) Coconut Eriophyid Mite- Issues and strategies. Proceedings of the international workshop on coconut mite held at Bangalore. Coconut Development Board, Kochi, pp 89–96

Koshy PK (1986a) Research on plant parasitic nematodes at the Central Plantation crops Research Institute, October 1972 to June 1986. CPCRl Regional Station, Kayangulam, p 45

Koshy PK (1986b) The burrowing nematode, *Radopholus similis* (Cobb 1893), Thorne, 1949. In: Swarup G, Dasgupta DR (eds) Plant parasitic nematodes of India problems and progress. IARI, New Delhi, pp 223–248

Koshy PK, Sosamma VK (1975) Host range of the burrowing nematode, *Radopholus similis* (Cobb, 1893) Thorne, 1949. Indian J Nematol 5:255–257

Koshy PK, Sosamma VK (1977) Identification of physiological race of *Radopholus similis* populations infesting coconut, arecanut and banana in Kerala, South India. Indian Phytopathol 30:415–416

Koshy PK, Sosamma VK and Sundararaj P (1980) Survey of plant parasitic nematodes of coconut in India. In: Proceedings, PLACROSYM II, UPASI, Ootacamaund, 26-29 Jun. 1979. (Eds.) CS Venkataram et al. p. 45–49

Koshy PK, Sosamma VK (1982) A simple method for inoculation and production of symptoms by *Radopholus similis* on coconut. Indian J Nematol 12:200–203

Koshy PK, Sosamma VK (1987) Pathogenicity of *Radopholus similis* on coconut (*Cocos nucifera* L.) seedlings under green house and field conditions. Indian J Nematol 17:108–118

Koshy PK, Sosamma VK, Nair CPR (1975) Preliminary studies on Radopholus similis (Cobb, 1893) Thome, 1949 infesting coconut and arecanut palms in South India. Indian J Nematol 5:26–35

Krishnakumar R, Maheshwari P (2003) Effect of cashew apple extracts on the mortality of red palm weevil, *Rhynchophorus ferrugineus* (Oliv.). Insect Environ 9:83–84

Krishnakumar R, Sudha G (2002) Indian tree pie *Dendrocitta vagabunda parvula* (Whistler and Kinnear) (Corvidae). A predatory bird of red palm weevil *Rhynchophorus ferrugineus* (Oliv.). Insect Environ 8:133

Krishnakumar V, Jerard BA, Josephrajkumar A, Thomas GV (2014) Four decades of CPCRI interventions on coconut-based Island ecosystem at Minicoy, Lakshadweep. Technical Bulletin No. 86. ICAR-CPCRI, Kasaragod, p 27

Kumar ARV (1997) Biology and management of arecanut white grubs, *Leucopholis* spp. (Coleoptera: Scarabaeidae) in Karnataka. Ph.D. thesis, Uniersity of Agricultural Science, Bangalore, p 230

Kurian C, Pillai GB, Abraham VA, Mathen K (1972) Record of a coreid bug (nut crinkler) as a new pest of coconut in India. Curr Sci 12:37

Kurian C, Pillai GB, Antony J, Abraham VA, Natarajan P (1983) Biological control of insect pest of coconut. In: Nayar NM (ed) Proceedings of international symposium of coconut research and development. Wiley Eastern, New Delhi, pp 361–375

Kurian C, Abraham VA, Ponnamma KN (1984) Attractants, an aid in red palm weevil management. In: Bavappa KVA et al (eds) Proceedings of PLACROSYM, vol V. Indian Society of Plantation Crops, Kasaragod, pp 581–585

Laborius A, Burgstaller H, Von Keyscrlingk N (1980) Storage pests in Tonga and West Samoa. Alfua Agric Bull 5(3):59–61

Latta R (1966) Attempts to relate nematodes to lethal yellowing disease of coconut palms in Jamaica. Trop Agric (Trinidad) 43:59–68

Lawson-Balagbo LM, Gondim MGC Jr, de Moraes GJ, Hanna R, Schausberger P (2008) Exploration of the acarine fauna on coconut palm in Brazil with emphasis on *Aceria guerreronis* (Acari: Eriophyidae) and its natural enemies. Bull Entomol Res 98:83–96

Leefmans S (1920) *Rhynchophorus ferrugineus* F. Inst Plantenziekten Meded 43:1–90

Lefroy HM (1906) Indian insect pests, Calcutta, p 318

Lever RJ (1969) Pests of coconut palm. Food and Agriculture Organization of the United Nations, Agricultural Studies, No. 77, p 190

Llacer E, Martinez de Altube MM, Jacas JA (2009) Evaluation of the efficacy of *Steinernema carpocapsae* in a chitosan formulation against the red palm weevil, *Rhynchophorus ferrugineus*, in Phoenix canariensis. Biocontrol 54:559–565

Llácer E, Negre M, Jacas JA (2012) Evaluation of an oil dispersion formulation of imidacloprid as a drench against *Rhynchophorus ferrugineus* (Coleoptera, Curculionidae) in young palm trees. Pest Manag Sci 68(6):878–882

Lloyd CJ, Hewitt PS (1958) Relative susceptibility to pyrethrin in oil of Coleoptera and Lepidoptera infesting stored products. Bull Entomol Res 49:177–185

Lomer CJAD (1986) Release of baculovirus Oryctes into *Oryctes monoceros* population in the Seychelles. J Invertbr Pathol 47(3):237–246

Longworth JF, Carey G (1980) The use of an indirect enzyme-linked immunosorbent assay to detect baculovirus in larvae and adults of *Oryctes rhinoceros* from Tonga. J Gen Virol 47:431–438

Mallik B, Chinnamade Gowda C, Jayappa J, Guruprasad H, Onkarappa S (2003) Coconut eriophyid mite in India – issues and strategies. In: Singh HP, Rethinam P (eds) Coconut Eriophyid Mite- Issues and strategies. Proceedings of the international workshop on coconut mite held at Bangalore. Coconut Development Board, Kochi, pp 27–34

Malumphy C, Moran H (2009) Red palm weevil, *Rhynchophorus ferrugineus*. Plant Pest Fact Sheet. http://www.defra.gov.uk/fera/plants/plant

Mani M (2010) Origin, introduction, distribution and management of the invasive spiralling whitefly *Aleurodicus dispersus* Russell in India. Karnataka J Agric Sci 23(1):59–75

Manjunath TM (1985) Coconut black headed caterpillar on banana and coconut. FAO Plant Prot Bull 33(2):71–72

Mariau D (1969) *Aceria guerreronis* Keifer: récent ravageur de la cocoteraie Dahoméenne. Oleagineux 24:269–272

Martin JH (2004) The whiteflies of Belize (Hemiptera: Aleyrodidae) Part 1 – Introduction and account of the subfamily Aleurodicinae Quaintance & Baker. Zootaxa 681:1–119

Martin JH (2008) A revision of *Aleurodicus* Douglas (Sternorrhyncha, Aleyrodidae), with two new genera proposed for palaeotropical natives and an identification guide to world genera of Aleurodicinae. Zootaxa 1835:1–100

Massa B, Lo Verde G (2008) Gli antagonisti naturali del Punteruolo rosso delle palme. In: La ricerca scientifica sul punteruolo rosso e gli altri fitofagi delle palme in Sicilia. Regione Siciliana-Assessorato Agricoltura e Foreste Dipartimento Interventi Infrastrutturali, Servizi allo Sviluppo, vol I, pp 73–78

Mathen K (1961) Storage pests of coconut. Ware Housing News Lett 4:3

Mathen K, Kurian C (1967) Insecticide trials against *Rhynchophorus ferrugineus* F., the coconut weevil (Curculionidae: Coleoptera). Indian J Agric Sci 37:521–523

Mathen K, Sathiamma B, Kurian C (1968) A mite to combat coconut caterpillar. Indian farming 9(5):32–33

Mathew TB, Ambily Paul, Beevi SN, Saradamma K (2000) Population dynamics of coconut mite, *Aceria guerreronis* Keifer on developing nuts. Paper presented at Entomocongress 2000, Perspective plan for the new millennium, 5–8, November, Thiruvananthapuram, p 195

Mayilvaganan M, Nair CPR (2002) Physicochemical changes due to coreid bug infestation on coconut. J Plantn Crops 30(2):61–63

Mayilvaganan M, Nair CPR, Shanavas M, Nair SS (2003) Field assay of locally synthesized ferrugineol for trapping *Rhynchophorus ferrugineus*. Indian Cocon J 33:8–9

Mazza G, Francardi V, Simoni S, Benvenuti C, Cervo R, Faleiro JR, Llacer E, Longo S, Nannelli R, Tarasco E (2014) An overview on the natural enemies of *Rhynchophorus* palm weevils, with focus on *R. ferrugineus*. Biol Control **77**:83–92

Menon KPV, Pandalai KM (1960) The coconut palm – a monograph, Indian Central Coconut Committee, Ernakulam, South India, p 384

Meyrick E (1905) *Nephantis serinopa*. J Bombay Nat Hist Soc 16:603

Mohamed UVK, Abdurahiman UC, Remadevi OK (1982) Coconut caterpillar and its natural enemies in Kerala, India. University of Calicut, Kerala, India, Zool Monogr No.2, p 162

Mohan P, Faizal MH (2004) Quantitative loss due to combined infestation of coreid bug, *Paradasynus rostratus* Distant and eriophyid mite, *Aceria guerreronis* in coconut. Pest Manag Hortic Ecol 10(2):191–195

Mohan KS, Pillai GB (1982) A method for laboratory scale mass cultivation of *Metarhizium anisopliae*. Folia Microbiol 27:281–283

Mohan KS, Pillai GB (1993) Biological control of *Oryctes rhinoceros* Linn. using an Indian isolate of *Oryctes baculovirus*. Insect Sci Applic 14(5–6):551–558

Mohan KS, Jayapal SP, Pillai GB (1983) Baculovirus disease in *Oryctes rhinoceros* population in Kerala. J Plantn Crops 11:154–161

Mohan KS, Jayapal SP, Pillai GB (1985) Diagnosis of baculovirus infection in coconut rhinoceros beetles by examination of excreta. Z Pflkranh Pflschutz 93(4):379–383

Mohan KS, Jayapal SP, Pillai GB (1989) Biological suppression of coconut rhinoceros beetle, *Oryctes rhinoceros,* Linn., in Minicoy, Lakshadweep by *Oryctes baculovirus*-Impact on pest population and damage. J Plantn Crops 16:163–170

Mohandas N, Remamony KS (1993) Emerging pest problems in coconut and their control. In: Nair K, Khan HH, Gopalasundaram P, Rao EVVB (eds) Advances in coconut research and development. Oxford and IBH Publishing Co. Pvt. Ltd, New Delhi, pp 493–503

Mohanty JN, Prakash A, Rao J, Pawar AD, Patnaik NC, Gupta SP (2000) Parasitization of the coconut black-headed caterpillar, *Opisina arenosella* Walk. in Puri district of Orissa by field release of hymenopterous parasitoids. J Appl Zool Res 11(1):17–19

Monty J (1978) The coconut palm rhinoceros beetle, *Oryctes rhinoceros* Linn. in Mauritius and its control. Revue agricole et Sucriere de I'iie Maurice 57(2):60–76

Moore D, Alexander L (1987) Aspects of migration and colonization of the coconut palm by the coconut mite, *Eriophyes guerreronis* (Keifer) (Acari: Eriophyidae). Bull Entomol Res 77:641–650

Moore D, Howard FW (1996) Coconuts. In: Lindquist EE, Sabelis MW, Bruin J (eds) Eriophyoid mites: their biology natural enemies and control. Elsevier, Amsterdam, pp 561–570

Moore D, Alexander L, Hall RA (1989) The coconut mite, *Eriophyes guerreronis* Keifer in St. Lucia: yield losses and attempt to control it with acaricide, polybutene and *Hirsutella* fungus. Trop Pest Manag 35:83–89

Moore D, Ridout MS, Alexander L (1991) Nutrition of coconuts in St Lucia and relationship with attack by coconut mite *Aceria guerreronis* Keifer. Trop Agric 68:41–44

Mullakoya P (2003) Status report on the incidence of coconut mite in Union territory of Lakshadweep. In: Singh HP, Rethinam P (eds) Coconut eriophyid mite- issues and strategies. Proceedings of the international workshop on coconut mite held at Bangalore. Coconut Development Board, Kochi, pp 97–98

Muralidharan K, Mathew J, Thampan C, Amarnath CH, Anithakumari P, Mohan C, Vijayakumar K, Sairam CB, Nair CPR (2001) Pilot sample survey on the incidence and yield loss due to eriophyid mite in coconut in Alappuzha district. Indian Cocon J 32:28–32

Murphy ST, Briscoe BR (1999) The red palm weevil as an alien invasive: biology and the prospects for biological control as a component for IPM. Biocont News Inform 20(1):35–46

Murthy KS, Gour TB, Reddy DDR, Ramesh Babu T, Zaheeruddeen SM (1995a) Host preference of coconut black headed caterpillar *Opisina arenosella* Walker for oviposition and feeding. J Plantn Crops 23(2):105–108

Murthy KS, Gour TB, Reddy DDR, Babu TR, Zaheruddeen SM (1995b) Pheromone baits for coconut black headed caterpillar, *Opisina arenosella* Walker. Insect Environ 1(2):3–4

Muthiah C, Bhaskaran R, Kannaiyan S (2001) Bioecology and control of eriophyid mite of coconut: an Indian experience. The Planter 77:255–263

Muthukrishnan P, Rangarajan M (1974) Laboratory studies on the control of black headed caterpillar *Nephantis serinopa* Meyr. by microorganisms. Labdev J Sci Technol Part B Life Sci 12(3):106–108

Nadarajan L, ChannaBasavanna GP (1980) Population dynamics of coconut black headed caterpillar, *Nephantis serinopa* Meyrick (Lepidoptera: Cryptophasidae) and its parasites. Mysore J Agric Sci 14(4):533–541

Nagarkatti S (1973) Biological control campaign against *Nephantis serinopa* M. in India in the light of modern concept. J Plantn Crops 1(1):28–31

Nair CPR (2002) Status of coconut eriophyid mite *Aceria guerreronis* Keifer in India. In: Fernando LCP, de Moraes GJ, Wickramanada IR (eds) Proceedings of international workshop on coconut mite *Aceria guerreronis*. Coconut Research Institute, Lunuwila, pp 9–12

Nair CPR, Daniel M (1982) Pests of arecanut. In: Bavapa KVA, Nair MK, Premkumar T (eds) The arecanut palm (*Areca catechu* Linn.). CPCRI, Kasaragod, pp 162–184

Nair CPR (1983) Effect of some insecticides on the control of cacao mealybug. Proceedings of the second annual symposium on plantation crops. Plant protection (entomology, microbiology, nematology, plant pathology and rodentology). PLACROSYM II 1979, p 461–465

Nair CPR, Nair SS (2003) Role of pheromones in the integrated management of coconut red weevil. Indian Cocon J 33:8–10

Nair CPR, Sathiamma B, Chandrika Mohan, Gopal M (1998) Newer approaches in the integrated management in coconut. Indian Cocon J 29(4):99–103

Nair MRGK (1986) Insects and mites of crops of India. Indian Council of Agricultural Research, New Delhi, p 83

Nair MRGK, Visalakshi A, Koshy G (1980) A new root infesting mealybug of coconut. Entomon 5(3):245–246

Nair SS, Abraham VA, Nair CPR (2000a) Efficiency of different food baits in combination with pheromone lures in trapping adults of red palm weevil, *Rhynchophorus ferrugineus*. Pestology 24:3–5

Nair, CPR, Rajan P, Chandrika Mohan (2000b) Distribution, ecology and management of the eriophyid mite, *Aceria guerreronis* Keifer affecting coconut. Paper presented at Entomo Congress 2000, Perspective plan for the new millennium 5–8, November, Thiruvananthapuram, p 180

Nair CPR, Chandrika Mohan, Rajan P, Sandhya SR, Gopinath A (2003) Technological advances in the management of coconut eriophyid mite, *Aceria guerreronis* Keifer. In: Singh HP, Rethinam P (eds) Coconut eriophyid mite- issues and strategies. Proceedings of the international workshop on coconut mite held at Bangalore. Coconut Development Board, Kochi, pp 99–103

Nair CPR, Rajan P, Chandramohanan R, Anithakumari P (2004) Current status in the management of root (wilt) disease and eriophyid mite affecting coconut. In: Abstracts of First Indian Horticulture Congress, 6–9, November, 2004, Horticultural Society of India, p 70

Nair CPR, Rajan P, Chandrika Mohan (2005) Coconut eriophyid mite, *Aceria guerreronis* Keifer – an overview. Indian J Plant Protec 33(1):1–10

Nair CPR, Rajan P, Chandrika Mohan, Gopal M (2008) Management of rhinoceros beetle, *Oryctes rhinoceros* Linn., in coconut gardens by biocontrol method. In: Singh HP, Thomas GV (eds) Organic horticulture: principles, practices and technologies. ICAR-CPCRI, Kasaragod, pp 342–347

Nalinakumari T (1989) Studies on the extent of damage caused by pests of stored copra and control of important pests. Ph.D. thesis submitted to Kerala Agricultural University, Thrissur, p 171

Nalinakumari T, Mammen KV (1996) Moisture content of copra- its effect on the extent of damage caused by insect pests. J Plantn Crops 24:339–341

Nalinakumari T, Mammen KV (1998) Varietal preference, growth and development of *Necrobia rufipes* De Geer and *Oryzaephilus surinamensis* Linn. in stored copra. J Trop Agric 36:37–40

Nalinakumari T, Mammen KV, Mohandas N (1992) Occurrence and nature of damage caused by pests of stored copra. Indian Cocon J 23(7):7–9

Nalinakumari T, Mammen KV, Mohandas N (1993a) Effect of moisture content of copra on the population build up of *Oryzaephilus surinamensis* Linn. and *Necrobia rufipes* De Geer. Fifth Kerala Sci. Congress, Kottayam, 28–30 January, 1993, pp 429–430

Nalinakumari T, Mammen KV, Mohandas N (1993b) Safer use of insecticides for the protection of copra from insect pests. J Plantn Crops 21:223–226

Narayanan K, Veenakumari K (2003) Nuclear polyhedrosis viruses from coconut black headed caterpillar *Opisina arenosella* Walker and sorghum spotted stem borer, *Chilo partellus* (Swinhoe). J Biol Control 17(1):97–98

Narendran TC, Abdurahiman UC, Mohamed UVK, Knutson L (1979) Field incidence of *Nephantis serinopa* Meyr. in the Malabar area of Kerala state. Indian J Plant Protec 6(2):91–92

Naseema Beevi S, Mathew TB, Bai H, Saradamma K (2003) Status of eriophyid mite in Kerala – Resume of work done. In: Singh HP, Rethinam P (eds) Coconut eriophyid mite- issues and strategies. Proceedings of the international workshop on coconut mite held at Bangalore. Coconut Development Board, Kochi, pp 64–75

Navia D, de Moraes GJ, Roderick G, Navajas M (2005) The invasive coconut mite *Aceria guerreronis* (Acari: Eriophyidae): origin and invasion sources inferred from mitochondrial (16S) and nuclear (ITS) sequences. Bull Entomol Res 95:505–516

Negloh K, Hanna R, Schausberger P (2008) Comparative demography and diet breadth of Brazilian and African populations of the predatory mite *Neoseiulus baraki*, a candidate for biological control of coconut mite. Biol Control 46:523–531

Nirula KK (1954) War against rats. Bull Indian Central Coconut Comm 6:243–247

Nirula KK (1955a) Investigations on the pests of coconut palm. Part I. The coconut palm and pests of coconut palm. Indian Cocon J 8:118–130

Nirula KK (1955b) Investigations on the pests of coconut palm. Part II. *Oryctes rhinoceros* Linn. Indian Cocon J 8:161–180

Nirula KK (1956a) Investigation on the pest of palm, Part IV, *Rhynchophorus ferrugineus*. Indian Cocon J 9:229–247

Nirula KK (1956b) Investigations on the pests of coconut palm. Part-III. *Nephantis serinopa* Meyrick. Indian Cocon J 9(2):101–131

Nirula KK (1958) Investigations on the pests of the coconut palm. Part V *Leucopholis coneophora* Burm. Indian Cocon J 12:10–34

Nirula KK, Menon KPV (1957) Insect pests of coconut palms in India. FAO Plant Protect Bull 6(1):1–5

Nirula KK, Antony J, Menon KPV (1951) Investigations on the pests of coconut palm, the coconut caterpillar, *Nephantis serinopa* Meyr. Life history and habits. Indian Cocon J 4(4):217–223

Nirula KK, Antony J, Menon KPV (1952) A new pest of coconut palm in India. Indian Cocon J 12(1):10–34

Nirula KK, Antony J, Sahasranaman KN, Menon KPV (1954) Observations on the biology and control of the coconut slug caterpillar, *Conthyla rotunda* H. Indian Cocon J **7**:142–155

Nirula KK, Radha K, Menon KPV (1955a) The green muscardine disease of *Oryctes rhinoceros* Linn. I Symptomatology, epizootology and economic importance. Indian Cocon J 9(2):3–10

Nirula KK, Radha K, Menon KPV (1955b) The green muscardine disease of *Oryctes rhinoceros* Linn. II The causal organism. Indian Cocon J 9(2):83–89

Nowell W (1918) Root disease of coconut palms in Grenada. Agric News (Bridgetown) 17:398–399

Oblisami G, Ramamoorthi K, Rangaswami G (1969) Studies on the pathology of some crop pests of South India. Mysore J Agric Sci 3:86–98

Oehlschlager AC, Chinchilla C, Gonzalez LM, Jiron LF, Mexon R, Morgan B (1993) Development of a pheromone-based trapping system for *Rhynchophorus palmarum*. J Econ Entomol 86:1381–1392

Ooi PAC (1988) Insects in Malaysian agriculture. Kuala Lumpur, Malaysia; Tropical Press Sdn. Bhd, p 106

Ortega CA, Rodriguez VJ, Garibay CV (1967) Investigaciones preliminares sobre el eriófido del fruto del cocotero *Aceria guerreronis* Keifer, en la Costa Grande Guerrero (Mexico). Oleagineux 6:371–372

Perera PACR (1984) Biological control of pests. Cocon Bull 1(2):49–50

Perera PACR (1987) Studies on *Opisina arenosella* Walker and its natural enemies in Sri Lanka. Ph.D. thesis, University of London, UK, p 399

Perera PACR (1993) Integrated control of *Opisina arenosella* Walker. In: Nair MK, Khan HH, Gopalasundaram P, Rao EVVB (eds) Advances in coconut research and development. Oxford & IBA Publ. Co. Ltd, New Delhi, pp 443–454

Perera PACR, Hassel MP, Godfray HCJ (1989) Population dynamics of the coconut caterpillar *Opisina arenosella* Walker (Lepidoptera: Xylorictidae) in Sri Lanka. COCOS **7**:42–59

Peter DM (1974) Report of the Govt. of Gilbert and Ellice Island colony on a survey of insect pests of crop. FAO UNDP Rep, p 35

Philip BM, Mathai S, Jacob A (1982) A nuclear polyhedrosis virus of the black headed caterpillar *Nephantis serinopa* (Meyrick) (Lepidoptera: Cryptophasidae). Curr Sci 51:6–11

Pillai NK (1919) Coconut, the wealth of Travancore. Agric J India 14:608–628

Pillai GB, Nair KR (1982) A note on *Trichospilus pupivora* Ferr., a parasite of *Nephantis serinopa* Meyr. Entomon **7**(2):225–226

Pillai GB, Nair KR (1983) *Xanthopimpla nana nana* Schulz. (Hymenoptera: Ichneumonidae) a new pupal parasitoid of *Opisina arenosella* Wlk. on coconut. J Plantn Crops 11:65–69

Pillai GB, Nair KR (1986) Additions to the natural enemy complex of the coconut caterpillar, *Opisina arenosella* Walker. J Plantn Crops 14(2):138–140

Pillai GB, Nair KR (1989) Observations on *Xanthopimpla punctata* F. (Hymenoptera: Ichneumonidae), a pupal parasitoid of *Opisina arenosella*. J Plantn Crops 16(Suppl):173–177

Pillai GB, Nair KR (1993) A checklist of parasitoids and predators of *Opisina arenosella* Walker on coconut. Indian Cocon J 3(9):2–9

Pillai GB, Nair KR (1995) Superiority of the solitary parasitoids over gregarious species in the biological suppression of the coconut caterpillar, *Opisina arenosella* Walk. J Plantn Crops 23(1):19–27

Ponnamma KN, Kamalakshyamma PG, Kurian C, Nambiar PTN (1982) Incidence of *Myllocerus curvicornis* (Fab.) in relation to nutritional status of coconut. Philippine J Coconut Stud 1&2(VII):40–41

Prathapan KD (1996) Outbreak of spiralling whitefly, *Aleurodicus dispersus* Russell (Aleyrodidae: Hemiptera) in Kerala. Insect Environ 2:36–38

Prathapan KD, Shameen KM (2015) Wallacea sp. (Coleoptera: Chrysomelidae) – a new spindle infesting leaf beetle on coconut palm in the Andaman and Nicobar Islands. J Plantn Crops 43(2):162–164

Prathibha PS (2015) Behavioural studies of palm white grubs, *Leucopholis* spp. (Coleoptera: Scareabaeidae) and evaluation of new insecticides for their management. Ph.D thesis, UAS Bangalore, p 119

Prathibha PS, Kumar ARV, Subaharan K (2013) Ethology of coconut root grub chafer, *Leucopholis coneophora* Burmeister (Melolonthinae: Scarabaeidae). Intl J Agric Food Sci Technol 4(2):24–28

Purrini K (1989) Baculovirus oryctes release into *Oryctes monoceros* population in Tanzania, with special reference to the interaction of virus isolates used in our laboratory infection experiments. J Invertbr Pathol 53(3):285–300

Pushpalatha NA, Veeresh GK (1995) Population fluctuation of coconut black headed caterpillar, *Opisina arenosella* Walker (Lepidoptera: Xylorictidae). J Plantn Crops 23(1):44–47

Puttarudriah M, Shastry SKS (1964) Incidence of the coconut caterpillar, *Nephantis serinopa* Meyr. in Bangalore. Curr Sci 33:340

Radhakrishnan B, Mathew TB, Premila KS, Mohan P (2003) New report of mealybugs occurring inside the perianth of immature nuts in coconut. Insect Environ 9(2):53–54

Rahalkar GW, Harwalkar MR, Rananavare HD, Shantaram K, Gopalayengar AR (1973) Laboratory studies on radiation sterilization of the red palm weevil (*Rhynchophorus ferrugineus* Oliv.) males. J Plantn Crops 1(Suppl):141–146

Rahalker GW, Tamhankar AJ, Shantaram K (1978) A artificial diet for rearing red palm weevil, *Rhynchophorus ferrugineus* Oliv. J Plantn Crops 6:61–64

Rai BK, Singh W (1977) Studies on insect pests of copra. Bull Grain Tech 15(3):59–61

Rajagopal V, Arulraj S (2003) Towards helping farming community. Central Plantation Crops Research Institute, Kasaragod, p 115

Rajamanickam K, Kennedy FJS, Lourduraj AC, Raveendran TS (1992) Attractants- an aid in rhinoceros beetle (*Oryctes rhinoceros* L.) management. Indian Cocon J 23(1):6–7

Rajan P, Nair CPR (2005) Coreid bug of coconut. Indian Cocon J 36(6):11–12

Rajan P, Chandrika Mohan, Nair CPR (2007) Technical folder on Coconut mite, *Aceria guerreronis* Keifer. CPCRI, Regional Station, Kayangulam

Rajan P, Chandrika Mohan, Nair CPR, Josephrajkumar A (2009) Integrated pest management in coconut. Technical Bulletin 55. ICAR-CPCRI, Regional Station, Kayamkulam, p 20

Rajan P, Josephrajkumar A, Sujatha P (2010a) Gradient outbreak of coconut slug caterpillar, *Macroplectra nararia* Moore in East coast of India. CORD 27(1):61–69

Rajan P, Josephrajkumar A, Chandrika Mohan, Subaharan K (2010b) Emerging pests of coconut. In: Thomas GV, Krishnakumar V, Jerard BA (eds) Improving productivity and profitability of coconut farming. Proceedings of international conference on coconut biodiversity for prosperity. ICAR-CPCRI, Kasaragod, pp 348–356

Rajan P, Nair CPR, Josephrajkumar A, Chandrika Mohan (2012) Another invasive pest, coconut leaf beetle, *Brontispa longissima* (Gestro), an imminent biosecurity threat at the doorsteps of India. J Plantn Crops 40(2):140–143

Rajkumar RKE, Rajesh MK, Sabana AA, Nagaraja NR, Shahin S, Subaharan K (2016) Molecular identification of entomopathogenic nematode isolate and its virulence to white grub, Leucopholis burmeisteri (Coleoptera: Scarabaeidae). Vegetos 29:4. https://doi.org/10.4172/2229-4473.1000187

Rama K, Dasgupta MK (2000) Population ecology and community structure of plant parasitic nematodes associated with coconut and arecanut in Northern West Bengal. Indian J Nematol 30(2):175–118

Ramachandran CP (1991) Effect of gamma radiation on various stages of red palm weevil, *Rhynchophorus ferrugineus* F. J Nuclic Agric Biol 20:218–221

Ramachandran CP (1998) Biotypical variability among four populations of red palm weevil *Rhynchophorus ferrugineus* Fab/Oliv. from different parts of India. CORD 14(1):26–41

Ramachandran CP, Kurian C, Mathew J (1963) Assessment of damage to coconut due to *Oryctes rhinoceros* Linn.-Nature and damage caused by the beetle and factors involved in the estimation of loss. Indian Cocon J 17:3–12

Ramani S (2000) Fortuitous introduction of an aphelinid parasitoid of the spiraling whitefly *Aleurodicus dispersus* Russell (Aleyrodidae: Homoptera) into the Lakshadweep Islands, with notes on host plants and other natural enemies. J Biol Control 14:55–60

Ramaraju K, Rabindra RJ (2002) Palmyra, *Borassus flabellifer* Linn. (Palmae): a host of the coconut eriophyid mite *Aceria guerreronis* Keifer. Pest Manag Hortic Ecosyst **7**:149–151

Ramaraju K, Natarajan K, Sundara Babu PC, Palanisamy S, Rabindra RJ (2000) Studies on coconut eriophyid mite *Aceria guerreronis* Keifer in Tamil Nadu, India. In: Proceedings of international workshop on coconut eriophyid mite. CRI, Sri Lanka, pp 13–31

Ramle M, Glare, T, Jackson, T, Mohd BM, Norman K, Ramlah AAS (2001) The sensitivity of polymerase chain reaction method in diagnosis in the presence of *Oryctes rhinoceros* virus.

In: Proceedings of PIPOC. International Palm Oil Congress, Kuala Lumpur, Malaysia, 20–22 August 2001, pp 594–601
Rao YR (1924) An outbreak of *Nephantis serinopa* at Mangalore. In: Report on the proceedings of the V entomology meeting Pusa. New Delhi, India, pp 92–98
Rao YR, Cherian MC, Ananthanarayanan KP (1948) Infestations of *Nephantis serinopa* Meyrick in South India and their control by the biological method. Indian J Entomol 10(2):205–247
Rao VP, Ghani MA, Sankaran T, Mathur KC (1971) A review of the biological control of insects and other pests in South East Asia and Pacific region. Commonwealth Institute of Biological Control. Technical Communications No. 6. Bureaux. Farham Royal, UK, p 149
Rao PR, Rao AG, Laxminarayana K (1979) New record of predatory spiders of coconut black headed caterpillar *Opisina arenosella* Walker (= *Nephantis serinopa* Meyrick) in Andhra Pradesh. APAU Res J 4(1–4):58
Reginald C (1973) Principal insect pests. In: Coconuts. Tropical agriculture series. Longmans, London
Rethinam P (2003) International efforts on coconut eriophyid mite, *Aceria guerreronis* Keifer, Problem – an overview. In: Singh HP, Rethinam P (eds) Coconut eriophyid mite- issues and strategies. Proceedings of the international workshop on coconut mite held at Bangalore. Coconut Development Board, Kochi, pp 10–21
Rethinam P, Singh SP (2004) Coconut hispine beetle, *Brontispa longissima* (Gestro) (Coleoptera: Chrysomelidae). CORD 20(1):1–20
Ridley HM (1919) Report on destruction of coconut palms of beetles, Govt. Press, Singapore
Rutgers AAL (1918) Report of the Director General Exptl. Station of the General Association of the Rubber Planters of East coast of Sumatra, p 44
Sadakathulla S, Ramachandran TK (1990) Efficacy of naphthalene balls on the control of rhinoceros beetle attacks in coconut. COCOS 8:23–25
Salama HS, Foda MS, El-Bendary MA, Abdel-Razek A (2004) Infection of red palm weevil *Rhunchophorus ferrugineus*, by spore forming bacilli indigenous to the natural habitat in Egypt. J Pest Sci 77:27–31
Santhosh Babu PB, Prabhu VKK (1987) How many larval instars are there in *Opisina arenosella*? Entomon 12(1):39–42
Saradamma K, Beevi SN, Mathew PB, Sreekumar V, Jacob AV, Vishalakshy A (2000) Management of coconut eriophyid mite in Kerala. Paper presented at Entomocongress 2000, Perspective plan for the new millennium, 5–8 November, 2000, Thiruvananthapuram
Sathiamma B, George MV, Kurian C (1973) Seasonal variation in the larval population of *Nephantis serinopa* Meyrick in the field as correlated with meteorological factors. J Plantn Crops 1(Suppl):161–163
Sathiamma B, Jayapal SP, Pillai GB (1987) Observations on spiders (Order Araneae) predacious on the coconut leaf eating caterpillar, *Opisina arenosella* Walker in Kerala: Biology of *Rhene indicus* Tikader (Salticidae) and *Cheiracanthium* sp. (Clubionidae). Entomon 12(2):121–126
Sathiamma B, Sabu AS, Pillai GB (1996) Field evaluation of the promising species of indigenous parasitoids in the biological suppression of *Opisina arenosella* Walker, the coconut leaf eating caterpillar. J Plantn Crops 24(1):9–15
Sathiamma B, Nair CPR, Koshy PK (1998) Outbreak of a nut infesting eriophyid mite, *Eriophyes guerreronis* (K.) in coconut plantations in India. Indian Cocon J 29(2):1–3
Selvaraj K, Sundararaj R, Venkatesan T, Chandish RB, Jalali SK, Gupta A, Mridula HK (2017) The invasive Rugose spiraling whitefly, *Aleurodicus rugioperculatus* Martin in India. J Biol Control https://doi.org/10.18641/JBC/0/0/108767
Shamseldean MM, Atwa AA (2004) Virulence of Egyptian steinernematid nematodes used against the red palm weevil, *Rhynchophorus ferrugineus* (Oliv.). Egypt J Biol Pest Control 14:135–140
Shanas S, Job J, Joseph T, Anju Krishnan G (2016) First report of the invasive rugose spiraling whitefly, *Aleurodicus rugioperculatus* Martin (Hemiptera: Aleyrodidae) from the Old World. Entomon 41(4):365–368

Shetty PK, Parida A, Swaminathan MS (2008) Biosecurity. National Institute of Advanced Studies:255

Simmons P, Ellington GW (1925) The ham beetle- *Necrobia rufipes* De Geer. J Agric Res 30:845–863

Singh SP (1999) Pest management – the ecofriendly approach. In: Ravi N (ed) The Hindu. Survey of Indian Agriculture National Press, Chennai, p 208

Soroker V, Blumberg D, Haberman A, Hamburger-Rishad M, Reneh S, Talebaev S, Anshelevich L, Harari AR (2005) Current status of red palm weevil infestation in date palm plantations in Israel. Phytoparasitica 33:97–106

Sosamma VK (1984) Studies on the burrowing nematode of coconut. Ph.D. thesis, Kerala University, Trivandrum, Kerala, India

Sosamma VK (1994) Interactions between Vesicular Arbuscular Mycorrhizae and Burrowing Nematode on Coconut. Final Report ICAR AP Cess Fund Scheme 1988–93. CPCRI, Kayangulam, Kerala, India

Sosamma VK, Koshy PK (1977) Additional hosts of the burrowing nematode *Radopholus similis* infesting coconut palm in South India. Plant Dis Rep 61:760–761

Sosamma VK, Koshy PK (1981) Other host records of *Radopholus similis*. Indian J Nematol 11:73–74

Sosamma VK, Koshy PK (1986) Survival of *Radopholus similis* in host free soil. Indian J Nematol 16:74–76

Sosamma VK, Koshy PK (1991) Etiology – nematodes. Coconut Root (Wilt) Disease Monograph Series No. 3. CPCRI, Kasaragod, India

Sosamma VK, Koshy PK (1998) Etiology – nematodes. In: Nampoothiri KUK, Koshy PK (eds) Coconut root (wilt) disease, Rev edn. CPCRI, Kasaragod, pp 21–27

Sosamma VK, Koshy PK, Rao EVVB (1980) Susceptibility of coconut cultivars and hybrids to *Radopholus similis* in field. Indian J Nematol 10:250–252

Sosamma VK, Koshy PK, Rao EVVB (1988) Response of coconut cultivars to the burrowing nematode, *Radopholus similis*. Indian J Nematol 18:136–137

Sosamma VK, Geetha SM, Koshy P K (1990) Effect of the fungus Paecilomyces lilacinus on the burrowing nematode, Radopholus similis infesting betelvine. Paper presented at the seminar on "Bioagents in nematode management" at IARI, New Delhi, July 13, 1990

Stocks I, Hodges G (2012) Pest Alert- DACS-P-01745. Florida Department of Agriculture and Consumer Services, Division of Plant Industry, p 6

Stride G (1977) Coconut palm rhinoceros beetle. Advisory leaflet. South Pacific Commission, p 4

Subaharan K, Ponnamma KN, Sujatha A, Basheer BM, Raveendran P (2005) Olfactory conditioning in *Goniozus nephantidis* (Muesbeck), a parasitoid of coconut black headed caterpillar, *Opisina arenosella* Walker. Entomon 30(2):165–170

Sujatha A (2001) Investigations on natural enemy complex of coconut leaf caterpillar *Opisina arenosella* Walker (Lepidoptera: Oecophoridae) with special reference to egg and early larval parasitoids and predators. Ph.D. thesis, University of Mysore, Karnataka, p 215

Sujatha A, Chalam MSV (2009) Status of coconut black headed caterpillar, *Opisina arenosella* Walker and evaluation of bio-agents. Ann Plant Protect Sci 17(1):65–68

Sujatha A, Singh SP (1999) Natural enemy complex of coconut leaf eating caterpillar *Opisina arenosella* Walk. (Lepidoptera: Xylorictidae). J Biol Control 13(1&2):51–58

Sujatha A, Singh SP (2002) Natural enemy complex of coconut leaf eating caterpillar, *Opisina arenosella* Walker (Lepidoptera: Oecophoridae) on palmyrah palm (*Borassus flabellifer* L.). J Plantn Crops 30(3):33–36

Talati GM, Butani PG (1988) A new host record for *Nephantis serinopa* Meyrick, the black headed caterpillar of coconut. Bull Entomol Res 29(2):142

Talati GM, Kapadia MN (1984) Influence of host plants on larval development, longevity and fecundity of *Nephanis serinopa* Meyr. (Lepidoptera: Cryptophasidae). Gujarat Agric Univ Res J 9(2):57–59

Tey CC, Ho CT (1995) Infection of *Oryctes rhinoceros* L. by application of *Metarhizium anisopliae* (Metsch.) Sorokin to breeding sites. Planter 71:563–567

Thomas GV, Jerard BA, Krishnakumar V, Chandran KP, Muralikrishna H, George J (2014) A decade of growth and achievements in CPCRI 2005–2014. Tech Bull No. 88. ICAR-CPCRI, Kasaragod, p 107

Thorne G (1949) On the classification of the Tylenchida, new order (Nematoda: Phasmidia). Proc Helm Soc Wash 16(2):3773

Valsala PG (1958) A white grub attacking roots of coconut palms. Cocon Bull 12:240–243

Van Hoff HA, Seinhorst W (1962) *Rhadinaphelenchus cocophilus* associated with little leaf of coconut and oil palm. Netherlands J Plant Pathol 68:251–256

Van Weerdt LG, Birchfield WR, Esser RP (1959a) Observations on some subtropical plant parasitic nematodes in Florida. Proceedings Soil Crop Sci Soc Florida 19:443–451

Van Weerdt LG, Martinez AP, Esser RP (1959b) Results of a survey designed to determine the etiology of lethal yellowing of *Cocos nucifera*. Proceedings Florida State Horticult Soc 72:421–430

Varadarajan MK, David PMM (2002) Population dynamics of the coconut mite *Aceria guerreronis* Keifer (Acari: Eriophyidae) and associated arthropods in Tamil Nadu, India. Insect Sci Appl 22:47–59

Veeresh GK, Vijayendra M, Reddy VN, Rajanna C, Rai PS (1982) Bio ecology and management of areca white grubs (*Leucopholis* spp.) (Melolonthinae: Scarabaeidae). J Soil Biol Ecol 2:78–86

Venkatasubban CS (1932) The coconut caterpillar in Cochin. Madras Agric J 20:192–198

Venkatesan T, Jalali SK, Murthy S, Rabindra RJ, Rao NS (2006) Field evaluation of different doses of *Goniozus nephantidis* (Muesebeck) for the suppression of *Opisina arenosella* Walker on coconut. CORD 22(Special issue):78–84

Viado GBS, Bigornia EA (1949) A biological study of the Asiatic palm weevil, *Rhynchophorus ferrugineus* Oliv. (Curculionidae: Coleoptera). Philippine Agric 33:1–27

Vidyasagar PSPV (1989) Scale insects and mealybugs of coconut. CPCRI Extension folder No. 16, p 4

Visalakshy A, Beevi SN, Pillai KS, Nair KKR, Das NM (1987) Control of the coreid bug on coconut. Entomon 12(4):395–396

Wakamura S, Tanaka H, Masumoto Y, Sawada H, Toyohara N (2007) Sex pheromone of the bluestriped nettle grub moth *Parasa lepida* (Cramer) (Lepidoptera: Limacodidae): identification and field attraction. Appl Entomol Zool 42:347–352

Watson GW, Adalla CB, Shepherd BM, Carner GR (2014) Aspidiotus rigidus Reyne: a devastating pest of coconut in Philippines. Agric Forest Entomol. https://doi.org/10.1111/afe.12074

Wattanapongsiri A (1966) A revision of the Genera Rhynchophorus and dynamis. Department of Agricultural Science Bulletin. Bangkok, Thailand

Way MJ, Cammell ME, Bolton B, Kanagaratnam P (1989) Ants (Hymenoptera: Formicidae) as egg predators of coconut pests, especially in relation to biological control of *Opisina arenosella* Walker (Lepidoptera: Xyloryctidae) in Sri Lanka. Bull Entomol Res 79(2):219–233

Weischer B (1967) UNDP report of plant parasitic nematodes. Report to the Govt. of India, No, T. A, 2332. FAO of United Nations, Rome

Wickramananda IR, Peiris TSG, Fernando MT, Fernanado LCP, Edgington S (2007) Impact of the coconut mite *Aceria guerreronis* Keifer on the coconut industry in Sri Lanka. CORD 23:1–16

Williams DJ (1981) New records of some important mealybugs (Hemiptera: Pseudococcidae). Bull Entomol Res 71(2):243–245

Williams DJ (1994) Distribution of the Pacific coconut mealybug, *Dysmicoccus cocotis* (Maskell) and of a new related species on coconut in southern Asia (Hemiptera: Coccoidea: Pseudococcidae). J Nat Hist 28(2):365–371

Wood BJ (1968) Pest of oil palms in Malaysia and their control. Incorporated Society of Planters, Kuala Lumpur

Zelazny B (1981) India – presence of the baculovirus of *Oryctes rhinoceros*. FAO Plant Protect Bull 29(3/4):77–78

Zelazny B, Alfiler AR, Crawford AM (1987) Preparation of a baculovirus inoculum for use by coconut farmers to control rhinoceros beetle (*Oryctes rhinoceros*). FAO Plant Protect Bull 35:36–42

Zipagan MR, Pecumbaba EP (1974) Important coconut pests of copra and coconut by products in the Philippines and factors that influence the occurrence. University Library, University of Philippines

Dr. A. Josephrajkumar is presently the Principal Scientist in the discipline of Agricultural Entomology at ICAR-CPCRI, Regional Station, Kayamkulam, India. He has refined IPM strategies for cardamom and black pepper and was instrumental in the release of three cardamom varieties. His current interest is in developing IPM of coconut pests and vector entomology of root (wilt) disease. He has published 50 research papers. He has received national awards including ICAR-Jawaharlal Nehru Award, ICAR-Lal Bahadur Shastri Young Scientist Award and AZRA-Young Scientist award. joecpcri@gmail.com

Dr. Chandrika Mohan is currently working as Principal Scientist (Agricultural Entomology) at the ICAR-CPCRI, Kerala, India. Her expertise is in coconut pest management. During her 32 years of research experience, she has contributed towards development and validation of biocontrol strategies for sustainable pest management of coconut palm. She has published 200 research papers and has contributed chapters for 10 books. Dr. Chandrika Mohan is a recipient of Fr. Gabriel Memorial Gold Medal. cmcpcri@gmail.com

Dr. P. S. Prathibha is a Scientist in entomology at the Central Plantation Crops Research Institute, Kasaragod, India. She has published five research papers. prathibhaspillai007@gmail.com

Dr. Rajkumar is a nematology Scientist at ICAR-CPCRI, Kasaragod, Kerala, India. He has also worked in ICAR-National Bureau of Agricultural Insect Resources (NBAIR), Bangalore. His specialization is phytoparasitic and entomopathogenic nematodes (EPN) and has over 10 years of research experience in bio-suppression of nematode disease associated with crops. He has published 25 research papers. rajkumarcpcri@gmail.com

Dr. T. Nalinakumari former Professor and Head, Entomology, Kerala Agricultural University, India, is currently serving as consultant to National Institute of Plant Health Management. She has specialized in storage entomology and integrated pest management. She has 28 years of teaching and research experience. She has published 75 research papers and contributed to 5 book chapters. achutnk@yahoo.co.uk

Dr. C. P. R. Nair former Principal Scientist and Head, ICAR-Central Plantation Crops Research Institute, Regional Station, Kayamkulam, Kerala, India, has more than 37 years of professional expertise in entomology of plantation crops. He had been the Principal Investigator of two International Projects on Integrated Pest Management in Coconut. He has more than 130 publications. cprnair_47@yahoo.com

Chapter 13
Harvest and Postharvest Technology

M. R. Manikantan, R. Pandiselvam, Shameena Beegum, and A. C. Mathew

Abstract This chapter enlists salient harvest and postharvest technologies of coconut. Various harvesting and postharvest gadgets, tools, machineries and equipments developed by different institutions are described. Technologies related to storage of nuts, processing of fresh coconut kernel and coconut water into different edible products, processing of dry kernel/copra, extraction and refining of coconut oil and processing of coconut sap (neera/sweet toddy) are explained in detail. The industrial and nonedible applications of coconut husk, coconut shell, coconut wood and coconut leaf are also described. Various sections that include harvesting and processing, technologies on food products as well as non-food products to future strategy in relation to harvest and postharvest technologies of coconut are concisely illustrated. A complete value chain information from harvesting to consumption and industrial utilisation from farm to fork is narrated. The need for refining some of the important equipments and technologies is indicated to make them more cost-effective and eco-friendly.

13.1 Introduction

Postharvest processing ensures effective utilisation of harvested produce and the quality of the end product, which ultimately tells upon the consumption and acceptance of the produce. It has been estimated that there is considerable loss in many of the agricultural products after harvest. This is either because the postharvest technology has not been developed to the desired extent or because of the special problems in the marketing of the produce.

Edible copra (milling copra) for the extraction of coconut oil, desiccated coconut powder, fermented sap and jaggery from the sap is the important edible products

M. R. Manikantan (✉) · R. Pandiselvam · S. Beegum · A. C. Mathew
ICAR-CPCRI, Kasaragod, Kerala, India
e-mail: manicpcri@gmail.com; anbupandi1989@yahoo.co.in;
shameena.pht@gmail.com; acmathew@yahoo.com

K. U. K. Nampoothiri et al. (eds.), *The Coconut Palm (Cocos nucifera L.) - Research and Development Perspectives*,
https://doi.org/10.1007/978-981-13-2754-4_13

of coconut palm. Coconut is an important source of vegetable oil used for both edible and industrial applications. India consumes 40% of coconuts for its culinary and religious purpose, 35% for copra, 6% for manufacture of value-added products (1% as virgin coconut oil, 1% as coconut milk/cream and 4% as desiccated coconut powder), 17% as tender nuts and 2% for seed purpose.

One of the main reasons for the fall in price of the coconut and its products is dependency of price of coconut oil which in turn depends on the cost of other vegetable oils and hence the importance of product diversification. The industry has been successful in evolving appropriate processing technologies for some of the value-added products of coconut. As a result, products like packed coconut milk, coconut creams, spray-dried coconut milk powder, virgin coconut oil, coconut chips, coconut water-based products like vinegar and *nata de coco*, coconut palm sap-based products like coconut jaggery, coconut sugar, etc., which have good commercial potential, have been developed. These products have become important as an agro-based raw material for many industries. As a result, it has become possible to encourage product diversification and development of value-added products in the industry.

13.2 Devices and Gadgets for Plant Protection and Anti-buckling

13.2.1 Telescopic Sprayer

The telescopic sprayer comprises of two coaxial pipes of ultra-lightweight (0.5 kg m^{-1}) which can be used to spray up to a height of 12.5 m. The pipe height can be locked at any desired level above 6 m. Marginal farmers could attach a rocker sprayer, whereas in large gardens, a power sprayer can be used. The telescopic pipe assembly developed by ICAR-Central Plantation Crops Research Institute (CPCRI) is very much useful if the garden is of uniform size and slope. They are lightweight and durable. This solves the constraint of height of palms in taking up plant protection measures.

13.2.2 Anti-buckling Device

Partial severing of the bunch stalk from the trunk is known as buckling, which is a serious problem often faced by coconut farmers. Heavy weight of bunches, long and less sturdy peduncle, wider angle between the leaf and the inflorescence and weak leaf petiole are some of the causes for buckling of bunches. ICAR-CPCRI has developed a mechanical gadget for bunch support (Baboo and Bhat 1980), which consists of a trunk-clamp, support-clamps and telescopic support-rods. The trunk-clamp is made in six different sizes to suit varying trunk girths. Support-clamps can be placed at any point on the trunk-clamp. Telescopic support-rods are made in two different sizes to suit bunches located at various heights and positions on the crown. The support can easily be fixed on the trunk of the palm. The fork end of the support-rod is inserted towards the tail end of the bunch (about one-fourth distance from the tip) to support the peduncle. The rod is slightly lifted up, and the other end of the rod is inserted in the support-clamp hole. One man can easily do the operation, and it is possible to support as many as six to eight bunches at a time. Harvesting of nuts can be done without removing the support-rod by severing the nuts first and the stalk last. After the nuts from the bunch are harvested, the support-rod can be used for another bunch that might require support. As the palm grows taller, the trunk-clamp has to be refixed at a higher level. Before removing the trunk-clamp, which is already fixed, another is fixed above the old clamp. The bunch already under support is then provided with another support-rod and the clamp, support fixed to the new trunk-clamp. The old ones can be removed and used for another palm. This operation may be necessary only once in two years.

13.3 Harvesting of Nuts

13.3.1 Stage of Maturity of Nuts at Harvest

Coconuts are harvested at different stages of maturity for specific uses. For tender nut purpose, harvesting is done when the nuts are 6–8 months old. For snowball tender nut and coconut chips purpose, 8–9 and 9–10 months old nuts are harvested, respectively. For the production of copra and other kernel-based products, only fully mature coconuts are harvested. The nuts reach full maturity in 11–12 months after the inflorescence is opened. At this stage, the output of copra and oil as well as brown fibre would be the maximum. In a study in India, it was found that compared to 12 month old nuts, the copra yield was less to the extent of 6% in 11 month old nuts, 16% in 10 month old nuts and 33% in 9 month old nuts. The corresponding reduction in the percentage of oil was found to be 5, 15 and 33%, respectively. In places where green husks are in demand for the production of white fibre, the usual practice is to harvest 11 month old nuts. The slightly low copra output at this stage would, however, be compensated by the additional income derived from the fibre and its products.

13.3.2 Harvesting

Though coconut palm produces an average of 12 inflorescences in a year, some of the inflorescences are likely to get aborted or fail to develop into fruit bunches due to environmental factors. Consequently, the number of bunches available for harvest is less than 12 in many areas. Similarly, the frequency of harvest also varies from country to country and also within the countries. In many areas, 6–12 harvests per year are the usual practice. In the properly managed gardens, harvest at monthly intervals is usually adopted. In the neglected gardens, bunches are not produced regularly, and, as such, not more than six harvests are possible in a year. In most of the coconut-growing countries, fully mature nuts of 12 months are harvested at bimonthly intervals.

13.3.3 Methods of Harvesting

Harvesting of coconuts is a skilled job and is generally performed by experienced climbers by climbing up to the top of the tree. Climbing is usually done with the help of a ladder, 3–4 m long. On reaching the top of the ladder, the climber invariably uses a rope ring at the feet or ankles for the next stage of climbing. The climber uses another rope ring for the hands also in some places. Climbing is also done by cutting notches on the trunk at convenient spaces, which is not a desirable practice because of the injury caused to the stem. Tied-on-husks are also used as steps for climbing. Another method involves the use of a rope loop which encircles the trunk and the climber. Various climbing devices have also been developed to reduce the drudgery involved in the climbing process for harvesting of coconuts (Thamban et al. 2011).

13.3.4 Harvesting Gadgets

13.3.4.1 Pole Harvesting

In Sri Lanka and also in some states in India, bunches are brought down by a knife attached to a long bamboo pole. But this will be very difficult to be used in very tall palms. In some countries such as Thailand and Malaysia, trained monkeys are also used to pluck the fruits.

13.3.4.2 Ultra-Lightweight Coconut Harvester

The harvester developed by CPCRI comprises of two coaxial pipes of ultra-lightweight (0.5 kg m^{-1}), which can be used to harvest up to a height of 12.5 m from the ground. The pipe height can be locked at any desired level above 6.25 m. On the top end of pipe, a specially designed knife is fitted using nuts and bolts. The harvesting knife could be fabricated by local craftsmen.

In Papua New Guinea, some of the African countries and other Pacific Islands, the fruits are allowed to ripe and fall down on their own, and the fallen nuts are collected at regular intervals.

In addition to harvesting, various other operations like cleaning the crown, pollination, tapping and insecticide application are being often neglected due to scarcity of climbers (Akyurt et al. 1839). Considering this scenario, many efforts have been made to ease the operation of palm climbing by developing gadgets and climbing devices. Davis (1961, 1963, 1964, 1977) developed three prototypes for palm climbing. Swamy and Patil (1975) developed a much simpler device which consists of movable supports for legs and hands and lifts alternately while the other one is gripped for climbing.

Various types of climbing devices like tractor-operated, self-propelled, manually operated and some robotic-type (electronic) devices have been developed and tested for harvesting coconut by both the government and private sector (Kolhe 2010; Kolhe and Jadhav 2011; Sial 1984; Shabana and Mohamad 1993). Among the manual types, one model was developed by an innovative farmer (Joseph model), another by TNAU (TNAU model) and the third by CPCRI (Singh et al. 2012) in India. Of all these manual devices, Joseph model is the only machine commercially available and used by professional climbers.

13.3.4.3 Paddle–Type Device

Chemberi Joseph Model Joseph model has got mainly two assemblies of similar construction. The steel rope wires of both top and bottom assembly need to be looped with the palm and locked. The user then climb on to the machine by placing one foot each on both the assemblies holding the handles provided. Standing on one assembly, the user lifts the other assembly to loosen the steel rope and raise it by hand. After attaining a comfortable height, he pushes back the assembly with foot so that it gets tightened to the palm. The user has to co-ordinate these two assemblies simultaneously by using hands and legs to climb on coconut palm. It does not require much skill and 2–3 days' training is enough to easily climb the palm.

ICAR-CPCRI has developed a safety attachment to this model of climbing machine. The safety device consists of a 6 mm steel rope of length slightly more than the circumference of the coconut trunk having an oval-shaped loop each at both ends. The only modification required in the climbing machine is providing two 6 mm metal (MS) rings at the bottom of the handle of the right leg unit of the climbing machine. The steel rope is taken through these loops around the palm, and one end of the steel rope is taken through the loop provided at the other end, making it a noose around the coconut palm. The free end is then connected to a commercially available body harness. The wire rope moves up and down along with the climbing machine during operation. In case of an eventuality of failure of the machine or accidental falling of the climber, the steel rope noose gets tightened to the coconut trunk and prevents the climber from further falling. The safety attachment is independent of the climbing machine and gives fool-proof safety to the climber from falling.

TNAU Model This is a sitting-type or push up-type model developed by Tamil Nadu Agricultural University (TNAU). The device has two MS frames – one upper and another lower one. They are connected by a belt while the equipment is on the coconut palm. The user has to sit on the seat which is provided on the upper frame and has to insert his foot between the rubber rollers available in the lower frame. The upper frame can be lifted by hands, and the lower frame has to be lifted by leg. In this type, the size can be adjusted as per the diameter of the palm. As both frames are positioned in an angle, it will get clung to the palm due to the friction by rubber bush, and the process has to be repeated for further climbing. Safety belt can be adjusted as per the body posture. Distance between the top and bottom frames can be adjusted by the belt as per convenience.

In Jamaica during 1976, Swiss equipment used in forestry for climbing, known as 'Baumvelo' (tree bicycle), was tried with some modifications for coconuts, but problems were faced when leaving the device near the crown of the tree. Sliding aluminium ladders are also being used as in Jamaica for coconut climbing. None of the above prototypes could reach the stage of full acceptability by the farmers and commercial production due to various reasons.

One farmer in Kerala has developed a successful model of tree climbing device, which is similar to the 'Baumvelo' tree bicycle. This device works on the principle of gripping the device through rings by self-weight with legs providing the motive power. The device has two stirrups and the rubber pad rings tightened or loosened on the trunk by wire rope connected to the sliding foot rest on the stirrups providing the gripping or loosening of the device. The alternate gripping action by hand and legs facilitates lifting and lowering of the gadget and thus enables the person to climb up or down. The unit is very cheap and handy and has become popular in India.

Attempts are also made by researchers at TNAU, ICAR – Central Institute of Agricultural Engineering (CIAE), Bhopal, and a local fabricator at Ratnagiri (India) to develop coconut harvesting machine based on self-propelled hydraulic mast. Efforts were made by local fabricators and innovative farmers also to make prototype of a 'robotic-type' coconut harvester. However, none of these machineries are yet fit for commercial use.

13.3.4.4 Tractor–Operated Multipurpose Hoist

TNAU has developed a tractor-operated multipurpose hoist that is amenable for fruit plucking, coconut harvesting, pruning, lopping and spraying tree crops. The equipment is attached to the back of a 45 hp agricultural tractor. Two persons can stand on the platform and do operations. Platform can reach a maximum height of 8.1 m.

13.3.4.5 Aerial Access Hoist for Coconut Harvesting

An aerial access hoist for coconut harvesting has also been developed by TNAU. The machine is the first of its kind in tractor-mounted form. A full-length chassis from front to rear of the tractor provides support. The entire weight of the hoist and moments is transmitted through the chassis to the stabilisers without transferring to the tractor chassis. Four palms can be accessed from a single position. The time required for positioning the unit and operating stabilisers is 1 min, and the time required for positioning against a palm of 10 m height is 2 min. The operator platform can be positioned by the operator himself. Lifting capacity of the machine is 120 kg.

13.3.4.6 Self–Propelled Hydraulic Multipurpose System

CIAE has developed a self-propelled hydraulic multipurpose system for orchard management operations which could be used for harvesting coconut also. The self-propelled hydraulic multipurpose system has a maximum vertical reach of 6 m, has a load carrying capacity of 200 kg and can be operated at maximum ground speed of 3 km h^{-1}. This machine was designed with dimensions of 2.2 m × 6.3 m × 1.89 m and can be used as a platform to reach fruits on trees for easy picking. It is hydraulically powered using 8.2 kW (11 hp) engine and has a low centre of gravity while still maintaining good ground clearance (290 mm) for good stability during operation. Lifting, lowering of the platform, forward and backward movement and steering of the machine are controlled by the operator from the platform itself. The hydraulic system of the machine has been designed to stop the movement of machine even if the engine needs to be shut off. The machine is easy to be operated and requires low maintenance.

13.4 Processing Machineries

The following devices were developed by ICAR-CPCRI (unless otherwise mentioned) to increase efficiency and reduce time required and drudgery involved.

13.4.1 Coconut Dehusking Machine

Coconut dehusking is the first postharvest operation in any coconut processing industry. Traditionally coconut is dehusked manually using a spike. Drudgery and risk of getting injured make the operation male dominated.

A power-operated semi-automatic coconut dehusking machine has been designed and fabricated. Dehusking is done by passing coconut over two spiked rollers rotating at different speeds in opposite directions. The coconut is pressed towards the rollers with the help of a pneumatic cylinder. Husk is removed by the rotating spikes. However, a small portion of the coir fibre would remain at the eye portion of the coconut. This would then fall on a pair of rollers rotating at the same speed but in opposite directions. The remaining coir also would be detached, and the fully dehusked coconut can be collected at the opposite side. The rollers are rotated by an electric motor. Feeding is done manually. Though the machine would not replace manual labour, the drudgery involved in dehusking is totally avoided. Anyone can use the machine even without any formal training. The machine has a capacity to dehusk 200 coconuts per hour. 'Keramitra', the coconut dehusking machine developed by Kerala Agricultural University (KAU), is a very popular device in households (Varghese and Jacob 2014). However, the extremely low output is a major constraint.

13.4.1.1 Coconut Dehusking Tool

An easy-to-operate simple coconut dehusking tool which is particularly useful for domestic purpose has been developed. It consists of mainly a stationary wedge, a movable wedge, a lever and a pedestal having a base. The new tool has the resemblance of a crowbar but has an additional mechanism for making the opening of the husk easier or effortless. The base of the tool is held firmly to the ground using the feet particularly with one foot on the side of the handle. The coconut is thrust on to the husking tool. On pulling the handle upwards with one of the hands, one sector of the husk is loosened from the shell. This is repeated three or four times for loosening the remaining sectors. It takes only about 8–20 s for husking a nut depending on the variety, maturity, etc. of the nut. It is light in weight, simple to use and handle and cheaper and can be used both indoors and outdoors.

13.4.1.2 Pedal–Operated Coconut Dehusker

This is a modified design based on the one developed by the Kerala Agricultural University (Indian Patent No.192670, 1995). About 100–120 nuts can be dehusked h^{-1}. The unit has been designed to withstand longer use. This unit has some advantages such as less requirement of energy for dehusking and reduced drudgery.

In India, three prototypes of mechanical dehuskers were developed in the 1980s. The first experimental prototype (Singh 1981) was provided with a set of three spring-loaded curved blades and a telescopic hand lever connected to the assembly. By pressing the hand lever manually, the blades pierce the husk, and on moving out in horizontal plane, the husk segments are removed. But the outturn was low and the breakage was to the extent of 5%. Moreover, it was not possible to dehusk the coconuts of all shapes and sizes in this prototype. This was modified by replacing the spring-loaded curved top blades with a set of straight blades fitted on a frame rotating freely in the shaft, to accommodate nuts of all shapes and sizes (Baboo and Bhat 1980). Here the breakage was reduced to 1%, but the outturn was further lowered to 500 coconuts in an 8-h operation as against 1500–2000 nuts in the traditional manual operation.

Another model with leg and hand operation was designed at ICAR-CPCRI in 1986. The device consists of a platform at waist level, with a frame. A shaft, which moves up and down on the platform, is operated by foot through linkages at the bottom end. The top end of the shaft is fixed with a circular plate on which a set of three fixed blades is provided. A rotating plate with guide slots is fitted in alignment and just below the bottom plate. The scooping blades move out radially when the rotating plate is rotated by a handle fixed on it. Another plate with three fixed blades is fixed at the top in such a way that the blades are faced in vertical alignment with guide slots by means of guide rods welded to them. The guide rods move freely in the radial guide slots in the bottom rotating plate. The plate can be rotated by pulling the handle. Coconut is kept inverted in between the top and bottom plates. The bottom plate assembly is lifted up by pressing the foot pedal; pressing it further, the bottom and top blades pierce into the husk. After piercing to the maximum, the handle is pulled out to rotate the free plate which in turn will actuate the scooping blades. Upon scooping, the husk gets detached from the nut into three portions. It involves less effort to dehusk the coconut. It can be fabricated locally and is simple in design and safe to operate. The outturn of this machine is 110 nuts h^{-1}, whereas the outturn by traditional way is about 160 nuts h^{-1}. But the main advantage of this machine is that an unskilled person can do this job without the risk of injury. It is 13–25% cheaper than the conventional method. Nijaguna (1988) has also developed another coconut dehusker, which works on similar principles.

A power-operated semi-automatic coconut husking machine has also been developed (CPCRI 1996). The machine consists of three main parts, viz. a rotating platform having a set of six coconut holders, a piercing system with shock absorber to avoid breakage of coconut due to excess load and a scooping blade set for splitting and loosening the husk. The required operating speed of various parts is achieved by using different types of gearboxes. The machine can dehusk 500–600 nuts h^{-1} much higher than the traditional spike method. Two unskilled labourers are required to operate this machine. It is economical and financially beneficial to farmers and copra manufacturers. The time required to dehusk one nut is 6 s.

Mechanical dehusking devices have been perfected and put to use in other countries as well. A portable spike has been developed at the IRHO for Côte d'Ivoire Region (Pomier 1984) which can dehusk 1200–1300 nuts per day. In Malaysia, a manually operated pneumatically powered dehusking device was developed (Alias and Hashim 1982) which gave an output of 75–80 nuts h^{-1}.

In Taiwan, a hydraulically operated coconut dehusking machine giving an output of 350–400 nuts per hour has been developed. The Tropical Development Research Institute (TDRI) has developed a power-driven coconut dehusker which has been commercially manufactured by M/s Jaffor Incorporated Ltd., UK. The capacity is reported to be 730 nuts h^{-1} (Coconis 1986).

The Caribbean Industrial Research Institute, Trinidad and Tobago (West Indies), has developed an electric motor-powered coconut dehusker (Rnam 1990). The capacity of the dehusker is 600–800 nuts h^{-1} and is powered by a 3 hp motor. The machine uses shearing force to strip the husk from the nut. The shearing action is produced by two spiked rollers which rotate in opposite directions and at different speeds.

In South Pacific Islands, dehusking of coconuts is not carried out; instead the whole nut is split into halves with an axe and the endosperm is gouged out in pieces, either while fresh or after preliminary drying. A machine for splitting coconuts in this way has also been developed, which splits the coconut into three parts.

13.4.2 Coconut Splitting Device

To solve the problem of splitting of coconuts by holding it in hand, a manually operated splitting device is developed. Knife of the machine is made of spring steel and is kept at a bevel angle of 25 degrees. As the cutting blade is made of spring steel, it does not require frequent sharpening. Nut is split manually by the impact force and the nut water is collected at the bottom.

13.4.3 Copra Dryers

The common practice of making copra is by sun drying the fresh coconut kernel on cement floor or on sand floor for 7–9 days. The endosperm of coconut is exposed while drying and so can be contaminated with dirt. Prolonged drying, especially during monsoon, also results in microbial infection. The energy-efficient dryers developed produce dust and microbial contamination-free copra in a short period (Vidhan Singh et al. 1999).

13.4.3.1 Small Holders' Copra Dryer

This is an indirect dryer that can dry up to 400 coconuts per batch, ideal for a farmer to make copra in his farm itself. Any agricultural waste could be used as fuel. It takes 36 h to get quality copra.

13.4.3.2 Shell-Fired Copra Dryer

This copra dryer works on indirect heating and natural convection principles using coconut shell as fuel. It requires less fuel, makes copra in short time and is less expensive too. Capacity of the dryer is 1000 nuts per batch. The copra obtained is light brown in colour which helps to fetch a better price in the market. The burner generates heat for 5 h without tending, and the residual heat is retained for one more hour. The average drying time is 24 h.

13.4.3.3 Forced Circulation Copra Dryer

Forced circulation dryers are more efficient and much faster than the dryers using natural convection to dry copra (Madhavan and Bosco 2002). Several types of such dryers are available in the market with varying capacities to suit the requirement of the copra industry.

13.4.3.4 Solar Tunnel Dryer

A solar tunnel dryer with a capacity of 1500 coconuts/batch is developed that produces good-quality copra in a shorter time (Madhavan and Bosco 2004). It consists of a semi-cylindrical shape tunnel structure having a transparent cover made from UV-stabilised polyethylene film of 200-micron thickness. The solar collector is the black polyethylene film of 250-micron thickness spread on the ground inside the dryer for better absorption of solar heat. The temperature inside the dryer is 20–25 °C higher than the ambient, and the RH value is 20–22% lower than the ambient. The copra, thus produced, is less infested by fungi and bacteria than the one produced by open sun drying. Drying time taken to dry copra is 32 sunshine hours. The dryer can also be used to dry arecanut, cocoa beans, cardamom, pepper and ginger.

13.4.3.5 Solar Cum Electrical Dryer with Agricultural Waste as Third Source of Energy

For cloudy and rainy days, a multisource dryer has been developed with solar energy as the main source of energy and electricity and biofuel as alternate sources of energy. The dryer consists of a semicircular parallel plate solar collector, electric heaters of 1000 W (six numbers), blower cum exhaust motor and the drying chamber. It is an auto-regulated dryer with temperature and humidity control. It is a batch-type dryer and the capacity of the dryer is 2000 coconuts batch^{-1}. The dryer can be used to dry other produce such as cardamom.

13.4.4 Coconut Deshelling Machine

A power-operated batch-type coconut deshelling machine has been developed to separate shell and copra after partial drying. Capacity of the machine is 400 half cups per batch. The optimum moisture content is 35% d.b for maximum deshelling efficiency of 92.16%. The optimum speed of the deshelling machine is 10 RPM, and the time taken for deshelling is 4 min batch^{-1} (Singh and Udhayakumar 2006).

13.4.5 Copra Moisture Meter

The quality of milling copra ultimately determines the quality of the oil and the residual cake. Good-quality copra will yield oil with free fatty acid (FFA) content of less than 1%. Moisture is the most important factor influencing the quality of copra. Copra with a moisture content of less than 6% is considered good quality as it is not easily damaged by insects, moulds or microorganisms. An electronic moisture meter was developed to determine the moisture content of copra, based on the electrical conductivity of the kernel. The instrument can read moisture content from 5% to 40%. It is very handy, and the accuracy is more than 94% in the lower levels of moisture content. It consists of an electronic unit, an output meter and a sensor. The sensor is made of a pair of steel rods of size 8 mm × 2 mm fixed 6 millimetres apart. The power supply is 9 V battery or equivalent. The instrument was calibrated against standard air-oven method. After switching on the instrument, adjust the Cal Port so that the meter needle rests on the 'Cal' position marked on the dial. Then, plug the sensor into the sensor socket. Probe the sensor into the inner side of copra. The reading now seen on the meter gives the percentage moisture content in the copra. The average moisture content from a minimum of three points is taken as the moisture content of copra. The instrument is handy and useful to copra processors and farmers. This unit is being marketed by M/s Kerala Agro Industries Corporation in view of its practical application (Madhavan 1987).

13.4.6 Tender Coconut Punch and Cutter

Tender nut punch and cutter are two simple devices to pierce and cut open the tender coconut. A clean hole sufficient enough to insert a straw is formed so that one can drink the fresh coconut water conveniently. The nut is placed on the wooden platform and cut open by pressing the lever attached to the blade. The efficiency is 20 tender coconuts h^{-1}.

13.4.7 Snowball Tender Nut Machine

Snowball tender coconut is a globular tender coconut containing tender coconut water inside. The ball is scooped out with the help of a specially devised tool after cutting the shell of tender coconut of 7–8 months maturity by using snowball tender coconut machine. The snowball tender coconut thereafter is made free of the adhering testa, packed hygienically to use either as fresh or after refrigerated storage. Output capacity is 250 coconuts in 8 h by one person.

13.4.8 Coconut Shell Removing Machine

Coconut shell removal is the second postharvest operation in a coconut processing industry using fresh coconut kernel as the raw material. Traditionally coconut shell is removed using a knife. Though women do this work, the labour-intensive process is not at all attractive to the processing industries. Coconut shell removing machine is intended to reduce both time and drudgery involved in the manual shell removal.

Two concentrically rotating multipointed circular blades and a stationary pointed blade on which coconut would be placed firmly are the major components of the shell removing machine. Coconut to be processed is pressed towards the rotating blades by firmly placing it on the stationary blade. Shell gets detached from the kernel due to the impact force of the rotating blade. The machine has a capacity to remove the shell of 150 coconuts h^{-1}.

13.4.9 Coconut Testa Removing Machine

Many high-value coconut products like coconut chips, virgin coconut oil, desiccated coconut powder, etc. require removal of testa. Manual removal of testa using potato peeler is a cumbersome and time-consuming process. Moreover, a sizable amount of coconut meat also would be lost along with the removed testa. The coconut testa removing machine developed reduces the drudgery and improves the efficiency and capacity of units that require removal of testa. The main component of the machine is a circular wheel covered with emery cloth or water paper attached to an electric motor. One person can remove testa of about 75 coconuts h^{-1}.

13.4.10 Coconut Slicing Machines

Slicing coconut kernel to produce chips of uniform thickness is the single most important unit operation in the coconut chips making process. Conventionally, this is done manually, and the process is very cumbersome and time-consuming. Moreover, quality of chips, especially uniformity of thickness, would depend on the skill of the operator. In order to make this operation simple and faster, two types of coconut slicing machines were developed.

13.4.11 Coconut Chips Making Machine: Electrical

The machine consists of two stainless steel slicing blades fixed on a circular blade supporting disc, a feeder to insert coconut endosperm for slicing, an exit guide to guide the sliced coconut chips towards the outlet and an electric motor as a prime mover. The electric motor rotates the blade supporting disc using a V-belt. When coconut endosperm comes into contact with the blades, it gets sliced. The sliced coconut chips are then guided towards the outlet by the exit guide and are collected in a container. Coconut chips of uniform and required thickness could be produced using this machine. Provision is made in the slicer to slice tuber crops, banana and vegetables. The capacity of the machine is 60 coconuts h^{-1}.

13.4.12 Manually Operated Coconut Slicing Machine

The machine is quite user-friendly especially for ladies who know operation of sewing machine. Approximately 25 coconuts can be sliced in 1 h using this machine.

13.4.13 Coconut Chips Dryers

The electrical dryer developed consists of a set of ten trays with wire-mesh screens for loading coconut slices. Temperature in the dryer is controlled automatically by a sensor and electronic control unit. Though the dryer is designed for 50 coconuts, the size could be enhanced to any desired capacity. Another prototype of chips dryers that uses agricultural waste as fuel has also been developed. Furnace of this indirect dryer is conveniently placed outdoor with the drying unit indoor. Temperature is controlled by a butterfly valve in the hot air inlet. Cost of drying is less than that of electrical dryer.

13.4.14 Coconut Grating Machines

Two coconut grating machines were developed to enhance the grating efficiency. First one is of single-user and the second one is of multiuser (four grating blades) type. These machines scrape off the coconut kernel into fine gratings with the help of stainless steel blade. The single-user machine has a capacity of 60 coconuts h^{-1}, and the multiuser has a capacity four times of the first one.

13.4.15 Coconut Pulveriser

The coconut pulveriser consists of a power-operated rotary blade. The coconut kernel pieces are fed into the chopper manually. Due to the impact of the rotary blade and the rubbing on the stationary blade, the coconut kernel turns into fine powder. The machine has a capacity of 250 coconuts h^{-1}.

13.4.16 Coconut Milk Extractors.

Four different coconut milk extractors have been developed to enhance the milk extraction efficiency. While two of them are manually operated, the other two are hydro pneumatic coconut milk extractors.

13.4.16.1 Manually Operated Milk Expellers.

Both the manually operated machines are similar to a hand-operated vertical screw press. The grated coconuts are kept in a perforated cylinder, and by rotating the handle provided at the top of the screw, the gratings are pressed. In one of the machines, the whole pressing process is done manually by rotating the handle. In the other one, an additional hydraulic jack is provided at the bottom. Gratings kept in the perforated cylinder are pressed initially by rotating the handle. Final pressing is done by operating lever of the hydraulic jack. The platform on which the perforated cylinder is kept moves up and exerts pressure on the coconut gratings thereby expelling coconut milk without much effort.

13.4.16.2 Screw-Type Coconut Milk Expellers

Two screw-type coconut milk expellers, single and double screw, with different capacities have been developed to extract coconut milk. The screw-type expellers have the maximum extraction efficiency among different types of coconut milk extractors. The single-screw expeller has a capacity of 300 coconuts h^{-1} and the double-screw has a capacity of 1000 coconuts h^{-1}.

13.4.17 Virgin Coconut Oil Cookers

The technique to produce virgin coconut oil (VCO) by hot processing method has been standardised and the protocol commercialised. Conventionally VCO is prepared by heating coconut milk in an open container at low flame with continuous stirring. It is done manually and the constant stirring is a laborious process. Many a times the milk gets charred when the stirring is not proper or when excess fuel is burnt. In order to overcome these limitations, two types of VCO cookers were developed to extract the VCO by hot processing (Mathew et al. 2014).

13.4.17.1 Virgin Coconut Oil Cooker: LPG/Biogas

This cooker consists of a double jacketed vessel filled with thermic fluid, which ensures efficient and uniform heat transfer to coconut milk in the cooker. Four Teflon-tipped stirrers are provided to stir coconut milk, which helps the cooker to distribute heat energy uniformly within the coconut milk. The stirrers are powered by an electric motor with a reduction gear. An outlet with a door attached to a lever is provided at the bottom of the cooker to take out the extracted oil. The cooker is supported by three legs with sufficient clearance from ground for easy collection of extracted oil. A thermometer is provided to measure the temperature of the thermic fluid so that it can be maintained at 100–120 °C. A safety valve is also provided for extra safety by releasing the pressure developed, if any, in the thermic fluid chamber. The cooker is heated by two burners provided at the bottom of the heating chamber. Biogas or LPG could be used as fuel. Virgin coconut oil cooker has a capacity of 125 litres which can be upscaled.

13.4.17.2 Agricultural Waste Fired Virgin Coconut Oil Cooker

In this cooker agricultural waste material is used as fuel. The design and construction features of the cooker are similar to the LPG/biogas model. A burning chamber is provided at the bottom of the double jacketed vessel. The cooker is heated by burning any agricultural waste, preferably coconut shell, in the burning chamber. An opening is provided at the front to feed the fuel. An exhaust to remove smoke is provided at the opposite side and is connected to a chimney kept outdoor for easy combustion of fuel. This can be fabricated as per the required capacity.

13.5 Storage of Coconuts

It is a common practice to store harvested nuts in heaps under shade for a few days, known as seasoning, before they are further processed. The advantages are:

1. Husking becomes easier.
2. Shelling is cleaner and easier, and the resulting shells are dry and hard and, when used as fuel, burn continuously and produce less smoke.
3. The moisture content of the meat decreases and the thickness of the meat layer increases. Consequently the yields of copra and oil increase.
4. The quality of copra produced is also superior to that of unstored nuts.

Studies have shown that storage of harvested nuts is beneficial only if the nuts are fully ripe and that good-quality copra can be obtained from nuts even immediately after harvest if the nuts are fully ripe. Though storage of comparatively immature nuts is helpful in obtaining a higher output of copra per nut, these are easily spoiled during storage. The results of a study conducted in India on storage of nuts are summarised in Table 13.1.

Table 13.1 Changes in copra content and nut spoilage during storage

	Changes in copra content (g) nut^{-1}				Nut spoilage (%)			
Maturity of nuts (months)	12	11	10	9	12	11	10	9
Storage								
Nil	161.1	150.0	128.2	96.8	–	–	–	–
30 days	162.5	150.4	135.2	117.6	2.24	3.16	22.02	58.33
60 days	162.2	156.8	138.9	127.0	7.62	13.10	31.55	78.57

Source: Rethinam and Bosco (2006)

The storage trials conducted at the Coconut Research Institute in Sri Lanka to study the quality of copra obtained from stored and unstored coconuts of different stages of ripeness showed that unripe green nuts when stored gave 25% more yield of copra with reduction of rubbery copra by 24% and the same without storage gave discoloured copra (Mathes and Marikkar 2010).

13.6 Food Products

Various products derived from the coconut palm are utilised as food either in their natural state or after processing into various products.

13.6.1 Products Derived from Wet Kernel

13.6.1.1 Wet Kernel

The kernel of 7–8 months old nut is consumed either as such or along with the sweet tender nut water. At this stage, the kernel will be very soft with maximum protein content and sugar which gradually decline as the nut matures. The kernel is considered ripe when the nut is 11–12 months old after fertilisation. The composition of unripe kernel is 90.50% moisture, 0.80% protein, 1.30% fat, 6.80% carbohydrates and 0.60% ash (Banzon et al. 1990).

As the nuts mature, the moisture, protein and ash content of the fresh endosperm slowly decreases, whereas the fat content increases. After the 10th month, the protein content undergoes practically no change, but the fat content continues to increase until it reaches its peak at 12th and 13th months. Subsequently, the fat content decreases until the 15th month (Hallexa and Seirra 1976). Table 13.2 shows the chemical composition of the kernel from three countries.

The ripe kernel derived from 11 to 12-months old nuts is used in different forms for edible uses both in the households and organised food industry. In the households, fresh kernel is used both in the grated form and in the form of milk or cream obtained by squeezing the gratings with or without the addition of water. Processed forms of kernel and milk find applications in the food industry.

Table 13.2 Composition of kernel from fresh coconuts in percentage

Source	Moisture	Protein	Fat	Crude fibre	Carbohydrates	Ash
U.K.[1*]	44.0	3.6	38.1	3.1	9.9	1.3
Sri Lanka[2]	44.0	4.6	38.2	2.3	9.7	1.2
India[3]	35.4	5.5	44.0	3.0	10.0	2.1

[a]Dendy and Timmins (1973); [2]Nathanael (1965); [3]Anonymous (1984)
bT.P.I. – Tropical Products Institute, UK

It is now possible to preserve kernel of 8–10-months old nuts. For this, the tender kernel is cut into strips of about 6 cm length and 2 cm width. These are put into cans, covered with hot syrup, exhausted at 80 °C, sealed and processed. The product has a shelf life of 4–6 months at ambient temperature and has been found to be microbiologically safe even after 1 year. In the Philippines, kernel from 8- to 9-months old nuts is first scooped out, the adhering testa is removed by using a sharp knife, and the pared meat after washing is cut into 0.5-cm-thick and 6-cm-long strips. The strips are put in cans to which 50 °C brix syrup with 0.01% sodium metabisulphite is added. The filled cans are then exhausted at 78 °C, sealed and processed at 110 °C for about 20 min. In the processing of jelly-like tender nuts of 6–7 months, the meat is scraped out, and an equal part of refined sugar is added. The mixture is cooked in low heat until the sugar is totally dissolved, hot packed in sterilised bottles and closed tightly.

Coconut kernel has been used as food since ancient times. Coconut kernel contains almost all essential vitamins and minerals, which are necessary for health. Coconut fat is made up of fatty acids of saturated, unsaturated and glycerol esters. The coefficient of digestibility of coconut oil is highest when compared with that of other fats. The coconut protein has a biological value of 71–77% and digestibility of 86–94%. Coconut contains the following carbohydrates: sucrose, raffinose, galactose, glucose, fructose, pentoses, cellulose, pentosans, starch, dextrin and galacton. The fresh kernel contains about 7% sugar though it is a poor source of minerals. It also contains 151 IU of thiamine, 1 mg of ascorbic acid (vitamin C), traces of vitamin A and 0.2 mg of tocopherol 100 g^{-1}. Coconut proteins are high in nutritive value and are fairly rich in lysine, methionine and tryptophane.

The fresh kernel of ripe coconut constitutes an essential ingredient in the recipes of diverse food preparations in the households as well as in food industries of different countries. In the household food preparation, fresh kernel is used extensively in the grated, paste or milk form. When the grating as such or in the form of ground paste is used, there is no loss of nutrients or wastage of kernel. Coconut is transported either as whole coconuts or after removal of a portion of husk to regions, where the palm is not cultivated.

By transporting the coconut kernel packages instead of whole or partially dehusked coconut, up to four times higher quantity (weight basis) could be transported. Predictive equations and regression models were developed to optimise the package parameters like respiration rate of kernel, permeability characters of polyfilm and partial gas pressures inside the package. Optimum gas composition of 5, 15 and 80% of O_2, CO_2 and N_2, respectively, was found to increase the shelf life of kernel under the storage temperatures of 15 °C and 25 °C.

Preservative mixture of butylated hydroxy anisole at 0.1% and propionic acid at 1000 ppm was found to increase the shelf life of kernel along with modified atmosphere package (MAP) system. At 15 °C storage temperature, low-density polyethylene film of thickness 0.041 mm and polypropylene film of 0.042 mm thickness and at 25 °C storage temperature low-density polyethylene film of 0.027 mm thickness were found to be the best for MAP. The shelf life of the kernel in MAP was increased to a minimum period of 90 and 25 days at 15 °C and 25 °C, respectively.

For vacuum package, 0.042-mm-thick polypropylene film was the best to store at both 15 °C and 25 °C. The shelf life of the kernel was increased to a minimum period of 60 and 20 days at 15 °C and 25 °C, respectively. The shelf life of the kernel stored in open condition at 15 and 25 °C was only 20 and 5 days, respectively (Bosco 1997).

The cost of production of one package of 0.15 kg at laboratory level was Rs. 0.49. A net benefit of 30 paise could be obtained by transporting and distributing the kernel at a distance of 1000 km. Further increase in the distance of transport increases the benefit. The loss of coconut due to crack development is reduced, indirectly increasing the benefit and profitability.

13.6.1.2 Sweet Coconut Chips

Dehydrated coconut chips can be prepared from coconut kernel with intermediate moisture (35–40% on wet basis). Moisture from mature coconut kernel is removed partially by osmotic dehydration by using osmotic mediums like sugar syrup. This osmotically dehydrated coconut slice is named as ‘sweet coconut chips’.

Artificially dried coconut slices by use of conventional tray dryers or vacuum dryers are wholesome, nutritious and palatable, but it has not been, in general, found popular because it does not have the flavour, colour and texture of the original material even after rehydration. Freeze-drying of coconut slices results in good quality of dried material with long storage stability, but cost of processing is very high. Sweet coconut chips are crispy and can be packaged and marketed in laminated aluminium pouches, which will have a shelf life of 6 months. Since it is in a ready-to-eat form, it could be used at any time, as snacks or just like fresh kernel after rehydration of the chips in hot water.

13.6.1.3 Coconut Milk and Related Products

Coconut milk is an emulsion of coconut oil in water into which some of the soluble components of the meat have already been passed. It is prepared from fresh mature coconuts. In this process, the white coconut kernel is ground into slurry from which coconut milk is separated by pressing. In continuous process, coconut milk is extracted in the screw press. The milk is then centrifuged and further processed to get milk concentrate, coconut cream and milk powder. The stability of the emulsion is better in the pH regions where coconut proteins are more soluble (Monera and del Rosario 1982). Apart from household culinary uses, the coconut milk is utilised as a substitute for dairy cream in beverage-type milk as evaporated and sweet condensed milk (Banzon 1978) in the preparation of white soft cheese (Sanchez and Rasco 1983), yoghurt (Sanchez and Rasco 1984) and many other food items.

The milk is pleasant and sweet with an agreeable flavour. A comparison of the coconut milk with cow’s milk has shown that white coconut milk is richer in fat but poorer in protein and sugar content. Diluted coconut milk is used as an adulterant

Table 13.3 Composition of coconut milk in percentage

Source	Moisture	Total solids	Ash	Fat	Protein (N × 6.25)	Total sugar as invert sugar
Philippines[a]	56.3	43. 7	1.2	33.4	4.1	5.0
Sri Lanka[b]	50.0	50.0	1.2	39.8	3.0	6.2
India[c]	51.0	37.0	1.8	26.0	4.5	5.5

[a]Jeganathan (1970); [b]Nathanael (1954); [c]Arumugham et al. (1993)

for cow milk in certain places. The proximate composition of coconut milk as reported by the Central Food Technological Research Institute (CFTRI) Mysore, India, is 41.0% moisture, 5.8% protein, 38–40% fat, 6.2% minerals and 9–11% carbohydrates. The composition of coconut milk is also subject to variation. The figures reported by other sources are given in Table 13.3.

Preserved Coconut Milk or Cream Preserved forms of coconut milk such as canned cream or milk and dehydrated whole milk are now becoming available in many countries. Commercial production of these products has been promoted in the Philippines, Thailand, Indonesia, Western Samoa, Sri Lanka and Malaysia.

The technology for the production and preservation of coconut milk has been perfected in India. In this process, pared kernels are first immersed in hot water for 10 min at 75 °C to 80 °C for microbial control. The kernels are then comminuted either using the hammer mill (3 mm) or the Krauss-Maffei disintegrator with 3 mm perforation. The optimum maturity of nut for achieving maximum yield of milk was found to be 11–12 months. The moisture content of the gratings is maintained at 40–45% for facilitating efficient extraction of milk. Pressure is applied on the comminuted kernel with or without the addition of water. The extraction of milk is done in a screw press. The extracted milk is passed through a vibratory sieve of 150 microns to ensure that the milk finally contain not more than 0.5% of the finely grated material. The milk is then pasteurised at 75 °C to 80 °C for 10 min using a plate heat exchanger. As an adjunct to heat treatment, sodium metabisulphite or the antibiotic nisin is used to enhance the shelf life of the finished product. After pasteurisation, stabilisers and emulsifiers are added to the milk for preventing the separation of the milk and for facilitating emulsion formation. Carboxy methyl cellulose (CMC) and guar gum were found to be good stabilisers. Among the emulsifiers tried, Tween 80 at 0.1–0.2% levels was found suitable in the presence of 0.5% CMC. The pasteurised milk is then passed through a colloidal mill and pressure homogeniser for the coagulum dispersal and the even distribution of emulsifiers and stabilisers. Pressure homogenisation at 4000 psi has been found to impart better stability to the finished product. For packing the product, hot filling and crown corking are presently followed. Other packing systems like flexible packing and canning are also found successful. The analytical data of the coconut milk produced in India are shown in Table 13.4 (Arumugham et al. 1993).

The technology developed in the Philippines has been patented under the Philippines Patent No. 5632. In this process, round mature nuts are selected and deshelled. After disintegration, the meat is mixed with one and a half to two times

Table 13.4 Composition of canned coconut milk products (in percentage)

Properties	Moisture	Protein	Fat	Carbohydrates
Canned cream	55.0	0.9	40.0	2.8
Canned whole milk	75.0	1.0	12.5	10.2

Source: NIIR Board of Consultants and Engineers (2012)

its weight of water and passed through a screw press or expeller to extract the milk. The extracted milk is centrifuged to separate the cream from the watery and solid portions of the extract. The cream is then mixed with water (half to two times its weight) and pasteurised for 15–30 min. The pasteurised cream is thoroughly mixed with sodium caseinate at 0.5% level to stabilise the cream and then homogenised. The homogenised mixture is then heated almost to boiling point and filled in cans or bottles and autoclaved at 115.5 °C for 40 min. The tin cans are cooled immediately by submerging in a cooling tank with running water before packing in carton boxes. In the process developed in the Philippines, the watery phase or skim milk is wasted as a result of centrifugation. In order to eliminate this wastage, a process to preserve the whole milk has been developed with the use of additional stabilisers. Both canned coconut cream and processed whole milk have been found to enjoy very high consumer acceptability. The composition of these products is given in Table 13.4.

The chemical composition and quality of coconut cream vary widely in different countries. The Tropical Products Institute, London, reported that the organoleptic qualities of a number of samples tested were inferior to those of freshly prepared cream, due to excessive heat treatment. As such, the TPI has standardised the processing technology with optimum heat treatments and judicious use of added emulsifiers and stabilisers. Here the coconut milk is first clarified at a temperature of about 45 °C in a centrifugal clarifier at 10,000 rpm. The clarified emulsion is fed to a centrifugal separator at a rate of 425 gph and a temperature of 38 °C by a centrifugal pump fitted with a 0.5 HP motor to separate the cream and the aqueous phases. The cream is then diluted with the aqueous phase to obtain three different concentrations (low, medium and high fat), and the creams are heated to 55 °C, and an emulsifier/stabiliser mixture of Triodan 55 (1% W/W) and bleached soya lecithin (0.1% W/W) is added with thorough mixing. The formulated creams are then subjected to a two-stage homogenisation, one at 3500 psi at 50 °C and the other 1 °C at 500 psi. The cream is then heated to 60 °C using a water bath arrangement, canned and sealed (Timmins and Kramer 1977).

In Thailand, coconut meat is soaked in sodium metabisulphite solution prepared by dissolving 1 gram of sodium metabisulphite in water for every coconut, for 1 or 2 h or overnight. After washing, the meat is grated and milk extracted using an expeller or screw press. The milk is evaporated at 60 °C to 70 °C in a steam-jacketed kettle equipped with a blade, rotating at 20 rpm with constant stirring for 4–6 h. About 600 ppm sodium metabisulphite and 240 g pectin for every 100 coconuts are then added to the milk concentrate. Finally the milk concentrate is pasteurised at

70 °C for 1 or 2 min. The concentrated cream is packed in sterilised lacquered tin cans or sterilised aluminium tubes at 70 °C, each containing 160–170 g net. The final product is greyish white in colour which turns white upon dilution with water. It has a proximate composition of 10–11% moisture, 76–80% fat and 6–7% protein. About 130 kg unhusked coconuts yield about 32 kg of processed cream.

In the process adopted in Malaysia, the pared kernels are sterilised in boiling water for 1–1.5 min prior to pulverisation and extraction of milk. The milk is concentrated in a steam-jacketed kettle at 80 °C for 12–14 h with continuous stirring. Sodium metabisulphite at 600 ppm is also added as a preservative. The product which is thick in consistency with 40–45% fat is packed both in cans and plastic pouches.

In Western Samoa, fresh nuts within 3 days of harvest are utilised for processing. The pared kernels are grated and the milk extracted in a screw press. The milk is diluted in the ratio of 2:1 and sterilised in a steam-jacketed kettle for about 30 min. The cream is subsequently homogenised while hot, canned and sealed. One hundred and eighty kg of unpared coconut kernel yields about 127 kg cream (Timmins and Kramer 1977).

In Singapore, the processed milk is packed in cans, tetra pack and combine packs. Besides coconut cream, various milk shakes like fruit juice, coconut water and to a small degree coconut cream are also produced in Singapore. In a study conducted in Indonesia, it was found that 160 g of cream was enough to satisfy the requirements of an average Indonesian family normally using four coconuts a day.

Coconut cream is mainly used as a fat source for the reconstitution of the skimmed dairy milk and as a component of infant milk powders. In the Philippines, the cream could be successfully included as a component for the production of recombined milk or filled milk into three types of milk products – beverage type, evaporated type and sweetened condensed type. Sanchez and Rasco (1983) found that about 60% of coconut milk can be used as cow's milk extender in processing white soft cheese. The product was comparable to that obtained from 100% cow's milk in flavour, aroma, texture and general acceptability. The shelf lives were 2 days and 1 week at ambient temperature (25 °C–30 °C) and refrigerated temperature (5 °C), respectively.

Coconut Cream Coconut cream is the concentrated coconut milk extracted from fresh coconut kernel. From 10,000 ripe coconuts, the yield is around 2500 kg of cream and a byproduct residue of 500 kg. This is an instant product, which can either be used directly or diluted with water to make preparations such as curries, sweets, desserts, puddings, etc. It can also be used in manufacture of bakery products and for flavouring foodstuffs. The technology for the preservation of coconut milk in the concentrated form has been standardised and its economic viability established through pilot testing. The product has also shown encouraging consumer acceptance in the domestic market (Ghosh 2015).

Bottled Coconut Milk The processing technology involves extraction of milk from finely grated unpared coconut with the addition of some water or coconut water, straining the milk in a cheese cloth into an aluminium kettle with 0.1% benzoic acid before placing the kettle in an autoclave at 117 °C for 3 min with steam injection. The temperature of the milk in the pot is then brought down to 80–85 °C, using running tap water. The milk is then homogenised for about 5 min and bottled at 70 °C to 80 °C in 230 ml sterilised soft drink glass bottles and sealed. The final product is as good as cow's milk and is highly nutritious (Seow and Gwee 1997). This is of high commercial utility to be used as a substitute for cow's milk and is being produced in many countries.

Dehydrated Coconut Milk This is produced on a commercial scale in the Philippines and Malaysia. In the Philippines, the fresh coconut milk is blended with small amounts of additives such as maltodextrin or casein and is spray dried. The final product is marketed in laminated foil bags. The powder is easily dissolved in water to form a milky white liquid obtaining the flavour and texture of coconut milk. It is suggested to mix or blend 100 g powder with 120 ml water to make coconut cream. About 1000 kg of kernel would yield 450 kg powder. The product contains 692 calories, 60.5% fat, 27.29% carbohydrates, 9.6% protein, 1.75% ash, 0.8–2.0% moisture and 0.02% crude fibre. In Malaysia, coconut milk powder is marketed in the domestic markets as 'Santan' and in overseas markets as coconut cream powder (Muralidharan and Jayashree 2011).

Coconut Jam from Coconut Milk Coconut jam is prepared using aqueous portion of the milk. In the process developed in the Philippines, coconut milk is extracted first after mixing coconut gratings with equal quantity of water. The milk is then mixed with brown sugar and glucose in the proportions of 10.25% and 5.5%, respectively, based on the weight of the milk. The mixture is cooked over a slow fire with constant stirring for about 20 min. The mixture is strained at this stage for removing suspended matter and again cooked over high heat. Before the mixture begins to thicken, citric acid at the rate of 0.25% of the original weight of the milk is added and cooking continued over low heat until the mixture thickens. The product is hot filled in sterilised containers and sealed hermetically. The jam so obtained has a rich creamy coconut flavour (Muralidharan and Jayashree 2011).

Coconut Syrup A commercial product known as coconut syrup is produced from coconut milk in the Philippines. For this, coconut milk is first extracted from the freshly grated pared coconut meat (Hagenmaier et al. 1975). After homogenisation, an equal quantity of sugar and 0.05% citric acid or 0.25% sodium phosphate are added and then steam-cooked to a total soluble solid content of 65–68%. The boiling hot syrup is poured into lacquered tin cans, sealed and cooled under running water. The coconut syrup contains 27.18% moisture, 3.11% protein and 63.47% total soluble solids. The syrup can be used for confectionery purposes. It gives a delicious instant drink, which is milk white in colour when mixed with water and is also an excellent bread spread (Sangamithra et al. 2013).

Coconut Honey This is another useful product prepared from coconut milk and is an excellent substitute for real honey. The milk is extracted from the gratings of unpared meat after adding an equal quantity of water. To this, 60% by weight of brown sugar and 30% by weight of glucose are added and then boiled in steam-heated containers until a thick consistency is reached. The product is then hot filled in lacquered tin containers or bottles and sealed. The final product is a golden-coloured thick paste with a nutty flavour. This can also be used as an excellent base for soft drink (Muralidharan and Jayashree 2011).

Coconut honey can be prepared from skim milk as well. One part of the skim milk is mixed with half part of refined sugar and half part of glucose and then blended with sodium alginate at 0.5% as a stabiliser. Coconut cream is sometimes added to improve the flavour of the product. The mixture is heated for 15 min, homogenised and cooked with constant stirring in steam-jacketed kettle to a total soluble solid content of about 75%. It is poured hot into sterilised containers and sealed hermetically (Banzon et al. 1990).

Coconut Cheese Fresh coconut meat is grated and pressed to extract milk. The milk is allowed to stand for 8 h, until the cream is collected at the top. The cream is slowly scooped out and the skimmed milk heated with vinegar to coagulate the proteins. This is mixed with cream and kneaded with salt. The process is simple and can be taken up as a household industry (Ghosh et al. 2000).

13.6.2 Coconut Skim Milk and its Products

Coconut skim milk is a solution of the soluble components of coconut after the cream is separated in a cream separator. This is one of the stages involved in the wet processing of coconut developed by the Texas A&M University. In this process, freshly prepared coconut milk from pared/unpared kernel is filtered through a 120 mesh vibrating screen, and the pH of the filtered milk is raised from 6.3 to 7.0 with the addition of sodium hydroxide. The milk is then pasteurised at about 60 °C for 1 h and subsequently centrifuged in a cream separator to yield the aqueous phase or the protein-rich skim milk besides the cream.

In the method suggested by Birosel et al. (1963), milk is extracted from the grated coconuts by hand without the addition of water. This is poured into clean wide-mouthed bottles up to half of the container's total volume and placed in the cold storage chamber at 17 °C. The bottles are taken out and then vigorously shaken until the coconut milk is separated into its different components. The fat forms one big white mass, whereas the skim milk remains in the liquid form. The fat and the coconut skim milk are stored in cold storage chamber.

The general composition of skim milk is 90.98% moisture, 1.956% oil, 2.641% protein, 0.324% reducing sugar with 0.013% acidity and 6.15 pH. Capulso et al.

(1981) investigated in detail the optimum conditions for maximum recovery of the proteins as well as their functional properties. Skim milk can be used for making a variety of products like spray-dried powder, coconut honey, coconut jam and sweetened condensed milk. In addition, it can also be used as a substitute for the preparation of fermented beverage concentrate and also as a source of vegetable casein. It is a good source of quality protein suitable for the preparation of many useful allied food products or as a supplementary protein source, especially in regions deficient in animal proteins.

13.6.2.1 Fermented Beverage Concentrate

The processing method has been developed by Sanchez and Rasco (1984). This is a type of cultured milk using skim milk as a substrate and *Lactobacillus bulgaricus* as a starter culture. For this process, the milk base should contain 10% total solids. On an average, coconut skim milk (CSM) is composed of 14.5% sugar and 85.5% water, whereas non-fat dry milk (NFDM) is composed of approximately 100% total solids. As such, different combinations of CSM and NFDM are formulated to maintain the 10% total solid content of milk base. This is then pasteurised at 90 °C for 30 min in a water bath, cooled to 40 °C and inoculated with 3% culture of *Lactobacillus bulgaricus*. The mixture is incubated at 37 °C–38 °C for 24 h. Curdled milk is homogenised for 5 min and heated to 60 °C. Sugar is then added in the ratio 1:1. The mixture is further heated to 80 °C and cooled down to 60 °C, and then 0.5% flavoured extract is added. The finished product is bottled and pasteurised in water bath at 70 °C for 30–60 s. Out of the different formulations of CSM and NFDM tried, a combination of 50% CSM and 50% NFDM was found to be optimal. The ready-to-drink fermented beverage (concentrate diluted at 1:3) contains 1.01% protein, 0.74% fat, 18.70% sugar and 79.25% water. The product is stable even after 2 months of storage, both at ambient and refrigerated temperatures. This is a highly nutritious drink suitable for kids and adults alike. Unlike carbonated drinks, the fermented beverage contains protein contributed by CSM and NFDM combination. It is non-fattening and easily digestible and is a perfect beverage for those suffering from digestive ailments. Similar products have been commercially prepared in other countries, for example, 'Calpis' in Japan and 'Bulgaricus milk' in Bulgaria.

13.6.2.2 Vegetable Casein

Skim milk is also a source of vegetable casein. In Brazil, the gastrointestinal disturbances were successfully treated in infants by feeding coconut milk, which shows that coconut skim milk having the same protein level (1.6%) as mother's milk is well-absorbed by infants. Both produce a soft curd when acted on by the gastric juice.

13.6.2.3 Low-Fat Coconut Jam

Skim milk is also utilised for the production of low-fat coconut jam. The whole milk is subjected to centrifugation for the separation of skim milk. For every 20 kg skim milk, 3.75 kg of brown sugar and 1.25 kg of glucose are added, and the mixture is boiled for 20 min, blended, strained and boiled again. Citric acid is added just before cooking is complete, and boiling is continued to an end point of 75–76% total soluble solid content as measured with a hand refractometer. The hot jam is poured in sterilised containers and sealed hermetically (Banzon 1978).

13.6.2.4 Sweetened Condensed Milk

Skim milk is also utilised as a base for the production of sweetened condensed milk. Corn oil, coconut cream and sugar are added to the pasteurised milk, and the mixture is passed through a colloid mill. It is then heated in a steam-jacketed kettle with constant stirring to a total soluble solid content of 68%. The finished product is packed hot in sterilised tin cans and cooled immediately in cooling tanks. It contains protein (5.53%), fat (10.35%), carbohydrates (60.68%), calcium (112.62 mg per cent) as well as magnesium, iron, zinc and manganese.

The protein-rich skim milk is spray dried to yield a product similar to non-fat dry milk (NFDM) containing about 70% of the original protein present in coconut. The powder is white and deliquescent having a fresh coconut flavour, the average composition being 24% protein, 6% fat and 5% moisture. Maltodextrin is added as an additive to improve body and fluidity of the final product. This is an excellent coconut-flavoured beverage base. The detailed composition is given in Table 13.5 (Banzon 1978).

An attempt has been made to explore the possibilities of spray drying of coconut skim milk powder to produce coconut skim milk powder (CSMP) (Ganesan and Gothandapani 1999). The critical operations involved to produce CSMP are centrifuging, evaporation and spray drying. The coconut milk could be successfully centrifuged in a cream separator to produce coconut skim milk. The percentage of recovery is 71%. The skim milk has a low-fat content of 0.27% with total solids of 5%. The density is 1.084 g cc^{-1}. The per cent of cream and sludge recovered were 22% and 7%, respectively. The concentrated skim milk after evaporation contains 40% solid and 2.70% fat with a density of 1.25 g cc^{-1}.

The coconut skim milk could be successfully spray dried in a spray drier. The characteristic of the product was comparable with commercially available ones. The skim milk had a bulk density of 0.5 g cc^{-1} with moisture content of 3% with an ash content of 8%. The composition of the skim milk powder was protein 13% and fat 2.70% with sugar and carbohydrate at the level of 9 and 7%, respectively. The product could be successfully stored for more than 2 months under vacuum and at refrigeration at 15–20° C.

The microstructure of coconut skim milk powder was observed using scanning electron microscopy (SEM). Small surface folds and cracks were noticed in some

Table 13.5 Composition of dehydrated coconut milk (in percentage)

Constituent	1[a]	2[b]
Protein		
Crude (N × 6.25)	25	30
Low molecular weight N (as % N)	8	7
Fat		
Crude fat	5	7
Free fatty acids (as % of oil)	3.2	1.4
Nonsaponifiables (as % of oil)	3.2	–
Iodine value	6.3	–
Carbohydrates		
Reducing sugars	2.8	2.0
Reducing sugar after inversion	45	37
Sucrose	33	–
Crude fibre	0.03	0.03
Minerals		
Phosphorous	0.50	0.50
Calcium	0.17	0.06
Magnesium	0.26	0.36
Potassium	3.60	3.30
Sodium	0.90	1.40
Chlorine	1.60	1.60

[a]Coconut water used in processing; [b]Tap water used in processing

particles, whereas many samples contained small depressions, resembling the surface of moon. The surface of particle resembled honeycomb structure. Some of the particles were more wrinkled and curl (Ganesan and Gothandapani 1999).

13.6.2.5 Edible Coconut Flour

Partially defatted edible coconut grating is becoming acceptable to different categories of consumers for use in households, bakeries, confectionery units and hotels. Appropriate production technology has been perfected, and its commercial viability also has been established. Because of the low content of fat and higher percentages of protein, sugars and minerals, the product possesses better water holding and thickening properties.

During the processing of coconut cream and other related products, the fibrous residue obtained after expelling the milk is dried and powdered to obtain a product called coconut flour. The flour so obtained typically contains 7–8% protein, 3–5% moisture and 17% oil. It can be used as an ingredient in dietary foods because of its high fibre content. The protein contained in the flour is identical to the original meat. Better grade coconut flour is made by powdering the pressed cake obtained after extracting oil from quick dried gratings of coconut meat. Such partially defatted coconut gratings can find use in bakery and confectionery preparations as well

Table 13.6 Composition of coconut flour and desiccated coconut(in percentage)

Constituent	Coconut flour	Desiccated coconut
Moisture	3.83	1.70
Fat	41.43	70.05
Protein	17.32	8.30
Crude fibre	7.00	4.65
Ash	3.26	1.62
Total carbohydrates	27.16	13.69

Source: Satyavathy Krishnan Kutty (1987)

as in nutrition programmes in schools. The Central Food Technological Research Institute (CFTRI), Mysore, India, has developed a process for the manufacture of edible flour. Here the wet kernel is first separated from the shell by crushing the cups in a hand-operated gadget. After removing the shell pieces by hand, the kernel is pared, washed free of dirt and passed through a pin-type disintegrator, and gratings are sieved. The gratings are then dried in less than 30 min in a crossflow drier to less than 3% moisture. Oil is partially extracted from these gratings using a hydraulic press in 10 kg batches. This oil possesses better quality than commercial samples, and the residual product is white and powdery with good flavour. The composition of the flour is given in Table 13.6.

NIIST (CSIR), Trivandrum, India, also perfected the technology for the partial extraction of oil from desiccated coconut in order to produce, besides oil, good-quality coconut flour. Here the pared kernel pieces are first washed and soaked in hot water at 80 °C for 10–15 min to reduce the microbial load and also to inactivate enzymes. Here also, the kernel pieces are comminuted into fine gratings using a pin mill and dried at 60 °C–70 °C. The dried gratings are charged into a perforated stainless steel cage, and pressure is applied from the top using a downstroke hydraulic press till the desired level of oil has been expelled. The oil is stored for 10–12 h and filtered using a filter press in the presence of 0.1% super cell as filter aid. The partially defatted gratings are removed and powdered using a cake-breaker and further dried in an electric drier to a moisture content of 2–3%. The fat content of the final flour is adjusted to 40–45%. The shelf life of the product stored in sealed aluminium foil pouches is 4–6 months at ambient temperature and more than 1 year under refrigerated conditions. About 12–15% of the fresh meat constitutes a byproduct, with an oil content of 60% on dry weight basis. About 90% of this could be recovered as commercial-grade coconut oil.

The product has a low content of fat and higher percentages of protein, sugars and minerals and has been found to possess better water holding and thickening properties. The oil extracted from the dried gratings is of superior quality, which could command premium price. Flour derived after removing 60% oil has been found to have higher consumer acceptability. The proximate composition of flour from pared and unpared coconut as reported from different sources is given in Table 13.7. It is also seen that 5% coconut flour can replace proportionate amounts of wheat flour and non-fat dry milk powder used in school nutrition programmes without affecting baking qualities and food value.

Table 13.7 Proximate composition of coconut flour (in percentage)

Constituent	From pared coconut[a]	From unpared coconut[b]
Moisture	5.69	5.4
Fat	7.18	2.0
Crude fibre	9.21	9.8
Protein	20.39	24.9
Ash	5.41	5.3
Carbohydrates	–	62.4

[a]Balakrishnamurthi (1979); [b]Abdon (1967)

Table 13.8 Nutrient composition of coconut flour compared with various seed flours (in percentage)

Nutrient	Soybean	Ground nut	Sesame	Cotton seed	Coconut	
Moisture	5.00	11.00	5.60	9.20	11.20	5.40
Protein (N × 6.25)	60.00	52.00	33.30	51.10	20.90	24.90
Fat	7.00	8.90	12.20	5.50	13.30	2.00
Carbohydrate	30.00	21.80	38.10	25.80	39.20	52.60
Crude fibre	2.50	1.00	4.80	1.50	10.50	9.80
Ash	5.50	4.60	6.00	5.90	4.90	5.30
Calcium	0.33	0.67	2.38	0.36	0.16	0.07
Phosphorous	0.62	0.50	0.63	0.82	0.49	0.47
Iron	20.00	2.90	19.30	12.00	5.70	8.10
Thiamine	0.7	0.95	1.05	0.99	0.17	0.09
Niacin	5.70	19.50	5.30	5.20	4.10	2.30
Riboflavin	0.38	0.20	0.3	0.30	–	0.08

Source: Banzon et al. (1990)

Nutritionally coconut flour compares favourably with most of the common cereal flours as could be seen from Table 13.8.

These flours can also be utilised in different food formulations such as extrusion, baking and confectionery. They can also be used for enriching the nutritional values of wheat flour, rice flour, etc. The oil extracted from the dried gratings is of superior quality, which could command premium price. Coconut flour is naturally low in digestible carbohydrates and high in fibre content and good proteins and hence is a health-promoting food. This has four times more fibre than oat bran and two times more than wheat bran. Technologies are available now to prepare coconut milk residue and virgin coconut oil (VCO) cake flour-based compressed bar, porridge, *laddoo*, *halwa* and noodles. Processes for preparation of extrudates, pasta, muffin cakes and ready-to-eat food items such as coconut pickle and coconut chutney powder have been developed and standardised by ICAR-CPCRI. These products can be prepared from the byproducts obtained while preparing the coconut milk, coconut milk powder and virgin coconut oil. It may have some health benefits and may encourage the industry to produce value-added products or functional foods which can help in the proper control and management of chronic diseases. This offers scope for utilisation of coconut flour as a dietary component for diabetes. Low-fat, high-fibre coconut flour, a unique product from 'sapal', is a good source of dietary fibre. It is comparable with other cereal flours in terms of carbohydrate, fat and energy content and is a good ingredient in nutraceuticals.

13.7 Coconut Oil

Coconut oil is the most important consumable product from coconut. Coconut oil extracted by dry process from copra is called coconut oil or copra oil, and the one extracted from wet kernel is known as virgin coconut oil. Coconut oil is extracted from milling copra using appropriate milling devices.

13.7.1 Extraction Methods of Coconut Oil from Copra

In most of the coconut-producing countries, copra crushing is a traditional industry. In rural areas, copra is still crushed in *chekkus* for domestic consumption. Power-driven *chekkus* or rotary mills, expellers and hydraulic presses are used for crushing on a commercial scale. In the modern industrial units, solvent extraction planks are linked with the expellers or hydraulic presses for the maximum recovery of residual oil from the copra cake.

As a cottage industry, oil extraction is still done in the *chekkus* in the rural areas of India and Sri Lanka. The *chekku* is a fixed wooden or stone mortar inside which wooden pestle is made to revolve. The pestle is attached to a long pole which is moved round by a pair of bullocks or even by men. Copra is crushed as a result of the friction caused by the revolving pestle. The pressure on the pestle is regulated by levers and weights. Usually only 100 kg of copra can be crushed with *chekkus* in 8 h.

Power-driven *chekkus* or rotary mills are used in medium establishments for oil extraction from copra. In principle, they are similar to *chekkus*, but here, the mortar itself revolves against the pestle. Both the pestle and mortar are made of cast iron. The rotaries are worked in pairs and are driven either by steam, oil engine or electric motor. The crushing capacity of a rotary mill varies from 200 to 300 kg of copra day^{-1} shift. About 1 kg of *gum acacia* is added to every quintal of copra as a binding agent to facilitate the extraction of oil.

Expellers are used in large milling establishments, which vary in number from 1 to 24, but two expellers usually constitute a unit since double crushing gives the maximum extraction. The extraction takes place within a steel cage or barrel by means of a hardened steel worm or screw so arranged on a revolving shaft as to produce increasing pressure as the copra is pushed from one end of the cage to the other. The oil escapes through the openings built on the barrel and the cake through an adjustable pressure orifice at the end of the barrel. In large installations, the clean copra is first passed over an electric magnet, by chain conveyor for removing extraneous material. Then it is passed to a disintegrator for conversion into a coarse meal. The disintegrated copra meal is fed continuously by special conveyors to the expeller from which residual oil and cake are constantly forced out through separate streams.

While passing through the first expeller, the pulped copra yields about 50% oil. The resultant cake is again ground, heated and pressed in a second expeller to extract the remaining oil. Generally expellers can squeeze out about 95% of the oil present in the copra. The crude oil collected in underground tanks is pumped to a filter press, and clean filtered oil is finally pumped to big storage tanks. Smaller expeller units, like those in operation in many states in India, have a crushing capacity of 2–3 tonnes of copra per shift of 8 h. A few bigger units also are available with expellers of 20–40 tonnes capacity for pre-pressing and 7–20 tonnes capacity for pressing per shift of 8 h.

The residual oil present in the rotary and expeller crushed cakes can be extracted using solvent extraction plants. In this process, the cake is broken into bits and fed to flaking rolls. The resulting raw material is then treated with suitable solvents in a countercurrent extraction process. The solvent containing the dissolved oil is drawn off from the extracted residue and filtered. The oil is then separated from the solvent by distillation. The common hydrocarbon solvents used in the solvent extraction plants in India are benzene and hexane.

13.7.2 Yield of Oil from Copra

In *chekkus*, the yield of oil varies from 58% to 60%, in rotaries 62–63% and in expellers 63–65%. The residual oil in the cake ranges from 12% to 17% in *chekkus*. The rotary and expeller cakes normally contain 10–12% and 8–10% residual oil, respectively. High-pressure expeller cakes hold about 6% oil, while solvent-extracted cakes have very low oil content ranging from 0.5% to 1.0%.

13.7.3 Ensuring the Quality of Oil

For obtaining a better quality product, it is necessary to crush copra having a moisture content of less than 6%. The crushing process should be done in clean surroundings and the oil collected in clean equipments. The crude oil should be filter pressed without much time lag and the storage tanks frequently cleared of sediments and other settled impurities. The oil should be free from moisture, for instance, under the British Standard specification, the moisture content must be under 0.25%. In the case of refined oil, it is treated and subsequently washed, filtered and/or heated to 110–120 °C to remove traces of moisture. The latter treatment also effects sterilisation. The oil must be stored as far as possible away from light and air. The receptacles may be filled to the maximum possible extent in order to reduce the amount of surface exposed to light and air. Small quantities of oil can be successfully stored in soldered kerosene tins and large quantities in storage tanks.

13.7.3.1 Physical Properties of Coconut Oil

Coconut oil is colourless to pale brownish yellow, which is fluid in the tropics and is a solid fat in the temperate climate. In the solid state, its melting point ranges from 23 to 26 °C. Among the common vegetable oils, this has the lowest turbidity and has excellent properties for electrical insulation. The oil also possesses high inductivity. The heat of combustion of coconut oil is estimated at 9285 cal g^{-1} gross or 8697 cal g^{-1} net. The vapour pressure at 202 °C is 0.054 mm, at 227 °C is 0.16 mm, and at 50 °C is 0.37 mm. The principal physical characteristics of coconut oil are specific gravity (0.926 at 15 °C and 0.9188 at 25 °C), saponification value (251–263), iodine number (8.0–9.6.0), Polenske value (15–18), melting point (23–26 °C), titre of fatty acids (20.4–23.5) and melting point of completely hydrogenated fat (44.5 °C).

13.7.3.2 Chemical Properties of Coconut Oil

Coconut oil is rich in saturated fatty acids, mainly lauric and myristic acids with notable proportion of still lower fatty acids. In comparison with the leading edible vegetable oils, coconut oil is low in unsaturated and polyunsaturated acids, particularly, linoleic acid. It has got the highest saponification value and the lowest iodine value. On account of its low iodine value, it is classified as nondrying oil. Coconut oil contains the largest percentage of glycerol compared with other oils and fats as shown in Table 13.9.

The glycerides of coconut oil have not been quantitatively separated. But the presence of certain mixed glycerides along with their quantity and melting point has been established. It comprises of large glycerides, caprylo-lauromyristin (melting point 15 °C) and dilauromyristin (melting point 33 °C). Small glyceride laurodimyristin (melting point 38.1 °C) and very small glycerides dimyristopalmitin (melting point 45.1 °C) and dipalmitostearin (melting point 55 °C) are also components of coconut oil. Coconut oil contains the largest percentage of glycerol compared with other oils and fats as shown in Table 13.10.

The composition of fatty acids in coconut oil changes with increasing stage of maturity. In a study reported by Banzon et al. (1990), 8-month old nuts gave oil containing an average 30% oleic acid, 16% lauric acid and 20% linoleic acid. This decreased rapidly as the nuts matured. Thus, 10-, 11- and 12-month old nuts yielded oil containing a stabilised value of 5% oleic acid, 1% linoleic acid and 49% lauric acid.

Table 13.9 Percentage of fatty acid composition of coconut oil compared with other oils

	Coconut	Palm kernel	Babassu	Palm oil	Soya bean	Safflower
Saturated fatty acids						
Caprylic	8.24	1.40	3.50	–	–	–
Capric	7.19	2.90	4.50	–	–	–
Lauric	47.31	50.90	44.70	0.30	–	–
Myristic	17.00	18.40	17.50	1.10	0.10	0.10
Palmitic	8.85	8.70	9.70	45.20	10.50	6.50
Stearic	2.27	1.90	3.10	4.70	3.20	2.40
Arachidic	–	–	–	0.20	0.20	0.20
Unsaturated fatty acids						
Palmitoleic	1.00	–	–	–	–	–
Oleic	6.27	14.60	15.20	38.80	22.30	13.20
Linoleic	1.87	1.20	1.80	9.40	54.50	77.70
Linolenic	–	–	–	0.30	8.30	–
Arachidonic	–	–	–	–	0.90	–
Percentage unsaturated	9.14	15.80	17.0	48.50	86.0	90.80

Source: Rethinam and Bosco (2006)

Table 13.10 Glycerol in oils of different saponification value

Type of oil	Glycerol content (%)	Saponification value
Coconut oil	13.84	253
Palm kernel oil	13.57	248
Olive oil	10.45	191
Groundnut oil	10.45	191
Gingelly oil	10.45	191
Cotton seed oil	10.83	198
Palm oil, tallow, etc.	10.83	198
Sun flower oil, olive oil, etc.	10.45	191
Rape seed oil	9.57	175

Source: Rethinam and Bosco (2006)

13.7.4 Uses of Coconut Oil

The main use of coconut oil whether raw or refined is for edible purposes. Oil obtained by direct processing of wet kernel or by crushing good-quality copra in clean surroundings is used for cooking without any further refining. In India, raw coconut oil is preferred for edible purpose as cooking and frying oil.

13.7.4.1 Edible Uses

Because of its high content of saturated fatty acids, coconut oil is highly resistant to oxidative rancidity and retains a pleasing flavour. Hence, it is preferred as a fat source in the preparation of filled milk, as an instant milk powder and also as a confectionary fat in the preparation of ice cream, imitation cream or whipped cream. Hardened coconut oil creamed with sugar in the proportion 40:60 is extensively used in the developed countries as biscuit and water fillings. Because of the easy digestibility of coconut oil, it has been used as an essential ingredient in many ghee substitutes. But with the introduction of hydrogenated oils, coconut oil has now lost its eminence as a solid vegetable fat. It is still used to some extent in Europe as a constituent of margarine and shortenings and also to fill the skim milk byproduct of butter manufacture (Rethinam and Bosco 2006). For details on nutritional aspects and health benefits of coconut oil, please see Chap. 15.

13.7.4.2 Nonedible Uses

One of the major nonedible applications of coconut oil is in soap industry. Because of its high saponification value, it combines with concentrated solutions of caustic soda in the cold, saponification being completely effected within 6–24 h. Marine soaps that lather in hard water and which are soluble in brine are made from coconut oil. Unlike ordinary soaps, coconut oil soap will not precipitate in weak brine. Coconut oil contains the maximum content of glycerol (13.5%) among the vegetable oils. In soap making with coconut oil as the major ingredient, glycerol is an important byproduct. This is not often recovered in unorganised soap units. The recovery of glycerol is possible with improved facilities. The glycerol is used in various industries, especially in pharmaceuticals, in food industry and in the manufacture of nitroglycerine.

One important chemical derivative of coconut oil is methyl esters which are produced by treating coconut oil with alcohol. The most commonly used alcohol in the process is methyl alcohol which produces the methyl esters of coconut oil fatty acids. Since they are more stable and are easier to separate by fractional distillation than the fatty acids, they can be converted into a very wide range of chemical intermediates or into alcohols. Both the methyl esters and mixed fatty alcohols constitute the basic raw material for the manufacture of a wide variety of detergent intermediates particularly in the ethoxylated and sulphated form.

Coconut monoethanol amide (CMA) prepared with coconut oil as the starting material can be used as foam stabilisers and super flattening and thickening agents, for detergent shampoos and cosmetics. A suitable technique has been devised for its preparation, which will enable the oil to command a much higher market value (Arida et al. 1979). The chemistry of surfactants derived from coconut oil was studied. In fact, coconut oil is the basis of a number of surfactants since their hydrophobic chains contain 12–14 carbon atoms. Thus, coconut oil derivatives can be used as raw material for the synthesis of polyolesters and imidazolines.

13.7.5 *Quality Standards for Coconut Oil*

For coconut oil, there is no uniform international quality standard. In India, the two common grades are unclean and clean oils; the former is *chekku* and rotary oil and the latter is expeller oil which is usually filtered. The Asian and Pacific Coconut Community has recommended five different grades of coconut oil. Grade I refers to refined and deodorised oil, whereas grade II refers to refined oil. Grade III oil is white oil obtained by wet processing. Grades IV and V are referred to as Industrial Oil No. 1 and 2 obtained by the process of mechanical and solvent extraction, respectively.

Storage of oil in air-tight brown-coloured containers along with either sodium meta bisulphate, citric acid or common salt will increase the shelf life of the oil for more than 6 months. So also the quality is better in copra produced in the solar dryer, followed by electrical dryer and then by sun drying on cement floor.

13.7.6 *Methods of Refining Coconut Oil*

Crude coconut oil is refined using methods like neutralisation, physical refining, etc. and more recently using hydrogen peroxide and oxalic acid. The free fatty acid content comes down to less than 1% as a result of refining. Neutralisation is done with caustic soda solution. Oil is mixed with solutions directly, washed and dried. The dry oil at 90–95 °C is then leached by the addition of adsorptive earth (fuller's earth, activated earth, activated carbon). In addition to the removal of some pigments, traces of remaining soap, gums and other impurities are removed, and a completely clear product is obtained (Rethinam and Bosco 2006).

Physical refining is also possible which is carried out by steam distillation. The removal of off flavours also takes place, so that oil may be neutralised and deodorised at the same time. Since free fatty acids are more volatile than triglycerides, separation is effected.

Brown coconut oil extracted from inferior-quality copra can be bleached to light yellow colour by pretreatment with hydrogen peroxide or oxalic acid followed by bleaching with earth and carbon. The bleaching of solvent-extracted oil is done by pretreatment with aqueous solutions of oxalic, citric and tartaric acids. The advantage is that bleaching retains the characteristic sweet aroma of the original oil.

13.7.7 *Rancidity in Coconut Oil*

Unrefined coconut oil is susceptible to rancidity due to the presence of certain proportion of free fatty acids. This is accelerated by the presence of initial moisture and by the action of air, light and fat-splitting enzymes like lipase and peroxidase. The first stage of rancidity is hydrolysis to produce the free fatty acids, known as 'hydrolytic' rancidity, the rapidity of which varies with the initial acidity and the amount of moisture present. This is followed by the oxidation of the fatty acids involving double bonds to form hydroperoxides, which are responsible for the off flavours and further reactions. This is known as 'oxidative' rancidity. In fact, rancidity is considered to be the objectionable smell that results from the accumulation of decomposition products of the oxidation reaction.

Most of the free acids and the accompanying bad odour and taste originate in the copra itself before the oil has been extracted. Oil derived from damaged copra favours the growth and multiplication of microorganisms resulting in poor-quality oil. During storage, the melting point, smoke point, refractive index (RI), peroxide value (PV) and saponification value (SV) show an increase, while iodine value (IV) decreases (Tribold and Aurand 1963; Palaniswami et al. 1989). Ghani oil has a higher percentage of residual moisture and thus a faster rate of peroxide formation.

Microorganisms under favourable humidity and temperature also induce spoilage of oil besides producing toxic substances such as aflatoxins. Mukherji (1967) reported the presence of *Aspergillus niger* in homemade coconut oil. Hoover et al. (1973) isolated lipase from a strain of *Aspergillus flavus* causing deterioration of coconut oil. Oil contaminated with *R. nigricans* and *B. theobromae* turned rancid more quickly than those contaminated with *A. flavus*. Presence of aflatoxins in samples of coconut oil affected by *A. flavus* Link Ex Fries and aflatoxin production in coconut oil was also reported elsewhere (Schindler and Eisenberg 1968; Arseculeratne and De Silva 1971; Samarajewa 1972).

13.7.8 Coconut Cake

Traditionally coconut is dried to produce copra which is then milled or solvent extracted to get the oil. Fresh coconut kernel contains about 4–4.5% protein. Major portion of the original protein passes on to the coconut cake or poonac, which is the residual product after oil extraction. It forms about 32–40% of the copra after the extraction of oil. The output of cake and its final composition depend upon the extraction methods employed (Table 13.11). The cake however is not considered suitable as a protein supplement because in the process of oil extraction, the original

Table 13.11 Composition of coconut cake

	Expeller cake	Solvent-extracted meal
Moisture	7.0	8.9
Fat	6.7	2.4
Protein (N × 6.25)	21.2	21.4
Nitrogen-free extract	47.4	47.4
Fibre	11.2	13.3
Mineral matter	6.5	6.6

Source: Rethinam and Bosco (2006)

protein gets discoloured and denatured due to the generation of very high temperature. Hence it is mainly used in ruminant feeding.

Coconut cake is also useful for feeding poultry. The cake easily absorbs atmospheric moisture while in storage and consequently is prone to mould attack. The equilibrium moisture content of the cake at 40–70%, 80% and 90% relative humidity of the atmosphere is 10%, 20% and 30%, respectively.

In India, a study on the shelf life of coconut cake revealed that the moisture content of the cake at 79% RH was 15.2% and this was the critical moisture content at which it can be stored free from moulds. The study also showed that the cake could be stored without any spoilage up to about 6 months if alkathene bags or alkathene-lined containers are used. The rancidity of the cake could be effectively checked if its moisture content is kept below the critical level of 15.2%.

Coconut cake is sometimes used as manure for field crops. But the cake has a high carbon nitrogen ratio and is, therefore, not considered suitable for manuring seasonal crops. Coconut cake contains 3% nitrogen, 1.9% phosphorus pentoxide and 1.8% potassium oxide. As the nitrogen, phosphoric acid and potash contents of coconut cake are not high, it is not considered as valuable organic manure (Thampan 1987).

13.8 Virgin Coconut Oil

The fresh meat can also be processed to yield oil, which is known as virgin coconut oil (VCO). The processing method adopted for obtaining oil and other products directly from the fresh kernel is known as wet processing. Various processes developed in this direction could successfully extract good-quality oil directly from the wet kernel without passing through the copra stage.

For production of VCO, coconut milk is filtered and concentrated, and then cream is separated by centrifugation. The cream is stirred vigorously to get the virgin coconut oil by a process called phase inversion. The oil thus obtained is very clear and nutritious and has a longer shelf life. VCO can also be prepared by hot fermentation and fresh dry process. In hot process, coconut milk is heated until the evaporation of water and coagulation of protein to recover the oil. In fermentation process, the milk is allowed to stand for 16–24 h fermentation, and then oil is separated from the skim milk and fermented curd by decanting and filtering. The fresh dry process involves screw pressing or hydraulic pressing of dried pulverised coconut meat to produce VCO and food grade full protein and medium fat coconut flakes (Marina et al. 2009).

Though oil recovery from wet processing normally does not exceed 90% as against 95–97% oil in dry processing, it is gaining renewed interest mainly because of the possible economic advantages associated with the production of nutritionally rich products. Though the efficiency of oil recovery is less than that of dry processing, the other products derived from the process such as coconut protein, coconut flour, etc. make wet processing economically viable. The processing technologies already developed and tried in various centres are briefly described below.

13.8.1 Krauss–Maffei CFTRI Process

M/s Krauss Maffei of West Germany developed a process in which the husked nuts are first steamed in an autoclave for 10 min at a pressure of 3 kg cm^{-2}. The nuts are then cut, kernels scooped out and the kernel pieces passed into a cutter. These are further comminuted by passing through a roller mill. The kernel mass is then fed into a screw press and milk extracted and is separated by centrifugation into cream and water phases. The cream phase is heated to 92 °C, centrifuged and filtered to get good-quality oil. The water phase is heated to 98 °C in a flow heater to coagulate the protein, which is separated by centrifugation and dried. The leftover whey is then concentrated under vacuum to a syrup called 'coconut honey',and the fluffy residue after extraction of the milk is dried in a drier and powdered to get an edible-quality coconut meal. The recovery of oil is about 89%, as compared to about 95% in the conventional expeller process.

To improve the recovery, the CFTRI has made some modifications in the process utilising the M/s Krauss-Maffei plant. The meat was scooped out without subjecting the nuts to heat treatment and milled. The milled meat was first squeezed in a dewaterer under reduced pressure (2:1) and then fed into the K.M. Press (12:1). With this two-stage extraction method, roughly 93% of the oil was extracted into the milk, and the final recovery of the protein was also high (roughly 85%). The residue after the extraction of milk was dried and fed into an expeller to extract the residual oil.

In one of the trials, 220 kg of wet meat yielded 73.5 kg of clean oil, 5.7 kg of residual oil, 8 kg of protein and 1 kg of coconut honey. In this modified process, the direct oil recovery efficiency was 90%, which was definitely better than the conventional oil milling process. The composition of the various products obtained in the process is given in Table 13.12.

The water phase concentrate or the coconut honey was used for the preparation of two types of processed food: (1) infant food and protein food and (2) cereal flakes (NIIR Board of Consultants and Engineers 2012).

Table 13.12 Composition of products from K.M./CFTRI process (in percentage)

Material	Moisture	Protein (N × 6.25)	Fat	Minerals	Carbohydrates
Coconut honey (60° brix)	40.0	15.6	2.0	6.0	35.6
Acid-coagulated proteins (by centrifugation at pH 3.8–4.0)	6.0	74.3	6.1	6.1	10.5
Heat-coagulated protein (at 82 °C)	8.4	66.1	3.4	8.2	13.9
Residual meal	9.7	5.2	8.7	1.6 (ash)	25.1 (crude fibre)

Source: Rethinam and Bosco (2006)

Table 13.13 Composition of products from Texas A&M University process (in percentage)

Material	Oil	Crude fibre	Ash	Crude protein (N × 6.25)	K_20	CaO	P_20_5	Fe_203	MgO	CI
Fibrous residue	22.0	23.0	0.6	3.4	–	–	–	--	–	--
Protein fraction of low solubility	19.0	3.0	7.3	57.0	0.62	0.39	1.9	0.07	1.0	–
Skim milk	5.0	–	9.0	31.0	3.3	0.07	0.5	0.004	–	1.6

Source: Rethinam and Bosco (2006)

13.8.2 Texas A&M University Process

This process known as the 'aqueous process' was developed by the Texas A&M University for the preparation of the coconut protein products (Samson et al. 1971). In this process, the scooped kernel is finely comminuted and mixed with heated coconut water. The mixture is then repeatedly pressed in a countercurrent system to remove the milk and other extractable components. The fibrous residue is dried. The milk is adjusted to pH 7.0 (from 6.3) with sodium hydroxide, pasteurised at 60 °C for 1 h and centrifuged while warm. Centrifugation yields three phases: (a) the cream or oil, (b) a protein fraction of low solubility and (c) protein-rich skim milk. The skim milk is spray dried to yield a tasty product, which resembles non-fat dry milk and can be a valuable commodity in food and beverage industry. The oil recovery has been reported to be 90%. The composition of the various products obtained in this process is given in Table 13.13.

The skim milk powder is rich in protein which contains good amount of important amino acids like lysine, leucine, phenyl alanine, threonine and valine.

13.8.3 The TPI Process

This process was developed by the Tropical Products Institute, London (Dendy and Timmins 1973). Here the fresh kernel is sliced, minced in a 5 mm plate of a wedge mill, mixed with half its weight of water and passed through a 1 mm plate assembly of a second wedge mill. The resulting slurry is then passed through the plate assembly of the second wedge mill. Then the emulsion is passed through a vibrating sieve to remove the coconut residue under a high-velocity overhead water spray. The residue is pressed again and the emulsions are pooled. The final residue is then dried and bagged to be used either as a ruminant feed or fed to an expeller mill to extract the residual oil.

The emulsion is adjusted to a pH 3.5–4.0 with acetic acid and allowed to stand for 6 h for the separation of cream and water phases. The lower water phase is removed either by gravity flow or by siphoning and discarded. The cream phase is centrifuged at 35 °C to separate the oil and protein. The protein component is vac-

Table 13.14 Composition of products from the TPI process (in percentage)

Produce	Fat	Crude protein	Moisture	Crude fibre	Ash	Carbohydrates
Protein isolate	7.2	82.0	4.9	0.6	4.95	0.35
Residue or meal	36.4	4.8	5.0	43.6	0.4	9.8

Source: Rethinam and Bosco (2006)

Table 13.15 Oil recovery from different wet processing methods

Processes	Products	Percentage of recovery
Krauss-Maffei	Oil, coconut protein, coconut honey and meal	89
Krauss-Maffei/ CFTRIprocess	Oil, residual oil, protein coagulate, coconut honey and meal	97
Texas A&M	Cream/oil, protein, skim milk and fibrous residues	90
TPI	Oil, protein isolate and meal	85

Source: Rethinam and Bosco (2006)

Table. 13.16 Composition of tender (TCW) and mature coconut water (MCW)

Constituents	TCW	MCW
Total sugar (per cent)	4.8	3.1
Total reducing sugar (per cent)	4.0	2.0
Total protein (mg/dl)	150	450
L-arginine (mg/dl)	30	150
Vitamin C (mg/dl)	25	15
Magnesium (mg/dl)	16	14
Potassium (mg/dl)	300	257
Calcium (mg/dl)	40	44

Source: Sandhya and Rajamohan (2008)

uum dried at 55–80 °C and the residual oil extracted with isopropanol. The final oil recovery is about 85% which is distinctly lower than that from the other two processes. In all these processes, the protein recovery was more or less the same, i.e. about 85%. The defatted protein is desolventised by vacuum drying for 1 h at 60 °C and milled to a powder. The product composition is shown in Table 13.14.

The products obtained and the percentage of oil recovery from the existing wet processing methods are compared in Tables 13.15 and 13.16.

The Industrial Technology Research Institute (erstwhile NIST), Philippines, initiated a programme for preparing edible oil for rural areas by wet processing of coconut. This was based on the 'substractive process' of Birosel et al. (1963). In the substractive process, the nuts are split and the meat dislodged from the shell and pared. The pared meat is conveyed to washing tanks and to the comminuter, which disintegrates and breaks the oil cells. The milk is extracted using a carver-hydraulic press, which is centrifuged to separate the coconut cream. This cream passes through the washing tanks six or more times to separate the cream from the other components except oil. On centrifugation of the cream, white oil free from odour and flavour is obtained.

Since this process requires specialised equipment, not ordinarily found in rural areas, a study was undertaken to simplify Birosel's procedure by using simple utensils that are readily available in the rural community. Here fresh-grated coconut meat from the local market was macerated and passed through a cheese cloth and pressed by hand to extract the milk. A second pressing was done on the pressed meal or 'sapal' after the addition of tap water, equal to the weight of the original coconut meat. The extracts were pooled and allowed to stand for gravity separation for 1 h in a plastic container with faucet. The milk is thus separated into skim milk and cream. The cream was filtered through a screen (140 mesh) to remove the last traces of residual meal and washed several times with tap water to remove any entrained salt. After 30 min, the cream was boiled at low heat, and the minimal dark fine solids were allowed to settle at the bottom. The oil was rapidly filtered in a dry cheese cloth or filter paper and stored in closed bottles. The oil yield was found to be 73.64%.

Considerable work has been carried out to recover the oil as well as high-grade protein from fresh coconut meat. Thus, Pratap Chakraborthy (1985) developed a process for the simultaneous extraction of oil as well as protein concentrate by ultrafiltration. Ultrafiltration yielded 15% more coconut protein than acid coagulation. The protein had superior functional properties and had higher chemical score due to better amino acid profile.

The technology for wet processing has reached to the extent that high-quality coconut-based food items are possible to be produced such as the natural oil, protein-rich skim milk powder or skim milk concentrate, coconut flour,coconut water concentrate,etc. Commercialisation of the same will lead to higher employment facilities and profit provided the products find consistent market demand.

A critical assessment of the various processes described above will bring into limelight some of the inherent limitations which discourage the commercial application of the technology. In all the processes, except the Krauss-Maffei/CFTRIprocess, the oil recovery is about 5–10% less compared to efficient milling process, which is a traditional industry in the major coconut-growing countries. In wet processing, the unit operations are more, and the processing cost may work out to four or five times that of normal processing. Moreover, the process is highly capital-oriented. Whether a ready market and favourable price relationship exists for the coconut protein and coconut flour and whether this will compensate for the high cost of production and justify the huge initial investment are still to be ascertained.

13.9 Desiccated Coconut

Desiccated coconut is the white kernel of coconut, comminuted and desiccated to a moisture content of less than 3%. The common grades have a particle size of less than 5 mm. It is a very important commercial product having demand all over the world in the confectionary and other food industries, as one of the main subsidiary ingredients of fillings for chocolate, candies and liquorice of all sorts. It is also used uncooked and as decoration for cakes, biscuits and ice cream and toasted for short eats.

Sri Lanka is the first country to manufacture desiccated coconut where the first mill was installed in the 1890s. The Philippines is another major producing country. Other countries producing small quantities of desiccated coconut are Malaysia, Indonesia, Fiji, Tonga, Côte d'Ivoire, Brazil and India. The major consumers of the product are the USA, West Germany, Australia, Canada, Japan, the Netherlands, Denmark, South Africa, Sweden, Belgium and the Middle East.

13.9.1 Processing of Desiccated Coconut

Deshelling and Paring Desiccated coconut is made out of fully mature nuts which have been stored for about a month to make the deshelling operation easier. The shell is removed by sharp taps on the shell by a small axe/chisel and the kernel separated from the shell. Shelling usually leaves the kernel including the testa intact with the water inside.

Next stage is paring the kernel to remove the brown testa using a special type of knife. The parings which amount to 12–15% of the kernel are usually sun dried, but sometimes over dried and pressed for oil, the yield being 60–62%. The resulting oil, known as paring oil, is of inferior quality and finds use in soap manufacture. The paring cake has approximately the same composition as the ordinary copra cake.

Washing The pared kernel is placed in tanks and sliced into two to release the coconut water. The sliced kernel pieces are then passed into other tanks through a two-stage washing. The coconut water is led into settling tanks, and the surface scums when formed are removed and boiled to get an oil of inferior quality. The resulting press cake has a high fertiliser value. This oil is not derived from the coconut water, but from the oil which has been washed away from the pared surface of the kernels.

Sterilising The kernel pieces after washing are sterilised by passing through large tanks containing boiling water. The pieces are carried in wire baskets continuously by mechanical device, and the movement of the baskets is adjusted in such a way that from one end of the tank to the other, a set period of 90 s is available for sterilisation. The oil layer that may form on the surface of the boiling water is removed at frequent intervals. In the Philippines, sterilisation is also effected by subjecting the disintegrated meat to live steam in stainless steel blanchers at about 80 °C for 5 min or 70 °C or 80 °C for 8–10 min. To stabilise the kernels, the sterilised pieces are immersed in a solution of sulphur dioxide.

Disintegrating and Desiccating The kernel pieces are transferred from the wire baskets straight to the disintegrator which shreds the kernel pieces to a fine wet meal. Disintegration is done by an attrition mill known as 'Devil Disintegrator' which shreds the kernels to pieces varying in size from 1 to 5 mm. When fancy grades are required, different cutting machines are used. The disintegrated product will have double the weight of the final product. This wet mass is subsequently desiccated or dried to a final moisture content of 2–2.5%. The dried product is allowed to cool on galvanised tables before being taken to the sifters for grading into coarse, medium and fine grades. For export, each grade is packed in plywood cases containing 50 kg. The product is available in other forms also such as extra fine shreds and threads and chips. In fully mechanised plants, as in the Philippines, the cooling system is integrated into the drying system. After drying, the product is sucked from the drier and blown along a pipe to a cyclone cooler mounted on a platform above the grading table and the packing section. At the time of packing, the temperature of the product will be about 40 °C, and the grading and packing are performed mechanically.

The following standard requirements and specifications for export-quality desiccated coconut have been recommended by the Asian and Pacific Coconut Community. Good desiccated coconut should be crisp and snow white in colour with a sweet pleasant fresh taste of the nut. The oil content range is from 68% to 72%, but the oil should not contain more than 0.1% free fatty acids. The moisture percent shall not exceed 3 for coarse, medium and fine grades and 3.5 for special grades such as thread and chips. The product shall also be free from all foreign matter including shell, coconut fibre, metal particles and textile fibre.

13.9.2 Yield of Desiccated Coconut (DC)

In India, 8000–9000 nuts yield 1 tonne of desiccated coconut. While the general figure in Sri Lanka is about 7000 nuts and in certain tracts, particularly in Chilaw District, 6000 nuts will give 1 tonne of DC. In the Philippines only 5500 selected nuts are needed to produce 1 tonne of desiccated coconut in most of the producing areas.

13.9.3 Composition

According to NIIR Board of Consultants and Engineers (2012), the main constituents of desiccated coconut are fat (67.5%), carbohydrates (5.9%), protein (9.30%) and pentosan (8.9%).

Good-quality desiccated coconut can be obtained by drying in the vacuum dryer at 100 °C in 200 mm of Hg vacuum. The drying time is reduced, in vacuum drying to 20 min instead of 40–50 min in conventional forced hot air drying. The colour and quality of the product are also very good. Due to vacuum, puffing of the desiccated coconut (about 40%) was also observed in vacuum drying.

13.10 Tender Coconut Water

Water from tender coconuts (7–8 months old) makes a refreshing and palatable drink particularly during summer. Tender nut water contains good amount of B group vitamins like nicotinic acid, pantothenic acid, ascorbic acid, riboflavin, biotin and minerals like Na, Ca, Mg, K, Fe, etc. The pH of the water varies from 4.8 to 5.3. The concentration of ascorbic acid ranges from 2.2 to 3.7 mg 100 ml^{-1}, which gradually diminishes as the kernel surrounding the water begins to harden. The percentage in respect of amino acids such as arginine, alanine, cystine and serine in the protein is higher than those in cow's milk. The tender nut water has a calorific value of 17.4 per 100g of water. When the nut is 7–8 months old, the nut water has the maximum concentration of sugar, and a large nut at this stage may contain over 28 g of sugar in solution. The tender nut water also contains various minerals of which potassium is the major constituent, and its concentration largely depends upon the nature of potash manuring.

13.10.1 Mature Coconut Water Vs Tender Nut Water

As the nut matures, the composition of the water especially the sugar content undergoes significant changes. During the early stages of development, the quantity of invert sugar present in the water increases and reaches a maximum at 220 days. After this stage, sucrose appears in the water and the concentration of total sugars falls and concentration of total solids also declines. Alanine, 2-aminobutyric acid and glutamic acid constitute about 75% of the free amino acids of the water from mature fresh nut. So, mature coconut water is not popular as a soft drink. The composition of tender nut water and mature nut water is given in Table 13.16.

Tender nuts are valued both for sweet water, which is a refreshing drink, and for its gelatinous kernel which is delicious. Moreover, the tender nut water has a number of medicinal properties, and it is an essential component in many of the Ayurvedic preparations. The use of tender coconut water is recommended in cases of gastro-enteritis and as a useful substitute to saline glucose in intravenous infusion. It is also prescribed in serious cases of diarrhoea and vomiting against dehydration of body tissues. It increases the blood circulation in the kidney and causes profuse dieresis. The values for the B group vitamins reported (Anon 1950) are nicotinic acid (0.64 g cc^{-1}), pantothenic acid (0.52 g cc^{-1}), biotin (0.02 g cc^{-1}), riboflavin (< 0.01), folic acid (0.003) and traces of thiamine and pyridoxine.

13.10.2 Products Derived from Tender Coconut

13.10.2.1 Snowball Tender Coconut

The soft tender kernel or solid endosperm of tender coconut is a delicious dessert. But the traditional method of its extraction is difficult, time-consuming and risky. In all the coconut-growing countries, a portion of the coconut is harvested at the tender stage (7–8 months) for using its water as a beverage. The kernel of the tender coconut is sometimes eaten or else thrown away, as it is difficult to remove it from the shell. The tender coconut kernel is good for convalescing patients. It contains good amount of nutrients. A technology for making snowball tender nut (SBTN) has been developed. SBTN is the tender coconut without husk, shell and testa which is in ball shape and white in colour. This white ball contains tender coconut water, which can be consumed by just inserting a straw through the top white tender coconut kernel. Coconut of 8-months maturity is more suitable for making SBTN. Making groove in the shell before scooping out the tender kernel with water is one of the important unit operations. For this, a suitable machine has been developed.

It is nutritive and is a drink and a snack at the same time. Since there is no refuse after the consumption, there is no littering of the premises. Since the snowball tender nut can be individually packaged and refrigerated under hygienic conditions, the shelf life of this product is prolonged, and therefore this ready-to-serve product is becoming popular.

Tender coconuts are usually transported and sold in their natural form, involving a lot of transportation cost due to the volume of the material. Further, the life of tender coconut water in the nut is short and therefore cannot be kept for a long period. A technology has been developed to extend the shelf life of the tender coconut water by packing in flexible pouches and aluminium beverage containers. Tender coconut water's characteristic flavour is contributed by heat-sensitive elements, and hence partial heat treatment combined with certain preservative is employed. To obtain a uniform taste, some of the sweetening agents are added. The product can be stored up to a period of 3 months under ambient condition and about 6 months under refrigerated condition. The flavour retention is better in the case of cans than flexible pouches.

The processing and packing of tender coconut water with a capacity of 10,000 tender nuts per day facilitates a direct employment potential of 30 personnel. The profitability after taking into account the prevailing prices of finished product is about 20% with a payback period of 3 years. Tender coconut water has a great potential as a health drink both in domestic and international market.

13.10.3 Products Derived from Mature Coconut Water (MCW)

Coconut water from mature nuts is mostly a waste product of the copra and desiccated coconut industries. Nevertheless, it has been recommended for the production of certain food products which could be developed on a cottage industry basis, for augmenting the income of coconut growers.

13.10.3.1 Bottled Coconut Water

Bottling of coconut water for use as a soft drink is gaining popularity. Coconut water can be marketed as a natural soft drink if preserved and packed. Although this product is already developed and marketed, limited shelf life is a constraint. Therefore, further investigation is required to extend the shelf life, to improve quality and to develop suitable packaging system for overcoming the problems of transportation and storage.

13.10.3.2 Coconut Water Concentrate

One of the problems for the use of coconut water as feedstock for the manufacture of beverage and other products is the cost of transportation and spoilage during transport of the material from the source to the beverage factory. A possible solution to this is to concentrate coconut water into a form that is easily rehydrated which will also help to reduce shipment weight, volume and cost, as well as improve product stability. Potential end users of the concentrate include not only the food and beverage industry but also the fermentation industry, hospitals, research laboratories and the beverage-consuming public.

Coconut water is adversely affected by extended processing at high temperature, so a nonthermal concentration process has to be used. One of the available techniques for producing coconut water concentrate is reverse osmosis. In the case of concentration of coconut water, the concentration is achieved by applying external pressure in order to overcome osmotic pressure and force the solvent (water) of all or most of the dissolved substances to evaporate in effect reversing the normal osmotic process. Concentration factor of five- to six folds can be obtained using 90 membranes at 4 MPa after 5 h of reverse osmosis with continuous retentate recirculation. Maximum concentrations of sugar and protein for the retentate, namely, 16.9 and 0.7%, respectively, are attained under these conditions.

13.10.3.3 Frozen Coconut Water

A brief description of the process is as follows. Fresh coconut water from newly opened coconuts is collected under hygienic condition. Suspended solids and oil in the samples are removed by means of three-way centrifuge. The removal of the solids and the oil is necessary in order to minimise fouling or clogging of the membranes. The salts present in coconut water may be removed if desired, prior to concentration, to produce a very sweet product. This is achievable by passing the centrifuged coconut water through a mixed-bed ion-exchange resin. However, additional costs are entailed, and problems dealing with re-generatability of the resin need to be overcome. The concentrate can be frozen or preserved in cans and after dilution to the desired strength can be used as a base for the production of carbonated and noncarbonated coconut beverages. The concentrated coconut water is also being used successfully in the brewery industry.

13.10.3.4 Coconut Vinegar

Coconut water can be converted into vinegar by using vinegar generators. The vinegar generator assembly comprises of a feed vat, an acetifier and a receiving vat for collection of vinegar. MCW consisting of about 3% sugar is concentrated to 10% level by fortifying with sugar, which is then fermented by inoculating the solution with yeast, *Saccharomyces cerevisiae.* After alcoholic fermentation for 4–5 days, the clear liquid is siphoned off and inoculated with mother vinegar containing *Acetobacter* bacteria. The alcoholic ferment obtained is then fed into a vinegar generator where the feed is uniformly sprayed over the surface of the porous packing medium (corn cobs). Here the alcoholic ferment is oxidised to acetic acid. The product is run out from the packing medium by gravity flow into the receiving vat from where it is recycled into the vinegar generator and the process of acidification is repeated until strength of 4% is attained. This acetified vinegar is then aged before bottling. The manufacture of coconut vinegar with a capacity of 500 litres day^{-1} involves a low capital investment with an employment potential of 10 persons. The profitability (taking into account the prevailing price) of the finished product is about 20%. Coconut vinegar enjoys a wide market as a preservative in pickling industry and as a flavouring agent for foodstuffs. A traditional popular drink, coconut lemonade, is prepared by boiling coconut water, sugar and lemon juice. A tangy coconut sauce can also be prepared from coconut water with red chilli, onion powder and little vinegar.

13.10.3.5 Coconut Honey from Coconut Water

Coconut honey is another product from coconut water containing many growth-promoting trace elements besides glucose, fructose and laevulose. Coconut water is filtered, evaporated and blended with a little golden syrup to produce coconut honey, a palatable, nutty-flavoured breakfast food, soft drinks additive and a sweetener.

13.10.3.6 Nata De Coco

Nata is a gelatinous substance formed by *Acetobacter* sp. on the surface of fermented coconut water. This product when preserved in sugar syrup is an excellent dessert and enhances the flavour and taste when mixed with other sweet preparations such as fruit salad. Nata de coco is prepared from MCW or coconut skim milk by mixing the medium with sugar, acetic acid and culture liquor. The recommended proportion of the ingredients is 1 kg of boiled, cooled and filtered coconut water, 65 g of sugar, 25 g of glacial acetic acid and 165 g of mother liquor. The mixture is allowed to stand undisturbed in wide-mouthed glass or plastic jars for 15–20 days. During this period, a white jelly-like thick surface growth is formed by the action of *Acetobacter xylinum.* The surface growth is harvested, sliced, acid washed, boiled in sugar syrup and preserved in either tin containers or bottles. The leftover liquid medium is used as the mother liquor for subsequent production. The gelatinous growth is composed mainly of polysaccharides, probably dextrose, and is cellulosic in nature. For the initial preparation of the starter liquor, three cups of pineapple residue left after the extraction of juice are mixed with six cups of water and one cup of sugar, and the mixture is left undisturbed in wide-mouthed glass jars covered with a thin cloth for approximately 2–3 weeks. The jelly-like growth which forms on the surface during this period is taken out, sliced into small cubes and used as the starter medium.

The product is served either mixed with other fruits or baked into a delicious cream pie or simply served with flavoured syrup. Since nata has no calorific value, it is a favoured delicacy. In the Philippines, where the nata is produced in large quantity, the demand is on the increase, both for domestic consumption and export.

13.10.3.7 Soft Drink

Coconut water has a potential use as a refreshing soft drink. In India, CSIR-National Institute for Interdisciplinary Science and Technology (NIIST), Thiruvananthapuram, has developed a technology for the formulation of noncarbonated beverage from coconut water of mature nuts. The process involves collection of water, upgradation, pasteurisation, filtration and bottling. The coconut water when collected from copra processing units is immediately filtered through a clean cheese cloth. Initial filtration at the collection centres reduces the bacterial load in the water. At the processing site, the pH of the coconut water is adjusted to 4.2 with citric acid and 0.1–0.15% sodium citrate. The addition of sodium citrate is to reduce the biting taste that may develop with the addition of citric acid. In some formulations, the use of 0.01–0.05% sodium chloride is found to improve the taste. The total soluble solid content is adjusted to 8–10% with refined sugar. Sodium benzoate is also added at the rate of 0.05% to increase shelf life. The formulation is immediately pasteurised at 94 °C for 25–30 min. The protein starts coagulating at 70 °C, and at this stage, super cell is added to aid sedimentation by co-precipitation. Excessive heat treatment is avoided as it imparts a cooked flavour to the finished product. Rigorous agitation is also avoided in order to prevent disintegration of the protein coagulation into fine particles. After pasteurisation, the formulation is passed through a pressure filtration system, and the filtered product is filled in sterilised returnable bottles and crown-corked under sterile conditions. The final temperature of the product in the bottle ranges between 72 °C and 75 °C. These can be stored for 3 months at ambient temperature without spoilage.

Processing technology has also been standardised for the preservation of tender coconut water obtained from 7- to 8-months old tender coconuts. During the processing, the precipitation after pasteurisation is much less, and it has been possible to obtain a clear product even without filtration. This also keeps well at ambient temperatures and is microbiologically safe for over 9 months.

13.10.3.8 Other Miscellaneous Uses

Even in the early 1890s, it was shown that coconut water is a medium for growing microorganisms like yeast for the production of which viable technology is available. The superiority of coconut water over molasses in yeast production is known. *Saccharomyces fragilis* can be grown on coconut water, yielding up to 0.54 g of dry yeast per g of total sugar at 40–45 °C. The resulting yeast which contains 45% protein is a rich source of amino acids and vitamins (Smith and Bull 1976).

For ensuring maximum yields, it is sufficient if 1 g of urea per litre of water is added as an additional nitrogen source. The yeast so produced after washing and drying will have a pale cream appearance, slight odour, a nutty taste and a good shelf life. In coconut complexes, where large number of coconut is processed, it would be feasible to grow the food yeast for the production of protein.

Quimio (1984) found that coconut water, both from mature nuts and tender nuts, can support the mycelial growth of edible mushroom. When incorporated with agar, coconut water performed almost equally well as the potato dextrose and malt extracts. Within 1-week incubation, 12 g dry weight of *Lentinus sajor-caju* mycelium can be harvested from 100 ml sterilised medium containing coconut water. Trials with other edible mushrooms such as *Volvariella volvacea*, *Auricularia polytrichia*, *Pleurotus sajor-caju*, *Tremella fuciformis* and *Agaricus bisporus* showed that the technique is successful.

Coconut water is also an excellent medium for culturing different organisms. Coconut water agar medium can be used as a routine laboratory agar medium for a number of plant pathogenic fungi such as *Fusarium*, *Colletotrichum*, *Pythium*, *Rhizopus*, *Botryodiplodia* and *Phytophthora.* Preliminary studies have shown that coconut water can also be a good sporulating medium for the non-sporulating fungi. It can even be dried into coconut water agar powder, ready to be rehydrated when needed, with the addition of peptone. The growth of *E. coronata* was faster when coconut water from either immature or mature nuts was included in the culture medium. Sugar-enriched coconut water can be used as a medium for the production of dextran by using *Leuconostoc mesenteroides*.

Coconut water also possesses growth-promoting activity. The water from tender nut shows better growth-promoting activity than that from mature ones. Gibberellin can be produced by cultivating *Gibberella fujikuroi* in coconut water. Apart from these, coconut water can be used for other miscellaneous purposes. It is mixed with lime to improve the adhesiveness of white wash. Fermented coconut water has been used successfully as a rubber coagulant to obtain good-quality rubber.

13.11 Toddy

Toddy is obtained by tapping the unopened spadix of the coconut palm. This is known as 'tuba' in the Philippines and 'tuvak' in Indonesia. The word tapping collectively connotes the extraction and the various manipulations for stimulating the palms to exude juice from a selected part. In the coconut palm, palmyra (*Borassus flabellifer* L.), and in the Indian sago palm (*Caryota urens* L.), it is the inflorescence that yields toddy on tapping, whereas in the date palm (*Phoenix dactylifera* Roxb), the juice exudes from the lateral portion of the young stem on being punctured.

13.11.1 Tapping Methods

In India and Sri Lanka, the spadix is considered ready for tapping when the female flowers within the unopened spathe cause a swelling at the base. In the process of tapping, the spathe is trained for a period of 3 weeks, which involves uniform beating all over the surface of the spathe daily with a tapping rod to rupture the cells and to induce the sap flow. During the process, the spathe gradually bends, and it is prevented from opening by tightly binding it with fibre. When the spadix is ready to yield the sap, about 4 or 5 cm is cut from the tip of the spathe, and an earthen receptacle is placed for sap collection. The interval between the beginning of tapping process and the commencement of sap flow is about 15 days extending to 21 days in summer months. The tapping is usually continued for a period of 6 months, and sometimes three spathes are continuously tapped. In Sri Lanka, the tapping continues for a period of 8 months, with a maximum toddy yield in the third month.

In Indonesia, the tapping begins when the spadix is about 1 month old. It is tightly wrapped in coconut leaves and the tip is cut about 1 cm long. The end is then lightly beaten with a round piece of wood, 30 cm long, till the flowers are bruised and the beaten part again wrapped with a piece of young leaf. This procedure is repeated several days until the flow of sap begins.

13.11.2 Yield of Toddy

The yield of toddy varies from palm to palm, season to season, day to day and even from spadix to spadix. There is also considerable variation between varieties. The yields of toddy obtained from different varieties of coconut in Sri Lanka are given in Table 13.17.

In India, an average of about 18 l of toddy spadix^{-1} has been recorded for a tapping period of 1 month. Palms which yield a large number of nuts have been found to yield up to 300 l of toddy during the 6 months of tapping with an average yield of about 50 l of toddy spadix^{-1}. On an average, the yield of toddy palm^{-1} day^{-1} is about 1.5 l.

In the Philippines, it was found that 40-years old palms under good management produce about 400 l of sap annually. In an experimental tapping of 100 palms, a daily yield of 1.38 l palm^{-1} was obtained. When calculated in a 12-h period, the flow during the day time was 0.64 l, and it was 0.74 l during the night (Gibbs 1911). The Laguna cultivars yield about 0.772 l day^{-1}, while the Coconino produces about 0.400 l day^{-1}. The annual yield reported for tall palm is about 370 l palm^{-1} (Fernandez 1978).

Table 13.17 Average toddy yields of tall, dwarf and hybrid coconut palms in Sri Lanka

	Average toddy yield (l)		
Particulars	Tall	Dwarf	Hybrid
Total yield for 365 days	577	110	792
Average yield per spadix (weighted average)	49.5	8.5	44.8
Range of yield per spadix	17–69	0.1–15	22–53
Number of spadices tapped	15	14	19

Source: NIIR Board of Consultants and Engineers (2012)

13.11.3 Kalparasa (Neera)

The trickled sap from the inflorescence is traditionally collected in earthen pots, and during this process it gets fermented. A coco-sap chiller technology has been developed by ICAR-CPCRI (Hebbar et al. 2015). A technology for production of hygienic and unfermented neera has been standardised by the Kerala Agricultural University also (Jayaprakash Naik et al. 2013). For details, please refer to Chap. 9.

13.11.4 Properties of Toddy

The fresh toddy obtained on tapping the coconut palm is mildly acidic and rich in sugar and vitamins. Apart from 10% to 15% sugar, it contains proteins, minerals and vitamins which makes it a nutritious drink and also an excellent fermenting medium. The fresh toddy, unless collected under sterile conditions, rapidly ferments, and the sugar is replaced by about 5–8% alcohol which on distillation yields strong liquor known as arrack. If the fermented toddy is kept further, it undergoes a process of acetic fermentation yielding the coconut vinegar containing about 4–7% acetic acid. The fresh toddy is also a source of baker's yeast. The composition of fresh as well as fermented toddy is given in Table 13.18.

Under natural conditions, toddy is fermented by the native microflora consisting of yeast and bacteria. *Saccharomyces* species of yeast has been identified to cause fermentation. Fully fermented toddy contains 6–7% (V/V) alcohol (Nathanael 1960). The fermented toddy may contain ethyl alcohol, aldehydes, higher alcohols or acetic acid. This is consumed as an alcoholic beverage in many coconut-growing countries. The composition of fermented toddy is given in Table 13.18.

Unson (1966) found that toddy can be preserved without fermenting by the addition of sulphanilamide or p-hydroxy-benzoic acid. Potty et al. (1978) attempted the preservation and bottling of coconut toddy. Here toddy was mixed with 5% activated carbon granules and pasteurised at 80 °C–82 °C. After centrifuging and autoclaving for 30 min with steam at atmospheric pressure, the product was bottled.

Table 13.18 Composition of fresh and fermented toddy

Fresh toddy		Fermented toddy	
Specific gravity	1.058–1.077	Specific gravity	0.998–1.033
Total solids	15.2–19.7 g 100 ml^{-1}	Acidity (as acetic)	0.32–0.67 (%)
		Water	98.2 (%)
Protein	0.23–0.32	Protein	0.2 (%)
Ash	0.11–0.41	Ether extract	0.1 (%)
Sucrose	12.3–17.4	Carbohydrates	1.3 (%)
		Mineral matter	0.01 (%)
Ascorbic acid	16.0–30.0	Calcium	<0.01 (%)
		Phosphorous	0.01 (%)
		B1	<51 μ 100 g^{-1}

Source: Browning and Symons (1916)

13.11.5 Effect of Tapping on Yield of Nuts

It has been found that tapping in medium- and poor-bearing palms improves the yield for about 4 years after discontinuation of tapping. In India, it was shown that tapping increases the yield of only poor bearers, while in the Philippines, the tapped palms fruited abundantly the year after the tapping but stopped for a few more years thereafter. However, a reduction in oil content of copra was seen ranging from 14.32 kg to 25.79 kg t^{-1} of copra. Similar results were reported from Malaysia also (Jack and Dennett 1925) indicating a reduction in the weight of meat (21 g), weight of copra (10.7 g), weight of oil (21.7 g) and oil content (3.3%) in the tapped palms.

13.11.6 Products Derived from Toddy

13.11.6.1 Jaggery

Unfermented toddy, when boiled to 118–120 °C and allowed to cool, solidifies. The solid mass is known as coconut jaggery or *gur*. During the production process, the scummy impurities along with froth are removed, and the saturated solution is poured out in moulds for hardening. Since the removal of scummy impurities involves considerable wastage of sugars, in certain places, frothing is avoided by adding a few drops of coconut oil or a little coconut grating. Before boiling, the juice is filtered through sand filters to remove the impurities, and a small quantity of alum is added to induce the precipitation of lime and magnesia. This will render the final jaggery much less deliquescent which gives a better colour and will enable to remain hard for a reasonable period. Twelve to fifteen percent jaggery is obtained from toddy. Jaggery contains 10.92% moisture, 68.35% sucrose, 6.58% reducing sugar, 2.17% ash, 1.64% protein and 8.72% of pectins and gums. The calorific value is 321.6100 g^{-1} (Anon 1984). The manufacture of coconut jaggery from coconut sap holds promise as a viable commercial venture.

13.11.6.2 Refined Sugar

The sugar content in sweet toddy tallies with that of cane sugar juice. On the basis of 175 palms to a hectare and 250 l of toddy palm^{-1} containing 15% crude sugar, the yield of raw sugar ha^{-1} would be about 6 tonnes. For recovering the raw sugar, toddy is treated with 2% lime to coagulate albuminous impurities. The limed toddy is then carbonated in two stages and filtered each time to remove excess lime. The clarified juice is evaporated to the extent of 75% sugar content, and the resultant syrup is concentrated in vacuum pans till crystallisation commences. The syrup is then discharged into crystallisers, and the crystalline sugar is separated by centrifugation. It is of nutritional value because of its low glycaemic index. But the process as a whole is not economically viable.

13.11.6.3 Vinegar from Coconut Toddy

Coconut toddy vinegar is formed when toddy is allowed to ferment for more than 24 h to yield acetic acid. Nathanael (1963) standardised the procedure for large-scale production of vinegar from coconut sap. Submerged fermentation using pure strain of *Acetobacter* and an isolated strain from jaggery were found to be suitable for vinegar production (Gupta et al. 1980). Besides this, a quicker method of vinegar production known as 'generator' process has been developed in Sri Lanka (Nathanael 1963). Its composition is density (30 °C, g cc^{-1}), acetic acid, total solids, potash and nitrogen (g 100 ml^{-1}) content of coconut vinegar range from 1.010 to 1.013, 4.7 to 5.4, 1.25 to 1.36, 0.16 to 0.22 and 0.025 to 0.033, respectively.

13.11.6.4 Treacle

This is the concentrated form of sweet toddy obtained by boiling the toddy. The final product is golden-coloured syrup. The recovery is about 16% of the toddy used. Treacle is considered a delicacy in many places, and its preparation is very common in Lakshadweep Islands.

13.11.6.5 Arrack

Arrack is the product obtained by the distillation of fermented toddy, the range of recovery being 17.5–25% of the original toddy. The reported ranges of analytical values of arrack are esters (164–258), total acids (116–158), fixed acids (6–13), volatile acids (105–152) and furfural (0.45–1.32). Double-distilled arrack is called coconut feni.

13.12 Other Food Products

13.12.1 Haustorium

When the nuts germinate, a spongy ball-like haustorium develops inside the nut. This is utilised for the preparation of various products like jam, marmalade, etc. It is sometimes sliced and preserved in sugar syrup for use as a constituent of fruit salad. Haustorium contains moisture (15.23%), ash (9.39%), protein (10.60%), fat (15.68%), carbohydrates (38.42%) and fibre (10.59%). The terminal bud or cabbage is also a delicacy. It is used raw or in salads and is sometimes pickled, canned or bottled as a preserve.

13.12.2 Dry Kernel or Copra

Copra and coconut oil are the principal products of coconut palm. The dried coconut endosperm is called copra. With an oil content of 65–70%, copra is the richest source of fat. Though there is variation in the oil content of copra derived from different regions, according to Indian Standards, 1 tonne of copra corresponds to about 680 kg of oil with an average extraction of about 645 kg of oil. There are two types of copra – edible copra and milling copra.

13.12.3 Edible Copra

Edible copra is available in two forms – ball copra and cup copra. Edible copra is used for sweet meat preparations in households and as an ingredient in the processed foods.

13.12.4 Ball Copra

Ball copra is made from fully mature (>12 months) whole unsplit nuts. The nuts are stored for about 8–12 months on a raised platform, usually made of bamboos, inside a shed. During this period, the coconut water slowly gets absorbed, and the kernel dries out and loosens itself from the shell. The nut at this stage is dehusked and the shell carefully removed to separate the whole dry kernel in the ball form. In some places, drying process is hastened by occasional heating of the nuts by a slow fire set under the platform.

Ball copra is soft, sweet, oily and cream coloured. Three grades of ball copra are available on weight basis, viz. 'large', 'medium' and 'small', depending on the number of copra required for a weight of 4 kg (< 20 for 'large', 20–40 for 'medium' and > 40 for 'small'). The moisture content shall in all cases be below 7%. This grading is being accepted generally in the marketing sector now.

13.12.4.1 Preparation of Ball Copra by Heat Treatment

In order to reduce the incubation period to the minimum, coconut is partially dehusked and heated in small holder's dryer at 55 °C–60 °C for 8 h daily for 3 days and stored in gunny bags for 10 days. This intermittent heating is repeated till all the nuts become ball copra. All the nuts under heat treatment become ball copra within 6 months, and the quality of the ball copra is also very good. In conventional method, it takes 11 months. In a similar way, puny nuts under heat treatment are transformed into ball copra within 3–4 months. Conversion of nuts into ball copra is faster during January to May compared to June to November period (Madhavan and Bosco 1991).

13.12.4.2 Cup Copra

Edible cup copra is prepared either from fully mature nuts or from stored nuts by cutting the dehusked nuts into halves and drying them in the sun. The cutting, drying and deshelling processes are done very carefully in order to get a good clean final product. In Sri Lanka and in some parts of India, hot air driers are used for the production of edible type of cup copra. In some copra markets in India, good-quality cups of uniform size are separated from bulk gatherings of commercial copra and marketed as edible copra at a premium price.

In Kerala, India, cup copra is classified into three grades. The major grade is 'Rajapur' copra. Here, the ball copra in which the kernel is completely dried out is cut into two and dried in the sun for 3 days. The second grade is called 'Malathi' where partially dried ball copra is cut and dried in the sun for 7 days. The third grade is known as 'Dil Pasand'. Nuts harvested and kept for 2–3 months are used for preparing this type of cup copra. Freshly harvested nuts also can be used. Nuts are split and dried in the sun for more than 7 days. Adequate care is taken in splitting the nuts and drying. Neat white copra cups separated from the bulk of milling copra are also sold as 'Dil Pasand' after trimming the edges.

13.12.5 Milling Copra

The conversion of fully mature coconuts into copra for milling purpose is the most common processing activity in the major coconut-producing countries. The unit operations involved in the process are dehusking, splitting the nuts into halves and drying the split halves. Drying is an important postharvest operation in the processing of coconut for the extraction of oil. For the efficient storage of copra and easy extraction of oil, fresh coconut meat, which contains 45–55% moisture, has to be dried to 5–6% moisture level to obtain the copra. The conventional system of copra drying is by spreading the split coconuts on an open surface for sun drying. This operation takes 6–8 days, and quality deterioration due to deposition of dirt and dust of wet kernel is unavoidable. Moreover, if the atmosphere is cloudy and the temperature goes down during the initial days of drying, the copra will get infested with mould. Copra obtained by this method is often of poor quality because of prolonged exposure to atmosphere resulting in microbial infection. During rainy season, with restricted sunshine, drying by artificial method is the only possible solution for making good-quality copra. The direct-type kiln dryers are not desirable as the copra becomes inferior in quality due to the smoking and improper drying. To obtain good-quality copra, particularly during rainy season, a suitable dryer using indirect heating for drying is essential. Drying must be carried out within 4 h of splitting since coconut kernel deteriorates very rapidly due to growth of mould and bacteria.

13.12.5.1 Dehusking

The first stage in the processing of coconut is the removal of the fibrous outer layer or the husk. In most of the coconut-growing countries, dehusking is done manually on a wooden spike provided with a sharp tip. The spike is fixed vertically into the ground with the sharp iron tip facing upwards, and the trained worker drives the husk against the sharp tip and, with three or four quick twists, tears it off. To make the dehusking operation more efficient and less tiresome, mechanical devices have been developed. Please see Sect. 13.4.1 for details.

13.12.5.2 Copra Drying

There is variation in the moisture and oil content from different parts of the endosperm. Studies conducted at the Coconut Research Institute, Ceylon (Sri Lanka), revealed that the moisture was more near the centre of the kernel and least in the outer region, 63.7 and 28.4%, respectively, while the oil content was the highest in the outer region (78.5%) and least in near the wet surface (49.2%). In a similar study conducted in the Regional Research Laboratory (CSIR) at Trivandrum, India, the variation in the moisture content was 61.9% in the inner region, 32.6% in the middle region, 18.1% in the outer region and 29.1% in the testa. The corresponding values for oil were 41.39%, 68.78%, 75.36% and 27.41%, respectively, on dry weight basis. Therefore the drying of copra has to be meticulously monitored for ensuring uniformity in drying.

Delay in drying of split nuts will cause deterioration of coconut kernel as it is a very good substrate for the growth of mould and bacteria. Microbial activity is related to the moisture content becoming much higher, say 20% or more (Nair 1984).

When drying is delayed, dipping the copra in dilute sodium carbonate and dilute sulphuric acid in conjunction with sun drying has been found to be successful to prevent deterioration of copra (Subramanyan 1966). Sreemulanathan et al. (1979) also confirmed that application of a coat of glacial acetic acid on the kernel surface prevented microbial growth during open sun drying. Treating the fresh meat with shell ash prior to drying (De Castro and Cora Nera 1978) and dipping the fresh kernel in 1000 ppm propionic acid for 60 min (Patil 1982) have also been found successful in preventing mould growth.

Drying Methods Different methods such as sun drying, solar drying, direct drying in kilns and indirect drying are used for copra drying. In many places, a combination of preliminary sun drying followed by kiln drying is followed.

Sun Drying This is the most popular and cheapest method of copra making, which is traditionally adopted in all the coconut-producing countries. In this method, the split nut halves (cups) are laid out in an open yard with the kernel portion facing the sun. After about 2 days of drying, the kernel gets detached from the shell. The partially dried kernel is then scooped out by means of a thin wooden lever. The detached kernel cups are again arranged in the yard for further drying. The drying process has to be continued for another 4–5 days to achieve the desired level of drying.

In Thailand, the preliminary drying of copra with shell is done only for 1 day after which deshelling and further drying are carried out. In Indonesia, the kernels are usually removed immediately after cracking and dewatering. In Sri Lanka, the drying time under the sun is about 9 days.

In India, sun drying takes 7–9 days (70–80 h) under the sun to obtain copra with 6% moisture content. Generally copra is dried on cement or concrete yards or on the bare soil surface. Patil and Nambiar (1982) found that black-painted palmyra mat surface reduced the drying time by 13%. Though good-quality copra could be produced by sun drying, the method is hindered during the monsoon. If proper care is not taken, the copra processed during the monsoon may have mould or get discoloured. Even though mould growth does not appear on the fresh meat until about 3 days, inadequate drying or intermittent drying develops sliminess on the surface of the meat resulting in subsequent discolouration. As such, copra drying has to be carried out without interruptions.

Solar Drying Solar energy can be more effectively utilised for copra drying by using solar heat collector-type drying systems. Patil (1984a) developed a polythene-covered drying gadget. Perforations are provided for air circulation on polythene sheet cover which is kept on a mild steel bar frame, and the gadget is kept on a black-painted palmyra mat surface. The cups are kept inside for drying. The temperature inside is above the ambient temperature, and 60 nuts per m^2 area can be placed and dried.

Solar cabinet dryer has been developed which is of chamber type having a direct heating and natural air convection arrangements (Patil 1984a). The dryer mainly consists of a drying cabinet provided with selectively coated solar aluminium absorbent sheet (sun sheet) as the drying surface, a trolley which supports the drying cabinet, sliding-type 4 mm transparent top cover, adjustable aluminium reflectors on hinges on three sides and transparent glass cover on all four sides. The drying surface capable of holding 100 nuts (200 cups) is kept at an inclination of 12.5° (equal to the latitude of the location). Thermocol insulation of about 25 mm thickness is provided between the sun sheet and plywood, for preventing heat loss. The aluminium reflectors with adjustable angle help in concentrating the solar radiation on the drying surface itself. The drying surface is so adjusted as to face the sun with the help of castor wheels, and an indicator rod is placed on top of the frame. For air inlet, a 10-cm-wide opening covered with expanded metal is provided at the bottom of the front side of the dryer. The exhaust provided at the rear top is also covered with expanded metal.

The split nuts are loaded on the drying surface with the cups facing up. The dryer is positioned to face the sun with the help of the indicator rod. The dryer is moved to track the sun twice during the day at 3-h intervals for the effective trapping of solar energy. Heat is generated inside the cabinet due to the absorption of solar radiation which promotes the rapid evaporation of moisture in the copra. The moisture-laden air escapes through the exhaust vent.

At the close of the day, the cups are covered with gunny cloth and the dryer is closed. On the second day, the kernel is detached from the shell and kept for further drying. The drying is continued every day from morning till evening to reduce the moisture content to less than 6%. The total drying time is 32 sunshine hours. Drying time is reduced by 50% compared to open sun drying. The quality of copra obtained is superior to that obtained by open sun drying. The dryer is versatile in design and can be adapted for drying other plantation crop produces also, such as cardamom, pepper, ginger, etc. It is easy to fabricate locally. It can be easily transported. It is very easy to operate and the maintenance cost is low. The cost of operation is negligible as there is no fuel requirement.

Direct Drying in Kilns The simplest direct heating system is the smoke drying. Here coconut husks and shells are fired in shallow pits over which a grill of wooden platform is erected on which the coconut cups are arranged in layers for drying. Firing is done either continuously or intermittently. Due to the uneven heat and heavy smoke, copra produced is of low quality and brownish with a smoky smell. The copra cake after extraction of oil is also not of good quality for use as animal feed.

In Kerala, India, kiln drying is combined with sun drying as it makes it less susceptible to mould infection. In the more common kilns, brick or mud walls are provided with tiled or thatched roof. In the Philippines, smoke dryers are called *tapahans* with a capacity of 400–4000 nuts batch^{-1}. Some of them are very large requiring larger area.

Improvements have been introduced in kiln drying in many countries in order to minimise the emission of smoke while drying. In the improved kilns, coconut shell is the commonly used fuel. With controlled burning of the fuel and adequate ventilation systems, excessive emission of smoke is avoided, and better quality of the final product is ensured. These kilns consist of a cabinet containing a rectangular grill 2–2.5 m above ground level. The bottom space is enclosed, where the firepit is made in serrated brick hearths. The kiln is roofed mostly of corrugated iron sheet and the roof is extended over a verandah. The four sides of the kiln are covered with brick walls.

The split coconuts to be dried are kept on the grill, with the kernel portion of the bottom layer facing upwards. The upper layers are arranged downwards up to a thickness of 30 cm. Coconut shells of even size are interlocked and laid loosely in single or double rows in the firepit. In order to avoid excessive emission of smoke, some shells are ignited outside the kiln and brought in while they are burning. The fire moves slowly burning the arranged shells without smoke. Six to eight firings are required with a drying time of 3–4 days. The drying temperature should not exceed 70 °C in the initial hours, and subsequent drying should be completed at 60 °C (Thampan 1987). Mostly the deshelling is done after two to four firings.

Various types of smokeless kilns are in current use. Cooke-type kilns are used in Malaysia and Sri Lanka. Here the hot air stream is more uniformly spread out from the firing tunnel outlet so that a traveling cone of heat is created below the drying platform. Varghese and Thomas (1952) modified and adopted the Malaysian kilns to suit Indian conditions. The drying chamber is made of red earth with bamboo grills which are loaded with coconut halves. These are placed one over the other on the drying platform. A perforated iron sheet hung 0.5 m below the grill platform serves as heat spreader. The drying time is 34 h at a temperature of 53–60 °C. The IST-NA Coco dryer developed by the NIST Philippines has a capacity of 1500–2000 nuts. In these large capacity dryers, fuel used is mostly charcoal or diesel oil.

Drying by Indirect Heat Here, the products of combustion do not come into contact with the coconut kernel, and as such high-quality smoke-free white copra can be produced. The hot combusted gases and the flame produced heat the clean air coming inside the dryer, through a heat exchanger, usually a metallic drum, without mixing with the clean air.

Indirect heating dryers are of two types based on the method of hot air circulation: (i) natural draft type and (ii) forced circulation type.

Natural Draft Dryer In these dryers, any type of fuel can be used, including the dry agricultural waste materials like coconut husk, shells, fronds, dry firewood, etc., available in the plantations. Copra is smoke-free with minimum deterioration during storage. But the higher consumption of fuel and higher cost of installation are the limitations.

These dryers essentially consist of a kiln and a heating unit. The kiln may be of a single compartment or several compartments, either vertical or laterally placed adjacent to a closed combustion system designed to obtain the maximum heat. The combustion unit consists of a furnace or firebox usually made of fire bricks from which flue pipes of suitable dimensions pass beneath the kiln along its full length and are connected with a damper chimney at the end opposite to the furnace. The flue pipes are inclined vertically up to 5° for easy flow of flue gases. In certain dryers, thermal efficiency is improved by providing additional furnace with damper controls. Sufficient inlets for the entry of cold air are provided on the walls below the level of flue pipes. The gases of burning fuel in the furnace pass through the flue pipes and heat the air around them. The hot air so produced moves into the kiln where the split nuts or copra are arranged. As the hot air passes through the layers of split nuts or copra, it removes the moisture, and the moisture-laden air is exhausted out by damper ventilators at the top of the kiln. For efficient drying, the drying bed thickness is not allowed to exceed 30–40 cm. Here the copra does not come into contact with smoke but only with the uncontaminated hot air because the flue pipes are connected to a chimney for the smoke to escape outside.

Simple smoke-free collapsible copra dryer has been developed to suit medium-sized plantation crop holdings and processing units (Vidhan Singh et al. 1999). The dryer is designed to hold 1000–1500 coconuts depending on the size of the nuts. The total drying time is 24 h. The dryer consists of a drying chamber, a unique burner, a heat exchanger and ventilation holes. This dryer was fabricated using locally available materials such as asbestos cement sheet, galvanised iron (GI) sheet, mild steel (MS) angle and fire-resistant plywood. Asbestos cement sheet has been provided only where the copra does not come into direct contact. The contact area (sides) with copra is provided with heat-resistant plywood. The shape of the heat exchanger is designed to avoid the flame coming into contact with the copra. The burner has thick fireclay lining. Broken shell pieces are used as fuel.

The use of indirect dryers has been promoted recently in India for making quality copra. Different models of low-cost indirect dryers have been designed and developed to suit small- and medium-sized coconut holdings (1 ha). These dryers have indirect heating and natural air convection arrangements.

The dryer uses low-cost agricultural waste as fuel. The dryer is comprised of (a) drying chamber, (b) plenum chamber, (c) burning cum heat exchanging unit and (d) chimney with regulators. The drying chamber is the upper portion of the dryer. It is made of asbestos cement sheets on sides and wire-mesh tray at the base supported on MS angle frame. The chamber just below the drying area is the plenum chamber. It is made of asbestos cement sheets supported on MS angle frame. An adjustable opening is provided at the bottom to facilitate entry of fresh air into the chamber. The burning cum heat exchanging unit is a 30-cm-diameter cylinder made of 22-gauge corrugated GI sheet. It is located in the centre of the plenum chamber at an inclination of 3° angle towards the exhaust side where it is connected to a chimney. The other end of the cylinder is covered by a damper with holes for entry of air for combustion. The fuel is burnt in a wire-mesh tray inside this cylinder. The outer surface of the cylinder at the points of contact on the plenum chamber sides is sealed with plaster of Paris. The chimney has a diameter of 10 cm and is made of GI sheet. Two butterfly valves are provided to the chimney, which regulate the escape of flue gases as well as the entry of air into the burning unit for combustion. By adjusting the valves, the drying air temperature is also controlled. The chimney is partially covered by asbestos rope cemented with plaster of Paris.

The small holder's dryer has a capacity to dry 400 coconuts (Patil 1984b). The fuel requirement is about 28 kg of coconut husk and shell. The fuel is fed as and when required to keep the fire burning. The drying air temperature is maintained at 60°–70 °C by adjusting the valves in the chimney. After 10–15 h of drying, the partially dried copra could be scooped out from the shell. The copra cups are to be raked every 2 h for uniform drying. The total drying period is over 4 days with overnight breaks till the moisture content of copra is reduced to 6%. The actual drying time is about 36 h.

The medium holder's dryer for holdings of above 1 hectare introduced by the CPCRI has more or less the same design as that of the small holder's dryer (Annamalai et al. 1989). It is designed to dry 1000–1200 coconuts batch^{-1}. The overall dimensions of the dryer are 2.5 m (L) × 1.2 m (B) × 1.75 m (H), and it requires a housing shed of 7 m^2 area. The drying chamber has an area of 3 m^2 and volume of 1.2 m^3. Three doors on hinges are provided on one side of the chamber to facilitate easier loading and unloading of the produce. The volume of the plenum chamber which is just below the drying chamber is 2.25 m^3. At the bottom of the plenum chamber, a 15-cm-wide opening is provided on either side along the length with baffles on hinges for fresh air flow into the dryer. The burning cum heat exchanging unit is a 45-cm-diameter cylinder made of 22-gauge galvanised iron sheet. It is housed at the centre of the plenum chamber longitudinally at a vertical inclination of 4° angle towards the exhaust side where it is connected to a 15-cm-diameter chimney. The volume of the cylinder is 0.40 m^3 and the surface area is 3.5 m^2. GI sheet fins are provided on the upper half of the cylinder for better heat transfer. The lower end of the cylinder is fixed on plenum chamber wall with a damper having holes for entry of air for facilitating combustion of fuel. The fuel is burnt inside the cylinder in a MS flat tray of size 100 cm × 43 cm × 15 cm. Two butterfly valves are provided in the chimney to regulate the escape of gases and entry of air for combustion which in turn controls the rate of combustion of fuel and the drying air temperature. The ratio of dryer width and burning cylinder width is 3:1. The fuel requirement is 110 kg of coconut husk and shell. At a drying air temperature of 60–70 °C, the dispelling could be done after 10–16 h. After about 12 h of continuous drying overnight, tempering is given to facilitate diffusion of moisture from the inner side. The copra cups are to be raked every 2 h for uniform drying. The total drying period is over 3 days with overnight breaks till the moisture content of copra is reduced to 5–6%. The actual drying time is 33–37 h.

A low-cost dryer for larger holdings has also been introduced during 1987. The capacity of the dryer is 3000–4500 nuts batch^{-1} and the drying time is 34–36 h. The fuel requirement is 800–1000 coconut shells and husks. By controlled fuel feeding, the drying air temperature could be maintained at about 55 °C to 70 °C. The overall dimension of the dryer is 4.6 m × 3 m × 2.6 m (Madhavan and Bosco 1991).

Indirect Heating with Forced Hot Air Systems These dryers are more efficient with reduced drying time. But the installation cost is high compared to the natural draft dryers. Here the hot air is forced into the drying system by means of a blower driven by diesel engine or electric motor. Husks, shells, firewood, electric power or oil can be used as fuel. This dryer usually consists of a furnace and an array of flue pipes with chimney, a drying chamber and a blower. The fresh air is heated by the flue gases passing through the flue pipes serving as heat exchanger. This hot air is forced into the drying chamber by the blower.

Different types of forced hot air driers are developed which use either solid fuel or automated diesel. They are Chula hot air dryers (BDO and BDI model and oil fired), Lister reversible airflow dryer, Pearson's patent dryer, Comessa dryer of Comoro Islands, ASP dryer in New Guinea, electrical dryer of CPCRI and agricultural waste-fuelled dryer of TNAU, India. Of these, Pearson's patent dryer is installed in many estates in Sri Lanka.

Electrical Drying The electrical dryer developed at the ICAR-CPCRI has a capacity of 1000 nuts per batch with drying time of 2 days at 60 °C. The dryer has been improved by modifying the drying systems. It is a tray dryer with mixed flow and forced hot air circulation. The main components of the dryer are drying chamber, plenum chamber, heating unit and blower unit. Drying chamber is made of jack wood planks lined with GI sheet inside. The air distribution chamber located vertically at the centre is made of GI sheet with perforations on both sides. The drying chamber can accommodate 10 trays of 92 cm × 45 cm size and is made of welded wire mesh. The trays are kept on aluminium angle runners on both sides of the air distribution unit. The top of the drying chamber is open with an adjustable lid to serve as exhaust. The heating unit consists of heaters of 6 KW capacity each and a blower equipped with 0.5 hp., 1440 rpm electric motor (Madhavan et al. 1998).

The temperature of inlet air is kept at 60 °C for copra. The dryer is operated continuously for 12 h initially and is switched off. The trays are taken out and the shells are separated from the kernel. Then the trays with the kernel cups are reloaded into the dryer with cups facing up, and the dryer is switched on again. The drying is continued till the desired moisture level of 6% is attained. Total drying time is 32 h. This dryer could be a feasible proposition for cooperatives, medium growers and copra processors. The dryer can be used in rainy season also when sun drying is not possible. The quality of copra is good, white and mould-free. Dryer design is simple and can be fabricated locally. A semi-skilled person can operate it. Mixed-type airflow provides uniform drying of the produce and hence no need for mixing.

Solar cum electrical dryer with agricultural waste as third source of energy has been designed for the capacity of 3000–3500 coconut batch^{-1}. It consists of a double-pass solar collector, fan, electric heaters and trolleys with trays to carry the drying materials. Electric heaters are provided before the fan so that hot air from the heater can also be drawn by it. The double-pass solar collector is of semicircular shape of 1.5 metre radius, and the drying chamber is the space below the collector where coconuts are kept in trays. The transparent cover of the collector is made of LDPE film of 200 micron size, and the absorber is black LDPE film of 250 micron size. The casing is provided with black-painted GI sheet of 26-gauge thickness. Hot air from the collector is sucked by the blower which is also fixed inside the drying chamber. Electronic control circuits are designed to switch on the heaters only when the temperature of the drying chamber is less than 50 °C. When the temperature of the drying chamber is higher than 70 °C, the electric heaters and the blower will be switched off. These temperature limits can be varied as per requirement.

13.12.5.3 Quality of Copra

The quality of milling copra ultimately determines the quality of the oil and the residual cake. The quality of copra is influenced by (i) moisture content, (ii) colour and cleanliness, (iii) microbial load, (iv) rubberiness, (v) case hardening and (vi) charring.

Moisture is the most important factor influencing the quality of copra. Copra with a moisture content of less than 6% is considered as good quality since it is not easily damaged by insects, moulds or microorganisms. An electronic moisture metre to determine the moisture content of copra was developed (Madhavan 1987), based on the electrical conductivity of the kernel. The instrument can read moisture content from 5% to 40%. It is very handy, and the accuracy is more than 94% in the lower levels of moisture content readings. Please see Sect. 13.4.5 for details.

Nuts from dwarf variety palms, unripe nuts as well as those from sulphur-deficient palms yield rubbery copra. This copra undergoes rapid deterioration, and good copra also becomes susceptible to microbial infestation when mixed with rubbery copra (Southern 1957).

Case hardening of copra is seen when nuts are dried in kilns and hot air dryers with no temperature control. If the initial drying temperature is too high, case hardening can occur, preventing the moisture from the interior of the meat from diffusing rapidly to the outside layers. Case-hardened copra develops a hard smooth surface covering a wet core. While drying copra, the temperature should not exceed 70 °C in the early stages and 60 °C subsequently, which will otherwise result in charring. The oil from charred copra will be turbid with a burnt odour.

Bio Deterioration of Copra Defective methods of processing and high moisture content of copra are the major factors responsible for copra deterioration. Coconut kernel is a favourable substrate for the growth of microorganisms. If drying is done under open conditions, copra spoilage due to fungal infection is very common. Usually bacterial action starts during the initial stages of drying, and later mould infection occurs. A gap of more than 4 h between splitting and drying facilitates the activities of bacteria. With relative humidity of 80% at temperatures above 30 °C, bacteria multiply rapidly, and within 4 h, a surface slime begins to develop on the wet kernel. The slime continues to develop, and it becomes more pronounced during the first and second days of drying. The bacteria are active at moisture levels above 20%. As a result of the bacterial growth, the copra loses white colour, turns red and becomes slimy. *Serratia marcescens*, *Staphylococcus aureus* and *Bacillus* sp. are the common bacteria, which cause discolouration, sliminess and softening of the copra (Rethinam and Bosco 2006).

The penetrating moulds make their appearance after the bacterial growth. *Rhizopus* sp. or the white mould thrives on wet meat and destroys a very high percentage of oil, and the oil from the infected kernel has a high percentage of free fatty acid. *Aspergillus niger* group or black mould has a lower moisture requirement. This mould causes considerable loss of oil up to 40%. The *Aspergillus flavus* group or brown mould is the most serious of all moulds. It flourishes at 8–12% of moisture, and the oil loss in this case also may be more than 40%. It causes maximum colour change and rancidity in the oil. The *Penicillium glaucum* group or the green mould is commonly found on copra and grows well on the copra with moisture content of 6%. This fungus does not penetrate deeper and causes only minimum reduction in oil content. In addition to these, only two other fungi, *Mucor hiemalis* and *Aspergillus tamarii* (yellow mould), are often found on copra causing considerable loss. Pure coconut oil is not a suitable medium for the growth of microorganisms. Moulds produce free fatty acid in the copra, and bacteria cause decomposition of albumin in the moist copra. Oil prepared from such spoiled copra becomes rancid quickly with bad taste and odour.

Fungi also cause deterioration of copra followed by bacteria *P. frequentans*,which is found to cause spoilage even at a moisture content of 4% (Nair and Sreemulanathan 1970). Rao et al. (1971) observed the presence of *Botryodiplodia theobromae* during the blackening of coconut kernel. Paul (1969) and Susamma and Menon (1981) isolated a number of fungi and bacteria from deteriorated copra. The fungi isolated were *R. stolonifer*, *R. oryzae*, *Mucor hiemalis*, *P. citrinum*, *Curvularia senegalensis*, *Cochliobolus lunatus*, *Paecilomyces lilanicus*, *Aspergillus oryzae* and *Aspergillus fumigatus*. Bacteria causing spoilage were identified as *R. subtilis*, *E. aerogenes*, *Pseudomonas fluorescens* and *Sarcina lutea*. The chemical change in the amino acid content due to fungal infection was also investigated (Sierra 1971; Susamma et al. 1981). Pests can also cause serious damage to stored copra. For details please see Chap. 12.

Storage of Copra The general observation is that the copra on storage is affected by excessive mould growth when the relative humidity is greater than 85% at room temperature or greater than 95% at 40 °C. In many of the establishments, it is a usual practice to dry copra under the sun for one or two days before bagging and storing. Painting of upper surface of the roof of the storage structure with white reflective paint has been reported to reduce temperature fluctuations within 10 °C thus preventing serious condensation effect. The walls also should be shaded from direct sunlight and should be provided with sufficient number of adjustable ventilators.

Sulphuring the copra is effective for storing for longer periods during rainy season. Fumigation of copra with methyl bromide at the rate of 3 kg 100 m^{-3} for 48 h with a gas-proof sheet is recommended when copra is stored in godowns. Fumigation with sulphur dioxide or lindane smoke or pyrethrum and piperonyl butoxide as a fog is also resorted to. Fumigation with a mixture of carbon dioxide and ethylene oxide (99:1) is effective against insect pests of copra. A chemical treatment of dipping fresh kernel in 1000 ppm propionic acid for 60 min to preserve it up to 49 days without further drying had been developed (Patil 1982).

Table 13.19 Standard specifications for copra

SI. No.	Characteristics	Grade 1	Grade 2	Grade 3
1.	Moisture content (per cent by weight, maximum)	6	6	6
2.	Oil content (on moisture-free basis) (per cent by weight, minimum)	70	68	66
3.	Free fatty acid (per cent as lauric, per cent by weight, maximum)	1	3	6
4.	Aflatoxin content (in parts per million, maximum)	20	20	20
5.	Impurities (per cent weight)	0.5	1	2
6.	Immature kernels (wrinkled) (per cent maximum)	Nil	5	10
7.	Mouldy cups (per cent by count, maximum)	Nil	4	8

Source: NIIR Board of Consultants and Engineers (2012)

Use of proper packing materials is also an important factor. Polythene/alkathene-lined gunny bags or netted polythene bags can be used for safe storage of copra even during the rainy season (Nalinakumari 1989). Multiwalled paper bags could keep copra free from insects up to 3 months, and those treated with pyrethrin could be kept up to 9 months. Studies at ICAR-CPCRI, during 1989, have indicated that copra stored in tin containers and polythene bags and fumigated with biogas, neem leaf gas, carbon dioxide and sulphur dioxide are effective in controlling microbial infestation during storage. Copra could be stored for more than 6 months by exposing it either to biogas for 1–3 days or to neem leaf gas for 15 days immediately after splitting prior to storage in any air-tight container.

Grades of Copra In all the coconut-growing countries, there are recognised grades of copra which form the basis of the trade transactions. Though each producing country has standards for grading, there is no common and uniform international standard for copra. Standard contract terms for milling copra are in force in India since 1949. In Sri Lanka, three grades of copra are defined based on moisture content, oil content and FFA level of pressed oil. In the Philippines, four classes and seven grades of copra are available based on type of moisture content and drying system used. In Western Samoa, grading is done based on the method of drying adopted.

Grade specifications for milling copra as suggested by the Asian and Pacific Coconut Community for adoption by member countries are shown in Table 13.19.

13.13 Non-food Products

Among the non-food products of coconut, coconut fibre, coconut pith and coconut shell assume commercial importance. Other parts of the palm especially coconut wood and leaves are also gaining attention.

13.13.1 Coconut Husk

The husk usually forms 35–45% of the weight of the whole nut when ripe. About 30% of the husk is fibre and 70% is coir dust. At present, only 35% of the total husk available is utilised by the industry, while there is scope for economically utilising at least 50% of the husk produced. When the nuts mature, the quantity of the fibre in the husk does not decrease, but it is the moisture in the fibrous mass of the husk which disappears. Thickness of the husk of an ordinary nut varies from 2.5 to 3.0 cm in the case of thin husked nuts and 4.0–5.0 cm for thick husked ones. The products of importance derived from coconut husk are coir fibre and coir pith.

13.13.1.1 Coir Fibre

There are two distinct varieties of coir, namely, white fibre and brown fibre. The white fibre is extracted from retted coconut husk and is used for making traditional coir products like mats, mattings, rugs and carpets. Brown fibre is extracted from unretted husk. It is mainly used for the manufacture of curled coir, which is used in the rubberised coir mattresses, sofa cushion, bolsters, pillows, carpet underlay, etc. In India, white fibre production has dominance, whereas in other countries, the production is confined to brown fibre. India and Sri Lanka contribute about 65% and 32%, respectively, of world coir fibre production.

Structure and Properties of Coir Fibre The chemical constituents of pure coir are cellulose (32–43%), lignin (40–45%), hemicellulose (0.15–0.25%) and pectin which makes it more extensible compared to other natural fibres. The fibre is weather-resistant and also resistant to fungal and bacterial decomposition which are attributed to the high lignin content. Lignin is the main constituent responsible for the stiffness of the coir and also partly responsible for the natural colour of the fibre. Generally, mature nuts have higher lignin and cellulose than tender nuts.

Fibre is polygonal to round in section. It is a multicellular fibre with a central pore called lacuna. The larger fibres have a length of up to 35 cm and can be 0.3–1.0 mm in diameter, being thickest in the middle. X-ray studies have indicated a spiral orientation of 45° to the fibrils (Bhowmick and Debnath 1984). There are three types of fibres, namely, mat fibre, bristle fibre and mattress fibre. Mat fibres are generally used for making yarn out of which ropes and floor coverings are made. Brushes, brooms, bags, net and twines are made from bristle fibre. Mattress fibre is mainly used for stuffing mattresses, pillows and cushions.

The white fibre or mat fibre is extracted from unripe husks by bacteriological process of retting, while the brown fibre is extracted from ripe dry husks by the mechanical defibering process. Husks from 10- to 11-months old nuts have been found to give superior-quality white fibre. Among the different states in India, white fibre production is concentrated in Kerala, whereas in other states such as Tamil Nadu, Karnataka and Odisha, the production is confined to brown fibre.

Extraction of fibre – Natural Retting Retting involves soaking the husk in water, preferably saline water, until the fibre becomes loose and soft. Water flows in and out of the soaking sites with the rise and fall of the tide. The efficiency of retting depends on many parameters like temperature, water quality, rate of removal of the foul water and stresses that husks are subjected to during the retting process.

Retting process essentially consists of a microbiological degradation of the pectic substances which hold the fibres together prior to soaking. The husk becomes soft, and a number of substances like carbohydrates, glucosides, tannins and nitrogen compounds are acted upon by a great variety of microorganisms which produce various organic acids and gas. When the fermentation progresses, the temperature of the husk increases, and water becomes turbid due to gas formation and frothing. At this stage, pectin in the middle lamella of the husk slowly dissolves. Subsequently the rate of fermentation slows down, and water becomes clear when the husk is ready for removal. The process is quicker during summer since heat is necessary for fermentation. The average retting period is 8–9 months in saline water and 4–5 months in freshwater. Saline water is preferred since the salts prevent over-fermentation and discolouration of the fibre. But investigations of Jayasankar (1966) and Bhat and Nampoothiri (1973) have indicated that under controlled conditions, salinity has no influence on the microflora and its activity.

It has been found that soaking of crushed husk reduces the retting period (Nagarajan et al. 1987). Crushing can be done by simple crushing rollers, similar to sugarcane crushers. Similarly, if the husks soaked for 1 month are removed, crushed and put back again, the retting period can be reduced to 3 months.

Mechanical and Chemical Methods of Retting Mechanical methods of retting are employed in areas where facilities for natural retting do not exist. Either dry or green husks are soaked in cement tanks or in specially dug pits for a period varying from a few hours to 6 weeks, and the fibre is extracted manually or mechanically. In medium to major coir units, cement tanks each measuring 8 m × 2.7 m × 1.8 m are built-in series with facilities for frequent change of water. In some units, the husk is first crushed in crushers and put in retting tanks where they undergo fermentation for a minimum period of 72 h. These processes however do not yield fibre of spinnable quality, but yield only bristle and mattress fibres.

Various chemical methods also have been developed for retting of husk. The advantages claimed are a higher yield of uniform-quality fibre and a considerable saving of time. But the economics of chemical retting compared to the natural process has not been fully investigated for commercial exploitation. In the Nanji process, the green or dry husks are partially crushed and treated under steam pressure of 5.6–7.0 kg cm^{-2} with sodium sulphate or sodium carbonate containing a trace of aluminium sulphate for 1–2 h. During this process, the pith is loosened from the fibre and removed by washing. The fibre obtained is of good quality, but slightly darker than that of natural retting. In the Elod and Thomas process, crushed husk is immersed in hot water twice. Slaked lime is added during the second immersion to avoid discolouration. Subsequently, the fibre is extracted mechanically. In the

Rowell process, crushed husk is subjected to a high steam pressure, and the fibres come out loose from the steaming chambers. In the Vander Jagt process, husks are first split into pieces. The pieces are boiled with a week solution of caustic soda and squeezed. The compressed fibres are reopened, softened and cleaned. Good-quality fibre is obtained within 2 h by this process. In the Hayes Gratze process, the husk is first split into sections by special cutting machines. The split pieces are then immersed in water, pressed or rolled and boiled in a solution of water and ionised oil of pH 17 maintained at 93 °C. Coconut oil is sometimes used for the preparation of ionised oil. Carren (1966) developed another method where the husk is fermented for 4 days at 37 °C with the aid of *Aspergillus* sp. previously isolated from partially retted husk. This process is reported to have given 37% fibre output.

Experiments carried out to improve the retting process reveal that inoculation of ret. microflora mass multiplied in the laboratory, collected from areas known for poor retting, leads to bettering the colour of the fibre, besides reducing the period of retting (Nagarajan et al. 1987). Coir fibre can also be extracted by decortication of dry husks. However, there is a need for developing a process requiring lesser retting period and yielding fibre of uniform quality irrespective of the seasonal and environmental variations.

Extraction of White Fibre After natural retting, the husk is taken out of water, washed and the outer skin is peeled off. The husk is then placed on wooden blocks and beaten with a wooden mallet to separate the fibre from the pith. It is then cleaned and spread in shade for drying. Occasionally, the fibre spread is beaten and tossed up for further cleaning and also for thorough mixing of long and short fibres. For making superior type of fibre especially for spinning, the fibre is combed in a specially designed combing or willowing machine, which consists of a number of knives with saw like teeth mounted on a wooden shaft set spirally and which is rotated by hand within a drum. The fibre from the retted husk is also extracted mechanically. The beaten husk is torn on rolling cylinders with nails on the cylinder casting. The raw coir fibre thus obtained is further cleaned by means of blowing fans. The machine helps to soften and remove the last traces of pith on the fibre, and the processed fibre is clean and more or less parallel. The fibres are then rolled into slivers which are used for spinning.

Extraction of Brown Fibre The extraction of fibre from the partially wet husk is carried out in specially constructed machines called drums. The drums are operated in pairs, and each pair consists of a breaker drum and a cleaner drum. The drum consists of an iron flywheel to the rim of which a number of wooden planks in two layers are fixed. A number of nails of about 6 cm long are fixed on the outer layer of the planks at intervals of about two with more than half the length of the nails projecting from the surface. For fibre extraction, the husk is held firmly between two rods fitted on the front side of the breaker drum so that one half of the husk is combed on the revolving drum. Then the second half is also combed similarly. In the breaker drum which is driven at a higher speed than the cleaner drum, the outer skins and some of the short fibres and pith are removed and get collected at the base of the drum. This is fed on to a sifter for separation of the fibre from the pith and dust. The fibres obtained are short and thin and are known as mattress fibre.

During the combing process, the longer fibres or bristle remain in the hands of the operator. The partially combed-out husk segments containing the bristle fibre are combed a second time by another operator in the cleaner drum for removing the remaining pith and smaller fibre. The long stiff fibres which remain with the cleaner or dresser drum operator are subsequently washed, dried and combed again for the final removal of the short fibres. The long fibres obtained after the final combing are called bristle fibres. In another method, mattress fibre is obtained from unrett husks using a husk bursting machine. The fibre so obtained is considered more suitable for the manufacture of rubberised coir.

The yield of fibre is subject to considerable variation depending upon the season, method of extraction and the quality of fibre produced. Yield from rett husk is more than that from unrett husk. The yield is less during monsoon as compared to summer season. Taking all these into consideration, it is estimated that the average yield of fibre from an average-sized fully mature nut weighing 1.1 kg is 80 gm of white fibre (George and Joshi 1961). In Sri Lanka, 1000 full husks yield an average of about 50 kg of bristle fibre and 100 kg of mattress fibre.

Varieties of Fibre and Grades The fibres are used for spinning into yarn for manufacturing mats and mattings, ropes, twines, etc. The bristle fibre which is long and stiff is preferred for brushes and brooms. The bulk of the fibre produced in India is mat fibre which is used for spinning into yarn and for manufacturing mats and mattings, ropes, twines, etc. The mattress fibre which is short and thin is generally used in mattresses, upholstery, cushions, etc. The bristle fibre which is long and stiff is used for brushes and brooms.

Coir is graded according to the colour and length of the fibre as also its refraction content. Four grades are recognised in India on the basis of the specifications drawn up by the Indian Standards Institution for coir fibre (Table 13.20).

Spinning of Coir Yarn The spinning of coir yarn is a traditional cottage industry in India, which is concentrated in backwater areas where natural retting is in vogue. Hand spinning, wheel spinning and machine spinning are the common methods adopted for this purpose.

In hand spinning, the fibre is rolled between the palms with a clockwise twist into strands of short length. When sufficient quantity is made, the strands are taken in pairs and twisted together in the opposite direction to form a two-ply yarn. The yarn is then held in position by the toes, and individual pieces of yarn are joined together one after another by continuing the counter twist using both the palms till the required length for a knot (6–18 m) is reached. The yarn is then reeled in the form of a hank and a knot made at the end. One worker is estimated to make about 2–2.5 kg of yarn day^{-1}.

In wheel spinning, two wheels are used. One wheel with two spindles is fitted to a stationary stand, and the other one with one spindle is mounted on wheels which

Table 13.20 Characteristics of coir fibre of various grades

Grade	Maximum impurities (%)	Length of fibre, proportion of long, medium and short fibres	Colour and utility
1	2.0	At least 70% by weight 'long' and the remaining 'medium' and/or 'short'	Making superior-quality double warp fancy fibre mats
2	3.0	At least 50% by weight 'long' and the remaining 'medium' and/or 'short'	Constitutes fibre of white lustrous colour
3	5.0	At least 80% by weight 'medium' and the remaining 'long' and/or 'short'	Slightly reddish or greyish coir containing little pith
4	7.0	At least 20% by weight 'medium' and/or 'long' and the remaining 'short'	Dark in colour containing more pith, used for making cheaper yarn known as beach yarn

Source: NIIR Board of Consultants and Engineers (2012)

can be moved forwards and backwards. In the actual working, one can rotate the wheel on the stationery stand by moving a handle fixed to its axis. Two persons make the strands by hooking short length of fibre strands onto the spindles of the stationary wheel and walking back deliver the fibre continuously to form strands of uniform thickness. The stationary wheel is made to rotate continuously to give the necessary twist to the strand. When they complete a length of about 15–18 m of strand each, the rotation of the stationary wheel is stopped. The two ends of the single strands are then joined together and hooked to the spindle of the movable wheel. One of the workers now takes charge of this movable wheel and rotates it to give the two-strand yarn a twist in a direction opposite to that of the single strands. The other worker now moves forwards towards the stationary wheel with a yarn guide in hand, which is a triangular block of wood, grooved on the sides. It regulates the counter twist, prevents knots and kinks and binds the strands very close. The spun yarn in lengths of 12–15 m is reeled into hanks. About 100 strands of 15 m length each and weighing about 15 kg can be produced day^{-1}.

Hand spun yarn is soft and has uniformity of twist and thickness. But the method is nowadays becoming more or less obsolete. Wheel spinning is the popular method of yarn spinning, but with some inherent defects. It requires long open yard and process is interrupted during monsoon. Mechanical spinning has been introduced to obviate these defects, but the output is very limited.

Coir yarn is graded according to colour, Scorage, moisture content and presence of sand, salt, etc. Bright golden-coloured yarns are considered to be the best, while the lower grades are comparatively dull and dark in colour. Scorage is another important factor in the classification of yarns. It represents one-20th the number of strands of yarns which could be held close to each other without overlapping in a span of 0.9 m.

Utilisation of Coir fibre and Yarn Coconut fibre is used in the preparation of hardboards, fibre boards and building boards. The hardboards prepared from coconut fibre by the Asplund process have outstanding flexibility. Fibre boards can be made from the husk of immature coconut by pressing them at 160 °C for 20 min. They are suitable for roofing, panelling and replacing plywood in tea chests. George and Joshi (1961) made hardboards and insulating boards from dry husk of mature coconuts after softening them with dilute sodium hydroxide solution and also from coir shearing waste softened with slaked lime solution. Building board was also developed by mixing the husk material with water and subjecting to elevated temperature and pressure (Totheringhan 1982). Semana et al. (1988) prepared standard-type hardboard by blending of coconut coir pulp and petiole pulp (50:50).

The coir yarn finds various domestic uses. For commercial purposes, the yarn is used for making ropes, mats, mattings, nets, bags, etc. Coir fibre of inferior quality and the mattress fibre are used for stuffing mattresses, cushions, upholstery, etc. The bristle fibre goes in the manufacture of brushes and brooms. The mattress fibre, decorticated fibre and bristle fibre and their mixtures are twisted into ropes to produce curled fibre which is used in the manufacture of rubberised coir. Coir and coir products have also been found useful in soil erosion control, road construction and as reinforcement in cement matrix.

Coir Fabrics for Groundwater Recharge Compared to other natural fibres,coir fibres degrade very slowly, and hence coir woven fabrics with loop construction retain moisture in the soil which can be used in water harvesting. Besides this, when the degradation of lignin starts, the acidic phenolic leaches may decompose even hard laterites and make the soil porous and help in downward seepage of water. It has been the experience of the farmers to condition the hard soil by stacking coconut husk in pits in the palm basins. This improves groundwater level in hard lateritic terrain. Infiltration trenches at suitable locations with bore holes at the bottom to a significant depth (i.e. rocky bed or water table) can be made. These bore holes can be lined with reinforced and treated coir felt, and metal chips can be loosely filled inside the trenches and bore holes. The top of the trenches can be covered with coir net with loop structure and fixed to a bamboo frame. This arrangement will ensure the collection of surface rainwater and encourage percolation enriching the collector wells, besides raising the groundwater level.

Coir Geotextiles Coir netting is an ideal material for prevention of soil erosion. Coir geo-fabrics are woven coir nettings or mesh mattings. They are inexpensive, ready-to-use and effective items for a variety of applications including control of soil erosion, control of landslides, slope stabilisation, seepage of water through canals and in other civil engineering applications like road embankments, etc. In these applications, coir is used because it is a natural hard fibre with high tensile strength, durability and moisture resistance.

Coir-Cement Composite Panel Reinforced composite panels are made using coir fibre of 50–350 mm long and 0.10–0.40 mm diameter. The fibres soaked in water initially for 1–2 h are thoroughly mixed with predetermined quantity of Portland cement and chemical admixture. The coated fibres are uniformly spread on a mould and pressed for 6–8 h to the required thickness. The pressed panel is moist cured for 10 days and dried under natural conditions for 4–6 days and trimmed to required size.

Husk Particle Boards Husk of mature coconut is a unique raw material to prepare particle boards, in view of the fact that wooden particle boards use 8–10% adhesives on weight basis, while coconut husk boards require only 0.25% adhesives. However, care has to be taken to see that the ingredients in pith are allowed to separate out from the fibre, while the chips should be of free-flowing nature. It should not interlock into bundles during handling and storage.

13.13.1.2 Coir Pith and its Uses

Coir pith constitutes as much as 70% of the husk and is now a waste product of the coir industry. Accumulation of this waste in industrial yards causes environmental pollution and fire hazard. Pith generally mixed with short fibres contains lignin, cellulose and hemicellulose as major constituents. Coir pith is open cell foam. The cells are of almost uniform size and cylindrical in shape. The walls are very thin and empty cavities (lumen) are comparatively large. Average lumen size of the pith is 50 mm. The maximum water holding capacity is 624%. The composition of coir pith is given in Table 13.21.

The main uses of coir pith are briefed below.

Manure Various studies have shown the advantages of application of coir dust to improve soil property and increase yield (Loganathan and Lakshminarasimhan 1979; Ramaswamy and Kothandaraman 1985; Clarson 1986). The continuous

Table 13.21 Composition of coir pith in percentage

	Fine coir dust	Coarse coir dust	Husk
Moisture	15.77	20.39	–
Organic matter	86.87	96.43	96.5
Ash	13.13	3.57	3.5
Organic matter			
Nitrogen	1.00	0.39	0.29
Potash	0.39	0.33	0.31
Lignin	37.71	43.65	45.45
Pentosan	11.95	13.10	19.15
Ratio of pentosan to lignin	0.32	0.30	0.42
Ash			
Sand	9.29	0.87	0.36
P_2O_5	0.07	0.06	0.08
Lime	0.79	0.87	0.94

Source: NIIR Board of Consultants and Engineers (2012)

application of coir dust influences reduction in bulk density and improves the water holding capacity and organic carbon status of soil. Coir pith absorbs over eight times its weight of water and parts with it slowly. It has been found that by incorporation of 2% by weight of coir dust with sandy soil, the water holding capacity of the latter is increased by 40%. It is an excellent organic mulch in all kinds of soils. Nagarajan et al. (1987) found that coir pith after inoculation with *Pleurotus sajor-caju* and treatment with urea showed a definite reduction in the cellulose and lignin contents on incubation for 26 days at room temperature. The total nitrogen and other nutrients increased, while the C:N ratio was narrowed from 112:1–24:1. The values of N, P_2O_5 and K (percentage) in cattle manure are 9.86, 4.82 and 1.81, respectively, whereas they are 4.42, 0.71 and 1.02 in coir dust.

The lignocellulosic components of coir pith can be acted upon by lignocelluloytic fungi (Uyenco and Ochoa 1984), converting it into a biomass product which can be used as an animal feed or fertiliser.

Composting of coir pith has the advantages of detoxifying phenolic compounds, which are deleterious to microbial growth, reducing the bulk of the material and converting plant nutrients to a form more readily available to plants. A technology has been developed by Tamil Nadu Agricultural University, India, to detoxify phenolics of coir pith and produce bio-polymerising enzymes using basidiomycetous fungus, *Pleurotus sajor-caju*. To compost 1 tonne of coir pith, five spawn bottles (one spawn bottle contains 350 g of *Pleurotus* fungus culture raised on sorghum or pearl millet grain) and 5 kg of urea are required. After 30 days of decomposition, coir pith turns into a black mass of compost with reduced lignin, cellulose, organic carbon and C:N ratio. The volume of the material is also reduced by 50% (Rethinam and Bosco 2006).

As a Source of Furfural, Oxalic Acid and Gypsum Coir dust contains pentosans which are the only source in the world for furfural production, involving a two-stage process (Thiagarajan 1987). About 90–95% of the weight can be eliminated at the original site of accumulation. The residue after the extraction of furfural can be used for the manufacture of oxalic acid by heating with caustic alkalis to a temperature of about 240 °C. Crude sodium oxalate is formed which can be converted to calcium oxalate by lime, regenerating the caustic alkalis for reuse. The calcium oxalate can be decomposed with waste sulphuric acid to produce oxalic acid and gypsum. Gypsum can be used as a fertiliser.

Aslam Ali and Bhaskar (1980) reported that when pith is treated with concentrated nitric acid containing vanadium pentoxide (0.05% by weight of the pith) and heated for 6 h at 80 °C, it gives oxalic acid crystals after filtration and evaporation. The recovery on weight basis is 31–33%. The nitrogen fumes can be recovered and the nitric acid can be reapplied. The oxalic acid finds varied uses as mordant in dyeing and cloth printing, as a bleaching agent, anodizer, in detergents and in metal polishes.

Biogas Production A mechanical mixer has been developed for proper mixing of coir pith and cow dung. The output is connected to biogas plant through a 2 inch pipe for direct delivery of mixture to the plant. Cow dung and coir pith are put in the required ratio on volume basis, and water is added in the ratio of 1:1 (cow dung + coir pith/water). For optimum biogas production, the C:N ratio should be in the range of 25–30, and in the case of 80% cow dung and 20% coir pith, the C:N ratio was 26.18, and the production of gas was high and was on par with 100% cow dung.

When temperature is more than 33 °C, the production of gas is constant, whereas gas production is low when the temperature falls below 30 °C. When a modified dual fuel engine was run with biogas, it was observed that in 8 h of operation, approximately 10 m^3 biogas is consumed, thereby saving 5 litres of diesel. But due to continuous operation of the engine with dual fuel, the rings and piston head were damaged indicating the presence of sulphur in the gas. So the maintenance cost is higher as compared to running the engine on diesel.

Coir pith can be used as a starting material for the production of gas by controlled combustion, which has been tried in industrial engines on a small scale. It is estimated that an energy equivalent to two million tonnes of coconut waste is available in the Philippines.

As Fluidised Fuel Dried coconut pith could be blown with air to burn in suspension using a blower. A fluid furnace using pith has been developed. The furnace was evaluated using multiple injection points. As the outlet pipe provided was of lesser diameter, burning of coconut pith was difficult, and also smoke formation was more. The modified furnace was provided with holes of larger diameter at different points to study the effect of injection of pith with air using a blower. It was observed that injecting at about 0.80 m from the ground level was suitable to burn 75% of the pith in suspension and the balance is burnt at the bottom where a heap of shell is in burning condition. The air keeps the pith rotating in circular motion, and the large particles are thrown outwards by the centrifugal force and burnt near the walls, while the smaller particles are burnt in suspension in the air. The exhaust temperature was about 380 ° C. The temperature around the furnace is about 85 °C, which is used to the dry pith.

A 400-nut copra dryer has been developed which can be connected to the fluid furnace. The highest furnace efficiency obtained was 80% at the pith feed rate of 10 kg h^{-1} and air flow rate of 87.6 cum h^{-1}. The temperature of the flue gas varied from 80 °C to 350 °C at various pith feed rates. The pith and air were introduced at various angles, and it was observed that the best results are obtained when the pith was introduced tangentially. About 400 nuts can be dried from the initial moisture content of 48% to less than 5% in 36 h.

Particle Board Using Coir Pith A process for the production of particle board from coir pith has been optimised. The process includes the preparation of the coir pith, mixing with resin (phenol-formaldehyde and urea-formaldehyde), mat formation and hot pressing.

Densification Varadaraju and Gothandapani (1998) developed a pelletiser for making pellets from decomposed coir pith. The unit is operated by a 5 hp. electric motor and has a capacity of 100 kg h^{-1} for extruding the pellets of 6–8 mm diameter and 10–12 mm length at 25% moisture content. The coir pith pellet with optimum durability, compaction and expansion ratios of 0.82, 3.14 and 1.33, respectively, is obtained.

Briquetting A continuous extruder-type briquetting machine, consisting of screw shaft, barrel housing, extruder die pipe and gear box, has been developed by Devadas (1996) with a capacity of 125 kg h^{-1}. Briquettes produced from coir pith using cow dung or molasses as binder having a calorific value of 3000–3200 kcal h^{-1} were used as an alternative source of fuel. Tender coconut husk with tender shell can be crushed into powder form. It can be converted into fuel briquette after bringing the moisture content to about 15%.

Building Materials A technology for the production of coir-cement corrugated roofing sheet has been developed (Viswanathan 1998). It is observed that the temperature is lowered by 1–4 ° C in coir-cement corrugated roofing than that in asbestos. Lightweight bricks using coconut pith in the proportion of up to 80:20 (clay and pith) by weight are found successful (Dan 1992). Medium-density (0.9 g cc^{-1}) particle boards of size 25 cm × 25 cm, using coir pith with phenol-formaldehyde and urea-formaldehyde as binder, have also been produced. The strength characteristics of the boards are on par with other commercial boards (Viswanathan and Gothandapani 1998). However, the water absorption and swelling characteristics are high which are not desirable for use in exteriors.

Miscellaneous Products from Coir Pith Many commercial products such as hardboards, insulators, expansion joint filters, etc. can be prepared using coir pith. A process developed in the Philippines during 1962 could convert coconut pith into charcoal briquettes by a continuous process. The gas can be used as domestic fuel and tar and liquor condensates as wood preservatives, and charcoal can be in the production of many chemicals like calcium chloride, carbon disulphide, silicon carbide, etc.

Medium-density particle board using coir pith has been produced, and the strength characteristics have been determined (Viswanathan and Gothandapani 1998). This board, made of phenol-formaldehyde resin, meets the strength requirements of the Bureau of Indian Standards. Particle boards of size 250 mm × 250 mm x 12 mm thickness with density 0.9 g cc^{-1} were produced from the coir pith of uniform particle sizes 2.1, 1.2, 0.80 and 0.40 mm. Particle boards were produced from mixed size of coir pith either with or without fibre. Phenol-formaldehyde and urea-formaldehyde resins of liquid type with hardeners are used for the purpose. Polymers and composites can be prepared from pith by copolymerising the lignin present in pith with either formaldehyde or phenols. Pith can also be used along with rubber to make composite flooring, ceiling floors and other similar products.

The pith in combination with cement has been found to be an excellent thermal insulating material. It is much lighter and easier to apply and gives much better thermal insulation at a cost equal to that of lime concrete. Coconut pith was successfully utilised in the production of a variety of lightweight high-strength bricks by the partial replacement of clay with pith. In India, cashew nut shell liquid filled pith composite has been used as joint filler between concrete slabs in roofs, roads and runway with a view to accommodating thermal movements. The pith joint fillers have been found resistant to alternate heating and wetting as well as to freezing and thawing. They have also been found resistant to termite and fungi and superior in qualities to those of bituminised fibre boards (Rethinam and Bosco 2006).

Mushroom cultivation is an economically feasible and eco-friendly process for bioconversion of coir pith into high-quality protein food. Oyster mushrooms are the ideal ones for the coconut sector due to their ability to utilise lignin-rich pith and due to the climatic conditions prevailing in plantations which are ideal for its growth.

The steps in oyster mushroom cultivation include development of spawn, substrate preparation, spawning, incubation for spawn running and opening and maintenance of beds for cropping. Polyethylene bag method is commonly followed for spawning. The bags are punched to facilitate cross ventilation. Spawning is done by multilayered technique using 3% spawn. Coconut bunch waste such as leaf stalk, mixtures of leaf stalk + coir pith (1:1) and coir pith + bunch waste (1:1) are suitable substrates.

Vermiculture technology involves the use of earthworms as versatile bioreactors for effective recycling of non-toxic organic wastes to produce high-quality manure. The byproducts from coconut plantations are converted to vermicompost with a nutrient content of 1.8% nitrogen, 0.2% phosphorus and 0.2% potassium using a locally isolated earthworm species *Eudrilus eugeniae.*

13.14 Coconut Shell

Coconut shell, the endocarp of coconut, is another important commercial product. Its main use is as a fuel. To a lesser extent, it is used as a raw material for the manufacture of hookah shells, various domestic utensils, curios, fancy items, etc. The commercial utilisation of coconut shell for the production of shell charcoal, activated carbon, shell flour, etc. is gaining importance in the producing countries with an expanding market demand. The coconut shell contains moisture (6.76%), ash (1.32%), pentosan (10.01%), lignin (32.22%) and cellulose (50.99%) (Narayanamurthy and Singh 1954).

13.14.1 Major Shell Products

Shell charcoal, activated carbon and shell flour are the main commercial products obtained from the shell. Major portion of the shell is used as fuel, for the production of copra and for domestic purpose. Much of the coconut shell is scattered in small quantities in individual farms. Setting up a viable industry based on coconut shell is likely to succeed only where large quantities of shell are available at one point as in large plantations, desiccated coconut factory, centralised copra processing plant, etc. The shell charcoal is already in high demand in the world market.

13.14.1.1 Coconut Shell Charcoal

Shell charcoal is manufactured by burning shell obtained from fully mature nuts in a limited supply of air, sufficient only for carbonisation, but not for complete destruction. Charcoal output is just under 30% of the weight of the original shell, which depends much on the efficiency of processing. In Sri Lanka and the Philippines, 17,000–24,000 whole shells for 1 tonne of charcoal is the usual requirement. In Malaysia, four times the number of nuts giving 100 kg of copra is required to make 100 kg of charcoal. This works out to 18,000–20,000 whole shells for 1 tonne of charcoal. In India, the average output has been found to be 35 kg of charcoal from 1000 whole shells equivalent to 30,000 whole shells for 1 tonne of charcoal. The popular methods adopted for the production of shell charcoal in different countries are discussed briefly below:

Covered Pit Method In this method, the carbonisation takes place in a pit of 1.25 m diameter and 2 m depth. The pit is excavated in such a way that the bottom tapers down to a point. The fire is started at the bottom, and the shells are heaped as the fire spreads. Towards the end, a thin mild steel plate is placed on the heap, and as the shells settle down, the lid comes to rest on the edges of the pit. The edges of the lid are covered with earth and left for 4–5 days to complete the carbonisation and further cooling. Then the lid is removed and charcoal collected. The charcoal produced is of uniform quality and is free from soil contamination. The yield ranges from 28% to 32% of the original weight of the shells used.

Modified Pit Method The pit is lined with fire bricks or GI sheets. The shells are then put in a thin layer at the bottom and ignited with kerosene. More shells are heaped during ignition to completely cover the pit. The pit is then covered with a corrugated iron sheet, leaving a gap of about 2 inches. The nature of the smoke that emits out is an indication of the progress of carbonisation. Towards the end of carbonisation, the colour of the smoke will change to pale blue from the initial dense white and steamy smoke. Now, the gap is closed and sides are sealed with mud. The pit is then left to cool for 4–5 days.

The Drum Method An ordinary 55 gallon gasoline drum is used for the carbonisation. The bottom of the drum is removed by cutting around the inside rim. The top is provided with two or three holes of 5 cm × 8 cm size equidistant from the two plug holes which are also kept open. The drum is placed on two 2.5 cm iron pipes with the open end up. The shells are then tightly stocked inside and fire lit between the base of the drum and the ground. As the charge burns, the contents of the drums are frequently shaken and fresh shells added in between. A restricted air circulation is allowed in the drum which is ensured by covering a portion of the space between the drum and the ground with soil. When the drum is filled with charcoal, the removed lid is replaced on the top tightly and the drum with the contents turned upside down on a bag. It is left standing until the smoke clears up. The holes are then covered with soil, and bottom is also piled up with sand around. The drum if properly sealed is ready to be discharged the next morning.

Properties of Coconut Shell Charcoal Good coconut shell charcoal is uniformly dark, snaps with a clean shining fracture and produces a metallic sound when dropped on hard ground. Underburnt shells do not give a metallic sound and a clean fracture, while overburnt ones are friable, and the surface of the fracture sounds dull when dropped and easily crumbles. Coconut shell charcoal contains the highest percentage of fixed carbon of all the ligneous charcoals. The average composition of good charcoal is moisture, 6.24%; volatiles, 5.46%; ash, 0.54%; and fixed carbon, 87.76%. The particle size should be such that less than 5% shall pass through 0.63 cm mesh sieve (Rethinam and Bosco 2006).

The charcoal has a high adsorption capacity for gases and colouring matter and can, therefore, be used as a refining agent both as a deodoriser and as a decolouriser. The shell charcoal also finds way to laundries, smitheries, etc. Well-powdered shell charcoal finds limited use as a dentifrice. The charcoal is used by gold smiths in melting gold and silver and for other metal works. The commercial value of shell charcoal lies in its use as the primary raw material for the production of activated carbon.

13.14.1.2 Activated Carbon

Activated carbon plays a very important part in solvent recovery processes, water and effluent treatment and the treatment of flue gas before its discharge into the atmosphere. Activated carbon is extensively used as agents for purifying, refining and bleaching of volatile oils and chemical solutions. They are also in demand as an adsorbent of gases. The common raw materials used for the manufacture of activated carbons are carbonaceous substances such as coal, lignite, wood and charcoal. Superfine coconut charcoal is one of the useful materials for activation. In the activation process, shell charcoal is fed continuously into a retort. The normal activation process involves the use of steam at selected temperatures for the selective oxidation of the material, resulting in the production of carbon with pores of molecular dimensions. Approximately 3 tonnes of shell charcoal is needed to produce a ton of activated carbon. Activation can be carried out with a variety of gases, including oxides of carbon, chlorine and mixtures of steam and air. After withdrawal from the retorts, the material is cooled and passed through a series of granulators and screens, thereby obtaining a carbon of known quality, in a variety of grade sizes to suit most of the applications.

A large number of plants based entirely on coconut shell charcoal have come up in the major coconut-growing countries. In India, the use of coconut shell for the manufacture of activated carbon is covered by the Indian Patent No. 109082. Here, coconut shells are crushed and treated with surface-active chemical followed by drying and carbonisation. It is then steam activated at 900 °C followed by air to facilitate oxidation and steam quenched to reduce the bed temperature. The material is then discharged in a receptacle and subjected to acid treatment to adjust the pH value. Finally the activated material is washed with water, dried and stored. Coconut shell-based activated carbon constitutes 10% of the total production of activated carbon in the world.

13.14.1.3 Coconut Shell Flour

It is prepared by grinding clean coconut shell to a fine powder. The shell is first pre-crushed in a beater-type disintegrator into 5 cm pieces, which are then conveyed to the first hammer mill. Suction in the conveying system draws the particles of flour into a cyclone, where they are separated into coarse and fine particles. Ultrafine particles are drawn away and collected separately. From the cyclone, the coarser particles pass to the second hammer mill, and the ground products are subjected to the same air separation as the particles from the first grinding. The fine particles from the cyclone are fed onto a vibrator-sieving unit and graded into the required mesh size. The shell flour should be a free-flowing powder and clear light brown in colour. The moisture and ash content shall not exceed 10 and 1.5%, respectively. The apparent density ranges from 0.6 to 0.7 g cc^{-1}. The particle size of shell flour which passes through 200 mesh sieve of British Standard should be between 6 and 10 microns (Rethinam and Bosco 2006).

Shell flour is used mainly as a filler, replacing wood flour either partially or wholly in the manufacture of phenolic moulding powders by the thermoplastic sector. It is used successfully with specialised surface finishes, liquid products, mastic adhesives, resin casting, mild abrasive products, hand cleaners and bituminous products. The shell flour gives a smooth and lustrous finish to moulded articles and improves their resistance to moisture and heat. Because of its higher resinous content and lower absorption properties, it can be used in higher concentrations than wood flour. Shell flour is used in the production of glues which is used in plywood industry. It is also used as a fuel and as a substitute for bunker oil in boiler operations. Shell flour has a variety of other uses. It is used as a filler for mosquito incense coils, filler in specialised surface finishes, resin castings, etc. As a mild abrasive, it is used as a soft blast to clean piston engines. It has been incorporated into hand cleaners and used as a diluent for potent insecticides.

Destructive Distillation of Shells In the ordinary method of charcoal making, about 70% of the shell weight is lost in the smoke. If the burning is done in retorts, heated from outside, and the vapour given off is condensed, apart from the recovery of good-quality charcoal, various chemical products of immense industrial value can also be obtained. This process is popularly known as destructive distillation of shells. The products include shell charcoal, which remains in the carbonising apparatus, pyroligneous acid, settled tar and uncondensable gases.

Experiments were conducted in Malaysia long back in 1929 on destructive distillation of coconut shell and the recovery of the byproducts besides charcoal. An experimental plant consisting of a retort 90 cm long and 60 cm in diameter holding 110 to135 kg of shell with a water-cooled vertical cooler was used. The distillate was treated for the byproducts. After the mechanical separation of the tar, the pyroligneous acid was redistilled from a still in order to eliminate dissolved tar and tarry condensation products, the residue in the still being completely freed from acetic acid by the injection of steam. The redistilled liquid which contained all the acetic acid and methyl alcohol was neutralised with lime and partially redistilled to recover the methyl alcohol leaving the acetate of lime in solution. This on evaporation dried up and gave the commercial product known as grey acetate of lime. The acetate of lime was treated in a special still with sulphuric acid and the crude acetic acid which distilled over was further purified by rectification. Out of 100 kg of shell, 40 kg of charcoal and 4.68 kg of acetic acid were obtained, besides smaller quantities of other byproducts such as tar, methyl alcohol and crude creosote oil from the tar.

In India, experiments were started in 1920, and product recovery could be established in subsequent years. It was reported that the average recovery of the products are charcoal (35–37%), pyroligneous acid (6–9%), settled tar (25–37%) and uncondensable gases (17–20%). The tar contained 70% phenols of which 32.6% was mono phenols, 21.3% diphenols and 46.1% triphenols.

In Sri Lanka, Nathanael (1964), on the basis of his pilot plant studies, reported that destructive distillation of the shell yields monomers like phenol, furfural, catechol, butadiene, etc. which could be used for the production of polymers. The shell tar contains about 30% phenols and furfural and croton aldehyde in lesser amounts, which are the potential raw materials for polymer development. Destructive distillation of 1 tonne of coconut shell was found to yield 300 kg each of uncondensable gases and charcoal, 350 kg pyroligneous acid (acetic acid 36%, wood spirit 13.5%) and 50 kg tar (5 kg phenol, 10 kg creosote, 10 kg pitch and natural oils in variable quantity).

The process of destructive distillation on an industrial scale could be divided into five different phases. (1) As a result of external heating to about 170 °C, the moisture in the shells evaporates, and no gas is formed. (2) When the temperature rises from 170° to 280 °C due to external heating, pyroligneous acid vapour, a little tarry vapour and gas consisting almost entirely of carbon monoxide and dioxide are evolved. (3) With no external heating, at about 290 °C, concentration of carbon in the charcoal takes place, and large quantities of hydrocarbon, acetic acid, wood naphtha and tar are produced when the temperature rises to 380–400 °C. (4) The charcoal is fully carbonised and the volatiles are completely driven off at about 600 °C–700 °C. Sometimes, a maximum temperature of 900 °C may be reached in retorts, and (5) the charcoal cools down in an atmosphere of hydrocarbons in the carbonising apparatus itself.

The shell charcoal obtained through this process is of the best quality attracting a high demand. The pyroligneous liquor, which is impure acetic acid, is a good

Table 13.22 Energy from coconut products

Components	Kg	KCal kg^{-1}	Energy KCal	% Total energy
Coconut oil	0.12	9000	1080	27.7
Carbohydrates and proteins	0.06	4000	225	5.7
Shell	0.18	5500	990	25.4
Husk	0.40	4000	1600	41.1
Total	0.76	22,500	3895	99.9

Source: Rethinam and Bosco (2006)

substitute for the coagulation of rubber latex, and the quality of smoked sheets was found to be on par with standard smoked sheets. The crude liquor may be further processed for the production of grey acetate of lime, which is also a marketable commodity. The carbolic acid and creosote present in the distilled tar may find useful applications. Destructive distillation will however be commercially viable only if sufficient quantity of shell could be centralised with minimum expense from the integrated processing areas.

13.14.1.4 Shell as an Energy Source

Coconut plantation has the advantage of producing energy besides food. The shell, husk and leaves have alluring prospects as energy source which needs to be exploited. Even if only 1% of the shell and husk is made use of, the energy produced is still in the order of 38.5 million litres of gasoline equivalent per day. 0.05 kg shell charcoal of 362 kcal and 0.086 kg of husk charcoal of 548 kcal can be obtained from 0.40 kg husk and 0.18 kg shell of one coconut, respectively. The energy obtained from the components of an average coconut is given in Table 13.22.

13.14.1.5 Other Uses of Coconut Shell

It is found feasible to use coconut shell in building construction in India (Balagopal and Menon 1987). In the constructions of load-bearing walls, the use of coconut shell cellular blocks made up of concrete masonry units (in which coconut shells were incorporated) was found to contribute to as much as 30% saving in material without loss in strength. The shells contribute substantially to the strength of the blocks by virtue of their inherent shell compressive strength. It was also found that the shell is capable of carrying a load of 100–400 kg at its top portion. The coconut shell cellular blocks could be produced by arranging the shells in the mould of a block making device prior to pouring the cement matrix (Cement sand-gravel of 1:3:6 by volume). After curing and drying for about 4 weeks, the blocks could be used in the construction of load-bearing walls. Adopting similar process, non-load-bearing partition walls having a thickness of 12 cm could also be produced. A single row of shells is accommodated in the 12-cm-wide block, whereas two interlocking rows of shells are accommodated in the 20 cm block.

13.15 Miscellaneous Products of Coconut

13.15.1 Coconut Wood

Coconut wood is currently a substitute to conventional wood. It is sold in lumber yards especially in coconut-producing areas. The timber can be used as materials for furniture items, novelties, wares and carvings. Since it is a highly dense material, it can be used in building components such as posts, trusses, doors, window frames, girders, etc. The top portion of the trunk is suitable for firewood, while the slabs and offcuts from saw milling can be converted into charcoal of good quality.

13.15.1.1 Preservation of Coconut Timber

Coconut wood is now available in plenty and will continue to be so for many years to come. Coconut wood belongs to the non-durable group of timber. When used in situations favourable to attack by decay fungi and wood-boring insects, the hard dermal portion of the trunk will last only for a period of 1–2 years. The soft inner portion will deteriorate in a few months when left exposed to the weather. Coconut wood should be properly treated to protect it against attack of wood-destroying organisms especially when used with ground contact and exposed to the weather. Attempts have been made to find the most suitable preservation technique and to select the best preservatives with respect to the various end uses of the treated wood. The methods tested for applying the preservatives are brushing or spraying, dipping, soaking/steeping, dip diffusion, double diffusion and hot and cold bath method. The wood preservatives used were copper chrome boron, copper chrome arsenate, zinc chloride + potassium dichromate, boric acid + borax, copper sulphate, potassium dichromate + arsenic pentoxide, cashew nut shell oil and creosote with bunker oil. The treated wood was tested for its shelf life by putting the wood for the following end uses: interior with ground contact, interior without ground contact, exposed with ground contact and exposed without ground contact. The following are the results obtained from the study. 1. Among the preserving methods dipping performed better than brushing. 2. For the end use exposed with ground contact, treatment with chemical creosote was found to be very effective. 3. All the treated as well as control samples were found to be intact for the end use interior without ground contact. Coconut wood is also found useful in the manufacture of particle boards. In a trial in the Tropical Products Institute, the timber was converted into flakes, dried to a moisture content of 5–8% and sprayed with a synthetic resin binder and a urea-formaldehyde/hardener mixture, at the rate of 0.8% by weight of the finished board. The sprayed articles were compressed at an elevated temperature and pressure for about 15 min in a hydraulic press to produce boards of 240 cm by 120 cm with 15, 22.5 or 30 cm thickness. The finished boards were found to be comparable to other quality products. Manufacture of pulp and paper from coconut stem is also a possibility.

13.16 Coconut Leaf

Coconut leaf is another product of importance for domestic use. The plaited and unplaited leaves are used for thatching houses, fencing and making baskets. The lifespan of coconut leaf is only 1–2 years. The leaves soaked in saline water before plaiting withstand climatic influences better than the unsoaked ones. The leaves being lignocellulosic in nature are affected by sunlight, rain and air and are susceptible to attack by fungi, insects, etc. The thick-walled sclerenchyma cells which impart mechanical strength to the tissues are relatively scarce in coconut leaf. Treatment with Bordeaux mixture (Pillai et al. 1983) is undertaken for increased longevity. Fibres can be extracted from fresh leaves. Here the fresh leaves are boiled in water and separated into upper and lower halves. Each half is made into strips of convenient width and again boiled in 5–8% sodium carbonate solution for 1–2 h. After thorough washing, they are immersed in a bleaching solution for 1–3 days with periodic stirring. Then they are washed and dried in shade. These strips, which form smooth, semi-transparent, waterproof threads, are excellent for making hats, bonnets, mats, bags and slippers.

The midribs of the leaves are used for stiff brooms, bird cages and lobster and fish traps. The petioles, bunch stalks, spathes, stipules, etc. are used as fuel. The roots have medicinal properties, and hence the decoction of the roots is used as mouthwash and gargle. The roasted roots can be used as a dentifrice.

13.17 Future Strategy

Coconut processing sector has to be strengthened by focusing attention on nontraditional products from coconut. Research undertaken in the past has generated viable technologies for the manufacture of diversified products from coconut. A few of them are still in their infancy requiring refinement in packaging and quality upgradation. There is an urgent need to undertake pilot-scale trials for commercialising some of the products such as coconut milk powder and coconut water-based beverages.

Various market promotional activities and consumer awareness campaigns are the need of the hour to strengthen the coconut-based economy. Strategic planning and implementation for free flow of proven technologies which have not reached the farmers/entrepreneurs; strong public-private-farmer partnership approach with needed technical support; strong market promotion through marketing networks; bilateral, regional and international collaborations with strong political will and commitment from government side; and determination of farmers for increased farm-level processing are essential to make the coconut industry competitive. Priority should be given for integrated processing of coconut, which is flexible enough to adjust to the market response.

References

Abdon IA (1967) Chemical and physical properties of the flour prepared from unpared dehydrated coconut (*Cocos nucifera Linn.*). Ph.D. thesis. University of Philippines, Diliman, Quezon City, Philippines

Akyurt M, Rehbini E, Bogis H, Aljinaidi AA (1839) A survey of mechanization efforts on date palm crown operations. World 2208:2645

Alias ZBHJ, Hashim AB (1982) A new coconut dehusking technique for Malaysian conditions. Paper presented in National Coconut Conference, 25–26 May, Kuala Lumpur, Malaysia

Annamalai SJK, Narayanaswamy KG, Patil RT (1989) Design and development of a natural convection type low cost dryer for medium holdings. J Plantn Crops 16:81–90

Anonymous (1950) Wealth of India, raw materials, vol 2. Council of Scientific and Industrial Research, New Delhi, pp 277–278

Anonymous (1984) Coconut cream, raw NDB no.12115. Nutrition data base, University of Minnesota, USA

Arida VP, Binlayo DL, Chua S (1979) Preparation of coconut monoethanol amide (CMA). Philippine J Cocon Stud 6(3):17–24

Arseculeratne SN, De Silva LM (1971) Aflatoxin contamination in coconut products. Ceylon J Med Sci 20:60–75

Arumugham C, Balachandran C, Sundaresan A (1993) Development of a process for coconut cream on commercial scale. J Food Sci Tech 30:408–412

Aslam Ali M, Bhaskar TD (1980) Utilization of coconut pith for the manufacture of oxalic acid. Coir 24(2):9–14

Baboo B, Bhat S (1980) Processing cocoa beans-some problems of the grower. Indian Arecanut Spices Cocoa J

Balagopal TS, Menon D (1987) Coconut shell and pith in building construction. Indian Cocon J 18(4):5–11

Balakrishnamurthi TS (1979) Correspondence with D.E. Peters cited in Tech. Data Handbook on the coconut. Philippine Coconut Authority, Philippines

Banzon JA (1978) Reconstitution of milk using coconut milk and non fat dry milk. Philippine J Cocon Stud 3:1–8

Banzon JA, Gonzalez ON, de Leon SY, Sanchez PC (1990) Coconut as food. Philippines Coconut Research Development Foundation, Philippines

Bhat JV, Nampoothiri AMD (1973) Effect of some of the environmental factors on quality of coir. Coir 17(2):23–25

Bhowmick BB, Debnath CR (1984) Coir fibre, part I. properties of coir fibre. Indian Cocon J 15(5):11–14

Birosel DM, Gonzales AL, Santos MP (1963) The nature and properties of the emulsifier system of oil globulins in coconut milk and cream. Philippine J Sci 92(1):1–15

Bosco SJD (1997) Optimization of modified atmosphere packaging parameters for coconut fresh kernel. Ph.D. thesis, Tamil Nadu Agricultural University, Coimbatore

Browning KC, Symons CT (1916) Coconut toddy in Ceylon. J Soc Chem Ind Lond 35:1138–1142

Capulso SA, Gonzales AL, Celestino VG (1981) Studies on the isolation and functional characteristics of protein from coconut skim milk. Philippine J Sci 110(1–2):25–32

Carren PA (1966) Fermentation of coconut husk by moulds. Sugar News 43:683–686

Chakraborthy P (1985) Functional properties of coconut protein isolate obtained by ultra filtration. J Food Sci Tech 22:248–254

Clarson D (1986) Coir dust for economic reclamation of saline-alkali soils and for better profit. Indian Cocon J 16(9):6–10

Coconis (1986) Dehuskers. Coconis Coconut Information Service 11:23–24

CPCRI (1996) Annual report. Kasaragod, India, pp 72–74

Dan TK (1992) Development of light weight building bricks using coconut pith. Res Indu 37(1):11–17

Davis TA (1961) Climbing the coconut. World Crops 13:486–470

Davis TA (1963) Ways of climbing the coconut palm. Cocon Bull 17:82–89
Davis TA (1964) Devising instruments for coconut research. In: Proceedings of FAO working party on coconut production, processing and marketing. Colombo, FAO, Rome, pp 70–77
Davis TA (1977) Attempts at mechanical climbing of palms with special reference to coconut palm. J Plantn Crops 5(1):1–35
De Castro MM, Cora Nera MC (1978) Comparative studies on the various methods of copra making and development of alternate technique. PCA-ARB devas research Centre, Annual Report, pp 7–8
Dendy DAV, Timmins WH (1973) Development of a wet coconut process designed to extract protein and oil from fresh coconut. Tropical Products Institute Report No. G78 pp 61
Devadas CT (1996) Densification and combustion studies on coir pith for utilization as a fuel briquette. Ph.D. thesis submitted to TNAU, Coimbatore, India
Fernandez WL (1978) The collection of tuba and lambanog toddy. Philippine J Cocon Stud 3(2):73–77
Ganesan V, Gothandapani L (1999) Microstructure of spray dried and partially defatted coconut skim milk powder. J Food Sci Tech 36(3):268–269
George J, Joshi HC (1961) Complete utilization of coconut husk –2. Hardboards from coconut fibre. Indian Pulp Paper 15(9):573–575
Ghosh DK (2015) Postharvest, product diversification and value addition in coconut. In: Value addition of horticultural crops: recent trends and future directions. Springer, New Delhi, pp 125–165
Ghosh DK, Hore JK, Sharangi AB (2000) Prospect of development of coconut based industry in West Bengal abstract of papers, VIIth W B State Science & Technology Congress, Bot-Agri 11
Gibbs HD (1911) The alcohol industry of the Philippine Island. Philippine J Sci 6:199–206
Gupta RC, Jain VK, Shankar G (1980) Studies on production of vinegar from 'neera'. Indian J Microbiol 20:119–121
Hagenmaier RD, Cater CM, Mattil KF (1975) Aqueous processing of coconuts: economic analysis. J American Oil Chem Soc 52(1):5–9
Hallexa CF, Seirra ZN (1976) Proximate analysis of the coconut endosperm in progressive stages of development. Philippine J Cocon Stud 1(2):37–44
Hebbar KB, Arivalagan M, Manikantan MR, Mathew AC, Thamban C, Thomas GV, Chowdappa P (2015) Coconut inflorescence sap and its value addition as sugar – collection techniques, yield, properties and market perspective. Curr Sci 109:1411. https://doi.org/10.18520/v109/i8/1411-1417
Hoover R, Laurentinus SF, Gunetileke KG (1973) Spoilage of coconut oil. Purification and properties of a fungal lipase that attacks coconut oil. J American Oil Chem Soc 50:64–67
Jack HW, Dennett JH (1925) The effect of tapping coconut palms for toddy on the copra and oil produced from subsequent fruiting. Malaysian Agri J 13:302–308
Jayaprakash Naik B, Suresh PR, Meera Manjusha, Balachandran TV, Madhu Subramonian, Balakrishnan PC (2013) Keraamritham- a healthy drink from coconut inflorescence sap. Indian Cocon J (Special issue) 9–10
Jayasankar NP (1966) Studies on aerobic microflora associated with retting of coconut husk. Ph.D. thesis. Indian Institute of Science, Bangalore, India
Jeganathan M (1970) Report of the officer-in-charge, Chemistry Division, Ceylon Coconut Research Institute. Ceylon Cocon Qtly 21(1/2):13–21
Kolhe KP (2010) Mechanized harvesting, the need of coconut growers in India. Indian Cocon J 73(2):15–19
Kolhe KP, Jadhav BB (2011) Testing and performance evaluation of tractor mounted hydraulic elevator for mango orchard. American J Engg Appl Sci 4(1):179–186
Kutty SK (1987) New products from coconut. In: processing and marketing of coconut in India. Coconut Development Board, Kochi, pp 70–76
Loganathan S, Lakshminarasimhan CR (1979) Physical properties of soil and yield of groundnut as influenced by application of amendments to soils. Coir 23(2):6–10
Madhavan K (1987) Design and development of copra moisture meter. J Plantn Crops 16:113–116

Madhavan K, Bosco SJD (1991) Coconut processing technology. Kisan World 11:19–21
Madhavan K, Bosco SJD (2002) Design of a forced convection solar cum electrical dryer. In: Sreedharan K et al (eds) Proceedings of the PLACROSYM, vol XV. Indian Society for Plantation Crops, Kasaragod, pp 690–696
Madhavan K, Bosco SJD (2004) Development of a solar tunnel dryer for plantation crops. J Plantn Crops 32:428
Madhavan K, Patil RT, George S (1998) Design of an electrical dryer for plantation crops. In: Proceedings of the XI Kerala science congress, India, pp 451–455
Marina AM, Che Mana YB, Amin I (2009) Virgin coconut oil: emerging functional food oil. Trends Food Sci Technol 20:481–487
Mathes DT, Marikkar JMN (2010) August) The impact of harvesting coconuts at monthly intervals. Cocos 16:56
Mathew AC, Madhavan KA, Arumuganathan T (2014) Development and performance evaluation of virgin coconut oil cooker. Agric Mech Asia Africa Latin America 45(1):56–59
Monera OD, Del Rosario E (1982) Physico-chemical evaluation of the natural stability of coconut milk emulsion. Ann Trop Res: Tech J Visayas State College of Agriculture
Mukherji KG (1967) *Aspergillus Niger* growing in coconut oil. J Gen Appl Microbiol 13:407–408
Muralidharan K, Jayashree A (2011) Value addition, product diversification and by-product utilization in coconut. Indian Cocon J 7:4–10
Nagarajan R, Rabindranath DA, Prabhu GN (1987) Studies on seeding of ret microflora in coconut husks. Coir 31(2):7–12
Nair RR (1984) Factors affecting the quality of copra, In: Proceedings of seminar on coconut processing and utilization, Trivandrum, India, pp 72–84
Nair MKC, Sreemulanathan H (1970) Spoilage of copra by *Penicillium frequentans*. Westling. Agric Res J Kerala 8:125
Nalinakumari T (1989) Studies on the extent of damage caused by pests of stored copra and control of important pests. Ph.D. thesis submitted to Kerala Agricultural University, Thrissur p 171
Narayanamurthy D, Singh T (1954) Moulding powders and moulds from coconut shells. Composite Wood 1:38–40
Nathanael WRN (1954) Proximate composition of coconut milk. Ceylon Cocon Qtly 5:108
Nathanael WRN (1960) Chemical and technological investigations on coconut products. Ceylon Cocon Qtly 11:31–54
Nathanael WRN (1963) Coconut toddy as an industrial raw material. Cocon Bull 18(5):145–151
Nathanael WRN (1964) Coconut products as industrial raw material. Cocon Bull 19(5):163–181
Nathanael WRN (1965) Some aspects of copra deterioration. Ceylon Cocon Qtly 16:111–120
NIIR Board of Consultants and Engineers (2012) The complete book on Coconut & Coconut Products (cultivation and processing). Asia Pacific Business Press Inc, New Delhi, p 293
Nijaguna BT (1988) Coconut dehusker. J Food Engg 8:287–301
Palaniswami A, Valliappan T, Neelakantan S (1989) Effect of fungal contamination of copra on the quality of coconut oil. Indian Cocon J 19(11):10–13
Patil RT (1982) Use of propionic acid for preservation of coconut kernel. In: George KV et al (eds) Proceedings of PLACROSYM III. Indian Society for Plantation Crops, Kasaragod, India, pp 173–179
Patil RT (1984a) Design and development of solar crop dryers. Agric Mech Asia Africa Latin America 15(2):59–62
Patil RT (1984b) Design and development of small holders copra dryer using agricultural waste as source of energy. In: Bavappa KVA et al (eds) Proceedings of PLACROSYM, vol V. Indian Society of Plantation Crops, Kasaragod, India, pp 324–334
Patil RT, Nambiar CKB (1982) Sundrying of copra. J Plantn Crops 10(1):26–32
Paul PG (1969) Studies on the microbial deterioration of copra and the utilization of certain vegetable oil by fungi. M.Sc. thesis, University of Kerala, Trivandrum
Pillai CKS, Venkitaswamy MA, Satyanarayana KG, Rohatgi PK (1983) Preserving coconut leaf thatch; a simple method. Indian Cocon J 14(8):3–6
Pomier M (1984) The portable husking spike. Oleagineux 39(6):321–323

Potty VP, Joseph KV, Menon KS, Jayasankar NP (1978) Processing and fermentation of coconut toddy. J Food Sci Tech 15(1):32–34

Quimio TH (1984) Coconut water as a growth medium for edible and plant pathogenic fungi. A research report of Department of Plant Pathology available at University Library, University of the Philippines at Los Banos, College, Laguna

Ramaswamy PP, Kothandaraman GV (1985) Role of coir pith on yield and uptake of rice in sodic soil. Paper presented in workshop on coir research held in Coir Board, Coir house, Kochi, Kerala, India

Rao KSN, Sreekanita KR, Ramachandra R (1971) Post harvest infection of coconut kernel by *Botryodiplodia theobromae* and a note on the hydrolytic enzymes secreted by the fungus. Indian Phytopathol 24(4):815–819

Rethinam P, Bosco SJD (2006) Coconut-harvest, processing and products. Asian and Pacific Coconut Community, Jakarta, Indonesia, p 206

Rnam (1990) Caridi-developed Dehusker. Rnam Newsletter 38:21

Samarajewa W (1972) Aflatoxins in coconut products. Ceylon Cocon Qtly 23:108–113

Samson AS, Khaund RN, Cater CM, Mattil XF (1971) Extractability of coconut proteins. J Food Sci 36:725

Sanchez PC, Rasco PM (1983) Utilization of coconut in white soft cheese production. Res Los Banos 2(2):13–16

Sanchez PC, Rasco PM (1984) Coconut milk (Gata) in yoghurt manufacture. Philippine J Cocon Stud 8:42–50

Sandhya VG, Rajamohan T (2008) Comparative evaluation of the hypolipidemic effects of coconut water and lovastatin in rats fed fat–cholesterol enriched diet. Food Chem Toxicol 46:3586–3592

Sangamithra A, Swamy GJ, Sorna PR, Chandrasekar V, Sasikala S, Hasker E (2013) Coconut: an extensive review on value added products. Indian Food Ind Mag 32(6):29–36

Schindler AE, Eisenberg WV (1968) Growth and production of aflatoxin by *A. flavus* on red pepper. J Assoc Off Anal Chem 51:911–912

Semana JA, Las Marias VB, Ballon CH (1988) Fibre board from coconut coir. Indian Cocon J 19(8):10–17

Seow CC, Gwee CN (1997) Coconut milk: chemistry and technology. Int J Food Sci Tech 32(3):189–201

Shabana HR, Mohamad S (1993) Mechanisation of date production. In: Proceedings of Ist Symposium on the Date Palm, KFU, Al-Hassa, pp 712–714

Sial FS (1984) Mechanisation of Saudi Arabian date industry, scope and limitation. Int J Dev Technol 2(1):317–325

Sierra ZN (1971) Proximate chemical composition of copra as influenced by fungal infection. M.S. thesis, UPLB College, Laguna, Philippines

Singh D, Kushwaha HL, Mathew AC, Manoj Kumar TS, Hebbar KB, Madhavan K, Thomas GV, Bhargav VK (2012) Performance evaluation of paddle type and push up type coconut climbing devices. J Plantn Crops 40(3):1–3

Singh J (1981) Coconut husking machine. Invent Intell 16(2):80–81

Singh TV, Udhayakumar R (2006) Development of a coconut de-shelling machine. Trop Agric Res 18:20

Smith ME, Bull AT (1976) Studies of the utilization of coconut waste for the production of the food yeast *Saccharomyces fragilis*. J Appl Bacteriol 41(1):81–95

Southern PJ (1957) Sulphur deficiency in coconuts, a widespread condition in Papua and New Guinea. II. The effect of Sulphur deficiency in copra quality. Papua New Guinea Agric J 19(1):38–44

Sreemulanathan H, Krishnamurthy S, Jainamma KM, Krishna Swamy C, Mathew AG, Subramanyan V (1979) Control of spoilage during sun drying of coconut. J Food Sci Technol 16(5):95–97

Subramanyan V (1966) Copra process gives better results. Cocon News 11(5):1–2

Susamma P, Menon MR (1981) Microorganisms associated with the deterioration of copra. Madras Agric J 68(10):686–687

Susamma P, Menon MR, Alice A (1981) Quantitative changes in the amino acids of copra due to fungal infection. J Food Sci Technol 18:254–255

Swamy ASK, Patil RF (1975) A simple device to help in climbing coconut palms. Curr Res 4:93–96

Thamban C, Mathew AC, Muralidharan K, Subramanian P, Singh VT, Madhavan K (2011) Coconut climbing methods and devices: a participatory analysis of constraints and strategies. J Plantn Crops 39:45–47

Thampan PK (1987) Handbook on coconut palm. Oxford & IBH Publishing Co. Ltd, New Delhi, p 311

Thiagarajan B (1987) The processing of fresh coconuts. Indian Cocon J 18(2&3):12–20

Timmins WH, Kramer EC (1977) The canning of coconut cream. Philippine J Cocon Studies 2(4):15–25

Totheringhan RH (1982) Process for making a building board (patent). ABIPC 52(10):12–21

Tribold HO, Aurand L (1963) Food composition and analysis. D. Van Nostrand Company Inc, London, pp 132–133

Unson CG (1966) The effect of some food preservatives on coconut toddy. Philippine J Cocon Stud 9(1):51–54

Uyenco FR, Ochoa JAK (1984) Microbial degradation of coconut coir dust for biomass production. Philippine J Cocon Stud 9(1):51–54

Varadaraju N, Gothandapani L (1998) Design and development of equipment for pelleting decomposed coir pith. Agric Mech Asia Africa Latin America 29(2):33–34

Varghese A, Jacob J (2014) A study on the KAU coconut husking tool. Annual International Conference on Emerging Research Areas: Magnetics, Machines and Drives. https://doi.org/10.1109/AICERA.2014.6908254

Varghese EJ, Thomas PK (1952) Copra kiln for small coconut plantations. Indian Cocon J 6(1):19–28

Vidhan Singh T, Swamy KGN, George SA (1999) Design and development of smoke free collapsible copra dryer for medium holdings. J Plantn Crops 27(2):112–120

Viswanathan R (1998) Coconut husk made products. In: Bosco SJD et al (eds) Harvest and post harvest technology of plantation crops. CPCRI, Kasaragod, pp 39–53

Viswanathan R, Gothandapani L (1998) Mechanical properties of coir pith particle boards. Bioresour Technol 67(1):93–94

Dr. M. R. Manikantan is a Principal Scientist in Agricultural Engineering at ICAR-CPCRI, Kasaragod, Kerala, India. He has published 64 research papers, 18 popular articles as well as 26 technical bulletins/manuals and presented 70 research papers in conferences. He has also edited 3 books and contributed 11 book chapters. He is the recipient of the Jawaharlal Nehru Award for postgraduate research and Distinguished Service Certificate from ISAE. manicpcri@gmail.com

Dr. R. Pandiselvam is a Scientist of Agricultural Process Engineering at the ICAR-Central Plantation Crops Research Institute, Kasaragod, Kerala, India. He has over 30 research papers, 5 books and 10 book chapters to his credit. anbupandi1989@yahoo.co.in

Dr. Shameena Beegum is a Scientist at the ICAR-Central Plantation Crops Research Institute, Kasaragod, Kerala, India. Her specialization is in post harvest technology of horticultural crops. She is a DST-Inspire awardee. She has authored 5 research papers, 10 popular articles and 4 book chapters. shameena.pht@gmail.com

Dr. A. C. Mathew Principal Scientist, ICAR-CPCRI, Kasaragod, Kerala, India, has 28 years of professional experience in agricultural engineering. He has developed 24 machineries/gadgets for coconut and arecanut processing and has three patents to his credit. He is the recipient of one international and two national awards. acmathew@yahoo.com

Chapter 14
Technology Transfer in Coconut-Global Scenario and Strategies

S. Arulraj, C. Thamban, and D. Jaganathan

Abstract The chapter on 'Technology Transfer in Coconut – Global Scenario and Strategies' covers the field-level experiences of technology transfer initiatives in major coconut-growing countries of the world. Components in technology transfer network, viz. technology generation system, extension system, support system and utilization system in India and other major coconut-growing countries including technology integration for inclusive development, are elaborately discussed. The chapter outlines multinational research organizations with coconut as their mandate crop, coconut research institutions in major coconut-growing countries and improved technologies on coconut varieties and hybrids, production system management and processing in coconut. Details of extension system comprising of extension organizations; innovative extension approaches using community-based organisations, participatory technology transfer module and promotion of self-help groups; farmer-participatory community extension; and farm field school are dealt along with case studies on successful extension methodologies and ICT initiatives in several coconut-growing countries. Adoption of improved technologies by farmers and constraints in adoption are also highlighted. Strategies for strengthening the transfer of technology programmes in coconut are suggested based on field-level experiences of different stakeholders. A rich documentation of literature supports the scientific analyses and the policy implications in each of the sections of this chapter, where different components of technology transfer intersect fruitfully, both at the national and global level.

S. Arulraj
ICAR-IIOPR, ICAR- CPCRI, Krishnagiri, Tamil Nadu, India
e-mail: aruldopr@gmail.com

C. Thamban (✉)
ICAR-CPCRI, Kasaragod, Kerala, India
e-mail: c.thamban@gmail.com

D. Jaganathan
ICAR-CTCRI, Thiruvananthapuram, Kerala, India
e-mail: djaganathtn@gmail.com

K. U. K. Nampoothiri et al. (eds.), *The Coconut Palm (Cocos nucifera L.) - Research and Development Perspectives*,
https://doi.org/10.1007/978-981-13-2754-4_14

14.1 Introduction

Coconut, one of the most important of all cultivated palms, provides livelihood security to several millions of people across the world. Through the systematic research conducted during the last century, a substantial number of viable technologies related to crop improvement, production, protection and post harvest processing have been evolved for enhancing coconut productivity and profitability. However, farmers are not able to exploit the production potential from these technologies to the extent desirable. Adoption of the recommended practices plays a crucial role in improving coconut productivity. Various organizations involved in the research and development of coconut are streamlining their activities to enable the farmers to make use of the technologies for enhancing the production and productivity.

Technology is defined as application of scientific knowledge for solving problems in a particular field. The term 'technology transfer' can be defined as the process of movement of technology from one entity to another (Souder et al. 1990; Ramanathan 1994). The transfer may be said to be successful if the receiving entity, the transferee, can effectively utilize the technology transferred and eventually assimilate it. The movement may involve physical assets, know-how and technical knowledge (Bozeman 2000). In a very restrictive sense, where technology is considered as information, technology transfer is sometimes defined as the application of information into use (Gibson and Rogers 1994).

The key elements in the technology generation, transfer, support and utilization system in coconut in India are presented as follows:

Technology generation system	Extension system	Support system
International Research Institutes	Front-line extension of research institutes	Input agencies
National Research Institutes	Ministry of Agriculture	Nurseries
Agricultural Universities	Coconut Development Board	Non-governmental organizations
Coconut Development Board	Department of Agriculture/horticulture private agencies	Farmers' organizations
	Non-governmental organizations	Marketing departments
		Financial institutions
Utilization system Coconut farmers, entrepreneurs and policy makers		

14.2 Technology Generation System

Technologies are generated mainly at the research institutes which are spread throughout the coconut-growing countries. Major research institutes which are working on coconut are indicated below.

14.2.1 Multinational Research Organizations

The three main agencies involved in the development of coconut industry, at global level, are Asian and Pacific Coconut Community (APCC), International Coconut Genetic Resources Network (COGENT) and Bureau for the Development of Research on Tropical Perennial Oil Crops (BUROTROP).

14.2.1.1 Asian and Pacific Coconut Community (APCC)

APCC is an intergovernmental organization of coconut-producing countries organized in 1969 under the aegis of the United Nations Economic and Social Commission for Asia and the Pacific (UN-ESCAP). The objectives of the Asian and Pacific Coconut Community are to promote, coordinate and harmonize all activities of the coconut industry which sustains the lives of millions of small farmers as well as those engaged in production, processing and marketing of coconut products.

14.2.1.2 International Coconut Genetic Resources Network (COGENT)

With the endorsement of the CGIAR and its donors, the International Plant Genetic Resources Institute (IPGRI) established the International Coconut Genetic Resources Network (COGENT) in 1992 to promote an international collaborative programme on coconut genetic resources conservation and use.

14.2.1.3 Bureau for the Development of Research on Tropical Perennial Oil Crops (BUROTROP)

BUROTROP is a non-profit-making association registered in France in January 1995. Its mandate is to assist, strengthen and further develop research on tropical perennial oil crops. At present, BUROTROP is non-functional. For details on international organizations, please see Chap. 2.

14.2.2 WTO and Its Impact on Coconut Research Priorities

With globalization, the trade and agriculture are governed by new rules with opportunities and threats which deal with liberalization in market access, export competition and domestic support. India needs improvement in productivity of coconut and its quality through proper policies, technologies, training and development of proper institutions. Exploration and development of value-added products and marketing is another area of concern (Naik 2005). Competitiveness would be determined by the competitive advantages in terms of climate suitability, management strategies and infrastructure. Sanitary and phytosanitary measures would become a barrier, if they are not harmonized and conditions are not made favourable to face the challenges. Value addition to each and every part of coconut palm is a way to make coconut cultivation profitable. In this era of free trade, comparative low cost of production and high-quality standards are important aspects to be competitive in the world market. To cut down cost, there is a need to emphasize on adoption of cost-effective inputs like cultural practices, recycling of farm biomass and free play of a host of biocontrol agents. There is also a need to synthesise traditional wisdom with modern scientific technologies. Coconut-based economy can expect a revival from the negative impact of liberalized imports only when the profitability of coconut farming is delinked from the price behaviour of coconut oil. This is possible through full utilization of the land under coconut and also the various products at the on-farm and community levels. As coconut farming has close linkage with other aspects of rural life, coconut is not to be treated in isolation but as a component of integrated rural development, and research strategies are to be modified and strengthened accordingly.

14.2.3 National Level Coconut Research Institutes

ICAR-central plantation crops research institute, state agriculture universities, central food technology research institute (CFTRI) and regional research laboratories under CSIR, etc., India.

Coconut research board-coconut research institute (CRI), Sri Lanka.

Indonesian Centre for Estate Crops Research & development, Bogor, Indonesia.

Coconut for agro-based industry (CABI), Bogor, Indonesia.

Philippine coconut authority (PCA), Davao research Centre, University of Los Banos, Philippines.

Horticulture research institute (HRI), Chumpon research Centre, Thailand.

Malaysia agriculture Research and Development Institute (MARDI), Kuala Lumpur, Malaysia.

Planning and science research department, oil Plant Institute of Vietnam, Ho Chi Minh City, Vietnam.

Chinese academy of tropical agriculture sciences (CRICATAS), Wenchang City, Hainan China and Wenchang coconut research institute, Hainan.
Centre de cooperation Internationale en Recherche Agronomique pour le development (CIRAD), Montpellier, France.
Centro de Investigacion Cientifica de Yucatan (CICY), Merida, Mexico.
Coconut Research Station, Marc Delorme, port-Bouet, Côte d'Ivoire.
Nigerian Institute for oil Palm Research, Benin City, Nigeria.
Coconut industry board, Kinston, Jamaica.
Mikocheni agricultural research institute, Dar Es Salaam, Tanzania.
Coconut Programme, OPRI, Sekondi, Ghana.
CRSC, Seme Podji, Benin.
EMBRAPA, Aracaju, Betume, Brazil.
BARI, Gazipur, Bangladesh.
RS, Islamabad, Pakistan.
Cocoa coconut research institute, Rabaul and Stewart Research Station, Madang, Papua New Guinea.
Ministry of Agriculture, Nuku'alofa, Tonga.
Saraoutou Research Station, Santo, Vanuatu.
Taveuni coconut research Centre, Fiji.
Olomanu coconut seed garden, RS, Samoa.
Research Station, Yandina, Solomon Islands.

These research institutes have evolved many varieties and hybrids and their cultivation technologies as well as developed technologies for product diversification, value addition and by-product utilization.

14.2.3.1 Coconut Research in India

Important organizations conducting research on coconut in India include ICAR-Central Plantation Crops Research Institute (CPCRI) under the Indian Council of Agricultural Research (ICAR) and the State Agricultural Universities.

ICAR-Central Plantation Crops Research Institute (ICAR-CPCRI) is the pioneering research organization in India conducting research on different aspects of coconut cultivation. CPCRI also coordinates research on coconut within the country and executes research programmes through the All India Coordinated Research Project on Palms. Technologies which are generated from ICAR-CPCRI are briefed here:

Crop Improvement Coconut research efforts in India, over a century, have recorded impressive achievements in terms of developing varieties and hybrids, which have the potential to give two-to threefold increase in yield. ICAR-CPCRI and other coconut research organizations in India have released a number of improved varieties and hybrids, capable of producing 2.79–6.28 tonnes of copra ha^{-1} year^{-1} in different parts of the country. Varieties such as Chowghat Orange Dwarf, Kalpa Sree, Kalpa Jyothi, Kalpa Surya and Kalpa Raksha were released for tender nut purpose.

Dual-purpose varieties such as Chandra Kalpa, Kera Chandra, Kalpa Dhenu, Kalpa Prathibha, Kalpa Mitra, Kalpatharu, Kalpa Haritha, Kera Keralam and Kalpa Shatabdi were released for different agroclimatic conditions. Six hybrids, viz. Chandra Sankara, Kera Sankara, Chandra Laksha, Kalpa Sankara, Kalpa Samrudhi and Kalpa Sreshta, were released which are capable of giving high yield and copra out turn. For further details, please refer to Chap. 4.

Crop Production The general recommended dose of fertilizer is 500:320:1200 g of N, P_2O_5 and K_2O palm^{-1} year^{-1}, to be applied in two split doses along with organic manures. Water requirement was worked out as 200 l palm^{-1} with an irrigation frequency of once in 5 days. Drip irrigation and fertigation practices have been standardised. Vermicomposting of coconut biomass using two species of earthworms, *Eudrilus eugeniae* and *Eisenia foetida*, was proved as an efficient method of composting. For maximizing economic returns, high-value medicinal and aromatic crops and flowering plants have been recommended as intercrops in the palm-based cropping system. A number of vegetables (snake gourd, ridge gourd, bottle gourd, amaranthus, coccinia, brinjal and bitter gourd), tuber crops (colocasia and greater yam), spices (ginger and turmeric), etc. are compatible for intercropping in coconut gardens. Various perennials like cocoa, clove, nutmeg, coffee, black pepper, mulberry, jack, bread fruit, mango, sapota, papaya and timber yielding trees were found to be suitable mixed crops in coconut garden. Mixed farming in coconut with various subsidiary enterprises such as dairy, poultry and sericulture integrating nutrient recycling with coconut cultivation was found to be quite advantageous. For further details on crop production technologies, please see Chap. 7.

Crop Protection Integrated pest management practices for insect pests, viz. eriophyid mite, rhinoceros beetle, red palm weevil, leaf-eating caterpillar, white grub, coreid bug, slug caterpillars, mealy bugs, scale insects and termite, have been standardised. Efficient disease management practices for diseases, viz. bud rot, leaf rot, stem bleeding, Thanjavur wilt/*Ganoderma*, leaf blight or grey leaf spot and root (wilt), have been recommended. Judicious cultural, biological, mechanical and chemical methods are emphasized in managing weeds, pests and diseases. For details, please see Chaps. 10, 11 and 12.

Post harvest Processing and Mechanization Agricultural implements/gadgets such as power-operated sprayer, copra dryers using different energy sources and capacities, coconut splitting device, coconut deshelling machine, coconut grating machine, tender coconut cutter, tender nut punching machine, coconut chips slicing machine and a simple and safe coconut palm climbing device have been developed by ICAR-CPCRI. Technologies for making value-added products like snowball tender nut, coconut chips with various flavours, virgin coconut oil, kalparasa, coconut sugar, coconut charcoal, etc. have been developed and are being promoted vigorously among prospective entrepreneurs. Details on these technologies are described in Chap. 13.

All India Coordinated Research Project on Palms In 1970, the ICAR initiated the All India Coordinated Research Project (AICRP) on Palms with its headquarters at CPCRI, Kasaragod. The project provides adaptive research support for coconut through collection, conservation, cataloguing and evaluation of germplasm, evaluation of new hybrids and high-yielding varieties of coconut, standardization of agro-techniques for various agroclimatic regions including development of appropriate farming systems as well as efficient pest and disease management strategies especially for pests like leaf-eating caterpillar and rhinoceros beetle and diseases like Tatipaka and *Ganoderma*/Thanjavur wilt. As in 2018, 15 centres are conducting research on coconut under AICRP on Palms.

State Agricultural/Horticultural Universities State agricultural/horticultural universities which are in the major coconut-growing regions undertake research programmes related to coconut through their academic programmes and also through AICRP on Palms. Results of the studies are regularly incorporated in the location-specific package of practices recommendations for coconut in the respective states. The state universities are implementing a large number of transfer of technology programmes through Directorate of Extension Education, Krishi Vigyan Kendra, Farm Clinic, Agricultural Technology Information Centre, etc.

Research Role of Coconut Development Board (CDB) In order to promote product diversification and by-product utilization of coconut and to create a vibrant coconut-based economy, a Technology Development Centre has been established at the headquarters of the Coconut Development Board. Technology has been developed for the production of nata de coco, coconut cream, spray dried coconut powder and coconut water-based vinegar, process for the preservation and packaging tender coconut water, etc., with the collaboration of CSIR (RRL), DFRL and CFTRI. Transfer of technology and other development programmes undertaken by CDB are described under Sect. 14.5.1.

14.3 Developmental Programmes for Coconut

The main developmental strategies for sustainability of the sector adopted by coconut-growing countries are coconut planting/replanting/rehabilitation, area expansion, technology adoption relating to product diversification, integrated farming system, provision of quality planting materials, institutional credit to farmers, skill training, information dissemination, organization of coconut farmers, market research and development, market promotion, product quality improvement, strengthening of coconut development organizations and extension system, reduction on dependence of copra as the major coconut product and introduction of coconut-based cottage industries, copra price support and strengthening of downstream processing sector.

14.3.1 Role of APCC

The Secretariat of APCC functions as the regional centre for collection, analysis and dissemination of coconut information, realizing the importance of sharing information in the development of the industry. Information is disseminated in various forms through journals (CORD, COCOMUNITY and Cocoinfo International), proceedings of meetings, country studies, technological sheets, directories, statistical yearbooks and video documentaries. APCC, from its inception, has promoted programmes for product and market diversification. Studies have been conducted to identify constraints and potential for expansion of markets for coconut products. APCC with the inputs provided by an expert group of food scientists drew up a set of uniform quality standards for aqueous coconut products which could serve as a basis for the countries to follow. APCC also frequently arranges training programmes to suit the varying needs of the member countries under the technology transfer exercise.

14.3.2 Coconut Development in India

Current Indian coconut production scenario presents a highly encouraging picture, with India reaching the top in coconut production in the world. India produced 23, 904 million nuts in the year 2016–17 from an area of 2.082 million ha. Bulk of the coconut production in the country comes from the Western plains and the Ghat region, followed by the Eastern Coastal Plains and hill regions. The islands of Andaman as well as Nicobar and Lakshadweep are the other traditional coconut-growing areas. Coconut cultivation has made progress, in recent years, in certain nontraditional areas also.

Coconut cultivation in the country is mainly in the hands of small and marginal farmers with more than 90% of the coconut holdings being less than 0.40 ha in size. Average productivity in India is 11,481 nuts ha^{-1}, but several coconut plantations in Tamil Nadu and Andhra Pradesh, which are well managed, have a higher productivity ranging from 13,000 to 15,000 nuts ha^{-1}. Thus, the challenge is to improve productivity which would in turn improve the economic returns to the farmers.

14.3.3 Yield Gap

Efforts taken by research institutes and extension agencies in the areas of technology generation and dissemination have immensely contributed to production and productivity of coconut. However, the productivity level remained low in India compared to potential yield. National average of coconut productivity is 11,481 nuts ha^{-1} $year^{-1}$, while that of the best-managed garden is 27,300 nuts ha^{-1} $year^{-1}$. A yield of 29,225 nuts ha^{-1} $year^{-1}$ has been obtained at ICAR-CPCRI. Globally, the estimated coconut technology gap in terms of nut and copra yield ranges from 33 to 84% (Batugal and Bourdeix 2005). Lack of adoption of scientific cultivation practices is one of the important reasons for such a low productivity. A comparison between the best-managed gardens and national average of productivity of coconut crop will reveal that there still exists a wide gap between the technologies generated and their utilization by the growers in coconut crop sector, especially in small holdings (Rajagopal et al. 2004a). Low level of technology utilization at farmers' fields calls for formulating effective extension strategies suitable to the heterogeneous farming situations.

14.3.4 *Technology-Wise Adoption*

14.3.4.1 Varieties, Nursery Management and Planting

Knowledge and adoption of nursery practices and correct method of planting in main field were found to be on a higher side, while the extent of cultivation of recommended varieties and hybrids was on the lower side. Bastine et al. (1991) found that only 6.5% of the farmers of Kannur district, Kerala, India, adopted coconut hybrid planting. Internationally 400 hybrids have been released for cultivation. But only 1–14% of farmers adopt planting of hybrids due to various reasons.

Yogananda et al. (1994) found that 50% of large holders and 61.67% of small farmers were not aware of the importance of copra content in selection of seed nut. But majority of them (more than 75%) had knowledge on all aspects of nursery management and seedling selection. Thampan (1999) opined that 95% of coconut farmers adopted planting of West Coast Tall (WCT). Ohler and Griffee (1999) stated that only 6% of the total coconut area was covered with hybrids.

The use of hybrid seedlings for new planting was the lowest adopted recommendation, whereas spacing and filling pits with top soil were highly adopted (Jnanadevan and Prakash 1994). Kalavathi and Anithakumari (1998) reported high level of knowledge and adoption of planting time, depth of planting, pit size for planting as well as seedling and seed nut selection with 80–96% knowledge and 62–96% adoption. However, adoption level of mother palm selection in root (wilt) disease-affected areas was 33%, while knowledge level was 90%. Similar was the trend for spacing (knowledge 65% and adoption 22%). Anithakumari and Kalavathi (2001) in their study in root (wilt) disease-affected areas reported high level of knowledge and adoption of nursery management and main field planting. Majority of farmers used either open well (62%) or bore well (32.7%) as the source of water for drip irrigation (Mathew and Thamban 2011).

14.3.4.2 Production System Management in Coconut

One of the major constraints identified for low productivity in coconut is the inadequate and irregular nutrient management. Studies reviewed indicated low to medium level of adoption of practices like organic manure application, chemical fertilizer application, green manuring, etc. (Thamban and Venugopalan 2002). Bastine et al. (1991) from their study reported low level in adoption of practices like split dose of fertilizer application (2.96%), cultivation of green manure and cover crops (2.96%), irrigation (16.3%), husk burial (nil) and fertilizer and manure application (25–30%).

Thampan (1999) opined that the farmers applied very small quantity of chemical fertilizers. Yogananda et al. (1994) found that more than 80% of the farmers lacked knowledge on farmyard manure application and quantity of chemical fertilizers to be applied. Jnanadevan and Prakash (1994) reported none of the farmers adopted application of fertilizers as per recommendation. However, 40–70% of the partner farmers of CDB programmes adopted recommended fertilizer dosage.

Kalavathi and Anithakumari (1998) reported 50% knowledge and 20% adoption regarding irrigation to coconut, medium level of knowledge (55%) and adoption (25%) for fertilizer application to adult palms, application of lime and salt (45% knowledge and 30% adoption) and mulching (35%). But fertilizer application to seedlings was found to be very low, i.e. 5%. The respondents were neither aware nor adopted the root (wilt) disease management technologies as a package. Level of knowledge (10%) and adoption (10%) of application of magnesium sulphate were low, and modified/improper dosage was adopted by 50% of the farmers. Anithakumari and Kalavathi (2001) reported low level of knowledge for nutrient management (20%) and moisture conservation (10%) with very few farmers adopting recommended nutrient management (12%) and moisture conservation.

Intercrops like cocoa, nutmeg, banana, black pepper, tuber crops, vegetables, fodder crops and lemon are grown by farmers in certain states in India depending on the agroclimatic and socio-economic factors. Mulching is generally practised using residues of coconut, banana, cocoa, etc. to conserve soil and water. Green manure crops like sunnhemp (*Crotalaria juncea*) and Kolinji (*Tephrosia purpurea*); green leaf manure crops like glyricidia, neem leaves, *Calotropis*, etc.; and leguminous crops like *Stylosanthes gracilis*, *Calopogonium mucunoides* and *Vigna unguiculata* are used as cover crops to prevent soil erosion and for enriching the soil fertility (Kalathiya et al. 2007; Jaganathan et al. 2013).

14.3.4.3 Plant Protection in Coconut

Lack of knowledge and practical difficulties resulted in low level of adoption of plant protection technologies as reported by Bastine et al. (1991), Yogananda et al. (1994) and Thampan (1999).

14.3.4.4 Postharvest Technologies

Knowledge and adoption level of palm-climbing device, copra dryer and copra moisture metre were found to be very low (Anithakumari and Kalavathi 2001).

14.3.4.5 Constraints in Improving Coconut Productivity

Many research workers (de Silva 1989; Thampan 1999; Anderson 2007; Kumar and Kapoor 2010; Pathiraja et al. 2010; Anithakumari et al. 2011, 2012; Kalavathi and Anithakumari 2011; Thamban et al. 2011; Anithakumari 2013; Jaganathan et al. 2013; Kerure et al. 2016) have analysed the constraints in adoption of new technologies, and a summary of their findings is presented here:

1. Nonavailability of root (wilt) disease-tolerant seedlings.
2. Lack of expertise on early identification of red palm weevil incidence.
3. Inadequate information base about technologies, limited ability of small enterprises to assess and acquire technology, inadequate technology acquisition, transfer and development to stimulate local firms to use indigenous technology as well as lack of targeted approach to technology development and technology transfer systems.
4. Inadequate investment in R&D.
5. Improper use of land, unscientific use of chemicals, micronutrient deficiencies.
6. Low level of investment due to increasing cost of cultivation, declining trend in yields and decrease in price of coconuts, affecting the profits.
7. Low level of awareness about the bio-intensive practices.
8. Small size of holdings restricting value addition of farmers' own produce.
9. High competition and low skills in marketing techniques, unorganized markets and lack of farmers' cooperatives for marketing.
10. Poor infrastructural facilities for storage and processing.
11. Labour problems (high-wage rates, scarcity of skilled workers and difficulty in community mobilization of labour).
12. High cost for transporting organic inputs to the farm due to the decline in livestock components in coconut-based farming system, lack of standard package of practices, nonavailability of biocontrol agents, biopesticides and organic manures, inadequacy of local certification agencies and lack of specialized markets for organic products.
13. Inadequate government support especially for initial operational funds and subsidies on electricity charges for small-scale units.

14.4 Transfer of Technology Programmes

Timely and sustainable transfer of technologies and extent of adoption of the recommended practices play a critical role in improving the productivity of the crop. It has been demonstrated that a fourfold increase in yield could be achieved by adopting proper cultivation practices as compared to the poor managed palms. Taking into cognizance the problems faced by farmers and the need to improve the coconut production scenario in the country, various development programmes are implemented by different agencies. In India, at national level, the front-line extension programmes of ICAR-CPCRI and programmes of the Ministry of Agriculture and Farmers' Welfare through the Coconut Development Board and Directorate of Extension and, at the state level, agriculture/horticulture departments, private agencies and NGOs play a critical role in the implementation of various transfer of technology programmes for coconut development.

14.4.1 Front-Line Extension Programmes of ICAR–CPCRI

Important front-line transfer of technology activities in coconut implemented by ICAR-CPCRI include training programmes; front-line demonstrations; information communication through mass media like radio, television, newspapers and farm magazines, extension pamphlets, CD ROMs, video cassettes, etc.; exhibitions; seminars; Krishi Melas; and group meetings, providing consultancy through field visits and replying postal queries (Fig. 14.1).

14.4.2 Conventional Extension Approaches

Training Programmes On-campus and off-campus training programmes for farmers and extension personnel on specific topics related to agro-techniques, integrated pest and disease management, nursery management, organic farming technologies, coconut-based cropping systems and post harvest technology form important components of TOT programmes to enhance the users' knowledge and skill for better technology utilization. A study conducted among extension personnel indicated that the training programmes were highly successful in achieving the objectives as reflected by the high Training Effectiveness Index (TEI) values.

Impact analysis of capacity building programme on coconut among subject matter specialists of KVKs during 2016 revealed that training had significant impact on enhancing knowledge level of the respondents in all subjects of coconut technologies. Average gain in knowledge was estimated to be 18.36%. Knowledge gain among youngsters as well as among respondents who had undergone training earlier was higher. Sixty-one per cent of the respondents graded training course as

Fig. 14.1 Transfer of technology programmes of ICAR-CPCRI

excellent. After the training, 38% of KVKs initiated on-farm trials and front-line demonstrations on new coconut technologies (Jaganathan et al. 2016).

Front-Line Demonstrations The institute has been organizing front-line demonstrations in farmers' fields on different coconut cultivation technologies such as coconut-based farming systems, soil and water conservation, management of root (wilt) disease-affected coconut gardens, mixed cropping with cocoa, etc. These demonstration programmes have proved to be effective in convincing the farmers about the technical feasibility and economic viability of the technologies. Through adoption of the proper management practice, average yield of the palms in the disease-affected gardens increased from 24.17 to 46.3 nuts palm^{-1} year^{-1} after 3 years, recording an improvement of 91.4%. Observations on yield of coconut revealed an increase in productivity of palms from a pre-demonstration yield of 95 nuts palm^{-1} year^{-1} under monocrop situation to 122 nuts palm^{-1} year^{-1} in coconut-based high-density multispecies cropping system.

14.4.3 Innovative Extension Approaches

Besides the conventional extension activities, a few innovative extension approaches have been pilot tested through action research with farmers' participation by ICAR-CPCRI for improving technology utilization. Several studies have indicated that knowledge and adoption of the coconut technologies need to be analysed in terms of risk management faced by the farming community, resources needed for the technology utilization, gaps in research and development needs and social process for evolving refined extension approaches. Alternative extension mechanisms for managing field problems of coconut need special emphasis while considering the following factors:

- Coconut being a perennial plantation crop, cultivated in a contiguous area in small and marginal land holdings (average holding size of 0.2 ha), provides congenial conditions for pest and disease incidence throughout the year.
- Approaches and strategies differ not only with social factors but also with technology features per se, nature of the crop, nature of pests/disease-causing organisms, incidence, severity and potential spread of pests or diseases as well as the observable nature of loss incurred influence the extension mechanisms/approaches required.
- Constraints such as input availability, technical/extension/economic factors, social components, and biophysical constraints require attention for evolving innovative extension approaches or mechanisms.

14.4.4 Paradigm Shift in Reaching Out to Stakeholders

Refinement of extension approaches paves the way for reaching out to the relevant stakeholders more efficiently and effectively. Farming is not an isolated activity by any means. It has the foundations in culture, heritage, experiences, tacit knowledge, knowledge innovations, innovation systems and social process. The possible approaches are:

1. Participatory technology transfer approach (PTTA) for coconut root (wilt) disease management.

2. Clustering coconut farmers – a successful extension approach for enhancing adoption and income from marginal and small holdings of root (wilt) disease-affected areas.
3. Area-wide community extension approaches (AWCA) in bio-management of rhinoceros beetle of coconut.
4. Farmer field Schools (FFS) in coconut.
5. Participatory community approaches in area-wide management of red palm weevil.
6. Information communication technologies (ICT) as interactive platform for technology transfer and field problem-solving for farming community.

Special features of innovative extension approaches are:

(i) Interactive and participatory approach involving relevant stakeholders.
(ii) Shift from individual farmers to farm family approach with gender concerns.
(iii) Inclusiveness incorporated for reaching out to all sections of society.
(iv) Area-wide interventions to overcome the challenges of fragmented holdings and resource base variability of farming community.
(v) Appropriate integration of extension techniques and methods for awareness building, knowledge dissemination and skill upgradation.
(vi) Technology-specific and problem-specific approaches for focused technology interventions, improving the impact of research.
(vii) Utilizing the digital literacy and advances in taking technologies and information, to different strata, of the society.

14.4.4.1 Community-Based Organizations

The livelihood of a substantial number of families in rural poor communities depends on coconut farming. Many a times, the income generated from coconut farming in small and marginal holdings does not provide enough for meeting their family requirements. The fragmented holdings don't render themselves viable for the optimum utilization of resources and adoption of improved technologies. Management of resources by groups helps to overcome the inherent weaknesses of the fragmented holdings and augment the production and productivity of these holdings. The effectiveness of organizing coconut farmers into community-based organizations (CBO) for efficient management of farmers' resources to reduce cost of cultivation and to increase productivity through integration of technologies even in very small farm holdings has been demonstrated by ICAR-CPCRI in selected localities (Thamban 2010).

A study was taken up for developing sustainable coconut-based income-generating technologies in poor rural communities in two selected coconut communities, one each from West Coast region and East Coast region in India. The three-pronged strategy for the project included growing suitable inter/mixed crops in coconut gardens and integrating animal husbandry and other subsidiary enterprises, cultivating high-yielding cultivars of coconut and diversification of coconut products. Implementation of the strategies, including microcredit, was routed through the CBO. A close linkage was developed between the CBO and scientists, and monitoring and evaluation of the interventions were done through CBO. Coconut clusters were found to possess better group characteristics and capacity development, and their performance is highly encouraging in terms of their increased knowledge and skills, improved behavioural changes and strong and responsible leadership. The coconut groups should be provided with organized markets, adequate long-term government support and a permanent establishment with infrastructural facilities for storage and processing just like the well-established commodity clusters (Thamban et al. 2016).

This innovative extension methodology was subsequently adopted for the implementation of other poverty alleviation and income enhancement projects. The cluster approach has been scaled up by other agencies like Coconut Development Board (CDB). The Board has initiated a massive programme for the formation of a large number of Coconut Producer Societies (CPS) by associating 40–100 coconut growers in a contiguous area with a consolidated minimum of 4000–5000 palms (Thamban 2010). Through 'cluster approach' in the root (wilt) disease-affected coconut area, average yield of coconut was doubled after technology package implementation for 3 years.

Technological interventions on soil and water conservation, soil health management, integrated nutrient management, inter/mixed cropping system and integrated pest and disease management implemented in farmers' gardens coupled with facilitating CBO of farmers and women self-help groups for effective integration of production and processing technologies in coconut holdings were found to be effective in obtaining substantially higher income.

Under the multi-country IFAD project on 'overcoming poverty in coconut-growing communities', participatory planning and implementation of diverse interventions notably intercropping and off-farm activities in small and marginal coconut homesteads were attempted through CBOs along with nutrition education. These efforts brought out significant improvements in the food and nutritional security as well as the income of the family members. Total annual income per homestead enhanced from Rs.25,617 to Rs.59,017 ha^{-1}over the project period. Income from coconut, increased by 50%, intercrops by fourfold, livestock rearing by sixfold and household level processing by 33-fold. At the end of the project, 96% of the members became completely food secure and 72% nutritionally secure (Thamban et al. 2016).

Based on the inferences drawn from evaluation and refinement of community approaches, an integrated model for coconut cluster was developed which was tested subsequently. An increase in knowledge index to the tune of 153% was recorded. Income from coconut recorded 2.4-fold improvement. The area under intercrops increased to double than that of the pre-project period, and the income from intercrops increased by 3.9-fold.

When community-based bio-resource management was adopted for sustaining production and livelihood security under coconut-based farming systems, the efforts resulted in ensuring quality bio-inputs to the farmers along with efficient use of land, water, sunlight and residue utilization, thereby contributing to sustainable production and productivity. An increase in knowledge level by 117% in case of bio-resource management and 76% in case of integrated nutrient management was recorded. Community coconut nursery established by the farmer groups produced seedlings with a recovery of 79.5% good-quality bio-primed dwarf coconut seedlings. Significant improvement in income from coconut and other intercrops resulted in 89.7% improvement in farm income (Thamban et al. 2016).

14.4.4.2 Farmer Empowerment Programmes

When PTTA was adopted for technology assessment with participation of farm families in a contiguous area affected by coconut root (wilt) disease, awareness, knowledge, attitude and adoption level of farmers towards the integrated root (wilt) disease management technologies were improved by 40–85%. Area-wide community extension approach (AWCA) was attempted in another cluster for management of coconut pests. Through this approach, more than 90% of the potential adopters were reached within 2 months, and post-intervention data indicated 75.8% reduction in fresh pest infestation (Thamban et al. 2016). Awareness and knowledge level of farmers improved by 100% when Farmer Field Schools (FFS) were implemented in 15 locations for the integrated pest management of rhinoceros beetle.

Entrepreneurship development was attempted in a group approach format for micro-level interventions on product diversification in coconut such as production of quality copra using copra dryers, coconut kernel-based food products, coconut candies, snow ball tender nut and coconut chips as well as production of oyster mushroom on coconut wastes and vermicompost using coconut leaves. Results revealed that women members of the CBO increased their income by three to five times through the production and marketing of value-added products compared to their previous income from copra. Equally important, the project intervention provided employment opportunities to formerly unemployed and underemployed rural women resulting in enhanced self-esteem and economic and social empowerment (Batugal and Oliver 2003).

Technology assessment and refinement increased the effectiveness of participatory approach in the adoption of various technologies related to high-yielding varieties, intercropping, nutrient management and crop protection (Arulraj et al. 2002; Thamban et al. 2004). In the interphase approach, researchers, extension personnel and farmers were brought together on a common platform which helped in enhancing the awareness and knowledge about the technologies (Rajagopal et al. 2004b).

Mera Gaon Mera Gaurav: This initiative by the Ministry of Agriculture and Farmers' Welfare, Govt. of India, envisages promoting the direct interface of agricultural scientists with the farmers to hasten the lab to land process. Cutting across all disciplines, farm problems are diagnosed and effective solutions delivered and showcased in farmer's fields. National priorities such as secondary agriculture, climate change, good agricultural practices and soil and health management of crops are given importance in this programme (Thamban et al. 2016).

Cyber extension programmes: Cyber extension include effective use of information and communication technology (ICT), national and international information networks, Internet expert systems, multimedia learning systems and computer-based training systems to improve information access to all the players. The video conferencing system installed at the Agricultural Technology Information Centre (ATIC) of ICAR-CPCRI, Kasaragod, facilitates interaction between various stakeholders for enhancing technology utilization in coconut through effective linkages with government institutes, commodity boards and farmers' organizations (Thamban 2010).

14.5 Developmental Agencies

There are various agencies under the government of India and states to transfer technologies at the appropriate level.

14.5.1 Coconut Development Board

The Coconut Development Board is a statutory body established by the government of India for the integrated development of coconut cultivation and industry in the country. The Board which came into existence in 1981 has its headquarters in Kerala state and regional offices and state centres in other states. Activities of the Board include implementing measures for the development of coconut industry; improving marketing of coconut and its products; imparting technical advice; providing financial and other assistance for expansion of area under coconut; encouraging adoption of modern technologies for processing of coconut and its products; taking steps to get incentive prices; recommending measures for regulating imports and exports; fixing grades, specifications and standards for coconut and its products; financing suitable schemes to increase the production of coconut and to improve the quality and yield of coconut; assisting, encouraging, promoting and financing agricultural, technological and industrial research on coconut and its products; maintaining the database on coconut industry; and undertaking publicity activities.

The Board brings out several publications, including a popular journal in several languages, to promote coconut-based technologies. CDB also participates in exhibitions, seminars, workshops and entrepreneurship development programmes, both in India and abroad, and organizes training programmes for coconut farmers and producers on improved methods of cultivation and latest technologies in coconut processing. Lack of availability of labour, especially skilled labour for coconut climbing and high-wage rate, is a serious problem faced by growers in adopting timely crop management practices. To tackle this, a professional group of youth is organized for harvesting and taking up plant protection operations. The Board, as a service organization, popularizes the latest development in coconut industry through mass media. The extension activities under the Coconut Development Board are planned and executed with the objectives of promoting scientific coconut cultivation and product diversification besides promoting coconut oil as a healthy cooking medium, so that the decreasing trend in the consumption of coconut oil could be arrested. Promotion of tender coconut consumption is also undertaken to divert an increased part of the production to the beverage sector.

Generation of information on wholesale and retail prices and arrivals and analysis of trends in various markets of the country help the farmers and traders. The Coconut Development Board is recognized as the National Information and Documentation Centre for coconut in the country, and it has a nationwide information network using modern information technology.

14.5.2 State Agriculture and Horticulture Departments

The agriculture and horticulture departments in the major coconut-growing states in the country implement various development schemes and extension programmes for the benefit of coconut farmers. Many of the states have been implementing separate programmes for the production and distribution of planting materials and fertilizers as well as distribution of plant protection chemicals and biocontrol agents against coconut pests. Financial incentives are also provided in the form of subsidies for irrigation infrastructure. Organizing coconut farmers at grass root level for group management of coconut production activities, conducting farmers training programmes and organizing other extension programmes are the major activities of these departments. Besides the above, local bodies like grama panchayats also implement location-specific development schemes under the technical guidance of the agriculture and horticulture departments.

14.5.3 District Level Agriculture Technology Management Agency (ATMA) Model

In a country like India where agroclimatic zones widely differ, besides significant variation in socio-economic status of farmers, uniform extension service cannot be the answer for all the regions. ATMA was formed as an alternate public extension institution to place the public extension system in a new decentralized institutional arrangement which is demand driven and farmer accountable adopting a bottom-up farming system approach. It is a registered society of key stakeholders (farmers, line/development departments, nongovernment organizations, input dealers, mass media, agribusiness companies, farmers' organizations, etc.) involved in agricultural activities for sustainable agricultural development in the district. Emphasis has been laid on providing flexible working environment and establishing effective integration of all the stakeholders at the district level thereby improving input into programme planning and resource allocation, especially at the block level thereby increasing accountability of stakeholders.

Development work plans at block/district level is based on a Strategic Research and Extension Plan (SREP), prepared through participatory appraisal techniques involving all the stakeholders. The SREP contains detailed analysis of the information on existing farming systems in the district and research-extension gaps required to be filled up. It also prioritizes the research-extension strategies within the district.

14.5.4 State Agricultural/Horticultural Universities (SAUs)

The SAUs are also implementing a number of transfer of technology programmes, in the concerned states through the Directorate of Extension Education, Krishi Vigyan Kendras and plant clinics. The basic extension education role of the university is to make available useful research information. The Directorate of Extension, the Krishi Vigyan Kendras (KVK) and the Village Adoption Programme are responsible for the transfer of technology programmes of the university.

Expert Centres and Village Resource Centres are established in collaboration with Indian Space Research Organisation (ISRO) for the interaction of farmers with experts on a variety of agricultural information. Mobile message services and Kisan Call Centres provide timely information to the farming community regarding agricultural technologies, weather data and market information.

14.5.5 Krishi Vigyan Kendra (KVK)

Krishi Vigyan Kendra (Farm Science Centre) is an innovative institution of ICAR established at district level. The first KVK was established during 1974 which has grown to 620. They play a vital role in conducting on-farm trials and front-line demonstrations in farmers' fields to identify location-specific agricultural technologies and demonstrate the production potential of new agricultural technologies in farmers' fields through front-line demonstrations. They also conduct need-based training programmes. Kerure et al. (2016) reported that ICAR-Krishi Vigyan Kendra, Chitradurga, could encourage many unemployed youth to use the coconut tree climber for harvesting and choose it to earn their livelihood. Critical and quality inputs like seeds, planting materials, organic products, biofertilizers and poultry strains are produced by the KVKs and made available to the farmers. Agricultural Knowledge and Resource Centres are set up at KVK to support the initiatives of public, private and voluntary sectors at district level. A number of successful case studies have emerged out of effective implementation of various technological and institutional interventions by KVKs.

14.5.6 Directorate of Extension

The Directorate of Extension functioning under the Ministry of Agriculture and Farmers' Welfare, Govt. of India, sponsors training programmes on various aspects of coconut production technologies for the benefit of extension personnel engaged in the development of coconut in different states. It also funds projects for production of audio and video programmes on coconut cultivation technologies for strengthening the TOT programmes on the crop.

14.6 Transfer of Technology Programmes in Other Countries

All the major coconut-growing countries have established their own research organizations supported by transfer of technology programmes so that the coconut farmers are fully benefitted by the technologies developed.

14.6.1 Indonesia

Coconut not only provides livelihood security to more than 3.5 million farmers but also plays its social role as the second important commodity after rice in Indonesia. Total coconut area is 3.571 million ha with a production of 14,804 million nuts, and smallholders own 98% of the area (APCC 2015). Coconut is largely consumed as fresh coconut. However, in manufacturing industries, coconut is processed into cooking oil, desiccated coconut, coconut milk, nata de coco, coir fibre, activated charcoal, etc. Coconut is an important source of foreign exchange for the country. As in the case of other coconut-producing countries, Indonesia experiences low international prices for its traditional export items, i.e. coconut oil and copra. The situation is aggravated by the defective marketing system in the country with the presence of long channels of marketing intermediaries especially in remote islands resulting in very low prices received by coconut farmers. Low productivity, due to palm senility and poor farm management, also poses as one of the industry's major problems. Recognizing these problems, the government implements the Smallholder Coconut Development Project.

A large number of improved technologies have been developed by several research institutes/university centres including the Indonesian Coconut and Palmae Research Institute (ICOPRI), Manado, which has the mandate for coconut research. Major developmental programmes are focused on enhancing production and productivity by adopting package of agricultural practices, rejuvenation of old and senile palms and regular replanting programme with selected high-yielding talls, dwarfs and hybrids, adoption of integrated farming system and encouraging farm level processing as a group approach (Novarianto 2004).

Transfer of technology programmes are implemented through the Ministry of Agriculture, National Centre for Agricultural Extension Development, Agricultural Universities, Provincial Agricultural Extension Coordination Offices, private companies, NGOs and farmers' organizations. Farmer's empowerment through agricultural technology and information project was started in 2007 for achieving sustainability.

14.6.2 Malaysia

Coconut industry being the oldest agro-based industry in Malaysia continues to be an important industry involving 70,000–80,000 smallholders, which accounts for 75% of the total area under coconut. Coconut is grown in an area of 82,001 ha with a production of 538 million nuts (APCC 2015). Major exports of coconut products are desiccated coconut, copra, coconut oil, coconut milk powder, activated carbon, etc. Extension programmes are implemented through the Ministry of Agriculture and agro-based industries, Malaysian Agricultural Research and Development Institute (MARDI), NGOs, private companies and farmer-based associations/cooperatives/societies (Fong 2004). Major developmental programmes for coconut in Malaysia are enumerated here under:

- Replanting old, uneconomic coconut palms with high-yielding hybrids like MATAG and MAWA with incentives for cleaning, building of drainage and irrigation infrastructure as well as making available quality seedlings and other agricultural inputs to all the needy farmers.
- Extension services to assist farmers to utilize the land resources.
- Enhancing income of smallholders through crop diversification and integration of cash crops and livestock component.
- Strengthening of technology transfer network through collaboration between the Malaysian Agriculture Research and Development Institute and Department of Agriculture under the Ministry of Agriculture.
- Collection, processing and marketing of coconut and its products through the Federal Agricultural Marketing Authority (FAMA) for the benefit of smallholders.

14.6.3 The Philippines

Coconut is grown in an area of 3.517 million ha with a production of 14,735 million nuts (APCC 2015). Productivity is only 4196 nuts ha^{-1} which is less compared to other countries mainly because the major area is under old/senile palms. Frequent hurricanes destroying large area of coconut is a constraint peculiar to the Philippines. Major coconut products are copra, copra meal, activated carbon, coco shell charcoal, coconut oil and desiccated coconut. In the Philippines, the Makapuno coconut (yielding nuts with creamy kernel) has a good market through ice cream industry.

Extension programmes are carried out through the Department of Agriculture under the Ministry of Agriculture and Food, Agricultural Training Institute, State Agricultural Universities, private companies, NGOs, farmers' organizations and Philippines coconut farmers' cooperatives. A major extension programme was funded by FAO in 2002 for strengthening agricultural extension through ICT applications for technology management service, farmers' information and technology services, Farmer-Scientist Bureau, Open Academy for Philippines Agriculture, Smart farmer Call Centre, school on the air and web portal as well as mobile Internet bus.

Developmental programmes for coconut include farm rehabilitation, planting, replanting and extension services for greater farm productivity with focus on the small coconut farms' development programme; skills training, information dissemination and organization of the coconut farmers for their economic upliftment and empowerment; intensification of agricultural and industrial research for better crop protection, high-yielding varieties and product processing; market research and development for the expansion of both domestic and foreign markets; and recovery of the multibillion coconut levy assets to be eventually used for the benefit of coconut farmers as well as for improvement of administrative and support services (Coronacion 2004).

14.6.4 Sri Lanka

Coconut, which is the oldest tree crop in Sri Lanka, continues to occupy a significant and eminent place in economic and social spheres in the lives of Sri Lankans. Coconut covers around one fourth of the cultivated area in the country which contributes 2% to the GDP of the country. Sri Lanka produces white edible copra, coconut cake, coconut cream, coconut milk, milk powder and defatted coconut and coconut shell products such as shell charcoal, shell flour and activated carbon. Shipping of fresh tender nuts in refrigerated conditions is a special feature of Sri Lankan market. Coconut Research Institute (CRI) of Sri Lanka develops technology and adapts them at experimental level, while Coconut Cultivation Board (CCB) is engaged in transfer of technology to coconut farmers up to the grass root level through extension officers at field level (Somasiri et al. 1993; Sugathadasa 2004).

Information on new technologies is disseminated to farmers using instruction leaflets and advisory circulars and also by organizing seminars and workshops both for extension workers and farmers. In addition, Coconut Development Authority in Sri Lanka brings out weekly bulletin, monthly bulletin (Coco Market Focus) and yearly bulletin (Sri Lanka Coconut Statistics). Transfer of technology programmes are implemented by the Department of Agriculture under the Ministry of Agriculture,

Agricultural Universities, Horticulture Crop Research and Development Institute, Coconut Cultivation Board, Agriculture Research Institute, Coconut Research Institute, Hector Kobbekaduwa Agrarian Research and Training Institute, NGOs, private companies, coconut triangle milk producers' union and agro-enterprise development and information service.

14.6.5 Thailand

Coconut is widely grown throughout the country, and it is one of the important crops in the southern and central regions of Thailand. Coconut plays an important role in Thai diets and domestic life of Thailand population. Major coconut products are coconut oil, palm sugar, coconut fibre, tender coconut, desiccated coconut, copra and activated carbon (Watanayothin 2004). Aromatic tender nut is unique to Thailand and finds a good market.

The Department of Agricultural Extension under the Ministry of Agriculture and Cooperatives, Kasetsart University, Agricultural Technology Transfer Centres, Lampang Agricultural Research and Training Centre, agricultural extension associations, NGOs and private companies are taking care of dissemination of technologies to the coconut growers. Coconut developmental programmes include replanting and rehabilitation schemes in Southern Thailand, value addition for small holder farmers' groups and Good Agricultural Practices (GAP), Good Manufacturing Practices (GMP) and the development of pre- and post harvest technologies to ensure increase in the production of safe, high-quality produce, making these technologies available to the small farmers and industrialists.

14.6.6 Vietnam

Vietnam has an area 0.162 million ha with a production of 1434 million nuts (APCC 2015). Major coconut products are coconut oil, desiccated coconut and shell charcoal (activated carbon). Coconut sugar is one of their unique products. Vietnam strengthened its coconut development programmes by establishing the Union of Vegetable Oil Manufacturers in 1984, having a wider scope covering all aspects of coconut research, development, processing and marketing (Nga 2004). The country also has in place an extension system at the national level and at the provincial level. In 1986, support was obtained from the USSR and East Germany for coconut development in the form of loans for the establishment of new plantations and rehabilitation of existing plantations.

Coconut developmental programmes include intercropping models through specific programmes/policies and mixed farming with shrimp, fish, livestock and honeybee, providing planting materials of new varieties to enhance productivity and training of farmers on scientific cultivation, harvesting and preservation.

14.6.7 Papua New Guinea

The executing agency of Papua New Guinea's national coconut research programme is the Cocoa and Coconut Research Institute (CCRI). Coconut research and development programme include a number of components designed to address the major problems faced by the coconut industry (including poor farm management, declining productivity and increasing number of senile trees), support price for copra, research on coconut breeding and embryo culture, encouragement of intercropping and value addition.

14.7 Technology Utilization Pattern in Coconut Sector

The process of research could be considered as successful only when the results reach the ultimate users. However, acceptance of a new idea by the members of a social system is rather slow and difficult. Each individual technology has to pass through different stages of adoption process. Various communication channels used by research and extension systems help the farmer to pass through these stages in a rapid and positive manner. Input agencies involved in the supply of fertilizers and plant protection chemicals, nurseries in public and private sector involved in the production and distribution of quality planting materials of coconut, financial institutions such as banks and cooperative societies who provide financial assistance to coconut growers and marketing organizations, who intervene in coconut marketing, are the major actors in the support system in coconut sector. Through their specific functions, they contribute to the strengthening of transfer of technology in coconut. Non-governmental organizations, farmers' organizations and input dealers in private and cooperative sector also play a critical role in the process of transfer of technologies.

It is often presumed that the innovations would sell themselves or the benefits of a new technology would be widely realized by its potential adopters, and therefore the technology would diffuse rapidly. Unfortunately, this is seldom the case because of a factor of uncertainty when the superior alternative is put forth. As a first step to understand the components of this complex situation, an effort was made to review the level of adoption of recommended coconut cultivation technologies, which would serve as a feedback to the coconut research and development community. Technologies which are successfully adopted and not adopted worldwide as reported by Hazelman (1994) are described below.

14.7.1 Successfully Adopted Technologies

- Nursery planting and management techniques have generally been adopted as long as they are coordinated or implemented by the national government or private plantations but are generally not followed if passed directly to small producers.
- Intercropping with different crops such as cocoa, coffee, banana, taro, vanilla and root crops and to some extent the use of cattle under coconut. Benefits from increased output per unit area is a favourable factor for adoption.
- Weed control where bioagents are relied upon (e.g. Lantana weed control in Niue and Solomon Islands and that for *Mimosa* in Cook Islands), as they are more self-sustaining with almost no further contact needed with researchers once bioagents are established, indicating the simplicity of the technology and its appropriateness to farmers' context.
- Pest control especially those which require minimum farmer input has largely been adopted.
- Processing of commodities that fit local eating habits and lifestyle such as coconut cream products and various cosmetics, which fit for both export and local market demands, has been readily adopted especially if market prices are favourable.
- Use of local cultivars which are already familiar to farmers and also have adapted to local conditions with low demand for management and environmental requirements.
- Use of various copra driers has been generally accepted since it ensures product acceptability in the market.
- Use of coconut wood for furniture, for fence posts and for carvings has positive acceptance due to good market potential.

14.7.2 Unsuccessful Technologies

- Planting of hybrids, since they more often require a higher level of management than the one most of the small farmers are accustomed to.
- Inorganic fertilizers, which have received much attention by researchers, are hardly adopted due to cost considerations.
- Replacement of senile palms, which is strongly advocated, but is resisted due to both economic and cultural considerations.
- Harvesting systems of various types have largely been avoided by farmers as they rely on traditional systems for collecting nuts after they are dropped.

The Asian and Pacific Coconut Community launched a project in the late 1980s to assess the farmers' receptivity to new technologies in coconut in several countries. The assessment covered a range of technologies that have been researched upon and recommended for implementation. By and large, adoption of new technologies has been very poor.

14.7.3 Country-Wise Status of Adoption of Technologies

14.7.3.1 Indonesia

The APCC conducted a study covering a range of technologies including improved planting material, fertilizer application and moisture conservation. Adoption was found to be very low, and the two main reasons for the disappointing level of adoption were that a great majority of farmers were not convinced about the technology. Awareness on the new technologies was also poor (Amrizal 1988).

14.7.3.2 Malaysia

The survey included two technologies, namely, the use of hybrids and intercropping in coconut farms. While there was some response to intercropping, the level of adoption of hybrids was dismal (Shahar 1988).

14.7.3.3 Papua New Guinea

The survey in Papua New Guinea covered a package of technologies including hybrids, fertilizers, cover crops, soil and moisture conservation and intercropping. Except for intercropping, none of the other technologies has been adopted by the farmers (Yarbro 1988).

14.7.3.4 The Philippines

The survey covered a range of new technologies including hybrids, fertilizers, agrochemicals, intercropping, soil conservation, cover cropping and scientific copra making. The survey revealed that only 4.2% of the farmers implemented any one or more of the above technologies. Reasons for non-adoption included lack of finance (66%), negative perception on relative advantage (12%), lack of awareness (4%) and non-conviction of the technology (18%) (Arancon 1988).

14.8 Strategies for Effective Transfer of Technology

Low level of technology utilization, at farmers' fields in coconut sector, calls for formulating effective extension strategies, suitable to the heterogeneous farming situations in coconut farming. Some of the strategies in this direction are discussed below:

14.8.1 Farmers' Participation in Research and Extension

Extent of adoption of technologies in coconut is not at a satisfactory level, as has been discussed earlier, and the main factor hindering adoption of technology is farmers' lack of knowledge regarding appropriate techniques for improving his farming. But in many other situations, the reason behind the low acceptance of technologies may be that the technologies are not economically viable, not technically feasible, not matching with the farmers needs and not compatible with the farmers' overall farming system. Recommending technology based on the results at the research station without adequate and proper testing in the farmers' fields under the farmers' resource and risk situation leads to inadequate technology integration. An active participation of the beneficiaries in the generation of technology can ensure that the technology is user-friendly. In the farmer participatory technology generation efforts, conducting on-farm trials is envisaged, wherein trials are conducted in the fields of participating farmers who will be allowed to manage the trials by themselves in collaboration with researchers and extension agencies. Farmer participation in agricultural technology development and dissemination should begin from diagnosis to planning and designing technological solutions, to implementation and to evaluation and feedback into dissemination. The experience in India endorses the view that the farmer participatory technology generation and dissemination in coconut could yield fruitful results, which will be worthy to be emulated by other agencies involved in the development of coconut.

14.8.2 Need-Based Training Programmes

Training programmes with appropriate methodologies are to be organized for farmers on various aspects of crop production, protection and post harvest processing technologies. Participatory methods could be effectively employed to unearth the areas of improved technologies in which farmers require exposure. Training programmes are also required for the extension personnel engaged in the development of coconut to keep abreast with the latest technological advances in coconut cultivation.

In the farmer empowerment programmes, special emphasis is to be given on the following aspects, as recommended by different workers.

- There is a need for popularizing production and use of organic manures including use of vermicompost and also encouraging the cultivation of green manure crops in coconut garden.
- To enhance the profitability of coconut cultivation especially in the present context of wide fluctuation in coconut prices, it is necessary to have programmes for popularizing optimum methods of inter/mixed cropping/mixed farming in coconut gardens. Suitable crop combinations are to be suggested depending on the farmers' preferences and his resource endowment and prevailing agroecological features of the locality.

Efforts are required to provide extension support to educate farmers about the scientific pest and disease management and also to implement location-specific schemes to provide incentives for the need-based plant protection measures.

Special training programmes are to be organized on new technologies such as climbing devices for the benefit of farmers.

14.8.3 Targeting the Extension Service

Often it is observed that the resource-rich farmers make use of the technologies in a better way than the resource-poor small and marginal farmers. Extension service obviously benefits these elite farmers better. Reaching resource-poor farmers with small and marginal holdings requires a different approach. The cropping system models such as high-density multispecies cropping system can be adopted by farmers who can afford to have a well laid-out irrigation unit. In the marginal holdings, which are often rainfed, choice of the cropping system and other production technologies would have to be different. The research and extension system is to work for the resource-poor farmers and low potential areas with appropriate technology, and active utiliser systems are to be generated. Transfer of technology model of innovation dissemination has to target heterogeneous situations.

14.8.4 Promoting Group Approach

Group approach helps to overcome the limitation of small size of holding in utilizing resources and improved technologies and also to provide opportunities for better social integration among farmers. Group approach programme could be implemented through local level farmers' associations. In the post harvest processing of coconut, there is immense potential for organizing cultivators with small and marginal holdings into viable groups to overcome the limitation of scale of cultivation. Apart from the efficient management of resources, these farmers' groups could be effectively utilized for organizing educational programmes for the benefit of member cultivators. Moreover, these groups could act as an effective link between various systems engaged in the technology generation and dissemination.

14.8.5 Producer-Driven Value Chain system

The study conducted by Jayasekhar et al. (2014) has integrated the concept of sectoral system of innovation and value chain theory to find the regional dynamics of an evolving commodity chain. It has captured the reflections of tender coconut sector from the comprehensive study conducted in Kerala, India. The tender coconut value chain of Kerala has been found middleman -driven, which offers only a meagre value share to the producer/farmer. For upgradation of position in the chain, the producers should proactively function in a group mode and should integrate the domestic value chain. According to the study, for ensuring a bargaining position to the producer, the chain should be restructured from the middleman-driven one to producer-driven. The study has argued that the support of an effective sectoral system of innovation is inevitable for the development of the sub-sectoral commodity chains.

14.8.6 Experiential Learning Through 'Satellite Farms'

Owing to the perennial nature of the crop, the scope for setting up and maintenance of demonstration plots in coconut is limited compared to field crops. In this context, encouraging farmer-farmer extension, by way of model farms/satellite farms, assumes significance. A farmer who is already maintaining his farm with the improved technologies of coconut farming who is cooperative and well accepted by his neighbouring farmers could be identified for group meeting, field visits, demonstration and discussion in his farm. His farm could act as a 'satellite farm' from where the concept of improved methods of farming radiates to neighbouring farmers.

14.8.7 On-Farm Adaptive Research Trials

These trials with the active participation of the growers would be very useful in assessing the suitability of technologies. Approaches may differ from country to country. Field trials are to be made very simple with minimal replicates. Guidance and close involvement of the concerned researcher are necessary during the establishment of the trials and subsequent monitoring. Grower too should contribute, at least by way of labour, other material inputs being provided by the researcher. In Sri Lanka, on-farm adaptive research trials to demonstrate the farming systems approach (with intercropping and animal husbandry) have been a priority activity in the research programme of the Coconut Research Institute of Sri Lanka. These have been becoming quite successful as centres of technology transfer and sources for planting material.

14.8.8 Participatory Demonstrations

Participatory demonstrations strive to demonstrate the technology with the active involvement of farmers and research stations. These should be easily accessible to the public and visible. Relevant information should be displayed in very simple language. In Sri Lanka, during the 1980s, over 200 demonstrations were established to popularize the use of fertilizers. This exercise was, by and large, successful with a significant number of smallholders being sensitized on the beneficial use of fertilizer, as was evident from the records available to the farmers on increased nut production (Rezania and Jayasekera 1992).

14.9 Future Strategy

Coconut research institutions have developed a number of technologies by the use of which remarkable improvement in the coconut productivity levels has been achieved within a short span of time. However, the wide gap that exists between the recommended practices and their actual level of adoption in various holdings indicates major barriers in improving the productivity levels. In order to change this scenario, the development efforts implemented by various research and development organizations are needed to be strengthened which should form one of the primary tasks towards achieving the targets for improving coconut productivity and profitability levels. As farmer participatory technology generation and dissemination in coconut could yield fruitful results, it will be worthy to be emulated by various agencies involved in the development of coconut. In the farmer empowerment programmes, special emphasis is to be given on popularizing production and use of organic manures and cultivation of green manure crops in coconut garden and optimum methods of inter/mixed cropping in coconut gardens depending on the farmers' preferences, his resource endowment and prevailing agroecological features of the locality. Efforts are required to provide extension support to educate farmers about the scientific pest and disease management and also to implement location-specific schemes to provide incentives for the need-based plant protection measures. Special training programmes are to be organized on new climbing devices for the benefit of farmers. The research and extension system is to work for the resource-poor farmers and low potential areas with appropriate technology, and active utiliser systems are to be generated. Transfer of technology model of innovation dissemination has to target heterogeneous situations. The producers should proactively function in a group mode and should integrate the domestic value chain coupled with creating storage facilities and market intervention to ensure a bargaining position for the producer.

Coconut plays a vital role in the agrarian economy of many countries. The present scenario of technology adoption in coconut calls for strengthening the technology generation and dissemination programmes based on a viable extension strategy with the active participation of beneficiaries. Effective linkage is to be established among different research and development agencies and coconut farmers through well-coordinated participatory research and extension programmes for ensuring a meaningful technology generation and transfer system in coconut.

References

Amrizal J (1988) Farmers' receptivity to new technologies in coconut: Indonesia. Asia and Pacific Coconut Community, Jakarta, pp 13

Anderson JR (2007) Agricultural advisory services. Background paper for the World Development Report 2008, Agriculture and Rural Development Department, World Bank, Washington, DC

Anithakumari P (2013) Labour availability in coconut cultivation and impact on technology adoption as perceived by coconut farmers. J Plantn Crops 41(2):186–190

Anithakumari P, Kalavathi S (2001) Knowledge and adoption of recommended practices by coconut cultivators of root (wilt) affected areas. Indian Cocon J 31(9):21–24

Anithakumari P, Rajeev MS, George J (2011) Improving income from coconut cultivation through farm level value addition- an analysis. J Plantn Crops 39(1):131–136

Anithakumari P, Muralidharan K, Kalavathi S (2012) Technology utilization among coconut farmers of root (wilt) disease affected areas. J Plantn Crops 40(3):174–180

APCC (2015) Asian and Pacific coconut community statistical year book 2015. APCC, Jakarta

Arancon RN (1988) Farmers' receptivity to new technologies in coconut: Philippines. Asia and Pacific Coconut Community, Jakarta, p 78

Arulraj S, Sairam CV, Thamban C, Rajagopal V (2002) Technology assessment through institution village linkage programme. CPCRI, Kasaragod, p 200

Bastine CL, Nair N, MP A (1991) Adoption rate and constraints of adoption of newer technologies by coconut farmers in Northern Kerala. In: Silas EG, Aravindakshan M, Jose AI (eds) Coconut Breeding and Management. Kerala Agricultural University, Thrissur, pp 251–254

Batugal P, Oliver JT (2003) Poverty reduction in coconut growing communities, Volume I: the framework and project plan. IPGRI-APO, Serdang

Batugal P, Bourdeix R (2005) Conventional coconut breeding. In: Batugal P, RaoV R, Oliver J (eds) Coconut genetic resources. International Plant Genetic Resources Institute, Regional office for Asia, the Pacific and Oceania, Serdang

Bozeman B (2000) Technology transfer and public policy: a review of research and theory. Res Policy 29:627–655

Coronacion DM (2004) Strategies for enhancing productivity and income of coconut farmers in Philippines. In: Rethinam P (ed) Proceedings of the XLI Cocotech meeting on "Strategies for enhancing productivity and income of coconut farmers". APCC, Jakarta, pp 54–58

de Silva S (1989) High yielding varieties of coconut – an interregional study on small farmers' experience. Asian pacific coconut community, Jakarta, p 171

Fong HL (2004) Strategies for enhancing productivity and income of coconut farmers in Malaysia. In: Rethinam P (ed) Proceedings of the XLI Cocotech meeting on "Strategies for enhancing productivity and income of coconut farmers". APCC, Jakarta, pp 46–50

Gibson DV, Rogers EM (1994) R and D collaboration on trial: the microelectronics and computer technology consortium. Harvard Business School Press, Boston

Hazelman SM (1994) Coconut technology needs and technology transfer experience in the pacific region. In: Arancon RN (ed) Proceedings of the XXXII Cocotech meeting on "technology transfer and application in relation to the coconut industry". Pacific Region, Kuala Lumpur

Jaganathan D, Thamban C, Jose CT, Jayasekhar S, Anithakumari P (2013) Analysis of organic farming practices in coconut in South India. J Plantn Crops 41(1):71–89

Jaganathan D, Thamban C, Chandran KP, Jayasekhar S, Muralidharan K (2016) Impact analysis of capacity building programme on coconut: the case of subject matter specialists of KVKs in South India. Abstract 3rd international symposium on coconut research and Development 10–12 December 2016. ICAR-CPCRI, Kasaragod, p 172

Jayasekhar S, Chann KP, Harilal KP (2014) Whether an effective innovation system is inevitable for sectoral value chain development? Evidence from coconut sector in Kerala. Agri Econ Res Rev 27:25–33

Jnanadevan R, Prakash R (1994) Adoption patterns of farmers in selected development programmes of coconut development board. Indian Cocon J 25(5):23–28

Kalathiya KV, Dave SK, Mehta (2007) Existing farming and cultural practices of coconut followed by the growers in Valsad District (South Gujarat). Agri Sci Digest 27(3):182–185

Kalavathi S, Anithakumari P (1998) Level of knowledge of coconut growers and level of adoption of technologies. Tree World 7(9):1–5

Kalavathi S, Anithakumari P (2011) Evaluation and refinement of community approaches in coconut technology transfer. J Plantn Crops 39(1):141–147

Kerure P, Channagouda RF, Salimath SB, Onkarappa S (2016) Study of coconut tree climber use efficiency over traditional methods in Chitradurga District of Karnataka. J Krishi Vigyan 5(1):101–103

Mathew AC, Thamban C (2011) Irrigation efficiency in coconut gardens- a comparative analysis of different systems of irrigation. J Plantn Crops 39(1):48–56

Naik G (2005) WTO, Indian agriculture and coconut sector. In: Thomas Mathew M, Kumar N (eds) Proceedings of the international seminar on 'global coconut industry-outlook for the farmers'. ICAR Research Complex, Goa, pp 200–209

Nga NT (2004) Strategies for enhancing productivity and income of coconut farmers in Vietnam. In: Rethinam P (ed) Proceedings of the XLI Cocotech meeting on "Strategies for enhancing productivity and income of coconut farmers". APCC, Jakarta, pp 76–78

Kumar N, Kapoor S (2010) Value chain analysis of coconut in Orissa. Agri Econ Res Rev 23:411–414

Novarianto H (2004) Strategies for enhancing productivity and income of coconut farmers in Indonesia. In: Rethinam P (ed) Proceedings of the XLI Cocotech meeting on "Strategies for enhancing productivity and income of coconut farmers". APCC, Jakarta, pp 28–39

Ohler J, Griffee P (1999) Modern coconut management, palm cultivation and products. Royal Tropical Institute, Amsterdam, p 39

Pathiraja PMEK, Fernando MTN, Abeysekara AWAD, Subasingha SDJN (2010) An account of labour availability in major coconut growing areas in coconut triangle. COCOS 19:13–26

Rajagopal V, Arulraj S, Thamban C (2004a) Research-extension interface for transfer of technology on coconut. CPCRI, Kasaragod, p 33

Rajagopal V, Thamban C, Arulraj S, Anithakumari P (2004b) Role of transfer of technology, training and demonstration in increasing income of coconut farmers. In: Rethinam P (ed) Proceedings of the XLI cocotech meeting on "Strategies for enhancing productivity and income of coconut farmers", APCC, Jakarta, pp 147–163

Ramanathan K (1994) The polytrophic components of manufacturing technology. Technological Forecasting & Social Change 46:221–258

Rezania M, Jayasekera KS (1992) FAO fertilizer programme: plant nutrition status of coconut palms and prospects for increasing production in the coconut small holding sector of Sri Lanka, GCPF/SRL/047/NET Field Document No.15, FAO

Shahar BS (1988) Farmers receptivity to new technologies in coconut: Malaysia. Asia Pacific Coconut Community, Jakarta, p 24

Somasiri LLW, Peiris TSG, Gunathilake HAJ, Fernando MTN, Pinto JLJG (1993) An appraisal of coconut growers' reactions and observations on CRI –Recommended cultural practices and other related issues- A diagnostic survey report, Coconut Research Institute of Sri Lanka, p 123

Souder WE, Nashar AS, Padmanathan V (1990) A guide to the best technology transfer practices. J Technol Transfer 15(1–2):23–26
Sugathadasa I (2004) Strategies for enhancing productivity and income of coconut farmers in Sri Lanka. In: Rethinam P (ed) Proceedings of the XLI Cocotech meeting on "Strategies for enhancing productivity and income of coconut farmers". APCC, Jakarta, pp 63–69
Thamban C (2010) Extension approaches to enhance technology utilization and income generation in coconut farming. In: Thomas GV, Krishna Kumar V, Augustine Jerard B (eds) Proceedings of international conference on "Improving productivity and profitability in coconut farming". CPCRI, Kasaragod, pp 443–456
Thamban C, Venugopalan M (2002) Extent of adoption of recommended package of practices for coconut cultivation. In: Rethinam P, Khan HH, Reddy VM, Mandal PK, Suresh K (eds) Research and development in plantation crops in the new millennium. Proceedings of PLACROSYM XIV, Indian Society for Plantation Crops, Kasaragod, pp 218–223
Thamban C, Arulraj S, Sairam CV (2004) Farmer participatory technology assessment and refinement in coconut through institution village linkage programme – experiences of CPCRI in India. J Plantn Crops 32(Suppl):459–464
Thamban C, Mathew AC, Muralidharan K, Subramanian P, Vidhan Singh T, Madhavan K (2011) Coconut climbing methods and devices: a participatory analysis of constraints and strategies. J Plantn Crops 39(1):148–152
Thamban C, Kalavathi S, Anithakumari P, Jaganathan (2016) Reaching Coconut Farmers and other Stakeholders. In: CPCRI Centenary Souvenir. ICAR-CPCRI, Kasaragod, pp 97–110
Thampan PK (1999) Enhancing the income and employment in the coconut sector through conservation and use of special coconut eco types in India. Report on farmer participatory survey. Peekay Tree Crops Development Foundation, Kochi
Watanayothin S (2004) Strategies for enhancing productivity and income of coconut farmers in Thailand. In: Rethinam P (ed) Proceedings of of the XLI Cocotech meeting on "Strategies for enhancing productivity and income of coconut farmers". APCC, Jakarta, pp 70–72
Yarbro S (1988) Farmers receptivity to new technologies in coconut: Papua New Guinea. Asia Pacific Coconut Community, Jakarta, p 22
Yogananda HG, Ramachandra KT, Manjula N (1994) Knowledge level of coconut growers. J Extn Edun 5(3):15–21

Dr. S. Arulraj served as Scientist (Agricultural Extension) and subsequently as Principal Scientist and Head of Division (Social Sciences); Project Coordinator, All India Coordinated Research Project on Palms at ICAR-CPCRI; as well as Director, ICAR-Indian Institute of Oil Palm Research in Andhra Pradesh, India. His major areas of research and development are participatory technology development, participatory technology transfer, documentation of indigenous technical knowledge, public-private sector partnership for agricultural development, policy research and use of information technology for transfer of technology. aruldopr@gmail.com

Dr. C. Thamban is working as Principal Scientist (Agricultural Extension) at ICAR-CPCRI, Kasaragod. He has also worked as Agricultural Extension Officer in the Department of Agriculture, Kerala State. His areas of interest include research on participatory technology transfer and community extension approaches in plantation crop sector with special emphasis on coconut. He has published 40 research papers and has edited 5 books. c.thamban@gmail.com

Dr. D. Jaganathan is a Scientist in agricultural extension at the ICAR-CTCRI, Trivandrum, Kerala, India. He has 8 years of experience in coconut while working at ICAR-CPCRI, Kasaragod, Kerala, India. He is a recipient of IARI Gold Medal for outstanding academic performance. His expertise is on human resource management, farmers' participatory research and innovation system analysis. He has published 20 research papers, 3 books and 10 chapters in books. djaganathtn@gmail.com

Chapter 15
Nutrition and Health Aspects of Coconut

T. Rajamohan and U. Archana

Abstract Coconut provides a wide spectrum of human consumable products such as oil, kernel, tender nut water, toddy, neera, coconut sugar, immature inflorescence and haustorium. Despite enormous beneficial utilities of coconut oil, many believe that it increases blood cholesterol level, thus promoting the risk of cardiac diseases. There are enough research results to indicate the health benefits of coconut oil because of its unique fatty acid composition. Wet processing of fresh coconut kernel yields virgin coconut oil (VCO) which is very nutritious and has important health-promoting properties. Coconut oil has antibacterial, anti-protozoan and antiviral properties because of the medium chain fatty acid (lauric acid) it contains. Apart from coconut oil, coconut kernel contains many beneficial factors. Coconut kernel protein possesses cardioprotective and antidiabetic properties. Dietary fibre isolated from coconut kernel has significant hypocholesterolemic effect, and its hemicellulose component is responsible for the cholesterol-lowering action. Research indicates that tender coconut water (TCW) and mature coconut water can reduce the occurrence of lifestyle diseases. TCW consumption reduces the risk of heart disease and has significant hypocholesterolemic, antioxidant and antithrombotic effects. Furthermore, hypoglycaemic effect of mature coconut water has been reported in diabetic rats. Neera tapped from coconut inflorescence enhances digestion and has a low glycaemic index. Studies indicate that immature coconut inflorescence can lower blood glucose in diabetic rats. Coconut haustorium possessing significant cardioprotective and antioxidant properties is underscored. This chapter draws attention on the health benefits of coconut products evidenced through systematic scientific approach.

T. Rajamohan (✉)
Coconut Research and Development Centre, Thiruvananthapuram, Kerala, India
e-mail: tr.mohan@yahoo.co.in

U. Archana
Westvest 233, Delft, The Netherlands
e-mail: archanaravind@gmail.com

K. U. K. Nampoothiri et al. (eds.), *The Coconut Palm (*Cocos nucifera *L.) - Research and Development Perspectives*,
https://doi.org/10.1007/978-981-13-2754-4_15

15.1 Introduction

Coconut is a major part of the diet for many people who live in tropical regions. The main consumable products of coconut are coconut oil, coconut kernel, tender coconut water and haustorium. Most important edible products derived from inflorescence include neera, toddy and coconut sugar.

15.2 Coconut Oil

Coconut oil is prepared both in households and commercial establishments. Oil extracted by dry process from copra is called coconut oil or copra oil, and the one extracted from wet kernel is known as virgin coconut oil. Please refer to Chap. 13 for details on processing.

Coconut oil is a light coloured, saturated and stable oil of natural origin, with a pleasant aroma and pleasing flavour. It can be used for cooking without refining as a frying and seasoning oil. The oil is biodegradable and highly resistant to oxidative deterioration during frying, which makes it ideal for high-heat cooking. The unique nutritional and medicinal qualities of coconut oil are easy digestibility, absorbability and readily oxidisable nature leading to reduced fat accumulation in the body. The oil is an excellent fat source in the preparation of filled milk and infant food formulae. It is used as massage oil due to its ready penetration into the skin (CDB 2005). Coconut oil exhibits antimicrobial and antiviral properties.

15.2.1 Composition of Coconut Oil

Coconut oil contains 90–92% saturated fatty acids and about 8–10% unsaturated consisting of oleic acid (C18:1) and linoleic acid (C18:2) as triglycerides. The saturated fatty acids in coconut oil are mostly medium chain fatty acids, and lauric acid (C12:0) is the most predominant among them. In Table 15.1, the fatty acid composition of coconut oil and VCO is compared.

15.2.2 Dietary Features of Coconut Oil

Among the different edible oils, coconut oil stands out as a triglyceride with unique physical, chemical and structural characteristics. Though coconut oil is basically a saturated fat, 63.5% of the total fatty acids consist of short and medium carbon chains of 12 or less carbon atoms. Consequently, coconut oil exhibits dietary properties which are specific to the group of short and medium chain fatty acids. On the

Table 15.1 Fatty acid composition of coconut oil and virgin coconut oil

Fatty acids	Composition	Percentage	
		Coconut oil	VCO
Caprylic acid	C 8:0	8.15	8.050
Capric acid	C 10:0	5.56	5.420
Lauric acid	C 12:0	43.55	45.510
Myristic acid	C 14:0	18.38	19.740
Palmitic acid	C 16:0	8.25	7.830
Stearic acid	C 18:0	2.65	3.140
Oleic acid	C 18:1	6.70	4.700
Linoleic acid	C 18:2	1.49	1.880
Arachidic acid	C 20:0	0.086	0.086

Source: Nevin and Rajamohan (2006)

other hand, the fatty acids present in other edible oils belong to the group of long chain fatty acids which are chemically different from the short and medium chain fatty acids. The only other major source of medium chain fatty acid is palm kernel oil (Babayan 1968).

The short and medium chain fatty acids (SMCFAs) have higher solubility in water and in other biological fluids than the long chain fatty acids (LCFAs). These features facilitate the action of various lipases leading to faster and complete digestion of glycerides composed of SMCFAs. The products of hydrolysis reach the liver directly via the portal venous system and preferentially are used for energy production. Studies have revealed that supplementation of coconut oil beneficially modulates fatty acid metabolism by reducing lipogenesis and enhancing the rate of fatty acid catabolism (Arunima and Rajamohan 2014). The only limitation of coconut oil as the sole source of dietary fat is its extremely low content of essential fatty acids. However, in the major coconut-consuming regions, the local populations have access to diverse food sources such as cereals, pulses, tubers, fish, etc. which provide substantial additions of essential fatty acids (EFAs) in their daily diet. Moreover, the presence of 8–10 carbon fatty acids which constitute 15% of the total fatty acids of coconut oil diminishes the requirement of EFAs in human nutrition (Kaunitz 1983).

15.2.3 Health Effects of Coconut Oil

Although coconut oil has many beneficial uses, the propaganda against the consumption of coconut oil is still continuing and is debated due to its high saturated fat content. It is generally believed that saturated fatty acids increase blood cholesterol and thus promote the risk of heart disease. Research findings now suggest that coconut oil can make a comeback because of its unique health-promoting properties.

15.2.3.1 Coconut Oil and Heart Disease

Coconut oil, being rich in saturated fatty acid, is frequently used for investigating the saturated fat hypothesis and generally maligned as hypercholesterolemic. High intake of saturated fats is considered as a risk factor for the occurrence of coronary heart disease (CHD). Several epidemiological studies however revealed that association between saturated fat and CHD was much weaker than that predicted by international comparisons (Keys 1980), because saturated fat increases HDL cholesterol which to a great extent compensates for its adverse effect on LDL cholesterol (Sreevastava 1994). Because of the high content of saturated fats in coconut oil, there are concerns that it could lead to more atherogenic lipid profiles (Ahrens 1957; Keys et al. 1957). It is argued that coconut oil supplementation increases adverse lipids, thus, promoting the risk of heart disease (Hegsted et al. 1965; Anderson et al. 1976). But Prior et al. (1981) showed that there is no harmful effect when islanders had a high intake of coconut oil. Blackburn et al. (1989) in a review concluded that coconut oil is a neutral fat in terms of atherogenicity. Thampan (1994) highlighted the fact that people consuming large amounts of fresh coconut kernel in a varied diet in different parts of the world did not demonstrate hypercholesterolemia and coronary atherosclerosis. Though coconut is a main ingredient in the diet of Polynesians and Filipinos, their blood cholesterol levels and incidence of heart diseases are reported to be lower (Dayrit 2000).

Research studies carried out in 258 human volunteers in Kerala population, India, revealed that coconut oil consumption does not cause increase in blood LDL cholesterol, but it increases HDL cholesterol. Consumption of coconut kernel along with coconut oil had a beneficial effect in that it reduced total cholesterol and more importantly raised HDL cholesterol levels and lowered LDL cholesterol/HDL cholesterol ratio (Rajamohan 1997). Comparative studies on CAD patients consuming coconut oil and sunflower oil did not show any significant difference in the blood lipid profile, vascular function and antioxidant levels (Vasudevan 2010, Vijayakumar et al. 2016).

Researchers of Harvard Medical School have shown that coconut oil reduces the risk of heart disease due to favourable alteration of blood lipid profile (Norton et al. 2004). Assuncao et al. (2009) conducted a study in women between the age of 20 and 40 with abdominal obesity given either soya bean oil or filtered coconut oil and regularly exercised. They reported that the subjects showed weight loss and only the coconut oil administered group experienced reduction in abdominal obesity and an improved LDL cholesterol/HDL cholesterol ratio, whereas the soya bean group actually had increased total and LDL cholesterol as well as decreased HDL cholesterol levels.

Studies by Feranil et al. (2011) in premenopausal women in the Philippines showed that coconut oil intake was positively associated with beneficial lipid profiles. A review by Eyres et al. (2016) does not support the view that use of coconut oil reduces the risk of heart disease. It has been pointed out that the quantity of cooking oil is more important than the type of cooking oil in contributing to the risk of CAD (Sabitha and Vasudevan 2010). The National Lipid Association also suggests that it is better to restrict the quantity of coconut oil, when used as part of a daily diet (Jacobson et al. 2015).

There are reports that in addition to hypercholesterolemia, free radical-mediated oxidative stress plays a key role in the occurrence of CVD (Singh and Devaraj 2006). Coconut oil is less susceptible to lipid peroxidation than unsaturated fatty acid-rich oils (Ergun et al. 2005). Free radical damage has long been believed to be a risk factor for the degenerative process which accompanies ageing. Comparative studies carried out in young and aged rats fed with coconut oil and sunflower oil indicate that the rate of tissue lipid peroxidation was less in coconut oil-fed rats when compared to sunflower oil-fed ones (Shalini and Rajamohan 2012). Feeding thermally oxidised fats increases the lipid peroxidation products in the tissues (Liu and Huang 1995). The products thus formed during oxidation have been suggested to have deleterious effects on the health of human beings and animals, since they can lead to atherosclerosis (Cohn 2002).

Studies carried out in rats showed that consumption of unsaturated fatty acid-rich cooking oil heated repeatedly at elevated temperatures is more deleterious to health compared to saturated fatty acid-rich oils such as coconut oil (Chacko and Rajamohan 2011). This study indicates that dietary oils heated repeatedly at elevated temperatures result in significant alterations in plasma lipids, peroxide levels, platelet aggregation and platelet function compared to fresh oils in rats and the deleterious effects were lower in heated coconut oil compared to heated mustard oil and sunflower oil. This observation suggests that unsaturated fatty acid-rich oils which are considered to be beneficial for consumption, when used as fresh oil, lose their beneficial effects on repeated heating.

Researchers and nutritionists are now beginning to recognise the need to distinguish features of dietary oils other than the degree of saturation. Apart from the structure and composition of triglycerides, the nature and quantity of non-glyceride components like polyphenol, tocopherols, carotenes, etc., all of which are essential factors, cannot be ignored.

15.2.3.2 Therapeutic Properties of Virgin Coconut Oil

VCO, extracted by wet processing, is very valuable because of its unique health-promoting properties. Most of the biologically active minor components are inactivated when coconut oil is extracted in the traditional method through extraction from copra which has been exposed to high temperatures. On the other hand, VCO extraction, by wet processing, retains most of the nonsaponifiable components. As in the case of coconut oil obtained from copra, VCO contains almost similar percentage of fatty acids, mainly medium chain fatty acids. In addition, it also contains higher amounts of biologically active components, viz. polyphenols, tocopherol and phytosterols with antioxidant and other beneficial properties (Nevin and Rajamohan 2004). Dia et al. (2005) revealed that VCO had higher total phenolic content compared to refined coconut oil. High total phenolic content in VCO has been reported by Marina et al. (2009) also.

Studies have been made by several investigators to determine the effects of VCO on blood lipids and lipid peroxidation and its effect on CVD. Animal studies showed that supplementation of VCO increased the HDL cholesterol compared to copra oil and groundnut oil (Nevin and Rajamohan 2006, 2009). Rats fed with VCO, copra oil, olive oil and sunflower oil showed that VCO feeding lowered the levels of total cholesterol, LDL + VLDL cholesterol, Apo B and triglycerides in serum, while HDL cholesterol and Apo A_1 were significantly higher in rats fed with VCO compared to other oil-fed groups. The increased Apo A_1 levels in VCO-fed rats were directly correlated with the increased HDL cholesterol levels (Arunima and Rajamohan 2012). Apo A_1 is the major protein component in HDL in serum, and its concentration is inversely correlated with the risk of premature atherosclerosis (Miccoli et al. 1996). These observations clearly indicate that consumption of VCO has significant beneficial effect on lipid metabolism than coconut oil extracted from copra.

Tissue lipid peroxidation has been linked to aetiology of various degenerative diseases (Halliwell 1997). The increased lipid peroxidation leads to the oxidative modification of LDL which plays a major role in pathogenesis of atherosclerosis (Dillon et al. 2003). Studies demonstrated that feeding VCO increases the antioxidant status, preventing the oxidative damage of lipids and protein oxidation (Arunima and Rajamohan 2013). It also possesses significant antithrombotic effect by inhibiting the activation of platelets and coagulation factors compared to rats fed with copra oil, olive oil and sunflower oil (Arunima and Rajamohan 2016). These findings indicate that supplementation of VCO has significant cardioprotective effects. The potential benefits of VCO in maintaining lipid levels, antioxidant status and antithrombotic effects are due to the presence of lauric acid and biologically active unsaponifiable components like polyphenols, tocopherols and phytosterols. Wet processing of VCO helps to retain higher amounts of polyphenols which is capable of reducing lipid levels and lipid peroxidation (Marina et al. 2009; Nevin and Rajamohan 2006; Mansor et al. 2012).

Administration of VCO ameliorates blood glucose and oxidative stress in diabetic rats (Iranloye et al. 2013; Akinnuga et al. 2016). Kochukuzhiyil et al. (2010) demonstrated that coconut oil-enriched diet helps to prevent accumulation of body fat and prevents insulin resistance. In type 1 diabetes, brain function can be improved by medium chain fatty acids found in coconut oil (Page et al. 2009). MCFA in coconut oil regulates blood sugar and increases insulin secretion and insulin sensitivity (Garfinkel et al. 1992; Han et al. 2003).

15.2.3.3 Coconut Oil and Antimicrobial Property

The antimicrobial properties of the medium chain fatty acids found in coconut oil have been known to researchers since 1966. Coconut oil is predominantly composed of medium chain fatty acids (MCFA), lauric, capric and caprylic acids. All the three of the medium chain fatty acids and medium chain triglycerides possess antimicrobial properties (Kabara et al. 1972). According to published reports, lauric acid is one of the best inactivating fatty acids, and its monoglyceride (monolaurin) is the most effective one (Kabara et al. 1972; Sands et al. 1979; Fletcher et al. 1985). Bacteria, yeast, fungi and enveloped viruses are inactivated by monolaurin. The lipid membranes of the microorganisms are inactivated by the medium chain fatty acids and their derivatives (Isaacs and Thormar 1991; Isaacs et al. 1992). Hornung et al. (1994) have shown that the antimicrobial effect in viruses is related to monolaurin's interference with virus assembly and viral maturation. However, Projan et al. (1994) reported that one of the antimicrobial effects in bacteria is related to monolaurin's interference with signal transduction/toxin formation. Ogbolu et al. (2007) indicated the possibility of using coconut oil in the treatment of fungal infections, in view of its antifungal activity against *Candida* spp. especially since some of the *Candida* species are becoming drug-resistant.

Monolaurin also inactivates the protozoan parasite *Giardia lamblia.*

15.2.3.4 Other Beneficial Properties

Studies demonstrated that VCO has potential beneficial effects in arthritis-induced rats (Vysakh et al. 2014). The anti-inflammatory and antioxidant effects of VCO which help to prevent CVD are due to its higher phenolic fraction. VCO has also been proved to have a positive effect on Alzheimer's disease (AD). Phenolic compounds of coconut oil may inhibit the aggregation of amyloid β peptide, a key step in the pathogenesis of AD (Porat et al. 2006). Coconut oil is easily absorbed and metabolised by the liver and can be converted into ketones. Ketone bodies may be beneficial to people developing or already suffering from memory impairment as in AD since they are important alternate energy sources in the brain (Fernando et al. 2015). Coconut oil has a beneficial role in colon cancer (Reddy and Maeura 1984) and breast cancer also (Cohen et al. 1984). Lauric acid is known to show anticancer activity by its ability to induce apoptosis (Fauser et al. 2013).

15.3 Tender Coconut Water (TCW)

TCW is a clear liquid found inside the tender coconut of 6–8 months' maturity. It has been a popular refreshing drink in the tropical countries, either served fresh, chilled or packed. Coconut water possesses several properties beneficial to human health (Pradera et al. 1942; Anurag and Rajamohan 2003a). Coconut water contains several bioactive components, viz. sugars, electrolytes, minerals, vitamins, amino acids, phytohormones, cytokinins and enzymes such as catalase, dehydrogenase, polymerases, etc. As it contains plant enzymes and sugar, TCW undergoes biochemical changes and loses its quality once the nuts are harvested. Various efforts have been made to preserve the quality and increase the shelf life of TCW through acidification, pasteurisation, etc. Technology is available to extend the shelf life of TCW by packing in flexible pouches and aluminium beverage containers (CDB 2017).

15.3.1 Chemical Constituents of Coconut Water

Coconut water contains minerals, sugars, vitamins and proteins besides growth-promoting factors and neutral fats. The composition of coconut water depends upon various factors like variety, nut maturity, soil features and climatic conditions (Msengi et al. 1985; Jayalakshmi et al. 1986). The major chemical constituents are sugars and minerals, the minor ones being fat, proteins and other nitrogenous substances (Chavalittamrong et al. 1982). Sugar content in the nut water increases from 1.5% to 5 to 5.5% during nut development and decreases to about 2% at full maturity (Child and Nathanael 1950). In the fully mature nut, sucrose content is approximately 50% of total sugar (Marar 1958).

The nitrogen and total protein concentration increases with maturity (Birosel et al. 1976; Jayalakshmi et al. 1986). The predominant proteins are globulin, albumin, glutelin and prolamine. Even though coconut water is not a good source of protein, it contains most of the amino acids (Suresh et al. 1968; Baptist 1956). In the ripening nut, the free amino acids increase from 4 mg to 16 mg 100 ml^{-1} (Shivashankar 1991). Among the free amino acids, L-arginine constitutes the major fraction. TCW contains 30 mg% of L-arginine (Sandhya and Rajamohan 2006). Fat is a minor constituent, and changes occur in fat content and fatty acid composition during maturity (Jayalakshmi et al. 1986). A comparison of the composition of TCW and mature coconut water (MCW) is presented in Table 15.2.

Coconut water is an abundant source of minerals. Composition analysis of TCW showed that it contains 300 mg/dl potassium. TCW is one of the best sources of electrolytes. The liquid endosperm containing nuclei as well as both positive and negative ions makes it a biological electrolyte (Bamunuarachchi and Ranaweera 2007). It contains almost the same level of electrolyte balance as in the blood (Suresh et al. 1968).

Tender nut water is a rich source of vitamin C (25 mg/dl) which is an important water-soluble antioxidant and acts as a radical scavenger. It also contains polyphenols (Bhagya et al. 2010a). Another useful component present in the water is phytohormones, namely, cytokinins (Kende and Zeevart 1997). Research studies suggest

Table 15.2 Composition of tender (TCW) and mature coconut water (MCW)

Constituents	TCW	MCW
Total sugar (%)	4.8	3.1
Total reducing sugar (%)	4.0	2.0
Total protein (mg/dl)	150	450
L-arginine (mg/dl)	30	150
Vitamin C (mg/dl)	25	15
Magnesium (mg/dl)	16	14
Potassium (mg/dl)	300	257
Calcium (mg/dl)	40	44

Source: Sandhya and Rajamohan (2008)

that cytokinins (e.g. kinetin and trans-Zeatin) show anti-ageing, anticarcinogenic as well as antithrombotic effects (Vermeulen et al. 2002). The natural pH of coconut water varies between 4.9 and 5.2. The acidity may be due to the organic acids and free amino acids present (Jayalakshmi et al. 1986). Dissolved CO_2 evolved during tissue respiration, and also fatty acids contribute to acidity.

15.3.2 Therapeutic Properties of Tender Coconut Water

Tender coconut water (TCW) is consumed all over the world not only because it is a refreshing drink but also because of its numerous therapeutic qualities. TCW has a unique composition of vitamins, minerals, amino acids, phytohormones and sugars due to which reason it is very useful in the upkeep of health (Yong et al. 2009). The most important use of TCW is as a rehydration medium (Pummer et al. 2001). TCW is also used as a blood plasma substitute and is readily accepted by the body (Anzaldo et al. 1975). The water is reported to be beneficial to bladder infections, kidney stones and sexual vitality (Macalalag Jr and Macalalag 1987). TCW is recommended as the best treatment for diarrhoea in children (Cooper 1986). It is also reported to contain substances capable of rapid proliferation of plant tissues (Tulecke et al. 1961) and has been used as a bacterial and plant tissue culture media (Smith and Bull 1976; Marquez et al. 1987).

Shah (1956) showed that intake of tender coconut water reduces the risk of heart disease. TCW has significant hypocholesterolemic, antioxidant and antithrombotic effects as evidenced from experimental rats with induced myocardial infarction (Anurag and Rajamohan 2003b). Feeding TCW showed improved activities of mitochondrial enzymes and provided protection against free radical-mediated damage, induced by isoproterenol (Anurag and Rajamohan 2003b). Pretreatment with TCW decreased the clotting and showed antithrombotic effects in rabbits with induced myocardial infarction. Studies using TCW and thrombolytic drug, streptokinase, showed similar thrombolytic effects, while the antioxidant effects were more with TCW (Prathapan and Rajamohan 2010). In another study, feeding TCW showed hepatoprotective and antioxidant effects in carbon tetrachloride-intoxicated rats (Anthony and Rajamohan 2003). TCW has a beneficial effect on blood pressure and lipid levels in fructose-fed hypertensive rats (Bhagya et al. 2010a). When such rats were treated with TCW, the blood pressure could be kept under control. The levels of serum lipids were higher in hypertensive rats compared to normal ones. Treatment with TCW decreased the levels of total cholesterol, triglycerides and free fatty acids both in serum and tissues. In fructose-fed hypersensitive rats, tender coconut water had a beneficial effect as far as aldosterone level, plasma renin activity and electrolyte imbalance are concerned (Bhagya et al. 2010b). Alleyne et al. (2005) also showed that consumption of TCW is effective in bringing about the control of hypertension.

15.3.3 Health Effects of Mature Coconut Water

While tender coconut water is mainly consumed as a natural drink, mature coconut water (MCW) is usually discarded. According to Sandhya and Rajamohan (2008), compared to TCW, MCW is superior in hypolipidemic action in cholesterol-fed rats. The quantitative difference in the composition of the active components in MCW is the reason for its superiority. Comparative studies using tender and mature coconut water indicated that administration of MCW had significant hypoglycaemic effect in diabetic rats and reduced the pancreas damage induced by alloxan and stimulated beta-cell regeneration (Preetha et al. 2013a). Comparative evaluation of lyophilised MCW and glibenclamide in alloxan-induced diabetic rats revealed that MCW has beneficial effect against diabetic-induced complications and its effects were comparable to those of the standard drug, glibenclamide (Preetha et al. 2013b). The defects of hyperlipidemia and lipid metabolism in diabetic rats were significantly reduced by MCW (Preetha et al. 2013c). Treatment with MCW and L-arginine exhibited significant antithrombotic activity in diabetic rats, which was evidenced from the reduced level of WBC, platelets, fibrin and fibrinogen. In addition, the activity of nitric oxide synthase in the liver and plasma arginine content and urinary nitrite was higher in MCW treated rats, and the effects were comparable with L-arginine-treated group (Preetha et al. 2015). This finding indicates that L-arginine in coconut water is one of the major factors which ameliorates the alterations in blood glucose, lipid levels and coagulation factors through the increased nitric oxide availability. Diabetes has a degenerative and destructive effect on the kidney which can be partially but significantly reversed by concomitant administration of coconut water (Nwangwa 2012).

15.4 Coconut Kernel

Since ancient times, coconut has been used as a vital source of food in many of the tropical countries. It is a major culinary ingredient in recipes in several parts of India, particularly in Kerala. Coconut kernel is a complete food rich in calories, vitamins and minerals. The kernel contains potassium, calcium, magnesium, manganese, iron, copper, phosphorus, sulphur and chlorine as well as vitamins, namely, thiamine, ascorbic acid, vitamin A, tocopherol, phenolic compounds and phytohormones. Fresh coconut kernel contains 7% dietary fibre and 5% proteins, in addition to coconut oil. Presence of these macro- and micronutrients in fresh coconut kernel is one of the reasons behind its biological effects. Studies indicate that consumption of coconut oil alone has less beneficial effects on blood cholesterol levels than that of coconut kernel along with coconut oil (Padmakumaran Nair et al. 1998a).

15.4.1 Coconut Kernel Protein (CKP)

The coconut protein is known to be of good nutritional quality. Seventy to eighty percent of the coconut kernel proteins are globulins, which are good protein from the point of view of digestibility and biological value. Amino acid analysis of the protein carried out by Salil et al. (2011) revealed that it has a much higher amount of L-arginine. The health benefits of coconut kernel protein are mostly due to the presence of L-arginine. Coconut kernel possesses significant hypolipidemic effect (Padmakumaran Nair et al. 1998b). Coconut kernel protein has a significant cardio-protective effect on isoproterenol-induced myocardial infarction in rats (Mini and Rajamohan 2002). Studies have shown that diet containing coconut kernel protein protects the heart by beneficially modulating endothelial nitric oxide synthase and antioxidant mechanism due to its ability to inhibit tumour necrosis factor-alpha and nuclear factor-kappa B activation in isoproterenol-induced myocardial infarction (Remya et al. 2013). Dietary supplementation of coconut kernel protein may have greater significance in reducing the extent of oxidative stress and inflammatory responses associated with myocardial infarction. The results show that the oxidative stress during isoproterenol administration in rats was alleviated effectively by coconut kernel protein, possibly due to the presence of L-arginine. The beneficial effects of coconut kernel protein in reducing glucose, enzymes and insulin facilitating carbohydrate metabolism in diabetic rats have been proved by the studies undertaken by Salil et al. (2011). Coconut kernel protein reversed the damage caused to the pancreas by alloxan as evidenced from the histopathological studies.

15.4.2 Coconut Kernel Fibre

Coconut kernel is an excellent dietary fibre source. It is prepared from finely ground, dried and defatted coconut kernel. It has the highest dietary fibre content than any other dietary fibre supplements and has all the benefits of other dietary fibres. Studies revealed that coconut fibre has significant hypocholesterolemic effect, and the results indicate that its hemicellulose component may be responsible for the cholesterol-lowering action (Sindhurani and Rajamohan 1998). Chemical composition of neutral detergent fibre (NDF) isolated from coconut kernel contains hemicellulose (45.10%), cellulose (43.28%), lignin (8.15%), cutin (3.28%) and silica (0.25%). Investigations suggest that inclusion of coconut fibre in the diet results in significant hypoglycaemic action (Sindhurani and Rajamohan 2000). Feeding dietary fibres isolated from coconut kernel and black gram showed significant decrease in serum and tissue lipids and lipid peroxides in high-fat diet-fed rats (Thampi et al. 1991). The fibre contained in the coconut kernel has a protective role in the prevention of

1,2-dimethylhydrazine-induced colon carcinogenesis (Manoj et al. 2001). Protective effect of coconut cake against colon carcinogenesis has also been reported (Nalini et al. 2004). In addition, the beneficial role of minor components such as phytohormones and phenolic compounds present in coconut kernel in preventing Alzheimer's disease has also been reported (Fernando et al. 2015).

15.5 Other Edible Coconut Products

15.5.1 Coconut Haustorium

Haustorium is a spongy part developed from the basal part of the embryo during germination, filling the entire cavity, in 8–16 weeks. During this period, it mobilises nutrients in the endosperm and nourishes the germinating embryo (Balachandran and Arumughan 1995). It is a nutritious and tasty tropical delicacy. Coconut haustorium contains about 66% carbohydrates in which 64% is sugar. It is a rich source of phenolics and has high antioxidant capacity (Arivalagan et al. 2018). The haustorium is a rich source of phosphorus and potassium and can be used as an additive in food for children (Konan et al. 2017). It is a good source of sucrose, fructose, glucose, phosphorous and potassium. Studies have shown that during germination, the total starch content of haustorium increased linearly, whereas reducing and soluble sugars decreased rapidly and remained in a steady state thereafter. The excess carbohydrates mobilised from the kernel are stored in the haustorium as starch (Balasubramaniam et al. 1973).

Studies using aqueous extract of coconut haustorium indicated cardioprotective and antioxidant properties during isoproterenol-induced myocardial infarction in rats (Chikku and Rajamohan 2012, 2013). In the myocardium of coconut haustorium pretreated rats, activities of antioxidant enzymes increased, and peroxidation products decreased. Histopathological examination of the heart of isoproterenol administered rats showed extensive cardiac damage, while rats pretreated with aqueous extract of haustorium showed minimal histological changes. These findings clearly show that coconut haustorium has significant cardioprotective property. Fresh coconut haustorium is enriched with phytoconstituents such as proteins, carbohydrates, terpenoids and flavanoids which possess strong anti-inflammatory and antioxidant activities (Abiraami Valli and Uma Gowri 2017). These workers reported the presence of squalene, a terpenoid showing strong antiulcer affinity against *Helicobacter pylori*, an ulcer-causing bacteria, and recommended coconut haustorium as a natural, economically potent food source for human health. Coconut haustorium can be used to improve foetal growth and development in polytocous animals by reducing foetal number (Sumiaty et al. 2017).

15.5.2 Coconut Inflorescence

Coconut inflorescence is used for production of diverse products for human consumption.

15.5.2.1 Value-Added Products from Coconut Inflorescence

Coconut Sap The coconut sap (neera, also called palm nectar) is the sap extracted from the inflorescence of coconut which is sweet, translucent and high in nutritional value. Due to its very good taste, it has attained popularity as a soft drink. But one of the issues is its short life. This has been overcome through technologies developed by ICAR-CPCRI and KAU (Kerala Agricultural University) for collecting fresh, hygienic and unfermented coconut inflorescence sap (Jayaprakash et al. 2013, CPCRI 2016). For details please see Chap. 13.

Neera is known to improve digestion and enable clear urination. Since its glycaemic index is only 35, neera can be consumed even by diabetic patients. Neera is rich in carbohydrates and sugar (10–15%), and its pH is almost neutral. It is a rich source of amino acids, antioxidants and vitamins such as nicotinic acid, vitamin C and vitamin B complex (Aalbersberg et al. 1997; Hebbar et al. 2015). The composition of neera is given in Table 15.3.

Coconut Toddy When fermented, neera becomes toddy due to conversion of sugar to alcohol (Iwuoha and Eke 1996). Coconut toddy, also known as palm wine, is a sweet alcohol beverage. It is known by many names in different regions and is a common drink in various parts of Asia, Africa, the Caribbean and South America. In small quantities, it cures intestinal disorders and insomnia. In some areas toddy is evaporated to produce unrefined sugar called jaggery.

Table 15.3 Composition of neera

Total sugar	18–20%
Vitamin C	1.3 mg/100 ml
p^H	6.8
Acidity	10.0 meq/l
Phenols	8.0 mg/1009 ml
Minerals	
Potassium	90.5 ppm
Calcium	60.0 ppm
Phosphorus	15.0 ppm
Iron	45.0 ppm
Sodium	9.5 ppm

Source: Jayaprakash et al. (2013)

Coconut Sugar Coconut sugar is produced from neera. Neera is boiled over moderate heat until most of the water is evaporated. As the water evaporates, it starts to transform into fine granules of sugar. Coconut sugar is subtly sweet, almost like brown sugar. It is used as a sweetener in many countries. Coconut sugar contains 70–79% sucrose and nearly 3% of both glucose and fructose (Purnomo 1992). Coconut sugar also contains several nutrients, viz. proteins, minerals, short chain fatty acids, polyphenols and antioxidants. According to the Philippine Department of Agriculture, coconut sugar has a low glycaemic index (GI) of 35 which is much lower than the value of table sugar (GI 60). Lower glycaemic index value of coconut sugar has also been reported by others (Trinidad et al. 2010; Srikae and Thongta 2015). Consumption of low glycaemic index diets is recommended because it decreases the risk of developing diabetes and cardiovascular disease (Jenkins et al. 2002). Palm sugars contain a fibre called inulin, which may decrease glucose absorption (Trinidad et al. 2010; Vayalil 2012).

Inflorescence Cream A creamy preparation obtained from young inflorescence is used to cure backache in Indian traditional medicine. Young coconut inflorescence (20% w/w)-containing diet has beneficial effect in diabetic rats. It improves glucose homoeostasis and antioxidant status in diabetic rats (Renjith and Rajamohan 2012b). Onset of hyperglycaemia resulting from alloxan administration was prevented by a lowered blood glucose level resulting from a treatment with inflorescence. In diabetic rats, activity of antioxidant enzymes declined, while their activities were reverted to normal range in inflorescence-supplemented animals. The young coconut inflorescence exhibited protective and ameliorative effects against alloxan-induced pancreative cytotoxicity and severe hyperglycaemia and also in repairing and rejuvenating the residual beta-cell population (Renjith and Rajamohan 2012a). These effects may be due to the presence of phenolic acids, flavanoids and other phytoconstituents which could act synergistically or independently in modulating the activities of glycolytic and gluconeogenic enzymes.

15.6 Future Strategy

It is advantageous to the coconut industry, if coconut moves out of the oil sector, to the extent possible, taking full advantage of its health and nutritional properties in view of the strong competition from other vegetable oils. Although the medicinal value of coconut and coconut products has been clearly indicated from experiments on animals, it is necessary to validate these by bringing in concrete evidence on its therapeutic properties and mode of action in curing specific diseases. With the growing demand for virgin coconut oil, it is important that the quality characteristics of the oil are studied and improved.

It is now necessary to follow up the traditional knowledge and the effects from animal studies to clinical outcomes for prevention and treatment of various diseases. In addition, studies are warranted using underutilised factors for exploring the medicinal benefits. The recognised medicinal properties of coconut oil by the traditional medical systems have to be validated through continued studies on dietary and medicinal qualities of coconut oil.

It is encouraging that many laboratories are undertaking research on coconut oil. Carefully controlled research is necessary before we can determine the long-term impact of coconut oil, especially in the case of long-term progressive dementias such as Alzheimer's disease or related disorders.

References

Aalbersberg B, Rohinee S, Paveen R (1997) Nutrient analysis of coconut toddy. Trop Sci 37(3):160–163

Abiraami Valli S, Uma Gowri S (2017) A study on the bioactive potential of fresh and dried coconut sprouts of *Cocos nucifera* L. an *in vitro* and in-silico approach. Intl J Pharm Sci 9(3):129–142

Ahrens EH (1957) Nutritional factors and serum lipid levels. American J Med 23:928

Akinnuga AM, Jejo SO, Bamidele O, Omu IE, Onum TB, Sunday VE, Odu OP (2016) Virgin coconut oil (VCO) supplemented diet ameliorates blood glucose, renal tissue and oxidant enzymes in diabetic rats. J Intl Res Medi Pharmaceu Sci 9(2):49–53

Alleyne T, Roacho S, Thomas C, Shirley A (2005) The control of hypertension by use of coconut water and mauby: two tropical food drinks. West Indian Med J 54:3–8

Anderson JR, Grande F, Keys A (1976) Independence of effects of cholesterol and degree of saturation of the fat in the diet on serum cholesterol in man. American J Clin Nutri 21:1184–1189

Anthony LL, Rajamohan T (2003) Hepatoprotective and antioxidant effect of tender coconut water on carbon tetra chloride induced liver injury in rats. Indian J Biochem Biophys 40:354–357

Anurag P, Rajamohan T (2003a) Effect of tender coconut water in experimental myocardial infarction. Plant Foods Hum Nutr 58:1–12

Anurag P, Rajamohan T (2003b) Beneficial effects of tender coconut water against isoproterenol induced toxicity on heart mitochondrial activities in rats. Indian J Biochem Biophy 40:27–28

Anzaldo FE, Kintanar QL, Recio PM, Velasco RU, Dela Cruz F, Jacaline A (1975) Coconut water an intravenous fluid. Philippine J Pediatr 24:143–166

Arivalagan M, Rakesh B, Sugatha P, Poonam S, Karteek H, Santosh RK (2018) Biochemical and nutritional characterization of coconut (*Cocos nucifera* L) haustorium during germination. Food Chem 238:153–159

Arunima S, Rajamohan T (2012) Virgin coconut oil improves hepatic lipid metabolism in rats-compared with copra oil, olive oil and sunflower oil. Indian J Exp Biol 50:802–809

Arunima S, Rajamohan T (2013) Effect of virgin coconut oil enriched diet on the antioxidant status and para oxonase1 activity in ameliorating the oxidative stress in rats – a comparative study. Food Funct 4:1402–1409

Arunima S, Rajamohan T (2014) Influence of virgin coconut oil enriched diet on the transcriptional regulation of fatty acid synthesis and oxidation in rats – a comparative study. British J Nutr 111:1782–1790

Arunima S, Rajamohan T (2016) Supplementation of virgin coconut oil compared with copra oil, olive oil and sunflower oil on thrombotic factors in rats and *in vitro* platelet aggregation. Intl J Curr Res Bio Sci Plant Bio 3(2):106–113

Assuncao ML, Ferreira HS, Dos Santos AF, Cabral CR Jr, Florencio TM (2009) Effects of dietary coconut oil on the biochemical and anthropometric profiles of women presenting abdominal obesity. Lipids 44(7):593–601

Babayan VK (1968) Medium chain triglycerides-their composition, preparation and application. J American Oil Chem Soc 45(1):23–25

Balachandran C, Arumughan C (1995) Triglyceride deposition in tissues of germinating coconut (*Cocos nucifera* Linn.). J American oil Chem Soc 72:1583–1586

Balasubramaniam K, Atukorala TMS, Wijesundera S, Hoover AA, De Silva MAT (1973) Biochemical changes during germination of the coconut (*Cocos nucifera*). Ann of Bot 37(3):439–445

Bamunuarachchi A, Ranaweera KKDS (2007) Coconut products in health and nutrition. In: Thampan PK, Vasu KI (eds) Coconut for rural welfare. APCC, Jakarta, pp 307–319

Baptist NG (1956) Gamma –amino- butyric acid and other free amino acids in the coconut. Nature 178:1403–1404

Bhagya D, Prema L, Rajamohan T (2010a) Beneficial effects of tender coconut water on blood glucose and lipid levels in experimental hypertension. J Cell Tissue Res 10(1):2139–2144

Bhagya D, Prema L, Rajamohan T (2010b) Tender coconut water maintains the level of electrolytes and renin in fructose-fed hypertensive rats. Intl J Biol Med Res 1(4):341–346

Birosel DM, de Oliviera FV, Holeberg IB, Pitelli AC (1976) Isolation of proteins in coconut water. Rev Farm Bioquin Univ Salo Paulo 14:35–42

Blackburn GL, Kates G, Macioli EA, Kovalchuk M, Babayan VK, Bistrain DR (1989) A revaluation of coconut oils effect on serum cholesterol and atherogenesis. J Philippine Med Assoc 65:144–152

CDB (2005) Coconut oil-unique and healthy. Coconut Development Board Publication No.133 p 6

CDB (2017) Preservation and Packing of Tender Coconut water. http:/www.coconutboard.nic.in/tender.htm. Downloaded, November, 2017

Chacko C, Rajamohan T (2011) Repeatedly heated cooking oils alter platelet functions in cholesterol fed rats. Intl J Biol Med Res 2(4):991–997

Chavalittamrong B, Pidatcha P, Thavisri U (1982) Electrolytes, sugars, calories, osmolarity and pH of beverage and coconut water. South East Asian J Trop Med Public Health 13:427–431

Chikku AM, Rajamohan T (2012) Dietary coconut sprout beneficially modulates cardiac damage induced by isoproterenol in rats. Bangladesh J Pharmacol 7(4):258–265

Chikku AM, Rajamohan T (2013) Coconut haustorium maintains cardiac integrity and alleviates oxidative stress in rats subjected to isoproterenol induced myocardial infarction. Indian J Pharm Sci 74(5):397–402

Child R, Nathanael WRM (1950) Changes in the sugar composition of coconut water during maturation and germination. J Sci Food Agri 1(11):326–329

Cohen LA, Thompson DO, Maeura Y, Weisburges JH (1984) Influence of dietary medium chain triglycerides on the development of N-Methyl nitrosourea induced rat mammary tumour. Cancer Res 44(11):5023–5028

Cohn JS (2002) Oxidized fat in the diet, post prandial lipaemia and cardio vascular disease. Curr Opin Lipidol 13:19–24

Cooper ES (1986) Coconut water. Lancet 2:281

CPCRI (2016) A century of service to the nation. Chowdappa P, Augustine Jerard B, Niral V, Jayasekhar S, Hebbar KB, Muralikrishna H (eds), ICAR-CPCRI, Kasaragod, India, Publication p 26

Dayrit CS (2000) Coconut oil in health and disease: its and Monolaurin's potential as cure for HIV/AIDS. XXXVII Cocotech Meeting Chennai, India, July 25, 2000

Dia VP, Garcia VV, Mabesa RL, Teeson–Mendoza EM (2005) Comparative physicochemical characteristics of virgin coconut oil produced by different methods. Philippine Agr Sci 88:462–475

Dillon SA, Burmi RS, Lowg GM, Billington D, Rehman K (2003) Antioxidant properties of aged garlic extract : an *in vitro* study – incorporating human low density lipoproteins. Life Sci 72:1583–1594

Ergun S, Yontem M, Yerlikaya A, Ozata A, Uysal K, Kurt H (2005) Influence of dietary oils on blood lipid peroxidation. Saudi Med J 26:442–446

Eyres L, Eyres MF, Chisholm A, Brown RC (2016) Coconut oil consumption and cardio vascular risk factors in human. Nutri Rev 74(4):267–280

Fauser JK, Mathews GM, Cummins AG, Howarth GS (2013) Induction of apoptosis by the medium chain length fatty acid lauric acid in colon cancer cells due to induction of oxidative stress. Chemotherapy 59(3):214–224

Feranil AB, Duazo PL, Kuzawa CW, Adair LS (2011) Coconut oil predicts a beneficial lipid profile in pre-menopausal women in the Philippines. Asia Pacific J Clin Nutri 20(2):190–195

Fernando WMADB, Martins IJ, Goozee KG, Brennan CS, Jayasena V, Martins RN (2015) The role of dietary coconut for the prevention and treatment of Alzheimer's disease : potential mechanism of action. British J Nutr 114:1–14

Fletcher RD, Albers AC, Albertson JN, Kabara JJ (1985) Effects of monoglycerides on mycoplasma pneumoniae growth. In: Kabara JJ (ed) Pharmacological effects of lipids II. American oil chemists' Society, Champaign, pp 59–63

Garfinkel M, Lee S, Opera EC, Akwari OE (1992) Insulinotropic potency of lauric acid: a metabolic rationale for medium chain fatty acids (MCF) in TPN formulation. J Surg Res 52:328–333

Halliwell B (1997) Antioxidants and human disease. A general introduction. Nutr Rev 55:554–549

Han J, Hamilton JA, Kirkland JL, Gorkey BE, Guo W (2003) Medium chain oil reduces fat mass and down regulates expression of adipogenic genes in rats. Obesity Res 11:734–744

Hebbar KB, Arivalagan M, Manikantan MR, Mathew AC, Thamban C, Thomas GV, Chowdappa P (2015) Coconut inflorescence sap and its value addition as sugar-collection techniques, yield, properties and marketing perspective. Curr Sci 109(8):1411–1417

Hegsted DM, Mc Gandy RB, Myers ML, Starc FJ (1965) Quantitative effects of dietary fat on serum cholesterol in humans. American J Clin Nutri 17:281–295

Hornung B, Amtmann E, Sauer G (1994) Lauric acid inhibits the maturation of vesicular stomatitis virus. J Gen Virol 75(2):353–361

Iranloye B, Oludare G, Olubiyi M (2013) Anti-diabetic and antioxidant effects of virgin coconut oil in alloxan induced diabetic male Sprague Dawley rats. J Diabetes Mellitus 3(4):221–226

Isaacs CE, Thormar H (1991) The role of milk and milk derived antimicrobial lipids as antiviral and antibacterial agents. In: Mesteky J et al (eds) Immunology of milk and the neonate. Plenum press, New York

Isaacs CE, Litov RE, Marie P, Thormar H (1992) Addition of lipases to infant formulas produces antiviral and antibacterial activity. J Nutri Biochem 3:304–308

Iwuoha CI, Eke OS (1996) Nigerian indigenous foods: their food traditional operation inherent problems- improvements and current status. Food Res Int 29:527–540

Jacobson TA, Maki KC, Orringer KC et al (2015) National lipid association recommendation for patient centred management of Dyslipidemia. Part. 2. J Clin Lipidol. https://doi.org/10.1016/j.jacl.2015.09.002 Epub 2015 Sep 18

Jayalakshmi A, Arumughan C, Narayanan CS, Mathew AG (1986) Changes in the chemical composition of coconut water during maturation. J Sci Food Agric 84:203–207

Jayaprakash Naik B, Suresh PR, Manjusha M, Balachandran TV, Subramonian M, Balakrishnan PC (2013) Keraamritham- a healthy drink from coconut inflorescence sap. Indian Cocon J (Special issue):9–10

Jenkins DJ, Kendall CW, Augustin LS, Franceschi S, Hamidi M, Marchie A, Jenkins AL, Axelsen M (2002) Glycemic index. Overview of implications in health and disease. American J Clin Nutri 76:266–273

Kabara JJ, Swieezkowski DM, Conley AJ, Truant JP (1972) Fatty acids and derivatives as antimicrobial agents. Antimicrob Agents Chemotherapy 2(1):23–28

Kaunitz H (1983) Coconuts today 1(2): 27–29 Cited in: Julian A Banzon et al. (1990) Coconut as food. Philippines Cocon Reseach and Development Foundation Dilman, Quezon City

Kende H, Zeevart J (1997) The five classical plant hormones. Plant Cell 9:1197–1210

Keys A (1980) Seven countries: a multi variate analysis of death and coronary heart diseases. Harward University press, Cambridge, MA

Keys A, Anderson JT, Grande F (1957) Prediction of serum cholesterol responses to man to change in fats in the diet. Lancet 2:959–966

Kochukuzhiyil BM, Devi K, Fattepur SR (2010) Effect of saturated fatty acid – rich dietary vegetable oils on lipid profile, antioxidant enzymes and glucose tolerance in diabetic rats. Indian J Pharmaco A2 3:142–145

Konan BR, Assa RR, Kouassi KN, Konan KJL, Et Amani NG (2017) Variation in physico–chemical parameters of coconut (*Cocos nucifera* L) haustorium during germination. Intl J Agron Agri Res 10(5):17–25

Liu JF, Huang CJ (1995) Tissue alpha–tocopherol retention in male rats is compromised by feeding diets containing oxidized frying oil. J Nutri 125:3071–3080

Macalalag EV Jr, Macalalag AL (1987) Bukolysis: young coconut water renoclysis for urinary stone dissolution. Int Surg 72:247

Manoj G, Thampi BSH, Leelamma S, Menon PVG (2001) Effect of dietary fiber on the activity of intestinal and fecal beta-glucuronidase activity during 1,2-dimethylhydrazine induced colon carcinogenesis. Plant Foods Hum Nutr 56(1):13–21

Marquez JC, Scorza JV, Anez N (1987) Coconut water as a monophasic liquid medium for cultivating Trypanosomatidae. Trans R Soc Trop Med Hyg 81:615

Mansor TST, Che Man YB, Shuhaimi M, Abdul Afiq MJ, Ku Nurul FKM (2012) Physicochemical properties of virgin coconut oil extracted from different processing methods. Intl Food Res J 19:837–845

Marar MMK (1958) Optimum stage of harvesting coconut for different purposes. Indian Cocon J 11:39–47

Marina AM, Che Man YB, Amin I (2009) Virgin coconut oil: emerging functional food oil. Trends Food Sci Technol 20(10):481–487

Miccoli R, Bestolotto A, Navalesir OL, Boni A, Wessling J, Funke H, Wiebush H, Eckardstein A, Assman G (1996) Compound heterozygosity for a structural apolipoprotein A1 null allele in patients with absence of HDL cholesterol, corneal opacifications and coronary heart disease. Circulation 9:1622–1628

Mini S, Rajamohan T (2002) Cardioprotective effect of coconut kernel protein in isoproterenol administered rats. Indian J Biochem Biophy 39:197–200

Msengi AE, Mbise RL, Msuya PM, Amsi DM (1985) The biochemistry of water from unripe coconuts obtained from two localities in Tanzania. East African Med J 62:725–729

Nalini N, Manju V, Menon VP (2004) Effect of coconut cake on the bacterial enzyme activity in 1,2 – dimethyl hydrazine induced colon cancer. Clin Chim Acta 342:203–210

Nevin KG, Rajamohan T (2004) Beneficial effects of virgin coconut oil on lipid parameters and *in vitro* LDL oxidation. Clinical Biol Chem 37:830–835

Nevin KG, Rajamohan T (2006) Virgin coconut oil supplemented diet increases the antioxidant status in rats. Food Chem 99:260–266

Nevin KG, Rajamohan T (2009) Wet and dry extraction of coconut oil: impact on lipid metabolic and antioxidant status in cholesterol co administered rats. Canadian J Physiol Pharmacol 87:1–7

Norton D, Angeeman S, Istfan N, Lopes S, Babayan VK, Putz MC, Steen SN, Black Burn GL (2004) Comparative study of Coconut oil, soybean oil and hydrogenated soybean oil. Philippine J Sci 29:1–2

Nwangwa EK (2012) The renoprotective effects of coconut water on the kidneys of diabetic wistar rats. J Health Sci 2(1):1–4

Ogbolu DO, Oni AA, Daini OA, Oloko AP (2007) *In vitro* antimicrobial properties of coconut oil on Candida species in Ibadan, Nigeria. J Med Food 10(2):384–387

Padmakumaran Nair KG, Rajamohan T, Kurup PA (1998a) Coconut kernel protein modifies the effect of coconut oil on serum lipids. Plant Foods Human Nutri 53:133–144

Padmakumaran Nair KG, Rajamohan T, Kurup PA (1998b) Changes in the metabolism of lipoproteins in rats fed coconut kernel protein. J Clin Biochem Nutri 25:159–168

Page KA, Williamson A, Namyi YU, Menay EC, Dzuria J, McCrimmon RJ, Sherwin RS (2009) Medium fatty acids improve cognitive function in intensively treated type I diabetic patients and support *in vitro* synaptic transmission during acute hypoglycaemia. Diabetics 58(5):1237–1244

Porat Y, Abramowitz A, Gazit E (2006) Inhibition of amyloid fibril formation by polyphenols: structural similarity and aromatic interactions as a common inhibition mechanism. Chem Biol Drug Des 67(1):27–37

Pradera ES, Fernandez F, Calberin O (1942) Coconut water, clinical and experimental and study. American J Dis Child 64:977–987

Prathapan A, Rajamohan T (2010) Antioxidant and antithrombotic activity of tender coconut water in experimental myocardial infarction. J Food Biochem 35(5):1501–1507

Preetha PP, Girija Devi V, Rajamohan T (2013a) Hypoglycemic and antioxidant potential of coconut water on carbohydrate metabolism and pancreatic pathology of alloxan induced diabetic rats. European J Integr Med 5:234–240

Preetha PP, Girija Devi V, Rajamohan T (2013b) Comparative effects of mature coconut water (*Cocos nucifera* L) glibenclamide on some biochemical parameters in alloxan induced diabetic rats. Brazilian J Pharmaco 23(3):481–487

Preetha PP, Girija Devi V, Rajamohan T (2013c) Anti hyperlipidemic effects of mature coconut water and its role in regulating lipid metabolism in alloxan induced experimental diabetes. Comp Clin Pathol 23:1331–1337

Preetha PP, Girija Devi V, Rajamohan T (2015) Mature coconut water exhibits antidiabetic and antithrombotic potential via L-arginine- nitric oxide pathway in alloxan induced diabetic rats. J Basic Clin Physiol Pharmacol 26(6):575–583

Prior IA, Davidson F, Salmond CE, Czochanska Z (1981) Cholesterol, coconuts and diet on Polynesian atolls: a natural experiment: the Pukapuka and Tokelau island studies. American J Clin Nutri 34(8):1552–1561

Projan SJ, Brown-Skrobot S, Schlievert PM, Vandenesch F, Novick RP (1994) Glycerol monolaurate inhibits the production of beta-lactamase, toxic shock toxin-1,and other staphylococcal exoproteins by interfering with signal transduction. J Bacteriol 176(14):4204–4209

Pummer S, Heil P, MaleckWand PG (2001) Influence of coconut water on hemostasis. Ann J Emerg Med 19:287–289

Purnomo H (1992) Sugar components of coconut sugar in Indonesia. ASEAN Food J 7(4):200–201

Rajamohan T (1997) Report of the study on the effect of consumption of coconut kernel and coconut oil on the serum lipid profile. Coconut Development Board, Government of India, Kochi

Reddy BS, Maeura Y (1984) Tumour promotion by dietary fat in azoxy-methane-induced colon carcinogenesis in female F344 rats; influence of amount and source of dietary fat. J Natl Cancer Inst 72:745–750

Remya S, Chikku AM, Renjith RS, Arunima S, Rajamohan T (2013) Coconut kernel protein in diet protects the heart by beneficially modulating endothelial nitric oxide synthase tumour necrosis factor – alpha and nuclear factor – kappa B expressions in experimental myocardial infraction. J Food Drug Anal 21:325–331

Renjith RS, Rajamohan T (2012a) Protective and curative effect of *Cocos nucifera* Inflorescence on Alloxan-induced pancreatic cytotoxicity in rats. Indian J Pharmacol 44(5):555–559

Renjith RS, Rajamohan T (2012b) Young inflorescence of *Cocos nucifera* contributes to improve glucose homeostasis and antioxidant status in diabetic rats. Intl J Diab Dev Ctries 32(4):193–198

Sabitha P, Vasudevan DM (2010) Lipid profile and antioxidant enzymes in coconut oil consumers. Indian Cocon J 42(12):3–6

Salil G, Nevin KG, Rajamohan T (2011) Arginine rich coconut kernel protein modulates diabetics in alloxan treated rats. Chem Biol Interact 189:107–111

Sandhya VG, Rajamohan T (2006) Beneficial effects of coconut water feeding in lipid metabolism in cholesterol fed rats. J Med Food 9(3):400–407

Sandhya VG, Rajamohan T (2008) Comparative evaluation of the hypolipidemic effects of coconut water and lovastatin in rats fed with fat-cholesterol enriched diet. Food Chem Toxicol 46:3586–3592

Sands JA, Auperin D, Landlin P, Reinhardt A, Cadden SP (1979) Antiviral effects of fatty acids and derivatives. In: Kabara JJ (ed) Pharmacological effects of lipids. American oil Chemists Soc, Champaign, p 75

Shah NJ (1956) Use of coconut water in treatment in congestive cardiac failure. Indian J Med Res 44:341–351
Shalini AN, Rajamohan T (2012) Age dependent variation in the antioxidant status and oxidative stress in rats fed diet containing coconut oil and sunflower oil. Intl J Biol Med Res 3(2):158–194
Shivashankar S (1991) Biochemical changes during fruit maturation in coconut. J Plantn Crops 19:102–109
Sindhurani JA, Rajamohan T (1998) Hypolipidemic effect of hemicellulose component of coconut fibre. Indian J Exptl Biol 36:786–789
Sindhurani JA, Rajamohan T (2000) Effects of different levels of coconut fibre on blood glucose, serum insulin and minerals in rats. Indian J Physiol Pharmacol 44:97–100
Singh U, Devaraj S, Gialal (2006) Oxidative stress and atherogenesis. Pathophysiology 13:129–141
Smith ME, Bull AT (1976) Studies of the utilization of coconut water waste for the production of the food yeast saccharomyces fragilis. J Appl Bacteriol 41:81–95
Sreevastava RA (1994) Saturated fatty acid but not cholesterol, regulates apolipoproteins a 1 gene expression of post transcriptional mechanisms. Biochem Mol Biol Intl 34:393–402
Srikaeo K, Thongta R (2015) Effects of sugar cane, palm sugar, coconut sugar and sorbitol on starch digestibility and physicochemical properties of wheat based foods. Intl Food Res J 22(3):923–929
Sumiaty A, Ajik S, Hera M, Wasmen (2017) The effects of coconut sprout administration during pregnancy in rats. Intl J Sci Basic Appl Res 33(3):105–119
Suresh VP, Hegde VR, Setty SV, Rangachar TR (1968) Fluid therapy by tender coconut water in veterinary practice. Indian Vet J 45:829–837
Thampan PK (1994) Facts and fallacies about coconut oil. Asian Pacific Coconut Community, Jakarta 10013
Thampi BS, Manoj G, Leelamma S, Menon VP (1991) Dietary fibre and lipid peroxidation: effect of dietary fibre on levels of lipids and lipid peroxides in high fat diet. Indian J Exp Biol 29(6):563–567
Trinidad TP, Mallillin AC, Sagum RS, Encabo RR (2010) Glycemic index of commonly consumed carbohydrate foods in the Philippines. J Funct Foods 2(4):271–274
Tulecke W, Heinstein LH, Rutucer A, Lausengot HI Jr (1961) The biochemical composition of coconut water (coconut milk) as related to its use in plant tissue culture. Contrib Boyce Thompson Inst 2:115–128
Vasudevan DM (2010) Lipid profile in blood and in plaque material from diseased coronary artery. Indian Cocon J 43(6):17–21
Vayalil PK (2012) Date fruits (Phoenix dactylifera Linn): an emerging medicinal food. Crit Rev Food Sci Nutri 52:249–271
Vermeulen K, Strned M, Krystof V, Havliuk L, Vanderla A, Lengou M, Ngis G, Rodrigus I, Stockman B, Van Onckelen H, Van Bockstaele DR, Bernemam ZN (2002) Antiproliferative effect of plant cytokinin analogues with inhibitory activity on cyclin dependent kinases. Leukemia 16:299–305
Vijayakumar M, Vasudevan DM, Sundaram KR, Krishnan S, Vaidyanathan K, Nandakumar S, Chandrasekhar R, Mathew N (2016) A randomised study of coconut oil vs sunflower oil on cardiovascular risk factor in patients with stable coronary heart disease. Indian Heart J 68(4):498–506
Vysakh A, Ratheesh M, Rajmohanan T, Pramod C, Premalal S, Gireeshkumar B, Sibi PI (2014) Polyphenolics isolated from virgin coconut oil inhibits adjuvant induced arthritis in rats through antioxidant and anti-inflammatory action. Intl Immuno Pharmacol 20(1):124–130
Yong JWH, Ge L, Ng YG, Tan SN (2009) The chemical composition and biological properties of coconut (*Cocos nucifera L.*) water. Molecules 14(12):5144–5164. https://doi.org/10.3390/molecules14125144

Dr. T. Rajamohan retired as Professor and Head, Department of Biochemistry, University of Kerala, India. He is presently the Director of the Coconut Research and Development Centre, Thiruvananthapuram, Kerala, India. He is a Biochemist by qualification and has 30 years of postgraduate teaching and research experience. His major areas of research are in atherosclerosis, diabetes, toxicology, nutrition and health, and he is especially interested in dietary coconut products in health and disease. He has published 75 papers and 1 book based on health aspects of coconut products and won the National Award for Best Research Worker 2010. tr.mohan@yahoo.co.in

Mrs. U. Archana is a nutrition specialist interested in the nutrition and health aspects of vegetable oils. She has an MPhil in Food and Nutrition and is a postgraduate in Bioinformatics (Radboud University, The Netherlands). She is presently working as Nutritionist at Sadhbhavana World School at Kozhikode, India. She has published five research papers on the subject. archanaravind@gmail.com

Chapter 16
Climate Change, Carbon Sequestration, and Coconut-Based Ecosystems

P. K. Ramachandran Nair, B. Mohan Kumar, and S. Naresh Kumar

Abstract Climate change, a key global environmental issue of the day, refers to the gradual increase in temperature of the Earth's atmosphere. It is believed to be caused by the increase in atmospheric concentration of carbon dioxide and other greenhouse gases (GHGs). Concerned by the serious consequences of anthropogenic climate change, several global initiatives have been launched to address the issue. They fall under two broad categories, climate-change mitigation and adaptation, aimed at reducing GHG emissions and their negative impacts, respectively. Carbon sequestration, the prominent mitigation strategy, refers to capturing atmospheric carbon and securing it in long-lived pools, such as through photosynthesis by plants. Climate-smart agriculture is the rallying theme for adaptation strategies, which is a combination of site-specific management activities. Most climate-change mitigation and adaptation studies in agriculture so far have focused on annual crops, with little attention being paid to perennials such as coconut. Coconut-based ecosystems offer good possibilities for enhancing carbon sequestration through crop combinations involving a variety of plants including food crops, tubers, vines, and tree crops. For climate-change adaptation, the annual intercrops planted under coconuts could be managed for optimum benefits for the whole system. A holistic approach focusing on the overall productivity and sustainability of the whole system rather than the palm alone is needed to make the coconut-based agroecosystems resilient to climate change.

P. K. R. Nair (✉)
School of Forest Resources and Conservation, University of Florida, Gainesville, FL, USA
e-mail: pknair@ufl.edu

B. Mohan Kumar
College of Forestry, Kerala Agricultural University, Thrissur, Kerala, India
e-mail: kau@gmail.com; bm.kumar@kau.in

S. Naresh Kumar
Centre for Environment Science and Climate Resilient Agriculture,
ICAR-Indian Agricultural Research Institute, New Delhi, India
e-mail: snareshkumar.iari@gmail.com

K. U. K. Nampoothiri et al. (eds.), *The Coconut Palm (Cocos nucifera L.) - Research and Development Perspectives*,
https://doi.org/10.1007/978-981-13-2754-4_16

16.1 Introduction

"Climate change" was probably unheard of a 100 years ago when coconut research was initiated in India. Although memorable climatic events such as floods and droughts may have been talked about, climate change may not have been thought of as an item of coconut research agenda. Today, climate change is a prime research agenda in all agricultural research endeavors including coconut research. After years of gridlock, a new United Nations' agreement on sustainable development bolstered the fight against environmental degradation, and nearly 200 countries signed a historic agreement on climate change. Thus, climate change and discussions and action plans surrounding it are happening in a big way.

Climate change, and the companion term global warming, refers primarily to a gradual increase in the average temperature of the Earth's atmosphere and its oceans, a change that is believed to be permanently changing the Earth's climate. Field investigations on the impacts of climate change on agricultural systems have mostly been limited to cereals and other short-duration species; such studies have not been extended in a significant manner to tree crops, in general, and to coconut in particular. Nevertheless, some projections can be made on the potential impact of climate change on coconut-based ecosystems, based on the current insights and understanding from other studies. The primary objectives of this chapter are to make some such projections and suggest some such strategies for minimizing the adverse effects.

It is important that the commonly used technical terms on the topic are explained and the causes, consequences, and strategies for climate-change mitigation and adaptation (M & A) are discussed in some detail at the outset. This chapter explains those details, presents the global spread of coconut-based ecosystems, briefly reviews the current state of knowledge on climate-change impacts and M & A strategies for agricultural systems, and makes some projections on climate-change scenarios for coconut-based ecosystems.

16.2 Climate Change: The Language, Extent, Causes, and Consequences

Some of the commonly used terms used in climate-change discussions are explained in this section. These details and explanations are compiled from various sources, the most significant being the Intergovernmental Panel on Climate Change (IPCC; www.ipcc.org). Additional reference sources include Simpson (2016) and FAO's (Food and Agricultural Organization; www.fao.org) Climate-Smart Agriculture (CSA) Sourcebook (http://www.fao.org/3/i3325e.pdf).

16.2.1 The Extent of Climate Change

The Earth continues to be warmer than it was several decades ago. According to NASA (the US National Aeronautical and Space Administration: https://www.nasa.gov/), all 10 warmest years in the 134-year documentation have all happened since 2000 (with the exemption of 1998); 2010 and 2005 ranked as the warmest years on record. The average global temperature has increased by about 0.8 °C (1.4 °F) since 1880 and 0.6 °C (1.0 °F) since 1970 as per the January 2014 assessment of NASA's Goddard Institute for Space Studies (GISS). The IPCC has projected that mean global temperatures could escalate between 1.4 and 5.8 °C by the year 2100. Studies conducted in India indicate that in the past 100 years, annual mean maximum temperatures over India increased by about 0.71 °C (INCCA 2010). Similarly, the minimum temperature has been rising at a rate of 0.29 °C/10 years. During 1871–2017 period, India faced 29 deficit and 20 excess monsoon years. Among these, as many as 17 deficit monsoon years and 6 excess monsoon years occurred in the post-1960 period.

It is worthwhile in this context to point out the difference between variability and change. *Variability* in weather is the difference between anything that occurs normally and that actually occurs. Variations can be observed between various time periods such as days, weeks, months, and years. *Change* is the trend in average conditions (of temperature, rainfall quantity and pattern, wind parameters, etc.) in one direction or another. Variability does not necessarily bring about increase or decrease from mean over a period of time; but change does.

16.2.2 Causes of Climate Change: Greenhouse Gases

Climate change (global warming) is thought to be caused by the rise in atmospheric concentration of greenhouse gases (GHGs). The key GHGs are carbon dioxide (CO_2), methane (CH_4), and nitrous oxide (N_2O). Among these, CO_2 is the major and most common one, but other gases are more powerful: methane is 25 times and N_2O nearly 298 times more potent than CO_2 (IPCC 2014). These are called GHGs because of their effect in the atmosphere, where they intercept the outgoing long-wave radiation from the Earth's surface. Solar energy (short-wave radiation) is absorbed by Earth's objects and reradiated back as long-wave radiation (heat). Trapping of GHGs in the atmosphere prevents the long-wave radiation escaping from the Earth back into space, in a manner analogous to the glass ceiling of a greenhouse that permits sunlight to pass in but traps the sun's heat within.

Atmospheric concentration of CO_2 has augmented from the preindustrial concentration of about 280 ppm to the current level of approximately 400 ppm, increasing at an average of 2 ppm per year (Fig. 16.1). This increase is thought to be caused by human activities, including burning of fossil fuels (about 60% of total) such as coal, gas, and oil for industrial and other purposes; agriculture, forestry, and other

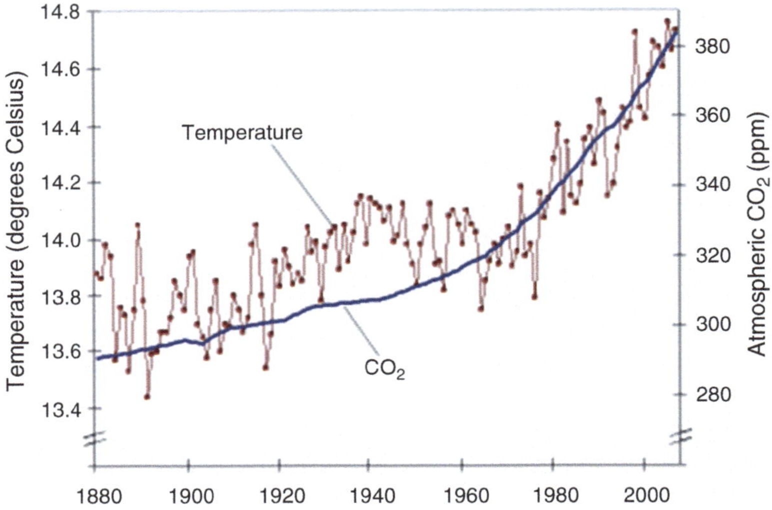

Fig. 16.1 Average global temperature and atmospheric carbon dioxide concentrations, 1880–2007. (Source: NASA GISS and NOAA/ESRL (USA))

land-use activities (AFOLU) including deforestation (approximately 24%); and transportation (about 14%). The full effect of warming is likely to continue into the foreseeable future as long as emission of GHGs continues unabated. Therefore, it is important to reduce GHG emission levels.

Climate-change "deniers" argue that the Earth's climate has always been changing; however, correlation is not the same as causation, i.e., although there is a correlation between increases in GHG concentration and atmospheric temperature, no direct cause-effect relationship has been established between the two, and that global temperature, in spite of its variability between years, has increased only by a mean of less than 1 °C in the past 130 years. The climate-change deniers also argue that most of the climate-change predictions are built on flawed computer models, and the so-called global consensus on climate change is generated more by political and executive pressures and coercion rather than solid scientific data. But the vast majority of the world and public opinion as well as scientific consensus are moving with the conviction that climate change is a real global issue and it needs to be addressed with deserving seriousness.

16.2.3 Impacts of Climate Change

Listed below are the key potential consequences of global temperature increase (climate change):

1. *Rainfall patterns*: Air temperature increase leads to increased evaporation of moisture (water) from soil and water bodies (rivers, lakes, and oceans). Moreover, warmer air holds more moisture than cooler air. Together, these cause more water to move through the global climate system leading to extra rainfall. The surge in rainfall, however, will be distributed unevenly: Less in drier areas and more in wetter areas. The total quantity of rainfall, its distribution, and the pattern (onset of rainy and dry seasons) will be unpredictable. All these effects are now of common occurrence throughout the world.
2. *Sea-level rise*: Similar to the phenomenon of warmer air holding more moisture, the volume of water expands with increase in temperature, leading to rise in sea levels. Ocean levels have levitated 20 cm over the past century, and higher levels of sea-level rises in the future are predicted. About a seventh of the world's human population live within about 10 m of sea level.
3. *Melting of glaciers and permafrost*: Increase in air and ocean temperatures causes melting of glaciers and ice caps, leading to further sea-level rise. As the temperature rises and permafrost melts, bacterial activity will increase leading to faster decomposition of organic matter and release of CO_2 and methane caught in a frozen condition in the Arctic Ocean floor.
4. *Inundation of low-lying islands*: Addition of large volumes of water to world's oceans will lead to inundation of low-lying islands (as has happened may 2016 in Solomon Islands in the Pacific Ocean: http://www.scientificamerican.com/article/sea-level-rise-swallows-5-whole-pacific-islands/).
5. *Ocean currents and El Niño/La Niña*: When ocean surface becomes much warmer than normal, an El Niño is said to occur; a La Niña happens when the reverse situation occurs resulting in lowering of sea surface temperatures. During an El Niño episode (as in 2014–2015), heavy rains and strong tropical storms may occur in some faraway areas and drought in others.
6. *Indirect effects*: These include effects on pollinators, pests and diseases, weeds and invasive species, and other ecosystem and environmental services.

16.2.4 Carbon Sequestration

Carbon sequestration is the important strategy for reducing atmospheric concentrations of GHGs, especially CO_2, and thereby global warming. The United Nations Framework Convention on Climate Change describes it "as the process of removing carbon (C) from the atmosphere and depositing it in a reservoir, or the transfer of atmospheric CO_2 to secure storage in long-lived pools" (UNFCCC 2007). Plants of all types, especially trees, and soil are important in C sequestration. Plants take up atmospheric C for photosynthesis and store the product of photosynthesis in plant parts. The soil is a major C sink; globally, soil to 1 m depth is estimated to contain 2300 Pg (1 petagram = 10^{15} g = 1 billion ton) consisting of 1550 Pg as soil organic C (SOC) and 750 Pg as inorganic C. This total soil C pool is threefold higher than the atmospheric pool (770 Pg) and 3.8 fold higher than the vegetation pool of 610 Pg.

16.2.5 Climate-Change Mitigation and Adaptation

Mitigation signifies the technological change and substitution that decrease GHG emissions through avoiding emissions and sequestering GHGs. Mitigation strategies include:

- Avoiding or reducing the emissions through increasing input-use efficiency by adopting sound management of nutrients and water and decreasing losses through appropriate soil and water conservation strategies.
- Sequestering CO_2 in terrestrial biosphere such as in biomass aboveground and belowground and soil C pools.

Adaptation refers to initiatives and actions to decrease the susceptibility of natural systems to climate change, by developing strategies to reduce the adverse consequences. These include:

- Enhancing soil resilience (augmenting SOC pool, restoring degraded lands).
- Adopting efficient land-use practices such as agroforestry and conservation agriculture.

16.3 Coconut-Based Ecosystems

Climate-change-related discussions pertaining to specific ecosystems or regions need to be contextualized because many of the manifestations of climate change and strategies for coping with them have to be specific according to the characteristics of the system and region in consideration. With that in view, it is important to review briefly the overall global distribution and management of coconut ecosystems.

16.3.1 Characteristics and Distribution

The coconut palm (*Cocos nucifera* L.), the most widely cultivated of all palms and one of the earliest among domesticated plants, has been described with various adulations and accolades, indicating the usefulness of its various parts and products. The palm has been and still is an intimate part of the bio-cultural legacy and economic well-being of the populations of its principal growing regions. As Purseglove (1972) states, coconut's uses are legion; every part of the palm is useful to man in one way or another.

Coconut is presently grown in over 93 countries (FAO 2018). The area under coconuts is difficult to estimate accurately because of the lack of standardized procedures for estimating areas when the species grows both naturally in scattered stands at varying densities and is planted and nurtured as a crop either alone (Fig. 16.2) or in combination with various other species (Fig. 16.3). This situation also raises doubts about the accuracy of repeatedly used area statistics. Equally unsatisfactory is the situation about its production statistics because a major share of the production is consumed locally in the households without the products entering any form of marketing channels. Nevertheless, statistics on area and production have traditionally been compiled by local, regional, national, and international agencies; the most widely cited and supposedly authentic are the FAO statistics (http://faostat.fao.org). Coconuts are common in countries of the tropical regions especially in coastal areas of Africa and the LAC (Latin America and the Caribbean), mostly in natural stands but also substantially in planted and managed stands. The vast majority of such planted and managed stands of the palm are in smallholder

Fig. 16.2 Sole stand of coconuts. Photograph of a sole stand of about 40-year-old coconut palms at the Coconut Research Institute, Lunuwila, Sri Lanka, 7.5 m × 7.5 m spacing, 175 palms per hectare, showing the high level of light (solar radiation) availability and low shading on the plantation floor, indicating the opportunities for intercropping. (Photo: P. K. R. Nair 1984)

Fig. 16.3 An intensively managed multi-species crop combination with coconuts at ICAR-Central Plantation Crops Research Institute in Kasaragod, India. The species other than the coconut palm include: Banana (*Musa* sp.), Black pepper (*Piper nigrum*), Pine apple (*Ananas comosus*), and nutmeg (*Myristica fragrans*). (Photo: M.A.Nair)

farms of less than 5 ha; the farms in Asia, the main coconut-growing region of the world, are much smaller.

16.3.2 Common Land-Use Features of Coconut-Based Ecosystems

Being a mono-stemmed woody perennial with no cambium, the principal stem (trunk) of the palm does not grow radially with age. In an even-aged (planted) stand of palms, this characteristic growth form permits substantial light infiltration into the understory, as the palm increases in height with age (Fig. 16.2). It is quite common to see younger coconut palms and a variety of shade-tolerant other species growing under tall (older) palms. Thus, smallholder farms of coconut that predominate the coconut areas comprise mostly of palms together with an array of other species of all types: herbs, shrubs, vines, and trees (Fig. 16.3). These multi-species, smallholder production systems, managed as family-farm enterprises, have been acclaimed as "an agroecological marvel" (Nair 2017). While the area estimates of coconut ecosystems are likely to be inaccurate as explained above, they could even be more inaccurate for intercropping systems involving coconuts. Exceptions to this general rule of smallholder farms are found in commercial holdings, as in large "coconut estates" that were developed in the Pacific islands by the European settlers during the twentieth century and a large number of smallholder plantations established later in those areas by local farmers (Bonnemaison 1996). In such situations, uniformity in spacing, mostly 7.5 m square (Fig. 16.2), and even-aged palms (with about 180 palms per ha) are the norm.

Intercropping under or between coconuts with a variety of other useful species is a common practice in most coconut-growing regions of the world. Nair (1979) has articulated the chronological sequence of the growth of coconut palms in Kerala (India) and the potential for growing a wide spectrum of intercrops at various growth stages of the palm. Reports on intercropping and the array of crops grown in different countries and regions abound. "The species so intercropped consist of food crops including roots and tubers, fruit trees, tree-plantation crops, medicinal plants, multipurpose trees, and others that provide multiple products such as food, fuel, fodder, timber, medicine, and such other basic necessities, and help meet the cash requirements of the growers" (Kumar 2007; Lamanda et al. 2006). Such integrated farming systems are generally characterized by higher productivity, better returns, and improved social functionality than many commercial farming systems (Tipraqsa et al. 2007).

16.3.3 Growth Habits of Coconut Palm in Relation to Multispecies Systems

The amenability of coconut stands for intercropping depends primarily on the growth stages of the palm. Based on the amount of light transmitted through coconut canopy during the palm's growth stages, Nelliat et al. (1974) divided the life span of the palm into three distinct phases from the perspective of intercropping in a sole stand of palms (see Chap. 7 for details).

Managed multi-species, multi-strata systems with the coconut palm and a wide variety of other economically useful plants are now quite common on the west coast of India (Fig. 16.3). The central hypothesis underlying the functioning of such agroecological marvels (Nair 2017) is the "niche complementarity hypothesis" (Harper 1977). It implies that "a larger array of species in a system leads to a broader spectrum of resource utilization making the system more productive and leads to better and more efficient use and sharing of resources." Some farmers, however, are apprehensive of the effect of competition for growth factors such as light, water, and nutrients between coconuts and associated plants. A compilation of the available reports, nonetheless, indicates that intercropping trees in the interspaces does not exert strong negative effects on the yield of coconut palms unless such trees grow taller and reduce light availability to coconut crown (Kumar 2007).

16.3.4 Integrated Animal and Fish Production Systems with Coconuts

Apart from the multi-species, multi-strata tree, shrub, and field crop systems, the generic coconut-based integrated farming popular in the Asian courtiers also involves animals such as cow, goat, poultry, duck, rabbit, pig, and aquaculture (production of shrimp, fish, prawn, etc.). Historically, considerable attention has been given to the integration of cattle with coconut, especially with the traditional Tall variety of palm in Southeast Asia and the Pacific (Plucknett 1979; Devendra and Thomas 2002). Reviewing the constraints to farmers' attitude to the acceptance of this traditional practice, Mack (1991) and UNESCO (1979) reported that grazing livestock in coconut stands also necessitates the agriculturalist to learn animal husbandry and pasture management techniques, which could sometimes dissuade farmers from adopting the practice.

16.3.5 Ecosystem Services of Coconut-Based Systems

Multi-strata, multi-species coconut ecosystems offer a variety of ecosystem services such as provisioning, regulating, supporting, and cultural functions as described by the UN Millennium Ecosystem Assessment Report (MEA 2005). In addition to the production of fruits, nuts, vegetables, spices, and medicinal plants (provisioning services), improvements in soil organic matter status and moisture holding capacity and consequential yield increases have been demonstrated (Kumar 2007). For example, soil moisture retention was better in the cacao + coconut systems than in the cacao + *Gliricidia sepium* system (Osei-Bonsu et al. 2002). Furthermore, global warming and the resultant accelerated soil organic matter oxidation could accelerate degradation of the nutrient-poor tropical soils; however, such problems are less likely to manifest in multi-strata production systems than in sole stands of coconuts.

Thus, coconut-based ecosystems around the world consist, in general, of predominantly small-to-medium-holder farm enterprises involving not only coconuts but a variety of other crops and trees grown in intimate associations and managed at various levels of intensity for multiple products and services. These systems are ecologically and structurally complex and economically and socially rather unique in terms of their history, structure, function, dynamics, outputs, and societal values and importance. As such, they present an interesting subject for climate-change mitigation and adaptation studies.

16.4 Climate-Change Mitigation and Adaptation Strategies for Coconut-Based Ecosystems

Consequent to the emergence of climate change as a major environmental issue on the global agenda about 20 years ago, numerous studies, reports, and action plans have emerged at all levels.

16.4.1 Impacts of Climatic Stresses and Projected Climate Change on Coconut Plantations

Coconut, being a perennial crop, lives through the climate change and may face several climatic stresses during its life span of 60–70 years. Although the effects of mean change in temperatures are difficult to predict, the consequences of extreme events on coconut palm have been demonstrated in a few studies. For example, Rajagopal et al. (1996) reported that coinciding dry spells with sensitive stages of inflorescence development such as primordial initiation, ovary development, and button-size nut stages will cause severe loss to coconut yield. Changes in temperature affect coconut yield mainly through the effect on phenological development process (Chmielewski and Rötzer 2001). In a cross-pollinated crop such as coconut, climate variability might influence phenology and peak flowering time. This is because the palm is susceptible to climatological changes throughout the period from primordium initiation to maturity of nuts (approximately 44 months). Consistent with this, Rajagopal et al. (1996) noted that under rain-fed farming situations, nut production was significantly affected by the duration of dry spells at sensitive phases and the dry spells during primordium and ovary development stages were particularly crucial.

Furthermore, such effects may manifest over the long term, given the long reproductive phase of the palm. For instance, drought affected coconut yield for the succeeding 4 years (Naresh Kumar et al. 2007). Consecutive drought years caused not only severe yield reduction but also left about 200,000 palms dead in Coimbatore district of Tamil Nadu and Tumkur district of Karnataka, India; productivity loss was approximately 3500 nuts ha^{-1} $year^{-1}$ (Naresh Kumar 2010). Likewise, the cyclone of 1995 reduced productivity by ~4100 nuts ha^{-1} $year^{-1}$ in Andhra Pradesh state of India (Naresh Kumar et al. 2008).

Naresh Kumar et al. (2008) developed a process-based simulation model for climate-change impact assessments for coconut in India, originally called "InfoCrop-coconut," now named "CoCoSim." It takes weather, soil, management, and varietal characteristics into account to simulate the growth and yield of coconut. Results of such simulation studies indicate that climate-change impacts on coconut yield in India will have considerable spatial variations (Naresh Kumar and Aggrawal 2013). According to these projections, coconut productivity may improve in the western coastal zone (consisting of Kerala, parts of Tamil Nadu, Karnataka, and Maharashtra), northeastern states, islands of Andaman and Nicobar, and Lakshadweep, if present levels of water and crop management are maintained. On the flip side, nut yield is predicted to diminish in Andhra Pradesh, Odisha, West Bengal, Gujarat, and parts of Karnataka and Tamil Nadu owing to climate change. Krishnakumar (2011), however, reported that productivity of a perennial crop like coconut would be more susceptible to climatic variability (e.g., summer droughts) rather than climate change, viz., rising temperature and declining rainfall. Nevertheless, such impact assessments may help adapt plantation crop management to climate change and maximize positive impacts while minimizing the negative impacts.

16.4.2 Mitigation and Adaptation Strategies

As far as the agricultural and other land-use issues pertaining to climate change are concerned, the most significant issues on which efforts are focused on are climate-change mitigation and adaptation (M & A). As explained earlier, mitigation refers to technological transformation and substitution that reduce GHG emissions through avoiding emissions and sequestering GHGs, whereas adaptation refers to "initiatives and measures to reduce the vulnerability of natural systems against climate change," by developing strategies to reduce the negative impacts. Although mitigation and adaptation are different concepts, the terms and action plans surrounding them are so related and intertwined that the terms are usually used together as M & A and sometimes synonymously for one another. Several strategies have been suggested for addressing the two issues individually as well as together, and the most prominent "rallying themes" are carbon sequestration for mitigation and "climate-smart agriculture (CSA)" for adaptation. These are discussed in detail here in relation to coconut-based ecosystems.

16.4.2.1 Carbon Sequestration

To reiterate, the fundamental concept of carbon sequestration in relation to land-use systems is that plants capture or absorb CO_2 from the atmosphere for photosynthesis and store the products of photosynthesis in their different parts. Depending on the nature of utilization of these parts (products), the C so stored may be retained in "long-lived" pools such as stem (timber). When plant parts fall to the ground either through natural processes such as leaf (and litter) fall or through management operations such as residue incorporation, mulching, and green manuring, they decompose, and some or most of the carbon could be "lost" back to the atmosphere, with a relatively small part becoming a part of soil organic matter (SOM).

General Situation in Land-Use systems Following the realization of the importance of C sequestration as a climate-change mitigation strategy, there has been a veritable explosion of such studies in land-use systems of all types. Numerous reports are available on C sequestration potential of various land-use systems, including tree-based systems, from different parts of the world. For example, extensive estimates of global forest biomass have been prepared, mostly as the product of estimated stem-wood volume and species-specific wood density and other "correction factors"—to calculate whole-tree biomass. Carbon content is taken as 50% of the whole-tree biomass, and the belowground components are generally ignored in such estimations. Although whole-tree harvesting, followed by determining the component weights and summing up the amount of harvested and standing biomass, has traditionally been used for obtaining more precise estimates of tree biomass stocks, the cumbersome nature of the method restricts its application to experimental studies only.

Allometric equations constitute yet another widely used method in forestry for estimating volumes of standing forests. It is based on biophysical attributes of trees and validated by measurements of destructive sampling occasionally. These are developed as regression equations linking easily measurable parameters such as tree diameter at breast height, total height or commercial bole height, and occasionally wood density as the independent variables and total dry weight as the dependent variable.

Based on such extensive studies reported from different forest- and other tree-based systems, it is well accepted that tree-based systems such as forestry and agroforestry have greater potential to sequester C compared to single-species crop or pasture systems (Nair et al. 2009, 2010). Primarily, this assumption and calculations are based on the higher amounts of aboveground biomass (and therefore carbon) stored in these systems, even if only a (small) part of such stored C could be considered as C stored in secure long-lived pools (which, by definition, is sequestered C). The estimates of C stocks in agroforestry systems (AFS) varied from 0.29 to 15.21 Mg ha^{-1} $year.^{-1}$ aboveground and 30–300 Mg C ha^{-1} for the top 1 m layer of the soil profile (Nair et al. 2010). Recent studies involving diverse AFS under varied ecological conditions also corroborate the fact that tree-based systems, compared to treeless systems, have higher soil C stocks especially in the deeper soil layers at close proximity to the tree than farther away from it, and species richness and tree density are major determinants of soil organic C content. Besides, C3 plants (trees) contributed more C in the silt + clay fractions of soil that represents the more stable C than C4 plants, in deeper soil profiles (Nair 2012a). However, environmental conditions and system management are two predisposing factors that determine the amount of C sequestered in AFS depends.

Carbon Sequestration Potential of Coconut-Based Ecosystems Reports on carbon sequestration potential of coconut palms and coconut-based ecosystems are, in general, scanty and superficial. Some reports suggest that tree plantation crops represent remarkable carbon pools based on the conjecture that trees retain much more carbon per unit land area than other categories of vegetation and coconut has great potential as a carbon sink (Ranasinghe and Silva 2007). Many of these reports, however, are too general in nature and not based on scientifically rigorous investigations. Indeed, detailed studies on the NPP (net primary productivity) of coconut palms and coconut-based ecosystems are rare. In one of the early reports on this subject, Nelliat et al. (1974) estimated the total annual biomass production of a sole stand of coconuts with 175 adult palms (in full bearing stage) to be in the range of 14.2 Mg ha^{-1} for an annual average production level of 60 coconuts per palm, 18.7 Mg ha^{-1} for 100 nuts per palm, and up to 35.5 Mg ha^{-1} for the very high production level of 250 nuts. These estimates appear to be too simplistic given that the differences in values among the three production levels are attributed solely to the nuts, the biomass production by other plant parts such as leaves, stem, and roots being assumed to be the same across all production levels, which is a questionable assumption. Navarro et al. (2008) reported that under optimal crop nutrition in a high fertility site without any drought problems, the total NPP of coconut + grass understory was 32 Mg ha^{-1} year.$^{-1}$. Simulation studies (Naresh Kumar et al. 2008) indicated a total dry matter production potential in the range of 52–62 Mg ha^{-1} year^{-1}, which was close to the potential dry matter production of coconut monocrop (51 Mg ha^{-1} year^{-1}) reported by Corley (1983).

Navarro et al. (2008) from published values of NPP deduced that coconut plantations exhibit high productivity typical of the tropical humid evergreen forest ecosystems. Nevertheless, biomass C stocks were lesser than for forests, because a major share of C allocation was reportedly going into labile or easily decomposable components such as fruits (nuts), leaves, peduncles, and fine roots. Moreover, many of these parts such as fruits and leaves, being economically useful, are removed from the farm so that their contribution to soil organic matter pool and soil carbon sequestration may not be substantial compared to that of stands of other trees. Kumar and Takeuchi (2009) and Kumar (2011) estimated the aboveground C stock of mixed tree species (>20 cm girth) in 839 homegardens of southwestern coast of India to be in the range of 16.3–35.2 (mean 24.3) Mg ha^{-1}; these included coconuts and a number of fruit trees, both of varying stand densities. Carbon storage by coconut palms in mixed stands will, understandably, be considerably higher than in sole stands of coconuts, especially when the mixed species are trees (e.g, compare Figs. 16.2 and 16.3).

Saha et al. (2010) reported one of the rare studies on soil carbon sequestration in coconut stands compared with other common land-use systems such as homegardens, natural forests, rubber (*Hevea brasiliensis*) plantation, and an agricultural field (rice paddy) in the humid lowlands of Thrissur district of Kerala, India (10°0′–10°47’ N latitude and 75°55′–76°54′E longitude; 2780 mm rainfall per year; soils: inceptisols). On the whole, SOC content up to 1 m depth decreased in the order forest > HGS = rubber ≥ HGL ≥ coconut > rice paddy (Fig. 16.4) [HGS and HGL refer to small and large homegardens, with the former less than 0.4 ha area and the latter more than 0.4 ha but usually less than 1.0 ha; both stands include intimate association of several types of plants including fruit trees, around the homestead; HGS has a higher density of plants than the HGL]. Comparing the SOC stock at different soil depth classes, forest had the highest stock of SOC and the rice paddy

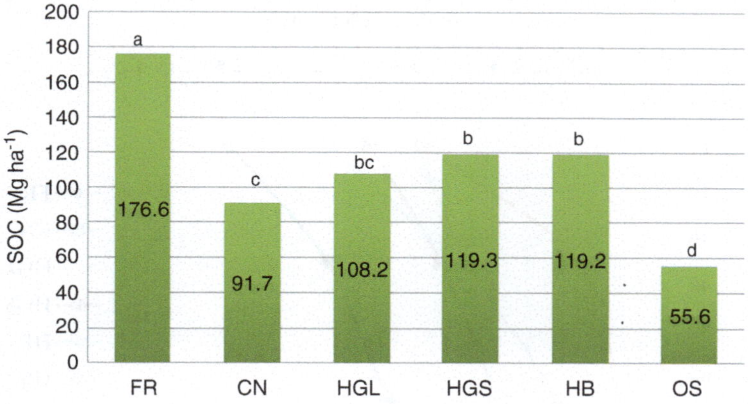

Fig. 16.4 Stock of soil organic carbon (SOC) down to 1 m depth six land-use systems in Madakkathara *Panchayat*, Thrissur, Kerala, India. The land-use systems are forest, FR; coconut grove, CN; large homegarden (> 0.40 ha in area), HGL; small homegarden (<0.4 ha in area), HGS; rubber plantation, HB; and rice paddy field (OS). The lowercased letters indicate differences ($p = 0.05$) in SOC among the systems to 1 m soil depth. (Source: Saha et al. (2010); reproduced with authors' permission)

the lowest among land-use systems at all depth classes, and tree-based systems (coconuts, rubber, and the homegardens) came in between the forest land and rice paddy fields (Fig. 16.5). Another aspect of this study examined soil carbon content in different soil fraction-size classes to assess the relative long-term nature of the sequestered carbon. It showed that coconuts and rice paddy, compared with forests and homegardens, had lower amounts of "recalcitrant" C, i.e., C that is strongly held (= sequestered) in the lower soil layers up to 1 meter depth. This suggests that soil carbon sequestration under coconut stands was relatively less compared with that of other tree-dominant systems such as forests, homegardens, and rubber plantation. It shows that in sole stands of coconuts, soil carbon sequestration, the predominant form of carbon sequestered in ecosystems, could be substantially lower than in mixed stands of coconuts especially when coconuts are combined with other tree species such as fruit trees and cacao. This ecological benefit of multi-species and multistoried crop combinations with coconuts is not adequately studied and, naturally, not appreciated.

It is also worth stating in this context that, in general, estimates of C sequestration in ecosystems are based on several assumptions such that the reported values may be lacking in scientific rigor and quality (Nair 2012b). For example, there is profound ambiguity in the perception of this concept of C sequestration when it comes to "long-lived" C pools. Most reports equate C stock to C sequestration, whereas sequestration refers to loading in "long-lived" pools. Furthermore, C content of plant biomass is 42–44%, whereas most calculations take it as 50% for the sake of convenience. The general assumption that all C in soil denotes sequestered C is also incorrect. Fresh C additions to the soil surface through litterfall and exog-

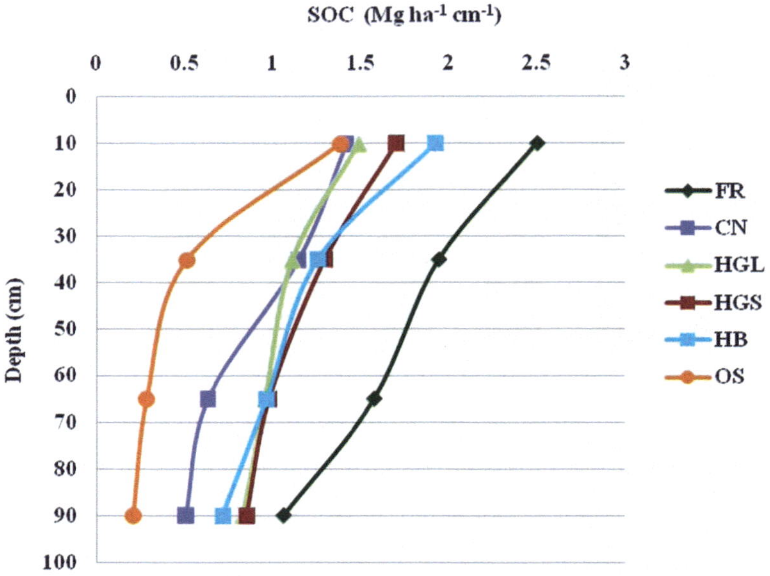

Fig. 16.5 Mean soil organic carbon (SOC) content at different soil depths under six land-use systems in Madakkathara *Panchayat*, Thrissur, Kerala, India. The land-use systems are forest, FR; coconut grove, CN; large homegarden (> 0.40 ha in area), HGL; small homegarden (<0.4 ha in area), HGS; rubber plantation, HB; and rice paddy field (OS). (Source: Saha et al. (2010); reproduced with authors' permission)

enous additions may undergo quick decomposition and release of CO_2, with only a small proportion of C entering the stable "long-lived" pools. If C stocks are augmented through time, that represents C sequestration. Yet another issue is that, in general, roots are believed to represent one-third of the total NPP. However, it is very difficult to measure this fraction, and it may not be true in the case of palms such as coconuts with high proportion of coarse roots than fine roots. Nair and Nair (2014) argue that the high degree of variability and lack of rigor in the reported results of available reports on C sequestration in agroforestry systems limit their potential use for arriving at widely applicable land-management decisions and recommendations.

16.4.2.2 Climate-Smart Agriculture

CSA is a collective rallying theme for various climate-change adaptation strategies. The concept was forged together by FAO in collaboration with a large number of organizations. It is not a solitary farming technology or practice that can be universally applied but is an approach that entails location-specific appraisals to recognize suitable agricultural production technologies and practices. It is comprised of three key objectives: "sustainably increasing agricultural productivity and incomes, adapting and building resilience to climate change, and reducing and/or removing greenhouse gas emissions, where appropriate" (FAO 2013).

The strategies for adaptation to climate change, obviously, are very situation-specific. Jat et al. (2016) discussed crop management technologies in relation to climate-change adaptation focusing on many agronomic management issues including conservation agriculture, pest management, climate-resilient genotypes, early climate warning systems, and crop insurance, all in terms of short-duration agronomic crops. Though the much-referenced FAO-CSA Sourcebook (FAO 2013) also describes similar climate-change adaptation strategies, there is only very limited information presented in the "module" on forestry.

Simulation analysis (Naresh Kumar and Aggrawal 2013) indicates that agronomic management such as soil and water conservation, summer irrigation, drip irrigation, and fertilizer addition can minimize the negative effects of climate change on coconut palms. These will also help to maximize any positive impacts of climate change (e.g., resulting from carbon nutrition) on nut yield. Consistent with this, low-cost practices such as mulching the palm basins with husk/plant detritus, application of composted coir pith, and husk burial have been recommended to conserve soil moisture in palm basins (Naresh Kumar et al. 2006). Drip irrigation was found to save the drought-affected gardens in Tamil Nadu, India, during the consecutive drought years of 1998–2002 (Subramanian et al. 2012). Along with drip irrigation and soil moisture conservation, fertigation will improve the water and nutrient use efficiency in coconut plantations. In addition to the agronomic management, adoption of improved and stress-tolerant hybrids is essential to improve coconut productivity in climatically vulnerable areas.

16.5 Future Directions: Climate-Change M & A and the Coconut-Based Ecosystems

Although climate change is now almost universally accepted as a reality, the existence of a strong and powerful group of "climate-change deniers" should not be ignored. But it cannot be denied that the world has been experiencing substantial climate variability lately even if that may not translate into significant changes when expressed as averages over longer periods of time such as decades. These inter- and intra-seasonal climate variations could have enormous impacts on agricultural production within a season as well as across seasons.

Research on coconuts and other tree crops are modelled for sole stands of species along the same lines as for short-duration agricultural crops. This is inappropriate for crops like coconuts, which exist in natural stands in association with a number of other species in many places. Accurate area estimate is essential for designing any regional management strategy such as climate-change M & A. The production statistics of coconuts are also of doubtful accuracy, which presents a problem in terms of calculating biomass and, therefore, carbon sequestration potential of such systems. These gaps and deficiencies in data gathering and reporting related to coconuts and coconut-based ecosystems will have to be rectified since they have a bearing on the appropriateness of climate-change M & A strategies.

As far as climate-change mitigation is concerned, enhancement of carbon sequestration in the ecosystem is well accepted as a viable strategy. Although some reports suggest that the coconut palm has "high" C sequestration potential, our analysis based on the growth habits of the coconut palm and the pattern of utilization of its products indicates that the palm by itself cannot be rated as a species of high C sequestration potential. On the other hand, "carbon farming" (Toensmeier 2016) through multi-species combinations with coconuts consisting of a variety of annual and perennial species offers tremendous opportunities for enhancing C sequestration in coconut-based ecosystems. This is a satisfactory situation in terms of climate-change mitigation given that such crop combinations are the order of the day in most coconut-based ecosystems. Considering the other ecological benefits of mixed species communities such as biodiversity and ecosystem sustainability, and above all possible economic gains, additional efforts in carbon farming present a "win-win" strategy for coconut-based ecosystems.

Adjusting the dates of planting and harvesting and such other management strategies as a strategy for climate-change adaptation are just not feasible for long-term crops like coconuts. What is feasible, though, would be to capitalize on the climate-change adaptation strategies for the annual crops that are interplanted with coconut. Crop substitutions of even perennial crops grown in association with coconuts could also be thought of as a strategy. For example, if coffee planted with coconuts is seen to be suffering under climate-change-induced higher temperature, it could be substituted by cacao that has better adaptability to higher temperatures than coffee, especially when replanting of old coffee bushes is on the management plan. The response of coconut palm itself to vagaries of climate in terms of various physiological processes and susceptibility to pest and disease incidence needs to be monitored and factored into the strategies.

Carbon substitution projects such as coconut biodiesel initiatives are also being discussed in the context of climate-change mitigation strategies. Coconut oil is used as a feedstock for biodiesel production in some countries such as the Philippines and the South Pacific. But research results on such seemingly novel initiatives are still too scanty that it is too early to include such experimental approaches.

Large-scale effects of climate change such as is sea-level rise will, naturally, have serious adverse impacts on the coconut palm that is traditionally adapted to coastal areas. Total submersion and disappearance of islands as has happened in May 2016 in Solomon Islands are catastrophic events for coping for which no local, short-term, climate-change adaptation strategies are available.

In conclusion, the best strategy for climate-change mitigation and adaptation in coconut-based ecosystems in the immediate future seems to be one of promoting multi-species combinations with coconuts by planting other species under or between coconuts. In order to implement such a strategy, a holistic outlook on coconut-based ecosystem is needed by all concerned: researchers, research administrators, and policy planners at all levels. The emphasis on looking at and planning solely for the coconut palm has to be replaced by one that considers coconut palm as a central component of an ecosystem. Research programs and policies need to be formulated with such a holistic ecosystem concept rather than for just one component of the system.

References

Bonnemaison J (1996) Gens de Pirogues et Gens de la Terre- Les fondements géographiques d'une identité, l'archipel du Vanuatu- Livre 1. Orstom (eds). Paris, 460p

Chmielewski F-M, Rötzer T (2001) Response of tree phenology to climate change across. Eur Agric For Meteorol 108:101–112

Corley RHV (1983) Potential productivity of tropical perennial crops. Exp Agric 19:217–237

Devendra C, Thomas D (2002) Crop–animal interactions in mixed farming systems in Asia. Agric Syst 71:27–40

FAO (2013) Climate-smart agriculture sourcebook. Food Agriculture Organization of the United Nations (FAO), Rome

FAOSTAT (2018). Food and Agricultural Organization of the United Nations-Statistics. [http://www.fao.org/faostat/en/#home]

Harper JL (1977) Population biology of plants. Academic Press, New York

INCCA (2010) Climate change and India: A 4 × 4 assessment- A sartorial and regional analysis for 2030s: chapter on agriculture, Ministry of Environment and Forestry. Under Indian Network for Climate Change Assessment. MoEF, Government of India, 160 pp

IPCC (2014) Climate change 2014: impacts, adaptation, and vulnerability. Summaries, frequently asked questions, and cross-chapter boxes. A contribution of working group II to the fifth assessment report of the intergovernmental panel on climate change [Field CB, Barros VR, Dokken DJ, Mach KJ, Mastrandrea MD, Bilir TE, Chatterjee M, Ebi KL, Estrada YO, Genova RC, Girma B, Kissel ES, Levy AN, MacCracken S, Mastrandrea PR, White L.L (eds)] World Meteorological Organization, Geneva, 190 pp

Jat ML, Dagar JC, Sapkota TB, Singh Y, Govaerts B, Ridaura SL, Saharawat YL, Sharma RK, Tetarwal JP, Jat RK, Hobbs H, Stirling C (2016) Climate change and agriculture: adaptation strategies and mitigation opportunities for food security in South Asia and Latin America. Adv Agron. https://doi.org/10.1016/bs.agron.2015.12.005

Krishnakumar KN (2011) Coconut phenology and yield response to climate variability and change. Thesis submitted in partial fulfilment of the requirements for the degree of doctor of philosophy in atmospheric science. Cochin University of Science and Technology Kochi, India. 216p

Kumar BM (2007) Coconut-based agroforestry for productive and protective benefits. In: Thampan PK, Vasu KI (eds) Coconut for rural welfare. Proceedings international coconut summit 2007, Kochi, India. Asian and Pacific Coconut Community, Jakarta, pp 87–98

Kumar BM (2011) Species richness and aboveground carbon stocks in the homegardens of Central Kerala. India Agric Ecosyst Environ 140:430–440

Kumar BM, Takeuchi K (2009) Agroforestry in the Western Ghats of Peninsular India and the Satoyama landscapes of Japan: a comparison of two sustainable land use systems. Sustain Sci 4:215–232

Lamanda N, Malézieux E, Martin P (2006) Structure and dynamics of coconut-based agroforestry systems in Melanesia: a case study from Vanuatu Archipelago. In: Kumar BM, Nair PKR (eds) Tropical homegardens: a time-tested example of sustainable agroforestry. Springer, Dordrecht, pp 105–122

Mack SD (1991) Livestock under tree crops in Africa. In: Proceedings of the Internat. Livestock-Tree Cropping Workshop, 5–9 December 1989. FAO/MARDI, Serdang, pp 1–15

MEA (2005) Millennium ecosystem assessment: ecosystems and human Well-being synthesis. Island Press, Washington, DC

Nair PKR (1979) Intensive multiple cropping with coconuts in India. Paul Parey, Berlin/Hamburg, p148

Nair PKR (2012a) Climate change mitigation and adaptation: a low hanging fruit of agroforestry. In: Nair PKR, Garrity DP (eds) Agroforestry: the future of global land use. Springer, Dordrecht, pp 31–67

Nair PKR (2012b) Carbon sequestration studies in agroforestry systems: a reality-check. Agrofor Syst 86:243–253. https://doi.org/10.1007/s10457-011-9434-z

Nair PKR (2017) Managed multi-strata tree + crop systems: an Agroecological marvel. Front Environ Sci 5:1–5. https://doi.org/10.3389/fenvs.2017.00088

Nair PKR, Nair VD (2014) 'Solid-fluid-gas': the state of knowledge on carbon-sequestration potential of agroforestry systems. Africa Curr Opin Environ Sustain 6:22–27. https://doi.org/10.1016/j.cosust.2013.07.014

Nair PKR, Kumar BM, Nair VD (2009) Agroforestry as a strategy for carbon sequestration. J Plant Nutr Soil Sci 172:10–23

Nair PKR, Nair VD, Kumar BM, Showalter JM (2010) Carbon sequestration in agroforestry systems. Adv Agron 108:237–307

Naresh Kumar S (2010) Climate change and Indian agriculture: current understanding on impacts, adaptation, vulnerability and mitigation. J Plant Biol 37(2):1–16

Naresh Kumar S, Aggarwal PK (2013) Climate change and coconut plantations in India: impacts and potential adaptation gains. Agric Syst 117:45–54

Naresh Kumar S, Rajagopal V, Thomas TS, Cherian VK (2006) Effect of conserved soil moisture on the source-sink relationship in coconut (*Cocos nucifera* L.) under different agro-climatic conditions in India. Indian J Agric Sci 76(5):277–281

Naresh Kumar S, Rajagopal V, Thomas TS, Cherian VK, Narayanan MKR, Ananda KS, Nagawekar DD, Hanumanthappa M, Vincent S, Srinivasulu B (2007) Variations in nut yield of coconut (*Cocos nucifera* L.) and dry spell in different agroclimatic zones of India. Indian J Hortic 64(3):309–313

Naresh Kumar S, Kasturi Bai KV, Rajagopal V, Aggarwal PK (2008) Simulating coconut growth, development and yield using InfoCrop-coconut model. Tree Physiol 28:1049–1058

Navarro MNV, Jourdan C, Sileye T, Braconnier S, Mialet-Serra I, Saint-Andre L, Dauzat J, Nouvellon Y, Epron D, Bonnefond JM, Berbigier P, Rouziere A, Bouillet JP, Roupsard O (2008) Fruit development, not GPP, drives seasonal variation in NPP in a tropical palm plantation. Tree Physiol 28:1661–1674

Nelliat EV, Bavappa KVA, Nair PKR (1974) Multi-storeyed cropping - new dimension of multiple cropping in coconut plantations. World Crops 26:262–266

Osei-Bonsu K, Opoku-Ameyaw K, Amoah FM, Oppong FK (2002) Cacao-coconut intercropping in Ghana: agronomic and economic perspectives. Agrofor Syst 55:1–8

Plucknett DL (1979) Managing pastures and cattle under coconuts. Westview press, Boulder

Purseglove J (1972) Tropical crops monocotyledons. Longman, London

Rajagopal V, Shivasankar S, Mathew J (1996) Impact of dry spells on the ontogeny of coconut fruits and its relation to yield. Plant Res Dev 3:251–255

Ranasinghe CS, Silva LRS (2007) Photosynthetic assimilation, carbohydrates in vegetative organs and carbon removal in nut-producing and sap-producing coconut palms. COCOS 18:45–57

Saha SK, Nair PKR, Nair VD, Kumar BM (2010) Carbon storage in relation to soil size-fractions under some tropical tree-based land-use systems. Plant Soil 328:433–446

Simpson BM (2016) Preparing smallholder farm families to adapt to climate change. Pocket Guide 1: extension practice for agricultural adaptation. Catholic Relief Services, Baltimore

Subramanian P, Dhanpal R, Mathew AC, Palaniswami C, Upadhyay AK, Naresh Kumar S, Reddy DVS (2012) Effect of fertilizer application through micro-irrigation technique on nutrient availability and coconut production. J Plantn Crops 40:168–173

Tipraqsa P, Craswell ET, Noble AD, Schmidt-Vogt D (2007) Resource integration for multiple benefits: multifunctionality of integrated farming systems in Northeast Thailand. Agric Syst 94:694–703

Toensmeier E (2016) The carbon farming solution: a global toolkit of perennial crops and regenerative agriculture practices for climate change mitigation and food security, vol 480. Chelsea Green Publishing, White River Junction

UNESCO (1979) Tropical grazing land ecosystems, Natural Resources Research Series no. 1. UNESCO, Paris

UNFCCC (2007) Report of the conference of parties on its thirteenth session, Bali, Indonesia. In United Nations framework convention on climate change. Geneva

Dr. P. K. Ramachandran Nair is presently the Distinguished Professor, University of Florida, USA. His main interest is in agroforestry and climate change. He has held various positions in ICAR-CPCRI, Kasaragod, India; ICRAF, Nairobi, Kenya; and University of Florida. He has authored 200 research papers and published 16 books. The honours received by Dr. Nair include Humboldt Prize, Humboldt Fellow and Doctor Honoris Causa from Japan, Canada, Ghana and Spain. pknair@ufl.edu

Professor B. Mohan Kumar is a leading researcher in agroforestry. He has held important positions in Nalanda University, Rajgir; Indian Council of Agricultural Research, New Delhi; and Kerala Agricultural University. He is a Visiting Professor to a number of overseas universities (Fulbright Scholar, Erasmus Mundus Visiting Professor, JSPS Visiting Professor, etc.). He has published 250 papers and edited 5 books.

He received Dr. KG Tejwani Award for Excellence in Agroforestry Research and Development in India. bmkumar. kau@gmail.com; bm.kumar@kau.in

Dr. S. Naresh Kumar is presently Professor and Principal Scientist at the Centre for Environment Science and Climate Resilient Agriculture, Indian Agricultural Research Institute, New Delhi, India. His contributions are mainly in climate change and simulation modelling research. He is a Government of India Expert for UNFCCC IPCC reports, Lead Scientist of Info Crop models and Member of the Global Steering Council, AgMIP. He has published 100 research papers and authored/edited 6 books. snareshkumar.iari@gmail.com

Chapter 17
World Coconut Economy: Sectoral Issues, Markets and Trade

C. V. Sairam and S. Jayasekhar

Abstract Liberalized marketing and trade regimes of coconut sector pose serious challenges for policymakers and planning experts of the coconut-growing countries across the world. This is particularly true, when the world coconut sector faces many daunting challenges such as low productivity, incidence of pests and diseases, fluctuating prices, volatile domestic marketing system, recession in trade sector, etc. This chapter depicts varied aspects like global scenario including the challenges faced by major coconut-producing countries, trade aspects, global competitiveness, price analysis, policy-level impediments, issues related to procurement and marketing, empowerment through agribusiness and future perspectives of global coconut economy.

17.1 Introduction

Coconut assumes considerable significance in the national economy of coconut-growing countries in view of rural employment and income generation. The traditional coconut farming in these countries is an integral part of their life, culture and identity. The livelihood of millions of people across the world relied on coconut farming, and hence the importance of coconut sector for the upliftment of rural economy in major coconut-growing countries cannot be undermined. Income- and employment-generating ability from kernel-based, husk-based and shell-based subsectors of coconut is significant. It is notable that more than 90% of coconut farmers in the world are asset poor with meagre size of holdings. Of late, apart from price-/market-related constraints, coconut growers face many other challenges such as

---

C. V. Sairam (✉)
ICAR-CIBA, Chennai, Tamil Nadu, India
e-mail: sairam_cpcri@yahoo.com

S. Jayasekhar
ICAR-CPCRI, Kasaragod, Kerala, India
e-mail: jaycpcri@gmail.com

K. U. K. Nampoothiri et al. (eds.), *The Coconut Palm (Cocos nucifera L.) - Research and Development Perspectives*,
https://doi.org/10.1007/978-981-13-2754-4_17

scarcity of skilled labour, high wage rate, low productivity of coconut, depletion of natural resources in coconut gardens, soil-related constraints and inadequate irrigation facilities as well as crop loss due to incidence of pests, disease and natural calamities. The post-World Trade Agreement (WTA) and ASEAN Treaty regime witnessed integration of plantation economies across the globe that resulted in fierce competition among producing countries (Jayasekhar et al. 2014; Lathika and Kumar 2009). A substantial number of technologies have been developed for enhancing productivity and income from coconut farming. These include planting high-yielding varieties and hybrids, irrigation techniques, integrated nutrient, pest and disease management as well as adoption of cropping/farming systems, and technologies for value addition through product diversification. However, due to several socio-economic factors, the adoption rates of these technologies are extremely low. Hence, it is highly imperative that a favourable pro-poor policy environment is evolved to protect the interests of coconut growers especially for small and marginal farmers, and appropriate development and extension interventions are implemented to enhance the efficiency of coconut sector. Moreover, unlike in the case of annuals, perennial crops pose varied challenges for the producers (farmers), scientists and developmental personnel, because the efforts taken by any of them either individually or in groups cannot be scaled up in the short run. In spite of several constraints stated above, coconut cultivation, especially in a coconut-based cropping/farming system, is a profitable venture (Kannan and Nambiar 1976; Krishnaji et al. 1976; Das 1984, 1991; Hegde et al. 1990). This chapter depicts in detail the global-level sectoral issues, marketing and trade aspects of coconut.

17.2 Global Scenario

In spite of the fact that coconut is broadly scattered in the vast majority of tropical tracts and grown in 93 countries across the world, 10 out of 12 million hectares under this crop are contributed by only 4 countries, namely, Indonesia, the Philippines, India and Sri Lanka (APCC 2016). Indonesia holds the largest area in coconut (30%) followed by the Philippines. India stands third in area of coconut and first in production with the share of 16.5% and 28%, respectively. As far as the productivity is concerned, Brazil holds the top position (11,630 nuts ha^{-1}) followed by India (10,119 nuts ha^{-1}). Eight percent of the coconut export market is in the hands of half a dozen countries.

17.2.1 Indonesia

Coconut farmers in Indonesia are predominantly small holders (98%) who are not well versed with the advanced production technologies. Hence, the cultivation practices and the productivity are functioning at suboptimal levels in the country. It is

Table 17.1 Export of coconuts and coconut products in Indonesia

Item	Value US$ 1000	Share (%)
Fresh coconuts	63,125	4.9
Copra	33,214	2.6
Coconut oil	811,981	63.1
Copra meal	46,494	3.6
Desiccated coconut	137,610	10.7
Coconut milk/powder	35,042	2.7
Shell charcoal	113,238	8.8
Activated carbon	34,270	2.7
Fibre and fibre products	12,204	0.9
Total	1,287,178	100.0

Source: APCC (2016)

also pertinent to mention that the coconut farms are located in far-off places from the main urban centres and the transportation and other infrastructural facilities are underdeveloped. With very limited resources, the coconut farmers are effortlessly lured by the private financial intermediaries and subsequently are trapped in the vicious debt trap. However, the export sector of the Indonesian coconuts and coconut products is well diversified both in terms of products and markets (FAO 2014). Table 17.1 depicts the extent of product diversification achieved in Indonesia. Coconut oil with 63% of total exports holds the major value share, followed by desiccated coconuts (11%) and shell charcoal (9%). The export destinations are also very well diversified across Europe and Asian countries.

The country has taken proactive efforts to adopt National Quality Standards of various coconut products in view of the proliferating international food safety standards (USDA 2017). Such a step has to be seen in the perspective of market competition as the stringency in food safety has already led to heavy consignment rejections of coconut products across the European markets. Conversely, the price movement of coconut oil in the country has shown topsy-turvy pattern, and in general, there is a tendency of price fall. In the recent times, there is a trend of price integration of domestic and international coconut oil (Fig. 17.1).

17.2.2 Philippines

The Philippine Coconut Authority has identified that coconut-based integrated farming systems and development of agribusiness have been promoting the two priority areas for income enhancement of coconut farmers in the country. Formation of coconut clusters is given utmost importance in the country. As of now, the consumption-export ratio is 25:75 (Pabuayon et al. 2009). Coconut oil accounts for the major share of coconut products export (64%) followed by desiccated coconut (12%). A substantial quantity of coconut water (in various flavours) finds the export

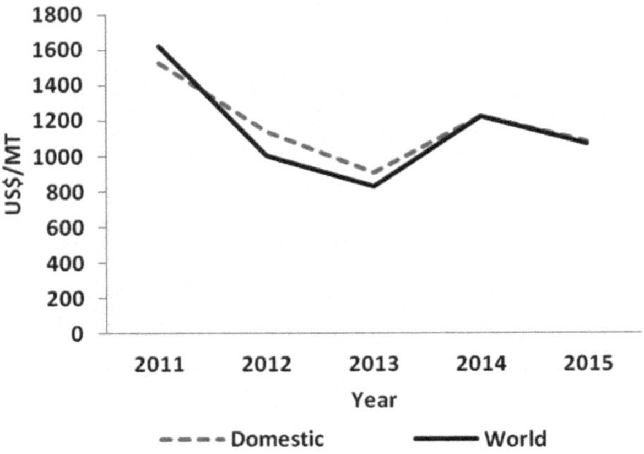

Fig. 17.1 Price movement of coconut oil – Indonesia

Table 17.2 Export of coconuts and coconut products in the Philippines

Item	Value (US$ 1000)	Share (%)
Fresh coconuts	1118	0.1
Coconut oil	843,160	63.7
Copra meal	70,424	5.3
Desiccated coconut	160,051	12.1
Coir products	13,810	1.0
Shell charcoal	32,343	2.4
Activated carbon	81,527	6.2
Coconut milk	17,093	1.3
Coconut water ('000 litres)	81,152	6.1
Others	22,767	1.7
Total	**1,323,445**	**100.00**

Source: APCC (2016)

destinations overseas (Table 17.2). Nevertheless, the break-up of export products categorically highlights the over-dependency on copra and coconut oil.

There exists a robust demand for conventional coconut products like desiccated coconut in the country. It is also remarkable that solid specialty markets are fast emerging for coconut-yoghurt, flavoured milk, coconut-based dessert, palm sugar and snacks. In spite of the fact that the country in general has a favourable business ambience for the coconut sector with excellent growth rate, the dominance of senile palms and overall decline in productivity are the grave concerns, which are to be addressed with immediate priority.

While looking at the coconut industrial structure of the Philippines, there are 68 oil mills with cumulative processing capacity up to 4.31 Mmt of copra year^{-1}, and 38 coconut oil refineries work with added up capacity of 1.44 Mmt of oil year^{-1}(FAO 2014). The capacity of desiccated coconut year^{-1} is 0.20 Mmt from 11 plants, and

Table 17.3 Coconut industry of the Philippines – a glance

Sl. No.	Particular	Total number	Capacity (Mmt $year^{-1}$)
1	Oil mills	68	4.30
2	Oil refineries	38	1.44
3	Desiccated coconut plants	11	0.21
4	Activated carbon plants	8	0.05

Source: Philippine Coconut Authority

that of activated carbon is 0.055Mmt $year^{-1}$ from 8 activated carbon plants (Table 17.3).

The coir industry in the Philippines has not yet developed to the fullest potential that the sector offers and hence warrants more focus and concerted effort for the full-fledged development of this aspect. It is also imperative to give more thrust on shell charcoal segment and activated carbon production, especially in the context of surging worldwide demand for the activated carbon. Having said that, it is also essential to address the newfound challenges, such as sanitary and phytosanitary requirements of the developed countries (food safety standards) and rising environmental standards.

Scarcity of resources was observed as the major reason behind the lack of adoption of innovative concepts on value addition and marketing in the coconut sector. The farmers of the country, in general, are vulnerable and incapable to upgrade their position in the coconut value chain. Therefore, the upstream end of the value chain, where the farmer concentration is the highest, receives the lowest revenue share of the global value chain of the coconuts/coconut products. Subsequently, the lack of price incentives in terms of revenue share has resulted in negligence of coconut gardens by the farmers and subsequent low productivity (Alma and Albert 2016). This partly explains the relatively higher poverty incidence among coconut farming communities in this part of the world.

In order to alleviate poverty in coconut farming communities, providing assistance to farmers for their market development and improvement of the present marketing system needs attention. Providing capital, technical assistance, training, equipment, tools and market information is the critical element to empower farmers to establish village-level enterprises for producing higher-value products and create necessary linkages with markets. They also need to increase their farm productivity to support the raw material needs of the new enterprises. Attractive domestic market for toddy (alcoholic beverage from coconut inflorescence sap) and VCO is prevalent even in the rural villages of the Philippines. Nevertheless, such opportunities are utilized only by the resource-rich farmers, entrepreneurs and other intermediaries, thereby marginalizing the majority of the coconut farmers. This fact is very well reflected in the widespread poverty incidence among the coconut farmers of many remote villages in the country. The authorities should urgently act on these issues by developing an action-oriented strategic plan for the inclusive growth of coconut farmers. The investment pattern in the coconut sector should be monitored by the

Fig. 17.2 Price movement of coconut oil – Philippines

coconut development authorities, wherein investment by farmers in cluster mode should be promoted. The production mode of the coconut value chain should be strengthened by creating appropriate linkages with research and development agencies and farmers and should propose strategies for addressing the issue of declining productivity. Although the price movement in the recent years is integrated with the world prices, an increasing price wedge is apparent (Fig. 17.2).

17.2.3 Sri Lanka

As in the case of other Asian countries, Sri Lanka is also confronting several sectoral challenges related to coconuts such as persisting adverse weather conditions, widely fluctuating farm-gate prices, escalating input costs, ineffective utilization of advanced technologies and encroachment of coconut farms for urbanization. In perspective of these, the Sri Lankan government chalked out a coconut revitalization agenda, and ministry of coconut development has been entrusted with the execution of programmes. According to the recent policy statement of the government of Sri Lanka, enhancement of coconut productivity and welfare of the farmer is the top most priority in coconut sector. The government has also devised special programmes to bring in more coconut area under mixed cropping/farming models to enhance the resource utility and income from unit area.

While analysing coconut product utilization and marketing, it was found that, out of the total production of nuts, more than 70% is used for household consumption. Two million nuts are required for annual domestic consumption. Depending on the global demand, the excess coconut is used for oil, copra and desiccated coconut production. Table 17.4 depicts the coconut industry outlook of Sri Lanka. It is noteworthy that the idle capacity in oil mills and DC units is huge.

Sri Lanka is the most specialized exporter of coir products in the world. In the past, they have come up with innovative coir products, and their export growth rate in the sector is impressive. Recently, apart from the traditional export destinations of coir products, they have diversified their markets, and China has become one of

Table 17.4 The glimpse of coconut industry – Sri Lanka

Industry	No.of units	Capacity (mt)	Utilization (mt)	Utilization (%)
Oil mills	1265	335,706	81,000	24.12
DC units	51	77,904	42,000	53.91
AC units	8	50,640	39,810	78.61

Source: CDA (2016)

the major importers of Sri Lankan coir products. Sri Lanka's coconut shell-based industries also have been experiencing rapid growth, and their market share in these products has improved in Germany, Japan, France and the USA.

The plight of small-scale coconut entrepreneurs in the villages of Sri Lanka is not different from other coconut-growing countries, in terms of low levels of awareness, know-how and marketing expertise on technologies of coconut-based value-added products. At the village level, farmers are not well organized into clusters or societies; thereby, they lack the bargaining capacity in the financial institutions, as their production capacities at the individual levels are not enough to reap the economies of size. There is an urgent need for capacity building in the rural coconut-growing tracts of Sri Lanka. The farmers and the small-scale entrepreneurs in the villages should be empowered, and their activities are to be promoted.

17.2.4 Pacific Island Countries

The export of coconut products (mainly copra, coconut oil and desiccated coconut) contributed 60% of the total merchandise export of the Pacific Island countries during the late 1960s and 1970s. In the mid-1970s onwards, there had been a rapid decline of the industry, caused by low prices that resulted in low investments. All forms of coconut products have fallen sharply between mid-1970s and 1987, and the share of coconut products fell to 24% of the total exports. In the recent times, the coconut sector of these countries plays an insignificant role in terms of productivity, competitiveness and export earnings. It is necessary to assess the current status of the coconut sector and identify the gaps and opportunities to provide a renewed focus for coconut sector of the Pacific Island countries.

Among the Pacific Islands, Papua New Guinea has the largest coconut area (37.5%), considering the nine major islands under consideration (Table 17.5). In the case of coconut production, again Papua New Guinea tops the list with 55.2% share of the total production of the islands under comparison. The low productivity of these countries is a matter of grave concern not only in view of the comparative disadvantage but also in terms of sustainability of the coconut sector. It is evident that the comparative picture of Pacific coconut growers on the aspects of coconut production is bleak as of now and requires adequate impetus to make it competitive and sustainable.

Table 17.5 Comparison of coconut production statistics

Country	Area (000'ha)	Production (million nuts)	Area (%)	Production (%)	Productivity (nuts ha^{-1})
F.S. Micronesia	18	59	3.1	2.2	2197
Fiji	62	200	10.5	7.4	2387
Kiribati	20	55	3.4	2.0	2730
Marshall Islands	8	35	1.4	1.3	4375
Papua New Guinea	221	1483	37.5	55.2	6710
Samoa	99	267	16.8	9.9	2697
Solomon Islands	38	100	6.5	3.7	2631
Tonga	31	75	5.3	2.8	2423
Vanuatu	92	415	15.6	15.4	4512
Total	589	2689	100.0	100.0	3407[a]

Source: APCC (2016)
[a]Average productivity of the nine islands

Table 17.6 Comparison of coconut product exports and its contribution

Country	Export (mt)							
	Coconut	Copra	Coconut oil	Copra meal	VCO	Cream	Desiccated coconut	% export contribution[a]
F.S. Micronesia	–	76	–	–	–	–	–	–
Fiji	74		1630	75	–	–	–	2.57
Kiribati	–	1332	2939	216	–	–	–	46.25
Marshall Islands	–		124	482	–	–	–	0.04
PNG	–	48,228	11,068	5250	25	–	–	0.43
Samoa	1311	NA	82	2094	–	100	13	4.66
Solomon Islands	–	1586	–	–	–	–	–	2.22
Tonga	1644		–	–	–	–	–	4.15
Vanuatu	–	25,194	9208	4786	–	–	0.1	43.98

Source: APCC (2016)
[a]Share of revenue from coconut exports in the total export revenue of the country

The share of revenue of a commodity/sector in the total merchandise trade of a country implies the importance of the particular commodity/sector to the country. In the case of coconut exports of the nine Island countries, the share of coconut export revenue in the total revenue is highest for Kiribati and Vanuatu (46% and 44%), respectively (Table 17.6). Tonga's coconut export revenue accounts for a meagre 4.15% of the total merchandise revenue of the country.

The World Bank report (World Bank 1990) rightly points out the importance of trade expansion, value addition and market access for a sustainable and profitable coconut sector of the Pacific Islands in the long run. But the worrisome factor is the lack of entrepreneurs in the sector. There are only very few coconut traders who are active at present. High labour costs (eight times higher than those in a competing country) like the Philippines remain a constraint. This makes returns to labour from copra about one tenth of the rural wage and the lowest of any recorded rural activity in the Pacific Islands.

It is evident from the analysis of data and qualitative research conducted among stakeholders that the present operating environment of coconut sector in the Pacific Islands has to be thoroughly restructured with the provision of adequate consideration of the economic incentives to the farmers. As far as international competitiveness is concerned, the Pacific Island countries have disadvantages on various facets. The production segment is characterized by very low productivity and dominance of senile palms. The contribution of coconut sector in the GDP and export earnings has been dwindling of late. It could be noticed that product diversification initiatives targeting export markets are unsatisfactory. The benefit of perfect competition and economy of scale is not realized because there are not many exporters in the sector. Although scientific replanting is necessary coupled with scientific package of practices, the return to investment in this regard has to be ensured. The return to investment is crucial, because as of now, the Pacific tract is lacking comparative advantage in terms of coconut exports.

17.2.5 India

In India, coconut production, processing and marketing sectors face an uphill task for achieving sustainable growth in the present international trade scenario, in which competitiveness through higher productivity is the goal for all the coconut-growing countries. At present, India is the largest coconut-producing country in the world, with second highest productivity of 10,119 nuts ha^{-1} (APCC 2016). The crop contributes Rs. 221,670 million to the gross domestic product (GDP) per $year^{-1}$ (CDB 2016). Export of coconut and coir products brings in foreign exchange earnings of Rs. 33,510 million.

The Philippines, Indonesia and Sri Lanka, which are the major coconut-producing countries, have lesser domestic demand and hence have competitive advantage of economies of scale of production for value-added production and trade sector.

In India, both at macro- and micro-level coconut production, processing and marketing sectors face many daunting challenges, for which several short- and long-term strategies need to be formulated and effectively implemented. At macro level, higher cost of production and pattern of domestic consumption places India at a disadvantageous position in the international trade map. At micro level, the problems like predominance of small and marginal holdings and consequent less marketable surplus, rainfed cultivation, higher cost of production and weak links in product diversification are the major challenges of Indian coconut sector.

According to the statistics of year 2015, the area under coconut cultivation in India is 2.08 million ha, from which 22,167 million nuts are produced. The productivity of coconut palms in India was estimated to be 10,614 nuts ha^{-1} (Table 17.7).

Table 17.7 Coconut – Indian scenario

State	Area (000 ha)	Production (m nuts)	Productivity (nuts ha^{-1})
Andhra Pradesh	104	1427	13,732
Karnataka	526	5129	9744
Kerala	771	7429	9641
Tamil Nadu	460	6171	13,423
Other states	227	2011	7295
Total	2088	22,167	10,614

Source: CDB (2016)

Majority of the coconut gardens in India fall under the category of marginal size of holdings. Such holdings are ineffective in view of generation of adequate income for livelihood support as well as for providing meaningful employment. The sustenance of coconut farming, in the recent times, is much more relied on the market prices of coconuts and its products, and essentially the practice of integrated farming with suitable crop diversification assumes greater importance. For brightening the future prospects of a sustainable coconut sector, it is necessary to delink the sector from the dependency on coconut oil and enhance production of diversified value-added products.

17.2.5.1 Area and Productivity Effects

The rate of increase of area in India is higher than that of productivity. The annual compound growth rate analysis for the period 1961–2014 (FAO 2017) indicates that the area under coconut has grown by 1.90% annum^{-1}, while the productivity has grown up by only 0.52%, whereas in countries like Brazil, Sri Lanka, Papua New Guinea and Micronesia, the productivity effect is more than that of area effect (Table 17.8).

Since the 1970s, the macro-level policy committees have concentrated primarily on increasing the production of coconut and have neglected the vital concepts of product diversification and product utilization. The detrimental effect of this has been realized in the post-WTO era, wherein small countries like Sri Lanka have proved competitiveness over India due to effective product diversification.

17.2.5.2 Trade Aspects

The total value of exports in the case of coconut products during the year 2015–2016 was found to be Rs. 14,502 million (here we have excluded the coir products), which is 11% higher than that of the export earnings of the year 2014–2015 (Table 17.9). On the other hand, the imports of coconut products to India in the year 2015–2016 (valued at Rs. 3833 million) were observed to be 11% less than that of the value of imports during 2014–2015 (valued at Rs. 4217 million). The major

Table 17.8 Compound growth rates of coconut area, production and productivity

	Annual compound growth rate (%)		
Country/region	Area	Production	Productivity
Brazil	1.04	5.28	4.20
India	1.97	2.50	0.52
Indonesia	2.18	2.75	0.55
Papua New Guinea	−0.30	0.77	1.07
The Philippines	1.61	1.94	0.33
Sri Lanka	−0.29	0.26	0.55
Oceania	0.25	0.45	0.20
Melanesia	0.49	0.62	0.13
Micronesia	0.02	0.70	0.69
World	1.50	2.00	0.49

Source: FAO (2017)

Table 17.9 Export/import of coconut products: India

Year	Export value (Rs. million)	Import value ((Rs. million)
2007–2008	690.1	559.3
2008–2009	1798.0	1030.8
2009–2010	2197.5	1071.6
2010–2011	4959.2	1207.7
2011–2012	9432.9	2098.8
2012–2013	10223.6	1919.0
2013–2014	11561.2	2311.1
2014–2015	13123.8	4216.6
2015–2016	14502.4	3832.6

Source: CDB (2016)

coconut product imported to India was copra meal (coconut oil cake), which accounts for 86% share of total imports, followed by the coconut oil (11%). Importantly, the import of coconut oil to the country has come down to around 1000 mt during the year 2015–2016 from 2660 mt imported during the year 2014–2015.

It was observed that the import intensity of the coconuts and coconut products is at low levels (Table 17.10) and thereby will not influence the domestic price behaviour of the product. Moreover, as far as the international trade is concerned, India can boast a robust domestic market in comparison with other competing counterparts. Furthermore, the coconut and coconut oil are in exclusion list of ASEAN-India Free Trade Agreement (AIFTA), which provides temporary immunity for the domestic coconut oil sector. Nonetheless, the commodities in the exclusion list are subjected to periodic revision, and in all probability, coconut oil will be included in the reduction list sooner or later. In such a scenario, the immunity of the coconut oil from the cheaper imports will be lost, and eventually there will be huge price crash in the domestic coconut oil sector.

Table 17.10 Coconut imports, production and import intensity

Year	Import (Rs. million)[a]	Production (million nuts)	Value output (Rs. million)	Import intensity[b]
2009–2010	1071.69	16918.40	203020.80	0.53
2010–2011	1207.77	16942.92	203315.04	0.59
2011–2012	2098.85	23351.22	280214.64	0.75
2012–2013	1908.52	22680.03	272160.36	0.70
2013–2014	2301.44	21665.19	259982.28	0.89

[a]HS Code: 0801, EXIM Data bank, Department of Commerce
Production NHB (Various years)
[b]Quantity of import as percentage of production

Table 17.11 Percentage share of world exports of coconut products

S. No	Product	Countries (percentage share)			
1	Coconut oil	Philippines (42)	Indonesia (35)	Malaysia (9)	India (0.30)
2	Copra meal	Philippines (64)	Indonesia (34)	Others (1.9)	India (0.004)
3	Desiccated coconut	Philippines (25)	Indonesia (20)	Sri Lanka (12)	India (1)
4	Coconut milk/cream	Indonesia (51)	Sri Lanka (44)	Philippines (4)	India (0.30)
5	Coconut shell charcoal	Indonesia (70)	Sri Lanka (20)	Philippines (7)	India (0.30)
6	Coir and coir products	Sri Lanka (42)	Thailand (12)	Indonesia (10)	India (25)

Source: APCC (2016)

17.3 Global Competitiveness

It is imperative to have a look at the international trade scenario of coconut value-added product exports. While comparing with other major global exporters, the share of India in coconut product exports is meagre (Table 17.11). Though it is an accepted fact that India holds a robust domestic market in the coconut sector, it is high time that India emerges as a major export player by upgrading its position in the global value chain of coconut exports. The Philippines and Indonesia together contribute the major world export share of coconut oil, copra meal and desiccated coconut. Sri Lanka too contributes substantially to the international exports of coconut milk, shell charcoal and coir products.

It is worth mentioning that a major proportion of coconut produced in India is consumed domestically itself (Fig. 17.3). On the other hand, the Philippines consumes only 25% of its coconut production domestically. The economic logic always points towards the correlation between the export growth and domestic consumption. In most of the cases when there is a market surplus developing outward market orientation and thereby in the long run, you will develop a robust export market for the product, and there will certainly have a first mover advantage as well. This is

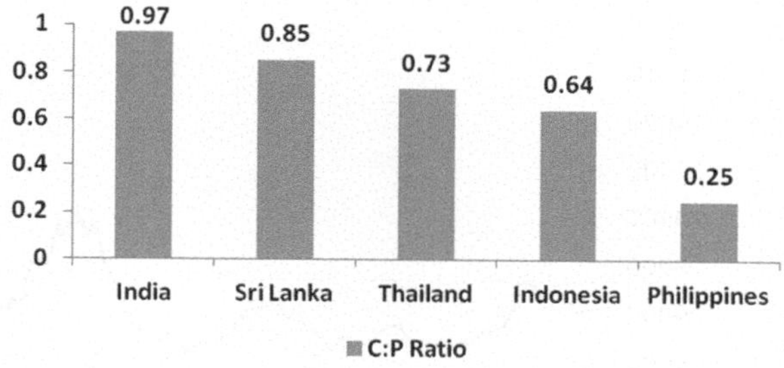

Fig. 17.3 Consumption-production ratio of coconut

Table 17.12 Revealed comparative advantage indices of major coconut-growing countries

Country	Coconut oil	Desiccated coconut	Activated carbon	Virgin coconut oil	Coconut milk
India	2.10	1.80	6.90	1.10	NA
Indonesia	21.20	4.90	26.20	16.10	NA
Malaysia	12.00	4.20	10.60	5.20	5.20
Philippines	32.30	29.10	28.10	38.00	NA
Thailand	1.60	5.60	12.60	8.10	39.20

Figures indicate revealed comparative advantage (RCA), *NA not available

exactly what happened with the Philippines, and now they are the most competent exporter with respect to coconut and coconut products. Nevertheless, India, of late, has been making concerted effort to penetrate their products in the high value export segments.

Competitiveness analysis of coconut and coconut products to export destinations through employing revealed comparative advantage (RCA) methodology indicates that comparative advantage of India is lower than major coconut-exporting countries like the Philippines, Indonesia and Sri Lanka (Table 17.12). The Philippines clearly dominates in most of the coconut value-added product lines. The analysis suggests the need to formulate plausible strategies to reach the overseas market and capture the optimal share in market segments. Though India has a strong domestic market base, it is an indubitable fact that in the near future due to the evolving trade agreements even in the domestic sector, India may confront fierce price competition from the overseas export. Hence, the country needs to chalk out modalities and execution plans to elevate our export competitiveness and comparative advantage.

17.4 Price Analysis of Mandate Crops in Value Chain Perspective

Beyond any degree of doubt, in the recent times, trade-related issues, market access and attractive prices are the major factors shaping up investment decisions in coconut farming enterprise. The above-mentioned factors assume much more

Fig. 17.4 Price movement of coconut (2004–2015)

importance than that of the increase in productivity, which otherwise conventionally considered as the single crucial component. The coconut farmers in India are so far concentrated in the upstream end of the coconut value chain, without any functional upgradation.

As a matter of fact, the confidence of coconut farmer can be elevated only when a stabilized price regime is experienced for a reasonable period. Analysis of coconut prices for over a decade (2004–2016) depicts the increasing price volatility, especially in the recent years (Fig. 17.4). From 2004 to 2015, the prices were keeping low with comparatively low price fluctuations. On the other hand from the year 2010 onwards, the price fluctuations are quite apparent wherein the prices started rising reaching peak levels during the mid-2011 after which it plummeted to low levels. But again from the beginning of 2013, the prices started improving, and the prices continued as attractive, and all over again, from 2015 onwards, the sector has been experiencing a price crash regime. Jnanadevan and Jayasekhar (2011) attempted to characterize the earlier price rise regime (during 2011). They have put forth the argument that the price rise regime experienced in the coconut sector is linked with the supply crunch of coconuts and copra coupled with huge industrial demand for processing and exports. They have provided corroborative evidence in the form of increasing export growth rate, inefficient copra procurement and low levels of supply and general supply deficit in edible oil sector. They have also rightly argued that the bubbles of price rise regime is not helpful for the sectoral prosperity, as these sort of price boom periods are not long-lasting enough to instil confidence in the coconut farmers to have a serious reorientation towards scientific farming approaches.

Besides, the analysis of demand-supply scenario using stock-use ratio revealed that there is a declining demand for coconut oil from 2012 to 2013 onwards and the

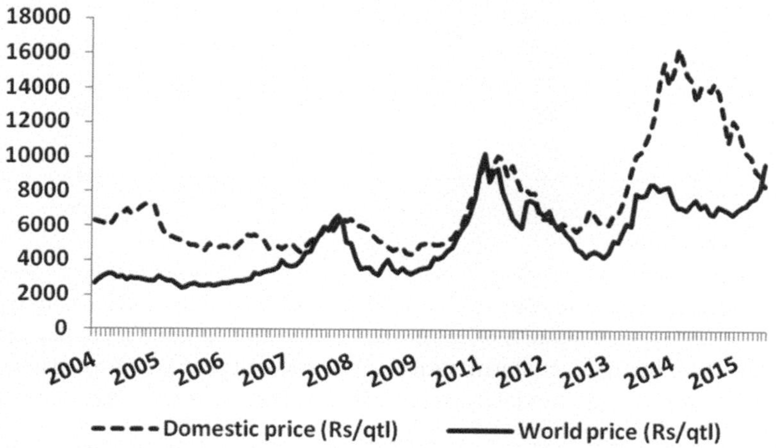

Fig. 17.5 Coconut oil: movement of domestic prices and international prices

wedge between demand and supply has been narrowed down. This of late has certainly reflected in realization of low prices for the commodity.

It was observed that there is huge price wedge between domestic and international prices (Fig. 17.5). As the prices will tend to integrate, there is a possible price crash in the near future.

17.5 Policy Level Impediments

For the past two decades, plantation sector in India has been confronting a commodity crisis, arguably as an off shoot of the ongoing trade liberalization. The regional trade agreements such as AIFTA have made the crisis even worse due to the adverse policy frame in the form of phased tariff reduction and fixation of import tariffs at extremely low level. In this context, it would be erroneous to view coconut sector in isolation, because the trade and tariff decisions on competing crops as well as edible oils in general would straight away affect the coconut sector as well. In Table 17.13, tariff reduction schedule of the special products is depicted, wherein the reduction commitment of palm oil (an immediate substitute of coconut oil) is notable. Unprecedented growth rate in palm oil imports in recent times is also a matter of concern in view of the domestic prices of the coconuts. The possibility of lowering the existing tariff structure of special products in the forthcoming review meetings of AIFTA is also bothersome.

With the ongoing liberalization process across the world, proliferation of regional free trade agreements (RTAs) has become inevitable. There will be a differential impact of such trade agreements on different sectors, and it is important to safeguard the plantation sector in general and coconut in particular in the forthcoming RTAs. In view of this, it is imperative to conduct studies on challenges faced by the

Table 17.13 Mandatory tariff schedule for special products

Tariff line	Base rate	2010	2015	2019
Crude palm oil	80	76	56	37.5
Refined palm oil	90	86	66	45
Coffee	100	95	70	45
Tea	100	95	70	45
Pepper	70	68	58	50

Source: Harilal (2010)

coconut sector at micro and macro levels to bring out plausible strategic action plans for the sectoral reorientation. It is also crucial to envisage appropriate policy options with regard to the trade and tariff structures of coconut sector and to ensure such sectoral details are appropriately represented in the national and international dialogues.

It is always better to have a floating import duty structure on edible oils, so that the tariffs can be adjusted in relation to the international prices of edible oils to stabilize the domestic price fluctuations. But in the case of palm oil in India, the import duty was always hovering around 5%, irrespective of the international price movements. The flawed tariff fixation of such pattern had detrimentally affected the domestic price scenario (and movements) of the coconut oil in the country. Therefore, it is vital to regulate the edible oil tariff structure, so that the state machinery can adopt flexible policy options to control the price fluctuations of coconut oil.

17.6 Issues Related to Procurement and Marketing

The studies on marketing margins and costs are important as they reveal many facets of trade, price structure and the efficiency of the system. The 'price spread' is associated with the movement of a commodity from the producer to consumer wherein the actual costs involved in the transaction at various nodes as well as the margins accrued to various actors at different nodes are accounted. In general, the term 'price spread' in agriculture implies the 'producer's share in consumer's rupee'.

The impact of risks is more severe in the case of perennials, in which heavy initial investments are made. Price-spread analysis of coconut marketing revealed that raw coconuts are sold through the village traders by almost 70% of the farmers.

Less marketable surplus due to small and marginal holding size is the major reason for the farmers for not undertaking copra or oil for sale. The marketing channel consists of village traders, whole sellers and retailers who in turn sell their products to oil millers and retailers and send some of their lots to upcountry markets as raw nuts and edible or ball copra. Predominant marketing channel identified is:

Producer➔Copra maker➔Oil miller➔Whole seller➔Consumer.

In Kerala (India) conditions, which are the same in many countries with predominantly small holder coconut gardens, the producer share in consumer rupee is around 64%, and the market chain consumes as much as 36% share in the total value chain. Higher price spread always indicates a lower share of the final price to the producer. In other words, it reflects the low marketing efficiency of the market channel. The price spread and marketing efficiency can be improved only through collective and constant efforts in terms of adoption of higher-value addition technologies at individual or group level.

Coconut sector in India has been experiencing a low-profit and low-income regime for quite long, and the impact is such that the farmers lost their interest in scientific coconut-based farming systems (Mani and Santhakumar 2011). As of now, the prices of coconuts are attractive, and this is the apt opportunity for creating awareness among coconut growers on integrated coconut-based farming systems, the adoption of which can act as shock absorber in the event of failure or price crash of the main crop. It is an experienced fact that the coconut prices are volatile and unpredictable and there are close substitutes available to replace the coconut oil in the event of any supply shock or price crash. In this scenario, it is wise to redefine the present coconut farming methods more towards high-density integrated farming, based on the agroclimatic specifications.

Coconut prices in India have been historically integrated with the coconut oil prices. Therefore, indubitably the coconut prices received by the farmers are integrated with the Minimum Support Price (MSP) of copra. In general, the farmer prefers to sell fresh coconut when the price of coconut is attractive, as he receives a remunerative sum as ready cash and he can avoid processing and transportation charges. Contrary to this, if the copra and oil prices are lucrative, farmer prefers to do at least primary-level processing which would augment farm-level copra production. Therefore, the MSP for copra fixed at higher levels would certainly influence and act as an incentive for the primary value addition in coconut.

Having said this, it should be mentioned that the copra procurement system in the country has been functioning always at suboptimal levels and never effective in lifting up the market prices to optimum levels. The National Agricultural Cooperative Marketing Federation of India Ltd. (NAFED) is the apex state machinery controlling the copra procurement. The major issue faced by the NAFED in the event of huge procurement was finding the appropriate market avenue to push the product with a reasonable margin, which in the past had resulted in market failures.

Minimum Support Price (MSP) should be in such a way that it ensures an incentive for processing to the coconut farmers when compared with that of selling fresh coconut. Other pertinent factors in this context of discussion are lack of effectiveness and efficiency in copra procurement by the agencies and inadequate infrastructural facilities for the storage of copra. It is noteworthy that for the most part of the year, copra is traded below MSP.

The effectiveness and efficiency of price support mechanism can be enhanced only by means of adequate quantity of procurement and by ensuring that the genuine farmers are benefitted by the system of procurement. It is also important to design the procurement pattern in such a manner that adequate quantity is procured throughout the year, without any seasonal restrictions.

The potential area of the coconut sector is the agribusiness, based on value-added products of coconuts. The breakthrough products developed from the coconuts have the export potential, and thereby in the long run, the price stabilization in the domestic coconut sector is also possible.

The coconut inflorescence sap called Kalparasa can be preserved up to 45 days under cold condition (in refrigerator) without addition of any preservatives and additives (with the bottling technology). This enterprise can bring very good earnings to farmers and tappers. For sustaining the value-added coconut sector, Women's Self-Help Groups like Kudumbashree can be equipped with technical know-how, and smooth functioning of the coconut value chain is to be ensured through continuous supply of value-added products to the downstream part of the chain.

Desiccated coconut (DC) export from India is only to the tune of less than 1% of the global demand, in spite of the fact that the country is the largest producer of raw coconut in the world. Nevertheless, during the year 2015–2016, India exported 4261 MT DC worth Rs. 526 million. In comparison with the export figure of the previous year, India achieved an increase to the tune of 63%, which is indeed remarkable. Due to the growing consumer demand for DC across the world, there exists an immense export potential for the product. The capital investment required to start up a DC production unit, of capacity to process 15,000 coconuts day^{-1}, amounts to Rs. 12.9 million. It is noteworthy that there are attractive export promotional schemes initiated by the Indian government under the new Foreign Trade Policy (2015–2020), wherein under merchandise export from India scheme, 5% export subsidy can be availed on free on board (FoB) prices. There is also a duty draw back scheme wherein up to 1% of the FoB prices is refunded for the service taxes paid for raw materials and other input services for the production of DC.

17.7 Empowerment Through Agribusiness

Complete package of practices is available for virgin coconut oil production (hot and fermentation process), coconut chips, coconut honey, jaggery and sugar (Sairam et al. 2008). Virgin coconut oil (VCO) has received much attention globally in the recent times. The popularity of VCO is growing among consumers in all the continents due to its myriad properties including potential health benefits. Feasibility analysis of the project on commercial production of VCO revealed a benefit cost ratio of 1.12 and an internal rate of return of 21.5%. Thereby, we may conclude that the commercial production of VCO could turn out to be a profitable venture. Though the market of VCO is expanding in the domestic and international front, as a matter of fact, coconut-growing countries are yet to realize the potential benefit that it holds in this segment. It is imperative to establish good quality, technically advanced VCO units across the countries so as to realize the competitive market share of VCO in the global market.

17.8 Future Strategy

In the evolving trade liberalization regime, processing and value addition in the coconut sector have to be scaled up manifold across the coconut-growing countries to sustain the coconut industry as a profitable enterprise. Establishment of coconut parks which can ensure production of value-added products in accordance with the sanitary and phytosanitary regulations will be a right step in this direction.

The coconut development authorities of the coconut-growing countries should assess the strengths, weaknesses, opportunities and threats of their respective countries and accordingly should formulate an action-oriented time-bound strategic framework for the sectoral development. Such a strategic framework should essentially consider the ongoing trade nuances, the food safety requirements and the market preferences of the major trade destinations of coconut products. It is desirable to give heuristic attention to the coconut value chain of each coconut-growing country and ensure that the upstream end players (mainly farmers) receive a reasonable share of the revenue generated in the entire value chain. In the ongoing competitive spectrum, it is advisable to work as farmer collectives for attaining the optimum economies of size as well as reducing the cost of marketing. Furthermore, the research organizations should highlight coconut-based farming systems which are economically viable and possibly forward linked with the processing units. The coconut sector should be placed as a priority crop in the agrarian policies and developmental agendas of each of the coconut-growing countries.

For better trade relations among the APCC countries, it is imperative to form a regional coconut trade agreement among the APCC countries. The modalities of such a commodity-specific trade agreement should be such that all the partners will be in a win-win position. In this respect, an unbiased tariff reduction schedule should be developed, considering the existing tariff structures of close substitutes/competing products as well, through a consensus.

References

Alma P, Albert A (2016) The long climb towards achieving the promises of the tree of life: a review of the Philippine coconut levy fund policies. FFTC agricultural policy article. Retrieved from http://ap.fftc.agnet.org/ap_db.php?id=577&print=1

APCC (2016) Coconut statistical year book-2015. Asia Pacific Coconut Community, Jakarta, p 351

CDA (2016) Coconut Development Authority, Sri Lanka Industry outlook. Retrieved from http://www.cda.lk/web/index

CDB (2016) Annual Report, Coconut Development Board-2015-16. Coconut Dev Board, Kochi, p 172

Das PK (1984) Economics of cocoa mixed cropping with coconuts in India. In: Sethuraj et al (eds) Proceedings of PLACROSYM – VI. Oxford and IBH Co. Ltd., New Delhi, pp 397–408

Das PK (1991) Coconut Intercropping With Cassava: an Economic Analysis. Cord 7(2):58–65

FAO (2014) Report of the FAO high level consultation on coconut sector development in Asia and the Pacific. FAO, Rome, p 177

FAO (2017) FAOSTAT-food and agriculture organization of the United Nations. Retrieved from http://www.fao.org/faostat/en

Harilal KN (2010) ASEAN-India free trade area- Noises of dissent from deep south, Occasional Paper No.2010:01 State Planning Board, Government of Kerala

Hegde MR, Gopalasundaram P, Yusuf M (1990) Intercropping in coconut gardens, Tech Bull No.23. CPCRI, Kasaragod, p 7

Jayasekhar S, Chandran KP, Thamban C, Muralidharan K (2014) Price stabilization through stakeholder synergy: the key to revitalize coconut sector. Indian Cocon J LVI (2):20–23

Jnanadevan R, Jayasekhar S (2011) Coconut sector experiencing a price rise regime. Indian Cocon J (4):26–30

Kannan K, Nambiar KPP (1976) Studies on intercropping coconut garden with annual crops. Cocon Bull 5(9):1–3

Krishnaji N, Mukerji C, Nelliat EV, Nair PKR, Mathew J, Varghese PT (1976) Economics of Inter and Mixed Cropping in the Coconut Gardens of Kerala. Some Preliminary Findings. Working paper No.41. Centre for Development Studies, Trivandrum, pp. 1–9 (Mimeographed)

Lathika M, Kumar A (2009) Indian stakes in the global coconut scenario by the turn of the century -An empirical investigation. South Asia Eco J 10(1):209–221

Mani S, Santhakumar V (2011) Diffusion of new technologies and productivity growth in agriculture: natural rubber vs coconuts. Econ Polit Wkly 46(6):58–63

Pabuayon IM, Cabahug RD, Castillo SV, Mendoza MD (2009) Key actors, prices and value shares in the Philippine coconut market chains: implications for poverty reduction. J ISSAAS 15(1):52–62

Sairam CV, Bosco SJD, Arulraj S, Rajagopal V, Prabhu Kumar S (2008) Prospects of coconut based agribusiness in India. Financing Agri pp, pp 28–33

USDA (2017) Indonesia oilseeds and products annual report. Gain Report Number: ID 1705, p 19. Retrieved from https://gain.fas.usda.gov

World Bank (1990) Kingdom of Tonga Agriculture sector strategy review, Report No-8554-TON, World Bank Document. Agriculture Operations Division, Asian Regional Office, Bangkok

Dr. C. V. Sairam is an Agricultural Economist at ICAR-CIBA, Chennai, India. His specialization is in farming system research and frontline extension programmes. He has 15 years of experience in plantation crops research as well as in technology transfer mechanisms. He has 160 publications to his credit which include 43 research articles, 51 conference papers, 1 book and 24 book chapters. sairam_cpcri@yahoo.com

Dr. S. Jayasekhar is a Senior Scientist in the discipline of Agricultural Economics of ICAR-CPCRI, Kasaragod. He has also served as Indian Technical and Economic Cooperation (ITEC) Expert to the Kingdom of Tonga and was awarded the certificate of merit from the Government of Tonga. His specific areas of interest include policy research in plantation crops and studies on global and domestic commodity chains. He has 87 articles to his credit including 36 research articles and 1 edited book. He has received the Best Researcher Award from the International Institute of Fisheries Economics and Trade (IIFET) at Montpellier, France. jaycpcri@gmail.com

Index

K. U. K. Nampoothiri et al. (eds.), *The Coconut Palm (*Cocos nucifera *L.) - Research and Development Perspectives*,
https://doi.org/10.1007/978-981-13-2754-4

N

Q

R

The manufacturer's authorised representative in the EU is Springer Nature Customer Service Centre GmbH, Europaplatz 3, 69115 Heidelberg, Germany. If you have any concerns regarding our products, please contact ProductSafety@springernature.com

Printed and bound by CPI Group (UK) Ltd, Croydon, CR0 4YY

30/09/2026

02225446-0019